The
Physiological
Ecology
of
Vertebrates

The Physiological Ecology of Vertebrates

A VIEW FROM ENERGETICS

Brian Keith McNab

with a foreword by James H. Brown

COMSTOCK PUBLISHING ASSOCIATES

a division of
Cornell University Press
ITHACA AND LONDON

First published 2002 by Cornell University Press

Printed in the United States of America

Library of Congress Cataloging-in-Publication Data

McNab, Brian Keith, 1932–
    The physiological ecology of vertebrates : a view from
energetics / Brian Keith McNab.
        p. cm.
    Includes bibliographical references (p.   ).
    ISBN 0-8014-3913-2 (cloth : alk. paper)
    1. Adaptation (Physiology) 2. Bioenergetics.
3. Animal ecology. 4. Vertebrates—Physiology. I. Title.
    QP82 .M38 2001
    596.14—dc21   2001002729

Cloth printing   10 9 8 7 6 5 4 3 2 1

To Roan and Derrick
May their choices in life be as
* exciting and rewarding as mine have been*

Much of the work that is done under the name of ecology is not ecology at all, but either pure physiology—*i.e.*, finding out how animals work internally—or pure geology or meterology, or some other science concerned primarily with the outer world. In solving ecological problems we are concerned with *what animals do* in their capacity as whole, living animals, [neither] as dead animals [nor] as a series of parts of animals. We have . . . to study the circumstances under which they do these things, and, most important of all, the limiting factors [that] prevent them from doing . . . other things. By solving these questions it is possible to discover the reasons for the distribution and numbers of . . . animals in nature.

CHARLES ELTON (1927, pp. 33–34)

[T]here is a great deal more value in having [views] corrected than there is in never stating them; the road to truth lies much through argument.

F. E. J. Fry (1947, p. 59)

It never occurs to people that the one who finishes something is never the one who started it, even if both have the same name, for the name is the only thing that remains constant.

JOSÉ SARAMAGO (1991, p. 37)

Publish *and* perish

Contents

7 Water and Salt Exchange in Terrestrial Vertebrates 174

PART V   Consequences

13 The Significance of Energetics for the Population Ecology of Vertebrates   431

14 Physiological Limits to the Geographic Distribution of Vertebrates   447

Foreword

This is a great book, a remarkable achievement. It gives us, at last, a grand synthesis of physiological ecology.

Physiological ecology emerged as a coherent discipline in the 1950s with the pioneering work of Bartholomew, Bogert, Irving, Kendeigh, Morrison, Pearson, Schmidt-Nielsen, Scholander, Hobart Smith, and others. It has remained vigorous into the 1990s, largely due to the application of new technologies for measuring the performance of free-living organisms in the field. But physiological ecology has never really had its own theoretical framework and empirical underpinnings. It has remained an orphan, healthy and growing, but at home neither in comparative physiology nor in ecology and evolution. It has been portrayed as the discipline that does some combination of (1) endlessly cataloging the weird and wonderful attributes of "oh my" organisms that live in far-off, stressful places; (2) trivially proving that an organism "can actually exist where it lives"; and (3) exemplifying the "adaptationist paradigm" by interpreting every trait as an adaptation to some feature of the environment. This impression has been justified to some extent, because there has been little conceptual unification to give context and meaning to the numerous, well-documented mechanisms of physiological tolerance and regulation that allow organisms to inhabit nearly all of Earth's diverse environments. Despite thousands of scientific papers and scores of books, the data and theory of physiological ecology have remained fragmented and specialized.

Now, in this book, Brian McNab provides a broad, synthetic overview of the physiological ecology of vertebrates. It is clear from the contents of this volume why it has taken so long. To pull it off required the dedicated effort of an exceptional scientist. In several decades of research on an amazing variety of organisms, from vampire bats and tree-kangaroos to frogmouths and Komodo dragons, McNab has distinguished himself as a keen naturalist and a creative "idea man" in physiological ecology. In this book, he also shows his exceptional capacity for scholarship and synthesis. The book is both encyclopedic and synthetic, factual and speculative. Drawing on more than 3100 references, the book encompasses the breadth of contemporary physiological ecology, the awesome diversity of vertebrates, and the impressive variety of mechanisms that enable them to live everywhere from Antarctica to the Sahara Desert, from tropical rainforests to the abyssal depths of the oceans. It would be easy to get so caught up in the special cases as to lose track of the big picture, but McNab has organized the factual material around the unifying framework of energetics.

The breadth of this framework is impressive. McNab recognizes that there can be no one theory of physiological ecology. On the one hand, the mechanisms of physiological tolerance and regulation must be explained largely in terms of the physical, chemical, and biological processes that govern the fluxes of energy and materials between organism and environment. So physiological ecology must be grounded in the fundamental principles of organismal biology, biochemistry, and biophysics. On the other hand, the influence of physiology on the abundance, distribution, and diversity of species must be interpreted in terms of the principles of ecology and the process of evolution. McNab sweeps back and forth over this vast terrain, using energetics as a unifying theme to explain how processes at the physical, chemical, cellular, and organismal level have been exploited by natural selection to solve problems of survival and reproduction in diverse, often stressful environments. The title *Physiological Ecology of Vertebrates* is almost a misnomer, because McNab's emphasis is as much evolutionary as it is ecological. Here we have a single treatise that integrates everything from energy balance equations, the Stefan-Boltzmann law, and the biochemistry of brown-fat metabolism to the fossil histories of therapsids and pterosaurs, the evolution of flightlessness in insular birds, and the role of temperature in setting the northern range limits of alligators, armadillos, and opossums. In between we do indeed learn how organisms are able to live where they do: how koalas survive on a diet of eucalyptus leaves; how normally marine sharks and rays adjust urea metabolism when they live far up the freshwater Amazon; how pocket-gophers cope with the low oxygen and high carbon dioxide concentrations in their burrows; and how some arctic seals manage to transfer 85% of the energy stores from mother to pup with 75% efficiency in just 4 to 6 days of lactation, by producing milk with a 50% fat content at a rate of 6 to 7 kg per day!

What I most admire about the book, however, is its point of view. The reader can visualize McNab struggling to organize the vast quantity of material into a coherent synthesis. And despite the success of that endeavor, McNab's unique creative insights, critical mind, and speculative approach come through clearly. In the Preface he confesses to having "worried throughout the writing of this book that it is either too confrontational or too bland." But then he reveals his true colors: "In my opinion being too conservative is a greater problem than being too speculative: conservative intellectual niches are fully occupied.... I wrote in part to 'push the envelope.'" We should be grateful. The book does much more than just compile the facts and place them in a conceptual framework. Critical commentaries call attention to data that may be biased or incorrectly interpreted. Queries and speculations highlight questions that remain unanswered.

McNab's book provides a masterful synthesis of the current state of physiological ecology. Its thorough, inspired scholarship summarizes what has been learned in the past. Its wide-ranging, speculative tone sets an ambitious, optimistic agenda for the future.

JAMES H. BROWN

Preface

This book is the first to collect and organize the observations of physiological ecology—a hybrid field within the biological sciences that integrates the physiology and behavior of organisms with the physical and chemical conditions in the environment. Its hybrid origins contribute to the general absence of an organizing theoretical framework, the creation of which has been my motivation for the more than 30 years it has taken to write this book. These origins are exacerbated by a massive amount of data on many physiological systems derived from a diversity of species that belong to a variety of families, orders, classes, and phyla.

To complicate further the task of building a theoretical framework for this field, many who think of themselves as physiological ecologists are interested in one class, one physiological problem, or one environment, and this impedes a broad philosophical viewpoint from being developed.

As Heinrich (1994, p. 486) stated, "Physiological ecology, the modern natural history, is a science rooted in mechanisms." The present work attempts to reflect the ecological and evolutionary contexts of physiological ecology and its mechanisms to demonstrate the intellectual cohesion potentially present in this field. Even when such contexts are available, no observations can be made intelligible if the gaps in information are too great, if the environmental context is unknown, or if the evolutionary relationships of the studied organisms are unclear. My solutions to these difficulties are (1) to restrict this book to vertebrates because many aspects of their environmentally relevant physiology have been examined, because we are often familiar with the environments in which they live, and because their evolutionary history is compara-

tively well known, and (2) to emphasize a common thread, that of energy expenditure, throughout the text. Energy expenditure is used to integrate individual topics, where possible, because all activities require energy, because energy availability may limit expenditures, which may require some species to maintain selected functions at the expense of others, and because energy units may be used as a currency common to all major components of an organism's performance. Furthermore, energy is a currency used in many ecological models.

A solution for the piecemeal state of physiological ecology that Sibly and Calow (1986) advocated in their book, *Physiological Ecology of Animals*, is the use of an intuitive-deductive (i.e., a priori) system of "theoretical" models to organize the conceptual framework and, presumably, the observations of physiological ecology. That choice is symptomatic of the times, when the search for theory in ecology outweighs a commitment to the collection of data. This approach has been most clearly demonstrated with respect to energetics by Kooijman (2000). In contrast, I prefer a more inductive method, what Sibly and Calow referred to as the comparative, correlational, a posteriori approach. The advantages of this method are that it (1) focuses on animals, which after all are the source of inspiration for most zoologists, and (2) makes no assumptions about how nature "should" be organized. It is based on a thorough knowledge of the natural history of vertebrates. The inductive approach, when effectively applied, not only leads to models that can be as instructive as any a priori model, but also is more sensitive to the unexpected results of many observations. In the long run, physiological ecology will profit from the use of both inductive and deductive models to organize data and to direct inquiries.

Thus here I examine the topics of scaling, temperature relations in ectotherms and endotherms, water and salt balance in terrestrial and aquatic vertebrates, gas exchange, energetics of locomotion, energy budgets, diet and nutrition, the significance of energetics for population ecology, and the physiological limits to the distribution of vertebrates.

A difficult decision involved determining the level at which this book should be written. Widespread interest in this topic by workers, often without an extensive familiarity with physiology, needs to be respected. But a full appreciation of the physiological ecology of vertebrates requires knowledge of their physiology, natural history, and evolution. I try here to supply the physiology; the rest I assume that readers will have. Therefore, I consider this book an essay intended especially for advanced undergraduates, graduate students, and professionals in physiological ecology, ecological physiology, environmental physiology, ecosystem process, population and community ecology, behavioral ecology, paleontology, evolutionary biology, conservation biology, and physical anthropology.

I wrote this book with the avowed intents of noting deficiencies in the current knowledge, of provoking discussion on the evolution of physiological characters, and of encouraging students to work on this fascinating aspect of natural history. When I found significantly different views on a topic, I tried to give them with balance, which is not to say that I always have a neutral position. When sufficiently moved by a topic, I included a boxed essay with a more partisan view. In my opinion being too conservative is a greater problem than being too speculative: conservative intellectual niches are fully occupied. But I have worried throughout the writing of this book that it is either too confrontational or too bland—and some of the many revisions have reflected this varying viewpoint. This book took a long time to write, in part because of the difficulty of organizing a very large (and growing) literature and in part because of the time I required to develop some unity of thought.

I wrote in part to "push the envelope" and to mark some of the boundaries and techniques for a new way of looking at products of natural selection. I wrote also to give student readers and biological researchers an opportunity to think in new ways about animal physiology and about the context for animal evolution. Essentially I offer this book as

• a synthesis of an emerging field of study within the biological sciences that until now has been represented in a disparate literature;
• a way to acquaint students of ecology, behavior, and vertebrate biology with the ecologically and evolutionarily relevant aspects of physiology;
• a means to show how to analyze physiological data; and
• an opportunity to apply analysis of physiological data to the study of the ecology, behavior, and evolution of vertebrates.

Finally, I hope that this book will bring more physiology into the mainstream of ecological and evolutionary thought and that it will be an intellectual stimulus to graduate students as they seek significant research topics. Too often, beginning graduate students believe that they are a generation too late, most of the interesting and profound observations having already been made—a conclusion that is far from the truth. Of necessity, this effort is compromised by limitations in the quality and quantity of available data, limitations in the conceptual framework in which the data are placed, and profound limitations in my ability to bring the data and concepts together into an integrated whole. Hopefully, this attempt will find its reward in stimulating others to extend the work that so many have started, to explore new aspects of this exciting field, and to succeed in the synthesis where I have failed.

BRIAN MCNAB
Gainesville, Florida
January 2001

Acknowledgments

This book represents a long struggle. It started in the late 1960s as a set of proposed readings, which I abandoned as a result of an inability to come to an agreement on which papers should be included. In the process, one reviewer suggested that I write a book on physiological ecology, an idea that I quickly rejected. However, the suggestion did plant the seed for this book. This project, which has gone through many revisions and tried the patience of many editors, profited greatly from the detailed reviews of many individuals, known and unknown, including John Anderson, David Evans, Leonard B. Kirschner, Carlos A. Navas, Terry Root, C. R. Tracy, P. Van Soest, and Bruce Wunder. The final draft of this book was thoroughly examined by C. R. Tracy and Marianne Preest, a yeomanly task that I greatly appreciate. I thank these "contributors" for their aid; this book would have been much poorer but for their efforts. Several editors, most notably Robb Reavill and the determined Peter Prescott, have greatly contributed to this book. Special thanks go to the several generations of graduate students that have been subjected to this book in various photocopied incarnations. I thank Peter Morrison, who gave me my first exposure to an analytical approach to comparative physiology, and Ernst Mayr for his insight and comments on the Introduction and his encouragement for this project. I also appreciate the early encouragement from João Balas and Greta Balas NcNab. I have nearly exhausted the patience of a series of artists with my fiddling with their computer graphics, including Grace Kiltie, Laurie Walz, Ulie Christman, and Susan Trammell. This book would have been bare without their contributions. I can only praise the Herculean effort of Mary Babcock, who copyedited this book: she saved me from many embarrassments. Vassiliki Smocovitis has been patient and encouraging in the face of this seemingly endless task. Only I am responsible for the errors that have crept into this book: they reflect my ignorance and intransigence.

Introduction

Physiological ecology as defined in the Preface, derives its theoretical basis from ecology and evolution, which is why I use *ecology* as the noun and *physiology* as the adjective. Physiological ecology potentially has the capacity to integrate the diverse approaches used in the study of the comparative biology of vertebrates. Biologists have segregated their approaches by technique and concept, but the boundaries among ecology, behavior, anatomy, and physiology are arbitrary and of no significance to organisms. The integration offered by physiological ecology is not derived from the examination of chemical and physical mechanisms, but from the application of energetics because all biological functions of significance require the expenditure of energy. An examination of energy expenditure permits a comparative estimate of the responses of organisms to the conditions in the environment and ultimately of their contribution to the evolution of biological diversity. The limits to geographic distribution are an ultimate expression of the limitation in the responses of organisms to challenges from the environment.

The techniques used in physiological ecology are often those used in comparative physiology. The principal difference between these fields lies in their theoretical bases, the latter being preoccupied with the physicochemical mechanisms by which organisms regulate their internal state and respond to the environment. The unifying principle of comparative physiology is found in the assumption that the overwhelming diversity in organisms can be reduced at the systemic, cellular, and molecular levels to a restricted set of widespread physicochemical mechanisms—that is, its "context" is ultimately physics and chemistry. This assumption minimizes the importance of the diversity that is at the heart of physiological ecology and evolutionary biology. In contrast, physiological ecology emphasizes the diversity of solutions (and evasions) to problems posed by the environment—that is, its "context" is proximately the evolutionary response to the environment and ultimately the environment itself. This environmental context has implications for conservation biology.

Physiological Ecology and Conservation Biology

Physiological ecology is relevant to conservation biology in at least two ways. One is in the general sense that all knowledge potentially is of value when decisions concerning the survival and protection of species and their environments are to be made. One can never know a priori which information will be "useful" and which will be "useless," under the assumption that some knowledge is useless or that we can distinguish between these categories. Anyone who has worked with threatened, endangered, or protected species has encountered the frustration associated with this false dichotomy. Knowledge is power. The more we know, the more likely the decisions required in conservation biology will be effective, or at least not destructive.

A second area of knowledge where physiological ecology is central to conservation is indicated by one of the ultimate goals of physiological ecology—to determine the factors responsible for setting the limits of geographic distribution of organisms. This topic is explored in Chapter 14. Recent evidence indicates that Earth's climate is changing (although we may be uncertain today whether this "global warming" is short or long term and whether it is or is not mediated by humans). Regardless of its duration or ultimate cause, global warming appears to be having an appreciable effect on the distribution and even the survival of some vertebrates. Physiological ecology is uniquely positioned by its broad knowledge base and approach to examine the factors responsible for changes in the distribution of organisms. It can help, therefore, to anticipate problems before they have an irretrievable impact on the survival of selected species. This approach may suggest geographic sites to which species endangered by climate change might be translocated for their conservation.

Physiological Ecology and the Evolutionary Synthesis

Between 1930 and 1950 morphology, paleontology, systematics, and genetics were brought together under the belief that a unifying principle existed in biology, the principle being evolution by means of natural selection. This unification was called the "evolutionary synthesis." The success of this synthesis derived from the unity of a quantitative mechanism given by population genetics, a temporal dimension drawn from paleontology, and a biological diversity identified by morphology and systematics. The process by which this synthesis was arrived at is described in a book edited by Mayr and Provine (1980) and in a later volume written by Smocovitis (1996).

Other disciplines, including developmental biology, ecology (at that time, not a coherent field), and physiology (a reductionist handmaiden of medicine), were not part of this synthesis. Their absence related to a fundamental requirement of the synthesis, and that was the acceptance of a populational viewpoint as derived from natural history and population genetics. Natural selection produces evolution as it operates on inheritable variation. All fields of biology, therefore, must emphasize the differential consequences of this variation in populations to be incorporated into the synthesis: the incorporation of systematics was possible when it moved from a typological to a populational approach.

Physiological ecology has as its technical basis classic physiology, a field that is predominantly preoccupied with a reductionist viewpoint whereby physiology at the individual level is broken down into a series of physicochemical mechanisms. Physiology is essentially typological in nature when it describes the (invariable) characteristics of species. From an ecological viewpoint, classic, reductionist physiology suffers from the absence of a context within which to examine the variability in physiological responses and mechanisms.

Although the ecological component of physiological ecology has been understood for some time, a realization has only just begun that the evolution of physiological systems is an important topic. Many earlier, and some contemporary, studies of physiology had, and still have, a simplistic phylogenetic standpoint. The physiological literature is replete with statements on the "primitive" or "advanced" nature of various regulatory processes without an adequate analysis of the comparative contributions of ecology, behavior, and phylogeny. As a result of recent developments in phylogenetic analysis, one can now place physiological, ecological, and behavioral observations into a phylogeny to separate homologous from convergent characters. But as we will see, the separation of historical from contemporary influences is difficult to determine because of factor interaction. A problem with most physiological studies is that they usually use a typological description of species performance, a viewpoint that hinders the contribution of physiology to the evolutionary synthesis.

Through the addition of comparative, ecological, and evolutionary approaches, the typological bias of physiology has diminished. The shift from a typological to a populational viewpoint required for a field to be included in the evolutionary synthesis reflects a stage in the evolution of a field. A field cannot be brought into the synthesis kicking and screaming: it must be "ready," ready in the sense of techniques whereby individual variation can be measured, ready in the sense of the description of the diversity of functions present, and ready in the sense of the ecological/evolutionary consequences of this variation.

Awareness that a populational component should be included in physiological ecology (see, for example, Feder et al. 1987) is increasing. But for all of the statements on the importance of examining individual variation, most work in this field remains typological, in part because so little of the physiological diversity of the animals on this planet has been studied. We have little understanding of the extent to which our present knowledge of almost any function that has been studied represents an adequate sampling of the diversity that is present. For example, the greatest diversity of birds is found in the tropical lowlands, but almost nothing is known of their physiological diversity, even from a typological viewpoint. Thus, much descriptive information is required before more than a few selective examples of physiological diversity can be incorporated into the evolutionary synthesis. We are light-years from any general theoretical bases for the physiological ecology of vertebrates. We should always remember that most successful theories of nature are derived from being acquainted intimately with the processes of nature, and are not superimposed in the belief of how nature "should" be organized.

The Approach and Organization of This Book

A serious concern I had in writing a book of this scope is organization. On one hand, I want to convey a great diversity of information in an organized, sequentially rational manner, being fully aware that many readers will not be familiar with the material. On the other hand, I want to convey the complex interactions among the factors under discussion. Thus, water exchange is associated with water balance, temperature regulation, gas exchange, and energy expenditure, and under some circumstances the limits of distribution. These, slightly conflicting approaches are addressed here first by summarizing detailed information on temperature relations, water balance, and gas exchange and then by integrating these relationships through the use of energy expenditure. Any attempt to integrate completely all of these phenomena in one seamless fabric is likely to run the risk of confusion, especially as the conceptual field attains the complexity encompassed here.

The topics I chose for this book directly contribute to our understanding of body maintenance in an environmental context. An introduction to the concept and complexities found in the term *adaptation* (Chapter 1) leads to the mathematical and physical bases of exchange and regulation (Chapter 2). A chapter on scaling (Chapter 3) follows. Then each major regulatory system is analyzed, namely, thermal biology (Chapters 4 and 5), water and salt balance (Chapters 6 and 7), and gas exchange (Chapter 8). Because activity is an important component of existence and energy expenditure in vertebrates, Chapter 9 examines the energetics of locomotion. Chapter 10 on energy budgets is an attempt to integrate the expenditures

for maintenance and activity, and Chapter 11 considers energy expenditure in the context of daily and annual periodicities in the environment. Nutritional complications inherent in balancing an energy budget with foods of various chemical compositions are explored in Chapter 12. Finally, examinations of the consequences of variations in energy expenditure for growth and reproduction (Chapter 13) and of the limits to geographic distribution (Chapter 14) round out the story.

Each chapter starts with a short synopsis to give an overview of the subject to be covered, followed by a brief introduction designed to lay the basis for the chapter's development. Then the data are arranged in a conceptual context, pointing out where discrepancies among measurements or between theory and observations may exist, and where critical observations are needed. The context is preferably ecological, but at times this cannot be rigidly separated from taxonomy because some groups of vertebrates occupy distinctive environments and have idiosyncratic responses to environmental conditions. On these occasions I organized the responses of vertebrates by taxonomic groups. Throughout, I prefer a quantitative analysis to examine the response of organisms to conditions in the environment. Each chapter concludes with a summary of the principal conclusions.

I make extensive reference to the sources of the data and ideas summarized in this book, and reject the use of "selected readings," which often turn out to be other summaries that make it difficult to find the original sources. I believe that (almost) all work on the topics addressed here can be discovered through an exploration of the more than 3100 references cited.

I have tried both to give the latest references on a particular topic (although this goal is a moving target, given the number of new articles that appear monthly) and to give a historical development of some major ideas; this is partly out of respect for the people who were responsible for the initiation of a line of thought and partly to demonstrate how ideas change with time, or how we repeatedly return to the same ideas. After all, we biologists live within a conceptual framework that has evolved with the accumulation of data, the invention of new techniques, and the development of thought. None of us is independent of such a context. I apologize to the contributors whose work has not been cited; this reflects either my ignorance or an attempt to limit the size of this book.

PART I | Foundations

1 The Limits to Adaptation

1.1 SYNOPSIS

This chapter defines the concept of adaptation, analyzes the complex problems associated with separating adaptation from other processes that determine character states, discusses the association of correlation with causation in intact organisms, and examines the view that many characteristics of an organism are determined by the interaction of other characters. Some changes in character states may open new evolutionary opportunities, whereas others can restrict future possibilities. Finally, a limitation to adaptation may issue from factor interaction, ecological inappropriateness or expense, or historical legacy.

1.2 CONCERNING ADAPTATIONS

Physiological ecology is concerned with the malleable features of physiology that when modified in relation to the physical or biological characteristics of the environment, contribute to the increased reproduction of a species, that is, to its inclusive fitness. Such modifications are referred to as *adaptations*, although the reproductive consequences of most presumptive adaptations are rarely known.

Slobodkin (1964) suggested that the principal property held constant by a species, when it adjusts to a changing environment, is the capacity for homeostasis (i.e., the ability to maintain a constant internal environment). He (pp. 352–353) argued that "... all ... responses [to an environmental change tend] to maximize the ability ... to respond to subsequent environmental change." The first response is likely to be behavioral, such as the selection of microclimate. If that response is not adequate or possible, the organism makes a rapid physiological adjustment, such as the increase in rate of metabolism shown by birds and mammals when exposed to a fall in ambient temperature (see Section 5.2). Acclimatization to a seasonal change in environmental conditions may replace an immediate physiological response. An example is found in the seasonal adjustment in the rate of metabolism to temperature in fishes and amphibians (see Section 4.5.3), and in the seasonal variation in insulation in birds and mammals (see Section 5.5.2). Presumably, the cost of facilitating the modified response to temperature in fishes (see Section 4.5.4), or of growing a thick fur coat in mammals, is less than the cost of not doing so.

If the change in environmental conditions is constant over a sufficient number of generations, either because of a climatic shift or because a species extends its distribution into new regions, then the appropriate response may well be genetic. Slobodkin suggested that such genetic changes restore and maintain the flexibility of short-term physiological adjustments. Genetic change, in this view, would be the response of last resort because a genetic change may reduce the capacity for further long-term change, or because a genetic change in one direction may preclude a response in another direction. Another difficulty with a genetic response is that mutation is random: mutations appropriate to the environmental challenge are not available when needed. Nevertheless, evolution is based on an accumulation of genetic changes, some of which may be "neutral," others deleterious, and some that contribute to the reproduction of organisms under specific environmental conditions (i.e., they represent an "adaptation" to those conditions).

The concept of adaptation has been attacked as being vague or representing circular reasoning (see Gould and Lewontin 1979). Without doubt the use of the term has often been inexact—in fact, rarely has an attempt been made to distinguish the character states in a species that are a direct response to conditions in the environment from those that result from a dependent coupling of characters with each other. Equally, the separation of derived characters in a phylogeny from those that represent an ancestral condition has often been difficult. Some enthusiasts for the concept of adaptation have assigned nearly every change in an organism as an adaptation—implying an independence of characters that has not been demonstrated. This error, however, does not discredit the concept so much as its application. Surely, no one familiar with the seasonal change in insulation found in polar mammals would doubt that it is a means of modifying heat loss and production that presumably contributes to the fitness of these species—call it by whatever name.

1.2.1 A partial solution. One way to test and potentially to falsify the validity of adaptation in a character is to map its various forms (states) on a cladogram describing the phyletic history of the species under study. This procedure theoretically will permit the separation of character states that represent the ancestral (plesiomorphic) condition from those that have been derived (apomorphic). If apomorphic character states can be shown to coincide with changes in the environment or behavior of a species, then these changes in character states may be adaptive. For example, Grismer (1988) described a cladogram for the geckkonid genus *Coleonyx*, based on morphological and meristic data. Dial and Grismer (1992) then mapped onto the cladogram (Figure 1.1) three physiological traits (standard rate of metabolism [see Section 3.3.2], rate of evaporative water loss [see Section 7.5.2], and "preferred" [or "selected"] body temperature [see Section 4.8.3]) in six of the seven species of *Coleonyx* and in species belonging to three of the five related genera (which formed their "outgroup" standard).

Dial and Grismer concluded that the ancestral condition (obtained from the outgroup) was characterized by a low standard rate of metabolism, high rate of evaporative water loss, and low temperature "preference," which also were the character states in the two tropical-forest *Coleonyx* species belonging to the earliest lineage. Derived species of *Coleonyx*, however, have different character states. Dial and Grismer suggested four hypotheses for their evolution, two of which are

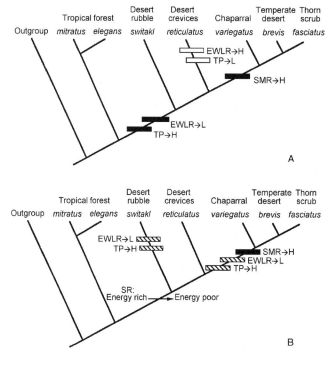

Figure 1.1 Character states in geckos belonging to the genus *Coleonyx* mapped onto a phylogeny in two hypothetical schemes. The characters include standard rate of metabolism (SMR), rate of evaporative water loss (EWLR), and preferred body temperature (TP), each of which may be either high (H) or low (L). A solid bar represents the evolution of a derived condition, a diagonal bar represents the independent evolution of a derived state, and a hollow bar represents a reversal to the ancestral state. A shift from an energy-rich to an energy-poor environment (selective regime), indicated by SR. The hypothetical schemes are (A), one which assumes that the character states in *C. reticulatus* represent a return to the ancestral condition typical of the outgroup that is characterized by a low BMR, high EWLR, and low TP; and (B), one where the conditions in *C. reticulatus* represent the unchanged ancestral state, which requires convergence between *C. switaki* and those species beyond *C. reticulatus*. Source: Modified from Dial and Grismer (1992).

indicated in Figure 1.1. One (Figure 1.1A) suggests that low evaporative water loss and high thermal preferences evolved between nodes 2 and 4 and that these changes were reversed in the evolution of *C. reticulatus*. Another (Figure 1.1B), which Dial and Grismer preferred and which incorporated distribution and natural history, suggests instead that *C. reticulatus* is a relictual species retaining the ancestral states, and therefore that a high temperature preference and low rate of evaporative water loss evolved convergently in *C. switaki* and in the *C. variegatus-brevis-fasciatus* clade in response to life in desert environments. The evolution of a high standard rate of metabolism occurred in the *variegatus-brevis-fasciatus* complex, apparently in association with an increase in activity, foraging area, or both. This example clearly shows the potential impact of a phylogenetic approach to adaptation, but even when only one phylogeny is considered, several interpretations of evolutionary events are always possible.

Baum and Larson (1991, p. 6) defined an adaptation as a change in a character that has an "... enhanced utility relative to its antecedent state, and the evolutionary transition must be found to have occurred within the selective regime of the focal taxon." Following Gould and Vrba (1982), Baum and Larson attempted to separate adaptations from "exaptations," which are defined as utilitarian traits that originally evolved for a use other than the current one. The value of such a distinction, however, usually appears slight, at best.

All phylogenies are hypotheses of evolutionary histories. The principal weakness of relying on them in an analysis of adaptation is the widespread absence of *reliable* phylogenies: obviously, as phylogenies change, so does the evaluation of which states are plesiomorphic and which are derived. The recent use of DNA to construct phylogenies shows how radical the changes may be: see the impact of the work by Sibley and Ahlquist (1990) on the classification of birds and, on a smaller scale, the reevaluation of the evolution of nectarivory in megachiropteran bats (Kirsch et al. 1995). What is unacceptable, however, is to argue that an evolutionary approach to the study of adaptation is required to account for phylogeny and then to maintain that "... we see no reason why our analyses [of the impact of phylogeny on basal rate of metabolism] should be systematically biased because they rest on unsatisfactory classifications" (Harvey et al. 1991, p. 558).

1.3 A COMMENT ON CORRELATION AND CAUSATION

The character states of organisms are commonly correlated with conditions in the environment or with other states in the same organism. Such observations lead to the caution that a correlation is not evidence, per se, of a causal relation. The paradox, however, is that nearly all "true" causal relationships appear first as correlations. The difficulty is in separating causal from incidental correlations.

In intact organisms, functional interactions among characters are complicated, and many of their physical and chemical bases are imperfectly understood. Furthermore, some ecologically significant correlations may involve "emergent" biological properties (see Mayr 1997). An effective way to determine whether a functional relationship exists between two variables is to examine how they vary in organisms. For example, is rate of metabolism correlated with food habits? If so, how is the correlation influenced by body mass, climate, and taxonomy? How does osmotic regulation vary with body size, food habits, and phyletic history? Fundamental to apparent character associations is that the residual variations in these characters be correlated. For instance, even though the correlation of brain mass in eutherians with body mass is similar to the correlation of basal rate of metabolism with mass, McNab and Eisenberg (1989) concluded that brain size and basal rate were not functionally connected, contrary to the conclusions of Martin (1981), Armstrong (1983), Harvey and Bennett (1983), and Hofman (1983), because the residual variations in the two scaling relationships were not correlated with each other.

Even when a functional relationship appears to exist between two variables, cause still may be difficult to separate from effect. This difficulty is illustrated in the positive correlation of rate of metabolism with body temperature. As environmental temperature increases, the body temperature of poikilotherms increases, and as a result of enzyme kinetics, the rates of chemical reactions increase (see Section 4.4). In this case rate of metabolism is said to depend on body temperature, an interpretation that has been extended to mammals: marsupials have been suggested to have lower rates of metabolism than eutherians *because* they have lower body temperatures (Dawson and Hulbert 1970), and by implication birds have higher rates of metabolism than mammals *because* of the higher temperatures of birds. Yet, the level of body temperature in endotherms shows little evidence of adaptive modification (Irving and Krog 1954), whereas the level of energy expenditure shows extensive evidence of such modification (see Section 5.4). Therefore, one can argue that mammals with high rates of metabolism have higher body temperatures because body temperature increases to a level at which heat loss equals heat production (see Section 5.6). A similar view can be applied to the differences in body temperature and rate of metabolism between birds and mammals (McNab 1970). Thus, in contrast to poikilotherms, which are characterized by low rates of metabolism in part because of their generally low body temperatures, the high body temperatures of some endotherms may be the effect, not the cause, of a high rate of metabolism. Another example of the difficulty of determining causal relations is found in the larger body mass of some endotherms in cold climates (Bergmann's rule; see Section 5.3.2), which is usually thought to be a response to cold temperatures, but such a cline also may represent a reduction in size in warm climates (especially in burrowing mammals; McNab 1966b, 1979b). These alternatives are better evaluated if the phyletic and geographic histories of the populations are known.

1.4 CONSEQUENCES OF ADAPTATIONS

Most organisms live in environments that pose a variety of physical and biological challenges. Each challenge normally has several potential solutions, the secondary consequences of which may be quite different. For example, an endotherm living in a polar or cold-temperate climate requires a large food intake to maintain its normal body temperature in winter. If an adequate food supply is not available, the endotherm may reduce heat loss by increasing thermal insulation, migrate to regions where food is available, or reduce energy expenditures by resorting to daily or seasonal torpor (hibernation). Some restrictions are placed on each of these alternatives: increased insulation cannot be tolerated if high temperatures are encountered; migration is possible only in species that have a means of long-distance locomotion; daily torpor will only work if food supplies are restricted but not eliminated; and hibernation requires an acceptable shelter and depresses reproduction. In

some species the benefits of these alternatives are so similar that some individuals use one solution and others use another. Free-tailed bats (*Tadarida brasiliensis*) that breed in Texas, New Mexico, and Oklahoma winter in Mexico, whereas those breeding in California, Louisiana, and Florida remain during winter. And many bird species in temperate climates show interpopulational, even interindividual, variation in their propensity or distance to migrate.

Nearly all solutions to particular challenges have secondary consequences reflecting character interaction. Such consequences often can be exploited to permit movement into new adaptive zones: the evolution of lungs permitted vertebrates to move onto land, whereas the use of gills or the posterior intestine for aerial gas exchange did not, except in the most marginal manner. In other words, some adaptive modifications may restrict organisms, whereas others may expand their opportunities.

1.4.1 Restrictive adaptations. Although an adaptation by definition is a change that contributes to enhanced reproduction in an organism in the short term, it also may constrain future possibilities. Restrictions occur because adaptive changes may decrease the potential to respond to other environmental challenges. The restrictive nature of several adaptations are illustrated here.

1. Some food resources, most notably the photosynthetic and woody parts of plants, require such radical morphological and physiological changes from a middle-of-the-road omnivorous diet (Section 12.6) that a specialized consumption of plants precludes the exploitation of other food resources. In the evolution of mammals from nontherian synapsids, clades of specialized herbivores always were evolutionary dead-ends (Hopson 1969).

2. Some organisms responded to environmental conditions by greatly reducing various structures or functions, which in turn often restricts these species to a narrow set of conditions. Such organisms include flightless birds, which are usually restricted to landmasses without eutherian predators (see Section 9.6.9); lungless salamanders (Plethodontidae), which cannot live in warm, stagnant bodies of water, especially in combination with a large body size (see Section 8.6.3); and hemoglobinless fish (Channichthyidae), which are restricted to nearly freezing water (see Sections 8.4.8 and 14.3.1). An extreme example of an evolutionary

cul-de-sac is found in the East African naked mole-rat (*Heterocephalus glaber*), which combines a very low rate of metabolism with a high thermal conductance and a small mass, producing such poor temperature regulation as to limit its geographic distribution (see Section 14.5.5). A significant decrease in rate of metabolism in many organisms reduces their ability to commit resources to reproduction and growth (see Sections 13.3 and 13.7).

3. Modern amphibians use their skin as an important organ for gas exchange in air, especially at low environmental temperatures (see Section 8.6.1). Dependence on integumental gas exchange is permissible only in species that are small (when total gas exchange is low and the surface area, relative to oxygen demand, is high) and live in moist microclimates (when rate of integumental water loss is low). Amphibians that live in hot, arid climates radically modify their behavior, often spending much of the time in aestivation (see Section 6.10.5), thereby evading the worst conditions in the macroclimate.

1.4.2 Expansive adaptations. Some adaptations, although a response to a particular challenge, appear in retrospect to open opportunities for new modes of existence. These are what Brown (1958) called "general adaptations," although they are not "general" in the sense of being a nonspecific response to the environment, but only in the sense of having widespread consequences. Whether a particular adaptation is likely to lead to new opportunities depends on the environmental context. Several examples can be cited.

1. The use by crossopterygians of lobed fins as an aid in movement within stagnant ponds permitted, in conjunction with other changes, the invasion of terrestrial environments: lobed fins were not an adaptation *designed* to move into terrestrial environments, but they *permitted* such movement.

2. The repeated evolution in vertebrates of homeothermy permitted vertebrates to undergo a rapid and extensive radiation into many heretofore unavailable niches, especially in cool to cold climates; such vertebrates include, at least, mammals and birds, and probably a few fishes, a marine turtle, some therapsids, saurischian and ornithischian dinosaurs, pterosaurs, and Mesozoic marine reptiles (see Sections 4.15 and 5.10.3).

3. The development by eutherian mammals of

the trophoblast from fetal and maternal tissues reduced the likelihood of immunological rejection of a fetus by the mother, while permitting high rates of gas and material exchange between fetus and mother and as a result, high fetal growth rates and extended fetal development in the uterus (see Section 13.4).

1.5 LIMITS TO ADAPTATION

Although a modification in physiology may facilitate survivorship and reproduction, such modifications may be limited. Many limits to adaptation may exist, including those that have functional, ecological, and historical bases.

1.5.1 Functional limits to adaptation. Each character of organisms is often treated as if it were determined independently of all others. On the contrary, few characters of biological importance are completely independent. For example, the ability to describe the cost of physiological regulation by an equation means that at most this system has $n - 1$ degrees of freedom, where n is the number of variable parameters in the equation: that is the essence of all equations. Far fewer degrees of freedom may actually be present. As we will see in Chapter 5, body temperature cannot vary independently of the rate of metabolism, thermal conductance, and air temperature: mammals and birds with low body temperatures normally have low rates of metabolism. Variation in these factors are intelligible only with regard to each other (see Section 5.6), and all vary in relation to body mass. In a similar manner, the amount of oxygen transported by blood depends on heart rate, the differential in oxygen loading and unloading tensions, stroke volume, and hematocrit (see Section 8.4.8). If the hematocrit is zero, as in ice-fish, oxygen transport can be maintained constant at a fixed oxygen differential only if the heart rate or stroke volume increases; otherwise, oxygen transport and rate of metabolism fall. To ask what determines body temperature or heart rate alone is to take these parameters out of the context in which they attain significance.

1.5.2 Ecological limits to adaptation. All modifications that are physiologically possible are not ecologically acceptable: what is acceptable in one environment may be unacceptable in another. A high rate of metabolism coupled with a thick fur coat and a large body mass in mammals would lead to overheating in warm, humid environments, but this combination is appropriate in a polar setting. Vertebrates living in waters high in dissolved carbon dioxide cannot use a respiratory pigment with a large Bohr shift for aquatic gas exchange because it would prevent significant amounts of oxygen from being transported in the blood (see Section 8.4.4), but fish living in waters with low carbon dioxide concentrations can use a large Bohr shift to unload oxygen during activity.

Another "ecological" limitation to adaptation is that the cost of some adjustments cannot be borne, given conditions in the environment. For example, bats and birds that feed exclusively on flying insects cannot remain active during winter in a cold-temperate environment because flying insects are not available. Furthermore, insectivorous bats, even in the tropics, do not generally maintain body temperatures and rates of metabolism as high as nectarivorous, frugivorous, or carnivorous species. Low rates appear to be imposed on insectivorous bats by a reduced availability of insects during a dry season. Similarly, saltwater and euryhaline teleosts maintain higher plasma concentrations in salt water than freshwater or than euryhaline teleosts do in fresh water (see Section 6.5.3). These plasma concentrations reduce the osmotic differential between the blood and the environment, thereby potentially reducing the cost of osmotic regulation, which may be required by the amount of energy that is dependably available.

Adaptation appears to be a graded response to conditions in the environment. Miller (1963), considering the desert adaptation of birds, emphasized that adaptation, even to extreme environmental conditions, usually occurs by the accumulation of many small quantitative responses, rather than by the acquisition of qualitatively distinct, extreme modifications of normal form and function. Physiologists too often are preoccupied with a search for a spectacular solution to a challenge, thereby neglecting the subtle, quantitative nature of most adaptations. Population genetics teaches that a particular character state may be selected, even if its fitness is only slightly greater than the fitness of some alternate state.

1.5.3 Historical (or evolutionary) limits to adaptation. Many biologists believe that phyletic (or historical) limits to adaptation also are important: mammals and birds do not breathe water and

turtles do not fly. It is one thing, however, to state that turtles do not fly and another to argue that turtles and their potential derivatives are constitutionally unable to evolve the capacity for flight. As Jacob (1977, p. 1166) so clearly wrote:

> Evolution does not produce novelties from scratch. It works on what already exists, either transforming a system to give it new functions or combining several systems to produce a more elaborate one. . . . It is always a matter of tinkering. . . . Yet living organisms are historical structures: literally creations of history. They represent, not a perfect product of engineering, but a patchwork of odd sets pieced together when and where opportunities arose. For the opportunism of natural selection is not simply a matter of indifference to the structure and operation of its products. It reflects the very nature of a historical process full of contingency.

The view that history limits adaptation implies that many or most characters are sufficiently conservative as not to depart from earlier norms, a conclusion that Westoby et al. (1995) noted should not be accepted without extensive evidence. As we know, many characters, especially those that are the concern of this book, are labile. The presence of endothermy in some insects, reptiles, fish, and even plants, as well as in birds and mammals, demonstrates the phyletic lability of energy expenditure and temperature regulation. Presumably, any (multicellular?) organism can be endothermic if it would gain a selective advantage from endothermy and if its cost is not prohibitive. Mammals do not breathe water and turtles do not fly possibly because no long-term selective gains are associated with these capacities, or because the commitments that mammals and turtles have already made have placed functional limits on them. Thus, gills in mammals possibly cannot be designed with an area for gas exchange sufficient to compensate for the low oxygen tensions in water and to meet the very high rates of gas exchange required by endotherms.

Although biochemists and regulatory physiologists often describe elegant mechanisms of regulation, they rarely show that an organism does not do something because it is prevented from doing so. Most functional limitations can be evaded by "tinkering" in the sense of Jacob: limitations ascribed to phylogeny upon detailed examination are often likely to be ecological limitations. *Evasion of functional limits is the name of the biological game.* Many taxonomic patterns in the distribution of physiological functions may result from ecological similarities among closely related organisms (Westoby et al. 1995). Again, cladistics may permit the separation of the ecological bases of adaptation from the historical consequences of phyletic legacy. Clearly, no sharp distinctions occur among functional, ecological, and historical limits to adaptation.

A particularly interesting example of the interaction of functional, ecological, and historical factors is found in the interdependence of reproduction and energetics in mammals. In eutherian mammals, growth rate increases with an increase in the basal rate of metabolism (see Section 13.3.2), whereas it shows little correlation with the basal rate in marsupials (see Section 13.5). This difference between eutherians and marsupials may be associated with differences in their means of reproduction. Eutherians and marsupials were derived from a common ancestor that is thought to have had a form of reproduction similar to that found in living marsupials (Lillegraven 1976, Lillegraven et al. 1987). In a sense, eutherians are therians that were freed from marsupial-like limits to reproduction, thereby expanding ecological opportunity: eutherians were able to exploit modes of life that required specialized use of the forelimbs, as is found in bats, ungulates, and whales, unlike marsupials, which require the forelimbs to be unspecialized to facilitate movement to a nipple at the time of birth (Lillegraven 1976, Lillegraven et al. 1987).

1.6 CHAPTER SUMMARY

1. Adaptations are modifications in organisms that increase reproduction, and therefore survivorship, in a particular environment.

2. A hierarchy of physiological responses to variation in the environment may exist, including responses that are short term, seasonal, and long term (climatic), all to some extent involving genetic changes.

3. Several potential solutions to an environmental challenge normally are available.

4. All solutions have side effects.

5. Some adaptations constrain future possibilities, whereas others open new opportunities.

6. Adaptation can be limited by the interaction of factors involved in the mechanisms of adaptation, by ecological opportunities, and possibly by the set of responses accumulated through time.

2   Patterns of Physical Exchange
with the Environment

2.1 SYNOPSIS

In this chapter the physical bases of material and energy exchange between organisms and the environment are described by equations that are used in later chapters to analyze the behavior of regulatory systems. Here biologically relevant examples are cited to illustrate particular principles of physics, or of their misunderstanding. First, the fundamental process of diffusion is discussed and the widely used Fick's law derived. Then the basic components of thermal exchange (radiation, conduction, convection, and evaporation) are described, after which they are unified in models to show the net energy exchange between an organism and the environment. The different models reflect varying complexities in the physical environment. Then osmotic exchange with the environment is discussed. Finally, the limits to gas exchange are described.

2.2 INTRODUCTION

All organisms are complex systems open to matter and energy. Material and energy exchange with the environment is described by the laws of physics. Under extreme environmental conditions these laws may limit the ability of organisms to regulate their internal physicochemical states. Examples of the biological application of these physical principles are examined in this chapter to clarify the physical limitations potentially faced by organisms, although many species evade simple physical limits through the use of behavior and a modification of complex physiological functions. Detailed physiological responses to physical condi-

tions in the environment are covered extensively in Chapters 4 through 8.

The most effective way of representing physical rules is to express them as equations. The power of equations is in their organization of our thoughts and in their application. Their object is to facilitate the analysis of biological phenomena: ignorance of physics impedes the understanding of ecology generally and physiological ecology specifically. (Note: equations are numbered if they are referred to later, but if an equation is given simply to make a point, or if it is described only as a step to obtain a more useful equation, it is not numbered.)

2.3 A NOTE ON UNITS

The International System of Units (SI) is used in this book with a few exceptions, all associated with reducing the use of multiplying factors ($\times 10^n$). One of these is for time, which in terms of SI units is a second (s); power thus is in units of watts, which equals joules per second (J/s). However, a second is not a meaningful unit of time for vertebrates. Instead, power is in units of joules per hour (J/h) when dealing with rate of metabolism, or kilojoules per day (kJ/d) when concerned with energy budgets. Velocities are given as meters per second (m/s) or as kilometers per hour (km/h). The SI unit for mass is kilogram (kg), whereas here the principal unit used is gram (g), mainly because most studied vertebrates weigh less than 100 g. Although the SI unit for temperature is K (absolute temperature), here temperature is usually expressed as °C, except when absolute temperatures are required (e.g., in radiation). Another SI unit that has been abandoned is the unit for pressure, pascals. Here pressure is represented by millimeters of mercury (mm Hg), which is what is usually measured and which avoids the factor 7.5×10^{-3}. (Torr is another unit of pressure; values in torr are equivalent to values in mm Hg.) In other words, the units used are those that are simple and easy to convert to SI units, if that is desired.

2.4 DIFFUSION

Diffusion, the physical process by which matter or heat moves from a region of high concentration to a region of low concentration, is one of the most ubiquitous of physical processes. It underlies many other processes, such as thermal conduction, thermal convection, evaporation, osmosis, and gas exchange. Diffusion may occur in solution, in air, or through porous solids; it may involve particles, atoms, molecules, or gases; and it may describe energy transfer by means of molecular collision.

The movement of matter by diffusion depends in part on the concentration gradient, dC/dX, and on the electrical charge (if any) of solutes, if diffusion occurs in an aqueous solution. In a solution, net solute flux ($\dot{u}$) is given by

$$\dot{u} = -AD\left[\left(\frac{dC}{dX}\right) + \left(\frac{zF}{RT}\right)C\left(\frac{dE}{dX}\right)\right], \quad (2.1)$$

where A is the surface area (cm^2); D is the diffusion coefficient; z is the charge on the solute; F, the Faraday constant (96,500 coulombs/mole); R, the universal gas constant (0.082 l·atm/[K·mole]); T, absolute temperature (K = °C + 273.16); and dE/dX is the electrical gradient (in V/cm) between the two solutions (Daniels and Alberty 1975). If the solute is uncharged, Equation 2.1 becomes

$$\dot{u} = -AD\left(\frac{dC}{dX}\right), \quad (2.2)$$

which is usually referred to as Fick's law. The diffusion coefficient D varies with the matter diffusing and the medium in which diffusion occurs; it has the units (cm^3·cm/[cm^2·s]), which is usually simplified to (cm^2/s).

2.5 THERMAL EXCHANGE—RADIATION

Radiation is the process by which energy is exchanged between two bodies at a distance by means of electromagnetic waves or particles

(photons, a mole of which is an einstein) that travel through space with the speed of light. Radiant energy can be characterized by the energy content of the particles, which is inversely related to the photon's wavelength. In the visible spectrum the energy content varies from 300 kJ/einstein at 0.4 μm (violet) to 171 kJ/einstein at 0.7 μm (red). When these waves or particles are not transparent to a body, impinging energy either is reflected or is absorbed and converted to heat.

All bodies at temperatures greater than 0 K radiate energy. As the temperature of a body increases, the total amount of energy radiated increases. This rule is called the *Stefan-Boltzmann law*. It can be written as

$$\dot{Q} = \sigma A T^4, \quad (2.3)$$

where $\dot{Q}$ is the amount of energy radiated in joules per hour, σ (Stefan-Boltzmann constant) equals 2.04×10^{-8} J/(cm^2h K^4), A is surface area of the body in square centimeters (cm^2), and T is its surface temperature in Kelvin. A body conforming to Equation 2.3 is called a *black body*.

Many real bodies emit less energy than is described by Equation 2.3, so a dimensionless quantity, called *emissivity*, is used to describe the degree to which a particular body conforms to black-body radiation. Emissivity (ε) varies from 0.0 to 1.0 and

$$\dot{Q} = \varepsilon \sigma A T^4. \quad (2.4)$$

If a body had an emissivity of 0.0, it would radiate no energy, and if it had an emissivity of 1.0, it would radiate the maximal amount of energy dictated by temperature (i.e., it would be a black body). Emissivity varies with the wavelength of the radiation—so knowledge of the emissivity of a body at one wavelength does not indicate the emissivity of the same body at some other wavelength. *Kirchhoff's law* says that a good radiator at a particular wavelength is also a good absorber at the same wavelength.

The net radiational exchange between two bodies is given by

$$\dot{Q} = \bar{\varepsilon}_2 \sigma A_2 T_2^4 - \bar{\varepsilon}_1 \sigma A_1 T_1^4,$$

a relation in which the emissivities $\bar{\varepsilon}_1$ and $\bar{\varepsilon}_2$ are means over a particular band of radiation. Or, if a small warm body with a surface temperature equal to T_s and a mean emissivity equal to $\bar{\varepsilon}$ is immersed in a large field penetrated by black-body radiation at ambient temperature T_a, then

$$\dot{Q} = \bar{\varepsilon}\sigma A(T_s^4 - T_a^4), \qquad (2.5)$$

in which A is the surface area of the body. As long as a temperature differential exists between two bodies, or between a body and its environment, a net radiative exchange of energy will proceed from the warmer to the cooler "body."

Radiation can be examined on Earth and in its environment (see, e.g., Gates 1962, 1980). The spectral distribution of energy radiated as a function of wavelength is given for the sun (Figure 2.1) and for Earth's surface and atmosphere (Figure 2.2). These "bodies" have the near equivalency of

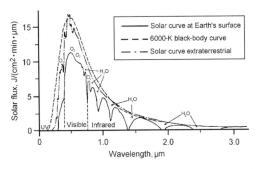

Figure 2.1 Solar flux as a function of wavelength at the surface of Earth, at the outer layer of the atmosphere, and as expected from an ideal body with a temperature of 6000 K. Absorption bands for UV, O_3, O_2, and water are indicated. Source: Modified from Gates (1962).

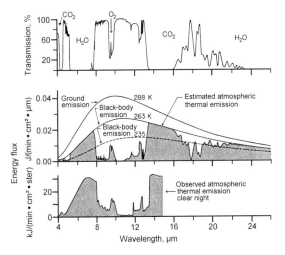

Figure 2.2 Energy flux and transmission as a function of wavelength in the atmosphere and at Earth's surface. Absorption bands for CO_2, O_2, and water in the atmosphere are indicated. Source: Modified from Gates (1962).

black-body radiation at 6000, 288, and 263 K, respectively. With an increase in the temperature of a body, the curve shifts to shorter wavelengths (*Wien's law*) and the area under the curve, which equals the amount of energy radiated, increases (Stefan-Boltzmann law). As a consequence, the curve for the sun encloses that of Earth, which in turn encloses that of the atmosphere, because of the differences in their mean temperatures.

The amount of solar flux reaching Earth's surface is attenuated by the atmosphere (Figure 2.1) because some of the impinging energy is either absorbed or reflected by the atmosphere. In fact, only slightly more than one-half of the quantity of energy impinging on the atmosphere reaches the surface of Earth. The attenuation of energy does not occur uniformly over the wavelength spectrum, but within specific spectral bands (Figure 2.1). These bands correspond to the absorption spectra of the molecules constituting the atmosphere. Ozone, oxygen, and water have the most important absorption bands for radiation at wavelengths from 0.1 to 3.0 μm, the range that includes the visible spectrum. The angle at which the sun's rays hit the atmosphere also dictates the proportion of the energy reflected: the lower the angle of the sun, the smaller the amount of energy impinging Earth's surface. This effect itself is affected by wavelength, as can be seen at sunset.

The atmosphere also has an effect on the transmission of energy in the infrared spectrum (Figure 2.2). Carbon dioxide (CO_2), water, and ozone absorb energy at wavelengths between 4.0 and 26.0 μm. For example, almost all of the energy radiated, or reflected, by the ground between 5.0 and 8.0 μm is absorbed by water in the atmosphere; that between 13.0 and 17.0 μm by CO_2; and that between 19.0 and 26.0 μm, by water. Most of the energy emitted, or reflected, at wavelengths between 8.0 and 13.0 and between 17.0 and 19.0 μm by the ground is unaffected by the presence of the atmosphere; these wavelength ranges are said to constitute atmospheric "windows." The absolute "size" of the window depends on the amounts of CO_2 and water in the atmosphere.

Because much of the energy in the visible spectrum freely moves through the atmosphere and is absorbed by the earth's surface, the earth's surface increases in temperature. Therefore, the surface radiates in the far infrared band, but not in the visible spectrum because it is far too cool, although it reflects some energy in the visible spectrum and therefore the earth can be "seen." Earth's radiation that is absorbed by the atmosphere is reradiated to the earth as well as to space. Because the earth absorbs nearly all of the impinging radiation that it does not reflect (an exception being cosmic radiation, which passes through the earth) and because much of the impinging radiation is transmitted by the atmosphere, the earth has a higher temperature than the atmosphere, which in turn has a higher temperature than space, differences that reflect the densities of these "bodies." The storage of heat on the earth resulting from the differential atmospheric transmission of energy is called the *greenhouse effect*, which, as the name implies, is similar to the thermal effect produced by the differential transmission of energy through glass. Glass has a high transmission in the visible wavelengths and a reduced transmission in the infrared wavelengths, which is why greenhouse atmospheres tend to be warmer than ambient temperatures when the sun is shining.

The mean global surface temperature of Earth has increased 0.3°C since 1975 and 0.7°C since 1910, although the latter increase included a stable period between 1950 and 1975 (Figure 2.3A). The increase since 1975 is correlated with an increase in atmospheric concentrations of methane and CO_2 (Figure 2.3B), molecules that absorb energy in the infrared spectrum and permit energy in the visible spectrum to reach the earth's surface. Whether the increase in global temperature results from a natural process or whether it reflects the human-induced increase in greenhouse gases is unclear (Mason 1995, Hayward 1997). Wandering ozone "holes" also have developed during winter over the South and North Poles; they permit a regional increase in ultraviolet radiation. An increase in the atmospheric concentration of chlorofluorohydrocarbons has been suggested to be responsible for the development of the polar ozone holes. These changes in the composition of the atmosphere may lead to a change in the climate of Earth, which, if continued, will have a striking impact on the survival and distribution of vertebrates (see Section 14.9).

Water in the atmosphere shows short-term and topographic variations in abundance. Most atmospheric water occurs near the ground because water vapor is derived from evaporation and because the density of the atmosphere is greatest at the earth's surface. An increase in the amount of atmospheric water intensifies the absorption of radiation between atmospheric windows and encroaches on the

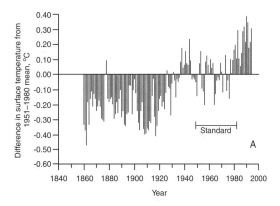

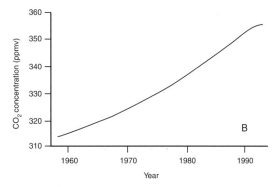

Figure 2.3 A. The difference in the combined land, air, and sea surface temperatures of Earth between 1860 and 1995, using the average of 1951–1980 as the standard. B. Atmospheric CO_2 content at the Mauna Loa Observatory, Hawaii, between 1958 and 1994. Source: Modified from Mason (1995).

edge of the windows, thereby decreasing the amount of energy lost directly to space by radiation from the ground. Deserts generally have more variable temperature regimes than other environments because they have dry air and large atmospheric windows, which means that a greater fraction of solar radiation reaches the ground during the day and that a greater fraction of ground emission is lost directly to space at night.

One of the clearest examples of how the laws of radiation have been misapplied in biology involves the dependence of emissivity on wavelength. Hesse et al. (1951, p. 616) argued that

> The warm-blooded animals of polar-regions contrast . . . with . . . poikilothermic forms in their pale or pure white coloration. . . . The poikilothermic forms are almost all dark, and thus absorb the greatest amount of heat during the brief period of their activity. *The white coloration of the homeotherms radiates less heat than the dark*, and prevention of heat loss is evidently of

greater importance than the absorption of the relatively small amounts of heat received from the sun. . . . (italics added)

The suggestion that the white color of homeotherms is a means of reducing the radiation of heat is incorrect because it erroneously assumes that an animal with an emissivity near zero in the visible spectrum (i.e., it appears white) has a similar emissivity in the infrared spectrum at which the homeotherm would be radiating. Indeed, the emissivities of Arctic mammals and birds in the infrared vary between 0.95 and 1.00, irrespective of coat color (Hammel 1956), which is compatible with observations that the heat loss of mammals with light and dark pelages is identical (Silva 1956). The color of an animal, therefore, has nothing to do with its rate of heat *loss*.

The color of animals may have an appreciable effect on the rate of heat *gain* because a large fraction of solar radiation is in the visible spectrum. Thus, poikilotherms increase heat gain by being dark (see Section 4.9.3). Studies of the energetics of endotherms also suggest that color may be thermally important. Dark birds, for example, reflect less (absorb more) energy in the visible and near-infrared wavelengths than do light-colored birds (Figure 2.4). This differential pattern of absorption produces higher surface temperatures in darker than in lighter mammals and birds (see Section 5.7.5). Øritsland (1970) showed in seals that dark fur absorbs much of the infrared radiation in the superficial layers, but light-colored fur permits a

deeper penetration of radiation, which leads to higher skin temperatures when seals are exposed to solar radiation. Hutchinson and Brown (1969) also found that the penetrance of radiation is greater in white than dark fur coats. In a comparison of two desert ground-squirrels, *Spermophilus teretricaudus* and *Ammospermophilus harrisii*, both of which have heavily pigmented dorsal skin, Walsberg (1988) demonstrated that the physical properties of the fur coat can greatly modify solar heat gain, but that the pigmented skin layer underneath the fur coat principally reduces ultraviolet transmission to deeper tissues.

Microclimate can greatly modify the radiant heat load. For example, Dawson (1972) showed in the central Australian desert that even small trees significantly reduced solar influx, although they had little effect on the long-infrared influx. At midday, when the shaded air temperature was 37.7°C, the highest radiant temperature in the sun was 78.3°C, whereas that in the shade varied between 52.5 and 57.6°C. In contrast, the radiant flux from the open sky away from the sun was 110 J/(cm²h), which if black-body radiation is equivalent to a radiant temperature of −2.2°C:

$$\dot{Q}/A = \sigma T^4$$

$$110 = (2.04 \times 10^{-8})T^4$$

$$T = 271.0\text{K} = 271.0 - 273.2 = -2.2°\text{C}.$$

The low sky temperature is due to the low water content of the desert atmosphere and consequently, a radiant flux dictated by large windows in the atmosphere and a low atmospheric temperature. In a similar manner, the Cape ground-squirrel (*Xerus inaurus*), a diurnally active resident of arid southern Africa, reduces the radiant heat load by positioning its tail above its back (Bennett et al. 1984). This behavior may reduce the radiant temperature by 5°C; the use of the tail as a parasol permits this squirrel to be active for much longer periods (7 h) than would be the case if the tail were not used (ca. 3 h). When the radiant temperature in the environment is low (i.e., early in the morning and late in the afternoon), free-living squirrels do not erect their tails.

Calder (1973a, 1974b) demonstrated the selective use of radiation in a study of the positioning of hummingbird nests. Incubating females at high elevations in summer may be exposed to radiant sky temperatures at night as low as −20°C (Calder

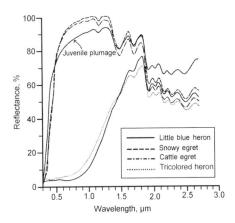

Figure 2.4 Energy reflectance as a function of wavelength in the plumage of four herons, little blue heron (*Egretta caerulea*), snowy egret (*E. thula*), tricolored heron (*E. tricolor*), and cattle egret (*Bubulcus ibis*), including juvenile (white) plumage of the little blue heron. Source: Modified from Ellis (1980).

1974b). Yet, the total estimated rate of energy expenditure for thermoregulation needed by the hummingbirds was within basal expenditures because the heavily insulated nest obscured three-fourths of an incubating female's surface and because the nest was placed under branches, thereby blocking radiation from its exposed back to the open sky. The predawn radiant temperatures of the branches varied from 3.5 to 6.5°C, some 25°C above those of the open sky.

A simple demonstration of radiative exchange is shown by the equilibrial body temperatures attained by cave bats (*Pipistrellus subflavus*) in torpor (McNab 1974a). The mean body temperature (T_b) of individuals hanging over a hay-covered floor was 19.5°C and of those roosting over a bare floor was 18.4°C. The radiant temperature of the hay was 19.0°C, that of the bare floor was 17.6°C, and the air temperature throughout the cave was 19.8°C. Body temperature here was determined principally by the radiant temperature of the surface the bats faced. Furthermore, water condensed on the fur of torpid bats on a side that faced a nearby wall because the humidity of the cave was 100%, the wall was 1.0°C or more below air temperature, and the tips of the bat's fur were in radiation equilibrium with the wall. Standing water in these caves, however, was warmer than cave atmospheric temperature. Consequently, water did not condense on the side of the torpid bat that faced a pond.

2.6 THERMAL EXCHANGE—CONDUCTION

Conduction is the process by which heat is transferred from one molecule to another without material exchange. It is an important means of heat transfer within bodies, depending on their thermal conductivity, and between bodies, depending on their area of contact.

For most terrestrial tetrapods conduction is the least important means of thermal exchange with the environment, mainly because the principal areas of contact with the environment are the small areas upon which tetrapods stand. At times of rest the area of contact and conductive exchange may be appreciably greater. Conductive heat transfer also occurs between a body and the boundary layer of the atmosphere at the solid-fluid interface, but heat transfer here is normally determined by the bulk movement of molecules in the fluid beyond the boundary layer, and thus is covered in the next section, on convection.

Conductive exchange is important in aquatic and burrowing vertebrates because of the large surface areas in intimate molecular contact with the environment. Such is the case in fish, many amphibians, and most snakes, although, again, most heat exchange in an aquatic environment is due to convective exchange because water is rarely still at the surface of a body. The conduction of heat across the surface is also much more important in aquatic than aerial environments because water is denser and has more intermolecular interactions than air (i.e., it has a higher heat capacity and thermal conductivity than air).

Thermal conduction can be described by *Fick's law of diffusion* (Equation 2.2): it varies with the surface area of contact between two bodies, the temperature gradient between the bodies, and a constant (thermal conductivity, k) that measures the ease of transfer of heat through a body:

$$\dot{Q} = -kA\left(\frac{T_2 - T_1}{X_2 - X_1}\right), \qquad (2.6)$$

where $(T_2 - T_1)/(X_2 - X_1)$ is the thermal gradient and X_1 and X_2 represent positions along a gradient. When written as a partial differential $\partial Q/\partial t = -k(\partial T/\partial X)$, Equation 2.6 is called *Fourier's law*. As in diffusion, heat moves from a region of high temperature to one of low temperature. Thermal conductivity is a constant reflecting the molecular makeup of the body, the units for which are J·cm/(cm²h°C), which are often reduced to J/(cm·h°C). Water has a conductivity that is about 23 times that of air; fat, a conductivity 8 times that of air; and fur pelts have a conductivity 0.9 to 2.4 times that of air (Gates 1980).

Few available studies demonstrate the importance of thermal conduction to animals. Cole (1943) showed that the temperature of the substrate (in the absence of a radiation load) is much more important in dictating the temperature of a lizard than is air temperature. Norris (1953) observed that desert, sand-dwelling lizards, such as the desert iguana *Dipsosaurus dorsalis*, often shuffle their feet and, after placing their body against the sand, move side to side when the sand is hot during midday. These behaviors cause them to break through the hot surface to cooler layers of sand, thereby reducing the conductive gain of heat.

The Andean lizard *Liolaemus multiformis* often basks exposed to the sun while lying flat on plant material, which greatly reduces the conduction of heat away from the animal. This behavior is especially important for the conservation of heat in a lizard living at an altitude of 4300 m (Pearson 1977).

2.7 THERMAL EXCHANGE— CONVECTION

Convective transfer of heat entails the bulk movement of heated molecules, and as such occurs only in fluids (gases and liquids), especially at an interface with a solid. This form of heat exchange is a function of several parameters, including surface area and the temperature differential between the fluid and the surface of the solid:

$$\dot{Q} = h_c A(T_s - T_a), \qquad (2.7)$$

where h_c is the convective coefficient, which has the units J/(cm²h°C).

Unlike thermal conductivity, which is a constant for particular materials, the convective coefficient is determined in an exceedingly complex manner (Gates 1962, 1980). It takes one of three forms, each of which corresponds to a unique set of physical conditions.

1. In the nearly stagnant layer of fluid adjacent to a solid, heat is transferred by conduction. Within this "boundary layer" h_c is a function of the thermal conductivity (k) of the fluid and the thickness of the boundary layer (d): $h_c = k/d$. The thickness of the boundary layer, which is but a few millimeters at most, depends on the size of the solid and on the velocity of the fluid beyond the layer.

2. Heat is transferred from (or to) the boundary layer by "free" or "natural" convection due to differences in the density of the fluid produced by temperature differences existing in the fluid. Under these conditions the convective coefficient depends on the temperature differential, the "characteristic dimension" of a solid (a linear measure of the size of an object, such as the diameter of a cylinder or the length of a plate), and certain dimensionless variables, such as the Nusselt, Prandtl, and Grashof numbers. If the movement in the fluid beyond the boundary layer is turbulent, rather than laminar, the convective coefficients are somewhat greater.

3. "Forced" convection depends on fluid movement produced by an external force, such as a current in water or wind in air. It is described by a dimensionless variable called the *Reynolds number*, which combines all physical influences on the thickness of the boundary layer (fluid velocity, viscosity, and density, as well as the dimension of the object). For various solids and fluids, the convective coefficient is usually estimated by a series of empirically derived approximations.

The importance of wind velocity to heat loss has been shown in mammals and birds. The form of the convective coefficient has been the source of disagreement. For example, Tregear (1965) showed in a series of mammal pelts that the thermal conductance was proportional (approximately) to $\dot{v}^{0.25}$, where $\dot{v}$ is wind velocity; he also showed that a dense fur coat greatly reduced the penetration of wind into the still layer of air of the coat. As a consequence, the effective thickness of the coat increased with an increase in fur density. In a study of convective heat loss from models of chipmunks covered with fur pelts, Heller (1972), however, found that the convective coefficient was proportional to $\dot{v}^{1.0}$. Tregear's and Heller's conclusions might be doubted because they were based on skins or models.

Rate of metabolism, which is proportional to heat loss when body temperature is constant, was proportional to $\dot{v}^{0.50}$ (Figure 2.5) in the white-crowned sparrow (*Zonotrichia leucophrys* [Robinson et al. 1976]), snowy owl (*Nyctea scandiaca* [Gessaman 1972]), and verdin (*Auriparus flaviceps*

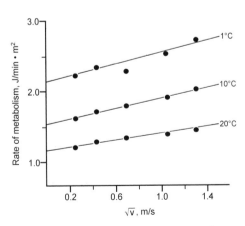

Figure 2.5 Rate of metabolism in white-crowned sparrows (*Zonotrichia leucophrys*) as a function of air temperature and the square-root of wind velocity. Source: Modified from Robinson et al. (1976).

[Webster and Weathers 1988]) in agreement with the observations of Church (1960) on heat loss in moths. Studies of convective heat loss from the ears of rabbits also suggest that h_c is approximately proportional to $\dot{v}^{0.50}$ (Wathen et al. 1971). Goldstein (1983), however, found that rate of metabolism was proportional to $\dot{v}^{1.0}$ in Gambel's quail (*Callipepla gambelii*) and noted that the recalculation of earlier data, including those from *Z. leucophrys*, was not proportional to $\dot{v}^{0.50}$. He nevertheless concluded that the most effective expression for the effect of wind on h_c was proportional to $\dot{v}^{0.50}$. The coefficient also depends on wind direction; the size, shape, and orientation of the body being studied; and ambient temperature.

The comparative importance of convective cooling can be shown in several examples. First, Marder (1973) studied the absorption of radiation by ravens (*Corvus corax*) under conditions similar to those existing in the Negev desert region. Superficial feather temperatures rose to 90 to 100°C in still air at a temperature of 28°C. With air movement the surface temperature of feathers on the raven's back fell, depending on wind velocity (Figure 2.6). If the decrease in surface temperature is assumed to be a linear function of wind speed (which is unlikely at high velocities), then a velocity of at least 8.3 m/s (30 km/h) would be required for the surface temperature to equal 28°C. Flight speeds of corvids often vary between 50 and 70 km/h, which should be sufficient to dissipate by convection all of the radiant heat load and much of the heat produced by metabolism. Ravens often use air currents to soar, thereby minimizing heat input from the work accomplished during flight.

Walsberg et al. (1978) developed a model of radiative heat gain in relation to plumage color, feather position, and wind velocity. Under extreme desert conditions ($T_b - T_a = -10°C$, solar flux = 360 J/[cm²h], radiant temperature = 91.5°C) the model suggests that wind velocity has little effect on heat gain by white plumage but greatly reduces heat gain by black plumage (Figure 2.7). If black plumage is erected (which increases the thermal resistance of the plumage), heat gain falls below that of white plumage when wind velocity is greater than 11 km/h (3.1 m/s) because of the penetration of white plumage by radiation. If, however, black plumage is compressed, the absorption and transfer of heat does not fall below that of the white plumage. Therefore, dark plumage in the desert may facilitate convective heat loss if (1) the wind velocity is high and (2) the plumage is ptiloerected.

Unlike most gulls, three species in the tropical eastern Pacific are dark as adults: Heermann's gull (*Larus heermanni*) from Baja California, lava gull (*L. fuliginosus*) from the Galápagos Islands, and gray gull (*L. modestus*) from the Atacama Desert of Chile. The position of the scapulars and wings, which can facilitate convective cooling in Heermann's gull during incubation, is strongly correlated with substrate temperature (Bartholomew and Dawson 1979). Unfortunately, no direct measurements of energetics have been made in these species, but Ellis (1985) showed that dark-colored seabirds, including terns, generally have lower rates

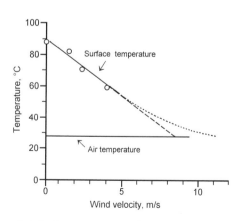

Figure 2.6 Surface temperature of ravens (*Corvus corax*) as a function of wind velocity when exposed to a 300-W tungsten lamp. Source: Modified from Marder (1973).

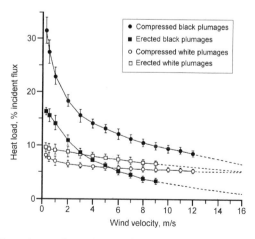

Figure 2.7 Radiative heat load as a function of wind velocity in black and white pigeons (*Columba livia*), depending on whether their plumage is erect or compressed. Source: Modified from Walsberg et al. (1978).

of metabolism than light-colored species, which raises the possibility either that a dark color might compensate for a low rate of metabolism or that a low rate of metabolism might compensate for a dark color (see Ohmart and Lasiewski [1971] on the energetics of the roadrunner [*Geococcyx californicus*], or Hennemann [1983b] on the anhinga [*Anhinga anhinga*]).

Chappell et al. (1989) showed the differential impact of convective cooling in the Adélie penguin (*Pygoscelis adeliae*) and the blue-eyed shag (*Phalacrocorax atriceps*) in Antarctica. At 20°C and a low wind velocity (2.6 m/s = 9.4 km/h), the smaller (2.6-kg) shag had a total rate of metabolism that was 14% greater than the rate in the larger (4.0-kg) penguin. When air temperature fell to −20°C at the same wind velocity, rate of metabolism in the penguin increased to a level that was 13% greater than that found in the shag. If at that temperature, wind velocity increased to 5.7 m/s (20.5 km/h), the rate of metabolism in the penguin was 31% greater than that in the shag, even though the shag's rate increased by 12%. The sensitivity of the Adélie penguin to convective heat loss issues from their poorly insulated flippers, whose design facilitates aquatic propulsion, not heat conservation. When the penguin is facing an ambient temperature below 0°C, heat loss through the flippers may be high, especially in the presence of relatively high wind velocities, because skin temperature in the flippers must be maintained above freezing. Consider the heat loss from the flippers when the penguin is swimming in water.

The last example concerns the thermal exchange of jackrabbits (*Lepus* spp.), which live in arid western North America. They have large ears (up to 20% of total surface area). Schmidt-Nielsen (1964), Schmidt-Nielsen et al. (1965), and Dawson and Schmidt-Nielsen (1966) suggested that the ears are a major conduit for heat dissipation by means of radiation to a cool sky. Even during midday in the Arizona desert, the radiant sky temperature (away from the sun) is usually less than 20°C. Jackrabbits spend much of the day in the shade cast by vegetation, thereby avoiding direct solar input to the ears. But the sky is only part of the radiant environment that a jackrabbit faces: the temperature of the vegetation and especially of the soil may be appreciably above that of the jackrabbit's body and ears (Wathen et al. 1971). Under these conditions jackrabbits gain heat. Heat loss, then, may depend principally on convective heat loss to the

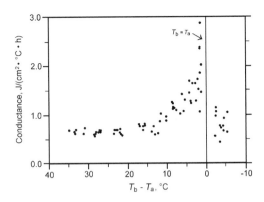

Figure 2.8 Thermal conductance in the antelope jackrabbit (*Lepus alleni*) as a function of the temperature differential between the body and environment. Source: Modified from Dawson and Schmidt-Nielsen (1966).

air. Wathen et al. (1971) estimated the heat dissipated from the ears by convection during a summer day in Arizona, assuming that ear temperature was equal to core body temperature. If the air temperature in shade is 30 to 35°C, up to 161% of the resting rate of metabolism can be dissipated through the ears by convection, depending on wind velocity. Convective loss becomes unimportant as shaded air temperature approaches 40°C, that is, as the difference between surface and air temperature (ΔT) approaches 0°C. At higher ambient temperatures the transfer of heat through the surface is generally reduced to impede a convective gain of heat by the animal (Figure 2.8). At these high environmental temperatures heat loss depends on evaporative cooling (Dawson and Schmidt-Nielsen 1966).

2.8 THERMAL EXCHANGE— EVAPORATION

The rate at which heat is transferred by evaporation equals $L\dot{E}$, where L is the latent heat of vaporization (ca. 2.40 kJ/g H_2O at 40°C) and $\dot{E}$ (g/h) is the rate at which water is vaporized. Like convection, the amount of heat transferred by evaporation varies with air movement. In still air, the rate of evaporation is limited by the rate at which water vapor diffuses to an external air mass from a boundary layer saturated with water vapor. The rate of water movement can be described by Fick's law (Equation 2.2): it depends on surface area, temperature of the surface (and boundary layer), and the differential in water vapor density between the boundary layer and external air mass:

$$\dot{E} = h_\mathrm{d} A(\rho_\mathrm{o} - \rho_\mathrm{a}), \qquad (2.8)$$

where h_d is a mass transfer coefficient; A, effective surface area; ρ_o, saturated water vapor density (g/cm^3) at surface temperature; and ρ_a, water vapor density of the air mass (g/cm^3). The saturated water vapor density increases with temperature according to the relation

$$\rho_\mathrm{o} = (9.16 \times 10^8)\,e^{-(5218/T)},$$

where e is the base of natural logrhythins, the constant 5218 is approximately H_evap/R, H_evap is the molar heat of vaporization, and R is the universal gas constant (Daniels and Alberty 1975).

If the air adjacent to the boundary layer moves, evaporation increases because convection is a faster process than diffusion. Theory suggests that rate of water loss is proportional to the square-root of wind velocity (Ramsey 1935). Evaporative heat loss, however, is difficult to separate from convective heat loss because both vary with $\dot{v}^{0.5}$. Tracy (1976) described in great detail the physical complexities associated with evaporative water loss and their implications for terrestrial amphibians.

Hutchinson (1955), followed by King and Farner (1964) and Salt (1964), described the coefficient in Equation 2.8 as a function of respiratory volume, a maneuver that converts the equation for passive evaporation into one for forced convection. Thus, Equation 2.8 becomes

$$\dot{E} = k\dot{V}A(\rho_\mathrm{o} - \rho_\mathrm{a}), \qquad (2.9)$$

where $\dot{V}$ is ventilation rate and k is the slope of the water loss–ventilation curve. In fact, rate of evaporative water loss in some lizards, birds, and mammals is proportional to the ventilation rate. Of course, many mammals also increase evaporative water loss by sweating. Equation 2.9 may not have a sound theoretical basis, but it is a convenient description of evaporative water loss, although integumental and respiratory water loss may each require a separate term.

2.9 NET THERMAL EXCHANGE WITH THE ENVIRONMENT

The net thermal exchange of resting animals with the environment is a sum of the heat exchanges by radiation, conduction, convection, and evaporation:

$$\begin{aligned}
\dot{Q}_\mathrm{net} &= \dot{Q}_\mathrm{rad} + \dot{Q}_\mathrm{cond} + \dot{Q}_\mathrm{conv} + \dot{Q}_\mathrm{evap} + (\dot{Q}_\mathrm{stor} + \dot{Q}_\mathrm{work}) \\
&= \varepsilon\sigma A_1(T_s^4 - T_a^4) - k_1 A_2(T_s - T_a) \\
&\quad + h_c A_3(T_s - T_a) + L\dot{E}, \qquad (2.10)
\end{aligned}$$

which assumes that the temperature of the animal is constant. If that is not the case, an additional term must be added for positive or negative heat storage ($\dot{Q}_\mathrm{stor}$) to represent an increasing or decreasing body temperature, respectively. Furthermore, if an animal is active, another term must be added for work done against the environment ($\dot{Q}_\mathrm{work}$). Sometimes an additional term is added for metabolism, but metabolism is already represented in Equation 2.10 by the heat exchange terms, which are the means by which the heat produced by metabolism is dissipated.

Several difficulties are associated with the use of Equation 2.10. One is that the body temperatures are surface temperatures because heat exchange with the environment occurs at the surfaces of a body, both integumental and respiratory. The problem with using surface temperatures is that (1) they are difficult to measure (compared to measuring core temperature), (2) a mosaic of surface temperatures at any particular moment may exist (see, e.g., Doncaster et al. 1990), and (3) they are not the temperatures being regulated if temperature regulation is occurring. The temperature differential between body and the environment ($T_\mathrm{b} - T_\mathrm{a}$) can be expressed as the sum of two differentials, one between core and surface and another between surface and environment:

$$T_\mathrm{b} - T_\mathrm{a} = (T_\mathrm{b} - T_\mathrm{s}) + (T_\mathrm{s} - T_\mathrm{a}).$$

The relative size of these differentials would be expected to vary with body size, as well as with environmental conditions. Veghte and Herreid (1965) showed that the differential $T_\mathrm{s} - T_\mathrm{a}$ is independent of T_a in several birds (Figure 2.9), and therefore that the differential $T_\mathrm{b} - T_\mathrm{s}$ increases with a decrease in ambient temperature. Furthermore, surface temperature is nearly equal to T_a in a large species like the raven, *C. corax* (see Figure 2.9). A small mass, as in the black-capped chickadee (*Parus atricapillus*), so restricts the thickness of the feather coat that the surface temperatures of feathers may be 10°C higher than in large species at the same air temperature. Still, the variable temperature differential is $T_\mathrm{b} - T_\mathrm{s}$, which permits T_b to replace T_s in Equation 2.10.

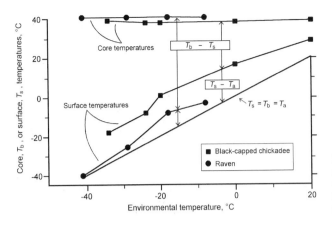

Figure 2.9 Core and surface temperatures of ravens (*Corvus corax*) and black-capped chickadees (*Parus atricapillus*) as a function of environmental temperature. Source: Derived from Veghte and Herreid (1965).

Representing the differential $T_s - T_a$ by $T_b - T_a$ is a simplification: if two species have the same $\Delta T = T_b - T_a$, but one has a higher T_s than the other, then the species with the higher T_s would lose heat at a higher rate and require a higher rate of energy expenditure; that is, it would have a higher thermal conductance (see Section 5.5.3). Although the replacement of T_s by T_b is a practical aid in analyzing energetics, especially when the internal generation of heat is important, surface temperatures cannot be neglected when an animal is subject to a radiant heat load that raises T_s above the value that is expected from heat loss at a particular T_a (see Ohmart and Lasiewski 1971).

Another problem with Equation 2.10 is that the areas A_1, A_2, and A_3, and the coefficients k_1 and h_c, not to mention the complexities incorporated into $\dot{E}$, are highly variable, and in turn are produced by changes in posture, ptilomotion or pilomotion (changes in the erection or depression of feathers or fur, respectively), peripheral circulation, and ventilation rate. Such complexity greatly reduces the value of Equation 2.10, except as a conceptual framework or as a way to estimate the comparative contributions to heat exchange made by various pathways. Because of these complexities, some simplified versions of Equation 2.10 are often used.

2.9.1 Burton and Scholander-Irving models.

A practical response to the complications in Equation 2.10 is to combine the four modes of heat exchange into two terms, one for "dry" heat exchange, which depends on a temperature differential and incorporates radiation, conduction, and convection, and another for evaporative heat loss. If core temperature is substituted for surface temperature,

which is appropriate when most of the temperature differential existing between the body and environment is found between the core and body surface, then

$$\dot{Q}_{net} = C'(T_b - T_a) + L\dot{E}, \qquad (2.11)$$

where C' is a coefficient (in J/[h°C]) called *thermal conductance* (or the coefficient of "dry" thermal exchange). Burton (1934) described Equation 2.11, which is applicable over all biologically relevant temperatures but cannot be used if wind velocity influences heat loss (see Tracy 1972), or if an appreciable external radiant heat load exists.

If the net thermal exchange between a body and the environment is studied at cool to cold temperatures, conditions in which the evaporation of water constitutes only about 10% of heat loss, Equation 2.10 can be further simplified to

$$\dot{Q}_{net} = C(T_b - T_a), \qquad (2.12)$$

where C is "total" thermal conductance (i.e., it incorporates evaporative as well as nonevaporative routes of heat dissipation; thus C has been called *"wet" thermal conductance*). The use of Equation 2.12 is even more restricted than is the use of Equation 2.11 when an animal is resting at cool temperatures in a burrow, in a nest, in shade, or in a metabolism chamber. This model of energy expenditure usually is used with data on oxygen consumption as a function of T_a, as presented in Figure 2.10A, where Equation 2.12 represents the curve at temperatures less than 22°C.

The "proper" name for Equation 2.12 has been a continuing controversy (e.g., see Strunk 1971, 1973; Kleiber 1972, 1973; Tracy 1972, 1973;

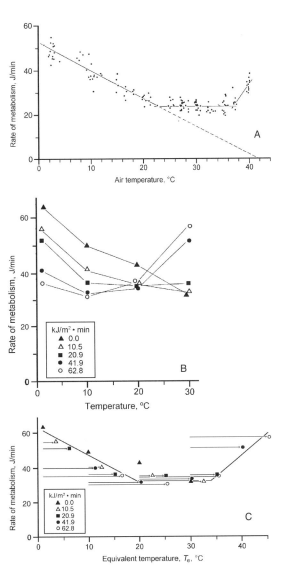

Figure 2.10 Rate of metabolism in white-crowned sparrows (*Zonotrichia leucophrys*) (A) as a function of air temperature, when the radiant temperature equals air temperature (modified from King 1964); (B) as a function of air temperature at various radiant heat loads (modified from DeJong 1976), and (C) as a function of equivalent temperature (modified from Robinson et al. 1976).

where ΔT_b is a change in body temperature, ΔQ is a change in the heat content of the body (J), c_p is specific heat of the body (J/[g°C]), and m is the mass of the body (g). Newton's law is a poor name for Equation 2.12, if only because Sir Isaac (1701) was concerned with the "cooling" of furnaces, not animals. An appropriate name for this model of energy exchange is the *Scholander-Irving model*, for these workers were the first to apply Equation 2.12 to a comparative study of animal energetics (Scholander et al. 1950b), although Bergmann (1847) had anticipated this application.

Other models of net thermal exchange with the environment have been suggested, all of which also involve a modification of Equation 2.10. They differ from the Burton and Scholander-Irving models by incorporating more complicated physical conditions. For example, Porter et al. (1973) predicted the proportion of time that the desert iguana (*Dipsosaurus dorsalis*) would be active during the year, and if active, the conditions to which it would be exposed. Porter and Gates (1969) graphically attempted to analyze heat exchange in the environment, especially with reference to ectotherms, and named these figures "climate spaces." They have been replaced by the use of equivalent and standard operative temperatures.

2.9.2 Equivalent temperature. This model of energetics lifts the restriction that the radiant temperature (T_r) equals air temperature (T_a). The equivalent temperature (T_e) is an estimate of the mean black-body temperature of the environment when a significant radiant heat load is present but no appreciable wind movement occurs. Winslow et al. (1937) described the model, and Bakken (1976, 1980), Robinson et al. (1976), and Mahoney and King (1977) recently applied it to vertebrates. In this model

$$T_e = T_a + \left(\frac{r_a}{\rho \cdot c_p}\right)(R_{abs} - \varepsilon \sigma T_b), \qquad (2.14)$$

where R_{abs} is the amount of radiant energy absorbed by the animal; r_a is the resistance of the boundary layer; ρ the density of air; and c_p, the specific heat of air. Under this circumstance, the total energy expenditure of the animal is given by

among others). Many biologists refer to it as "Newton's law of cooling." Much of the argument centers around the semantic nit of whether *cooling* refers to the loss of heat or to the fall in temperature. The fall in temperature, however, is related to the loss of heat by the relation

$$\Delta T_b = \Delta Q / c_p \cdot m, \qquad (2.13)$$

$$\dot{Q}_{net} = K_e (T_b - T_e) + L\dot{E}, \qquad (2.15)$$

where K_e is the equivalent thermal conductance. When $R_{abs} = \varepsilon\sigma T_b$ (i.e., when the radiant heat load equals the rate of radiant heat loss from the animal), $T_e = T_a$ and Equation 2.15 becomes Equation 2.11.

The application of effective temperature to rates of metabolism in white-crowned sparrows (*Z. leucophrys*) under a radiant heat load can be seen in an analysis of Robinson et al. (1976) by comparing Figures 2.10B and 2.10C. When the data of DeJong (1976) are plotted as a function of T_a (Figure 2.10B), rates of metabolism are low at cold temperatures and high at high temperatures, compared to the data obtained from birds without a radiant heat load (Figure 2.10A). These data suggest that a radiant load may substitute for a high rate of metabolism at cold temperatures but lead to overheating at high temperatures. When these data are plotted against T_e, however, they follow one curve (Robinson et al. 1976; Figure 2.10C). That is, T_e adequately incorporates the radiant heat load to account for the differences in measured rates of metabolism at a given T_a. The correlation in great frigate-birds (*Fregata minor*) of postures to reduce the radiant heat load with T_e demonstrated the utility of incorporating the radiant heat load into "temperature" (Mahoney et al. 1985).

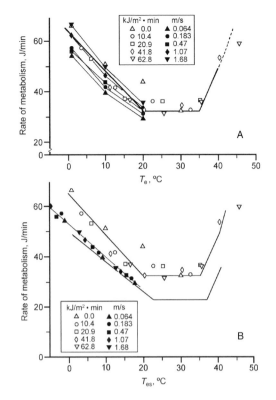

Figure 2.11 Rate of metabolism in white-crowned sparrows (*Zonotrichia leucophrys*) (A) as a function of equivalent temperatures at various radiant heat loads and wind velocities and (B) as a function of standard operating temperature. Source: Modified from Bakken (1980).

2.9.3 Standard operative temperature.

Gagge (1940) and Gagge and Hardy (1967) described another integrated "temperature" that takes both the radiant heat load and wind velocity into consideration when describing environmental conditions. Bakken (1976, 1980) also advocated this analysis. The temperature is called the *standard operative temperature*, T_{es}. It is defined by

$$T_{es} = T_a + \left[\frac{4\varepsilon\sigma(T_a)^3 A_1(T_r - T_a)}{4\varepsilon\sigma(kT_a)^3 A_1 + kA_3 \dot{v}^{0.5}} \right], \quad (2.16)$$

where T_r is the radiant temperature in the environment. Under these circumstances, a standard operative conductance, K_{es}, can be defined in which the equation for rate of energy exchange becomes

$$\dot{Q}_{net} = K_{es}(T_b - T_{es}) + L\dot{E}. \quad (2.17)$$

When $\dot{v} = 0$, $T_{es} = T_e$, and Equation 2.17 becomes Equation 2.15, and when $\dot{v} = 0$ and $T_r = T_a$, $T_{es} = T_a$, and Equation 2.17 becomes Equation 2.11.

The usefulness of T_{es} is seen in Bakken's analysis (1980) of data on the rates of metabolism in white-crowned sparrows at various heat loads and wind velocities (Figure 2.11A): when the data are plotted as a function of T_{es}, the variation in rate of metabolism at a given T_a disappears (Figure 2.11B) and the data agree with the measurements in still air made by King (1964). Bakken et al. (1991), by plotting rate of metabolism as a function of T_{es}, also showed in the junco (*Junco hyemalis*) that the effects of wind velocities disappear between 0.1 and 3.0 m/s (0.4 and 10.8 km/h). Although they are not expressed in terms of T_{es}, both Wood and Lustick (1989) on titmice (*Parus*) and Rogowitz and Gessaman (1990) on *Lepus* species showed experimentally the reciprocal interactions among rate of metabolism, wind velocity, and radiation.

Goodfriend et al. (1991) used standard operative temperatures to analyze the comparative thermal biology of two Middle Eastern gerbils, *Gerbillus allenbyi* and *G. pyramidum*, in the Negev desert region of Israel. They showed that *G. pyramidum* encountered a higher T_{es} under all environmental conditions, which may have reflected its larger body size. Bakken (1992), however, pointed out that the T_{es} calculated by Goodfriend et al. was always greater than T_e, which should not be the case because

$$T_{es} = T_e - \delta(T_b - T_e),$$

where δ is always positive. As long as T_b is greater than or equal to T_e, which was the case in this study, T_{es} is less than or equal to T_e. Nevertheless, this study properly focused attention on the thermoregulatory cost faced by small desert endotherms that avoid high daytime temperatures by nocturnal activity.

The differences among the models of total energy expenditure can be simply summarized. The Burton and Scholander-Irving models generally use a simple measure of an animal's performance—rate of metabolism—and correlate it with a simple parameter of the environment—air temperature. Both effective and standard operative temperatures use the same simple index of performance but use an increasingly complicated measure of the physical environment, radiation and air temperature for the former and radiation, convection, and air temperature for the latter. With the use of these derived "temperatures," values for rate of metabolism essentially, are moved along an ambient temperature scale to compensate for the presence of radiation, or of radiation and convection. Further adjustment of "temperature" could be made to account for the influence of humidity on the rate of evaporative water loss (see Gagge 1937, Winslow et al. 1937, Bakken 1980).

2.10 OSMOTIC EXCHANGE

Osmosis is the movement of water through a differentially permeable membrane that separates two solutions with different concentration of solutes. Concentration, then, is of primary importance to osmotic exchange in the characterization of a solution.

Concentration can be represented in many ways. The thermodynamic expression for concentration compares the number of dissolved particles to the number of solvent molecules. Then concentration is expressed in terms of *molality*, which is defined as the number of moles of solutes per *kilogram* of water (at a fixed temperature). A mole, or gram-molecular weight, of solute has 6.02×10^{23} particles (Avogadro's number) of a non-ionizing molecule. The significance of molal solutions is found in their colligative properties, that is, the numerical properties independent of the chemical nature of the solutes. For example, the "pressure," or more accurately the energy per mole of solute (π, in atm or in J) exerted on a membrane that permits only water to move through is given by

$$\pi = iCRT, \qquad (2.18)$$

where R is the gas constant (0.082 l·atm/[K·mole], or 8.31 J/[K·mole]), T is absolute temperature, and C is the molal concentration of a solute. The coefficient i indicates the mean number of particles produced from a mole of solute when in solution. For non-ionizing molecules, such as simple sugars, $i = 1.0$, but for ionic complexes, such as salts, i is greater. Sodium chloride (NaCl) ionizes into two particles; therefore, i would be expected to be 2.0. Measurements of the decrease in freezing point of NaCl solutions suggest that i actually is about 1.8 and approaches 2.0 only in very dilute solutions. For calcium chloride ($CaCl_2$), i is expected to be 3.0 but actually is about 2.6 (again, depending on the molal concentration of the solution). Because all biological solutions contain salts, the concentration of a solution is convenient to describe in osmolals ($c = iC$), which takes the molal concentration of solutes into consideration along with the number of particles produced when the solutes are dissolved. In Equation 2.18, $\pi = 22.4$ atm when the osmolal product $iC = 1.0$ and $T = 273.16$ K (= 0°C).

Osmotic "pressure" in this (conventional) view is the result of a partial pressure produced by the bombardment of solutes against a membrane, just as a partial pressure is exerted by a gas in a mixture. Scholander (1966, 1967, 1971) with his coworkers (1965, 1968) extended this view to argue that solute pressure is always positive (because of a solute's restricted freedom to go through a membrane), and that when a free water surface is present, the solvent (water) has a pressure equal and opposite to that of the solutes. Scholander (1971, p. 217) suggested that " . . . the solvent is coupled to the solute pressure via the free surface." Hence arises the concept of negative "pressure" in the solvent, and the concept of the osmotic potential of water, which is given by

$$\pi_w = -iCRT, \qquad (2.19)$$

which is equal and opposite to the positive osmotic "pressure" of the solutes (Equation 2.18). The negative sign in Equation 2.19 indicates that a solution having a concentration greater than 0.0 osmole will absorb pure water. Under these circumstances, π_w is defined as a component of the water potential, which is a measure of the chemical potential of water compared to that of pure water: all solutions that contain solutes have a negative water potential. This view has been expanded to derive the colligative properties of a solution from the enhanced solvent tension produced by the presence of solutes (Hammel 1976a).

Concentrations are often represented as *molar* solutions, which are defined as the number of moles of solutes per *liter* of solution. A liter of water weighs a kilogram only at 3.98°C, which means that a molar solution is slightly more concentrated than a molal solution at higher temperatures because fewer water molecules are present in a liter of water at temperatures above 4°C. This difference increases with temperature but is quite small. For example, at 20°C, osmolal concentrations would be 99.83% of the corresponding osmolar concentration.

The value of molal over molar solutions is readily seen in the effect of concentration on the freezing point depression of solutions. The freezing point of an aquatic solution is given by the relation

$$f_p = -1.86iC = -1.86c. \qquad (2.20)$$

Irrespective of solute composition or molecular weight, a given osmolal concentration has a fixed depression on freezing point, because at a given temperature a solution of a particular molal concentration has a fixed number of solute particles relative to solvent molecules. In molar solutions, however, freezing point depression varies with molecular weight, because a gram-molecular weight of a large molecule takes more volume than a gram-molecular weight of a small molecule, and thus less water is present in the former solution. The concentration of a solution measured by freezing point depression always is an osmolal concentration as calculated from Equation 2.20, not as is sometimes reported as an osmolar concentration. The concentrations of molal and molar solutions are often expressed as milliosmoles per kilogram (mOsm/kg) or milliosmoles per liter (mOsm/L), respectively, because most biological solutions are rather dilute.

If two solutions are separated by a differentially permeable membrane, water moves from the solution with the lower concentration to the solution with the higher concentration (Fick's law, Equation 2.2). The flux of water ($\dot{w}$) between the two solutions depends on the differential in osmotic potential, $\Delta\pi$:

$$\dot{w} = AP\Delta\pi$$
$$\dot{w} = AP(\pi_2 - \pi_1),$$

which, using Equation 2.18, permits

$$\dot{w} = APRT(c_2 - c_1),$$

where A is the surface area of the membrane separating the solutions, c is the concentration in osmolals, and P is the coefficient of osmotic permeability.

Solutes, as well as water, may move across membranes, the rate varying with the solute, its charge, and the concentration gradient (Equation 2.1). In fact, the flux of solutes may be greater or lesser than expected from the passive movement implied by Fick's law. One way that net flux may be less than expected from the concentration gradient is if the solute is actively transported from a solution of low concentration to a solution of high concentration. Movement against a gradient requires energy to be expended. The minimal amount of energy needed to transport actively a mole of solute is estimated from the change in Gibb's free energy (ΔG) associated with this movement:

$$\Delta G = RT\ln(c_2 - c_1), \qquad (2.21)$$

where ΔG is a *minimal* estimate of the work that must be done to transport a mole of solute from c_1 to c_2 when c_2 is greater than c_1. The actual cost of transporting a mole of solute against a concentration gradient may be much higher, depending on the efficiency of the transport system.

2.11 CONSTRAINTS ON GAS EXCHANGE

Vertebrates exchange gas with their environments because all species use aerobic metabolism, which requires molecular oxygen and produces CO_2 as a waste product. The acquisition of oxygen and the elimination of CO_2 occur in both aerial and aquatic environments, although the absolute abundances of gases differ greatly in these environments.

The atmosphere is a mixture of gases. Most compartments of the atmosphere accessible to vertebrates have a similar gas mixture, the principal exceptions being the compartments in which the composition is actively modified by organisms, and then only if these compartments are cut off from free exchange with the atmosphere as a whole. Gas abundance can be expressed either as partial pressures (mm Hg) or as concentrations (cm^3/L). The advantage of using concentrations is that they are absolute, requiring no conversion factors, whereas the advantage of using pressures and partial pressures is that they are familiar and measured directly.

Three major physical factors modify the dry atmospheric gas abundances given in Table 2.1:

1. Total atmospheric pressure falls with an increase in altitude (Table 2.2), so even though the relative composition of gases remains the same, the partial pressure of individual gases decreases proportionally to the fall in total pressure.

Table 2.1 Major gas components of the dry atmosphere

Gas	Abundance (vol%)
N_2	78.03
O_2	20.99
A	0.94
CO_2	0.03

Source: Condensed from Hodgman et al. (1954).

Table 2.2 Barometric pressure at various altitudes

Altitude		Barometric pressure	
m	ft	mm Hg	% Sea level
0	0	760	100
500	1,641	715	94
1,000	3,281	673	89
1,500	4,921	633	83
2,000	6,562	594	78
2,500	8,202	558	73
3,000	9,842	524	69
4,000	13,123	461	61
5,000	16,404	405	53
6,000	19,685	354	47

Source: Condensed from Hodgman et al. (1954).

2. The capacity of the atmosphere to hold water is set by atmospheric temperature: it increases with temperature (Table 2.3). The amount of water in the atmosphere, measured in terms of mass (g/L), volume (cm^3/L), or pressure (mm Hg), is referred to as the *absolute humidity*. *Relative humidity*, represented as a percentage, compares the amount of water present to the capacity of the atmosphere at a particular temperature to hold water.

3. The gas having the most variable content in the open atmosphere is water vapor. If the atmosphere at sea level has a relative humidity of 50% at a temperature of 20°C, the amount of water contained in the atmosphere exerts a pressure of $17.5 \times 0.5 = 8.8$ mm Hg, which means that the sum of the pressures exerted by the other gases is $760 - 8.8 = 751.2$ mm Hg, assuming that the total barometric pressure is at the "standard" value. The pressure exerted by oxygen, then, is $0.2099 \times 751.2 = 157.4$ mm Hg. The presence of water vapor obviously diminishes the abundance of oxygen and other gases in the atmosphere when the total pressure is fixed by conditions in the environment, such as altitude and weather events.

The gas content of water is more complexly determined. It depends on the solubility of the individual gases and the abundance of each gas in the atmosphere with which the water is in equilibrium. For example, the content of oxygen in freshwater can be calculated in the following manner:

$$C_{O_2} = \alpha_{O_2} \cdot P_{O_2}, \qquad (2.22)$$

Table 2.3 Maximal pressure and content of water in the atmosphere as a function of temperature

Temperature (°C)	Pressure (mm Hg)	Content (g/cm^3)
0	4.6	4.9
5	6.5	6.8
10	9.2	9.4
15	12.8	12.8
20	17.5	17.3
25	23.7	23.0
30	31.7	30.4
35	42.0	39.6
40	55.1	31.9
45	71.7	65.6
50	92.3	83.2

Source: Condensed from Hodgman et al. (1954).

Table 2.4 Solubility coefficients $(cm^3/cm^3 H_2O \cdot mm\,Hg)$ of various gases in water

Temperature (°C)	Nitrogen	Oxygen	Carbon dioxide
5	0.0209	0.0429	1.424
10	0.0186	0.0380	1.194
15	0.0169	0.0342	1.019
20	0.0155	0.0310	0.878
25	0.0143	0.0283	0.759
30	0.0134	0.0261	0.665
35	0.0126	0.0244	0.592

Source: Condensed from Hodgman et al. (1954).

where C_{O_2} is the oxygen content of water in cm^3O_2/cm^3H_2O, α_{O_2} is the solubility coefficient of oxygen in water $(cm^3O_2/[cm^3H_2O \cdot mm\,Hg])$, and P_{O_2} is the partial pressure of oxygen in air. The coefficient of solubility varies with the gas and decreases with an increase in water temperature (Table 2.4).

A body of pure water in equilibrium with air at 20°C and 760 mm Hg has 9.2 (= 0.0155 × 760 × 0.7803) cm^3 of N_2, 4.9 cm^3 of O_2, and 0.20 cm^3 of CO_2 per liter of water. In seawater the gas contents decrease by about 20%. Note that although the solubility of CO_2 is much greater in water than oxygen (see Table 2.4), the partial pressure of CO_2 in the atmosphere is so small, relative to the partial pressure of oxygen, that the amount of CO_2 in water in equilibrium with the atmosphere is still (normally) small relative to the amount of oxygen (but see Section 8.4.4). Nevertheless, in the atmosphere oxygen is 700 (= 20.95/0.03) times as abundant as CO_2, whereas in freshwater this ratio is 24.5, the decrease reflecting the difference in solubilities between these gases.

The low solubilities of gases in water imply that the amount of oxygen, nitrogen, and CO_2 dissolved in water is appreciably less than the amount found in the atmosphere. A particularly graphic illustration of this difference is indicated by the following calculation. About 32.3 times (= $1/\alpha_{O_2}$) as much oxygen is present in the atmosphere at 20°C as in water equilibrated with the atmosphere. Because water has a density 806.7 times that of dry air (at 20°C), an animal must move approximately 32.3 × 806.7 = 26,056 times the mass of water as air across its respiratory surfaces to extract a given volume of oxygen (assuming that the extraction efficiency is equal in the two media). At higher temperatures even more water relative to air must be moved to satisfy oxygen requirements because of the decrease in gas solubility in water with an increase in temperature. Consequently, the energy expended for ventilation is much greater in an aquatic than in an aerial environment.

Gas exchange by diffusion occurs in both air and water and can be described by Fick's law (Equation 2.2) in which D is the Fick coefficient of diffusion. At least four values for D are important for animals, two for oxygen (in air [0.196 cm^2/s at 20°C] and in water [0.183 × $10^{-4}cm^2$/s at 20°C]) and two for CO_2 (in air [0.159 cm^2/s at 20°C] and in water [0.177 × $10^{-4}cm^2$/s at 20°C]). With an increase in temperature, the Fick coefficient increases according to the relation

$$D = D_0 \left| \frac{T}{273} \right|^n, \qquad (2.23)$$

where D_0 is the coefficient at 0°C, T is temperature in Kelvin, and n is a power equal to 1.75 for oxygen in air and to 2.00 for CO_2 in air.

Because the thickness of a gradient is difficult to measure, it is often replaced with a differential, expressed either in terms of concentration (ΔC, cm^3/L) or partial pressure (ΔP, mm Hg). For diffusion in an aerial environment, Equation 2.2 becomes

$$\dot{s} = D_g A \cdot k \Delta P, \qquad (2.24)$$

where $\dot{s}$ is the amount of gas transported by diffusion, D_g is the Fick diffusion coefficient for oxygen or CO_2 in air, and k is a factor for converting units, which can be eliminated if concentrations, rather than pressures, are used:

$$\dot{s} = D_g A \cdot \Delta C.$$

In an aquatic environment the diffusion of a gas must take the solubility of that gas into consideration:

$$\dot{s} = D_w A \cdot k \cdot \alpha \cdot \Delta P,$$
$$\dot{s} = K \cdot A \cdot \Delta P, \qquad (2.25)$$

where α is the solubility coefficient of oxygen or CO_2, D_w is the Fick coefficient of diffusion for oxygen or CO_2 in water, k is a factor converting units, and $\alpha \cdot k \cdot D_w = K$ (cm³/[cm²·atm]) is Krogh's diffusion constant (Dejours 1975). Notice that Equations 2.24 and 2.25 lack the negative sign found in Equation 2.2 because a gradient has been replaced by a differential.

Gas exchange in organisms often involves diffusion across a water-air boundary. The K in Equation 2.25 is derived from the boundary membranes between the air and the organism (Dejours 1975). But as Dejours pointed out, confusion over Krogh's constant makes the application of these equations difficult. Ultsch (1974c) estimated K in a live, intact salamander (*Siren intermedia*) at the critical P_{O_2} below which resting metabolism depends on P_{O_2} (see Section 8.4.6). By estimating the thickness of the skin and the loading tension of oxygen in the blood at the skin, Ultsch estimated K to equal 0.14 cm³$O_2 \cdot \mu m/(cm^2 \cdot min \cdot atm)$, a value that is essentially identical to those found by Krogh (Steen 1971) for other biological tissues. Thus, the relationship described by Equation 2.25 may limit gas exchange in animals facing low oxygen tensions because diffusion, a passive means of gas exchange, is normally limited in effectiveness to distances of a few millimeters (Harvey 1928, Alexander 1971).

Most animals, and all vertebrates, depend on the convective transfer of oxygen and CO_2. Rahn et al. (1971b) described equations that incorporate convection in terms of ventilation rate ($\dot{V}_g$, L/min):

$$\dot{s} = \dot{V}_g \cdot k \cdot \Delta P \quad \text{(in air)}$$
$$(\text{or} \quad \dot{s} = \dot{V}_g \cdot \Delta C)$$

$$\dot{s} = \dot{V}_g \cdot k \cdot \alpha \cdot \Delta P \quad \text{(in water)}$$
$$(\text{or} \quad \dot{s} = \dot{V}_g \cdot \alpha \cdot \Delta C).$$

Total gas exchange equals the sum of gas transported by diffusion and convection, although the amount transported by convection is greater because $\dot{V}_g$ is greater than $D \cdot A$. The use of convectional exchange tends to shear the boundary layer, which increases diffusional exchange.

Vertebrates that live in the soil are exposed to an atmosphere with low oxygen and high CO_2 concentrations, both when freely moving through soil, as in some snakes, lizards, and caecilians, and when living in burrows. Convection or diffusion can transfer oxygen to burrowing mammals, depending on whether the burrows are open or closed, respectively. Vogel and Bretz (1972) and

Vogel et al. (1973) described the bulk flow of gas through the open burrows of the prairie-dog (*Cynomys ludovicianus*). If wind is present at the surface and if a burrow has two openings, then the burrow atmosphere is coupled to wind movement on the surface by viscous entrainment (Bernoulli's principle). Gas flow through the burrow can be enhanced and given a principal direction by placing mounds around the openings: air enters a burrow through the lower of two mounds and exits through the higher. Furthermore, mounds with sharp rims are better exits for air than mounds with rounded rims. Vogel et al. (1973) used a scale model to estimate that 5 min is required to replace the air in a 20-m burrow with air from the external atmosphere if the external flow is 1.30 m/s (4.7 km/h), or 10 min if the flow is 0.50 m/s (1.8 km/h). Obviously, CO_2 will not accumulate and oxygen will not be depleted in such a burrow, except possibly in blind side passages.

All burrows are not open to the external atmosphere; consequently, burrow atmospheres may have a gas composition that is markedly different from that of the external atmosphere. Gas exchange between closed burrows and the external atmosphere depends principally on diffusion through the soil (McNab 1966b). From the Fick diffusion equation Penman (1940a) derived the following equation for gas diffusion through porous soil:

$$\dot{s} = \left[\frac{0.66 D_0 \cdot S \cdot A (T/273)^2 (c_2 - c_1)}{l} \right], \quad (2.26)$$

where D_0 is the diffusion constant for the gas at 1 atm pressure and 0°C; S is the fractional porosity of the soil; A is the burrow surface area; $(T/273)^2$ is the temperature correction factor for D_0 to compensate for burrow temperatures above 0°C, which in the case of oxygen might be better represented by $(T/273)^{1.75}$ (see Equation 2.23); c_2 and c_1 are the concentrations of oxygen in the atmosphere and burrow, respectively; and l is the depth of the burrow. Equation 2.26 can be used to estimate the minimal surface area of a burrow needed to supply the amount of oxygen consumed by an animal in the burrow. For example, McNab (1966b) estimated that a pocket-gopher, *Geomys pinetis*, requires about 450 cm² of burrow surface, assuming that (1) the gopher's rate of oxygen consumption is 180 cm³/h, (2) burrow temperature is 25°C,

(3) depth of the burrow is 30 cm, (4) the differential oxygen concentration is $0.21 - 0.15 = 0.06$, (5) net soil porosity in sandy soil is 0.40, and (6) the Fick coefficient of diffusion for oxygen is $0.178 \text{cm}^2/\text{s} = 640.8 \text{cm}^2/\text{h}$. If the burrow is 8 cm wide, a burrow 56 cm long is required for oxygen exchange: this distance is much smaller than actual burrow lengths. Withers (1978), however, suggested that burrow depth does not have an appreciable impact on burrow gas composition. A similar analysis can be made for the diffusion of CO_2 in soil (Penman 1940b).

The minimal surface area (and therefore the minimal length) of the burrow needed for gas exchange may depend on the activity of the gopher and on soil type through its influence on soil porosity. A soil with a low net porosity, whether produced by low total porosity or (as in clays) a high total porosity and a high water content, may exclude fossorial mammals that plug their burrows because the rates of diffusion would be inadequate to supply the demands of gas exchange in these species. Thus, in the example cited here, a net soil porosity equal to 0.05 would require a burrow 8 times as long (i.e., 4.5 m). Pocket-gophers are not found in clay soils, whereas moles (*Scalopus aquaticus*) often are. This difference is related to the shallow feeding burrows of moles, which often have cracked ceilings that presumably permit bulk gas exchange with the external atmosphere. Clearly, the limitations imposed by diffusion may have significance for the ecology and distribution of some vertebrates under field conditions. Olszewski and Skoczén (1965), however, showed that 26% of the variation in burrow atmosphere velocity in the closed burrow of a mole, *Talpa europaea*, was correlated with surface air movement. They eliminated this correlation by covering the closed mounds with mud or clay and enhanced burrow air movement by opening the burrows. Burrowing habits may expose mammals to unusually low oxygen tensions when these habits are found at high altitudes (see Section 8.8.5).

2.12 CHAPTER SUMMARY

1. Radiation is a major means by which vertebrates exchange energy in terrestrial environments.

2. The color of a vertebrate influences its heat gain from, but not heat loss to, the environment.

3. In terrestrial vertebrates, thermal conduction is normally of little importance as a means of heat exchange with the environment, principally because of the small area of contact with surfaces, but it is an important process in aquatic species.

4. Convection is an important and highly complex form of heat exchange, most of the complexities depending on the shape and size of the vertebrate and on the presence or absence of movement in the fluid external to the vertebrate.

5. Evaporation is similar to convection in its complexity, but it can be selectively increased by raising ventilation rate or sweating.

6. The net thermal exchange between a vertebrate and its environment is the sum of the exchanges by radiation, conduction, convection, and evaporation; it can be represented by several approximations, each of which uniquely relies on the physical conditions to be included.

7. The physical conditions in the environment may be represented simply by temperature, or they can be increased in complexity by including radiation, convection, and evaporation.

8. Osmotic exchange depends on diffusion and active transport, the balance between these processes varying with the organism and the environment.

9. Most impediments to gas exchange result from the variable abundances of gases (at high altitudes, in terrestrial burrows, or in aquatic environments) or from high resistances to diffusion in organisms.

PART II | Thermal Exchange with the Environment

3 The Scaling of Metabolism and Thermal Relations

3.1 SYNOPSIS

Body size has a fundamental impact on the thermal biology of vertebrates, a pattern referred to as *scaling*. Total rate of metabolism increases with body size. This relationship usually is expressed as a power function. The impact of a power function, especially in terms of the numerical value of the power, which is always less than 1.0, is examined. As a result of this scaling, mass-specific rate of metabolism decreases with an increase in mass. The comparative merits of total and mass-specific units for rate of metabolism are discussed. The level at which scaling occurs is compared in various vertebrate groups, which leads to the conclusion that whether an organism is a poikilotherm or a homeotherm principally reflects its level of energy expenditure. A relationship between metabolism and body mass separates strict endothermy from facultative endothermy and ultimately from classic ectothermy, especially at small masses, although these differences are less marked at large masses.

3.2 INTRODUCTION

Two fundamental characteristics of organisms are their body size and whether they regulate body temperature by the generation of heat through metabolism. These two characteristics are intimately intertwined and are the subjects of this chapter. A discussion of the thermal relations of vertebrates and the impact of body size on these relations is basic to our understanding of the ecology of vertebrates.

Thermal relations probably have received the most attention of all physiological functions of vertebrates. At one extreme, some vertebrates have body temperatures that are nearly identical to, and vary with changes in, environmental temperature: they are called *poikilotherms*. They are characterized by low rates of heat production and poor insulation, which means that the principal source for body heat is external to their bodies. Consequently, they are called *ectotherms* (see Chapter 4). At the other extreme, some species precisely control body temperature over a range in environmental temperature by regulating the rates at which heat is generated and lost. Such animals are called *homeotherms*, because they have a constant body temperature, and *endotherms*, because the principal source of body heat depends on high rates of metabolism (see Chapter 5).

Many intermediate states in thermal biology exist between strict ectothermic poikilothermy and strict endothermic homeothermy. For example, some endotherms do not regulate body temperature precisely but permit body temperature to fall selectively, and some ectotherms maintain body temperature rather independent of environmental temperature, usually by controlling heat gain from the environment. These intermediate states led some vertebrates to be called *heterotherms*, which is a nearly useless term given that it represents a heterogeneous grab bag of species that do not conform to either extreme.

In this chapter the transition between the extreme thermal states will be shown to be graded; the factor most important in determining where a species falls along this continuum is the relationship between rate of metabolism and body mass. That is, the thermal characteristics of a species are related to the manner by which rate of metabolism is scaled to body mass. Another factor important to the distinction between ectotherms and endotherms is the rate at which heat is lost to the environment. Like heat production, heat loss is also scaled to body size. To avoid undue complications at this point, a discussion of heat loss is deferred to Sections 4.13 and 5.5.

3.3 SCALING RATE OF METABOLISM

The scaling of a function refers to the manner in which the function is quantitatively correlated with body size, which here usually is represented by body mass. Because large animals expend more energy than small animals, if only because they have more structure to maintain, energy expenditure might be assumed to be proportional to body size. That presumption is not correct.

3.3.1 Scaling basal rate in mammals. The first function to be considered in terms of scaling was the *basal rate of metabolism* in mammals, which is the minimal rate of a thermoregulating endotherm during rest when it is not digesting a meal and is not exposed to temperatures that require an increase in metabolism (see Section 5.2). Early work on the energy expenditure of mammals led Sarrus and Rameaux (1839) to conclude that (1) heat production is proportional to oxygen consumption, (2) heat production equals heat loss as long as body temperature is constant, (3) heat loss is proportional to surface area, and therefore (4) oxygen consumption is proportional to mass raised to the $\frac{2}{3}$ power. Bergmann (1847) independently came to the same conclusions and to the logical corollary that "...a gram of a large animal must, in general, respire less than a gram of a small animal" (p. 605). This analysis set the stage for all future speculations on the relation between rate of metabolism and body size in mammals and all other organisms.

Bergmann and Leuchart (1852), Rubner (1883), and Richet (1885, 1889) concluded that heat production is dictated by heat loss through the surface of the body. For example, Rubner (1883) showed that heat production varied more in dogs when expressed per unit of mass than when expressed per unit of surface area: in dogs that weighed 3.2 to 31.2 kg, heat production varied from 369 to 49 kJ/kg and from 5073 to 4337 kJ/m², respectively. He concluded that heat production is more closely associated with surface area than

with body mass. This conclusion is often called the *surface law*.

This interpretation of the relation between metabolism and body size remained dominant for about a century. Although doubts about this conclusion had been expressed earlier (von Hoesslin 1988, Benedict 1915), the first effective challenges to the surface law were mounted with the use of a power model of growth described by Julian Huxley (1932). It describes a variable y, organ size in the case of Huxley or rate of metabolism here, as a function of body mass (m) raised to a power (b), a pattern referred to as *allometry*:

$$y = a \cdot m^b, \qquad (3.1)$$

where a is a coefficient. Kleiber (1932) originally examined rate of metabolism in eight species, three birds (dove, pigeon, and chicken) and five mammals (rat, dog, sheep, human, and cattle), five of which were represented by two or more measurements. He concluded that the power of this relationship was 0.73 to 0.74, depending on whether ruminants were included. He rounded this value to 0.75, partly out of calculating convenience. Brody and Procter (1932) demonstrated that the fitted power for domesticated birds and mammals was 0.73. Benedict (1938) included a greater size range of mammals, from mice to elephants, which led his curve to be called the *mouse-to-elephant curve*. He committed himself to no particular scaling power, concluding that "... it seems unjustifiable to apply mathematics to the pooled end result of the activities of millions of cells, each highly differentiated, with different energy potentialities, and actuated by different stimuli" (p. 179). This view perplexed Kleiber (1961, pp. 202–203), raising in his mind the question of how Benedict could tolerate calculating averages, which themselves involved pooling millions of cells! Brody (1945) in his monumental book *Bioenergetics and Growth* concluded that the power of mass is 0.73. These analyses led to the downfall of the surface law and to the establishment of the "$\frac{3}{4}$ rule."

Much discussion has occurred as to the "proper" value for the power b, including the continuing belief that the most appropriate value is 0.67, which Heusner (1982a, 1982b) most strongly advocated. Heat production at temperatures where basal rate is measured is compensated for by changes in insulation. That is, the basal rate is not dictated by heat loss, even though they remain balanced. A second difficulty, at least as it relates to heat production, is that a similar scaling relation occurs in ectotherms and presumably is not dictated by heat loss.

Other explanations for the value of b are as follows:

1. The power reflects the scaling factor for some other component of the body, such as the surface area for gas exchange (Ultsch 1973a) or mitochondrial numbers (Else and Hulbert 1985a, 1985b). For example, Ultsch argued that if rate of oxygen consumption is proportional to m^b and if the area for gas exchange is proportional to m^d, then $b \le d$, which is generally correct among a wide range of vertebrates (Table 3.1). The difficulty with this explanation is that either metabolism or surface area can be the independent variable. A similar analysis could be made for mitochondrial number.

2. Blum (1977) maintained that 0.75 can be derived directly from assuming that an organism can be represented by a fourth dimension, which as Speakman (1990) noted, was never defined.

3. Spatz (1991) argued that 0.75 resulted from a constraint on metabolism arising from the increase in resistance to blood flow with an increase in mass and from size-dependent parameters related to the unloading of oxygen.

4. A power less than 1.0 might result from the accumulation of inert body matter with an increase in body size. Such matter might include body fat (Scott and Evans 1992, but see McNab 1968, Scott et al. 1972, Goyal et al. 1981) and the skeleton. Ultsch (1974b) examined scaling of the skeletal

mass of mammals, but the increase in mass was proportional to $m^{1.13}$, which is not sufficient to account for the decrease from $m^{1.00}$ to $m^{0.75}$, $m^{0.71}$, and least of all $m^{0.67}$. Spaargaren (1992), however, renewed this suggestion by including connective tissues, body fluids, and energy stores.

Two persistent suggestions for the scaling of basal rate of metabolism in mammals were based on dimensional analyses. von Hoesslin (1888) and Lambert and Teissier (1927), based on geometric similarity, concluded that rate of metabolism should be proportional to $m^{0.67}$. Heusner (1982a, 1982b, 1991) maintained that this is the correct relationship (Figure 3.1) and concluded that powers for mass other than 0.67, such as 0.75, are statistical artifacts produced by interspecific regression analysis. When interspecific comparisons are made, especially over a wide range in body mass, species that have different body proportions and compositions are being compared, thereby violating the assumption in a regression that all species have a common a. In fact, he stated that the only cases in which geometric similarity is held constant is in intraspecific studies. Further, he noted that a increases with body mass, which results in an artificially high interspecific power (Figure 3.1). When Heusner (1982a) fitted 11 intraspecific curves for seven species of mammals, b varied from 0.48 and 0.91, which gives little security to the fact that their

Table 3.1 Scaling of rate of metabolism and gas exchange*

Group	Metabolism (b)	Gas exchange (d)
Anurans	0.67–0.86	0.85
Lunged salamanders	0.86	0.86
Lungless salamanders	0.72	0.74
Fish	0.70–0.81	0.78–0.94
Reptiles	0.60–0.86	0.75
Birds	0.72–0.73	0.94–0.95
Mammals	0.71–0.75	0.75–1.06

*Where $\dot{V}_{O_2} \approx m^b$ and $\dot{V}_g \approx m^d$.
Source: Condensed from Ultsch (1973a).

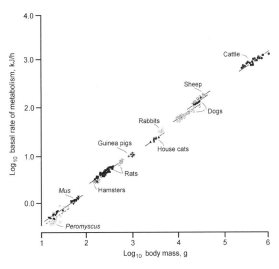

Figure 3.1 Log_{10} basal rate of metabolism within and between various species of mammals as a function of log_{10} body mass. Source: Modified from Heusner (1982a).

mean happens to be 0.67. Although he adheres to 0.67, Heusner (1982b) maintained that it is not associated with heat loss through the surface, because "... this explanation [is] not sufficiently general to account for the same relationship in poikilotherms which do not regulate their body temperature" (p. 14). Exactly.

An alternative hypothesis for the magnitude of b, proposed originally by Rashevsky (1960) and most recently by MacMahon (1973, 1975), is based on dimensional analysis and on the elastic similarity of skeletons and other support structures. From certain assumptions, one of which leads to the conclusion that mass is proportional to the fourth power of a linear dimension, MacMahon concluded that the basal rate of metabolism should be proportional to $m^{0.75}$. In response to the view of Heusner that "$\frac{3}{4}$" is a statistical artifact, Feldman and MacMahon (1983) recalculated Heusner's data, using a modified covariance model, and argued that both exponents are valid, 0.67 for intraspecific scaling and 0.75 for interspecific scaling.

The conclusion that both 0.67 and 0.75 are appropriate scaling powers has not met with acceptance. For example, Schulz (1988) maintained that both geometric and elastic similarities were acceptable bases for scaling, but that empirical evidence indicates that the scaling of body functions is generally proportional to elastic similarity (i.e., $b = 0.75$). Platt and Silvert (1981) used dimensional analysis to conclude that 0.67 is the appropriate power in aquatic animals, because gravity has little impact on species that are neutrally buoyant with the environment, and that 0.75 is appropriate in terrestrial species, when "... gravity imposes an additional metabolic cost on land animals and plants" (p. 858). In accord with that conclusion, Economos (1979) earlier noted that among mammals the exponent is 0.75 for terrestrial species and 0.67 for marine species. From this view rate of metabolism in snakes should scale proportionally to 0.67, but an extensive analysis of intraspecific and interspecific scaling in boids (Chappell and Ellis 1987) showed that intraspecific powers were less than the interspecific power (0.732 vs. 0.806, respectively), but both were greater than 0.67. Finally, the body proportions of terrestrial mammals appear to conform more closely to geometric than to elastic similarity, at least at masses between approximately 10 g and 100 kg (Prothero 1992), unlike the predictions of

MacMahon (1975). Alexander and Jayes (1983) and Beiwener (1983) also found widespread geometric similarity in mammals.

West et al. (1997) derived the allometric equation from the branching transportation networks of organisms, as is found in the circulatory system. They argued that a branching pattern of vessels is required to supply substrates and to eliminate wastes, that the final branch of the network is size invariant, and that the energy required for the distribution of materials is minimized. "Scaling laws arise from the interplay between physical and geometric constraints implicit in these three principles" (p. 122). From these assumptions they derived a variety of allometric exponents, including the $\frac{3}{4}$ rule for metabolism. These authors further suggested that the 3 in the "$\frac{3}{4}$ rule" reflects three-dimensional existence, which raises the question of whether two-dimensional organisms, bryozoans or flatworms, might scale to the "$\frac{2}{4}$" power. They later (West et al. 1999) argued that fluid flow is not necessarily involved in the derivation of the quarter power, but that it can be derived from the geometry of circulatory networks: "The vast majority of organisms exhibit scaling exponents very close to "$\frac{3}{4}$" for metabolic rate and to "$\frac{1}{4}$" for internal times and distances" (p. 1679). Banavar et al. (1999) similarly concluded that the "$\frac{3}{4}$ rule" can be derived from network structure independent of dynamics or geometry.

Charnov (1993) and Kozłowski and Weiner (1997) proposed a radically different approach to the scaling of physiological and ecological functions. They recommended constructing models based on life-history optimization, especially with regard to fitness as measured by the lifetime production of offspring. For example, Kozłowski and Weiner (1997) suggested that the amount of energy available for production (P) in an individual equals the amount that it assimilates (A) minus the amount it expends for maintenance (i.e., respires, R):

$$P(m) = A(m) - R(m) = am^b - \eta m^\beta = cm^d, \quad (3.2)$$

where all factors vary with body mass (m); a, η, and c are coefficients; and both assimilation and respiration are power functions of mass.

In this analysis, several assumptions were made: (1) production is growth until the age of maturity in species with determinant growth; (2) beyond this age production is represented by reproduction; (3)

the parameters of production and mortality are species specific; and (4) natural selection has optimized body size in each species. From these assumptions Kozłowski and Weiner derived interspecific allometries for assimilation, respiration, production, age at maturity, and life expectancy from repeated simulations of intraspecific Equation 3.2. Thus, when they set the intraspecific $b = 0.67$ and $\beta = 0.75$ with 5% coefficients of variation, they derived interspecific $b = 0.76$ and $\beta = 0.70$. Furthermore, they derived a frequency distribution of body sizes that is similar to that observed (see Brown et al. [1993], who also derived a model of fitness based on energetics and predicted a body size distribution, but assumed a power equal to 0.75 for the relationship between energy intake and body mass). Kozłowski and Weiner concluded that interspecific allometries are by-products of the optimization of body size and of the distributions of the parameters of intraspecific production and mortality. They also concluded that the model Charnov (1993) described, which assumed a 0.75 power for metabolism, is a special case of their more general model. These probabilistic models derived from population biology and selection theory differ fundamentally from engineering models, which depend on the mechanical properties of skeletons, the surface area of isometric bodies, or the structure of branching networks.

These analyses proceed on the assumption that a uniform relationship exists between rate of metabolism and body mass. Actually, this relationship shows much residual variation (Figure 3.2), especially if other than domesticated species are considered. At any given mass, the maximal range in basal rate is at least 3 or 4:1 and is as great as 10:1! In the face of this variation, how can the question of the power b be resolved, unless the factors responsible for the variation in rate at a particular size are addressed? From a statistical viewpoint, this scatter is "noise," but it is not noise if the residual variation has a biologically significant pattern. Most students of energetics agree that patterns exist in this variation, although they might not agree on what these patterns are: some suggest that standard or basal rate usually is associated with phylogeny, whereas others maintain that it is associated principally with the ecological and behavioral characteristics of the organisms. (This discussion is delayed to Sections 4.6 and 5.4.) Whatever the source of the variation, it means that the basal rate in mammals cannot be uniquely pre-

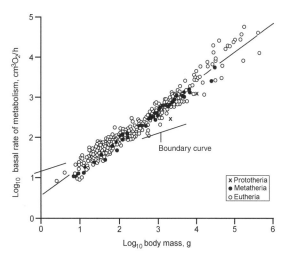

Figure 3.2 Log_{10} basal rate of metabolism in mammals as a function of log_{10} body mass, along with the all-mammal mean curve. The boundary curve for endothermy is also indicated. Source: Modified from McNab (1988a).

dicted by body mass, and therefore if these additional factors are not distributed uniformly along the mass axis, they will affect the power b. Consequently, nearly all empirically derived values of b will be contaminated by factors that are correlated with mass. For example, if the basal rate of mammals is correlated with food habits and family affiliation, and if food habits are correlated with family affiliation and body mass, all three factors will be incorporated into an empirically derived b.

The fitted equation for the mammals ($n = 321$) in Figure 3.2 is given by

$$\dot{V}_{O_2} = 3.45\,m^{0.713}, \qquad (3.3)$$

where $\dot{V}_{O_2}$ is rate of oxygen consumption (cm^3/h) and m is mass (g). This is a more appropriate standard for mammals than the Kleiber curve, even in a later reincarnation (Kleiber 1947), which dealt principally with domestic species, a highly biased, production-oriented standard (McNab 1988a). The power in Equation 3.3 is significantly different from both 0.67 and 0.75. Hayssen and Lacy (1985), in a study of 293 species, found a scaling power equal to 0.693. Heusner (1991), using a larger ($n = 391$) but somewhat less narrowly defined set, obtained a fitted power of 0.710.

A doubling of body mass leads to a 64% ($2^{0.713} = 1.64$) increase in basal rate in mammals, which means that basal rate increases proportionally less

BOX 3.1
Total or Mass-Specific Rates of Metabolism?

The unequal impact of mass on total and mass-specific rates of metabolism has encouraged biologists to be selective in their use of units. Do large mammals have greater or lesser rates of metabolism than small species? Both, depending on the units used. The question remains whether total or mass-specific units are biologically most meaningful (McNab 1999). Total units might refer to ecologically relevant limitations to metabolism, such as food quality or quantity, whereas mass-specific units might reflect physiologically based limitations, such as mitochondrial number or enzyme content, but they too increase with body size (Else and Hulbert 1985b), unless they are expressed in mass- or volume-specific units. The fact is that only total expenditures are measured, and either both units have the same meaning because they are conceptually and mathematically tied to each other, or one is the result of mathematical manipulation without biological significance.

The dichotomy between reasoning based on total and mass-specific units can be seen in the comparative energetics of temperate and Arctic endotherms: does a large body mass in cold climates (Bergmann's rule) increase or reduce energy expenditure? A widely accepted view suggests that energy expenditure in a cold climate can be decreased by reducing the surface-to-volume ratio (see Section 5.3.2). Yet, large individuals have higher total rates of metabolism (and thermal conductances) under all environmental conditions (Figure 3.3). An increase in size, presumably, cannot occur unless an animal can increase its total food intake. Nearly every quantitative justification for a large mass derived from mass-specific rationales can be replaced by another analysis that uses total rate of metabolism (see Section 5.3.1). Although mass-specific units of metabolism may be convenient to use, the only animals that live on a per-gram or per-kilogram basis weigh exactly 1 g or 1 kg, respectively.

Does this analysis mean that mass-specific rates of metabolism are meaningless (as Kleiber [1970] once suggested), simply the result of mathematical manipulation? Not necessarily. Because mass can

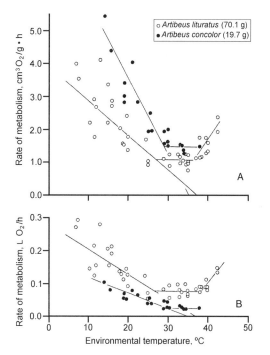

Figure 3.3 Rate of metabolism in two fruit bats, the larger (70.1-g) *Artibeus lituratus* and the smaller (19.7-g) *A. concolor*, as a function of environmental temperature. The data on metabolism are expressed in (A) mass-specific and (B) total units. Notice that a higher rate of metabolism is found in the larger or in the smaller species at all temperatures, depending on the units of metabolism. Source: Modified from McNab (1982).

be represented as having an equivalency in energy, a mass-specific rate with the units of joules per gram·hour (J/g·h) can be expressed with the units joules per joules·hour (J/J·h), which equals J per hour (1/h). In other words, a mass-specific rate of metabolism is equivalent to a turnover constant (Kleiber 1975). This constant is inversely related to mass: small species have high turnover constants and large species, low turnover constants. The significance of this constant and its relation to body size is addressed in Section 5.3.1.

than the increase in body mass. That is, as a consequence of scaling basal rate proportional to mass raised to a power less than 1.0, mass-specific rates are proportional to mass raised to a negative power, which is shown by dividing both sides of Equation 3.3 by mass:

$$\frac{\dot{V}_{O_2}}{m} = \frac{3.45m^{0.713}}{m} = 3.45m^{-0.287}. \qquad (3.4)$$

A doubling of body mass leads to a 18% reduction in mass-specific basal rate (i.e., $2^{-0.287} = 0.82$). Thus, an increase in mass results in an increase in total basal rate and a reduction in mass-specific basal rate: dividing rates by body mass does not make them independent of body mass. To obtain a mass-independent measure of basal rate would require total rates to be divided by m^b, then the values would equal a:

$$\frac{\dot{V}_{O_2}}{m^{0.713}} = \frac{a \cdot m^{0.713}}{m^{0.713}} = a,$$

a value that will characterize each species relative to the mean curve (for a discussion of this problem, see Ultsch [1995]).

Various complications exist in analyzing the influence of body size on rate of metabolism. One is the unequal distribution of species by body mass, an inequality that differentially weighs points along the curve. Of the sample of 321 species reported by McNab (1988a), 124 had a mass between 10 and 100 g, whereas only 21 had a mass between 10 and 100 kg. Another complication is associated with how data are segregated. They can be divided by family, class, and order. An alternative division is by ecological or behavioral categories, such as food habits or climatic distribution. In this sample of mammals, the mean power for basal rate is 0.712 ± 0.013 ($n = 11$) when grouped by orders or 0.684 ± 0.028 ($n = 16$) when grouped by food habits (using only those groups that have six or more species). These differences raise further doubts that the power of mass is a fundamental property, especially if it is estimated from a heterogeneous array of species. The level at which scaling occurs, as is reflected in the coefficient a, however, is important, but unlike b, no one has suggested that a particular value of a is "fundamental." Indeed, within mammals a and b are reciprocally correlated (McNab 1988a). Marsupials have basal rates that average 71% those of eutherians (McNab 1988a),

but their basal rate scales in a manner similar to that in eutherians (MacMillen and Nelson 1969, Dawson and Hulbert 1970).

The assumption that scaling relationships can be fundamentally represented by a power function has been questioned. Smith (1984, p. R158) argued that this equation is usually justified "... by two poor criteria: 1) the highly distorted visual appearance of points on a log-log plot and 2) correlation coefficients." Its conceptive justification is that functions (e.g., growth, metabolism) are multiplicative and not additive, but Smith maintained that the power function has "... falsely simplified a complex problem." As a (partial) remedy he recommended comparing species within a narrow size range, which is essentially equal to examining the residual variation around a fitted allometric curve, what Smith called a "narrow allometry." Indeed, Heusner (1982a, 1982b) noted that much more attention should be paid to a than heretofore has occurred, and argued that its "... explanation ... is, and remains, the central question in comparative physiology" (Heusner 1991, p. 34). This view seems to be exaggerated, but an analysis of a would be of value and has been started (see Section 5.4).

3.3.2 Scaling rate of metabolism in other vertebrates. The pattern described for mammals by Equations 3.3 and 3.4 is widely applicable to other organisms (Table 3.2). Birds other than passerines have slightly higher basal rates than eutherians, and passerines generally have still higher basal rates, with basal rates in both groups scaling proportionally to $m^{0.73}$ (Lasiewski and Dawson 1967, Aschoff and Pohl 1970). Some doubt has existed that passerines truely have higher basal rates than other birds. Thus, alcids (Johnson and West 1975, Gabrielsen et al. 1991) and shorebirds (Kersten and Piersma 1987) have high basal rates, and many birds other than passerines with low basal rates (e.g., hummingbirds, swifts, nightjars, etc.) have distinctive habits often associated with low basal rates (see McNab 1988b). Furthermore, a size difference is often found between passerines and other birds (few passerines weigh more than 500 g, and only one more than a kilogram [*Corvus corax*]). These observations and additional measurements on the energetics of birds led Prinzinger and Hänssler (1980) to conclude that no difference was found in the relationship between basal rate and body mass in passerines and other birds, a

Table 3.2 Scaling of standard rate of metabolism in vertebrates*

Group	n	T_a	a	b	r^2	Source
Fishes						
All freshwater	266	20	0.297	0.81	0.94	Winberg 1956
Acipenseridae	33	20	0.391	0.81	0.99	Winberg 1956
Salmonidae	30	20	0.498	0.76	0.96	Winberg 1956
Cyprinidae	50	20	0.336	0.80	0.87	Winberg 1956
Cyprinodontidae	25	20	0.192	0.71	0.93	Winberg 1956
All marine	123	20	0.321	0.79	0.94	Winberg 1956
All fishes	389	20	0.316	0.78	0.95	Winberg 1956
Amphibians						
Caudata	43	5	0.020	0.81	0.85	Gatten et al. 1992
	100	15	0.045	0.81	0.83	Gatten et al. 1992
	39	20	0.068	0.80	0.90	Gatten et al. 1992
	61	25	0.095	0.80	0.91	Gatten et al. 1992
Anura	26	5	0.049	0.81	0.84	Gatten et al. 1992
	66	15	0.104	0.79	0.80	Gatten et al. 1992
	98	20	0.103	0.82	0.86	Gatten et al. 1992
	71	25	0.174	0.84	0.92	Gatten et al. 1992
Gymnophiona	3	20	0.050	0.93	0.98	Gatten et al. 1992
	4	25	0.067	1.06	0.91	Gatten et al. 1992
Reptiles						
Lizards	24	20	0.096	0.80	0.64	Bennett & Dawson 1976
	26	30	0.240	0.83	0.74	Bennett & Dawson 1976
	19	37	0.424	0.82	0.67	Bennett & Dawson 1976
Diurnal lizards	26	Preferred	0.327	0.82	0.54	Text
Snakes	35	20	0.120	0.86	0.64	Bennett & Dawson 1976
	13	30	0.280	0.76	0.83	Bennett & Dawson 1976
Boidae	14	20	0.032	0.83		Chappell & Ellis 1987
	14	30	0.083	0.83		Chappell & Ellis 1987
	14	34	0.122	0.83		Chappell & Ellis 1987
Varanidae	10	25	0.117	0.87	0.98	Thompson & Withers 1997
	10	35	0.209	0.92	0.99	Thompson & Withers 1997
Turtles	10	20	0.066	0.86	0.61	Bennett & Dawson 1976
All reptiles	73	20	0.102	0.80	0.69	Bennett & Dawson 1976
		20	0.103	0.80		Andrews & Pough 1985
	44	30	0.278	0.77	0.83	Bennett & Dawson 1976
		30	0.248	0.80		Andrews & Pough 1985
Birds						
Nonpasserines	72	TN	4.60	0.72		Lasiewski & Dawson 1967
Passerines	48	TN	7.50	0.72		Lasiewski & Dawson 1967
Nonpasserines$_\rho$	17	TN	4.00	0.73		Aschoff & Pohl 1970
Nonpasserines$_\alpha$	17	TN	4.95	0.73		Aschoff & Pohl 1970
Passerines$_\rho$	14	TN	6.60	0.73		Aschoff & Pohl 1970
Passerines$_\alpha$	14	TN	8.10	0.70		Aschoff & Pohl 1970
All birds	254	?	6.96	0.67	0.96	Reynolds & Lee 1996
Mammals						
All mammals	(13)	TN	3.42	0.75		Kleiber 1932
All mammals	293	?	4.33	0.69	0.75	Hayssen & Lacy 1985
All mammals	293	?	4.33	0.75	0.91	Hayssen & Lacy 1985
Metatheria	42	?	2.50	0.70	0.74	Hayssen & Lacy 1985
Eutheria	248	?	4.46	0.71	0.78	McNab 1988a
All mammals	321	TN	3.45	0.74	0.92	McNab 1988a
Metatheria	46	TN	2.49	0.74	0.92	McNab 1988a
Eutheria	272	TN	3.53	0.72	0.77	McNab 1988a

*All equations are in the form $\dot{V}_{O_2} = a \cdot m^b$, where $\dot{V}_{O_2}$ is in units of cm³O₂/h and body mass is in grams. In birds, ρ indicates measurements made during the resting period; α, measurements made during the active period; n, number of species in a sample; TN, that endotherms were measured in thermoneutrality; ?, questioning whether all measurements were made in thermoneutrality (see McNab 1997).

conclusion followed by Reynolds and Lee (1996). The difficulty with completely accepting the analysis of Reynolds and Lee is that one cannot tell whether strictly defined basal rates alone were used. In any case, a question exists as to whether a common standard should be used in birds, which is parallel to whether a single mammal curve should be used, or whether marsupial and eutherian standards should be used. The answer in both cases probably depends on the questions being asked. If one is interested in whether this or that marsupial, vole, or woodpecker has a high or low basal rate, then a narrower curve might be used as a standard, but if a broader context is appropriate, then a general standard would best be used.

Bennett and Dawson (1976) and Andrews and Pough (1985) demonstrated that a power relationship also describes the resting rates of metabolism in reptiles, except that the level of the relationship varies with ambient and body temperature (Table 3.2). When resting rates of metabolism in inactive, postabsorptive ectotherms are measured at a particular temperature, they are referred to as *standard* rates of metabolism; they differ from the basal rates of endotherms by not including a minimal cost of endothermy (see Section 5.2). In both studies the power by which mass is raised is approximately 0.80. Andrews and Pough (1985) indicated that the intraspecific b is less than the interspecific b. In an extensive study of boid snakes Chappell and Ellis (1987) showed that the interspecific b equaled 0.81, similar to the values found in other reptiles, whereas the mean intraspecific b was 0.73, significantly less than the interspecific mean. At least one of the boids, *Epicrates cenchria*, has a significantly distinguishable scaling power (0.56). As Heusner (1982a) suggested, intraspecific estimates of a increase with mass, which is presumably why interspecific b is higher than the mean intraspecific b. However, Thompson and Withers (1997) showed in *Varanus* species that standard rate of metabolism scaled interspecifically proportionally to $m^{0.92}$, whereas intraspecific b values varied between 0.43 and 1.20 (pooled mean, 0.97).

An extensive survey of the scaling of amphibian standard rates (Table 3.2) indicated that in caudates and anurans the power b is approximately 0.81, anurans having rates 1.5 to 2.5 times those of caudates (Gatten et al. 1992). Caecilians have still lower rates, although the data are few. The data on fishes are characterized by a power equal

to approximately 0.80, the rates generally being higher than those found for amphibians and most reptiles (Winberg 1956, 1961). The data reported by Winberg represent a rather narrow taxonomic distribution with a heavy emphasis on salmonids. Clarke and Johnston (1999) recently reported data on a broader diversity with a mean power equal to 0.79.

Hemmingsen (1960) maintained that unicellular organisms, multicellular poikilotherms, and homeotherms have resting rates that are proportional to approximately $m^{0.75}$ over a wide range in body mass (Figure 3.4). Such a pattern precludes heat loss from dictating heat production because neither unicellular organisms nor multicellular poikilotherms maintain a temperature differential with the environment by the internal generation of heat. The principal differences among these groups are in the level (a) at which such scaling occurs: the difference between homeotherms and multicellular poikilotherms undoubtedly reflects the cost of endothermy, and the difference between multicellular poikilotherms and unicellular organisms may reflect the cost of multicellularity.

Many physiological functions other than energy expenditure are scaled to body size, summaries of which are found in the works by Peters (1983), Calder (1984), and Schmidt-Nielsen (1984). Many of these relations are addressed in the appropriate

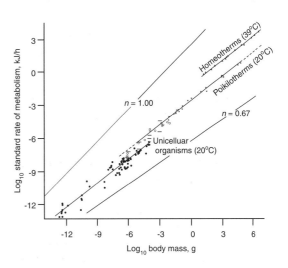

Figure 3.4 Log_{10} standard rate of metabolism in unicellular organisms (at 20°C), multicellular poikilotherms (at 20°C), and homeotherms (with a body temperature of 39°C) as a function of log_{10} body mass. Source: Modified from Hemmingsen (1960).

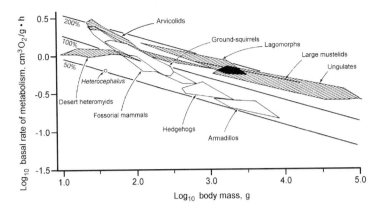

Figure 3.5 Log_{10} mass-specific basal rate of metabolism in various groups of eutherian mammals as a function of log_{10} body mass. Source: Modified from McNab (1983a).

chapters. One such function is thermal relations, which is addressed next.

3.4 SCALING THERMAL RELATIONS

Another scaling relation between rate of metabolism and body mass has been described (McNab 1983a). In this relation, basal rate shows a much shallower increase with mass than a Kleiberian relation shows:

$$\dot{V}_{O_2} = 15.56\,m^{0.33}, \tag{3.5}$$

where mass is in grams (see Figure 3.2). This relation is much steeper on a mass-specific basis ($m^{-0.67}$). The curve separates rigid endothermic homeotherms from endotherms that enter daily torpor and from ectothermic poikilotherms. It was called the *boundary curve for endothermy* because it appears to define the minimal conditions for continuous endothermy.

The significance of the boundary curve can be seen by a detailed examination of thermoregulation in relation to this curve.

1. Some mammals that never enter torpor, such as microtid rodents, have high basal rates that are scaled parallel to the mean mammal curve down to a mass of 50 g, below which they follow the boundary curve (Figure 3.5; McNab 1992b). Mammals that have low basal rates at large masses often follow the boundary curve at even larger masses (see fossorial mammals, ground-squirrels, hedgehogs, and armadillos; Figure 3.5).

2. The consequence in mammals of having basal rates that fall below those expected from the

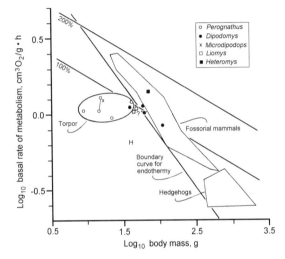

Figure 3.6 Log_{10} mass-specific basal rate of metabolism in fossorial mammals, heteromyid rodents, and hedgehogs as a function of log_{10} body mass. The data in heteromyid rodents are examined with respect to their propensity to enter daily torpor and to the boundary curve for endothermy (see text). The question mark represents uncertainty as to whether a particular species enters torpor. The all-mammal and 2× all-mammal curves are indicated. Source: Modified from McNab (1983a).

boundary curve is seen in heteromyid rodents (Figure 3.6) and marsupials (Figure 3.7): small species readily enter torpor, and the division between those that enter torpor and those that do not coincides with the boundary curve. The superficial "reason," then, why the small species that belong to these groups of mammals enter daily torpor is that they do not have basal rates that are high enough to ensure continuous endothermy.

Similar reasoning accounts for daily torpor in white-toothed shrews (Crocidurinae), which fall below the boundary curve, and for the general absence of torpor in red-toothed shrews (Soricinae), which have basal rates above the boundary curve (McNab 1983a, 1991). This relationship emphasizes the connection that exists among the level of metabolism, temperature regulation, and body size.

3. A similar pattern in the occurrence of daily torpor is found in small birds: species with basal rates that fall on, or below, the boundary curve often enter torpor (Figure 3.8). Such species include small goatsuckers, small pigeons, mouse-birds, some swifts, and all hummingbirds.

4. Precisely regulating endotherms conform to one of two equations, depending on whether their mass is greater or less than approximately 50 g, a condition suggesting that conformation to a single scaling relation at a small mass leads to the sacrifice of endothermy.

Another consequence of the existence of the boundary curve (in continuous endotherms) is that the standard rate of metabolism is not described by one power curve, but at least by two curves that intersect at approximately 50 g (McNab 1983a). No obvious reason exists as to why a single power function should describe the scaling relation in all vertebrates—its justification is practical rather than theoretical. Only Brown et al. (1993) also suggested that a nonlinear (or bilinear) relationship exists for the allometry of the life-history, ecological, and by extension, physiological characteristics of vertebrates. The implication of bilinear scaling relationships simply would be that different rules apply to large and to small species, which is quite reasonable in terms of the scaling of thermal biology.

The boundary curve has weaknesses, which are found in species that have basal rates falling below this curve but show no evidence of entrance into reversible torpor. For instance, some small tropical cave swiftlets (*Collocalia*) have basal rates that are only 51% to 63% of the values expected from the boundary curve, but show no evidence of a spontaneous entrance into torpor (McNab and

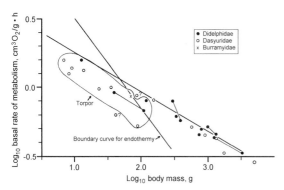

Figure 3.7 $\log_{10}$ mass-specific basal rate of metabolism in marsupials as a function of $\log_{10}$ body mass. The propensity of marsupials to enter into torpor is examined relative to the boundary curve for endothermy. Question marks represent uncertainty as to whether a particular species enters torpor. The all-mammal curve is indicated. Source: Modified from McNab (1983a).

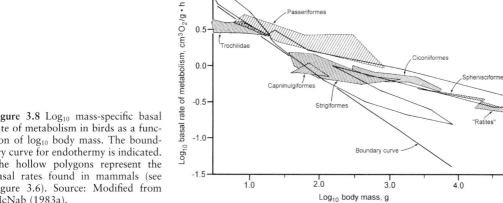

Figure 3.8 $\log_{10}$ mass-specific basal rate of metabolism in birds as a function of $\log_{10}$ body mass. The boundary curve for endothermy is indicated. The hollow polygons represent the basal rates found in mammals (see Figure 3.6). Source: Modified from McNab (1983a).

Bonaccorso 1995a). The absence of torpor does not necessarily mean they will not enter torpor, but it may reflect either that adequate stimuli are absent or that the boundary curve was established with temperate and Arctic species and that the endothermy of small tropical species is different. Another difficulty is that some mammals with basal rates that fall above the boundary curve, especially ground-squirrels (*Spermophilus*, *Marmota*) and large carnivores (e.g., *Taxidea*, *Ursus*), enter torpor. The torpor of these species, however, is seasonal (i.e., hibernation or aestivation) and represents an actively regulated state: it is not an inability to thermoregulate, as tends to be the case in those species with basal rates that fall below, especially far below, the boundary curve.

The boundary curve also has significance for poikilotherms (Figure 3.9). Most small to moderately sized poikilotherms have such low standard rates of metabolism, even at relatively high ambient and body temperatures, as to fall far below the boundary curve. These vertebrates include lizards, most snakes, most turtles, all amphibians, and most fishes. Only a few ectotherms cross the boundary curve (e.g., the largest turtles, snakes, and presumably the largest fishes) and do so because of a large mass, not because they have high resting rates of metabolism. These species cross the boundary curve without acquiring endothermy. This observation points out the asymmetrical nature of the boundary curve: continuous endothermy normally cannot be maintained below this curve, but endothermy need not develop above the curve.

Some vertebrates belonging to taxa that typically are ectothermic show an intermediate form of endothermy. This behavior coincides with a rate of metabolism that equals or exceeds the rates described by the boundary curve. Thus, tuna are endothermic to varying degrees (see Section 5.9.2), and the skipjack (*Katsumonus pelamis*) has resting rates that fall above the boundary curve (Figure 3.9). Female Indian pythons (*Python molurus*) incubate their eggs through the internal generation of heat (see Section 5.9.3); their energy expenditure during incubation exceeds the level described by the boundary curve (Hutchison et al. 1966, Van Mierop and Barnard 1978), but falls below this rate when the python is not incubating at the same ambient temperature, that is, when the python is strictly poikilothermic (Figure 3.9). The marine leatherback turtle (*Dermochelys coriacea*) may be endothermic (Frair et al. 1972, Mrosovsky 1980), and given its large mass (up to 500–750 kg), probably has a rate of metabolism that exceeds the values expected from the boundary curve (Paladino et al. 1990, Lutcavage et al. 1992). At large masses, endothermy is clearly compatible with rates of metabolism that are well below the levels typical of birds and mammals. This condition may well have applied to large dinosaurs (see Section 4.15).

The boundary curve for endothermy suggests that the difference between strict endothermy and

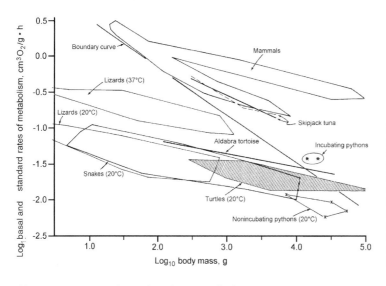

Figure 3.9 $\log_{10}$ basal rate of metabolism in mammals and standard rate of metabolism in skipjack tuna (*Katsuwomus pelamis*), lizards at 37 and 20°C, snakes and turtles at 20°C, Aldabra tortoise (*Geochelone gigantea*), nonincubating pythons (*Python molurus*) at 20°C and incubating pythons, as a function of $\log_{10}$ body mass. The boundary curve for endothermy is indicated. Source: Modified from McNab (1983a).

classic ectothermy is quantitative, not qualitative. Any organism theoretically can be an endotherm at any mass if it has a sufficiently high rate of metabolism, or at any rate of metabolism if it is sufficiently large. Thus, the periodic endothermy found in many insects during activity and found in the spadices of some plants occurs when rates of metabolism equal or exceed the values expected from the boundary curve. So the question as to which vertebrates are endotherms and which are ectotherms is likely to be answered in terms of "economics," that is, whether the energy to pay for such expenditures is available and whether a sufficient benefit accrues to organisms by being endothermic (see Pough 1980a and Section 4.16).

3.5 CHAPTER SUMMARY

1. Vertebrates with small body masses are found along a continuum from ectothermic poikilotherms, which have low rates of metabolism, to endothermic homeotherms, which have high rates of metabolism.

2. Total resting rates of metabolism scale proportionally to mass raised to a power usually between 0.65 and 0.80, a consequence of which is that mass-specific resting rates of metabolism decrease with an increase in body mass. Great variety exists in the level at which this scaling occurs.

3. The factors determining the power of mass in the scaling of standard rates of metabolism are complex, but intraspecific powers tend to be less than interspecific powers.

4. Given the complexity with which standard rates are determined, a great emphasis on the scaling power b is probably unwarranted.

5. A greater emphasis should be placed on the factors responsible for determining the level of energy expenditure of vertebrates, that is, in determining the scaling coefficient a.

6. Total rates of metabolism have the greatest relevance in ecological and evolutionary contexts; mass-specific units at best represent turnover constants. Total and mass-specific units of metabolism must lead to the same conclusions because they are mathematically derived from each other.

7. A second scaling relation describes the lowest rate of metabolism that is compatible with continuous endothermy at a particular mass; its scaling power is approximately 0.33.

8. In theory, any organism can be endothermic if it has a sufficiently high rate of metabolism or if it is sufficiently large; the absence of endothermy can be traced to restrictions in energy availability, the absence of a gain associated with endothermy, or a small body size.

9. The difference between ectotherms and endotherms may be blurred at large body masses.

4 Adaptation to Temperature Variation: Poikilothermy-Ectothermy

4.1 SYNOPSIS

Poikilotherms collectively tolerate body temperatures from −2 to 45°C, although all individuals and species have much narrower limits of tolerance. Poikilotherms are ectothermic in that their heat content is principally determined by the environment. Within the acceptable range in body temperature, physiological rates increase with temperature and show thermal acclimatization. Poikilotherms that are exposed to freezing temperatures have a few options: they can avoid low temperatures by seeking shelter, by accumulating compounds that decrease the freezing point of body fluids, by living in a supercooled state, or by selectively permitting extracellular fluids to freeze; the use of these strategies depends on whether the species is aquatic or terrestrial. Some species that live in thermally complex environments behaviorally control body temperature by the selective absorption of radiant energy. Most physiological functions are maximized in the range of body temperatures preferentially maintained in the field. Dinosaurs probably were thermally diverse, given their great range in body size; most large species likely were homeothermic as a result of their large mass, and probably were ectothermic. Some small, highly active predatory dinosaurs may have been endothermic homeotherms. Ectothermy, often coupled with behavioral temperature regulation, is an effective means of balancing energy budgets at small body masses, in species that use seasonal or rare foods, and in physically harsh environments.

4.2 AN INTRODUCTION TO TERMINOLOGY

One of the most pervasive factors in the environment is temperature, which in terrestrial environments is highly variable. Organisms, in so far as they are elaborate chemical "machines," are affected by this variation. Various relevant temperatures, including radiant, operative, air, soil, and water temperatures, exist in the environment.

All vertebrates do not respond to a change in temperature in the same manner. Those in which body temperature equals ambient temperature, or ambient temperature plus a small differential, over an appreciable range in ambient temperature (Figure 4.1) are called *poikilotherms*. As a result, the rates of chemical reactions (collectively referred to as *metabolism*) in poikilotherms are low at low environmental and body temperatures and high at high environmental and body temperatures (Figure 4.2). Most poikilotherms, then, tend to be sluggish at cool to cold temperatures and active at high temperatures, a correlation that, as we will see, has its limits and is subject to adjustment.

In contrast to poikilotherms, *homeotherms* tend to maintain body temperature constant over a considerable range in ambient temperature (Figure 4.1). Homeotherms maintain body temperature independent of, and usually greater than, ambient temperature by producing enough heat to replace heat lost to the environment, which means that most homeotherms are characterized by relatively high rates of metabolism. Carl Bergmann (1847) first suggested the terms *poikilotherm* and *homeotherm* (= homoiotherm).

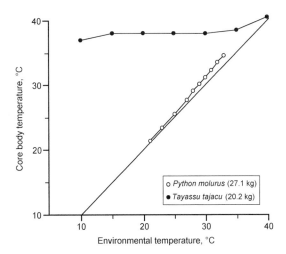

Figure 4.1 Body temperature in poikilotherm (*Python molurus*) and a homeotherm of similar mass (the collared peccary [*Tayassu tajacu*]) as a function of environmental temperature. Sources: Data taken from Zervanos (1975) and Van Mierop and Barnard (1978).

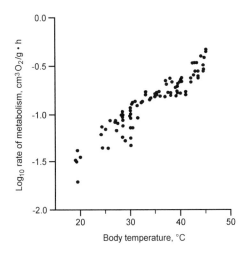

Figure 4.2 Log$_{10}$ rate of metabolism in the desert iguana (*Dipsosaurus dorsalis*) as a function of environmental temperature. Source: Modified from Dawson and Bartholomew (1958).

Because poikilotherms tend to have body temperatures dictated by ambient temperatures, whereas those of homeotherms are usually determined by the rate at which heat is produced through metabolism, Cowles (1962) suggested that the terms *poikilotherm* and *homeotherm* be replaced by the terms *ectotherm* and *endotherm*, respectively. Whereas the older terms refer to the constancy of body temperature, his terms refer to

the source of the body's heat: external or internal to the animal's body. Cowles's terms, therefore, are complementary, not equivalent, to those of Bergmann. Thus, although most homeotherms are also endothermic, an ectothermic homeotherm is conceivable—for example, a fish living at great depths in the ocean; a giant tortoise living in a benign, cloudy, tropical climate; or (on a periodic basis) a lizard basking in the sun. An endothermic poikilotherm is somewhat more difficult to imagine unless incompetent endotherms, like the naked mole-rat (*Heterocephalus glaber*), are included (Buffenstein and Yahav 1991).

Some additional terms describing the thermal biology of vertebrates are occasionally encountered. For example, Cowles broke ectotherms into *thigmotherms* and *heliotherms*, depending on whether the external heat source is a substrate or the sun, respectively. *Heterotherm*, as noted, is a vague term for species that do not fall into one clear category or another, such as mammals that enter torpor or fish that marginally thermoregulate with heat produced by metabolism. The difficulty with all performance categories is that they attempt to place each organism exclusively into one compartment or another, but these "intellectual boxes" are of human invention and nature is discontinuously continuous.

This chapter is limited to the thermal characteristics and adaptations of ectothermic poikilotherms, whether they are thermal conformers or behavioral thermoregulators; discussion of the states intermediate to ectothermy and endothermy is delayed until Chapter 5, which concerns endothermy.

4.3 THE LIMITS OF TEMPERATURE TOLERANCE

The range of temperatures found on Earth is much greater than the range tolerated by poikilotherms in an active state. The coldest recorded air temperature is −88.3°C (Antarctica) and the warmest recorded air temperature is 58.0°C (Libya), but poikilotherms collectively tolerate (body) temperatures only from about −2 to about 45°C, although some thermophilic bacteria inhabit boiling water.

The lower limits of temperature tolerance in poikilotherms are often associated with the temperature at which water freezes because their composition is 60% to 80% water. For example, intercellular fluids freeze at −2 to 0°C, although some vertebrates tolerate lower body temperatures by selectively accumulating solutes that depress the freezing point of body fluids, or by permitting some fluid to supercool (i.e., to be at temperatures below the freezing point without ice crystal formation) (Section 4.7). The freezing of body fluids has two consequences: structural damage to cells, because of intercellular ice crystal formation, and the osmotic movement of water from cells to intercellular space, which makes cellular enzymes precipitate. The lower limits of temperature tolerance in some poikilotherms, however, are well above 0°C, as has been shown in some tropical, freshwater fish that have become established in southern Florida (Shafland and Pestrak 1982). Thus, the South American cichlid *Astronotus ocellatus* stops feeding at 14.5°C, loses equilibrium at 13.6°C, and dies at 12.9°C.

The upper limits to temperature tolerance also are imperfectly understood. Many poikilotherms have lethal body temperatures between 40 and 45°C; desert pupfish (*Cyprinodon* spp.) tolerate temperatures as high as 44°C (Elliott 1981). No vertebrates are known to tolerate core body temperatures as high as 50°C. A common explanation is that upper lethal temperatures cause irreversible denaturation of the enzymes critical for survival. Such denaturation has been rarely shown. Furthermore, reaction rates can be reduced severely at temperatures below those at which enzymes are irreversibly damaged (Hochachka and Somero 1984). Results of studies on the temperature tolerance of organisms living in the runoff from hot springs suggest that tolerance to high temperatures is reduced in species with a high level of morphological complexity, especially as reflected in terms of membrane or neural structures. Mammals at a lethal body temperature often show brain damage, such as a malfunctioning of the middle ear (e.g., walking in circles), before any other apparent difficulty. Whether the susceptibility to high body temperature reflects directly a biochemical limitation to adaptive modification, or whether it simply demonstrates that organisms have not developed biochemical mechanisms of tolerance to high temperatures because they never encounter them, is unclear.

4.4 THE INFLUENCE OF TEMPERATURE BETWEEN LETHAL LIMITS

Rate of metabolism, like all rates of chemical processes, increases with an increase in temperature (Figure 4.2). To make effective comparisons of such a variable character as rate of metabolism, we use *standard* rates, which are measured at a particular temperature when the animals are quiescent, postabsorptive, and adult. Arrhenius (1912) first suggested that the velocity of a chemical reaction, $\dot{V}$, is given empirically by the following relation:

$$\dot{V} = Ne^{-Ea/RT},$$

where N is a frequency factor (which measures the number of molecules colliding), e is the base of natural logarithms, E_a is the energy of activation, R is the gas constant, and T is temperature in Kelvin. From this view an increase in rate with temperature is due to an increase in the kinetic energy of the substrates of a reaction and therefore the increased fraction of the molecular population that has the energy of activation. Two reactions, or two organisms, can have the same upper and lower limits of tolerance but still differ in their sensitivity to a change in temperature. A difficulty with using the Arrhenius relation is that it involves E_a, which may be an appropriate characterization for a chemical reaction, but its significance for an intact organism is unclear. It has been suggested to represent the energy of activation for the limiting step in a chemical pathway. But the inability to demonstrate the correctness of this or any other interpretation has led biologists to seek a less complicated means of describing temperature sensitivity in intact organisms.

The sensitivity of an organism to a change in temperature is usually described by Q_{10}, which is the ratio of a rate at one temperature to the rate at a temperature 10°C less:

$$Q_{10} = \frac{\dot{V}_{T+10}}{\dot{V}_T}. \qquad (4.1)$$

This relation has as its basis the assumption that the rate is proportional to e^{KT}. Thus, when the log of the rate is plotted as a function of temperature, the curve is linear (Figure 4.2), at least over a limited range in temperature. When the difference in temperature is not 10°C, Q_{10} can be calculated from

$$Q_{10} = \left(\dot{V}_2/\dot{V}_1\right)^{\left[\frac{10}{T_2-T_1}\right]}. \qquad (4.2)$$

If the rate is desired at some temperature T_2, it can be calculated easily from

$$\dot{V}_2 = \dot{V}_1 Q_{10}^{\left[\frac{T_2-T_1}{10}\right]}. \qquad (4.3)$$

Q_{10} has no theoretical pretense. This estimate of temperature sensitivity is, at best, approximate, because Q_{10} is itself a function of temperature, which means that reactions, or sets of reactions, do not precisely conform to the relation $\dot{V} = k \cdot e^{K \cdot T}$. At high ambient and body temperatures, poikilotherms increasingly depart from an exponential relation, presumably because of an increase in enzyme inactivation. Q_{10} normally has a value between 2 and 3.

4.5 THERMAL ACCLIMATIZATION AND ACCLIMATION

The general relationship described in Figure 4.2 holds for all poikilotherms, although no single rate-temperature curve applies to all species or even to all individuals in a population. In fact, these curves usually show a reversible change as a result of thermal experience. When this change occurs in nature, it is referred to as *acclimatization*, implying that it is a response to the complex seasonal change of conditions in the environment. A similar change in a rate-temperature curve is often produced in laboratories by exposing animals to selected factors, such as temperature or photoperiod. This response is usually called *acclimation*, to distinguish it from the response to a potentially more complex, natural signal. Significant variations between species or individuals in different seasons occur in the lethal limits of tolerance (the "critical" thermal maxima and minima), in Q_{10}, and in the rates of reaction at any given temperature.

4.5.1 Limits of thermal tolerance. *Lethal limits of temperature* are defined as the ambient temperatures at which 50% of exposed animals die. These "critical" temperatures, however, are not absolute: the time to death is an inverse function of the lethal temperature (Figure 4.3). One way around this

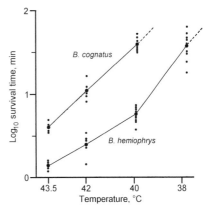

Figure 4.3 Log_{10} survival time in toads (*Bufo cognatus* and *B. hemiophrys*) as a function of the temperature to which they were exposed. Source: Modified from Schmid (1965b).

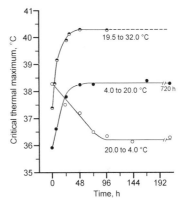

Figure 4.4 Critical thermal maximum in the salamander *Diemictylus viridescens* as a function of time of exposure to various temperatures and thermal experience. Source: Modified from Hutchison (1961).

complication is to use a fixed time period and measure the lethal temperature associated with that period. (The "end point" need not be death itself but can be some distinctive, temperature-dependent behavior, such as the inability of an animal to turn over or to maintain its position in a water column; these measures are often ecologically more meaningful than death itself, and they have the added advantage of conserving animals.) As noted, critical thermal maxima or minima vary with the thermal experience of the animal. For example, Hutchison (1961) showed that salamanders require 2 d to shift acclimation at 4°C to acclimation at 20°C, and 4 d to shift acclimation from 20°C to 4°C (Figure 4.4). His results consistently showed, as has the work of others, that acclimation to a cold temperature takes longer than acclimation to a warm temperature.

Independent of thermal experience, a correlation of critical thermal temperatures and geographic distribution has been demonstrated. For example, Brett (1970) demonstrated that tropical fish have higher upper and lower lethal limits than polar species (Figure 4.5). Hutchison (1961) showed that the critical thermal maximum in salamanders is correlated with habitat and geographic distribution, as have Brattstrom (1968) and Snyder and Weathers (1975).

The correlation of temperature tolerance with habitat and distribution raises the question of cause and effect. Do warm-tolerant salamanders live in warm environments *because* they are heat tolerant, or are they heat tolerant *because* they live in warm environments? Both factors, experience (= acclimatization) and potential (= genetics), appear to operate. Thus, even though thermal acclimatization does occur, it is limited (Brett 1944, Hutchison 1961): an exposure to high or low temperatures does not always lead to a change in the thermal limits to tolerance. The polygon of acceptable ambient temperatures bounded by the upper and lower limits of tolerance represents the set of all possible temperatures tolerated by a species, and is referred to as a *tolerance polygon*.

It presumably represents the genetic capacity of the individuals tested, although their tolerance also may be influenced by their thermal experience during development. The tolerance polygon for other individuals or other populations may be different, most likely due to a different genetic makeup. Because of practical problems with measuring an end point at low temperatures (i.e., poikilotherms become more sluggish with a fall in body temperature), most of the studies on thermal tolerance have examined only maximal temperatures, except in fish where the capacity for swimming is a low-temperature end point.

A tolerance polygon is not used equally over its temperature range. Thus, lethal temperatures in the brown trout (*Salmo trutta*) at which 50% of the individuals die vary inversely with time of exposure, so that 7-d or 1000-, 100-, or 10-min limits sequentially expand the polygon (Figure 4.6). Even within the 7-d tolerance zone, feeding occurs at only about one-half of the acceptable temperatures, so the area between the feeding limit and the 7-d boundary is not indefinitely tolerable. An even smaller area describes the region of maximal energy intake and growth.

Tolerance polygons vary greatly among species (Figure 4.7): the goldfish (*Carassius auratus*) has a wide temperature tolerance, whereas the brown

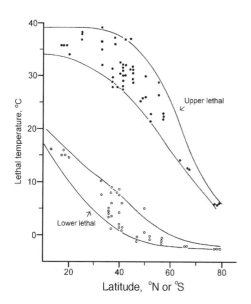

Figure 4.5 Upper and lower lethal temperatures in various fishes as a function of latitude. Source: Modified from Brett (1970).

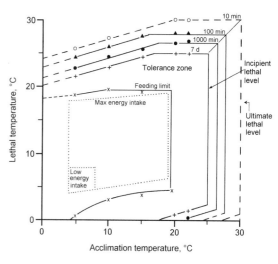

Figure 4.6 The thermal limit to feeding and lethal temperature in brown trout (*Salmo trutta*) as a function of acclimation temperature and tolerance time. Source: Modified from Elliott (1981).

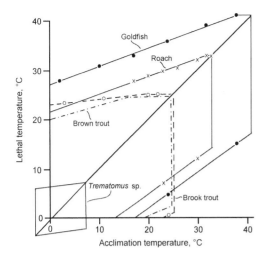

Figure 4.7 Lethal temperature as a function of acclimation temperature in teleosts: goldfish (*Carassius auratus*), roach (*Rutilus rutilus*), brown trout (*Salmo trutta*), brook trout (*Salvelinus fontinalis*), and an antarctic fish (*Trematomus* sp.). Sources: Modified from Brett (1970) and Elliott (1981).

(*S. trutta*) and brook (*Salvelinus fontinalis*) trouts have much narrower tolerances. A measure of the thermal tolerance of fish can be given by the areas of the tolerance polygon (in °C²): they vary from 1220°C² for the goldfish to 770°C² for the roach (*Rutilus rutilus*), 625°C² for the brook trout, and 582°C² for the brown trout. Some Pacific salmon (*Oncorhynchus*) have tolerance polygons as small as 450°C² (Brett 1956), which is only 37% of the area for goldfish. An antarctic *Trematomus* has a polygon that is maximally 83°C², or 7% of that for goldfish!

Differences among tolerance polygons led Elliott (1981) to divide freshwater teleosts into several thermal categories, including *stenotherms*, which are limited to a narrow range in temperatures; *mesotherms*, which tolerate an intermediate range in temperatures; and *eurytherms*, which tolerate a wide range in temperatures. By his classification, Elliott designated members of the Coregonidae, Thymalidae, and Salmonidae as cold-water stenotherms; Esocidae, most Cyprinidae, Gasterosteidae, Percidae, and Anguillidae as mesotherms; and some Cyprinidae—carp (*Cyprinus carpio*), goldfish, and grass-carp (*Ctenopharyngodon idella*)—as warm-water eurytherms, although the boundaries to these categories are not sharp. A caution should be noted that this work (as is so often the case) may give a

reasonably good picture of north-temperate fish, but almost no data are available from species native to the tropical lowlands.

These polygons can be used potentially as an analytical tool to study the thermal behavior of free-living poikilotherms. For example, if the tolerance polygon of a poikilotherm were known, or even if only the critical thermal maxima were known, the thermal experience of poikilotherms in the field could be estimated by collecting animals in the field and *immediately* measuring their critical temperatures in the laboratory. Field thermal experience could be examined as a function of behavior, season, and geography.

4.5.2 Q_{10}. The question of whether Q_{10} also is modified by thermal acclimatization has been debated. Scholander et al. (1953), in their classic study of thermal adaptation in arctic and tropic poikilotherms, argued that Q_{10} showed no acclimatization when these groups are compared *at normal environmental temperatures*, because the same tangent to the curve of metabolism on temperature applies to both sets of data (Figure 4.8). They concluded that habitat selection and thermal acclimatization of rates are more important than acclimatization of Q_{10}. Paradoxically, Rao and Bullock (1954) used the data of Scholander et al. to conclude that Q_{10} indeed shows acclimatization! They suggested that warm (tropic) and cold (arctic) acclimatized animals must be compared *at the same temperature*: then adaptation in Q_{10} can be shown. The difficulty with accepting Rao and Bullock's view is that by using a fixed temperature, one or both groups may be considered at ecologically inappropriate temperatures. The question of whether Q_{10} is adaptive or not indirectly addresses a more fundamental question, namely, whether Q_{10} is a "real" property of organisms or whether it is a mathematical fiction resulting from comparing a rate at two temperatures, each rate being adjusted independently of the other as a result of acclimatization (or acclimation). If that happens, Q_{10} will be said to have changed; nevertheless, it is not an "entity" but remains a quantitative abstraction.

4.5.3 Rates of reaction. Compared to temperate and tropical species, polar fishes have rate curves that are displaced to lower temperatures (Figure 4.8). This correlation holds even when the taxonomic affiliation of the fishes is taken into account (Edwards et al. 1970). As a consequence of this

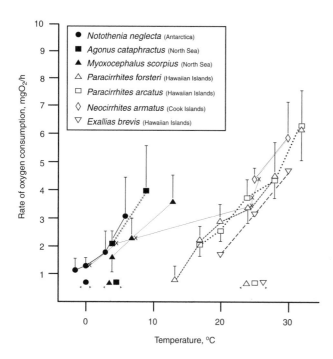

Figure 4.8 Rate of metabolism in sedentary marine fishes as a function of temperature. The ranges of ambient temperatures are indicated by arrows. The x's indicate rates for 50-g individuals. Source: Modified from Johnston et al. (1991).

adjustment, the upper and lower tolerance are shifted, as is the so-called optimal temperature, which is simply the temperature at which the maximal rate occurs. A similar pattern is seen in thermal acclimation when cold-acclimated individuals have a higher rate than warm-acclimated individuals at an intermediate temperature (Bullock 1955), a pattern also seen in climatically acclimatized plethodontids (Figure 4.9). These adjustments tend to make the rates of activity in poikilotherms independent of habitat temperature.

Temperature acclimatization has behavioral consequences other than setting the limits to temperature tolerance or changing the rate of metabolism. For example, frogs in the Northern Hemisphere, such as *Rana sylvatica*, tend to have earlier breeding periods at southern latitudes and later periods at northern localities, which means that these species reproduce at a narrower range in water temperature than is implied by their range in latitude. All frog species, however, do not follow the same acclimatization curve, for species at a given locality breed at different times in spring and summer depending on whether they have a northern or southern geographic distribution (Moore 1939, 1949). This thermal distribution of breeding frogs at a locality has the ecological consequence of temporally partitioning breeding sites and rep-

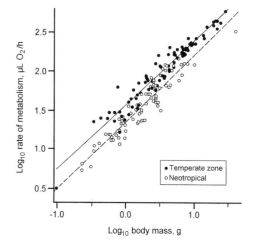

Figure 4.9 Log$_{10}$ rate of metabolism of temperate and tropical plethodontid salamanders at 15°C as a function of log$_{10}$ body mass. Source: Modified from Feder (1976b).

resents, presumably, a genetic or a geographic component to acclimatization, or both.

A striking example of thermal acclimatization is found in the activity and rate of metabolism of poikilotherms living in polar waters, which would be essentially uninhabitable if no thermal acclimatization occurred. Scholander et al. (1953),

Wohlschlag (1960, 1962, 1963, 1964), Yalph and Emerson (1968), Hemmingsen et al. (1969), and DeVries and Eastman (1981) described thermal acclimatization of the rate of metabolism in arctic and antarctic fish (Figures 4.8 and 4.10). They demonstrated that most arctic and antarctic fish have standard rates that are higher than would be expected in temperate fish at 0°C by extrapolation along a Q_{10} curve. One antarctic zoarchid (*Rhigophila dearborni*), however, had a standard rate that showed no evidence of thermal acclimatization. At −1.5°C, Holeton (1974) measured in arctic fish standard rates that ranged from those typical of cold acclimatization to those found in *R. dearboni* (Figure 4.10); he suggested that most of these differences were produced by spontaneous activity and doubted the existence of cold acclimatization in marine fish. Holeton, unfortunately, did not measure complete metabolism-temperature curves, so his judgement relied exclusively on data taken at one temperature. He used minimal values but presented no quantitative data on the level of activity under these conditions. In contrast, Somero et al. (1968) showed that brain tissue slices in *Trematomus bernacchii*, an antarctic nototheniid that has a high standard rate (Wohlschlag 1960), have high rates of oxygen consumption at low temperatures, indicating cold acclimatization at the tissue level. Although activity is a contaminant in the assessment of cold acclimatization, it is unlikely to account completely for the differences seen among fish (and other poikilotherms) originating from disparate climates.

Part of the problem in studying thermal acclimatization is the (understandable) desire to find a "general" pattern, when in fact we are dealing with a tremendous diversity in fishes. For example, largemouth bass (*Micropterus salmoides*), a temperate species, showed activity almost independent of temperature above 7°C, below which activity was greatly reduced (Lemons and Crawshaw 1985). These results were interpreted to mean that the central nervous system showed thermal acclimation, whereas other systems, such as the liver and muscles, showed little acclimation, judged by food intake. At temperatures below 7°C, the bass were nearly inactive, suggesting that they had entered a torpid-like state. Furthermore, rate of metabolism in mesopelagic fishes decreased with depth in both Californian and antarctic waters (Torres and Somero 1988), but the decrease was twice as great in California, which was interpreted to mean that the antarctic fishes showed thermal acclimatization.

A complication for the view that acclimatization reduces the temperature dependence of poikilotherms is found in the observation that some species show an "inverse" acclimatization in which rate of metabolism decreases with cold exposure (for reviews, see Precht 1958, Bennett and Dawson 1976). A few other species have rates of metabolism that are independent of cold exposure. These differential responses to a prolonged exposure to cold possibly reflect conditions in the environment, namely, that inverse acclimatization is a means of preparing individuals living in an environment with a cold winter for hibernation and that the absence of acclimation is typical of species living in a seasonally constant environment (Bennett and Dawson 1976, Gregory 1982, Tsuji 1988a). Tsuji (1988a, 1988b) tested these hypotheses in a study of rate of metabolism in the lizard *Sceloporus occidentalis* in Washington (cold winter with hibernation) and southern California (cool winter with year-around activity) and *S. variabilis* in lowland Costa Rica (temperature mild and constant). The Washington *occidentalis* in fact showed a decreased rate of metabolism in winter preparatory to and

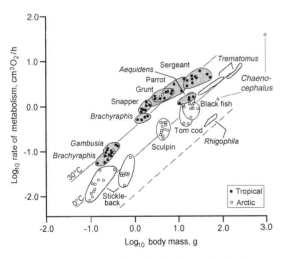

Figure 4.10 Log_{10} rate of metabolism in tropical and arctic fishes at their normal habitat temperatures (30°C for tropical species and 0°C for arctic species) as a function of log_{10} body mass. The dashed curve represents the rates expected from tropical species when extrapolated to 0°C. Source: Modified from Scholander et al. (1953). Additional data on the antarctic teleosts *Trematomus bernacchii* (Wohlschlag 1962), *Rhigophila dearborni* (Wohlschlag 1963), and *Chaenocephalus aceratus* (Ralph and Emerson 1968, Hemmingsen et al. 1969) are indicated.

during hibernation, whereas the California *occidentalis* had higher rates in winter, when this lizard is active, and lower rates in summer during a period of reduced activity. The Washington population had a higher rate of metabolism than was found in the California population at all seasons, except during winter. In both populations the variable rates of metabolism were associated with the variable thermal environments that they faced. In contrast, the Costa Rican *variabilis*, which lives in a seasonally constant environment, showed no seasonal acclimatization of metabolism. The critical thermal minimum, as might be expected, was lowest in Washington, intermediate in California, and highest in Costa Rica, and in each case decreased with cold exposure (Tsuji 1988b).

The concept of acclimatization implies that an individual changes its response to temperature as a result of its exposure to higher or lower environmental temperatures. The pattern noted in *Sceloporus* may be widespread. John-Alder et al. (1988) showed in hylid treefrogs that subtropical and tropical species had higher thermal minima for activity than did temperate species. These authors concluded that the low-temperature performance in temperate hylids is the derived condition (in ancestors that came out of the tropics). A comparison of 11 species of anurans that belong to 4 genera (*Atelopus*, *Colostethus*, *Hyla*, and *Eleutherodactylus*), each of which had species at high (2900–3550 m) and low (90–350 m) altitudes, demonstrated that none of these species showed evidence of individual thermal acclimatization, but the rate of metabolism in high-altitude *Atelopus* and *Colostethus* species was greater than that in low-altitude species at low ambient and body temperatures (Navas 1996). Furthermore, activity metabolism (see Section 9.4.1) was greater in high-altitude species at nearly all body temperatures,

except in the genus *Hyla*. Among natricine snakes, those living at high latitudes were less temperature sensitive than those living at lower latitudes, and even at the latitude of Spain, a nocturnal, stream-dwelling *Natrix* species was less sensitive to low temperatures than a diurnal terrestrial one (Hailey and Davies 1986). Notice that these differences are usually intrageneric and interspecific, which counteracts the argument that species within a genus are evolutionarily redundant.

As noted, the net effect of acclimatization is to increase the independence of poikilotherms from temperature: they are not sluggish in polar regions and hyperactive in the tropics. They permit body temperature to vary with ambient temperature while maintaining their rates of metabolism independent of ambient and body temperature. Endotherms, in contrast, maintain body temperature constant by varying their rates of metabolism with ambient temperature. In terms of energetics, endotherms are thermal slaves to their environments.

4.5.4 Mechanisms of thermal acclimatization. Thermal acclimatization, whereby a cold-acclimatized poikilotherm has a higher rate than a warm-acclimatized poikilotherm at intermediate to cold temperatures, undoubtedly reflects an adjustment at the molecular level. For example, the enzyme lactate dehydrogenase in the bluefin tuna (*Thunnus thynnus*), which operates at warm body temperatures, has an activity that is only 76% of that found in the cold-temperature nototheniid *Pagothenia borchgrevinki*, when both are measured at 5°C (Somero and Siebenaller 1979). In fact, the concentration of pyruvate at which the reaction reaches 50% of the maximal rate, K_m or *Michaelis constant*, is similar in vertebrates at their normal body temperatures over a very wide range in body temperature (Figure 4.11). Because the substrate

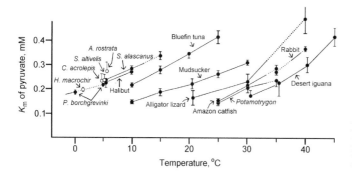

Figure 4.11 The Michaelis constant of pyruvate in a variety of vertebrates as a function of temperature. Source: Modified from Hochachka and Somero (1984).

concentrations of pyruvate are comparatively independent of body temperature, the adjustment of K_m maintains the realized $\bar{V}$ generally independent of body temperature.

Such independence occurs even over small temperature ranges. For example, Graves and Somero (1982) showed in eastern Pacific barracudas (*Sphyraena*) that the K_m of pyruvate is nearly identical when four species are compared at normal environmental and body temperatures (Figure 4.12). Cold acclimatization in the north-temperate *S. argentea* and south-temperate *S. idiastes* produced nearly identical functioning lactate dehydrogenases. Two species of rockfish (*Sebastes*) showed a similar thermal acclimatization in enzyme activity (Wilson et al. 1974), although here the species living in the warmer but more thermally unstable environment showed a quantitatively greater acclimatization of rate in response to a change in temperature. Hochachka and Somero (1984, p. 399) concluded that ". . . body temperature, not phylogenetic status, is the dominant component in establishing the relation of enzymes to temperature."

The molecular means by which such adaptation occurs is unclear. The maintenance of K_m could be produced either by a qualitative change from one form of an enzyme (isozyme) to another or by a change in the amount of enzyme present. Little hard evidence exists for a switch in the isozyme.

First of all, such a switch may be from an enzyme produced by one locus to that produced by another locus, the so-called multiple locus isozyme system. This system has been described for acetylcholinesterase in rainbow trout (*Salmo gairdnerii* [Baldwin and Hochachka 1970]), but if correct seems limited to species that are tetraploid (Somero 1975), that is, in species that do not have a limiting amount of genetic material. Another possibility is to have allelic (one locus) isozymes (= allozymes), an alternative that has been rarely demonstrated. One example was two alleles of the enzyme lactate dehydrogenase B in *Fundulus heteroclitus* along a 15° latitude and 15°C mean annual temperature cline (Place and Powers 1979). The allele with the higher rate constant (K_m) at colder temperatures is most abundant in the north, whereas the allele with the higher rate constant at higher temperatures is most abundant in the south (Figure 4.13).

A more subtle means of changing rates of reaction appears to be a change in the relative abundances of enzymes that are present at all temperatures, a pattern that has been described for *Taricha* (Feder 1983b). Similar conclusions have

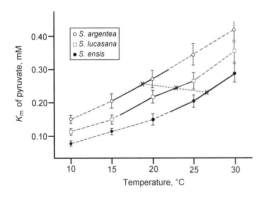

Figure 4.12 The Michaelis constant of pyruvate as a function of temperature in three species of Pacific barracudas (*Sphyraena*). A fourth species, *S. idiastes*, is identical to that of the other temperate species, *S. argentea*. Notice that the K_m for each of the species is nearly identical when evaluated at their normal water temperatures (indicated by the x's). Source: Modified from Graves and Somero (1982).

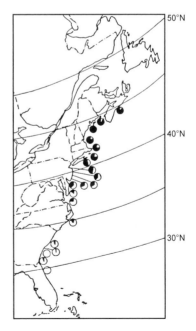

Figure 4.13 The relative frequencies of two alleles of lactic acid dehydrogenase in *Fundulus heteroclitus* along the East Coast of North America. Source: Modified from Place and Powers (1979).

been drawn from studies of thermal acclimation in *Lepomis* (Sidell 1977), *Xenopus* (Tsugawa 1976, 1980), *Discoglossus* (DeCosta et al. 1981), and *Rana* (Enig et al. 1976). A change in enzyme amount, however, is difficult to distinguish from the multiple isozyme or allozyme systems, especially if evidence of thermal acclimatization is demonstrated solely by a change in rate with temperature exposure, or by the constancy of K_m with respect to temperature. A detailed examination of the green sunfish (*Lepomis cyanellus*) suggested that the principal means by which it acclimates to temperature is based on a change in enzyme concentration (Shaklee et al. 1977), but even that conclusion was inferred only from the absence of electrophoretic evidence of a qualitative change in the analyzed enzymes. An increase in enzyme levels with cold acclimation was also described for chain pickerel (*Esox niger* [Kleckner and Sidell 1985, Sidell and Johnston 1985]), striped bass (*Morone saxatilis* [Jones and Sidell 1982]), flounder (*Platichthys flesus* [Johnston and Wokoma 1986]), and three-spined stickleback (*Gasterosteus aculeatus* [Guderley and Blier 1988]).

Some structural changes occur in poikilotherms with cold acclimatization: mitochondrial densities in striped bass increase in both red (oxidative) and white (glycolytic) muscle fibers, and red muscle fibers show a dramatic increase (13.2 times) in intracellular lipid droplets (Egginton and Sidell 1989). Intermitochondrial distances decrease with cold acclimatization, which compensates for the reduction in the rate of diffusion produced by a fall in body temperature. Lipid droplets may accelerate oxygen flux because of the greater solubility of oxygen in lipid than in water, and they likely provide an intracellular store of oxygen, a function that may be important in ice-fish (Channichthyidae), which have no myoglobin (see Section 8.4.8).

The use of the oxidative red muscle fibers for sustained swimming in fish requires an adequate oxygen supply, so their use increases with temperature to a maximum, beyond which it decreases. Burst swimming, which principally uses the anaerobic white fibers, however, is more independent of temperature (Guderley and Blier 1988). Sustained swimming at low temperatures is facilitated by thermal acclimatization, whereby the proportion of red fibers in muscles increases. This adjustment is most well developed in eurythermal species and apparently is not found in stenothermal species living in cold water.

In a review and analysis of the literature on the biochemistry and microstructural adjustments in fish to low temperatures, through either acclimatization or evolution in a polar environment, Dunn (1988) noted that mitochondrial density increased. This response may partially offset the decrease in the rate of diffusion at low temperatures but has the secondary consequences of an increase in diameter of muscle cells and a decrease in myofibrillar volume. Dunn argued that the reduction in myofibrillar volume reduces the capacity of the cold-adapted muscles to use ATP and therefore acts as a limit to cold adaptation. For this reason, Dunn suggested, most antarctic fishes are predominantly sluggish (also, see Hochachka 1988).

In summary, acclimatization occurs at the cellular level, but the molecular bases for such changes are complex and vary with the thermal conditions that vertebrates face in the environment. When acclimatization reflects changes in enzyme quantity, parallel changes in enzyme activity often occur in several enzymes coupled in the same metabolic pathway (Shaklee et al. 1977).

4.5.5 Doubts on thermal acclimatization. On a regular basis the "reality" of acclimatization is doubted. Newell measured rate of oxygen consumption in intertidal invertebrates (Newell 1966, Newell and Pye 1970) and questioned whether thermal acclimatization actually occurs. He suggested that its appearance stems from the integration of activity. Rates of metabolism during pumping of water across gills had a Q_{10} of about 2, whereas the rates during periods of quiescence were nearly temperature independent (i.e., with a Q_{10} near 1). He argued that resting rate of metabolism in intertidal invertebrates is independent of season. Tribe and Bauler (1968) doubted that Newell's interpretation was correct. They maintained that the intertidal animals used by Newell, in fact, may have used anaerobiosis during low tide and aerobiosis during high tide, so that the apparent "quiescent" rate at low tide cannot be measured by oxygen consumption. Holeton (1974), as noted before, also suggested that the apparent cold adaptation of polar fish results from the contamination of measurements by activity. And most recently Clarke (1993, p. 159) maintained that "... the traditional view of latitudinal and seasonal acclimatization of metabolic rate to temperature has no useful biological meaning." He maintained that much of the seasonal change in physiology is associated with activity, growth, or reproduction, not a seasonal change in the standard rate of metabolism. Johnston et al. (1991) reported data on the metabolism of sedentary marine fishes (Figure 4.8), which they surprisingly concluded shows no evidence of cold acclimatization. Yet, the cold climate species have a higher rate at colder temperatures, even when corrected for body mass. In these fishes, therefore, the cold acclimatization is not "perfect" because the rate, corrected for size, is not flat, but it is much less steep than the intraspecific metabolism-temperature curves (Figure 4.8). Acclimatization is most likely to occur in seasonally variable temperate environments.

Given the molecular evidence of thermal acclimatization, its occurrence in most poikilotherms seems assured, although every effort should be made to ensure that the animals are studied under controlled conditions. Variables that may erroneously imply acclimatization, besides activity, include the impact of feeding, fasting, duration of exposure, and the behavioral response of organisms to respirometers (Feder et al. 1984).

4.6 THE INFLUENCE OF ECOLOGICAL FACTORS ON RATE OF METABOLISM

Factors other than body mass and body temperature also may influence the rate of metabolism in ectotherms. One potential factor is the type of food used, especially if foods differ in quality and in spatial and temporal availability. Few attempts have been made to analyze the influence of food habits on standard rate of metabolism in ectotherms, in part because comparative measurements of energy expenditure are available only for a narrow spectrum of fish, and because adult amphibians, the tuatara, crocodilians, snakes, and amphisbaenids are strictly carnivorous.

Although all adult amphibians are carnivorous, with the possible exception of sirenid salamanders, some diversity in standard rate of metabolism may be correlated with food habits. For example, Taigen and Pough (1983), in a study of three dart-poison frogs (Dendrobatidae) and a leptodactylid, concluded that ant-eating specialists have high standard rates. In a follow-up study, Pough and Taigen (1990), as a result of adding another dendrobatid and dropping the leptodactylid, noted that the ant-eating specialist *Dendrobates auratus* has a low standard rate! In either case, the strength of the correlation of standard rate with ant eating was

low: the most important ecological correlates of metabolism in amphibians occur between the use of aerobic or anaerobic metabolism and foraging mode (see Section 9.4.2). A lesson from this analysis is that the selection (or availability) of data has an extemely important effect on its conclusions.

Among vertebrate ectotherms, only lizards have both some diversity in food habits and many measurements of rate of metabolism. Pough (1983) and Pough and Andrews (1985) examined the correlation of standard rate of metabolism with food habits in lizards. An analysis of this relationship is complicated by the influence of mass and temperature on rate of metabolism, and on the correlation of food habits with body mass: small adult lizards are insectivorous, whereas those weighing over 300 g tend to be carnivorous, omnivorous, or herbivorous (Pough 1973b).

After the influence of mass is taken into consideration, rate of metabolism in lizards remains correlated with food habits (Pough 1983): herbivorous lizards have standard rates below those of insectivorous, omnivorous, or carnivorous species, which are similar to each other (Figure 4.14). These differences, however, are small compared to those described in mammals (see Section 5.4.1) because energy intake in ectotherms is low (Pough 1983). Ant-eating *Phrynosoma* species, unlike ant-eating mammals, have rates no lower than those of other insectivorous species, possibly because the low energy expenditures of ectotherms are not limited by the availability of ants as a food resource, whereas the energy requirements of mammals may

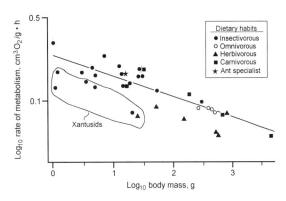

Figure 4.14 Log_{10} mass-specific standard rate of metabolism in lizards at a body temperature of 30°C as a function of log_{10} body mass and food habits. Source: Modified from Pough (1983).

be influenced by low availability and digestibility of ants.

Lizards that belong to the family Xantusiidae have especially low rates of metabolism (Figure 4.14), irrespective of food habits (Mautz 1979, Pough 1983). These lizards are secretive, they feed on restricted resources, and some species are cavernicolous, all characteristics that appear to contribute to low rates of metabolism. Associated with these low rates are low body temperatures, low growth rates, late maturity, and low reproductive rates (Mautz 1979). Thus, standard rate of metabolism also may be correlated with "behavior," and in fact, Andrews and Pough (1985) showed a correlation of standard rate in lizards with several behavioral-ecological categories (i.e., the rates of day-active species were greater than those of reclusive or fossorial species). In this analysis membership in an ecological category accounted for more residual variation (45%) in standard rate (after the effect of body mass had been accounted for) than did family affiliation (14%). Herbivory, however, was associated with intermediate rates. Pough (1983, p. 160) suggested that a comparison of "... low energy-flow endotherms and high-energy ectotherms ... should be especially instructive in defining the differences between the ectothermal and endothermal modes of life."

A thorough examination of the factors setting energy expenditure in 34 species belonging to 18 genera of boas and pythons indicated that 97% of the variation in $\log_{10} \dot{V}_{O_2}$ could be accounted for by $\log_{10}$ body mass and ambient temperature (Chappell and Ellis 1987). Nevertheless, other factors showed a significant correlation with rate of metabolism, as was demonstrated by the following correlations: (1) boas (subfamily Boinae) have lower standard rates than pythons (Pythoninae); (2) some genera (e.g., *Python* and *Eryx*) showed large interspecific differences, whereas others (e.g., *Epicrates* and *Liasis*) showed few interspecific differences; (3) significant differences in standard rate were found among four ecological categories (fossorial, arboreal, semiaquatic, terrestrial), the lowest rate occurring in arboreal species and the highest in semiaquatic species; and (4) significant differences were found among 6 fossorial and 23 terrestrial species. Snakes belonging to the family Achrochordidae also have low standard rates of metabolism, apparently in association with sluggish behavior and prolonged submergence (Seymour et al. 1981, Lillywhite 1987).

If comparable sets of data were available on other ectotherms, a much better picture of the impact of ecological, behavioral, and morphological factors on the energetics of ectothermic vertebrates would be attained. A group worthy of study are sea turtles, especially given that the green turtle (*Chelonia mydas*) is herbivorous, one of the few vertebrates known to feed on the seagrass *Thalassia*, whereas the other species are highly specialized carnivores. They include the leatherback (*Dermochelys*), which feeds on free-swimming medusae; the hawksbill (*Eretmochelys*), which feeds primarily on sponges (Meylan 1985); and the loggerhead (*Caretta*) and ridley (*Lepidochelys*), which feed on molluscs and crustaceans (Mortimer 1982). The leatherback is especially intriguing because it may be endothermic to some degree (see Section 5.9.4), presumably in association with its large mass, even though it feeds on a food that has a very low energy density.

Finally, many more measurements on energetics should be made over a phylogenetic and ecological diversity of fish. Not only do all "primitive" fish (agnathans, chimeras, lungfish, sturgeons, polypterids, bowfin, gars, etc.) need to be measured to examine the question of whether phylogeny is a factor influencing the evolution of fish energetics, but also a diverse array of species should be measured to determine whether climate or food habits, or both, impact energy expenditure. For example, Goulding (1980, Tables 13.1 and 13.2) examined the fishes found in the Rio Machado, a tributary of the Rios Madeira and Amazonas in Brazil. He found that of 33 species, 10 were specialized to feed on fish, 10 on fruit and seeds, 5 on a combination of fruit and leaves, 4 on fruits and invertebrates, and 1 each on invertebrates, leaves, detritus, and the combination of fish and crustaceans. Clearly, fruit is an important food for some tropical, freshwater species, but we have no idea whether it has an impact on fish energetics. Other foods used in this fauna included semidecomposed wood, flowers, algae, and the feces of mammals.

4.7 LIFE NEAR 0°C

The responses of poikilotherms to environmental temperatures near 0°C are as varied as are the thermal environments in which they live. A principal dichotomy is whether vertebrates are found in terrestrial or aquatic environments: aquatic vertebrates do not face ambient temperatures below −2°C, whereas terrestrial species in temperate or polar environments potentially face temperatures that are much lower. When poikilotherms encounter temperatures at or below 0°C, their responses are limited to (1) avoiding these conditions by seeking shelters in which these temperatures do not occur (see Section 11.6.1), (2) reducing the freezing point of body fluids, (3) permitting supercooling, or (4) tolerating the freezing of extracellular body fluids. The use of these responses reflects their environments: aquatic vertebrates can either avoid temperatures near 0°C, reduce the freezing point, or tolerate supercooling, whereas terrestrial vertebrates principally avoid freezing temperatures or selectively tolerate freezing of intercellular fluids.

Several temperatures associated with the freezing of liquids, namely, the freezing point, melting point, and supercooling point, should be defined. The *freezing point*, a term used with some ambiguity, is the temperature at which a liquid will theoretically freeze based on the concentration of the solution. For pure water this is 0°C, a temperature that will decrease as solutes are added to water as a result of their colligative (numerical) properties. Pure water and other liquids, however, do not normally freeze at the freezing point but rather at a somewhat lower temperature, the *supercooling point*. A liquid is said to be supercooled if it is at a temperature between the freezing point and the supercooling point. The difference between these two temperatures is not fixed but subject to modification by the addition of selected molecules that as a result of their concentration, will lower the freezing point and lead to a greater reduction in the supercooling point. The freezing point can be operationally defined in two ways: (1) The freezing point is equal to the *melting point*, which is defined as the highest temperature of a fluid in which ice crystals can be maintained on a stable basis. (2) The freezing point is the temperature to which a supercooled liquid increases as a result of the release of the heat of fusion (the *exotherm*) when the liquid freezes. The difference between the supercooling point (sometimes confusingly referred to as the "freezing point") and the melting point of a liquid is called *thermal hysteresis*.

In aquatic environments a decrease in temperature is limited by the formation of ice, which acts as an insulator that reduces the further loss of heat to the external environment. Temperate poikilo-therms in freshwater are usually exposed to temperatures no lower than 0°C, which means that most temperate fish can survive in a lake covered by ice because the freezing point of their tissues is approximately −0.5°C and because the water they occupy is slightly above 0°C. The only threat in freshwater would be if the lake were shallow and froze solid; then fish cannot live in the lake or they must seek shelter in the warmer mud on the lake bottom (e.g., the arctic blackfish [*Dallia pectoralis*]; DeVries 1971b).

4.7.1 Reduction in the freezing point. In marine environments ice formation occurs at approximately −1.8°C, a temperature well below the freezing point of most marine teleosts (which is usually about −0.8°C, the 0.3°C difference with freshwater fish being due to the higher plasma concentration of marine teleosts; see Section 6.5.3). Scholander et al. (1957) examined the freezing point of plasma in relation to its molar concentration of salt (NaCl) (Figure 4.15). Most of the fish and invertebrates examined had a seasonal varia-

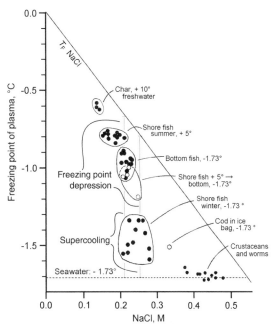

Figure 4.15 Freezing point of plasma as a function of the plasma sodium chloride (NaCl) concentration in fishes and marine invertebrates. A standard freezing point curve for NaCl is included. The difference between the standard curve and the freezing point of plasma is due to either supercooling or the accumulation of organic molecules. Source: Modified from Scholander et al. (1957).

tion in plasma NaCl concentration and had a freezing point lower than dictated by the NaCl concentration of the plasma. Only part of the tolerance of low marine temperatures, then, depends on high plasma NaCl concentrations.

The accumulation of organic molecules also reduces the freezing point of a liquid (see Fletcher et al. 1986). All marine animals have some unbound organic molecules, but they are most prevalent in certain species of fish, crustaceans, and worms. The organic molecules in antarctic fish are, principally, glycoproteins with molecular weights from 2600 to 33,700 (DeVries and Wohlschlag 1969, DeVries 1970, DeVries et al. 1970, DeVries 1971a, Duman and DeVries 1975, DeVries 1982, DeVries and Eastman 1981). Up to 60% of the freezing point depression may be due to these molecules; most of the remaining depression is due to NaCl, which suggests that the impact of the organic molecules is not produced by a colligative mechanism (DeVries 1984), that is, by the number of such molecules. Arctic and North Atlantic fish, in contrast, use a diversity of molecules as "antifreeze," most of which are peptides, although northern cods (Gadidae) use glycoproteins (Osuga and Feeney 1978, van Voorhies et al. 1978, Hew et al. 1981, Fletcher et al. 1987). The only antarctic fish known to use peptides is the eelpout *Rhigophila dearborni* (DeVries 1980). Umminger (1969) reported that the killifish *Fundulus heteroclitus* uses high levels of serum glucose. The use of antifreeze is phylogenetically widespread among teleost fishes, which in conjunction with the diversity of molecules used, suggests that it has evolved repeatedly (Duman and DeVries 1975).

The molecular identity of antifreeze proteins in marine teleosts has an impact on the difference between the supercooling point and the melting point of fish blood, that is, on its thermal hysteresis. Thermal hysteresis increases with the concentration of proteins in the blood and with their molecular weight (Kao et al. 1986), although the effect of an increase in molecular weight decreases as it surpasses 6000. The small peptides typically found on a seasonal basis in arctic teleosts have higher "activities" than do glycoproteins of the same mass. Large glycoproteins with a molecular weight of 10,000 or more have the greatest effect on thermal hysteresis, which may be why they are principally found in species that face a continuous

threat of freezing, especially among antarctic notothenids.

In many antarctic fishes antifreezes are permanently present, which has contributed to the evolution of aglomerular kidneys (see Section 6.6.3), but in subpolar fishes in the Northern Hemisphere, these molecules are present only on a seasonal basis (DeVries 1982). This difference reflects the seasonal constancy or variability of sea temperatures. For example, Fletcher et al. (1987) showed that the "freezing point" (= supercooling point) and thermal hysteresis are inversely related to water temperature in the Atlantic cod (*Gadus morhua*) (Fig. 4.16). The seasonal appearance of antifreeze glycoproteins that are responsible for the change in thermal hysteresis in this species is not associated with photoperiod but depends directly on the presence of cold (0°C) water. This dangerous strategy is permitted by the highly mobile habits of this species, which usually ensure but do not guarantee the avoidance of freezing temperatures. Widespread winter die-offs have been recorded for this species (Woodhead and Woodhead 1959, Templeman 1965). In contrast, winter flounder (*Pseudopleuronectes americanus*) synthesize antifreeze proteins before they encounter freezing temperatures, possibly because they inhabit shallow inshore waters and encounter ice on a dependable basis (Fletcher et al. 1987).

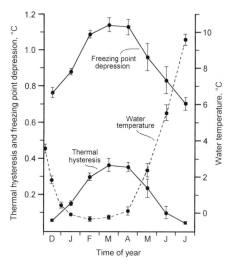

Figure 4.16 Water temperature and the freezing point and thermal hysteresis of the body fluids of Atlantic cod (*Gadus morhua*) as a function of the time of the year. Source: From Fletcher et al. (1987).

4.7.2 Supercooling. Supercooling occurs when a fluid attains a temperature below the freezing point without solidifying (see Lowe et al. 1971). Supercooling is facilitated in invertebrates by polyhydroxyl alcohols, such as glycerol, because they lower the supercooling point more than twice as much as they reduce the freezing point (Hochachka and Somero 1984). As noted, fish use proteins to reduce the supercooling point. The difficulty with the use of supercooling is that it exposes an organism to the possibility of freezing, which can be reduced by avoiding ice crystals in the environment, a goal that may be difficult in polar environments. The supercooled state thus has been referred to as being *metastable*. The marine sculpin *Myoxocephalus octodecemspinosus*, which commonly supercools, migrates in winter from shallow water, where ice formation occurs, to deep water (Leim and Scott 1966), which is ice free, thereby avoiding freezing body fluids.

4.7.3 Tolerance of freezing. Unlike aquatic poikilotherms, some terrestrial species are known to tolerate a controlled freezing of extracellular body fluids. These species include some amphibians and reptiles. The first direct evidence of this behavior, long ignored, can be traced back at least to Samuel Hearne, a polar explorer, who found frozen *Rana sylvatica* in moss at 61°N latitude on an expedition to northern Canada between 1769 and 1772 (Storey 1990). The first biologist to report freeze tolerance in a vertebrate was Weigmann (1929), working on the European wall lizard (*Podarcis muralis* [Claussen et al. 1990]). Recent interest in this subject stems from a paper by Schmid (1982). In it he showed that anurans that hibernate at or near the ground surface, often covered in winter only by leaf litter and a snow layer, face microclimate temperatures that fall to less than −7°C. Under these circumstances the body fluids of *R. sylvatica*, *Hyla versicolor*, and *Pseudacris crucifer* will freeze: they can be kept at a body temperature of −6°C for 5 to 7d and still recover when permitted to thaw. Schmid showed that *H. versicolor* accumulated large quantities of glycerol during freezing. The size of the exotherm produced by freezing indicated that approximately 35% of body water had been converted into ice crystals. When *H. versicolor* was placed at −30°C, 58% of body water was converted into ice, but none of the animals recovered. Schmid found,

however, that two species of anurans, *R. pipiens* and *R. septentrionalis*, could not tolerate freezing: both hibernate under water.

This study set off a rush to examine freeze tolerance in these and other terrestrial ectotherms, most notably by Janet and Kenneth Storey (see reviews by Storey 1990, Storey and Storey 1988a, 1988b, 1992). In addition to confirming the observations of Schmid, they showed that (1) *Pseudacris triseriata* also was capable of freeze tolerance (Storey and Storey 1986; see also MacArthur and Dandy 1982); (2) all anurans studied, except for *H. versicolor*, accumulate glucose as a cryoprotectant; (3) glucose and glycerol are derived from glycogen sequestered in the liver (Storey and Storey 1985a); (4) the production of cryoprotectants is initiated by ice crystal formation (Storey and Storey 1985b); (5) cryoprotectants operate by increasing the colligative properties of intracellular fluids, thereby counteracting the withdrawal of water from cells by the freezing of intercellular fluids (which theoretically should occur because the freezing of intercellular fluids takes water out of a liquid state, thereby relatively "drying" the intercellular spaces); (6) just as some proteins facilitate supercooling, others induce and control intercellular ice crystalization; (7) the tissues of anurans are not frozen, although blood circulation completely stops; (8) anurans rely on anaerobiosis for cell maintenance when frozen (Storey and Storey 1985a); and (9) glycogen is resynthesized from glucose and glycerol after the anuran thaws.

An attempt to find freeze tolerance in *Plethodon cinerea*, *Ambystoma laterale*, and *Bufo americanus* (Storey and Storey 1986) was unsuccessful, although later observations (Storey and Storey 1992) suggested that a marginal capacity to tolerate freezing is present in *A. laterale*. The occurrence or absence of freeze tolerance in amphibians appears to correlate principally with their behavior in winter: if they hibernate in exposed sites, such as under leaf litter in terrestrial settings, they are freeze tolerant, but if they burrow into the soil (toads), descend into crevices or rodent burrows (salamanders), or (as noted) hibernate under water, they are not. Berman et al. (1984) made a striking observation: the Siberian salamander *Hynobius keyserlingi* is capable of freeze tolerance down to temperatures of −35 to −40°C(!), their shelter temperatures are as low as −10 to −32°C, and during freeing they accumulate appreciable amounts of glycerol derived from liver glycogen.

The capacity of a terrestrial ectotherm to tolerate microclimatic temperatures below 0°C is also correlated with their geographic distribution. Thus, the very low temperature tolerance of the Siberian salamander is found in its northern limit of distribution on tundra above the Arctic Circle (67°N). *Hyla versicolor* from Minnesota can tolerate freezing at body temperatures as low as −9°C (Schmid 1982), whereas this species in Indiana can only tolerate −3°C (Layne and Lee 1989). Unlike Minnesota and Ontario populations of *H. versicolor*, Indiana populations do not have substantially elevated glycerol levels during freezing.

Using ice crystal formation as the trigger for cryoprotectant synthesis permits glycogen to be preserved until cryoprotectants are needed. This system occurs in species that live in environments in which the likelihood of freezing is uncertain from year to year and uncertain from one hibernating site to another. The synthesis of simple cryoprotectants, like glucose and glycerol, can be delayed because it starts within only 5 min from the release of the exotherm (Storey 1987). Freeze tolerance is a much more reliable response to subfreezing temperatures than is supercooling because supercooling is always threatened by the presence of ice crystals in the environment, a threat that is especially important for amphibians because they must hibernate in a moist environment to prevent dessication. Freeze tolerance, however, is limited: ambient temperatures must be higher than −10°C, and less than 50% to 60% of body water is tolerated as ice (depending on the species [Storey and Storey 1992]). *Hynobius keyserlingi*, however, appears to be a marked exception to these limits, a conclusion that has been challenged (K. Storey, pers. comm., 2000).

Some reptiles are now known to tolerate the freezing of intercellular fluids. Originally seen in *Podarcis* by Weigmann (1929) and reconfirmed by Claussen et al. (1990), it has also been found in garter snakes (*Thamnophis sirtalis* [Constanzo et al. 1988, Churchill and Storey 1992b]), hatchling pond sliders (*Trachemys scripta* [Churchill and Storey 1992a]), hatchling painted turtles (*Chrysemys picta* [Storey et al. 1988]), and box turtles (*Terrapene carolina* [Costanzo and Claussen 1990, Storey et al. 1993]). Of these species, only turtles are truely capable of recovering from a long-term, deep freeze: hatchling painted turtles remain in the shallow nest cavity throughout winter, often being exposed to temperatures as low as −10°C, and box

turtles hibernate in shallow burrows in which temperatures often fall below 0°C. Although Packard and Packard in an earlier report (1990) seemed to suggest that painted turtle hatchlings could tolerate freezing, later (1995) they pointed out that hatchlings usually tolerate subfreezing temperatures by supercooling, not by tolerating freezing; in fact, freezing led to death (see also Packard and Packard 1993). Box turtles accumulate a cryoprotectant during freezing, glucose, whereas *Th. sirtalis* did not do so. *Chrysemys picta* is the only freeze-tolerant species that is known not to accumulate cryoprotectants. The two reptiles that show only a limited capacity for freeze tolerance, *Th. sirtalis* and *P. muralis*, usually evade low ambient temperatures, the first by denning in winter (see Section 11.6.1) and the second by movement into shelters.

4.8 BEHAVIORAL TEMPERATURE REGULATION

Poikilotherms can control body temperature by the differential use of the thermal complexity in the environment. In most aquatic environments thermal complexity usually is limited to interfaces between large water masses of relatively uniform temperatures, whereas in terrestrial environments, poikilotherms can control body temperature by the controlled absorption of radiant energy, evaporation of water, and convection of heat. The control of body temperature by behavior thus can occur both in aquatic and in terrestrial environments, but it is most prevalent in terrestrial environments, where the thermal complexity is greatest.

4.8.1 Aquatic environments. The simplest form of behavioral temperature regulation involves the selection of a limited temperature range from the range of temperatures available in the environment (see Elliott 1981). Fish show a thermal preference in the laboratory, the actual temperature that they select depending on the species and their thermal experience (Figure 4.17). Massman and Pachecho (1957) and Ferguson (1958) found general agreement between the thermopreferenda of fish in the field and laboratory: cold-water fish, as expected, selected lower temperatures than warm-water species having the same thermal experience. At one extreme, antarctic ice-fish (*Chaenocephalus aceratus*) selected temperatures between 1.5 and 3.0°C, avoiding "warm" temperatures of 3.0 to 7.5°C

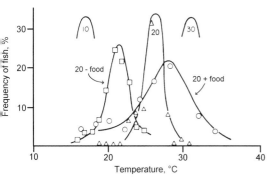

Figure 4.17 The distribution of carp (*Cyprinus carpio*) in a temperature gradient as a function of temperature, thermal acclimation, and the presence or absence of food. The numbers associated with the distributions indicate acclimation temperatures. Source: Modified from Elliott (1981).

(Crawshaw and Hammel 1971), whereas at the other extreme, desert pupfish (*Cyprinodon macularius*) selected temperatures of 38 to 40°C (Lowe and Heath 1969) rather than cooler temperatures.

Brett (1971a) and Crawshaw (1977) suggested that fish select temperatures at which they maximize various functions. For instance, sockeye salmon (*Oncorhynchus nerka*) selected an ambient temperature (15°C) at which feeding rate, digestive rate, active metabolism, cardiac work, growth rate, and sustained swimming velocity are maximized (see Section 4.8.5). With the cessation of feeding, young sockeyes selected a colder temperature, thereby reducing the cost of maintenance. Elliott (1975) showed a similar dependence of feeding rate, maintenance cost, and growth rate on temperature in the brown trout (*Salmo trutta*), the optimal temperature for growth increasing with the frequency of feeding. Hammel and Crawshaw also showed temperature preferences for a wide variety of teleosts (e.g., Crawshaw et al. 1973, Crawshaw and Hammel 1974) and for at least one elasmobranch, *Heterodontus francisci* (Crawshaw and Hammel 1973). Such preferences minimize the changes in rate of metabolism and electrolyte and acid-base balances that are associated with even minor changes in body temperature (Crawshaw 1977), as well as maximize growth rates.

Licht and Brown (1967) studied the temperature selection of the newt *Taricha rivularis* along a thermal gradient in the field. They showed that aquatic larvae up to 16 weeks old had wide preferenda, from 20 to 26°C; upon metamorphosis, the larvae transformed into terrestrial juveniles and then into terrestrial adults, but the preferendum remained approximately the same. Preparatory to entering cold water at the beginning of the breeding season, however, the thermopreferendum of adults fell to 8 to 17°C. Adult bullfrogs (*Rana catesbeiana*) tend to avoid shallow water, which in warm climates has high temperatures at midday and low temperatures at night (Lillywhite 1970). Bullfrogs, upon being disturbed at the edge of a pond, seek shelter on land rather than enter shallow water during the afternoon. Hammel et al. (1973) suggested such temperature sensitivity and selection to be characteristic of all vertebrates.

4.8.2 Amphibians in terrestrial environments. In terrestrial environments amphibians may selectively use radiant energy: Lillywhite (1970) showed that bullfrogs on land shuttle between sun and shade, thereby balancing heat gain from insolation with heat loss by evaporation. As a result, body temperature was approximately equal to air temperature, except at the highest air temperatures, when evaporation may lead to a lower body temperature, all but 8% of the evaporative water loss being cutaneous (Lillywhite 1975). In a study of three European anurans, Sinsch (1984) showed that the northern species, *Rana temporaria*, thermoregulates by shifting microhabitats and activity periods to compensate for high or low ambient temperatures, whereas the southern species, *R. lessonae*, selects higher temperatures than the other species and shows extensive basking behavior. The third species, *R. rudibunda*, thermoregulates principally by moving between water and land. Brattstrom (1963) found behavioral thermoregulation also in *Acris crepitans* and *R. pipiens*.

Amphibians, because of the high permeability of their integuments to water in association with a limited supply of body water (see Section 6.10.2), are limited in their ability to tolerate high ambient and body temperatures (for a summary of the literature on amphibian temperature regulation, see Hutchison and Dupré [1992]). The environmental temperatures most important in determining body temperature in terrestrial amphibians are those of the air and substrate (Tracy 1976). Because of high rates of evaporative water loss, the greatest difference in core temperature between a fully shaded frog weighing 60 g and one fully exposed to the sun is only about 2.5°C. Some few amphibians, the so-called waterproof frogs, greatly reduce integumental permeability to water (see Section 6.10.4): among them, *Chiromantis* maintains a body temperature 2 to 4°C below the environmental temperature by increasing the discharge of cutaneous mucous glands (Kaul and Shoemaker 1989).

Behavioral temperature regulation in anurans may have a greater impact at high altitudes, where ambient temperatures are often quite low. Thus, in an Andean toad, *Bufo spinulosus*, body temperature at 3200 m exhibits great scatter; during the day it usually is above substrate and shaded air temperatures and below unshaded air temperatures (Sinsch 1989). Toads extend their period of exposure to the sun by replacing water lost from the skin with water stored as urine in the bladder (see Section 6.10.2). At night, body temperature tends to remain elevated above air temperature, with the toads obtaining heat by conduction from higher surface temperatures, a behavior that prolongs the nocturnal period of food acquisition. Body temperatures in *B. spinulosus* at 4300 m are similar to those at 3200 m. In contrast, the Andean lizard *Liolaemus multiformis*, when exposed to the sun, has a body temperature higher by 10 to 15°C than the toad. Even at 4300 m toads retreat to the shade to reduce evaporative water loss (Pearson and Bradford 1976).

A modest increase in body temperature may have an impact on amphibians: in a laboratory experiment, body temperature in *Hyla cinerea* increased 4.1°C by basking, which permitted increased feeding and digestion rates and an increase in growth rate (Fried 1980). The impact of temperature on growth rate, however, is not uniform: Lillywhite et al. (1973) showed that growth rate in *Bufo boreas*, like other aspects of "metabolism" in poikilotherms, follows a bell-shaped curve with a maximum near 27°C, so that higher temperatures will normally be avoided.

4.8.3 Lizards in terrestrial environments.

Heliothermic lizards differentially apportion their time between the sun and shade, thereby maintaining body temperature between the radiant and shade temperatures. The daily pattern of body temperature in a heliothermic lizard is illustrated in Figure 4.18. In the morning a lizard leaves its nighttime shelter with a low body temperature. During basking, body temperature increases more rapidly than air temperature; if the radiant temperature is high enough, the increase will continue until the *maximal temperature of voluntary tolerance* is

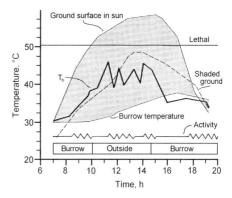

Figure 4.18 Environmental temperatures, body temperature, and activity in the lizard *Dipsosaurus dorsalis* as a function of the time of day. Source: Modified from McGinnis and Dickson (1967).

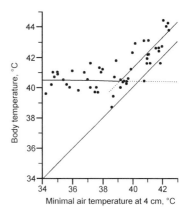

Figure 4.19 Body temperature in *Dipsosaurus dorsalis* as a function of shaded air temperature in the field. Source: Modified from DeWitt (1967).

attained, whereupon the lizard will move into shade. If body temperature is greater than shaded air temperature, heat is lost from the lizard and the lizard's body temperature falls. If the shaded air temperature is low, the lizard will attain the *minimal temperature of voluntary tolerance*, when it will commence basking in the sun. Such behavior tends to bring body temperature within a range independent of time of day (Figure 4.18) and independent of ambient temperature (Figure 4.19), as long as an external source of energy is available. Cowles and Bogert (1944) provided one of the first descriptions of this behavior.

The existence of a temperature differential between a lizard and air in the presence of a radiant heat source is not sufficient proof that thermoregulation is occurring. Heath (1964a) demonstrated that any inanimate object (such as a rock or beer can) exposed to the sun will attain a temperature above air temperature. An object exposed to the sun will attain a steady-state temperature, which simply means that radiant input equals heat loss. For a dry object (see Equation 2.12),

solar radiant heat gain
= (radiative + conductive + convective) loss

$$\dot{Q}_{abs} \approx C'(T_b - T_a).$$

Therefore, if $\dot{Q}_{abs} > 0$, then $\Delta T > 0$ because $C' \neq 0$. Heath noted that the frequency distribution of temperatures in inanimate objects (filled beer cans) exposed to the sun is truncated at the upper end, just as is found in lizards, because the maximal differential that can be attained is determined by $\dot{Q}_{abs}$, given a particular reflectance. Temperature differentials as great as 31.5°C have been measured in the Andean lizard *L. multiformis* tethered in the sun (Pearson 1954a). Such large differentials are only acceptable in cold environments (e.g., T_a for *L. multiformis* was −0.5°C) because body temperature would attain lethal levels at high ambient temperatures. At high ambient temperatures lizards avoid thermal death by spending time in the shade. Evidence of behavioral temperature regulation requires the existence of the maximal and minimal temperatures of voluntary tolerance (Cowles and Bogert 1944).

Heath (1965) most clearly described the temperature limits to specific behaviors for North American lizards belonging to the genus *Phrynosoma* (Figure 4.20) and Bradshaw and Main (1968), for Australian lizards belonging to the genus *Amphibolurus*. The thermal characteristics of various behaviors are what give the capacity for, and precision to, the thermoregulation of lizards.

The precision of behavioral thermoregulation has been rarely quantified, but an estimate can be made. DeWitt (1967) obtained a frequency distribution of body temperatures in desert iguanas (*Dipsosaurus dorsalis*), when they were placed in a thermal gradient in the laboratory: the modal temperature was 38.5°C and the central 50% of the values was between 37.0 and 39.5°C, although trunk temperatures were higher (Figure 4.21). A similarly sized, endothermic fruit bat, *Carollia perspicillata*, at ambient temperatures between 6 and 33°C had a mean body temperature of 36.6°C and the central 50% of the values was between 36.0 and 37.5°C (Figure 4.21). Thus, this endotherm has a lower temperature than this ectotherm—which is not "cold-blooded"—but body temperature was maintained with about twice the precision by the endotherm. Nevertheless, an ectothermic "homeotherm" can approach the precision of an endothermic homeotherm, even at a small mass.

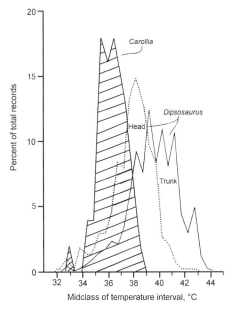

Figure 4.21 Frequency distributions of cephalic and trunk temperatures in desert iguanas (*Dipsosaurus dorsalis*) confined in a circular temperature gradient and rectal temperatures in the fruit-bat *Carollia perspicillata* held at room temperature. Sources: Modified from DeWitt (1967) for *Dipsosaurus* and personal observations (1970) on *Carollia*.

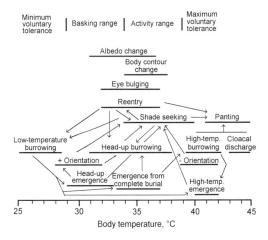

Figure 4.20 Body temperature limits to various behaviors in the field in the lizard *Phrynosoma coronatum*. Source: From Heath (1965).

Lizards collectively show a great range in preferred (= "selected"; Pough and Gans 1982) body temperatures. They range from 21°C in *Elgaria multicarinata* to 40°C in *Cnemidophorus tigris* and *D. dorsalis* (Table 4.1). Some of this variation is related to taxonomic affiliation, although this explanation of the variation is superficial because so little is known of other factors that may influence body temperature, including food habits, predatory style, and microclimate. For example, not all lizards are heliothemic: nocturnal forms, such as *Xantusia* species and most geckos, are active at low body temperatures. In a study of 26 species of Australian lizards, all 9 agamids selected significantly higher temperatures in a thermal gradient than did 9 species of scincids, whereas 5 species of gekkonids were thermophilic with temperatures intermediate to those of agamids and skinks and 3 were cool temperate with temperatures equal to those of the coolest skinks (Licht et al. 1966). The difference among geckos does not correlate with taxonomic affiliation. Field measurements of body temperature in these lizards are usually similar to those measured in the laboratory, as long as conditions in the field permit the preferred body temperatures to be attained. Such conditions in the field include low air temperatures, cloudy weather, and wind (see also Muchlinski et al. 1990). Nevertheless, some lizards, like *Conolophus pallidus* in the Galápagos Islands (Christian et al. 1983), show a seasonal shift in preferred body temperature to maximize the period

Table 4.1 Field body temperatures during activity in lizards

Species	Mean °C	Species	Mean °C
Agamidae		Phrynosomatidae	
Amphibolurus barbatus	34.8	*Callisaurus draconoides*	38.0
Amphibolurus minor	34.0	*Phrynosoma coronatum*	34.9
Amphibolurus inermis	36.7	*Phyrnosoma m'calli*	37.4
Anguidae		*Sceloporus gracilis*	33.6
Elgaria multicarinata	21.2	*Sceloporus magister*	34.8
Ophisaurus attenuatus	32.0	*Sceloporus occidentalis*	35.0
Chamaeleonidae		*Sceloporus undulatus*	34.8
Chamaeleo dilepis	31.2	*Sceloporus woodi*	36.2
Chamaeleo namaquensis	33.5	*Uma scoparia*	35.7
Gekkonidae		*Urosaurus ornatus*	35.5
Coleonyx variegatus	24.7	*Uta stansburiana*	35.4
Gerrhosauriidae		Polychridae	
Gerrhosaurus flavigularis	33.3	*Anolis allisoni*	33.0
Helodermatidae		*Anolis sagrei*	33.1
Heloderma suspectum	27.2	Lacertidae	
Corytophanidae		*Eremias namaquensis*	38.5
Basiliscus vittatus	35.0	*Lacerta vivipera*	29.9
Crotaphytidae		Scincidae	
Crotaphytus pallidus	33.8	*Eumeces fasciatus*	33.0
Crotaphytus collaris	37.2	*Lygosoma laterale*	28.8
Crotaphytus wislizeni	38.3	*Mabuya striata*	35.8
Iguanidae		*Tiliqua rugosa*	32.6
Amblyrhynchus cristatus	32.9	*Tiliqua scincoides*	32.6
Conolophus pallidus	33.8	Teiidae	
Dipsosaurus dorsalis	40.0	*Ameiva quadrilineata*	37.6
Iguana iguana	33.3	*Cnemidophorus sexlineatus*	40.4
Sauromalus obesus	37.9	*Cnemidophorus tigris*	40.4
		Xantusiidae	
		Klauberina riversiana	23.5
		Varanidae	
		Varanus gouldii	37.5

Sources: Derived from Lee and Badham (1963), Brattstrom (1965), Bradshaw and Main (1968), Dawson (1975), and Van Damme et al. (1990).

over which a relatively fixed temperature can be maintained.

Other, "internal" factors may influence the preferred body temperature in lizards. After feeding, various lizards, including *Sceloporus, Lygosoma, Elgaria,* and *Xantusia* species, maintain higher body temperatures than before feeding or after defecation (Regal 1966), apparently to facilitate the mechanical and enzymatic breakdown of prey. Similar observations have been noted in snakes (Regal 1966, Kitchell 1969). Pregnancy in viviparous lizards also may affect body temperature (Beuchat 1986b): in some lizards (*Lacerta, Sceloporus*) pregnant females have lower body temperatures than males or nongravid females, whereas others (*Hoplodactylus, Elgaria*) have higher body temperatures. A higher body temperature is also found in pregnant garter snakes (*Thamnophis sirtalis*) (Stewart 1965, Gibson and Falls 1979). These data hint that reptiles with high preferred temperatures tend to lower their temperature during pregnancy, whereas those with low preferred temperatures tend to increase their temperature during pregnancy. These alternatives may vary with the conditions encountered in the environment (Beuchat 1986b) and may represent a trade-off between the requirements of the mother and those of the developing young (Beuchat and Ellner 1986).

The impact of altitude on the behavior required for temperature regulation can be seen in heliothermic lizards. Lizards belonging to the genus *Anolis* on Hispaniola increased the proportion of time spent in the sun to compensate for the decrease in ambient temperature and an increase in overcast weather associated with an increase in altitude from 5 to 2200 m (Figure 4.22). This behavior permitted a shift from the occupancy of shaded habitats at low altitudes to the occupancy of open habitats at high altitudes and an increase of mean ΔT (Figure 4.23). At low altitudes *Anolis* selected shaded habitats to avoid overheating (Hertz and Huey 1981).

One of the greatest environmental challenges faced by heliothermic lizards occurs at still higher altitudes. Above 3000 m a high body temperature can be attained only when direct exposure to the sun is possible because shaded air temperatures are low. In northern Chile, two species of *Liolaemus* found at 3500 to 4000 m are active for about 4 more hours per day than are two others found at altitudes between 4000 and 4500 m (Marquet et al.

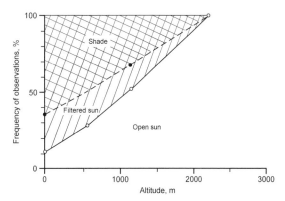

Figure 4.22 Frequency of observations in full sun, in filtered sun, and in shade in *Anolis* as a function of altitude. Source: Derived from Hertz and Huey (1981).

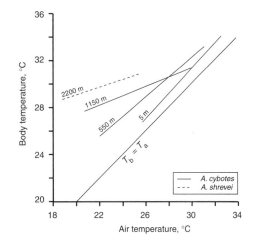

Figure 4.23 Body temperature in *Anolis* as a function of environmental temperature at various altitudes. Source: Modified from Hertz and Huey (1981).

1989). At 4300 m in Peru ambient temperature was cool enough during the day to ensure that overheating in *Liolaemus* would not occur, but the activity period only lasted about 4 to 5 h/d because of cloud formation, rain, and hail in the afternoon (Pearson and Bradford 1976). To reduce conductive heat loss *Liolaemus* species basked on insulated substrates, namely, mats of plant matter (Pearson 1977), even though this behavior increases the lizard's conspicuousness to predators. The local absence of plant mats may prevent lizards from occupying otherwise acceptable sites (Pearson and Bradford 1976). By implication altitudes exist above which the basking period is so reduced and ambient temperature is so low that acceptable activity temperatures cannot be attained. This limit to altitude undoubtedly would be modified by cloud cover and the availability of food and shelter.

In temperate latitudes low air temperatures may be encountered by active lizards at both low and high altitudes. For example, the Eurasian lizard *Lacerta vivipara* is found from southern Europe to the Arctic Circle and from sea level to over 2000 m in altitude. Field measurements of body temperature were 3.2°C higher at an altitude of 25 m than at 2000 m; air temperatures at the low site were 4.2°C higher, which means that the mean temperature differential was slightly greater at the higher site (Van Damme et al. 1990). The greater differential was accomplished in montane individuals by spending much more time basking.

4.8.4 Other reptiles in terrestrial environments.
The thermal physiology of reptiles other than lizards is poorly understood. The tuatara (*Sphenodon punctatus*), which belongs to a sister group of the †Rhynchocephalia,[1] lives in cold, damp environments in New Zealand; its temperature during nocturnal activity in summer varies from 6.2 to 13.3°C (Bogert 1953). Similarly Werner and Whitaker (1978) found low body temperatures in the tuatara at night, but they also found that it attained body temperatures as high as 24°C while basking during daylight hours. A trogonophid, *Trogonophis wiegmanni*, which occupies damp microclimates in North Africa, selects low temperatures (20–25°C) in a gradient, which is typical of fossorial species (Gatten and McClung 1981). These include legless lizards (*Anniella* [Bury and Balgooyen 1976]) and burrowing snakes (Uropeltidae [Gans 1973]).

Goodman (1971) made one of the first comparatively complete studies of behavioral temperature regulation in a snake. He showed that the water snake *Nerodia taxispilota* in Florida spends much of its time in water at the edge of lakes, but it will emerge to bask in the sun in winter (Figure 4.24). In the laboratory Goodman was able to define the combination of air and water temperatures that determined whether this snake was to be found basking in the sun, lying in shade, or submerged in water (Figure 4.25): it did not leave the water if air temperature did not exceed 16°C or if water tem-

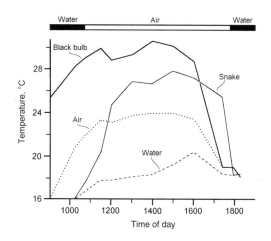

Figure 4.24 Various environmental temperatures and body temperatures in the semiaquatic snake *Nerodia taxispilota* as a function of the time of day. Source: Modified from Goodman (1971).

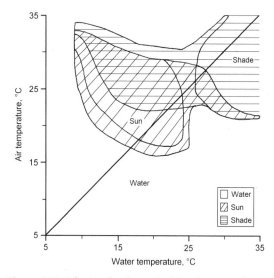

Figure 4.25 Selection by the snake *Nerodia taxispilota* of water, shaded air, or exposure to artificial sun depending on water and air temperatures. Source: Modified from Goodman (1971).

perature did not exceed 10°C. Studies similar to that of Goodman would be of great value to determine the extent to which these "behavioral polygons" vary, how such variations are correlated with the habits and distribution of the species, and whether they contribute to the partitioning of resources in geographically sympatric species. The northern water snake (*N. sipedon*) shows a similar temperature-dependent selection of water or air

---

1 The names of all extinct organisms are preceeded by †.

(Robertson and Weatherhead 1992), but this result is difficult to compare directly with Goodman's study because of the different techniques used.

Few other snakes have been studied in detail. In a comparative study of the thermal biology of seven species of Australian elaphid snakes, Lillywhite (1980) showed that (1) adults in a gradient selected mean temperatures between 32.2 and 34.5°C with the precision of heliothermic lizards, (2) juveniles and newborn selected lower temperatures, and (3) the lowest critical thermal minima were found in species restricted to a temperate distribution. An analysis of the physics of heat exchange in montane (2600 m) populations of *Thamnophis elegans* (Scott et al. 1982) showed that it could maintain body temperature within the optimal range of 26 to 32°C by behavioral means for about 6 h/d in May only on sunny days, for 2 to 4 h/d in June through August on both cloudy and sunny days, but for 4 h/d in early September only on sunny days. This species became inactive after 15 September at this altitude.

The ability of *Th. elegans* to maintain body temperature within the preferred range at an altitude of approximately 700 m depends directly on air temperature: when air temperature was less than 15°C, the snakes did not emerge from shelters and body temperature depended directly on shelter temperature, which is below the preferred range (Peterson 1987). At ambient temperatures between 15 and 30°C, body temperature oscillated around the preferred range as a result of microhabitat selection and postural changes, whereas on warm days ($T_a > 30°C$) body temperature was maintained constant by basking for extended periods. The temperatures maintained on warm days were similar to the temperatures selected in a laboratory gradient, which indicates that the low and unstable body temperatures attained on cooler days reflected limitations to heating imposed by conditions in the environment. Studies on *Th. elegans* indicated that some (heliothermic) snakes can be as precise thermoregulators as the most precise (heliothermic) lizards (Lillywhite 1987, Peterson 1987), contrary to the conclusion of Avery (1982), although snakes tend to chose lower temperatures than do most thermophilic lizards. In fact, most snakes, independent of taxonomic affiliation, have preferred temperatures between 28 and 34°C (Lillywhite 1987), a narrower range than is found in lizards, in which preferred temperatures range from 20 to 40°C. Nevertheless, snakes tend to select temperatures in a thermal gradient that reflect habitat temperatures: semiaquatic snakes (*Nerodia*) select lower temperatures; upland snakes (*Coluber*, *Heterodon*), higher temperatures; and snakes with an intermediate distribution (*Thamnophis*), intermediate temperatures (Kitchell 1969).

Body size is another factor that may influence the thermal behavior of snakes. Most species are rather small (i.e., < 0.5 kg) and thus reach thermal equilibrium with the environment rather rapidly (see Sections 4.13 and 4.14). Diamond pythons (*Morelia spilota*), at 1 to 4 kg, cool slowly when coiled and may retain a 2 to 4°C change in temperature after a night of cooling (Slip and Shine 1988b). This time lag results from the thermal inertia associated with a large mass, the reduction in surface area (and conductance) associated with coiling, and a propitious microhabitat selection. Such behavior is appropriate in sit-and-wait predators, thereby potentially extending the time and temperature range over which predation can occur (Slip and Shine 1988b). This combination of factors (a large size, reduced cooling rates, and ambush predation) may permit some snakes to extend their geographic ranges into otherwise unacceptably cool environments (e.g., *Python molurus*).

Little is known of the thermal behavior of turtles, except that many bask. A quantitative study by Auth (1975) of *Chrysemys picta* showed that (1) the frequency and duration of basking increase with light intensity; (2) basking frequency increases with water temperature up to 28.5°C, above which basking frequency decrease; (3) basking increases in duration from August through December (in Florida); (4) larger turtles tend to bask for longer periods than smaller individuals; and (5) some individuals never bask. Alligators and crocodiles also bask.

Basking in aquatic reptiles, especially turtles, has many explanations (e.g., Boyer 1965, Pritchard and Greenhood 1968, Spray and May 1972, Auth 1975). It may increase body temperature to facilitate digestion, which would explain why large individuals require longer periods of basking than small individuals (because they require longer to reach a high temperature and because larger individuals have larger guts and gut contents) and why the duration of basking increases at low air temperatures or low solar inputs. This explanation also may account for why the turtles that bask tend to be omnivorous or herbivorous: the digestion of cellulose is facilitated by high temperatures because most animals require assistance from an intestinal microflora that requires high, stabile temperatures to digest and ferment food materials. With feeding the preferred body temperatures increased by 4.5°C in *Trachemys scripta* and 1.5°C in *Terrapene ornata*, and the latter species showed a great increase in the precision of thermoregulation (Gatten 1974b). Feeding and basking may decrease at low water temperatures because digestion is not ensured under these conditions. Most strictly carnivorous turtles, like *Macroclemys* and *Chelydra*, do not bask. The only marine turtle known to come to shore and bask is *Chelonia* (A. Carr, pers. comm., 1980), which, unlike other marine turtles, is herbivorous as an adult.

4.8.5 Maintenance of a preferred temperature range.
Because rates of chemical reaction increase with temperature, behavioral temperature regulation might be expected to increase T_b to the highest level that is compatible with environmental constraints (i.e., the concept of maxithermy; see Hamilton 1973). As in some fish, the selection of warm temperatures by newly metamorphosed toads (*Bufo boreas*) accelerates growth, as long as sufficient quantities of food are available, but if the food supply is limited, the toads select lower temperatures (Lillywhite et al. 1973), as was seen in trout. Diurnal lizards reduce the cost of maintenance by selecting low ambient temperatures at night (Regal 1967).

By maintaining a high body temperature at high altitudes and latitudes, lizards increase the velocity of movement, forage distance, and food intake, in spite of the increased fraction of time spent basking (Avery et al. 1982). Predation by lizards at high body temperatures is associated with rapid movement and decreased prey-handling time, which permits a wider range of prey types to be consumed. In *Lacerta vivipara* growth rate increases with time spent thermoregulating, body temperature, and food consumption (Avery 1984), all of which are interconnected. Similarly, snakes operating at preferred body temperatures are more likely to succeed in capturing prey than snakes with a lower body temperature (Greenwald 1974). Conversely, predation on lizards is reduced when they have high body temperatures than when they have low temperatures (Christian and Tracy 1981). As a result of operating at lower body temperatures, montane *L. vivipara* ran at lower velocities (van

Damme et al. 1990), which may be acceptable if their prey is slow and the risk of predation is reduced. But selection for the highest possible temperatures seems generally not to be the case. Instead, as noted, most reptiles, especially heliothermic lizards, tend to maintain body temperature within a preferred range as long as conditions in the environment permit.

Why does a terrestrial poikilotherm bother to maintain body temperature within some narrow range? Part of the answer is that some functions, like maintenance of acid-base balance and respiration, are less disrupted if body temperature is kept within narrow limits. Dawson (1975) showed that several aspects of lizard physiology reach their peak rates at or near their preferred body temperatures. These aspects include (1) scope in aerobic metabolism for activity, (2) auditory sensitivity, (3) digestion and egestion, (4) immunological response, (5) secretion and action of certain hormones, and (6) renal function. A similar pattern has been seen in sockeye salmon (*Oncorhynchus nerka*) (Brett 1971a). This congruence among several physiological functions and preferred body temperature in *Dipsosaurus dorsalis* is seen in Figure 4.26 (Huey and Kingsolver 1989). In some reptiles, such as the garter snake *Thamnophis elegans*, the congruence in maximal rates of various functions with body temperatures is not precise (Figure 4.27), principally because of great differences in the thermal sensitivities of the functions. Therefore, the impact of the daily variation in body

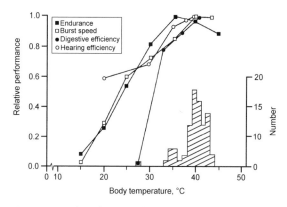

Figure 4.26 The relative performance of endurance, burst velocity, digestive efficiency, and hearing efficiency in the desert iguana (*Dipsosaurus dorsalis*) as a function of body temperature, and the frequency distribution of body temperature of *Dipsosaurus* in the field. Source: From Huey and Kingsolver (1989).

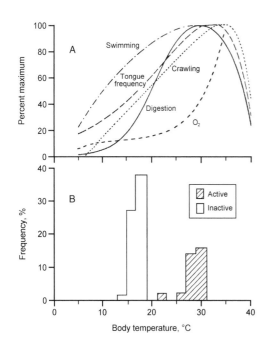

Figure 4.27 A. The relative performance of tongue-flicking frequency, swimming velocity, crawling velocity, digestion, and rate of metabolism in the garter snake (*Thamnophis elegans*) as a function of body temperature. B. The bimodal frequency distribution of body temperature of *Th. elegans* in the field, depending on whether the snake was active or inactive. Source: Derived from Stevenson et al. (1985).

temperature is highly varied in this snake (Stevenson et al. 1985). Lizards of the genus *Anolis* that experience more variable body temperatures in the field sprint quickly at a wider range in body temperatures, compared to species with a more narrow range in body temperatures (van Berkum 1986). Other characteristics of poikilotherms, such as the anaerobic scope for metabolism, have great thermal independence (Bennett 1980). Like other lizards, the leopard gecko (*Eublepharis macularius*) from Pakistan shows a maximal growth rate at a body temperature (30°C) near the preferred range (29°C), in spite of being nocturnal, and thus rarely encounters these temperatures, unless they occur at night in burrows (Autumn and DeNardo 1995). A dependence on high nocturnal temperatures restricts nocturnal lizards to low latitudes.

4.9 BEHAVIORAL CONTROL OF HEAT EXCHANGE

The behavioral thermoregulation of heliotherms relies on the behavioral control of the physics of heat exchange, especially with respect to such parameters as (1) time exposed to the sun and shade, (2) surface area exposed to the sun, and (3) reflectance of the integument.

4.9.1 Time in the sun. The maintenance of a temperature differential by a heliotherm requires exposure to the sun. What regulates the apportioning of time between sun and shade? Surprisingly, this basic parameter of behavioral temperature regulation has been little studied, even though time spent by a heliotherm in the sun is equivalent in the maintenance of temperature differential to rate of metabolism in endotherms. Several predictions follow from this analogy:

1. At a given body mass, a lizard must spend a greater fraction of time in the sun to maintain a larger temperature differential with the environment.

2. A small heliotherm would have to spend a greater fraction of its time exposed to the sun to maintain (not attain) the same temperature differential as a large heliotherm. Conversely, if the fraction of time spent in the sun is independent of mass, then body temperature should increase with body mass.

3. The time spent in the sun should be inversely correlated with ambient temperature for body temperature to be independent of T_a.

Some of these predictions are substantiated by observations. Walter Auffenberg (pers. comm., 1975) noted that small Komodo monitors (*Varanus komodoensis*) spend more time in the sun than large individuals; body temperature in the field is independent of size in these monitors (McNab and Auffenberg 1976). Bradshaw and Main (1968) showed that smaller species of *Amphibolurus* have lower body temperatures than do larger species, even though the smaller species spend a much greater fraction of their time in the sun. Hillman (1969) found that small species of the lizard genus *Ameiva* maintain approximately the same body temperature as large species and tend to live in areas characterized by a higher incidence of sun, and juveniles belonging to the largest species (*A. leptophrys*) forage in more sunlit areas. Trial observations (pers. observ., 1978) on lizards in the laboratory indicated that smaller lizards spend more time exposed to an artificial "sun" and that a larger differential is established with longer time exposed to the "sun."

The relation that exists among time exposed to the sun, body mass, and the temperature differential maintained with the environment is an ecologically and evolutionarily important relation that requires study. As has been seen, a modification of the time exposed to the sun is a means of compensating for low air temperatures at high altitudes (Hertz and Huey 1981, Marquet et al. 1989) and during cold seasons (Muchlinski et al. 1990). The allocation of time between the sun and shade also may be associated with food acquisition. That is, the allocation of time between the sun and shade may be as associated with foraging behavior, as with the necessities of temperature regulation: the high, precisely regulated temperature of *Cnemidophorus* may be associated with its pursuing form of predation, whereas the lower, variable temperatures found in *Sceloporus* and *Anolis* may reflect a sit-and-wait predatory behavior.

4.9.2 Surface area.

Control of the surface area exposed to the sun is a potentially important means of regulating temperature differential. Many qualitative reports indicate that basking postures increase or decrease heat exchange. Heath (1965), however, gave the most concrete illustration of the magnitude of its control in lizards of the genus *Phrynosoma* (Table 4.2). Although some of the capacities of *Phrynosoma* to modify the effective surface area may be due to its flattened body, the capacity to change surface area by a factor of six shows its remarkable behavioral control of heat input. Comparative studies on other species are needed to determine the differential abilities of various lizards to regulate effective surface area. Snakes have a great capacity to regulate the surface area for heat exchange with the environment as a result of their high surface areas and ability to coil. Cooling rates in large constrictors can be reduced by at least one-half by coiling (Lillywhite 1987). This behavioral reduction in heat exchange is augmented by aggregation in populations of snakes seasonally exposed to cool temperatures either at high altitudes, for example, constrictors (Myers and Eells 1968), or at high latitudes, for example, colubrids (Lillywhite 1987) and *Crotalus* species (Klauber 1956). The importance of reducing heat exchange by reducing surface area is found in female pythons, which form a tight coil when regulating body temperature through the internal generation of heat during incubation (Van Mierop and Barnard 1978; see Section 5.9.3).

4.9.3 Reflectance.

Another physical factor that can control heat exchange is integumental *albedo*, the fraction of incident radiant energy that is reflected. The amount of energy reflected or absorbed *in the visible spectrum* varies with the color of the integument. Because dark objects absorb more energy in the visible spectrum than do light-colored objects, heliotherms living in cold climates (such as high latitudes and altitudes) might be expected to be darker. In general, these expectations hold for lizards (Hutchison and Larimer 1960, Pearson 1977) and butterflies (Watt 1968). The color of the integument, however, is not determined simply, or even primarily, by concern for heat exchange. The absorption of energy in the visible spectrum by a lizard's integument tends to match the absorption spectrum of its background, which suggests that camouflage is an important factor determining a lizard's color. Nevertheless, small chuckwallas, *Sauromalus varius*, reflect less visible and infrared radiation than do large individuals (Norris 1967), which agrees with the suggestion that small individuals must absorb more energy to maintain the same temperature differential as a large lizard (see Section 4.9.1). Furthermore, most heliothermic lizards are darker when they have a low body temperature, thereby facilitating heat absorption to reach their preferred body temperature, and are a lighter color at a high body temperature, thereby reducing heat absorption and the probability of overheating. Sometimes lizards find themselves in an environment where the demands for crypsis conflict with those for heat absorption: *Liolaemus multiformis* at an altitude of 4300 m in the Andes will bask on insulative plant mats in spite of its conspicuously dark coloration

Table 4.2 Behavioral control of exposed surface area by *Phrynosoma*

Species	Control (%)	Surface area Orientation (%)* +	Surface area Orientation (%)* −	Ratio +/−
P. m'calli	100	160	27	5.9
P. platyrhinos	100	173	28	6.2

*Positive orientation occurs when the lizard augments the exposure of its surface area to the sun, whereas negative orientation occurs when the lizard remains exposed to the sun but minimizes its surface exposed to the sun.

Source: Condensed from Heath (1965).

required for heat absorption (Pearson 1977), which might be tolerated because of a lower level of predation.

4.10 THE COST OF BEHAVIORAL TEMPERATURE REGULATION

The regulation of body temperature by ectotherms requires an expenditure of energy and time. Even though much discussion of the costs and benefits of regulation has occurred (e.g., Huey and Slatkin 1976), few measurements or estimates of these costs are available. For example, Huey and Slatkin (1976) constructed a simple quantitative model for lizards and concluded that (1) optimal body temperatures reflect ambient temperatures; (2) lizards regulate precisely only if the cost is low, which for a heliothermic species is most likely to occur in open habitats; (3) many lizards, especially those living in lowland tropical forests, will turn out to be thermal conformers with little commitment to precise thermoregulation; and (4) the precision of thermoregulation reflects the productivity of the environment.

Withers and Campbell (1985) attempted estimate the cost of thermoregulation for the desert iguana (*Dipsosaurus dorsalis*). They trained lizards to turn on heat lamps (by breaking a light beam) for various periods of time (from 30 to 180 s). Body temperature in this species increased with shuttle rate (number of movements/h) and the length of time that the lamp was on (Figure 4.28). The precision of temperature regulation increased with ambient temperature and with shuttle rate. A fixed expenditure for thermoregulation, as represented by a fixed shuttle rate, leads to a lower body temperature at lower ambient temperatures. A maximal expenditure for temperature regulation appears to exist in this species, but it is only 1.8% of the standard rate at 30°C and 5.4% at 20°C. The body temperatures of lizards, therefore, vary with the environment as the cost of thermoregulation varies (see also Ruibal and Philibosian 1970, Huey and Slatkin 1976). The stenothermy of *D. dorsalis* may reflect the high temperatures and high solar input of its environment. Yet, the eurythermy found in *Elgaria* species is not associated with a high body temperature variance, as might be expected. The increase seen in body temperature with an increase in ambient temperature reflects a fixed expenditure for temperature regulation (Campbell 1985).

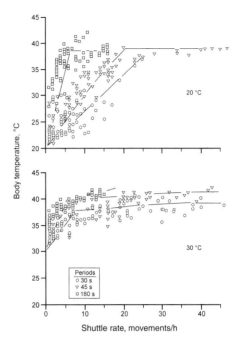

Figure 4.28 Body temperature in the desert iguana (*Dipsosaurus dorsalis*) as a function of the shuttle rate at two ambient temperatures when exposed to a heat lamp for periods that lasted from 30 to 180 s. Source: Modified from Withers and Campbell (1985).

As long as models of behavioral temperature regulation are difficult to test critically, any particular model should be considered tentative. Penn and Campbell (1981) argued that models other than those based on cost and benefit should be considered, including those that emphasize the mechanisms by which temperature regulation is accomplished coupled with an assessment of the environmental conditions under which they operate.

4.11 BEHAVIOR AS A COMPLICATION IN SCALING RATE OF METABOLISM

Chapter 3 described the general pattern by which rate of metabolism scales with respect to body mass. This pattern is given by the relation $\dot{V}_{O_2} = am^b$, where b is a power less than 1.0. Bennett and Dawson (1976) described this pattern in detail for lizards, snakes, and turtles, as well as for reptiles in general, as have Pough (1983) and Andrews and Pough (1985). Because the rate of metabolism in reptiles reflects body temperature, several curves usually are drawn, each representing measurements at a different fixed body temperature (e.g., 20, 30, or 37°C). Although these curves are often compared with those for birds and mammals (see, e.g., Figure 3.9), the basal rates of mammals and birds correspond to their normal body temperatures, not to a fixed body temperature. For example, some mammals, like shrews, arvicolid rodents, weasels, and rabbits, have body temperatures near 38°C, whereas others, like tenrecs, armadillos, sloths, and anteaters, have body temperatures between 30 and 34°C. This pattern in mammals and birds suggests that the standard rates in reptiles should be compared at their preferred body temperatures.

The data summarized on rates of metabolism in diurnal lizards at their preferred body temperatures in Table x of Bennett and Dawson (1976) give the following equation:

$$\dot{V}_{o_2} = 0.32 m^{0.83} (n = 19, r = 0.626),$$

where mass is in grams, or if augmented by additional data on *Sauromalus*, *Tiliqua*, *Varanus*, and *Amphibolurus*:

$$\dot{V}_{o_2} = 0.327 m^{0.82} (n = 26, r = 0.733). \qquad (4.4)$$

As expected, these curves are intermediate to those fitted by Bennett and Dawson for lizards at 30 and 37°C (Table 3.2), and approximately those expected from the mean body temperature and a Q_{10} of 2.1. (That is, the coefficient $a = 0.240$ for lizards at 30°C, mean $T_b = 34.2$°C for the sample of 26 species; therefore, the expected $a = 0.240 \cdot 2.1^{[(34.2-30.0)/10]} = 0.240 \cdot 2.1^{0.42} = 0.328$.) Behavioral temperature regulation in lizards has no effect on the scaling power of metabolism (at least as long as small, nocturnal lizards are not included), but it does influence the level at which scaling occurs. Behavioral thermoregulation should be taken into account when analyzing the scaling of physiological functions in species that have this behavior because of the influence of Q_{10}. The use of a fixed body temperature is arbitrary and biologically unrealistic, convenient though it may be.

4.12 THE EVOLUTION OF BEHAVIORAL TEMPERATURE REGULATION

Nearly all vertebrates are sensitive to changes in the environmental temperatures that they encounter. The two principal responses to such changes are a passive acquiesence or an active selective use of the thermal complexities in the environment to maintain a preferred body temperature. The evolution of thermal preference in vertebrates appears to be widespread.

Although vertebrates other than reptiles show a diversity in thermal preferences, the evolution of preferences has been little studied. One exception is the thermal tolerances of desert pupfish in the Death Valley region of California and Nevada (Brown and Feldmeth 1971). Populations and species of *Cyprinodon* that live in hot springs have high lethal thermal maxima, but most of these differences have developed in the last 30,000 y and can be accounted for by thermal acclimatization alone without invoking genetic changes. However, a comparison of two populations of *C. nevadensis*, one of which lives in a constant-temperature spring and the other in a thermally variable river, suggested that their differences in critical thermal maxima and minima involve a genetic difference (Hirshfield et al. 1980).

Most of the discussion on the evolution of behavioral temperature regulation has focused on reptiles, which have a great range in preferred body temperatures (Table 4.1). As indicated, many factors affect the preferred body temperatures of ectotherms. Bogert (1949a) suggested that phylogeny is one such factor. He showed that species within the genera *Sceloporus* and *Cnemidophorus* are more similar to each other in body temperature than they are to species of the other genus. He concluded that body temperature is evolutionarily conservative. In a study of Mexican *Sceloporus* populations, Bogert (1949b) argued that mean body temperature in the field is not correlated with any environmental parameter, thus furthering his view that body temperature is conservative. Brattstrom (1965), however, showed that Bogert's data were inversely correlated with altitude (and presumably reflected ambient temperature). Bogert's analysis also ignored the striking differences in foraging behavior between *Sceloporus* and *Cnemidophorus*.

In a study of the behavior and temperature regulation of Cuban *Anolis* species, body temperature varied with habits and habitats (Ruibal 1961). An analysis of the phylogeny of these anoles led Ruibal to the conclusion that perch selection was evolutionarily conservative and that body temperature was a dependent character. Later, Ruibal and Philibosian (1970) noted that many *Anolis* species have body temperatures that conform to ambient conditions, and Huey (1974) showed that body temperature in *A. cristatellus* was significantly higher in an open habitat than in an adjacent forest (Figure 4.29). Huey and Slatkin (1976) concluded from these and other observations that precise thermoregulation occurs only when the benefits accruing from the regulation exceed the cost. Thus, *Cnemidophorus* may have a higher body temperature than *Sceloporus* because the former has food and foraging habits that result in higher body temperatures.

The data on lizards are sufficient to conclude that several factors influence the level of body temperature. In species belonging to the genera *Sceloporus*, *Cnemidophorus*, *Lacerta*, and *Agama*, body temperature appears to be evolutionarily conservative (e.g., King 1980, Hertz et al. 1983, Crowley 1985, van Damme et al. 1990), whereas in lizards belonging to the genus *Anolis* it appears to be highly flexible (e.g., Ruibal 1961, Ruibal and Philibosian 1970, Huey and Webster 1976, van Berkum 1986, Wilson and Echternacht 1987). What is most likely is that some groups or types of lizards have a flexible response to environmental conditions and others do not. More important would be a discussion on why such differences exist, rather than debating which of these two extreme views is "correct" for all species. For example, are "flexible" or "labile" genera and species more likely to be tropical rather than temperate or desert dwelling? Or are thermally "flexible" taxa more likely to be sit-and-wait predators than "inflexible" or "static" species, which might more likely be widely foraging? (For a discussion of some of the correlates of foraging mode in lizards, see Huey and Pianka 1981.)

One way to circumvent some of the difficulties associated with assigning a "direction" to the evolution of characters is to place the characters in a phylogenetic context. This can be accomplished by mapping the character states of a related group of species on a cladogram of this group. As noted in Section 1.2.1, a phylogenetic analysis of geckos belonging to the genus *Coleonyx* (Figure 1.1) suggested that a high temperature preference has been repeatedly derived (Dial and Grismer 1992). Huey and Bennett (1987) mapped preferred body temperatures on a phylogeny of eight genera of Australian skinks and calculated presumptive preferred temperatures at various evolutionary nodes (Figure 4.30), although justification for the calculated nodal temperatures is more out of mathematical convenience than out of biological reality. They concluded that a low temperature preference in skinks is derived and that the coupling ("coadaptation") of the optimal temperature for sprinting and critical thermal maximum to preferred temperatures is tight only in genera with high temperature preferences. With a modified phylogeny, the use of species rather than genera as units, and a modification of statistical techniques, Garland et al. (1991) concluded that the only evidence of coadaptation in these skinks is between the critical thermal maximum and preferred body temperature. The difficulty fundamental to the use of phylogenetic analyses is the assumption that a particular phylogeny is correct; as a phylogeny changes, so too does the interpretation of the plesiomorphic or apomorphic nature of character states. But, as seen in Section 1.2.1, even with a constant (correct or erroneous) phylogeny, one

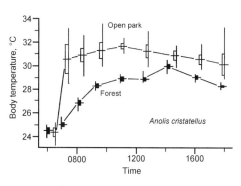

Figure 4.29 Body temperature in a lizard (*Anolis cristatellus*) as a function of the time of day in an open park habitat and in a closed forest. Source: Modified from Huey (1974).

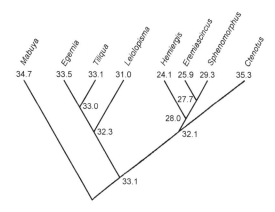

Figure 4.30 A phylogeny of Australian skinks (Scincidae) and a suggested pattern for the evolution of preferred body temperatures according to Huey and Bennett (1987).

unambiguous interpretation of the evolution of the physiological characters is often not possible. More often the problem to be faced, however, is the absence of detailed phylogenies.

4.13 PHYSIOLOGICAL CONTROL OF HEAT EXCHANGE

Many heliothermic lizards control the rate of heat exchange within their bodies and between their bodies and the external environment by controlling the pattern and rate of blood circulation. A consequence of restricting the pattern of circulation is the establishment of temperature differentials within the body. For example, DeWitt (1967) found a difference in the frequency distribution of body temperatures of the lizard *Dipsosaurus dorsalis* in the laboratory, depending on whether its abdomen or head was over the heat source in a thermal gradient (Figure 4.21), which implies that heat was not freely transferred by blood between the head and abdomen.

The influence that circulation has on the thermal behavior of lizards is shown by the "head-up" burrowing of *Phrynosoma* and *Holbrookia* (Heath 1964b). These lizards burrow in sand but at high ambient temperatures will often remain just under the surface with their heads protruding from the sand. Bogert (1959) suggested that this behavior will warm the entire body, but Stebbins (1954) argued that it would only warm the head. The measurements of Heath show that at first the head alone is heated, producing a temperature differential between the head and cloaca. When this differential exceeds 2.0°C, *Phrynosoma* shows what has been termed *eye-bulging*—that is, the eyes appear to protrude from the orbits. This protrusion occurs because of increased blood pressure in the orbital venous sinuses that results from contraction of the internal jugular constrictor muscle. The constriction of this sphincter prevents the return of blood to the heart from the head by means of the internal jugular vein, which is part of a countercurrent heat exchanger that maintains the cephalic-abdominal temperature differential. Eye-bulging, thus, reflects a changed pattern of blood circulation that reduces the temperature differential between the head and abdomen by reducing the increase in cephalic temperature and increasing the abdominal temperature. Eye-bulging diminishes the likelihood that cephalic temperatures will attain lethal levels

and reduces the necessity for the lizard to evade solar input.

Modifying the pattern of circulation influences heat exchange with the environment, as Bartholomew and Tucker (1963) implied in their studies on the dynamics of temperature in the lizard *Amphibolurus barbatus*. They measured heating and cooling rates in live and dead lizards (see Box 4.1) and found that (1) live animals have higher heating and cooling constants than dead animals, and (2) heating constants in live animals were about 33% greater than cooling constants (Figure 4.31).

The reason that live lizards heat and cool faster (i.e., have higher rates of heat exchange with the environment) than dead lizards is that heat is transported by convection with the circulation of blood. Heating constants in live *A. barbatus* are higher than cooling constants because its heart rate and, presumably, the rate of circulation are higher at a given body temperature when the lizard is heating than when it is cooling (Figure 4.33). A high heating constant at the beginning of the day coupled with a low cooling constant at the end of the day extends the period of time over which preferred body temperatures are maintained.

Reptiles other than *A. barbatus* have been examined to see whether they also differ in heating and cooling rates. Observations on lizards in the genera *Amblyrhynchus* (Bartholomew and Lasiewski 1965), *Tiliqua* (Bartholomew et al. 1965), and

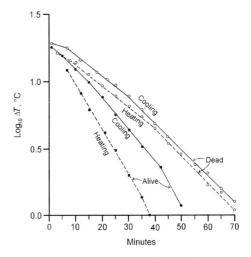

Figure 4.31 Log_{10} temperature differential between the body and environment in a 480-g lizard (*Amphibolurus barbatus*) when cooling or heating, and when alive or dead. Source: Modified from Bartholomew and Tucker (1963).

Derivation of the Cooling Curve Constant

A cooling curve for the lizard *Varanus varius* is shown in Figure 4.32A. A slope, dT_b/dt, cannot be used to characterize the curve because it varies with the temperature differential:

$$dT_b/dt = a(T_b - T_a), \qquad (4.5)$$

where a is the cooling constant (1/time). The value for a can be obtained from the slope when dT_b/dt is plotted as a function of ΔT (Morrison and Tietz 1957). Such a procedure is cumbersome because of the need to estimate dT_b/dt by taking repeated tangents to the curve in Figure 4.32A. This difficulty is eliminated by integrating Equation 4.5:

$$dT_b/(T_b - T_a) = a\,dt$$

$$\int dT_b/(T_b - T_a) = a\int dt$$

$$\ln(T_b - T_a) = \ln\Delta T = at + c, \qquad (4.6)$$

where c is an integration constant equal to $\ln\Delta T$ at $t = 0$. When $\ln\Delta T$ is plotted against time, the slope of the curve equals a, which is either positive or negative, depending on whether the animal is heating or cooling, respectively. The data presented in Figure 4.32A, after transformation, appears in Figure 4.32B. Equation 4.6, when taken out of logarithms, is $\Delta T = k_o e^{at}$, where k_o is the antilog of c. This equation is invaluable in the analysis of heating and cooling curves. Robertson and Smith (1981) explored some of the complications in a, expressed as a thermal time constant $t = 1/a$.

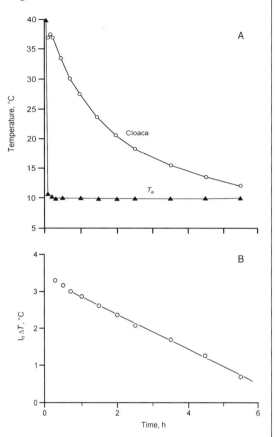

Figure 4.32 A. Cooling curve in a varanid lizard (*Varanus varius*). B. The ln temperature differential as a function of time. Source: Derived from Brattstrom (1973).

Dipsosaurus (Weathers 1970), among others, and on the alligator (*Alligator mississippiensis*) (Smith 1976, 1979) show that many reptiles heat more rapidly than they cool. The circulatory bases of this thermoregulatory behavior have been studied in crocodilians (Grigg and Alchin 1976, Robertson and Smith 1979, Turner and Tracy 1983).

The information available on turtles is inconsistent. Field measurements on the daily temperature cycle in the Galápagos giant tortoise (*Geochelone elephantopus* [MacKay 1964]) and laboratory data on a soft-shelled turtle (*Trionyx* [Smith et al. 1981]) indicate that they too heat more rapidly than they cool. Spray and May (1972) showed a similar pattern in the aquatic basking turtles *Chry-*

semys and *Trachemys*, but measured faster cooling than heating rates in the terrestrial *Gopherus* and *Terrapene* species. Weathers and White (1971), however, found that although aquatic turtles heated more rapidly than they cooled in water, this pattern was reversed in air. These contradictions remain unresolved. In any case, the difference in heating and cooling rates in turtles reflects variation in blood flow to the carapace and skin (Weathers and White 1971).

Further complications in heating and cooling rates were found in three species of South African tortoises (Perrin and Campbell 1981). All species heated faster than they cooled, except under certain conditions. These included a body size in *Geoche-*

lone pardalis larger than 10 kg, when the heating and cooling rates are identical; a reduction in the heating rate of *Cherdina angulata* when body temperature exceeds 30.5°C; and a reduction in the heating curve of *Homopus areolatus* as body temperature exceeds 32°C and a reduction in the cooling curve as body temperature falls below 14°C. The sequence of species just given is also a sequence from large (a maximum of 11.1 kg) to intermediate (0.9 kg) and small (0.2 kg), which suggests that the behavioral control of body temperature is most effective at small masses and most passive at large masses. Limited data available on Galápagos giant tortoises (65–150 kg)

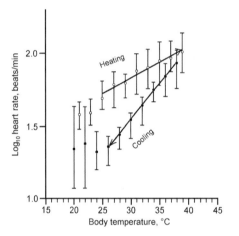

Figure 4.33 Log_{10} heart rate in *Amphibolurus barbatus* during heating and cooling. Source: Modified from Bartholomew and Tucker (1963).

show a limited control over changes in body temperature (MacKay 1964). As always, our knowledge of ectotherms with the largest masses is most limited.

Nearly all reptiles that show a difference in the rates of heating and cooling share a rather sluggish life-style, which poses the question of whether an active predator too would show circulatory control over the rates of heat gain and loss. Bartholomew and Tucker (1964) studied heating and cooling rates in predatory lizards of the genus *Varanus*. They, like other lizards, heat more rapidly than they cool, but the magnitude of the difference was reduced when the heating and cooling rates were corrected for rate of metabolism. (Note that Strunk [1971] criticized this technique.) Unlike *A. barbatus*, however, heart rate is not influenced by heating and cooling.

Endogenous heat production may be more important in varanids than other lizards:

1. Varanids can attain temperature differentials up to 2°C, even when tied to a restraining board, whereas similarly sized *Amphibolurus* can maximally attain differentials of 1°C.

2. Varanids have higher *aerobic scopes*, a measure of the maximal ability of an organism to increase aerobic energy expenditure above the resting rate, than other lizards, so that their augmented rates of metabolism (even under the restraining conditions of measurement) may exceed the basal rates of many mammals of the same mass (Table 4.3).

Table 4.3 Aerobic scope in lizards in comparison with mammals

| Species | Mass (g) | T_b (°C) | Rate of metabolism* | | Scope (max/min) |
			Min	Max	
Tiliqua scincoides	493	32.6	13.7	37.7	2.7
Amphibolurus barbatus	373	34.9	15.8	33.0	2.1
Bradypus variegatus	3790	33.0	41.8	53.1	1.3
Varanus gouldii	714	35.5	26.6	82.9	3.1
Choloepus hoffmanni	4250	34.5	44.6	90.2	2.0
Heterocephalus glaber	39	32.4	48.5	257	5.3
"Standard" mammal		37.0	100	~600	~6.0
Lepus americanus	1500	39.0	178	>450	?

*Rate of metabolism expressed as a percentage of the basal rate expected in mammals of the given mass.
Sources: Data taken from Bartholomew and Tucker (1963, 1964), Bartholomew et al. (1965), Hart et al. (1965), and McNab (1978b).

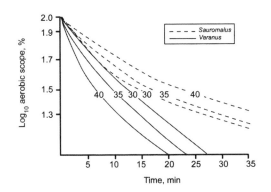

Figure 4.34 Log$_{10}$ aerobic scope after activity in *Sauromalus obesus* and *Varanus gouldii* at three body temperatures. Source: Modified from Bennett and Licht (1972).

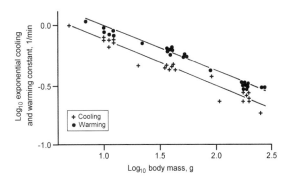

Figure 4.35 Log$_{10}$ exponential cooling and warming constant in brown trout (*Salmo trutta*) as a function of log$_{10}$ body mass when heating or cooling. Source: Modified from Elliott (1981).

3. Varanids have a high capacity for oxygen transport and therefore incur a small *oxygen debt*, a measure of the extent to which an organism has used anaerobic metabolism (usually measured in terms of the accumulation of lactic acid), and rapidly recover from a period of activity. *Sauromalus*, an iguanid, takes about twice as long as *Varanus* to recover from activity (Figure 4.34). A high body temperature facilitates the recovery from activity in *Varanus*, whereas it delays recovery in *Sauromalus*.

Anaerobic metabolism, unlike aerobic metabolism, is nearly independent of body temperature, which permits short-term activity to be independent of temperature (Bennett and Licht 1972); dependence on anaerobiosis, however, limits the period of time over which continuous activity can occur because of the accumulation of lactic acid. Lizards that actively pursue their prey cannot rely on anaerobiosis to maximize the periods of sustained activity; they require high body temperatures and high rates of circulation to ensure adequate oxygen transport for aerobiosis. Evidence of the physiological distinctiveness of *Varanus* is that a free-living *V. komodoensis* has field energy expenditures that are approximately twice those expected from iguanids (Green et al. 1991).

The conclusion that the differential heating and cooling rates in reptiles tend to reflect their lifestyle, subject to modification by body size, must address the observation (Stevens and Fry 1974, Elliott 1981) that heating rates in fishes are 1.23 times those of cooling rates, and that both rates decrease with body size (Fig. 4.35). Most fish face small temperature changes, so the differential in heating and cooling rates may simply demonstrate a physiological difference of no adaptive significance. The impact of body size, however, may be appreciable: Elliott calculated that a brown trout (*Salmo trutta*) encountering a 2 to 10°C temperature change would require 3 to 6 min to reach thermal equilibrium if it weighed 10 g, 7 to 15 min if 100 g, and 23 to 35 min for cooling and 18 to 27 min for heating if 1000 g.

The impact of body mass on the cooling constant in varanid lizards can be derived from Bartholomew and Tucker (1964):

$$a = 0.46\,m^{-0.39},$$

where a has the unit of 1/min and mass is in grams. A 150-g *Varanus* would require 6.3 min to cool from 35°C to 25°C in an environment of 20°C (based on Equation 4.5), whereas a *Varanus* weighing 150 kg would require 93 min. That is, a 1000-fold increase in size would extend the cooling period by a factor of 15. An increase in size, therefore, potentially contributes through the time lag for cooling to thermoregulation at a large body size (see Stevenson 1985 and Section 4.14) to thermal constancy, which has been called *inertial homeothermy* (McNab and Auffenberg 1976). Varanid cooling constants are only one-fourth those of brown trout of the same mass, presumably because the conductivity of air is only 4% that of water.

4.14 BODY SIZE AND THE THERMOREGULATION OF LIVING REPTILES

Bartholomew and Tucker (1964) noted that standard rate of metabolism and thermal conductance in varanid lizards are a function of body mass: standard rate (in cm³O₂/h) at 30°C is

$$\dot{V}_{O_2} = 0.23m^{0.82}, \qquad (4.7)$$

where mass is in grams, and thermal conductance, which is derived from the cooling constant, a, is

$$C = 4.71m^{0.63},$$

where C is in units of cm³O₂/h°C. Because the minimal temperature differential between an animal and its environment, maintained by the internal generation of heat, is defined by the ratio $\dot{V}_{O_2}/C$ (Equation 2.12), in varanids

$$\Delta T = \left[\frac{0.23m^{0.82}}{4.71m^{0.63}}\right] = 0.05m^{0.19}. \qquad (4.8)$$

The minimal temperature differential expected in varanids, thus, increases with mass, although the estimated differential is small due to the very small coefficient (0.05). A 1-kg varanid would be expected from Equation 4.8 to have a minimal differential equal to 0.19°C, whereas a 100-kg varanid would have a 0.45°C differential. According to this equation, a varanid would have to weigh 1.1×10^6 metric tons to attain a differential equal to 10°C! Bartholomew and Tucker (1964, p. 353), not surprisingly, concluded: "Obviously homeothermy evolved some other way [than through an increase in body size]."

This conclusion seems at first glance to be mathematically reasonable, but on closer inspection it requires adherence to the set of assumptions contained in Equation 4.8. They include the following: (1) rate of metabolism corresponds to a body temperature of 30°C, (2) rate of metabolism measured at small masses can be extrapolated to very large masses, (3) thermal conductances estimated at small masses can be extrapolated to large masses, (4) activity is minimal, and (5) temperature differential will reach an equilibrium. Each of these assumptions deserves reexamination before the conclusion that a large mass makes no contribution to the evolution of endothermy is accepted.

A body temperature equal to 30°C is not appropriate because most field measurements for varanids fall between 32 and 39°C and usually between 35 and 38°C at ambient temperatures greater than 30°C. However, if Equation 4.4 is used instead of Equation 4.7, little effect is made on these calculations because the power is the same (0.19) and the coefficient is almost as low as it is in Equation 4.8 (= 0.07). At present, the second assumption must be accepted because no standard measurements of metabolism are available in varanids weighing more than 4.4 kg, unless the field values of Green et al. (1991) on *V. komodoensis* are used. Activity may be appreciable, especially in varanids, which have a large aerobic scope for metabolism. An increase in body temperature and the addition of activity will elevate the rate of metabolism and thereby increase the temperature differential.

The assumption that conductances of large reptiles can be extrapolated from the conductances of small reptiles is incorrect. McNab and Auffenberg (1976) measured cooling constants of Komodo monitors (*V. komodoensis*) weighing from 6.7 to 35.0 kg and estimated cooling constants from other reptiles weighing up to 150 kg. Thermal conductances were calculated from the cooling constants. The conductances of reptiles weighing less than 3 kg conform to the varanid curve, but at greater masses, reptiles, including *V. komodoensis*, have conductances that are less, often much less, than those expected by extrapolation from small reptiles (Figure 4.36).

When a rate of metabolism corresponding to Equation 4.4 and a lower thermal conductance are used, the estimated temperature differential is much greater than that suggested by Bartholomew and Tucker. For example, at 100 kg a temperature differential approaching 10°C is expected. Even if the rate of metabolism of reptiles at 100 kg is only 50% of the value expected from Equation 4.4, the differential would still be approximately 5°C, which is 10 times the value expected from Equation 4.8. Field measurements of body temperature in a 45.2-kg *V. komodoensis* varied between 1.3 and 1.8°C above the ambient (Green et al. 1991), although how much of that was due to metabolism is unclear.

A large body size in reptiles appears to be associated with marked temperature differentials after all, at least in active species: they may even approach those of some mammals having the same

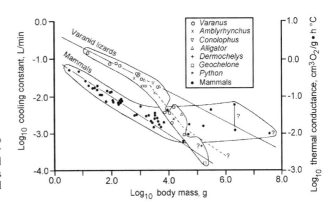

Figure 4.36 $\log_{10}$ cooling constant and $\log_{10}$ thermal conductance as a function of $\log_{10}$ body mass in various reptiles and mammals. A standard curve for both mammals and varanid lizards is included. Source: Modified from McNab and Auffenberg (1976).

size. This does not mean that large reptiles are endothermic, even though most large reptiles probably have rates of metabolism that exceed the minimal rates required for endothermy (Figure 3.9). A fundamental difference remains between most reptiles and mammals: as ambient temperature falls, temperature differential decreases in reptiles because body temperature decreases and, consequently, rate of metabolism falls, whereas temperature differential increases in mammals because body temperature remains constant, or nearly so, and rate of heat production increases. Large reptiles tend toward homeothermy (defined as the constancy of body temperature) because of their large mass and resulting thermal inertia (Spotila et al. 1973). Whereas most mammals are endothermic homeotherms, large reptiles tend to be inertial homeotherms (McNab and Auffenberg 1976).

Although the assumptions made on the level of body temperature, rate of metabolism, and thermal conductance can be questioned, the principal difficulty with the analysis of Bartholomew and Tucker (1964), as pivotal in the development of our understanding of the impact of heat exchange as it was, was that it ultimately relied on the assumption that reptiles reached a thermal equilibrium. In fact, the principal influence of a large mass is on the dynamic state (also see Section 5.3.1). Large reptiles potentially are inertial homeotherms because their cooling constants are small and the length of a night is too short for a large reptile to reach thermal equilibrium. That is, assumption 5 mentioned earlier is not correct in large reptiles. Such nonequilibrial conditions have been demonstrated during the night in the Galápagos tortoise by MacKay (1964) and in the Komodo monitor by Auffenberg (1981).

Stevenson (1985) developed a nonequilibrial heat transfer model to evaluate the variation and lag in the body temperature of terrestrial ectotherms under the assumption that body mass is divided into a core and a shell. He compared his predictions with data taken on insects and reptiles. He concluded that (1) the daily fluctuation in body temperature is greatest at masses between 10 and 1000 g (Figure 4.37A), (2) the time lag of the daily increase in body temperature increases with mass above 10 g (Figure 4.37B), and (3) the extreme difference between body and air temperature increases with mass. This analysis suggests that small ectotherms can easily take advantage of short periods of favorable weather as might be found in temperate zones, tropical highlands, and even polar environments, whereas large ectotherms (e.g., >10 kg) need high ambient temperatures to attain the same activity temperatures, this difference reflecting the impact of body size on rates of heating and cooling. Large ectotherms must avoid extended periods of low ambient temperature because of their effect on body temperature, which may explain the restriction of large terrestrial ectotherms to warm temperate and tropical environments.

4.15 THE THERMOREGULATION OF DINOSAURS

The analysis of the thermoregulation of living reptiles is relevant to the recent controversy about whether dinosaurs were ectotherms or endotherms (e.g., Bakker 1971, 1972a, 1986; Bennett and Dalzell 1973; Spotila et al. 1973; Feduccia 1973, 1974; Case 1978a; Thomas and Olson 1980; Weaver 1983; Reid 1987; Farlow 1990; Lambert 1991; Barrick and Showers 1994; Millard 1995;

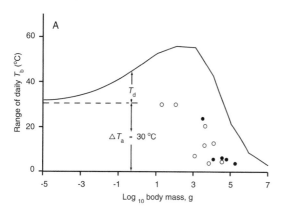

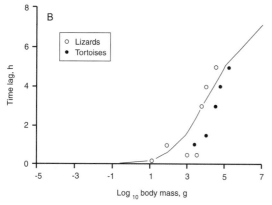

Figure 4.37 A. Maximal temperature differential attained in ectotherms as a function of $\log_{10}$ body mass. Source: Modified from Stevenson (1985). B. Time lag for an increase in body temperature in terrestrial ectotherms.

Farlow et al. 1995; Ruben et al. 1996). Most dinosaurs probably maintained appreciable temperature differentials with the environment, if for no other reason than that most had large masses, low thermal conductances, and body temperatures that did not reach an equilibrium during night. For example, if a 3-metric-ton tyrannosaurid had resting rates of metabolism that fell on the 37°C lizard curve (Table 3.2) and if it had a thermal conductance equal to 2.5 times the value expected from a mammal of the same mass, then the equilibrial temperature differential calculated from an adjusted version of Equation 4.7 would equal about 20.0°C. This means that an air temperature of 15°C would be required for the body temperature of the tyrannosaurid to be 35°C. Even if the appropriate rates of metabolism corresponded to a body temperature of 30°C, the equilibrial temperature differential would be 13.2°C. Furthermore, if

the tyrannosaurid were active, if rate of metabolism during activity increased to 3 times the resting value, and if thermal conductance doubled (due to an increased convective loss of heat and an increased circulation of blood to the skin), the differential temperature would, using the 30°C metabolism curve, be approximately equal to 19.8°C. These approximations suggest that dinosaurs were inertial homeotherms (i.e., they had a constant body temperature provided by a large mass), but that they were most likely characterized by low rates of metabolism by mammalian standards, a conclusion similar to that of Spotila et al. (1991). Indeed, dinosaurs may have had rates of metabolism above the boundary curve (see Section 3.4), thereby enhancing homeothermy, but well below the rates found in mammals. However, some large reptiles, like the pelycosaur [†]*Dimetrodon* (Bramwell and Fellgett 1973) and the ornithosuchian [†]*Stegosaurus* (Farlow et al. 1974), had dorsal sails or plates that may have been important for heat exchange, most likely for the acquisition of radiant heat from the environment rather than for the dumping of heat produced by metabolism.

Variation in the body temperature of a [†]*Tyrannosaurus rex* has been estimated by the oxygen isotopic composition of its skeletal phosphate (Barrick and Showers 1994). Evidence indicated that the $^{18}O/^{16}O$ ratio in dense bone was not affected by diagenic (metamorphic) processes. Although the absolute temperature could not be estimated because the isotopic composition of body water was not known, differences in body temperature were estimated: the highest temperatures were found in dorsal vertebrae, a rib, and a proximal caudal vertebra, all of which were within 1.3°C of each other. Leg elements (femur, tibia, metatarsal, phalange) were 2.9 to 3.5°C lower than the warmest dorsal vertebra, and a distant caudal vertebra was 4.2°C lower than the warmest dorsal vertebra. More importantly, Barrick and Showers (1994) concluded that core body temperature in this 6000-kg [†]*T. rex* fluctuated no more than 3°C, which is about the daily fluctuation found in hydrated camels (Schmidt-Nielsen et al. 1967). Millard (1995), however, argued that this estimate should be 5.7 times as great (i.e., 17.1°C), which represents a poikilothermic condition, although this analysis has been disputed (Barrick et al. 1995). Even if this dinosaur had a fluctuation in core temperature of less than 5°C, that does not mean that this species was characterized by a high rate of metabolism (compared to those found in living mammals and birds). The conclusion of Barrick and Showers (1994, p. 222) that the "... maintenance of homeothermy implies a relatively high metabolic rate that is similar to that of [living] endotherms ..." ignores the impact of body size and confuses homeothermy with endothermy.

Although various kinds of evidence have been examined to determine the energetics of dinosaurs, including fossil predator-prey ratios (which are highly unreliable), the most reliable characters probably are anatomical, at least because they are properties of the dinosaurs themselves and not some pseudoproperty divined by "theoretical" arguments (see Farlow et al. 1995). For example, de Ricqlès (1974) examined the structure of dinosaur bones and found that they had well-developed haversian canal systems, as did therapsids and as do mammals and birds, unlike living reptiles. He concluded that these structures indicate that dinosaurs (and therapsids) were endotherms. Later (1980) he softened that view to suggest that they probably were homeotherms, possibly with an intermediate level of energy expenditure. Reid (1987) surveyed bone structure and concluded that the distribution of bone type is much more complicated than previously thought, and therefore that bone structure is not correlated with the thermal biology of dinosaurs. The discovery that some dinosaurs had young that may have remained in a nest (Horner 1982) raises the possibility that these young were altricial, which might imply that their parents were homeothermic and that the young accelerated growth rates by being endothermic (Lambert 1991), although the maternal behavior of living crocodilians should caution against jumping to the conclusion of endothermy. An examination of the cross sections of the nasal passages of dinosaurs (Ruben et al. 1996) indicated that they were similar to those found in living reptiles, which is about one-fourth of the cross sections found in living birds and mammals (Figure 4.38). This observation led to the conclusion that dinosaurs did not have the ventilation rates found in living endotherms, but rather rates that were similar to those found in living ectotherms.

The large body size of herbivorous dinosaurs (e.g., †*Diplodocus*, †*Apatosaurus*) may have permitted a thermal constancy in the absence of a high rate of metabolism, its importance being the maintenance of an intestinal flora for cellulose digestion. In contrast, mammals attain thermal constancy by

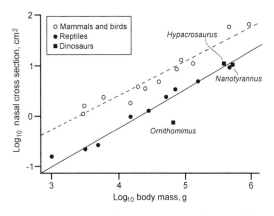

Figure 4.38 $\log_{10}$ nasal cross section as a function of $\log_{10}$ body mass in mammals, reptiles, and dinosaurs. Source: Modified from Ruben et al. (1996).

means of high rates of metabolism, which may have prevented mammals from becoming larger, owing to the prospect of overheating. The huge body masses of some whales, which greatly exceed those of most large dinosaurs, are compatible with a high rate of metabolism only because the threat of overheating is diminished by living in an environment characterized by a high heat capacity and thermal conductivity. The suggestion of Bakker (1971, 1972a, 1986) and of Barrick and Showers (1994) that dinosaurs were endotherms with rates of metabolism equal to those of mammals seems highly unlikely from this view. In fact, their large masses in a terrestrial environment may be indirect evidence of low rates of metabolism.

The largest dinosaurs may have avoided lethal heat storage in the warmest environments by having low rates of metabolism, which may have permitted the generally larger body size not found in terrestrial mammals. After all, large tropical mammals face a problem of dumping heat when exposed to high radiant heat loads in open environments or during intense activity. That is why they tend to be naked or nearly so in warm environments. Woolly mammoths (†*Mammuthus*) and woolly rhinoceroses (†*Coelodonta*) had cold-temperate distribution. On the other hand, large terrestrial ectotherms are unlikely to have tolerated cool to cold-temperate climates because their equilibrial body temperatures would have been too low (Stevenson 1985, contra Spotila 1980). If dinosaurs tolerated cool environments, they may have attained some control of body temperature through an intermediate rate of energy expenditure that was low by present mammalian standards (see McNab 1983a). Yet, some persistent reports (e.g., Sloan et al. 1986) indicate that the disappearance of some dinosaurs at the end of the Cretaceous may have been associated with a fall in ambient temperatures. In this context, the little that is known of the thermal biology of the few living, large, terrestrial reptiles (e.g., *Geochelone* [Galápagos, Aldabra] and *Varanus* [Australia, Komodo]) is unfortunate.

A difficulty with most analyses of the biology of extinct species is that they usually are forced to conform to patterns found in living species. The suggestion of Spotila et al. (1991, p. 203) that "... Arctic dinosaurs of moderate to large size probably migrated away from cold winter conditions, but small dinosaurs would have hibernated ..." may not be necessary. Even though some dinosaurs were found at high latitudes (Rich and Rich 1989, Clemens and Nelms 1993, Molnar and Wiffen 1994), they may have only encountered seasonally dark and cool conditions (Farlow et al. 1995). A unique pattern in energetics may have existed in dinosaurs, especially in species that were terrestrial, huge, and scaled. They may have had a rather constant core temperature and a level of energy expenditure below the level found in mammals but above the boundary curve. In fact, Farlow (1990) predicted that a diversity in thermal biology would be found in dinosaurs in relation to body size, latitude, and season.

Some species had heat-exchange structures, which suggests a degree of ectothermy, whereas others, like "coelurosaurs," were small (down to 5–10 kg, as in †*Compsognathus*) and highly active predators (Ostrom 1969), possibly endotherms with high rates of metabolism. At least one relative of †*Compsognathus* was recently found in China with what was described as a feather tract down its spine (Gibbons 1996), which may or may not be the case. Ostrom (1976) believed that small theropods were the source from which birds evolved, a view that has been challenged (Feduccia 1996), in part because these theropods lived long after the first birds.

Farlow et al. (1995, p. 465) clearly stated the intellectual value of analyzing dinosaur biology in the context of this book: "Dinosaurs push the envelope of what it means to be a large, terrestrial vertebrate. Their very existence poses questions about how body size is related to locomotion, reproduction, growth, metabolism, and trophic ecology that are broader than would be asked if land animals of the modern world were the only terrestrial creatures we knew."

4.16 THE ADVANTAGES OF ECTOTHERMY

Pough (1980a, 1983) addressed the common bias that ectothermy is an inferior state compared to endothermy. He argued that the low energy demand by amphibians and reptiles is in part responsible for their diversity and abundance and that it permits the exploitation of habits and resources that otherwise would not be possible. This conclusion is based on several lines of evidence:

1. Small masses are almost exclusively found in ectotherms: much less than 1% of mammals and birds weigh less than 5 g, whereas 36% of lizards, 50% of anurans, and 65% of salamanders weigh less than 5 g. As seen in Section 3.4, the cost of continuous endothermy is prohibitively high at small masses because of the scaling factor associated with the boundary curve for endothermy (Figures 3.3 through 3.9).

2. Ectotherms have a greater morphological plasticity than do endotherms: no mammals have the wormlike shape found in snakes, legless lizards, amphisbaenids, caecilians, and salamanders. Brown and Lasiewski (1972) showed how the cost of endothermy is markedly higher in weasels because of their large surface areas, even though their areas are not nearly as extreme as those found in salamanders and snakes.

3. Ectothermy permits the use of food supplies that are highly seasonal or patchy in distribution. For example, Gila monsters (*Heloderma*) and some snakes (*Dasypeltis*) are specialist feeders on vertebrate eggs, which means that these reptiles fast for months until the next breeding season of their prey, a behavior that is impossible in committed endotherms. Some geckos (*Coleonyx*) can store enough energy in 4 d to last 9 months of fasting; this behavior is rivaled in mammals only by temperate bats, which enter torpor on a seasonal basis. *Sauromalus obesus*, the chuckwalla, is a lizard that feeds on green vegetation; it may stop eating for 8 months if the water content of its food decreases enough in summer to put the chuckwalla into a negative water balance (Nagy and Shoemaker 1975).

4. The low level of energy expenditure for maintenance by ectotherms means that a greater fraction of available energy can be converted into mass, which may account for the finer partitioning of resources by ectotherms, as is found in the huge diversity of cichlids in the rift lakes of East Africa (e.g., 500+ species in Malawi, 350+ in Victoria, and 170+ in Tanganyika). Such a diversity of endotherms is unimaginable on the resource base available.

5. Ectotherms often evade harsh environmental conditions by withdrawing into an inactive state. Although a similar behavior occurs in some endotherms, it reaches an extreme in ectotherms because they do not periodically arouse; for example, hibernating mammals expend energy at a rate two orders of magnitude greater than is found in hibernating garter snakes (*Thamnophis sirtalis*) (Costanzo 1985). Spadefoot toads (*Scaphiopus*) are reproductively active in a desert after a summer rain, only to retreat underground and remain there for 9 to 10 months until the summer rains come again (see Section 6.10.5); in some areas local rains may not come, so the individuals living in these areas are known to wait an additional year (Mayhew 1965b)! Some reptiles can withstand anoxia for long periods (e.g., under water, in sand, or in compacted soils), in part because their minimal aerobic requirements are met by low rates of diffusion.

Endotherms first appeared in the terrestrial fauna in the early Mesozoic, and came to be dominant at body masses greater than 50 to 100g only since the beginning of the Paleocene, some 120 million years later. Prior to the Paleocene, large terrestrid vertebrates, namely, pelycosaurs, therapsids, ornithischians, and saurischians, were ectotherms, even if they were rather homeothermic. Some few large ectothermic tetrapods continue to the present, but they generally fall into one of two categories: (1) specialized predators (e.g., leatherback, hawksbill, and loggerhead turtles, crocodilians, and various large snakes) with food habits that cannot support a continuously high rate of metabolism; or (2) terrestrial herbivores (*Geochelone*) and carnivores (*Varanus komodensis*) that are isolated on island refugia. In all cases these ectotherms are limited in distribution to tropical or warm-temperate environments.

Most other habits that remain in the domain of ectotherms because they were, and remain, unsuited to the high levels of energy intake required by endotherms. In marine environments a large size is as prevalent in ectotherms as in endotherms. Some habits occur in ectotherms under particular conditions and in endotherms under others. For example, specialist ant and termite eating among mammals occurs only in the tropics and subtropics, environments in which these foods are highly abundant and clumped. In warm-temperate North America ants and termites are less abundant and they are exploited mainly by lizards of the genus *Phrynosoma*. Active flower pollination is effected principally by insects in polar or temperate communities, whereas birds and bats are important vectors of pollen in many tropical communities. Large herbivores and carnivores are ectotherms in the tropics only on islands, not only because equivalent endotherms are not present but also because the resource base is so limited on small islands that populations of large endotherms cannot be sustained (McNab 1994b). These differences in resource exploitation reflect the quantity and seasonal availability of resources relative to latitude, the rarer and more seasonal resources being exploited exclusively by ectotherms.

4.17 CHAPTER SUMMARY

1. Poikilotherms, collectively, tolerate environmental and body temperatures from about −2 to 45°C, although individual species usually have appreciably narrower temperature limits.

2. Between these limits physiologically important rates vary according to the Q_{10} function; Q_{10} usually is between 2 and 3.

3. Rate-temperature curves show thermal acclimatization, whereby cold acclimatization decreases the upper and lower lethal temperatures. Warm acclimatization has the opposite effects.

4. A tolerance polygon is a measure of the capacity of individuals to tolerate a range in temperature as modified by thermal acclimatization: it may vary in a species from one population to another.

5. The comparative importance of the molecular mechanisms of thermal acclimatization is only partially understood.

6. Some evidence indicates that lizards have standard rates of metabolism that vary with food habits; a similar pattern also might be expected in other groups of ectothermic vertebrates with diverse food habits, namely, turtles and teleosts.

7. Poikilotherms survive at ambient temperatures near 0°C in various ways, including increasing cellular salt concentrations, synthesizing antifreeze molecules, supercooling, permitting extracellular fluids to freeze, or avoiding freezing temperatures.

8. Many poikilotherms select body temperature in thermally complex environments; this behavior is most marked in terrestrial environments where poikilotherms can selectively absorb radiant energy.

9. Behavioral temperature regulation by reptiles involves maximizing radiant heat gain in the morning, so that body temperature increases to a range appropriate for activity; moving forth and back between sun and shade to maintain body temperature within a preferred range; remaining in the sun late in the afternoon to minimize the fall in body temperature; and remaining cool overnight in a shelter.

10. Behavioral temperature regulation may approach but rarely equals the precision of endothermic regulation.

11. Body temperature is maintained within a range, which may be principally determined by the habits of the animal and by conditions in the environment.

12. Thermoregulation occurs through control of the duration and surface area exposed to the sun and control of heart rate and of circulation.

13. The temperature differential maintained by a reptile with the environment increases with body mass, which does not lead to overheating in large terrestrial reptiles because of their low rates of metabolism.

14. Large dinosaurs probably were inertial homeotherms but characterized by low rates of metabolism, whereas some small, predatory theropods probably were endothermic homeotherms.

15. Ectothermy is advantageous over endothermy in animals with small body masses, in animals with high surface-volume ratios, in species using highly limited or seasonal food supplies, and in species living under harsh ambient conditions.

16. The few large ectothermic tetrapods in existence today have food habits that will not sustain endotherms, or they live on islands limited in resources and protected from displacement by endotherms.

5 Adaptation to Temperature Variation: Homeothermy-Endothermy

5.1 SYNOPSIS

Endothermy is the state in which body temperature is regulated by balancing the heat lost to the environment with that generated by metabolism. The rate of heat production in endotherms is much greater than that in ectotherms and is influenced by many factors, including the temperature differential maintained with the environment, body size, and thermal conductance. The basal rate of endotherms is the lowest rate compatible with endothermy. It varies with body size, food habits, climate, body composition, and phylogeny. When body size is correlated positively with latitude, it principally reflects a positive correlation of resource availability with latitude, although the increase of fasting endurance facilitated by a larger size may contribute to survival in unpredictably variable environments. Thermal conductance varies with body size and climate. The level at which body temperature is regulated reflects variations in body size, basal rate of metabolism, and minimal conductance. At low ambient temperatures some tropical endotherms conserve energy by permitting peripheral temperatures to approach ambient levels, thereby establishing a large core-shell temperature differential. Heat gain in mammals is derived from shivering, nonshivering thermogenesis, and basking; birds show no evidence of nonshivering thermogenesis. Although endothermy is most highly developed in mammals and birds, it is found to a limited extent in some insects and in large sharks, tunas, pythons, and the leatherback turtle. Endothermy may evolve either through an increase or a decrease in body size, although the characteristics of endothermy may vary with its pathway of evolution.

5.2 INTRODUCTION TO THE ENERGETICS OF ENDOTHERMY

Endothermic homeothermy is the condition in which the core temperature of an organism is maintained independently of environmental temperature by regulating the rates of heat production and loss. It is most clearly demonstrated in mammals and birds but also is found to varying degrees in other organisms. Endothermy usually requires high rates of metabolism, distinguishing it from behaviorally based homeothermy in terrestrial environments, which is principally based on the selective absorption of solar radiation. Compared to the other forms of homeothermy, endothermy has the advantage of being portable, although as seen in Section 3.4, it is very expensive at small masses, when some compromise of endothermy often occurs (see Sections 11.5.3 and 11.8.3). The high energy expenditures associated with endothermy are translated into high rates of food intake and extended periods of food acquisition. A high rate of food intake turns out to be the Achilles' heel of endothermy: under many environmental circumstances this cost is more than what can be paid or more than the benefits warrant.

Endothermic homeothermy, like all forms of homeostasis, is limited: a low ambient temperature exists below which body temperature can no longer be maintained at its "normal" (= normothermia) level, and a high ambient temperature exists above which body temperature increases (Figure 4.1) and ultimately is limited by a lethal level. Maintenance of a fixed body temperature over a range of environmental temperatures implies variation in the temperature differential (Figure 4.1), which itself implies variation in the rates of heat loss and gain. Thus, endothermic temperature regulation cannot be studied meaningfully without examining its energetics.

At moderate to cool temperatures, heat exchange between a body and its environment mainly varies with the temperature differential (ΔT) (Equation 2.12): for body temperature to remain constant, all heat loss from the animal must be replaced. In the absence of an external heat load, the heat must be produced by metabolism. Therefore,

$$\dot{V}_{O_2} = C(T_b - T_a), \qquad (5.1)$$

where metabolism is measured as the rate of oxygen consumption ($\dot{V}_{O_2}$) and C is thermal conductance. This equation is referred to as the Scholander-Irving model (see Section 2.9.1). The units for rate of oxygen consumption are normally cm^3O_2/h, and those for thermal conductance are $cm^3O_2/h°C$. Mass-specific units, usually $cm^3O_2/g·h$ for rate of metabolism and $cm^3O_2/g·h°C$ for thermal conductance, can be used in Equation 5.1 with the appropriate symbols, namely, $\dot{V}_{O_2}/m$ and C/m. (For a comparison of total and mass-specific units, see Section 3.3 and Box 3.1.) Rate of metabolism also can be expressed in energy, such as joules per hour, which is derived from oxygen consumption by assuming appropriate energy equivalents of the food stuffs metabolized (ca. 20.09 J/cm^3O_2). The conversion of oxygen consumption to energy is most relevant when energy budgets are considered (see Section 5.4 and Chapters 10 and 11).

Equation 5.1 indicates that rate of metabolism is proportional to the temperature differential between an endotherm and its environment (Figure 5.1). The slope of the metabolism-ΔT curve equals C when the curve extrapolates through the origin. If evaporative water loss is included, Equation 2.11 is appropriate and the slope of the curve of $\dot{V}_{O_2}-L\dot{E}$ on ΔT equals C', if $\dot{V}_{O_2}-L\dot{E} = 0$, when $\Delta T = 0$. (Parallel analyses can be made for models that incorporate equivalent temperature [Equation 2.15] and standard operative temperature [Equation 2.17].)

Rate of metabolism never equals zero but attains a minimal value when ΔT falls below a value equal to T_b-T_l, where T_l is the lower limit of thermoneutrality. This means that rate of metabolism is independent of ΔT over a range in ΔT and, by implication, over a range in T_a's, because all variation in ΔT reflects a variation in T_a when T_b is constant. The temperature independence of rate of metabolism can be seen when it is plotted against environmental temperature (Figure 5.2). At cool to cold temperatures, rate of metabolism is high, reflecting large temperature differentials (i.e., increased rates of heat loss from the body to the environment). Thermal conductance, C, is the slope of the metabolism-T_a curve at cool temperatures, if (and only if) the curve extrapolates to $T_b = T_a$ when $\dot{V}_{O_2} = 0$. At higher temperatures, the rate of metabolism is constant within a range in temperatures, which is called the *region* or *zone of thermoneutrality* (Figure 5.2). The lower limit of thermoneutrality in Figure 5.2 corresponds in Figure 5.1 to the break in the metabolism-ΔT curve. An upper limit to the region of thermoneutrality is produced by an increase in forced evaporative cooling and by an increase in body

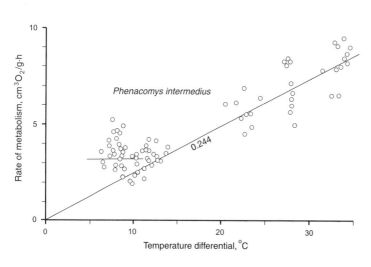

Figure 5.1 Rate of metabolism in *Phenacomys intermedius* as a function of the temperature differential maintained with the environment. The slope of the curve, 0.244 $cm^3O_2/g·h°C$, is an estimate of thermal conductance. Source: Data taken from McNab (1992b).

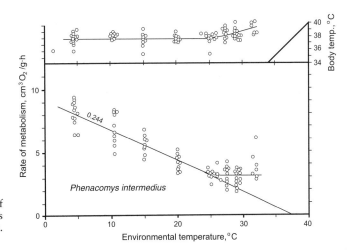

Figure 5.2 Body temperature and rate of metabolism in *Phenacomys intermedius* as a function of environmental temperature. Source: Modified from McNab (1992b).

temperature (Figure 5.2). The rate of metabolism in the zone of thermoneutrality, when an endotherm is regulating body temperature, is postabsorptive, and is quiescent during the period of inactivity (Aschoff and Pohl 1970, Daan et al. 1989), is the basal (= "standard") rate of metabolism (see McNab 1997). (Note that the use here of "standard" rate of metabolism is confusing because it implies that it is similar to the "standard" rate in ectotherms; only endotherms have a *basal* rate. The basal rate, of course, is a standard rate in that it conforms to a series of conditions, some of which, however, do not apply to ectotherms.)

How can the region of thermoneutrality exist— that is, how can rate of metabolism remain constant when the temperature differential changes in this region? Heat production (and loss) remains constant because thermal conductance decreases in a manner that the product $C\Delta T$ remains constant. That is, with a decrease in ambient temperature and a constant T_b, ΔT increases and in compensation C decreases as a result of a postural reduction in exposed surface area, an increase in piloerection or ptiloerection, or an increase in peripheral vasoconstriction. In other words, each measurement of basal rate in thermoneutrality corresponds to a thermal conductance dictated by ΔT. Furthermore, T_l divides those lower ambient temperatures at which an endotherm responds to a changing ΔT by modifying heat production (region of chemical regulation), from those higher ambient temperatures at which it responds by modifying heat loss (region of physical regulation).

Many observers (e.g., King 1964, Packard 1968, MacMillen and Lee 1970, Baudinette 1972, Gettinger 1975, McNab 1980a) noticed that the curve below thermoneutrality often extrapolates to an ambient temperature above the mean body temperature. These observations led to the conclusion that the Scholander-Irving model is inadequate, which is correct in that this model (and all others) does not include all behavioral and physiological complications potentially found in endotherms. But the assumptions of this model relate principally to the physical conditions in the environment (see Section 2.9.1). The curve below thermoneutrality does not extrapolate to body temperature when endotherms do not sharply separate the regions of physical and chemical regulation. That is, some endotherms respond to a fall in ambient temperature both by reducing thermal conductance and by increasing rate of metabolism. In these species T_l is not determined by minimal thermal conductance.

In conclusion, the energetics of endothermy, under conditions when the radiant and air temperatures are similar and when air velocity is low, are generally described by Equation 5.1 as long as evaporative cooling is not an important means of heat loss. An examination of this equation indicates that a finite set of factors describes endothermy. They include basal rate of metabolism, minimal thermal conductance, body temperature, and (under certain conditions) rate of evaporative water loss. The first three factors are examined in this chapter; the last is examined in Chapter 7. Fundamental to this analysis is the observation that most of the parameters in Equation 5.1 vary with body mass.

5.3 IMPLICATIONS OF THE CORRELATION OF BASAL RATE WITH BODY MASS

Since Sarrus and Rameaux's 1839 article, the basal rate of metabolism in mammals has been recognized to be fundamentally correlated with body mass raised to a power less than 1.0 (see Section 3.3). This conclusion has been reinforced repeatedly (Bergmann 1847, Rubner 1883, Brody and Procter 1932, Kleiber 1932, Benedict 1938, Heusner 1982a, Hayssen and Lacy 1985, McNab 1988a). (For a short history of the study of the energetics of endotherms, see McNab [1992c].) A mathematically trivial consequence of this relationship is that mass-specific rates decrease with an increase in mass (Equation 3.4), which, as noted in Chapter 3, led to confusion as to whether total, or mass-specific, rates have the greatest biological significance. Only total rates of metabolism are measured; mass-specific rates are always derived by division. From an ecological viewpoint, total rates are basic (see Box 3.1).

5.3.1 Turnover time. The dependence of rate of metabolism on mass has implications for a class of time-limited phenomena that involve energy use (Calder 1974a, Lindstedt and Calder 1981, Lindstedt and Boyce 1985). These phenomena include the time period over which an endotherm can tolerate starvation without giving up endothermy, the duration an organism can live off its accumulated fat stores in hibernation or in migration, and the time that an organism requires to reach thermal equilibrium with the environment. For example, Morrison (1960) suggested that mammals can survive torpor for a period that is proportional to the size of the energy (fat) store and inversely proportional to the rate at which the store is used (i.e., the rate of metabolism):

$$t = \frac{Q}{\dot{Q}_{met}}, \qquad (5.2)$$

where t is survival time in hours, Q is energy store in kilojoules, and $\dot{Q}_{met}$ is rate of metabolism in kilojoules per hour. He concluded that if the size of the energy store is proportional to body mass and if

rate of metabolism is proportional to $m^{0.75}$ (as approximations), then

$$t = \left(\frac{k_1 m^{1.00}}{k_2 m^{0.75}} \right) = k_3 m^{0.25}, \qquad (5.3)$$

where k_1, k_2, and k_3 are appropriate coefficients. That is, survival time is proportional to $m^{0.25}$, which means that a doubling of mass increases survival time by 19% ($2^{0.25} = 1.19$). At small masses, then, the frequency of feeding is of necessity higher: from 10 times/d in shrews, 5 to 6 times/d in voles and mice, twice daily in ungulates, and one long period/y in large cetaceans (Millar and Hickling 1990). If this pattern is prohibited by a commitment in small species to a limited portion of the daily light cycle (e.g., in endothermic insects or hummingbirds), then a circadian abandonment of strict endothermy occurs (see Sections 5.9.1 and 11.5.3).

The relationship between t and $m^{0.25}$ in Equation 5.3 can be stated in another manner: survival time is inversely proportional to the mass-specific rate of metabolism. Note, however, that mass-specific rates were not used in the derivation of this relation. In fact, most justifications for the importance of mass-specific rates have this fate—they disappear on close analysis (McNab 1999). If rate of metabolism is proportional to mass raised to a power less than 0.75, which is the case at temperatures below T_l (see Section 5.5.1), then t is proportional to a power greater than 0.25, as long as fat deposits are proportional to $m^{1.00}$. Empirical evidence suggests that total fat deposits in terrestrial mammals scale proportionally to $m^{1.19}$ (Calder 1984), superficial fat deposits in carnivores scale to $m^{1.10}$ (Pond and Ramsay 1992), and intraspecific scaling of fat stores in small rodents generally equals or exceeds 1.00 (Millar and Hickling 1990). These scaling factors further enhance the ability of a large species to tolerate extended periods of starvation. Fasting endurance indeed is more closely proportional to $m^{0.40}$ than to $m^{0.25}$ (see Calder 1974a), although responses to a harsh, variable climate other than by an increase in mass are possible, including food cashing, migration, and entrance into torpor (see Chapter 12).

5.3.2 Bergmann's rule. A widely recognized factor thought to influence the body size and energetics of endotherms is latitude. This relationship has been codified as Bergmann's rule, which states that endotherms living in cold climates tend to be larger than those living in warm climates (also see the discussion in Section 5.5.2). Carl Bergmann first described this concept in 1847 in his study of the heat economy of animals in relation to their body size. Bergmann's description compared species belonging to the same genus, whereas Mayr (1956) modified the "rule" to apply to populations within the same species. Fundamental difficulties with this rule are whether it exists as a generality, which would demand a general explanation, or whether it occurs sporadically in individual cases, which would demand individual explanations for each case, and whether the rule simply describes a geographic pattern or whether it includes an explanation. Few surveys of endotherms have been made to determine its prevalence. Among the mammals of North America (McNab 1971b), (1) either most species with a wide latitudinal distribution did not follow the rule, or if they did, the conformation occurred only over a limited latitudinal range; (2) a decrease of size with latitude was approximately half as common as an increase in size; (3) those species that showed an increase in size often were carnivores or granivores; and (4) the pattern in body size appeared to be a likely result of resource availability and character displacement. Later Dayan et al. (1991) found that a minority of mammalian Carnivora studied worldwide conformed to Bergmann's rule. In contrast, Ashton et al. (2000) showed with some ambiguity (i.e., with the dismissal of species that demonstrated no correlation of size with latitude or temperature and the inclusion of nonsignificant positive correlations) that a (small) majority of mammals conformed to Bergmann's rule. An examination (Hamilton 1961) of latitudinal size trends in birds demonstrated the complexity of this relationship, and James (1970) showed that wing length, as a measure of body size, in 12 species of birds in eastern North America increased at northern latitudes, at western longitudes, and at high altitudes. That some species conform to the rule, however, has never been disputed. What has been disputed is the interpretation of the rule.

The nature of Bergmann's rule, when it occurs, is often obscure. The correlation with latitude usually is thought to represent an inverse correlation of body size with ambient temperature. For example, Brown and Lee (1969) demonstrated a strong negative correlation of body mass in woodrats (*Neotoma*) with the mean annual temperature of the locality. James (1970), however, found a better correlation of the size of the birds she studied with wet-bulb temperature and atmospheric water vapor pressure, and thus concluded that body size in these species principally reflects the combination of ambient temperature and humidity, a conclusion followed by others, including Wigginton and Dobson (1999) in their analysis of body size in bobcats (*Lynx rufus*). Blackburn et al. (1999, p. 169), in an attempt to free the rule from any specific intrepretation, defined Bergmann's rule as ". . . the tendency for a positive association between the body mass of species in a monophyletic higher taxon and the latitude inhabited by those species."

A persistent problem with most explanations of the correlation of body mass with latitude has been the use of mass-specific energy expenditures. In otherwise worthy studies, including those by Brown and Lee (1969) and James (1970), a mass-specific rationale for size trends was given. Thus, Brown and Lee (1969, p. 337) concluded that ". . . thermal conductance . . . [was] positively correlated with environmental temperature and inversely correlated with body size." But that is only because mass-specific units for conductance were used; indeed, total thermal conductance in wood-rats increases with a *reduction* in environmental temperature (because of the increase in mass). Ashton et al. (2000) also based their analysis on a mass-specific rationale. Steudel et al. (1994) most clearly stated this mass-specific view when they argued (pp. 74–75) that "Bergmann's rule might very well be a consequence of thermal considerations in animals . . . *if mass-specific, rather than total, metabolic rate is being selected*" (italics added). No evidence exists that mass-specific rates are being selected, and such selection probably is impossible (McNab 1999).

Equation 5.3 gives a possible physiological justification for Bergmann's rule. In cold climates animals may have to live on stored resources if food availability is temporally unreliable or if the length of the hibernal period is unusually long or highly variable. Originally Morrison (1960) proposed this view, and later Lindsey (1966), Brodie (1975), Boyce (1978), and Lindstedt and Boyce (1985) supported it. This explanation may also account for poikilotherms that conform to Bergmann's rule (Ray 1960, Lindsey 1966).

Dunbrack and Ramsay (1993) criticized the possibility that starving endurance might contribute to the occurrence of Bergmann's rule. Although their analysis is plagued by a reliance on mass-specific rates, they argued that other responses to a cold climate (i.e., other than an increase in mass), including an entrance into torpor, the storage of food, and the selective use of mild microclimates, are more likely. Of course, all of these "evasions" occur, but that does not deny that some mammals use none of these adjustments. An analysis that suggests that the tolerance of fasting is an explanation for a latitudinal increase in body size, like all other analyses, states or implies that all conditions other than those considered are held constant. In any case, some mammals and birds are larger in a cold climate, and this pattern needs an explanation.

To maximize the ability to withstand long periods of energy shortage, energy storage and body mass must be maximized, but that is possible only if food availability permits the higher cost of maintenance inherent in a large size. Thus, resource availability suggests another basis for a correlation of body size with latitude: in Norway a higher-quality forage grows in cooler and drier climates at high latitudes, and red deer (*Cervus elaphus*) grow larger when they feed on higher-quality foods for longer periods in summer (Langvatn and Albon 1986). The mass of moose (*Alces alces*) increases with latitude in Sweden (Sand et al. 1995), reflecting a growth period that is extended by 2 y at northern laitudes, which possibly is permitted by an increase in the quality of the food. Geist (1987) generalized this view by suggesting that the duration of the annual pulse of plant production increases with latitude to 50 to 60°N, beyond which the duration of production decreases. Body size in (at least large) mammals follows the duration of production, so that the largest individuals coincide with 50 to 60°N, thereby generating a correlation with temperature at latitudes less than 50°N. Purdue (1989) concluded that white-tailed deer (*Odocoileus virginianus*) showed size variation during the Holocene in Illinois that reflected food availability. This pattern is seen in deer (Figure 5.3) and in the wolf (*Canis lupus*). Furthermore, the blind mole-rat (*Spalax ehrenbergi*) shows a 30% decrease in mass in populations from the Negev desert region compared to northern Israel, although local reverses of this trend occur depending on local climate: the two most impor-

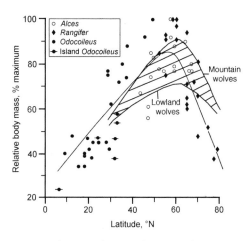

Figure 5.3 Body mass relative to the maximal mass attained in moose (*Alces alces*), caribou (*Rangifer* spp.), and white-tailed deer (*Odocoileus virginianus*) as a function of latitude. Source: Modified from Geist (1987).

tant factors affecting mass are temperature and food resources (Nevo et al. 1986). Murid rodents belonging to the genus *Rattus* in New Guinea showed negative interspecific and intraspecific size correlations with altitude (Taylor et al. 1985), which may be a response to a higher resource availability at lower altitudes. Resource availability also was suggested to be principally responsible for determining body size in mammals and birds limited in distribution to oceanic islands (McNab 1994b).

The factors responsible for setting body size clearly are complicated: fasting endurance may favor a large size, but only if resources are available, whereas resource availability may favor a large or small size. This complexity can be seen in red-winged blackbirds (*Agelaius phoeniceus*), where males are larger than females, the larger size permitting males to defend territories effectively (Searcy 1979). However, a limit to an increase in size occurs because the largest males face a limit to resources. This was shown by the observations that (1) males lost mass in the spring, (2) large males reduced their territorial defense compared to small males, and (3) males supplied with supplementary food spent more time defending their territories. Searcy (1980) generalized this pattern to include the effect of ambient temperature, when he argued that the optimal body size is that at which the difference between energy gain from foraging and the cost of maintenance is maximized. This difference increases with body size as ambient temperature decreases because the cost of maintenance increases as ambient temperature falls, especially in small individuals. Kendeigh (1969a) suggested that larger birds in cold climates have physiological advantages, such as a lower limit of thermoneutrality and a greater tolerance of low temperatures, although he pointed out the increase in energy expenditure associated with an increase in mass.

An increase in the availability of food resources with latitude can occur either if the resources themselves increase with latitude or if fewer species are using the resources. For example, among weasels belonging to the genera *Mustela* and *Martes*, smaller species get larger beyond the northern limits of distribution of the next larger species (McNab 1971b). In a similar manner, pumas (*Puma concolor*) are smallest in the neotropical lowlands where they coexist with the larger jaguar (*Panthera onca*), but increase in mass in North America beyond the northern limits of the jaguar and in South America beyond the southern limits of the jaguar (Figure 5.4A). Iriarte et al. (1990) showed that mean size of the prey used by the puma increases with latitude (Figure 5.4B). Erlinge (1987) and Dayan et al. (1989, 1994) showed a similar pattern and interpretation in mustelids. The interaction among food size, predator size, and latitude is compatible with observations on the food habits of two Chilean foxes (Fuentes and Jaksic 1979) and with the size structure in communities of coexisting predators (e.g., Storer 1966, Ashmole 1968, McNab 1971a). It is also compatible with the temporal response of goshawks (*Accipter gentilis*) in Finland over a 30-y period to a reduction in grouse abundance: males (which are smaller than females) have shifted to smaller prey (ducks, corvids, young hares) and become smaller, whereas females have switched to mountain hares (*Lepus timidus*), a larger prey, and have increased in size (Tornberg et al. 1999).

As repeatedly occurs in a discussion of Bergmann's rule, the Langvatn-Albon-Geist hypothesis (that the correlation of body size with latitude reflects resource availability) generated as much heat as light. Paterson (1990) criticized Geist severely but failed to give alternative explanations

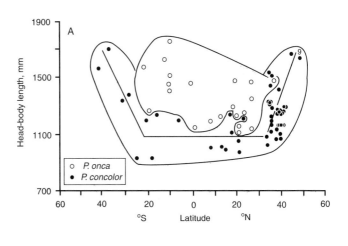

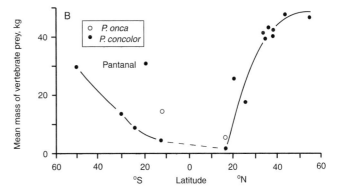

Figure 5.4 A. Head-body length of the puma (*Puma concolor*) and jaguar (*Panthera onca*). B. Mean mass of vertebrate prey as a function of latitude in the Americas. Sources: Data taken from McNab (1971b) and Iriarte et al. (1990).

for a positive correlation of body size with latitude and failed to understand the unacceptability of a mass- or surface-specific explanation (see Section 5.5.2). Paterson insisted on the heat-conservation interpretation for *Odocoileus* and *Alces*, dismissing the pattern found in *Rangifer* as "aberrant" and throwing out inconvenient data on *Canis*. Geist (1990) noted various problems with Paterson's view and reemphasized that many large terrestrial mammals other than deer, including *Ovis dalli, O. nivicola, Ovibos moschatus, Alopex lagopus,* and *C. lupus,* decrease in size at latitudes greater than 60°N. This disagreement is reminiscent of the earlier conflict that stemmed from an article by Scholander (1955), the essence of which was that a change in insulation was the principal (only?) means of responding to a cold climate, a change in body size being either inappropriate or too expensive. Mayr (1956) and Newman (1956) challenged this view, but Irving (1957) supported it. A reply by Schölander (1956) was a classic example of hyperbole: when noting the small size of the northernmost populations of muskrats (*Ondatra*), he suggested that it resulted from subtle selection for a size compatible with small holes in lake ice, and that that explanation should be accepted until a better one is found!

Several entirely different explanations for an increase in body mass with latitude have been given. One by Kalela (1957, p. 18), resulting from his work on rodents, was that in ". . . the evolution of greater body sizes in cooler climates, natural selection has operated . . . principally through [an increase in] litter size." Another stemmed from the study of the energetics of migration in shorebirds: Piersma et al. (1996) showed that maintenance metabolism in knots (*Calidris canutus*) is correlated with body mass and composition. Piersma and coworkers showed that body mass is flexible, varying among subspecies and between seasons within subspecies. Larger subspecies winter in colder climates and smaller subspecies are larger on their arctic breeding grounds than on their tropical wintering grounds. These correlations are explained by the positive correlation of energy expenditure with body mass. The seasonal and geographic variation in mass and body composition permits an adjustment of energy expenditure to match environmental requirements: low in the tropics and high during temperate winters. A similar pattern was seen (Castro et al. 1992) in the related sanderling (*C. alba*). Whether this analysis,

or any other, applies to all endotherms, or even to other shorebirds is unclear. As suggested earlier, each example of a positive correlation of body mass with latitude may require an individual explanation, which may or may not conform to a general pattern. The most general explanation for Bergmann's rule appears to be that body mass increases with latitude because food availability, quality, or size, or a combination there of, increases with latitude. An increase in energy storage and starvation time probably is, at best, a secondary contributor to a correlaton of mass with latitude. Care must be taken, therefore, when using historic changes in body size to estimate climate temperatures (see Klein 1986, Klein and Scott 1989).

An examination of climatic variability is a potential tool in the attempt to understand the effects of climate and latitude on the body size of mammals. Post and Stenseth (1999) showed that plant phenology and ungulate body size and fecundity varied with the North Atlantic Oscillation, an interannual and decadal variation in the North Atlantic air mass that influences the distribution of wintertime temperatures and rainfall in North America and Europe. In the face of this variation, inland populations of ungulates decreased in body mass and increased in fecundity during a warm winter, whereas maritime populations increased in body mass and decreased in fecundity. Winter weather appears to act on ungulates through its cumulative effects on reproductive females, with the effects on population abundance occurring with a delay of 2 to 3 y. The effects of climatic variability on ungulate size and fecundity are very complex and probably derived from the action of plant phenology and production on the nutrition condition of the ungulates.

The determination of body mass in endotherms obviously is infuenced by many factors. This complexity may be even greater than expected: in an examination of the structure of neotropical bird communities, Blackburn and Gaston (1996) showed that (1) mean body mass is lowest at the equator and increases with latitude in both hemispheres, (2) species richness is greatest at equatorial latitudes and smallest at high latitudes, and therefore (3) species diversity is correlated inversely with mean body mass. Blackburn and Gaston suggested that a functional relationship may exist between diversity and species richness: if a large mass is favored at high latitudes (for whatever reason), a low primary production rate may require a low species richness, although why many small species are found at low latitudes is unclear (unless they reflect the resources being used).

5.3.3 Dehnel's phenomenon.
A conflict potentially exists between the influence of chronic food shortages (which may require a small size) and of periodic or short-term food shortages (which through starvation time may encourage a large size). Hanski (1985), for example, argued that although an acute food shortage might select for a larger body mass among shrews, chronic low food availability would select for small individuals as a result of their smaller energy demands. In fact, some continuous endotherms that live in cold climates show a reduction in body mass during winter (Dehnel's phenomenon). These include some meadow voles (*Microtus* [Iverson and Turner 1974]) and red-toothed shrews (*Sorex* [Dehnel 1949; Pucek 1963, 1964; Mezhzherin 1964; Mezhzherin and Melnikova 1966; Genoud 1988; Hyvärinen 1984; Merritt and Zegers 1991]). The propensity of a species to demonstrate a decrease in mass in winter is an alternative to the use of seasonal torpor and is most likely to occur in species that do not conform to Bergmann's rule (in that it is an opposite response). Indeed, *Microtus pennsylvanicus*, which conforms to Dehnel's phenomenon, has its smallest body size in the coldest parts of its geographic range (Snell and Cunnison 1983). This reduction in mass is not simply a reduction in body fat but involves a decrease in skeleton, brain case, and brain tissue, that is, a reduction in active tissues. Caribou (*Rangifer tarandus*) have small-bodied populations characterized by a compact body form, thick fur coat, and "lethargic" behavior, all of which reduce energy expenditure, when isolated in environments with few resources, for example, Svalbard (Cuyler and Øritsland 1993). A similar decrease in mass occurred in Holocene mammoths (†*Mammuthus primigenius*) on Wrangel Island in the Siberian Arctic (Vartanyan et al. 1993; also see Section 10.10). This pattern is compatible with the interpretation of Geist (1987) for the impact of resource availability on body size.

5.3.4 Heat storage.
Equation 5.2 also can be used to analyze the size dependency of heat storage. Heat storage equals the product of the increase in T_b, body mass, and the body's specific heat ($c_p = 3.43 \, J/g°C = 4.186 \, J/cal \times 0.82 \, cal/g°C$). To simplify the calculation, metabolism is assumed to be the only source of heat. Then storage time is given by

$$t(h) = \frac{c_p \cdot \Delta T_b \cdot m^{100}}{69.31 m^{0.71}} = 0.049 \Delta T_b \cdot m^{0.29}. \quad (5.4)$$

If the body temperature of a 300-g mammal increases 6°C (e.g., from 35 to 41°C), Equation 5.4 estimates that t would be about 1.5 h, whereas t in a 300-kg mammal would equal 11.4 h. A 300-kg camel (*Camelus dromedarius*) lengthens this time period with excellent insulation, thereby reducing heat gain from the environment, and with a basal rate that is only 70% of the standard value (Schmidt-Nielsen et al. 1967). The reduction in basal rate alone is sufficient to increase storage time from 11.4 to 16.3 h. Heat storage occurs in other large mammals living in savannahs or deserts at low latitudes (where they encounter high heat loads), including black (*Diceros bicornis*) and broad-lipped (*Ceratotherium simum*) rhinoceroses, giraffe (*Giraffa camelopardalis*), and Asiatic (*Elephas maximus*) and African (*Loxodonta africana*) elephants (V. A. Langman and M. F. Rowe, pers. comm., 2001).

Most desert mammals are nocturnal; the few diurnal species either are large or seek shade to evade heat loads. Both strategies reduce reliance on evaporative water loss in environments where water availability is low. Yet, some small diurnal ground-squirrels (e.g., *Ammospermophilus leucurus*) store heat for short periods, after which they enter cool burrows to dump heat. Later they may be active again outside the burrow. The consequence for behavior in storage time in a ground-squirrel and a camel is shown in Figure 5.5 (Bartholomew 1964). The very short storage time in *A. leucurus* results from its small mass, poor insulation, and significant environmental heat load. Storage time in this squirrel, at 80 g, is expected from Equation 5.4 to be 1 h, but it actually can only tolerate 9 to 13 min because it has a rate of metabolism that is 3 times the basal rate during activity (which should lower t to 20 min) and faces a standard operative temperature (T_{es}) between 50 and 60°C (Chappell and Bartholomew 1981). Heat loads in *A. leucurus* are modestly decreased when its tail is used to shade the back (e.g., T_{es} fell maximally 3°C). Bennett et al. (1984) made similar observations on the African ground-squirrel *Xerus inauris*. This species has a longer tail than *A. leu-*

curus and under equally harsh conditions reduces T_{es} maximally 8.3°C. *Xerus inauris* may be active as much as 4 h daily but not continuously.

5.4 RESOLVING RESIDUAL VARIATION IN BASAL RATE

Although endotherms in the field are rarely under standard conditions, basal rates are valuable for comparative studies because they are relatively easily measured, are equivalent in all endotherms (see McNab 1997), and appear to be correlated with maximal rates of metabolism (see Section 5.7.4) and with energy expenditures in the field (see Section 10.5). The principal factor setting basal rate in endotherms is body mass. However, all scaling relations are mean curves with a residual

variation unaccounted for by mass (Figure 3.2). Therefore, using the most recent all-mammal curve (McNab 1988a) as a standard, the basal rate of a particular mammal can be written as

$$\dot{V}_{O_2} = 3.45 \cdot f_m \cdot m^{0.713}, \qquad (5.5)$$

where f_m is a dimensionless decimal coefficient that describes the degree to which the observed basal rate agrees with the value expected from the standard curve: when $f_m = 1.00$, the measured basal rate equals the expected rate.

A statistical analysis of the factors correlated with the basal rate in mammals has indicated how complicated the determination of basal rate is (Table 5.1). The residual variation around the mean curve might be considered to be an "error" term, but not if it correlates with biologically relevant variables. For example, in the Xenarthra, the only significant correlate of $\log_{10}$ basal rate is $\log_{10}$ body mass. However, among the Carnivora $\log_{10}$ basal rate is correlated with $\log_{10}$ body mass, food habits, and activity level. Family affiliation is not a significant correlate of $\log_{10}$ basal rate in carnivorans, unless food habits are deleted from the analysis, thereby implying a correlation between family affiliation and food habits. Indeed, felids are strict vertebrate eaters, most mustelids are vertebrate eaters, and most procyonids and viverrids have mixed diets or are frugivores. Furthermore, among the Rodentia, if family affiliation is eliminated from the analysis, entrance or not into torpor becomes significant, suggesting that the use of torpor is correlated with family affiliation: for example, arvicolids do not enter torpor, whereas heteromyids,

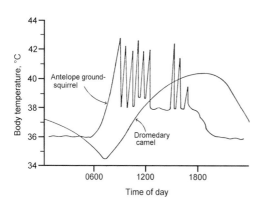

Figure 5.5 Body temperature in the antelope ground-squirrel (*Ammospermophilus leucurus*) and the dromedary camel (*Camelus dromedarius*) as a function of the time of the day. Source: From Bartholomew (1964).

Table 5.1 Analysis of basal rate of metabolism in mammals*

Taxon	n	$\log_{10} m$	Subgroup	Food	Climate	Activity	Torpor	Reproduction	r^2
Xenarthra	15	0.0001	Genus NS	NS	NS	NS	NS	NS	0.954
Carnivora	30	0.0001	Family NS	0.0001	NS	0.0069	NS	*	0.989
	30	0.0001	Family 0.0490	—	NS	0.0396	NS	*	0.983
Rodentia	106	0.0001	Family 0.0001	0.0384	0.0001	*	NS	NS	0.980
	106	0.0001	—	NS	0.0001	*	0.0001	NS	0.964
Mammalia	297	0.0001	Family 0.0001	0.0001	NS	NS	0.0001	NS	0.993
	297	0.0001	—	0.0001	0.0001	0.0001	0.0003	0.0001	0.977

*This analysis results from ANCOVA, where $\log_{10}$ basal rate was the dependent variable and the independent variables were $\log_{10}$ body mass, taxonomic subgroup, food habits, climate, activity level (high, intermediate, or low), torpor (yes, no), and type of reproduction. A factor arbitrarily eliminated, is indicated by —; a factor not relevant is indicated by *; NS means that $\log_{10}$ basal rate is not correlated with the factor; r^2 is for the correlation of $\log_{10}$ basal rate with all significant factors; n is the number of species; if a factor is significant, it probably is given.
Source: Condensed from McNab (1992a).

glirids, and zapodids do. When all 297 studied eutherians are examined (McNab 1992a), basal rate is correlated with food habits, climate, activity level, the use of torpor, and type of reproduction, but only when family affiliation is deleted from the analysis. In other words, the basal rate of metabolism in mammals is correlated with a complex array of interacting factors. Many fewer data are available on birds so that an extensive analysis is not presently possible. The data on birds thus will be compared to those available on mammals.

5.4.1 Food habits. Many data suggest that the basal rates of mammals are correlated with food habits (for a general review, see McNab 1986a). The first, and still one of the clearest, examples was found in bats (McNab 1969) because of their diversity in food habits: species that are specialized feeders on fruit, nectar, or vertebrates have basal rates that are similar to those expected from the mammalian standard and significantly higher than those measured in insectivorous species of the same mass (Figure 5.6). Vampire bats and some species with a mixed diet have intermediate basal rates. The only significant change in this picture has been recent observations that a few small, tropical, insectivorous species belonging to the Emballonuridae (Genoud and Bonaccorso 1986, Genoud et al. 1990) and Mormoopidae (Bonaccorso et al. 1992) have average to high basal rates. The correlation of basal rate with food habits obviously occurs only in families (or orders) that have diverse food habits: Phyllostomidae ($P \leq 0.0001$), Chiroptera ($P = 0.0004$), Carnivora ($P \leq 0.0001$), and Rodentia ($P = 0.0384$).

The procyonid with the narrowest diet, the kinkajou (*Potos flavus*), which principally feeds on fruits and insects, has the lowest basal rate (compared to that expected from mass), whereas the species with the greatest diversity in food habits, including eating vertebrates, the North American raccoon (*Procyon lotor*), has the highest basal rate (Mugaas et al. 1993). All three other procyonids studied, including the crab-eating raccoon (*Procyon cancrivorous*), have food habits of an intermediate diversity and intermediate basal rates of metabolism.

A general summary of the interaction between food habits and basal rate in mammals is shown in Figure 5.7 A complex pattern can be seen. At masses less than 100 g, most mammals have rather high basal rates independent of food habits (two exceptions being those species that feed on flying insects [bats] and those that feed on seeds in a desert [heteromyid rodents]). Basal rates at masses greater than 100 g are highly variable: mammals that eat grass, vertebrates, and (possibly) seeds have high basal rates, whereas those that eat invertebrates, fruit, or the leaves of trees have low rates (although see Section 5.4.3). Mammals with mixed diets have basal rates that fall between the rates of mammals that are specialized to feed on the components of the mixed diet (McNab 1986a).

An explanation for a correlation of basal rate with food habits is difficult to establish without ambiguity, although suggestions can be made. In insectivorous bats a low basal rate may be associated with variation in the seasonal availability of flying insects. Bats with other habits, especially

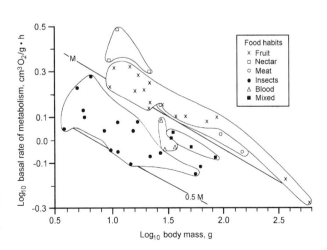

Figure 5.6 Log$_{10}$ mass-specific basal rate of metabolism in bats as a function of log$_{10}$ body mass and food habits. The all-mammal curve (M) and 50% of the all-mammal curve (0.5 M) are indicated. Source: Modified from McNab (1980b).

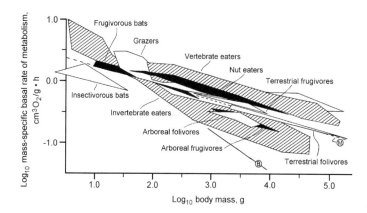

Figure 5.7 $\log_{10}$ mass-specific basal rate of metabolism in eutherian mammals as a function of $\log_{10}$ body mass and food habits. The all-mammal (M) and boundary (B) curves are indicated. Source: Modified from McNab (1986a).

those that feed on fruit, nectar, and vertebrates, either are less committed to a particular food or feed on foods having a greater energy and nutritional density and a greater seasonal reliability than flying insects. This difference is reflected in the seasonality of fat deposits of bats: strictly insectivorous species in the tropics, especially those that feed above or beyond the forest, have the greatest annual amplitude in body fat content, whereas frugivorous and nectarivorous bats tend to have the smallest amplitudes (McNab 1976). As further evidence of the differential seasonality of foods in tropical localities, insectivorous bats generally have shorter breeding seasons than do nectar-, fruit-, or blood-eating bats in the same area (McNab 1969, Mares and Wilson 1971). The high basal rates in some small tropical insectivorous bats in part are permitted by the small (total) energy requirements associated with a small mass and (at least in some emballonurids) a seasonal change in foraging area (Bradbury and Vehrencamp 1976).

In mammals generally the principal differences between foods that are associated with high basal rates and those that are associated with low rates appear to reflect the properties of the foods, namely, digestibility, energy and nutritive content, seasonality, and content of secondary compounds. A large body mass in consumers multiplies the impact of food habits on basal rate because large mammals have higher energy and nutritional requirements than do small species, and therefore are more likely to encounter limits to energy and nutrient availability imposed by the food.

The data available on basal rates in birds cover such a small trophic and taxonomic diversity that the impact of food habits is best compared to that in mammals (McNab 1988b). Passerines generally

are recognized to have higher basal rates than other birds, which in turn have higher basal rates than mammals of the same mass (see Chapter 3), but when mass and food habits are considered, mammals and most birds have similar basal rates of metabolism. For example, arvicolid rodents and soricine shrews have the same basal rates as passerines of the same mass (McNab 1988b). The lower basal rates of many eutherians compared to passerines are correlated with a much greater proportion of small eutherians entering torpor. This analysis does not deny that differences in energetics exist between mammals and birds: large invertebrate-eating birds have higher basal rates than do invertebrate-eating mammals of equal mass, and large passerines have high basal rates compared to mammals with the same mass and food habits. Birds, like mammals, appear to have basal rates that reflect food habits, although many more data are needed before any general analysis can be made of avian energetics. Data are especially needed on species indigenous to the tropical lowlands, where a great unexplored ecological and taxonomic diversity exists. A step in this direction was taken by comparing the basal rates of tropical, fruit-eating toucans (Ramphastidae), pigeons (Columbidae), and flying foxes (Pteropodidae): although an appreciable interspecific variation in their basal rates was demonstrated, no difference among these families, or between these birds and mammals, was found (McNab 1994b, 2001).

At small masses, unlike mammals, birds can avoid the worst environmental conditions and food shortages through seasonal migration, which may explain why Yarbrough (1971) found that temperate sparrows, flycatchers, and warblers showed no correlation of basal rate with food habits.

Migratory birds, from this view, are less likely than sedentary mammals to encounter food shortages. Yet, the birds Yarbrough studied were small and therefore, like small mammals, are less likely to show a correlation between food habits and rate of metabolism because a low (total) rate of expenditure is unlikely to exceed energy availability.

Food habits have a greater impact on resting rates of metabolism in mammals and birds than on the standard rates of reptiles (Pough 1983, Andrews and Pough 1985). This difference reflects the high rates of energy turnover in endotherms and the consequent necessity to harvest food in large quantities. Depending on the ecological and nutritional characteristics of a food, some endotherms might be required to reduce their expenditures and therefore to compromise the precision, level, and continuity of their endothermy. As suggested, this is the Achilles' heel of endothermy.

5.4.2 Climate. Scholander et al. (1950a, 1950b, 1950c), in three pioneer articles, presented data showing that arctic mammals and birds have higher basal rates than tropical species, although these authors were reluctant to draw this conclusion. Subsequent measurements generally have shown that arctic and cold-temperate terrestrial mammals have high basal rates, whereas desert, fossorial, and many tropical species have low basal rates (Figure 5.8). The difficulty with this conclusion is that the basal rate in mammals is correlated with many factors that interact with climate. For example, in Figure 5.8, the arctic/cold-temperate species, which are vertebrate eaters belonging to the Mustelidae and the Canidae or are grazers and browsers belonging to the Lagomorpha, Arvicoli-dae, or Artiodactyla, have basal rates that average 54% higher than those for tropical species, which are ant/termite-eating or folivorous Xenarthrans, frugivorous or mixed-diet procyonids and viverrids, and herbivorous or mixed-diet hystricognath rodents. These data raise the question as to whether the difference seen in Figure 5.8 is due to climate, food habits, or phylogeny. Thus, the few tropical carnivores and grazers studied have high basal rates (Taylor and Lyman 1967, Taylor et al. 1969, McNab 2000b).

Endotherms in aquatic environments face a thermal conductivity that is 24 times that of air, which may explain why mammals living in cold water, such as whales, pinnipeds, the sea otter (*Enhydra lutris*), otters (*Lutra*), beavers (*Castor*), and the muskrat (*Ondatra zibethica*), have high basal rates. Lavigne et al. (1985), however, argued that seals and whales do not have high basal rates, but that those reported by earlier investigators (Scholander 1940, Irving et al. 1941b, Irving and Hart 1957) resulted from measurements on stressed individuals. That may well be the case, but recent measurements by Ochoa (1999) indicated that cold-water seals indeed have high basal rates. The high basal rates in aquatic carnivores are similar to those of terrestrial carnivores, which may indicate that the high rates of seals are fundamentally connected with food habits. The low basal rates of herbivorous manatees (*Trichechus* [Gallivan and Best 1980, Irvine 1983]) may reflect food habits as much as life in warm water. Unfortunately, no warm-water carnivores, such as the monk seal (*Monachus*) or Amazonian porpoises (*Inia* and *Sotalia*), have had their basal rates measured, but they are unlikely to have low basal rates.

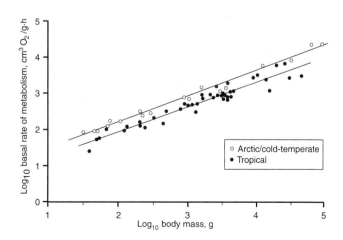

Figure 5.8 Log$_{10}$ basal rate of metabolism in mammals as a function of log$_{10}$ body mass and climate. Source: Data derived from McNab (1988a).

A study along a 3000-km transect from northern Finland to southern Bulgaria showed no correlation in the bank vole (*Clethrionomys glareolus*) of basal rate with latitude (Aalto et al. 1993), probably because habitat selection varied with locality (e.g., altitude was 110m in southern Finland and 1700m in Bulgaria), thereby overriding any effect of latitude on climate. The basal rate of *Peromyscus maniculatus* increased at high altitudes independently of temperature (Hayes 1989b), possibly as a means of maintaining a high maximal rate of metabolism.

Another example of the apparent influence of climate is the low basal rates found in many desert mammals (e.g., McNab and Morrison 1963), especially in species that feed heavily or exclusively on seeds, a food having a low water content. Low rates of metabolism may be required of seed-eating rodents to balance their water budgets and by the ephemeral nature of seeds as a food supply. Within the rodent family Heteromyidae, desert species have low basal rates, whereas species belonging to the genus *Heteromys*, which live in tropical rainforests, have high basal rates (McNab 1979a, Hinds and MacMillen 1985). Desert lagomorphs, which are browsers, have high basal rates (Schmidt-Nielsen et al. 1965; Dawson and Schmidt-Nielsen 1966; Hinds 1973, 1977), possibly reflecting a food with a high water content. The desert canid *Fennecus* at a mass of 1.1kg has a low basal rate (Noll-Banholzer 1979); it feeds principally on invertebrates, a habit that is associated with low basal rates in all climates in mammals that weigh more than 100g.

A potential example of a "climatic" influence is found in the low basal rates of all burrowing and fossorial mammals that weigh more than 100g (McNab 1966b, 1979b; Contreras and McNab 1990). Low basal rates in these species have been ascribed to the gaseous composition of burrow atmospheres (Baudinette 1972, Arieli et al. 1977, Arieli 1979, Arieli and Ar 1981); to the difficulty in dissipating heat in a still, warm atmosphere saturated with water vapor (McNab 1966b, 1979b; Nevo and Shkolnik 1974; Contreras 1986; Contreras and McNab 1990); or to a limited energy supply (Vleck 1979, 1981). The greatest reduction in the basal rate of burrowing mammals occurs in the largest species and those living in the warmest environments (McNab 1966b, 1979b; Nevo and Shkolnik 1974).

Birds also show evidence of a correlation of basal rate with climate, again complicated by food habits. For example, desert-dwelling birds also tend to have low basal rates. These include pigeons (Dawson and Bennett 1973, Withers and Williams 1990), quail and partridges (Goldstein and Nagy 1985, Frumkin et al. 1986), sandgrouse (Thomas and Maclean 1981, Hinsley et al. 1993), and starlings (Dmi'el and Tel-Tzur 1985), all of which feed on seeds, as do desert rodents with low basal rates. More generally, basal rate in birds is correlated with climate as represented by latitude (Weathers 1979, Hails 1983, Ellis 1985) (Figure 5.9), but at low latitudes species that face high levels of solar radiation have high rates of metabolism only if they have light-colored plumage, whereas those exposed to solar radiation and have dark plumage have low basal rates of metabolism (Ellis 1980, 1985; see Section 2.5).

The impact of climate on the basal rate of some endotherms may be seasonal (for a general review of the physiological basis of temperature acclimation in endotherms, see Chaffee and Roberts [1971]). The available data indicate a highly complicated picture in rodents (Hart 1971): some species show no seasonal variation in basal rate, some have higher rates in winter, whereas others have higher rates in summer. Djungarian hamsters (*Phodopus sungorus*), lagomorphs, and peccaries (Hinds 1973, 1977; Zervanos 1975; Heldmaier and Steinlechner 1981b) have higher basal rates in winter, whereas deer have low basal rates in

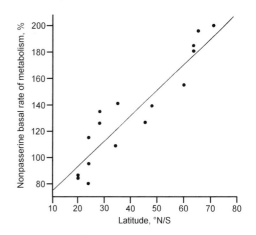

Figure 5.9 Basal rate of metabolism in nonpasserine birds, expressed as a percentage of the basal rate expected from body mass, as a function of latitude. Source: Modified from Ellis (1985).

winter (Silver et al. 1969, Weiner 1977), as do desert-dwelling dingos (*Canis familiaris* [= *lupus*] [Shield 1972]), coyotes (*C. latrans* [Golightly and Ohmart 1983]), and wolves (*C. lupus* [Afik and Pinshow 1993]). Birds are equally variable (Irving et al. 1955, Hart 1962, Veghte 1964, West 1972). Thus, Root et al. (1991) found that house finches (*Carpodacus mexicanus*) living in cold climates have higher basal rates in winter than do house finches that come from mild climates. Dark-eyed juncos (*Junco hyemalis*), however, showed no seasonal variation in basal rate (Swanson 1991a, 1993).

An explanation for this variable response to seasonality is not readily available. The physiological response to winter may vary with latitude: an increase in basal rate may be least likely in polar species and most highly developed in temperate species (i.e., the response may be a function of the length and severity of winter conditions). Small species might be more prone to increase basal rate because the maximal rate of metabolism varies with basal rate of metabolism (Hemmingsen 1960, Jansky 1962, McNab 1980b; see Section 5.7.4) and because the response to low temperatures by increasing the thickness of the feather or fur coat is physically limited. Some of the response to winter conditions may reflect a shift in diet (Macari et al. 1983), for example, in deer from grass to browse.

A particularly striking example of the seasonal acclimatization in basal rate of metabolism is found in king penguin (*Aptenodytes patagonicus*) chicks (Barré 1984), which hatch in summer and remain on land fasting through winter until the following summer. A lower basal rate in winter combined with a lower thermal conductance and huddling permits winter-acclimatized chicks to fast for periods twice as long as summer-acclimatized chicks (84 vs. 44 d, respectively). This behavior and a large mass (10–20 kg) permit king penguins to overwinter on subantarctic islands.

5.4.3 Body composition. Another factor that is associated with basal rate in mammals is body composition. McNab (1978b) and, more generally, S. Thompson and T. Grand (pers. comm., 1985) showed that basal rate in eutherian mammals is directly related to the proportion of body mass that is muscle. That is, eutherians that are highly active have a large muscle mass and a high basal rate. Konarzewski and Diamond (1995) found that

much of the variation in the basal rate of domestic house mice can be accounted for by variation in the size of the heart, kidney, liver, small intestine, or their correlates. The reason why (larger) arboreal mammals have low basal rates may be because they have small muscle masses and are relatively sedentary (Grand 1977), rather than because they feed on leaves or fruit (as previously proposed by McNab [1978b, 1986a]). Body composition, however, does not account for the lower basal rates found in marsupials (McNab 1986b), and it accounts for little of the variation in basal rate in *Peromyscus maniculatus* (Koteja 1996b). Among birds, Kersten and Piersma (1987) maintained that basal rate reflected variations in "metabolic machinery," and Daan et al. (1990) noted that 50% of the variation in basal rate was accounted for by the mass of heart and kidneys in 22 species of birds. Most flightless birds have lower basal rates than their flighted relatives, the reduction being correlated with a reduction in the pectoral muscle mass (McNab 1994a; see Section 9.6.9). However, Burness et al. (1998) found no correlation between body composition and rate of metabolism in tree swallows (*Tachycineta bicolor*).

5.4.4 Taxonomic affiliation. An analysis of the factors setting the level of basal rate of metabolism in mammals and birds is greatly complicated by factor interactions. Food habits are correlated with body size and climate; body size is correlated with climate; activity level, with food habits and body size; torpor, with body mass, food habits, and climate; and so on. And many of these factors are correlated with taxonomic affiliation. The ability to separate cause from effect consequently is exceedingly difficult, if not actually impossible. Understandable then are disagreements on the factor interactions that set this or that phenotypic character. Furthermore, the residual variation in basal rates, after the influence of mass was considered, is much greater in eutherians than in marsupials (see Section 13.5). Such observations led Hayssen and Lacy (1985) and Elgar and Harvey (1987) to argue that most of the residual variation in mammalian basal rates reflects taxonomic affiliation, although Elgar and Harvey concluded that vertebrate eaters indeed have higher basal rates than what is expected from mass and taxonomic affiliation. Bennett and Harvey (1987) maintained that most of the residual variation in the basal rate of birds is also associated with phylogeny.

The view of Harvey and his colleagues has been bolstered by use of the method of "contrasts," which attempts to correct for the influence of phylogeny in the evolution of character states, here basal rate of metabolism. Derrickson (1989), however, pointed out several defects with the original analysis, including the retention of the effect of taxonomic affiliation, the bias against discovering significant differences in another food habit by the retention of food habits that have already been shown to be different, and the unwillingness to consider variation within a taxonomic assemblage. For example, Elgar and Harvey (1987) argued that variation in basal rate among species belonging to a genus is irrelevant because all species belonging to a genus are identical in terms of diet and habitat. This view ignores the many examples of ecological and physiological differentiation among species belonging to a genus (e.g., McNab and Morrison 1963; Shkolnik and Borut 1969; Sparti 1990; McNab 1992b, 1994b; Mugaas et al. 1993) and even among subspecies belonging to a species (e.g., McNab and Morrison 1963, Nevo and Shkolnik 1974). Another approach to "correct" for the impact of phylogeny on basal rate is to map

changes in basal rate onto a cladogram of the phylogeny of a group (see Section 1.2.1). As reasonable as this suggestion appears, its implementation depends on a reliable phylogeny, and even then no unique mapping is possible. A notable example of the difficulty in attempting to analyze the factors responsible for setting basal rate of metabolism is seen in pteropodid bats (McNab and Bonaccorso 1995b, Bonaccorso and McNab 1997). Within this family at least 161 species belong to 41 genera. Nectarivorous species are small (<40 g); have low basal rates; enter daily torpor; live in the tropics from the Solomon Islands to the Philippines, Southeast Asia, and Africa; and usually have been assigned to one subfamily, the Macroglossinae. Their assignment to one subfamily implied that the evolution of these characteristics occurred once in the history of this family. But a recent study of DNA hybridization (Kirsch et al. 1995) indicated that nectarivory has evolved independently in the Pteropodidae at least 5 times. Consequently, most of the genera of nectar feeders may have evolved a small size, low basal rate, and poor temperature regulation independently of each other.

BOX 5.1
On the Significance of the Correlation of Physiological Characters with Taxonomic Affiliation

The quantitative variation among species in physiological characters is often associated with taxonomic affiliation. For example, the basal rate of metabolism independent of body mass is greater in eutherians than in monotremes and marsupials, as Martin (1902) originally observed. The difference between eutherians and marsupials, however, really is not that marsupials have low basal rates—the lowest basal rates are found in eutherians—but that no marsupial has a high basal rate (McNab 1988b; Section 13.5). What does this statement mean?

Even if basal rate is correlated with taxonomic affiliation, it is correlated also with food habits, activity level, and climate, which raises the question of what determines basal rate. Thus, all ant and termite eaters that weigh over 1 kg, including such eutherians as the aardwolf (*Proteles cristata*) and the sloth bear (*Ursus ursinus*), as well as tamanduas, aardvarks, pangolins, and the numbat (*Myrmecobius fasciatus*), have low basal rates

(McNab 1980c, 1984; Anderson et al. 1997). The red panda (*Ailurus fulgens*), an arboreal folivore belonging to the order Carnivora, has a basal rate as low as a tree-sloth (McNab 1988c). The insectivorous phyllostomid bat *Macrotus californicus* has a much lower basal rate than frugivorous and nectarivorous phyllostomids (Bell et al. 1986), but similar to the rates found in bats belonging to the strictly insectivorous families Molossidae and Vespertilionidae.

If affiliation is important in determining characters like the basal rate, then the basis for this impact should be specified. For example, I have argued that the basal rate of marsupials is distinctive, but in this case the association is with the marsupial form of reproduction (McNab 1986b). The absence of high basal rates in marsupials may occur because of the inability of pregnant marsupials to augment maternal-fetal exchange by means of a high basal rate, owing to the presence of egg-shell membranes surrounding the develop-

---

ing embryo (see Section 13.5). The absence of high basal rates in marsupials undoubtedly reflects some factor that distinguishes marsupials from eutherians, rather than reflecting some vague view that marsupials have basal rates "frozen" at a low level derived from an ancestor. After all, eutherians "thawed" basal rates in association with their evolution from an ancestor common with marsupials (see Lillegraven et al. 1987). In this analysis the difference between the attribution of a character state to "ecology," "behavior," or "morphology" versus "phylogeny" becomes unclear and ultimately artificial.

Another fundamental problem exists with the analysis that emphasizes the importance of phylogeny. Westoby et al. (1995, p. 531) argued that the phylogenetic correction advocated by many "... is not in fact a 'correction,' i.e., an adjustment to remove errors. Rather, it is a conceptual decision to give priority to one interpretation over another." The fundamental problem with the attempt to dissect the factors influencing character states is the interaction that occurs between "phylogeny" and "ecology." The nature of this interaction is shown in Figure 5.10. In this example the combination of phylogeny, which is represented by taxonomic affiliation, and ecology, which would include reactions with the physical environment, food habits, and behavior, accounts for 80% of the variation in the character, the remaining 20% being unresolved. A 40% overlap occurs between the influences of phylogeny and ecology. If an analysis starts with the assumption that ecology is the primary factor (assumption A), then the conclusion is that 60% of the variation in the character is due to ecological factors, 20% is due to phylogeny, and 20% remains unresolved. This has been the analytical procedure used by McNab and Morrison (1963), Bartholomew et al. (1964), McNab (1966b, 1969, 1984, 1988b), Hulbert and Dawson (1974), Shkolnik and Schmidt-Nielsen (1976), Contreras and McNab (1990), Bonaccorso et al. (1992), and Genoud (1993), among others. If, however, the assumption is made that phylogeny is primary (assumption B), then the conclusion is that phylogeny accounts for 60% of the variation, ecology for 20%, and 20% remains unresolved. That is, "... only after the correlation with phylogeny has been extracted, is residual variation analysed for its relationship to present day ecology. [Phylogenetic correction] therefore allocates the maximum possible variation in a trait to phylogeny, considering only the residual as potentially

DETERMINATION OF CAUSES OF TRAIT VARIATION

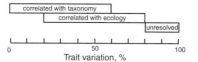

ANALYTICAL APPROACHES

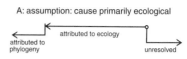

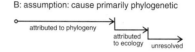

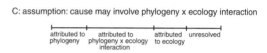

Figure 5.10 The comparative impact of "phylogeny," "ecology," and "unknown factors" on the variation in phenotypic characters, such as rate of metabolism. Three types of analyses are recognized: (A) under the assumption that ecological factors are primary, which incorporates the overlap between ecology and phylogeny into ecology; (B) under the assumption that phylogeny is primary, which incorporates the overlap between ecology and phylogeny into phylogeny; and (C) in which the determinate factors are phylogeny only, interaction between phylogeny and ecology, ecology only, and unknown. This scheme follows that of Westoby et al. (1995).

attributable to ecology" (Westoby et al. 1995, p. 531). This has been the approach of Bennett and Harvey (1987), Elgar and Harvey (1987), Harvey et al. (1991), Harvey and Pagel (1991), Garland et al. (1992), and Konarzewski (1995), among others. Their approach, which often has been justified by the claim that data from species belonging to the same genus are redundant and represent pseudoreplication (Felsenstein 1985, Elgar and Harvey 1987, followed by many others), is often erroneous and represents a convenience for calculations or an ignorance of biological diversity. The most appropriate analysis is seen in assumption C, when 20% of the variation is due (uniquely) to phylogeny, 20% is due (uniquely) to ecology, 40% reflects an interaction between phylogeny and ecology, and 20% remains unresolved. Even then, the conclusion that 20% of the variation in a char-

Continued

acter "... is due (uniquely) to phylogeny ..." has no clear meaning. It would be more precise to state that 20% is *correlated* with phylogeny, leaving open the possibility that some future study will indicate a correlation of the character state with some previously unsuspected aspect of the natural history of the organism.

Both assumptions A and B assume that phylogeny and ecology are mutually exclusive explanations, which is surely incorrect. Westoby et al. concluded (p. 534) that "... phylogenetic constraint asserts that a trait has resisted selection over tens of millions of years, and very strong evidence should be required to support so implausible a proposition." What is needed is the realization that the interaction term cannot be allocated exclusively either to "phylogeny" or to "ecology." How

this interaction term should be handled, however, is not clear. Can one seriously consider the possibility that a species could be characterized by ecologically unacceptable characteristics dictated by phylogeny alone? Such species are usually referred to as "extinct." More likely, energy expenditure might be determined by food habits or other behavioral characteristics, which themselves are dictated or limited by phylogeny. That is, the relatives of most species have similar characteristics and in part that is why they are classified together. Thus, nearly all cats (Felidae) feed exclusively on vertebrates and nearly all cats have high basal rates (McNab 2000b), which means that separating the effects of "felidness" and carnivory on basal rate of metabolism is difficult, except as they are examined in other vertebrates eating mammals.

5.5 FACTORS AFFECTING THERMAL CONDUCTANCE

Thermal conductance measures the ease with which heat leaves or enters the body. It is influenced by environmental factors, such as radiant temperatures (see Section 2.5) and wind velocity (see Section 2.7), and by behavioral factors, such as posture, piloerection and ptiloerection, and peripheral vasomotion. In fact, the concept of "thermal conductance" is a great simplification of a highly complex set of physical relationships between an endotherm and its environment, which has led some biologists (e.g., Tracy 1972) to dismiss the use of conductance as being too imprecise. Its value is that it represents a practical description of heat exchange under the conditions in which rate of metabolism is usually measured. Minimal thermal conductance is most convenient to analyze because environmental factors that modify heat exchange have been eliminated and the behavioral component in conductance has been standardized. Surface area and surface-specific conductance are important components of total thermal conductance. The first is determined mainly by body mass and shape; the second, by mass and climate. Because minimal thermal conductance corresponds to maximal insulation, thermal conductance also can be analyzed in terms of insulation. Scholander et al. (1950a, 1950b, 1950c), Hammel (1955), and Hart (1956, 1957) extensively studied the insulative properties of fur coats.

5.5.1 **Body mass.** Body mass is the one most important determinant of thermal conductance. Minimal total conductance (C_m, in cm^{3}O$_2$/h°C) in mammals is

$$C_m = 1.00\,m^{0.50},\qquad(5.6)$$

where mass is in grams. Morrison and Ryser (1951) first implied this relation, and McNab and Morrison (1963) first described it. Herreid and Kessel (1967) and Bradley and Deavers (1980) derived an essentially identical relation. Total minimal conductance of birds is

$$C_m = 0.85\,m^{0.49}\qquad(5.7)$$

(Lasiewski et al. 1967). The thermal conductances of birds are 15% lower than those of mammals of the same size (Figure 5.11) because feathers trap air better than fur.

Total thermal conductance is equal to the product of surface area (A) and the surface-specific thermal conductance. In mammals

$$C_m\,(\text{J/h°C}) = A(\text{cm}^2)\cdot C_m/A(\text{J/cm}^2\text{h°C})$$
$$20.09\,\text{g}^{0.50} = (10m^{0.67})\cdot C_m/A$$
$$C_m/A = 2.01\,m^{-0.17}.$$

Consequently, surface-specific insulation (I_a) increases with body mass:

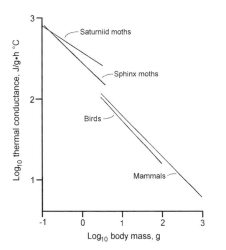

Figure 5.11 Log₁₀ mass-specific thermal conductance in saturniid moths, sphinx moths, birds, and mammals as a function of log₁₀ body mass. Source: Modified from Bartholomew and Epting (1975).

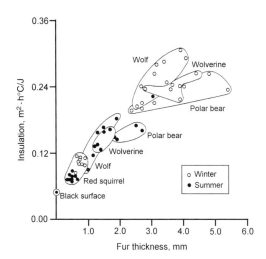

Figure 5.12 Insulation in mammals as a function of fur thickness and season. Source: Modified from Hart (1956).

$$I_a \left(cm^2 h°C/J \right) = 0.50 m^{0.17}. \qquad (5.8)$$

Although the major factor determining surface-specific insulation is the length of the fur (Figure 5.12), differences in the density of the coat also contribute to insulation. The principal reason why large mammals have better insulation, then, is that they have longer fur than do smaller species.

Another morphological factor related to size that influences thermal conductance and insulation is body shape. Most mammals have similar shapes,

but some, such as weasels, have large surface areas. Brown and Lasiewski (1972) showed that long-tailed weasels (*Mustela frenata*) have higher thermal conductances than do similarly sized wood-rats (*Neotoma* sp.), a difference that results from a larger surface area and a thinner fur coat. The very high basal rates of weasels may be viewed as compensatory for their large surface areas, as well as being associated with vertebrate-eating habits.

5.5.2 Climate. The correlation of thermal conductance with body mass has been suggested to be the basis for the correlation of body size with latitude and climate in endotherms. Bartholomew (1977, p. 364) gave the clearest statement of this position: ". . . large birds and mammals . . . have lower *mass-specific* conductances than smaller ones. Therefore, it should be less energetically expensive *per kilogram* for a large homeotherm to live in a low temperature than it would be for a small one to do so. This is the rational basis for Bergmann's rule. . . ." (italics added). This suggestion suffers from two difficulties. One is that some poikilotherms conform to the same cline in body mass (Ray 1960, Lindsey 1966); these species surely are not reducing heat loss because they do not maintain a temperature differential with the environment through the internal generation of heat. More important, thermal conductances decrease with an increase in body mass only if conductances have mass-specific units. Such a mass-specific justification for Bergmann's rule persists in the analysis of Steudel et al. (1994). Total conductances, however, are greater in larger than in smaller endotherms (Equations 5.6 and 5.7). Large endotherms expend more energy to live at a low temperature than do small species (see Figure 3.3), except on a per-gram basis, which has little ecological significance. As seen (see Section 5.3.2), the most appropriate explanation for Bergmann's rule is a geographic correlation of body size with food availability.

Climate has a marked influence on conductance. Scholander et al. (1950a, 1950b, 1950c) and Hart (1956) showed that tropical mammals have poor insulation because they have a shorter and less dense fur coat (Figure 5.12), whereas arctic mammals have high insulation because they have a long, dense fur coat (the musk-ox [*Ovibos moschatus*] being an extreme example). The difference in thermal conductance between winter and summer within a species may be appreciable (Figure 5.13A), but some species, especially those having a small mass, usually show little difference in thermal conductance between winter and summer (Figure 5.13B) because they cannot carry a long fur coat. Small desert rodents with low basal rates, however, tend to have low thermal conductances (McNab and Morrison 1963). Lovegrove et al. (1991a) examined this observation in some detail; they suggested that the low conductances in two Kalahari rodents, *Thallomys paedulcus* and *Aethomys namaquensis*, permit these rodents to tolerate low environmental temperatures in spite of low basal rates.

Large mammals living in warm climates have high thermal conductances (Figure 4.36) to prevent overheating. Overheating occurs at large masses because heat production increases proportionally to about $m^{0.75}$, whereas body surface area, through which much of heat loss occurs, is proportional to $m^{0.67}$. Thus, most naked mammals are large and have a tropical distribution. Nakedness in humans (*Homo sapiens*) may have reduced a heat load produced by a combination of a tropical distribution, moderate size, and a high level of activity (while running down injured prey?). It also conforms to a

scaling relationship in primates whereby the density of the fur coat decreases with surface area (Schwartz and Rosenblum 1981) and therefore with mass. Great apes larger than humans (e.g., the gorilla [*Gorilla gorilla*], orangutan [*Pongo pygmaeus*]) have fur coats of low density and are sedentary and vegetarian. Naked mammals of intermediate size (e.g., aardvark [*Orycteropus*], armadillos, giant pangolins [*Manis*]) have burrowing habits, overheating being a threat in warm, humid burrows. The only small mammals that are naked as adults live in the tropics, either in a warm, humid burrow (the naked mole-rat [*Heterocephalus glaber*] from East Africa) or in large colonies (the naked bat [*Cheiromeles torquatus*] from Borneo). On the other hand, tropical mammals with low basal rates, such as tree-sloths and some monkeys, anteaters and termite eaters, and viverrids, often have in compensation excellent fur coats (McNab 1984).

Many aquatic mammals use a specialized form of insulation to reduce heat loss to water. The insulative properties of fur and feathers depend principally on a layer of still air trapped in the coat. A few marine mammals, such as fur seals (*Callorhinus*, *Arctocephalus*) and the sea otter (*Enhydra lutris*), have a well-developed fur coat, but most rely instead on a layer of blubber for insulation. A steep thermal gradient is established in the layer, heat loss being inhibited by a low capillary density in the blubber (see Scholander et al. 1950a, Irving and Hart 1957). As a consequence, the insulation provided by blubber in water is usually higher than that given by a fur coat, although in baleen whales blubber may principally be an energy store (Hokkanen 1990). The largest whales, such as the blue (*Balaenoptera musculus*) and humpback (*Megaptera novaeangliae*), have a bipolar distribution, which indicates that they are more likely to face overheating in warm waters or during periods of intensive activity than to encounter excessive heat loss in cold waters. In winter they reside in tropical waters.

The variation in minimal thermal conductance in mammals, independent of mass, can be represented by the dimensionless factor, f_c:

$$C_m = 1.00 \cdot f_c \cdot m^{0.50}, \qquad (5.9)$$

so that the minimal thermal conductance in mammals equals the value expected from mass when $f_c = 1.00$.

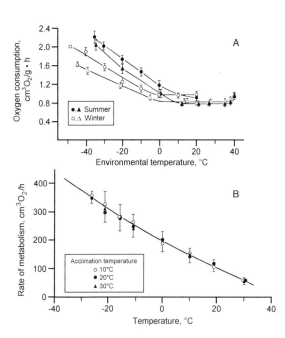

Figure 5.13 Rate of metabolism as a function of environmental temperature in (A) snowshoe hares (*Lepus americanus*) in winter and summer and (B) white-footed mice (*Peromyscus leucopus*) acclimated to 10, 20, and 30°C. Source: Modified from Hart et al. (1965) and Hart and Heroux (1953).

5.5.3 The control of peripheral temperatures.
Thermal conductance is an integrated measure of the ability of heat to move across the surface between an organism and its environment. This "leakage," as noted, is a function of surface area, the thickness and density of the insulation, and the temperature of the tissues at and below the surface. The maintenance of high peripheral temperatures increases heat loss from endotherms exposed to low ambient temperatures and is directly related to the supply of blood in the skin. Conversely, heat conservation is facilitated by permitting peripheral temperatures, including those of appendages, to fall below core body temperature and is accomplished by the restriction of blood flow to the skin. Irving and Krog (1955) showed that arctic endotherms at ambient temperatures from 17 to −55°C usually have abdominal skin temperatures between 29 and 35°C (Figure 5.14), whereas the temperatures of extremities often are between 1 and 20°C. The unequal loss of heat through the surface of endotherms exposed to low ambient temperatures is clearly demonstrated by infrared photography: a raven (*Corvus corax*) at −18°C had head surface temperatures above 12°C, whereas the crown, back, and much of the body had temperatures between −6 and −9°C (Veghte and Herreid 1965), although much of this difference was due to the presence or absence of feathers and

did not reflect a difference in skin temperature. A similar dependency of surface temperature on ambient temperature was found in the gerbil *Meriones unguiculatus* (Klir et al. 1990). Naked mammals have skin temperatures between these extremes (Figure 5.14). When tropical and subtropical species, like the nine-banded armadillo (*Dasypus novemcinctus*), are exposed to cold, their skin temperatures decrease with air temperature but always remain above freezing (e.g., Johansen 1961). Armadillos have very high thermal conductances, usually between 133% and 225% of the values expected from mass (McNab 1980c).

The limit to which a commitment to energy conservation is achieved by peripheral vasoconstriction can be seen in birds and mammals in winter. Peripheral tissues cannot be permitted to freeze: at a water temperature of 0°C, harbor seals (*Phoca vitulina*) have skin temperatures between 1 and 2°C (Irving and Hart 1957), and at an air temperature of −16°C, the foot webbing of glaucous-winged gulls (*Larus glaucescens*) standing on ice is between 0.0 and 4.9°C (Irving and Krog 1955). The hoofs of reindeer (*Rangifer tarandus*) equal 3°C when air temperature is −31°C. These appendage temperatures reflect a thermal gradient along the axis of the appendages. The gradient is produced by countercurrent heat exchangers that are inserted into the circulatory system, whereby much of the heat carried by arterial blood leaving the core is transferred to the cold venous blood returning from the appendage (Irving and Krog 1955, Steen and Steen 1965). In birds these exchangers are located in the distal part of the tibia above the termination of the feather coat. The efficiency of such heat exchangers appears to vary with the distribution of species (Midtgård 1989): the arctic Iceland gull (*L. glaucoides*) has a more efficient heat exchanger than the temperate herring gull (*L. argentatus*). The melting point of fats deposited in the appendages decreases distally (Irving et al. 1957), which permits the appendages of arctic endotherms to function at low tissue temperatures.

At warm ambient temperatures, peripheral tissues can aid the dumping of heat to the environment. Thus, shorebirds can lose heat via their legs (Steen and Steen 1965); the wood stork (*Mycteria americana*) and some New World vultures (Cathartidae) spray urine on their legs to facilitate evaporative cooling (Kahl 1963); and elephants

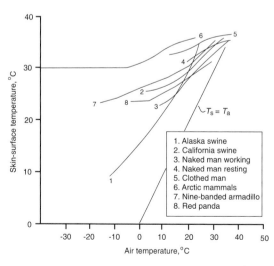

Figure 5.14 Skin temperature in mammals as a function of environmental temperature. Source: Modified from Hart (1956) with additional data from Johansen (1961) and McNab (1988c).

(*Loxodonta*) can lose a large proportion of their heat production by convection through their large pinnae (Wright 1984, Phillips and Heath 1992). Recent measurements on African elephants (Rowe 1999) indicate that pinnae movement increases at ambient temperatures greater than 21°C, and pinnae surfaces can lose nearly 100% of elephants' heat production (not including heat gain from the environment) at an ambient temperature near 30°C.

5.5.4 A consequence of low peripheral temperatures.

The distribution of temperatures at the surface of an endotherm reflects its pattern of heat exchange with the environment. Some mammals and birds of intermediate size so radically reduce peripheral circulation that the body becomes divided into a warm "core" and a cool "shell," which results in such a dramatic decrease in thermal conductance that rates of metabolism at low ambient temperatures are equal to or less than the thermoneutral rate of metabolism without interfering with the maintenance of core body temperature. This behavior has been observed in the red panda (*Ailurus fulgens*; Figure 5.15), a tree-kangaroo (*Dendrolagus matschiei* [McNab 1988c]), lemurs (*Lemur, Varecia* [Daniels 1984, McNab pers. observ., 1988]), several viverrids (*Arctictis, Arctogalidia, Nandinia* [McNab 1995]), and the trumpeter (*Psophia crepitans* [McNab 1989b]). These species share several characteristics: (1) an intermediate size, (2) tropical habits, (3) a sedentary life-style, and (4) a short-term exposure to cold. This capacity to reduce rate of metabolism

at cool to cold temperatures by restricting circulation of blood to the periphery, thereby establishing a low temperature shell, is limited by the necessity to protect peripheral temperatures, and requires rate of metabolism to increase at ambient temperatures near or below freezing (Figure 5.15). Thus, *Ailurus fulgens*, in spite of a thick, dense fur coat, permits abdominal skin temperatures to fall to 20 to 25°C and occasionally to 15°C at an ambient temperature of 0°C, unlike arctic carnivores in which abdominal skin temperatures usually exceed 30°C at ambient temperatures as low as −50°C (Figure 5.15)! Nevertheless, when ambient temperatures approach 0°C, phasic relaxation of vasoconstriction occurs (Irving et al. 1956) to prevent tissue injury; consequently, rate of metabolism increases (Figure 5.15).

This behavior is not found at small masses because a circulatory separation between core and shell is not possible. The lower size limit of this behavior varies: flying foxes of the genus *Pteropus* that show this behavior weigh more than 450 g (McNab and Bonaccorso 1995b), lemurs that show this behavior weigh more than 1 kg (McNab pers. observ., 1988), and a toucan that showed this behavior (*Ramphastos toco*, the largest species) weighed 582 g (McNab 2001). The largest endotherm to show this behavior is the binturong (*Arctictis binturong*) at 15 kg. This is one of the few physiological phenomena that is restricted to an intermediate body size; most others, if limited, occur in small (e.g., daily torpor [see Section 11.5.3]) or large (e.g., heat storage [see Section 5.3.4]) species.

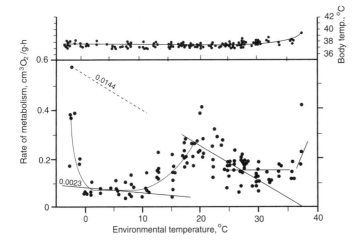

Figure 5.15 Body temperature and rate of metabolism in the red panda (*Ailurus fulgens*) as a function of environmental temperature. Source: Modified from McNab (1988c).

5.5.5 The impact of huddling on energy expenditure. Thermal conductance can be modified by behavior through a change in posture (with the consequent modification of exposed surface area), ptiloerection and piloerection, and vasomotion. Such modification permits thermal conductance to vary between maximal and minimal values in response to ambient temperature. Social behaviors, such as huddling, also modify thermal conductance.

When several individuals huddle at ambient temperatures below the lower limit of thermal neutrality, total rate of metabolism for the group is greater than that of a single individual but less than the sum of the rates for the same number of solitary individuals. This occurs because the surface area exposed to the environment is less than the sum of the areas of the huddled individuals, which means that the surface areas not exposed to the external environment face the warm surface temperatures of adjacent individuals. Many studies demonstrated this reduction in rate of metabolism at low temperatures in mammals (e.g., Pearson 1960b, Herreid 1963, McNab 1969, Canals et al. 1989) and birds (e.g., Le Maho 1977, Chaplin 1982, Prinzinger 1988, Brown and Foster 1992, McNab and Bonaccorso 1995a). The impact of huddling increases with the number of individuals huddling; in flying-squirrels (*Glaucomys volans*) the number of animals huddling together in a shelter varied with season and inversely with environmental temperature (Stapp et al. 1991).

The observation of Brown and Foster that the speckled mousebird (*Colius striatus*) shows a reduction in rate of metabolism with huddling in the zone of thermoneutrality raises questions. This reduction was not seen in the glossy swiftlet (*Collocalia esculenta*) because the swiftlets did not aggregate in warm temperatures (McNab and Bonaccorso 1995a). A decrease in the thermoneutral rate of metabolism is difficult to understand because at these temperatures all variations in temperature differential are compensated for by a change in thermal conductance, unless the depression in metabolism simply reflects a decrease in thermal conductance without a change in temperature differential. If that is the case, what limits the extent to which metabolism can be depressed, and what is the basal rate?

5.6 SETTING THE LEVEL OF BODY TEMPERATURE IN ENDOTHERMS

The regulated level of body temperature varies among mammals and birds. Mammals generally have temperatures between 36 and 38°C, although some species, such as anteaters and sloths, have body temperatures as low as 28 to 31°C. Birds usually have body temperatures between 39 and 42°C, but kiwis and penguins have temperatures as low as 37 to 38°C. Most mammals and birds thus regulate body temperatures between 35 and 40°C, although a circadian variation in body temperature exists when it is highest during the activity period. As Aschoff (1981) and Daan et al. (1989) demonstrated, this range is negatively correlated with body mass. Why does such uniformity in body temperature exist in endotherms? Why do some mammals have higher temperatures than others? And why do birds usually have higher body temperatures than mammals?

The remarkable uniformity of body temperature in endotherms is difficult to explain. Is this range intrinsically valuable? Does the range result from selection for the precision of temperature regulation? Or is this range produced passively by selection for a range in rate of energy expenditure? The realized range in body temperature may result from the interaction of many factors. For example, precise temperature regulation may be difficult to maintain at low levels (e.g., 10–20°C) because on Earth this would generally require a great dependence on evaporative cooling; body temperatures between 40 and 50°C, in contrast, would require inordinately high rates of energy expenditure in temperate and polar latitudes. In fact, Calloway (1976), Dunitz and Benner (1986), and Paul (1986) argued that endotherms have body temperatures near 36°C because of the physical properties of water. McArthur and Clark (1988) maintained that the heat and water balances of endotherms dictate that their temperatures fall between 35 and 43°C, but these authors ignored the effect of body size, except as the units of heat exchange are area specific.

The argument that body temperatures cannot be higher because enzymes would denature is unlikely; enzymes may simply denature at temperatures above normal (see Section 4.3). If body temperatures were higher, denaturation would probably occur at still higher temperatures; note that with their higher body temperatures, birds have higher lethal body temperatures, approximately 46°C, than do mammals, about 42.5°C. Why should an organism produce enzymes that denature at body temperatures well beyond those that will be encountered?

Body temperature is potentially adaptive to climate. If it were, polar mammals and birds might be expected to have low temperatures (by lowering temperature differential, they reduce heat loss), and tropical mammals and birds might have high temperatures (to reduce their dependence on evaporative cooling in humid environments). Actually, Scholander et al. (1950c), Morrison and Ryser (1951), and Irving and Krog (1954) showed that no consistent difference in body temperature exists among arctic, temperate, and tropical mammals. Indeed, "cool" mammals, like armadillos, anteaters, and sloths, live in the tropics, whereas arctic species, like arvicolid rodents, hares, carnivores, and ungulates, have high body temperatures.

A fundamental question should be asked at this point: can the various parameters of endothermy, namely, body temperature, thermal conductance, rate of metabolism, and body mass, be modified independently of each other, or does Equation 5.1 mean that all (or most) adaptive modifications of these parameters are limited by factor interdependency?

From Equation 5.1 the interaction of basal rate of metabolism and minimal thermal conductance defines (under ideal conditions, see Section 5.2)

the temperature differential at the lower limit of thermoneutrality:

$$\Delta T_1 = \dot{V}_{O_2}/C_m.$$

The basal rates and minimal conductances of mammals are determined primarily by body mass (Equations 5.5 and 5.9, respectively). Therefore,

$$\Delta T_1 = \frac{3.45 \cdot f_m \cdot m^{0.71}}{1.00 \cdot f_c \cdot m^{0.50}} = 3.45 \left(\frac{f_m}{f_c} \right) m^{0.21} = 3.45 \cdot F \cdot m^{0.21},$$

(5.10)

where $F = f_m/f_c$, which means that the temperature differential between the body and environment at the lower limit of thermoneutrality in mammals increases with an increase in body mass and in basal rate, or with a decrease in minimal thermal conductance (McNab 1970, 1974b, 1980c; also see Lovegrove et al. 1991b). For example, compared to high-intensity species (such as soricine shrews and arvicolid rodents), anteaters and most marsupials have smaller temperature differentials for their size principally because they have low basal rates (Figure 5.16). Armadillos, pangolins, and other naked mammals have especially small differentials because they combine low basal rates with high thermal conductances. Note that the two "naked" mammals with high ΔT_1 values, the Tasmanian subspecies of the common echidna (*Tachyglossus aculeatus*) and the New Guinean mountain echidna (*Zaglossus bruijni*), are furred. The size of

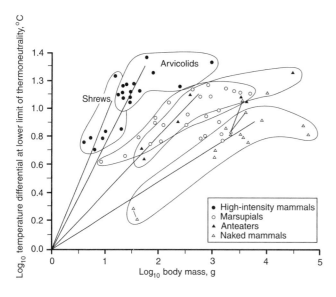

Figure 5.16 Log$_{10}$ temperature differential at the lower limit of thermoneutrality in arvicolid rodents, marsupials, anteaters, and naked mammals (echidnas, armadillos, pangolins, and *Heterocephalus*) as a function of log$_{10}$ body mass. Data on the same species are connected. Sources: Derived from McNab (1978c, 1984, 1992b) and Contreras and McNab (1990).

the differential, therefore, is set by the interaction of body mass, food habits, and climate as they operate through basal rate and minimal thermal conductance. Criticizing this analysis, Calder and King (1972) argued that it reflected circular reasoning. However, when the data on body mass, T_b, T_l, basal rate, and minimal thermal conductance are examined, the suggested relations generally hold and measurements of the first four parameters are independent of each other.

The ability to account for variations in ΔT_l does not necessarily account for variations in body temperature because a decrease of ΔT_l with a decrease in mass may mean either that body temperature decreases or that the lower limit of thermoneutrality increases. If small mammals regulate body temperature precisely, they have high lower limits of thermoneutrality, which is why soricine shrews must be measured at temperatures of at least 30°C to obtain basal rates. Variations in ΔT_l at small masses usually reflect variations in body temperature but at large masses reflect variations in T_l (McNab 1970). Above a particular mass ΔT_l is independent of mass (Figure 5.16). If this analysis is correct, body temperature is to a great extent a dependent character (also Lovegrove et al. 1991b). Furthermore, the conditions associated with a high body temperature—a large size, high basal rate, and low thermal conductance—are those that lead to precise temperature regulation, which means that the level and precision of thermoregulation are intertwined (McNab 1969).

The analysis of variation in ΔT_l and T_b in mammals can be used to examine the difference in body temperature between mammals and birds (McNab 1966a). The ΔT_l maintained by birds other than passerines should equal

$$\Delta T = \frac{4.00 \cdot f_m \cdot m^{0.73}}{0.85 \cdot f_c \cdot m^{0.50}} = 4.71 \left(\frac{f_m}{f_c} \right) m^{0.23},$$

which means that birds other than passerines usually have a larger ΔT than do mammals of the same mass. Therefore, birds either have a higher T_b and/or a lower T_l than do mammals. Generally, the former is correct. That is, birds have higher body temperatures than mammals because birds have higher rates of metabolism and lower thermal conductances. Furthermore, unlike mammals, which have body temperatures that are independent of body mass (Morrison and Ryser 1951), birds tend to have temperatures that are correlated inversely

with mass (McNab 1966a), especially during activity (Prinzinger et al. 1991).

The adaptive significance of variation in ΔT_l can be seen in fossorial rodents. These mammals greatly vary in mass, which according to Equation 5.10 should be reflected in ΔT_l. Rodents living in closed burrows potentially overheat, especially during activity, because convective and evaporative cooling are minor pathways for heat loss. To prevent overheating in warm burrows, body mass and basal rate decrease, which with an increase in thermal conductance markedly reduces ΔT_l. A reduction in ΔT_l accomplished by a reduction in mass alone is clearly inadequate at high burrow temperatures (Figure 5.17). The extreme to which such modifications can be carried is shown in the naked mole-rat (*Heterocephalus glaber*), which lives in humid (100%), high-temperature (30°C) burrows. Not only does it have a small mass (39 g) but also has a basal rate that is 48% of expectations and a thermal conductance that is 242% of expectations (i.e., $F = f_m/f_c = 0.48/2.42 = 0.20$). When these data are used in Equation 5.10, the calculated ΔT_l is 1.5°C, whereas the standard value expected from mass is 8.5°C. Given the small mass of this species, body temperature is very low (ca. 32.5°C). The reduction in body temperature (e.g., from ca. 37.0 to 32.5°C) is more than accounted for by the reduction in ΔT_l from 8.5 to 1.5°C, which means that the lower limit of thermoneutrality increased to approximately 31°C.

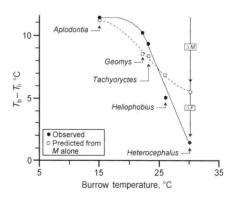

Figure 5.17 Temperature differential between the body and environment at the lower limit of thermoneutrality in burrowing rodents as a function of burrow temperature. The dashed line represents the change in temperature differential due only to a change in mass (ΔM), the remainder reflecting a change in basal rate or thermal conductance (ΔF) (where $F = f_m/f_c$, see text). Source: Modified from McNab (1979b).

5.7 MECHANISMS OF HEAT ACQUISITION

If endotherms maintain a constant body temperature, they must compensate for an increase in ΔT and heat loss at low ambient temperatures by increasing rate of metabolism, unless some supplementary heat source is used. Rate of metabolism may be increased at temperatures below thermoneutrality by muscle contraction during shivering and activity or by nonshivering thermogenesis.

5.7.1 Shivering thermogenesis. An important source of heat production at low environmental temperatures in many endotherms is the nonsynchronous contraction of skeletal muscle called *shivering*. Unless an endotherm is exposed to a very low temperature, shivering is difficult to perceive, but its presence can be demonstrated by measuring electrical activity in muscles. For example, in birds the amount of shivering, measured in terms of microvolts, increases approximately linearly with a reduction in ambient temperature (Figure 5.18); rate of oxygen consumption is directly proportional to the amount of shivering. Similar results have been obtained in other birds (Steen and Enger 1957, Hart 1962, West et al. 1968) and in mammals (Pohl and Hart 1965, Pohl 1965). The heat produced by a given amount of shivering depends on the size of the bird (Figure 5.18): 1 mV of activity in a 390-g crow (*Corvus*

brachyrhynchus) corresponds to 2.2 times as much heat produced as in the 125-g common grackle (*Quiscalus quiscula*) and 7.8 times as much as in the 14-g redpoll (*Carduelis flammea*). These differences disappear when rates of metabolism are expressed on a surface-area basis, suggesting that muscle activity at temperatures below thermoneutrality is attuned to the rate of heat loss, which is a surface-dependent phenomenon.

The correlations among electromyographic activity, rate of metabolism, and ambient temperature are complicated by cold acclimation: electromyographic activity is reduced at low temperatures with cold acclimation (Figure 5.19). Nevertheless, Swanson (1991b) showed that winter-acclimatized juncos (*Junco hyemalis*) have greater tolerance of low temperatures than do

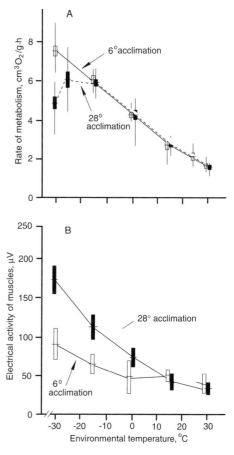

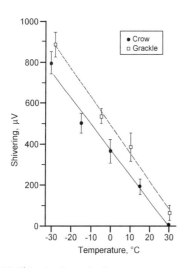

Figure 5.18 Shivering intensity in common crows (*Corvus brachyrhynchus*) and common grackles (*Quiscalus quiscula*) as a function of environmental temperature. Source: Modified from West (1965).

Figure 5.19 Acclimation to 6 and 28°C in the golden hamster (*Mesocricetus auratus*), as shown by a correlation of (A) rate of metabolism and (B) electrical activity of the muscles with environmental temperature. Source: Modified from Pohl (1965).

summer-acclimatized juncos, a difference that is correlated with a greater pectoralis muscle mass in winter. Shivering in mammals may be an emergency means of increasing heat production at temperatures below those normally encountered.

5.7.2 Nonshivering thermogenesis. Mammals respond to low environmental temperatures by mechanisms other than shivering, especially after cold acclimatization (Hart 1971). Such mechanisms are collectively referred to as *nonshivering thermogenesis* and include various cellular mechanisms under the control of catecholamines (e.g., norepinephrine) and the sympathetic nervous system. Nonshivering thermogenesis involves heat production by many organ systems, including muscles, liver, and brown fat (Hart 1971, Jansky 1973), although brown fat appears to be a tissue specialized for this heat production (Cannon et al. 1978). The question of whether nonshivering thermogenesis applies only to the increase in rate of metabolism above the basal rate (Himms-Hagen 1978), or to all processes that do not involve shivering (Jansky 1973, Wunder 1984), is the subject of some disagreement.

The importance of nonshivering thermogenesis to small rodents is seen in the golden hamster (*Mesocricetus auratus*) and shown in Figure 5.19 (Pohl 1965). Cold- and warm-acclimated hamsters have similar rates of metabolism at ambient temperatures from 30 to −15°C, even though warm-acclimated individuals have much higher electrical activity in their muscles. That is, most of the increase in rate of metabolism below T_1 in cold-acclimated hamsters is due to nonshivering thermogenesis. Furthermore, the maximal rate of oxygen consumption is much greater in cold-acclimated than in warm-acclimated individuals, presumably because the increase in nonshivering thermogenesis with cold acclimation is greater than the decrease in shivering thermogenesis.

The effect of cold acclimatization on nonshivering thermogenesis is seen in *Clethrionomys gapperi*, where it reached a maximum in winter and a minimum in summer (Merritt and Zegers 1991). The seasonal response of nonshivering thermogenesis contrasts with the seasonal insensitivity of its resting rate of metabolism in thermoneutrality. The red-backed vole survives winter through a behavioral change in microhabitat and food resources, a seasonal reduction in body mass, and an increased capacity for nonshivering thermogen-

esis, which is used at the low ambient temperatures often encountered in winter. The value of nonshivering thermogenesis, as Wunder (1992) noted, is that it is a much cheaper response to low ambient temperature than is an increased maximal rate of metabolism, which may depend on an increase in basal rate (see Section 5.7.4). An increase in basal rate is a payment made even in thermoneutrality, whereas nonshivering thermogenesis is used only on demand at temperatures below thermoneutrality.

Nonshivering thermogenesis is most important in young mammals and in the adults of small species (Jansky 1973): brown fat, an important site for nonshivering thermogenesis, is most prevalent in small species and in the young, such as newborn rabbits (Cannon et al. 1978). The influence of body size can be seen in the increment in oxygen consumption produced by the injection of norepinephrine (Figure 5.20). In an analysis of the correlates of nonshivering thermogenesis in rodents, Haim and Izhaki (1993) demonstrated that its development is greater in diurnal than in nocturnal species and greater in arid than in mesic species. Furthermore, a negative correlation was found between basal rate and nonshivering thermogenesis. Thus, desert rodents often have low basal rates and low normothermic body temperatures but increase heat production and body temperature at low ambient temperature by turning on nonshivering thermogenesis. Yet, nonshivering thermogenesis is well developed in arvicolids (Feist and Morrison 1981, Wunder 1984, Klaus et al. 1988,

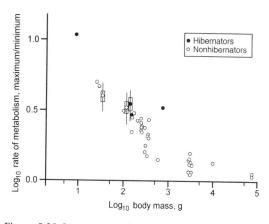

Figure 5.20 Log_{10} maximal rate of metabolism produced by the injection of norepinephrine relative to the minimal rate of metabolism in various mammals as a function of log_{10} body mass. Source: Modified from Hart (1971).

Merritt and Zegers 1991), which are characterized by high basal rates (McNab 1992b, Koteja and Weiner 1993).

Little evidence indicates that birds use nonshivering thermogenesis (Hart 1962, West 1965, Chaffee and Roberts 1971). Thus, cold acclimation produces no change in the electrical activity of bird muscles (Hart 1962), and no effect is seen in birds injected with catecholamines (Chaffee et al. 1963, Hissa and Palokangas 1970, Koban and Feist 1982). Furthermore, birds usually are thought to lack brown fat (Johnston 1971), although Oliphant (1983) described its presence in the ruffed grouse (*Bonasa umbellus*) and black-capped chickadee (*Parus atricapillus*). Olson et al. (1988) and Saarela et al. (1989) reexamined this claim in parids and other birds and concluded that no evidence demonstrated the presence of brown fat in birds. This apparent difference between birds and mammals is not surprising, given that the evolution of endothermy in birds was independent of that in mammals.

5.7.3 Thermogenesis during activity. All endotherms increase heat production during activity, which raises the question of whether heat produced in this manner can be used for thermoregulation at low ambient temperatures. Heat produced during activity does not normally reduce the cost of temperature regulation because thermal conductance and heat loss increase with activity. For example, collared lemmings (*Dicrostonyx groenlandicus*) have higher rates of metabolism during activity than at rest at all temperatures, but body temperature is greater during work only at temperatures above −10°C. Hart and Heroux (1955) interpreted the decrease in body temperature at cold temperatures to mean that heat loss is greater during activity than at rest, so heat storage is negative. The energy expended during activity also is added to that used for temperature regulation in small birds (West and Hart 1966), although Pohl and West (1973) and Paladino and King (1984) found evidence of substitution at very low temperatures. The difficulty with these analyses, as Zerba and Walsberg (1992) noted, is that convective heat loss may have contributed to a higher heat loss in active birds and mammals. They showed in Gambel's quail (*Callipepla gambelii*) that when a wind equivalent in velocity to that of a running bird passed over a resting bird, the resting bird had the same rate of metabolism as a

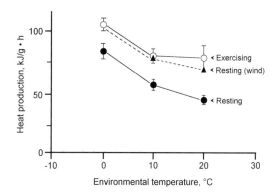

Figure 5.21 Heat production in Gambel's quail (*Callipepla gambelii*) as a function of environmental temperature when resting, when exercising, and when resting with a wind velocity equal to that encountered when exercising. Source: Modified from Zerba and Walsberg (1992).

running bird at the same temperature (Figure 5.21). The heat produced by exercise completely substituted for the heat required for temperature regulation, at least at temperatures from 0 to 20°C. Therefore, at temperatures when an increase of metabolism is required for temperature regulation, moderate levels of activity are "free."

Thermal conductance is higher during activity for many reasons: the boundary layer is sheared, the fur or feather coat is no longer intact, the surface area is no longer minimal, and blood flow to the appendages and skin is greatly increased. Because many of these factors are influenced by body mass, the conclusion that substitution does not occur, or is only partially effective, may apply principally to small species, and then only if convective heat loss is ignored. Most studies on the energetics of locomotion in large mammals have been at high environmental temperatures, thereby offering no information on the possibility that heat production during activity may be used by large species for temperature regulation at cold temperatures.

5.7.4 Maximal rates of metabolism. The capacity of an endotherm to tolerate low environmental temperatures while body temperature remains constant depends on its ability to increase rate of metabolism above the basal rate. In small mammals the maximal steady-state rate of metabolism maintained during cold exposure, or during activity, varies from about 5 to 7 times the basal rate and also scales proportionally to about $m^{0.75}$

(Hemmingsen 1960, Jansky 1962, Hart 1971, Lechner 1978, Peterson et al. 1990, Bozinovic 1992, Hammond and Diamond 1997). Kirkwood (1983) suggested that the maximal metabolizable energy intake (the amount after the losses in feces, urine, and fermentation are subtracted) equals about 1713 kJ/kg$^{0.72}$·d in mammals and birds, or about 6 times the mammal basal rate. Short-term bursts of activity, such as running and flight, can exceed this limit and may involve a draw down of energy reserves. The similarity of Kirkwood's values to those of maximal steady-state rates of metabolism and field expenditures (Nagy et al. 1978, Karasov 1981) implies that free-living mammals and birds often live near the limit for energy expenditure (Weiner 1992). The limit to sustained maximal rate of metabolism may depend on the maximal rate at which the digestive tract can assimilate nutrients (Karasov 1981, 1990; Weiner 1987, 1989, 1992; also see Sections 12.5 and 12.6.1).

Koteja (1987) found no correlation between the residual variation in (short-term?) maximal and basal rates in 18 species of mammals, even though both parameters are correlated in a similar manner with body mass. He concluded that these rates are independent. This finding contrasted with the correlation of the residual variations in basal and maximal rates in 29 species of rodents belonging to 42 populations (Bozinovic 1992). The difference between these studies may have reflected the use of uniform methods in the latter study and the heterogeneous data of the earlier study. Hayes (1989a) showed by an analysis of covariance that maximal rate of metabolism in two populations of *Peromyscus maniculatus* at 4 times of the year was correlated with basal rate of metabolism.

Evidence of the selective independence of the basal and maximal rates of metabolism is found in some mammals. Active terrestrial mammals with basal rates of metabolism down to 50% of the standard value often have maximal rates of metabolism that are about 4 to 7 times those of the measured basal rates, whereas arboreal mammals, such as tree-sloths, anteaters, and some prosimians, all of which are characterized by slow movements and small muscle masses, have low basal rates and maximal rates that are only 1.6 to 2.8 times the basal rate (Figure 5.22). These data indicate that the maximal rate of metabolism can be modified independently of the basal rate.

A low basal rate may be produced in a variety of ways in mammals, including a decrease in

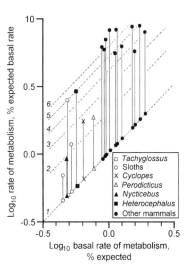

Figure 5.22 Log$_{10}$ measured basal and maximal rate of metabolism in mammals, expressed relative to the expected basal rate, as a function of the basal rate, also expressed relative to the expected basal rate. The parallel curves represent maximal rates that vary from 1 to 6 times the measured basal rate. Source: Modified from McNab (1980b).

thyroid activity (Hudson and Wang 1968, Lemaire et al. 1969, Yousef and Johnson 1975) and a reduction in muscle mass (S. D. Thompson and T. I. Grand 1990; McNab 1978a, 1994b). A marked reduction in scope for metabolism may be produced by a reduction in muscle mass, which also reduces the ability to produce heat by shivering. In other words, a reduction in muscle mass in arboreal mammals leads to a reduction in basal rate and scope for metabolism, which in turn leads to a sedentary life-style (e.g., *Cyclopes*, *Bradypus*, and *Choloepus*). The terrestrial genera *Tachyglossus* and *Heterocephalus*, however, have low basal rates and scopes for metabolism that are similar to those in species with high basal rates (Figure 5.22). A decrease in basal rate produced by a reduction in thyroid activity, in combination with a normal scope, only marginally decreases the tolerance of low ambient temperatures.

5.7.5 Basking. Even though endotherms depend principally on metabolism to generate the heat used in temperature regulation, some species supplement this source with the selective absorption of solar radiation. As seen in Section 2.9.2, the absorption of solar radiation leads to a reduction in the rate of metabolism at temperatures below thermoneutrality and to an increase in the rate of metabolism at high ambient temperatures (Figure 2.10). Similar observations have been made repeatedly in birds (Hamilton and Heppner 1967; Lustick 1969; Ohmart and Lasiewski 1971; Hennemann 1982, 1983b). For example, a roadrunner (*Geococcyx californicus*) basking in artificial sunlight with an intensity of 3.35 J/cm²·min maintains body temperature equal to 38.4°C with a rate of metabolism equal to approximately 75% of the value required by metabolism alone (Ohmart and Lasiewski 1971). Such a reduction in rate of metabolism in the dark corresponds to a body temperature of 34.3°C, a 4.1°C reduction. Sometimes basking is used for other functions, such as wing drying: this occurs in vultures after a rain and in cormorants, including paradoxically the flightless cormorant (*Phalacrocorax harrisi*) after fishing.

Because dark birds absorb more light than light-colored birds (Lustick 1969, 1971; Heppner 1970; Ellis 1980), dark species potentially show a greater decrease in rate of metabolism than light-colored birds in the presence of a radiant load (Hamilton and Heppner 1967, Lustick 1969). The difference in energy absorption also may be reduced by increasing the angle of incidence to solar radiation (Lustick et al. 1978, 1980). The difference between dark and light-colored species may be diminished because radiant energy penetrates deeper into the feather or fur coat of light-colored species, leading to high skin temperatures, whereas the energy absorbed by dark species occurs principally at the surface, where it can be convected away (Øritsland 1970, Walsberg et al. 1978; see Section 2.5).

Several explanations have been given for the decrease in rate of metabolism that occurs with the absorption of radiation: (1) insulation is increased—that is, thermal conductance is decreased (Lustick 1969); (2) the absorbed energy is used as a source of heat in the maintenance of a temperature differential with the environment (Hamilton and Heppner 1967, Ohmart and Lasiewski 1971); and (3) ΔT is reduced or even reversed (Cowles 1967). The appropriate ambient temperature in the presence of a large radiant heat load is T_e, not T_a. By equating Equations 2.11 and 2.14, T_e is higher than T_a as long as the radiant heat load is greater than zero, and therefore the effective conductance K_e is *greater*, not less, than the standard conductance C. Because the surface temperature of birds increases with the absorption of solar radiation (Heppner 1970, Lustick et al. 1970), the thermal gradient is reduced at low

ambient temperatures and the gradient is reversed at thermoneutral temperatures (Figure 5.23). These observations substantiate the view of Cowles (1967) and explain why rate of metabolism diminishes with the absorption of solar radiation at temperatures below thermoneutrality and increases within thermoneutrality (Figure 2.10).

Endotherms living in environments with high solar heat loads adjust their behavior to maximize the absorption of heat at cool temperatures and minimize its absorption at high temperatures. At low air temperatures, anhingas (*Anhinga anhinga*) increase the frequency of spread-winged behavior (Hennemann 1982), which exposes pigmented areas of the apteria (featherless tracts). At high temperatures, anhingas show little or no spread-winged behavior and spend much time gular fluttering to increase evaporative water loss (Hennemann 1982). As radiant temperature increases, cactus wrens (*Campylorhynchus brunneicapillus*) spend less time in the sun, which reduces their heat load and the amount of water that they need for temperature regulation (Ricklefs and Hainesworth 1968). Herring gulls (*Larus argentatus*) modify solar inputs by changing the angle of their body with the sun and by differentially exposing white or dark surfaces to the sun (Lustick et al. 1978).

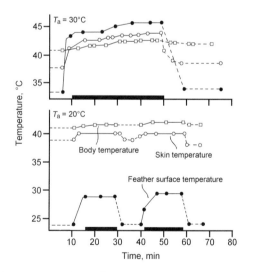

Figure 5.23 Feather surface temperature, skin temperature, and core body temperature in redwing blackbirds (*Agelaius phoeniceus*) at ambient temperatures 20 and 30°C with and without a radiant heat load. The presence of a radiation load is represented by the dark bars along the time axis. Source: Modified from Lustick et al. (1970).

In hot climates many birds adjust to high solar heat loads. Tropical birds that forage in the sun have lower basal rates than do those that forage in the shade (Weathers 1979). Ellis (1980) showed among four species of tropical and subtropical herons that (1) the (dorsally) dark tricolored heron (*Egretta tricolor*) has a high basal rate compared to the standard curve and preferentially nests in the shade, (2) the white snowy (*E. thula*) and cattle egrets (*Bubulcus ibis*) have low basal rates compared to the standard and nest without regard to the presence of sun and shade, and (3) the completely dark little blue heron (*E. caerulea*) has the lowest basal rate and often nests in the sun. The basal rate in these herons is correlated negatively with the maximal solar heat load encountered at the nest (Figure 5.24). The low basal rate in the little blue heron, however, is not completely compensatory for high solar inputs because it starts gular fluttering at lower ambient temperatures than does the cattle egret, and it spends only one-half the time as the cattle egret brooding eggs in the sun at high temperatures (Figure 5.25). The immature plumage of the little blue heron is completely white, which may well be a means of reducing the heat load in young exposed to the sun during late stages of development in the nest. Dark, tropical birds that are exposed to the sun often are characterized by low basal rates, whereas white birds that are exposed to the sun have high basal rates (Ellis 1980, 1985). A question that remains is whether a dark plumage then is compensatory for a low basal rate, or whether the low basal rate is compensatory for the dark plumage: chicken and eggs. The extent

to which this pattern in color, basal rate, and radiant heat load occurs is unclear; it does not apply to corvids or to Heermann's gull (*Larus heermanni*), a gull that is dark as an adult (Ellis and Frey 1984), but see Marder (1973).

5.8 DEVELOPMENT OF ENDOTHERMY

Many endotherms are born blind, naked, and helpless and therefore depend completely on parental care for survival. Others can see, run, and feed within a few minutes or hours after birth

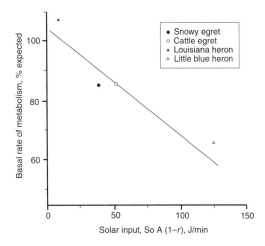

Figure 5.24 Basal rate of metabolism in the little blue heron (*Egretta caerulea*), snowy egret (*E. thula*), tricolored heron (*E. tricolor*), and cattle egret (*Bubulcus ibis*), expressed relative to the basal rates expected from mass, as a function of the maximal solar heat load encountered when nesting. Source: Modified from Ellis (1980).

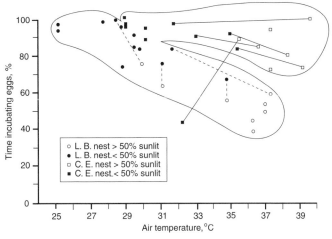

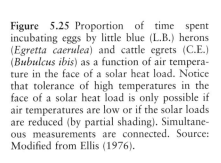

Figure 5.25 Proportion of time spent incubating eggs by little blue (L.B.) herons (*Egretta caerulea*) and cattle egrets (C.E.) (*Bubulcus ibis*) as a function of air temperature in the face of a solar heat load. Notice that tolerance of high temperatures in the face of a solar heat load is only possible if air temperatures are low or if the solar loads are reduced (by partial shading). Simultaneous measurements are connected. Source: Modified from Ellis (1976).

or hatching; at an extreme "... newly hatched [megapodes] have been observed landing aboard ships well offshore" (Jones et al. 1995, p. 22). The former species are called *altricial*; the latter, *precocial*. Altriciality and precociality are extremes at the ends of a continuum in development. Precocial mammals include hares, histricognath rodents, cetaceans, and most ungulates; ducks and geese, shorebirds, gulls and terns, and grouse are precocial birds.

5.8.1 Altricial endotherms. Many studies on the development of temperature regulation have involved altricial mammals and birds. Depending on the species, effective endothermic temperature regulation usually takes 1 to 3 weeks after birth or hatching to develop. Most small passerines develop effective endothermy within 7 d after hatching (Dawson and Evans 1957, Maher 1964); larger altricial birds (e.g., cattle egrets [Hudson et al. 1974]) may require 12 to 14 d. Small mammals take a slightly longer period to develop endothermy than do small birds, some 12 to 14 d (Morrison et al. 1954).

One of the slowest rates for the development of endothermy is found in the opossum *Didelphis virginiana* (Morrison and Petajan 1962). Effective temperature regulation takes between 85 and 95 d from birth. Many factors, including a large mass, a low basal rate, and (especially) marsupial development (see Section 13.5), contribute to this delay. In contrast, some precocial eutherians at a slightly larger mass, such as the porcupine *Erethizon dorsatum*, a histricognath rodent, have effective thermoregulation at low temperatures only 3 d after birth (Morrison and Petajan 1962). The difference between the opossum and porcupine, however, is deceptive because of the porcupine's long gestation period. The period from conception to effective endothermy in the opossum is about 105 d; in the porcupine it is 115 d, 112 of which represent the gestation period. What is so surprising is that a precocial mammal as small as a guinea pig (*Cavia porcellus*) takes almost 100 d from conception to attain endothermy. One of the shortest periods from conception to endothermy at larger masses is the 70 or so days found in the domestic cat and dog, considerably shorter than that in the opossum, their short periods possibly reflecting high rates of metabolism (see Section 13.4).

The importance of using the time of conception as the baseline against which to make comparisons is illustrated by a comparison of two similarly sized rodents (ca. 50 g), one having altricial young (*Gerbillus perpallidus*) and the other precocial young (*Acomys cahirinus*). Waldschmidt and Müller (1988) showed that *G. perpallidus* required approximately 19 d after birth to regulate body temperature, when exposed to an ambient temperature of 15°C for 15 min, whereas *A. cahirinus* required only 7 d after birth. However, this difference is not nearly so dramatic when the difference in gestation period is included, 20 d for *G. perpallidus* and 40 d for *A. cahirinus*, so the total period from conception to attain temperature regulation is 39 and 47 d, respectively. The view that altriciality permits energy to be transferred selectively from maintenance to growth is only marginally demonstrated here, given the similar periods required to develop endothermy and the similarities in the gompertzian growth constants, 0.045 and 0.041 1/d for *G. perpallidus* and *A. cahirinus*, respectively.

The altriciality of mammals and birds probably was derived from the "precocial" development in reptiles (Hopson 1973). (Reptiles are precocial at least in terms of being independent of parental care at the time of birth.) Hopson suggested that altriciality evolved in mammals to facilitate the evolution of the very small body mass that developed in the transition from therapsid reptiles to the earliest mammals (see Section 5.10.3). Altriciality may permit energy to be channeled to growth and development, rather than to maintenance (Dawson and Evans 1957; Ricklefs 1968, 1973b), although as seen, this difference may not always be great. The high body temperatures needed for effective growth can be maintained more efficiently by parental brooding than by requiring young to expend energy, especially during periods when the young are poorly insulated. The nakedness of young altricial endotherms can be viewed not as a liability but as a means of facilitating heat transfer from brooding adults to thermally dependent offspring.

5.8.2 Precocial endotherms. More information exists on the development of endothermy in precocial birds than in precocial mammals. The time at which endothermy is acquired in precocial birds varies greatly. Gallinaceous birds (Ryser and Morrison 1954, Koskimies 1962) and gulls may take 20 to 30 d before they can maintain a high body temperature at an ambient temperature of 10°C, whereas the newly hatched young of some arctic ducks can tolerate an ambient temperature of 0°C for at least 15 h without body temperature falling. Intermediate to these extremes are some wading birds, like great snipe (*Gallinago media*), which as hatchlings permit core temperature to fall as low as 29°C without an appreciable reduction in rate of metabolism or, more importantly, in mobility (Steen et al. 1991b).

The excellent study of 10 species of European ducks by Koskimies and Lahti (1964) deserves special attention. These species show an appreciable variation in cold tolerance: the young of diving ducks, including the common eider (*Somateria mollissima*), common golden-eye (*Bucephala clangula*), velvet or white-winged scoter (*Melanitta fusca*), and red-breasted merganser (*Mergus serrator*), are much better regulators at the time of hatching than are surface-feeding ducks, such as the mallard (*Anas platyrhynchus*) and common teal (*A. crecca*). These differences in tolerance are related to differences in insulation. Koskimies and Lahti calculated that eider hatchlings could tolerate −10°C at an expenditure 4 times the basal rate, whereas mallard and teal hatchlings at 4 times the basal rate could only tolerate 15°C, a difference of 25°C. Cold-hardy species only relinquish temperature regulation when they run out of energy reserves, which at hatching are derived completely from egg yolk. These differences are associated with the thermal independence of the young and the breeding ranges of the adults—the most cold-tolerant species have an exclusively northern distribution. One of the surface-feeding ducks, the European wigeon (*A. penelope*), has a strictly northern breeding range, and for the *Anas* genus, is a cold-tolerant species.

The evolution of precociality in endotherms presumably was derived secondarily and repeatedly from an altricial condition. In mammals precociality preferentially occurs in large species and is associated with a reduced litter size. By extending development in the uterus, a mammal greatly reduces the period of lactation and potentially the maternal investment, measured in terms of kilojoules, used in raising the young to independence. Yet, as seen, the total period from conception to complete endothermy in precocial young may not be greatly shortened. Whether precociality reduces the cost of producing a young is not at all clear (estimates of the comparative costs of reproduction would be of great value). Precociality may simply

increase the probability of young surviving to the age of reproduction.

The implications of precociality may be rather different in birds. Female mammals retain precocial young for longer periods in utero than altricial young, and the incubation period of birds is longer than for altricial species. In fact, precociality occurs at much smaller masses in birds than in mammals (e.g., in small charadriiform birds). Why precociality occurs in some small rodents and not in others is unclear. One factor influencing the presence of precociality in birds is that precocial young must use food resources that do not require flight; that is, they are ground feeders (Galliformes and Charadriiformes) or are inshore aquatic species (ducks and some murrelets) (Ricklefs 1974).

5.8.3 The energetics of development. The scaling of basal rate in adult mammals and birds at small masses is connected intimately with the effectiveness of endothermy (McNab 1983a; see Section 3.4); a similar pattern occurs during development. After immature mammals acquire endothermy, their basal rates tend to overshoot the standard curve and, if the mammal is small, to follow the boundary curve for endothermy (Figure 5.26); this pattern occurs even in mammals that as adults have low basal rates of metabolism (e.g., *Baiomys*, *Perognathus*). A similar pattern is generally seen in passerines (Figure 5.27) and in chinstrap (*Pygoscelis antarctica*) and gentoo (*P. papua*) penguins (Taylor 1985). Such high rates during

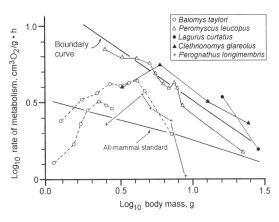

Figure 5.26 Log$_{10}$ mass-specific rate of metabolism in thermoneutrality in rodents during the period from birth to adulthood as a function of log$_{10}$ body mass. The boundary curve and the all-mammal standard curves are indicated. Source: Modified from McNab (1983a).

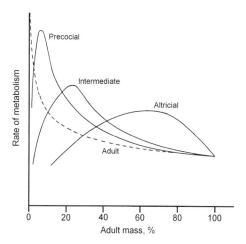

Figure 5.27 Schematic representation of the mass-specific standard rate of metabolism in birds that conform to a precocial, intermediate, or altricial form of development as a function of the proportional attainment of adult mass. The curve for basal rate in (nonpasserine) birds is indicated by the dashed curve. Source: Modified from Ricklefs (1974).

development raise doubts on the intraspecific scaling of metabolism obtained from an ontogenetic series in small species: the high cost of endothermy at small masses would be incorporated as a reduced power for metabolism.

High rates of metabolism ensure effective endothermy at a small mass (see Section 3.4) and high postnatal growth rates (see Section 13.3.2). High growth rates are borne in mammals by the lactating mother and in birds by the foraging parents: the "value" of high growth rates to parents is a significantly shortened period that the young depend on the parents. Reproduction occurs at a time when food availability is not limiting, and therefore when high rates of consumption associated with reproduction and growth can be sustained. In contrast, immature marsupials do not adjust basal rate at small masses and have a delayed attainment of effective temperature regulation (McNab 1986b).

King and Farner (1961) pointed out that the concepts of precociality and altriciality are two extremes along a continuum in behavior and physiology. Most of the cost of thermoregulation in precocial birds switches at the time of hatching from parental investment to investment by the young, modified to some extent by parental brooding. In fact, the switch from parental to juvenile investment in precocial species appears to occur a few days before hatching when the developmental increase in rate of metabolism stabilizes, apparently while sensory, neuromuscular, and thermoregulatory systems mature (Vleck et al. 1979). In altriciality the developmental increase in metabolism continues through hatching and the switch to juvenile investment is much more gradual, parental investment continuing at least until the birds fledge, and often beyond. In mammals, this transition is delayed by lactation.

5.9 INTERMEDIATE FORMS OF ENDOTHERMY

Endothermy in its various forms has a wide phylogenetic distribution. Mammals and birds are the only two classes of vertebrates dedicated to endothermic temperature regulation. The few apparent exceptions to this generalization, namely, temperate insectivorous bats and hibernating mammals, do not represent an inability to regulate but rather strategic deviations to balance energy budgets during periods of environmental inhospitality, and as such are discussed in Chapter 11. Some organisms other than mammals and birds, even including some plants, are endothermic to varying degrees (Nagy et al. 1972, Knutson 1974, Prance and Arias 1975, Schroeder 1978, Seymour et al. 1984). A short description of some of these intermediate forms of endothermy is given here.

5.9.1 Insects. Many insects are endothermic, especially large, nocturnal moths belonging to the family Sphingidae (Heath and Adams 1967; Hanegan and Heath 1970a, 1970b; Heinrich and Casey 1973), noctuid moths of the subfamily Cuculiinae (Heinrich 1987), honey bees (Esch 1960, Verma 1970, Heinrich 1981a, Southwick 1983), bumblebees (Heinrich 1972a, 1972b), euglossine bees (May and Casey 1983), beetles (Bartholomew and Casey 1977a, 1977b), vespid wasps (Heinrich 1984), and dragonflies (May 1976a, Heinrich and Casey 1978, May 1979a). A general review of thermoregulation in endothermic insects is found in the works by Heinrich (1974, 1981b) and May (1979b). The study of insect endothermy is relevant to vertebrate energetics because it demonstrates the difficulties encountered in attaining endothermy at a small mass. Thoracic and to a lesser extent, abdominal temperatures are regulated in insects.

The endothermy of insects is maintained only when insects are "active," although activity here is stretched to include wing movement dissociated from flight, often called *shivering*, which is associated with a warm-up preparatory to flight. Activity is used to maintain a temperature differential because the resting rates of metabolism of insects are very low: at rest most insects have body temperatures nearly identical to ambient temperature. When they are thermoregulating, frequency of wing beat, rate of metabolism, and differential temperature increase as ambient temperature falls. The thermal conductances of moths are similar to those of mammals and birds when adjusted for body mass (Figure 5.11), mainly because moths are covered with hairlike scales that conserve heat, as do the analogous structures in mammals and birds. Low conductances permit larger temperature differentials than would otherwise be obtained (Church 1960). This system can be so effective that winter-active cuculiine moths in northeastern North America are able to fly at air temperatures near 0°C, when thoracic temperatures are 30 to 35°C (Heinrich 1987), a feat accomplished by their ability to shiver at low muscle temperatures, effective insulation, and heat exchangers that reduce heat loss to other parts of the body.

The expense of maintaining continuous endothermy at a small mass can be shown by a comparison of shrews and endothermic moths. Soricine shrews are the smallest continuous endotherms. According to Equation 3.5, a 3.3-g *Sorex cinereus* must have a basal rate that is at least 2.9 times that expected from the all-mammal curve, which means that this shrew must have a nearly continuous feeding schedule, given its small mass and consequent inability to rely on fat stores to support metabolism for any appreciable period. A strictly nocturnal moth of this (thoracic) mass would have total lipid stores that are enough to sustain such high rates for only about 3 h (Hanegan and Heath 1970a), which means that it cannot maintain continuous endothermy. Smaller insects are even less likely to have continuous endothermy. Similar reasoning suggests that all hummingbirds, unlike most shrews, enter torpor (Pearson 1950, Morrison 1962, Lasiewski and Lasiewski 1967, Lasiewski et al. 1967, Hainsworth and Wolf 1970, Beuchat et al. 1979, Krüger et al. 1982, Schuchmann and Prinzinger 1988) because they feed only during daylight.

5.9.2 Fishes.
Fish generally are not endothermic because most of the heat produced in muscles is lost through the body surface or at the gills during gas exchange (Davis 1955; Stevens and Fry 1970, 1974; Stevens and Sutterlin 1976; Erskine and Spotila 1977; Kubb et al. 1980; Elliott 1981). The high rates of ventilation required by the small quantities of oxygen dissolved in water (see Section 2.11) ensure heat loss because the rate of thermal diffusion is about 10 times the rate of molecular diffusion. The resulting temperature differential between the lateral muscles (usually the warmest part of a fish) and water increases with body mass but normally is 1°C or less in sluggish fishes.

Many active fishes, such as tuna (Scombridae) and mackerel sharks (Lamnidae), are endothermic to varying degrees (Barrett and Hester 1964; Carey and Teal 1969a, 1969b; Carey et al. 1971, 1982; Dizon and Brill 1979; Smith and Rhodes 1983; Holland et al. 1992). For example, bluefin tuna (*Thunnus thynnus*), which may weigh 200 to 350 kg, establish temperature differentials up to 20°C, whereas smaller tuna maintain differentials of 4 to 8°C (Figure 5.28), still well above those found in sluggish fishes. Lamnid sharks maintain temperature differentials with the water that are 6 to 8°C. The lateral muscle masses of many fishes are dark red because of high myoglobin concentrations, which promote aerobic expenditures during activity. In endothermic fish these muscle masses lie near the vertebral column, whereas cold-bodied sharks have a smaller mass of red muscles and they are placed under the skin. Large temperature differentials result from combining a large size, high rate

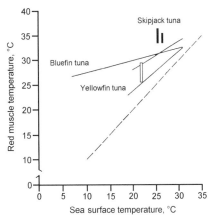

Figure 5.28 Maximal temperature of the lateral red muscles in skipjack (*Katsuwonus pelamis*), bluefin (*Thunnus thynnus*), and yellowfin (*Th. albacares*) tuna as a function of sea surface temperature. Source: From Dizon and Brill (1979).

of metabolism (see Figure 3.9), and a circulatory system having a countercurrent heat exchanger inserted between the lateral muscle masses and the body surface, heart, and gills (Figure 5.29). Cold-bodied sharks have no countercurrent heat exchangers inserted between the red muscle masses and the body surface. Several thresher sharks (*Alopias*) have these retia, which suggests that they too are somewhat endothermic (Bone and Chubb 1983), although their temperatures have not been measured. Similarly, mobulid (e.g., manta) rays have retia that suggest some degree of endothermy (Alexander 1995, 1996).

The heat produced by muscle contraction warms the cold arterial blood entering the muscle mass from the lateral cutaneous artery, thereby keeping the muscle mass warmer than the surrounding tissues. Carey and Gibson (1983) calculated that up to 99% of the heat carried by venous return in the bluefin tuna is transferred to the arterial supply to the muscles. In spite of such high efficiencies, about 50% of heat loss in active species occurs across the gills (Neill et al. 1976, Brill et al. 1978), whereas only some 20% is lost across the gills in sluggish fishes (Stevens and Sutterlin 1976). This difference reflects differences in cardiac output and ventilation rates. Holland et al. (1992) showed that the heat exchangers in bigeye tuna (*Th. obesus*) can be turned off when entering warm, shallow water and turned on when entering cold, deep water, thereby expanding foraging space.

The control of body temperature in large, active fishes is not limited to the lateral musculature. Lamnid sharks have a warm liver, stomach, and spiral valve, the temperature of which is maintained by local generation of heat, which is

conserved by a superhepatic rete (Carey et al. 1981): in free-swimming mako sharks (*Isurus oxyrhinchus*) stomach temperatures often are 6 to 8°C greater than water temperatures. Block and Carey (1985) showed that porbeagle (*Lamna nasus*) and mako sharks have brain temperatures up to 5.3°C higher than water temperature. A similar temperature differential is found between the eyes in these sharks and water temperature. These temperature differentials are slightly smaller than the maximal differentials found between the lateral musculature and water. Similar brain and eye temperature differentials also have been seen in tuna (Stevens and Fry 1971, Linthicum and Carey 1972).

The presence of cephalic differentials is correlated with the presence of orbital retia mirabilia in sharks and carotid retia in tuna, heat exchangers that conserve heat produced by metabolism. Sharks that do not have orbital retia do not maintain brain and eye temperatures above water temperatures. The principal source of the heat that maintains cephalic temperature differentials in sharks is the lateral abdominal red muscles (Wolf et al. 1988), the heat being transported to the head by two "red muscle veins," one from each lateral muscle mass. They carry blood anteriorly, enter the neural canal, and join the myelonal vein, which transports blood forward to join a venous plexus covering the brain. Some of the heat that contributes to the cephalic differentials is contributed by the metabolism of the brain and the eye musculature, although these sources may be secondary.

A similar system may be present in some large mobulid rays, especially in *Manta birostris* and *Mobula tarapacana*, which have cranial retia (Alexander 1996), large amounts of red muscle, and in the case of *M. tarapacana*, a countercurrent heat exchanger on the ventral surface of the pectoral fin (Alexander 1995). Alexander suggested that these structures, including the brain, viscera, and lateral musculature, indicate that the rays are warm-bodied, although no body temperatures are available.

Another means of controlling brain and eye temperatures is found in swordfish (Xiphiidae) and billfish (Istiophoridae). Unlike warm-bodied sharks and tuna, the swordfish (*Xiphias gladius*), white marlin (*Tetrapturus albidus*), and sailfish (*Istiophorus platypterus*) have warm brain temperatures, even though they have cool muscle and visceral temperatures (Carey 1982). These species

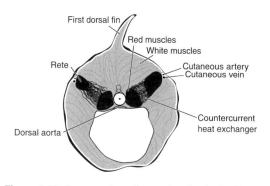

Figure 5.29 Cross section of a porbeagle shark (*Lamna nasus*) showing the countercurrent heat exchangers (retia). Source: Derived from Carey and Teal (1969a).

have structures below the brain that act as brain heaters (Block 1986). They are derived from the superior rectus eye muscles. In these fishes the heater tissue, brain, eyes, and eye muscles are encased by a layer of insulating fat. Brain-water temperature differentials in live, active swordfish are as great as 10 to 14°C, a difference that may facilitate the maintenance of a constant brain and eye temperature as these predators pursue prey through the water column and a thermal isocline. A similar structure, derived from the lateral rectus eye muscle, is found in the scombrid *Gasterochisma melampus*, which implies that this species also has warm brain temperatures. This diversity of morphology and taxonomic affiliation of warm-bodied fishes attests to the repeated evolution of (partial) endothermy in fishes.

The difference between warm- and cold-bodied fishes can be seen in free-swimming sharks (Carey et al. 1981, Carey and Scharold 1990). Cold-bodied blue sharks (*Prionace glauca*) encountered water temperatures between 10 and 20°C as a result of vertical movements in the water column up to 600 m. Their muscle temperatures remained between 15 and 21°C when water temperatures varied between 7 and 27°C (Figure 5.30A). Blue

shark muscle temperatures heated faster than they cooled when water temperatures varied, this hysteresis giving the shark a degree of temperature stability, but when water temperature remained constant for several hours, muscle temperature was within 0.3°C of the water temperature (Carey and Gibson 1987). In contrast, when mako sharks encountered temperatures between 15 and 24°C, their stomach temperatures were usually 6 to 8°C warmer than water temperature, except when they made a short excursion in cold water by diving, when temperature differential increased, or when they entered warm water at the surface, when the differential decreased (Figure 5.30B). Stomach temperatures in the mako usually were between 23 and 27°C when water temperatures ranged from 13 to 23°C.

Field measurements of body temperature in tuna are difficult to use as direct evidence of active temperature regulation because of the thermal inertia associated with a large mass (Neill and Stevens 1974, Neill et al. 1976, Dizon and Brill 1979). Yet, evidence of behavioral and physiological control of body temperature exists (Dizon and Brill 1979, Carey and Gibson 1983). Endothermy permits the body temperature of fast-moving, predatory fishes

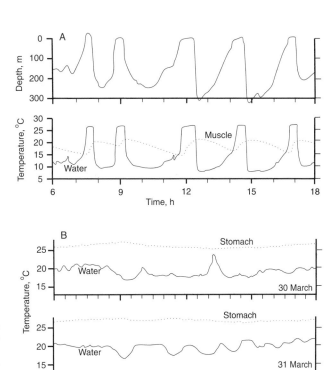

Figure 5.30 The response of (A) lateral muscle temperatures to a change in water temperature in a cold-bodied blue shark (*Prionace glauca*) and (B) stomach temperature in a warm-bodied mako shark (*Isurus oxyrhinchus*). Sources: Modified from Carey and Scharold (1990) and Carey et al. (1981).

to be relatively independent of water temperature, especially if they have a large mass: the large bluefin is the only tuna to enter cold waters, but it does so for such extended periods that thermal inertia alone cannot account for a constant body temperature and a large temperature differential. In the case of billfish, warm temperatures permit predatory fishes to pursue active prey during deep dives, when water temperature may decrease by 10 to 20°C.

5.9.3 Pythons.

Facultative endothermy has been described for *Python molurus* (Hutchison et al. 1966, Vinegar et al. 1970, Van Mierop and Barnard 1978), *Morelia spilota* (Harlow and Grigg 1984), and pythons belonging to the genera *Aspidites*, *Liasis*, and *Morelia* (Charles et al. 1985) and *Chondropython* (Van Mierop et al. 1983). Pythons usually are ectothermic behavioral thermoregulators (e.g., Cogger and Holmes 1960, Slip and Shine 1988c), but brooding *P. molurus* and *M. spilota* females attain temperature differentials up to 7.3°C (Figure 5.31A). Brooding females form a tight coil around the eggs, the temperature differential resulting from an increase in heat production with a fall in ambient temperature (Figure 5.31B). Rate of metabolism is increased by increasing the rate of body contractions, so that at an ambient temperature of 24°C the body pulsates at a frequency of 35 contractions/min (Van Mierop and Barnard 1978). This form of endothermy is similar to that found in birds in that heat generation appears to be exclusively due to muscle contraction ("shivering"). The switch from ectothermy to endothermy in a female python seems to anticipate the laying of eggs (Van Mierop and Barnard 1978), which may mean that it is hormone mediated. This behavior does not occur in males or with behavior other than reproduction, including feeding (Slip and Shine 1988a).

Slip and Shine (1988b) studied the thermoregulation of *M. spilota* in the field near Sydney, Australia. They showed that peritoneal temperatures were more independent of ambient temperatures in brooding females than in nonbrooding females or males. This difference was greatest on cloudy days, when brooding females used shivering; brooding females usually maintained body temperatures between 28 and 33°C, whereas those in nonshivering females fluctuated between 20 and 31°C. On sunny days brooding females often stopped brooding late in the morning to bask. Basking, coupled

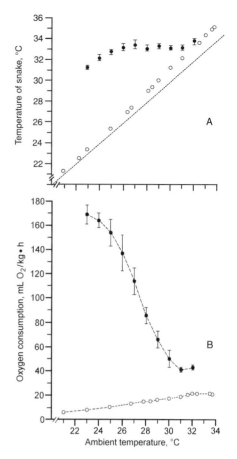

Figure 5.31 (A) Body temperature and (B) rate of metabolism in a python (*Python molurus*) during incubation (closed circles) and when nonreproductive (open circles). Source: Modified from Van Mierop and Barnard (1978).

with shivering, permitted the body temperature of brooding females to remain above ambient temperature (Figure 5.32). Although males also bask on sunny days, their body temperatures do not remain much above air temperature (Figure 5.32). The retention of a tightly coiled position and the use of a well-insulated nest permit brooding females to maintain large differentials with the environment, presumably at a minimal cost. Nevertheless, the cost of incubation is quite high, as shown by a 15% loss in the mass of females during brooding, a period when they do not feed. The opportunistic use of basking thus appears to be a means of reducing their dependence on shivering, an exceedingly expensive form of thermoregulation that at low temperatures may involve a 22-fold increase in rate of metabolism. This mixture of behavioral and physiological thermoregulation

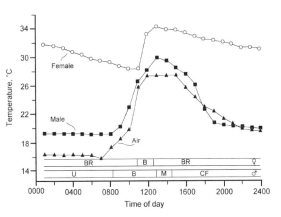

Figure 5.32 Body temperature of a female diamond python (*Morelia spilota*) when brooding (BR) or basking (B) and of a male diamond python when under cover (U), basking (B), or coiled under a filtered cover (CF). Source: From Slip and Shine (1988b).

permits this python to maintain adequate incubation temperatures, to shorten the incubation period, and therefore to enter a temperate environment with its attendant short feeding season, during which the young must grow and deposit fat to survive until the next summer.

The propensity of pythons to show an intermediate level of endothermy is correlated with their oviparous habits (unlike the viviparous boas, which warm their developing young by temperature selection), a distribution that enters suboptimal conditions, and a large mass (which may attain 60 kg in *P. molurus*, but only 1–2 kg in *M. spilota*). In the smallest (<2 kg) species of *Python* (the ball python, *P. regius*) incubation may increase the rate of metabolism at cool ambient temperatures, but only in a transient manner (Ellis and Chappell 1987). Egg temperature is not maintained much above the surroundings and shivering was never observed. The poorly developed capacity for endothermic brooding in *P. regius* may reflect a small size and its tropical distribution in Africa. The occurrence of endothermic brooding in the smaller (<1 kg) *Chondropython viridis* may be associated with the insulation provided by hole nesting (Van Mierop et al. 1983, Ellis and Chappell 1987).

5.9.4 Dermochelys. The leatherback turtle (*Dermochelys coriacea*), one of the largest living reptiles (>900 kg), is the only marine turtle regularly found in cold water, which suggests that it, like bluefin tuna, has some control of its body temperature. Mrosovsky and Pritchard (1971) showed that leatherbacks lay warmer eggs than other marine turtles. Frair et al. (1972) made the striking observation that an active leatherback in Nova Scotia had a temperature differential of 18°C with the environment. Furthermore, Greer et al. (1973) demonstrated that leatherbacks have countercurrent heat exchangers in the flippers, and Goff and Stenson (1988) reported brown adipose tissue in this species (although see Section 5.7.2).

The difficulty with concluding that leatherbacks are endotherms is in defining the border between endothermy and ectothermy, especially at large masses and with respect to rate of metabolism (see Figure 3.9). Two estimates of resting rate of metabolism have been obtained from leatherbacks, one being 69% of the basal rate expected for mammals at 358 kg (Lutcavage et al. 1992) and the other 78% of the expected mammalian basal rate at 353 kg (Paladino et al. 1990). These rates are well above those required for continuous endothermy by the boundary curve. (However, Paladino disavowed his rates [Ruben 1995].) Standora et al. (1982) called *Chelonia mydas* endothermic in that they could maintain a 8°C differential with water when *vigorously swimming*, but the differential fell to 1 to 2°C when inactive. At best, the green turtle at 100 to 150 kg is transitional between ectothermy and endothermy, which points out the limited ability of a large mass to contribute passively to endothermy through thermal inertia alone. The basic difference between endotherms and ectotherms is that endotherms increase their rate of metabolism with a fall in environmental temperature, whereas ectotherms show a decrease in metabolism with a fall in environmental temperature. This criterion cannot yet be applied to marine turtles: their temperature differentials may simply reflect activity, as in *Chelonia*.

Although the data on leatherbacks are incomplete (Mrosovsky 1980), they undoubtedly are more endothermic than smaller marine turtles. The occurrence of endothermy in *Python* and possibly *Dermochelys* raises many questions about the evolution of endothermy: How readily can endothermy evolve? What conditions in the animal and environment lead to endothermy? At present these questions have few answers; the best hope for some insight comes from an examination of the evolutionary history of endothermy in the phylogeny of mammals and of some fishes.

5.10 THE EVOLUTION OF ENDOTHERMY

Interest in the evolution of endothermic homeothermy has been widespread, but the analyses have often told us more about the analyst's area of interest than they have about the evolution of endothermy. This topic can be addressed in several related subquestions: (1) What value is derived from endothermic homeothermy? (2) What mechanisms were involved in the evolution of endothermy? (3) What was the historic pathway by which endothermy evolved?

5.10.1 What value is derived from endothermy?

Three main answers have been suggested: endothermy produces (1) high body temperatures, (2) constant body temperatures, and (3) high levels of sustained activity based on aerobic metabolism. For example, Hamilton (1973) speculated that high body temperatures were selected to maximize rate of metabolism and growth, the constancy of body temperature being a by-product of selection for a high temperature. He also suggested that a high basal rate in eutherians may lead to an increase in the rate of reproduction and r_{max}, where r_{max} is the maximal population growth constant (also see McNab 1980b and Section 13.4). Heinrich (1977) maintained that a high body temperature permits enzymes to specialize for high activity rates, especially when body temperature is fixed, and that the high set points used by mammals and birds are necessary to dissipate the heat produced by activity. A high body temperature also may reduce response time for behavior (Hulbert 1980). But Hammel (1976b) argued that the constancy of body temperature is more important than its level because a constant temperature permits enzyme activity to specialize on a narrow temperature range and because a constant body temperature permits activity over a wide range in environmental conditions. Block et al. (1993) and Block and Flinnerty (1994) proposed a similar view. The level of energy expenditure needed for a constant body temperature is similar to that required to produce a high body temperature, at least at small masses (McNab 1983a), which may make these characteristics difficult to separate.

High rates of aerobic activity are a correlate of endothermy: maximal rates of aerobic metabolism fall between 5 and 10 times the resting rate in both ectotherms (Bennett and Ruben 1979, Taigen 1983, Bennett 1991, Ruben 1995) and endotherms

(see Section 5.7.4), and endotherms have resting rates that are 7 to 9 times those found in ectotherms. Duncker (1991, p. 348) went so far as to state that "... homeothermy is a secondary result of the evolution of the aerobic capacity of the locomotor apparatus"; that is, "... evolving birds and mammals finally acquired a higher metabolism as an *unavoidable side effect* of the increasing upper limit of their aerobic capacity ..." (p. 347, italics added).

The principal difficulty with this analysis, as always in comparative biology, is one of cause and effect. Taigen (1983) addressed this issue when he noted that a proportionality between resting and maximal rates does not exist in certain "special cases." Among them are varanid lizards and snakes and some amphibians, which do not show a tight coupling between scope and resting rate. The scincid lizard *Chalcides ocellatus* also shows no correlation between activity and standard rates (Pough and Andrews 1984). As noted, tree-sloths and some other arboreal mammals have low basal rates and low factorial scopes, whereas the naked mole-rat *Heterocephalus* has a low basal rate and a normal scope (McNab 1980b). Furthermore, Koteja (1987) denied a connection between maximal and minimal rates of metabolism in mammals, which he concluded contradicts the "aerobic capacity" model of the evolution of endothermy. Bozinovic (1992), however, found a correlation between maximal and basal rates and within their residual variations in mammals (see Section 5.7.4). The observation of Gatten et al. (1992, p. 377) that "... the low resting metabolic rate of salamanders [as compared to anurans] is accompanied by a low aerobic capacity and a low aerobic scope, which has been interpreted as an adaptation for an economic life style (Feder 1983[a], Pough 1983), ... supports the arguments for the 'aerobic capacity' model for the evolution of endothermy (Taigen 1983)" is doubtful, especially if the low aerobic capacities and scope reflect "an economic life style." Amphibians tell us nothing about endotherms and even less about evolution of endothermy.

The ultimate rationale for the evolution of endothermy may include several factors, including the independence of body temperature from ambient temperature, the constancy of body temperature, an increase in aerobic scope for activity, and an increase in reproductive output. Endothermy may accrue many benefits to an organism, *as long as it can be afforded*. But under many conditions ectothermy is a more appropriate response to the environment (see Section 4.16), irrespective of any potential advantages associated with endothermy, and in fact many endotherms have come to the same "conclusion" as demonstrated by the repeated evolution of daily and seasonal torpor (see Chapter 11).

BOX 5.2

**Was the Evolution of Endothermy a Means of Attaining
Thermal Stability, of Increasing Aerobic Scope, or Both?**

The question as to the "value" of endothermy in evolution has often generated more heat than insight. The two principal explanations are that it increased the stability of body temperature and that it increased aerobic scope. Some authors have felt the necessity to chose one explanation and to attack the other. For example, Bennett (1991) noted that intermediate levels of energy expenditure would not confer thermal stability to a line evolving endothermy. Hillenius (1994, p. 225) following this logic stated that "[g]iven the long time period apparently involved in the evolution of endothermy, it is unlikely that homeothermy was a significant selective factor in this process." What was not said, however, is that the "gradual" shift from a "low" to a "high" rate of metabolism in the process of converting an ectotherm to an endotherm was associated with a *decrease* in body size (McNab 1978a), so that all stages of the conversion of a large ectotherm to a small endotherm could have been homeothermic. The sharp dichotomy in rate of metabolism between ectothermy and endothermy that is so often emphasized exists only at small body sizes. No reason presently exists to exclude either thermal stability or increased aerobic scope as factors in the evolution of endothermy in both birds and mammals. The evolution of endothermy in association with a large mass, as has been found in various marine fishes, may have been associated primarily with the stability of brain and visceral temperatures (Block et al. 1993). On close inspection, most naturally occurring biological phenomena are complicated, not simple. Bennett (1991, p. 16) concluded that "[u]nifactorial explanations for the evolution of any complex character are almost certainly incomplete." I agree.

5.10.2 What was the mechanism by which endothermy evolved? Stevens (1973) and Hulbert (1980) addressed the cellular mechanisms involved in converting ectothermy to endothermy. Both emphasized the importance of thyroxine. Stevens, noting that thyroxine stimulates the sodium (Na^+) pump in membranes, suggested that endothermy evolved by means of an increase in Na^+ transport, which increased the rate of heat production. Sodium transport may have increased in association with an increase in the Na^+ permeability of cell membranes (Hammel 1976b). Most of the differences in the rate of resting energy expenditure between ectotherms and endotherms are related to mitochondrial enzyme concentrations, and these are influenced by thyroxine levels (Hulbert 1980; also see Hulbert and Else 1981 and Else and Hulbert 1981, 1987). Thyroxine and its influence on Na^+ transport may well have been a cellular means by which rate of metabolism was increased. That is, "... in mammal[s] ... the increased leaki-ness of the cell membrane [was] the *driving force* behind an increased level of energy consumption" (Else and Hulbert 1987, p. R5, italics added). "The evolutionary advantage that favored this increased membrane permeability to ions in mammals ... is increased thermogenesis and the resultant high and constant body temperature ..." (p. R6). These factors are unlikely to have *directed* the development of endothermy but may have permitted it. The directing and determining factors are more likely to have been associated with temperature independence, reproductive output, and resource availability.

Another approach is to ask what sequence of changes occurred in the evolution of endothermy. Whereas a discussion of the biochemical changes that may have been involved might address whether a mechanism is general to all instances in which endothermy evolved, the changes at the organismal level might well be particular for each instance in which endothermy evolved. The evolution of endothermy is now examined in both mammals and fishes.

5.10.3 How did endothermy evolve historically in the class Mammalia? Because of the extensive fossil record of mammals, the evolution of endothermy in mammals can be documented more thoroughly than for any other group of organisms. Several "scenarios" have been suggested. For example, Heath (1968) concluded that the change in posture and gait that accompanied the transition from amphibians to mammals was associated with an increase in muscle tone, which in turn contributed to the increase in the resting rate of metabolism. Hulbert (1980) doubted this contribution; an upright posture may actually reduce, rather than increase, rate of metabolism (McNab 1978a). Hammel (1976b) gave the first detailed analysis of the physiological changes that may have occurred in the evolution of mammalian endothermy. He thought that (1) all reptiles, therapsids, and early mammals sensed and behaviorally responded to ambient temperatures; (2) these vertebrates were homeothermic to varying degrees; (3) advanced therapsids and mammals evolved during a warm period; (4) body temperatures of large reptiles were between 25 and 30°C, if they lived near coastal seas; (5) higher body temperatures were possibly an adaptation to arid environments; (6) rate of metabolism remained low in the phylogeny of mammals until the late Tertiary, when Earth cooled; and (7) high rates only developed in small to intermediately sized species after the development of a fur coat. This analysis argues, then, that mammals had a reptilian level of energy expenditure (and ther-moregulation?) until the late Tertiary. It is based partly on the erroneous assumption that body mass remained between 1 and 20 kg throughout the Mesozoic and into the later Tertiary (Figure 5.33).

Crompton et al. (1978), after observing that diurnal mammals often have higher basal rates than nocturnal species, argued that the evolution of endothermy occurred in two steps: the first permitted diurnal reptiles to move into a nocturnal insectivorous niche, and the second was associated with the movement of some mammals into a diurnal habit. The original nocturnal mammals were said to have weighed 30 to 40 g, to have a body temperature of 20 to 30°C, and to have a low (reptilian? or intermediate?) resting rate of metabolism. Subsequently, some mammals evolved diurnal habits, high rates of metabolism, and high body temperatures. In this view the fundamental dichotomy in energetics among mammals is nocturnality versus diurnality, rather than in relation to burrowing, climate, or food habits (see Section 5.4). They justify this view by comparing the running energetics of two tenrecs (*Tenrec*, *Setifer*), a hedgehog (*Erinaceus*), an opossum (*Didelphis*), and an echidna (*Tachyglossus*). Paradoxically, the echidna and opossum, a monotreme and a marsupial, are concluded to have "mammalian" energetics, whereas the eutherian tenrecs and hedgehog have "reptilian" energetics! Basal rates, which they dismiss, are not correlated with period of activity. The low basal rates measured in hedgehogs and tenrecs are correlated with their burrowing and invertebrativorous habits, and the especially low basal rate in a desert hedgehog (*Paraechinus*) is

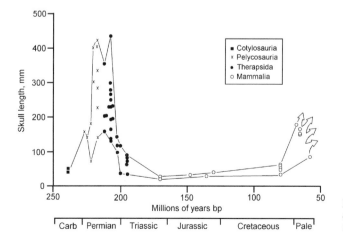

Figure 5.33 Skull length in cotylosaurs, pelycosaurs, therapsids, and early mammals as a function of time. Source: From McNab (1978a).

related to a desert distribution (Shkolnik and Schmidt-Nielsen 1976).

An examination of the fossil record led McNab (1978a) to a different description of the evolution of endothermy in mammals. A striking feature in the evolution of mammals was a great reduction in body mass. From skull length (Figure 5.33), body mass was estimated to have decreased by a factor up to 1000:1, roughly from 20 kg to 20 (or possibly 40) g! Such a radical decrease in mass must have had significant consequences for the energetics of the earliest mammals.

Most therapsids probably were characterized by fairly constant body temperatures, given their rather large masses (to 250 kg). The most persuasive evidence for their thermal constancy, however, comes from bone structure (de Ricglès 1974, 1980): living vertebrates show a correlation between homeothermy and the occurrence of haversian canals (but see Reid 1987), and most (all?) therapsids had these canals. Therapsids, then, may well have been inertial homeotherms, but no evidence suggests that they had mammalian rates of metabolism. This picture suggests that the evolution of endothermy was intimately related to the radical decrease in mass because endothermy is the only means by which homeothermy can be transferred from a large to a small mass.

Morphological evidence supports this hypothesis. First of all, a decrease in mass is compatible with a shift from reptilian to mammalian rates of metabolism. Because mammals have higher total rates than reptiles at all masses (Figure 5.34), a large ectothermic reptile can be converted to a small endothermic mammal if total rate of metabolism scales by a factor less than $m^{0.71}$. Given the difference in the coefficients of the power curves, the smallest change in mass compatible with the switch from ectothermy to endothermy is by a factor of 15:1, which is well within the decrease implied in the fossil record, as is the (250:1) decrease if the therapsid-mammal line had followed the boundary curve for endothermy (Figure 5.34).

The evolution of another morphological trait, the secondary palate, corroborates the view that the evolution of mammalian endothermy was associated with a decrease in mass. Ventilation rate increases both with a decrease in mass and with an increase in rate of metabolism (Figure 5.35). Consequently, the evolution of a small endotherm from a large ectotherm must have involved a very large

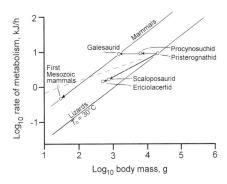

Figure 5.34 Log_{10} basal rate of metabolism in mammals and standard rate of metabolism in lizards at 30°C as a function of log_{10} body mass, illustrating a possible scenario as to the evolution of endothermy through a reduction in body mass. The shift in rate of metabolism in the evolution of mammals from cynodonts is given in two extreme patterns: the smallest decrease in mass compatible with the conversion to endothermy (solid curve to the galesaurid), or a gradual shift to endothermy by following the boundary curve (dashed curve). Also indicated is a potential shift in metabolism in the evolution of ericiolacertid theriocephalians. Source: Modified from McNab (1978a).

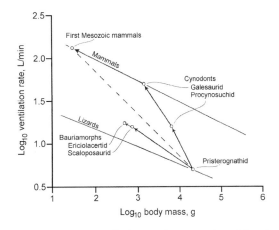

Figure 5.35 Log_{10} ventilation rate in mammals and lizards as a function of log_{10} body mass, illustrating the probably large increase in ventilation rate associated with the evolution of a small, endothermic mammal from a large, ectothermic cynodontid therapsid. Source: From McNab (1978a).

increase in ventilation rate. For example, if the decrease in mass was by a factor of 1000, the ventilation rate would have increased about 29 times! A large increase in ventilation rate apparently had an impact: a secondary palate evolved with a decrease in mass in cynodonts, the therapsid suborder that gave rise to mammals. For example, cynodonts of the family †Pristerognathidae sequentially gave rise to the †Procynosuchidae and the

†Galesauridae, which gave rise to the earliest meso-zoic mammals (Hopson 1969, Hopson and Cromp-ton 1969, Hopson and Barghusen 1986). The first appearance of the secondary palate in this sequence was in the †Procynosuchidae, where it was incom-plete; the palate was complete in the †Galesauridae. The secondary palate permits ventilation and mas-tication to occur simultaneously. Furthermore, a palate also developed in the other therapsid subor-der that showed a marked decrease in body mass, the suborder †Therocephalia (including the families †Ericiolacteridae and †Scaloposauridae). Endo-thermy thus appears to have evolved independently in cynodonts and therocephalians.

Furthermore, the evolution of complex nasal maxilloturbinates in mammals, which reduce the evaporative water loss associated with high venti-lation rates, also is associated with the evolution of endothermy (Figure 4.38; Hillenius 1992, Rubin et al. 1996). It occurred both in the phylogeny of cynodonts and therocephalians, which confirms the suggestion that the evolution of endothermy was associated with the reduction in body size. Homeothermy developed in therapsids long before endothermy, their homeothermy being associated with the thermal inertia of a large mass, possibly aided by the presence of a fur coat. Endothermy evolved in conjunction with the evolution of a small mass, which presumably was associated with a shift from a carnivorous to an insectivorous diet. These changes may have been instrumental in permitting the earliest mammals to occupy active nocturnal habits (Hopson 1973).

5.10.4 The evolution of endothermy in verte-brates other than mammals. Because of the diversity seen in the physiology of endothermy in various vertebrates, the endothermy of vertebrates other than mammals probably evolved in a variety of ways. The means by which the endothermy of birds evolved is unclear because little is known of the early evolution of birds. Ostrom (1976) argued that they evolved from "coelurosaur" dinosaurs, whereas others (e.g., Feduccia 1996) maintained that the morphological similarity between early birds and therapods is convergent, the first birds long predating the evolution of the coelurosaurs. If birds evolved from dinosaurs, they would appear to be the only small descendents of this line. The evolution of a small body mass in birds too may have been an integral element in the evolution of endothermy. Or, conversely, the small, active

coelurosaurs that may have given rise to birds already could have been endothermic.

Ruben (1991) suggested that the first birds, including †Archaeopteryx, were characterized by "reptilian" physiology, that is, with a low rate of metabolism and a low, unstable body temperature. His conclusion (p. 6) that "[a] fully feathered Archaeopteryx could have been either ectothermic or endothermic . . ." is difficult to believe, as is his view (p. 9) that †Archaeopteryx could have main-tained horizontal flight with ". . . an ectothermic, anaerobically powered . . ." pectoral muscle mass. Yet, some recent work (Chinsamy et al. 1994) described Cretaceous birds with what may have been annual rings in leg bones, which if correct implies an ectothermal life-style. An analysis of the allometry of the long bones of †Archaeopteryx, nevertheless, indicated that it was endothermic (Houck et al. 1990). †Archaeopteryx need not have had the level of energy expenditure found in living birds.

With respect to reptiles, observations by Brioli (1941) and Sharov (1971) indicated that some pterosaurs may have had a "fur" coat. If so, pterosaurs may well have been endothermic to some degree. The earliest pterosaurs were small; all large pterosaurs lived late in their history. Unfortunately, nothing is known of the origin of pterosaurs. Recent evidence indicates that the wings of pterosaurs had structural fibers in the wing membranes (Martill and Unwin 1989, Padian and Rayner 1993, Unwin and Bakhurina 1994); these fibers may be what was identified as "fur."

The evolution of endothermy in fishes has been examined in relation to phylogeny (Block et al. 1993, Block and Finnerty 1994). For example, a molecular phylogeny of the suborder Scombroidei, which includes tunas (Scombridae), swordfish (Xiphiidae), and billfishes (Istiophoridae), was based on a mitochondrial cytochrome *b* gene. This phylogeny, which shows some inconsistencies with contemporary taxonomy, indicates that billfishes and the swordfish have a common shared charac-ter (A) in the modification of the superficial rectus eye muscle into a thermogenic organ (Figure 5.36). In addition, tuna (tribe Tunnini) evolved systemic endothermy (C), which includes the maintenance of warm temperatures in the lateral musculature (D), viscera, and brain, and independently the but-terfly mackerel (*Gasterochisma melampus*) evolved a cranial thermogenic organ (B) from the superior rectus eye muscle (Figure 5.36). Thus, two types of

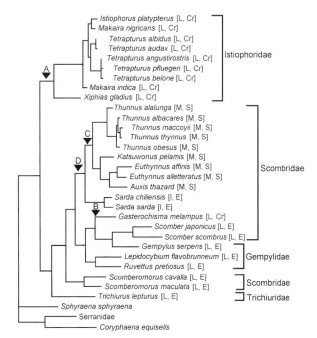

Figure 5.36 The phylogeny of the Scomroidei with several evolutionary steps: A, modification of the superior rectus (eye) muscle into a thermogenic organ; B, modification of the lateral rectus (eye) muscle into a thermogenic organ; C, systemic endothermy using countercurrent heat exchangers in the lateral musculature, viscera, and brain; and D, internalization of red muscle along the horizontal septum. Each species is followed by two letters, the first indicating red muscle position and the second, thermal strategy. Red muscle positions are lateral (L), medial (M), and intermediate (I). Thermal strategies are cranial endothermy (Cr), systemic endothermy (S), and ectothermy (E). Source: Modified from Block et al. (1993).

endothermy (cranial and systemic) have evolved in this suborder, and it has occurred 3 times (twice for cranial endothermy). "All three endothermic scombroid lineages have expanded their ranges . . . into cool temperate waters (50°N and S); this supports the primacy of niche expansion over increased aerobic capacity as a selective force" (Block et al. 1993, p. 213). An analysis of the phylogeny of living sharks suggests that endothermy probably evolved twice, once for lamnids and once for threshers (Block and Finnerty 1994). Consequently, endothermy evolved at least 5 times in marine fishes (not counting the mobulid rays, which if endothermic would constitute a sixth time) and niche expansion would be effective in all cases.

Several stages may have characterized the evolution of endothermy in fishes (Block and Finnerty 1994). First of all, a large body mass may have been plesiomorphic for all groups: the large size reduces the cooling rate (i.e., it contributes to thermal inertia). Furthermore, fat deposits around the body and brain reduce thermal conductance, and thermal hysteresis permits the reduction of heat loss to the environment. Behavioral temperature regulation may have preceded the evolution of endothermy, which required the shift of red muscles to the interior. This intermediate state, internal red muscles and no peripheral countercurrent retia, is seen today in bonitos (*Sarda*). Countercurrent heat exchangers were then added to

complete the evolution of endothermy. A shift of the lateral red muscles to the interior was associated with a change in swimming style to the stiff form seen in tuna and lamnid sharks. Although ". . . the high aerobic capacity of tuna red muscle is important for endothermy, . . . it did not evolve under selection for endothermy. Rather, the evolution of high red muscle oxidative capacity predates endothermy in the lineages leading to tunas" (Block and Finnerty 1994, p. 300).

Endothermy may evolve through an increase in mass, which is likely in sharks, tuna, pythons, and *Dermochelys*, or through a decrease in mass, as appears to be the case in mammals. The endothermy derived from a decrease or an increase in mass may not be identical. A decrease in mass may be required for the development of an endothermy characterized by the great precision found in most birds and mammals because such endothermy cannot rely on the thermal inertia offered by a large mass to buffer fluctuations in the external environment. Endothermy that evolved in association with an increase in mass may depend on thermal inertia for the thermal stability of body temperature.

5.11 CHAPTER SUMMARY

1. Endothermy provides a form of homeothermy that is based on high rates of heat

production and reduced rates of heat loss; it is most highly developed in mammals and birds.

2. Rate of heat production varies directly with the product of thermal conductance and the temperature differential maintained between the body and the environment.

3. Rate of metabolism is independent of ambient temperature in the region of thermoneutrality, which is produced by changes in thermal conductance that compensate for changes in the temperature differential; the rate of metabolism in thermoneutrality is the basal rate when an adult animal is quiescent, postabsorptive, and thermoregulating.

4. As a consequence of scaling basal rate proportional to about $m^{0.73}$, many biologically important time periods, such as the time for fat use or heat storage, are proportional to approximately $m^{0.27}$.

5. The principal factor responsible for a positive correlation of body size with latitude is a correlation of resource availability with latitude.

6. Food habits are an important determinant of basal rate in mammals and birds, especially at masses greater than 100 g.

7. Climate and activity level (muscle mass) also appear to influence the basal rates of endotherms.

8. Body mass and climate influence minimal thermal conductance.

9. Because basal rate and minimal conductance are unequal functions of body mass and because they each reflect the influence of food habits and climate, the temperature differential maintained with the environment and body temperature varies with mass, food habits, and climate.

10. The principal sources of heat for thermoregulation at low ambient temperatures are shivering, nonshivering thermogenesis, and basking, although nonshivering thermogenesis is not found in birds; activity increases thermogenesis and may contribute to temperature regulation at cold temperatures.

11. Small altricial and precocial endotherms follow the boundary curve for endothermy during growth to ensure endothermy; high rates of metabolism, which also are found in the young of large mammals, increase growth rate, thereby shortening the period of time that young depend on their parents.

12. Some animals other than mammals and birds including many insects, tuna, billfishes, some sharks, some pythons, and the leatherback turtle, are endothermic to varying degrees.

13. Endothermy, although expensive to maintain, confers advantages, including the stability of body temperature, the independence of body temperature from ambient temperature, an increase in the aerobic scope for metabolism, and an increase in reproductive output.

14. The evolution of endothermy may occur through either an increase or a decrease in body mass, but the latter may be the only way in which the precise thermoregulation of mammals and birds may be attained.

PART III | Material Exchange with the Environment

6 Osmotic Exchange in Aquatic and Transitional Vertebrates

6.1 SYNOPSIS

Aquatic vertebrates collectively tolerate water with a wide range of solute concentrations and compositions, the highest concentrations occurring in seawater and in some inland lakes with internal drainage. All vertebrates regulate the ionic concentration of body fluids, and all vertebrates except hagfishes osmoregulate. Chondrichthyans usually store urea in their plasma and tissues, which makes them slightly hyperosmotic with the ocean but increases their osmotic differential with freshwater. The enzymatic impact of urea storage is counteracted by the storage of other compounds, principally methylamines. The only truely freshwater chondrichthyans are South American rays belonging to the family Potamotrygonidae: they do not store urea. Most other "freshwater" chondrichthyans are short-term residents that breed in salt water. Some bony fishes use urea as a waste product, usually in association with air-breathing habits, but the living coelacanth, like chondrichthyans, stores urea and is approximately isosmotic with ocean water. Teleosts do not store urea, are hyperosmotic in freshwater, and are markedly hyposmotic in seawater. As a result, marine teleosts lose water osmotically, which they replace by drinking seawater and excreting the excess monovalent ions against the concentration gradient. Teleosts in freshwater excrete the excess water acquired by osmosis, replacing the salts lost in the urine by active transport from the environment. The transport of ions against a concentration gradient requires an expenditure of energy. Whether the cost of osmotic regulation is ecologically significant is unclear, but it should be proportional to the gill surface area and to the osmotic differential maintained by a fish with the environment. Vertebrates probably originated in salt water, the glomerulus originally being a structure used for ionic regulation, only later developing its volume regulation function with an entrance into freshwater. The evolution of chondrichthyan osmoregulation and the use of urea in combination with methylamine possibly occurred in a marine environment as a means of reducing the impact of salts on the activity of enzymes. The use by marine teleosts of hyposmotic regulation usually is thought to be derived from a freshwater ancestry, but the evolution of hyporegulation in a marine environment remains a possibility. Aquatic amphibians are generally limited to dilute waters and use ammonia as their principal nitrogenous waste product, whereas most terrestrial species use urea as their principal waste product. To reduce water gain, aquatic amphibians usually have a lower integumental permeability to water than do terrestrial amphibians. Arboreal frogs generally have rates of water loss that are about one-half those of terrestrial species, although a few have rates of water loss that are as low as lizards, often coupled with the use of uric acid as a waste product, at least in species that weigh more than 10 g. Terrestrial amphibians conserve water in many ways, including the reuse of bladder water, entrance into burrows, and seasonal inactivity. The reduction of water loss is accentuated in many hibernating and aestivating amphibians by the development of a cocoon that isolates the amphibian from the enclosing soil when the soil has a high water tension. This pattern also is found in some aestivating bony fishes.

6.2 INTRODUCTION

Vertebrates originated in an aquatic environment, where they were required to balance salt and water exchange to maintain acceptable concentrations of fluids in the various body compartments. These aquatic vertebrates are collectively referred to as *fishes* and constitute more than one-half of all living vertebrates. In analyzing the water and salt exchange of aquatic vertebrates in this chapter, most of the attention is given to the diverse array of fishes. Although some fishes use aerial gas exchange and some move onto land for short periods, no fish are totally terrestrial. In contrast, amphibians include some species that are completely aquatic, some of which have a dual existence (i.e., are "αμφβιοσ"), and some that are completely terrestrial. Amphibians are included in this chapter to represent the transition from aquatic to terrestrial habits. Some reptiles have completely aquatic habits, but they, like aquatic birds and mammals, are terrestrial vertebrates that have "returned" to aquatic habits, carrying with them physiological characteristics that evolved in association with a terrestrial existence (e.g., uricotelism). Some reptiles have made such a readjustment to aquatic habits that they have reevolved the capacity for aquatic gas exchange (see Section 8.9). The water and salt balance of these "aquatic" vertebrates is considered in Section 7.9.

Most chemical reactions underlying life's processes occur in dilute aqueous solutions and are sensitive to the concentration of these solutions. Biological fluids contain a variety of solutes, including ionized salts, amino acids, and various other organic compounds, collectively referred to as *osmolytes*, that is, particles that contribute to the osmotic concentration of a fluid. Intracellular and intercellular fluids are maintained within an acceptable range of concentration by control of the exchange of water and salts with the external environment. Water freely moves across many boundary surfaces, especially those associated with gas exchange. Most salt exchange with the environment occurs across surfaces specialized for exchange with the environment, such those found in gills, intestine, and kidneys.

As shown in Section 2.10, salt and water exchange are proportional to the differential concentrations in fluids between an organism and its environment. Two terms are often applied to this exchange. *Ionic regulation* refers to the maintenance of the concentration of an ion in a compartment like blood plasma. *Osmotic regulation* is defined as the maintenance of the total concentration in a fluid compartment. These definitions make ionic and osmotic regulation appear rather different: osmotic regulation, however, can be viewed as the process by which a differential concentration of the solvent molecule, water, is maintained between an organism and its environment. The solutes are not necessarily ions, particularly in chondrichthyan fishes, but the solutions are still characterized by concentration gradients across differentially permeable membranes.

A note on the concentration of fluids should be made. As seen in Section 2.10, the thermodynamically appropriate measure of concentration is molals or osmolals (see Sweeney and Beuchat 1993). Many other units are often used, including molar or osmolar solutions, percentage, and permillage (‰). Here all concentrations have been converted to osmolals (or mOsm/kg). Noteworthy is the observation that many publications report plasma concentrations as milliosmoles per liter when they have been measured by freezing point depression, which is proportional to milliosmoles per kilogram (Equation 2.20) and should have been reported as such. In any case, osmolar concentrations can be called osmolal concentrations with maximally a 1% error.

6.3 THE CHEMICAL COMPOSITION OF AQUATIC ENVIRONMENTS

Aquatic vertebrates live in the ocean, inland waters, and estuaries, thereby facing a wide range in physical and chemical conditions. The chemical composition of a body of water dictates the adjustments required of species in such an environment. This section surveys the ionic and total concentrations that vertebrates potentially face.

Ocean waters are essentially uniform in the relative abundances of dissolved ions (Sverdrup et al. 1942), but they may vary in total concentration (Clarke 1924). Most of this variation is found in embayments and semi-isolated seas, due to regional differences in the amount of rainfall, the rate of evaporation, and the amount of runoff funneled into the embayment. Ocean embayments that have a high rate of evaporation, small amounts of rainfall, and little direct runoff (such as the Mediterranean [1130 mOsm/kg] and Red Seas [1160 mOsm/kg]) have higher concentrations than do oceanic embayments with low rates of evaporation, high rainfall, and more direct runoff (such as the Baltic [210 mOsm/kg] and White Seas [80 mOsm/kg]). These factors also influence the surface salinity of the open ocean: it is low near the equator, maximal at 25°N and 25°S latitudes, and minimal beyond 40° latitude (Sverdrup et al. 1942), but usually is about 1000 mOsm/kg.

Rivers have a limited range in concentration, nearly all being very dilute by oceanic standards. River water may vary in ionic composition, depending on the mineral composition of sediments in the drainage basin. A spatially fluctuating region at the mouth of rivers, the estuaries, is characterized by complex daily and seasonal variation in salinity due to the mixture and stratification of fresh and marine water.

The greatest total concentrations and the greatest range in ionic compositions occur in lakes that have no external drainage to the ocean (Clarke 1924). The total concentration in large lakes of this class can be as high as 4000 mOsm/kg (Caspian Sea, Russia), 5800 mOsm/kg (Great Salt Lake, Utah), and 6500 mOsm/kg (Dead Sea, Palestine). Small ponds on salinas in arid regions may attain high concentrations in the process of drying. Depending on the petrographic characteristics of the drainage basin, lakes belonging to a closed basin may have waters whose predominant ion is chloride (Great Salt Lake [55% of total ions]; Dead Sea [66%]), sulfate (Little Manitou, Saskatchewan [48%]; Redberry Lake, Saskatchewan [71%]), or carbonate (Lake Nakuru, Kenya [62%]; Malheur Lake, Oregon [45%]) (Livingstone 1963).

Further complications in the distribution of ions in inland waters occur when springs discharge "salt" waters into rivers or lakes, thereby producing local pockets of water having a distinctive distribution of ions. Deep-sea upwelling also may influence the local composition of ocean water.

The variation in composition and concentration of naturally occurring waters has implications for the distribution of aquatic vertebrates because aquatic vertebrates have unequal capacities to withstand the range of osmotic and ionic concentrations found in environments.

6.4 TOLERANCE OF EXTERNAL CONCENTRATIONS

The tolerance of organisms to external concentrations is limited. Such limitations are produced either by a limited tolerance to a variation in the concentration of body fluids or by a limited ability to osmoregulate. Some marine invertebrates tolerate plasma concentrations that are slightly higher than the concentrations of "standard" ocean water (ca. 1000 mOsm/kg), but the only vertebrates that tolerate plasma *salt* concentrations equal to ocean water are the hagfishes (Myxinoidea). Elasmobranchs have plasma concentrations that are slightly hyperosmotic to seawater, but plasma salt concentrations are only 60% of seawater; the remainder is due to the accumulation of organic molecules, especially urea. In contrast, most marine bony fishes have plasma salt concentrations that are about 35% that of seawater and do not store urea: they are hyposmotic to the sea. At the other extreme, no vertebrate tolerates body fluids as dilute as freshwater, the most dilute vertebrates living in freshwater being a few "primitive" bony fishes and some amphibians. These species have plasma concentrations that are about 200 mOsm/kg. Aquatic vertebrates other than hagfishes, then, collectively have plasma salt concentrations between 20% and 60% that of seawater, usually between 20% and 35%.

6.5 OSMOTIC REGULATION IN AQUATIC VERTEBRATES

Tolerance to external concentrations depends principally on the capacity of aquatic vertebrates for osmotic regulation. Many tolerate a wide range in external concentration; that is, they are *euryhaline*. Others, however, tolerate only a narrow range in external concentrations, beyond which osmotic regulation fails: these species are called *stenohaline*. Stenohaline species may live either in freshwater or salt water. From the view of osmotic regulation, the most variable environments are estuaries, where the osmotic differential maintained by vertebrates with the environment may reverse several times a day.

Aquatic vertebrates usually have plasma and cellular salt concentrations that differ from those in the environment. In freshwater sodium and chloride tend to be lost to the environment, and in seawater these salts tend to be gained. In both cases, these ions must be transported against the concentration gradient to ensure constant plasma and cellular concentrations. As noted in Section 2.10, the transportation of salts against an osmotic gradient requires an energy expenditure. Important exchange mechanisms include those between Na⁺ and H⁺, Na⁺ and NH₄⁺, and Cl⁻ and HCO₃⁻ (Evans 1975, 1980, 1984, 1993), which are prominent in freshwater teleosts but are also found in hagfishes, sharks, and marine teleosts. The distribution of these mechanisms suggests that they evolved before vertebrates entered freshwater under demands for acid-base regulation and nitrogen excretion in a marine environment (Evans 1984).

The most abundant and diverse group of aquatic vertebrates includes a complex array of fusiform species: the Agnatha (jawless hagfishes and lampreys), Chondrichthyes (sharks, rays, and chimaeras), and Osteichthyes (bony fishes). The osmotic exchange of each of these groups is examined in turn.

6.5.1 Cyclostomes. Living agnaths are divided into two orders: the exclusively marine hagfishes (Myxinoidea) and the marine and freshwater lampreys (Petromyzontia). Both groups are limited in distribution to temperate waters in the Northern and Southern Hemispheres (see Section 14.8.1).

Hagfishes are the only vertebrates known to be osmoconformers (Figure 6.1), even though they have glomerular kidneys. The relative abundances of ions in their plasma differ from those found in

the ocean (Table 6.1), showing that hagfishes, like all vertebrates, regulate ionic concentrations. However, hagfish blood pressure is so low, about 5 to 6 cm H₂O (Riegel 1978), that its kidney may function principally as a secretory organ (Alt et al. 1981). Furthermore, urine flow rates equal the filtration rate (Munz and McFarland 1964), which implies that no net fluid resorption occurs in hagfish kidneys. In the absence of an osmotic differential with the environment, the urine of hagfishes is isosmotic both with the plasma and with seawater (Fänge 1963, Munz and McFarland 1964). Hagfishes do not enter freshwater, which reflects a limited tolerance to low plasma concentrations. Cholette et al. (1970) had no apparent difficulty in acclimating *Myxine glutinosa* to 630 mOsm/kg, within which the hagfishes remained in good condition for many weeks. But Strahan (1962) could not maintain *Paramyxine atami* at 760 mOsm/L. *Eptatretus stouti*, the Pacific hagfish, could tolerate external concentrations between 80% and 122% that of seawater; much of its inability to tolerate a greater range in concentration was associated with a rapid gain or loss of water and an inability to regulate sodium chloride exchange with the environment (McFarland and Munz 1965). The modest volume control that occurs in hagfish may result from changes in the glomerular filtration rate (Alt et al. 1981). The very low standard rates of metabolism in *Eptatretus* (Munz and Morris 1965) and other hagfish (Smith and Hessler 1974, Forster 1990) may prevent them from being effective osmoregulators.

Unlike hagfishes, lampreys are not confined to marine waters. Some species spend much of their life in the sea, coming into freshwater only to breed; others breed in freshwater and spend the rest of their life in estuarine waters; whereas the remainder are restricted to freshwater. Lampreys are hyperosmotic at ambient concentrations less than 300 mOsm/kg and hyposmotic at higher concentrations (Figure 6.1). The osmotic differential that lampreys maintain in freshwater leads to water gain in spite of a thick mucous coat over their bodies; most water gain probably occurs through their gills. Lampreys consequently produce a copious urine and have appreciably more glomeruli than hagfishes have. As expected, the volume of urine produced increases with a fall in the concentration of the external medium (Figure 6.2)—that is, with an increase in the osmotic differential—but approaches zero at the isosmotic

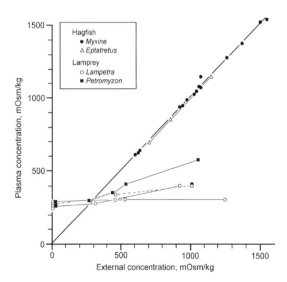

Figure 6.1 Plasma concentration in hagfish and lampreys as a function of the external concentration. Data obtained on a species in the same study are connected. Sources: Data derived from Fontaine (1930), Galloway (1933), Morris (1956, 1958), Bellamy and Chester Jones (1961), McFarland and Munz (1965), Cholette et al. (1970), and Pickering and Morris (1970).

Table 6.1 Plasma and urine composition of various aquatic vertebrates

Species	Total (mOsm/kg)	Na	K	Ca	Mg	Cl	Urea	TMAO
				(mEq/L)				(mM/L)
A. Hagfishes								
Eptatretus stouti								
Seawater	1029	496	10.3	21.8	103.1	543	0	0
Plasma	1031	570	7.0	9.0	24.0	547	—	—
Myxine glutinosa								
Seawater	898							
Plasma	980	439	5.9	7.2	19.5	455	—	—
B. Lampreys								
Lampetra fluviatis (fw)	>239	120	3.2	4.0	4.2	96	—	—
Petromyzon marinus								
Plasma (fw)	>236	124	—	3.6	—	105	0.2	—
Plasma (sw)	>396	212	—	7.0	—	173	0.2	—
C. Elasmobranchs								
Marine (species = 5)	1026	—	—	—	—	270	393	66
Squalus acanthus								
Seawater	925–935	440	9.1	10	51	492–500	0	0
Plasma	1001–1036	254–320	4.4–7.0	2.6	3.9	239–250	350	—
Urine	754–780	327–352	2.0	4.0	50	170–286	150	—
Rectal gland	1001–1036	520–580	5.6–8.4	0.1	1.0	490–562	0	—
Carcharhinus leucas (plasma)								
Seawater (Florida)	956	288	6.1	5.7	3.8	288	356	47
Freshwater (Nicaragua)	652	245	6.4	4.5	1.4	219	169	—
Freshwater rays (plasma)								
Daysatis (Congo)	550	153	11.4	—	—	—	212	—
Potamotrygon (Amazon)	308	150	6	7	4	149	0	—
Dasyatis sabina (plasma)								
Seawater	1034	310	7.0	6.2	—	300	395	—
Freshwater	621	212	5.2	8.6	—	208	196	—
D. Chimeras								
Chimaera monstrosa (sw)	>1020	362	10.2	—	—	380	265	—
Hydrolagus colliei								
Seawater	892							
Plasma	898	300	3.4	4.3	7.9	306	245	5.5
Urine	844	162	7.8	16.7	69	268	52	—
E. Coelacanth								
Latimeria chalumnae								
Seawater	1035							
Plasma	931	197	5.8	4.9	5.3	187	377	122
F. Lungfish								
Protopterus aethiopicus (fw)	238	99	8.2	4.2	—	44	0.6	—
G. Sturgeon								
Acipenser sturio								
Plasma (fw)	318	156	4.3	4.6	3.0	120	—	—
Plasma (sw)	343	164	4.7	4.2	3.1	126	—	—
H. Bowfin								
Amia calva (fw)	>276	133	2.0	10.6	0.8	120	—	—
I. Gars								
Lepisosteus osseus (fw)	>289	140	2.7	12.2	0.6	118	—	—
J. Teleosts								
Shallow marine (species = 14)	434	—	—	—	—	177	4	14.4
Midwater marine (sp = 6)	462	—	—	—	—	220	0.7	3.7
Lophius piscatorius (sw)	>424	198	7.4	5.8	11.6	186	—	—

Table 6.1 *Continued*

Species	Total (mOsm/kg)	Na	K	Ca	Mg	Cl	Urea	TMAO
		(mEq/L)					(mM/L)	
Gadus callarias (sw)	>368	180	4.9	10.0	7.6	158	—	—
Thunnus thynnus (sw)	437	190	26.8	—	—	181	—	—
Scomberomorus maculatus (sw)	386	188	9.8	—	—	167	—	—
Sphyraena barracuda (sw)	476	215	6.4	—	—	189	—	—
Mycteroperca bonasi (sw)	461	228	7.9	—	—	208	—	—
Coregonus clupoides (fw)	>297	141	3.8	5.3	3.4	117	0.7	—
Salmo trutta								
Plasma (fw)	326	150	2.7	—	—	133	—	—
Plasma (sw)	356	166	3.5	—	—	138	—	—
Salmo gairdnerii (fw)	>306	144	6.0	5.3	—	151	—	—
Onchorhynchus kisutch (fw)	295	146	—	—	—	132	—	—
Onchorhynchus tschawytscha								
Plasma (fw)	>288	161	0.3	5.4	2.4	114	10.3	—
Plasma (sw)	>332	179	1.0	2.0	1.8	139	—	—
Anguilla anguilla								
Plasma (fw)	328	150	2.8	—	—	105	—	—
Plasma (sw)	377	175	3.2	—	—	155	—	—
Cyprinus carpio (fw)	274	130	2.9	4.2	2.5	125	—	—

fw, a freshwater species or a euryhaline species measured in freshwater; sw, a saltwater species or a euryhaline species in salt water; —, measurement not made; TMAO, trimethylamine oxide. When plasma osmotic concentration was not indicated, its minimal concentration was estimated by the sum of ionic and plasma molecular concentrations.

Sources: Data derived from Smith (1929, 1930), Robertson (1954), Becker et al. (1958), Gordon (1959), Burger (1962), Fänge and Fugelli (1962), Magnin (1962), Fromm (1963), Munz and McFarland (1964), Sharratt et al. (1964), Conte (1965), Thorson et al. (1967, 1973), Urist and Van de Putte (1967), Houston and Madden (1968), Read (1971a), Griffith et al. (1974), Thorson and Watson (1975), Alt et al. (1981), Griffith (1981), and Piermarini and Evans (1998).

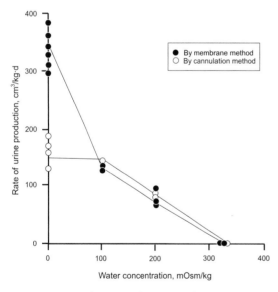

Figure 6.2 Rate of urine production in a lamprey (*Lampetra fluviatilis*) as a function of the external concentration. Source: Modified from Morris (1956).

point (ca. 300 mOsm/kg). Most of the change in the glomerular filtration rates of lampreys that occurs in association with a change in ambient concentration is produced by a change in the filtration rate of individual nephrons (McVicar and Rankin 1985), a pattern that appears to be shared only with mammals; most other vertebrates, including teleosts (Brown et al. 1983), modify filtration rate by changing the proportion of nephrons used.

Lampreys are difficult to obtain in a marine environment, so most observations on "marine" lampreys have been made on those caught at their entrance into freshwater. Galloway (1933) and Morris (1956) showed that the responses to external concentration in adult (freshwater) and "fresh-run" lampreys are different. The highest external concentration that mature *Lampetra* can tolerate is one-third that of seawater, and then only through an increased tolerance to a high plasma concentration. The permeability of body surfaces to water decreases with entrance into freshwater (Morris 1956). Lampreys probably replace water lost to a marine environment by swallowing seawater, absorbing water from the intestine by means of the active transport of monovalent ions, and excreting excess ions at the gills (Morris 1958, Pickering and Morris 1970), although few observations justify this conclusion, except in analogy with marine teleosts.

6.5.2 Chondrichthyans.

Cartilaginous fishes are much more diverse than cyclostomes; this class contains six orders, including sharks, sawfish, and rays in the subclass Elasmobranchii, and chimaeras, or rat-fish, in the subclass Holocephali. The vast majority of the chondrichthyans are exclusively marine, although some sharks, sawfish, and a few rays can be found in large rivers.

Marine sharks and rays normally are slightly hyperosmotic to the environment (Figure 6.3). Unlike hagfish, but similar to lampreys, marine elasmobranchs have a plasma salt concentration below that of seawater (Table 6.1); in the case of elasmobranchs, the salt content is about 60% of seawater, apparently much higher than that found in marine lampreys. The total plasma concentration in sharks and rays is slightly greater than that of the ocean because urea and trimethylamine oxide (TMAO) are stored. Urea production occurs by means of two principal pathways: all vertebrates synthesize urea at low rates from the breakdown of the purines (uricolysis), but the high-volume pathway, when present, is based on the ornithine-urea cycle, which is found in chondrichthyans and (as we will see) in the coelacanth, lungfish, selected teleosts, amphibians, and mammals. Elasmobranchs reduce urea loss by having gills that are relatively impermeable to urea (Smith 1936) and by the active reabsorption of urea in the kidney tubules (Kempton 1953).

The storage of urea by elasmobranchs is surprising in that this molecule at high concentrations disrupts enzyme function. Elasmobranchs have adjusted to the presence of urea in various ways. One is that the affinity for substrates of some enzymes increased, thereby requiring urea for "normal" enzyme function (Yancey and Somero 1978). Some species have hemoglobin that is insensitive to the presence of urea (Bonaventura et al. 1974). Most elasmobranchs accumulate TMAO. Evidence indicates that this oxide counteracts the impact of urea on enzyme function (Yancey and Somero 1979, 1980; Yancey et al. 1982). One measure of the facility with which an enzyme catalyzes a reaction, the Michaelis constant (K_m), is the concentration of a substrate needed to achieve one-half of the maximal rate. This constant for pyruvate kinase in the stingray (*Urolophis halleri*) is 0.30 mM ADP in the absence of urea (Figure 6.4). In the presence of urea, K_m increases, whereas in the presence of TMAO, K_m decreases, in both cases depending on the concentration of these

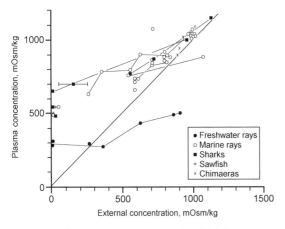

Figure 6.3 Plasma concentration in chondrichthyans as a function of the external concentration. Data obtained on a species in the same study are connected. Sources: Data derived from Smith (1931b), Smith and Smith (1931), Fänge and Fugelli (1962), Price (1967), Price and Creaser (1967), Thorson (1970), Thorson et al. (1967, 1973), de Vlaming and Sage (1973) and Bedford (1983).

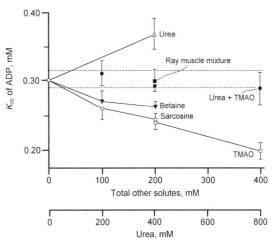

Figure 6.4 Michaelis constant for pyruvate kinase in the presence of various organic osmolytes in the stingray (*Urolophis halleri*) as a function of external concentration or the concentration of urea. The 95% confidence limits of the control value of K_m are enclosed by dashed lines. Source: From Yancey et al. (1982).

molecules. Notice that the molar impact of TMAO on enzyme function is twice that of urea, which accounts for the observation that when urea and TMAO are combined in a 2:1 ratio, K_m remains unchanged from the condition in which both urea and TMAO are absent.

Molecules other than urea and TMAO, including other methylamines, such as betaine and sarcosine, and some free amino acids, affect enzyme function, which often leads these molecules to be accumulated as a quantitative mixture. For example, free amino acids increase enzymatic K_m values, as with TMAO, but they decrease in the presence of arginine. Only proline appears to be a neutral amino acid. The ray muscle mixture indicated in Figure 6.4 contained 400 mM urea + 65 mM TMAO + 55 mM sarcosine + 50 mM β-alanine + 30 mM betaine, which assuming a 2:1 impact is a "balanced" solution: 400 = 2(65 + 55 + 50 + 30). Furthermore, many ions, such as K^+, Na^+, and Cl^-, and the polyhydric alcohols glycerol and sucrose also affect the catalytic rates and K_m of many reactions (Yancey et al. 1982). Therefore, exposure to high salt concentrations has the smallest impact on cellular function either if the cellular concentrations are maintained hyposmotic to the environment or if they contain appropriate organic solutes to balance an increased salt content (also see Ballantyne et al. 1987 and Section 6.8).

This use of a "balanced" set of interacting solutes to preserve enzyme function is most appropriate in organisms exposed to variable environments. Otherwise, species can respond to constant extreme environmental conditions by a massive alteration of protein structure through amino acid substitution. This solution, as found in the highly salt-tolerant bacterium *Halobacterium*, however, produces an obligatory dependence on high environmental salt concentrations: environmental flexibility has been lost (Yancey et al. 1982). A potential exception to this dichotomy is found in the accumulation of urea in some aestivating and hibernating vertebrates, for example, *Protopterus* and *Scaphiopus* (see Section 6.10.5), in which the counteracting solutes apparently do not compensate for the presence of urea, a condition that may aid in reversibly inhibiting some enzymes and facilitate the depression of energy metabolism during torpor (Hand and Somero 1982, Yancey et al. 1982).

Urea retention permits elasmobranchs to maintain a small, positive osmotic differential with the environment and consequently, to draw water into body fluids. As a result, they produce a dilute urine (Table 6.1), even though they live in the ocean. Urine is the pathway by which marine elasmobranchs eliminate excess water and urea, whereas the rectal gland is important for ionic regulation (Burger and Hess 1960, Burger 1962) (Table 6.1). Sharks and rays that feed on teleosts, which are hyposmotic to seawater, have low salt intakes. Thus, their prey pay some of the cost of osmoregulation for elasmobranchs. A few elasmobranchs feed exclusively on marine invertebrates (e.g., Port Jackson shark [*Heterodontus*], whale shark [*Rhincodon*], basking shark [*Cetorhinus*], manta [*Manta*]); they undoubtedly have much greater salt loads because marine invertebrates usually are isosmotic with the ocean.

Chimaeras, or rat-fishes, also are invertebrate-eating chondrichthyans. Little is known of their osmoregulation, except that their body fluids are slightly hyperosmotic (Figure 6.3), but they have a higher plasma salt concentration and a lower plasma concentration of urea and TMAO than sharks and rays (Table 6.1). Whether the higher plasma salt concentrations of chimaeras are associated with smaller, grapelike rectal glands imbedded in the wall of the intestine (Fänge and Fugelli 1962, 1963), a proximate explanation, or with their invertebrate-eating habits, an ultimate explanation, is unknown.

Some sharks and rays enter large rivers, especially in tropical and subtropical latitudes. One species prone to enter freshwater is the bull shark (*Carcharhinus leucas*). This species, or a close relative, is found in the Perak River in Malaysia (Smith and Smith 1931), in Lake Nicaragua (Thorson et al. 1966a), in Lake Izabal in Guatemala (Thorson et al. 1966b), 4000 km up the Amazon River (Thorson 1972), 2800 km up the Mississippi River (at least one record) (Thomerson et al. 1977), and 500 km up the Zambezi River in Africa, among other localities. Note that most of these records are tropical, which may mean that a low water temperature depresses rate of metabolism and the capacity for osmotic regulation, thereby limiting the entrance of *C. leucas* into freshwater. The record for Illinois was in September, when water temperature was estimated to have been 23 to 28°C. The movement of marine teleosts into freshwater is also most prevalent in tropical and subtropical environments (see Section 14.6.1).

Entrance of marine elasmobranchs into freshwater presumably is hindered by the storage of urea and TMAO (which in marine waters comprise up to 35% of total plasma solutes) because these compounds accentuate the osmotic differential. Elasmobranchs in freshwater reduce plasma concentrations of both salt and urea (Figure 6.3, Table 6.1), principally as a result of increased renal urea clearance and excretion (Payan et al. 1973), although a reduction in the rate of urea synthesis also may contribute to the fall in urea concentration (Goldstein and Forster 1971). The decrease in osmotic pressure in freshwater sharks is estimated to vary from 32% (Thorson et al. 1973) to 45% (Smith and Smith 1931) of the values found in marine species. The smaller decrease found by Thorson et al. may reflect movement of their animals between freshwater and salt water, or the Smiths' difference may be exaggerated by comparing different species in the two environments.

The decrease in plasma concentration in elasmobranchs with an entrance into freshwater raises a question of terminology. In a study of the dogfish (*Squalus acanthias*), Bedford (1983) found that plasma concentration fell from 987 to 849 mOsm/kg when the external medium decreased from 950 to 750 mOsm/kg. She concluded that this shark is an osmoconformer, as have Ballantyne and Moon (1986) generally for elasmobranchs. The dogfish and elasmobranchs generally, however, are not osmoconformers in the sense that hagfish are (i.e., plasma concentration equals environmental concentration and $\Delta p/\Delta m = 1.0$), because the change in plasma concentration (Δp) is less than the change in medium concentration (Δm) ($\Delta p/\Delta m = 138 \, mOsm/200 \, mOsm = 0.69$). In fact, the slope of the plasma-environmental concentration curve is concave upward (Figure 6.3), which indicates control in the plasma concentration. In dilute environments the slope of this curve is likely to be even more shallow. As Bedford (1983) noted, most of the decrease in plasma concentration (113/138 mOsm/kg = 82%) in the dogfish was due to a reduction in urea concentration. The response of the plasma to a decrease in the concentration of the medium in this elasmobranch is not osmoconformity, but it does reflect plasma osmotic flexibility.

Because of the presence of urea, freshwater elasmobranchs would be expected to have a copious, dilute urine. Smith (1931b) showed that the ray *Raja* has a marked decrease in urine concentration when it is placed in dilute seawater: in full seawa-

ter the mean concentration of the urine was 995 mOsm/kg, but at an external concentration of 587 mOsm/kg, the urine concentration was 237 mOsm/kg. Unfortunately, no indication of urine volume was given, but Goldstein and Forster (1971) demonstrated in *R. erinacea* that with a 50% reduction in the concentration of seawater, urine volume increased 6.5 times and glomerular filtration increased 4.3 times. This flexibility in renal clearance is important for the entrance of elasmobranchs into freshwater. A marked reduction in the size of the rectal gland also occurs with the entrance of elasmobranchs into freshwater (Oguri 1964). These "freshwater" elasmobranchs really are euryhaline species: they apparently do not reproduce in freshwater and as shown by tagging sharks in Lake Nicaragua (Thorson 1971), move between freshwater and salt water.

Truely freshwater rays belonging to the family Potamotrygonidae live in the drainages of the Amazon, Orinoco, and Paraguay Rivers of South America: they reproduce and spend their entire lives in freshwater. Thorson et al. (1967) showed that Amazonian species of *Potamotrygon* have plasma with an osmolality of approximately 300 mOsm/kg with no accumulation of urea (Table 6.1). Their enzyme systems do not require urea for normal function (Yancey and Somero 1978), unlike marine elasmobranchs. Furthermore, these rays are hyposmotic at high external concentrations (Figure 6.3). *Potamotrygon* does not accumulate urea even at high external concentrations (Thorson 1970). The tolerance of *Potamotrygon* to high external concentrations may reflect age: Thorson had fish (2.2 kg) that tolerated 650 to 1000 mOsm/kg, whereas Griffith et al. (1973) used juveniles (0.2 kg), only one of which could tolerate 650 mOsm/kg. A limited ability of *Potamotrygon* to tolerate high external concentrations is not surprising, given that these rays apparently have no rectal glands (Goldstein and Forster 1971) (see Griffith et al. 1973, pp. 309–310).

In Africa and Asia freshwater rays also exist, but they may be recent immigrants or transients, as judged from their morphology (including *Dasyatis* of the marine family Dasyatidae) and their retention of urea (Thorson and Watson 1975). *Dasyatis sabina*, a species widespread along the Atlantic coast of North America from Chesapeake Bay (Maryland) south to Central America, has a freshwater population that breeds and completes its life cycle in the St. John's River, Florida. When in the river (38 mOsm/kg), it maintains a plasma concentration of 621 mOsm/kg and urea concentration of 196 mM/L, whereas in salt water (ca. 1000 mOsm/kg) it has a plasma concentration of 1034 mOsm/kg and a urea concentration of 395 mM/L (Piermarini and Evans 1998). Even though this population has made some adjustments of life in freshwater, it has not made the complete commitment made by *Potamotrygon*.

The picture of the osmotic regulation of chondrichthyans presented here is spliced together from incomplete measurements made on many species. In fact, no complete study has evaluated the variation in total and ionic plasma concentrations in a single species of euryhaline elasmobranch as a function of the concentration of the external medium, and little is known of the osmotic regulation of chondrichthyans that feed on marine invertebrates (chimaeras and some elasmobranchs).

6.5.3 Osteichthyans. Included within the class of bony fishes is a morphologically and ecologically diverse set of species that is classically divided into four subclasses (Nelson 1994): Dipneusti (lungfishes), Crossopterygii (coelacanths), Brachyopterygii (polypterids), and Actinopterygii, the latter including chondrosteans (sturgeon and paddlefish), holosteans (bowfin and gar), and the array of teleosts.

A general picture of osmotic and ionic regulation in bony fish was developed through the extensive efforts of Homer W. Smith (summarized in 1959). Most freshwater species maintain plasma concentrations near 300 mOsm/kg (Figure 6.5). As a result, these fish gain water through the gills, which in turn is eliminated as a copious, dilute urine produced by high rates of filtration at the glomeruli. Some salt is lost in the urine, but it is replaced from freshwater by active transport at the gills. In marine species the osmotic differential is reversed, the plasma concentration being approximately 350 mOsm/kg (Figure 6.5). Under such conditions water diffuses out of the fish, again mainly through the gills. Water loss in marine teleosts is reduced by increasing the concentration of urine (although it is never more concentrated than the plasma), by resorbing much of the fluid that was filtered, and by reducing the glomerular filtration rate, which is produced in part by reducing the size and number of glomeruli. Water is replaced by drinking seawater and by absorbing water in the gut in association with the active transport of monovalent ions. These salts are returned to the ocean by active secretion at the gills (see, e.g., Karnaky et al. 1976, Karnaky 1980). Divalent ions are eliminated in marine teleosts principally via the gut or kidney (Hickman 1968, Evans 1980). Euryhaline teleosts, however, are not perfect osmoregulators: plasma concentration weakly increases with external concentration because of salt gain, in spite of an increase in body water, as shown by the hematocrit (Naiman et al. 1976). Differences in salinity tolerance between populations within a species are often present but usually are not marked, except as they reflect a slightly lower tolerance to high external concentrations in freshwater populations (Nordlie et al. 1992, Jordan et al. 1993).

Griffith et al. (1973) questioned why marine teleosts have not evolved urea retention in marine

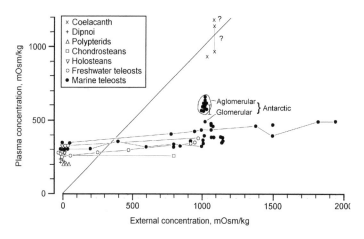

Figure 6.5 Plasma concentration in osteichthyans as a function of the external concentration. Data obtained on a species in the same study are connected. Some data on the coelacanth are questioned because the animals were moribund. Sources: Data derived from Smith (1930), Pickford and Grant (1967), Lutz and Robertson (1971), Potts and Rudy (1972), Griffith et al. (1974), Dobbs and DeVries (1975a), Naiman et al. (1976), Griffith (1981), and Bedford (1983).

waters parallel to the pattern seen in marine elasmobranchs. They suggested that this is not due to the absence of the ornithine-urea cycle, for all of its constituent enzymes have been found in most teleosts examined, even if at low concentrations (Huggins et al. 1969). They argued, somewhat vaguely, that the absence of urea retention may be associated with the presence of variable environmental salinities, which led them to the suggestion that bathypelagic and bathybenthic teleosts facing stable environmental salinities might be prone to use urea as the principal nitrogenous waste product and to store urea (like *Latimeria*). This seems not to be the case (Griffith 1981).

One teleost that stores urea is the tilapia *Oreochromis alcalicus grahami* that lives in Lake Magadi, Kenya. In this lake water has an osmolality equal to $525\,mOsm/kg$, a pH of 10, total CO_2 concentration equal to $180\,mM/L$, and a temperature between 30 and 36.5°C. Nearly all of this tilapia's nitrogenous wastes are excreted as urea, whereas a related species, *O. nilotica*, which lives nearby in freshwater streams and rivers, excretes 85% of its nitrogen as ammonia and 15% as urea (Wood et al. 1989). The Magadi tilapia stores urea in its plasma and body tissues at concentrations up to $17.1\,mM/L$, which however is only 6% to 7% of the value found in elasmobranchs. In contrast, *O. nilotica* maintains a plasma concentration of urea that is only 20% to 25% of that found in the Magadi tilapia and 1% to 2% that of elasmobranchs. Urea synthesis in *O. a. grahami* is accomplished by the ornithine-urea cycle (Randall et al. 1989), whereas *O. nilotica* depends on uricolysis. The difference in nitrogen metabolism between these species appears to reflect the high bicarbonate concentration of Lake Magadi waters, which minimizes the ammonia gradient between the tilapia and water, thereby shifting nitrogen metabolism to the ornithine cycle (Walsh et al. 1990) in spite of the high cost of urea synthesis (i.e., ca. 4–5 moles of ATP/mole of urea). This observation on two closely related species points out the flexibility of nitrogen excretion in teleosts when environmental conditions are appropriate.

Studies on the nitrogen excretion of marine toadfish (*Opsanus*) also have demonstrated the use of ureotelism. Thus, *O. tau* has an active ornithine-urea cycle (Read 1971b), although it usually excretes most of its nitrogenous wastes as ammonia. In contrast, the gulf toadfish (*O. beta*), which also uses the ornithine cycle to synthesize urea, excretes most of its nitrogen as urea (Walsh et al. 1990), depending on environmental conditions. The proportion of nitrogen excreted as urea increases whenever ammonia cannot be excreted (e.g., when ammonia concentrations in the environment are high). In toadfish, this condition occurs when they are trapped in environments in which water flow is restricted, especially during short periods out of water. Under these circumstances toxic ammonia, which cannot be stored, has to be converted into less toxic urea, which can be stored. Under similar conditions *O. tau* may rely on urea synthesis. Other conditions that also may lead to the switch from ammonia to urea excretion include exposure to exogenous ammonia and exposure to alkaline conditions (Mommsen and Walsh 1992).

The excretion of urea may be widespread in fish faced with ambient ammonia, but it is a weak response in rainbow trout (*Oncorhynchus mykiss*) and a strong response in goldfish (*Carassius auratus*) (Olson and Fromm 1971). Available data on the distribution of urea production and the presence of enzymes that constitute the ornithine-urea cycle (Huggins et al. 1969, Mommsen and Walsh 1989) indicate that the production of urea as a nitrogenous waste product by the ornithine cycle is a fundamental characteristic of vertebrates found at least as early as the evolution of Chondrichthyes (and possibly placoderms) and has been selectively lost or suppressed in some derivatives, most notably in some teleosts, most reptiles, and all birds. As Watts and Watts (1966, p. 795) noted, the absence of a functional urea cycle does not necessarily "... mean that the genetic potential for [the] metabolic pathway is also absent"; it may simply reflect "... the physiological demands of the environment. ..."

Another factor associated with the excretion of urea in osteichthyan fishes is the evolution of amphibious behavior. Such fishes include African (*Protopterus* [Brown et al. 1966, Janssens and Cohen 1966]) and probably South American (*Lepidosiren*) lungfish; *Channa gachua* and *Anabas scandens* (Ramaswamy and Reddy 1983); *Heteropneustes fossilis* (Saha and Ratha 1987); Chilean clingfish (*Sicyases sanguineus* [Gordon et al. 1970]); *Periophthalmus* (Gordon et al. 1969, 1978); and *Monopterus cuchia*, *Clarias batrachus*, and *Anabas testudineus* (Saha and Ratha 1989). When exposed to air, most of these species increase the excretion of urea, many by means of the ornithine-urea cycle.

These studies indicate that ureotelism is more widespread among teleosts than often recognized, although Huggins et al. (1969) argued that the fish that produce urea by the ornithine cycle would be better classified into groups based on the fate of urea. They proposed the term *ureogenic* for species that produce urea (in any quantity), which would probably hold for most vertebrates; *ureotelic* for species that eliminate most nitrogenous wastes as urea; and *ureosmotic* for species that retain urea to maintain osmotic equilibrium. Such diversity in nitrogen excretion indicates that the type of nitrogenous compounds used by aquatic vertebrates may be as sensitive to the physical conditions under which they presently live and to their behavioral response to these conditions as they are to the conditions their ancestors faced during earlier periods of evolutionary history. Huggins et al. (1969) maintained that fishes were at least primitively ureogenic, the degree to which the ornithine cycle is used depending on the conditions faced in the environment.

Knowledge of osmotic and ionic regulation in bony fishes is very incomplete because so few species have been studied thoroughly. Some of the known complications are worthy of mention. For example, stenohaline freshwater bony fishes maintain lower plasma concentrations than euryhaline teleosts do in freshwater, and stenohaline marine bony fishes maintain higher concentrations than euryhaline teleosts do in salt water (Figure 6.5). Thus, some antarctic teleosts belonging to the families Nototheniidae, Bathydraconidae, and Zoarchidae have unusually high plasma concentrations (ca. 550–650 mOsm/kg) (Dobbs and DeVries 1975a). These high osmolalities are principally due to sodium and chloride, although up to 10% is produced by the accumulation of organic molecules (Table 6.1). Midwater and benthic marine teleosts may have higher plasma osmolalities than shallow-water teleosts (Griffith 1981), but this difference may have reflected the stress of capture or death in deep-water species when brought to the surface. Freshwater fish other than teleosts, such as polypterids (*Polypterus* and *Erpetoichthys*) and lungfishes (*Neoceratodus*, *Lepidosiren*, and *Protopterus*) have a more dilute plasma (ca. 200–250 mOsm/kg) than do freshwater teleosts (Figure 6.5), although chondrosteans (*Polyodon* and the euryhaline *Acipenser*) have

somewhat higher plasma concentrations (ca. 300 mOsm/kg) (Potts and Rudy 1972).

The only completely marine, nonteleost bony fish, the coelacanth (*Latimeria chalumnae*), has a plasma nearly isosmotic with the ocean (Figure 6.5). The first values, obtained from frozen specimens (Pickford and Grant 1967, Lutz and Robertson 1971), indicated a hyposmotic plasma, but a later measurement taken from a live (but moribund) individual was slightly hyperosmotic (Griffith et al. 1974). All measurements show that the coelacanth stores urea and TMAO in concentrations equal to or greater than those found in marine elasmobranchs (Table 6.1), which means that the coelacanth tolerates internal salt concentrations similar to those of marine elasmobranchs and some antarctic teleosts. *Latimeria chalumnae* synthesizes urea principally with the ornithine cycle. In association with the storage of urea, this species, like elasmobranchs, has a rectal gland.

A particularly interesting group of euryhaline osteichthyans are those that live in one environment and breed in another: they include species that live in salt water and breed in freshwater (anadromous species, including acipenserids and salmonids) and those that live in freshwater and breed in the ocean (catadromous species, including anguillid eels and some mullets). McDowell (1987, 1988), following Myers (1949), defined a third class in which migration between freshwater and salt water is not associated with breeding (amphidromous species, including some galaxiids, southern graylings [Prototroctidae], and several eleotrids and gobiids). He combined these categories under the term *diadromy*. In his survey of the 20,000 species of fish, McDowell estimated that only about 160 species are diadromous, with 86 (54%) being anadromous, 40 (25%) catadromous, and 34 (21%) amphidromous species.

The osmotic regulation of diadromous teleosts has received some attention, although individuals living at sea have been difficult to obtain, so knowledge of these stages is usually poor. The most prominent anadromous teleosts are salmon belonging to the genera *Salmo* (Atlantic salmon) and *Oncorhynchus* (Pacific salmon). Parry (1961) performed one of the most complete studies of salmon, on *S. salar*. She showed that plasma concentration, after hatching, remains rather constant and is slightly higher in salt water than in freshwater (Figure 6.6). A similar pattern seems to apply to other anadromous salmonids, such as *S. trutta*

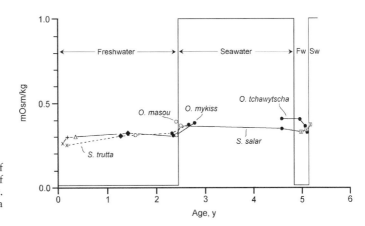

Figure 6.6 Concentration of the plasma of various anadromous salmonid fishes and of their environments through their life cycle. Fw, freshwater; Sw, seawater. Source: Data derived from Parry (1961).

(brown-sea trout), *O. mykiss* (rainbow-steelhead trout), *O. masou* (cherry salmon), and *O. tschawytscha* (Chinook salmon), and to catadromus eels belonging to the genus *Anguilla*. The plasma osmotic concentrations of eels are about 15% higher in salt water than in freshwater (Sharratt et al. 1964).

Diadromous fishes are often referred to as euryhaline in the sense of tolerating a wide range of external salinities (c.f., Evans 1984). However, a distinction can be made between species that can tolerate a wide range in external salinities at any stage in their life history (euryhaline) and those that make such movements only at specific times during their life history (*amphihaline*). McDowell (1987, 1988) believed that many fishes referred to as euryhaline in fact will be recognized to be amphihaline.

The ability of marine or estuarine teleosts to enter freshwater is influenced by the abundance of calcium ions in freshwater, due to the addition of calcium from springs or watershed erosion. Carrier and Evans (1976) showed that the pinfish *Lagodon rhomboides* can tolerate low-salinity water in some freshwater lakes in the Bahamas if calcium is at a sufficiently high concentration, but only if the fish is acclimated to brackish water for several days (Carrier 1974). The propensity of marine teleosts to invade freshwater is most prevalent in tropical and subtropical regions with rather "hard" freshwater (Neill 1957).

6.6 OSMOTIC EXCHANGE IN TELEOSTS

A complex set of parameters regulate net osmotic exchange with the environment. As a first approximation, ignoring salt flux and assuming that mass is constant, water gain must equal water loss:

$$\dot{f}_{\text{drink}} + \dot{f}_{\text{osm}} = \dot{f}_{\text{urine}},$$

where $\dot{f}_{\text{drink}}$ is the rate at which water is swallowed and absorbed in the gut, $\dot{f}_{\text{osm}}$ is the rate at which water is gained (in freshwater) or lost (in the ocean) by osmotic exchange, and $\dot{f}_{\text{urine}}$ is the rate of water loss in the urine (Motais et al. 1969). All three parameters vary with external concentration (Table 6.2): $\dot{f}_{\text{urine}}$ is high in freshwater but not zero in the ocean; $\dot{f}_{\text{drink}}$ is high in the ocean but not zero in freshwater; and f_{osm} is proportional to the osmotic differential established between the plasma and external medium. These data confirm the pattern Smith (1959) described: a teleost in freshwater has a high rate of urine production because of a high rate of osmotic gain of water, whereas marine teleosts have high drinking rates because of the high rate of osmotic loss of water. These fluxes are examined in turn.

6.6.1 Osmotic flux. Evans (1969) showed that the total water fluxes (in this case somewhat differently defined than $\dot{f}_{\text{osm}}$) are proportional to $m^{0.88}$, mainly because the majority of water exchange is across the gills (also see Evans 1980), and gill surface area is proportional to $m^{0.88}$ (see Section 8.4.2). Nagy and Peterson (1988) found a similar pattern between water flux and body mass. At a given mass, water flux is higher in freshwater than in marine species (Figure 6.7). Water fluxes might be expected to be low in freshwater species to reduce salt loss, but the osmotic differential maintained between plasma and the environment in teleosts is 2 to 3 times greater in the ocean than in freshwater. This perspective gives a rationale for the low permeabilities found in marine species. Euryhaline species have water fluxes similar to

Table 6.2 Osmotic exchange in teleosts

Genus	Environment (mOsm/kg)	$\dot{f}_{urine}$	$-$ $\dot{f}_{drink}$ (μL/100 g·h)	$=$ $\dot{f}_{osm}$	$C_p - C_o$ (mOsm/kg)
Carassius	10	1445	51	1394	250
Anguilla	10	538	135	403	255
Anguilla	1150	31	325	−294	−815
Platichthys	10	287	37	250	265
Platichthys	1150	47	192	−145	−825
Serranus	1150	70	277	−207	−820

Source: Data derived from Motais et al. (1969).

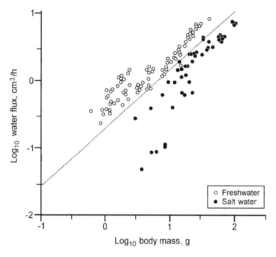

Figure 6.7 Log_{10} water flux in freshwater and saltwater fishes as a function of log_{10} body mass. Source: Modified from Evans (1969).

those of marine species, which further emphasizes the necessity to reduce the cost of osmotic regulation in salt water. Some stenohaline freshwater fishes may be restricted to low external concentrations because they have inflexibly high water permeabilities.

6.6.2 Drinking flux. One of the perplexing features of drinking in fish is that it occurs in freshwater. Drinking increases the water load that must be eliminated by the kidney and therefore contributes to an increased urinary salt loss. The rate for drinking freshwater appears to be greater in euryhaline than in stenohaline species: for example, in goldfish only 4.5% of the volume of the urine comes from drinking (Table 6.2), whereas in the euryhaline flounder and eel, 12.9% and 25.1%, respectively, comes from drinking.

The reason why drinking occurs in freshwater is unclear. It may permit fish to exchange divalent ions with the environment via the intestine (Hickman 1968, Dall and Milward 1969, Evans 1980). The higher drinking rate in euryhaline species may be part of the behavioral repertoire required for the return to salt water. The retention of high drinking rates by euryhaline species, in fact, may be the price paid in freshwater for the ability to return to salt water.

6.6.3 Urinary flux. The rate of urine production in teleosts is high in freshwater and low in the ocean. In freshwater the rate is higher in stenohaline species than it is in euryhaline species (Table 6.2). The question remains, why do marine teleosts produce any urine at all, given that urine production represents a loss of water to a salt-rich environment? The production of urine in a marine environment, like drinking in a freshwater environment, is important in the exchange of solutes, for example, divalent ions (Mg^{2+}, $SO_4^=$) and various organic acids and bases, that otherwise might be difficult to handle (Hickman 1968).

The rate at which urine is produced is determined in part by the number and size of the glomeruli in the kidneys. The product of the number of glomeruli and the area of Bowman's capsule, which can be roughly estimated from the glomerular diameter, is an estimate of the total area available for filtration. This area is proportional to body mass raised to a power of about 0.85; that is, it is proportional to gill surface area and to osmotic flux. In freshwater teleosts filtration area is about 5.2 times that of marine species of the same mass (Figure 6.8).

A difference in renal structure is seen between a flounder (*Pseudopleuronectes*) and a hagfish (*Myxine*) of the same size. These species have

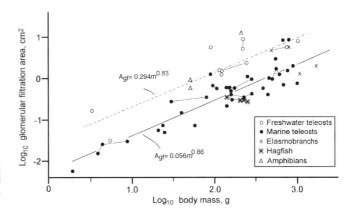

Figure 6.8 Log_{10} glomerular filtration area (A_{gf}) in hagfish, elasmobranchs, teleosts, and amphibians as a function of log_{10} body mass. Source: Data derived from Nash (1931).

similar filtration surface areas ($0.37 cm^2$/kidney in the flounder and $0.10 cm^2$/kidney in the hagfish), but they have very dissimilar numbers of glomeruli and glomerular sizes: 4696 glomeruli/kidney and 50-nm diameter in the flounder, and 40 glomeruli/ kidney and 500-nm diameter in the hagfish (Nash 1931). A particular filtration area can be attained by having either many small or a few large glomeruli. The solution used is perhaps related to the mechanical properties of the circulatory system: hagfishes have very low blood pressures compared to teleosts (Alexander 1975), which again may mean that the hagfish kidney is mainly a secretory organ (Alt et al. 1981). Low blood pressures require a low resistance to blood flow at the glomerulus, which may explain its large size.

The small area available for filtration in marine teleosts mainly results from a smaller glomerular size (diameter ca. 46 nm) than is found in fresh-water species (ca. 84 nm; Nash 1931), although some marine species have kidneys without glomeruli. Aglomerular kidney tubules have a blind end without any Bowman's capsules; urine forma-tion in these species is secreted into the nephrons.

The occurrence of aglomerularity raises several questions. Marine species may have aglomerular kidneys, partly glomerular kidneys, or completely glomerular kidneys, even though all three types of fish can live under similar environmental condi-tions. How is it that such fishes, often closely related, can live side by side? Dobbs and DeVries (1975b) suggested that aglomerular kidneys in some antarctic fish prevent the filtration of the small glycoproteins used as antifreeze (colloidal osmotic pressure; Section 7.10.2). Indeed, aglomerular antarctic fishes have the highest concentrations of organic molecules in plasma,

whereas glomerular species are nearly free of these compounds (Dobbs et al. 1974, Eastman et al. 1979). The presence of antifreeze molecules, however, can explain only some aglomerularity in marine fishes.

An unstudied aspect of aglomerular fishes is that some invade freshwater, especially in the tropics and subtropics. These include pipefish (*Syng-nathus*) and soles (*Trinectes*) in Florida, *Batra-choides* in Central America, and *Thalassophryne* in Amazonia. Aglomerular fish could live in fresh-water if they gained little water, which would be the case if they had (a) small gill surface areas, (b) low permeabilities, or (c) low drinking rates. Pipefish at least have a rigid, scaly coat and very small gill surface areas.

6.7 THE ENERGETICS OF OSMOTIC REGULATION

Unlike the study of temperature regulation, where energetics plays a prominent role (see Chapter 5), it has been nearly neglected in the study of osmotic regulation. This neglect is partly due to difficulties in measuring rate of metabolism in excitable fish and to the belief that the cost of osmotic regulation is so low as to make a minus-cule contribution to energy expenditure. Fish move salt against a concentration gradient in both the ocean and freshwater, which means that most species expend energy to maintain a constant solute composition and concentration in their body fluids.

Potts (1954) described a model of the energetics of osmotic regulation in fresh or brackish water (i.e., when the plasma concentration exceeds that of the external medium). Potts and Parry (1964) repeated the model in a modified form. If an animal is in salt balance, lost salt must be replaced, which therefore permits a *minimal* estimate of the cost of osmotic regulation to be derived from thermodynamics (Equation 2.21). Salt loss can occur either through the body surface or in the urine. The amount of salt lost varies with the various surface areas (A_i), their permeabilities (P_i), the volume of urine (V_u), and the concentrations of salt in the body (B), urine (U), and environment (E). The volume of urine is related to the osmotic differential and to the permeability of the body, a relation that can be expressed as

$$\dot{V}_u = P \cdot A(B - E), \qquad (6.1)$$

so that if many different surface areas exist, each characterized by its own permeability, total $P \cdot A = P_1 A_1 + P_2 A_2 + \ldots + P_n A_n$. Nevertheless, "... the major site of osmotic and ionic permeability is the branchial epithelium..." (Evans 1980, p. 94), whose high $P \cdot A$ is required for gas exchange.

The principal loss of salt in freshwater vertebrates occurs in the urine. The minimal work expended for osmotic regulation in a hyposmotic environment equals the sums expended to absorb salt in the nephron and to replace that lost in the urine. The value of extracting salts from the filtrate is that the filtrate has a higher salt concentration than freshwater, and therefore, less energy need be expended for salt extraction because the filtrate constitutes a smaller differential with the plasma than freshwater. This expenditure is approximated by

$$\dot{W}_n = RT\dot{V}_u U\left[\left(\frac{B - U}{U}\right) \cdot \ln\frac{B}{U}\right], \qquad (6.2)$$

(see Equation 2.21). The amount of energy expended to replace the salt lost in the urine from the environment is given by

$$\dot{W}_u = RT\dot{V}_u U \cdot \ln\left(\frac{B}{E}\right). \qquad (6.3)$$

Total expenditure for osmotic regulation in a hyposmotic environment is given by the sum of Equations 6.2 and 6.3, which with the use of Equation 6.1, gives

$$\dot{W}_t = RTPA(B - E)\left[U \cdot \ln\left(\frac{U}{E}\right) + B - U\right]. \qquad (6.4)$$

The mathematics of osmotic regulation deserves little emphasis because so few data are available, but an examination of Equations 6.2, 6.3, and 6.4 illustrates the consequences of variation in water permeability, plasma concentration, and urine concentration. A reduction in the cost of osmotic regulation can be attained in freshwater by (1) reducing permeability, (2) decreasing the surface area for exchange, (3) decreasing plasma concentration, and (4) reducing urine concentration. These conclusions generally correspond with observations made in euryhaline fish: (1) all euryhaline fishes, including lampreys, elasmobranchs, and teleosts, have lower plasma and urine concentrations in freshwater than in marine water, and (2) freshwater teleosts have smaller gill surface areas than do marine teleosts of the same mass. The cause for the reduction in gill area, however, is complicated because gill area is also correlated with activity (see Section 8.4.2). Milton (1971) suggested that active fish may avoid freshwater (because of a high cost of osmoregulation or because less food is available in freshwater?). Yet, in spite of smaller gill surfaces, freshwater teleosts have larger water fluxes than do marine teleosts (Figure 6.7), which implies that marine teleosts have lower permeabilities than freshwater species.

Potts (1954) estimated that the minimal expenditure for osmotic regulation in two freshwater invertebrates, a crayfish (*Potambius*) and a crab (*Eriocheir*), was less than 1% of the total energy expenditure. These calculations, however, are made under the clearly erroneous assumption that the efficiency of osmotic regulation is 100%. If, for example, the system is 25% efficient, then the cost of osmotic regulation would be 4 times the calculated values, but that still would be only about 4% of the energy expenditure.

Because the *estimated* thermodynamic cost of osmotic regulation is so low, any correlation of oxygen consumption with environmental concentration usually is assumed to represent a change in the activity of an animal or a transitory cellular response (Potts and Parry 1964, Styczynska-Jurewicz 1970). Yet, oxygen consumption in some euryhaline teleosts varies with external concentration even after a period of acclimation. For example, Nordlie and Leffler (1975) acclimated juvenile (16-g) mullet (*Mugil cephalus*) to 20- to 1570-mOsm/kg solutions. Measurements of resting rates of metabolism showed a marked correlation

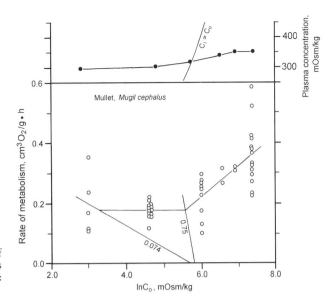

Figure 6.9 Plasma concentration and rate of metabolism in juvenile mullet (*Mugil cephalus*) as a function of ln external concentration. Source: Data derived from Nordlie and Leffler (1975).

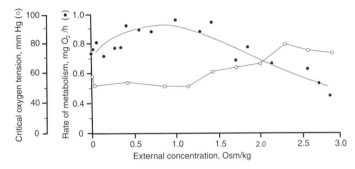

Figure 6.10 Rate of metabolism and critical oxygen tension in the teleost *Cyprinodon variegatus* as a function of the external concentration. Sources: From Nordlie et al. (1991) and Haney and Nordlie (1997).

with solution concentration, except over a range from about 30 to 300 mOsm/kg (Figure 6.9). This range is demonstrated most clearly when the rate of oxygen consumption is plotted against ln E, a tactic that is justified by Equations 2.21, 6.3, and 6.4. The apparent cost of osmotic regulation is minimal in this region, which (by analogy with temperature regulation) might be called the "region of osmoneutrality." Nordlie (1978) showed a similar pattern between rate of oxygen consumption and concentration in the Bornean euryhaline teleost *Ambassis interrupta*, but only at a small mass (1.2 g). At 5.0 g, rate of oxygen consumption was independent of salinity, except possibly in the most dilute solutions.

Teleosts that tolerate hypersaline environments do not conform to the pattern seen in juvenile mullet. For example, *Cyprinodon variegatus*, a euryhaline teleost, had a lower rate of metabolism in an environment of high salt concentrations than

in freshwater (Barton and Barton 1987). Nordlie et al. (1991) found a similar pattern in this species: rate of metabolism was independent of ambient salinities from 5 to 363 mOsm/kg, above which rate of metabolism increased up to salt concentrations between 1000 and 1400 mOsm/kg; at still higher concentrations rate of metabolism fell (Figure 6.10). At these external salinities plasma concentration increased slightly. This species is known to tolerate external concentrations as high as 4100 mOsm/kg, which is about 4 times the concentration of open-sea water! The reduction of its rate of metabolism at high concentrations may have resulted from a reduction in exchange through the gills in the face of a very large negative osmotic differential (Nordlie et al. 1991). Indeed, the critical oxygen tension, below which rate of oxygen consumption declines with oxygen tension (see Section 8.4.6), increased at salinities greater than 1145 mOsm/kg (Figure 6.10). The

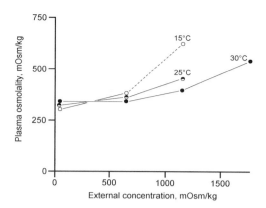

Figure 6.11 Plasma concentration in the desert-spring teleost *Cyprinodon salinus* at various temperatures as a function of external concentration. Source: Modified from Hillyard (1981).

reduction in metabolism reflected a reduced level of activity and an "emaciated" condition when these fish were maintained at external saline concentrations greater than 2030 mOsm/kg (Nordlie and Walsh 1989). The plasma concentration of a freshwater subspecies of *C. variegatus*, *C. v. hubbsi*, was lower in freshwater and higher at external salt concentrations up to twice that of seawater (Jordan et al. 1993).

Temperature also affects osmoregulation. In a pupfish, *C. salinus*, from Death Valley, California, the rate of metabolism increased with salinity, but the effectiveness of osmoregulation is greatest at higher temperatures (Figure 6.11), which may mean that the increase in metabolism required to pay the cost of osmoregulation is only possible when the rate of metabolism is boosted by an increase in temperature (see Section 6.5.2). The arctic char (*Salvelinus alpinus*), a freshwater salmonid, can tolerate external salinities as high as 950 mOsm/kg if water temperataure is 8°C but not if 1°C (Finstad et al. 1989).

These data pose problems for the conclusion of Potts. One is that the higher rates of metabolism at some salinities are not simply due to activity. Nordlie and Leffler (1975) pointed out that activity in the mullet was *diminished* at concentrations that correspond to the highest rates of metabolism. Secondly, the greatest rates of metabolism may occur at the highest external concentrations in accord with the greatest osmotic differentials (Figure 6.9). The low permeabilities found in marine teleosts, as shown by low water fluxes

(Figure 6.7), can be viewed as an adjustment to reduce energy expenditure in the face of a large differential. Another difficulty in applying the model of Potts is that it was derived only for animals that are hyperosmotic to the environment, and therefore is not applicable to marine teleosts, which have a reversed differential.

Kirschner (1993), using another approach to estimating the cost of osmoregulation, summed the amount of urea lost in marine elasmobranchs, multiplied that by the cost of urea synthesis, and added it to the sum of the cost of urea reabsorption in the kidney and the cost of sodium secretion in the rectal gland. He concluded that the cost of osmoregulation in the dogfish (*Scyliorhinus canicula*) is approximately 15% of its standard rate of metabolism. In the skate *Raja erinacea* the cost is 11% to 25% of the presumed standard rate. Marine teleosts (*Salmo* and *Platichthys*) by his estimates spend only 8% to 17% of their standard rates for hyposmotic regulation. These estimates for marine elasmobranchs and marine teleosts are opposite to the conclusion of Griffith and Pang (1979), who estimated that the cost of osmoregulation in marine teleosts is 20 times that of chondrichthyans (and the coelacanth).

Kirschner (1995) estimated that a freshwater salamander (*Necturus*) spends slightly more energy for osmoregulation than a teleost (*Oncorhynchus*), but because the trout has a standard rate that is about 3 times that of salamanders, the trout spends proportionally less (3%) for osmoregulation than the salamander (7%). Most of the cost in salamanders is associated with the absorption of sodium by the kidneys, which reflects the higher urine volume and permeability of salamander integuments, whereas most of the cost in teleosts reflects the absorption of sodium at the gills. Nevertheless, teleosts spend a greater fraction of their expenditures for osmotic regulation in a marine environment, where the osmotic differential is greatest.

Osmotic regulation, like thermal regulation, may be best understood through the study of energetics as a holistic response to the environment. At present, however, so few data on the energetics of osmoregulation exist that the evolution of ionic and osmotic regulation in aquatic vertebrates must be examined without the aid of energetics.

6.8 OSMOTIC REGULATION AND THE ORIGIN OF VERTEBRATES

A difficulty with reconstructing the evolution of osmoregulation in the context of the origin of vertebrates is that the evolutionary relationships existing among the major groups of fish are unclear. Furthermore, controversy exists with respect to the environments faced by fish in the geological record.

The conclusion that the earliest vertebrates evolved in freshwater was long accepted. Chamberlin (1900), arguing against the earlier presumption of a marine origin for vertebrates, maintained that the presence of eurypterids (phylum Crustacea) in deposits with early fishes and the fusiform morphology of early vertebrates suggested a freshwater origin. This view was sustained by two principal lines of evidence:

1. The earliest vertebrates (rather small, heavily armored jawless "fishes" called *ostracoderms*) were found in freshwater deposits, or if the deposits were marine, they were thought to have been fluviatile in estuarine conditions (Barrell 1916, Romer and Grove 1935).
2. All vertebrates have glomeruli, or have secondarily lost them, which was interpreted to mean that the earliest vertebrates lived in freshwater because the principal function of the glomerulus was believed to be to void excess body water (Smith 1932).

These conclusions with modification continue to have support. For example, Kirschner (1967) argued that the (marine) environment in which protochordates gave rise to vertebrates (as represented by hagfish) should be distinguished from the (freshwater) environment in which the predecessors of all other living fish lived, as is suggested by the low plasma and ionic concentrations in all fishes other than hagfish. These progenitors lived so long in freshwater that they became "genetically fixed" on low plasma salt concentrations.

Several paleontologists and physiologists challenged these arguments, thereby reasserting an earlier view that vertebrates originated in marine waters. For example, Denison (1956), Robertson (1957), and White (1958) concluded that most of the sediments that contained the early ostracoderms were marine. This conclusion was based on the occurrence of ostracoderms with eurypterids, which these authors argued were marine or estuarine in distribution, and on geological evidence of the marine nature of the sediments. Denison (1956) comprehensively reviewed the paleoenvironments of various taxa. Later discoveries (e.g., Lehtola 1973) found ostracoderms in strictly marine limestone deposits. Janvier (1985) maintained that osteostracan ostracoderms, at least, included both freshwater and marine species.

The glomerulus may not have evolved as a means of eliminating excess water: hagfish, which are isosmotic with the sea, have well-developed glomeruli. Robertson (1957), Morris (1960, 1965), Munz and McFarland (1964), and Pickering and Morris (1970) maintained that the glomerulus originally involved ionic regulation, only later assuming the role of volume regulation with movement into brackish water and freshwater. For example, Munz and McFarland (1964) presented evidence that hagfish (*Eptatretus stouti*) lacked a renal system to reabsorb sodium, which is fundamental to the invasion of freshwater. These studies further suggested that hagfish, in their marine distribution, ionic regulation, isosmoticity, absence of osmotic regulation, and presence of the glomerulus, represent the earliest condition in vertebrates. Many authors, including Bray (1985) and Halstead (1985), have followed this view.

Vertebrates are generally thought to have been derived from protochordate-like ancestors. All living protochordates are strictly marine, are isosmotic with the sea, and lack a glomerulus. The earliest ostracoderms (†Heterostraci) are found in marine deposits in the Ordovician (Lehtola 1973). Even in the Silurian, ostracoderms belonging to the †Heterostraci, †Thelodontida, and †Galeaspida were limited to marine waters, whereas freshwaters and brackish waters were occupied by the †Osteostraci and the †Anapsida. The earliest ostracoderms probably evolved in marine waters from protochordates, and the glomerulus developed for filtration and ionic regulation, possibly in association with a circulatory system characterized by the higher blood pressures needed for activity in animals having a mass much greater than that of protochordates. Early ostracoderms probably retained the isosmoticity and the absence of osmotic regulation found in protochordates and still present in their hagfish descendents. The movement of some ostracoderms into freshwater in the Silurian necessitated the development of osmotic

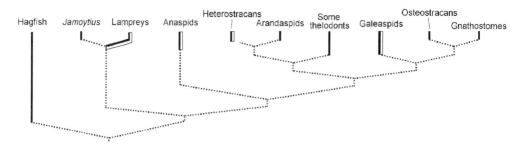

Figure 6.12 The phylogeny of the earliest vertebrates with records, or estimates, of their occurrence in marine water (dark bars) or freshwater (hollow bars). Source: Modified from Forey and Janvier (1994).

regulation, presumably based on a slight redirection of glomerular function (from ionic regulation to a combination of ionic and volumetric regulation) and on the development of the capacity of the kidney to resorb sodium. The numerous anatomical and osmoregulatory differences between lampreys and hagfishes are evidence that they evolved from different ostracoderm lines, a view Stensiö (1927, 1968), Jarvik (1964), and Hoy-Thomas and Miles (1971) advocated, lampreys from freshwater or brackish-water species and hagfishes from marine species.

In a review of early vertebrate evolution, Forey and Janvier (1994) concluded that hagfishes are the most primitive living or extinct vertebrates, that lampreys are more closely related to jawed vertebrates than to hagfishes, and that among ostracoderms, anaspids were the most primitive and osteostracans the most advanced. They also concluded that various groups occupied both marine and brackish-water/freshwater environments (Figure 6.12), but all of the earliest groups either were exclusively marine or had marine representatives. Forey and Janvier further proposed that the low plasma salt concentrations found in most marine vertebrates evolved in brackish environments characterized by highly variable salinities and large food supplies. Ballantyne and Moon (1986) noted that methylamines are common organic solutes (to counteract high salt concentrations) in many marine invertebrates and in hagfishes, but that neither group stores urea. They concluded that methylamine retention preceded urea retention. Indeed some marine teleosts accumulate TMAO without storing urea (Griffith 1981).

The relationships among jawed vertebrates are also uncertain. At present, most evidence suggests that chondrichthyans are related to placoderms, a group of uncertain origin that was mainly marine. The observation that chondrichthyans have a plasma salt concentration only 60% of seawater led Thomson (1971) originally to suggest that elasmobranchs (but not chimaeras [rat-fish]) had a freshwater stage in their evolution. Potts (1985, p. 320) stated, "[T]he presence of ... [a high urea concentration in blood and tissues] ... in the selachians, the coelacanths, and the holocephalians (rat fish) suggests that they too are of fresh water origin." Thomson (1980, p. 196), reversing his earlier position, concluded that "... ionic concentrations were reduced in all lines of early fishes as a result of an adaptation of the earliest vertebrate stock (excluding the craniate hagfishes and their allies as more primitive in this respect) ... and that this occurred when this stock was still living in the sea." Chondrichthyans may have evolved the urea-TMAO system through the addition of urea retention, or this system may have been a "primitive gnathostome character" (Bray 1985, p. 304) found in chondrichthyans, dipnoans, and crossopterygyans—the ancestors of terrestrial vertebrates—and possibly acanthodians and placoderms. Elasmobranchs and chimaeras, thus, may have attained a lower plasma salt concentration by the imposition of rectal glands and the storage of urea in an estuarine or marine environment, the lower plasma salt concentrations facilitating enzymatic function. South American freshwater rays demonstrate the plasticity in the osmotic regulation of elasmobranchs with respect to the environment. Indeed, if the type of osmotic regulation reflects the history of fish evolution, how can the osmoregulation of *Potamotrygon* be explained?

The history of osmotic regulation in bony fishes is equally ambiguous. Osteichthyans may have been derived from the earliest known jawed vertebrates, the acanthodians, which were first found in

BOX 6.1
Is the Evolution of Physiological Characters Conservative?

The assumption repeatedly implied in discussions of the evolution of osmoregulation in vertebrates is that this system is phylogenetically conservative and much of the pattern seen in living vertebrates reflects their evolutionary history, not their contemporary distributions. This conclusion appears in the form that (1) marine elasmobranchs and teleosts have low plasma salt concentrations, so they must have had a freshwater stage in their evolution, or (2) elasmobranchs and *Latimeria* produce urea as the principal waste product, which is evidence that they had a freshwater stage in their evolution. Yet, such a view does not seem (with good reason) to require that *Potamotrygon* was derived from "primitive" freshwater elasmobranchs. Presumably, these rays were derived from marine ancestors, and the absence of urea retention in these species is *adaptive* to the environment (i.e., *not* phylogenetically conservative). No other physiological function is assumed to be as phylogenetically conservative as ionic-osmoregulation. Although a controversy exists on the comparative contributions of the history and ecology of the level of body temperature in endotherms, that system is usually analyzed in terms of the interactions among body size, rate of metabolism, and thermal conductance (see Box 5.1). The view that phylogenetically conservative mammals have low body temperatures and low rates of metabolism simply because of their particular phyletic histories (Martin 1902, Kayser 1961) is now generally discarded. Phyletic history may make a contribution to endothermy, but it is only one among many factors. A more profitable way to examine the diversity in ionic and osmotic regulation in aquatic vertebrates is to combine an analysis of the fossil record (even in its incompleteness) with an inquiry into the parameters that produce the level of regulation and how these parameters vary with conditions in the environment.

Silurian marine deposits. The earliest known bony fish were lungfishes and rhipidistian crossopterygians. They have two characters that must be accounted for: the presence of lungs and the use of urea as the principal nitrogenous waste product. The earliest known fossil lobe-finned fishes, including the earliest coelacanths, the first rhipidistians, and the first lungfishes, were marine (Denison 1956; Thomson 1969, 1971). Urea production and retention may be an adaptation to a marine environment: ". . . the physiological adaptation of urea synthesis and retention, which is common to lobe-finned fishes and tetrapods, is primitively a marine adaptation . . ." (Thomson 1980, p. 196), a view that is compatible with an analysis of the impact of salt and other solutes on enzyme activity (Yancey et al. 1982). This analysis does not account for the development of lungs. Packard (1974) raised the question of whether lungs evolved in vertebrates living in shallow marine lagoons, although this view was challenged by Graham et al. (1978) but sympathetically viewed by Thomson (1980). In any case, as Thomson noted (1969), the comparative importance of lungs in lungfish increased with time, culminating with the present-day lepidosireniids, a South American freshwater family.

Brachyopterygians and actinopterygians probably evolved in freshwater from freshwater acanthodians, or from some ancestor in common with the acanthodians. Little is known of nitrogen metabolism in these species (Huggins et al. 1969). The observation that the bowfin (*Amia*) produces and may even store urea (Matter 1966) can be interpreted as a response to aerial gas exchange, which raises the question of whether other "primitive" actinopterygians (such as sturgeon, paddlefish, and gar) and *Polypterus* store urea in relation to the use of aerial gas exchange. As seen earlier (in Section 6.5.3), many amphibious and air-breathing teleosts increase the use of urea as a nitrogenous waste product during exposure to air.

Because osmoregulation often differs among fish living in the same environment, the evolution of osmoregulation has been assumed to reflect the environmental history of fishes. Thus, Smith (1932, 1959) argued that marine teleosts have plasma salt concentrations that are one-third that of seawater because they are descended from teleosts that evolved in freshwater. In this view, teleosts carried a dependence on a low internal salt concentration to the sea. Griffith persistently maintained that vertebrates originated in freshwater environments (1985), or that the earliest vertebrates had anadromous habits (1987, 1994), which contrasts with his earlier view that ". . . the retention of high levels of urea and, perhaps, TMAO is a primitive feature of jawed vertebrates, secondarily lost in actinopterygian fishes and most tetrapods upon the invasion of freshwater environments, but retained in an unmodified state in those groups such as the elasmobranchs, holocephalians, and coelacanths [that] remained in the sea" (Griffith et al. 1974). He retuned the earlier Romer-Smith view that the earliest fossils were in freshwater and that the glomerulus originated as a volume regulator. He concluded that Na^+/H^+ and Cl^-/HCO_3^- pumps facilitated Na^+ and Cl^- uptake in freshwater, that the absence of osmotic regulation in hagfish is a degenerate character derived from an osmoregulating ancestor, and that chondrichthyans had a freshwater/estuarine stage in their evolution, as did teleosts and tetrapods (Griffith 1994). He dismissed a marine ancestry for vertebrates, arguing that such an explanation requires a belief that the characters were "preadaptations" to a freshwater existence, rather than having these features explained as responses to life in a marine environment (see Box 6.1).

The question remains, however, whether the low plasma concentrations of marine teleosts and the intermediate plasma salt concentrations of marine elasmobranchs might have a functional, rather than a historical, explanation. Marine teleosts clearly do not have low plasma concentrations to reduce energy expenditure, because the most inexpensive regulation would have plasma isosmotic to seawater (Griffith and Pang 1979, but see Kirschner 1993). Thomson (1980), Boucot and Janis (1983), and Bray (1985) variously suggested that the loci of early vertebrate evolution and differentiation were shallow marine, continental platforms. The origin of the glomerular kidney then may have been associated with a control of divalent ions, linked to a change from small, sluggish, benthic, filter-feeding protochordates to larger, active, predatory vertebrates, that is, from species with low to those with high resource requirements. The low plasma ionic concentrations also may have been reduced in a marine environment (Thomson 1980, Ballantyne et al. 1987), possibly in association with an active life-style (see also Bray 1985). The observation that a high cellular salt concentration often disrupts enzymatic function (see

Section 6.5.2) and that this disruption can be cancelled through a reduction in cell salt concentration, or by the substitution of a balanced set of organic solutes, may account for the evolution in a marine/estuarine environment of the chondrichthyan and possibly even the osteichthyan osmoregulatory systems without involving a period of freshwater residency. Furthermore, a reduction in blood solutes, coupled with a reduction in skeletal size and an increase in water content, increases buoyancy in fish (Blaxter et al. 1971), which is most important in midwater fish that lack a swim bladder.

One of the greatest weaknesses in most analyses of the evolution of physiological function is the tendency to ignore the possibility of convergence. For example, Thomson (1969, 1971) and Griffith et al. (1974) raised the question of whether the presence of ureotely indicates a common relationship between elasmobranchs and lungfish or crossopterygians; Bray (1985) thought that the use of urea and TMAO was plesiomorphic to all vertebrates; and Ballantyne and Moon (1986) suggested that the use of methylamines may be plesiomorphic to vertebrates and (some?) invertebrates. Lutz and Robertson (1971) and Pang et al. (1977), however, maintained that the use of ureotely reflects convergence. Recent observations that several unrelated lines of arboreal frogs use uric acid as the principal waste product during dehydration (see Section 6.10.1) emphasizes that the use of nitrogenous waste products is evolutionarily pliable.

6.9 AMPHIBIANS AS AQUATIC VERTEBRATES

With the evolution of amphibians from rhipidistian crossopterygians, vertebrates emerged from water and moved onto land, and thus amphibians have a transitional position between water and land. The impetus for terrestrial locomotion may have been to escape drying ponds in the search for water where an aquatic existence could be continued (Romer 1947). The earliest known tetrapod, †*Ichthyostega*, a labyrinthodont from the Upper Devonian of Greenland, in fact retained a fishlike tail (Jarvik 1952).

All living amphibians are small, most are terrestrial, and most of these return to water to breed. Those that lay their eggs in water have gilled larvae, which remain in water until metamorphosis, when the gills are generally lost. Some amphibians remain in water as adults, and a few salamanders retain gills as adults. Thus, living amphibians as a group occupy a truly transitional position between life in water and life on land. While resident in water, amphibians, as either larvae or adults, face the same problems of water and salt balance as fish.

6.9.1 Osmotic regulation in anurans.

Most amphibians are limited to freshwater during their aquatic phase. This limitation is strictest during early development: for example, abnormal development always occurs in *Rana pipiens* at ambient concentrations greater than 150 mOsm/kg, and often above 100 mOsm/kg (Ruibal 1959). The eggs of *Bufo marinus* can develop normally in water at concentrations to 150 mOsm/kg (Ely 1944). In general, these data mean that the reproduction of amphibians is limited to waters having a concentration equal to, and preferably less than, 10% to 15% that of seawater.

Adult amphibians tolerate higher external concentrations than their developing embryos. Adults in freshwater typically maintain plasma concentrations of 200 to 250 mOsm/kg (Table 6.3), which are somewhat more dilute than is typical of freshwater teleosts. Some of this variation in plasma concentration is interspecific: for example, aquatic species tend to be more dilute than terrestrial species (Schmid 1965a, Mullen and Alvarado 1976). Furthermore, plasma concentration often decreases with acclimation to, or hibernation at, low temperatures (Zamachowski 1977b, MacArthur and Dandy 1982, Jørgensen 1991), but not always (Sinsch 1991). In some anurans a sexual difference in plasma concentration occurs (Zamachowski 1977a), possibly in relation to differential behavior or location of the sexes. With an increase in external concentration to 100 or 150 mOsm/kg plasma concentration shows a small increase (Figure 6.13). Ruibal (1959) observed that *R. pipiens* can tolerate water of 174 mOsm/kg without difficulty for months at a time. Plasma concentration increases markedly at external concentrations greater than 250 mOsm/kg and is lethal at an external concentration of 250 to 280 mOsm/kg. Adult anurans are hyperosmotic to the external medium, so that a lethal external concentration of 275 mOsm/kg corresponds to a lethal plasma concentration of about 300 mOsm/kg.

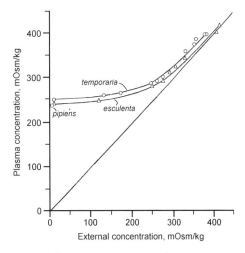

Figure 6.13 Plasma concentration in three *Rana* species as a function of the external concentration. Source: Modified from Adolph (1933).

Some populations of *Rana* tolerate salt water better than others. Neill (1958) pointed out that *Rana* species in coastal regions often have a larger body size, are more pigmented, and tolerate salt water better than those from freshwater populations. For example, *R. sphenocephala* in coastal Florida has been collected in water at a concentration of 432 mOsm/kg (Christman 1974), although that exposure may have been transient. In the laboratory some coastal individuals could live at least 4 d at 370 mOsm/kg and for 29 h at 440 mOsm/kg. Individuals from inland, freshwater populations, however, could only tolerate 370 mOsm/kg for about 14 h and 440 mOsm/kg for 6 h. This difference in tolerance may be related to population differences in permeability and body size, acclimatization, or even genetics. Ruibal (1959) did not find any such differentiation in *R. pipiens*: the exposure to salt water in the populations that he studied was periodic and not chronic, as was faced by coastal populations of *R. sphenocephala*. The ecological consequences to anurans of population differentiation in the tolerance to salt water are not understood.

Some anurans tolerate even higher external concentrations. Ruibal (1962a) showed in the leptodactylid genus *Pleuroderma* that species living in arid or saline habitats in northern Argentina (*P. nebulosa* and *P. tucumana*) can tolerate an external concentration of approximately 350 mOsm/kg,

Table 6.3 Plasma concentrations in amphibians in freshwater

Species	Total (mOsm/kg)	Na	K	Cl	Urea (mM/L)
			(mEq/L)		
Caudata					
Sirenidae					
Siren lacertina	230	119	3.8	78	8
Anura					
Ascaphidae					
Ascaphus truei	172	106		81	
Ranidae					
Pyxicephalus adspersus	243	102	4.4		13
Rana pipiens	193	112		68	
Rana cancrivora	290	125	9	98	40
Hylidae					
Hyla pulchella	221	108	4.3	77	13
Hyla regilla	218	110		78	
Pachymedusa dacnicolor	221	108	5.5	87	8
Agalychnis annae	265	115	5.8	86	15
Phyllomedusa pailona	213	95	4.6	71	9
Phyllomedusa iherengi	208	121	4.7	72	5
Phyllomedusa sauvagei	221	114	2.7	82	4
Bufonidae					
Bufo boreas	235	109		78	
Bufo viridis	247	129	9	86	35

Sources: Derived from Gordon et al. (1961), Gordon (1962), Shoemaker and McClanahan (1975), Mullen and Alvarado (1976), Loveridge and Withers (1981), and Etheridge (1990a).

whereas *P. cinerea*, which lives in freshwater environments, can only tolerate one of 300 mOsm/kg. The tolerance of xeric species is based on an ability to withstand high plasma concentrations (Figure 6.14). Thus, *P. nebulosa* maintains the largest osmotic differential, osmotically draws more water into its body than the other species, and produces the most dilute urine, whereas *P. cinerea*, which is essentially isosmotic with the external medium, produces a urine that is isosmotic with the plasma and medium. *Pleuroderma tucumana* is intermediate.

Amphibians, such as *P. nebulosa*, cannot tolerate high external salt concentrations by maintaining a dilute plasma because they cannot produce a urine more concentrated than the plasma and because they have a large integumental area across which salt and water exchange can occur (as a result of using their skin for gas exchange; see Section 8.6). Furthermore, the tolerance of adult *P. nebulosa* for moderate salinities is not found in its eggs, which do not develop at concentrations above 175 mOsm/kg.

Little is known of the method by which an adult *Pleuroderma* tolerates high internal concentra-

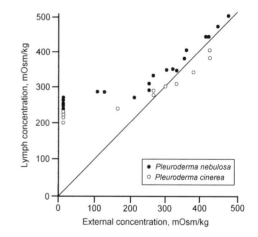

Figure 6.14 Lymph concentration in two anurans, *Pleuroderma nebulosa* and *P. cinerea*, as a function of external concentration. Source: Modified from Ruibal (1962a).

tions; more is known of these mechanisms in other anurans. For example, about 50% of cane toads (*B. marinus*) can tolerate exposure to an external concentration of 420 mOsm/kg for at least 7 d. At that concentration, plasma concentration increases to 425 mOsm/kg, 70% of the increase being due to

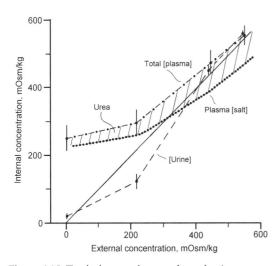

Figure 6.15 Total plasma, plasma salt, and urine concentrations in *Bufo viridis* as a function of the external concentration. The plasma urea concentration is indicated by the shaded area. Source: Modified from Gordon (1962).

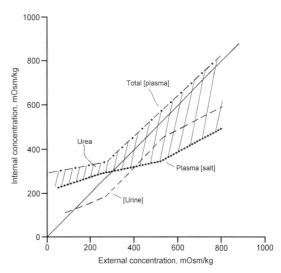

Figure 6.16 Total plasma, plasma salt, and urine concentrations in *Rana cancrivora* as a function of the external concentration. The plasma urea concentration is indicated by the shaded area. Source: Modified from Gordon et al. (1961).

sodium and chloride (Liggins and Grigg 1985). Freshwater populations of the green toad (*B. viridis*) of Europe and the Middle East can withstand 425 mOsm/kg, or after acclimation, 500 mOsm/kg; most individuals from Naples and saltwater Romanian populations, after acclimation, can tolerate external concentrations of 625 to 825 mOsm/kg (Stoicovici and Pora 1951, Gordon 1962). Like the cane toad, *B. viridis* tolerates high plasma salt concentrations: 84% of the increase in plasma concentration after the transfer of a toad from freshwater to an environment of 550 mOsm/kg is due to an increase in sodium and chloride; only 8% of the increase is due to urea (Figure 6.15).

Xenopus laevis, an aquatic frog from Africa, can tolerate external concentrations of 250 mOsm/kg indefinitely, and some individuals can tolerate 400 mOsm/kg for up to 82 h (Munsey 1972a). When it is placed in a 290-mOsm/kg solution, glomerular filtration rate falls to one-third of the rate in freshwater (McBean and Goldstein 1970), and as a result, plasma urea concentration increases 15.2-fold. Thus, in contrast to *B. viridis*, the ability of *X. laevis* to withstand concentrated solutions is related to its tolerance of high plasma urea concentrations.

Rana cancrivora is a euryhaline frog that lives in the mangrove swamps of Southeast Asia and carries urea tolerance to an extreme. All individuals of this species tolerate an external concentration of 525 mOsm/kg, and with acclimation most survive one of 800 mOsm/kg (Gordon et al. 1961), similar to the capacity of *B. viridis*. Unlike the green toad, *R. cancrivora* adults tolerate high external concentrations mainly through the accumulation of urea in the plasma (Figure 6.16): in the transfer from freshwater to an environment of 525 mOsm/kg, 30% of the increase in plasma concentration is due to sodium and chloride, and approximately 70% is due to urea (Gordon et al. 1961). As in *X. laevis*, exposure of *R. cancrivora* to concentrated solutions greatly reduces the glomerular filtration rate, thereby increasing urea storage (Schmidt-Nielsen and Lee 1962).

The capacity of *R. cancrivora* to withstand high external concentrations does not mean that this frog can freely enter seawater. At 800 mOsm/kg (ca. 80% the concentration of seawater), the plasma salt concentration is about 490 mOsm/kg, whereas in freshwater it is about 230 mOsm/kg. In a sense, then, *R. cancrivora* reaches a lethal external concentration because the increase in plasma concentration is not limited to the accumulation of urea. Urea accumulation is not greater because the integumental permeability to urea and urea loss increases at external concentrations greater than 400 mOsm/kg (Figure 6.17). A high urea concentration could be maintained in the face of a high rate of urea loss only if the rate of protein catabolism were high. Rate of metabolism, however, remains constant or decreases at high external concentrations (Figure 6.17).

Unlike *R. pipiens*, the tadpoles of *R. cancrivora* can tolerate higher external concentrations than adult frogs (to 1200 mOsm/kg; Gordon et al. 1961)! This remarkable ability depends on the maintenance of an internal concentration below that of the environment at solutions greater than 300 mOsm/kg (Figure 6.18). Even using hyposmotic regulation, tadpoles face high plasma concentrations: at an external concentration of 1000 mOsm/kg, the plasma salt concentration is 523 mOsm/kg, which is similar to the maximum tolerated by adults. In saline waters, *R. cancrivora* tadpoles osmoregulate like marine teleosts; they produce little urine, presumably swallow water, and excrete salt either at the gills or through the

integument (Gordon and Tucker 1965). This form of regulation depends on the use of ammonia as the principal nitrogenous waste product. Tadpoles are able to continue this regulation until developmental stage XIX, after which plasma concentration gradually accumulates urea (Figure 6.19) as they switch from ammonia to urea excretion. By metamorphosis, osmotic regulation has attained the adult form. Early embryological development and metamorphosis, however, will still not occur at external concentrations greater than 200 mOsm/kg, a limit similar to that found in *R. pipiens*, *B. marinus*, and *P. nebulosa*. The dependence of *R. cancrivora* on dilute solutions for reproduction (presumably occurring during the monsoon) prohibits it from making a complete commitment to a marine existence.

The shift in *R. cancrivora* from hyposmotic to hyperosmotic regulation at metamorphosis, which reduces its tolerance to high external concentrations, is correlated with two changes that occur at metamorphosis, both of which are correlated with the shift to terrestrial life. One, as noted, is the shift from ammonia to urea excretion. The second is the loss of gills, which may reduce the capacity to excrete salt. The use of urea may not only permit but also require urea storage because the kidney is the main pathway by which urea is lost in a terrestrial environment, and with entrance into salt water, anuran renal function is shut down. The integument appears to be quite impermeable to urea, except at high external concentrations (Gordon and Tucker 1968). As in the case of elasmobranchs, low glomerular filtration rates in *R. cancrivora* are coupled with a relatively low permeability to urea, which leads to urea storage.

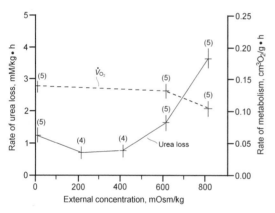

Figure 6.17 Rate of urea loss and rate of metabolism in *Rana cancrivora* as a function of the external concentration. Source: Modified from Gordon and Tucker (1968).

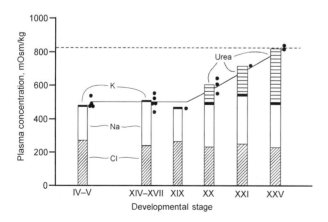

Figure 6.18 Plasma concentration in *Rana cancrivora* tadpoles at various developmental stages while residing in water having a concentration equal to 820 mOsm/kg. Source: Modified from Gordon and Tucker (1965).

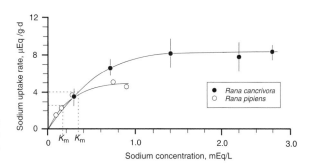

Figure 6.19 Rate of sodium uptake in *Rana pipiens* and *R. cancrivora* as a function of the sodium concentration in the environment. Source: Modified from Greenwald (1971, 1972).

The crab-eating frog has the greatest known capacity of any amphibian to tolerate marine water, but this tolerance is not sufficient for it to be truly marine. This condition has implied that some amphibian "limitations" (e.g., permeability, reproduction) prohibit this frog from making a successful transition to the sea. Another interpretation of this incomplete conversion is possible. In moving from a freshwater or semiterrestrial habitat to an estuarine mangrove environment, a frog moves from an area with adequate or modest food supply to a one with a superabundant food supply (especially in terms of crustaceans; see Elliott and Karunakaran 1974). Movement into a completely marine environment would likely have reduced the amount of available food. *Rana cancrivora* may have made the minimal adjustments required for entrance into a food-rich environment, the physiological remnants of a previous freshwater existence simply being those that are compatible with life in a mangrove swamp, not evidence of physiological or phylogenetic incompetence. A move from an estuarine mangrove habitat to an open-ocean environment may not have been "prohibited" physiologically; it just may not have had an ecological rationale.

6.9.2 Osmotic regulation in other amphibians.
Little is known of osmotic regulation in salamanders. Aquatic species are limited to freshwater. For example, Ireland and Simons (1977) showed that larval *Ambystoma mexicanum* maximally withstand an external concentration of 300 mOsm/kg, 93% of the attendant increase in plasma concentration (from 168 to 305 mOsm/kg) being due to an increase in salt and 7% to urea. With an increase in external concentration, *Dicamptodon* and *Ambystoma* larvae markedly decrease glomerular filtration rate (Stiffler and Alvarado 1974), thereby reducing water loss, although an increase in urinary water resorption also occurs. Differences in glomerular filtration rate between larval salamanders may reflect a difference either in exposed surface areas or in their permeabilities (Stiffler and Alvarado 1980).

Other salamanders tolerate higher concentrations: populations of *Salamandra salamandra* from semiarid environments tolerate plasma concentrations of 500 to 530 mOsm/kg at external concentrations up to 440 mOsm/kg, whereas those from mesic environments only tolerate plasma concentrations to approximately 360 mOsm/kg at environmental concentrations up to 270 mOsm/kg (Degani 1981). Licht et al. (1975), in a study of salamanders belonging to the genus *Batrachoseps*, demonstrated that *B. relictus* were often found under salt-encrusted logs only a few meters from high tide along the California coast. Adults from this setting can tolerate plasma osmolalities at 600 to 700 mOsm/kg for approximately 140 h with few deaths, when ambient osmolality is 490 mOsm/kg, which is similar to the tolerance found in *Bufo viridis*. Although significant variations in salt tolerance occurred among populations in both *Batrachoseps relictus* and *B. attenuatus*, the most tolerant populations of *B. attenuatus* are less tolerant of saline solutions than the least tolerant populations of *B. relictus*. Licht et al. (1975) concluded that *B. relictus*'s tolerance of high saline solutions involved high plasma concentrations that were not based on either salt or urea accumulation. Jones and Hillman (1978), however, concluded that *Batrachoseps* species store urea.

Little is known of osmotic regulation of caecilians (Apoda): they are limited in distribution to terrestrial and freshwater habits. The aquatic caecilian *Typhlonectes compressicauda* has a lower osmotic permeability than the terrestrial caecilian *Ichthyophis kohtaoensis* (Stiffler et al. 1990), which presumably reduces water gain in a freshwater species. (Note: a similar difference is found in anurans; see Section 6.10.3.) As expected, *T. compressicauda* produces a more dilute urine and has less water reabsorption after filtration than is found in *I. kohtaoensis*. Plasma concentration, as is typical of amphibians, is dilute, namely, 196 mOsm/kg in the former and 211 mOsm/kg in the latter species when exposed to freshwater. The response of caecilians to high external concentrations is unknown.

6.9.3 Salt balance in amphibians. Because amphibians are hyperosmotic in all environments, they osmotically gain water when immersed and lose salt in the urine. All lost salt must be replaced if amphibians are to remain in salt balance. Salt replacement requires active transport by enzymatically catalyzed reactions. In Figure 6.19 the rate of sodium uptake is given as a function of the sodium concentration in pond water for *R. cancrivora* and *R. pipiens*. *Rana pipiens* has a "sodium pump" with a higher affinity (affinity being defined as $1/K_m$) than is found in *R. cancrivora*. Greenwald (1972) demonstrated that animals that live in water with a low salt content, irrespective of taxon, have a higher affinity for sodium transport than do species that live in salt-rich environments (Table 6.4). This difference extends even to the small differences in water availability seen among anurans living in mountain streams, for example, *Ascaphus truei*; semiterrestrial species, *R. pipiens*; or terrestrial species like *Hyla regilla* and *Bufo boreas* (Mullen and Alvarado 1976).

6.9.4 Marine amphibians. No contemporary amphibians are truly marine, but they represent a biased sample of all species that have existed. Many labyrinthodonts were large (up to 2 m or more in length) and were partly scaled. One group

of rhachitome amphibians, the long-snouted trematosaurs, are found in Lower Triassic marine sediments (Romer 1966, Olson 1971). These may well have been marine in distribution, where they fed on fish. Some related rhachitome amphibians that lived in freshwater were known to have gilled larvae, an observation that led Romer (1947) to suggest that trematosaurs may have been anadromous, the gilled larvae being restricted to freshwater. Thomson (1980) indeed noted that a diversity of amphibians lived in marine waters during the Devonian. Given that lungfish, the living coelacanth, and most amphibians use urea as the principal nitrogenous waste product, trematosaurs may have lived in estuarine or marine waters in a manner similar to elasmobranchs today but returned to freshwater for reproduction. Whatever the habits of trematosaurs, caution should be used when assuming that the physiological and ecological limitations of living amphibians were shared by all amphibians (for a further exploration of this problem, see Section 8.7).

6.10 AMPHIBIANS AS TERRESTRIAL VERTEBRATES

With the movement of vertebrates from freshwater onto land, excretion and evaporation became important avenues of water loss. Amphibians have no capacity to produce a urine more concentrated than the plasma, so most species dramatically reduce the glomerular filtration rate when they move onto land. Furthermore, living amphibians, unlike other terrestrial vertebrates (see Chapter 7), generally do not have reduced integumental water loss, mainly because they breathe through their skin. Therefore, amphibians are excellent subjects for studying how vertebrates with high rates of water loss adjust to a terrestrial existence.

6.10.1 Nitrogenous waste products. The excretory water loss of terrestrial animals is required by the excretion of excess salts obtained in food and of nitrogenous waste products derived from protein metabolism. The regulation of an amphibian's ionic composition occurs in the kidney. Adult amphibians feed on vertebrates or invertebrates, neither of which have high salt contents. The principal molecules that must be voided in the urine are nitrogenous waste products. With movement onto land, body water must be conserved and conse-

Table 6.4 Sodium affinity of amphibians

Species	Environment	Affinity (mM Na)
Bufo americanus	Terrestrial	>1.0
Rana cancrivora	Fresh to salt water	0.4
Ambystoma gracile	Freshwater larvae	0.3–0.5
Rana pipiens	Freshwater/terrestrial	0.2
Amphiuma meansi	Freshwater	~0.2
Xenopus laevis	Freshwater	<0.05

Source: Derived from Greenwald (1972).

quently nitrogenous waste products often must be stored, which means that toxic products like ammonia cannot be used. The classic picture is that amphibians as aquatic tadpoles excrete ammonia, and with metamorphosis adults switch to urea as the principal waste product, urea being less toxic.

Detailed observations on the nitrogen excretion of amphibians have shown a much more complex pattern (Table 6.5). *Rana*, *Bufo*, and *Hyla* species mainly excrete urea as adults, but terrestrial species even in these genera have higher plasma urea concentrations than do aquatic species (see Schmid 1965a, Rovedatti et al. 1988), and aquatic species generally excrete a greater proportion of nitrogen as ammonia than do terrestrial species (Schmid 1968), although the ratio of ammonia to urea depends on whether the anurans are in water or air (Shoemaker and McClanahan 1980; Table 6.5). Amphibians that are completely aquatic as adults mainly excrete ammonia: for example, Amazonian *Pipa pipa* adults excrete almost 93% of nitrogen as ammonia, and the neotenic larvae of *Ambystoma mexicanum* excrete approximately 62% of nitrogen as ammonia (Cragg et al. 1961). Unfortunately, the urine of few salamanders has been analyzed: in a study of nitrogen excretion in large aquatic salamanders, Suhr (1976) showed that *Necturus* apparently is ammonotelic, that *Cryptobranchus* is normally ammonotelic but becomes ureotelic when exposed to high salinities, and that *Amphiuma* and *Siren* are ureotelic under most external conditions. Aestivating *S. lacertina* stores urea in its plasma (Etheridge 1990a).

The shift from ammonia to urea synthesis can be accelerated as well as delayed or postponed, especially in association with a shift in the mode of reproduction. The larvae of *Leptodactylus bufonius* excrete about 85% of the nitrogen as urea during the first 8 d of development (Shoemaker and McClanahan 1973), a time when most amphibian larvae excrete ammonia. This frog nests in mud burrows, where the only moisture is provided by a proteinaceous foam deposited and whipped by amplexing parents. The restricted water supply available to these larvae places a premium on the early conversion to urea synthesis. Spadefoot toads (*Scaphiopus*), which often breed in ephemeral ponds and have a greatly shortened period of larval development, shift to urea synthesis before metamorphosis (Jones 1980a)—a behavior that enables these toads to survive out of water. Similarly, the pouch embryos of the marsupial frog (*Gastrotheca*

riobambae) from South America excrete urea (Alcocer et al. 1992). The urodele *Salamandra salamandra* is ovoviviparous, the gestation period lasting 12 to 14 months, a period when the water available to the developing larvae is limited. During uterine development the larvae are predominantly ureotelic (80%), whereas the ensuing free-living aquatic larvae produce at least 50% ammonia, only to revert to ureotelism (67%) at the time of metamorphosis (Schindelmeiser and Greven 1981).

The extreme modification of nitrogen metabolism is shown in some arboreal frogs in xeric environments: they excrete more than half of their nitrogen as urate (Table 6.5). These frogs belong to the families Rhacophoridae (*Chiromantis* [Loveridge 1970, Drewes et al. 1977]) in Africa and Hylidae (*Phyllomedusa* [Shoemaker et al. 1972, Shoemaker and McClanahan 1975]) in South America. Because of the low solubility and low toxicity of uric acid (see Section 7.6.2), the use of uric acid permits a reduction in water turnover, but as we will see (in Section 6.10.4), the consequent water parsimony has significance only if water loss by other pathways, especially evaporation, is also reduced.

The data on *Leptodactylus*, *Gastrotheca*, *Scaphiopus*, *Salamandra*, *Phyllomedusa*, and *Chiromantis* suggest that much remains to be learned about the ecological and phylogenetic variation in nitrogen metabolism in amphibians: specifically, an examination should be made of nitrogen excretion in Australian anurans, some of which may be shown to excrete uric acid. At present, nitrogen metabolism appears to be readily modified to meet the exigencies of the environment independent of phylogeny.

6.10.2 Tolerance to dessication. A problem faced by amphibians in a terrestrial environment, given their high rates of water loss, is dehydration. The most widespread response of terrestrial amphibians is, simply, tolerance. Some anurans can tolerate a loss of nearly 50% of the original body mass and 60% of the total water content. The ability to tolerate water loss is greatest in terrestrial and fossorial species and least in aquatic species (Thorson and Svihla 1943). Schmid (1965a) and Zamachowski (1977c) confirmed this conclusion for anurans and Littleford et al. (1947) and Ray (1958), for salamanders, although salamanders generally have less tolerance for water loss than anurans. Furthermore, small species tolerate a

Table 6.5 Nitrogenous waste products in amphibians

Species	NH$_3$ (%)	Urea (%)	Uric acid (%)
Caudata			
Salamandridae			
Salamandra salamandra	4.4	81.9	
Triturus cristatus	3.3	79.6	0.0
Ambystomidae			
Ambystoma mexicanum	42.5	28.3	
Anura			
Pipidae			
Pipa pipa	92.5	7.5	0.0
Xenopus laevis	62.2	37.8	0.0
Ranidae			
Rana blythi			
Water	21	79	0.0
Air	24	76	0.0
Rana esculenta	7.1	76.9	0.0
Rana kuhli			
Water	65	35	0.0
Air	54	46	0.0
Rana limnochaeris			
Water	51	49	0.0
Air	42	58	0.0
Rana temporaria	7.4	78.1	0.0
Hylidae			
Agalychnis annae	4.7	88.5	7.1
Hyla arborea	4.1	86.0	0.1
Hyla pulchella	5.9	94.1	0.0
Hyla regilla	1.1	98.6	0.3
Pachymedusa dacnicolor	4.8	88.8	6.8
Phyllomedusa iherengi	7.3	37.2	55.6
Phyllomedusa pailona	7.1	39.7	52.9
Phyllomedusa sauvagei	4.8	10.7	84.5
Leptodactylidae			
Leptodactylus ocellatus	8.9	91.1	0.1
Bufonidae			
Bufo arenarum	4.3	95.3	0.1
Bufo arunco	5.0	95.1	
Bufo bufo	5.3	90.4	
Bufo calamita	5.9	86.7	0.5
Bufo fernandezae	4.6	95.3	0.1
Bufo quadriporcatus			
Water	91	9	0.0
Air	43	57	0.0
Rhacophoridae			
Chiromantis xerampelina	1–8	20–35	60–75
Chiromantis petersi	2.0	1.0	97.0
Rhacophorus leucomystax			
Water	37	63	0.0
Air	17	83	0.0
Hyperoliidae			
Hyperolius nasutus	3.6	96.3	0.1

Sources: Data derived from Cragg et al. (1961), Shoemaker et al. (1972), Shoemaker and McClanahan (1980), and Rovedatti et al. (1988).

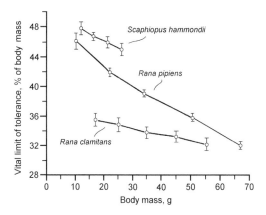

Figure 6.20 Vital limit of tolerance to water loss in three anurans as a function of body mass. Source: Modified from Thorson (1955).

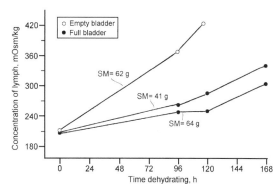

Figure 6.21 Change in the concentration of lymph in dehydrating toads, *Bufo cognatus*, depending on whether their bladders were empty or filled. Standard mass (SM), that is, hydrated mass with an empty bladder, for each toad is indicated. Source: Modified from Ruibal (1962b).

proportionally larger water loss than do large species (Figure 6.20). Yet, even though small individuals start out with proportionally more water and tolerate a greater proportional loss of water, small individuals lose water so rapidly that they approach death by dehydration sooner than large individuals do (see Geise and Linsenmair 1988).

The complex nature of tolerance to dehydration is demonstrated in frogs belonging to the genus *Eleutherodactylus* (Beuchat et al. 1984). Of three Puerto Rican species, the species restricted to the hot, dry lowlands, *E. antillensis*, tolerated more dehydration at high temperatures, as reflected by jump distance, than did species restricted to high, cool altitudes or species found at both high and low altitudes. (A similar pattern has been seen in *Bufo americanus* as a function of their state of hydration [Preest and Pough 1989].) At a water loss equal to 25% of the hydrated mass (with an empty bladder), jumping distance decreased in all three species. The widely distributed species *E. coqui* selects microenvironments that buffer it from high rates of water loss and thus, permit it to be a habitat generalist.

Water storage by an amphibian delays an approach to its dessication limit. Water is stored in the plasma, as shown by the low plasma concentrations found in amphibians (i.e., down to 200 mOsm/kg). Another water reservoir is urine stored in the bladder. Steen (1929) originally showed that water can be reabsorbed from an amphibian's bladder. Ruibal (1962b) subsequently demonstrated in *Bufo cognatus* that this reabsorbed water can replace water lost by evapora-

tion, as did Shoemaker and Bickler (1979) in the waterproof frog *Phyllomedusa sauvagei*. Thus, a hydrated toad with an empty bladder, which is defined as the "standard" mass, has a more rapid increase in lymph concentration than a toad of the same standard mass with a full bladder (Figure 6.21). The use of bladder urine as a water store may be the principal reason why no amphibian is known to be aglomerular, although two species of the central Australian frog genus *Cyclorana* have a reduced number and size of glomeruli (Dawson 1951). Other water stores may also be used: *Cyclorana platycephalus* stores fluid in the "... subcutaneous tissues and peritoneal cavity; when it is fully charged with water [the frog] is almost spherical ..." (Buxton 1923). Australian aborigines used *Cyclorana* as a source of drinking water.

The amount of water available in the bladder for delaying death by desiccation depends in part on the concentration of the urine, as can be demonstrated in the three species of *Pleuroderma* studied by Ruibal (1962a). At an external concentration of about 300 mOsm/kg, *P. nebulosa* has a urine concentration of 200 mOsm/kg; if the lethal plasma concentration is about 450 mOsm/kg, approximately 56% (= [(450 − 200)/450]·100) of bladder water is available for replacing the water lost by evaporation. If *P. tucumana* also has a lethal plasma concentration of 450 mOsm/kg, 42% (= [(450 − 260)/450]·100) of its bladder water is available because of the higher urine concentration (260 mOsm/kg). *Pleuroderma cinerea* has a lethal plasma concentration equal to about 300 mOsm/kg and produces a still more concentrated urine (ca. 290 mOsm/kg); consequently, only 3% (= [(300–290)/300]·100) of the water stored in the bladder is available to replace the water lost by evaporation. These calculations demonstrate the value of producing a dilute urine and the significance of tolerance to high plasma concentrations for life in a desiccating environment.

Shoemaker (1964) extended these observations quantitatively by analyzing the increase in plasma sodium concentration with water loss. He reasoned that because the original concentration of a fluid, C_o, equals $K/[H_2O]_o$, where K is the amount of solutes (in this case sodium) and $[H_2O]_o$ is the original amount of water, then $C_o[H_2O]_o = K$ and

$$C_o[H_2O]_o = C_t[H_2O]_t = C_t([H_2O]_o - d)$$

for any time t, assuming no loss of solutes and that $[H_2O]_t$ equals $[H_2O]_o$ minus the water lost (d, or deficit). Consequently,

$$1 - \left(\frac{C_o}{C_t}\right) = \left[\frac{d}{[H_2O]_o}\right] = \left[\frac{1}{[H_2O]_o}\right]d.$$

When $1 - (C_o/C_t)$ is plotted as a function of d (see Figure 6.23), a linear curve is obtained, the slope of which is $1/[H_2O]_o$. Thus, the original water content of an anuran can be estimated without killing the animal and the curve is displayed along the d axis depending on the amount of usable urine stored in the bladder (Figure 6.22).

The amount of bladder water that can be used to replace water lost by evaporation depends on urine volume, as well as urine concentration, which is why bladder capacity is greatest in terrestrial amphibians. For example, bladder capacity in some Australian anurans, such as *Notaden*, *Neobatrachus*, and *Cyclorana*, can be as much as 50% of the standard mass, whereas the completely aquatic African toad, *Xenopus*, has a bladder volume that maximally holds only about 1% of the standard mass (Bentley 1966). The proportion of bladder water available in *Pleuroderma* can be converted to amount of water if the volume is known; this consideration may further emphasize the difference in tolerance to dessication between *P. nebulosa* and *P. cinerea*. Terrestrial salamanders have larger bladders than aquatic species (Spight 1967), but they are smaller than those of anurans with similar habits.

The ability of bladder urine to extend survival times of amphibians in terrestrial environments raises the question as to when the bladder appeared in the evolution of amphibians from bony fish. No fish have a "true" (mesodermal) urinary bladder, although many teleosts have a slight enlargement of the posterior end of the fused ureters; half of filtered water is reabsorbed in these teleosts, most in the "bladder." If the earliest labyrinthodonts were almost exclusively aquatic, as indicated by the finned tail of †*Ichthyostega* (Jarvik 1952), they were probably also bladderless. The development of a bladder, however, must have been critical to the invasion of land because it is found not only in living amphibians but also in reptiles, which were derived from amphibians that are only distantly related to those that gave rise to anurans and caudates.

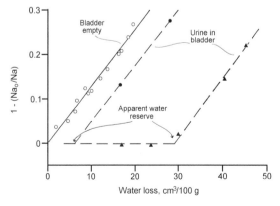

Figure 6.22 The effect of dehydration on the concentration of sodium in the plasma of the toad *Bufo marinus* depending on whether its bladder was empty or contained urine. Source: Modified from Shoemaker (1964).

6.10.3 Rates of water exchange. Amphibians generally have very high rates of evaporative water loss: for example, most frogs lose water by evaporation at rates that are 40 to 50 times those of lizards having the same mass (Thorson 1955, Claussen 1969, Wygoda 1984), which suggests that the variation in rate found among most amphibians is minuscule, at least compared to the rates found in lizards. Amphibians still might be expected to lose and gain water at rates that reflect their habits and the habitats in which they live, but the integument in amphibians must be kept moist both for its protection and for its use of gas exchange. Some terrestrial amphibians even increase integumental water loss in an apparent attempt to control body temperature when basking, especially at high air temperatures or in the face of high radiant temperatures. The water lost through this behavior can be derived from physical contact with the soil and transported to the dorsal skin by capillarity in channels on a scuptured integument, or it can be supplied by discharges of dilute secretions from mucous glands. Capillarity is found in bufonids and secretion occurs in terrestrial species with a smooth skin independent of phylogeny (Lillywhite 1970, 1971; Lillywhite and Licht 1974, 1975). The use by bufonids of a scuptured integument means that most of the water used for behavioral temperature regulation is derived directly from sources external to the body, a behavior that may be required in species having a xeric distribution. In the species that use mucus, the frequency of discharge increases with body temperature (Lillywhite 1970, Lillywhite and Licht 1975). This mechanism is used by the "waterproof" frog *Chiromantis* (see later) to regulate body temperature at high environmental temperatures and radiant heat loads through controlling the rates of evaporative water loss (Shoemaker et al. 1987). Such a response to high ambient temperatures is similar to the use of sweating by mammals. In contrast, *Phyllomedusa* species at ambient temperatures greater than 38°C open and protrude their eyes and then close and retract their eyes. This behavior accelerates the evaporation of water, cooling the surface of their eyes by 2 to 4°C and their adjacent brain by 0.5 to 1.0°C (Shoemaker and Sigurdson 1989).

Many amphibians, especially those that are nocturnal, fossorial, or aquatic, do not regulate body temperature to any appreciable extent. Their rates of water loss usually are dictated by the environmental conditions under which they live. If dehydrated, amphibians must replace lost water. The minimal rate of water loss and the maximal rate of water gain have been suggested to be adaptive to the environment. Almost every possible correlation of these rates with the environment has been claimed:

1. The rate of evaporative water loss is greatest in species or populations from aquatic or humid environments (Cohen 1952, Warburg 1965, Gehlbach et al. 1969); the rate is greatest in species from dry environments (Dumas 1966); or the rate has no correlation with the environment (Littleford et al. 1947, Thorson 1955, Ray 1958, Schmid

1965a, Spight 1968, Claussen 1969, Heatwole et al. 1969).

2. The rate of rehydration is greatest in terrestrial or arid-dwelling species (Cohen 1952; Bentley et al. 1958; Main and Bentley 1964; Schmid 1965a; Warburg 1965, 1971; Dumas 1966; Spight 1967; Christensen 1974; van Berkem et al. 1982), or has no correlation with the environment (Thorson 1955, Bentley et al. 1958, Gehlbach et al. 1969, Claussen 1969, McClanahan and Baldwin 1969, Fair 1970).

3. Skin structure does (Cohen 1952, Ray 1958, Schmid and Barden 1965, Warburg 1965, McClanahan and Baldwin 1969) or does not (Littleford et al. 1947, Thorson 1955, Heatwole et al. 1969) influence the rates of water loss or gain.

This confusion results from variation in techniques (see Warburg 1971), species, and environments.

A recourse in this confusion, other than collecting additional data, is to rely on the comparative studies that seem to be most complete. For example, Wygoda (1984) showed the importance of body mass to rate of evaporative water loss (Figure 6.23):

$$\dot{E} = 0.177\,m^{0.559}, \tag{6.5}$$

where $\dot{E}$ is the rate of water evaporation (cm³/h) and mass is in grams. Large individuals have higher total rates of water loss than small individuals, but on a mass-specific basis, small individuals have a higher rate of water loss. Thorson (1955), Ray (1958), Schmid (1965a), and Claussen (1969) showed similar results. The conclusion that evaporative water loss is not correlated with mass (e.g., Cohen 1952) stems from a small range in body mass.

Some studies of water exchange in amphibians (Schmid 1965a, Spight 1967, 1968) concluded that rate of rehydration is correlated with the environment, whereas no correlation is found between rate of water loss and the environment. This dilemma may be resolved by the observations (McClanahan and Baldwin 1969, Roth 1973, Christensen 1974, Drewes et al. 1977, Geise and Linsenmair 1986) that the ventral pelvic integument is highly specialized for water uptake. The differential development and exposure of this area may account for some of the differences between the rates of water gain and loss heretofore described. Sherman and

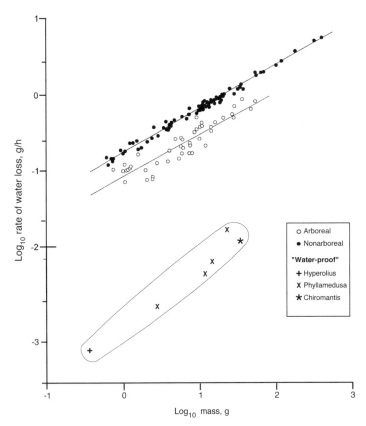

Figure 6.23 $\log_{10}$ rate of cutaneous evaporative water loss in arboreal and terrestrial frogs as a function of $\log_{10}$ body mass. Source: Modified from Shoemaker and McClanahan (1975), Withers et al. (1982), Wygoda (1984), and Stinner and Shoemaker (1987).

Stadlen (1986) suggested another complication: the rate of rehydration is more rapid in lungless than in lunged salamanders.

The Puerto Rican anuran *Eleutherodactylus antillensis* lives in dry, tropical lowlands and rehydrates more rapidly than another species, *E. coqui*, which lives in a variety of habitats (van Berkum et al. 1982). Rates of rehydration in *E. coqui* collected in the lowlands and mountains showed little difference. One of the principal responses of this species to a dry atmosphere is to retreat during the daytime to moist microclimates, few of which are apparently present in the dry lowlands, where the normally solitary coqui is forced to use communal roosts.

A study by Schmid (1965a) deserves special attention because of the ecological diversity of anurans studied. He showed that the greatest rate of water uptake after dehydration was found in terrestrial anurans, mainly because aquatic species have lower integumental permeabilities (also see Mullen and Alvarado 1976, Zamachowski 1977b). A similar difference was found between aquatic and terrestrial caecilians (Stiffler et al. 1990). Schmid and Barden (1965) demonstrated that the

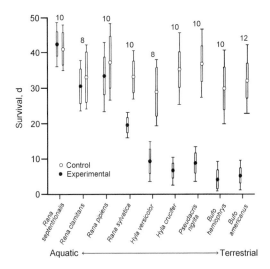

Figure 6.24 Survivorship of a series of anurans depending on whether they sat in distilled water that was 6 mm deep (control) or in distilled water that covered the entire anuran, except for the top of its head (experimental). Means, confidence, intervals, ranges, and sample sizes are indicated. Source: Modified from Schmid (1965a).

low permeability of aquatic anurans results from a high integumental lipid content. A complex seasonal difference in skin permeability to water has been shown in several European anurans, such that it is lowest during hibernation and during the breeding season (when the adults are in water), except in *Bombina bombina*, which spends most of its life in water and has a continuously low skin permeability (Zamachowski 1977b). These data suggest that plasma dilution in anurans sitting in a pond is a greater threat than plasma concentration in a terrestrial environment. Thus, Schmid (1965a) showed that anuran survivorship in distilled water is reduced in terrestrial species (Figure 6.24). From this view, then, anurans are truly terrestrial vertebrates that must reduce integumental permeability to remain in water for extended periods.

6.10.4 "Waterproof" frogs.

Some amphibians have unusually low rates of evaporative water loss. Wygoda (1984) demonstrated that arboreal frogs belonging to the genera *Hyla* and *Osteopilus* have rates of evaporative water loss that are about one-half those of terrestrial or aquatic frogs (Figure 6.23). He expressed the rates of evaporative water loss as a series of resistances to water loss, including a skin resistance (r_s) and a boundary layer resistance (r_b): $1/r = (1/r_s) + (1/r_b)$. Most of the difference in evaporative water loss between terrestrial and aquatic species was due to the resistance of skin to water transfer. Wygoda (1988) later showed that skin resistance increases in *Hyla* when exposed to low vapor densities, whereas *Bufo* has little skin resistance to water loss at all vapor densities.

Loveridge (1970), Drewes et al. (1977), and Withers et al. (1984) described an even greater reduction in the rate of evaporation, down to those typical of lizards, for *Chiromantis*; Shoemaker et al. (1972), Shoemaker and McClanahan (1975), and Withers et al. (1984), for *Phyllomedusa*; Loveridge (1976) and Withers et al. (1982a, 1982b, 1984), for *Hyperolius*; Withers et al. (1984), for *Litoria*; and Lillywhite et al. (1997), for *Polypedates* (Figure 6.23). For example, water evaporates in *Phyllomedusa* at only 1.0% to 2.1% of the values expected from Equation 6.5, whereas in its relatives *Pachymedusa* and *Agalychnis*, both of which live in more mesic environments, water evaporates at 20.6% and 25.7% of the expected rates, respectively (i.e., at rates that are one-half the rates of "typical" tree frogs). Low rates of evaporative water loss have evolved, apparently inde-

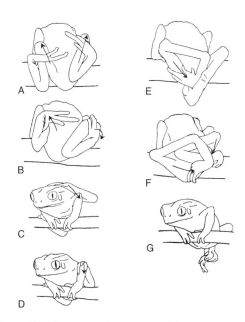

Figure. 6.25 Sequence of postures and leg movements in *Phyllomedusa sauvagei* when spreading lipids over its integument preparatory to assuming a posture that minimizes evaporative water loss. Source: From Blaylock et al. (1976).

pendently, in Africa (twice: *Chiromantis* and *Hyperolius*), South America (*Phyllomedusa*), Australia (*Litoria*), and India (*Polypedates*). The convergent nature of this evolution is shown in part by its taxonomic distribution in at least three families (Rhacophoridae, Hyperoliidae, and Hylidae) and in part by the different means used to reduce integumental water loss.

The very low rate of evaporation in *Phyllomedusa* is produced by a thin lipid layer (mainly of triglycerides), derived from integumental glands and applied over the skin by means of an elaborate pattern of wiping (Figure 6.25). The layer's effectiveness requires that the animal remain inactive so that the "seal" will not be broken. If disturbed, *Phyllomedusa* must rewipe its integument. A similar behavior is seen in the Indian rhacophorid *Polypedates maculatus* (Lillywhite et al. 1997). The anatomical basis for the low permeability in *Chiromantis* and *Hyperolius* is uncertain (Drewes et al. 1977; Withers et al. 1982b, 1984); these genera do not show the elaborate wiping behavior of *Phyllomedusa* and *Polypedates*. Withers et al. (1984) suggested that some of the skin's resistance to water loss involves a lipid barrier in *Hyperolius* and *Chiromantis*, whereas the barrier in *Hyperolius* may be due to a combination of dry mucus and the presence of iridophores in the skin (Geise and

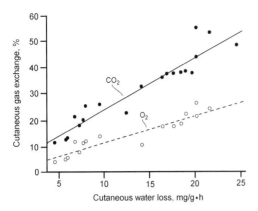

Figure 6.26 Cutaneous gas exchange of CO_2 and O_2 in *Phyllomedusa sauvagei* as a function of the rate of cutaneous water loss. Source: From Stinner and Shoemaker (1987).

Linsenmair 1986, Kobelt and Linsenmair 1986). The iridophores are integumental structures that contain purine (principally guanine) platelets that reflect a substantial proportion of incident solar radiation.

Integumental gas exchange is correlated with cutaneous evaporative water loss (Figure 6.26): at the lowest rates of cutaneous water loss, only 4% of total CO_2 and 2% of oxygen exchange occurred through the skin in *Phyllomedusa sauvagei* (Stinner and Shoemaker 1987). Under such conditions nearly all gas exchange is pulmonary (Geise and Linsenmair 1986). When faced with ambient temperatures above 39°C, *Chiromantis* increases evaporative water loss by increasing the discharge of cutaneous mucous glands, thereby maintaining body temperature 2 to 4°C lower (Kaul and Shoemaker 1989). In fact, the ability of *Chiromantis* and *Phyllomedusa* to maintain body temperature below high ambient temperatures depends on their bladder reserves (Shoemaker et al. 1989). Frogs without bladder reserves and without access to water move to cooler ambient temperatures. *Chiromantis* further facilitates the regulation of a low body temperature by increasing the reflective properties of its integument to short-wave radiation.

Phyllomedusa and *Chiromantis* can be considered "ideal" frogs from the standpoint of water conservation: not only do they have low rates of water loss, but also they hydrate very rapidly, which in both genera involves the ventral abdominal skin. In the daytime they perch in such a manner as to shield the ventral abdomen to reduce evaporative water loss. Furthermore, these genera, as noted, excrete urate as their principal nitrogenous waste product. Uricotelism undoubtedly requires a low evaporative water loss, because, as Blaylock et al. (1976, p. 294) pointed out, "[t]he amount of water conserved through uricotelism would in fact be of little significance to amphibians with a freely evaporative skin [because] evaporative water loss would far exceed the savings yielded by uricotelism."

The use of uric acid in *Phyllomedusa* is connected with its renal function: when out of water, glomerular filtration rate decreases, but even after 27 d filtration rate remains at 29% of the rate when *Phyllomedusa* sits in water (Shoemaker and Bickler 1979). This situation contrasts with what is found in "typical" anurans: for example, *Bufo boreas* within 12 h of being out of water shows a reduction in glomerular filtration rate to only 3% of the rate in freshwater. This difference is associated with the use of uric acid by *Phyllomedusa*, presumably because of its low solubility (which as in reptiles is deposited in the bladder).

Shoemaker and McClanahan (1975) illustrated the impact on a water budget of combining a low rate of evaporative water loss with the use of uric acid in *Ph. sauvagei*. Given its protein intake, use of urea to excrete nitrogen would require about 60 cm³H₂O/kg·d, which far exceeds the 8.6 cm³H₂O/kg·d that the frog obtains preformed in an insectivorous diet and from aerobic metabolism. Approximately 3.8 cm³H₂O/kg·d is needed to excrete uric acid, which if coupled with 0.5 cm³H₂O/kg·d lost in the feces, means that 4.3 cm³H₂O/kg·d is available for evaporation, or at least one-third of the amount of water needed, assuming the low rates of evaporation typical of *Ph. sauvagei*. Its water budget might be balanced by selecting humid or protected microclimates. This calculation assumes no drinking or water adsorption. If *Ph. sauvagei* lost water as predicted by Equation 6.5, it would have only 0.4% of the water needed for evaporation. Clearly, the combination of a low rate of evaporation with the use of uric acid permits some anurans to approach a balanced water budget under harsh environmental conditions.

Withers et al. (1982b) noticed that the connection between uricotelism and high skin resistance varies with body mass: uric acid is used as the principal nitrogenous waste product in anurans with high skin resistances only if they are large (i.e., if they weigh more than about 10 g) (Figure 6.27).

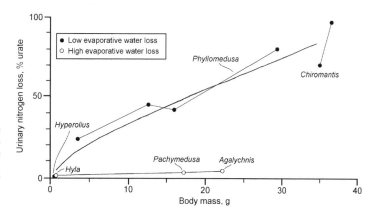

Figure 6.27 Proportion of urate in the nitrogenous waste products of anurans with low and high rates of evaporative water loss as a function of body mass. Sources: Data obtained from Shoemaker and McClanahan (1975), Drewes et al. (1977), and Withers et al. (1982b).

Water turnover is so great at small masses that uricotelism is of little value. Thus, *Hyperolius nasutus*, which weighs only (450 mg), excretes only 0.1% of its nitrogen waste as uric acid (Withers et al. 1982b). This relationship may explain why many other tree frogs do not use uric acid, even though they have integuments that are somewhat resistant to water movement. In the case of *H. nasutus*, however, dry-season acclimated individuals store a large proportion of their nitrogenous wastes as another purine, guanine, which is stored in the iridophores both as an osmotically neutral nitrogen waste and as a means of reflecting the radiant heat load from the environment (Schmuck et al. 1988).

The ability of *H. viridiflavus* to survive a dry season lasting several months is limited by its small mass (Geise and Linsenmair 1986). At 400 mg adult grass frogs can survive 60 to 80 d, if they have the lowest rates of evaporative water loss and lowest rates of metabolism characteristic of inactive individuals acclimatized to the dry season. Such extended periods are limited by the inability of frogs to tolerate a loss greater than 50% of body water. Individuals weighing less than 300 mg cannot tolerate extended periods of dessication because of high water turnover and limited energy reserves. Geise and Linsenmair (1988) concluded that the time limit for dessication increases with mass, and therefore after metamorphosis energy preferentially is shunted to growth to attain a mass equal to 400 to 500 mg, which is sufficient to permit *H. viridiflavis* to tolerate dehydration for up to 4 weeks. After the attainment of this mass, energy is allocated to fat storage, the source of energy used during periods of dehydration. Feeding cannot be used during the dry season because it leads to an imbalance in the water budget as a result of protein metabolism and the consequent accumulation of urea.

6.10.5 Seasonal withdrawal. Some amphibians evade dry conditions by behaviorally withdrawing to an acceptable microhabitat. For instance, Bentley et al. (1958) showed that the anuran *Heleioporus* digs deep burrows in sandy soil, thereby evading the most dessicating conditions, which may account for why its rate of water uptake is independent of climate. *Neobatrachus*, however, lives on clay soils, which prevents it from digging deeply; its rate of rehydration is correlated with environmental aridity. In some environments conditions are so hostile at certain times of the year that anurans may remain in the soil for months at a time in a state of extended inactivity. For example, *Scaphiopus* in California (Mayhew 1965b, McClanahan 1967) and in Arizona (Ruibal et al. 1969, Shoemaker et al. 1969) generally spends September to July continuously in soil, only to emerge for breeding during the sporadic rains of summer. During an unusually dry summer, toads may not emerge but wait in the soil until the following summer for emergence (i.e., some 21–23 months after burrowing into the soil)! Because this withdrawal occurs during dry summers, it is sometimes referred to as *aestivation*, although in *Scaphiopus* it extends through a cold, wet winter, when the term *hibernation* would be more appropriate.

Amphibians have two different approaches to an extended residence in soil (Katz 1989). One permits extensive water exchange with soil, the amount and direction depending on variations in soil water content. The direction of water movement between a buried toad and the surrounding soil depends on the osmotic differential between

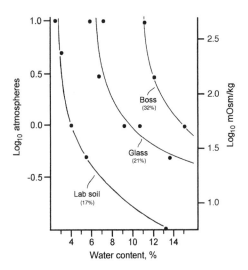

Figure 6.28 Log_{10} water tension in soil, as measured in terms of atmospheres and milliosmoles per kilogram, as a function of log_{10} soil water content and soil composition (% silt and clay). Source: Modified from Ruibal et al. (1969).

the toad and soil. The osmotic concentration of soil is related to soil water tension (1 atm = 44.6 mOsm/kg), which in turn depends on the water content of soil and soil particle size (Figure 6.28): as mean soil particle size decreases, soil water tension radically increases, even though water content remains fixed (Ruibal et al. 1969). At the beginning of the aestivational period, both toad and soil are hydrated, the toad having a slightly higher water tension. Water, therefore, moves into the toad and is stored in the bladder as a dilute urine. During the winter rainy period, soil and toad remain hydrated (Figure 6.29A). But during the dry spring and early summer, water is lost via evaporation from the soil, soil water tension increases, the osmotic differential reverses, the toad loses water to the soil, and the toad plasma concentration increases (Figure 6.29A). Much of the increase in plasma concentration in *Scaphiopus* is due to urea (Figure 6.29B). An aestivating *Scaphiopus* thus is an osmometer that depends for survival on its tolerance of high plasma urea concentrations. In fact, an osmotically stressed *Scaphiopus* shifts to protein metabolism, thereby increasing urea production (Jones 1980b). Similar to *Scaphiopus*, *Bufo viridis* in drying soil for 3 months had a plasma concentration equal to 1320 mOsm/kg, of which 900 mM was urea (Degani et al. 1984), a different response from when the green toad was exposed to seawater.

Some salamanders also find refuge from high

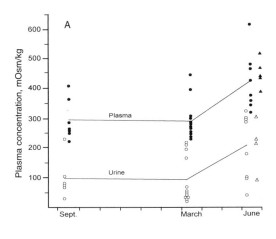

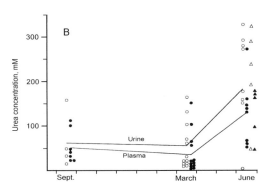

Figure 6.29 A. Plasma and urine concentrations in field *Scaphiopus hammondii* living in soil as a function of the time of year. B. Concentration of urea in plasma and urine in *S. hammondi* as a function of time of year. Source: Modified from Shoemaker et al. (1969).

rates of water loss by burrowing into soil. *Ambystoma tigrinum*, some populations of which live in the deserts of North America, can withstand dehydration in soil for periods up to 9 months. Under these circumstances plasma concentrations increased to over 550 mOsm/kg, much of the increase (ca. 220 mM) being due to urea (Delson and Whitford 1973). *Salamandra salamandra* tolerated 1.5 months in soil with a plasma concentration of 636 mOsm/kg, of which urea accounted for 223 mM (Degani 1981).

In the second approach, a barrier is placed between an amphibian and soil, greatly reducing water exchange. The barrier is derived from cornified skin, as is sometimes the case in spadefoot toads, especially at localities where rainfall was insufficient for 2 y (Mayhew 1965b). These membranes, however, are incomplete and probably reduce only modestly water exchange with the soil. Other amphibians, including leptodactylid anurans in Australia (e.g., *Cyclorana, Neobatrachus* [Lee and Mercer 1967, Withers 1995]) and Argentina (*Lepidobatrachus, Ceratophrys* [McClanahan et al. 1976]); the bullfrog *Pyxicephalus adspersus* in southern Africa (Loveridge and Withers 1981); hyperoliids in Africa (*Leptopelis* [Loveridge 1976]); two burrowing hylids, one in southwestern United States and Mexico (*Pternohyla fodiens* [Ruibal and Hillman 1981]) and one in Central America (*Smilisca baudini* [McDiarmid and Foster 1987]); and the North American sirenids *Siren intermedia* (Reno et al. 1972, Etheridge 1990b), *Pseudobranchus striatus* (Etheridge 1990b), and presumably *S. lacterina*, construct a complete cocoon during aestivation. Disagreement apparently exists as to whether *Lymnodynastes spenceri* does (Lee and Mercer 1967) or does not (Withers 1995) make a cocoon. These cocoons are made of many layers of squamous epidermal cells, which completely encase the amphibian except for the opening of the nares. Reno et al. (1972) believed that the cocoon of *Siren* was secreted by skin glands, but this is not correct (McClanahan et al. 1976, Etheridge 1990a). The thickness of the cocoon (i.e., the number of cell layers alternating with intercellular material) depends on the length of time that the amphibian has been aestivating and the rate at which skin epidermal layers are shed (Withers 1995). The cocoon often consists of 40 to 60 layers.

The cocoon of *Lepidobatrachus llanensis* from Argentina reduced evaporative water loss to one-tenth of the rate found in individuals without cocoons (McClanahan et al. 1976), and will retard water exchange with the soil. Thus, in a comparative study of *L. llanensis* (which forms complete cocoons) and *Scaphiopus couchi* (which does not) after 44 d of desiccation in the same dry soil, the plasma concentration of *L. llanensis* was 363 mOsm/kg and that of *S. couchi* was 691 mOsm/kg. Most (70%) of the difference in plasma concentration between these anurans at the end of the experiment was due to urea: 34.5 mOsm/kg in *L. llanensis* and 263 mOsm/kg in *S. couchi*. *Siren intermedia* encased in cocoons showed no increase in plasma Na$^+$ and K$^+$ concentrations after aestivating for 36 to 63 d in dry soil (Asquith and Altig 1986). Dormant *Pyxicephalus* individuals that are completely encapsulated in a cocoon have rates of evaporative water loss reduced to about 5% of the rates found in bullfrogs resting without cocoons. Enclosed *Cyclorana* and *Heleioporus* from Aus-

tralia evaporate water at rates that are 6% to 13% of the values found in nonenclosed individuals (Lee and Mercer 1967).

The contribution to long-term survival of a cocoon in combination with water stored as urine in the bladder can be seen in the African bullfrog (Loveridge and Withers 1981). If the original hydrated mass is 500 g, if it can lose 38% of the original mass as water, and if it loses water at the minimal rate of 0.03 g H$_2$O/h, then the bullfrog could last 316 d before dying. Furthermore, if *Pyxicephalus* stores 32% of body mass as bladder water, another 100 d of tolerance would be added. The extended time for dehydration in *Pyxicephalus* reflects its large mass. In *S. lacertina* time to 50% dehydration in aestivation increases 2.6-fold with an increase in mass from 20 to 500 g (Etheridge 1990a).

McClanahan et al. (1976) suggested that the use of a barrier is correlated with soil type. *Scaphiopus* generally burrows in friable sands, where the water tension is usually low due to large soil particles and interstitial spaces. *Lepidobatrachus* and *Siren*, in contrast, burrow into mud (i.e., clays and silts), which when dry is characterized by high water tension because of the small soil particles and interstitial spaces. Such soils, upon drying, would dehydrate amphibians, thereby requiring amphibians to impose a structural barrier to water exchange or to abandon the use of these soils for aestivation.

The essential significance of a seasonal withdrawal in amphibians is as a means of avoiding unacceptably harsh environmental conditions, the most important of which is the shortage of water. This withdrawal, however, implies a great reduction in all activities, which is clearly demonstrated by a great reduction in energy expenditure (see Section 11.8.5).

6.11 AESTIVATING FISHES

The behavior of the African lungfish *Protopterus* during periods of low water is similar to that of *L. llanensis* and *S. intermedia*. Unlike amphibians, the cocoon of *Protopterus* consists of mucus secreted by epidermal cells (Kitzan and Sweeney 1968). This cocoon also retards water loss (Smith 1930). During aestivation this lungfish synthesizes urea (Smith 1930, 1935a, 1935b). *Protopterus*, therefore, shares with some amphibians a nearly identical solution to the problems posed by a periodically desiccating environment.

Other fish that use aerial gas exchange, including the bowfin (*Amia* [Neill 1950]), *Clarias* (Das 1927, Bruton 1979), *Heteropneustes* (Das 1927, Singh and Hughes 1971), *Channa* (Das 1927), *Electrophorus* (Carter 1935b), *Synbranchus* (Carter and Beadle 1930, Johansen 1966, Bicudo and Johansen 1979), and the South American lungfish *Lepidosiren* (Carter and Beadle 1930), often live in swamps subject to dessication. Carter and Beadle (1930) stated that the swamps of the Paraguayan Chaco, where *Lepidosiren* and *Synbranchus* are found, are underlaid by compact clays. By burrowing into this impervious substrate, these fishes encounter a standing pool of water, or at least a moist burrow, even when the external conditions are unusually dry. Although *Lepidosiren* is generally thought to aestivate, a question remains as to whether it builds a cocoon. Evidence is available, however, to suggest that indeed it builds a cocoon, or at least lines its burrow with a "... slimy, sticky, gelatinous fluid ..." (Kerr 1898, p. 43; see also Sawaya 1946). The extent to which aestivation occurs in this species and whether it builds a cocoon needs to be examined (although I saw living *Lepidosiren* species in Professor Paulo Sawaya's laboratory in São Paulo, Brazil, enclosed in cylinders of "dry" mud).

Aestivation has also been described in some Southern Hemisphere fishes, including *Neochanna burrowsius* (Eldon 1979) from New Zealand and *Galaxius cleaveri* (Fulton 1986) from Tasmania, both members of the Galaxiidae, and *Lepidogalaxius salamandroides* (Pusey 1986, 1989) from Australia, a member of the Lepidogalaxiidae. These species apparently secrete a large amount of mucus when burrowed into a substrate (e.g., see Pusey 1989), although as Bruton (1979) noted with reference to *Clarias* in Africa, the production of copious amounts of mucus does not a cocoon make. Unfortunately, little is known of the physiological ecology of aestivating fishes.

Another environment in which fishes potentially face high rates of evaporative water loss is in the marine intertidal zone. Some fishes are trapped in tidal pools, are stranded in moist microclimates under boulders and among algae, or freely move onto sand bars in the intertidal zone. In a study of five species of intertidal stichaeoids, species that were found at high intertidal levels tended to survive longer out of water and had lower rates of evaporative water loss than did those found at low intertidal levels (Horn and Riegle 1981). One of the principal factors dictating survival out of water was body size: large (92-g) *Cebidichthys violaceus* could tolerate up to 37h of emersion. Small juveniles of these species have to reduce water loss in the intertidal microenvironments where evaporative water loss is reduced. Nevertheless, no intertidal fish is known to show the seasonal withdrawal shown by some freshwater fishes.

A solution to a seasonally drying environment is found among some cyprinodontid fishes. These species survive for long periods in mud only as fertilized eggs. With the return of rain, the eggs hatch to produce larvae and adults that belong to a generation that does not overlap with the previous generation. These fishes are collectively referred to as *annual fishes* (Wourms 1967). Their "diapause" may last for up to a year and has been found in a variety of genera and species in Africa and South America.

6.12 CHAPTER SUMMARY

1. Aquatic vertebrates face appreciable variations in the concentration and ionic composition of naturally occurring waters.

2. Most vertebrates have plasma salt concentrations between 20% and 35% of that in seawater.

3. The only vertebrates known to be osmoconformers are hagfishes. Lampreys have effective osmoregulation, being hyperosmotic in freshwater and hyposmotic in salt water.

4. Chondrichthyans generally are hyperosmotic in all environments; their plasma contains moderate amounts of salt and comparatively large amounts of urea and TMAO. Salt regulation is accomplished principally by rectal glands. South American freshwater rays differ from other freshwater chondrichthyans by having low plasma concentrations, even when exposed to seawater, and they do not store urea.

5. TMAO biochemically compensates for the impact of urea on enzyme function.

6. Teleosts are effective osmotic regulators: they are hyperosmotic in freshwater, where they produce a dilute urine and replace lost salt by active transport at their gills; they are hyposmotic in the ocean, where they replace lost water by swallowing seawater and excrete salt by active transport at their gills.

7. The living coelacanth is isosmotic with the sea, stores urea and TMAO, and like elasmobranchs has a rectal gland.

8. In teleosts, osmotic flux with the environment is proportional to the osmotic differential and with the surface area of the gills.

9. Some fluxes between a teleost and its environment, namely, drinking in freshwater and urine production in seawater, are unexpected. These behaviors are most marked in euryhaline species, and because they entail an increased expenditure for osmotic regulation, they represent a "cost" of euryhalinity.

10. Little attention has been given to the energetics of osmotic regulation, but the cost of regulation is likely to be greatest when the osmotic differential is greatest, that is, in the ocean for teleosts and in freshwater for elasmobranchs.

11. The cost of osmotic regulation can be reduced by a decrease in the osmotic differential, in permeability, and in surface area for exchange; it also can be reduced by the production of a dilute urine in freshwater and a less dilute urine in salt water.

12. Vertebrates probably originated in salt water; the earliest species were likely to have been osmoconformers with glomeruli. They acquired osmotic regulation before entering freshwater, possibly in brackish environments, at which time the glomerulus went from being an ionic regulating device to one concerned with both ionic and volume regulation.

13. The reduction in plasma salt concentration may have evolved in marine (or estuarine) environments to facilitate enzyme function.

14. Most amphibians cannot tolerate water with concentrations above 300 mOsm/kg because they cannot tolerate high plasma salt concentrations. Two notable anurans tolerate fluids up to 80% of the concentration of seawater: *Bufo viridis*, by tolerating high plasma salt concentrations, and *Rana cancrivora*, by storing urea. Larval *R. cancrivora* tolerate 100% seawater by hyposmotic regulation, a behavior that is lost at the time of metamorphosis.

15. Most amphibians use ammonia as their principal nitrogenous waste product when they are confined to water, but use urea as terrestrial adults. A few arboreal frogs use uric acid, but only if they weigh more than 10 g and have low rates of evaporative water loss.

16. Bladder urine in amphibians can be used to replace the water lost by evaporation, especially if the urine stored is more dilute than the plasma.

17. Terrestrial amphibians tolerate dessication better than aquatic species and appear to gain water more rapidly than aquatic species.

18. Anurans generally are terrestrially adapted; to move into aquatic habits they reduce skin permeability to prevent an unacceptable dilution of body fluids.

19. Some terrestrial amphibians evade harsh conditions by burrowing and by making a cocoon out of epidermal skin layers. Cocoon formation reduces water exchange with the soil and is most likely to occur in species occupying soils with a high water tension. A similar behavior occurs in some air-breathing fishes.

7 Water and Salt Exchange in Terrestrial Vertebrates

7.1 SYNOPSIS

Terrestrial vertebrates must balance their water budgets. The water content of food is an important determinant of vertebrate behavior: if they consume food with a low water content and are not parsimonious with water, they must drink water and therefore usually are limited in distribution to regions with available water. The most important pathways of water loss are respiratory and urinary, whereas integumental water loss is usually minor, or at least can be reduced by behavioral means. Respiratory water loss is a by-product of gas exchange but can be modulated. Birds and terrestrial reptiles use uric acid as their principal nitrogenous waste product, whereas aquatic reptiles use urea and ammonia. Reptiles cannot produce a urine more concentrated than the plasma, but they reabsorb water from the urine in the cloaca, thereby precipitating uric acid. Birds maximally produce urine that is 4 times as concentrated as the plasma, the concentration being limited because only a few nephrons have a loop of Henle. Birds also reabsorb water in their cloaca, which may limit the capacity of birds to produce a concentrated urine. Many reptiles and birds supplement kidney function by using cephalic glands to excrete excess salts. Mammals use urea as their nearly exclusive nitrogenous waste product. Their ability to produce a concentrated urine depends on an osmotic gradient that is established in the medulla of the kidney and on the ratio of long to short loops of Henle. Aquatic mammals have few long-looped nephrons and therefore cannot produce a urine much more concentrated than the plasma. Desert species tend to have more long-looped nephrons, especially if they have a large fluid intake, whereas those that eat dry foods and have a small fluid intake have relatively few long-looped nephrons. Some species can produce a urine that is more than 25 times the concentration of the plasma. The tendency for anuria in desert mammals with a small fluid intake is also found in some desert birds.

7.2 INTRODUCTION

Reptiles were the first vertebrates to make an unambiguous commitment to a terrestrial life, even though some, such as many turtles, crocodilians, some snakes, and an occasional lizard, have to varying degrees "returned" to aquatic habits. Birds and mammals evolved from reptiles, and some species in both groups evolved aquatic habits. Because the characteristics of most (all?) aquatic reptiles and all aquatic birds and mammals are derived from species that were committed to a terrestrial existence, the characteristics of the water and salt balance of these aquatic species must represent readjustments of the characteristics of terrestrial species. For that reason this chapter, which emphasizes the adjustments reptiles, birds, and mammals made to life on land, also deals with the readjustments that have been made to aquatic habits.

Life requires a balanced water budget: water intake must equal water loss. Water is obtained preformed in food, is synthesized through the oxidation of food, and is gained by drinking. In terrestrial environments water is lost by evaporation from the integument and during respiration, through the excretion of salts and nitrogenous waste products, and in feces. Although all pathways of exchange are important in a balanced water budget, ingestion of food and consumption of fluids are quantitatively the most important routes by which water is gained, whereas respiration and excretion are the most important pathways by which water is lost. The difficulty of balancing a water budget reaches its extreme on land in desert environments because they often combine high ambient temperatures with low water availability. Then, integumental and renal water loss must be minimized. Variations in the major components of a terrestrial water budget are examined.

7.3 WATER FLUX IN TERRESTRIAL VERTEBRATES

Water flux has been studied in a variety of terrestrial vertebrates, including reptiles, birds, and marsupial and eutherian mammals, both in captivity and in the field (for a comprehensive survey, see Nagy and Peterson [1988]). In all cases, water flux increases as a power function of body mass (Figure 7.1A). Field water fluxes in endotherms are generally similar to each other, 15 to 20 times those of reptiles, but only one-tenth to one-twentieth those of freshwater fishes (compare Figures 6.7 and 7.1).

Much residual variation in water flux remains after body mass is accounted for. Passerines have field fluxes that are 3 times those of other birds. Among eutherians, herbivores have water fluxes that are about 3 times those of carnivores, whereas the fluxes in omnivores are intermediate (Figure 7.1B). In contrast, marsupial herbivores have field water fluxes that are less than one-half those of marsupial carnivores (Nagy and Peterson 1988). Desert eutherians have lower fluxes than eutherians living in mesic environments, a difference that might reflect food habits as well as environmental conditions. Desert birds and reptiles have water fluxes that are about one-half those of birds that live in mesic environments. Thus, water flux in terrestrial vertebrates varies with their mass and with water availability, as reflected in their food habits and the climate in which they live. These topics now will be explored in some detail.

7.4 PATHWAYS OF WATER INCOME

The water content of food is the most important factor dictating the distribution and behavior

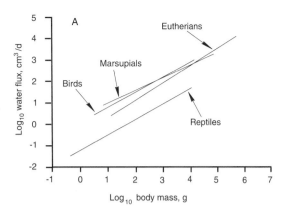

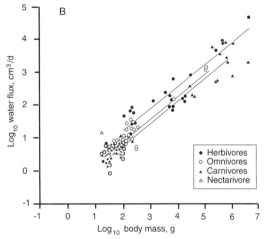

Figure 7.1 $\log_{10}$ field water fluxes as a function of $\log_{10}$ body mass in (A) vertebrates generally and (B) eutherian mammals in relation to food habits. Source: Modified from Nagy and Peterson (1988).

of terrestrial vertebrates relative to surface water. Dry foods, such as seeds, which often have a water content of 10% or less, normally can be used only when surface water is available. But hydrated foods, such as insects, vertebrates, and fruit, which usually have a water content of 60% to 70%, contain enough water to permit some vertebrates to have a geographic distribution independent of surface water, as long as these foods are available. The photosynthetic parts of plants, in spite of a high water content, often require water for processing, either because they contain high salt loads or more frequently because of the requirements of fermentation, the presence of indigestible molecules (like lignin), or the presence of plant secondary compounds (see Section 12.9).

7.4.1 Preformed water. Because birds are relatively easy to observe in the field, more is known of their food habits, behavior, and distribution in relation to water than of the habits of other terrestrial vertebrates. Most passerines that are widely distributed in deserts feed on insects, for example, rock wrens (*Salpinctes obsoletus*) in North America. Others, like the black-throated sparrow (*Amphispiza bilineata*), have a flexible diet: they feed on green vegetation and insects from March through June, when they have a distribution independent of surface water; in July, when flying insects are scarce and green vegetation is unavailable, this sparrow shifts to feeding on seeds and begins to visit water holes (Smyth and Bartholomew 1966a). Other small passerines, such as the Namibian larks *Spizocorys starki* and *Eremopterix verticalis* (Willoughby 1968) and the North American sage sparrow *Amphispiza belli* (Moldenhauer and Wiens 1970), have a mixed diet and do not depend on surface water. At great distances from surface water, like in the Great Victoria Desert of Australia, no seed-eating birds are included in the 15 most abundant species (Ford and Sedgewick 1967), but when water becomes available, usually through the agricultural practices of humans, seed-eating birds become very abundant.

7.4.2 Drinking. When surface water is available, the amount consumed by a bird is correlated with body mass (derived from Table 1, Bartholomew and Cade 1963):

$$\dot{D} = 1.46m^{0.387}, \tag{7.1}$$

where $\dot{D}$ is the rate of water consumption (cm³/d) and mass is in grams. This curve has great residual variation ($n = 21$, $r^2 = 0.264$): some desert birds (e.g., *A. belli* [Moldenhauer and Wiens 1970], *Zenaida macroura* [Bartholomew and MacMillen 1960], *Zenaida asiatica* [MacMillen and Trost 1966]) have unusually high rates of water consumption, whereas others (e.g., *Callipepla californicus* [Bartholomew and MacMillen 1961], *Melopsittacus undulatus* [Cade and Dybas 1962], *Taeniopygia castanotis* and *Estrilda troglodytes* [Cade et al. 1965]) have low rates. Some salt-marsh subspecies of savannah and song sparrows have very high rates of water consumption, whereas others have low rates (Bartholomew and Cade 1963).

Because water consumption is proportional to mass raised to a power less than 1.0, large birds tend to be more independent of surface water than small species. That is, large individuals can tolerate longer periods of time without drinking than can small individuals. This difference is enhanced in some moderately sized species, especially parrots, doves, and quail, by having a crop in which a relatively large volume of water can be stored (Fisher et al. 1972).

Body size and environmental temperature influence the frequency and temporal pattern with which desert birds visit a water hole. Strong fliers, like parrots and doves, visit a water hole early in the morning and (often) again late in the afternoon or early in the evening (Figures 7.2A, B), whereas finches and small doves are limited to the proximity of water holes and usually visit throughout the day (Figure 7.2C). Thus, the mourning (*Z. macroura*) and white-winged (*Z. asiatica*) doves can be found up to 10 km from surface water,

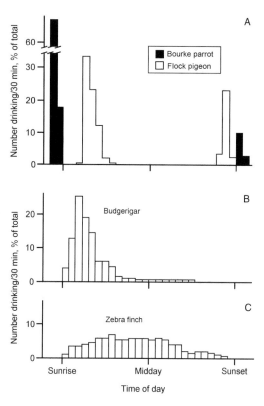

Figure 7.2 Daily pattern of drinking at water holes in central Australia: (A) Bourke parrot (*Neophema bourkii*) and flock pigeon (*Phaps histrionica*), (B) budgerigar (*Melopsittacus undulatus*), and (C) zebra finch (*Taeniopygia castanotis*). Source: Modified from Fisher et al. (1972).

whereas the smaller ground (*Columbina passerina*) and Inca doves (*Scardafella inca*) are restricted to the vicinity of surface water (Miller and Stebbins 1964, MacMillen and Trost 1966, Willoughby 1966). The mourning dove has peak visitation between 0900 and 1300 (Smyth and Coulombe 1971), often after feeding has occurred. In contrast, Bourke's parrot (*Neophema bourkii*) in Australia visits water holes almost exclusively before sunrise and after sunset (Figure 7.2A), a behavior that diminishes exposure to predators and reduces congestion at a shared water hole (Fisher et al. 1972). Some parrots increase their frequency of visitation to water holes at high temperatures to compensate for an increase in evaporative cooling.

Any bird that depends on surface water faces special problems during the breeding season, when water must be provided to their young. Pigeons and doves produce "pigeon's milk," which is nutritious and serves as a water source. The most distinctive means of supplying water to young far from a water source is found in sandgrouse (*Pterocles*) from Asia and Africa (Cade and Maclean 1967, Thomas and Robin 1977). They nest up to 40 km from water. Males, but not females (Cade and Maclean 1967), carry water adsorbed to specially modified breast and belly feathers. They can transport 12 to 22 mg of usable water per milligram of dry feathers (Thomas and Robin 1977). Cade and Maclean estimated that a male can load 25 to 40 g of water and that 10 to 18 g can be transported 30 km. This behavior markedly increases the area of a desert that is available to sandgrouse for nesting.

Vertebrates other than birds rely less on drinking because they are less mobile and therefore (must?) use moist foods. All reptiles feed on moist foods, such as insects, vertebrates, or succulent vegetation; most, however, will drink water when dehydrated (e.g., *Malaclemys terrapin* [Dunson 1970], *Varanus semiremax* [Dunson 1974]). The importance of drinking to some reptiles living in deserts is demonstrated by two examples. Many land tortoises have flared shells. This flaring is constructed such that when the tortoise maintains a stance that inclines the shell headward, rainwater is funneled to the head, where it can be drunk (Auffenberg 1963). The skin of the Australian agamid lizard *Moloch horridus* adsorbs water from the soil, after which it moves by capillary along fine keratinized channels to the corners of the mouth, and drinking occurs (Bentley and Blumer 1962).

Mammals may be abundant at great distances from surface water, which precludes all but the largest species from drinking. Small mammals balance a water budget either by using moist foods or by being parsimonious with water if they feed on dry foods like seeds. A reduction in water turnover is greatly aided by nocturnal habits and an extended residence in closed burrows. The amount of water consumed by small mammals in the laboratory is correlated with body mass (Adolph 1949), but that does not mean that they often drink in the field; rather, it is a convenient comparative standard for the rate of water turnover. Species from xeric environments drink less water than those from mesic environments (Lindeborg 1952, Glenn 1970). Some small desert mammals so restrict water loss through the behavioral evasion of dehydration, a low rate of metabolism, and the ability to produce a concentrated urine that they drink little or no water (see Section 7.10.3).

Large mammals, such as deer, peccaries, sheep, antelopes, elephants, rhinoceroses, and camels, often traverse great distances in arid environments to locate water. These species may disperse up to 50 km from a water hole (Schmidt-Nielsen 1964, Dorst 1970), returning every few days to drink. Upon returning, they can consume large amounts of water; for example, dromedaries drink water that is up to a third of their body mass in a few minutes, in one case 104 L (Schmidt-Nielsen 1964)! Even species that normally attain sufficient moisture from their food may show seasonal shifts in local distribution related to rainfall, as rainfall in turn affects the availability of acceptable food (Newsome 1965).

The local distribution of large mammals varies relative to water requirements, water availability, and behavior. Xeric-tolerant zebu cattle drink less water than Herefords, and the dry bushland oryx (*Oryx beisa*) requires less water than the grassland wildebeest (*Connochaetes taurinas*) (Figure 7.3). The smaller Thomson's gazelle (*Gazella thomsoni*) surprisingly is more parsimonious with water than the larger Grant's gazelle (*G. granti*). Cape buffalo (*Syncerus caffer*) depend more on surface water than eland (*Taurotragus oryx*), even though these species have similar mass-independent water requirements (Figure 7.3), because eland selectively feed on succulent food, seek shade at midday, produce dry feces, and permit body temperature to fluctuate (Taylor 1968, 1969). The greater depen-

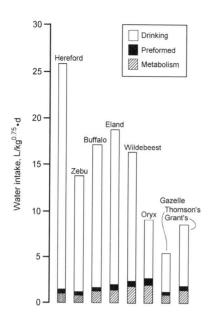

Figure 7.3 Minimal rates of water intake in various East African ungulates. Source: Modified from Taylor (1968).

Table 7.1 Water production from the oxidation of various foodstuffs

Foodstuffs	g H_2O/ g food	L O_2/ g food	g H_2O produced/ L O_2 used
Starch	0.56	0.80	0.70
Fat	1.07	2.01	0.53
Protein	0.40	0.95	0.42

Source: Derived from Schmidt-Nielsen and Schmidt-Nielsen (1950).

dency of the buffalo on surface water reflects its greater mass and the use of drier food.

7.4.3 Metabolism. Water is a by-product of aerobic metabolism. The amount of water produced from the metabolism of common food stuffs is indicated in Table 7.1. Aerobic oxidation of fats produces more water per gram of food stuff than the oxidation of other food stuffs, which led to the conclusion that desert mammals store fat to balance their water budgets. Because water is produced by the reduction of molecular oxygen, the production of water is tied to oxygen consumption and gas exchange. Schmidt-Nielsen and Schmidt-Nielsen (1950) suggested that the best measure of water gain from aerobic metabolism is the ratio of water produced to oxygen consumed, that is, the ratio of water gained to water lost by respiration. From this criterion, the oxidation of carbohydrates produces the most water, followed by fats and by proteins, although these differences are small. Fat storage in desert species is used as an energy store probably because it has the greatest energy density (McNab 1968), not because it produces the most water.

7.5 EVAPORATIVE WATER LOSS

Terrestrial vertebrates lose water principally by evaporation from the integument and respiratory tract. Respiratory water loss is usually greater, but integumental water loss may be very important in certain species.

7.5.1 Integumental water loss. Reptiles, evolutionarily the first truly terrestrial vertebrates, are characterized by dry scaly skin that was long thought to be impermeable to water. Measurements in reptiles suggest, however, that 40% to 70% of total water loss may occur through the skin at moderate ambient temperatures (Bentley and Schmidt-Nielsen 1966). The magnitude of this pathway depends on many factors, including body size (and surface area), wind speed, relative humidity, and environmental temperature, as might be expected from Equations 2.8 and 2.9. For example, the rate of evaporative water loss in the snake *Elaphe* is proportional to $m^{0.60}$ and its external surface area is proportional to $m^{0.63}$ (Gans et al. 1968), a correlation that is complicated by respiratory water loss. Water loss increases with wind velocity in lizards that have little skin resistance to water movement (e.g., *Anolis*), but not in those in which most of the resistance to water loss is in the skin (e.g., *Uta* [Claussen 1967]). The differential sensitivity of reptiles to air movement, along with differences in technique, may account for the inability of some studies to demonstrate an effect of wind velocity (e.g., Gans et al. 1968). Cutaneous water loss also depends on temperature (e.g., Crawford and Kampe 1971), presumably because the vapor density deficit between the skin and air varies with air temperature.

Cutaneous water loss in reptiles is correlated with water availability: species that live in dry environments lose little water through their skin, whereas species from moist to wet environments have an appreciable water flux (Bogert and Cowles

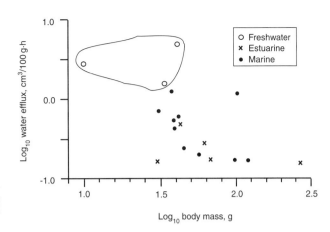

Figure 7.4 Log_{10} water loss in freshwater, estuarine, and marine snakes, when held in seawater, as a function of log_{10} body mass. Source: Modified from Dunson (1978).

1947, Bentley and Schmidt-Nielsen 1966, Claussen 1967, Gans et al. 1968, Krakauer et al. 1968, Prange and Schmidt-Nielsen 1969, Elick and Sealander 1972, Munsey 1972b, Hillman et al. 1979a). Some desert lizards lose less than 1% of body mass as water per day, whereas caiman (*Caiman sclerops*) may lose 11.5% per day. Marine snakes have much lower integumental permeabilities than freshwater or brackish-water species of the same size (Figure 7.4), thereby reducing water loss in a hyperosmotic environment, a pattern also seen in teleosts. Among amphisbaenid reptiles, *Amphisbaena* from the humid tropics loses water at a rate up to 130% of that expected from Equation 6.5 for amphibians of equal mass; *Rhineura* from humid, subtropical Florida, where it may occupy sandy soils (and thus lose appreciable amounts of water), loses water at rates up to 55% of the rate expected from mass, and *Monopeltis* from arid South Africa loses water at rates that are only 1% to 3% of the expected rates (Gans et al. 1968, Krakauer et al. 1968).

In spite of the variation demonstrated in integumental water loss in reptiles, little is known of the integumental barrier to water loss. Measurements of cutaneous water loss in mutant, scaleless snakes (Licht and Bennett 1972, Bennett and Licht 1975) showed that the rate of water loss is unaffected by the absence of scales. The doubling of integumental water loss during shedding may well be due to activity, rather than to an increased permeability per se (Gans et al. 1968).

Few studies of cutaneous water loss are available for terrestrial vertebrates other than reptiles, although it may be greater than often thought. For example, more than 50% of water loss in some

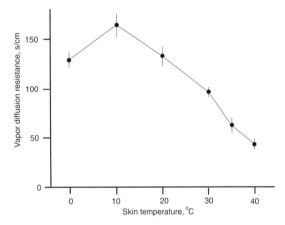

Figure 7.5 Vapor diffusion resistance in rock pigeons (*Columba livia*) as a function of skin temperature. Source: Modified from Webster et al. (1985).

birds may occur through the skin in thermoneutrality (Bernstein 1971). Using a modification of Equation 2.8 in which the total body diffusion coefficient for water vapor is replaced by total body resistance, Marder and Ben-Asher (1983) demonstrated that body resistance to the diffusion of water vapor decreases at high ambient temperatures (Figure 7.5) and that pigeons and doves have lower resistances at high temperatures than do quail. The body resistance in some pigeons and doves is so low that they can tolerate very high ambient temperatures (i.e., 45–52°C), as long as ambient humidity is low: then between 52% and 98% of heat production in columbids may be lost by the evaporation of water through the skin at these temperatures. The use of evaporative cooling at high environmental temperatures, coupled with

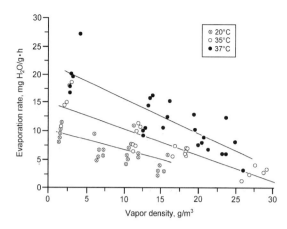

Figure 7.6 Rate of evaporative water loss in Anna's hummingbird (*Calypte anna*) as a function of the vapor density of air and air temperature. Source: Modified from Powers (1992).

the ability of columbids to fly long distances rapidly, permits them to have a widespread distribution in xeric environments (Marder 1983).

Total body resistance to water loss equals the sum of the resistances associated with the skin, feather layer, and boundary layer. Webster et al. (1985) showed that 6.2% to 25.8% of the resistance to water loss in the rock dove (*Columba livia*) is due to the feather and boundary layers. The remaining resistance is in the skin. Skin resistance falls at high temperatures (Figure 7.5), potentially because of a change in skin lipid conformation and an increased peripheral blood flow, the former explanation preferred by Webster et al. (1985) and the latter by Edwards and Haines (1978) and Marder (1983).

The rate of evaporative water loss in Anna's hummingbird (*Calypte anna*) varies directly with the air water vapor density (ρ_a) and ambient temperature (Figure 7.6): at low ρ_a and high T_a values, evaporative water losses are high, as expected from Equation 2.9. The sensitivity of water loss to ρ_a increases at temperatures greater than 33°C, an increase that may well reflect the incorporation of respiratory water loss. Except at the highest temperatures to which the hummingbird is exposed (40°C), evaporative water loss is less than the excretory water loss, presumably because of the high fluid intake associated with feeding on nectar. At no temperature is this hummingbird able to dissipate by evaporation more than about 60% of the heat produced by metabolism.

Among mammals, water loss occurs passively through the integument, although it may be augmented by sweating. In the absence of sweating, cutaneous water loss may be appreciable: 46% of the total evaporative water loss at 27°C is integumental in *Peromyscus maniculatus* (Chew 1955); this proportion surely depends on temperature and body size. The occurrence of sweat glands is highly variable. In some species they are widely distributed, in others they are topographically restricted (e.g., at the base of the tail [deer], about the lips [rabbits, some rodents], on the muzzle [pigs, sheep], and on the feet [some rodents, cats, hyraces]), and still others (especially aquatic species) are reported to have no sweat glands.

One of the few comparative studies of sweat gland function examined eight species of African bovids (Robertshaw and Taylor 1969). The importance of cutaneous water loss (mainly due to sweating), relative to respiratory water loss, increases with body mass: sweating is important in large bovids, such as buffalo and eland, but relatively unimportant in small species, such as duiker (*Silvicapra*) and Thomson's gazelle, the smaller species preferentially using panting. Panting may be too expensive a means to dissipate sufficient heat in large bovids. Rate of sweating shows little correlation among bovids with distribution in mesic or arid environments, which may reflect the tendency of bovids to follow seasonal variations in water availability (Lamprey 1963).

Another source of variation in sweating is the condition under which it is used. In most cases sweating takes place with either an internal (metabolism) or an external (environmental) heat load. In kangaroos sweating occurs only when they are running, that is, when the heat load is internal (Dawson et al. 1974; Figure 7.7); stationary red kangaroos (*Macropus rufus*) will pant and spread saliva when exposed to ambient temperatures of 55°C for 4 h, but will not sweat. Considering the high environmental heat loads often faced by *M. rufus*, it would be expected to sweat to dissipate such a load. Is such forbearance an evolutionary oversight? Or is it a means of conserving water? But if the latter, at what cost? Presumably spreading saliva is less efficient in dissipating a heat load than is sweating because saliva wets the fur coat, which destroys the insulative resistance to heat gain from the environment (see Schmidt-Nielsen et al. 1957), whereas sweating that occurs under a coarse coat will cool the skin surface.

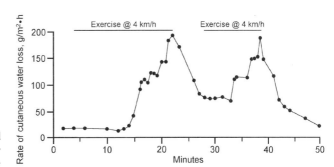

Figure 7.7 Rate of cutaneous water loss in red kangaroos (*Macropus rufus*) in relation to activity. Source: Modified from Dawson et al. (1974).

A specialized use of sweating occurs in hyraces, which sweat on the soles of their feet. At times such sweating may be used for evaporative cooling: in the rock hyrax (*Heterohyrax brucei*) up to 22% of evaporative heat loss is derived from the feet (Bartholomew and Rainey 1971). Sweating on the feet of this hyrax also increases the coefficient of static friction (Adelman et al. 1975), thereby permitting the hyrax to scale inclined rock surfaces without slipping.

Humans (*Homo sapiens*), as a "naked ape," represent a fascinating case for the study of integumental heat dissipation (see Section 5.5.2). Nakedness in mammals facilitates dissipating an internal heat load, especially in large species, such as warm-climate elephants and rhinoceroses, which are characterized by high rates of metabolism and relatively small surface areas (McNab and Auffenberg 1976). (Cold-climate proboscidians [†*Mammut*, †*Mammuthus*] and rhinoceroses [†*Coelodonta*, †*Elasmotherium*] were "woolly.") The few small mammals that are naked live in confined spaces with high temperatures and humidities (e.g., the naked mole-rat [*Heterocephalus glaber*; McNab 1966b] and possibly the "naked" bat [*Cheiromeles torquatus*]). To some extent this category includes armadillos, the aardvark, and some pangolins. Humans are too small to fall in the first category and too large to fall in the second. Our nakedness may stem from being a slow, but persistent pursuing predator (like Kalahari bushmen), capable during running of generating large internal heat loads that are principally dissipated by sweating. Larger apes, like the orangutan (*Pongo*) and the gorilla (*Gorilla*), are sedentary vegetarians that do not attain sustained internal heat loads, and they are furred. Wheeler (1984, 1985), however, related nakedness to a bipedal posture that reduces solar input and increases evaporative heat loss. Nakedness clearly increases the rate at which integumental evaporative water loss can occur, by reducing the thickness of the boundary layer.

7.5.2 Respiratory water loss. Water is lost during gas exchange, the amount of which depends on the ventilation rate (and rate of metabolism), lung temperature, ambient temperature, and water content of the external atmosphere. An animal will even lose water to an atmosphere saturated with water as long as body (lung) temperature is greater than atmosphere temperature. These factors are incorporated into Equation 2.9.

Respiratory water loss in reptiles increases with ambient temperature, at least because of the increase in body temperature and the consequent increase in gas exchange. At high temperatures the increase in water loss can be reduced to conserve water or enhanced to facilitate temperature regulation. Lizards reduce respiratory water loss by having low nasal temperatures. Murrish and Schmidt-Nielsen (1970) calculated that *Dipsosaurus dorsalis*, at an environmental temperature of 30°C and a core temperature of 42°C, saves up to 31% of the amount of water lost by evaporation, if nasal temperature is reduced from 42 to 35°C. If, however, temperature regulation is paramount, water loss can be increased by panting. Some small, heliothermic lizards can lose more heat by evaporation than is produced by metabolism and thus maintain body temperature below environmental temperature (as long as the radiant temperatures in the environment are not too high).

At moderate ambient temperatures (ca. 25°C) rate of evaporative water loss in birds (most of which is derived from the lungs and air sacs) varies with body mass (Figure 7.8). When studying the rate of water loss, Crawford and Lasiewski (1968) differentiated between passerines and other birds, but the few data on passerines make the curve for other birds more useful:

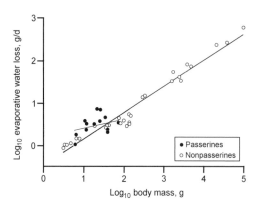

Figure 7.8 $\log_{10}$ rate of evaporative water loss in birds as a function of $\log_{10}$ body mass. Source: Modified from Crawford and Lasiewski (1968).

$$\dot{E} = 14.6m^{0.613}, \qquad (7.2)$$

where $\dot{E}$ is in milligrams of water per hour and mass is in grams. At moderate temperatures birds also reduce respiratory water loss through a reduction in nasal temperatures, both at rest (Schmidt-Nielsen et al. 1970) and in flight (Berger et al. 1970).

At high environmental temperatures birds markedly increase the rate of respiratory water loss through an increase in respiratory frequency (e.g., Lasiewski and Seymour 1972) and in tidal volume (e.g., Calder and Schmidt-Nielsen 1966). Lasiewski et al. (1966b) showed that nearly all birds can dissipate more heat by forced evaporative cooling at high temperatures than what is produced by metabolism, if the humidity of the inspired air is low. Therefore most birds (with the possible exception of small species, such as hummingbirds [Powers 1992]) can maintain body temperatures below ambient temperature. Earlier studies of evaporative water loss were unable to demonstrate this capacity because the air velocities in experimental chambers were too low, and as a result, chamber moisture was sufficiently high to jam evaporative water loss.

Two major problems are associated with increasing the frequency of respiration to facilitate water loss at high temperatures. One is that it involves an expenditure of energy and thus contributes to the heat that must be lost. Second, an increase in frequency eliminates appreciable quantities of CO_2, which leads to respiratory alkalosis (e.g., Calder and Schmidt-Nielsen 1966). To avoid respiratory alkalosis, some birds at high ambient temperatures

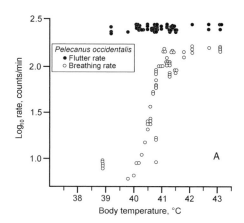

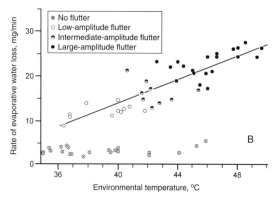

Figure 7.9 (A) $\log_{10}$ gular flutter rate and $\log_{10}$ breathing rate in white pelicans (*Pelecanus occidentalis*) as a function of body temperature and (B) rate of evaporative water loss as a function of environmental temperature and the amplitude of gular fluttering used by poorwills (*Phalaenoptilus nuttallii*). Sources: Modified from Bartholomew et al. (1968) and Lasiewski (1969).

restrict an increase in rate and volume of respiration to the buccopharyngeal region through the oscillation of the hyoid apparatus, a behavior referred to as *gular fluttering*. This behavior partially separates the evaporation of water from the exchange of respiratory gases. It is found in many species of birds but is not known from any passerine. Gular fluttering may occur at a frequency synchronous with breathing (roadrunner, owls, doves) or at a (nearly) temperature-independent frequency (pelicans [Figure 7.9A], cormorants, egrets, caprimulgids). The birds increase evaporative water loss at high temperatures by increasing the breathing rate (Figure 7.9A), the proportion of time when fluttering is used, or the amplitude of gular movement (Figure 7.9B). Gular fluttering apparently adds little to the heat load of a bird (Lasiewski 1969), because it attains a resonance frequency dictated by the elastic properties and mass of the gular region (Crawford and Kampe 1971). Yet, the increased air flow across gular surfaces is not completely uncoupled from pulmonary gas exchange, as is shown by the occurrence of respiratory alkalosis during gular fluttering (Calder and Schmidt-Nielsen 1966). Just as sweating is the most prominent means of facilitating cooling in large mammals, gular fluttering is most marked in large birds, especially in species that are exposed to high ambient thermal loads.

As in other terrestrial vertebrates, mammals increase water loss at high environmental temperatures by varying the frequency of respiration and the temperature of exhaled air as an emergency means of temperature regulation. For instance, the ratio of evaporative water loss to oxygen consumption is nearly constant at temperatures below thermoneutrality in the Australian murid *Notomys*, but at higher temperatures water loss increases exponentially, mainly due to an increase in ventilation rate, whereas the increase in oxygen consumption is small (MacMillen and Lee 1970). MacMillen and Grubbs (1976) argued, contrary to the suggestion of Schmidt-Nielsen and Schmidt-Nielsen (1950), that little evidence exists for a difference in the ratio of water loss to oxygen consumption between desert and mesic rodents. However, MacMillen and Grubbs used total rates of evaporative water loss in the calculation of this ratio, whereas the ratio, as described by the Schmidt-Nielsens, compared respiratory water loss and oxygen consumption.

Many intermediate to large mammals use panting to facilitate heat loss at high ambient temperatures and during strenuous activity. For example, when resting at 41°C, domestic dogs (*Canis*) and the African hunting dog (*Lycaon*) lose more heat by panting than they produce by metabolism, and therefore are able to maintain body temperature below air temperature (Taylor et al. 1971a). A striking difference, however, occurs between these species during activity: when running at velocities greater than 10 km/h, *Lycaon* loses only one-half as much heat by evaporation as *Canis*; that is, *Lycaon* stores more heat and relies more on convective and radiative losses of heat than does *Canis*. Heat storage permits the hunting dog to endure an extended pursuit of prey, often in hot, dry environments, and gives it a greater range from water sources for predation than would otherwise be attained.

The response to high temperature in two similarly sized bovids is different: one (zebu steer) uses sweating and the other (wildebeest, *Connochaetes taurinus*) uses panting (Taylor et al. 1969). As ambient temperature rises, the wildebeest has a greater increase in respiratory frequency than the zebu. Only at high temperatures does the zebu attain a high rate, but even then it has a tidal volume only 63% that of the wildebeest. Much of the increased ventilation volume in the wildebeest is associated with a large anatomical dead space, large nasal passages, and large openings into the paranasal sinuses; these spaces permit an increased evaporative water loss without an increased loss of CO_2. Thus, the wildebeest showed an increase in arterial pH (alkalosis) only at environmental temperatures near 50°C. However, when the zebu was forced to pant (by placing a plastic sheet over the animal), the arterial pH increased at ambient temperatures as low as 43°C. Panting as a means of emergency temperature regulation is obviously limited by the development of respiratory alkalosis, unless some adjustments are made to separate air flow for water loss from the air flow associated with gas exchange.

7.5.3 Minimizing evaporative water loss. Diurnal vertebrates in xeric environments face high temperatures, low relative humidities, and a water shortage. They potentially face a conflict between the demands of water conservation and those of temperature regulation. Diurnal reptiles and birds may reduce water loss by being inactive at midday and by seeking shelter in moderate microclimates (e.g., Cade and Dybas 1962, Cade et al. 1965). In mammals the means by which water loss is reduced in xeric environments depends on body size.

Most small mammals are nocturnal, spending the day in closed burrows, where humidity is high and temperature is moderate (Schmidt-Nielsen and Schmidt-Nielsen 1950, Kennerly 1964, McNab 1966b), conditions that reduce evaporative water loss. As burrow temperature falls, rate of metabolism increases, thereby increasing water production (MacMillen 1972). As a result, the ratio of water produced to water evaporated increases: this ratio equals or exceeds 1.0 at temperatures below 17°C in small rodents (see Section 7.10.3.)

Some large species reduce their dependence on evaporation by heat storage (see Section 5.3.4). Because most mammals have a similar upper lethal body temperature (ca. 42.5°C), any flexibility in heat storage independent of mass must come from variation in the level to which body temperature falls at night. Bligh and Harthoorn (1965) and Taylor (1970a, 1970b) found that the circadian variation in body temperature of East African ungulates (1) increased with body mass, (2) was greater in some large species (e.g., camel, oryx, eland, and rhinoceroses) than in others (e.g., cattle, sheep, giraffe, and hippopotamus), and (3) increased in some species with dehydration (e.g., oryx, buffalo, camel, and some small gazelles). Species that used heat storage have the most xeric distributions and thus are most subject to solar heat inputs.

A large species need not show a great variation in body temperature for heat storage to be important: a small increase in temperature in a large mammal is thermally equivalent to a large increase in a smaller species (Equation 5.5). For example, a 3-ton broad-lipped rhinoceros (*Ceratotherium*) shows a daily temperature variation of about 3°C (Allsbrook et al. 1958), which is equivalent to a storage time of about 11 h, if heat is stored at the expected basal rate (see Section 5.3.4). A similar storage time in a 274-kg mammal requires a 6°C increase in body temperature. A 300-g mammal stores enough heat from metabolism to increase body temperature by 3°C in just 46 min.

Factors other than size and dT_b also contribute to a reduction in evaporative water loss. In dromedaries (*Camelus dromedarius*), the thick fur coat is an effective barrier to heat gain from the environment: shorn camels lose approximately 50% more water by evaporation than unshorn camels because of the increased environmental heat load (Schmidt-Nielsen et al. 1956). Consequently, a shorn camel attains the maximal permissible heat storage sooner than an unshorn camel. The fur coat, coupled with a low basal rate of metabolism (see Section 5.3.4), heat storage, and a large mass, permits camels to tolerate heat exchange with the environment at a minimal expenditure of water, and accounts for their remarkable capacity to make long voyages through severe deserts without drinking (Schmidt-Nielsen 1964).

7.6 RENAL WATER LOSS AND EXCRETION IN REPTILES

Renal water loss is highly flexible. It reflects variation in glomerular number and size, filtration rates, rates of solute secretion and resorption along the tubules, type of nitrogenous waste products,

and presence or absence of postrenal and extrarenal handling of salt and water.

7.6.1 Reptilian kidney structure. In reptiles relatively little is known of renal morphology. Detailed descriptions of their renal anatomy are not available, and only slightly more information is available on the structure of their nephrons. Reptilian nephrons are small without a loop of Henle. Although snakes and lizards have many aglomerular tubules, no reptile is known to have aglomerular kidneys. The glomeruli of snakes and lizards have a small diameter (40–70 nm) and are few in number, whereas those of aquatic turtles are larger (55–90 nm) and more abundant (Huber 1917, Marshall and Smith 1930, Marshall 1934). Consequently, snakes and lizards have a smaller total area for filtration than do aquatic turtles. The extent of the ecological and taxonomic diversity in reptilian kidney structure is unknown.

7.6.2 Nitrogenous waste products. Reptiles use three principal nitrogenous waste products: ammonia, urea, and uric acid. The greatest variety in waste products occurs among chelonians: aquatic turtles mainly excrete ammonia and urea, whereas tortoises living in arid or desert regions excrete most nitrogen as uric acid or as a salt of uric acid (Table 7.2; Moyle 1949). Crocodilians have been reported to excrete at least 50% to 65% of nitrogen as ammonia and 20% to 30% as

Table 7.2 Nitrogenous waste products in reptiles

Environment	NH$_3$ (%)	Urea (%)	Uric acid (%)
Crocodylus porosus			
Freshwater	86.5	12.1	1.4
Estuarine	83.0	15.1	1.9
Salt water	49.2	39.4	11.4
Acrochordus granulatus			
Freshwater	86	12	2
Estuarine	81	17	2
Salt water	64	22	14
Turtles			
Aquatic	20–25	20–25	5
Semiaquatic	6–15	40–60	5
Terrestrial			
Hydrophilic	6	30	7
Xerophilic	5	10–20	50–60
Pseudemys scripta	4–44	45–95	1–24
Gopherus agassizi	3–8	15–50	20–50

Sources: Data derived from Moyle (1949), Dantzler and Schmidt-Nielsen (1966), Grigg (1977), and Lillywhite and Ellis (1994).

uric acid (Cragg et al. 1961), although Khalil and Haggag (1958) claimed that the uric acid fraction of crocodilian urine is greater in the Nile crocodyle (*Crocodylus niloticus*). However, freshwater *C. porosus* excreted more than 85% as ammonia, and uric acid was the least important nitrogenous waste product in all environments (Grigg 1977, 1981). Uric acid is a somewhat more abundant waste product in salt water (11%) than in fresh water (1%), an adjustment that may conserve water by excreting nitrogen in an insoluble form (Grigg 1977) (Table 7.2). A similar pattern was seen in the sea snake *Acrochordus granulatus* (Lillywhite and Ellis 1994). Lizards and snakes are thought to use uric acid as the nearly exclusive means of disposing of nitrogen (Khalil 1948a, 1948b, 1951), but very few species have been studied. Marine snakes (as noted) excrete ammonia as the principal nitrogenous waste product (Dunson 1975, Lillywhite and Ellis 1994). The tuatara (*Sphenodon*) principally voids uric acid, although urea is often a significant compound in its urine (Hill and Dawbin 1969).

The analysis of reptilian urine composition is complicated: the proportion of uric acid can be exaggerated because it readily precipitates, whereas the more soluble compounds ammonia and urea are recycled, thereby permitting the accumulation of uric acid (Khalil and Haggag 1955). This accumulation occurs in the cloaca or bladder, which are the sites at which urine is usually sampled. A second problem, especially among turtles, is that the relative abundances of the nitrogenous compounds vary with the degree of hydration, but such variation has not been demonstrated in species for which ureteral urine has been analyzed (Dantzler and Schmidt-Nielsen 1966).

Two principal explanations for the excretion of particular nitrogenous compounds have been used.

1. Toxicity. Ammonia is very toxic (lethal plasma values in mammals are as low as 3×10^{-5} M/L), urea is moderately toxic (see the analysis in Section 6.5.2), and uric acid is essentially nontoxic. Ammonia and (to a lesser extent) urea therefore must be excreted as dilute solutions, which means that species facing a restricted water income use less toxic nitrogenous waste products such as uric acid. Needham (1931) modified this explanation to suggest that the critical period for the toxicity of nitrogenous wastes is during embryological development. He noted that reptiles and birds, among vertebrates, and insects and snails, among terrestrial invertebrates, use uric acid. These groups deposit eggs in dry environments; uric acid is used because water cannot be spared for nitrogen excretion and therefore the developing embryo stores nontoxic compounds within the egg shell.

2. Solubility. Ammonia and (to a lesser degree) urea are highly soluble in water, whereas uric acid is insoluble. Ammonia and urea thus make an appreciable contribution to plasma and cell osmolality, whereas uric acid makes no contribution to the concentration of fluids and is excreted as crystals. The low solubility of uric acid salts has been suggested to be an important factor in the excretion of bound sodium and potassium in reptiles (Minnich 1972).

Neither toxicity nor solubility alone completely account for the pattern of nitrogen excretion. First of all, toxicity correlates with, and may depend on, solubility. Second, toxicity cannot explain why mammals, many of which face severe water shortages, do not use uric acid. This discrepancy may be related to the structure of the mammalian kidney: uric acid may precipitate in the renal pelvis and clog the ureter, or it may require the postrenal modification of urine. (Of course, here again one is faced with the chicken-and-egg problem: does the structure of mammalian kidneys prevent the use of uric acid, or does the use of urea permit the structure of mammalian kidneys?) Third, developing embryos of uricotelic snakes and lizards excrete uric acid only near the time of hatching, so that at earlier stages of development nitrogen is mainly lost as ammonia or urea (Clark 1955, Packard et al. 1977). The principal nitrogenous waste product of an adult cannot be explained by the protein metabolism of a developing embryo if the embryo does not use the same waste product. However, an embryo grows with time and therefore when it is at its greatest mass, highest total rate of metabolism, and highest rate of nitrogen excretion, it uses uric acid as its principal product. Yet, if toxicity to embryos were the principal problem with nitrogenous waste products, the lizards and snakes that are viviparous would be expected to use a different proportion of waste products than those that are oviparous, which seems not to be the case (Packard et al. 1977). Furthermore, many eggs of terrestrial reptiles absorb water during development, thereby diluting the accumulated urea produced.

The interaction of toxicity and solubility accounts for much of the distribution of nitrogenous waste products among vertebrates, although on occasion "phylogeny" is used as an explanation (Minnich 1972). For example, Schmidt-Nielsen and Skadhauge (1967) suggested that the use of uric acid in crocodiles may reflect a terrestrial ancestry secondarily adapted to an aquatic environment. This explanation ignores the extensive use of ammonia and semiterrestrial habits or life in salt water versus freshwater that may contribute to the use of uric acid (see Grigg 1977, Lillywhite and Ellis 1994). Thus, the use of ammonia by some aquatic reptiles (e.g., turtles, snakes, and crocodilians) suggests that some factor countervailing water conservation, such as the high cost of synthesis, selects against uric acid.

The principal nitrogenous waste product used by the "dominant" terrestrial vertebrates has not always been the same. Amphibians may have been in some sense "dominant" in the late Paleozoic; they surely excreted urea. In the early Mesozoic, the dominant (in size, diversity, or "impact") terrestrial vertebrates were probably the line of reptiles that included pelycosaurs and the therapsids, which ultimately gave rise to the mammals. Campbell et al. (1987) argued that this lineage probably evolved the use of urea as its principal waste product at its origin. They in turn were replaced by the sauropsid line that gave rise to the lepidosaurians (i.e., lizards and snakes) and to the archosaurians, which include crocodilians, dinosaurs, and birds, all of which principally (or exclusively) use(d) uric acid as their main waste product. (Although as seen above, this does not apply—secondarily?—to crocodilians.) Campbell et al. (1987, p. 357) proposed that one reason for this replacement was that it occurred during a period, the Permian-Triassic transition, characterized by "... an increase in ... environments with high year-round temperatures and at least one seasonal drought. ... Under such conditions, the selective pressure for water-conserving mechanisms ... was very high. ... The advantages of uricotely in water conservation have been proposed by Robinson (1971) to have been the major factor in the success of the archosaurs. ..." This shift in the use of urea to the use of uric acid occurred in spite of the fact that the "... cost of converting α-amino-N to uric acid is greater than the cost of converting it to urea (direct utilization of nine pyrophosphate bonds vs four for urea)" (p. 357). Ascribing the use of uric

acid to be an important contributor to the replacement of therapsids by dinosaurs is an exaggeration. It ignores the capacity of most mammals to produce a urine more concentrated than the plasma. In any case, the subsequent replacement of dinosaurs by mammals is also unlikely to have involved nitrogenous waste products.

7.6.3 Reptilian kidney function. Few studies of kidney function in reptiles are available, but some generalizations can be made. Glomerular filtration rate is low in reptiles, compared with amphibians, because of small glomerular size and reduced glomerular number (Roberts and Schmidt-Nielsen 1966). Reptilian, like amphibian, glomeruli appear to behave in an all-or-none manner, so that reptiles modify filtration rate by changing the number of glomeruli in which filtration occurs (Forster 1942, Schmidt-Nielsen and Forster 1954, LaBrie and Sutherland 1962, Dantzler and Schmidt-Nielsen 1966). An increase in body temperature leads to an increase in the glomerular filtration rate and the rate of urine production in reptiles (Figure 7.10). The temperature dependence of glomerular filtration is due to a decrease in blood pressure at low body temperatures and not to variations in the number of functioning glomeruli.

The filtrate collected in Bowman's capsule contains all plasma constituents, except cells and large proteins. Therefore, the filtrate has essentially the same osmolality as the plasma. When urine has an osmolality or molecular composition different from the plasma, it is due to active processes occur-

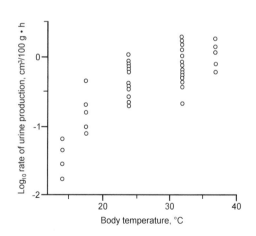

Figure 7.10 Rate of urine production in the lizard *Tiliqua rugosa* as a function of body temperature. Source: Modified from Shoemaker et al. (1966).

ring after filtration, either along the nephron or at postrenal sites, such as in a bladder or cloaca. Thus, if urine is hyposmotic to the plasma, filtered solutes must have been differentially reabsorbed compared to water. Because most (60%–93%) of the water filtered at the glomerulus is reabsorbed (LaBrie and Sutherland 1962), a very high proportion of salt must be reabsorbed to produce a dilute urine. A urine hyperosmotic to the plasma, in contrast, represents a selective removal of water or a selective secretion of solutes, or both. In fact, the production of a urine more concentrated than the plasma depends on a special structure in the nephron, the loop of Henle. No reptile is known to have the loop or is known to produce a hyperosmotic urine.

The concentration of reptile urine tends to conform to one of two patterns (Schmidt-Nielsen and Skadhauge 1967). In chelonians, especially those living in freshwater, the urine is hyposmotic to the plasma; urine concentration, however, is highly variable due to variations in the glomerular filtration rate, in the rate of water reabsorption in the kidney tubule, and in the water or salt loads (Dantzler and Schmidt-Nielsen 1966). These animals generally use urea as the principal nitrogenous waste product. In contrast, lizards, snakes, and crocodilians (at least *Crocodylus acutus* [Schmidt-Nielsen and Skadhauge 1967]) produce hyposmotic (most species) or isosmotic urine; in either case urine concentration shows little variation because the rates of glomerular filtration and water reabsorption remain comparatively constant, irrespective of salt or water loads (Roberts and Schmidt-Nielsen 1966).

7.6.4 Selective abandonment of osmotic regulation.
Because reptiles are not able to produce a urine more concentrated than the plasma, some are forced to store a salt load in the absence of an adequate water intake. Under these conditions reptiles reduce the glomerular filtration rate. For example, the Australian lizards *Trachydosaurus* (Bentley 1959) and *Amphibolurus* (Bradshaw and Shoemaker 1967, Braysher 1976) store salt in their body fluids during summer and are almost anuric. In *Amphibolurus*, the increase in plasma sodium and potassium concentrations is derived from its food, ants. During summer ants also provide *Amphibolurus* with water, thereby permitting it to remain hydrated. However, some populations of *A. ornatus* are polymorphic in response to the salt

loads contained within ants (Bradshaw 1970): slowly growing individuals continue to feed (maintaining body mass and fat stores), whereas rapidly growing individuals first switch from ants of the genus *Iridomyrmex* to those of the genus *Campanotus* (thereby reducing the salt load?) and then stop feeding altogether (leading to a reduction in fat stores and ultimately to death). *Amphibolurus* readily drinks water with a heavy rain and immediately produces urine that rapidly reduces plasma concentrations to normal.

The extracellular storage of electrolytes by *Amphibolurus* may be atypical of reptiles, but very few species have had their plasma osmolalities studied in the field. Minnich (1970) demonstrated that *Dipsosaurus dorsalis* maintains a constant plasma concentration during the summer, even though it obtains a salt load from the plants it eats. It excretes some of its salt load in the urine, especially potassium, as a monourate salt (Minnich 1970, 1972). The excretion of electrolytes by a salt gland (see Section 7.8.3) supplements kidney excretion in the desert iguana, and is undoubtedly indispensable for the maintenance of a constant plasma concentration when a dietary salt load is combined with an inadequate water supply. Given the widespread taxonomic distribution of salt glands, the maintenance of a constant plasma osmolality appears to be more common in lizards than is the tolerance of high plasma concentrations found in *Amphibolurus*, which is not known to have such a gland. The discovery that *Trachydosaurus* has a salt gland (Braysher 1971) raises the question, in light of the observations of Bentley, of whether some species with salt glands use them on a selective basis. This discrepancy may also reflect the dependency of kidney function on body temperature (Figure 7.10).

The desert tortoise (*Gopherus agassizii*) shows a selective abandonment of osmotic constancy during periods of drought (Peterson 1996). When hydrated, this tortoise produces a dilute urine, which is stored in the bladder. With dehydration water is removed from the stored urine until the urine reaches the concentration of the body fluids, after which all body fluids increase in concentration with a continued loss of water. During prolonged droughts, the tortoises may lose up to 40% of initial body mass: plasma concentrations, when the tortoise is hydrated, are approximately 300 mOsm/kg, whereas during the peak of drought they could reach concentrations that exceed

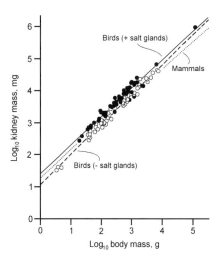

Figure 7.11 $\log_{10}$ kidney mass in birds with and without salt glands and in mammals as a function of $\log_{10}$ body mass. Sources: Modified from Hughes (1970b) and Stahl (1965).

500 mOsm/kg. These high concentrations are eliminated with the first heavy rainfall, when the tortoise drinks rainwater, voids concentrated bladder urine, and stores dilute urine in the bladder.

7.7 RENAL WATER LOSS AND EXCRETION IN BIRDS

7.7.1 Kidney structure. Little is known of the structure of bird kidneys and of its significance for renal function. The available data suggest that body size may be an important parameter of renal function. For example, Hughes (1970b) showed that kidney mass in birds that depend exclusively on the kidney for excretion can be described by the equation

$$K = 0.042m^{0.928}, \qquad (7.3)$$

where K is the mass (in grams) of two kidneys (Figure 7.11). Equation 7.3 means that small species have slightly larger kidneys per unit of body mass than larger species. Marshall (1934) demonstrated that the number of glomeruli in birds increases with body mass, but the allometric relation is unknown. Starlings (*Sturnus vulgaris*), however, have 57% more nephrons per kidney than Gambel's quail (*Callipepla gambelii*), which lives in dry environments and weighs twice as much as the starling but has the same kidney mass (Braun

1978). These observations suggest that climate may be an important factor determining kidney structure in birds, as it is in mammals (see Section 7.10.3). The tolerance to dehydration may be greater in the white-winged dove (*Zenaida asiatica*, 140 g) than in the mourning dove (*Zenaida macroura*, 105 g) and the Inca dove (*Scardafella inca*, 41 g) because of the larger number of nephrons and loops of Henle assumed to be associated with a larger body mass (MacMillen and Trost 1966). In birds, glomerular filtration rates scale proportionally to $m^{0.73}$, but the capacity to produce a concentrated urine relative to plasma is proportional to $m^{-0.10}$ (Calder and Braun 1983); that is, it is greater in smaller species (for a similar relation in mammals, see Sections 7.10.2 and 7.10.3).

Kidney tubules in birds are of two types (Huber 1917, Marshall 1934, Braun and Dantzler 1972, Braun 1978). One has a small glomerulus (diameter of 65–75 nm, volume of 0.032 nL), short proximal and distal convoluted tubules, and no loop of Henle, whereas the other has a large glomerulus (diameter of 100–150 nm, volume of 0.245 nL), long convoluted tubules, and a loop of Henle (Figures 7.12 and 7.13). The former has been referred to as a "reptilian"-type nephron, the latter as a "mammalian"-type nephron. Mammalian nephrons vary in the length of the loop of Henle from 0.7 to 3.7 mm in Gambel's quail (Braun and Dantzler 1972). In this quail approximately 90% of the nephrons are of the reptilian type, whereas 10% are of the mammalian type. In the starling 68% of the nephrons are of the reptilian type and 32%, of the mammalian type (Braun 1978). Thus, the starling not only has more nephrons per kidney than the quail but also has more of the mammalian type.

Avian kidney organization is shown in Figure 7.13. The nephrons are grouped into lobes, each of which has a cortex and a medulla. The cortex contains the glomeruli, proximal and distal convoluted tubules, a (central) efferent vein, and (peripheral) afferent veins and collecting ducts. The medulla is a compact bundle of vessels in the shape of a cone comprising several concentric layers, the innermost being capillaries and descending limbs of the loop of Henle, surrounded by collecting ducts, which in turn are surrounded by ascending limbs of the loops and capillaries (Poulson 1965, Johnson and Mugaas 1970a). Medullas attain a cone shape by the sequential fusion of collecting ducts into a branch of the ureter and by the return of the

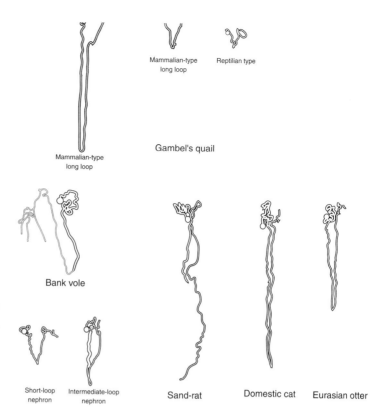

Figure 7.12 Kidney tubules in a bird, Gambel's quail (*Callipepla gambelii*), and in several mammals, bank vole (*Clethrionomys glareolus*), sewellel (*Aplodontia rufa*), sand-rat (*Psammomys obesus*), domestic cat (*Felis sylvestris*), and Eurasian otter (*Lutra lutra*). Sources: Drawings derived from Sperber (1944), Nungesser and Pfeiffer (1965), and Braun and Dantzler (1972).

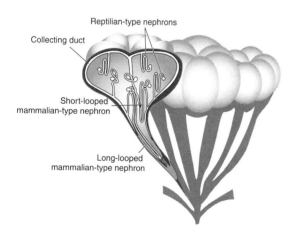

Figure 7.13 Kidney structure in Gambel's quail (*Callipepla gambelii*). Source: Modified from Braun and Dantzler (1972).

loops of Henle to the cortex; only the longest loops extend along the entire length of a cone. Mammalian-type nephrons in a lobe are centrally located, which permits their loops to enter the medulla.

The number of medullary cones in bird kidneys is highly variable (Poulson 1965, Johnson and Mugaas 1970b). Such variation may reflect either the proportion of nephrons with loops of Henle or the proportion of mammalian-type nephrons with long loops. In some species the medullary cones may be very long, so much that the cones may be J-, U-, or W-shaped, a condition that does not necessarily reflect a long loop (Johnson and Mugaas 1970b). The proportion of kidney mass found in the medulla in birds increases with the proportion of nephrons having a mammalian structure.

7.7.2 **Kidney function.** Uric acid is the principal nitrogenous waste product of birds, although no general survey of birds has been made. Lonsdale and Suter (1971) demonstrated that it normally occurs as a dihydrate, and not as a monobasic salt or as a simple acid. Birds apparently eliminate quantities of cations in association with uric acid in the form of negatively charged dihydrate colloids (McNabb and McNabb 1975b). Braun (1978) estimated that up to 98.5% of sodium is lost in association with uric acid in starlings without a salt load; at high salt loads, approximately 50% of sodium is excreted with uric acid. Nitrogen may also be lost as ammonia in domestic ducks and chickens (Shoemaker 1972). For example, McNabb and McNabb (1975a) showed that

Table 7.3 Nephron function in the Gambel's quail (*Callipepla gambelii*)

| Condition | Single-nephron filtration rate (nL/min) | | | Nephrons (% open) | |
	Long mammalian	Short mammalian	Reptilian	Mammalian	Reptilian
Control (diuresis)	15.8	10.9	6.4	100?	71
Salt load	13.4	12.4	0.0?	100?	16
Glomerular volume (nL)	0.247	0.237	0.032		

Source: Derived from Braun and Dantzler (1972).

chickens excrete from 55% to 72% of their nitrogenous wastes as uric acid, 11% to 21% as ammonia, and 2% to 11% as urea, independent of protein or water load. Whether these proportions are capable of ecologically significant variation in birds is unknown. Preest and Beuchat (1994) provided a hint. They showed that Anna's hummingbird (*Calypte anna*) excretes more than 55% of nitrogen in urine as urate and less than 30% as ammonia when water restricted, and less than 40% as urate and more than 50% as ammonia when hydrated.

Glomerular filtration rates are less variable in birds than in most other vertebrates (Shoemaker 1972), and most of the variation in filtration rate occurs in the reptilian-type nephrons (Braun and Dantzler 1972). The difference in behavior between the mammalian- and reptilian-type nephrons is shown in Table 7.3 for Gambel's quail. Mammalian-type nephrons have a high filtration rate, both during diuresis and following a salt load; reptilian-type nephrons, however, have a filtration rate only one-half that of the mammalian type during diuresis, and that rate falls to near zero during salt loading (Braun and Dantzler 1972). Morphological evidence showed that during diuresis approximately 71% of the reptilian-type nephrons were open, whereas after a salt load only 16% were open (and functioning?). During a salt load the blood supply to the reptilian-type nephron is shunted to a circulatory bypass that permits the arterial supply to enter the capillary net surrounding the convoluted tubules without first entering the glomerulus. With an osmotic challenge, most reptilian-type nephrons are consequently turned off, thereby saving water, whereas the mammalian-type nephrons continue to function, which maximizes urine concentration (Dantzler 1970). The total glomerular filtration rate (GFR) in this quail can be estimated from the sum of the filtration rates

from each type of nephron (reptilian or mammalian), each of which equals the product of the total number of nephrons (N), the proportion of reptilian or mammalian nephrons, the proportion of nephrons functioning (p), and filtration rate per nephron (f, nL/min). During diuresis

$$
\begin{aligned}
GFR &= (N)(\text{reptilian})(p)(f) + (N)(\text{mammalian})(p)(f) \\
&= (46780)(0.90)(0.71)(6.4) \\
&\quad + (46780)(0.10)(1.00)(13.3) \\
&= 0.253 \text{cm}^3/\text{min},
\end{aligned}
$$

and after a salt load

$$
\begin{aligned}
GFR &= (46780)(0.90)(0.16)(6.4) \\
&\quad + (46780)(0.10)(1.00)(12.9) \\
&= 0.103 \text{cm}^3/\text{min},
\end{aligned}
$$

if the reptilian nephron filtration rate did not change with a salt load.

Measured values were 0.209 and 0.126 cm³/min, respectively, a 43% reduction in filtration rate. House sparrows (*Passer domesticus*) during dehydration reduce urine flow rates by 85% (Goldstein and Braun 1988). This decrease is correlated with a 54% reduction in the glomerular filtration rate, which accounts only for 64% of the reduction in urine flow. The remainder of the reduction in urine flow presumably is due to a reabsorption of water in the nephrons.

The potential impact of a change in glomerular filtration rate produced by the differential contributions of the two nephron types can be seen in the ability of the quail to modify the concentration of the urine. As an approximation, the mean concentration of the urine over some particular time period may equal $[\Sigma(n_t)(f_t)(r_t)(c_t)]/[\Sigma(n_t)(f_t)(r_t)]$, where c_t is the concentration of the urine associated with a particular nephron type (t). The concentration can be expressed either as an absolute measure of concentration (i.e., mOsm/kg) or relative to the plasma. Unfortunately, urine-plasma (U/P) ratios for individual nephron types were not given in the quail, but the differential shift from the use of reptilian and mammalian nephrons alone would lead to an increase in the U/P ratio with a salt load even if no change in the individual ratios occurred: thus, *if* the mammalian nephron U/P ratio were fixed at 3.0 and the reptilian nephron U/P ratio were fixed at 0.8, then the U/P ratio during diuresis would be

$$\frac{U}{P} = \frac{(46780)(0.90)(0.71)(6.4)(0.8)}{0.253}$$
$$+ \frac{(46780)(0.10)(1.00)(13.3)(3.0)}{0.253}$$
$$= 1.34,$$

and after a salt load

$$\frac{U}{P} = \frac{(46780)(0.90)(0.16)(6.4)(0.8)}{0.103}$$
$$+ \frac{(46780)(0.10)(1.00)(12.9)(3.0)}{0.103}$$
$$= 2.09.$$

Therefore, a simple differential shift in the use of reptilian and mammalian nephrons can (partially) account for a change in the overall U/P ratio, which can be accentuated by variation in the U/P ratios of the nephrons with the exposure to a water or salt load.

In the starling, unlike Gambel's quail and house sparrow, urine flow increased with a salt load, even though total filtration rate remained constant, because the reabsorption of water was reduced (Braun 1978). Starlings increased both urine flow and urine concentration to eliminate the salt load, whereas the quail reduced urine flow and kept urine concentration constant, preferring to conserve water and to permit plasma osmolality to increase. The house sparrow is intermediate in that it decreased urine flow and increased urine concentration. A complete analysis of excretion in birds, however, cannot be given in terms of kidney function because the urine is modified further in the cloaca. Yet, the diversity seen among the few birds examined indicates that many more data on avian kidney function are needed (for a review see Skadhauge 1981).

An osmotic gradient is found along the length of medullary cones in birds. Skadhauge and Schmidt-

Nielsen (1967) found such gradients in chickens and turkeys. The lowest concentrations during salt loading were in the cortex and the highest were in the base, or tip, of the medullary cones. Most of the gradient was due to sodium chloride (NaCl), less than 0.5% being due to the accumulation of urea. During diuresis the gradient was reversed. A refined technique permitted Emery et al. (1972) to measure the gradient present in the budgerigar (*Melopsittacus undulatus*) and a saltmarsh savannah sparrow (*Passerculus sandwichensis rostratus*). In dehydrated budgies the cortex had a fluid concentration of 370 mOsm/kg, which in the medulla increased to 1200 mOsm/kg (Figure 7.14). The fluid in the collecting duct corresponded in concentration to that found in the surrounding tissues, except near the tip of the medulla, where a retrograde mixture of urine may have occurred. A bird produces a concentrated urine by sending the urine through a looped nephron, the function of the loop being to establish and maintain the osmotic gradient. The gradient permits water to diffuse out of the collecting duct and requires that the duct have a position parallel to the loop of Henle (this system is called a *countercurrent multiplier* and is discussed in more detail with reference to mammals [see Section 7.10.2]).

The ability to produce a urine more concentrated than the plasma permits birds, such as the savannah sparrow (Poulson and Bartholomew 1962a), to control plasma concentration over a wide range of salt and water loads. Consequently, U/P ratios increase with a salt load, but only up to some maximum that reflects the capacity of the kidney to produce a concentrated urine. Maximal U/P ratios in birds generally vary from about 1.3 to 2.8 (Table 7.4). The highest U/P ratios have been found in a saltmarsh savannah sparrow (*P. s. beldingi*): mean maximal osmolal U/P ratio equaled 4.5 and mean maximal chloride U/P ratio equaled 5.5. In the field, however, this subspecies had a mean U/P ratio equal to 1.65, whereas a migratory, upland subspecies that did not face a salt load but may have been subject to dehydration had a mean U/P ratio equal to 1.44 (Goldstein et al. 1990). Birds may attain somewhat higher anion than osmolal U/P ratios (e.g., Poulson and Bartholomew 1962a, 1962b), although these ratios may be underestimated if anions are bound to precipitated uric acid as are cations.

The effectiveness of the avian excretory system is shown by the ability of some species to maintain

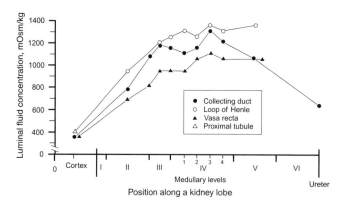

Figure 7.14 Concentration of luminal fluids in the kidney of dehydrated budgerigar (*Melopsittacus undulatus*) as a function of the location in the kidney. Source: Modified from Emery et al. (1972).

Table 7.4 Maximal urine concentrations and maximal urine-plasma (U/P) ratios in birds

Species	Maximal [urine] (mOsm/kg)	Maximal U/P
Dromaiidae		
Dromaius novaehollandiae	459	1.4
Phasianidae		
Colinus virginianus	643	1.6
Callipepla californica	669	2.0
Callipepla gambelii	884	2.5
Columbidae		
Zenaida macroura	544	1.3
Streptopelia senegalensis	661	1.7
Geophaps lophotes	655	1.8
Psittacidae		
Cacatua roseicapilla	982	2.5
Cacatua sanguinea		1.9
Platycercus zonarius		2.6
Neophema bourkii		2.5
Alcedinidae		
Dacelo gigas	944	2.7
Meliphagidae		
Meliphaga virescens	925	2.4
Anthochaera carunculata	917	2.4
Emberizidae		
Amphispiza belli	674	1.9
Passerculus sandwichensis	2010	4.5
Fringillidae		
Carpodacus mexicanus	770	2.1
Estrildidae		
Taeniopygia castanotis	1005	2.8
Ploceidae		
Passer domesticus	826	2.1

Sources: Derived from Poulson and Bartholomew (1962a, 1962b), Smyth and Bartholomew (1966a), McNabb (1969), Moldenhauer and Wiens (1970), Skadhauge (1974a), and Goldstein and Braun (1988).

body mass while drinking concentrated salt solutions, and by the ability of other species to live under laboratory conditions without drinking water. Sparrows of the North American genus *Amphispiza* living in xeric environments (black-throated sparrow, *A. bilineata* [Smyth and Bartholomew 1966a]) can drink more highly concentrated solutions than those living in less xeric conditions (sage sparrow, *A. belli* [Moldenhauer and Wiens 1970]). Saltmarsh savannah sparrows (*P. s. beldingi* and *P. s. rostratus*) can drink solutions up to 0.7 M NaCl and maintain body mass (seawater is about 0.55 M NaCl [Cade and Bartholomew 1959, Poulson and Bartholomew 1962a, Johnson and Ohmart 1973a]). Zebra finches (*Taeniopygia castanotis*) from Australia also can drink concentrated solutions, in part because they are able to produce a concentrated urine (ca. 1000 mOsm/kg; maximal osmolal U/P ratio equals 2.7 [Skadhauge and Bradshaw 1974]). The zebra finch, however, cannot produce a urine more concentrated than 0.5 M NaCl ($\cong$ 1000 mOsm/kg), so that at solutions of greater concentration drinking rate is greatly reduced (Figure 7.15). At a concentration of 0.8 M NaCl, saline water intake in the zebra finch is only about 15% of the total water turnover.

Some birds are so parsimonious with water that they can survive in the laboratory on a dry-seed diet without drinking. These species include the zebra finch (Cade et al. 1965), black-throated sparrow (Smyth and Bartholomew 1966a), budgerigar (Cade and Dybas 1962), Brewer's sparrow (*Spizella breweri* [Ohmart and Smith 1970]), and vesper's sparrow (*Pooecetes gramineus* [Ohmart and Smith 1971]).

Variations in renal function are probably linked to variations in kidney structure, but this interdependency has been difficult to demonstrate in birds because of few data. Presumably, the ability to produce a concentrated urine (in analogy with mammalian kidneys) depends on the length of the loop of Henle, on the length of an osmotic gradient, and therefore on the length of the medullary cones. Anatomical data collected by Johnson (Johnson and Mugaas 1970b; Johnson and Ohmart 1973a, 1973b; Johnson 1974), however, did not demonstrate a correlation between the concentration of a solution that a bird can drink and the number or length of medullary cones. Part of the problem is that little morphological differentiation occurs in medullary cone length in small birds (Johnson 1974), possibly because of uniformly high rates of water turnover and a limitation to kidney size. Among large birds, aquatic species have shorter medullary cones than terrestrial species (Johnson 1974). Because kidney mass and glomerular filtration rate in birds are correlated with body mass raised to a power less than 1.0, relative kidney function, as expected by the maximal U/P ratio in xeric species, correlates negatively with body mass ($m^{-0.09}$), although this ratio is only one-sixth the value in mammals of the same size (Calder and Braun 1983).

The extent to which kidney function in birds is generally correlated with environmental conditions is unclear. The clearest examples are found in sparrows belonging to the genera *Passerculus*, *Amphispiza*, and *Spizella*. Some structural modifications of the kidney may be more radical. Thomas and Robin (1977, p. 246) suggested that the smaller kidneys of desert species of sandgrouse (*Pterocles*) in Africa, compared to those of steppe species, are "... equivalent to a permanent and adaptive antidiuresis...." Sandgrouse, in addition, have long medullary cones compared to other birds of similar size. Sandgrouse have low estimated

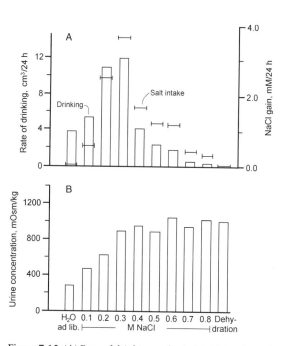

Figure 7.15 (A) Rate of drinking and salt (NaCl) intake and (B) urine concentration in the zebra finch (*Taeniopygia castanotis*) as a function of the concentration of the fluid ingested. Source: Modified from Skadhauge and Bradshaw (1974).

rates of water turnover (Thomas and Robin 1977, Thomas and Maclean 1981). The arid-dwelling Gambel's quail also has small kidneys (and low filtration rates [Williams et al. 1991]). In contrast, salt-marsh savannah sparrows (*P. s. beldingi*) have kidneys that are 1.62 times the mass of upland, migratory savannah sparrows, this difference appearing to reflect the higher salt loads faced by the salt-marsh population. Thus, kidneys may be reduced in size if a species is faced with a water shortage, but not if they face a high salt load. As a result, birds with salt glands, which typically face high salt or protein loads, have larger kidneys than do species without salt glands (see Section 7.8.3). Associated with the larger kidneys in *P. s. beldingi* is a urine flow rate 4 to 6 times that found in the upland population (Goldstein et al. 1990).

Guiliano et al. (1998) examined the response of two quail to water deprivation. They compared female northern bobwhites (*Colinus virginianus*), a resident of mesic eastern North America, with female scaled quail (*Callipepla squamata*), which live principally in the Chihuahuan Desert. With a restriction of water consumption to two-thirds and to one-third of ad libitum intake, bobwhites lost body mass and kidney mass and had an increase in serum osmolality, whereas scaled quail lost mass only at one-third water intake but showed no change in kidney mass or serum osmolality. During these experiments, females of both species continued to lay eggs, but at the intermediate level of water intake, bobwhites had larger ovaries and a higher rate of egg production than did scaled quail. In response to water deprivation, scaled quail, which live in an environment "... where sufficient rainfall needed for successful reproduction is infrequent and unpredictable..." (p. 785), preferentially allocated resources to body maintenance, whereas bobwhites preferentially channeled resources to reproduction.

A simple correlation of kidney function with bird distribution may not exist (Skadhauge 1974a) because birds are highly mobile and adjust their use of water and food to reflect environmental conditions. Such correlations are further complicated by the postrenal modification of urine and by extrarenal salt excretion in birds. A correlation of kidney size and function with environmental conditions in fact may be present, but principally in species that do not have salt glands (e.g., passerines and *Pterocles*).

7.8 EXTRARENAL SALT AND NITROGEN EXCRETION IN REPTILES AND BIRDS

The excretion of electrolytes and nitrogenous wastes in reptiles and birds is not the exclusive activity of the kidneys. Many reptiles modify urine stored in a bladder, reptiles and birds modify the composition of urine in the cloaca, and some reptiles and birds use cephalic glands to excrete electrolytes.

7.8.1 Bladder. The bladder is a storage site for urine produced by the kidneys. Among birds it is found only in the ostrich, but among reptiles bladders have a complex pattern of distribution. A bladder is found in *Sphenodon* (where it is small) and in turtles but is absent in crocodilians. Snakes lack bladders, although Hoffmann (1890) described swellings in the ureters of some species anterior to their entrance into the cloaca; he called these swellings "bladders." Whether they store urine is unknown. Bladders have been found or implied (Cope 1898, Beuchat 1986a) in most lizard families with the probable exception of the Varanidae and Teiidae.

A comparative study of excretion in the freshwater turtle *Pseudemys scripta* and the desert tortoise *Gopherus agassizii* (Dantzler and Schmidt-Nielsen 1966) demonstrated the ecological significance of urine storage in the bladder. *Pseudemys scripta*, when hydrated, produces a urine hyposmotic to the plasma, which is stored in the bladder, where urine retains its hyposmoticity. When moderately dehydrated, *P. scripta* produces an isosmotic urine. With continued dehydration, this turtle stops filtration. *Gopherus agassizii*, in contrast, produces hyposmotic urine, even when it is markedly dehydrated. During storage in the bladder, urine increases in concentration to reach isosmoticity with the blood as a result of water being withdrawn from the bladder. Although the bladders of both species are permeable to water, the permeability of the bladder in *G. agassizii* appears to be much greater.

The difference in bladder function between these turtles is related to the use of uric acid by *G. agassizii* and its near absence in *P. scripta*. In *G. agassizii* the uric acid filtered by the kidney precipitates in the bladder with the diffusion of water out the bladder. Urea, which is highly soluble, also diffuses out of the bladder, returns to general circulation, and is converted to uric acid, which is excreted by the kidney and sent to the bladder, where it precipitates. In this manner, waste nitrogen is taken out of circulation and stored as a precipitate in the bladder. The accumulation of uric acid in the bladder produces the highly variable ratios of uric acid to urea found in terrestrial tortoises (e.g., Khalil and Haggag 1955), and warns against the assumption that a sample of bladder urine accurately represents the composition of ureteral (or kidney) urine. Continued filtration during dehydration has little value for *P. scripta* because only a small fraction of its nitrogenous waste is uric acid and therefore can be taken out of circulation.

Dantzler and Schmidt-Nielsen (1966) noted that the desert tortoise does not store urine in the bladder when it is feeding, a time when the tortoise, a herbivore, has a large intake of K^+ ions. If a urine rich in K^+ were stored in the bladder, K^+ would freely diffuse out of the bladder and accumulate in the body and plasma. The tortoise only eats succulent plants, which ensures that enough water is ingested for K^+ elimination to occur.

The relevance of this picture of bladder function to other reptiles is uncertain, especially given the spotty taxonomic distribution of bladders. Can a very small lizard have as functional a bladder as a large lizard? If two related species differ by the presence or absence of a bladder, what does that mean in terms of their water relations or their nitrogenous wastes?

Beuchat et al. (1986) partially answered these questions in a study of bladder urine in neonatal lizards belonging to the species *Sceloporus jarrovi*. They showed that bladder urine at birth is very dilute (36 mOsm/kg) and acts as a source of water that can be drawn upon to delay dehydration. No urine is added to the bladder after birth: the filled bladder at birth is 14% of body mass but gradually decreases to a vestigal condition. Why the bladder degenerates is unclear, especially because it occurs without apparent regard for habitat.

7.8.2 Cloaca. The contribution of cloacal function to osmotic regulation is only slightly better understood than the contribution of the bladder. The cloaca in reptiles and birds is undoubtedly a major site of water conservation. Junqueira et al. (1966), for example, showed that snakes (*Xenodon*) have a highly vascularized cloaca; if the ureters are permitted to open to the exterior, rather

than into the cloaca, *Xenodon* loses body mass at rates 3 to 4 times those of controls. Death occurs in these snakes after about 5 d of dehydration, whereas control snakes show no death within 20 d.

The relation of cloacal water conservation to electrolyte excretion is unclear. Urine enters the cloaca directly from the ureters, whereupon it may move anteriorly as far as the large intestine (Ohmart et al. 1970, Shoemaker 1972, Skadhauge 1976). In reptiles, the concentration of urine produced by the kidney is maximally isosmotic with the plasma. Water is usually thought to be passively withdrawn from the cloaca by the active transport of sodium from the urine, thereby conserving water and precipitating uric acid and various monobasic urates (Schmidt-Nielsen et al. 1963). The active transport of Na^+ may require an extrarenal pathway for the elimination of electrolytes. Yet, all vertebrates with a cloaca do not use extrarenal salt excretion (e.g., most snakes, many lizards, and most birds do not have salt glands). If cloacal reabsorption of sodium occurs in these species, sodium excretion is difficult to understand unless sodium is lost as insoluble urates (Minnich 1972).

Shoemaker (1972) suggested that the rates of water and solute exchange are so low in the cloaca that they can influence urine concentration only during dehydration, that is, when glomerular filtration rates are reduced. Shoemaker concluded that the cloaca does not enhance the concentration capacity of the kidney; the cloaca in fact may simply permit the precipitation of uric acid, which cannot occur in the kidney because of possible clogging of the ureters and their branches. However, measurements of renal and cloacal function in the lizard *Amphibolurus maculatus*, a species that inhabits dry lake beds in South Australia, indicate that the cloaca may indeed be a site of urine concentration (Braysher 1976). When given a salt load, this lizard produces a urine in the kidney that is hyposmotic to the plasma (osmolal U/P ratio = 0.74), which when voided is hyperosmotic to the plasma (osmolal U/P ratio = 1.48). Samples of urine collected from the cloaca are intermediate (osmolal U/P ratio = ca. 1.08). In some manner, either water is withdrawn relative to the electrolytes in the cloaca, or the cloacal walls in this species are relatively impermeable to water and electrolytes are secreted into the cloaca.

A comparison of renal and cloacal function in a crocodile (*Crocodylus porosus*) and the American alligator (*Alligator mississippiensis*) showed that both species can easily tolerate exposure to hyposmotic salt water (ca. 200 mOsm/kg), but they accomplished this tolerance in radically different ways (Pidcock et al. 1997). *Alligator mississippiensis* in freshwater produced urine with low concentrations of Na^+ and K^+, which remained nearly the same when the urine was retained in the cloaca. With exposure to dilute salt water, the alligator had high urinary concentrations of Na^+ and Cl^-, which decreased during retention in the cloaca. The crocodile showed little change in the urine with exposure to dilute salt water and even less of a change in the urine during retention in the cloaca. These two crocodilians, then, differ in their response to osmotic and ionic environments: the alligator depends principally on a renal response (which may be undercut in the cloaca), whereas the crocodile uses a combination of renal, postrenal, and extrarenal reponses.

The ability of birds to produce a urine more concentrated than the plasma is a potential complication for cloacal function. If the cloacal walls were permeable to water, water would be drawn from the plasma to the urine, thereby undoing the concentrating work of the kidney, or if the cloaca was not permeable to water, all of the water excreted by the kidney would be lost, an amount that is appreciable given the inability of most birds to produce a high U/P ratio. Skadhauge (1974b) investigated cloacal water exchange in an Australian parrot, the galah (*Cacatua roseicapilla*), which can produce an osmolal U/P ratio equal to 2.5 when dehydrated. Skadhauge concluded that a hyperosmotic urine will gain water from the plasma, but isosmotic solutions in the cloaca will lose water in association with the active transport of ions. The postrenal absorption of water does not necessarily involve the cloaca: hydrated house sparrows (*Passer domesticus*) permit urine to enter the ileum (from the cloaca), where much of the water is reabsorbed, but when dehydrated (and urine is more concentrated than the plasma), house sparrows do not permit urine to leave the cloaca (Goldstein and Braun 1988). Dehydrated house sparrows absorb water in the ileum only from food that is being processed. The emu (*Dromaius novaehollandiae*), which can maximally produce a U/P ratio equal to 1.4 to 1.5, can reabsorb in the rectum most of the water coming from the ileum and from the urine. In this species the functioning of the lower gut is exceedingly important in tolerating a

desert climate, aided by the emu's tendency to feed on plants containing sufficient water to excrete the contained solutes (Skadhauge et al. 1991). The use of a cloaca for urate precipitation may limit the capacity of birds to produce a concentrated urine, and may explain why no birds are known to produce the concentrated urines typical of most mammals. The question remains, do birds capable of higher U/P ratios have lower water permeabilities in the cloaca?

Water movement through the cloacal walls need not involve the active transport of cations (Murrish and Schmidt-Nielsen 1970). In *Dipsosaurus dorsalis* water moves along an osmotic gradient produced by the colloidal osmotic pressure of plasma proteins, which is counteracted by the negative hydrostatic pressure of the fecal and urine mixture present in the cloaca (see the Scholander-Hammel concept of negative pressures in Section 2.10). Thus, water will be removed from a cloacal fluid by the colloidal osmotic pressure of the blood circulating through the cloacal walls until the force keeping water in the cloaca is equal and opposite to that of the plasma. Much more work needs to be done to clarify the significance of a cloaca to osmotic regulation in reptiles and birds.

7.8.3 Salt glands. Many reptiles and birds use an extrarenal pathway for salt excretion. All known cases of such excretion involve cephalic glands. The distribution of salt glands is taxonomically patchy, and the glands used are varied. For example, about half of the lizard families contain some species that have salt glands, all of which are derived from nasal glands. No terrestrial reptiles other than lizards have salt glands. Salt glands occur generally in large families that are ecologically diverse, whereas the families that apparently do not have salt glands, with the exception of the Gekkonidae, have few species and often have burrowing habits. Most snakes do not have salt glands. The two snake families that use sublingual glands for salt excretion (Elaphidae and Acrochordidae) have a marine or estuarine distribution (see Dunson and Taub 1967; Dunson et al. 1971; Dunson and Dunson 1973, 1974). The only chelonians known to have salt glands, all of which are derived from lachrymal glands, are sea turtles (Chelonidae and Dermachelidae) and the estuarine diamondback terrapin (*Malaclemys*, Emydidae). *Sphenodon* does not have a salt gland. Taplin and

Grigg (1981) and Taplin et al. (1982) showed that crocodiles, but not caiman and alligators, have lingual salt glands.

The salt glands of birds are derived from the nasal region, which because of the structure of a bird's beak are displaced to an orbital or supra-orbital position. These glands are widespread in about one-half of avian orders, mainly in those that live near marine environments. The only terrestrial birds known to have nasal glands modified for salt excretion are the ostrich, some hawks, a partridge, and the roadrunner. Although some passerines face appreciable salt loads, especially in salt marshes, none is known to have salt glands (e.g., see Rounsevell 1970, Paynter 1971). As noted, passerines rely exclusively on the renal excretion of salt loads, which may be why passerines have the highest known U/P ratios.

Most detailed studies of salt gland structure have been made in birds. In marine species each gland is constructed of a series of lobes, most of which run the length of the gland (Figure 7.16). Each lobe has a central canal into which empties a series of blind, branched tubules arranged in a radial fashion around the central canal. Between the tubules is a capillary network that carries blood in a direction countercurrent to fluid movement in the tubules (Schmidt-Nielsen 1960). The tubules are lined with cuboidal (presumably secretory) epithelium. The size of the avian salt gland reflects the experience of an individual with salt water (Schildmacher 1932). Thus, Schmidt-Nielsen and Kim (1964) showed that ducklings raised with dilute chloride solutions have larger salt glands than those raised with distilled water.

In marine and estuarine snakes a great variation in salt gland size occurs. Gland size is related to body size: species with large glands show a decrease in relative gland size with an increase in body mass, whereas in species with small glands relative gland size is independent of body size (Figure 7.17). The factors other than body mass that determine sublingual gland mass are not clear. It may be associated with variations in either renal function or skin permeability (Dunson and Dunson 1974, Dunson 1975). For example, the small size of the salt gland and its independence from body size in *Hydrophis elegans* reflects a low skin permeability.

Unlike urine, salt gland solutions only contain significant concentrations of salts; no ultrafiltration (i.e., filtration under pressure) occurs (Schmidt-

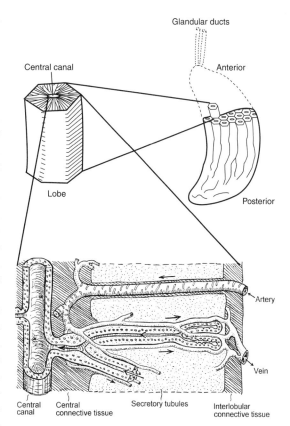

Figure 7.16 The structure of the salt gland in the herring gull (*Larus argentatus*). Source: Modified from Fänge et al. (1958).

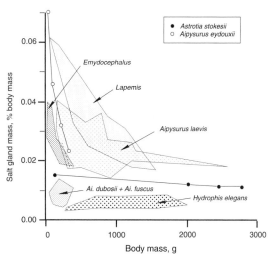

Figure 7.17 Salt gland mass in marine snakes as a function of body mass. Source: Modified from Dunson (1975).

Table 7.5 Salt gland secretion concentrations

Species	Food	Condition	Cl⁻ (mEq/L)	Na⁺ (mEq/L)	K⁺ (mEq/L)	[Secretion] (mOsm/kg)
Reptiles						
Crocodilians						
Crocodylus porosus	A	Lab	512	509	11.2	1085
		Field	632	663	20.8	>1316
Crocodylus acutus	A	Lab maximum		498	12.4	
Crocodylus johnstoni	A	Lab maximum		386	10.3	
Alligator mississipiensis	A	Lab maximum		186		
Turtles						
Lepidochelys olivacea	A	Average	782	713	28.8	>1524
Lizards						
Sauromalus obesus	P	Average	827	150	1102	>2097
Ctenosauria pectinata	P	Lab	477	439	253	>1379*
		Field	487	78	527	>1215*
Amblyrhynchus cristatus	P	Average	1256	1434	235	>2925
Conolophus subcristatus	P	Average	486	692	214	>1392
Snakes						
Pelamis platurus	A	Average	627	607	16	>1240
Aipysurus laevis	A	Average	791	798	28	>1617
Lapemis harwickii	A	Average	704	676	23	>1403
Hydrophis elegans	A	Average	520	509	20	>1049
Acrochordus granulatus	A	Average	492	483	15	>990
		Seawater	548	470	10	~1140
Birds						
Procellariiformes						
Oceanodroma leucorhoa	A	Range		900–1100		
Sphenisciformes						
Spheniscus humboldtii	A	Range	635–805	726–840	21–29	
Pelecaniformes						
Phalacrocorax auritus	A	Average	517	529	12	>1058
Pelecanus occidentalis	A	Average	698	722	13	>1433
Gruiformes						
Fulica americana	P	Average	542	530		>1072
Gallirallus owstoni	O	Average	785			
Falconiformes						
Micronisus gabar	A	Average	983	1023	45	>2051
		Maximum	2441	2316	128	>4885
Buteogallus meridionalis	A	Average	455	433	16	>904
Terathopius ecaudatus	A	Average	256	259	9	>524
Charadriiformes						
Larus glaucescens	A					
		Drinking seawater	872	785	68	>1725
		Na/K load = 400/100	677	646	61	>1384
		Na/K load = 250/250	633	545	94	>1272
Larus argentatus	A	Average	720	718	24	>1462

A, animal; P, plant; O, omnivorous.
*In *Ctenosaura* the anion HCO_3 was found: 210 mEq/L in lab animals and 123 mEq/L in field animals.
Sources: Derived from Schmidt-Nielsen and Fänge (1958), Schmidt-Nielsen et al. (1958), Schmidt-Nielsen and Sladen (1958), Schmidt-Nielsen (1960), Templeton (1964, 1967), Cade and Greenwald (1966), Dunson (1969), Carpenter and Stafford (1970), Hughes (1970a, 1975), Dunson et al. (1971), Nagy (1972), Dunson and Dunson (1973, 1974), Taplin and Grigg (1981), and Taplin et al. (1982).

Nielsen 1960). The principal ions concentrated are those most commonly encountered in the environment or in food, namely, Na^+ in marine species, or K^+ in terrestrial herbivorous lizards, and Cl^- (Table 7.5). In some terrestrial lizards the nasal gland and kidney can excrete bicarbonate, thereby regulating the acid-base balance (Norris and Dawson 1964, Templeton 1967). Bicarbonate is excreted when the cation loads (usually K^+) are high and the chloride loads are low (Templeton 1967). The relative abundances of Na^+ and K^+, or of Cl^- and HCO_3^-, vary widely, even within a given species (Table 7.5).

With a salt load the flow rate and concentration of salt gland secretions increase to a maximum, but the quantitative importance of nasal gland secretions stems mainly from their concentration rather than from their volume. The maximal concentration of these secretions is always greater than that of urine, and is often greater than the concentration of ocean water: the maximal osmolal U/P ratio in birds is normally 1.3 to 3.0 (Table 7.4), whereas the ratio of salt gland secretion to plasma concentration (G/P ratio) is usually about 5 or 6 (Carey and Schmidt-Nielsen 1962). Double-crested cormorants (*Phalacrocorax auritus*), for example, require twice as much water to excrete a given amount of salt via the kidneys as via the nasal glands (Table 7.6; Schmidt-Nielsen et al. 1958), as is to be expected from the ability of nasal glands to produce a secretion twice as concentrated as that produced by the kidney.

A great variation occurs in both the secretion rates (ca. 10-fold) and the maximal concentrations (ca. 2-fold) of avian nasal glands. Although the maximal concentrations produced by the salt glands of sea snakes are similar to those of the nasal glands of birds, the total capacity for salt exchange is about 10 times greater in birds because of the higher secretory rates in birds. The rate of

fluid excretion by avian salt glands is about 3 times that of reptilian salt glands (Dunson and Dunson 1974). This difference is principally due to temperature, as shown by the following calculation: if a marine reptile has a body temperature equal to 25°C, if the temperature of an avian salt gland is 38°C, and if the Q_{10} of nasal gland secretion is 2.5, then the factorial difference in rates between these glands should be (Equation 4.4)

$$\frac{\text{avian}}{\text{reptile}} = 2.5^{\frac{38-25}{10}} = 3.2,$$

which is what is observed. In fact, Staaland (1967b) reduced the volume of nasal gland secretion in ducks by local cooling.

The relative importance of the salt gland and kidney for the excretion of electrolytes varies greatly. Approximately 50% of NaCl in the double-crested cormorant load is lost by each pathway (Table 7.6) (Schmidt-Nielsen et al. 1958). Birds other than the cormorant, however, have a greater dependence on nasal glands; they account for 70% to 80% of salt loss in the gull *Larus glaucescens* (Hughes 1970a), for 80% in the pelican *Pelecanus occidentalis* (Schmidt-Nielsen and Fänge 1958), and for 90% in the penguin *Spheniscus humboldti* (Schmidt-Nielsen and Sladen 1958). In birds, potassium is lost mainly through the renal-cloacal pathway.

The nasal gland in birds varies in size, not only as a function of experience but also as a function of body size. Peaker and Linzell (1977) described nasal gland size (N, in grams for the two glands) in marine birds as a function of body size (in grams):

$$N = 0.0024m^{0.92} \tag{7.4}$$

(Figure 7.18A). Hughes (1968) showed a similar relation in the common tern (*Sterna hirundo*). Birds with salt glands have two kidneys (K, in grams), the sizes of which are given by

$$K = 0.052m^{0.879} \tag{7.5}$$

(Hughes 1970b). They are approximately 25% larger than kidneys found in species without salt glands (see Equation 7.3 and Figure 7.11). In absolute size, kidneys have 20 times (= 0.052/0.0024) the mass as nasal glands. Thus, salt glands supplement, not replace, kidney function.

Variation in gland function depends on variation

Table 7.6 *Phalacrocorax* water and salt balance*

Condition	Cl	Na	K	H$_2$O
	(mEq)			(g)
Salt and water ingested	54	54	4	50
Cloacal elimination	27.5	25.6	2.66	108.9
Nasal elimination	26.1	23.8	0.31	51.4
Total elimination	53.6	49.4	2.97	160.3

*Response of a 1.82-kg cormorant to ingesting 60 g of meat loaded with 3 g of NaCl.

Source: Derived from Schmidt-Nielsen et al. (1958).

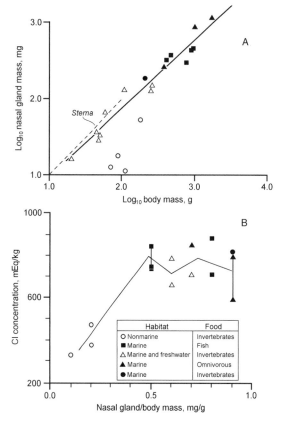

Figure 7.18 A. Log₁₀ mass of nasal gland in marine birds and in the common tern (*Sterna hirundo*) as a function of log₁₀ body mass. B. Chloride concentration of the secretion of salt glands in charadriiform birds as a function of the relative salt gland size. Sources: Modified from Staaland (1967a) and Hughes (1968).

what more concentrated secretion (600–800 mEq/L); the great black-backed gull (*L. marinus*), which is more strictly marine than the herring gull, has a still more concentrated secretion (700–900 mEq/L); and Leach's petrel (*Oceanodroma leachii*), which feeds extensively on marine plankton, has a highly concentrated secretion (1000–1200 mEq/L). Marine invertebrates, unlike marine teleosts, are generally isosmotic with the sea, and therefore birds that feed on marine invertebrates have much higher salt loads. Marine teleosts, then, have contributed to the osmoregulation of their predators.

The correlation of nasal gland size and function with the habits and habitats of birds can be seen in five species of Australian cormorants (*Phalacrocorax*) (Table 7.7). Nasal gland mass reflects both the environment and food habits: the glands are larger in individuals found in marine or estuarine environments than in individuals of the same species found in freshwater environments, are larger in species restricted to marine or estuarine environments, and are larger in invertebrate-eating species. That is, gland mass is larger in species and individuals that have higher salt loads. The observation that individuals collected in freshwater environments had smaller nasal glands than individuals of the same species collected in marine or estuarine environments implies either that little movement must have occurred between these environments, or that hypertrophy and atrophy occur very rapidly with movement and exposure to new conditions. Such information is required to understand the functioning of salt glands in birds (and reptiles).

The failure of nasal glands in birds presumably would tend to the collapse of osmotic regulation. Cooch (1964), in a study of migratory ducks on saline lakes in Nebraska, suggested that avian botulism is a disease that results from the action of a neurotoxin produced by the bacterium *Clostridium botulinum* on the nasal glands of ducks, thereby leading to an osmoregulatory failure. Such a collapse may occur because function of the avian salt gland is controlled by the parasympathetic nervous system (Fänge et al. 1958). Botulism is most prevalent on alkaline and saline lake basins, where functional salt glands are required. Ducks showing mild symptoms of botulism may be restored to health by drinking freshwater. The ducks most susceptible to botulism have small nasal glands; that is, they are migrants to alkaline or saline regions from areas with waters of low osmolality. Bang (1964) noted

in gland dimensions (Staaland 1967a). Chloride concentration in salt gland secretion is determined in part by the radius of the lobes and thus by the size of the nasal gland (Figure 7.18B), that is, by the length of the tubules. Volume of secretion is related mainly to gland size in birds (Staaland 1967b), as it is in reptiles (Dunson and Dunson 1974). The variation in the chloride concentration of the nasal gland secretion correlates with a bird's habits (Staaland 1967a; Figure 7.18B).

The influence of food habits in determining variation in nasal gland function has been shown in birds (Schmidt-Nielsen 1960). For example, the double-crested cormorant (*Ph. auritus*), a strict fish eater, produces a gland secretion that has a relatively low maximal Na⁺ concentration (500–550 mEq/L); the herring gull (*L. argentatus*), which feeds on invertebrates and fish, has a some-

Table 7.7 Nasal gland size and food habits of cormorants

Character	Species				
	Black (*carbo*)	Little Black (*sulcirostris*)	Pied (*varius*)	Little Pied (*melanoleucus*)	Black-faced (*fuscescens*)
Marine/estuarine environments					
Body mass (g)	2626	1030	1557	683	1515
Abundance (%)	18.9	43.8	15.8	20.4	1.2
Stomach contents					
Fish (%)	99.3	96.6	93.1	82.2	100.0
Crustacea (%)	7.3	10.2	17.2	36.6	0.0
Nasal gland (mg/100 g body mass)	14.4	29.4	26.7	26.6	30.2
Freshwater environments					
Body mass (g)	2018	811		697	
Abundance (%)	43.0	30.9	3.8	22.3	0.0
Stomach contents					
Fish (%)	92.6	81.2	90.9	46.5	
Crustacea (%)	13.2	20.0	9.1	64.0	
Nasal gland	5.5	10.9		13.3	

Source: Modified from Thomson and Morley (1966).

that vultures are very resistant to botulism and have very well-developed nasal glands.

Factors other than high salt loads also stimulate the excretion of salt by nasal glands. Many hawks initiate secretion upon feeding, the maximal concentrations reported being surprisingly high (Table 7.4). Falconiform birds have appreciable protein loads as a result of their carnivorous habits; these loads may limit the capacity of kidneys to handle ingested salts. A similar argument may apply to marine fish-eating birds, which because they feed on fish that are only slightly hyperosmotic to bird plasma, normally do not have high salt loads (Hughes 1968). Cade and Greenwald (1966) described a bateleur eagle (*Terathopius ecaudatus*) chick that upon exposure to the sun, started to pant and within a minute copiously excreted from its nasal glands. An ostrich produced nasal gland secretions upon exposure to the sun (Schmidt-Nielsen et al. 1963).

If some birds that face a high protein load use salt glands to void salts and kidneys to void nitrogenous wastes, why then do so many carnivorous birds not use (have) nasal glands? All accipterids studied by Cade and Greenwald showed some salt excretion by nasal glands, but falconids did not. For example, the American kestrel (*Falco sparverius*) showed no such secretion. Is this behavior related to a small body mass? And why have not owls shown functional salt glands? Could their absence be related to nocturnal habits and the evasion of large solar loads? Or, are these differences simply phylogenetic correlates?

The spotty taxonomic distribution and diversity of glands used for salt excretion suggest that the evolution of extrarenal excretion occurred repeatedly. All vertebrates, except mammals, use structures other than kidneys for significant levels of salt excretion: gills (fishes, amphibians), rectal glands (elasmobranchs, chimaera, coelacanth), and cloaca (reptiles, birds), as well as the various cephalic glands found in reptiles and birds. Peaker and Linzell (1977) suggested that nasal glands, which are found in all tetrapods (Bang and Bang 1959), were originally cephalic salt glands. These glands secondarily lost their salt secretory function in many lizards (why?), in snakes (in relation to burrowing?), in crocodilians, and in mammals (where they became tear glands?). Chelonians that live in a marine or estuarine environment secondarily converted lachrymal glands to salt glands. Dunson and Dunson (1973) concluded that sea snakes of the Elaphidae and Acrochordidae independently evolved a salt secretory role from the posterior sublingual gland. Crocodilians independently evolved lingual salt glands.

Whether the use of nasal glands for salt excretion by birds was obtained from reptiles, or was evolved de novo, is unknown. The property shared by birds and reptiles that is likely to be responsible for the presence of salt glands is an excretory system that uses uric acid in conjunction with a cloaca. The near restriction of salt glands in birds to aquatic species is difficult to understand, but they were established at least as early as the Cretaceous in †*Hesperornis* and †*Ichthyornis*, as seen by impressions of glands in the skulls of these birds (Marples 1932). Whether dinosaurs had salt glands is unknown: Whybrow argued that hadrosaurs that were terrestrial, or semiterrestrial, had salt glands, and Osmólska (1979) suggested that most, if not all, large herbivorous dinosaurs had salt glands located (unusually for a vertebrate) in the nares to compensate for their large potassium intake.

7.9 THE RETURN TO AQUATIC HABITS IN REPTILES

Several groups of reptiles have evolved aquatic habits, presumably from terrestrial ancestors. Aquatic reptiles today include crocodilians and some turtles, snakes, and lizards, as well as various extinct reptiles, including mosasaurs, pleisiosaurs, and ichthyosaurs. Unlike aquatic birds and mammals, aquatic reptiles have appreciable rates of integumental exchange of water and salt with the environment (e.g., Lillywhite and Ellis 1994). Some of this exchange, at least in sea snakes and some freshwater turtles, is associated with integumental gas exchange (see Section 8.9).

The integumental water and salt exchange of crocodilians may be associated with their differential tolerance to saline waters. For example, Bentley and Schmidt-Nielsen (1966) found that the (freshwater) caiman (*Caiman sclerops*) had a rather high integumental permeability to water. Doubt exists, however, as to whether they adequately controlled for water exchange through buccal membranes (Dunson and Mazzotti 1988). Calculations by Taplin (1984b, 1985) suggested, and measurements by Dunson and Mazzotti indicated, that the integuments of *Crocodylus porosus*, *Caiman crocodillus*, and *Alligator mississippiensis* have rather low permeabilities, similar to those of other freshwater and estuarine reptiles, but that buccal and nictitating membranes have very high permeabilities. The high permeability of buccal surfaces is associated in *Crocodylus* with the excretion of sodium by lingual salt glands (Taplin 1985), but a high buccal permeability to water is also found in alligators

(Dunson and Mazzotti 1988), which do not have salt glands.

In contrast to water exchange, integumental sodium exchange in crocodilians is very low (Taplin 1985, Dunson and Mazzotti 1988). Most sodium exchange in crocodiles occurs in association with the lingual salt glands, whereas in alligators it occurs through a renal-cloacal pathway that requires an adequate supply of water obtained either from food or from drinking freshwater. Unlike sodium, most potassium is excreted via the renal-cloacal pathway in crocodiles (Taplin 1985). As a consequence of their exchange system, estuarine and freshwater crocodiles have plasma osmotic concentrations that are independent of ambient concentration, although the Australian freshwater crocodile (*C. johnstoni*) living in hypersaline waters may require periodic access to hyposmotic water (Taplin et al. 1993). A similar situation may also exist in the "estuarine" American crocodile (*C. acutus* [Dunson 1982, Dunson and Mazzotti 1988]). Some evidence exists that *A. mississippiensis* cannot regulate plasma electrolytes at high external salinities (Lauren 1985). An examination of the broad-snouted caiman (*Caiman latirostris*), which is found in estuaries in southern Brazil, demonstrated that individuals weighing less than 500 g could tolerate external concentrations only up to about 90 mOsm/kg and larger individuals (for a limited period) could tolerate external concentrations up to 650 mOsm/kg, at least as long as they have access to freshwater (Grigg et al. 1998).

The occurrence of lingual salt glands in all freshwater and saltwater crocodiles, including the genera *Crocodylus* and *Osteolaemus*, is compatible with the suggestion that extant crocodiles were derived (recently?) from marine/estuarine crocodiles (Densmore and Dessauer 1982; Taplin et al. 1982, 1985; Taplin 1984a). Thus, even the predominantly freshwater crocodile *C. johnstoni* and the Nile crocodile *C. niloticus* (Taplin and Loveridge 1988) have salt glands that function when these crocodiles are exposed to salt water. The Southeast Asian freshwater crocodilians *Tomistoma schlegelii* and *Gavialis gangeticus* have lingual salt glands (Taplin et al. 1985), although apparently not as well developed as in *Crocodylus* and *Osteolaemus*. In contrast, *Alligator* and *Caiman*, both of which belong to the family Alligatoridae, appear to have had a long-term freshwater ancestry.

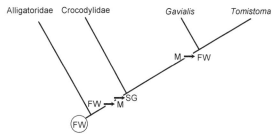

Figure 7.19 Phylogeny of living crocodilians and its implication for the evolution of freshwater and marine habits. FW, freshwater; M, marine water; SG, salt gland. Source: Modified from Poe (1996).

This pattern of gland distribution in crocodilians led Taplin et al. (1985) and Taplin and Grigg (1989) to suggest that alligators fundamentally are freshwater in distribution, but that all other living crocodilians are either marine/estuarine or are freshwater species that were derived from marine/estuarine species. A difference in renal and cloacal function that has also been found substantiates this evolutionary dichotomy between *Alligator* and *Crocodylus* (Pidcock et al. 1997). Furthermore, the worldwide distribution of *Crocodylus* primarily resulted from transoceanic dispersal, a behavior that was facilitated by an impermeable integument and lingual salt excretion. As a result, Taplin and Grigg (1989) offered a cladogram of eusuchians indicating the possible points at which lingual and buccal modifications for saltwater habits evolved. A modified cladogram, based on morphology and on DNA fragments and sequences (Poe 1996), is shown in Figure 7.19: it suggests that freshwater is plesiomorphic for all living crocodilians, but reaffirms that marine habits are pleisomorphic for all crocodilians other than alligators and caimen, but that some of this "marine" lineage has secondarily invaded freshwater habitats. Mazzotti and Dawson (1989), however, suggested that other hypotheses for the evolution of crocodilian osmotic regulation may exist, including the possibility that lingual salt glands in freshwater crocodilians are a response to seasonal droughts.

Almost nothing is known of the salt excretion of extinct marine reptiles. At least four major groups were at one time or another important predators in the sea: mosasaurs (lizards related to varanids), sauropterygians (including nothosaurs and plesiosaurs), placodonts, and ichthyosaurs. Some of the mosasaurs, most plesiosaurs, and all ichthyosaurs fed exclusively on marine fish. These reptiles might have handled salt loads with their kidneys, because the fluid concentrations of their prey would have been nearly isosmotic to the reptiles' plasma. Yet, estuarine crocodiles, which also are vertebrate predators, use lingual salt glands. Whybrow (1981) suggested that at least one ichthyosaur, †*Opthalmosaurus*, had nasal glands. Placodonts (Romer 1956) and some mosasaurs (Russell 1967), however, fed on marine molluscs. They surely must have had extrarenal salt excretion, but no evidence of cephalic salt glands has been found in the structure of their skulls.

7.10 RENAL WATER LOSS AND SALT EXCRETION IN MAMMALS

Unlike reptiles and birds, mammals depend almost exclusively on renal function for salt and water balance, except for some minor salt losses through sweating and salivary secretion. The kidney is also the principal means by which nitrogenous wastes are voided in mammals, the nearly exclusive waste product in mammals being urea. The egg-laying mammals *Tachyglossus* and *Ornithorhynchus* use urea as their predominant nitrogenous waste product (Denton et al. 1963). Renal function is closely tied to renal structure. An extensive comparative study of the structural organization of mammalian kidneys by Sperber (1944) laid the basis for all subsequent studies of the relation between kidney structure and function and among kidney structure, food habits, and environment. Much of the anatomical information that follows was derived from this source.

7.10.1 Kidney structure. The structure of mammalian nephrons is similar to the structure in other vertebrates (Figure 7.12), except that all mammalian nephrons have loops of Henle (Schmidt-Nielsen 1964). A loop consists of the distal portion (pars recta) of the proximal convoluted tubule and the thin and thick segments of the nephron. Some nephrons have short loops and others have long loops; Sperber (1944) also distinguished cortical nephrons, short-looped nephrons that lie completely within the cortex. Depending on the length of the loop, the hairpin turn may occur either in the thick segment (short-looped) or in the thin segment (long-looped).

The kidney consists of an outer layer, the cortex, and an inner region, the medulla. The cortex contains the glomeruli, Bowman's capsules, most of the proximal convoluted tubules, the distal part of the thick segments, the distal convoluted tubules, and part of the collecting ducts. The medulla contains the loop of Henle for most nephrons and most of the collecting ducts. Because of nephron alignment, the medulla is often divided into two zones: the inner zone and the outer zone. The only part of the nephrons contained within the inner zone is the thin segments of the long-looped nephrons. The outer zone contains thin and thick segments of the loop of Henle.

Species differences in kidney structure reflect the variable proportion of long- to short-looped nephrons, which is most noticeable in the "simple" kidneys found in small mammals (Figure 7.20A).

At one extreme, all (beaver, *Castor castor* [Sperber 1944]) or nearly all (ca. 98%, sewellel, *Aplodontia rufa* [Nungesser and Pfeiffer 1965]) of the nephrons are short-looped. At the opposite extreme, all the nephrons in some small carnivores (*Mustela*) and the sand-rat (*Psammomys*) are long-looped (Sperber 1944). In most other mammals, only a fraction of the nephrons have long loops (e.g., in the kangaroo-rat *Dipodomys merriami* approximately 27% of its nephrons are long-looped). Thus, desert seed-eating rodents, such as *D. merriami*, *Notomys alexis*, and *Cricetulus griseus*, have a long, narrow papilla that extends into the ureter (Sperber 1944, MacMillen and Lee 1969, Trojan 1979; Figure 7.20A).

Superimposed on this renal organization is an

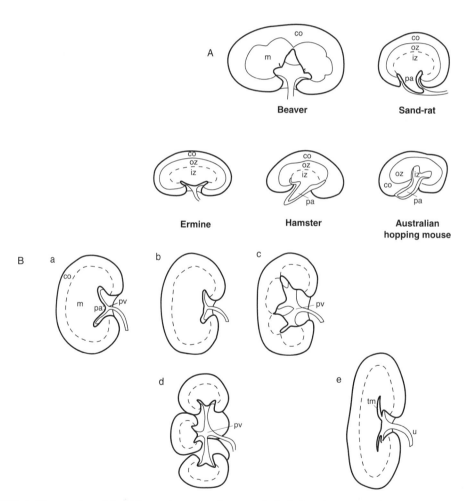

Figure 7.20 A. Cross section of the kidneys of some small mammals, including the Eurasian beaver (*Castor castor*), sand-rat (*Psammomys obesus*), ermine (*Mustela erminea*), hamster (*Cricetulus griseus*), and Australian hopping mouse (*Notomys alexis*). B. Types of mammal kidneys: (a) simple with papilla, (b) crested, (c) with several papillae, (d) renculi, and (e) with tubi maximi. co, cortex; m, medulla; pa, papilla; pv, pelvis; oz, outer zone of medulla; iz, inner zone of medulla; tm, tubus maximus; u, ureter. Sources: Modified from Sperber (1944), MacMillen and Lee (1969), and Trojan (1979).

afferent arterial blood supply to the glomerulus and an efferent arterial drain, which carries blood into the medulla, where the arterioles break up into capillaries. The capillaries form a network, the vasa recta, that surrounds the loops of Henle and collecting ducts and carries blood countercurrent to the flow in the loops; the capillaries then recombine into venules that empty into the renal vein (Plakke and Pfeiffer 1964). In species with a well-developed zonation in the medulla, a complementary zonation occurs in the blood vessels of the medulla, but in species with little or no zonation of the nephrons, none is found in the vasa recta.

As in most other functions, renal function in mammals varies with body size, and this correlation has an anatomical expression. Rytand (1938) showed that kidney mass (in grams) in mammals varies with body mass (in grams) according to the relation (adjusted to two kidneys)

$$K = 0.016 m^{0.85}, \qquad (7.6)$$

which is nearly identical to the relation Brody (1945) described for mammals and similar to the relations described for birds (Equations 7.3 and 7.5; Figure 7.11). Rytand further showed that both glomerular volume and number are a function of kidney mass, which when combined with the relation incorporated in Equation 7.6, means that total glomerular volume is proportional to body mass raised to the 0.85 power (also see Holt and Rhode 1976, Calder and Braun 1983). Glomerular filtration rate, as measured by inulin clearance, is proportional to $m^{0.72}$ and urine volume is proportional to $m^{0.75}$ (Edwards 1975). These renal functions clearly scale to body mass in a manner similar to that of basal rate of metabolism. In spite of this similarity, the glomerular filtration rate of birds, which scales identically with mammals, is only about 37% that of mammals (Calder and Braun 1983), even though birds generally have higher basal rates. This difference in filtration rate between birds and mammals may reflect the use of uric acid and urea, respectively, as the principal nitrogenous waste products, and may contribute to the greater capacity to produce a concentrated urine by mammals (by about 6.5-fold; Calder and Braun 1983).

Body mass also affects renal structure and function through its influence on the length and diameter of the kidney tubules. The length of the proximal convoluted tubule is correlated with kidney size, at least in kidneys less than 80 mm. A consequence of increasing nephron length is that the resistance to flow within the tubule increases because resistance is proportional to the surface area in the tubule and hence to its length. Resistance can be decreased by increasing the diameter of the tubule, and this is exactly what happens: the diameters of the proximal convoluted tubule, thin segment, thick segment, and distal convoluted tubule are correlated with the length of the proximal convoluted tubule, at least at lengths less than 100 mm. As we will see, the increase in nephron diameter influences kidney function.

Kidney form varies with kidney size (Sperber 1944; Figure 7.20B). Mammalian kidneys having a dimension less than 2 cm are simple in form, with either one papilla, or in the case of aquatic species, none. Other kidneys have a more complicated structure. Kidneys that are 3 to 6 cm have a short or moderate crest, which in form is somewhat like an elongated, low papilla. In larger species the kidneys are subdivided in various ways, presumably to maintain a surface-volume ratio that ensures exchange and drainage: they may have several papillae (with subdivided medullary areas); they may be so subdivided that the cortex and medulla are allocated to separate papillae (the so-called renculi kidney); or the kidney may be simple in form, but with two large collecting ducts entering the pelvis from opposite sides (tubi maximi). Kidneys with multiple papillae are found in pigs and the pigmy buffalo (*Anoa*); renculi are in panda, bears, cetaceans, elephants, and rhinoceroses; and tubi maximi are in capybara (*Hydrochaerus*), chevrotain (*Tragulus*), cervids, and the horse.

7.10.2 Kidney function. The structure of the mammalian nephron is intimately related to its function, although its interpretation has been changing (compare Gottschalk and Mylle [1959] with Kokko and Rector [1972] and Jamison and Robertson [1979]). Some views are summarized in Figure 7.21. Ultrafiltration, the movement of small molecules under arterial pressure through a differentially permeable membrane, occurs at the glomerulus because the small diameter of the efferent arteriole maintains a high blood pressure in the glomerular capillaries and because this hydrostatic pressure exceeds the colloidal osmotic pressure of blood. The filtrate, which is essentially isosmotic with the plasma, is collected by Bowman's capsule

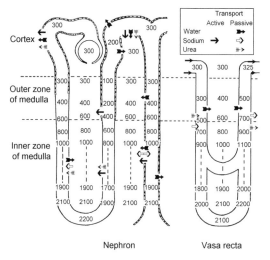

Figure 7.21 Schematic diagram of a mammalian nephron and its associated vasa recta. Concentrations are in mOsm/kg. Source: Modified from Gottschalk and Mylle (1959).

and conducted to the proximal convoluted tubule. Most of the filtered water passively diffuses out of the proximal tubule with the active reabsorption of sodium. As the remaining filtrate moves down the descending limb of the loop, the filtrate moves through tissues of increasing concentration, and water passively diffuses out of the tubule, leaving a concentrated salt solution in the tubule. The ascending thin segment is impermeable to water, but NaCl diffuses out of the tubule; some urea may diffuse into the tubule in this region. In the ascending thick segment, NaCl is actively transported out of the tubule. The walls of the distal convoluted tubule and the collecting duct are permeable to water (especially under the influence of the antidiuretic hormone [ADH]), whereas the walls of the collecting duct found in the inner medulla are permeable to urea as well. Consequently, most of the osmotic gradient in the outer medulla is due to the accumulation of NaCl, whereas urea is an important component of the gradient in the inner medulla. As it passes down the collecting duct, urine is concentrated by the passive diffusion of water to the surrounding tissues.

As Schmidt-Nielsen (1958) pointed out, the countercurrent configuration of the descending and ascending limbs of the loop of Henle maintains an osmotic gradient in the surrounding medullary tissues, but the active transport of ions is needed to establish such a gradient. The total medullary gradient established within the mammalian kidney

is well beyond the differentials produced across a membrane by secretion alone. In fact, the medullary gradient is the sequential product of the small osmotic differentials (maximally about 200 mOsm/kg) established by the active transportation of NaCl across the walls of the ascending limb of the loop. Because the sodium and chloride removed from the ascending limb contribute to the concentration of the tissues external to the limb, these ions contribute to a higher base against which active tubular transport can occur. The modest osmotic differential maintained across the wall of the ascending limb, thus, is "multiplied" into a much larger corticomedullary gradient external to the nephron. This behavior of the loop of Henle has been referred to as *countercurrent multiplication*.

To understand the ecological consequences of kidney function in mammals, selected features of nephron behavior must be examined in detail.

1. Rates of filtration at the glomerulus and resorption along the nephron vary greatly. Mammals and birds have filtration rates that are 30 to 100 times those of marine fishes and 4 to 20 times those of freshwater fishes and amphibians (Schmidt-Nielsen 1964). These high rates are due to the high body temperature and the high blood pressure of endotherms (see Shoemaker et al. 1966). For example, the filtration rate of a single glomerulus, F_{sg} (nL/s), depends on the difference between the differential in hydrostatic pressure (P, in mm Hg) and the differential in colloidal osmotic pressure (π, in mm Hg) existing between the glomerular capillaries (g) and the kidney tubule (t):

$$F_{sg} = k \cdot A(\Delta P - \Delta \pi)$$
$$= k \cdot A[(P_g - P_t) - (\pi_g - \pi_t)], \qquad (7.7)$$

where k is the hydraulic permeability coefficient of the capillary (nL/s·mm Hg·cm²) and A is surface area (cm²) of the glomerular capillaries (Brenner et al. 1976). In the salamander *Necturus maculatus*

$$F_{sg} = k \cdot A[(17.7 - 1.5) - (10.4 - 0.0)]$$
$$= 5.8(k \cdot A)$$

at the arteriole afferent to the glomerulus (data from White 1929). At the afferent arteriole in the squirrel monkey (*Saimiri sciureus*)

$$F_{sg} = k \cdot A[(48.5 - 12.6) - (24.4 - 0.0)]$$
$$= 11.5(k \cdot A)$$

(Maddox et al. 1974). *S. sciureus* has a total glomerular filtration rate that is about 4.5 times that of *N. maculatus*, which means that nearly half of this difference (2.0 = 11.5/5.8) is related to a greater blood pressure in the monkey, the remainder being due to variations in k, A, and the number of functional glomeruli.

If the filtration rate near the afferent arteriole is compared with that near the efferent arteriole at the circulatory exit from the glomerulus, insight is given into the functioning of the glomerulus. Because much water is lost by filtration in the glomerulus and because the proteins are not filtered, the concentration of protein in blood increases as the blood moves through the glomerulus. In the squirrel monkey at the efferent arteriole

$$F_{sg} = k \cdot A[(48.5 - 12.6) - (36.6 - 0.0)]$$
$$= -0.7(k \cdot A)$$

which indicates that the blood entering the vasa recta will absorb water from the surrounding tissues because the differential in colloidal osmotic pressure is greater than the differential in hydrostatic pressure. Filtration, clearly, is a self-limiting process (Starling 1899).

Equation 7.7 implies that filtration is limited by the protein content of blood: as protein concentration increases, filtration decreases, unless a compensatory increase in blood pressure occurs. This is one reason why hemoglobin is contained within cells in vertebrates (see Section 8.4.3), and why marine teleosts that use glycoproteins to reduce the freezing point of plasma in antarctic waters tend to be aglomerular (see Section 6.6): filtration possibly could not occur even if a glomerulus were present.

Estimates have been made in humans of the amount of fluid that is filtered at the glomerulus and the amount of filtrate that is reabsorbed along the tubule. About 10% of blood volume is filtered at the glomerulus, 60% to 70% of the filtered volume being reabsorbed in the proximal convoluted tubule in association with the active transport of ions. Only about 1% of the filtered volume (i.e., 0.1% of the blood entering the glomerulus) is excreted as urine under normal conditions of hydration. All of the water reabsorbed from the nephron is returned to blood, so that the blood leaving the kidney has 99.9% of its original volume. These values are variable: during maximal diuresis about 10% of the filtered volume in mankind is excreted as urine; during maximal

water conservation about 0.2% of the filtered volume is excreted. These parameters of excretion appear to be independent of mass in mammals (Calder and Braun 1983).

2. Hargitay and Kuhn (1951) first suggested that the loop of Henle was a countercurrent multiplier. They described a model in which the concentration of the fluid within a tubule, C_x, depends on the position along the loop, x; on the original concentration of the fluid entering the tubule, C_o; on the diameter of the tubule, d; and on a coefficient k that incorporates permeability, pressure, and velocity of the filtrate:

$$\frac{C_x}{C_o} = \frac{1}{1 - k(x/d)}. \tag{7.8}$$

(Kuhn et al. [1963] described a similar countercurrent multiplier system in the gas gland of the swim bladder [see Section 9.3].)

Equation 7.8 indicates that the concentration along the loop increases with loop length and decreases exponentially with an increase in nephron diameter. Notice that the two morphological parameters x and d are inversely related to each other, so that the concentration capacity of the loop is related to the ratio x/d. When this ratio increases, the concentration at any point x along the loop, C_x, increases. Equation 7.8 indicates why the large kidneys found in large mammals are not automatically capable of producing a concentrated urine, even though they tend to have longer loops of Henle. As Sperber (1944) showed, nephrons in large kidneys are longer and have a greater diameter than nephrons in small kidneys. Therefore, large kidneys cannot normally produce as concentrated a urine as small kidneys because the increase in x is not sufficient to compensate for the increase in d. Sperber therefore argued that the best measure of kidney capacity was a relative index of loop length (relative medullary thickness), and not the absolute dimensions of the tubules themselves (see later). Schmidt-Nielsen and O'Dell (1961; Figure 7.22) confirmed this conclusion.

Beuchat (1990) reexamined the correlation of urine concentration capacity with body size. She showed that although medullary thickness increases with body mass (Figure 7.23A), relative medullary thickness decreases with mass (Figure 7.23B), and because maximal urine osmolality increases with relative medullary thickness (Figure 7.23C), maximal urine osmolality decreases with

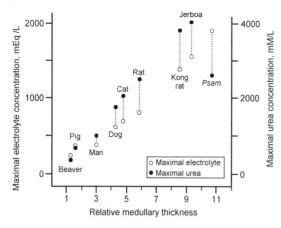

Figure 7.22 Maximal electrolyte and urea concentrations produced by various mammalian kidneys as a function of the kidney's relative medullary thickness. Source: Modified from Schmidt-Nielsen and O'Dell (1961).

mass (Figure 7.23D). As Beuchat clearly pointed out, however, the appreciable variation in the relation between urine osmolality and relative medullary thickness means that other factors must also play a part in determining the maximal concentration of urine, potentially including nephron heterogeneity, the structure of the collecting ducts and renal vasculature, the development of medullary zones, and so on.

3. The concentrations of luminal fluids along the nephron are those expected from the countercurrent multiplier hypothesis. For example, Gottschalk and Mylle (1959) and Ullrich et al. (1963) showed that the fluid in the proximal convoluted tubule is isosmotic to the plasma, and that the fluid in the proximal end of the distal convoluted tubule is hyposmotic to the plasma, whereas the fluid in the distal section is, again, isosmotic, independent of diuresis or antidiuresis. Furthermore, Ullrich and Jarausch (1956) and Gottschalk et al. (1963) demonstrated that the concentration of urine is correlated with the sodium concentration of the interstitial fluids of the papilla (Figure 7.24).

4. Tubular fluid is concentrated as it moves down the collecting duct through tissues of increasing concentration. Water passively leaves the collecting duct, depending on the permeability of its walls. The pituitary hormone ADH increases the permeability of the collecting duct to water and therefore contributes to the production of a concentrated urine (Berliner et al. 1958). Water that leaves the collecting duct and nephron enters the

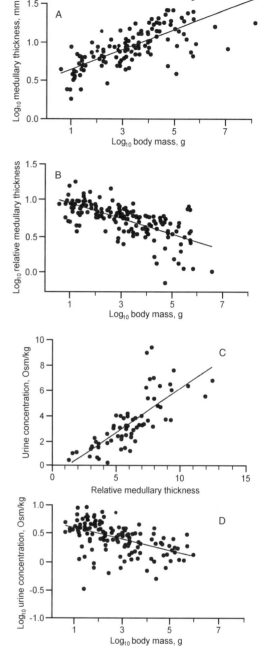

Figure 7.23 The structure and function of mammalian kidneys: (A) $\log_{10}$ medullary thickness as a function of $\log_{10}$ body mass; (B) $\log_{10}$ relative medullary thickness as a function of $\log_{10}$ body mass; (C) urine concentration as a function of relative medullary thickness; and (D) $\log_{10}$ urine concentration as a function of $\log_{10}$ body mass. Source: Modified from Beuchat (1990).

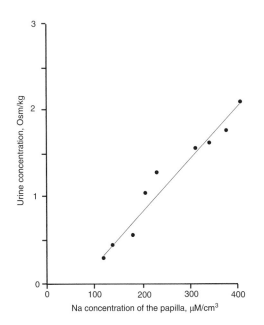

Figure 7.24 Concentration of urine as a function of the sodium concentration of the interstitial fluids of the kidney papilla. Source: Modified from Ullrich and Jarausch (1956).

surrounding tissues and the vasa recta. Water presumably enters the vasa recta from both the duct and the surrounding tissues because concentration of the plasma is greater, mainly due to the colloidal osmotic pressure of plasma proteins. That is, the differential hydrostatic pressure between the plasma and tissue fluids, which forces water out of the vasa recta, is less than the opposite differential in colloidal osmotic pressure at each level in the kidney (see Equation 7.7, and Wirz 1953), which thus draws water from the tissue fluids. Control of these processes permits the osmotic gradient in the medulla to be maintained and regulated.

5. Urea contributes to the medullary osmotic gradient in most mammals, its relative importance being related to their ability to concentrate urea. Most mammals can concentrate urea more than electrolytes, but some (e.g., beaver, pig, *Psammomys*) may concentrate these substances equally (Schmidt-Nielsen et al. 1961). Some of these differences may be related to species-specific permeabilities of the collecting duct to urea. Pfeiffer (1968) suggested that the capacity to concentrate urea may be related to renal morphology. Thus, urea may be recycled from the renal pelvis, thereby contributing to high urea concentrations, especially if the pelvis has many folds and secondary pyramids (which increase the surface area for

exchange); if the papillae are covered with squamous, rather than stratified, epithelium; and if many thin loops are present in the papilla. Urine formation is not complete until the urine enters the ureter.

7.10.3 Ecological correlates of renal function. Marked variations in kidney form and function occur in mammals in relation to habits and climate (see Section 7.10.1). For instance, the proportion of nephron types within a kidney varies greatly: cortical nephrons are found almost exclusively in herbivores, although they are also present in monotremes; some carnivores have only long-looped nephrons; and in some mammals that have very low protein and salt loads and that live in aquatic or wet terrestrial environments, from 0% (*Castor*) to 2% (*Aplodontia*) of the nephrons have long loops (e.g., Pfeiffer et al. 1960). The width of the papilla is related to the ratio of long to short nephrons and, ultimately, to the amount of fluid in the diet (Schmidt-Nielsen 1964). Seed eaters, like in the genus *Dipodomys*, have a low water intake and a long, thin papilla, but species, such as in the genus *Psammomys*, that feed on succulent plants with high salt loads excrete a large volume of urine, have a high fraction of long-looped nephrons, and consequently have a wide papilla. MacMillen and Lee (1969) found a similar dichotomy between two Australian murids, *Notomys alexis* and *N. cervinus*, as did Carpenter (1969) for some desert bats (*Eptesicus fuscus* and *Pizonyx viviesi*). In each case the second species had an appreciable fluid intake and a wide papilla.

Sperber (1944) provided the classic description of the ecological significance of kidney structure. (Note that Sperber defined kidney size as the cube root of the product of a kidney's length, width, and height.) He showed that a relative medullary index, which is defined as the thickness of the medulla divided by the kidney size index, correlates with environment: desert-dwelling seed eaters have large indices and aquatic species, small indices. This index depends on mass, as Sperber first recognized and Greegor (1975) and Blake (1977) formally described: the relative medullary thickness decreases with an increase in mass (Figure 7.23B). This inverse relation may reflect water conservation compensatory for the high rates of water turnover typical of small species. Greegor inappropriately referred to a "weight-specific RMT": relative medullary thickness (RMT) is already mass specific because it incorporates kidney thickness in the denominator, which is a function of body mass. Medullary thickness per se is a "total" index, but it is a poor measure of kidney capacity because, as already noted, an increase in medullary thickness produced by an increase in size is accompanied by an increase in nephron diameter.

Relative medullary thickness correlates with the environment (Figure 7.25). Greegor (1975) described two relations, one for mesic mammals:

$$RMT = 5.17m^{-0.090}, \quad (7.9)$$

which is nearly identical to the equation described independently by Blake (1977) and recently by Beuchat (1990), and one for xeric species:

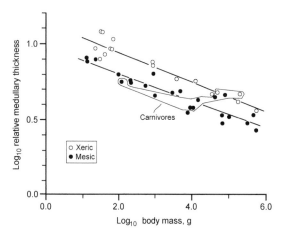

Figure 7.25 Log_{10} relative medullary thickness in mammals as a function of log_{10} body mass and distribution in xeric or mesic environments. Notice that the correlation of relative medullary thickness with climate is found even in carnivores in spite of the high dietary protein loads, although that may explain why some species living in mesic environments have a high medullary thickness. Source: Modified from Greegor (1975).

$$RMT = 7.04m^{-0.097}, \quad (7.10)$$

where mass is in grams. Xeric species, thus, tend to have a relative medullary thickness that is approximately 36% greater than that in mesic species of the same mass (ignoring the stastistically insignificant difference in the exponent of mass; Figure 7.25). The correlation of urine concentration with relative medullary thickness (Schmidt-Nielsen and O'Dell 1961, Heisinger and Breitenbach 1969, Blake 1977, Trojan 1977, Beuchat 1990) (Figure 7.23C) means that the capacity to produce a concentrated urine is correlated with a xeric or mesic distribution. In eutherians both the maximal U/P ratio and the relative medullary thickness scale approximately proportional to $m^{-0.9}$, the maximal U/P ratio being about 32% greater in xeric than in mesic species (Calder and Braun 1983). That is, the functional differences in the kidney with respect to habitat type reflect the morphological differences in the kidney. Some relative renal indices other than relative medullary thickness have been proposed to describe the concentrating capacity of mammalian kidneys (Heisinger et al. 1973, Brownfield and Wunder 1976).

The correlation of renal capacity with climate and food habits is clear. Heisinger et al. (1973) showed that not only does the maximal urine concentration in rodents, even at the subspecies level, vary with medullary thickness, but also this index varies with mean annual rainfall (Figure 7.26). The granivorous-insectivorous *Reithrodontomys* and *Peromyscus* have higher renal indices and higher maximal urine concentrations than does grazing *Microtus*, a correlation that reflects the water and protein contents of the foods used (Figure 7.27). Some desert, seed-eating rodents have carried the capacity to produce a concentrated urine so far as to have maximal osmolal U/P ratios as high as 26.8

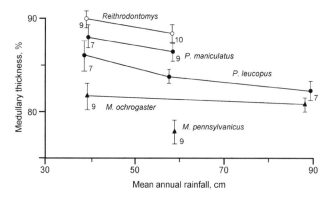

Figure 7.26 Renal medullary thickness in some rodents as a function of the mean annual rainfall encountered by these rodents. Source: Modified from Heisinger et al. (1973).

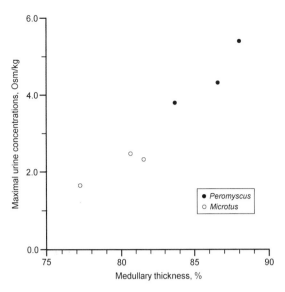

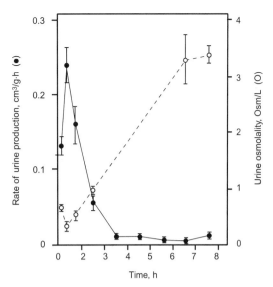

Figure 7.27 Maximal urine concentration in *Peromyscus* and *Microtus* as a function of the renal medullary thickness. Source: Data derived from Heisinger et al. (1973).

Figure 7.28 Urine flow and urine concentration in the vampire bat (*Desmodus rotundus*) as a function of time since the beginning of a meal. Source: Modified from Busch (1988).

(Table 7.8)! Similar correlations between urine concentration and renal morphology and between renal morphology and environments were found in three species of chipmunks (*Tamias* [Blake 1977]), in four species of Eurasian hamsters (Trojan 1977), and in many insectivorous bats (Geluso 1975, 1978; Bassett and Wiebers 1979; Bassett 1982, 1986; Happold and Happold 1988).

The convergent nature of the ability to produce high U/P ratios by the kidney in mammals distributed in arid regions is illustrated by the data summarized in Table 7.8. This ability is found in various rodent families, including Sciuridae, Heteromyidae, Cricetidae, Muridae, and Dipodidae, and on various continents, including Eurasia, Africa, Australia, and North America. Although some cricetid rodents in South America belonging to the genera *Calomys* and *Phyllotis* are thought to be somewhat independent of drinking water (Koford 1968, Mares 1977, Meserve 1978, Cortés et al. 1988), none are known to live without access to free water. Bozinovic et al. (1995) used the data of Cortés et al. (1990) to show that some small rodents living in the Mediterranean climate of coastal Chile are capable of producing rather high U/P ratios (Table 7.8) but have much lower ratios of water produced by metabolism to evaporative water loss than is found in some heteromyid rodents. Chilean rodents are not as good urine concentrators as Australian rodents or as effective at

surviving without drinking water as water-independent heteromyids. Bozinovic and colleagues argued, however, that a better comparison would be between Chilean rodents and a *balanced* sample of Australian and Californian rodents living in a Mediterranean climate. In this comparison, the Chilean rodents would likely appear to be quite effective at water conservation.

Urine concentration in mammals is further influenced by other factors:

1. Urea and electrolyte concentrations in the urine reflect the amount of protein and salt in the diet.

2. A temporal difference in the consumption and excretion of water and solutes may occur, so that a greater renal capacity is needed than would be expected from the diet. For example, the vampire bat (*Desmodus rotundus*) feeds on the blood of endotherms, a food that is isosmotic to the vampire's plasma. The vampire, however, excretes much of the water in consumed blood within the first half hour of feeding (Figure 7.28) to lighten the load for a return flight to a roost. Most protein metabolism and urea synthesis therefore occur after much of the ingested water is excreted. Consequently, a marked capacity for the production of a concentrated urine is required (McFarland and Wimsatt 1969, McNab 1973, Busch 1988).

3. As indicated, food habits, as they reflect various combinations of water and solutes, place

Table 7.8 Maximal urine concentrations and maximal U/P ratios in mammals

Species	Maximal [urine] (mOsm/kg)	Maximal U/P*
Chiroptera		
Pteropodidae		
Rousettus aegyptiacus	556	1.9
Vespertilionidae		
Pipistrellus hesperus	5000	(12.8)
Antrozous pallidus	4550	(11.7)
Euderma maculatum	4000	(10.3)
Lasionycteris noctivagans	4325	(11.1)
Myotis lucifugus	3700	(9.5)
Myotis yumanensis	3350	(8.6)
Desmodontidae		
Desmodus rotundus	4656	13.1
Carnivora		
Canidae		
Fennicus zerda	4022	(10.3)
Canis familiaris	2425	7.4
Rodentia		
Aplodontidae		
Aplodontia rufa	820	2.7
Arvicolidae		
Microtus pennsylvanicus	1663	(4.3)
Microtus ochrogaster	2544	(6.5)
Sciuridae		
Spermophilus lateralis	2425	(6.2)
Ammospermophilus leucurus	3900	9.5
Tamias striatus	2591	(6.6)
Tamias alpinus	3871	(9.9)
Tamias speciosus	3613	(9.3)
Tamias amoenus	4263	(10.9)
Tamias minimus	4312	(11.1)
Heteromyidae		
Dipodomys merriami	5540	14.0
Dipodomys spectabilis	4090	10.4
Cricetidae		
Peromyscus leucopus	3839	(9.8)
Peromyscus maniculatus	5465	(14.0)
Peromyscus truei	4750	(12.2)
Peromyscus crinitus	3430	9.0
Phyllotis darwini	4468	(11.5)
Abrothrix olivaceus	4468	(11.5)
Oryzomys longicaudatus	4168	(10.7)
Onychomys leucogaster	4250	(10.9)
Neotoma albigula	2670	6.7
Cricetus cricetus	3840	(9.8)
Mesocricetus auratus	3440	(8.8)
Cricetulus griseus	4590	(11.8)
Phodopus sungorus	4572	(11.7)
Castoridae		
Castor canadensis	550	(1.4)
Gerbillidae		
Gerbillus gerbillus	5500	14.0
Psammomys obesus	4400	(11.3)

Table 7.8 *Continued*

Species	Maximal [urine] (mOsm/kg)	Maximal U/P*
Muridae		
Acomys cahirinus	4700	12.1
Acomys russatus	4800	12.3
Notomys alexis	9370	24.6
Notomys cervinus	4920	14.2
Leggadina hermannsburgensis	8970	26.8
Dipodidae		
Jaculus jaculus	6500	16.0
Octodontidae		
Octodon degu	4443	(11.4)
Spalacopus cyanus	3272	(8.4)
Bathyergidae		
Heterocephalus glaber	1521	4.6
Leporidae		
Sylvilagus aquaticus	1920	6.6
Sylvilagus floridanus	3000	8.6
Sylvilagus auduboni	3650	9.8
Camelidae		
Camelus bactrianus	3170	(8.1)

*Values in parentheses are calculated under the assumption that the maximal plasma concentration is about 390 mOsm/kg.
Sources: Data derived from Schmidt-Nielsen et al. (1956), Schmidt-Nielsen and O'Dell (1961), Dolph et al. (1962), MacMillen and Lee (1967), Heisinger and Breitenbach (1969), McFarland and Wimsatt (1969), Shkolnik and Borut (1969), Abbott (1971), Heller and Poulson (1972), Heisinger et al. (1973), Bradford (1974), Blake (1977), Trojan (1977), Geluso (1978), Noll-Banholzer (1979), Urison and Buffenstein (1994), and Bozinovic et al. (1995).

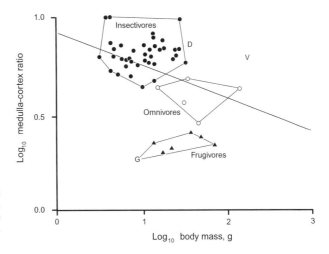

Figure 7.29 Log_{10} medulla-cortex ratio in bats as a function of log_{10} body mass and food habits. D, the vampire *Desmodus*; V, the carnivore *Vampyrum*. Source: Modified from Studier et al. (1983).

differential demands on the renal system of mammals. This impact has been shown best in an examination of kidney morphology and function in bats (Carpenter 1969, Studier and Wilson 1983, Busch 1988, Happold and Happold 1988, Arad and Korine 1993). For example, Studier et al. (1983) showed that insectivorous, carnivorous, and sanguivorous bats have high renal indices, irrespective of taxonomic affiliation, whereas frugivo-

rous and nectarivorous species have low indices, and omnivorous species are intermediate (Figure 7.29). These differences correspond to having high protein and intermediate water loads, low protein and high water loads, and intermediate protein and water loads, respectively.

The temporal separation of water and urea excretion in vampires emphasizes the importance

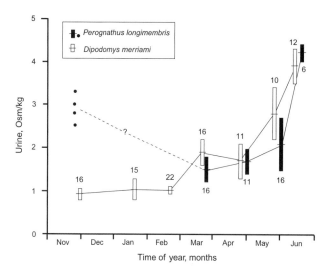

Figure 7.30 Urine concentration in free-living desert rodents, *Perognathus longimembris* and *Dipodomys merriami*, as a function of the time of year. Means, confidence intervals, and sample sizes are indicated. A dashed line and question mark indicate possible urine concentrations of *P. longimembris* during winter. Source: Modified from MacMillen (1972).

of field measurements of renal function. Unfortunately, few such data are available. The urine concentrations of the red kangaroo (*Macropus rufus*) and the euro (*M. robustus*) in the field increased during the dry season, whereas their plasma showed little seasonal variation (Dawson and Denny 1969b). These marsupials apparently did not face sufficient heat or salt loads to require maximal urine concentrations. Actually, the red kangaroo generally had a slightly more concentrated urine than the euro, in spite of the greater medullary thickness in the euro. The difference in urine concentration between these species may reflect different foods and microclimates (Dawson and Denny 1969a).

The Schmidt-Nielsens (1950) made one of the first attempts to describe a "complete" water budget of mammals in the field. They demonstrated that some desert *Perognathus* and *Dipodomys* species could balance their water requirements while feeding on seeds, by producing water through metabolism, reducing water loss by taking shelter during the day in burrows with humid atmospheres, and reducing urinary water loss through a high concentrating capacity of their kidneys.

MacMillen (1972) and MacMillen and Christopher (1975) performed the most extensive study to date on renal function of mammals in the field. They measured plasma and urine concentrations of various rodents in the Mohave Desert of North America. In the laboratory heteromyids can maintain body mass without drinking water; the cricetids *Neotoma* and *Onychomys* cannot live

without drinking water; and the sigmodontid *Peromyscus crinitus* is intermediate, that is, capable of mass maintenance without drinking water under some conditions. In the field the granivorous heteromyids had low urine concentrations during winter, when environmental temperatures and evaporative water losses are low and when the water content of seeds may be high. During summer the urine concentrations in the heteromyid *Dipodomys merriami* increased markedly, presumably due to the increased rate of water loss and the reduction in water content of the food (Figure 7.30). The carnivorous-insectivorous grasshopper mouse (*Onychomys torridus*) had high urine concentrations throughout the year, in spite of eating food with a high water content owing to the high protein content of the food and the resultingly high urea load in the urine (Figure 7.31). *Peromyscus crinitus* also had high urine concentrations throughout the year, in summer because it fed on dry seeds (like heteromyids) and in winter because it ate insects. In contrast, the wood-rat *Neotoma lepida* fed on succulent foods low in salt and protein and had dilute urine throughout the year (Figure 7.31). The evasion of an osmotic load by *N. lepida* may, however, have its cost, because by eating some cacti *N. lepida* has a high intake of the poison oxalic acid (Schmidt-Nielsen 1964).

In an attempt to give an integrated view of the water economy of heteromyid rodents, MacMillen and Hinds (1983) examined the production of water by metabolism, rate of water loss, chemical composition of the seeds eaten, urine osmolality, and entrance into torpor with respect to life in an

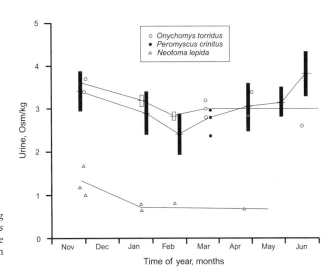

Figure 7.31 Urine concentration in free-living desert rodents, *Peromyscus crinitus*, *Onychomys torridus*, and *Neotoma lepida*, as a function of the time of year. Source: Modified from MacMillen (1972).

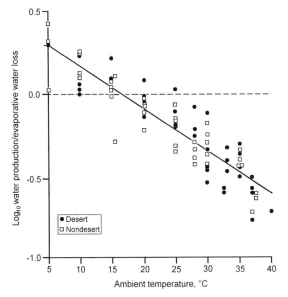

Figure 7.32 The ratio of rate of water production by metabolism to rate of evaporative water loss in desert and nondesert rodents as a function of environmental temperature. Source: Modified from MacMillen and Grubbs (1976).

arid or desert environment. Both the production of water by metabolism and the evaporation of water vary with ambient temperature, the first increasing and the second decreasing with a fall in temperature. Therefore, some temperature should exist at which water production equals the loss of water by evaporation, as has been shown repeatedly (e.g., Figure 7.32). Furthermore, given the scaling relations between the rates of metabolism and of evaporative water loss and body mass, MacMillen and

Hinds predicted that the ratio of water production to evaporative water loss would decrease with an increase in mass and that the temperature at which water production equals evaporation decreases with an increase in mass. When 13 species of heteromyids were examined in the laboratory, a decrease in the "equivalent" temperature was shown relative to body mass. The data indicated that *Perognathus*, which weighed between 8 and 32 g, established a balanced water budget at temperatures up to 26°C, whereas *Dipodomys*, at masses from 35 to 105 g, usually balanced its water budget at temperatures below 20°C.

These interactions have an effect on kidney function in heteromyids: when heteromyids are held at temperatures of 22 to 25°C in the laboratory, *Perognathus* appeared to be in better water balance than did *Dipodomys* and as a result had a more dilute urine (1.2–1.7 Osm/kg) than *Dipodomys* (3.2–3.8 Osm/kg). The urine concentration in heteromyids is also a function of the chemical composition of the food used: for example, as the protein content of seeds increased by a factor of 3.1, the urine concentration of *Perognathus penicillatus* increased by a factor of 3.7. Measurements of the urine concentation of *P. longimembris* and *D. merriami* in the field, however, were essentially identical, increasing as might be expected with environmental temperature (Figure 7.30) because of the decrease of metabolism and the increase of evaporation with an increase in temperature. This observation led MacMillen and Hinds to suggest that the two genera used different food supplies: high-protein seeds by *Perognathus* and high-carbohydrate seeds by *Dipodomys*. These authors (p. 156) argued that "... the urinary system [is] a fine-tuning device..." used to balance the water budget. Other differences between these two genera are that *Perognathus* often enters daily or seasonal torpor, probably in relation to its small size (i.e., in relation to a diminished period of tolerable starvation), and uses quadrupedal locomotion, whereas *Dipodomys* rarely enters torpor and often uses bipedal locomotion. MacMillen and Hinds concluded, "... *Perognathus* opted for a small size, increased water regulatory efficiency, and decreased absolute energy needs, while trading off locomotor efficiency; *Dipodomys* opted for larger size and greater locomotor efficiency, trading off a fixed intermediate level of water regulatory efficiency, and increased absolute energy and water needs" (p. 161).

Aside from its water content and the amount of water produced by aerobic metabolism, the composition of food has an impact on the water budget of mammals. Withers (1982) showed that as ingested food decreases in digestibility, fecal water loss increases (as expected from the increase in feces production). Body mass consequently falls as the digestibility of the food decreases when a mammal is placed on a water-restricted diet. For example, the pocket-mouse *Perognathus parvus*, which can maintain body mass without water on a diet of seeds, a food that has a low water content and a high digestibility, cannot maintain mass without water on a diet of soybeans or bran (Figure 7.33). The fiber content of millet seeds is only about 1%, whereas that of soybeans is 6% and bran is 32%; the assimilation efficiencies of these diets are approximately 90%, 83%, and 66%,

respectively. Another complication is that soybeans have a high protein content, which leads to increased urea production and urinary water loss, as well as a lower yield of water produced by metabolism. Withers concluded that desert mammals cannot survive for long periods as long as the assimilation efficiency is less than 85% because of the marked fecal water loss. Much plant matter in deserts, thus, cannot be used because it combines a high fiber content with a low water content.

As the measurements on *Neotoma* show, one way to survive a potentially high osmotic load is to evade the most hostile conditions. Such evasion takes several forms. A particularly interesting case is found in the kangaroo-rat *Dipodomys microps*, which uses its chisel-shaped incisors to remove hypersaline vesicles from the leaves of the halophyte *Atriplex confertifolia* (Kenagy 1972). Similar behavior is found in the murid *Psammomys obesus* in North Africa and the octodontid *Tympanoctomys barrerae* in South America, both of which also feed on *Atriplex* (Mares et al. 1997). This behavior ensures a large water and low salt intake, which explains why *D. microps* has the lowest known capacity to produce a concentrated urine of all *Dipodomys* species. An evasion little used by mammals is the abandonment of osmotic regulation. Baverstock (1976) showed that of four species of Australian *Rattus*, only one, *R. lutreolus*, tolerated high plasma concentrations; this species cannot produce urine as concentrated as that of the species that are osmotic regulators. Finally, some mammals evade high rates of water turnover by entering torpor, which has been shown in the laboratory for *Peromyscus truei* (Bradford 1974), and in both the laboratory and the field for *P. eremicus* (MacMillen 1964, 1965). This evasion undoubtedly occurs in various ground-squirrels; in contrast, the antelope ground-squirrel (*Ammospermophilus leucurus*), which does not enter torpor, can produce a highly concentrated urine (Hudson 1962).

The details of water conservation cited here can be integrated into the rate of water turnover and compared to conditions in the environment. Macfarlane et al. (1971) measured water turnover with tritiated water in pastured ruminants and concluded that the principal factor influencing water turnover other than body mass was rate of metabolism: species with high rates of metabolism had high rates of water turnover independent of the environment in which they lived. This may account

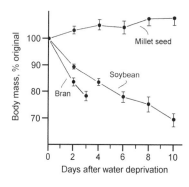

Figure 7.33 Body mass in the rodent *Perognathus parvus* as a function of time without water and diet composition. Source: Modified from Withers (1982).

for the observation that many desert-dwelling mammals have low basal rates of metabolism (e.g., McNab and Morrison 1963, Shkolnik and Schmidt-Nielsen 1976). Furthermore, Macfarlane and coworkers suggested that marsupials have lower turnover rates than eutherians that relate to the lower rates of metabolism found in marsupials. Nicol (1978) showed the correlation of the rate of water turnover under standard laboratory conditions with body mass, and argued that turnover rate, independent of mass, is principally correlated with the aridity of the habitat. He suggested that most differences between marsupials and eutherians actually reflect a difference in habitat. The different interpretations of Macfarlane and Nicol probably can be integrated into a unified picture of water turnover and may result in part from the use of different species and techniques.

Evidence of the influence of food habits, renal function, and behavior on water turnover in the environment is found in the study of eight species of rodents by Yousef et al. (1974). These rodents, all of which came from desert or desert-mountain environments, showed a 10-fold variation in water turnover. It was highest in species using moist foods, having diurnal habits, living in moist environments at high elevations, and having the least capacity to produce a concentrated urine. The lowest water turnovers were found in species using dry seeds, having nocturnal habits, living in the driest environments at low elevations, and having the greatest capacity to produce a concentrated urine.

Although most attention has been directed to the widespread problems associated with excreting excess salt loads, some food habits and environments have inadequate supplies of electrolytes. For example, many ungulates that eat food with a low salt content use salt licks. In an extreme case, the Snowy Mountains of Australia are deficient in sodium, and what little is present is often lost due to leaching during winter. As a result, the sodium content of grass available in this region in spring and summer is low. Blair-West et al. (1968) demonstrated that the urine and plasma concentrations of sodium in rabbits, kangaroos, and wombats in this region are low at this time. Low levels of sodium in the plasma and urine were correlated with hypertrophied adrenal glands and with high plasma levels of aldosterone (an adrenal hormone that increases sodium reabsorption in the distal convoluted tubule). Wild rabbits, in fact, selectively fed on sticks impregnated with sodium salts. In Zimbabwe, the abundance of elephants (*Loxodonta africana*) at salt licks correlates with the sodium content of the water (Weir 1972). And porcupines (*Erethizon*) in North America are known to eat urine-splashed boards in outhouses!

7.11 CHAPTER SUMMARY

1. Life in a terrestrial environment requires a balanced water budget, the principal pathways of water gain being drinking and ingestion of food (preformed water), and the principal means of water loss being respiration and excretion.

2. Water content of food is the most important determinant of the distribution and behavior of terrestrial vertebrates in relation to water: species that feed on food with a high water content are often independent of surface water; those that feed on dry food are not, unless they behaviorally evade high rates of water turnover.

3. The rate of water loss through the integument is relatively low in terrestrial vertebrates living in dry environments, except in mammals that thermoregulate at high environmental temperatures by sweating.

4. Respiratory water loss is a by-product of aerial gas exchange: it may be facultatively increased to facilitate temperature regulation at high ambient temperatures.

5. Many birds increase evaporative water loss at high temperatures by gular fluttering, in an attempt to separate heat dissipation from CO_2 exchange.

6. Terrestrial vertebrates may minimize evaporative water loss by selecting mild microclimates if body mass is small, or by storing heat if mass is large.

7. Reptiles produce a variety of nitrogenous wastes, including uric acid in terrestrial environments, urea and ammonia in aquatic environments, and a mixture of uric acid and urea in intermediate conditions.

8. No reptile is known to produce a urine more concentrated than the plasma because reptilian nephrons lack a loop of Henle.

9. The principal nitrogenous waste product of birds is uric acid.

10. Bird kidneys have two kinds of nephrons, a "reptilian" type with a small glomerulus and no loop of Henle, and a "mammalian" type with a larger glomerulus and a loop of Henle.

11. Differences in the behavior between the two types of nephrons account for much of the variation in glomerular filtration rates in birds.

12. The production by birds of urine more concentrated than the plasma is permitted by the establishment of an osmotic gradient in the medullary cones. The gradient consists mainly of NaCl.

13. The maximal osmolal urine-plasma (U/P) ratio found in birds is 4.5:1.

14. Reptiles and birds use extrarenal mechanisms for the conservation of water and the excretion of salt, including such structures as the bladder, cloaca, and salt glands.

15. The bladder and cloaca permit the precipitation of uric acid with the withdrawal of water, thereby furthering the conservation of water.

16. The reabsorption of water in the cloaca by birds may limit their ability to produce a concentrated urine.

17. Cephalic salt glands in reptiles and birds permit the secretion of salt solutions that are 2 or more times the maximal concentration of urine.

18. Salt glands in birds supplement kidney function and are especially well developed in species with high salt loads; species that produce the most concentrated urine lack salt glands.

19. Some reptiles, including crocodilians, some snakes, and most turtles, have returned to aquatic habits.

20. Aquatic reptiles often have high rates of gas, water, and ion exchange with the environment through their integument; marine species depend on cephalic salt glands to maintain a plasma concentration hyposmotic to the environment.

21. Mammals excrete urea as the only important nitrogenous waste product; they do not use extrarenal salt excretion to any appreciable extent.

22. All mammalian nephrons have a loop of Henle, although they may be short in aquatic species.

23. An osmotic gradient is established in the kidney medulla by the active transport of NaCl and the accumulation of urea.

24. A relative medullary morphological index describes the ability of a mammalian kidney to produce a concentrated urine better than an absolute measure of the thickness of the medulla because a relative index compensates for the impact of body size on kidney morphology and function.

25. The U/P ratio in mammals is correlated with environment, food habits, and body mass; the maximal osmolal ratio recorded in mammals is 26.8:1.

8   Adaptation of Gas Exchange

8.1 SYNOPSIS

Gas exchange maintains an adequate oxygen tension at the mitochondria, voids CO_2 produced by metabolism, and ensures an appropriate base ratio in body fluids. The maintenance of oxygen tension requires a high ventilation rate in water breathers and permits a low rate in air breathers, the rate reflecting the shortage of oxygen dissolved in water and its abundance in air. Gas exchange is facilitated in aquatic environments by the presence of gills with a large surface area, thin epithelium, and countercurrent exchange. Most of oxygen transported in blood is bound to hemoglobin in erythrocytes. Its affinity for oxygen reflects the demand for, and the abundance of, oxygen in the environment and is controlled by the organophosphate concentration in erythrocytes. The only vertebrates that have no hemoglobin are antarctic ice-fishes, which can transport only 10% as much oxygen in blood as other fishes. The elimination of erythrocytes reduces blood viscosity at low temperatures but in compensation requires a high cardiac output. Ventilation rates fell and plasma bicarbonate levels increased with the evolution of aerial gas exchange. Bimodal gas exchange evolved in some fish and all amphibians. In bimodal exchange CO_2 is mainly lost through the gills or skin, whereas most of oxygen uptake occurs through the lungs or other internal, ventilated structures. Gas exchange limits body size in aquatic amphibians; lungless salamanders are limited to an even smaller body size and are generally excluded from living in warm water. The earliest amphibians could not have depended on their skin for gas exchange because of their large size. Gas exchange

in reptiles, birds, and mammals is almost exclusively pulmonary. Birds have lungs that permit a unidirectional movement of gas, a high arterial oxygen tension, and a low arterial CO_2 tension. In compensation for a less efficient lung structure, mammals have larger lungs than birds, and bats have by far the largest lungs among mammals. Burrowing mammals have hemoglobin with high affinities and are relatively insensitive to high CO_2 levels. Birds and mammals that live at high altitudes have responded to low atmospheric pressures by adjusting hemoglobin affinity, erythrocyte concentration, myoglobin concentration, cardiac outputs, and ventilation rates. A few aquatic turtles and snakes have returned partially to aquatic gas exchange.

8.2 INTRODUCTION

All vertebrates expend energy for maintenance and activity. The energy is derived from the oxidation of reduced organic molecules obtained in food. If this extraction occurs under anaerobic conditions, the metabolism of glucose is given by

$$C_6H_{12}O_6 + 2ADP + 2P_i = 2CH_3COOH + 2CO_2 + 2ATP,$$

where ADP is adenosine diphosphate, P_i is inorganic phosphate, ATP is adenosine triphosphate, and CH_3COOH is lactic acid. Aerobic respiration is more efficient than anaerobiosis in converting the potential energy of substrates into ATP because the substrates are completely oxidized:

219

$$C_6H_{12}O_6 + 6O_2 + 36ADP + 36P_i = 6CO_2 + 6H_2O$$
$$+ 36ATP.$$

Aerobiosis produces 18 times as many moles of ATP per mole of glucose as glycolysis does. Vertebrates use both anaerobic and aerobic metabolism, anaerobiosis usually as a short-term supplement during activity (Ruben and Bennett 1980); no vertebrate depends exclusively on anaerobiosis. Gas exchange is essential to aerobic metabolism because molecular oxygen receives electrons from the electron transport system, thereby forming water as a waste product, and the excess carbon residues are discarded as CO_2.

The transition of vertebrates from aquatic to terrestrial environments requires an adjustment in vertebrates gas exchange because of striking differences in the composition of the atmosphere and the gases dissolved in water (see Section 2.11). This chapter explores ecologically significant aspects of this transition.

8.3 THE REGULATION OF GAS EXCHANGE

The regulation of a system generally means that certain parameters are held constant at the expense of others. In endothermic temperature regulation, body temperature is held constant at the expense of energy expenditure. Thermal acclimatization in ectotherms, however, tends to keep rate of metabolism constant as body temperature varies. Which parameters of gas exchange are held constant and which vary?

Gas exchange involves three major structures: an external gas exchanger (e.g., gill, lung, swim bladder, or skin); a circulatory system consisting of blood, vessels, and a pump; and the tissues where oxygen is consumed and CO_2 produced. Oxygen and CO_2 diffuse between the external exchanger and blood and between the blood and tissues. The greatest resistances to gas transport exist at these diffusional barriers, and they can be adjusted by varying the rate of ventilation, rate of circulation, hemoglobin content of blood, and arterial and venous CO_2 tensions (P_{CO_2}) and oxygen tensions (P_{O_2}).

At least two parameters of gas exchange are regulated. One, presumably, is the minimal oxygen tension at the mitochondria because aerobic respiration cannot proceed without an adequate oxygen supply. Rahn (1966a, 1966b) suggested that this

tension may be about 2 mm Hg, but it might be somewhat higher. The other factor that appears to be regulated is the ratio of the concentration of hydroxyl ions to hydrogen ions ($[OH^-]/[H^+]$) in blood (Rahn 1966a, 1966b) or the charge on blood proteins (Reeves 1976).

Mammals appear to regulate blood pH, but when a broader view is taken, this constancy reflects a constant body temperature. Blood pH in poikilotherms varies inversely with body temperature parallel to the ionization constant of water. For example, blood pH in the toad *Bufo marinus* increases with a fall in body temperature parallel to the increase in the neutrality of water (Figure 8.1). Water at neutrality has pN = pH = pOH and pH + pOH = pK_w, where K_w is the ionization constant of water. Thus, pN = $pK_w/2$. At 3°C, K_w = $10^{-14.8}$ and pN = 7.4; at 25.5°C, K_w = 10^{-14} and pN = 7.0; and at 37°C, K_w = $10^{-13.6}$ and pN = 6.8. That is, a pH of 7.0 in pure water is neutral only at 25.5°C. So mammals have a low, constant pH mainly because they are warm and thermally constant.

The blood of vertebrates is not neutral (Figure 8.1), that is, pH ≠ pOH, but the pH and pOH

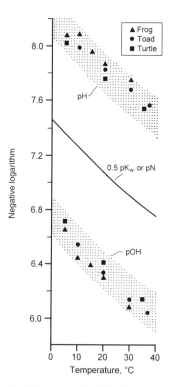

Figure 8.1 pH, pN, and pOH of blood in frogs, toads, and turtles as a function of temperature. Source: Modified from Howell et al. (1970).

curves are parallel to the pN curve, which means that a fixed difference exists between these curves at all temperatures. Therefore, the ratio $[OH^-]/[H^+]$ is independent of temperature because the curves describing the concentration of these ions are parallel. In *B. marinus* the distance between the pH and pOH curves is 2(0.6) log units and the ratio $[OH^-]/[H^+]$ equals $10^{1.2} = 16$. (The ratio is expressed with $[OH^-]$ in the numerator to give the ratio a value greater than 1.0; the reason why $[OH^-]$ is greater than $[H^+]$, even though it is plotted below the $[H^+]$ curve, is that p represents a negative logarithm, so that a larger p represents a smaller number.) Other species may defend a different ratio, but in no poikilotherm that has been studied has this ratio failed to remain constant. One striking exception is found in hibernating mammals, which do not maintain a constant $[OH^-]/[H^+]$ ratio, but instead maintain a constant pH independent of body temperature (Howell et al. 1970, Bickler 1984), almost as if endotherms regulate their $[OH^-]/[H^+]$ ratio by regulating pH at a constant body temperature, but with the secondary evolution of hibernation were locked on to a regulatory system that does not protect the $[OH^-]/[H^+]$ ratio.

Another parameter that remains constant with a change in temperature is the distribution of Cl^- ions across the membrane of red blood cells, even in mammals, where local cooling of blood may occur along an extremity (Reeves 1976). This constancy results from the maintenance of a fixed charge on red blood cells and on serum proteins. The charge comes from peptide-linked histidine, which has an ionization constant nearly identical to that of water and responds to temperature as does pH. Regardless of the "ultimate" regulatory factors in gas exchange, the pattern of regulation and its ecologically significant variants can be described in terms of a fixed $[OH^-]/[H^+]$ ratio (Rahn 1966a, 1966b).

Variation in temperature, of course, will affect the rate of metabolism of a poikilotherm and, thus, the rate of CO_2 production. Dissolved CO_2 contributes to the hydrogen ion concentration of blood because of the association of CO_2 and water and its subsequent dissociation (see Box 8.1). As with K_w, the dissociation constants K' and K'' are temperature dependent. The temperature dependence

---

BOX 8.1
The Henderson-Hasselbalch Equation

When CO_2 dissolves in water,

$$CO_2 + H_2O \overset{K'}{=} H_2CO_3 = H^+ + HCO_3^- \overset{K''}{=} 2H^+ + CO_3^=.$$

The first dissociation can be written as

$$K' = \frac{[H^+][HCO_3^-]}{[H_2CO_3]}$$

or,

$$[H_2CO_3] \cdot K' = [H^+][H_2CO_3^-],$$

where the dissociation constant K' represents the balance between the coefficients of forward and reverse reactions at equilibrium. Therefore,

$$\log K' + \log[H_2CO_3] = \log[H^+] + \log[HCO_3^-],$$

$$\log K' = \log[H^+] + \log\frac{[HCO_3^-]}{[H_2CO_3]}$$

$$-\log[H^+] = -\log K' + \log\frac{[HCO_3^-]}{[H_2CO_3]}.$$

Because carbonic acid (H_2CO_3) readily dissociates, $[H_2CO_3]$ can be replaced by the force driving the production of carbonic acid, which equals the product of α_{CO_2} ($cm^3/L \cdot mm\,Hg$), the solubility coefficient for CO_2, and P_{CO_2} ($mm\,Hg$), the tension (or pressure head) of CO_2 in plasma. As a consequence,

$$pH = pK' + \log\frac{[HCO_3^-]}{[\alpha_{CO_2}][P_{CO_2}]} \qquad (8.1)$$

which is called the *Henderson-Hasselbalch equation*. At 25°C, the first dissociation constant (K') is 8000 times larger than the second constant (K''), for example, 3.5×10^{-7} versus 4.4×10^{-11}, so that if CO_2 becomes associated with water, most CO_2 will remain as H^+ and HCO_3^-.

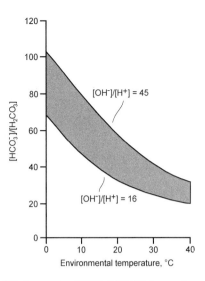

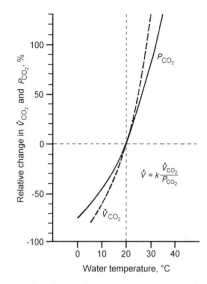

Figure 8.2 Base ratio, $[HCO_3^-]/[H_2CO_3]$, as a function of temperature when the $[OH^-]/[H^+]$ ratio varies between 16 and 45. Source: Modified from Rahn (1966b).

Figure 8.3 The relative change in P_{CO_2} and rate of CO_2 production in fish as a function of water temperature. Source: Modified from Rahn (1966b).

of pH and pK′, when applied to the Henderson-Hasselbalch equation, implies that the base ratio

$$\frac{[HCO_3^-]}{[H_2CO_3]} = \frac{[HCO_3^-]}{[\alpha_{CO_2}][P_{CO_2}]}$$

decreases with an increase in temperature (Figure 8.2). Because the bicarbonate concentration in blood is independent of temperature and because the solubility of CO_2 in a fluid decreases with an increase in temperature, the base ratio decreases because P_{CO_2} increases (Figure 8.3). P_{CO_2} is influenced by the ventilation rate ($\dot{V}$):

$$P_{CO_2} = k\left(\frac{\dot{V}_{CO_2}}{\dot{V}}\right); \quad \text{or} \quad \dot{V} = k\left(\frac{\dot{V}_{CO_2}}{P_{CO_2}}\right), \quad (8.2)$$

where k is an arbitrary coefficient. The increase in the production of CO_2 by metabolism ($\dot{V}_{CO_2}$) is greater than the increase in P_{CO_2} (Figure 8.3), which "results in," or at least is associated with, an increase in ventilation rate ($\dot{V}$) with temperature. Therefore, the ratio $[OH^-]/[H^+]$ remains constant with an increase in the temperature of a poikilotherm as a result of an appropriate reduction in pH produced by an increase in plasma P_{CO_2}. Plasma P_{CO_2} in turn is regulated by ventilation rate.

Finally, the ventilation rate varies with the abundance of oxygen in the environment. Ventilation rate must be higher in environments having a low oxygen tension to ensure an adequate supply of oxygen to the tissues. A striking contrast in oxygen abundance exists between aquatic and terrestrial

environments: oxygen is about 20 times more abundant in air than in water at 0°C; this ratio increases to 40:1 at 40°C (see Section 2.11). Therefore, in the evolution of aerial from aquatic gas exchange, ventilation rate decreases markedly, which results in an increase in the P_{CO_2} of blood (Equation 8.2). Rahn (1966a, 1966b) estimated that 17 cm³ air/min must be moved over lung alveoli, or 480 cm³ H_2O/min over gills, to maintain an oxygen tension of 100 mm Hg at the surface of the exchanger when the environment oxygen tension is 150 mm Hg. Then, the alveolar or branchial CO_2 tensions are 50 or 2 mm Hg, respectively. In other words, fish have such high ventilation rates in water that they blow off the CO_2 produced by metabolism, whereas reptiles, birds, and mammals have such low ventilation rates that they accumulate CO_2 in their blood.

An accumulation of CO_2 in the plasma of terrestrial vertebrates would lead to a reduction in pH at a fixed temperature, unless some compensatory modification is made (Equation 8.1). Two solutions to this problem are possible. One is to lose CO_2 by other than a pulmonary pathway, which as we will see, occurs in air-breathing fishes and amphibians. The second is to have an increase in the concentration of bicarbonate in the plasma, so that the base ratio at a fixed temperature remains constant. This solution is used by reptiles, birds, and mammals, and to a lesser extent by amphibians.

In a review of the data on pH and temperature in poikilotherms, Ultsch and Jackson (1996) found general agreement with the models proposed by Rahn (1966a, 1966b) and Reeves (1976). They did find, however, that some patterns were unaccounted for, such as the blood pH of marine fishes being consistantly lower than those of freshwater fishes, and reptiles having a lower blood pH than freshwater fishes and amphibians, which are similar to each other.

Because homeotherms have a rather constant body temperature, the correlation of blood pH and body temperature is unlikely to be seen, but limited observations indicate that blood pH in the echidna (*Tachyglossus aculeatus*) and the platypus (*Ornithorhynchus anatinus*) is temperature independent (Parer and Metcalfe 1967a, 1967b). A ground-squirrel (*Spermophilus tereticaudus*) shows an increase in arterial pH and a decrease in P_{CO_2} with a decrease in body temperature (Bickler 1984). Bickler suggested that the change in acid-base balance in torpid *S. tereticaudus* may contribute to the suppression of metabolism (see also Malan 1988 and Section 11.8.2).

8.4 GAS EXCHANGE IN AQUATIC ENVIRONMENTS

The only classes of vertebrates that depend exclusively on aquatic gas exchange are cyclostomes and chondrichthyans. Most osteichthyans depend exclusively on aquatic exchange, but many facultatively supplement aquatic with aerial exchange; a few are obligatory air breathers. Aquatic amphibian larvae rely heavily on aquatic exchange, whereas most adults in water preferentially use aerial exchange. Most aquatic reptiles and all aquatic birds and mammals are strict air breathers; a few reptiles secondarily reevolved aquatic exchange.

Hughes (1973) pointed out that a minimal set of conditions is required for adequate gas exchange to occur, irrespective of whether the exchange is aquatic or aerial. These conditions include the following: (1) the layer separating the blood and external fluid must be very thin (ca. 1–3 μm); (2) the surface area for exchange must be large, which is normally accomplished by an intricate folding of the exchange surface; (3) the respiratory surface must be moist because gases diffuse through membranes in solution; and (4) adequate rates of renewal at the exchange surface must exist in

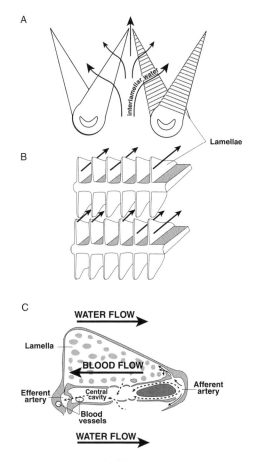

Figure 8.4 Diagrams of gill structures in teleosts. A. Transverse section through two adjacent gill arches. B. Two adjacent gill filaments. C. A cross section of a gill lamella. Sources: Modified from Steen and Kruysse (1964) and Hughes (1966).

both blood and external medium to ensure that exchange attains a steady state and does not approach an equilibrium. The adequacy of exchange rates can be further ensured by the subdivision of the blood supply to the surface so that diffusional exchange with erythrocytes occurs on all sides, a condition that requires capillaries with dimensions similar to those of corpuscles.

8.4.1 Structure of gills. The gills of fish have a complicated structure (Figure 8.4). They consist of vertical columns of narrow, horizontal shelves, the *filaments*, over which water passes on its way from the buccal cavity to the exterior via the gill slits. Most gill arches have two sets of filaments, one directed anteriorly to a gill slit and the other posteriorly to an adjacent gill slit. Gill slits thus have two sets of filaments, one from each adjacent arch, thereby producing a screen across the slit (Figure 8.4A, B). Each filament is given structural rigidity by possessing a cartilaginous gill ray. Most of the water passes between the filaments belonging to the same arch (Figure 8.4B). An arch that carries two sets of filaments is called a *holobranch*; one that has only one set of filaments (as might be found on the anterior side of the first slit or on the posterior side of the last slit) is called a *hemibranch*.

The surface area of gills is greatly increased by the presence of a set of thin, tall, vascular plates called *secondary lamellae* along the length of a filament. These lamellae are oriented at right angles to the filaments and are located on both the upper and the lower surfaces of the filaments (Figure 8.4B). With the lamellae of adjacent filaments, they form small pores through which the water flows. The lamellar surface area can be modified by changing the width and the spacing of the lamellae, but most effectively by changing the length of the filaments and therefore the number of lamellae attached to a filament.

Furthermore, the secondary lamellae are roughly triangular (Hughes 1973) and are so positioned that the leading (with reference to water movement) edge of a filament is taller than the trailing edge (Figure 8.4C). Blood enters the trailing edge of a filament by the afferent artery. Blood may pass either through a lamella or through a central cavity in the filament: blood flow is countercurrent to water flow. Blood leaves the filament via the efferent artery, which is found along the leading edge of the filament.

8.4.2 Functioning of gills. Hughes (1973) described the rate of gas exchange at the gills:

$$\dot{V}_{O_2} = \frac{K \cdot A \cdot \Delta P_{O_2}}{d}, \qquad (8.3)$$

where $\dot{V}_{O_2}$ is the rate of oxygen uptake ($cm^3 O_2$/min), K is Krogh's diffusion constant ($cm^3 O_2 \cdot nm/cm^2 \cdot min \cdot mm\,Hg$), A is surface area for exchange (cm^2), ΔP_{O_2} is the differential in oxygen tension between water and blood (mm Hg), and d is the distance between water and blood in the gills (nm). The variation in exchange at the gills can be analyzed in terms of Equation 8.3.

By far the greatest factor influencing rate of oxygen uptake is the surface area over which gas exchange occurs, which in turn is principally correlated with body mass (Figure 8.5). The relation

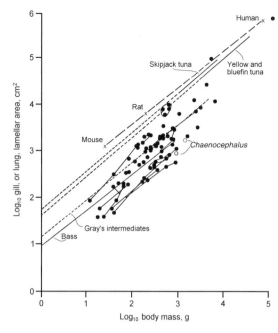

Figure 8.5 Log_{10} gill lamellar area in skipjack tuna (*Katsuomus pelamis*), yellowfin tuna (*Thunnus albacares*), bluefin tuna (*Th. thynnus*), and other fishes and log_{10} lung area in mammals as a function of log_{10} body mass. Sources: Modified from Muir (1969) and Hughes (1972a).

of gill surface area to body mass is described by $A = am^b$, where a is scaling coefficient and the power of body mass (b) varies from 0.7 to 1.0 (Muir 1969, Muir and Hughes 1969, Hughes 1973, Rombough and Moroz 1990, Palzenberger and Pohla 1992), the best estimate being about 0.85. In marine fish of intermediate activity

$$A = 13.92 m^{0.82}, \qquad (8.4)$$

where mass is in grams (Gray 1954). Most other fishes, irrespective of whether they are active or sluggish, freshwater or marine, or whether the comparison is made ontogenetically or interspecifically, have a similar power function to describe gill area (Figure 8.5). For example, intermediate-activity species of freshwater fish have a similar but slightly lower relationship (Palzenberger and Pohla 1992):

$$A = 9.85 m^{0.76}.$$

Great variation in gill area is present at a given mass (Figure 8.5). The coefficient a may vary by a factor of 22 : 1 (Muir 1969). Much of this variation is related to activity (Gray 1954, Hughes 1966, Wells 1987): active fish have gill surface areas that average 10 times those of sluggish fish of the same mass. For instance, a 1.6-kg goosefish (*Lophius piscatorius*) has a gill surface area equal to $0.22\,m^2$ (Hughes 1966), whereas a 1.7-kg skipjack tuna (*Katsuwonus pelamis*) has a gill area of $3.07\,m^2$ (Muir and Hughes 1969), a factor of 14:1. The implication is that inactive fish have low rates of metabolism and therefore low rates of gas exchange with the environment. A similar pattern exists in freshwater fishes, where the range in gill surface area at a fixed mass is 6:1 (Palzenberger and Pohla 1992), although no freshwater fish have the large surface area of the endothermic tunas.

Deep-sea fishes have small gill surface areas (Hughes 1976). Smith and Hessler (1974) showed that bathypelagic fish (at a depth of 1230 m) have much lower rates of metabolism than do related littoral fishes of the same mass. Torres et al. (1979) made similar observations, which Somero (1982) and Hochachka and Somero (1984) reviewed (Figure 8.6). Hughes (1976) found that at 1 kg fish tend to show little variation in the ratio $\dot{V}_{O_2}/A$ (although this assumption must be used with caution because the variation in this ratio is much greater at other masses [Hughes 1977]). By combining this assumption with measurements of gill surface area, Hughes (1972b, 1976) estimated that the bathypelagic coelacanth *Latimeria chalumnae* has a very low rate of metabolism. Smith and

Hessler (1974, p. 73) suggested that some of the reduction in metabolism may ". . . be a synergistic function of food availability, pressure, and temperature" (see Section 8.4.7). Barham (1971) suggested that bathypelagic fishes may reduce energy expenditure (i.e., enter a kind of "torpor") by migrating into colder waters at depths below 200 m.

As Barham (1971) implied, one explanation for the low rate of metabolism found in bathypelagic fish is that it is depressed by cold temperature. To separate the influences of depth and temperature, Torres and Somero (1988) compared the metabolism-depth profiles of fish off the California coast with one in antarctic waters. These two regions differed in the temperature-depth profiles: in California the temperatures were approximately 10°C at depths above 100 m and 5°C between 100 and 1000 m, whereas in antarctic waters the temperature was −1.8 to +2.0°C above 100 m and 0.3 to 2.0°C at 1000 m. That is, no appreciable decrease in water temperature occurred with depth in antarctic waters. Yet, whereas the California fishes showed a 20-fold decrease in rate of metabolism at 500 m, compared to the surface, antarctic fishes showed a 10-fold decrease over the same depth range. Correlated with the decrease of rate of metabolism in antarctic fishes was a decrease in skeletal lactate dehydrogenase and citrate synthetase levels, representing a decrease in anaerobic and aerobic metabolism with depth. Torres and Somero drew two conclusions: (1) rate of metabolism of mesopelagic fishes is correlated with depth, independent of water temperature; and (2) antarctic fishes show cold-water acclimatization.

The low rates of metabolism and the sluggishness of bathypelagic fishes may reflect a reduced quantity of food available at great depths in the sea, just as cave fish, salamanders, and crayfish have lower rates of metabolism than epigean species, apparently in relation to the reduced food availability in subterranean environments (Poulson and White 1969, Dickson and Franz 1980). A similar situation may exist in deep-water fishes in Lake Tanganyika (Coulter 1967). A low rate of metabolism in bathypelagic fishes might also be related to a reduction in muscle mass (Childress and Nygaard 1973, Torres et al. 1979) and a smaller osmotic differential with the environment (see Griffith 1981).

Gill surface area is modified in several ways. One is a change in the number of gill arches: most fish

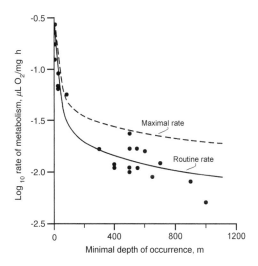

Figure 8.6 Maximal and routine rates of metabolism in midwater teleost fishes as a function of the minimal depth of occurrence. Source: Modified from Hochachka and Somero (1984).

have three holobranchiae arches with or without a hemibranchia arch, but some benthic genera (e.g., *Ogcocephalus*, *Monopterus*) have only two holobranchiae, and air-breathing fishes (e.g., synbranchids) may have as few as one holobranchia (Hughes 1973). Among lungfishes, the obligate water-breather *Neoceratodus* has four holobranchiae, whereas the air-breathing *Lepidosiren* and *Protopterus* have three and two holobranchiae, respectively. As a result of these modifications, air-breathing and bathypelagic fishes have a small total filamental length, whereas active surface fish have a long filamental length. In freshwater fish the increase in gill surface area with mass reflects an increase in both filament length and lamellar area (Palzenberger and Pohla 1992).

Another factor important in gas exchange is the ΔP_{O_2} established across the gills between the water and blood. This differential decreases only partially along the length of the gill lamellae. Under some conditions, the efferent blood leaving the lamellae and gills may have oxygen tensions that exceed the tensions found in the excurrent water (Steen and Kruysse 1964), which can only occur because the exchange at the lamellae is countercurrent. Hughes (1966, 1972a) argued that the triangular shape of the lamellae, with the leading edge being greater than the trailing edge (Figure 8.4C), ensures a larger area for gas exchange where ΔP_{O_2} is likely to be smallest.

Finally, the rate of oxygen uptake is inverse to the thickness of gill tissue through which oxygen must diffuse (Equation 8.3). The thinnest barriers are found in the most active species (Hughes 1973): marine species may also have a thinner barrier than freshwater species of the same mass and activity.

Ionic as well as gaseous exchange occurs at the gills (Steen and Kruysse 1964). Ionic exchange can be reduced by shunting blood through the central cavity of the gill filament, rather than through the lamellae (Figure 8.4C). Under these circumstances, however, gas exchange is reduced. During rest fish usually maintain a low arterial P_{O_2}, thereby minimizing ionic exchange. With exercise, arterial P_{O_2} and (presumably) ionic exchange increase. Some fish, such as *Cyprinodon variegatus*, can tolerate very high external salinities by reducing exchange with the environment (see Section 6.7). The differential use of lamellae or the central cavity of the filament as the pathway for branchial blood apparently depends on whether acetylcholine or adrenaline is present; in the presence of adrenaline, blood

is shunted through the lamellae and arterial oxygen P_{O_2} tension increases. Adrenaline also increases ionic exchange (Keys and Bateman 1932). Holeton and Stevens (1978) found that the cost of swimming in a characin was similar in "black" and "white" waters from the Amazon basin, but the "whiteness" or "blackness" of these waters relates to the presence or absence of organic compounds and silt, not to ionic concentrations.

8.4.3 Oxygen transport. Vertebrates must transport gas between the exchanger and the tissues where aerobic metabolism occurs. Oxygen is transported by blood, either dissolved in solution or bound to hemoglobin located in erythrocytes.

The amount of oxygen transported in solution depends on many factors (see Section 2.11), the most important being temperature. As temperature increases, the amount of oxygen in solution decreases: water at 20°C can hold only 63% of the oxygen held at 0°C. At any temperature, however, little oxygen is held in water (e.g., only $0.031 \, cm^3/cm^3 \, H_2O$ at 20°C), which when coupled with a small blood volume (2%–8% of total volume) means that the absolute amount of oxygen transported in solution is quite small. All vertebrates, except ice-fish (Channichthyidae), use hemoglobin to increase oxygen transport by blood. Hemoglobin is the only respiratory pigment found in the blood of vertebrates, and it is always located in erythrocytes. Except in the cyclostomes, in which hemoglobin is a monomere, vertebrate hemoglobin occurs as a tetramere, which imparts a curvilinear relationship to the correlation of the oxygen carried by hemoglobin with the amount of oxygen in the environment.

Blood picks up oxygen from an oxygen-rich environment and loses oxygen to an oxygen-poor environment, the tissues. Such association-dissociation curves for the blood of three fishes are shown in Figure 8.7. The amount of oxygen carried by blood is the sum of the amounts bound to hemoglobin and in solution. At saturation, 95% to 98% of the oxygen carried in these species is bound to hemoglobin. As can be seen in Figure 8.7A, great differences in the amount of oxygen transported by blood exist among species. These differences reflect (1) the amount of hemoglobin, (2) the affinity of hemoglobin for oxygen, (3) the amount of organophosphates in the erythrocytes, (4) the amount of CO_2 in the blood, (5) the sensitivity of hemoglobin to CO_2, and (6) the temperature of

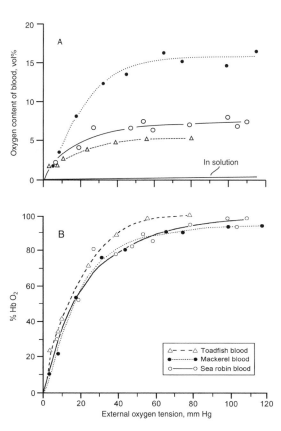

Figure 8.7 Oxygen transport by blood in three fishes, the mackerel (*Scomber scombrus*), sea robin (*Prionotus carolinus*), and toadfish (*Opsanus tau*), as a function of the external oxygen tension, (A) in terms of the total amount of oxygen transported and (B) in proportion to the maximal amount of oxygen capable of being transported. Source: Modified from Root (1931).

blood. Each of these factors will be examined in turn.

The amount of hemoglobin contained in an erythrocyte is approximately constant, so the amount of hemoglobin in blood varies mainly with hematocrit, the fraction of total blood volume that is cells. Hematocrit and oxygen capacity generally are large in vertebrates that have high rates of metabolism (e.g., in active fish) but are small in sluggish fish (Root 1931). In a survey of 80 species of south Atlantic fishes, Wilhelm Filho et al. (1992) showed that active species have higher hematocrits, that hematocrits increased from rays to sharks and to teleosts, and that elasmobranchs had erythrocytes 3 times the size found in teleosts. These authors suggested that the large erythrocytes of elasmobranchs reduce cardiac output, thereby contributing to a reduced energy expenditure. Hematocrit is nearly independent of body mass (Burke 1966).

Variation in the amount of hemoglobin obscures the influence of the other factors affecting gas transport. To examine these factors, the amount of oxygen carried by blood at a particular P_{O_2} can be expressed as a percentage of the amount transported when the blood is saturated with oxygen. Such a transformation of the data in Figure 8.7A is found in Figure 8.7B. Notice that after transformation the curves are *deceptively* similar. The transformed data, however, show with clarity the differential affinities of hemoglobin for oxygen. *Affinity* is quantitatively defined as the reciprocal of the oxygen tension at which the blood is 50% saturated (i.e., $1/P_{50}$). Thus, the toadfish (*Opsanus tau*) has the hemoglobin with the highest affinity (Figure 8.7B), although this is difficult to see in Figure 8.7A because of the low oxygen content of its blood at saturation.

Various factors influence the affinity of hemoglobin for oxygen, one being the conditions in the environment. It is usually high in fishes living in water with low oxygen tensions and low in species that are active, or that have high rates of metabolism and live in water with high oxygen tensions (Powers et al. 1979, Powers 1980, Wilhelm Filho and Reischl 1981). Fish obviously require a high affinity to extract oxygen from water with a low oxygen content, but why should affinity be inversely correlated with rate of metabolism?

A high P_{50} (low affinity) permits high unloading tensions, large blood-tissue differentials in oxygen concentration, and consequently high rates of oxygen diffusion (see Equation 2.22) from blood to the tissues where oxygen is used. A relation between affinity and rate of metabolism implies that affinity varies with body size. Unfortunately, body mass is often unknown in the fishes that have had their oxygen dissociation curves studied, so as a substitute, the P_{50} of Amazonian fishes was compared with body length: P_{50} is not correlated with body size but remains correlated with activity. Wood et al. (1979), however, showed that P_{50} in the piranha (*Serrasalmus rhombius*) is inversely related to body mass.

Variations in P_{50} result from structural differences in hemoglobin and from variations in the amount of organophosphates in the erythrocytes (Johansen et al. 1978a). The ability to control the internal environment of erythrocytes is undoubtedly the principal reason why vertebrate hemoglobins are located in cells and not free in the plasma. As the concentration of organophosphates increases in red blood cells, P_{50} increases and affinity falls (Johansen et al. 1978b, Riggs 1979). In *Fundulus heteroclitus* two alleles found at one locus control the ratio of ATP to hemoglobin in the erythrocytes and therefore the affinity of hemoglobin (Powers et al. 1979. Short-term changes in the affinity of hemoglobin for oxygen, often in response to a change in temperature (Grigg 1969), are controlled by changes in the concentration of organophosphates in erythrocytes (Weber 1982). Such adjustments may be common to all vertebrates (Wells et al. 1989).

Plasma has a low concentration of organophosphates, and therefore, hemoglobin dissolved in plasma is characterized by a high affinity and a low unloading tension (Krogh and Leitch 1919). Such hemoglobin would be suited to animals with low rates of metabolism living in hypoxic environments, which is the case for invertebrates that have plasma hemoglobin, such as burrowing polychaetes. Plasma hemoglobin would be unacceptable to most vertebrates because of their high rates of metabolism (i.e., larger body size) and unnecessary because most vertebrates live in waters that have a higher oxygen content than that encountered by burrowing polychaetes. Plasma hemoglobin would also contribute to the colloidal pressure of the plasma and thereby potentially interfere with ultrafiltration at the glomerulus (see Sections 6.6 and 7.10.2).

The affinity for oxygen of myoglobins, monomeric heme proteins found in heart tissue and red skeletal muscles, strongly depends on temperature (Nichols and Weber 1989). The affinities of the myoglobin of the sperm whale (*Physeter catodon*), yellowfin tuna (*Thunnus albacares*), albino rat, buffalo sculpin (*Enophrys bison*), and coho salmon (*O. kisutch*) are similar when they are compared at normal body temperatures.

8.4.4 Oxygen transport in the presence of carbon dioxide. Carbon dioxide in blood influences oxygen transport, mainly, but not exclusively, owing to a change in pH. An increase in P_{CO_2} causes hemoglobin to dump oxygen, which is often described as a shifting of the oxygen association curve "to the right" (Figure 8.8). This shift is called the *Bohr effect*, which can be defined as an increase in P_{50} (reduction in affinity) of hemoglobin with an increase in P_{CO_2}. This response aids the unloading of oxygen at tissues where the CO_2 concentrations are high (Black 1940).

All fish do not have the same sensitivity to CO_2 (Figure 8.8). Those that are active and live in water with high oxygen and low CO_2 tensions, such as the mackerel (*Scomber scombrus*), have hemoglobins with a large Bohr effect; those that are inactive and live at low oxygen and high CO_2 tensions, such as the toadfish, have small Bohr effects. Because high CO_2 tensions are usually associated with low oxygen tensions in naturally occurring waters, a low Bohr effect is often found in hemoglobins with high affinities (Black 1940). Contrary to expectations, however, suckers (*Catastomus*) that live in pools or sluggish water (e.g., *C. insignis*) showed a Bohr effect, and a fast-water species (*C. clarkii*) had some hemoglobin molecules that had no Bohr effect (Powers 1972).

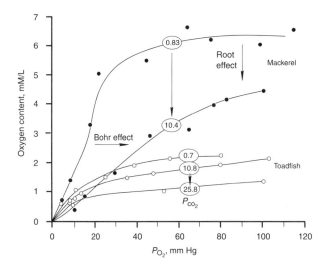

Figure 8.8 Oxygen transport by blood in two fishes, the mackerel (*Scomber scombrus*) and toadfish (*Opsanus tau*), as a function of the external oxygen tension and the CO_2 tension in blood. Source: Modified from Root (1931).

High CO$_2$ tensions also reduce the oxygen storage capacity of blood in many fishes (Figure 8.8). This reduction is called the *Root effect* after the pioneer work of R. W. Root (1931). Some authors (e.g., Riggs 1979, Farmer et al. 1979) maintained that the Root effect is simply an exaggerated Bohr effect. A large Root effect is most often found in fishes with a gas bladder (Scholander and van Dam 1957, Fänge 1966, Steen 1970), or in species living in fast-flowing water (Willmer 1934). The Root effect was most prominent in Amazonian fishes with a gas bladder and with a choroid rete in the eyes (Farmer et al. 1979). Farmer and colleagues suggested that a choroid rete is an "ancestral" (plesiomorphic) character, being found as it is in charachoids, culpeoids, and salmonoids, most of which also have a gas bladder rete. The original function of the Root effect may have been to deliver oxygen to a retina that did not have a capillary network (Riggs 1979). Dafré and Wilhelm (1989) confirmed this analysis. In fact, Wittenberg and Wittenberg (1962, 1974) showed that the oxygen tension in the eyes of teleosts and *Amia* is correlated with the presence of a choroid rete. The occurrence of this rete is usually associated with the presence of a pseudobranch, a modified gill, which may permit the accumulation of oxygen without accumulating CO$_2$ (Wittenberg and Haedrich 1974). Gas secretion into a bladder for buoyancy regulation (Ball et al. 1954, Kuhn et al. 1963) thus probably was a secondary use of the Root effect (see Section 9.3).

8.4.5 Oxygen transport and temperature.
Another factor influencing oxygen transport by hemoglobin is temperature: P_{50} increases (affinity falls) with an increase in temperature (Figure 8.9), although thermal acclimation may partially compensate for this effect (Grigg 1969). This relation reflects the fall in pH with an increase in temperature (Dejours 1975) (Figure 8.1) and the increase in P_{50} with a fall in pH (increase in P_{CO_2}) (Figure 8.8). As a result of these interactions, the affinity of poikilotherm hemoglobins (at normal body temperatures) is higher than that of homeotherms, but when fish have the same (high) temperatures as homeotherms, homeotherms have a hemoglobin with a higher affinity (i.e., the hemoglobin of homeotherms has been "corrected").

Some fish have hemoglobin with a reduced sensitivity to temperature. For example, the bluefin tuna (*Th. thynnus*), which makes extensive sea-

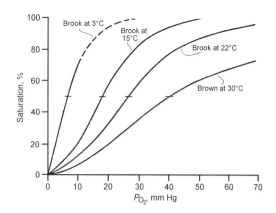

Figure 8.9 Oxygen saturation in the blood of two species of trout, the brook (*Salvelinus fontinalis*) and brown (*Salmo trutta*), as a function of the external oxygen tension at various temperatures. Source: Modified from Irving et al. (1941a).

sonal migrations between north Atlantic and tropical waters, has a temperature-insensitive hemoglobin (Johansen and Lenfant 1972; also see Powers 1980). (Even though this fish is semiendothermic [see Section 5.9.2], it regulates lateral muscle, not gill, temperatures.) Conversely, stenothermic fish often have a hemoglobin with a high sensitivity to temperature (Grigg 1967). Yet, some fishes that live in a thermally variable environment (e.g., *Fundulus heteroclitus*) have temperature-sensitive hemoglobin (Powers and Powers 1975).

The diversity of fish hemoglobins in response to temperature reflects a variety of adjustments made by fish, including a structural modification of hemoglobin, the evolution of multiple hemoglobins with different temperature sensitivities, and a change in the chemical environment within erythrocytes (Powers 1980). Indeed, *F. heteroclitus* kept P_{50} constant with an increase in acclimation temperature by reducing the molar ratio of ATP to hemoglobin in erythrocytes (Figure 8.10).

8.4.6 Critical oxygen tension.
Because many factors influence the oxygen dissociation curve, oxygen transport is best compared under conditions that a fish normally faces. Unfortunately, few data are available on the conditions faced in the environment and found in the blood of fish. This limitation is partially evaded by studying the response of an intact animal to the amount of oxygen present in the environment. Most vertebrates are "oxygen regulators" in the sense that

they maintain a resting rate of oxygen consumption independent of P_{O_2} down to a "critical" oxygen tension (P_c) (Figure 8.11), below which oxygen consumption decreases with P_{O_2}. Ultsch et al. (1981) showed an often-quoted example of the absence of oxygen regulation in the toadfish (*Opsanus tau*) to be erroneous; this fish, like all others examined, is a regulator. (Verheyen et al. [1994], however, concluded that two African cichlids are conformers.) $\dot{V}_{O_2}$ remains independent of P_{O_2} because most fish (1) increase ventilation rate and (2) increase cardiac stroke volume with a decrease in the oxygen content of water. Holeton and Randall (1967) showed that some fishes may also respond to low P_{O_2} by a partial shift to anaerobic metabolism and by an increase in hematocrit.

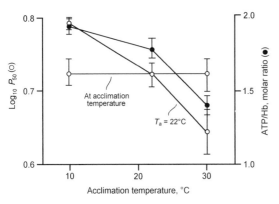

Figure 8.10 Ratio of ATP to hemoglobin (Hb) and $\log_{10} P_{50}$ in the teleost *Fundulus heteroclitus*, both at 22°C and at the acclimation temperatures of 10, 22, and 30°C, as a function of temperature. Source: Modified from Powers (1980).

Marine fishes that burrow into substrates face low oxygen tensions. Some species, for example, *Typhlogobius californicus*, a blind goby living commensally with a ghost shrimp in intertidal burrows, have low critical oxygen tensions, in this case 9 to 16 mm Hg (Congleton 1974). In contrast, other gobies, such as *Gillichthys mirabilis*, which may take refuge in burrows, and *Coryphopterus nicholsii*, which does not enter burrows, have higher critical oxygen tensions, namely, 16 to 25 and 19 to 28 mm Hg, respectively. *Typhlogobius californicus* has other characteristics that facilitate life in an anoxic burrow, including a very low rate of metabolism (20%–30% that in the other two gobies), the use of oxygen stored in the swim bladder during anoxic periods, and the extensive use of glycolysis during periods of anoxia (which corresponds to low tide). Other burrowing teleosts, such as *Cepola rubescens* and *Lumpenus lampretaeformis*, have higher critical oxygen tensions (i.e., 50–70 mm Hg), possibly reflecting higher field oxygen tensions encountered by these species than encountered by gobies (Pullin et al. 1980, Pelster et al. 1988). The Pacific sandlance (*Ammodytes hexapterus*) burrows into intertidal sediments, especially during winter. This species has a lower critical oxygen tension in winter (15.7 mm Hg) than in summer (31.1 mm Hg), possibly in association with lower water temperature in winter (Quinn and Schneider 1991).

Below P_c the resting rate of oxygen consumption falls because the amount of oxygen in the environment is inadequate to load hemoglobin and because the adjustments of circulation and respira-

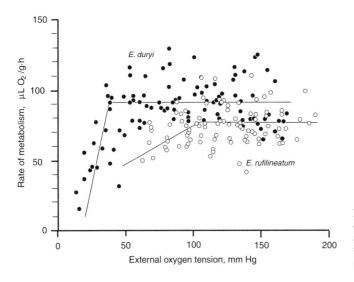

Figure 8.11 Rate of metabolism in two darters, *Etheostoma duryi* and *E. rufilineatum*, as a function of the external oxygen tension. Source: Modified from Ultsch et al. (1978).

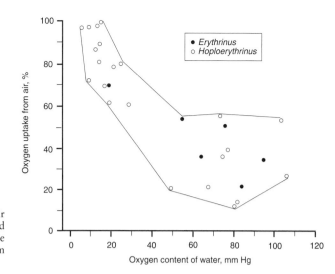

Figure 8.12 Proportion of oxygen uptake from air in two erythrinid fishes, *Erythrinus erythrinus* and *Hoploerythrinus unitaeniatus*, as a function of the oxygen tension in water. Source: Modified from Stevens and Holeton (1978).

tion are not adequate to compensate for the low P_{O_2}. The critical oxygen tension varies within and among species.

1. Large individuals and species have higher critical tensions than small individuals and species, which again emphasizes the importance (i.e., the "reality") of total over mass-specific rates of metabolism.

2. Species that depend exclusively on aquatic respiration usually have a lower P_c than species that mix aerial and aquatic respiration (see also Wakeman and Ultsch 1975).

3. Species that facultatively use both aerial and aquatic respiration depend increasingly on aerial gas exchange as the oxygen tension of water falls (Figure 8.12).

4. Aquatic species with a high P_c may be limited in distribution to rapidly flowing, highly oxygenated water (e.g., the darter *Etheostoma rufilineatum* [Ultsch et al. 1978]; Figure 8.11).

5. Activity can be depressed by low oxygen tensions in the environment (Figure 8.13) for the obvious reason that more oxygen is needed during activity.

6. An increase in temperature appears, generally, to increase P_c, but this trend may not be universal (Ultsch et al. 1978, Ott et al. 1980).

A systematic examination of P_c as a function of body size, type of respiration, morphology of the gas exchanger, temperature, distribution, and taxonomic affiliation would contribute to our understanding of the response of vertebrates to low

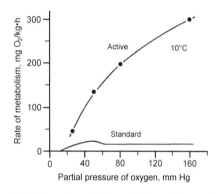

Figure 8.13 Rate of metabolism during activity and at rest at 10°C in carp (*Cyprinus carpio*) as a function of the external oxygen tension. Source: Modified from Beamish (1964).

oxygen tensions in the environment. Unfortunately, few such data are presently available: the one available study (Verheyen et al. 1994) showed much variation in P_c but no apparent correlation with other factors.

8.4.7 The impact of hydrostatic pressure. Although some vertebrates live in environments with low oxygen tensions, often coupled with high CO_2 tensions, and some face low barometric pressures at high altitudes (see Section 8.8.5), the only vertebrates that face uniformly high pressures on a chronic basis are fish that live deep in the ocean. At sea level the pressure by definition is 1 atm (= 760 mm Hg); the pressure increases with depth by 1 atm for each 10 m. On the edge of the continental shelf at a depth of 200 m, this hydrostatic pressure is 21 atm, and in the abyssal plain at 4000

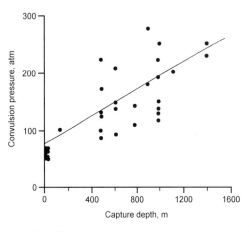

Figure 8.14 Hydrostatic pressure at convulsion in mesopelagic fishes as a function of the depth of capture. Source: Modified from Brauer et al. (1984).

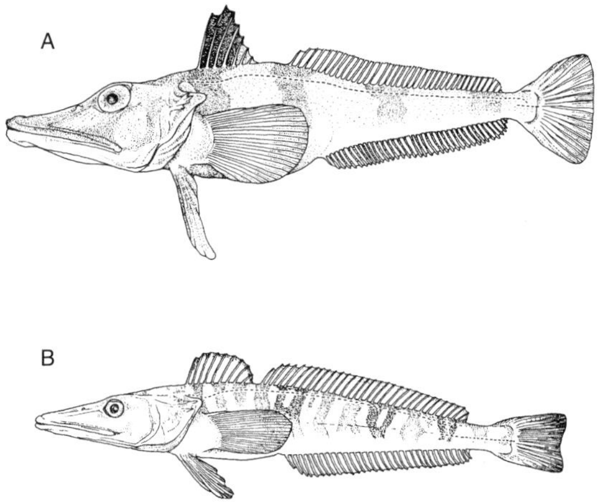

Figure 8.15 Ice-fishes. A. *Chaenocephalus aceratus*. B. *Champsocephalus esox*. Source: Modified from Iwami and Kock (1990).

to 5000 m it is 401 to 501 atm (Macdonald et al. 1987).

Fish that live deep in the ocean obviously have adjusted to the depth at which they live, but cannot withstand a sudden change in depth (i.e., pressure). For example, deep-water fishes, even those without swim bladders, often become moribund when brought to the surface, in part because of gas bubble formation in the blood, as a result of the great reduction in pressure. At the surface these fishes regain normal activity if they are artificially subjected to high pressures (Macdonald et al. 1987). Midwater fish, however, enter a convulsive state when they are subjected to pressures that exceed those that are normally encountered. The hydrostatic pressure at which convulsion occurs is directly correlated with the depth at which the fishes are captured (Figure 8.14). In Lake Baikal, Siberia, cottids without swim bladders that were captured at 15 to 75 m became convulsive at a mean pressure of 83 atm, whereas abyssocottids caught at 415 to 1400 m became convulsive at a mean pressure of 193 atm, which is equivalent to 1930 m. One exception of interest is that the cottid *Batrachocottus nikolskii* was collected at a depth of 600 to 1000 m, which is deep compared to the depth for the other cottids studied; it became convulsive at 119 atm, appropriate for its depth. Brauer et al. (1984, p. 699) concluded that hydrostatic pressure may have a role "... as a selection factor in the evolution of deep-water faunas." The evolution of a differential response to hydrostatic pressure opens new ecological opportunities,

thereby permitting an increased diversity in the Lake Baikal fish fauna.

As seen, fishes that live at high pressures have low rates of metabolism (Figure 8.6). The reduction in rate of metabolism with an increase in depth is probably not due to high pressures (Meek and Childress 1973, Gordon et al. 1976, Torres et al. 1979), low oxygen tensions (Meek and Childress 1973, Gordon et al. 1976, Torres et al. 1979), or even low temperatures (Torres et al. 1979), given that the rates in Figure 8.6 were all measured at one temperature (5°C). The low rates in mesoplagic fishes appear to reflect a reduction in muscle mass (Childress and Nygaard 1973) and lower activity levels (Torres et al. 1979), but ultimately the low rates are due to a great reduction in the resource base (which rains down sparsely from the surface).

8.4.8 Ice-fish: a special problem in gas transport. Ice-fishes (Channichthyidae, Figure 8.15) are unique among vertebrates in that they lack hemoglobin and erythrocytes; they have very low concentrations of myoglobin in heart muscle, but none in striated muscles (Douglas et al. 1985). All oxygen transported in these fish is dissolved in the plasma. Much interest has been shown in the mechanism by which ice-fish survive without respiratory pigments since the first reports of Rudd (1954, 1958), but little attention has been given to the question of why they lost erythrocytes.

Ice-fish are mainly distributed in the cold waters of Antarctica, where water temperature often remains between −1.4 and 0°C throughout the

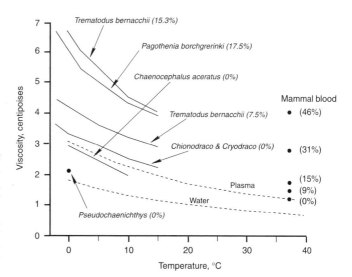

Figure 8.16 Viscosity of water and plasma of several antarctic fishes as a function of temperature. Data on the viscosity of mammalian blood as a function of hematocrit at a body temperature of 37°C are also given. Hematocrits are indicated as percentages enclosed in parentheses. Sources: Derived from Schmidt-Nielsen and Taylor (1968), Hemmingsen and Douglas (1972), and Macdonald and Wells (1991).

year. At these temperatures oxygen has its highest solubility in water, and water has its highest viscosity. A high solubility means that blood at this temperature can carry a relatively large amount of oxygen (although only one-tenth of the amount carried by fish with a full complement of hemoglobin; Rudd 1954). An increase in plasma viscosity is probably the major reason why ice-fish have eliminated their erythrocytes.

The relations that exist among viscosity, temperature, and hematocrit are indicated in Figure 8.16. With a decrease in temperature, the viscosity of water and all aqueous solutions, such as blood (Snyder 1971), increases. Schmidt-Nielsen and Taylor (1968) showed that viscosity (in mammals) at 37°C is greater in plasma than in water and that viscosity increases with hematocrit. The viscosity of blood greatly increases if hematocrit is high at a low temperature (e.g., Langille and Crisp 1980, Macdonald and Wells 1991). If some maximal limit to viscosity is dictated by the ability of the cardiovascular system to pump blood through the capillaries at a reasonable cost, a fall in blood temperature may require a compensatory reduction in hematocrit.

Even though ice-fish are the only fish to be completely erythrocyte free (which may not be technically correct; see Hureau 1966), other antarctic (Kooyman 1963, Everson and Ralph 1968) and arctic fishes (Scholander and van Dam 1957) have reduced numbers of erythrocytes. All channichthyids apparently are erythrocyte free (but see *Pseudochaenichthys*; Everson and Ralph 1968), including the only species, *Champsocephalus esox*,

that is found north of the Antarctic Convergence in Patagonia and the Malvinas (Falkland) Islands (Rudd 1954; see Section 14.3.1). A reduced hematocrit among marine fishes is mainly found in sluggish species that live in constantly cold water. Fewer subarctic than antarctic fishes show a reduced hematocrit, possibly because the water in which the subarctic species live is more subject to seasonal variations in temperature. Everson and Ralph (1968) raised the question of whether nototheniid fishes north of the Antarctic Convergence have higher hematocrits than those living south of the convergence; such data are presently unavailable, as are data from the high arctic.

Significant adjustments in gas transport are required in ice-fish to compensate for the 10-fold reduction in oxygen capacity of their blood. One adjustment is that ice-fish generally have low rates of oxygen consumption (Hureau 1966, Hemmingsen et al. 1969) due to low water temperatures and to a reduction of standard rates to one-half of expected values (Hemmingsen and Douglas 1970; Figure 4.10). The depression in rate of oxygen consumption is related to the decrease in hematocrit, as measured by the decrease in oxygen capacity of blood (Hemmingsen et al. 1969): ice-fish represent an extreme condition (Figure 8.17). No evidence exists of any unusually large reliance on anaerobic metabolism by ice-fish (Hemmingsen and Douglas 1970, 1972). Some large ice-fish, such as *Chaenocephalus aceratus*, which may attain a meter in length and weigh 3 kg, are sluggish bottom-dwellers (Ralph and Emerson 1968), a behavior that further reduces oxygen demand.

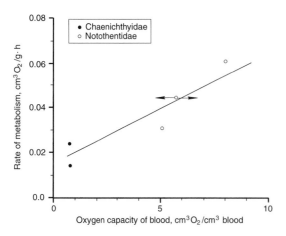

Figure 8.17 Rate of metabolism at 0 to 1°C in (hemoglobinless) ice-fish (Chaenichthyidae) and (hemoglobin-bearing) nototheniid fish, corrected to a mass of approximately 200 g (using $b = 0.78$), as a function of the oxygen capacity of blood. Sources: Derived from Hemmingsen et al. (1969), Hemmingsen and Douglas (1970), and Holeton (1970).

Channichthyids have modest to small gill areas for gas exchange (Steen and Berg 1966, Hughes 1972a; Figure 8.5), as befits sluggish fish. Walvig (1960), however, showed that ice-fishes are scaleless with a dermis rich in capillaries; he suggested that the pelvic fins may be important sites for gas exchange. At least 8% of the oxygen consumed is derived from the tail fin in *Ch. aceratus*; up to 40% of the total oxygen consumption may be obtained through the integument, which is unusual for a large fish (Hemmingsen and Douglas 1970). Other scaleless antarctic fishes, including the Bathydraconidae, Harpagiferidae, and Zooarchidae, may also depend on cutaneous gas exchange (Wells 1987): *Rhigophila*, a zoarchid, obtains about 34% of its oxygen uptake through its skin (Wells 1986). Scaled notothenids get 8% to 17% of their oxygen through their skin (Wells 1987). Cutaneous respiration appears to be most important in species with low rates of metabolism.

One of the principal compensations in channichthyids for the reduced oxygen capacity of blood is an increased rate of blood circulation. The rationale for this increase can be seen in the following relation (Holeton 1970, Hughes and Morgan 1973):

$$\dot{V}_{O_2} = \dot{Q} \cdot \alpha_b \cdot Pa_{O_2}, \qquad (8.5)$$

where $\dot{Q}$ is cardiac output (cm³ blood/min), α_b is absorption coefficient for oxygen in blood (cm³/L·mm Hg), and Pa_{O_2} is the arterial oxygen tension (mm Hg). A reduction in the oxygen capacity of the blood ($\alpha_b \cdot Pa_{O_2}$) by a factor of 10 (as occurs in ice-fish with the loss of hemoglobin) would require a 10-fold increase in cardiac output to transport the same amount of oxygen. Actually, a standard rate of metabolism in ice-fish that is 40% of the rate expected from mass means that $\dot{Q}$ needs to increase only fourfold to compensate for the fall in α_b.

Ice-fish have large hearts and cardiac outputs that are 3 to 5 times (Holeton 1970, Hemmingsen et al. 1972) and blood volumes that are 2 to 3 times (Hemmingsen and Douglas 1970) those of hemoglobin-bearing teleosts. Even though the cardiac output of ice-fish is high, their blood pressure is low (Holeton 1970; Hemmingsen and Douglas 1972; Hemmingsen et al. 1972, 1973), presumably because of a low vascular resistance produced by a large capillary bore and a high capillary density (Hemmingsen et al. 1969).

These circulatory adjustments are effective enough in *Ch. aceratus* to permit it to tolerate water temperatures as high as 4 to 5°C with some activity (Hemmingsen and Douglas 1972). It maintains rate of metabolism independent of P_{O_2} down to 40 to 50 mm Hg (Holeton 1970, Hemmingsen and Douglas 1970), and *Pagetopsis macropterus*, another chaenichthyid, can maintain standard rates at P_{O_2} down to 35 mm Hg (Hemmingsen et al. 1969). In contrast, the red-blooded *Notothenia gibberifrons* (Nototheniidae) maintains its standard rate at P_{O_2} down to 14 to 20 mm Hg (Holeton 1970). Clearly, hemoglobin is not required to regulate oxygen transport. The circulatory adjustments made by ice-fish appear to be costly: blood circulation in *Ch. aceratus* may require up to 27% of the standard rate of metabolism (Hemmingsen et al. 1969), whereas in most fish the cost is only about 5% (Garey 1970). The sum of the costs of gill ventilation and circulation may account for 50% of the standard rate of metabolism in ice-fish. These values are greater than the values that can be accounted for by a reduction in standard rate, and therefore probably represent an increase in the absolute cost of ventilation and circulation.

Much more work on the gas exchange of channichthyids obviously is required to understand the ecological and physiological limitations that they face; one of the most important species to study is *Champsocephalus esox* at its northern limits to distribution in Patagonia (see Section 14.3.1). The use of integumental gas exchange may pose difficulties for osmotic regulation: do ice-fish have higher plasma concentrations than other marine fish?

8.4.9 Carbon dioxide transport. Carbon dioxide is transported in vertebrates as gas dissolved in blood (see Section 2.11), as bicarbonate (see Section 8.3), and bound to hemoglobin in the erythrocytes. Because CO_2 has a higher solubility than oxygen, a greater proportion of CO_2 than oxygen is carried in solution. The amount of CO_2 transported in blood increases with external P_{CO_2}, although the absolute amount may vary appreciably with the species; few comparative studies of CO_2 transport have been made, but one of the important differences among species is the differential transport of CO_2 by hemoglobin. The amount of CO_2 bound to hemoglobin varies inversely with the amount of oxygen bound to hemoglobin (the Haldane effect), so that the oxygenation of hemoglobin in the gills facilitates the unloading of CO_2.

As a result of increasing blood P_{CO_2}, plasma pH falls, although fish show great variation in the pH associated with a particular P_{CO_2}. For example, two freshwater teleosts from Guiana, the bom-bom (*Pterodoras granulosus*) and the haimara (*Hoplias malabaricus*), have a low pH, whereas marine teleosts, such as the mackerel (*Scomber scombrus*), sea robin (*Prionotus carolinus*), and toadfish (*Opsanus tau*), have a high pH at the same P_{CO_2} (Root 1931, Willmer 1934), a difference that may reflect buffering by hemoglobin.

8.5 THE TRANSITION FROM AQUATIC TO AERIAL GAS EXCHANGE IN FISHES

A combination of low oxygen and high CO_2 tensions in the environment constitutes the greatest challenge to gas exchange in vertebrates, especially at high environmental temperatures (i.e., when rates of metabolism are high). These conditions are often encountered in temperate swamps during summer and throughout much of the year in tropical swamps (Carter and Beadle 1931, Carter 1935a, Dehadrai and Tripathi 1976, Ultsch 1976b, Kramer et al. 1978, Val and Almeida-Val 1995). For example, in swamps covered with water hyacinths (*Eichhornia*) P_{O_2} may approach 0 mm Hg and P_{CO_2} may reach 80 to 100 mm Hg (Ultsch 1973b). Such a combination is produced by many factors: swamp waters tend to be stagnant, thereby

reducing the bulk exchange of oxygen and CO_2 with the atmosphere; high temperatures lead to increased rates of decay (and high rates of CO_2 accumulation) and reduced solubility of oxygen; and swamp water is often turbid, shaded, and covered with floating vegetation.

Few fish tolerate such extreme conditions. Ultsch (1976b) showed that of nine species of teleosts found in Florida ponds, only two (*Gambusia affinis* and *Fundulus chrysotus*) were found in association with thick mats of hyacinths. These species live in the uppermost layers of water, where they use the surface film, which has more oxygen and less CO_2 than is found under the hyacinths.

Fish widely use the two solutions to the hostile conditions described for Florida waters: frequenting the superficial layer and switching to aerial exchange. Carter and Beadle (1931) showed that of the 20 most common fishes in Paraguaian swamps, 6 species use the uppermost layer of water and 8 are air breathers. Kramer et al. (1978) emphasized that in Amazonia, in spite of a diverse array of air-breathing fishes, the majority of fishes depend on aquatic exchange, and many of these species exploit the higher oxygen tensions near the surface without any marked morphological specialization.

Strictly speaking, fish that use the superficial layer of water are not bimodal exchangers, but like air breathers, they evade hostile conditions in the water column. The diffusion of oxygen from the atmosphere is effective in raising the oxygen tension of still water only to a depth of less than 1 mm (Lewis 1970). Most fish that use this layer as a source of oxygen have upturned mouths and flattened heads, both of which aid in maintaining a horizontal position near the surface with a minimal energy expenditure. These modifications are found most notably among cyprinodontoids, such as *Gambusia*, *Poecilia*, and *Fundulus*. A small size permits effective use of the surface layer of water (Carter 1935a, Lewis 1970) because it reduces demand for oxygen. Some mochokid catfishes that belong to the genus *Synodontis* in Central Africa use aquatic surface respiration (Chapman et al. 1994a): one species, *S. nigriventris*, exchanges gas at the surface while swimming upside down and has reversed shading, whereas *S. afrofischeri* exchanges gas at the surface while taking a vertical position relative to the surface and has normal counter-shading. The difference in position relative to the surface may indicate a lower cost of surface gas exchange in *S. nigriventris*. A body larger than a few grams can tolerate hypoxic waters only if aerial gas exchange is used.

The diversity in the tolerance of East African fishes to conditions in freshwater has implications for their local distributions and for their ability to survive human-mediated changes in the environment, partly as a result of the introduction of Nile perch (*Lates niloticus*) into Lake Victoria. This introduction led to the decline or disappearance of hundreds of haplochromine cichlids and other indigenous fishes. Some survivors have depended on their ability to tolerate the high P_{CO_2} and low P_{O_2} in adjacent swamps. Cichlids (Chapman et al. 1996, Rosenberger and Chapman 2000), mormyrid electric fishes (Chapman and Chapman 1998, Chapman and Hulen 2001), and cyprinids (Chapman and Liem 1995, Olowo and Chapman 1996) that tolerate these conditions rely on aquatic surface respiration, often in association with large gill surface areas, low rates of metabolism, and low critical P_{O_2}. Surface respiration is sometimes facilitated by holding air bubbles in the buccal cavity, which may both facilitate buoyancy and increase the P_{O_2} of the water passing over the bubble (Burggren 1982, Gee and Gee 1991, Chapman et al. 1994b). This differentiation occurs not only among species but also among populations belonging to the same species: longer gill filaments (ca. 50%) are found in populations of the cyprinid *Barbus neumayeri* inhabiting papyrus swamps, compared to populations living in streams and rivers, the gill filament length increasing with a decrease in ambient P_{O_2} (Figure 8.18). An increase

in gill area also occurs in swamp-dwelling *Pseudocrenilabrus multicolor*, a cichlid; this increase results from developmental plasticity and may have a genetic component (Chapman et al. 2000). Mormyrid fishes that live in hypoxic waters have larger gill areas and smaller brains, which led Chapman and Hulen (2001) to conclude that large gill areas permit survival at a low P_{O_2} but are inadequate to support a large brain size. Under similar environmental circumstances, *Colossoma macropomum*, a serrasalmid in Amazonia, increases its hemoglobin content, erythrocyte number, and gill surface area (Saint-Paul 1984) in response to seasonal hypoxic waters.

In cold-temperate climates lakes freeze over in winter. If the lakes are both shallow and eutrophic, they are often characterized by a winter kill of fishes due to the development of anoxic and hypercarbic conditions by springtime. Such lakes have a distinctive fish fauna, usually consisting of various species of minnows or mudminnows. The fishes that survive these hostile conditions tolerate or avoid hypoxia by moving to water inlets or to the ice-water interface. Mudminnows (*Umbra limi*) use oxygen contained within air bubbles trapped under the ice. A combination of a small mass, low rate of metabolism, tolerance to low oxygen levels, reduced activity, head shape, and use of local pockets of oxygen usually permits these fish to survive winter (Klinger et al. 1982; Magnuson et al. 1983, 1985).

Intertidal fishes, like those in freshwater swamps, often use aerial respiration (Martin 1995). Some species spontaneously move in and out of water

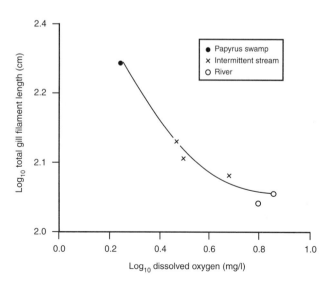

Figure 8.18 Log_{10} total gill filament length in populations of the East African cichlid fish *Barbus neumayeri* as a function of log_{10} dissolved oxygen content of the water in which the populations reside. Oxygen content depends on whether the water occurs in papyrus swamps, intermittent streams, or rivers. Source: Modified from Chapman et al. (1999).

(the so-called skippers, including, e.g., *Periopthal-amus*); some species remain in place, often under boulders or among algae, when the tide goes out (the "remainers," including *Blennius* and *Sicyases*); and the species that are trapped in tide-pools when the tide goes out, although they may emerge from the pools when the water increases in temperature, become anoxic, or become hypercarbic (the "tide-pool emergers," including various sculpins). Unlike many freshwater fish that show a decrease in rate of metabolism upon emergence from water, some intertidal fishes, most notably in the genera *Periopthalamus* (Gordon et al. 1969, 1978) and *Clinocothus* (Martin 1991), maintain their rates equal to those found in water, although anaerobiosis may be an important pathway during activity after emergence (Martin 1995).

8.5.1 Physiological adjustments to aerial respiration.
Many physiological adjustments are made to air-breathing habits.

1. In marine species hematocrit is correlated mainly with activity, whereas in freshwater species hematocrit is correlated with conditions in the environment (Willmer 1934, Johansen 1966). Thus, freshwater fishes that live in hypoxic swamps, especially those that use air breathing, have high hematocrits, which increase oxygen transport and the buffering of CO_2: *Clarias* and *Saccobranchus* have an oxygen capacity of blood equal to 18.0% and 17.5%, respectively (Singh and Hughes 1971).

2. Water breathers generally have blood with a higher affinity for oxygen than that found in air breathers (Figure 8.19; Johansen and Lenfant 1972), although two surveys of Amazon fishes (Johansen et al. 1978b, Powers et al. 1979) questioned this conclusion. These surveys found no correlation between P_{50} and breathing habits when the bloods were compared at one pH, but they found a significant difference at in vivo pHs between air breathers (mean P_{50} = 12.9 mm Hg) and water breathers (mean P_{50} = 5.3 mm Hg) (Johansen et al. 1978b). This difference in P_{50} occurs because air breathers are more sensitive to a change in pH than are water breathers, which emphasizes the importance of making measurements under realistic conditions. The difference between water and air breathers is shown dramatically by a comparison of *Arapaima* and *Osteoglossum* (Johansen et al. 1978a, Powers et al. 1979): if *Arapaima* were

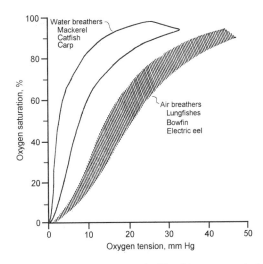

Figure 8.19 Oxygen transport by blood in water- and air-breathing fishes as a function of the external oxygen tension. Source: Modified from Johansen and Lenfant (1972).

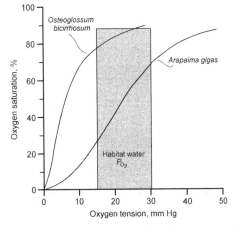

Figure 8.20 Oxygen transport by blood in two osteoglossid fishes, *Osteoglossum bicirrhosum* and *Arapaima gigas*, as a function of the external oxygen tension. The range in oxygen tension encountered in their natural habitats is indicated by the shaded area. Source: Modified from Johansen et al. (1978a).

limited to aquatic exchange, its blood would be maximally 70% saturated, whereas *Osteoglossum*, a water-breathing relative, has blood that would be minimally 80% saturated with oxygen (Figure 8.20). Nevertheless, Riggs (1979, p. 263) concluded that ". . . the major differences [in P_{50}] occur in comparison of fish of slow- and fast-moving waters and that the comparison between the water- and air-breathing fish yields almost no difference. The fish from rapid waters tend to have bloods with P_{50} values about 50% higher than found for the bloods of fish from slow waters; the P_{50} values for the air breathers average less than 10% higher than those from water breathers."

The observed differences in affinity of the blood of Amazonian fishes mainly reflect the molar ratios of organophosphates (ATP and GTP) to hemoglobin in the erythrocytes (Johansen et al. 1978b): air breathers have relatively greater concentrations of organophosphates, which have the effect of reducing the affinity (increasing the P_{50}) of hemoglobin. For example, these changes occur in *Synbranchus* as it shifts from aquatic to aerial respiration (Figure

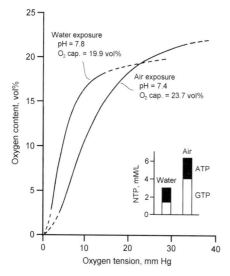

Figure 8.21 Oxygen content in blood of the teleost *Synbranchus marmoratus* in air and water as a function of the external oxygen tension and nucleotide triphosphate (NTP) levels in the erythrocytes. Source: Modified from Johansen et al. (1978b).

8.21). The difference in affinity for oxygen existing between the hemoglobins of *Arapaima* and *Osteoglossum*, however, is not related to organophosphate concentrations or to temperature sensitivity, but may be related to hemoglobin structure.

3. With a switch from aquatic to aerial exchange, ventilation rates fall and arterial P_{O_2} and P_{CO_2} increase (Rahn 1966a, 1966b; see Section 8.3). In compensation for the increased arterial P_{CO_2}, air breathers show an increase in plasma bicarbonate concentration proportional to the increase in P_{CO_2} (Figure 8.22).

4. Whether the Bohr effect varies with the mode of gas exchange is unclear. Carter (1931) and Willmer (1934) suggested that air breathers are less sensitive to a change in arterial P_{CO_2} than water breathers. Johansen et al. (1978b) concluded that Amazonian water-breathing fishes had smaller Bohr effects than air-breathing species. Yet, Powers et al. (1979) found no such correlation in Amazonian fishes. Data on the Bohr effect (and oxygen affinity) are complicated by the different methods used to evaluate oxygen association-dissociation curves (Riggs 1979).

8.5.2 The morphology of bimodal gas exchange in fishes. Air-breathing fishes use a variety of structures for exchange, including gills and branchial "trees," pharyngeal lungs, swim bladders, branchial "lungs," buccal cavities, stomachs, lower intestines, and to some extent, skin (Table 8.1). The eight air breathers from Paraguaian swamps listed by Carter and Beadle (1931) represent at least six independent lines of air-breathing habits, judging from the structures used (including three that use

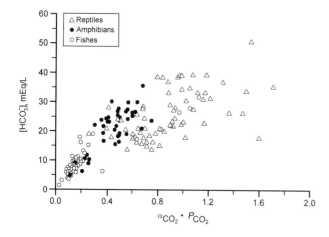

Figure 8.22 Blood bicarbonate concentration in fishes, amphibians, and reptiles as a function of the product $\alpha_{CO_2} \cdot P_{CO_2}$. Source: Modified from Ultsch (1996).

Table 8.1 Accessory gas exchange structures in fishes

Accessory organ	Examples	Habitat	Region
Pharyngeal diverticula			
Channidae	*Channa*	Tropical pools	Asia, Africa
Gobiidae	*Periophthalmus*	Muddy shores	Africa, Pacific
Synbranchidae	*Synbranchus*	Swamps	S. America
Branchial chamber diverticula			
Clariidae	*Clarias*	Tropical pools	India, Africa
Heteropneustidae	*Heteropneustes*	Tropical pools	India
Anabantidae	*Anabas*	Tropical pools	Asia
Belontiidae	*Betta, Macropodus*	Tropical pools	S.E. Asia
Osphronemidae	*Osphronemus*	Freshwater	S.E. Asia
Branchial chamber filled with air			
Hypopomidae	*Hypopomus*	Swamps	S. America
Synbranchidae	*Synbranchus*	Swamps	S. America
Pharyngeal epithelium			
Synbranchidae	*Monopterus*	Freshwater	S.E. Asia
Electrophoridae	*Electrophorus*	Freshwater	S. America
Air bladder			
Amiidae	*Amia*	Rivers, swamps	N. America
Lepidosteidae	*Lepisosteus*	Rivers	N. America
Osteoglossidae	*Arapaima*	Swamps	S. America
Gymnarchidae	*Gymnarchus*	Rivers, swamps	Africa
Erythrinidae	*Erythrinus*	Swamps	S. America
Umbridae	*Umbra*	Swamps	N. America, Eurasia
Lungs			
Ceratodontidae	*Neoceratodus*	Rivers	Australia
Lepidosirenidae	*Lepidosiren*	Swamps	S. America
Protopteridae	*Protopterus*	Rivers, swamps	Africa
Polypteridae	*Erpetoichthys, Polypterus*	Rivers, swamps	Africa
Stomach and intestine			
Cobitidae	*Misgurnus, Cobitis*	Rivers	Eurasia
Doradidae	*Doras*	Rivers, swamps	S. America
Callichthyidae	*Callichthys, Hoplosternum*	Swamps	S. America
Loricariidae	*Hypostomus, Ancistris*	Swamps	S. America

Skin
Many fishes that are regularly exposed to air exchange a significant proportion of gas through their skin. These include *Erpetoichthys, Anguilla, Clarias, Heteropneustes, Electrophorus, Blennius, Periophthalmus, Dormitator, Monopterus, Chaenocephalus, Rhigophila,* and *Synbranchus,* among others.

Sources: Derived from Carter and Beadle (1931) and Graham (1997).

intestines, two that use gills, and one each that use swim bladder, lungs, and stomach). Unlike many freshwater air breathers, intertidal air breathers rarely have special structures for gas exchange but tend to rely either on gills or on skin (Martin 1995). But even here subtle anatomical differences may occur among gobies: the completely aquatic *Gobius* has the largest gill surface area; the highly terrestrial *Periopthalamus*, the smallest gill area; and the intermediate *Boleopthalamus*, an intermediate gill area (Martin 1995, after Schottle 1932).

The diversity of structures used by fishes for aerial gas exchange is an excellent example of the opportunistic nature of natural selection. The use of one or another structure has its particular consequences. For example, gills tend to collapse in air, a condition that can be minimized by increasing the breadth of the secondary folds (Carter and Beadle 1931). Gill surface area tends to be small in air-breathing fish, but in no case have gills been completely replaced, because (as we will see) they are the principal means by which CO_2 is excreted. Some air breathers use their integument for gas exchange, but it requires that fish leave water because the use of skin for gas exchange is compromised in hypoxic water by the loss of oxygen.

One of the most peculiar structures used for exchange is the intestinal tract. Stomachs and intestines used for gas exchange are thinly walled and nearly transparent (due to the absence of muscles), which raises the question, unanswered at present, of how two separate functions, respiration and digestion, can occur in the same tube at the same time, albeit separated along the length of the tube. Surely, food must enter the anterior end of the digestive tube and (rapidly?) pass through the respiratory section. Most intestinal breathers apparently swallow and void air through the mouth (Carter and Beadle 1931), although some species void gas via the anus (Gee and Graham 1978). Gee and Graham observed that air in the intestine provides about 75% of the lift needed for neutral buoyancy in *Hoplosternum* and *Brochis*. The use of periodic air breathing in *Umbra limi*, even at high water P_{O_2} values, may be required to maintain the appropriate volume in its swim bladder and the correct buoyancy (Gee 1980).

8.5.3 Comparative importance of structures for exchange.
In all bimodally exchanging fish the gills (and skin, when used) are mainly used for CO_2 excretion, whereas the accessory organs are used principally for oxygen uptake: approximately 65% of oxygen requirements is obtained from air and about 75% of CO_2 is lost to water (Rahn and Howell 1976). Gills then are not eliminated in air-breathing fishes, even though their area is reduced to about one-half that found in water-breathing species (Dubale 1951, Dehadrai and Tripathi 1976). In the reedfish (*Erpetoichthys* [= *Calamoichthys*] *calabaricus*), approximately 32% of the uptake of oxygen is through the skin; 28%, through the gills; and 40%, through the lungs (Sacca and Burggren 1982). The surface area of the accessory organs does not completely compensate for the reduced gill surface area (Rahn et al. 1971a): in *Anabas* the total area for exchange is equivalent to the gill surface area of an inactive fish, only some 30% of which is contributed by the accessory organs (Hughes et al. 1974). The various areas for gas exchange do not make proportional contributions to exchange because of differences in the thickness of the diffusion pathway. In the case of *Anabas*, the pathway is thinnest in the suprabranchial gas exchanger, greatest in the skin (ca. 60 times as thick), and intermediate in the gills (3–50 times as thick). Consequently, although gill area in *Anabas* is 70% of the total area, it accounts for only 14% of the diffusion capacity. In *Heteropneustes* (= *Saccobranchus*), gills account for 20% of the area for gas exchange but only 17% of the diffusion capacity. Another factor affecting the comparative contributions to gas exchange by the organs involved is the comparative rate at which the organs are perfused.

A very important factor determining the comparative balance between aquatic and aerial exchange is the physical-chemical conditions in the water. Willmer (1934), for example, showed that the Guianian yarrow (*Hoplerythrinus*) breathes water at high oxygen and low CO_2 tensions but uses air when the water has low oxygen and high CO_2 tensions (Figure 8.23). Dehadrai (1962) showed a similar pattern in the Indian teleost *Notopterus*. An increase in temperature also shifts a bimodal exchanger from aquatic to aerial

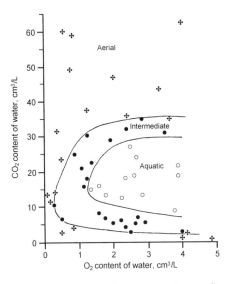

Figure 8.23 The use of aerial, aquatic, or intermediate gas exchange in a teleost (*Hoplerythrinus unitaeniatus*) as a function of the CO_2 and oxygen content of water. Source: Modified from Willmer (1934).

exchange: *Amia* obtains about 92% of oxygen through its gills at 10°C, 63% at 20°C, but only 26% at 30°C, the remainder obtained via the open air bladder (Johansen et al. 1970). Rahn et al. (1971a) demonstrated a similar temperature sensitivity in the gar *Lepisosteus*: the gar must use air at warm temperatures, but in winter it can live continuously under ice without breathing air. The African brachyopterygian *E. calabaricus* shifts from 66% of oxygen uptake from water at 25°C to 54% at 33°C (Pettit and Beitinger 1985), the remainder being obtained from air through its paired lungs. The influence of high temperature reflects both an increase in oxygen demand and a reduced solubility of oxygen in water. The use of lungs for aerial gas exchange in *E. calabaricus* permits it to tolerate 6 to 8h out of water, to increase its total rate of metabolism, and to be active, even to capturing insects (Sacca and Burggren 1982). The transition from a dependence on branchial to aerial respiration in *U. limi* occurs at a low water P_{O_2}, the dependence on aerial respiration increasing with a further reduction in P_{O_2} (Figure 8.24). This transitional P_{O_2} increases with temperature from about 10 to 15 mm Hg at 5°C to 40 to 50 mm Hg at 30°C (Gee 1980).

Another factor influencing the balance between aquatic and aerial exchange is body size. Bimodal fishes that weigh more than a few hundred grams depend more on aerial exchange than smaller species: such large fish as *Arapaima gigas*, *Electrophorus electricus*, *Lepidosiren paradoxa*, and *Protopterus aethiopicus* are obligatory air exchangers. The impact of body size is shown during ontogeny in *Anabas* (Hughes et al. 1974): at masses smaller than 45 g, *Anabas* depends mainly on aquatic exchange, whereas at larger masses aerial exchange is most important. A similar condition exists in *Clarias* and *Heteropneustes* (Das 1927). The large mass of *A. gigas* may explain why it depends more on aerial respiration than the smaller *Osteoglossum*.

Air breathing is found in some marine fishes, where it plays a different role than it does in freshwater species. Only a few freshwater fish are truely amphibious (e.g., *Clarias*, *Heteropneustes*, *Anabas*, *Synbranchus*). Most freshwater fishes use air breathing during periods of water stagnation and hypoxia, or during drought, that is, during periods of *reduced* activity. Aerial respiration permits some marine species, especially those that belong to the families Gobiidae and Blennidae, to exploit amphibious habits, mainly on mudflats in mangroves (Graham 1976), that is, during periods of *increased* activity. Consequently, amphibious fish have normal or high rates of metabolism when in air; *Periophthalmus australis* even has bradycardia upon (forceably?) entering water (Garey 1962)! Unlike freshwater species, most marine air breathing fish use their gills and skin for oxygen uptake, yet no marine species is known to be an obligatory air breather. Marine amphibious fish, as might be expected, shift to ureotelism in air (Gordon et al. 1969, 1970).

8.5.4 Bimodal exchange in lungfishes. The gas exchange of lungfishes is of special interest because they use lungs (two in *Lepidosiren* and *Protopterus*, one in *Neoceratodus*) that are homologous with the lungs used by tetrapods. Two of the three genera (*Lepidosiren* and *Protopterus*) are swamp dwelling; *Neoceratodus* is river dwelling.

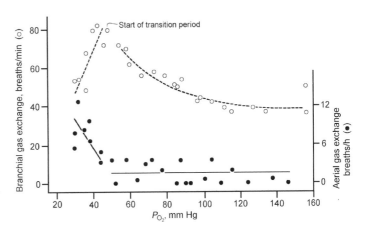

Figure 8.24 Branchial and aerial gas exchange in the mudminnow *Umbra limi* as a function of the oxygen content of water when ambient temperature was 25°C. Source: Modified from Gee (1980).

The swamp dwellers are obligatory air breathers, whereas *Neoceratodus* normally is a water breather that uses air breathing facultatively at night during activity (Grigg 1965). Both *Protopterus* and *Lepidosiren* aestivate during droughts (Smith 1931a, Sawaya 1946).

Lepidosiren and *Protopterus* separate aerial from aquatic gas exchange by differentially shunting blood through four gill arches. Systemic blood, which is rich in CO_2 and poor in oxygen, is sent through the posterior gill arches, which are supplied with gill filaments that permit CO_2 to be lost to ambient water. The blood, low in oxygen (ca. 20 mm Hg) and in CO_2 (26 mm Hg), is then sent to the lungs via the pulmonary artery. Pulmonary venous blood, which is richer in oxygen (34 mm Hg) and poorer in CO_2 (22 mm Hg), is selectively sent by the heart to the first two branchial arches, which because they are devoid of filaments, does not permit oxygen to be lost to the ambient water (Johansen et al. 1968, Lenfant and Johansen 1968). *Neoceratodus*, in contrast, has a full set of filaments on all gill arches, and relies principally on aquatic exchange for the loss of CO_2 and the acquisition of oxygen. Circulation of blood to the lung in *Neoceratodus* has little effect on the oxygen content of blood (e.g., 39 mm Hg in the pulmonary artery and 36 mm Hg in the pulmonary vein; Lenfant et al. 1966).

A little-studied observation in *Lepidosiren* is that a series of filaments develop on the pelvic fins in males during the breeding season. Cunningham and Reid (1932) suggested that the filaments oxygenated the eggs in the nest during hypoxic periods. Foxon (1933) argued to the contrary, that the filaments are more likely to be used for oxygen uptake. More data are needed to know the direction of the oxygen differentials that exist between the water and *Lepidosiren* in the field; the South American is the least studied of the lungfish.

8.5.5 The comparative costs of bimodal exchange.

Kramer (1983) explored the comparative costs of using aerial and aquatic exchange. The costs include those of energy, time, materials, and risk of predation. The balance of these factors may indicate the environmental conditions under which aquatic, aerial, or bimodal gas exchange would be preferred. Thus, if water were continuously shallow and hypoxic, energetics, time, and reduced risk of predation would favor unimodal aerial exchange; if water were continuously deep and

normoxic, unimodal aquatic exchange would be favored; but if water levels fluctuated and if the oxygen and CO_2 content varied greatly, bimodal exchange would be selected for. The cost of respiratory exchange may be minimal for bimodal exchange only in a radically fluctuating environment, whereas in a stable environment one modal exchange may interfere with another. One teleost, *Dormitator latifrons*, which has a respiratory epithelium on top of its head, reduces the cost of aerobic gas exchange by using perches under the surface of water or by inflating a swim bladder (Todd 1973), although use of the swim bladder preferentially occurs in the absence of disturbances (i.e., without the threat of predation).

Aquatic vertebrates that use the surface for gas exchange subject themselves to increased predation from aerial and aquatic predators. For example, the survival time of water-breathing fish was reduced when the oxygen tension of water was lowered in the presence of a predatory green heron (*Butorides striatus*), but when P_{O_2} is high, water breathers survived better than bimodal breathers (Figure 8.25). Aquatic predators increase their capture rates of bimodal fishes (Wolf and Kramer 1987) and tadpoles (Feder 1983c) when these prey are forced by low water P_{O_2} to breathe air.

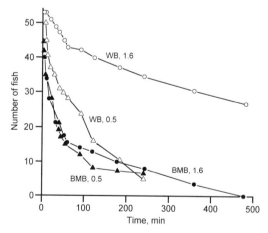

Figure 8.25 Survivorship of fish in the presence of a predatory green heron (*Butorides striatus*) depending on dissolved oxygen concentration (in mg/L) and on whether the fish were water breathers (WB) or bimodal breathers (BMB). The numbers following the respiratory mode indicate oxygen content of water (in mg/L). Source: Modified from Kramer et al. (1983).

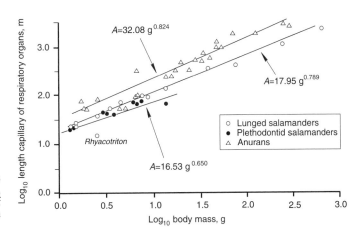

Figure 8.26 Log_{10} capillary length for gas exchange in anurans, lunged salamanders, and lungless salamanders as a function of log_{10} body mass. Source: Data derived from Czopek (1965).

8.6 BIMODAL GAS EXCHANGE IN AMPHIBIANS

The problems faced by amphibians are similar to those faced by bimodally exchanging fishes, the principal difference being that most (but not all) amphibians move freely between water and land, whereas only a small minority of fishes are truely amphibious. In addition, most amphibians reproduce in water and, thus, during their ontogeny shift from water to land. Like bimodal fish, amphibians use a combination of structures for gas exchange.

8.6.1 Bimodal exchange.
The principal sites for gas exchange in amphibians are the skin and lungs; August Krogh (1904) first demonstrated gas exchange through the skin. Lungs are more "efficient" for gas exchange than the skin because they have greater capillary density and a thinner epithelium (Czopek 1965). They tend to be a minor site for CO_2 loss because of the ease with which it diffuses through tissues, due to its high solubility, whereas pulmonary gas exchange is limited by the ventilation rate. Gills in adult amphibians have small surface areas (Szarski 1964), although a thin epithelium and high perfusion rates (compared to skin) make them more important than their proportional area would indicate (see Section 8.6.2).

The combined pulmonary and cutaneous areas for gas exchange in anurans scale in the following manner (Hutchison et al. 1968):

$$A = 13.52m^{0.58}, \qquad (8.6)$$

where A is surface area in square centimeters and m is body mass in grams. Surface area in anurans,

then, is similar to that found in small fish of similar mass (Equation 8.4), but increases less in anurans with an increase in body mass. This conclusion conflicts with the observation of Tenney and Tenney (1970), who estimated that lung mass and volume in amphibians are proportional to body mass raised to the 1.00 and 1.05 powers, respectively.

The total capillary length exposed for gas exchange in amphibians is approximately proportional to $m^{0.80}$ (Figure 8.26). The same mass relationship occurs within the ontogeny of a species (Czopek 1965). Anurans have capillary lengths that are about 1.8 times those of salamanders; toads tend to have slightly larger areas than frogs (Czopek 1965, Tenney and Tenney 1970). This difference is correlated with the propensity for activity metabolism to be aerobic in toads and anaerobic in frogs (see Section 9.4.2) Plethodontid salamanders, which are lungless, have capillary lengths that are proportional to $m^{0.65}$, which results in comparatively low capillary lengths at large masses. The lowest capillary length relative to body size is found in *Rhyacotriton*, which has rudimentary lungs and low cutaneous capillary density (Figure 8.26).

Other differences in amphibians relative to the medium in which gas exchange occurs are found in the characteristics of their blood:

1. The oxygen capacity of amphibian blood increases with movement onto land (Lenfant and Johansen 1967), which along with larger surface areas for exchange and longer capillary lengths, indicates higher rates of gas exchange in a terrestrial setting. Some aquatic amphibians, however, have high hematocrits (Hillman 1976).

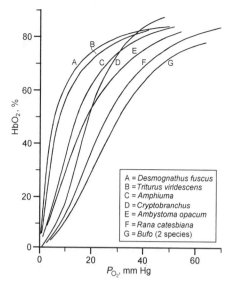

Figure 8.27 Hemoglobin oxygen dissociation curves in the blood of amphibians as a function of the external oxygen tension. Source: Modified from McClutcheon and Hall (1937).

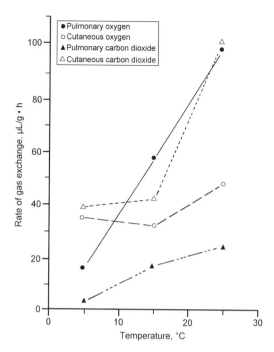

Figure 8.28 The exchange of oxygen and CO_2 through the skin and lungs of *Hyla gratiosa* as a function of environmental temperature. Source: Modified from Hutchison et al. (1968).

2. Terrestrial species have hemoglobins with low affinities (Figure 8.27; McCutcheon and Hall 1937, Lenfant and Johansen 1967). A similar decrease in affinity occurs ontogenetically (McCutcheon 1936). The problem faced by tadpoles and aquatic adults is to load hemoglobin in an oxygen-poor environment; that faced by terrestrial adults is to supply sufficient oxygen for the higher rates of metabolism and higher rates of activity associated with terrestrial life (McCutcheon 1936, Hillman 1976). In fact, a somewhat linear association-dissociation curve in larvae shifts to a sigmoidal curve in adults, which further increases the unloading tensions of blood.

3. Either adult frogs have a large Bohr effect, whereas their larvae have a "reversed" Bohr effect (McCutcheon 1936), or no correlation of the Bohr effect exists with habits (Lenfant and Johansen 1967).

4. Terrestrial amphibians have an increased buffering capacity compared to aquatic species (Lenfant and Johansen 1967), as befits animals that have low ventilation rates.

8.6.2 Partitioning of gas exchange. Victor Hutchison and his students first described quantitative measurements of the partitioning of gas exchange in amphibians. At temperatures near 5°C the skin of anurans is the most important site for oxygen uptake and CO_2 loss (Figure 8.28; Vinegar and Hutchison 1965, Hutchison et al. 1968). With an increase in ambient temperature, lungs take an increasingly larger fraction of the oxygen consumed: in anurans oxygen uptake through the lungs is usually greater at temperatures above 10°C and in salamanders it is greater at temperatures above 15°C. The augmentation of pulmonary exchange reflects an increase in ventilation rate and ventilation volume with an increase in temperature (Hutchison et al. 1968). In contrast, CO_2 at high temperatures is lost mainly via the skin. Hutchison et al. (1968) found that *Bufo boreas*, the northern toad, used its lungs for most gas exchange even at 5°C, which may be evidence of climatic adaptation. Depending on the temperature, blood from the skin and blood from the lungs differ in their content of oxygen and CO_2. Because amphibian hearts have only one ventricle, concern has been expressed (Foxon 1955) that the bloods are mixed, thereby undercutting the rationale for the "double circulation." Johansen and Ditadi (1966), however, showed that such a mixture does not occur to any appreciable extent.

Crowder et al. (1998) examined the change in critical oxygen tension in water during the devel-

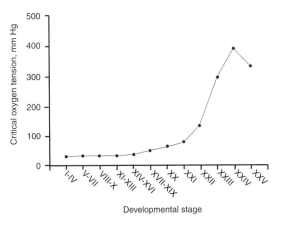

Figure 8.29 Critical oxygen tension of the anuran *Rana catesbeiana* as a function of the stage of development. Source: Modified from Crowder et al. (1998).

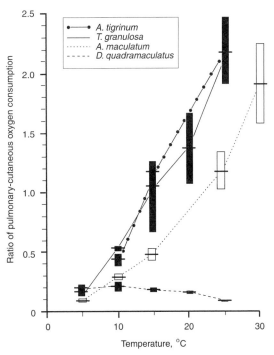

Figure 8.30 The ratio of pulmonary to cutaneous oxygen exchange in four amphibians, *Ambystoma tigrinum, A. maculatum, Taricha granulosa,* and *Desmognathus quadramaculatus,* as a function of environmental temperature. Sources: Modified from Whitford and Hutchison (1963, 1965).

opment of bullfrog (*Rana catesbeiana*) tadpoles. They showed that it remained low and rather constant at 29 to 27 mm Hg until stage XVI, gradually increased to 77 mm Hg at stage XXI, and abruptly increased to 160 mm Hg at XXII and to 372 mm Hg at stage XXIV (Figure 8.29). These higher tensions are well beyond any that would be encountered; they are predicated on the exclusive use of aquatic respiration. The abrupt shift of P_c at stage XXII is correlated with an increased number of deaths at stage XXI (25% compared to 1.4% at stage XX) and a shift in the preference for being on land from 7.5% at stage XXI to 74.0% at stage XXII, where aerial gas exchange is used. Obviously stage XXII is pivotal in the development of bullfrog tadpoles, when they shift from being primarily aquatic to being primarily terrestrial.

Salamanders collectively show a much greater diversity in the partitioning of gas exchange than do anurans. Oxygen uptake by lungs greatly predominates over integumental exchange at high temperatures (Figure 8.30). For example, at 15°C or higher *Taricha granulosa* and *Ambystoma tigrinum* obtain 50% or more of their oxygen requirements from pulmonary exchange. Other salamanders conform to this pattern (Whitford and Hutchison 1965). At all temperatures *A. tigrinum* relies on its lungs for gas exchange more than does *A. maculatum* (Figure 8.30), a difference that is correlated with the larger mass of *A. tigrinum*. Lungless salamanders breathe mainly through their skin, the remainder of gas exchange being buccopharyngeal, which by experimental protocol is lumped with pulmonary exchange. The fraction of oxygen consumption in plethodontids that is buccopharyngeal is nearly independent of temperature (Figure 8.30).

Some of the most interesting salamanders from the viewpoint of gas exchange are aquatic: *Necturus* supplements cutaneous gas exchange with exchange through the gills (Guimond and Hutchison 1972); at 25°C its gills become flushed with blood and are waved through water, apparently to break the boundary layer of water and to enhance convective and diffusive exchange. The gills of *Necturus* accounted for 54% of the aquatic uptake of oxygen at 5°C, 61% at 15°C, and 60% at 25°C; they accounted for up to 61% of CO_2 loss. The lungs of *Necturus*, however, seem mainly to have a hydrostatic function: less than 5% of oxygen uptake occurs in the lungs (Figure 8.31). *Cryptobranchus*, which has no gills, uses its skin as the principal organ for gas exchange; its lungs also account for only 10% of oxygen uptake (Figure 8.31) and function as a hydrostatic device. Cutaneous gas exchange in this salamander is enhanced

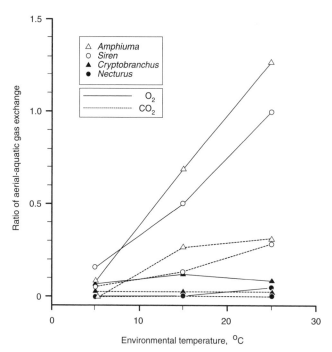

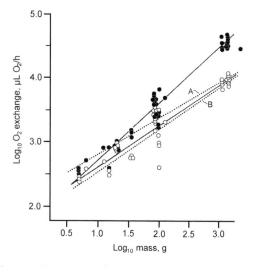

Figure 8.31 The ratio of aerial to aquatic exchange of oxygen and carbon dioxide in the aquatic salamanders *Necturus maculosus, Cryptobranchus alleganiensis, Siren lacertina,* and *Amphiuma meansi* as a function of temperature. Sources: Derived from Guimond and Hutchison (1972, 1973a, 1973b, 1974).

by large lateral folds of the trunk skin and by swaying, a behavior that presumably also breaks the boundary layer of water (Guimond and Hutchison 1973b). In both salamanders almost all CO_2 exchange is aquatic (Figure 8.31).

In contrast to these salamanders, the aquatic salamanders *Siren* and *Amphiuma* obligatorily depend on their lungs for gas exchange (Guimond and Hutchison 1973b, 1974). Although most CO_2 is lost to water through the skin during bimodal exchange, the proportion of pulmonary oxygen uptake increases with temperature (Figure 8.31). The importance of lungs in these salamanders is reflected in the few capillaries found in their skin compared to the number found in their lungs (Czopek 1965), an adjustment that is important in species living in hypoxic water. The gills of *Siren* maximally accounted for 5% of oxygen uptake, although they accounted for up to 18% of CO_2 loss.

Aquatic salamanders have made another adjustment to a nearly continuous life in water: they have a low rate of metabolism compared to other amphibians (Guimond and Hutchison 1974, Wakeman and Ultsch 1975). This reduction is accentuated in *Siren* when forced to live completely under water (Ultsch 1976a); at this time the surface area available for gas exchange is reduced (i.e., the lungs are not available). The depression of rate of

Figure 8.32 Log_{10} total oxygen consumption and log_{10} aquatic oxygen consumption in *Siren lacertina* as a function of log_{10} body mass. Curves for log_{10} oxygen exchange capacity as a function of log_{10} body mass (curve A) and log_{10} rate of metabolism when forced to use aquatic gas exchange only (curve B) are shown. Sources: Modified from Ultsch (1973a, 1976a).

metabolism with submersion increases with mass (Figure 8.32), which suggests that body size is limited when *Siren* exclusively uses aquatic exchange. This limit can be roughly estimated from the critical oxygen tension, P_c (Ultsch 1974a), which increases with mass (curve B in Figure 8.32).

An extrapolation of this curve to a P_c equal to 155 mm Hg (i.e., when water is in equilibrium with the atmosphere at sea level) indicates that the largest mass compatible with aquatic exchange is about 2090 g. Another, more precise estimate of maximal body size is given by comparing the rate of metabolism of submerged sirenids (which is proportional to $m^{0.65}$) to what Ultsch (1973a) called the *exchange capacity* (curve A in Figure 8.32), which equals the product of maximal permeability, surface area, and the differential in oxygen tension between the environment and animal. The exchange capacity in sirenids is estimated to be proportional to $m^{0.54}$. Because metabolism shows a greater increase with mass than the exchange capacity, a maximal body size is established, which by interpolation is approximately 950 g. Observations on *Siren* at 20°C suggest that the maximal mass compatible with complete submersion is about 800 g.

A comparative study of these giant North American salamanders (Duke and Ultsch 1990) demonstrated that the air breathers are (*Amphiuma*), or usually are (*Siren*), oxygen conformers, whereas water breathers (*Cryptobranchus* and *Necturus*) are oxygen regulators (in that they show clearly a critical oxygen tension). Furthermore, the air breathers had a lower cutaneous exchange of oxygen with the water than did water breathers, a difference that facilitates containing the oxygen obtained from air and not losing it to oxygen-poor water.

In reality, *Siren* and other bimodal salamanders use both aerial and aquatic respiration and can compensate for a low water P_{O_2} by reverting to aerial respiration. An increase in aerial respiration at a low P_{O_2} coincides with the critical oxygen tension in water (Figure 8.33). As in some swamp-dwelling fishes, the balance between aerial and aquatic exchange in aquatic salamanders depends on the P_{CO_2} and P_{O_2} of water. Of three salamanders, the most aquatic species (*Pseudobranchus striatus*) is relatively insensitive to P_{CO_2}, whereas the most terrestrial species (*Taricha torosa*) is sensitive to both P_{CO_2} and P_{O_2} (Wakeman and Ultsch 1975). These differences were related to the presence or absence of gills, body shape, and body size. *Diemyctulus viridescens* is intermediate in habits, form, and sensitivity to P_{CO_2} and P_{O_2}. Aquatic salamanders can compensate for a large body size by the use of aerial gas exchange: *Siren lacertina* attains a mass of 1700 g, which is approximately

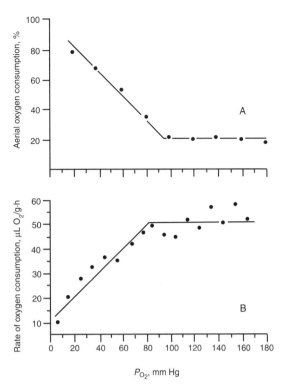

Figure 8.33 A. Proportion of total oxygen consumption derived from aerial gas exchange (when permitted to breathe both air and water) in *Pseudobranchus striatus* as a function of the oxygen tension in water. B. Aquatic rate of oxygen consumption (when no air is available). Source: Modified from Ultsch (1976a).

twice the mass maximally permitted by aquatic exchange alone.

Aquatic amphibians live in an environment characterized by a low P_{O_2}. With an increase in altitude, P_{O_2} falls, and therefore aquatic amphibians at high latitudes live in an environment with an especially low oxygen content. This is the problem faced by the Lake Titicaca frog (*Telmatobius culeus*). Lake Titicaca, which is situated on the border of Bolivia and Peru, is at an altitude of 3812 m, has an atmospheric P_{O_2} of 100 mm Hg and has a water temperature of 10°C. *Telmatobius culeus* is distinctive in several ways: the lungs are one-third those of equally sized ranids; large, highly vascularized loose folds are in the skin; its erythrocytes are small and high in number; it has blood with a very high affinity for oxygen and a very low rate of metabolism. All of these characteristics facilitate existence at high altitudes, low environmental P_{O_2} values, and low temperatures (Hutchison et al. 1976). This frog apparently has a low critical oxygen tension and can extract oxygen from water with a very low

P_{O_2}. It is not known to breathe air in the field but can be made to do so in the laboratory if P_{O_2} falls below 62 mm Hg, which may simply mean that this reduction does not (often?) occur in Lake Titicaca, except possibly deep in the lake.

Few data are available on the gas exchange of the third group of living amphibians, the caecilians (Gymnophiona). The skin is important for CO_2 loss in *Siphonops* (Mendes 1945). Sawaya (1947) showed that about 94% of oxygen uptake in *Typhlonectes* is pulmonary and about 60% of CO_2 loss is cutaneous. Based on these meager data, caecilians appear to be similar to other amphibians, but an examination of the fossorial *Boulengerula taitanus* from Kenya showed that it had blood with a high affinity for oxygen, a high oxygen capacity, and a low Bohr effect (Wood et al. 1975), characteristics that are to be expected from its burrowing habits.

The gas exchange of living amphibians, depending as it does on cutaneous exchange, especially at high temperatures and large masses, limits aspects of their life-style (Czopek 1965): maximal body size may be limited by their dependence on cutaneous exchange, and tolerance to hyperosmotic solutions may be limited by the high permeability of the skin.

8.6.3 An evolutionary cul-de-sac: lunglessness in salamanders.

One of the most distinctive modifications of gas exchange in tetrapods is the loss of lungs by some salamanders and their marked reduction in a few anurans. All members of the predominately New World family Plethodontidae are lungless, as are the European salamanders in the genus *Chiloglossa*. Other salamanders, namely the European salamandrid *Salamandrina* and members of the subgenus *Euproctus* of the genus *Triturus*, have rudimentary lungs (Lonnberg 1899, Wilder and Dunn 1920, Noble 1925). *Rhyacotriton* and *Ambystoma opacum*, both of which are members of the Ambystomidae, have rudimentary lungs (Lonnberg 1899, Noble 1925). Although no anurans are lungless, some have greatly reduced lungs, such as *Trichobatrachus* (= *Astylosternus*, the West African "hairy" frog), *Ascaphus* (from northwestern North America), tadpoles of some Asian *Rana* (= *Staurois*) (Noble 1929), and *Telmatobius* from Peru. No caecilians are known to have reduced lungs, although like other vertebrates with a long, thin body, such as snakes, one lung (here, the right) tends to be larger than the other.

Most amphibians accentuate pulmonary gas exchange in response to an increase in temperature and body size, which raises a question of the response of lungless amphibians. Whitford and Hutchison (1967) suggested one answer. They measured rates of oxygen consumption in lunged and lungless salamanders at 15°C and concluded that plethodontids have lower rates than lunged salamanders at all masses and that this difference increased with mass; that is, resting rate scales proportionally to $m^{0.86}$ in lunged salamanders and proportionally to $m^{0.72}$ in plethodontids. This difference is similar to that found in the scaling of capillary length used for gas exchange between these groups (Czopek 1965). The measurements of Feder (1976a, 1976b), however, did not show a difference in rate of metabolism between plethodontids and lunged salamanders. He suggested that the principal difference between these groups occurs during activity. Aerobic activity may be reduced in plethodontids compared to lunged amphibians, a difference that may be greatest in large individuals (Hillman et al. 1979b). Nevertheless, amphibians collectively have such low rates of metabolism (Feder 1983b) that they may have facilitated the evolution of lunglessness, especially in small species that live at cool temperatures.

Plethodontids and small-lunged anurans compensate for the reduction or absence of lungs in several ways:

1. These species have a very thin epithelium covering the capillary network in the skin (Noble 1925), thereby reducing the barrier to diffusion.
2. Many small-lunged amphibians have enlarged cutaneous surface areas produced either by an extensive set of papillae (*Trichobatrachus*) or by lateral skin folds (*Telmatobius*).
3. They tend to be terrestrial, unless they live in cold water: among species of *Ambystoma*, the most terrestrial, *A. opacum* (Salthe 1965), has rudimentary lungs (Lonnberg 1899), whereas the other species of the genus have normally developed lungs and more aquatic habits.
4. Most important of all perhaps, lungless salamanders are small, which reduces the demand for gas exchange and ensures a high surface-volume ratio, especially given their elongate shape.

Plethodontids and salamanders with rudimentary lungs have hemoglobin with a high affinity (Figure 8.27) and have lactic acid dehydrogenases that show little substrate inhibition (Salthe 1965). These characteristics are similar to those found in amphibians living in standing water. Both conditions, standing water and lunglessness, reduce oxygen availability, which explains why plethodontids do not live in warm, standing water (i.e., at low elevations): they would be doubly denied oxygen (see Feder 1983b). Even in cool to cold flowing water, plethodontids have a problem obtaining an adequate supply of oxygen because the boundary layer offers the greatest resistance to oxygen uptake (Booth and Feder 1991): at a low P_{O_2} *Desmognathus quadramaculatus* may behaviorially depress the rate of metabolism or depend on anaerobic metabolism. Beckenbach (1975) demonstrated the effectiveness of these adjustments: the critical oxygen tensions of plethodontids are generally below those of lunged salamanders and approximately equal to those of teleosts.

The question remains, why have some amphibians lost (or nearly lost) their lungs? Lonnberg (1899) suggested that the loss permitted aquatic species to walk on substrates and to crawl among boulders without floating. Wilder and Dunn (1920) embraced this suggestion, noting that mountain-brook salamanders have a reduced buoyancy as a result of having small or no lungs, just as mountain-brook teleosts have small swim bladders (Nelson 1961, Gee and Northcote 1963). Noble (1925, 1929), however, maintained that the loss of lungs was a response to a more complex set of conditions than buoyancy in a rheotic environment. He argued that the loss of lungs is intimately connected to the low temperature and high oxygen content of streams. He maintained that cold temperature alone might be enough to shift the balance from pulmonary to cutaneous exchange, which is correct in some aquatic and terrestrial amphibians (Figures 8.30 and 8.31), but whether cold is sufficient alone to bring about the evolution of lunglessness is doubtful.

Ruben and Boucot (1989) proposed another view of the evolution of lunglessness. They claimed that if plethodontids evolved in Mesozoic Appalachia, as most biologists have assumed, it could not have been a response to life in mountain streams because this area then was an erosional peneplain. As a countersuggestion, they pointed to the inverse correlation in ambystomids and *Rhyacotriton* between the proportion of oxygen consumption that is cutaneous and head width, and concluded that the selection for narrow head restricted the capacity for forced ventilation and led to selection for a sedentary life-style and the loss of pulmonary exchange. Reagan and Verrell (1991) maintained that the key factor leading to the evolution of lunglessness was the evolution of terrestrial courtship and mating, possibly coupled with the reduction of buoyancy. Beachy and Bruce (1992) reemphasized the significance of life in mountain streams as the basis for the evolution in plethodontids. They noted that other lungless or reduced-lunged salamanders also live in, or near, stream environments and that most larval aquatic plethodontids are benthic. They suggested that the evolution of lunglessness in plethodontids promoted larval survival in streams by reducing buoyancy in an environment rich in oxygen. An open season obviously exists for suggestions as to the factors that led to a lungless condition in plethodontids!

Most plethodontids are small: although a few weigh more than 20 g, the majority weigh less than 5 g, and many in the neotropics weigh less than 1 g. Szarski (1964) proposed that their small size is an adaptation to a lungless condition. Although a close connection between a lungless condition and small size undoubtedly exists, the causation may be reversed: lunglessness might have resulted from a small body size, whose significance is the exploitation of small nooks and crannies. How small can a vertebrate be and still rely on the forced ventilation of lungs? The resistance to free ventilation increases with a reduction in the diameter of pulmonary branchiae, which undoubtedly decrease with body size. Whatever the answer, a lungless condition sets a maximal body size (Beckenbach 1975): because P_c increases with mass in plethodontids (Figure 8.34) and because the maximal P_{O_2} in the environment is about 155 mm Hg, the maximal body mass is about 31.7 g (at 22.5°C). If the maximal P_{O_2} is 120 mm Hg (at 2000-m altitude), the maximal mass would be about 24.5 g. The largest plethodontids should be at intermediate altitudes, high enough to face cool temperatures but low enough to be exposed to high barometric pressures, which is exactly what is seen in Central America (Figure 8.35).

In spite of the conclusion of Salthe (1965, p. 405) that the only environments acceptable for plethodontids are "...in or near cold mountain streams, or in cool, damp terrestrial situations," plethodontids are the only salamanders to have invaded the

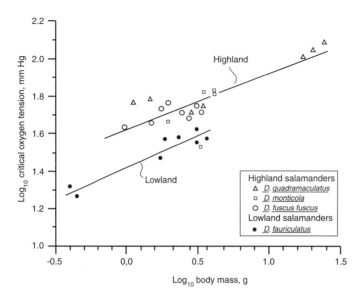

Figure 8.34 Log_{10} critical oxygen tension in lowland and highland plethodontid salamanders (*Desmognathus*) as a function of log_{10} body mass. Source: Modified from Beckenbach (1975).

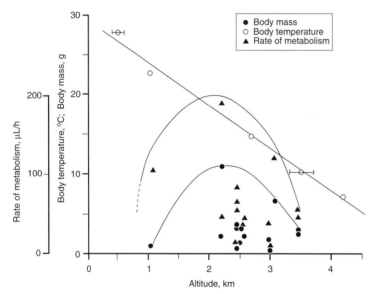

Figure 8.35 Body temperature, body mass, and rate of metabolism in plethodontid salamanders as a function of altitude in Central America. Source: Derived from Feder (1976b).

tropics (Brame and Wake 1963, Wake 1970, Wake and Lynch 1976). Feder (1976b) showed that tropical plethodontids have lower resting rates of metabolism than do temperate species and that standard rates at a given temperature increase with altitude in tropical species, which may be why montane species have higher critical oxygen tensions than lowland species (Figure 8.34). A low rate of metabolism can be viewed as an adaptation to compensate for high ambient temperatures in the tropical lowlands: nevertheless, ambient and body temperatures increase so markedly in the lowlands that total rate of metabolism increases with

a decrease in altitude (Figure 8.35). If a limit to total rate of metabolism is imposed by the capacity for gas exchange, then body mass would have to decrease to compensate for an increase in temperature. *Bolitoglossa occidentalis* from lowland Guatemala has a maximal mass of about 1 g; a similar size is found in the Amazonian *B. altamazonica*. These lowland species are not aquatic, as is to be expected, but are arboreal. Wake (1970) argued that the invasion of the tropical lowlands by plethodontids, indeed, was associated with terrestrial habits. A nagging question remains, why have lunged salamanders not invaded the lowland tropics? Is the answer so simple as the absence of opportunity?

8.7 THE EVOLUTION OF AERIAL GAS EXCHANGE

"Migration from the water to the land has occurred very frequently in the evolution of animals, perhaps more frequently, than any other change so far-reaching in its effects upon the organisation of the animal . . ." (Carter 1931, p. 1). The means by which vertebrates switched from aquatic to aerial exchange is difficult to describe, because the structures used for gas exchange are usually poorly preserved in the fossil record. All pictures drawn of this evolution are necessarily speculative.

Aerial gas exchange is almost as old as vertebrates themselves. This conclusion stems from a series of observations:

1. The placoderm †*Bothriolepis* from the Upper Devonian had what appears to have been a pair of ventral lungs attached to the pharynx (Denison 1941).

2. †*Bothriolepis* was found in sediments that also contained two rhipidistian crossopterygians and the lungfish †*Scaumenacia*.

3. Lungfish are known to have aestivated during the Devonian, Carboniferous, and Permian (Romer and Olson 1954, Carroll 1965, Thomson 1969).

4. The first amphibians were also found in the Upper Devonian.

5. A marine coelacanth from the Upper Cretaceous, †*Macropoma*, had an air bladder (Thomson 1969).

6. Among living vertebrates, lungfish, the brachyopterygians *Polypterus* and *Erpetoichthys*, and tetrapods have lungs, whereas the coelacanth *Latimeria* and most actinopterygians have bladders.

However, no evidence exists that any agnathans or chondrichthyans ever developed lungs or other aerial gas exchangers, or that any of their immediate ancestors had these structures.

These data lead to the following interpretation of the evolution of gas exchange in vertebrates. Aquatic exchange, presumably by means of gills, was the first form of gas exchange in vertebrates, given that vertebrates evolved from marine protochordates. Lungs were well established among various lines of vertebrates by the Upper Devonian.

Crossopterygians, or their ancestors, developed a pair of ventrally attached lungs, as are found today in the South American and African lungfishes, *Polypterus*, and tetrapods. From this condition, *Neoceratodus* developed a single, dorsal lung; *Latimeria*, a single, fat-filled bladder; and many bony fishes, a single, dorsal swim bladder. Many structures other than lungs or their derivatives must have been used for aerial exchange early in vertebrate history, just as they are today. A detailed description of the evolution of the morphology of gas exchangers is found in the article by Gans (1970a).

Thomson (1969) described a balance between aquatic and aerial exchange in the phylogeny of lungfish. He noticed that *Protopterus* and *Lepidosiren* have very small opercula compared to *Neoceratodus*. As Dipnoi are traced backward to the Devonian, the operculum becomes proportionally larger, which Thomson suggested indicates a greater reliance on aquatic respiration than is found today in the Lepidosirenidae. Some lungfish in the Devonian were capable of aestivating, and therefore of using aerial respiration, large operculum or not. The Permian lungfish †*Gnathorhiza* had a large operculum and has been found aestivating in burrows (Carlson 1968). No pattern in opercular size is present in crossopterygians and no evidence is known of them entering aestivation.

Air breathing usually is thought to have evolved in vertebrates principally in relation to life in freshwater. Packard (1974), however, suggested that air breathing arose in saltwater marshes and estuaries in response to low oxygen tensions in water at high temperatures and salinities: in fish under these conditions ventilation rates increase by 25% to 30% to attain the same rates of oxygen consumption as in cool freshwater. He further noted that the earliest jawed vertebrates were found in tropical marine water (Denison 1956) and suggested that lungs appeared prior to the movement of vertebrates into freshwater. Graham et al. (1978) disagreed, noting that no marine fish are obligate air breathers. Air breathing permits marine species to exploit new habits, whereas in freshwater, it is a method of survival during periods when water has high CO_2 and low oxygen tensions, or when water levels fall during a drought. The minor differences in oxygen content between freshwater and salt water can be handled by an adjustment of hemoglobin content, as occurs in salmon (Vanstone et al. 1964). Graham et al. (1978) concluded that the evolution of air breathing in the line leading to tetrapods occurred in freshwater, to increase survival during critical periods of aquatic gas exchange.

Sarcopterygians (lungfish, coelacanths, "rhipidistian" crossopterygians) gave rise to the earliest amphibians in the Devonian, although the identity of the specific group has been hotly debated: the early suggestion of lungfishes has been revived by Rosen et al. (1981) and Stiassny et al. (1996), a view strongly opposed by Jarvik (1981), Holmes (1985), and Carroll (1988), who advocated "crossopterygians" as the source of amphibians. Jarvik (1968) maintained that "rhipidistian" crossopterygians are really two groups, the †Porolepiformes and the †Osteolepiformes. Some recent interpretations have designated a Devonian family related to †Osteolepiformes, the †Panderichthyidae, to be the sister group of amphibians and to tetrapods generally (Vorobyeva and Schultze 1991, Ahlberg and Milner 1994, Ahlberg 1995, Ahlberg and Johanson 1998). But problems still abound.

Two major groups of amphibians were established early, one of large size, with labyrinthodont teeth, otic notch in the skull, and complex centra (the "labyrinthodonts"); and a second group of small size, with simple teeth, no otic notch, and simple centra (the Lepospondyli) (Carroll and Gaskell 1978). The labyrinthodonts, represented by †*Acanthostega* and †*Ichthyostega*, had finned tails that were used for locomotion in water, heavy ribs, a length up to a meter, and a scaled body covering, at least on the venter (see Romer 1972). The second group, the "Lepospondyli," have an uncertain relationship with the Lissamphibia, a collective term for living amphibians, including caecilians, anurans, and urodeles. Whether the origin of amphibians was polyphyletic (Jarvik 1968, Carroll and Gaskell 1978, Carroll 1988) or monophyletic (Thomson 1968, Duellman and Trueb 1986, Ahlberg and Johanson 1998) has been contested.

Packard (1976) argued that the gas exchange of early amphibians occurred principally through the lungs and that it was characterized by low ventilation rates, high arterial P_{CO_2} values, and high bicarbonate levels (i.e., like the gas exchange system of living reptiles), evidence of which were a large size, heavy ribs (which probably reflected costal ventilation of the lungs), and the presence of scales (which reduced or prohibited integumental gas exchange). Life in hypoxic and hypercarbic waters would have facilitated the development of an impermeable integument to reduce oxygen loss to, and CO_2 gain from, the environment. The ". . . evolutionary transition of acid-base status from that typical of water-breathing fishes to that typical of air-breathing amniotes may well have been completed before the protoamphibians left the water, obviating the need for a gas-permeable integument" (Ultsch 1996, p. 21). The presence of some tetrapods from the Upper Devonian suggests that this transition had already been made (see Daeschler et al. 1994).

The gas exchange used by living amphibians is unique to their small size. Living amphibians cannot be considered to represent an intermediate stage in the evolution of the gas exchange system used by reptiles, birds, and mammals (Cox 1967; Gans 1970b; Romer 1972; Ultsch 1987, 1996), as Rahn and Howell (1976) suggested. In any case, no living amphibian is closely related to the line that gave rise to reptiles. Nevertheless, living amphibians are functionally intermediate to aquatic gas exchangers that use gills and terrestrial air breathers that depend solely on lungs.

8.8 GAS EXCHANGE IN TERRESTRIAL ENVIRONMENTS

Reptiles, birds, and mammals depend almost exclusively on lungs for gas exchange. Pulmonary exchange reduces the rate of evaporative water loss compared to that found in amphibians, thereby permitting a terrestrial distribution that is more independent of surface water than is usually found in amphibians, especially during activity.

8.8.1 Aerial exchange. The structure of lungs in terrestrial vertebrates varies greatly. Most lizards have lungs that are simple sacs, with the respiratory area confined to the anterior section, although varanids have intrapulmonary bronchi. Varanids use the entire lung for gas exchange (Bennett 1972), which correlates with their use of aerobic metabolism during activity. This difference in lung structure between varanids and other lizards explains how *Varanus* during activity can have a rate of metabolism twice that of *Sauromalus*, whereas both have the same ventilation rate. Snakes have only one lung, on the left, to permit a reduction in body diameter. Mammals have larger lungs than reptiles and these lungs are highly subdivided, forming alveoli, to enhance gas exchange. Nevertheless, oxygen uptake by lizards is about as efficient as in mammals: the complex lungs of mammals simply permit a greater rate of oxygen uptake and of CO_2 loss (Bennett 1972).

Birds have a much more complicated respiratory system than either reptiles or mammals: it consists of two lungs, various air sacs, and air spaces located in many of the larger bones. The air sacs and spaces are a system used for the one-way ventilation of the lungs. On inhalation, air enters the bronchi, whence it moves in sequence to the posterior air sacs, lungs, anterior air sacs, and again the bronchi, whereupon the air is exhaled (Bretz and Schmidt-Nielsen 1972). This unidirectional flow of air permits a cross-current exchange of gas with the circulatory system (Scheid 1979). As a result, birds maintain a higher arterial oxygen tension (ca. 100 mm Hg) and a lower arterial CO_2 tension (ca. 28 mm Hg) than mammals (95 and 40 mm Hg, respectively), even though birds generally have higher rates of metabolism than mammals, ventilation rates one-third of mammalian rates, and minute volumes three-fourths of mammalian values (Lasiewski and Calder 1971). As a consequence of their highly efficient gas exchange system, birds have minimal rates of evaporative water loss that are only two-thirds those of mammals. Mammals compensate for less "efficient" lungs by having larger lung volumes (by 27%) than birds (Maina et al. 1989). The bat *Epomorphorus wahlbergi* has a lung volume that is 3.6 times that expected from mammals generally (Maina et al. 1982), and bats generally have lung masses twice those of nonvolant mammals (Jürgens et al. 1981), presumably an adjustment for flight.

Daniels and Pratt (1992) developed a mathematical model of breathing in terrestrial vertebrates to analyze gas exchange in the 30-ton sauropod dinosaur †*Mamenchisaurus hochuanesis*, which had a trachea 11 m long. They used four combinations of characters, a "mammalian" or "reptilian" level of metabolism and a "mammalian" or "avian" respiratory system. They concluded that the mammalian bellows lung and a high rate of metabolism were incompatible with a long trachea: the only acceptable combination was an avian air sac system with a low rate of metabolism. Bakker

(1971) indeed showed that many saurichians had air sacs in their long bones.

8.8.2 Oxygen transport. The transport of oxygen by blood depends on the rate of blood circulation, which depends on cardiac output and heart size. Birds have heart masses 40% larger than those found in most mammals; this permits a greater cardiac output commensurate with the high activity of birds during flight (Laswieski and Calder 1971). Bats have heart masses equal to those of birds (Hartman 1963, Snyder 1976, Jürgens et al. 1981).

The oxygen capacity of blood is also adjusted. Endotherms have high blood oxygen capacities (usually 20–25 volumes %; Bartels et al. 1963, Sealander 1964); reptiles have low oxygen capacities (snakes 10.2, lizards 8.4, and turtles 5.6 volumes %; Pough 1976). The low oxygen capacities of reptilian blood reflect a reduced hematocrit and hemoglobin concentration; the low oxygen capacities do not simply represent the absence of a high demand for oxygen, but also represent a means of attaining a low fluid resistance in a low-pressure vascular system (ca. 30–50 mm Hg in reptiles vs. 80–200 mm Hg in mammals) (Pough 1980b).

Oxygen capacity varies with body temperature in reptiles (Figure 8.36). Like many other functions (see Section 4.8.5), oxygen transport is greatest at the normal body temperatures of reptiles (Pough 1976): maximal oxygen capacities occur at high

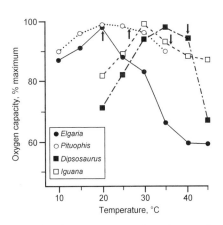

Figure 8.36 Blood oxygen capacity in various reptiles as a function of body temperature. Preferred body temperatures for each species are indicated by an arrow. Source: Modified from Pough (1976).

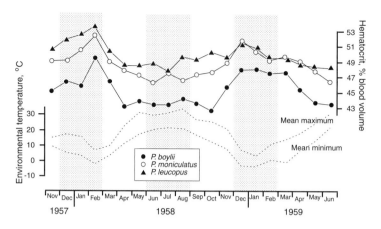

Figure 8.37 Hematocrit in the blood of deer mice (*Peromyscus* spp.) and mean air temperature as a function of the time of year. Source: Modified from Sealander (1964).

body temperatures in *Dipsosaurus*, at lower temperatures in *Elgaria*, and still lower temperatures in snakes and turtles. In lizards (*Sceloporus, Uma*), turtles (*Chrysemys*), and snakes (*Thamnophis, Pituophis*) that have broad thermal tolerances, oxygen capacity shows little variation with respect to temperature. Short-term changes in oxygen capacity involve changes in hematocrit through the mobilization of erythrocytes stored in the spleen.

Oxygen transport in blood also may be related to body size in terrestrial vertebrates: oxygen capacity increases with body size in *Thamnophis*, due to an increase in hematocrit and in hemoglobin concentration (Pough 1977a, 1977b). A similar increase occurs in turtles (Payne and Burke 1964). In small mammals, however, oxygen capacity tends to increase with a decrease in mass (Sealander 1964). Bartels (1964) and Bartels et al. (1969) argued that small mammals have relatively low hematocrits but very high hemoglobin concentrations in the erythrocytes; this combination permits a high blood oxygen capacity without an increase in blood viscosity, thereby reducing the blood pressure required for circulation. At present the relative importance of various factors on blood oxygen capacity, hematocrit, and hemoglobin concentration is unclear. Bartels (1964), for example, proposed that hemoglobin concentration is unaffected by the intensity of metabolism in mammals.

Rate of metabolism in small endotherms usually increases in fall and winter with the decrease in environmental temperature. This variation is reflected in the seasonal capacity to transport oxygen: hematocrit is inversely related to mean ambient temperature and follows a seasonal cycle in small mammals, at least in Arkansas (Figure 8.37), where small mammals cannot avoid exposure to low temperatures in winter. *Clethrionomys* in Alaska, however, showed no seasonal variation in hematocrit (Sealander 1966), possibly because the vole spends most of the winter in tunnels under the snow, where temperature remains stable. These data raise the possibility that a study of oxygen transport in mammals can be used to measure the microclimate faced by mammals. Northern species of mammals and birds that face the rigors of a harsh winter, such as in the genera *Dicrostonyx, Sciurus, Parus,* and *Corvus,* would also be expected to demonstrate a marked seasonal adjustment of oxygen transport.

8.8.3 Hemoglobin affinity. As in aquatic respiration, the affinity of hemoglobin for oxygen is highly variable in aerial respiration: in reptiles it falls (P_{50} increases) with an increase in temperature (Figure 8.38); a similar pattern is seen in thermally variable mammals like the naked mole-rat (*Heterocephalus glaber* [Johansen et al. 1976]) and hedgehogs (*Erinaceus* [Bartels et al. 1969]). Temperature sensitivity is greatest in snakes and turtles and least in lizards, but in all reptiles is less than in mammals. In spite of the intraspecific thermal sensitivity of the oxygen dissociation curve, P_{50} is nearly independent of preferred body temperature when the comparison is made interspecifically (Figure 8.38).

A slight but significant decrease in P_{50} occurs with an increase in mass in reptiles (Pough 1969, 1977a, 1980b), birds (Lutz et al. 1974), and mammals (Schmidt-Nielsen and Larimer 1958), the

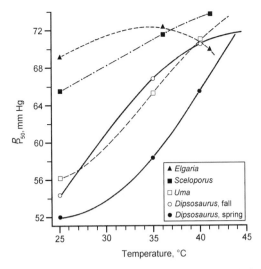

Figure 8.38 The partial pressure of oxygen at which the blood of lizards is 50% saturated, P_{50}, as a function of body temperature. Source: Modified from Pough (1980c).

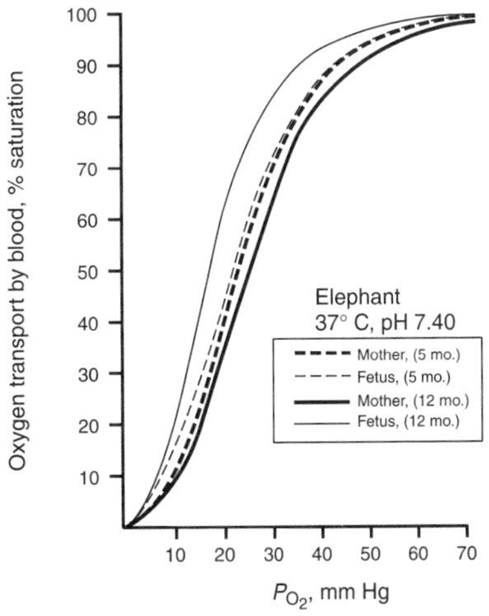

Figure 8.39 Oxygen dissociation curves in a fetal elephant (*Loxodonta africana*) at ages 5 and 12 months and of its mother as a function of external oxygen tension. Source: Modified from Riegel et al. (1967).

P_{50} values of mammals and birds at a given body mass being nearly identical (Lutz et al. 1974). Schmidt-Nielsen and Larimer's explanation for this correlation, which had been accepted by most subsequent workers, is that P_{50} is high in small mammals to compensate for their high mass-specific rates of metabolism (but see Box 3.1). A dissociation curve with a high P_{50} has a high unloading tension, which leads to a larger differential in oxygen tension between the capillaries, where oxygen is unloaded, and the tissues, where oxygen is consumed. A large differential facilitates high rates of oxygen diffusion to the tissues (Equation 2.2), thereby permitting high rates of oxygen consumption.

This explanation has at least two difficulties, ignoring the issue of whether mass-specific rates have any "reality." One is related to scaling: in mammals P_{50} scales to the −0.054 power and mass-specific basal rate scales approximately to the −0.25 power. At best, the change in P_{50} could satisfy only 20% of the increase in mass-specific rates with a decrease in mass. Steen (1971) spotted the second difficulty. Most comparisons of P_{50} used a fixed P_{CO_2}. Because ventilation rate is inversely related to body mass (Stahl 1967, Lasiewski and Calder 1971), Steen suggested that the renewal of alveolar gas is more complete in a small than in a large animal. Consequently, small species, because of a higher ventilation rate, would have a higher

arterial P_{O_2} and a lower P_{CO_2} than would large species. Lahiri (1975) indeed found that small mammals had a P_{CO_2} equal to 20 to 31 mm Hg, whereas in larger species it equaled 35 to 40 mm Hg. When the dissociation curve of a mammal is measured at its normal arterial P_{CO_2}, P_{50} and affinity are independent of mass.

This generalization has an exception. Pough (1977a) showed that the P_{50} of snake blood increases with body mass. All measurements of the dissociation curve in snakes were made at one temperature (25°C) and at a fixed P_{CO_2}, so if snakes behave like other vertebrates, an adjustment of these factors would further increase the dependence of P_{50} on body mass. Pough suggested that this unusual correlation may be related to the single lung present in snakes: small species and individuals have higher affinities to compensate for a low ventilation rate. These data raise the question of whether P_{50} in plethodontid (lungless) salamanders might also be positively correlated with body mass; it is surely lower than the P_{50} found in other amphibians (Figure 8.27). In contrast, the affinity of mammalian fetal hemoglobin, which is always greater than that of the mother, increases with the growth of the fetus, that is, as the fetal requirements for oxygen increase (Figure 8.39).

Much of the variation in P_{50} among mammals is related to the amount of 2,3-diphosphoglycerate (DPG) contained within the erythrocytes: as DPG within a species increases in concentration, P_{50} increases (Harkness 1972). This pattern, however, is much more complex than usually recognized: at least two separate groups of mammals follow this pattern (Kay 1977). One group is characterized by a high P_{50} and a low concentration of DPG; it includes species (e.g., domestic cats, bovids, many cervids) that can move rapidly, have little endurance, and generally are limited in distribution to low elevations. The second group (e.g., camels, pocket-gophers, armadillo, whales, and ground-squirrels) has a low P_{50} and a high DPG concentration, has endurance or faces hypoxia, and includes hibernators. *Heterocephalus*, in contrast, has hemoglobin with a high affinity based on its structure, not on a high DPG concentration in its erythrocytes (Johansen et al. 1976). These patterns need to be clarified, but the similarity of mammals facing low oxygen tensions in burrows and at high altitudes appears significant.

8.8.4 Adaptation to burrow atmospheres.

The atmospheres of burrows, as seen in Section 2.11, often have low oxygen and high CO_2 pressures. Burrowing lizards make few physiological adjustments to these conditions (Pough 1969), which may simply reflect their low rates of gas exchange. Equally, amphibians buried in soil rarely face oxygen tensions sufficient to affect their rates of oxygen consumption, mainly because they dig in soils of high porosity and have low rates of metabolism (Seymour 1973b). Seymour suggested that a reduction in the affinity for oxygen in *Scaphiopus* blood during the active period represents a means of supplying oxygen for activity.

Many mammals face especially low oxygen pressures and high CO_2 pressures in burrows (Kennerly 1964, McNab 1966b; Studier and Baca 1968). These atmospheres are modified by the metabolism of the burrow residents themselves, as was clearly shown in golden hamsters (*Mesocricetus auratus*) as they cycled into and out of torpor (Figure 8.40). Such an impact is pronounced in clay soils, due to their high water and low air contents (McNab 1966b, Arieli et al. 1977, 1984; Arieli 1979) and in areas with high rainfall (Arieli et al. 1984). Blind mole-rats (*Spalax*) from moister soils tolerate a lower P_{O_2} than do mole-rats from drier soils (Arieli and Nevo 1991). Burrow atmospheres faced by rodents may have oxygen concentrations as low as 6% (Kennerly 1964) and CO_2 concentrations as high as 6.2% (Studier and Procter 1971). These

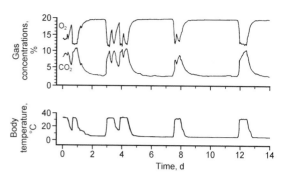

Figure 8.40 Gas composition of a burrow atmosphere and body temperature of a golden hamster (*Mesocricetus auratus*) occupying the burrow as a function of time. Source: Modified from Kuhnen (1986).

extreme atmospheric compositions are found in species that have closed burrows; species with burrows having two unobstructed openings have bulk flow when the wind is blowing as a result of viscous entrainment of the burrow atmosphere with the external atmosphere (Vogel et al. 1973; see Section 2.11).

Burrowing mammals are surprisingly variable in their capacity to transport oxygen in their blood. Many have high hematocrits, including *Talpa* (Quilliam et al. 1971), *Thomomys* (Lechner 1976), and *Spalax* (Ar et al. 1977), but not *Ctenomys* (Busch 1987) and some burrowing squirrels (Hall 1965). Some have erythrocytes with high hemoglobin concentrations, for example, *Talpa*, *Thomomys*, and *Ctenomys*, but not *Heterocephalus* (Johansen et al. 1976), *Marmota*, and other ground-squirrels (Hall 1965, Boggs et al. 1984). Most burrowing mammals have blood with a high affinity for oxygen (Figure 8.41), including all of the above and the burrowing pangolin *Manis pentadactyla* (Weber et al. 1986). Some fossorial mammals have high myoglobin levels (Nevo 1979). Curiously, some burrowing mammals (*Talpa* [Quilliam et al. 1971]) have blood with much higher affinities than others, such as *Thomomys* (Lechner 1976) and *Spalax* (Ar et al. 1977); this difference may reflect soil characteristics. This variability in dissociation curves demands further study of gas transport in burrowing mammals. Boggs et al. (1984) and Nevo (1999) reviewed these and other characteristics of the respiration of burrowing mammals.

Burrowing mammals have a reduced sensitivity to a high P_{CO_2} and to a low P_{O_2} compared to non-burrowing species. Pocket-gophers (*Thomomys*)

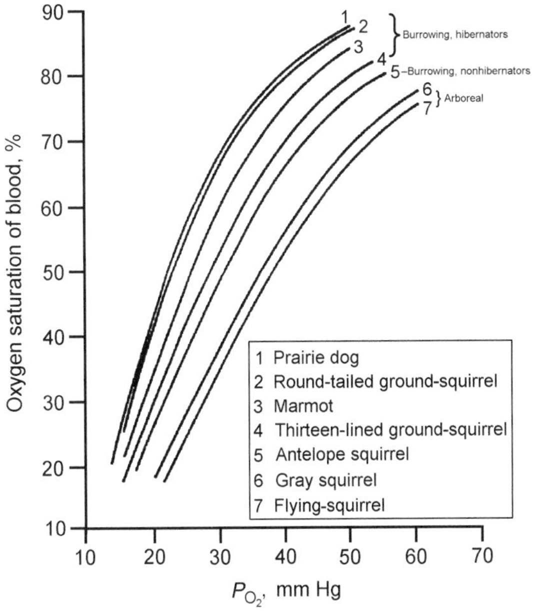

1 Prairie dog
2 Round-tailed ground-squirrel
3 Marmot
4 Thirteen-lined ground-squirrel
5 Antelope squirrel
6 Gray squirrel
7 Flying-squirrel
Burrowing, hibernators
Burrowing, nonhibernators
Arboreal
Oxygen saturation of blood, %
P_{O_2}, mm Hg

Figure 8.41 Oxygen dissociation curves in rodents as a function of external oxygen tension and whether the squirrels burrow or are arboreal and whether they hibernate or remain normothermic. Source: Modified from Boggs et al. (1984).

and armadillos (*Dasypus*) can breathe air that contains 4% CO_2 with only a 20% increase in minute volume, whereas humans breathing such a gas have a 200% increase; many burrowing mammals are intermediate (Darden 1972). Arieli et al. (1977) showed that the fossorial mole-rat *Spalax* can maintain normal rates of metabolism at lower environmental oxygen tensions than white rats of equal size, which is exactly what is to be expected from the observations of Hall (1966) that P_c in rodents is positively correlated with P_{50}.

The presence of a high burrow P_{CO_2} means that an inhabitant is rebreathing CO_2 and therefore faces a problem voiding CO_2 to its environment. Some mammals excrete divalent cations such as Mg^{2+} and Ca^{2+} in the form of bicarbonate and carbonate salts. Many of these mammals are herbivorous, often with caecal fermentation (Shirley and Schmidt-Nielsen 1967, Louw et al. 1972, Cheeke and Amberg 1973, Haim et al. 1985a). The excretion of bicarbonate by the mole-rat *Cryptomys* reflects both its diet and subterranean habits (Haim et al. 1985b), whereas the nonburrowing rodent *Otomys* only excretes bicarbonate when its food contains an appreciable amount of calcium (Haim et al. 1985c). The use of bicarbonate, however, is not limited to herbivores: the mole *Talpa* excretes significant amounts of bicarbonate even though it eats a carnivorous diet (Haim et al. 1987).

One problem that deserves greater consideration than it has received is the limit burrowing places on body size. Body size may be limited by the mechanics of digging a burrow of large diameter and by the demand for sufficient rates of gas exchange. Wilson and Kilgore (1978) addressed this question and concluded that the principal physical factor limiting body size is the porosity of the soil in conjunction with the lethal P_{CO_2}. They estimated that the upper limit to body size is approximately 0.9 kg. Burrowing mammals, however, often exceed a kilogram, the largest burrowers being the aardvark (*Orycteropus*), giant pangolin (*Manis*), and giant armadillo (*Priodontes*), each of which may attain 50 to 60 kg! Of course, some of these large burrowers live in open burrows, but *Priodontes*, at least, will spend the daylight hours in a closed burrow (pers. observ., 1980). Dhindsa et al. (1971), Darden (1972), and Wilson and Kilgore (1978) argued that the low rates of metabolism found in burrowing and fossorial mammals (McNab 1966b, 1979b; Contreras and McNab 1990) reduce the dependency of these species on a high oxygen tension. If so, such a reduction may permit a larger body mass than would otherwise occur, but it will not account for a mass of 60 kg.

Some birds, including shearwaters, petrels, kingfishers, bee-eaters, some barbets, and some swallows, nest and shelter in burrows and many more do so in tree holes. The few studies of the atmospheres of burrows occupied by birds indicated that these species, like burrowing mammals, often encounter low oxygen (usually >14%) and high CO_2 (usually <7%) pressures (White et al. 1978, Wickler and Marsh 1981, Birchard et al. 1984). In burrow atmospheres the decrease in P_{O_2} is correlated with an increase in P_{CO_2} (White et al. 1978, Wickler and Marsh 1981). The circulation of air in these burrows, especially when the burrow opens on a vertical exposure, is correlated with the wind speed at the mouth of the burrow and the angle at which the wind strikes the mouth of the burrow (White et al. 1978). The concentration of CO_2 in the burrow increases with the mass of the birds living in the burrow and with its depth (Wickler and Marsh 1981).

The atmospheres encountered by birds that live in tree holes appear to be more like the external atmosphere (e.g., oxygen usually is >17% and CO_2 <2%–3%) than the atmospheres in burrows (White et al. 1978, Howe et al. 1987). This similarity occurs because the holes are not plugged, are often exposed to wind (and thus coupled by Bernoulli's principle to the external atmosphere), and are subjected to mass air movement into and out of the hole owing to temperature fluctuations in the cavity and heat production by the occupant. These conditions even are found in hornbill (family Bucerotidae) nests; the males seal the breeding female into a tree cavity with mud, leaving only a slit for food to be offered by the male. In this case gas is exchanged with the external atmosphere through the slit because of air currents established by the heat provided by the female and growing young (White et al. 1984).

8.8.5 Adaptation to high altitudes. All organisms living at high altitudes face low oxygen pressures because barometric pressure declines with altitude: the barometric pressure and therefore oxygen tension at about 5600 m is one-half that at sea level. Reptiles living at high altitudes, like those living in burrows or sand, show few modifications of gas transport, at least at altitudes up to 2500 m (Dawson and Poulson 1962, Pough 1969). Weathers and McGrath (1972) showed, though, that low-altitude *Dipsosaurus*, facing a simulated altitude of 5500 m, had a 22% increase in hematocrit and a 19% increase in hemoglobin content. They also noted that *Sceloporus* individuals living at 2750 to 3200 m have 16% more erythrocytes than individuals at sea level, so a reduction in atmospheric pressure must be at least one-third for any response to be seen in these poikilotherms.

Endotherms respond strongly to high altitudes because of their high rates of metabolism. Although endotherms native to high altitudes tend to have low to normal hematocrits, hemoglobin concentrations, and blood capacities (Hall 1937; Morrison et al. 1963a, 1963b; Bullard et al. 1966; Carpenter 1975; Jürgens et al. 1988), they have blood with a high affinity (Figure 8.42), high myoglobin concentrations (Reynafarje and Morrison 1962, Lechner 1976), and high ventilation rates (Rosenmann and Morrison 1975). The high affinity found in camelids generally and especially in the high-altitude vicuña (*Vicugna vicugna*) reflects the structure of their hemoglobins and their differential sensitivities to the concentrations of Cl⁻ and DPG in the erythrocytes (Jürgens et al. 1988). High-altitude

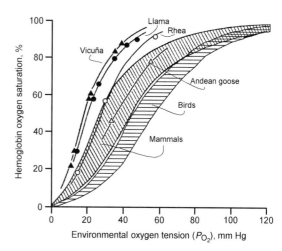

Figure 8.42 Oxygen dissociation curves for eight lowland mammals and the highland vicuña (*Vicugna vicugna*) and llama (*Lama glama*) and for six lowland birds and the highland rhea (*Rhea pennata*) and Andean goose (*Choephaga melanoptera*) as a function of external oxygen tension. Source: Modified from Hall et al. (1936).

endotherms also have high cardiac outputs, as has been inferred from their large hearts (Norris and Williamson 1955, Johnston 1963, Morrison 1964, Jürgens et al. 1988). As a result of these adjustments, the critical oxygen tension is lower in high-altitude mammals than in lowland species and subspecies (Morrison 1964, Rosenmann and Morrison 1975) at all temperatures and at all levels of energy expenditure. For example, a lowland mammal like *Microtus oeconomus* has a critical oxygen tension that ranges from 100 to 145 mm Hg (Rosenmann and Morrison 1974), depending on the environmental temperature, whereas at thermoneutral temperatures some high-altitude mammals (e.g., South American rodents belonging to the genus *Akodon*) have critical tensions as low as 18 mm Hg, which is the P_{O_2} at an altitude of 15 km! At temperatures low enough to raise the resting rate of metabolism to twice the basal rate, high-altitude species of *Akodon* still have critical oxygen tensions as low as 50 to 60 mm Hg, which is equivalent to an altitude of 7700 m (Morrison 1964).

The adjustment of *Peromyscus maniculatus* to high altitudes has been extensively examined (e.g., Hock 1964). In spite of the low P_{O_2} values encountered at 3800 m, deer mice at this altitude have higher maximal rates of oxygen consumption and higher field rates of metabolism than they do when living at 340 m (Hayes 1989a). This tolerance to high altitudes is associated with blood having a high affinity for oxygen. Both affinity and maximal rate of metabolism are directly linked to a genetic polymorphism in the alpha-globin units of hemoglobin molecules at two, closely linked loci (Chappell and Snyder 1984, Snyder et al. 1988). One homozygous genotype leads to a low affinity and a high maximal rate at low altitudes, and it is nearly "fixed" at altitudes less than 1750 m. Another homozygous genotype is associated with a high affinity and a high maximal rate at altitudes above 2750 m, where it predominates (Chappell and Snyder 1984). Other genotypes have intermediate consequences and occur principally at intermediate altitudes. The frequency of the genotypes is accounted for both by the altitude at which the mice are found and by their subspecies affiliation, the latter presumably reflecting the level at which genetic exchange occurs (Snyder et al. 1988). Furthermore, the differences seen in the maximal rate of metabolism and field energy expenditures between high- and low-altitude populations of *P. maniculatus* principally reflect the ambient temperatures (Hayes 1989b). Clearly, some mammals native to high altitudes can easily withstand low atmospheric pressures at low ambient temperatures.

Mammals that live at low altitudes, unlike those native to high altitudes, respond to high altitudes by increasing ventilation rates and by increasing the oxygen capacity of blood by increasing hematocrit (Hall 1937, Kalabuchov 1937). Humans, for example, show a striking increase in hematocrit with an increase in altitude, which may reflect their historically short period of residence at high altitudes (Hall et al. 1936). The difference between "native" and "immigrant" species at high altitudes is seen in the difference in South America between camelids and humans: in camelids P_{50} ranges from 17.6 to 20.3 mm Hg (Jürgens et al. 1988), whereas in humans P_{50} is 31.2 mm Hg at high altitudes (4500 m) and 29.2 mm Hg at sea level (Winslow et al. 1981). The hematocrit of Peruvians living at 3800 m is 55% (Whittembury et al. 1968), whereas the hematocrit of the alpaca (*Lama pacos*) at the same altitude is only 33%.

The problem with responding to a low barometric pressure by increasing hematocrit is that the viscosity of blood increases (Morrison et al. 1963b), which requires the heart to exert a greater force to circulate the blood and can lead to circulatory failure. The ideal solution to a low barometric pressure is used by South American camels:

they combine a high hemoglobin concentration with small erythrocytes.

Birds are able to withstand high altitudes better than mammals because they have much a higher cardiac outputs than mammals of the same size (e.g., house sparrows have a heart mass 2.7 times that of house mice) and because they are also more efficient in the exchange of gas than mammals because of the one-way ventilation of bird lungs. Consequently, at high altitudes birds maintain a higher arterial P_{O_2} and lower arterial P_{CO_2} than mammals, the lower P_{CO_2} increasing the affinity of the blood for oxygen via the Bohr shift (Lutz and Schmidt-Nielsen 1977). Few mammals are found above 6000 m, whereas some birds nest at altitudes as high as 6500 m and others migrate at elevations above 9000 m. This difference was seen in a comparison of house sparrows (*Passer domesticus*) and house mice (*Mus musculus*): house sparrows behave normally at a simulated altitude of 6.1 km, whereas house mice are immobile at the same altitude (Tucker 1968a). The mice had an arterial oxygen tension that was only 24% saturated (unacclimated humans become unconscious at this tension), whereas the sparrow can fly at 40% saturation.

The ability of birds to adjust to high altitudes is readily seen. A survey of four South American passerines, two from low altitudes (600 m) and three from intermediate altitudes (2100 and 3000 m, one species found both low and high), showed no reduction in rate of metabolism at P_{O_2} values equivalent to 10,000 m, even though the measurements were made at an ambient temperature of −5°C, that is, when these species must increase metabolism to maintain temperature regulation (Novoa et al. 1991). A detailed comparison of two North American cardueline finches, the house finch *Carpodacus mexicanus* (which is limited to low altitudes) and the rosy finch *Leucosticte arctoa* (which is generally found at 3000 to 4000 m in summer at the southern limits to its distribution), showed that the rosy finch had blood with a higher affinity for oxygen and maintained a moderate hematocrit, whereas the house finch had a lower affinity and responded to high altitudes by markedly increasing hematocrit, which may have contributed to a lower survivorship at high altitudes (Clemens 1988, 1990). With exposure to low temperatures at low altitudes, both finches responded by increasing rate of metabolism and tidal volume, whereas they responded to an increase in altitude by increasing respiratory frequency (Clemens 1988); a similar pattern was seen in chukars (*Alectoris chukar* [Chappell and Bucher 1987]). No change in critical oxygen tension was found between low (250 m) and high (4540 m) populations of rufous-collared sparrows (*Zonotrichia capensis*) in Peru; the principal problem that this species faces at high altitudes is an adjustment to low temperatures (Castro et al. 1985). The most dramatic example of high-altitude tolerance in birds is the often-quoted migration of bar-headed geese (*Anser indicus*) over the high Himalayan Mountains at altitudes up to 9200 m. This species showed no behavioral effects of simulated high elevations until they exceeded 10,668 m (Black and Tenney 1980). It has blood with a high affinity for oxygen and shows no increase in erythrocyte count.

As noted, burrowing mammals and those living at high altitudes face low oxygen tensions in the environment and respond in similar ways. Some mammals that live at high altitudes also burrow (e.g., pocket-gophers of the genus *Thomomys*, many ground-squirrels, tuco-tucos [*Ctenomys*]), which poses the question of how these species survive being doubly denied oxygen. At 3150 m barometric pressure is about 515 mm Hg; if the burrow atmosphere is 15% oxygen, the burrow oxygen tension would be 77.3 mm Hg, which corresponds at sea level to an atmosphere having only 10.2% oxygen. Lechner (1976) performed the one study on high-altitude burrowers: low-altitude (250 m) *Thomomys bottae* has blood with the same affinity as high-altitude (3150 m) *Th. umbrinus*, which may indicate that the adjustment in affinity made to burrowing habits was carried as far as possible, so that no further increase in affinity is made in response to high altitude. High-altitude *Th. umbrinus* did have a slightly higher blood oxygen capacity (24.2 vs. 22.8 volume %) and a 27% increase in myoglobin concentration. Lechner suggested that myoglobin may facilitate the diffusion of oxygen to tissues, rather than as an oxygen storage molecule per se. Further studies of gas exchange in high-altitude burrowers would be of value to determine the limits of adaptation to a low oxygen tension.

8.8.6 Adaptation to diving. Many reptiles, birds, and mammals have secondarily acquired aquatic habits, and many of these species dive for prolonged periods in search of food or as a means of evading predators. With the exception of a few reptiles (see next section), none of these divers is able to extract oxygen from water. The ability to tolerate prolonged periods of diving, then, depends on the ability of these vertebrates to store oxygen, on the rate at which these stores are used, or on the use of anaerobiosis.

Voluntary dives in most vertebrates are relatively short, which reflects the limits inherent in aerobic metabolism. Aquatic snakes like *Acrochordus* (Pough 1973a), *Cerberus* (Heatwole 1977b), and *Farancia* and *Nerodia* (Jacob and McDonald 1976) have voluntary dives that last 25 to 35 min, although the green sea turtle (*Chelonia*) may have voluntary dives that last up to 50 min, and sea snakes may dive for periods of over an hour (Heatwole 1977b). The leatherback turtle (*Dermochelys*) tends to have short (usually 7–14 min), shallow (to 35–110 m) dives, although it occasionally dives for periods up to 37 min and to depths greater than 1000 m (Eckert et al 1989). These depths are as great as those found in sperm whales (*Physeter catodon*) and elephant seals (*Mirounga*). Most of the leatherback's dives are to the deep scattering layer, where it encounters prey, namely, medusae, siphonophores, and salpae. The loggerhead turtle (*Caretta*) and Kemp's ridley (*Lepidochelys kempi*), unlike pelagic leatherbacks, are found principally in shallow marine water, but they are known to spend time on the sea floor for extended periods (to 8 h).

Many animals show bradycardia during diving, but Belkin (1964) and Heatwole (1977a) argued that this low heart rate may be the normal rate. From this view, divers have a breathing tachycardia to ensure rapid equilibration of blood with air at the water's surface, thereby reducing their vulnerability to predation. Nevertheless, most diving reptiles and all diving birds and mammals are air breathers so that their fundamental heart rate should occur at rest when breathing air. Observations of diving using radiotelemetry in caiman (Gaunt and Gans 1969), alligators (Smith et al. 1974), birds (Butler and Woakes 1979, Kanwisher et al. 1981), and mammals (Smith and Tobey 1983) indicate that much of the experimentally induced bradycardia (Scholander 1940, Irving et al. 1941b) is a "fear" response; some shallow dives by free-living animals occur without any appreciable reduction in heart rate. "Natural dives are usually short and aerobic, involving swimming effort that is not energetically costly" (Kooyman et al. 1981, p. 353). However, Culik (1992) found that diving Adélie penguins (*Pygoscelis adeliae*) have bradycardia during a dive and tachycardia after a dive ended. Free-diving Weddell's seals (*Leptonychotes weddelli*) showed a marked bradycardia during dives, although the extent of the decrease in heart rate was highly variable (Hill et al. 1987). So some ambiguity remains in our knowledge on the occurrence of bradycardia during normal dives in aquatic vertebrates. Recent observations with video cameras and data recorders attached to diving seals, sea lions, and whales have indicated that deep dives start with a few powerful strokes followed by an extended glide and bradycardia, thereby reducing the cost of diving (Hill et al. 1987, Costa 1988, Le Boeuf et al. 1989).

Body size influences diving times, as is most clearly shown in mammals. Calder (1969) suggested that diving times can be estimated by Equation 5.4; that is, it is proportional to $m^{0.25}$. Very small divers, such as the water shrew (*Sorex palustris*), have much shorter times than expected from mass because of high rates of heat loss and activity, both of which raise the rate of metabolism. Large divers, at least, have reduced rates of metabolism when diving and large oxygen stores, which extend their diving times beyond the periods expected from mass.

All divers are not equal. Some, such as porpoises, most baleen whales, sea lions, the walrus, and sea otter, are shallow divers that make dives of short duration, whereas others, namely, the sperm whale, bottlenose whales, and phocid seals, make deep, prolonged dives (Scholander 1940, Lenfant et al. 1970). Many physiological differences correlate with this dichotomy: deep divers have large oxygen stores (mainly due to high myoglobin levels and to high oxygen capacities in the blood), small lungs, high blood bicarbonate levels, and high buffering capacities. Shallow divers, the most extreme being the sea otter (*Enhydra*), have the opposite set of characteristics, including large lungs. The small lungs and propensity of alveoli to collapse in deep divers reduce the rates at which oxygen and nitrogen gases are forced by hydrostatic pressure (also see Section 8.4.7) into solution in the blood and other tissues. The difficulty with forcing gas into solution is that it returns to a gaseous state during the ascent to the surface (Scholander 1940),

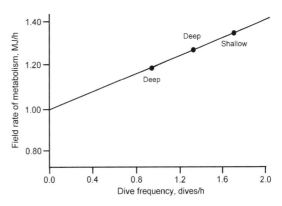

Figure 8.43 Field energy expenditure in northern fur seal (*Callorhinus alascanus*) as a function of dive frequency and depth. Source: Modified from Costa (1988).

thereby causing the "bends," or Cassion's disease. Even though adult sea lions (*Eumetopias*) belong to the shallow-diving group, fetal sea lions have the characteristics of deep, prolonged divers, which is what they are during gestation. Marine snakes and turtles, unlike diving mammals and birds, are not affected by the bends, apparently because of a venous shuttling of blood past the lung and the permeability of skin to dissolved gas (Seymour 1974).

Costa (1988) started an analysis of the energetics of diving behavior. Some female northern fur seals (*Callorhinus alascanus*) make deeper (mean 185 m) and some make shallower (50–60 m) dives. Deep-diving females make half as many dives as, and make shorter trips than, shallow-diving females. Therefore, deep-diving females expend less energy per trip, and the amount of energy expended is directly related to the dive frequency (Figure 8.43). Furthermore, deep divers consume fewer, but larger and more energy-dense prey than do shallow divers (i.e., pollock vs. squid, respectively). Shallow diving is a better strategy only if the probability of capturing prey is much greater in shallow than in deep water.

An important factor dictating the length of the diving period in reptiles is water temperature. As expected of poikilotherms, the period that *Chrysemys* can survive immersion increases with a fall in water temperature (Figure 8.44): this turtle can only survive 2 d submerged in air-equilibrated water at 26°C, but it can survive about 118 d at 1.5°C! Part of this increase is due to the effect of temperature on metabolism, but some other factor,

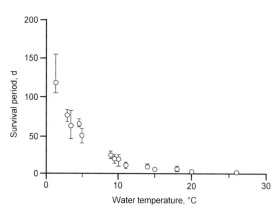

Figure 8.44 The survival of the turtle *Chrysemys picta* submerged in water as a function of water temperature. Source: Derived from Musacchia (1959).

such as aquatic gas exchange, must be important, because the Q_{10} (= 5.3) of this increase is well above what is expected from the effect of temperature alone.

Belkin (1963) showed that freshwater turtles may survive forced dives for periods that vary from 360 min to indefinite, whereas sea turtles can tolerate dives of 114 to 126 min, and other reptiles 20 to 118 min. The long dives of turtles in some species may result from their large masses, whereas those in others may result from their dependence on anaerobic metabolism (Belkin 1968): for example, at 22°C untreated *Sternothaerus odoratus* in nitrogen-equilibrated water can survive for about 12.2 h, but those poisoned with iodoacetate (which blocks glycolysis) can only survive for 0.32 h.

Ultsch and Jackson (1982a) examined the physiological basis of long-term tolerance of submergence in cold water by *Chrysemys*. They established that *Chrysemys* survives in anoxic water for more than 130 d at 3°C, although it develops severe acidosis due to the accumulation of lactic acid in the plasma. When the water is aerated, survivorship was higher, plasma acidosis due to lactic acid was lower, and arterial P_{O_2} higher, all of which indicate that *Chrysemys* is able to support a significant proportion of its metabolism by extracting oxygen from water.

8.9 THE RETURN TO AQUATIC EXCHANGE

The only vertebrates of a terrestrial ancestry that have (partially) returned to aquatic respiration are a few reptiles, although some completely aquatic amphibians, such as sirenids and pipids, undoubtedly have reemphasized aquatic gas exchange. No endotherms have evolved aquatic gas exchange because their high rates of metabolism demand such large surfaces for gas exchange that a reinvention of a gill-like structure would be required. Reptiles operating at the temperature of most naturally occurring waters have rates of metabolism sufficiently low that a limited surface area of high permeability will make a significant contribution to gas exchange.

Significant rates of aquatic gas exchange in reptiles have been found only in snakes and turtles. The snakes include sea snakes (Elaphidae [Graham 1974; Heatwole and Seymour 1975, 1978]), sea kraits (Elaphidae [Heatwole and Seymour 1978]), file snakes (Acrochordidae [Standaert and Johansen 1974; Heatwole and Seymour 1975, 1978]), and homalopsine snakes (Colubridae [Heatwole and Seymour 1978]). Among turtles, significant aquatic gas exchange is known in *Trionyx* (Gage and Gage 1886; Dunson 1960; Girgis 1961; Stone et al. 1992), *Sternothaerus* (Root 1949; Belkin 1963, 1968; Stone et al. 1992), and *Kinosternon* (Stone et al. 1992). Low rates of aquatic exchange have been found in *Trachemys* (Jackson et al. 1976), *Chrysemys* (Ultsch and Jackson 1982a), *Malaclemys* (Jackson et al. 1976), and *Chelydra* (Gatten 1980), which may indicate that aquatic exchange is a general property of freshwater turtles, especially in small individuals at low temperatures.

The site of aquatic gas exchange in reptiles varies. Most aquatic exchange in snakes occurs through the integument. Some of the exchange in turtles may occur through the exposed skin of the legs and head, but most occurs in the buccopharyngeal cavity in *Trionyx* (Gage and Gage 1886, Dunson 1960, Girgis 1961) and *Sternothaerus* (Belkin 1968). In *T. sinensis* 68% of aquatic gas exchange was buccopharyngeal and 32% was integumental; this ratio was independent of water temperature between 16 and 30°C (Wang et al. 1989). In *Trionyx* buccopharyngeal exchange occurs through the dense villiform processes that cover the wall of the pharynx (Girgis 1961). High rates of buccopharyngeal exchange are facilitated by gular movements. The high rates of aquatic gas exchange found in *Trionyx* and *Sternothaerus* are associated with large surface areas and the propensity to live in shallow, well-mixed riverine habitats, unlike *Kinosternon*, which has a low rate of aquatic gas exchange and a small surface area and tends to live in stagnant ponds (Stone et al. 1992). Nonpulmonary exchange of CO_2 is more important in aquatic species, terrestrial species being forced by evaporative water loss to reduce integumental gas exchange (Jackson et al. 1976). Gas exchange in the cloaca of *Trionyx* is of little importance, most cloacal exchange being ionic (Dunson 1966).

The use of aquatic exchange by reptiles varies with several factors. Firstly, the oxygen tension of water is important to turtles that use aquatic exchange: survivorship at 22°C markedly increases in *Sternothaerus* with an increase in oxygen tension, whereas the survivorship of *Pseudemys*, which depends little on aquatic exchange at such high temperatures, is unaffected (Belkin 1968). The survival of *Sternothaerus* in anoxic water, however, is only one-half that of *Pseudemys*, a condition that may reflect a high integumental permeability with the subsequent loss of oxygen to the water. When *Sternothaerus* was given the opportunity to choose aerial or aquatic exchange, it obtained 33.4% of its oxygen requirements from water; under similar conditions *Pseudemys* obtained only 4.2% of its oxygen from water. Secondly, aquatic exchange diminishes as body mass increases (Belkin 1968, Graham 1974, Heatwole and Seymour 1978).

The only large reptiles that apparently depend on aquatic gas exchange are some sea turtles (*Chelonia*, *Caretta*) that have been shown to spend weeks (months?) partly buried in marine shallows during the coldest part of the year (Felger et al. 1976, Carr et al. 1980), when metabolism is depressed by low water temperatures. In each case, tropical sea turtles were trapped by cold weather in warm-temperate waters. At water temperatures between 10 and 15°C, these turtles dive to the bottom of estuaries, or other shallow bodies of seawater, and remain partially or completely buried in sand. They remain in situ for several months, or at least long enough to obtain an algal and slime growth on the exposed carapace. Presumably, all gas exchange during this period is aquatic. These observations raise questions on the temperate limits to distribution of tropical turtles and contrast strikingly with the presumptively endothermic behavior of the largest sea turtle, *Dermochelys*, which has been found as far north as Nova Scotia (Frair et al. 1972). Whether alligators remain underwater in winter, or whether they shelter in burrows containing an air pocket is unclear.

A third factor to influence the balance between aquatic and aerial exchange is the type of gas. Aquatic exchange is more important for CO_2 loss than for oxygen uptake. Thus, in the sea snake *Pelamis* CO_2 loss is 55% to 94% aquatic, whereas oxygen uptake is 30% to 62% aquatic (Graham 1974). Heatwole and Seymour (1975) made similar observations on marine snakes and Stone et al. (1992), on turtles.

Aquatic gas exchange may have a different significance for snakes and turtles. Girgis (1961) and Belkin (1968) suggested that aquatic exchange in turtles is adequate only during periods of inactivity, especially at low temperatures; aerial exchange is more important during activity. In this, turtles are similar to bimodal-breathing fishes. Graham (1974) and Heatwole and Seymour (1975, 1978), in contrast, stated that cutaneous exchange in aquatic snakes increases with activity, especially in sea snakes that feed underwater, when extended periods of struggling with prey may occur. Aquatic exchange may permit sea snakes to have an increased scope for activity.

The relative importance of aquatic exchange in snakes is related to their habitats (Heatwole and Seymour 1978). High rates of aquatic exchange occur in sea snakes, which are typically found in deep water with high oxygen and low CO_2 tensions. The homalopsine *Cerberus*, however, has aquatic exchange rates that are only one-third to one-fourth those of sea snakes; this snake lives in shallow mangrove swamps that are characterized by water with a high temperature, low P_{O_2}, and high P_{CO_2}, conditions that are hostile to gas exchange. Sea kraits occupy habitats intermediate to sea snakes and homalopsines, often spending much time on land. They have rates of aquatic exchange as low as those for *Cerberus*. Acrochordids have the lowest rates of aquatic oxygen consumption; they live in freshwater, in estuaries, or in mangrove swamps. High rates of aquatic exchange may be difficult to tolerate in species that move from freshwater to mangrove swamps because of the coupling that likely occurs between ionic and gaseous exchange.

In fish and amphibians the affinity of blood for oxygen is greater in water than in air breathers. Do reptiles that use aquatic gas exchange have a higher affinity than those that are strictly air breathers? These data are not available.

8.10 CHAPTER SUMMARY

1. The parameters that are regulated in gas exchange are P_{O_2} at the mitochondria, plasma $[OH^-]/[H^+]$ ratio, and protein charge.

2. In mammals and birds pH appears to be regulated principally because body temperature is constant; in poikilotherms pH decreases with an increase in body temperature.

3. As temperature increases, pH is adjusted to ensure a constant $[OH^-]/[H^+]$ ratio by an increase in P_{CO_2} and ventilation rate.

4. Ventilation rate is principally determined by the amount of oxygen available in the environment: it is high in water and low in air.

5. In the evolution of air breathing, the bicarbonate level of blood increases to compensate for the higher P_{CO_2} that results from lower ventilation rates.

6. Gills facilitate gas exchange in water by having a large surface area, thin epithelium, and countercurrent exchange.

7. The gill surface area in fish varies with body mass and activity level.

8. Oxygen is transported by blood in solution (ca. 5%) and bound to hemoglobin (95%), which is located in erythrocytes in all vertebrates.

9. The affinity of hemoglobin for oxygen varies with its structure and with the amount of organophosphates in the erythrocytes: affinity is high in inactive fishes that live in oxygen-poor water and low in active fishes that live in oxygen-rich water.

10. High CO_2 concentrations in water reduce the capacity of fish hemoglobin to bind oxygen, unless the fish live in swamps, when their hemoglobin is relatively insensitive to the presence of CO_2.

11. Ice-fish are the only vertebrates that lack hemoglobin and erythrocytes; this condition reduces the viscosity of blood at constant, low temperatures. All oxygen transport in ice-fish occurs in solution. Ice-fish compensate for the low oxygen capacity of their blood by having low rates of metabolism and high cardiac outputs.

12. Some fish respond to a low P_{O_2} and high P_{CO_2} in water by combining aerial and aquatic exchange.

13. Compared to water-breathing fishes, air-breathing species have high hematocrits, low affinities for oxygen, and low ventilation rates.

14. The relative importance of aerial exchange in bimodal species varies with water temperature and body size.

15. Amphibians use both aquatic and aerial gas exchange. They use the integument and gills in water, and the integument and lungs in air. Terrestrial species tend to use the integument principally for CO_2 loss and lungs for oxygen uptake, the partitioning of exchange being influenced by temperature.

16. Terrestrial amphibians have higher hematocrits, lower affinities for oxygen, and higher plasma bicarbonate levels than do aquatic species.

17. Some aquatic salamanders use their lungs as hydrostatic devices, whereas others depend on them for aerial exchange.

18. Salamanders limited to aquatic exchange have an upper limit to body mass.

19. Some salamanders are lungless, a condition that may represent an adaptation to a small mass or to life in moving water. It restricts these species to a small mass and prevents them from living in warm water.

20. Air breathing evolved early in the history of vertebrates, principally in freshwater with low oxygen and high CO_2 tensions and high temperatures, as well as in areas with periodic droughts. Air breathing permits amphibious habits in some marine fishes.

21. The dependence of living amphibians on cutaneous gas exchange represents a specialization associated with small body masses and is not representative of the gas exchange of the amphibians that gave rise to reptiles.

22. Aerial gas exchange in reptiles, birds, and mammals occurs almost exclusively in the lungs.

23. Birds have a respiratory system that permits unidirectional air flow through the lungs via the air sacs. This system permits a higher arterial P_{O_2} and lower arterial P_{CO_2} than the system in mammals permits.

24. Variations in oxygen transport correlate with levels of metabolism, rate of activity, body temperature, and climate in terrestrial vertebrates.

25. Burrowing mammals have a low sensitivity to CO_2 and have blood with a high affinity for oxygen.

26. Birds that live in burrows or in tree holes face atmospheres that are less extreme than those found in mammalian burrows because these atmospheres are usually open to the external atmosphere.

27. Birds and mammals native to high altitudes generally have high affinities for oxygen, high myoglobin concentrations, high cardiac outputs, and high ventilation rates; lowland species exposed to high altitudes respond by increasing hematocrit.

28. Birds are able to tolerate high altitudes better than mammals because of the more efficient gas exchange mechanism associated with the structure of avian lungs and because of the higher cardiac outputs of birds.

29. Air-breathing vertebrates dive in water for periods that depend on the size of oxygen stores, the rate of metabolism, and the ability to use anaerobic metabolism. In general these periods increase with mass. In poikilotherms these periods increase at cold temperatures.

30. A few reptiles use aquatic gas exchange. It is most important in turtles during inactive periods at low temperatures but may increase the scope for activity in marine snakes.

PART IV | Ecological Energetics

9 The Energetics of Locomotion

9.1 SYNOPSIS

Most vertebrates are highly mobile; they move by swimming, walking, running, flying, gliding, or brachiating. The swimming velocity of active fish varies with temperature because it is powered by aerobic metabolism, subject to thermal acclimatization, whereas sit-and-wait predatory fishes principally rely on anaerobic metabolism, which is nearly temperature independent. Amphibians that actively forage move slowly, use aerobic metabolism, and are usually protected by poisonous secretions, a large size, and burrowing habits; amphibians that are sedentary use anaerobiosis to move rapidly and avoid predation. Activity levels are the greatest in anurans and the least in lungless salamanders, a difference related to oxygen transport. The anaerobic and aerobic scopes for metabolism in reptiles are greatest in active species at their preferred body temperatures. Most activities in the field are powered by aerobiosis, anaerobiosis being used principally during emergencies. Power output during flight in birds and bats is curvilinear with velocity, the minimal output at an intermediate velocity corresponding to the minimal cost per unit of distance. Flight velocity and cost are highly variable and influenced by body mass, wing span, and wing beat and amplitude. The flight apparatus of birds and bats reflects their diversity in ecology and behavior, although the diversity in bats is much less than that in birds. Pterosaurs had powered flight, but large species relied principally on soaring. The movement of terrestrial quadrupeds is described in terms of gaits, the shift from one gait to another being a means of reducing the cost of transport. The cost of transport increases with body mass, but at a given mass is least in swimmers, intermediate in fliers, and greatest in walkers-runners. Long-distance movement occurs most commonly in swimmers and fliers because of their low costs of transportation and high velocities, or in large walkers because of their high velocities.

9.2 INTRODUCTION

Physiology, like other experimental sciences, facilitates the analysis of complex natural phenomena by sequentially controlling variables, thereby converting a complex phenomenon into a series of comparatively simple relationships. In the study of energetics, animals usually have been studied at rest, activity being considered a contaminant to be eliminated because its presence, for example, confuses the relationship between rate of metabolism and body size. The exclusion of activity has led to progress in the study of resting metabolism (see Chapters 3 through 5). Activity metabolism, however, deserves attention because most vertebrates spend many of their waking hours moving from one place to another in search of food, shelter, and mates, or in the evasion of predators. The energy expenditures associated with locomotion comprise a large fraction of the daily energy budget in most vertebrates (see Chapter 10). Within the last two decades the energy expenditures of locomotion in vertebrates have been measured under controlled conditions. These studies used swimming flues for aquatic vertebrates, wind tunnels for flying species, rope or ladder mills for brachiators, and treadmills for terrestrial vertebrates, the results of which are summarized in this chapter preparatory to the examination of energy budgets.

9.3 SWIMMING IN FISHES

The extent and intensity of activity in fish are highly variable (Beamish 1978). Swimming is often divided into three categories: sustained, prolonged, and burst swimming. Sustained swimming occurs over long periods of time (>200 min) without muscular fatigue. It occurs during migration or "routine" activity or may be associated with the maintenance of hydrostatic equilibrium. Sustained activity is powered by aerobic metabolism. Prolonged swimming occurs over shorter periods than sustained swimming (20 s–200 min), ends in muscular fatigue, and relies on both aerobic and anaerobic metabolism. One example of prolonged activity is defined as the "critical" velocity, which is the maximal velocity maintained for a precisely defined period of time. Because of the inclusion of muscle fatigue in the definition of prolonged activity and because of the implied use of controlled water velocities in the definition of critical velocity, the concept of critical velocity and possibly even prolonged activity is nearly inapplicable in the field. Burst swimming occurs only over very short periods of time (<20 s) and involves high velocities. It is powered mainly by anaerobic metabolism and is used to capture prey, escape from predators, and swim against rapid currents.

The cost of swimming incorporates maintaining a position in a vertical column, forward motion, or a fixed position in the face of a current. Maintenance of a position in a water column can be accomplished by at least two methods:

1. Lift must be generated to balance gravity if a fish has a density greater than that of the water in which it is found. This is noticeable in elasmobranchs and in bony fishes with ganoid scales (bichurs) or with bony plates (sturgeons). For example, dogfish sharks (*Scyliorhinus*) have a specific gravity of about 1.075, whereas full seawater has a specific gravity of about 1.026 (Alexander 1975). The force that the shark exerts against gravity results from a combination of lift produced during forward motion by the pectoral fins and that produced by a heterocercal tail in which the dorsal lobe is appreciably larger than the ventral lobe (Alexander 1965, 1968). This solution

requires a continuous expenditure of energy (also see Magnuson 1970).

2. Gravity can be combatted passively by having a body that is neutrally buoyant. Fish may become neutrally buoyant either by reducing the specific gravities of muscle tissue, which usually has a specific gravity of 1.06 to 1.09, and of skeletons and scales, which have specific gravities up to 2.00, or by including a volume that has a low specific gravity to compensate for denser tissues. Some bathypelagic fishes without a swim bladder maintain neutral buoyancy by reducing the density of muscle, skeleton, and scales (Denton and Marshall 1958, Eastman and DeVries 1981, Eastman 1988). The effectiveness of these modifications is seen in a 32.2-kg *Dissostichus mawsoni*, an antarctic notothenioid, which weighed only 14.5 g in seawater. Otherwise, high-density tissues can be compensated for by the addition of fat (as in the skull and swim bladder of *Latimeria* [Alexander 1975, Hughes 1976], in the large oily livers of sharks [Alexander 1965, Corner et al. 1969], and in subcutaneous and intramuscular lipid deposits in antarctic notothenioid fish, such as *Pleuragramma antarcticum* and *D. mawsoni* [Eastman 1988]). The most common response is the inclusion of air in the form of respiratory structures or swim bladders.

The gas that fills a swim bladder is derived indirectly from the environment and directly from the blood at the gas gland through a reduction of the solubility of gas in the blood (Hall 1924, Scholander 1954, Scholander and van Dam 1954, Ball et al. 1955, Steen 1963). The pressure of gas is multiplied by a countercurrent exchanger at the "rete mirabile" associated with the gas gland (Kuhn et al. 1963). The release of gas is produced by the Root effect (see Section 8.4.4) with the secretion of lactic acid and H^+.

The volume of a swim bladder (or a fat deposit) required to give a fish neutral buoyancy depends on the density of the fish and the environment:

volume of fish at neutral buoyancy = volume of fish without bladder + volume of bladder

$$\left[\frac{m}{\rho'}\right] = \left[\frac{m}{\rho}\right] + x,$$

where m is the mass of the fish, ρ is its specific density, ρ' is the specific gravity of the medium, and *x* is the volume of the bladder (Alexander 1966). Freshwater teleosts with swim bladders have specific gravities close to 1.000, whereas those that live in seawater have specific gravities near 1.026. This is theoretically accomplished by having swim bladders that are 5.7 to 8.3 volume % in freshwater species and 3.1 to 5.7 volume % in saltwater species (Alexander 1966). For example, if a teleost weighed 1 kg and had a specific gravity (without a swim bladder) of 1.08, it would require a 74-cm^3 swim bladder to be neutrally buoyant in freshwater and 49-cm^3 swim bladder in seawater. These volumes are commonly found (Jones and Marshall 1953). Fish that live on the bottom of a water body need to make no adjustments of their buoyancy: they lack swim bladders and large lipid deposits and often are sedentary. These are the characteristics of many notothenioid fishes, a diverse radiation of bottom-dwelling fishes, in the southern ocean, some few of which have evolved a pelagic habit, gaining neutral buoyancy by the storage of lipids and the reduction of the skeleton (Eastman 1988).

9.3.1 Swimming velocities. Many factors influence swimming velocities. When a body moves through water, a force, called *drag*, is exerted against the body by water. Drag is proportional to the square of the velocity of the body pushing against water. That is, the velocity of the body is proportional to the square root of drag. Drag itself varies with the drag coefficient and the cross-sectional area of the body. Velocity increases with body mass because muscle mass increases with body mass. D'Arcy Thompson (1917) argued that the velocity of aquatic organisms is proportional to the square root of their length. This relation holds in the critical velocity of sockeye salmon (Brett 1965).

Prolonged swimming velocity increases with temperature (Figure 9.1), as expected of a behavior based on aerobic metabolism, which usually has a Q_{10} of 2 to 3. Eurythermal species, such as goldfish (*Carassius*) and bass (*Micropterus*), have maximal prolonged swimming velocities at temperatures between 25 and 30°C; cool stenothermal species, such as lake trout (*Salvelinus namaycush*) and various Pacific salmon (*Oncorhynchus*), attain maximal velocities between 15 and 20°C; and the antarctic stenothermal *Trematomus borchgrevinki* has a maximal velocity at −0.8°C (and can swim at temperatures only up to 2°C). The peak in the

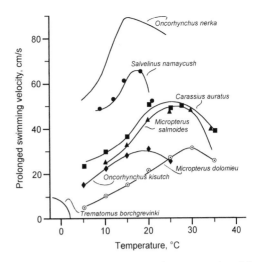

Figure 9.1 Prolonged swimming velocity in various fishes as a function of environmental temperature. Source: Modified from Beamish (1978).

scope for activity occurs at the upper end of the range of acceptable temperatures. Swimming velocity shows thermal acclimation: cold-acclimated individuals usually have higher prolonged swimming speeds than do warm-acclimated individuals at the same temperature (Beamish 1978). Burst swimming velocities, however, are temperature independent (Beamish 1978), reflecting the low temperature sensitivity of anaerobiosis ($Q_{10} \sim 1.2$).

Environmental factors other than temperature may influence swimming velocities. The oxygen content of water limits the prolonged swimming velocities of fish below some "critical" oxygen content (Figure 9.2; see also Figure 8.13 and Dahlberg et al. 1968). In this case the cold-water *Oncorhynchus* is less tolerant of low P_{O_2} than the warmer-water *Micropterus*. High concentrations of CO_2 also limit prolonged periods of swimming at high velocities. The influence of variation in salinity on swimming has been little studied; it seems to have little effect if fish are capable of osmoregulation (Beamish 1978). Holeton and Stevens (1978) showed that a low concentration of electrolytes or a low pH does not affect the energetics of swimming in an Amazonian characin.

9.3.2 Biochemical energetics. The muscles of fish are constructed of two kinds of fibers, white and red (Driedzic and Hochachka 1978). White fibers have poor vascularity, little (if any) myoglobin, and low rates of oxygen consumption. These fibers are principally anaerobic. Red fibers, in contrast, have high concentrations of myoglobin and

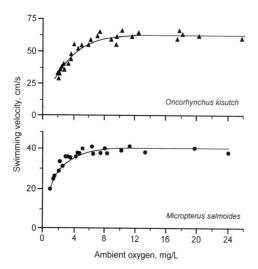

Figure 9.2 Swimming velocity in the teleosts *Oncorhynchus kisutch* and *Micropterus salmoides* as a function of the ambient oxygen. Source: Modified from Beamish (1978).

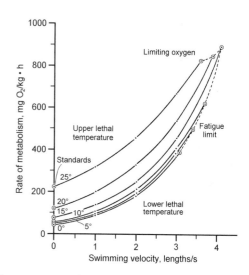

Figure 9.3 Rate of metabolism in yearling sockeye salmon (*Oncorhynchus nerka*) as a function of swimming velocity at various temperatures. Limitations by high and low temperatures, oxygen availability, and fatigue are indicated. Source: Modified from Brett (1964).

citric-acid-cycle enzymes and have large lipid stores; they also have a good vascular supply. Red fibers have high rates of oxygen consumption. At low velocities, fish mainly use red fibers and aerobic metabolism; at higher velocities both red and white fibers are used and the energy expended for muscle contraction is derived from both aerobic and anaerobic sources. Active fishes, such as tuna, have a greater proportion of red fibers, which tend to be located in discrete, superficial bands (Bennett 1978).

Because activity in fish may be powered by both aerobic and anaerobic metabolism, only with caution can oxygen consumption be used as a measure of the energy expenditure during prolonged swimming at high velocities. The rate of oxygen consumption at sustained and prolonged velocities increases with velocity (Figure 9.3) up to 10- to 15-fold. Fry (1947) defined the difference between the maximal and standard rates of metabolism as the "scope for activity," which is a measure of the range of energy expenditure for a normally active animal. In sockeye salmon oxygen consumption increases exponentially with an increase in velocity, often conveniently measured in terms of body lengths per second, but the increase is limited by lethal temperature, the oxygen tension in water, and fatigue. This relation is modified by thermal acclimation: the relation between rate of metabolism and acclimation (Figure 9.4) describes an activity polygon that is limited at high velocities

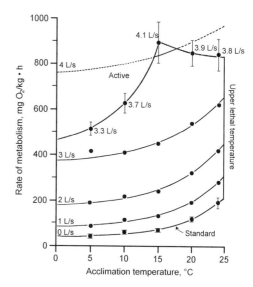

Figure 9.4 Rate of metabolism in yearling sockeye salmon (*Oncorhynchus nerka*) as a function of acclimation temperature and swimming velocity. L, lengths. Source: Modified from Brett (1964).

by a low solubility of oxygen at higher water temperatures (Brett 1964). The maximal capacity to consume and transport oxygen limits velocity (Bennett 1991).

Carbohydrates (glycogen) and lipids are the energy stores usually used in swimming, but proteins may be important in the long-distance migrations of salmonids (Bilinski 1974). King salmon

(*Oncorhynchus tshawytscha*) lost up to 30% of muscle mass during migration upstream, whereas fat content fell from 20% to 1% or 2% (Greene 1926). In the Frazier River of British Columbia, sockeye salmon (*O. nerka*) used 30% to 40% more fat than protein from muscle tissues for the first 400 km of migration upstream, but during the last 750 km sockeye used twice as much protein as fat (Idler and Bitners 1959).

Lactate is a by-product of anaerobic metabolism in vertebrates. In mammals 80% to 90% of the lactate that enters the blood is withdrawn by the heart, muscles, liver, and kidney. At these sites lactate is aerobically metabolized to CO_2 and water. This conversion is significant because energy is discarded if lactate is excreted. The fate of blood lactate is unclear in fish, but it probably is metabolized as it is in mammals (Driedzic and Hochachka 1978); at least, lactate is not excreted (Bennett 1978). Although the maximal rates of lactate production in fish are similar to those in amphibians and reptiles, fish appear to be more sensitive to its presence, which at high concentrations may lead to death (Bennett 1978). Death may be produced by a reduced oxygen capacity in blood, a decrease in blood bicarbonate level, or a breakdown in circulation (Black 1958). Even if a fish survives strenuous activity, it often requires a long time to recover from the accumulation of lactate.

The differential use of anaerobiosis and aerobiosis depends on the life-style of the fish. The creek chub (*Semotilus atromaculatus*), which lives in swift-flowing streams and is an active forager, was compared to the central mudminnow (*Umbra limi*), which is found in and near marshes and is a sit-and-wait predator (Goolish 1991). *Umbra limi* has no superficial red-muscle tissue, whereas 9% of the caudal peduncle in *S. atromaculatus* is red muscle. A swimming velocity of 2 body lengths/s is required in *U. limi* to start accumulating lactate in its muscles; this velocity is 5 body lengths/s in *S. atromaculatus*. At a 4-min exhaustion period, the mudminnow had accumulated more lactate than the chub, 9.82 versus 6.46 μmol/g, respectively. Complete glycogen restoration after exhaustion is shorter in the mudminnow (6 h) than in the creek chub (14 h). Goolish (1991) suggested that the reliance on anaerobiosis during swimming is characteristic of sit-and-wait predatory fish, at least at small masses.

Ruben and Bennett (1980) found that the use of anaerobiosis during activity is widespread, including in hagfish (*Eptatretus stouti*) and lampreys (*Lampetra pacifica*), but that it is poorly developed in the cephalochordate *Branchiostoma caribaeum*. The expanded capacity for anaerobic-based burst activity may have been associated with the development of cephalization and the evolution of vertebrates from invertebrate chordates with an increase in mass. Further study indicated that the brittle star *Ophioderma* and an appendicularian tunicate *Oikopleura* used glycolysis during activity, which suggests that this pattern originated in the earliest deuterostomes (Ruben and Parrish 1990).

9.4 AMPHIBIAN LOCOMOTION

The study of amphibian locomotion here is divided into the cost of locomotion and the biochemical basis of energy expenditure.

9.4.1 The cost of locomotion. Gatten et al. (1992) summarized the cost of locomotion in amphibians. During sustainable locomotion (i.e., during locomotion fueled by aerobic metabolism), the steady-state rate of oxygen consumption is correlated with the velocity of locomotion (Figure 9.5). At any particular velocity, the rate of metabolism is greater in anurans and lower in salamanders. The maximal steady-state velocity is lowest in lungless salamanders (Figure 9.5), presumably because of a lower maximal rate of cutaneous and buccopharyngeal gas exchange. Other factors, such as temperature and body size, may influence the cost of locomotion, but they either have not been studied to an appreciable extent (e.g., temperature) or have been examined over too small a range in mass (Gatten et al. 1992).

The cost of transport is obtained from dividing the rate of metabolism during activity by the velocity of movement and has the units of energy per mass times distance (J/g·m). It varies with velocity of movement, reaching a minimal value at high velocities because the *y*-intercept in the graph of expenditure on velocity (Figure 9.5) is not at the origin and consequently is disproportionately incorporated into the cost at low velocities (see Sections 9.7.3 and 9.9; Full 1986). The minimal cost of transport in amphibians is similar to that found in other terrestrial vertebrates (John-Alder et al. 1986, Gatten et al. 1992, Walton 1993).

The ability of frogs to sustain locomotion

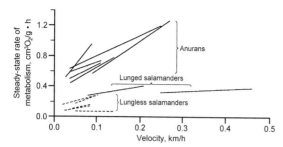

Figure 9.5 Rate of metabolism in amphibians as a function of velocity. Source: Modified from Gatten et al. (1992).

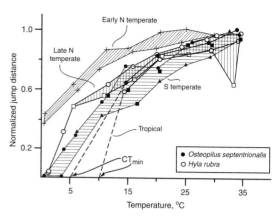

Figure 9.6 Normalized jump distance in North (N) Temperate hylids that breed early and late, in southern (S) North Temperate hylids, and in tropical hylids as a function of environmental temperature. CT_{min} is critical thermal minimum. Source: Modified from John-Alder et al. (1988).

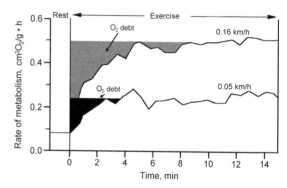

Figure 9.7 Rate of metabolism in the plethodontid salamander *Plethodon jordani* as a function of time and velocity. Oxygen debts accumulated at the beginning of a bout of activity are indicated by shaded areas. Source: Modified from Gatten et al. (1992).

depends on the cost of locomotion and the maximal capacity for aerobic energy expenditure. In hylids the maximal rates of metabolism are correlated with their resting rates of metabolism and independent of body size and phylogeny (Walton 1993). Because North American hylids are thought to have evolved from neotropical species (Duellman and Trueb 1986) and because tropical species belonging to the genera *Agalychnis*, *Pternohyla*, and *Osteopilus* have low resting and maximal rates, the movement into temperate environments appears to have been associated with an increase in both rates (Walton 1993).

A climatic adjustment is seen in the jump distance of hylids: early-breeding northern North Temperate hylids jump farther at low temperatures than similar hylids that are late breeders, which jump farther at these temperatures than southern North Temperate hylids (Figure 9.6). Tropical hylids are similar to southern North Temperate species, except that they are immobilized at higher cold temperatures (John-Alder et al. 1988). At warm temperatures all hylids jump about the same distance (Figure 9.6). A similar pattern is found between South American hylids in the high cold-tolerance of high-altitude species compared to those limited to low altitudes (Navas 1996).

Such adjustments facilitate activity at low ambient temperatures: ectotherms from high latitudes and altitudes have higher rates of metabolism both at rest and during activity at lower temperatures than do low-latitude and low-altitude species at the same temperatures. These adjustments have permitted hylids to invade temperate environments.

9.4.2 Biochemical energetics. The energy expended by amphibians for activity can be esti-

mated by the sum of the increment in oxygen consumption and the amount of lactate produced (Bennett and Licht 1973, 1974). Anaerobic metabolism during activity can also be estimated by the increment in oxygen consumption after activity, the so-called oxygen debt. Most of the expenditure in the first 2 min of activity depends on anaerobic metabolism (Figure 9.7; see Gatten et al. 1992).

Bennett and Licht (1973, 1974) suggested that the sum of aerobic and anaerobic expenditures is roughly constant in amphibians, so that species that rely on aerobic metabolism for activity have a low anaerobic scope, and species that rely on anaerobiosis have a small aerobic scope. Toads

(Bufonidae) tend to be comparatively inactive and depend principally on aerobic metabolism for activity, whereas frogs (Ranidae) are more active and mainly use anaerobic metabolism. Bennett and Licht (1974) and Bennett (1978) further suggested that the inactivity of toads was correlated with their use of passive defenses, such as the secretion of various toxins, and that the activity of frogs reduced their need for such defenses, because their principal defense is escape. The distinction between aerobic and anaerobic sources of energy, however, occurs during activity and immediately thereafter: in the final analysis, all activity in vertebrates is paid for by aerobic metabolism as long as lactate is not excreted. Anaerobiosis in vertebrates is simply aerobiosis on the installment plan.

This picture of amphibian energetics was challenged on technical grounds. In the studies of Bennett and Licht, oxygen consumption was usually measured with a manometer, and activity was produced by an electrical shock, often while the amphibian was sitting on a moist paper towel (to prevent the amphibian from becoming dehydrated). Hillman et al. (1979b) showed that this technique diminished the measurements of oxygen consumption through the combination of short-term temperature transients and the electrolytic generation of oxygen gas from water. Maximal rates of oxygen consumption are much greater when measured with an oxygen analyzer and when activity is stimulated by mechanical means (rotation of the chamber). These changes affect the conclusion as to the relative importance of aerobic and anaerobic metabolism during activity. For example, Hillman and coworkers, using the measurement of anaerobic scope by Turney and Hutchison (1974), concluded that *Rana pipiens* derived only 22% of the energy used in activity from anaerobic metabolism when the frog was manually stimulated, whereas when it was electrically stimulated 87% of the expenditure was described to be anaerobic. This analysis in no way denies the importance of anaerobic metabolism in lunged amphibians during activity, for even Hillman and coworkers concluded that anaerobic metabolism, depending on the species, may directly contribute from 21% to 64% of the energy used during activity (Table 9.1).

A conclusion of Bennett (1978) remains: great diversity in the relative importance of aerobic and anaerobic metabolism during activity exists among amphibians (Table 9.1). A high maximal consumption of oxygen during activity is correlated with a large cardiac stroke volume (as measured by ventricular mass) and with a large oxygen store (due to either a high corpuscular hemoglobin concentration or a high hematocrit) (Hillman 1976). The circulatory rate only doubled when oxygen consumption increased by 10-fold during activity. The remaining oxygen used during activity was drawn from blood stores, and the high rates of oxygen consumption after activity permitted the replenishment of these stores. Terrestrial anurans had higher aerobic scopes and higher oxygen stores in blood than did aquatic species. This difference permitted terrestrial species to increase foraging time, either by increasing the time to fatigue or by shortening the period required to retire the oxygen debt. On the other hand, lungless salamanders face a limited capacity to increase oxygen consumption (Figure 9.5), and may have a marked dependence on anaerobic metabolism during high levels of activity (Table 9.1). Activity in lungless salamanders, at least in *Plethodon jordani*, does not appear to be limited by oxygen availability at low-velocity movement, although it may limit aerobic scope and thereby limit maximal aerobic velocities (Full 1986).

Taigen et al. (1982) reexamined the energetics of activity in anurans. They showed that aerobic and anaerobic scopes are independent of each other and are independent of body size, habitat, and phylogeny. The dependence of anurans on aerobic metabolism is high in burrowers and in species that are active foragers, but low in jumpers. Burrowing and active foraging are usually found in the same species, so it is not presently possible to determine if one factor is responsible for the use of aerobiosis. In any case, burrowing anurans (toads) actively forage for extended periods, depend primarily on aerobic metabolism to power activity, move slowly, and seek protection in burrows, in a large body size, or by the use of chemical defenses. Hopping anurans (frogs) tend to be sit-and-wait predators and move rapidly for short periods to avoid predation, when they depend mainly on anaerobic metabolism.

The use of anaerobiosis as a supplement to aerobic metabolism suggests that a threshold in intensity of activity exists above which lactate formation occurs. Taigen and Beuchat (1984) showed that such is the case in anurans (Figure 9.8). This threshold occurs at rates of oxygen consumption

Table 9.1 Aerobic and anaerobic scopes in amphibians and reptiles

Species	T_a (°C)	Aerobic scope* mM ATP/g·h	%	Anaerobic scope* mM ATP/g·h	%
Amphibians					
Caudates					
Batrachoseps attenuatus	23	0.22	36	0.39	64
Plethodon jordani	15	0.03	24	0.11	76
	25	0.04	34	0.08	66
Anurans					
Hyla cadaverina	23	0.34	79	0.09	21
Hyla regilla	23	0.32	61	0.20	39
Hyla chrysoscelis	20	0.25	59	0.17	41
Hyla versicolor	20	0.27	71	0.11	29
Rana pipiens	23	0.17	78	0.05	22
	25	0.13	41	0.18	59
	26	0.21	55	0.17	45
Rana sylvatica	20	0.19	53	0.17	47
Xenopus laevis	26	0.35	62	0.21	38
Bufo americanus	20	0.29	73	0.11	27
Bufo boreas	26	0.36	79	0.10	21
Reptiles					
Turtles					
Pseudemys scripta	30	0.18	39	0.28	61
Terrapene ornata	30	0.10	25	0.30	75
	40	0.19	45	0.23	55
Lizards					
Dipsosaurus dorsalis	25	0.11	17	0.56	83
	35	0.44	40	0.66	60
	40	0.66	42	0.91	58
Sceloporus occidentalis	30	0.32	27	0.96	73
	35	0.38	33	0.84	67
	40	0.36	38	0.65	62
Amblyrhynchus cristatus	25	0.05	9	0.50	91
	30	0.10	15	0.55	85
	35	0.24	29	0.61	71
Iguana iguana	35		36		64
	35		33		67
Trogonophis wiegmanni	25	0.15	26	0.42	74
Ophiosaurus ventralis	25	0.36	27	0.96	73
Annellia pulchra	25	0.12	22	0.43	78
Snakes					
Lichanura roseofusca	32	0.06	48	0.06	52
Coluber constrictor/ Masticophis flagellum	35	0.28	44	0.35	56
Thamnophis butleri	25	0.12	23	0.40	77
Crotalus viridis	35	0.13	40	0.19	60

*Under the assumption that the ATP equivalencies for aerobic and anaerobic scopes are 0.29 mM/cm³O₂ and 0.0167 mM/mg of lactate, respectively (Bennett and Licht 1972).

Sources: Data derived from Bennett and Dawson (1972), Bennett and Licht (1972), Gatten (1974a), Bennett et al. (1975), Bennett and Gleeson (1976), Ruben (1976a), Hillman et al. (1979b), Hillman and Withers (1979), Taigen and Beuchat (1984), Kamel and Gatten (1983), Kamel et al. (1985), Gatten (1987), and Stefanski et al. (1989).

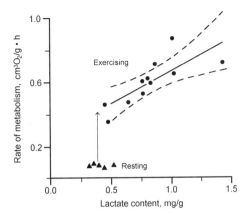

Figure 9.8 Rate of metabolism in *Rana sylvatica* in relation to lactate production. Lactate production increases above the resting value only after a threshold in intensity of activity, as measured by an increase in rate of metabolism, is attained. Source: Modified from Taigen and Beuchat (1984).

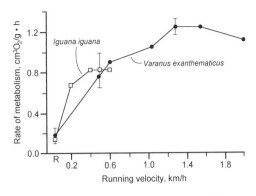

Figure 9.9 Rate of metabolism in two lizards, *Iguana iguana* and *Varanus exanthematicus*, as a function of running velocity. Source: Modified from Gleeson et al. (1980).

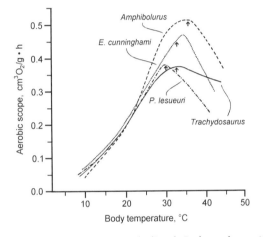

Figure 9.10 Aerobic scope in the lizards *Podarcus lesueuri*, *Eremias cunninghami*, *Trachydosaurus rugosa*, and *Amphibolurus barbatus* as a function of body temperature. The preferred body temperature for each species is indicated by an arrow. Source: Modified from Wilson (1974).

that vary from 45% to 63% of the maximal rates of oxygen consumption in the anurans studied. Yet, in some amphibians the use of anaerobic metabolism occurs at the beginning of activity before the increase in gas exchange is mobilized (Figure 9.7). The picture that we have of the energetics of amphibian activity clearly is incomplete and pertains almost exclusively to anurans.

9.5 LOCOMOTION IN REPTILES

Relatively few data are available on the comparative energetics of locomotion in reptiles. They indicate that rate of oxygen consumption increases with the velocity of locomotion to a maximum (Figure 9.9), as demonstrated in *Iguana* (Moberly 1968), *Varanus* (Gleeson et al. 1980), and *Tupinambis* (Bennett and John-Alder 1984). Unlike mammalian locomotor energetics, the scaling of locomotor energetics in reptiles essentially has not been examined, with the exception of the work by Bakker (1972b), who demonstrated in five species of lizards that the increment of metabolism with velocity is equal to or less than that of mammals of the same mass (see Section 9.9). Bennett and John-Alder (1984) showed that the cost of locomotion in lizards also increases with body temperature. Most studies of reptilian locomotion, however, have concerned its biochemical bases.

9.5.1 Aerobic expenditures. The increase in rate of oxygen consumption with locomotion in reptiles

is limited (Figure 9.9). Aerobic scope is a measure of the maximal increase in oxygen consumption with activity and can be defined either as a ratio (>20 in some lizards), wherein the maximal rate of oxygen consumption is *divided* by the minimal (or resting) rate, or as a differential, wherein the minimal rate is *subtracted* from the maximal rate. However aerobic scope is defined, *Varanus* has a larger aerobic scope than *Iguana*.

Aerobic scope in reptiles reaches a maximum near the preferred body temperature (Moberly 1968, Bennett et al. 1975, Bennett and Gleeson 1976; Figure 9.10). The reduction in aerobic scope at body temperatures above the preferred level stems from an increase in the resting rate, whereas the maximal rate remains constant, or declines (Wilson 1974). The correlation of maximal aerobic scope with preferred body temperature does not imply that all reptiles operate at a preferred body temperature in the field. Thus, the marine iguana (*Amblyrhynchus*) operates at two distinct body temperatures: its aerobic scope is maximal at 35°C, which is close to its preferred body temperature on land, but far above its temperature when foraging in the sea (Dawson et al. 1977). The marine iguana simply tolerates oceanic and body temperatures of about 25°C while grazing on algae (Bartholomew 1966). The aerobic scope of natricine snakes was less temperature sensitive in *Natrix maura*, which is nocturnal and semiaquatic, and in higher-latitude populations of *N. natrix*, a pattern that diminishes the effect of temperature on activity compared to low-latitude populations (Hailey and Davies 1986).

Thermoregulation by lizards in the field ensures that the available aerobic scope is usually greater than 85% of maximal scope, which means that they can attain aerobic walking velocities at least 75% to 80% of their maximal aerobic velocities (Wilson 1974). The rate and duration of activity in *Amphibolurus barbatus* are independent of body temperature in the field as long as the body temperature is in the preferred range, which includes 94% of all observed measurements during the day (Lee and Badham 1963). Body temperatures are low at emergence in the morning, when the aerobic scope is only 40% to 53% of the maximum; at this time, the aerobically based walking velocities are only 25% to 35% of the maximum (Wilson 1974). Then emergencies are met with anaerobic expenditures.

All lizards do not maximize aerobic scope in the same manner. At rest and during modest activity, *Iguana* has a high cardiac output and a small arterial-venous differential in oxygen content, whereas *Varanus* has a low cardiac output and a large differential (Gleeson et al. 1980). The product of cardiac output and the oxygen differential equals rate of oxygen consumption, which is equal in *Iguana* and *Varanus* at rest and at velocities up to 0.5 km/h. With a further increase in velocity, *Varanus* increases its cardiac output to the maximum found in *Iguana*. As a consequence of the larger arterial-venous differential in oxygen content, *Varanus* attains higher levels of oxygen consumption at peak activity (Figure 9.9). About 80% of the increase in cardiac output associated with peak activity in *Varanus* is due to an increase in heart rate, whereas the remaining 20% is due to an increase in stroke volume.

9.5.2 Energetics of legless locomotion.

A legless condition has evolved independently in several groups of lizards, amphisbaenians, and snakes. Gans (1975) argued that the evolution of limblessness, usually in association with body elongation, was associated with entrance into crevices and burrowing habits. Shine (1986b) supported this view in a comparison of the Australian/New Guinean lizard family Pygopodidae, which has no front limbs and only remnants of hind limbs, with their close relatives, the Geckonidae. Less obvious is the view that leglessness in some reptiles is an adaptation to cold climates (Shine 1985).

The question arises as to whether the evolution of leglessness has had an impact on the energetics of locomotion. A potential complication is found in the observation that most legless lizards and amphisbaenians are fossorial, whereas many snakes are not. For example, in a study of the fossorial, legless lizards *Anniella pulchra* and *Ophiosaurus ventralis*, the fossorial amphisbaenian *Trogonophis wiegmanni*, and the (nonfossorial) snake *Thamnophis butleri*, the three fossorial species showed low resting rates of metabolism (Kamel and Gatten 1983). During activity, however, *A. pulchra*, *T. wiegmanni*, and *Th. butleri* had aerobic and anaerobic scopes similar to those found in legged reptiles of similar mass. *Ophiosaurus ventralis*, in contrast, had a very high scope for activity. Limblessness thus is not correlated with unusual levels of activity metabolism, although fossoriality may be associated with low resting rates of metabolism.

In a comparison of the energetics of locomotion in four species of snakes, Ruben (1976a, 1976b) found no difference in resting rate of metabolism but a great difference in maximal rate: active snakes belonging to the genera *Coluber* and *Masticophis* had aerobic scopes twice that of the low-active *Crotalus* and 5 times that of the sedentary rosy boa (*Lichanura*). These differences in aerobic scope are correlated functionally with the pulmonary area for gas exchange and the amount of myoglobin present in muscles. Thus, the saccular lung is 40% to 45% of the total length of *Coluber* and *Crotalus*, whereas it is only 11% of the length of *Lichanura*. In addition, the number of air cells per cross section of lung is much greater in *Coluber* than in *Crotalus*. Consequently, the area for gas exchange is much greater in *Coluber* (and presumably in *Masticophis*) than it is in *Crotalus*, in which the area is much greater than it is in *Lichanura*. Furthermore, myoglobin is 4.9 times as concentrated in the skeletal muscles of *Coluber* as it is in *Lichanura*, a ratio that is similar to the ratio of aerobic scopes (i.e., $5.0:1$). High concentrations of myoglobin may not simply be a passive store for oxygen but may facilitate oxygen transfer to muscle tissue. Ruben (1977) further showed that constrictors like *Lichanura* have muscle units that extend over 20 to 30 vertebrae, whereas the coachwhip *Masticophis* has muscle units that extend over only 9 vertebrae. As a result, *Masticophis* expends little energy to overcome lateral muscle resistance compared to *Lichanura*. Constriction and high velocity movement may be functionally incompatible.

9.5.3 Anaerobic expenditures.
Bennett and Licht (1972) summarized the use of anaerobiosis by lizards during activity. Lizards have low total body levels of lactate during rest; with activity, especially during the first 30 s, rate of lactate formation greatly increases. The main value of anaerobiosis, other than that it does not require high rates of gas exchange and effective circulation, is that it has a low thermal dependence (Q_{10} is about 1.2). It permits the marine iguana to be active at cool marine and body temperatures (Bartholomew et al. 1976, Dawson et al. 1977). Small lizards would be expected to rely preferentially on anaerobiosis during activity and large lizards on aerobiosis, if an important determinant is the stability of body temperature.

Lactate produced by anaerobic metabolism persists for 30 to 60 min after activity, the time for complete recovery being shorter at higher temperatures. The period of recovery depends on body temperature (Q_{10} is about 3; Hailey et al. 1987); the shortest period may be at, or near, the preferred body temperature (Moberly 1968). Lactate persists longer than the oxygen debt, which suggests that some of the debt is not related directly to lactate metabolism but rather to the establishment of oxygen and creatine phosphate stores. The principal disadvantage of anaerobic metabolism—that the accumulation of lactate disrupts various physiological functions through a change in pH—is outweighed by its potential for supplying energy rapidly for short periods at nearly all temperatures.

The capacity for anaerobic metabolism undoubtedly reflects the underlying biochemical pathways. Ruben (1976b) showed that the snake genera *Coluber*, *Crotalus*, and *Lichanura* had concentrations in muscle tissue of the enzymes phosphofructose kinase and lactic acid dehydrogenase, both of which catalyze rate-limiting steps in glycolysis, in the proportions $4.9:1.3:1.0$, respectively. He measured anaerobic scopes in these snakes and found that they varied in the following order: *Coluber* (5.8), *Crotalus* (3.2), *Lichanura* (1.0). The similarity of these ratios suggests that *Lichanura* has a small anaerobic scope *because* it has rate-limiting enzymes in low concentrations, but this answer begs the question as to why these enzymes are found in low concentrations. They may simply be the means by which anaerobiosis is reduced, and not the "cause" in itself. Most "ultimate" causes of physiological limits, upon close inspection, are likely to have an ecological basis.

9.5.4 Total energetics.
All reptiles use a mixture of aerobic and anaerobic metabolism during activity. Aerobic and anaerobic scopes are apparently positively correlated in lizards (Bennett 1978). This pattern is also seen in snakes (Table 9.1): active snakes, such as *Coluber* and *Masticophis*, have high aerobic and anaerobic scopes; the sluggish rosy boa (*Lichanura*) has small scopes; and the intermediate *Crotalus* is intermediate in both scopes. That is, *Coluber* and *Masticophis* rapidly pursue their prey and flee predators, whereas *Lichanura* moves slowly, is a constrictor, and voids noxious secretions from the cloaca when disturbed. *Crotalus*, the rattlesnake, is intermediate in both its defensive and predatory behaviors. (This pattern is similar to the one seen in anurans.)

The energy used by reptiles during intermediate to intense activity is estimated to be 50% to 90% derived from anaerobic pathways (Moberly 1968, Bennett and Dawson 1972, Bennett and Licht 1972, Bennett and Gleeson 1976, Ruben 1976a). These estimates suffer from many difficulties. One is that the choice of biochemical pathways is sensitive to the intensity of activity and to the temperature at which it occurs, so the ratio of aerobic to anaerobic metabolism is not constant. For example, *Iguana* supplements aerobic metabolism with anaerobiosis at lower velocities compared to *Varanus*, because of the limited aerobic scope of *Iguana* (Gleeson et al. 1980). This difference reflects the sedentary, herbivorous habits of *Iguana* and the pursuing predatory habits of *Varanus*. The balance between aerobic and anaerobic metabolism also appears to reflect body temperature: aerobiosis generally makes a greater contribution at higher body temperatures (Table 9.1). A nagging question, originally raised for amphibians and not yet addressed in reptiles, is whether the technique of measurement during activity affected the aerobic scope and therefore the perceived balance between aerobic and anaerobic metabolism during activity. For example, some studies of "activity" concerned lizards that were tied to a platform—this "activity" in reality was struggling. Treadmill studies would appear to be the best way of obtaining quantifiable degrees of activity.

As Wilson (1974) clearly stated, long-term field studies are needed to determine whether most activity by reptiles is aerobic or anaerobic, which can be partly judged from the velocities at which they spontaneously move. Although all reptiles depend on anaerobic metabolism when running near their maximal velocities, they actually may run at these velocities only rarely (e.g., when evading herpetologists, fortunately a rare event). The few measurements of field lactate levels in lizards indicate that they increase with territorial defense (Bennett et al. 1981, Pough and Andrews 1985), routine activity, distance moved, and the size of prey captured (Pough and Andrews 1985). Furthermore, the type of metabolism during activity may depend more on aerobiosis at high body temperatures because of the high rates of blood circulation and oxygen transport, whereas anaerobiosis may be much more important at low body temperatures (Hailey et al. 1987).

9.6 FLIGHT

The analysis of the energetics of flight has proceeded further than for any other means of locomotion because of its application to aviation.

9.6.1 Theoretical energetics. A fundamental relation in the mechanics of gliding flight is

$$L = 0.5\rho \cdot C_l \cdot \dot{v}^2 \cdot A,$$

which can be rearranged as

$$\dot{v} = \sqrt{\frac{2m \cdot g}{\rho \cdot C_l \cdot A}}, \quad (9.1)$$

where L (in newtons, or N) is the lift force (which equals the weight [w = mass × gravity = $1000\,g$ × $9.8\,m/s^2$] of the flier, m is mass (g), $\dot{v}$ is velocity of flight (m/s), C_l is the dimensionless lift coefficient, ρ is air density (g/cm^3), and A (cm^2) is wing surface area (Lighthill 1977). Equation 9.1 describes the gliding velocity relative to the aerodynamically important factors of wing loading (m/A) and aspect ratio (wing span^2/A). Wing loading is explicitly included in Equation 9.1 and the aspect ratio is proportional to C_l^2. Thus, gliding velocity increases with mass and wing loading (Greenewalt 1975, Brower and Veinus 1981) but decreases with aspect ratio. A glider in still air sinks with a velocity $\dot{z}$ (m/s), which is given by:

$$\dot{z} = \left(\frac{D}{L}\right)\dot{v}, \quad (9.2)$$

where D (in newtons, or N) is the drag force.

The minimal power requirements ($\dot{P}$, watts, or W = J/s) for horizontal flight can be estimated from the relation

$$\dot{P} = D \cdot \dot{v}, \quad (9.3)$$

which, because $L = m \cdot g$, can be rewritten

$$\dot{P} = \left(\frac{D}{L}\right)m \cdot g \cdot \dot{v}, \quad $$

or

$$\dot{P} = \dot{z} \cdot m \cdot g. \quad (9.4)$$

Equation 9.4 is a first approximation of the minimal expenditure for steady-state, horizontal flight but does not take acceleration or ascending flight into consideration, and it does not consider factors other than aerodynamic drag.

Pennycuick (1969) performed the first comprehensive analysis of the cost of flapping flight in vertebrates. Flapping flight simultaneously produces lift and forward thrust, the total power output consisting of the sum of three components: (1) induced power, (2) parasite power, and (3) profile power. Induced power is the rate at which energy is expended to overcome weight. It is produced by the

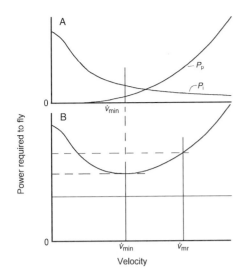

Figure 9.11 Power output during flight for (A) induced power (P_i) and parasite power (P_p), and (B) total power output, which is the sum of P_i, P_p, and profile power, P_o. Also indicated is $\dot{v}_{min}$, the velocity at which the cost of flight is minimal, and $\dot{v}_{mr}$, the velocity at which the maximal flight range is attained. Source: Modified from Pennycuick (1969).

flapping of wings, which accelerates air downward, thereby producing an upward force on the wings. Induced power ($\dot{P}_i$) is given by

$$\dot{P}_i = \frac{K_1 w^2}{s_d \cdot \dot{v}}, \qquad (9.5)$$

where K_1 is a coefficient and s_d, the area swept by the wings. At moderate to high velocities, induced power decreases with an increase in velocity (Figure 9.11A), but as velocity approaches zero (i.e., as a flier approaches hovering flight), induced power becomes exceedingly high. Parasite power ($\dot{P}_p$) is the output needed to overcome the drag of the body, that is, when $D = L$. It is the relationship incorporated into Equation 9.3, which can be rewritten as

$$\dot{P}_p = D_b \cdot \dot{v} = K_2 \cdot A \cdot \dot{v}^3, \qquad (9.6)$$

where D_b is body drag; K_2, a coefficient; and A, the flat-plate equivalent area of a flier. Parasite power increases steeply with an increase in velocity (Figure 9.11A) because it is proportional to $\dot{v}^3$. Finally, profile power ($\dot{P}_o$) is the power output required to overcome the drag of the wings; it is highly complicated because of variations in the shape of wings during flight. Pennycuick suggested

that profile power is approximately twice the minimal expenditure for flight calculated from the sum $\dot{P}_i + \dot{P}_p$.

Total power output ($\dot{P}_t$) for flight according to Pennycuick therefore equals

$$\dot{P}_t = \frac{K_1 \cdot w^2}{s_d \cdot \dot{v}} + K_2 A \cdot \dot{v}^3 + 2(\dot{P}_i + \dot{P}_p)_{min}. \qquad (9.7)$$

Equation 9.7, when represented by a graph, has the shape of a parabola with respect to the velocity of flight (Figure 9.11B), which means that at the velocity $\dot{v}_{min}$ the cost of flight is minimal ($\dot{P}_{min}$). The velocity that permits the maximal flight range for a given energy expenditure ($\dot{v}_{mr}$), however, is greater than $\dot{v}_{min}$ because the power output curve above $\dot{v}_{min}$ is rather flat with respect to velocity (Figure 9.11B). If this curve were flat, $\dot{v}_{mr}$ would occur at the highest velocity possible. Greenewalt (1975) suggested that $\dot{v}_{mr}$ is 1.32 times $\dot{v}_{min}$, but Pennycuick maintained that this ratio is a minimal difference, the ratio in some cases equaling 2.0.

Tucker (1973) and Greenewalt (1975) proposed models for the energy expenditure required for flight similar to that of Pennycuick. Tucker argued that Pennycuick's model did not accurately account for the influence of body mass. He incorporated the influence of Reynold's number on parasite and profile powers, and of wing span on induced power, and he considered the cost of internal maintenance. Greenewalt also included the influence of Reynold's number and wing span, and added the effect of aspect ratio, but neglected the contribution made by profile power. In spite of these differences, all three authors are in substantial agreement that the power-velocity curve for flight is a parabola produced by the sum of at least two terms, one (induced power) that controls power output at low velocities and shows an inverse correlation with velocity, and the other (parasite power) that controls power output at high velocities as the result of being proportional to the cube power of velocity.

Rayner (1979) described another model of the energetics of horizontal flight, which unlike the models of Pennycuick, Tucker, and Greenewalt was derived from an analysis of the vortices present in the wake of flight. Rayner argued that such a model would be free of the earlier, erroneous assumptions that wings operate under steady-state conditions, that lift and thrust are generated separately, and that the downward wake left by a bird's forward velocity can be ignored. Unfortunately, the size, strength, and geometry of the vortices are complicated and cannot be simplified easily. As in earlier models, the power output for flight is expressed as a sum of induced, parasitic, and profile powers. Fortunately, the estimates and the U-shaped pattern of expenditures in relation to velocity derived from this model are similar to those obtained from the earlier models, and his graphical representation of the expenditures (Rayner 1988) is similar to that described by Greenewalt. Rayner explored some of the consequences of variations in avian morphology, especially stroke period, stroke amplitude, and wing span, for the energetics of flight. He concluded that slow flight at a minimal expenditure is facilitated in birds by a low stroke period and a large wing span, whereas fast flight is facilitated by a small wing span and high aspect ratio.

9.6.2 Measurement of flight energetics.

The near impossibility to measure the power output *for* flight in vertebrates means that the models describing the energetics of flight are difficult to evaluate. What in fact is measured is the expenditure *during* flight. The relationship between the power output for flight and the expenditure during flight is in part the efficiency by which energy expenditure is converted into mechanical work, under the erroneous assumption that no work other than flight is occurring during flight. If this efficiency is only 20% to 25%, as appears to be the case for individual muscles, a calculated expenditure must be multiplied by 4 or 5 to estimate total expenditure, assuming that efficiency is not a function of velocity. Such a huge "correction" negates the ability to apply the intellectual rigor incorporated into Equation 9.7 with precision (also see Rayner 1988). In an attempt to address this problem directly, Masman and Klaassen (1987) measured the energy expenditure of Eurasian kestrels (*Falco tinnunculus*) in the field by using doubly labeled water; calculated the cost of flight estimated from the equations of Pennycuick, Greenewalt, and Tucker; and concluded that the muscular efficiency was approximately 15%.

Using a wind tunnel, Tucker (1966b, 1968b) obtained the first effective measurements of power output during flight, in the Australian budgerigar (*Melopsittacus undulatus*). Tucker concluded that his measurements agreed rather well with calculations based on his equations, which incorporated a 20% efficiency and maintenance metabolism (Figure 9.12). As expected, the cost of flight decreases when descending and increases when

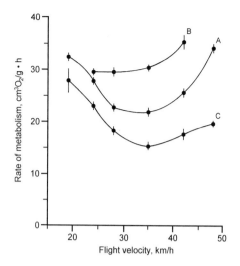

Figure 9.12 Rate of metabolism in flying budgerigars (*Melopsittacus undulatus*) during (A) level flight, (B) ascending flight, and (C) descending flight. Source: Modified from Tucker (1968b).

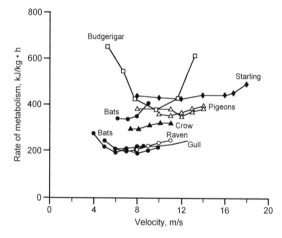

Figure 9.13 Rate of metabolism of birds and bats during flight as a function of velocity. Source: Modified from Ellington (1991).

ascending. Power estimates from the models are less sensitive to velocity than are the measurements in the budgy. In other species, however, measurements of rate of metabolism in flight show less dependence on velocity than expected: in most birds and bats the curves representing these measurements are relatively flat (Figure 9.13). The difference may be due to weaknesses in theory, changes in attitude and wing-beat amplitude in flight (Torre-Bueno and Larochelle 1978), or a trade-off between lift and drag (Withers 1981). The unusually high dependence of power output on flight velocity in the budgerigar may reflect the use

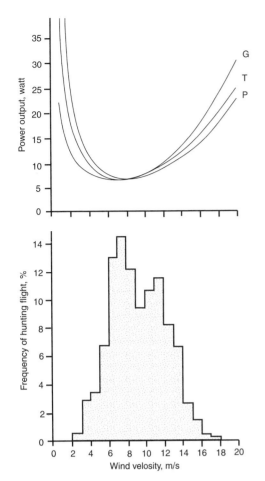

Figure 9.14 Frequency distribution of wind velocities at which Eurasian kestrels (*Falco tinnunculus*) hovered compared to the total cost of flight estimated by the Pennycuick (P; 1969), Tucker (T; 1973), and Greenewalt (G; 1975) models. Source: Modified from Masman and Klaassen (1987).

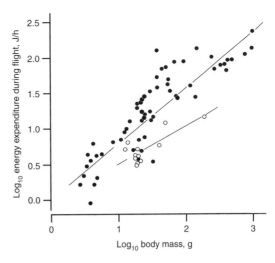

Figure 9.15 Log_{10} mass-specific rate of metabolism of birds during flight as a function of log_{10} body mass. Hollow symbols are for swallows and a swift (*Apus apus*); solid symbols are for other birds. Source: Modified from Hails (1979).

of hooded birds (Greenewalt 1975), or the absence of a marked dependence of power on velocity in other species is related to a larger body mass: notice in Figure 9.13 that larger birds and bats usually are represented by flatter curves than are smaller species (Thomas 1975, Ellington 1991). Whatever the cause of these flat curves, they have a clear consequence: the energy expenditure during flight decreases with an increase in flight velocity because under these circumstances the principal determinant of the expenditure is the time spent in flight. If the observed connection between curve shape and body size is correct, large species fly long distances more economically than do small species.

The theoretical estimates for the cost of avian flight, however, do coincide with field observations in at least one way (Masman and Klaassen 1987):

the Eurasian kestrel often hovers while searching for prey but does so only against a wind, which energetically is equivalent to flying horizontally at the velocity of the wind. The frequency distribution of the wind velocities in the field at which a kestrel hovered had a nearly normal distribution, with the mode occurring at a velocity that corresponded to the lowest cost of flight as estimated by the models of Tucker, Greenewalt, and Pennycuick (Figure 9.14), although with much variance. Masman and Klaassen cautioned that the measurements of energy expenditure made in wind tunnels that involved training birds to use a hood to collect respiratory gases gave values for expenditure that averaged 50% greater than those obtained by other means (especially $D_2{}^{18}O$).

Direct evidence demonstrates that the morphological characteristics of birds influence energy expenditure in flight. For example, Utter and Le Febvre (1970), Hails (1979), and Flint and Nagy (1984) showed that the energy expenditures of swallows, swifts, and the sooty tern are only one-half to one-fourth those of other birds of the same mass (Figure 9.15), principally because they have large aspect ratios (Figure 9.16) and spend a large proportion of the day in flight (Dol'nik 1982, Masman and Klaassen 1987). These "aerial species" are characterized by long wings (increasing the area swept by the wings, thereby reducing induced power), reduced cross-sectional area associated with streamline contouring (reducing para-

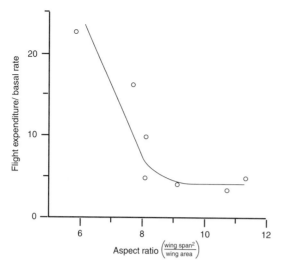

Figure 9.16 Energy expenditure of birds during flight, expressed relative to the measured basal rate, as a function of aspect ratio (wing span²/wing area). Source: Data from Masman and Klaassen (1987).

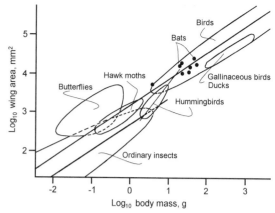

Figure 9.17 Log_{10} wing area in butterflies, hawk moths, "ordinary" insects, hummingbirds, bats, ducks and gallinaceous birds, and other birds as a function of log_{10} body mass. Source: Modified from Kokshaysky (1977).

site power), and an increased aspect ratio (reducing profile drag). Masman and Klaassen gave a revised equation to predict the energy expenditure in flight by including body mass, wing span, and wind area. They noted that aerial feeders, such as swallows, have a low flight cost in part because they mix powered flight with gliding.

Pennycuick (1972) estimated that the incremental cost of gliding over resting in *Gyps africanus* is only about 50% of the basal rate. Baudinette and Schmidt-Nielsen (1974) measured the rate of oxygen consumption in herring gulls (*Larus argentatus*) gliding in a wind tunnel. When gliding, this gull expends energy at a rate that is twice the resting rate (i.e., a 100% increase). The cost of flapping flight in the laughing gull (*L. atricilla*), which is much smaller than the herring gull, is about 15 times the predicted standard rate (Tucker 1972), or approximately 9 times the resting rate (assuming that the resting rate is 1.7 times the standard rate, as in the herring gull). In gulls, then, gliding is one-fourth to one-fifth as expensive as flapping flight. This difference may be even greater in soaring species (e.g., Pennycuick [1972] estimated soaring to be 1/23rd of the cost of flapping flight in the white stork [*Ciconia ciconia*]). Body size appears to affect the increment by which power increases during gliding.

9.6.3 The flight of birds.
A fundamental parameter of flight in birds is velocity, which (as noted)

is influenced by body mass, wing area, and wing span (Equation 9.1). In general, $\dot{v}_{min}$ increases with mass: small species fly at low velocities and large birds that are dimensionally similar fly at higher velocities. Many large birds, however, increase maneuverability by having large wing areas and wing spans, both of which reduce $\dot{v}_{min}$. This modification is especially important in species that live in closed habitats, that use pinpoint landing, or that carry a heavy load (e.g., hawks, vultures). Such species often have blunt wing tips with well-developed slots, the slots being used to distribute lift and thrust (Kokshaysky 1973) and to reduce induced drag (Rayner 1988). In open habitats, birds such as shorebirds and ducks have pointed wings, make fewer adjustments of dimensional relations, require less maneuverability, and fly at high velocities. Most variation in flight speed thus is related to body size, the means of feeding, the need to carry a load, habitat, and the necessity for precision landing, but not to long-distance flight (Greenewalt 1975).

Some of the variation among birds in the velocity of flight is related to wing area and wing-beat frequency (Kokshaysky 1977, Rayner 1988). Wing area increases with marked residual variation with body mass (Figure 9.17). Small wing areas require swift, direct flight, whereas large areas permit slow, erratic flight. For example, the large wing surface areas of butterflies are associated with low thoracic temperatures, low wing-beat frequencies, and erratic flight. This prevents most birds from feeding on butterflies, with the exception of some long-tailed flycatchers (e.g., paradise flycatcher [*Terp-*

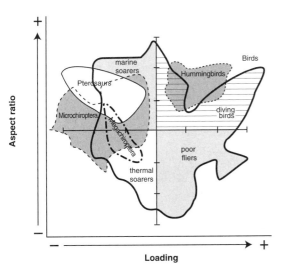

Figure 9.18 Aspect ratio as a function of wing loading, both factors normalized for body size, in birds generally, hummingbirds, microchiropterans, megachiropterans, and pterosaurs. Sources: Modified from Rayner (1988) and Hazlehurst and Rayner (1992).

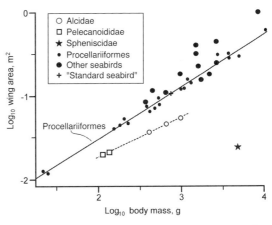

Figure 9.19 $\log_{10}$ wing area of members of Procellariiformes, alcids, diving-petrels, and a penguin as a function of $\log_{10}$ body mass. Source: Modified from Pennycuick (1987).

sipone paradisi] from Asia), which swing their tails to change flight direction sharply (Kokshaysky 1977). Gallinaceous birds and tinamous, in contrast, have small wing areas and high wing-beat frequencies and specialize in swift departures but have low flight ranges. Power output for flight also correlates with wing-beat frequency (see Rayner 1979, 1988): the minimal output during flight of a crow is only about 76% of the expenditure of a similarly sized laughing gull, and the crow has a wing-beat frequency that is about 77% of the frequency of the gull (Bernstein et al. 1973).

Rayner (1988) graphically analyzed bird flight in terms of wing loading and aspect ratio: when these two parameters, normalized for body size, are plotted against each other, the plane is broken into four segments (Figure 9.18). One represents birds with high wing loading and high aspect ratios (i.e., diving birds); another, those with low loads and high aspect ratios (marine soarers); a third, those with low loads and low aspect ratios (thermal soarers); and the last segment, those with high loads and low aspect ratios ("poor fliers"). Larger aerial predators, such as nightjars, kites, skuas, and falcons, tend to have low loading coupled with an intermediate aspect ratio, whereas smaller aerial predators, such as swallows and swifts, have higher aspect ratios. Forest-dwelling birds often are characterized by intermediate loading and aspect ratios, a response to living in a "cluttered" environment.

Rayner concluded that flight morphology represents a compromise among the various factors that influence flight performance and that unrelated species having a broadly similar flight morphology group together; that is, behavior and ecology are more important than phylogeny. Although Rayner (p. 26) maintained that "... scaling is unreliable and of limited validity ... for predicting the performance of an individual bird [and species?]," body size accounts for 97.6% of the variation in wing morphology. He (p. 31) went further to say that "... the use of allometry carries the implication that outlying species are maladapted." Not so: scaling has a value as a standard against which the characteristics of individuals and species can be compared and evaluated (see Chapter 3). Distinctive habits often call for distinctive solutions.

The impact of scaling is seen in an analysis of seabird flight (Pennycuick 1987). Pennycuick defined a "standard seabird" that weighted 700 g, with a wing span of 1.09 m, a wing area of 0.103 m², and a wing loading of 66.6 N/m². Most other seabirds, especially those belonging to the order Procellariiformes, from 20-g petrels, through shearwaters, to albatrosses that weighed up to 9 kg, conform to a geometrically similar series that differ principally in mass (Figure 9.19). This pattern is independent of the feeding methods used, with the exception of the Southern Hemisphere diving-petrels (Pelecanoididae). Diving-petrels, unlike other birds in the order Procellariiformes, use their wings for underwater propulsion, which leads to a reduction in wing span and area associated with the high density of water (>800 times that of air). Diving-petrels have wing areas similar to those in the Northern Hemisphere alcids (Figure 9.19), which also use their wings for underwater propulsion. Wing areas in these groups are intermediate to those in other procellariiform birds and in penguins, which are committed to aquatic propulsion by their wings. The wing area of diving-petrels and auks is a compromise determined by the demands of "flight" in air and in water. Thus, the razorbill (*Alca torda*), an alcid, has a wing span that is 60% that of a "standard" procellariiform bird of the same mass, which according to Equation 9.1 means that its velocity of flight is 29% greater and would require 2.1 times (Equation 9.7) the power output for flight, compared to Procellariiformes. The increase in power output can be accomplished by an increase in wing-beat frequency: one of the distinctive field characters of auks and puffins is a high wing-beat frequency. In theory, the macaroni penguin (*Eudyptes chrysolophus*) could fly in spite of its small wing area and high wing loading, but it would have to have a cruising velocity of 40 m/s (= 144 km/h!) by increasing its wing-beat frequency to 50 Hz (= 3000 cycles/min!), that is, equivalent to that of a hummingbird. Unlikely.

Various factors have been claimed to set the maximal flight velocity. One view is that it is determined by the intersection of the power-velocity curve and the maximal rate of metabolism, although some birds show no increase in rate of metabolism during flight near their maximal flight velocities (Figure 9.13; Greenewalt 1975). Based on observations of starlings, Torre-Bueno and Larochelle (1978) suggested that the maximal velocity may be set by a limit to an increase in wing-beat amplitude or a limit to wing-beat frequency, or a limit to both.

Measurements of bird flight velocity in the field (Schnell and Hellack 1979) indicated that the mean velocities of gulls and terns are a compromise between those predicted by a minimal expenditure and those predicted by a minimal cost of transport. These results may have been influenced by observations of flight near a nesting colony and by the relatively flat shape of many power-velocity curves, wherein an appreciable increase in velocity is attained with a relatively small increase in power output. In contrast, ducks that breed in the high arctic have smaller, more pointed wings and higher wing loadings than do warm-temperate or tropical species (Rayner 1988). This pattern, which has not been found in other birds, was explained as a means of increasing flight velocity, which reduces migratory time and increases the ability of arctic species to avoid inclement weather or to counter head winds.

Observations of bird flight in the field indicate that a complex interaction exists between behavior and aerodynamics:

1. The angle of descent in a glide, especially at high velocities, is increased by lowering the feet, which in *Gyps* can reduce the glide ratio (the ratio of the horizontal distance moved when gliding in still air to the decrease in altitude) from 15.5 : 1 to 10.3 : 1 (Pennycuick 1971b). At low velocities a reduction in wing area (by folding the wings) can additionally increase the angle of descent.

2. Some birds in flight can turn in a small arc. This behavior is especially important for soaring birds, which may need a small arc to remain within a thermal or to use the strong lift near the center of a large thermal. The radius of the smallest arc is proportional to the ratio w/A, where w is weight and A is wing surface area (Pennycuick 1971a). Thus, reducing wing loading reduces the turning radius, although the radius also can be reduced if the banking angle is increased or if forward velocity is reduced. The difficulty with using morphological adjustments to reduce the turning radius is that they also lead to low flight velocities and a low glide ratio, characteristics that are ineffective for movement from one thermal to another. For example, an albatross has the wing loading and aspect ratio that permit it to fly faster and have a lower sinking speed than a vulture, but it also has a 50% greater turning radius, requires higher air velocities to take off, and has poor maneuverability compared to a vulture.

3. The mass that can be carried by a bird in flight is limited because the extra mass requires an increase in power output to fly. This limit is easily seen in vultures, which because of their feast-and-famine regimen, eat as much food as possible when it is available. An empty white-backed vulture (*Gyps africanus*), which weighs about 5.5 kg, may maximally attain only sufficient power for $\dot{v}_{mr}$ (Pennycuick 1969). An individual that had consumed 1.14 kg of carrion, however, could not fly; that is, it could not expend energy at a rate sufficient to attain $\dot{v}_{min}$, even after a running attempt at takeoff. A larger vulture, the Andean condor (*Vultur gryphus*), may weigh 12 kg when empty.

McGahan (1973) concluded that this species is incapable of sustained level flight in still air, even when flying empty and flapping; it needs a head wind to remain aloft. A full crop increases wing loading by about 10%, which further increases forward and sinking velocities. The kori bustard (*Ardeotis kori*), which may also approach 12 kg, can barely fly even without an additional load, in spite of a wing surface area that may exceed that of a similarly sized vulture (see Greenewalt 1975). In contrast to the condor, the bustard has a small crop and feeds nearly continuously in the field. A large crop would require flight to be completely abandoned. Birds larger than 12 kg could potentially fly, especially if they used soaring flight, but takeoff and landing would be difficult (Pennycuick 1969). Windy regions, or cliffs, might be required. (Note that albatrosses at smaller masses [ca. 8–9 kg] have this difficulty, mainly because they have a wing loading twice that of a vulture; they require a head wind to take off.) Some extinct "vultures" (e.g., †*Teratornis* in North and South America) may have weighed 20 kg or more, which surely required soaring flight.

The reduction in wing span described by Pennycuick (1987) as a response to the use of wings for propulsion under water in auks and diving-petrels impacts the ability of these birds to forage for food and to transport food back to dependent young. Storm-petrels (Hydrobatidae) can carry 15% of their body mass as food, whereas puffins, which are alcids, can only carry 2% to 7% of their mass. Consequently, alcids preferentially feed within a few kilometers of their nests and make several rapid feeding trips, which implies that alcids have higher energy expenditures for foraging than do storm-petrels.

4. The distance that birds can fly without resting and refueling is limited because of the limited size of fat deposits. When flying long distances, birds may fly at or near velocities that maximize the distance that can be attained for a given expenditure. This distance varies with the lift-drag ratio, which itself varies with the aspect ratio (Greenewalt 1975), and is proportional to body mass and to the size of the fat store (Pennycuick 1969). Large birds can fly farther because they have larger fat stores and because the cost of flight increases with a power of mass less than 1.0: the specific range (km/g of fat) is inverse to mass at a given lift-drag ratio. Greenewalt (1975) suggested that migratory birds tend to have a 35% larger fat mass than

required for migration; this margin may be greater in small species, such as hummingbirds (e.g., a 2.8-g lean, premigratory ruby-throated hummingbird [*Archilochus colubris*] can carry up to 2.1 g of fat [Odum and Connell 1956]). (Irving [1960] explored the ecological significance of such margins in energy stores in arctic birds.)

The distance over which continuous migration occurs is also influenced by flight strategy and atmospheric conditions. Many large species, such as eagles (2 kg), storks (3–4 kg), and cranes (4 kg), regularly use soaring during migration, whereas most small birds use flapping flight, which is why many small birds accumulate proportionally large fat deposits preparatory to migration. A few large migrants, like geese and swans, do not soar. They may use flapping flight because they have reduced drag due to a reduced cross-sectional area and an increased aspect ratio, and therefore have a reduced cost for flight. Furthermore, the presence of a head or tail wind affects distance flown, especially in small species because they fly at low velocities (Pennycuick 1969). Consequently, small birds often use favorable tail winds at the edge of a storm front to reduce the cost of migration.

Another potential "resource" that might limit flight is water availability, which for a nonstop migrant reflects a balance between the water contained in the body and produced by metabolism and that lost by excretion and evaporation. This balance will be negative during nonstop flight (because no food or water intake occurs), which along with the metabolism of fats will lead to a decrease in mass. Consequently, the velocity and cost of flight will change during migration (Carmi et al. 1992). Carmi et al. (1992) concluded that a small passerine migrating over the Sahara is likely to become dehydrated before it runs out of fuel, which might well require migrating passerines to be selective with regards to the route taken (also see Miller 1963). Klaassen (1995), taking into consideration the effect of a reduction in mass during migration, proposed that the problem posed by Carmi et al. is not as dire as predicted, especially if migration occurs at higher altitudes than assumed. In any case, water balance may be an important factor for small birds migrating across deserts.

5. Some large birds, such as geese, swans, and pelicans, fly in a V-formation in which one bird normally is in the lead and subsequent individuals are staggered sequentially to the side and rear.

Others, like cranes, often fly in "disorganized" V-formations. One explanation for this pattern is that it permits birds to orient with each other and to their destination (Gould and Heppner 1974, Williams et al. 1976, Badgerow 1988), whereas another suggests that the V-formation reduces energy expenditure during flight (Lissaman and Schollenberger 1970; Hainsworth 1987, 1988; Hummel and Beukenberg 1989). A V-formation may permit a following bird to have a lower expenditure that is obtained from lift derived from vortices produced by the preceding bird. To obtain the maximal energy savings, adjacent birds in the formation may have to have a lateral overlap of their wing tips (Lissaman and Schollenberger 1970, Badgerow and Hainsworth 1981, Hainsworth 1987). Pink-footed geese (*Anser brachyrhynchus*) at best saved 2.5% of flight cost (Cutts and Speakman 1994) and greylag geese (*A. anser*), 4.5% to 9% of flight cost (Speakman and Banks 1998), based on analyses of photographs of formations. The potentially larger savings in the larger greylag geese may indicate why V-formations are found only in birds that weigh more than a few kilograms: small species may not be able to benefit from a V-formation because of small vortices, downdrafts, and a requirement that their short wings would have to overlap.

6. Because a flying bird obtains lift from forward motion (Equation 9.5), the cost of flight increases as the velocity of flight decreases. This finding led to the conclusion that hovering flight must be very expensive. Pearson (1950) made the first attempt to measure the cost of hovering, in hummingbirds. He found that in hovering flight Anna's hummingbird (*Calypte anna*) expended energy at 1.36 kJ/g·h, which is 6 times the "resting" rate, or about 17 times the basal rate of metabolism recorded by Lasiewski (1963). Later, Lasiewski (1963), Wolf and Hainsworth (1971), Berger and Hart (1972), and Bartholomew and Lighton (1986) measured rates of metabolism in hummingbirds during hovering that minimally equaled 0.84 to 0.88 kJ/g·h, or about 10 to 12 times the basal rate of a 3- to 6-g hummingbird. The nectarivorous bat *Glossophaga soricina* when hovering spends energy at a rate (0.54 kJ/g·h; Winter 1998) that is approximately 12 times its basal rate (McNab 1989d), a ratio that is similar to that in hummingbirds, but at a lower absolute level. At a fixed size, sphingid moths have higher rates of metabolism while hovering than do hummingbirds, which have a higher rate than hovering glossophagine bats: restated, a hovering expenditure of 66 J/min corresponds to a mass of 3 g in sphingids, 4 g in hummingbirds, and 7 g in glossophagine bats (Voigt and Winter 1999). These values are appreciably higher than the minimal cost of powered flight in most birds (see Winter and von Helversen 1998), except for the smallest, nonaerial species (Figure 9.16).

Another approach to the energetics of hovering, in analogy with helicopter power requirements, involves the concept of wing disc loading. This loading factor equals the ratio of body weight (= mass × gravity) to wing disc area, this latter factor equaling $\pi(b/2)^2$, where b is the wing span (Pennycuick 1969, Epting and Casey 1973). Measurements of energy expenditure in hovering hummingbirds increased with wing disc loading (Epting 1980), although the range in this loading factor was small (1.7 : 1).

Some variation in wing disc loading, and therefore in the cost of hovering, is correlated with the behavior and distribution of hummingbirds. For example, territorial species have shorter wings, which permit a greater maneuverability, the resulting higher cost of hovering being supplied by a high density of food resources (Feinsinger and Chaplin 1975). Nonterritorial species reduce the cost of flight and hovering by having a lower wing disc loading, which is appropriate when foraging on dispersed food resources. Wing disc loading also tends to decrease with altitude in partial compensation for the decrease in atmosphere density, with the consequence that the power output for hovering falls within a rather narrow range independent of altitude (Feinsinger et al. 1979).

9.6.4 The morphology of flight muscles in birds.
The size and composition of the breast muscles reflect the flight capacities of birds. For example, the maximal load that can be lifted by birds with powered flight is directly correlated with the mass of flight muscles (Figure 9.20), and the maximal induced power output, calculated from the maximal load and wing length, is directly proportional in flying animals to $m^{1.08}$ (Marden 1987, 1990). If the flight muscles of birds are equal to or less than 16% of total mass, they have a marginal capacity to take off (Marden 1987). As Pennycuick (1969) pointed out, Kori bustards (*Ardeotis kori*)

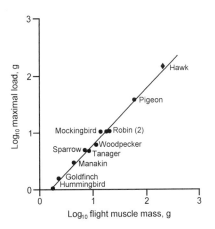

Figure 9.20 Log$_{10}$ maximal load that can be lifted by birds with powered flight as a function of log$_{10}$ pectoral muscle mass. Source: Modified from Marden (1987).

have only 16.4% of their total mass as flight muscles, which gives them a marginal capacity to fly and prevents them from increasing mass to any appreciable extent through food intake. In an examination of 425 species of birds, many of the data coming from Hartman (1961), Marden (1987) showed that only 12 (3%) have flight muscle masses less than 16% (Figure 9.21), and they either are aquatic and semiaquatic species (grebes [10%–11%], rails [8%–10%], gallinules [11%–12%], coots [9%]) or are reclusive species living in dense foliage (e.g., squirrel cuckoo [*Piaya*, 9%–12%], tody flycatcher [*Todirostrum*, 12%–13%], plain wren [*Thryothorus*, 11%–13%]). Having a high proportion of the body as flight muscles is of little value during flight at the minimal power or maximal flight range, but of great value in species that need maximal power for lifting loads, capturing prey, and avoiding predators. For example, woodpigeons (*Columba palumbus*) were more likely to be captured by goshawks (*Accipiter gentilis*) if they had small pectoral muscle masses (Kenward 1978). The largest pectoral muscles are found in species that have a rapid vertical takeoff, namely, tinamous (22%–32%), grouse (21%–29%), and pigeons (24%–37%) (Rayner 1988). These considerations may explain why so few birds have folivorous habits: they need such a large increase in the size of the gut that the ratio of flight muscle mass to total mass, especially when the gut is filled, either does not permit flight or makes it marginal (Dudley and Vermeij 1992).

Two principal flight muscles are found in birds: the pectoralis and the supracoracoideus. The pec-

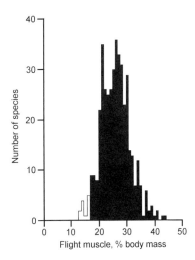

Figure 9.21 Frequency distribution of bird species as a function of the proportion of total mass that is pectoral muscles. Species with a proportion equal to or less than 16% are indicated by hollow bars. Source: Modified from Marden (1987).

toralis is principally responsible for producing the downward stroke of the wings, thereby giving lift and propulsion. The supracoracoideus is partly responsible for the upward stroke of the wings, but because most of the upward movement of the wings is produced by aerodynamic forces, the supracoracoideus is usually much smaller than the pectoralis. The supracoracoideus is most active during takeoff, when power output is maximal, and at slow flight velocities, when the contribution by aerodynamics is least (Rayner 1988). These conditions define which birds have the largest supracoracoideus (Figure 9.22):

1. Hummingbirds, in which the supracoracoideus is about 47% of the mass of the pectoralis to facilitate a powered upstroke during hovering when the wing is inverted
2. Tinamous and Galliformes, which have an explosive takeoff
3. Diving-petrels and alcids, which "fly" underwater in a medium that does not passively elevate the wings during propulsion

The size of the pectoral muscle varies with the flight pattern of birds. Thus, among ducks and geese, dabbling ducks, which become airborne with a quick, nearly vertical ascent, have large pectoral and supracoracoid muscles, whereas diving ducks, which become airborne after a long run on the

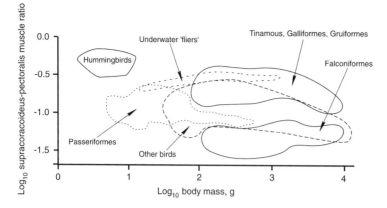

Figure 9.22 Log_{10} flight-muscle ratio (supracoracoideus-pectoralis) in birds generally and in hummingbirds and "underwater fliers" as a function of log_{10} body mass. Source: Modified from Rayner (1988).

surface of the water, have small pectoral muscles (Bethke and Thomas 1988). Geese are intermediate in behavior and relative muscle mass. Heart size, however, is not correlated with muscle mass or take-off pattern but rather with foraging depth. These two groups of ducks tend to inhabit different habitats: dabblers inhabit small, shallow lakes and ponds, where vertical take-off is required, and diving ducks, large, deep lakes, where they have long runways, thereby minimizing the cost of becoming airborne.

Birds also differ from one another in terms of muscle fiber composition (Rosser and George 1986a). Pectoral muscles are made of slow tonic and fast-twitch fibers, the latter including white (large-diameter fibers, low mitochondrial density, glycolytic), intermediate (intermediate-diameter fibers and mitochondrial density, oxidative-glycolytic), and red (small-diameter fibers, highest mitochondrial densities, oxidative-glycolytic) fibers. The slow tonic (glycolytic) fibers usually are located deep in the pectoralis and appear to be used for the maintenance of posture, and possibly in soaring (Rosser and George 1986b). Fast-twitch fibers (with varying amounts of myoglobin) constitute most of the pectoral muscles and occur in a variety of combinations. In most species red fibers predominate, although red fibers are very much more abundant in the deep layer of the muscle than superficially (Rosser and George 1986a). Long-distance migratory passerines have in their pectoral muscles the smallest myofibers, higher capillary densities, and shorter diffusion distances compared to those in nonmigratory or short-distance migratory passerines (Lundgren and Kiessling 1988). This morphology facilitates aerobic metabolism by the predominant fast-twitch oxidative-glycolytic

fibers. Migratory species, however, tend to have lower oxidative, higher glycolytic, and higher anaerobic capacities during the breeding season (Lundgren and Kiessling 1985).

Species that depend on flapping flight (e.g., hummingbirds, geese, herons) tend to have a predominance of red fibers, whereas those that use gliding (e.g., buteos, owls) tend to have a majority of intermediate fibers and those that are "reluctant" fliers (e.g., chicken, grouse) have a majority of white fibers. Tinamous (Tinamidae), which have an explosive start to flight, fly rapidly for short distances, land, and hide, have very large pectoral muscles made of white fibers, have small hearts, and apparently use anaerobic metabolism to fuel the "getaway" flight.

Although many factors other than type of flight may affect fiber composition, the impact of flight style can be illustrated by a comparison of three members of Falconiformes: the sharp-shinned hawk (*Accipiter striatus*), which does some soaring and much flapping pursuit, has a ratio of the relative abundance of intermediate to red fibers in the superficial layer of the pectoralis equal to 30:70; the northern harrier (*Circus cyaneus*), which soars most of the time, has a ratio equal to 43:57; and the red-tailed hawk (*Buteo jamaicensis*), which soars most of the time when flying but spends much of the time sitting, has a ratio equal to 54:25, with the remaining 21% of the fibers being white. In these three species, red fibers constitute 90%, 76%, and 30%, respectively, of the deep layer of the pectoralis; only the red-tailed hawk has fibers other than red fibers as the most abundant, those being intermediate fibers (67%). The turkey vulture (*Cathartes aura*), which also spends most of its time soaring, has a ratio equal to 66:34 in the superficial layer; the deep layer of the pectoralis is entirely slow tonic fibers, a composition that may facilitate the maintenance of the dihedral position of the wings in flight (also see Rosser and George 1986b).

9.6.5 The flight of bats. The cost of flight in bats is similar to that in birds (Thomas 1975, Carpenter 1986; Figure 9.13) but see Winter and von Helversen (1998). As in large birds, large bats show little variation in power output in relation to velocity. Thomas showed that the minimal power output in flight scales proportionally to $m^{0.78}$ in both birds and bats, and Carpenter found it was proportional to $m^{0.74}$. The power output of hovering nectarivorous phyllostomids is proportional to $m^{0.95}$ (Voigt and Winter 1999), although only over a narrow mass range (7–18 g). Rate of oxygen consumption and heart rate during flight are similar in birds and bats, the smaller stroke volume of bats being compensated by a greater hematocrit (ca. 60%) than that in birds (ca. 50%) (Thomas and Suthers 1972). Bats appear to be more sensitive to high ambient temperatures during flight than do birds because of a limit to evaporative water loss (Carpenter 1986).

Many bats face a limit to the load that can be carried (McNab 1973). Vampires (*Desmodus*) avoid this limit by excreting much of the water ingested with a blood meal while feeding, to reduce the inert mass that is transported (Figure 7.28). Pregnant vampires may feed twice in a night because the meal that can be transported is reduced by fetal mass. Some bats may be limited in their ability to take off from a horizontal surface (Carpenter 1986) because they have small pectoral muscle masses (Hartman 1963), although Marden (1987) questioned whether the dissections on which these measurements were based were complete (also see Carpenter 1986). Small muscle masses might require bats to start flight by dropping from a height, which often occurs.

The interaction between wing loading and aspect ratio in bats is similar to that found in some birds (Figure 9.18), although no bats have high wing loading (the vast majority being less than 20 N/m²) and none have as extreme an aspect ratio as albatrosses, tropicbirds, or swifts (Norberg and Rayner 1987). Most (97.4%) of the variation among bats in wing area (Figure 9.17) and shape and in wing loading is accounted for by the variation in body mass. The remaining variation in these characters

among microchiropterans is associated with food habits and behavior: insectivorous species that pursue high-flying insects have small pointed wings, high wing loading, high agility, and high velocities, whereas those that glean surfaces or feed among vegetation have short broad wings, good maneuverability, and slow velocities. This difference is clearly shown between molossids and vespertilionids, which, respectively, are similar to swifts and flycatchers. Fish-eating bats have long wings, low power requirements for flight, and low cost of transport. Species that eat vertebrates other than fish, however, have large wing areas, which permits them to carry heavy loads and to be maneuverable. Nectarivores have relatively small wing areas, which may reflect the necessity to navigate within "cluttered" environments. Megachiropterans are similar, although their mean wing loading is somewhat higher and their range is narrower than in microchiropterans. The narrow range in the wing morphology of bats undoubtedly reflects the narrower range in habits found in bats compared to birds. For instance, no truely marine bats exist, so the absence in bats of the wing loading and aspect ratios of albatrosses is not surprising.

9.6.6 The flight of pterosaurs. The third group of vertebrates to evolve powered flight belonged to the order †Pterosauria. Nothing is directly known of their flight because they have been extinct since the end of the Cretaceous (65 million years ago). Nevertheless, an examination of their fossil remains permits an analysis of their flight mechanics and energetics. Pterosaurs belonged to two suborders, the earlier †Rhamphor-hynchoidea, generally characterized by a small size and long tails, and the later †Pterodactyloidea, of small to huge size with very short tails, often with crests. These two morphotypes represented fundamentally different solutions to flight maneuverability and therefore to the energetics of flight. These groups coexisted in the Jurassic.

Pterosaurs are often thought to have been gliders, mainly because of their low wing loading and small pectoral muscle masses. Lawson (1975) discovered a huge pterosaur, †*Quetzalcoatlus*, which he estimated to have had a wing span of 15.5 m! Greenewalt (1976) estimated that its wing span was 5.3 m and mass was 30 kg; McMasters (1976), a wing span of 11 m and mass of 43.8 kg; and Brower and Veinus (1981), a mass of 66 to 86 kg! This pterosaur was very large, which again raises the question of whether it had powered flight.

One of the difficulties in analyzing the energetics of pterosaur flight is related to the structure of the wings. An early suggestion was that the wings were similar to those found in bats, with the trailing wing membranes attached to the ankle. Padian (1983) and Pennycuick (1988) reexamined the wing structure of pterosaurs and speculated on their terrestrial locomotion and flight. Padian's reconstruction indicated that pterosaurs were more like birds than bats in terms of their terrestrial locomotion and wing structure. He proposed that pterosaurs were bipedal, that the trailing edge of the wings attached to the body, and that the wings had stiffening fibers that ran anteroposteriorly (also see Padian and Rayner 1993). These fibers were functionally analogous to the shafts of the primary and secondary feathers in a bird's wing. Pennycuick disputed most of these suggestions, arguing that the fossils themselves, especially those preserved in the Jurassic Solnhoven limestones of Germany, indicate that (1) the pelvis was splayed (and could not have supported a pterosaur); (2) the trailing edge of the wing membranes, at least in a few cases, suggested attachment to the foot; and (3) the wing had no stiffening fibers, but that the parallel ridges seen in the wings were folds produced by elastic fibers in the distal part of the wing. Pennycuick concluded that the wing of pterosaurs was in form rather like the wing of a bat and thus suited for long-distance flight. Unwin and Bakhurina (1994) provided a similar interpretation for †Sordes pilosus, although Kellner (1996) criticized a parallel analysis for an unnamed pterosaur (Martill and Unwin 1989). Wellnhofer (1996) took an intermediate view, suggesting that some diversity in flight membrane size and attachment occurred, at least between the two suborders. He concluded, however, that pterosaurs were quadrupedal.

Several authors used Equation 9.4 to estimate the minimal power output for flight in †Pteranodon ingens. Bramwell and Whitfield (1974) suggested that it had a mass of 16.6 kg, a stall velocity of 6.7 m/s, and a sinking velocity of 0.42 m/s, where the sinking velocity equals the stall velocity times the drag-lift ratio. Then, the minimal power output for level flight would be

$$\dot{P}_{min} = (0.42)(16.6)(9.8)$$
$$= 68.3 \, \text{W} = 245.9 \, \text{J/h}$$

Stein (1975) estimated that P. ingens weighed 15 kg and that it had a stall velocity of 4.5 m/s. The power output of †P. ingens flying near its stall velocity therefore would equal $(0.56)(15)(9.8) = 82.3 \, \text{W}$. Brower (1983) calculated that the minimal power output for flight equaled $(0.53)(14.9)(9.8) = 77.4 \, \text{W}$.

These estimates of the power output required by †P. ingens for level flight are surprisingly similar. Bramwell and Whitfield concluded that if the wing muscles weighed 3 kg, they could have maximally generated 78 W, or 1.14 times the minimal power output required for powered flight, implying that †Pteranodon could have been capable of powered flight but could not have made steep, powered climbs. Stein maintained that †P. ingens did not primarily use soaring or gliding because the elbow was flexible, and estimated the power output for flight to be about 30% greater than the minimum needed. Brower argued that the steady-state power output by muscles was only 70% to 88% of the requirement for horizontal flight, but that the short-term burst expenditure was 2.1 times $\dot{P}_{min}$. He concluded that †P. ingens was not a continuous flapper but depended on soaring. In his view, a smaller pterosaur (†Nyctosaurus, 1.9 kg) was capable of continuous flapping flight; its muscles produced 1.4 to 1.5 times the minimal requirements for flight under steady-state conditions.

The morphology of pterosaur wings has been compared to that of bats and birds in an attempt to obtain insight into the ecology of pterosaurs (Rayner 1989, Hazlehurst and Rayner 1992). As noted, the complication in this analysis is that the mass and wing surface area have to be estimated from fragmentary fossils, and these calculations invariably involved an assumption as to whether the posterior edge of the wing membrane attached to the body adjacent to the femur (Brower 1983, Padian 1983, Rayner 1989), to the knee (Wellnhofer 1987, 1988), or extended to the ankle (Brioli 1938, Pennycuick 1988, Unwin and Bakhurina 1994). A question also exists as to the extent of a uropatagium, a membrane between the legs that adds to the flight surface membrane. The wing loading for pterosaurs used in Figure 9.18 is from model 2 of Hazlehurst and Rayner, which attaches the wing to the knee, but in which the tail membrane is ignored, so that their estimates of surface area are likely to be minimal. The resulting calculations, corrected for body mass, indicate that pterosaurs were characterized by a low wing loading and high aspect ratio. This pattern may have been most similar to that of frigate-birds (Alexander 1994).

Based on their analysis of wing morphology, Hazlehurst and Rayner (1992) drew several conclusions on the flight and ecology of pterosaurs.

1. They had low flight velocities and high maneuverability as a result of low wing loading. This would be most marked in species belonging to the genus †*Pterodactylus*, which were thought to have weighed between 8 and 96 g.

2. Larger pterosaurs like †*Pteranodon*, with its low wing loading, would have had low glide speeds and may well have depended on thermal soaring similar to that seen in frigate-birds.

3. The combination of low wing loading and high aspect ratio suggests that known pterosaurs did not live in "cluttered" environments such as forests.

4. The structure of the lower jaw in †*Doryganthus* and †*Rhamphorhynchus* is similar to that found in the extant skimmer (*Rhyncops*), which along with stomach contents argues that these pterosaurs scooped fish from the water surface.

5. †*Pteranodon* probably was a fish eater.

6. The numerous, closely spaced bristles in †*Ctenochasma* and †*Pterodaustro* indicate that they probably were filter feeders, but whether these "flamingo" pterosaurs stood in water, as depicted by Wellnhofer (1996), or whether they fed in flight (as do the much smaller prions [*Pachyptila*]), is unknown.

7. The apparent absence of pterosaurs with broad short wings may have reflected either a limitation to their morphology or a biased sample of species that were preserved.

8. The observation that individuals belonging to †*Pterodactylus kochi* showed a great range in body size suggests that they, unlike birds and bats, had indeterminate growth. This may well be evidence that pterosaurs were ectothermic. Maybe the apparent limitations in wing shape in pterosaurs reflected an inability to generate a high power output, as would be required of species with high wing loading (also see Section 5.10.4), but possible only in endotherms. Wellnhofer (1996), however, thought that most (all?) pterosaurs were endothermic based on their capacity to fly, the presence of air sacs and pneumatic bones, and reports of fur (Brioli 1941, Sharov 1971), which may turn out to be structural fibers.

9.6.7 The evolution of flight in vertebrates.
Powered flight evolved in at least three different times, assuming that bats are monophyletic (but see Pettigrew 1986, 1995). Many other vertebrates use gliding flight, including the anuran *Polypedates*; the lizards *Draco*, *Ptychozoon*, and *Cosymbotus*; the snake *Chrysopelia*; and various mammals (gliding opossums [*Petauroides*, *Petaurus*, *Acrobates*], flying "lemurs," or colugos [*Cynocephalus*], flying-squirrels [*Petaurista*, *Pteromys*, *Glaucomys*, etc.], and scaly-tailed squirrels [*Anomalurus*, etc.]). Ostrom (1974), Caple et al. (1983), Pennycuick (1986), and Feduccia (1996) considered the evolution of powered flight and reached conflicting conclusions. Ostrom maintained that bird flight was derived from a predatory, cursorial, bipedal form of terrestrial locomotion; Caple and coworkers suggested that all forms of powered flight were derived from a cursorial form of locomotion (terrestrial in birds and from cliffs or trees in bats); Pennycuick and Feduccia maintained that all forms of powered flight in vertebrates were derived from arboreal gliders. This diversity of opinion in each species was backed by anatomical, paleontological, or physical analyses. Nevertheless, the view that the origin of bird flight was substantially different from that found in bats and pterosaurs is suggested by observations that (1) all birds are bipedal, (2) few arboreal vertebrates (other than birds) are bipedal, and (3) bats and pterosaurs use membranes stretched between the limbs, in varying degrees similar to most arboreal, gliding mammals. Bird flight most likely was derived from terrestrial, cursorial predators, which is appropriate if birds were derived from coelurosaurian dinosaurs, as Ostrom (1976) suggested, or from stem archosaurs, a suggestion Feduccia (1996) preferred. Bats and pterosaurs may have been derived from either arboreal or cliff-dwelling ancestors.

9.6.8 A reduced capacity for flight in birds.
Some birds have a chronically limited capacity for flight. These include the hoatzin (*Opisthocomus hoazin*) and trumpeters (*Psophia* spp.) from South America; tapaculos (Rhinocryptidae) from Central and South America; the wrenthrush (*Zeledonia coronata*) from Central America; the kokako, or blue-waddled crow (*Callaeas cinerea*) and fernbird (*Bowdleria punctata*) from New Zealand; and mesites (*Mesitornis* and *Monias*) from Madagascar. These species have been little studied; they may be

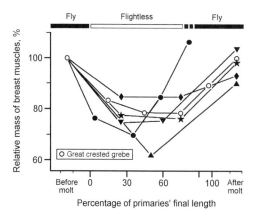

Figure 9.23 Relative loss and gain of breast muscles in various anserids and the great crested grebe (*Podiceps cristatus*) as a function of the molt cycle. Source: Modified from Piersma (1988).

characterized by small pectoral muscle masses and high wing loading (Feduccia 1996). For example, high wing loading in the hoatzin is associated with processing of leaves by pregastric fermentation (Grajal 1995). The limited flight in most of these birds reflects small pectoral girdles, small keels on the girdles, and small pectoral muscle masses. Some are limited to islands, which may represent an intermediate stage in the evolution of flightlessness. Unfortunately, the natural history of these birds is little studied, so that the ecological significance of a limited commitment to flight is not understood.

Other flighted birds, especially members of the families Podicepedidae (grebes) and Anseridae (ducks, qeese), develop a flightless condition on a seasonal basis as a result of the simultaneous molt of their primary feathers. This flightless condition is usually associated with a reversible atrophy of the pectoral flight muscles (Figure 9.23): the breast muscles of the great crested grebe (*Podiceps cristatus*) were 10% of total mass during winter, whereas this proportion dropped to 7.9% when the grebes were molting (Piersma 1988). The decrease was associated with an 80% decrease in food intake, which suggests that muscle atrophy might be due to the metabolism of muscle protein, a phenomenon that has been described during starvation in the willow ptarmigan (*Lagopus lagopus* [Grammeltveldt 1978]), a weaver finch (*Quelea quelea* [Kendall et al. 1973]), and the savannah sparrow (*Passerculus sandwichensis* [Swain 1992]). Yet, Piersma argued that the decrease in muscle mass in the grebe was principally due to muscle disuse. In defense of the disuse explanation,

Piersma cited data on muscle atrophy in western grebes (*Aechmorphus occidentalis*) during the breeding season while feeding occurs but when flight is rare. Seasonal flightlessness in Wilson's phalarope (*Phalaropus tricolor*) occurred in late August and early September in a small proportion (<1%) of individuals as they rapidly deposited fat preparatory to fall migration from Mono Lake, California, to northern South America (Jehl 1997). The largest fat load compatible with flight in the phalaropes is about 45% of total mass; flightless individuals had a fat load up to 54% of total mass. Flightless birds can resume flight by stopping feeding, thereby burning off excess fat.

Whereas the molting period in the great crested grebe lasts 2.5 months in the Netherlands, in eared grebes (*P. nigricollis*) it may last 2 to 4 months in breeders and juveniles and up to 8 months in non-breeders on Mono Lake, California (Gaunt et al. 1990). During the flightless molting period, eared grebes increase body mass by about 25% through fat storage, whereas flight muscle mass decreases by 50%. (With the decrease in muscle mass, myofibers atrophy.) Within 2 weeks of a failure in their food supply (brine shrimp), the grebes completed molt, their flight muscles hypertrophied in association with exercising their wings, and they flew to lakes with adequate food supplies (Gaunt et al. 1990). The reacquisition of flight capacity depends on the depletion of their accumulated fat stores. If, indeed, disuse is responsible for the atrophy of flight muscles, then bristle-thighed curlews (*Numenius tahitiensis*), which simultaneously molt their primaries on the wintering grounds of Central and South Pacific islands and are flightless for 2 to 4 weeks (Marks 1993), may show an atrophy of their pectoral muscle masses. Over 50% of the adult population of curlews become flightless, a condition that is rare in shorebirds; it is tolerated because this species winters only on islands that have no mammalian predators. Further evidence of the use-disuse explanation is found in migratory passerines that are characterized by large fat stores, including the yellow wagtail (*Motacilla flava* [Fry et al. 1972]) and the gray catbird (*Dumetella carolinensis* [Marsh 1984]): pectoral muscle masses hypertrophied during the migratory period.

9.6.9 The evolution of flightlessness in birds.
Although a few birds temporarily attain a flightless condition in an annual cycle, others have evolved

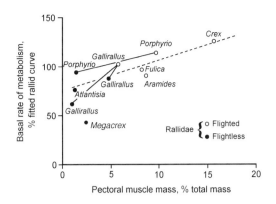

Figure 9.24 Relative basal rate of metabolism in flighted and flightless rails as a function of relative pectoral muscle mass. The dashed curve is the mean rail curve. Connected points indicate species belonging to the same genus. Source: Modified from McNab (1994a).

a permanently flightless condition on oceanic islands that lack mammalian predators. Flightlessness has evolved repeatedly in Podicipediformes (grebes), Anseriformes (ducks, geese), Ciconiiformes (ibises), Columbiformes (dodo, solitaire), Gruiformes (rails, kagu [*Rhynochetos jubatus*], †*Aptornis*), and Passeriformes in New Zealand (acanthisittid wrens, Chatham Island fernbird [Bell 1991]), and at least once in ratites, Sphenisciformes (penguins), Pelecaniformes (a cormorant), Psittaciformes (kakapo), and Strigiformes (owls).

The explanation for this phenomenon has been rather diffuse (e.g., flightlessness was produced by the absence of mammalian predators, it prevents island birds from being blown out to sea, or it is associated with phylogeny). Flightlessness more specifically has been suggested to be a means of energy conservation (Olson 1973, Fleming 1982). McNab (1994a) examined this suggestion and showed that (1) flightless rails have lower basal rates of metabolism than their flighted relatives, (2) the reduction of basal rate in rails is correlated with the reduction in pectoral muscle mass (which may be as small as 1.0%–1.4% of body mass; Figure 9.24); and (3) kiwis also have low basal rates in association with very small pectoral muscle masses (ca. 0.13%), and thereby conform to the relationship seen in rails. These observations led to the conclusion that the Galápagos cormorant (*Phalacrocorax harrisi*), the New Zealand flightless parrot (the kakapo, *Strigops habroptilus*), and the flightless grebes found in Peru (*Rollandia microptera*, *Podiceps taczanowskii*) and now extinct in Guatemala (*Podilymbus gigas*) are prob-

ably also characterized by low basal rates in association with their very small pectoral muscle masses (1.2%, 1.5%, 3.3%, 7.6%, and 4.3% of total mass, respectively; Livezey 1989, 1992a, 1992b).

The evolution of a flightless condition in penguins and ducks appears to differ from that in rails and kiwis (McNab 1994a). The absence of a low basal rate of metabolism in flightless ducks, however, raises the possibility that a reduction in energy expenditure in these species occurs through a reduction in activity level and in field expenditures. That is, the reduction in energy expenditure associated with the evolution of flightlessness is not all-or-none but is graded, as implied by the interpretation that reduced flight activity (see Section 9.6.8) of birds on islands leads to a reduced field expenditure.

The evolution of flightlessness in birds appears to have been driven by the necessity to reduce energy expenditure on islands characterized by a limited resource base (McNab 1994b), but its evolution is possible only in the absence of mammalian predators. Adjustments other than flightlessness that facilitate long-term persistence on small islands involve a reduction in resource requirements, including a small body size, a reduction in activity level, and possibly the abandonment of endothermy (see Section 10.10).

9.7 TERRESTRIAL LOCOMOTION IN BIRDS AND MAMMALS

Terrestrial locomotion in birds is bipedal and in mammals is bipedal or quadrupedal. Most of the studies of terrestrial locomotion have concerned mammals, where it reflects their diversity in morphology and the mechanics of movement.

9.7.1 Morphology. Mammals show great variation in body proportions, which is clearly shown when arboreal and terrestrial mammals are compared (Grand 1977, 1978, 1983). Mass is limited in arboreal species, which also differ from terrestrial species in having wrist and ankle joints that are laterally displaced. Active arboreal species grasp branches, which means that an appreciable muscle mass is located distal from the central body mass in the limbs and tail. Terrestrial species, in contrast, are supported on small surfaces at the end of thin legs, a condition that permits great acceleration and deceleration of the limbs and a high velocity. The terrestrial agouti (*Dasyprocta aguti*) has about 75% of total mass in the trunk and head, whereas active arboreal species, such as galagos (*Galago*) and cebid monkeys (*Cebus, Ateles*), have head and trunks that comprise only 50% to 65% of their total mass. Sedentary arboreal mammals such as sloths (*Choloepus, Bradypus*) and lorisids (*Perodicticus, Nycticebus*), like terrestrial mammals, have head and trunks that are 75% to 80% of total mass; they are usually tailless. Sedentary species, furthermore, have a great reduction in the proportion of total mass that is muscle, down to 25%, whereas in terrestrial species the proportion ranges from 43% to 55% of total mass. Correlated with the reduction in muscle mass is a reduction in basal rate (McNab 1978b; S. D. Thompson and T. Grant, pers. comm, 1985). An equal mass does not guarantee an equal expenditure of energy.

9.7.2 Mechanics of locomotion. Significant variation in the morphology of the locomotor system has an effect on the mechanics of locomotion. Terrestrial locomotion at a constant velocity occurs as a cycle of steps or strides during which gravitational potential energy and kinetic energy oscillate with the vertical displacement, acceleration, and deceleration of body mass (Cavagna et al. 1977). The energy expenditure for locomotion is reduced by an exchange (1) between gravitational potential energy and kinetic energy and (2) between the mechanical energy stored in the elastic elements of muscles and recovered as kinetic and gravitational energy.

The pattern of steps, or strides, is specified by a series of variables: length of step or stride, stepping frequency, and phase differences in footfall. A continuous set of uniformly patterned steps is called a *gait*, of which at least three major types exist (Pennycuick 1975):

1. *Walk* is an asymmetrical gait during which the body is continuously supported. At each step energy is exchanged between vertical potential energy and horizontal kinetic energy. Variation in velocity during a walk is due to a change in stepping frequency and a change in the number of feet touching the ground at a particular instant (which may vary from 2 to over 3 in a quadruped).

2. *Trot* is a symmetrical gait in which the diagonal pairs of legs move in unison. Some anatomists (e.g., Hildebrand 1965) consider the pace, in which the ipsilateral (same side) legs move together, different from the trot, although Pennycuick considers it to be mechanically equivalent to a trot. In the trot (and pace) energy is exchanged between potential energy and elastic energy stored in the ligaments of the legs. A change in velocity during a trot depends mainly on a change in the mean number of legs in contact with the ground (which may vary from 1.2 to 2 in quadrupeds), rather than due to a change in stepping frequency. (Note that the *mean* number of legs touching the ground is the mean over a cycle of steps.) Some mammals, especially those with long thoracic neural spines (i.e., with sloping backs), namely, the wildebeest (*Connochaetes*), giraffe (*Giraffa*), and spotted hyena (*Crocuta*), do not trot. With an increase in velocity, these species change directly from a walk to a canter. Mammals with horizontal backs, such as gazelles (*Gazella*), dogs, buffalo (*Syncerus*), lion (*Panthera*), and rhinoceroses (*Diceros, Ceratotherium, Rhinoceros*), use the trot at velocities intermediate between those of a walk and a canter.

3. *Canter* is an asymmetrical gait that uses all 4 legs in turn. The mean number of legs in a quadruped touching the ground during a canter is low, varying from 0.6 to 1.8. A fast canter is called a *gallop*; at these high velocities an unsupported period occurs in each stepping cycle.

Heglund et al. (1974) concluded that with an increase in velocity a recruitment of storage elements is found with a shift in gait: little energy is stored during walking, energy is stored in elastic elements of the limbs while trotting, and energy is stored in the entire trunk during galloping.

The range of velocities exhibited by an animal varies with body mass, the gait used, and the species. For example, the maximal running speed in mammals generally increases with body mass, although the largest species are not the fastest (Garland 1983). Among rodents and lagomorphs, hoppers have a higher maximal velocity than runners. The variation in velocity of movement relates to the individual characteristics of a species. This is illustrated by the frequency distribution of velocities in adult Thomson's gazelles (*Gazella thomsoni*) and the wildebeest (*Connochaetes taurinus*) in the field (Figure 9.25). While walking, stride length is greater, velocity slightly higher, and stepping frequency lower in the wildebeest than in the gazelle. These differences are related to shoulder height (i.e., leg length). Similar differences in

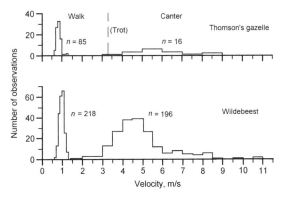

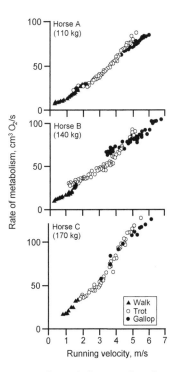

Figure 9.25 Frequency distribution of the velocity of movement by Thomson's gazelle (*Gazella thomsoni*) and wildebeest (*Connochaetes taurinus*) in the field. Source: Modified from Pennycuick (1975).

Figure 9.26 Rate of metabolism in three horses (*Equus caballus*) as a function of running velocity and gait. Source: From Hoyt and Taylor (1981).

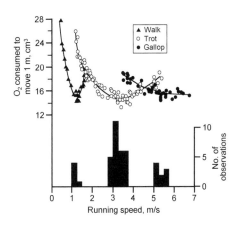

Figure 9.27 Cost of movement and frequency of walking, trotting, and galloping in horses (*Equus caballus*) as a function of the velocity of movement. The frequency distribution of gaits was determined when horse B was free in a pasture. Source: Modified from Hoyt and Taylor (1981).

velocity are found between calf and adult wildebeest, which also relate to shoulder height. Over a wide range in body size, stepping frequency is inversely proportional to shoulder height raised to the −0.5 power during a walk, trot, or canter (Pennycuick 1975).

The significance of the shift from one gait to another for mechanics and energetics is dramatically illustrated in measurements on the horse (*Equus*) by Hoyt and Taylor (1981). In horses the rate of metabolism varies in a curvilinear manner with velocity in all three gaits (Figure 9.26). The transition from one gait to another occurs at velocities where the curve for each gait intersects the metabolism-velocity curves of the adjacent gaits, although for experimental purposes the horses were trained to use gaits at velocities that are typical of adjacent gaits. The minimal cost of transport (i.e., expenditure per unit of distance) is similar for all three gaits (Figure 9.27); switching from one gait to another maintains a minimal cost of transport, although not a minimal expenditure because the velocities vary among the gaits (Figure 9.26). An optimal velocity or range of velocities exists for each gait, and these velocities correspond to those used by free-ranging horses (Figure 9.27). Thus, free-ranging quadrupeds tend to have a discontinuous distribution of locomotor velocities (Figures 9.25 and 9.27), apparently to minimize the expenditure of energy associated with activity.

9.7.3 Energetics of quadrupedal locomotion.

In quadrupedal locomotion aerobic energy expenditure is a linear (but see Figure 9.26) function

of velocity. This relationship has been demonstrated in many mammals, including marsupials (Baudinette et al. 1976) and eutherians (Taylor et al. 1970; Figure 9.28). It is described by

$$\dot{V}_{O_2} = a \cdot \dot{v} + y, \tag{9.8}$$

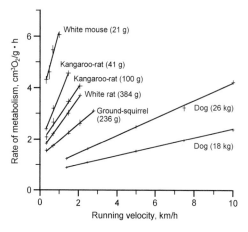

Figure 9.28 Rate of metabolism in mammals as a function of running velocity and body mass. Source: Modified from Taylor et al. (1970).

where a (cm³O₂/km) is the slope of the curve, $\dot{v}$ is velocity (km/h), and y (cm³O₂/h) is the intercept at zero velocity (note: $\dot{V}_{O_2}$ in Figure 9.28 is in cm³O₂/g·h). The coefficient a and the intercept y vary with body size: the best estimate in mammals for a is $0.0049m^{0.68}$ and for y is $0.0024m^{0.70}$ (Taylor et al. 1982), where mass is in grams.

A complication in this expression has been the observation that y in Equation 9.8 is about 1.7 times the resting rate (Taylor et al. 1970). Part of the problem has been that y usually is compared to the rate *expected* from the Kleiber curve, out of convenience, rather than to a *measured* rate. Mammals show appreciable variation in the basal rate independent of the influence of body mass: indeed, Paladino and King (1979) showed that y is only 1.2 times the measured resting rate. Schmidt-Nielsen (1972) suggested that the difference between the extrapolated y-intercept and the expected (or measured) rate reflects the cost of changing from a resting to an active posture, but this interpretation is difficult to prove.

The observation that the relations between rate of metabolism and velocity (one for each gait) are curvilinear in the horse (Figure 9.26) and in humans (Margaria 1938), whereas they appear to be linear in other mammals (Figure 9.28), raises the question of whether the apparent linearity of much of the data reflects scatter in the measurements and masks a "true" curvilinearity in the data. Taylor (1977) argued that once an animal has attained a constant velocity, energy expenditure should be

proportional to $\dot{v}^2/2$, if the variation in velocity were due to variation in stride length, or to $\dot{v}^2 \cdot \dot{v}/2 = \dot{v}^3/2$, if the variation in velocity were accomplished by variation in the number of strides. The distribution of mass in the limbs might influence energy expenditure because the kinetic energy required to rotate the limb is equal to $m \cdot r^2$, where m is the limb mass and r is the distance between the pivot of the leg and its center of mass. But in a comparison of the cheetah (*Acinonyx*), goat (*Capra*), and gazelle (*Gazella*), which had r values equal to 18, 6, and 2 cm, respectively, no difference occurred in the relationship between rate of metabolism and running velocity (Taylor et al. 1974). Nevertheless, Chassin et al. (1976) showed that 55-kg lions (*Panthera leo*) reached their aerobic maximum at about 8 km/h; any further increase in velocity required anaerobic expenditures. The lion combines a low anaerobic scope with a high cost of movement. The maximal velocity of an animal may reflect a combination of anaerobic and aerobic scopes, the latter being potentially limited by the ability to increase the resting heart rate.

A serious difficulty with attempting to estimate the amount of work required for locomotion from mechanics involves the question of mechanical efficiency, that is, the fraction of chemical potential energy used that is converted to mechanical work (also see Section 9.6.2). This efficiency is generally assumed to be about 25% because individual striated muscles in vertebrates maximally attain this efficiency (Taylor 1980). Taylor, however, concluded that the efficiency of terrestrial locomotion increases with velocity, and in humans may exceed 70%; efficiency may also vary with body mass. Mechanical efficiency cannot be assumed to be constant.

The cost of locomotion was claimed to be lower in some quadrupeds, such as lizards (Bakker 1972b, Taylor 1977), marsupials (Baudinette et al. 1976), and monotremes and insectivores (Taylor 1981), than in (most) eutherian mammals. As seen, the cost of locomotion at a particular velocity depends on two parameters: the increment in expenditure with velocity and the intercept. Quadrupeds show few differences in the incremental cost of running, other than those associated with mass, most differences being associated with the y-intercept. Paladino and King (1979), however, disputed this conclusion; they stated that no significant differences occur in the cost of locomotion or in the cost of "maintaining a running posture" among these groups. Taylor et al. (1982) agreed with this conclusion. Consequently, the principal variable dictating the cost of locomotion is the velocity of movement: the cost of walking is very low in the slow loris (*Nycticebus*) because the maximal velocity in this species is about 1.5 km/h (Parsons and Taylor 1977), not because it has a relation between energy cost and velocity that differs from that in other species.

The cost of quadrupedal locomotion usually is measured on a horizontal treadmill. In reality, much locomotion involves ascending and descending slopes. As an extreme, tree squirrels and tree shrews rapidly scurry up and down vertical tree trunks. Taylor et al. (1972) found little difference in energy expenditure in white mice (30 g) at velocities up to 1.7 km/h on +15, 0, and −15° inclines, but substantial differences in chimpanzees (*Pan*, 17.5 kg) at velocities from 1.5 to 7 km/h. The difference between these species may be related to body size, because the cost of transport of each unit of mass per kilometer is so high in small species that the effect of an inclined plane is proportionally small.

Wunder and Morrison (1974) found that rate of energy expenditure in the red squirrel (*Tamiasciurus hudsonicus*) is positively correlated with velocity and angle of inclination:

$$\dot{V}_{O_2} = \alpha(\dot{V}_{O_2 rest}) + a \cdot \dot{v} + k_1 \dot{v} \cdot \sin \Theta, \qquad (9.9)$$

where α is the "postural" factor (ca. 1.2), $\dot{V}_{O_2 rest}$ is the resting rate of metabolism (which varies with body size and ambient temperature), and Θ is the angle of the plane. Wunder and Morrison suggested that a may fall at temperatures below thermoneutrality, that is, that some of the heat produced with running can be used for temperature regulation. For a horizontal plane (when $\Theta = 0°$) Equation 9.9 becomes Equation 9.8, whereas it would equal $\alpha(\dot{V}_{O_2 rest}) + (a + k_1)\dot{v}$ for climbing on a vertical plane (i.e., $\Theta = 90°$).

9.7.4 Bipedal locomotion. Nearly all birds (some partial exceptions being grebes, loons, shearwaters, etc.) and some mammals use bipedal locomotion on land. Much has been written on the question of whether bipedal locomotion is less expensive than quadrupedal locomotion. Several early papers indicated that this was indeed the case, at least at larger masses (Taylor et al. 1970, 1971b; Fedak et al. 1974), but a detailed statistical analysis of the available data (Paladino and King 1979) and the addition of more data at large (Fedak and Seeherman 1979) and at small masses (Thompson et al. 1980) revealed that no consistent difference in the energetics of locomotion occurs between quadrupeds and bipeds. In primates no difference exists in the costs of quadrupedal and bipedal locomotion in the same species (Taylor and Roundtree 1973a).

Bipedal locomotion may offer some advantages over quadrupedal locomotion. In small species, particularly those living in open habitats, bipedal locomotion may enhance predator avoidance (Thompson et al. 1980) because it permits rapid changes in direction. Similarly, although brachiation may be at least as expensive as walking in primates (Parsons and Taylor 1977), its principal value may be a reduction of energy expenditure through the distance moved (Grand 1977). The factors selecting for the evolution of locomotor diversity are complex; they are not necessarily based on energetics, or at least not on energetics alone. Nevertheless, Baudinette (1991, 1994) reiterated the view that macropodid marsupials are more efficient during hopping at high velocities than during quadrupedal locomotion at the same velocities, at least at body masses greater than 5 kg: the rate of metabolism in a wallaby (*Macropus eugenii*) is independent of velocity above approximately 3 m/s (Figure 9.29), although as Baudinette (1994) noted, the measurements should be corrected for wind resistance. The difference between bipedal and quadrupedal locomotion may be due to a greater storage of energy in the tendons of the hindlegs. This uncoupling of energy expenditure and velocity does not occur in species that weigh less than 1.5 kg, even among macropodid marsupials, which may be the reason why Thompson et al. (1980) could not find it in small, bipedal rodents.

Leonard and Robertson (1997) summarized data that indicate humans have lower energy expenditures as a function of velocity, or at least when walking, than expected from the descriptive model of quadrupedal locomotion by Taylor et al. (1982). The relatively low cost of bipedal movement is further amplified in comparison with nonhuman primates (monkeys and apes), which usually move

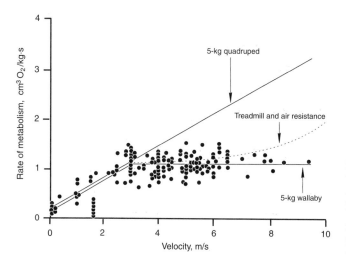

Figure 9.29 Rate of metabolism in the tammar wallaby (*Macropus eugenii*) as a function of running velocity. The dashed line represents a correction for wind resistance. The relation between rate of metabolism and running velocity for a quadruped of equal size, as calculated from Taylor et al. (1982), is indicated. Source: Modified from Baudinette (1994).

less than 5 km/d, whereas human hunter-gatherers (!Kung in Botswana and Ache in Paraguay) may move between 10 and 20 km/d. Although some of this difference is related to differences in body size, Leonard and Robertson (p. 308) concluded, "[E]ven allowing for the fact that the earliest hominids were less efficient than modern humans, because bipedality is *so much less* costly than either generalized mammalian and primate quadrupedality it is still likely that there were energetic benefits associated with the locomotor shift."

9.7.5 Heat storage as a limit to locomotion.

Only a fraction of the energy used during activity is used for work against the environment. Most energy is converted to heat, which either is lost to the environment or is stored as an increase in body temperature.

Heat storage, which in effect represents a delayed heat loss to the environment, is limited by the lethal body temperature, and therefore by the size of the animal (Phillips et al. 1981; see Section 5.3.4). A cheetah (*Acinonyx*) can attain a velocity of 110 km/h and a gazelle (*Gazella*), 90 km/h. For a few hundred meters a cheetah can multiply its resting rate of metabolism 54 times, but it stops running when body temperature reaches 40.5 to 41°C, that is, when the increase in body temperature has reached 1.0 to 1.5°C (Taylor and Roundtree 1973b). Thomson's gazelles (*G. thomsoni*), in contrast, may store enough heat to increase body temperature by 4.5°C during a short run (11 min): the increase in core temperature is greater at the highest velocities, when little of the heat is lost to the environment. Gazelles can toler-

ate body temperatures above 42°C because they have a countercurrent heat exchanger inserted in the carotid artery, so that during a 7-min run at 40 km/h, the carotid artery temperature increases 4.9°C, whereas the brain temperature increases only 1.2°C (Figure 9.30). Gazelles, then, are able to run farther than cheetahs when pressed. Baudinette (1994) argued that the low body temperatures of resting marsupials, especially macropodids, permit them to increase heat storage during activity.

9.7.6 The cost of burrowing.

Although burrowing by mammals might be treated most appropriately as an activity other than locomotion, some burrowers, such as moles, mole-rats, and pocket-gophers, forage by this method, and it thus can be considered locomotion through a dense medium. For example, 80% of the burrow system of *Thomomys bottae* may be feeding tunnels (Miller 1957). Vleck (1979) measured and modeled the cost of burrowing in pocket-gophers. In a horizontal burrow the cost equals the sum of two expenditures, one from shearing the soil from the end of the tunnel and the other from moving the loose soil out of the tunnel. Both expenditures involve the mass of the soil and the length of the burrow constructed. Later, Vleck (1981) added the cost of pushing the expelled soil to the surface, which involves counteracting gravity.

The expenditure for burrowing varies with soil type: as soil becomes more compacted, the cost of burrowing increases (Figure 9.31). Thus, a 150-g *Th. bottae* expends about 3.3 kJ to construct a burrow 1 m long with a 3.5-cm radius in sand,

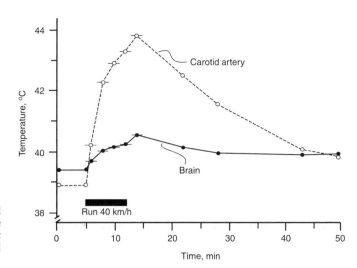

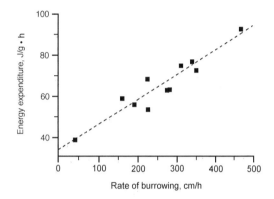

Figure 9.30 The temperature of the carotid artery and the brain in Thomson's gazelle (*Gazella thomsoni*) after a 12-min run at 40 km/h. Source: Modified from Taylor and Lyman (1972).

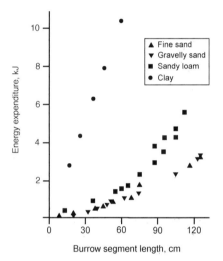

Figure 9.31 Oxygen consumption in the pocket-gopher *Thomomys bottae* as a function of burrow length and soil type. Source: Modified from Vleck (1979).

Figure 9.32 Rate of metabolism in the Cape mole-rat (*Georychus capensis*) as a function of the rate of burrowing. Source: Modified from du Toit et al. (1985).

9.8 AQUATIC LOCOMOTION IN BIRDS AND MAMMALS

The study of the energetics of aquatic locomotion has included a variety of mammals, including cetaceans, humans, mink, muskrats, seals, sea lions, and sea otters (for a summary, see Fish 1992, 1994) and a few birds, including ducks and penguins (Baudinette and Gill 1985). For example, the rate of energy expenditure of little penguins (*Eudyptula minor*) increases almost as an exponential function of swimming velocity, so the cost of swimming a unit of distance is not constant. As in horses (Figures 9.26 and 9.27), a unique minimal cost of transport exists at higher velocities. The little penguin reduced oxygen requirements by 40% when swimming completely submerged in water compared to swimming at the surface, which

whereas the same burrow in cohesive clay would require an expenditure of some 30 kJ. In *Cryptomys damarensis* and *Heterocephalus glaber* the rate of metabolism when digging was 4.3 to 5.3 times the resting rates, and in both species the rate in wet sand was 1.5 to 3.7 times that in dry sand (Lovegrove 1989). Vleck (1979) estimated that the cost of burrowing for 1 m requires 360 to 3400 times as much energy as is required to move the same distance on the surface. A similarly high cost of burrowing has been found in the Cape mole-rat (*Georychus capensis*; Figure 9.32), where the increase in cost is directly related to the rate of burrowing (Du Toit et al. 1985).

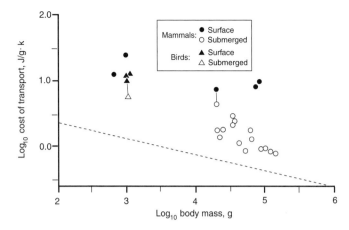

Figure 9.33 $\log_{10}$ mass-specific cost of aquatic transport in birds and mammals as a function of $\log_{10}$ body mass and whether birds and mammals swim submerged or at the surface. The mean cost of (submerged) transport in fishes is indicated by the dashed line. Source: Modified from Fish (1994).

itself was less than one-half the cost of running on land. The velocity at which Adélie penguins (*Pygoscelis adeliae*) chose to swim in a tank (ca. 2 m/s) corresponds to the velocity at which the cost of transport was minimal (Culik and Wilson 1991).

When the aquatic cost of transport is examined, two points are clear (Figure 9.33): (1) the cost of transport is appreciably greater in aquatic mammals and birds than in fish and (2) swimming at the surface is more expensive than swimming submerged. Vertebrates that swim at the water surface, like ducks (Prange and Schmidt-Nielsen 1970), mink (*Mustela vison*), and humans, have very high costs of locomotion. Thus, mink spend 40% more energy than humans to swim a body length, but do so at 2.6 times the velocity, and humans spend 4.7 times as much energy as seals, which swim at 7.4 times the velocity (Videler and Nolet 1990). The high cost of swimming at the surface is related to the cost of making wakes and to the fact that little of the body is involved with exerting a force against the environment. Fish, in contrast, use nearly their entire body to push against the environment.

9.9 THE COMPARATIVE COST OF LOCOMOTION

The "cost of transport" (i.e., the cost of moving a unit of mass a unit of distance) has been widely used to compare the various forms of locomotion (Weis-Fogh 1952, Tucker 1970, Taylor et al. 1970, Schmidt-Nielsen 1972). Taylor et al. (1970) and Schmidt-Nielsen (1972) defined this estimate as the rate of metabolism at a particular velocity divided by that velocity:

$$\text{cost of transport (J/g·km)}$$
$$= \frac{\text{rate of metabolism (J/g·h)}}{\text{velocity (km/h)}}. \quad (9.10)$$

This definition has several difficulties:

1. In terrestrial vertebrates, at least, the correlation of rate of metabolism with velocity is usually linear as described by Equation 9.8. Therefore,

$$\text{cost of transport} = \frac{a \cdot \dot{v} + y}{\dot{v}} = a + \left(\frac{y}{\dot{v}}\right). \quad (9.11)$$

This estimate of the cost of transport is not independent of velocity unless $y = 0$, which is never the case. A better estimate of the cost of transport is a (= $[\dot{V}_{O_2} - y]/d\dot{v}$), that is, the slope of the metabolism-velocity curve, as Schmidt-Nielsen (1972) clearly saw.

2. Rate of metabolism in fish (Figure 9.3), penguins, horses (Figure 9.26), and seals is exponentially correlated with velocity. Consequently, the cost of transport in these species is not constant but varies with velocity (Figure 9.27). Therefore, the minimal cost of transport is normally compared.

3. The cost of transport, defined by Equation 9.10, actually is mass specific, a/m, which in terrestrial locomotion is proportional to $m^{-0.32}$ (Taylor et al. 1982). Using mass-specific units does not "correct" for body mass because cost of transport is not proportional to $m^{1.00}$, and it obscures the ecologically significant observation that large animals expend more energy to move than do small animals (Kleiber 1975, Baudinette 1991).

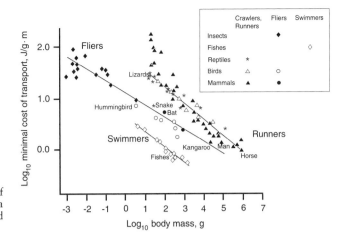

Figure 9.34 $\log_{10}$ mass-specific minimal cost of transport in runners, fliers, and swimmers as a function of $\log_{10}$ body mass. Source: Modified from Tucker (1975).

4. The cost of transport is sensitive to temperature in both ectotherms and endotherms (Paladino 1985), which therefore complicates the question raised in Section 9.7.3 on the mechanical "efficiency" of locomotion. For example, bipedal locomotion in the white-crowned sparrow (*Zonotrichia leucophrys*) varies with temperature in a manner that implies that some of the heat produced by running can be used for temperature regulation and that this dual use is greatest at the lowest temperature (also see Section 5.8.3).

A comparison of the major means of vertebrate locomotion, namely, walking or running, flying, and swimming, leads to several conclusions:

1. The total cost of transport increases with body mass, approximately proportional to the factors $m^{0.68}$, $m^{0.78}$, and $m^{0.69}$, respectively. Whether the mass factor for flying, $m^{0.78}$, truly is greater than the factors for walking-running or swimming is unclear.

2. The differences in mass-specific cost of transport (a/m) among the forms of locomotion are seen in Figure 9.34: walking-running has the highest cost of transport; flying, a cheaper cost; and swimming, the cheapest cost of transport (Tucker 1970, 1975; Schmidt-Nielsen 1972). For example, a 10-g bird expends only 3.5% as much energy to fly as a mouse does to walk the same distance.

3. Among birds, these differences can be refined (Butler 1991): energy expenditure during flight is 2.5 times that during running and swimming, except that in cursorial birds energy expenditures

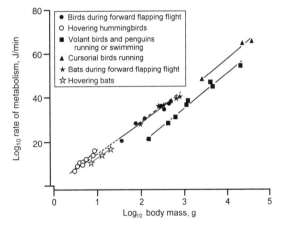

Figure 9.35 $\log_{10}$ rate of metabolism during activity in flying birds and flying and hovering bats, hovering hummingbirds, volant birds and penguins running and swimming, and cursorial birds running, as a function of $\log_{10}$ body mass. Sources: Modified from Butler (1991) and Voigt and Winter (1999).

during running are similar to those found in flight when adjusted for body mass (Figure 9.35).

4. Although hovering flight is thought to be more expensive than forward flight (Rayner 1979), in fact it appears to be similar to that expected from mass in species that show forward flight, both in hummingbirds and in nectarivorous bats (Figure 9.35).

The questions remain, why is swimming cheaper than flying and why is flying cheaper than walking or running? Most swimmers are neutrally buoyant and need not pay the cost of supporting their mass; in addition, they are completely surrounded by water, which facilitates exerting a force against their environment. The wings of fliers are highly efficient in transferring energy from downward to forward motion. Runners, in contrast, have only a small surface area with which they exert a force against the environment. Furthermore, as Grand (1977) pointed out, great variations occur in the amount of muscle that vertebrates have: some fishes may be composed of 85% muscle, whereas terrestrial mammals may have between 25% and 55% muscle. The range of expenditures in vertebrates for locomotion would be diminished by expressing the expenditures as a function of muscle mass, rather than as a function of total mass, total mass being a heterogeneous mixture of tissues, many of which make no direct contribution to movement. Snakes have lower costs of transport than runners and walkers (Figure 9.34), probably because they, like fish, need not work against gravity, but they must overcome the friction encountered with the substrate.

A comparison of the cost of transport throws some light on the question of which vertebrates are likely candidates for long-distance migration. Migration does not occur in small walkers and runners, such as mice, but is frequent in large species, like ungulates. A large mass is compatible with migration because velocity of movement increases with mass and because energy storage is proportional to $m^{1.0}$, whereas the cost of locomotion is proportional to $m^{0.7}$. Small birds, however, often migrate because the cost of transport is lower in fliers than in walkers and the velocity of migration is high in flight and low in walking. Tucker (1975) suggested that a mass-specific cost of transport of 8.4 kJ/kg·km separates the larger species capable of migration (many of which do not migrate) from the smaller species that are unable to migrate. Thus, many fish of intermediate size migrate long distances, whereas equivalently sized mammals do not migrate (unless they fly).

9.10 CHAPTER SUMMARY

1. Vertebrates use many kinds of locomotion, including swimming, walking, running, flying, gliding, and brachiating.

2. Maximal aerobic swimming velocities in fish are correlated with normal body temperatures, although high velocities, which depend on anaerobic metabolism, tend to be temperature independent.

3. Fish muscles are made of red (aerobic) and white (anaerobic) fibers, the most active fishes having a larger proportion of red fibers.

4. The relative dependence in fish on aerobic or anaerobic metabolism to fuel activity depends on life-style: pursuing predators use principally aerobic metabolism, whereas sit-and-wait predators use principally anaerobic metabolism.

5. Rate of metabolism increases with velocity in amphibians, but at a given velocity the rate is greatest in anurans and least in lungless salamanders, the difference reflecting their abilities to acquire and transport oxygen.

6. Amphibians use both aerobic and anaerobic metabolism during activity: toads tend to forage actively, move slowly, use aerobic metabolism, and find protection in burrowing, large size, and the secretion of noxious compounds, whereas frogs are sit-and-wait predators, are inactive, move rapidly to avoid predation, and use anaerobic metabolism.

7. Aerobic scope in lizards is maximal at their preferred body temperature and is usually greater than 85% of the maximum at field body temperatures.

8. Aerobic scope is similar in legged and legless reptiles, although in snakes it is correlated with the surface area for gas exchange and the amount of myoglobin found in muscles.

9. All reptiles use both aerobic and anaerobic metabolism during locomotion; these scopes are positively correlated with each other and are greatest in the most active species.

10. The use of anaerobiosis by amphibians and reptiles in the field may only occur occasionally, most species making appropriate adjustments, such as increasing heart and ventilation rates, to permit their routine activities to be powered by aerobic metabolism.

11. When activity is integrated over a long period, all activity is aerobic because lactate is aerobically recycled after activity has finished.

12. Power output during flight is represented by a parabola with respect to velocity, the minimum corresponding to a velocity at which flight per unit of distance is minimal. Another, slightly higher velocity corresponds to the maximal distance obtained from a given energy expenditure.

13. For some birds and bats in flight, the power-velocity curves are comparatively flat, possibly in

relation to a large body mass; the distance that can be attained on a fixed energy expenditure increases if the curves are relatively flat.

14. Many factors, including body mass, wing span and area, and wing-beat amplitude and frequency, influence flight velocity.

15. Body size, wing loading, aspect ratio, and muscle composition collectively determine the flight pattern in birds and reflect their behavior and ecology. Flight pattern reaches its extreme among Procellariiformes, of which albatrosses have a very high aspect ratio and flap only rarely, and diving-petrels have very short wings and flap vigorously, the wings being short because of their use for propulsion under water.

16. The amount of mass that a flying bird can lift, either as food or as fat, is limited by body mass, wing loading, and pectoral muscle mass.

17. The energetics of bat flight is similar to that of birds, except that bats have a much narrower range in flight morphology and ecological roles.

18. An analysis of the morphology of pterosaurs suggests that they were capable of powered flight and soaring, but their flapping flight probably was slow and maneuverable. Like bats, pterosaurs had a much narrower range in habits than birds.

19. Powered flight evolved independently at least 3 times in vertebrates, although gliding evolved many more times, including in an amphibian, several reptiles, and many mammals.

20. Flighted birds have repeatedly evolved a flightless condition as a response to a limited resource base, but this occurs normally only on oceanic islands in the absence of mammalian predators.

21. Terrestrial quadrupeds have three kinds of gaits: walking, trotting, and cantering. Each can be characterized by length of step, frequency of stepping, and phase of footfall.

22. The shift from one gait to another minimizes energy expenditure per unit of distance, which leads to a discontinuous distribution of velocities in the field, the peaks corresponding to minimal expenditures associated with each gait.

23. Rate of metabolism during terrestrial locomotion is correlated with velocity up to the rates that correspond with the maximal aerobic scope, beyond which anaerobic metabolism is used.

24. Bipedal locomotion has the same cost as quadrupedal locomotion in small mammals but may be cheaper in species that weigh more than 5 kg.

25. Heat storage in warm environments may limit activity.

26. The cost of burrowing is much higher than movement on the surface, especially when the soil is compacted.

27. The cost of swimming in birds and mammals is higher than it is for fish, but is reduced if swimming is under water rather than at the surface, where wake formation occurs.

28. The cost of locomotion increases with body mass and is low in swimmers, intermediate in fliers, and greatest in walkers-runners.

29. Swimmers and fliers can move long distances because of the low cost and high velocities of these forms of locomotion and in large species, because of their high velocities and large energy stores.

10   Energy Budgets

10.1 SYNOPSIS

An energy budget is a quantitative summary of the energy expenditure of an organism. Energy budgets have been estimated in many ways involving both laboratory and field techniques, the most effective estimates being derived from isotopic measurements of energy exchange in free-living individuals. An energy budget measured in the field reflects the behavior of the individuals and the environmental conditions under which it is measured, which makes a comparison of species, of the impact of environments, and of time periods difficult because of its complexity. An energy budget is most valuable when subdivided into its component expenditures, so that equivalent expenditures can be compared. Field energy expenditures increase with body mass and are especially high if reproduction or extensive activity occurs, or if an endotherm faces conditions in which heat loss is elevated; field expenditures usually are 2 to 5 times the standard rates of metabolism. The allocation of energy to production (reproduction and growth) is influenced by many factors, including body size and whether the young are precocial or altricial. Among mammals lactation is more expensive than pregnancy. Whether field expenditures are limited by physiological performance, by the capacities of species to process food, or by the availability of resources in the environment is unclear, although they may be correlated with variations in body composition. The spatial area required by a vertebrate to harvest the energy expended, home range, reflects in part the level of energy expenditure, which varies with body size, behavior, and energy availability in the habitat, and ultimately with climate and food habits. Estimates of population energy expenditure are derived from combining the energy budgets of individuals with the demographic parameters of populations. The production of a population is correlated with its maintenance expenditure, as long as its thermal behavior is fixed, but varies inversely with maintenance expenditure when thermal behavior varies.

10.2 INTRODUCTION

Body maintenance, activity, and production require expenditures of energy. Expenditures tend to be high for temperature regulation in endotherms, especially at low ambient temperatures (see Section 5.2); for water and salt balance in aquatic species at high external concentrations (see Section 6.7); for high levels of activity, especially involving extensive periods of flight (see Section 9.6.1) or running (see Section 9.7.3); and for reproduction (see Section 10.6) and growth (see Section 10.7). At these times vertebrates require larger amounts of energy in the form of food than they would if they maintained smaller differentials with the environment, were less active, or were not reproducing. If energy availability is limited (but see King and Murphy 1985), the expenditures for these behaviors may be restricted, the behaviors may be staggered with respect to time, or energy stores accumulated during periods when energy availability was high must be drawn on (see Section 11.5.1). The amount of energy that is harvested and the allocation of this energy to various functions can be summarized in a numerical statement that is called an *energy budget*. How such budgets are estimated, how energy is partitioned within a budget, and how various components of a budget are interconnected are the subjects of this chapter.

10.3 ENERGY BUDGETS DESCRIBED

In its simplest form an energy budget is a number that indicates the amount of energy that is expended over a given period of time; it should not to be confused with the "thermal" budget that describes the magnitudes and directions of energy exchange in terms of radiation, conduction, convection, and evaporation-condensation (see Section 2.9). A simple summed energy budget masks complexity. Of greater interest is a statement that describes the amount of energy obtained by an organism and allocated to maintenance, activity, growth, and reproduction. Such a budget can be described by the following equation (Brody 1945):

$$\dot{E} = \dot{M} + \dot{P} + \dot{U} + \dot{F}, \qquad (10.1)$$

where $\dot{E}$ is the total amount of energy expended; $\dot{M}$, the energy expended for maintenance and activity (or metabolism); $\dot{P}$, the energy used for production (growth and reproduction); $\dot{U}$, the energy lost in the urine (mainly in the form of nitrogenous wastes); and $\dot{F}$, the energy lost in the feces (predominately as undigested food). Although each term could be either an amount of energy (e.g., kJ) or a rate (e.g., kJ/h), rates are usually used (and consequently a dot is placed over each term). $\dot{E}-\dot{F}$ is defined as *assimilated* energy, which reflects the ability of the digestive system to process the consumed food (see Chapter 12), and $\dot{E}-(\dot{U}+\dot{F})$ is defined as *metabolized* energy, the amount of energy that can be used for metabolism and production. The terms $\dot{M}$ and $\dot{P}$ in Equation 10.1 can be further broken down into their constituent parts: maintenance broadly defined can be subdivided into temperature regulation, other maintenance, and activity; production can be subdivided into reproduction and growth. The boundaries of these categories are not clear: should the cost of territorial defense or of food gathering to feed young be included under reproduction or activity? Clearly, the categories of maintenance and activity are heterogeneous. Furthermore, energy budgets often reflect energy availability in the environment: the tendency of many vertebrates to stagger in time costly activities (see Section 11.6.2), such as reproduction, fat storage, molting, and migration, has

been used as evidence of constraints on energy expenditure placed by energy availability or nutrient processing.

As interesting as energy budgets are in theory, they have severe shortcomings in practice. Their principal difficulties lie in the near impossibility of estimating the confidence limits of a summed budget (but see Travis 1982) and in our ignorance of the energy equivalencies of all behaviors seen in the field (McNab 1989b). Some estimates of the energy equivalencies of behavior are little more than educated guesses; others may not be that good! Nevertheless, of greatest value is a budget that is dissected into the expenditures associated with each frequent behavior. This dissection permits equivalent components of an energy budget to be compared in different species. Species may differ in their total budgets because they differ in size, emphasize different components of a budget, or live under different environmental conditions, and because the costs of various activities vary with species. Some components of a budget may be inflexible, whereas others may be modified in relation to climate, food habits, body size, and social system. Such complexity prohibits the rigorous testing of ecological theories that incorporate energetics.

10.4 ESTIMATING ENERGY BUDGETS

The energy budgets of vertebrates have been estimated by many techniques. Some estimates are derived from the amount of (1) food consumed, (2) fluctuation in body mass, (3) oxygen consumed, or (4) isotope turnover in a fixed period. Others are derived from a sum of component expenditures, which usually depend on time budgets, although they might apply to continuous measurements of oxygen consumption.

10.4.1 Food consumption. Kleiber and Dougherty (1934) reported one of the first attempts to use food intake to measure energy expenditure, but S. Charles Kendeigh (1949, 1969b, 1970) and his students obtained many of the early comparative data based on this technique. They developed a series of equations relating daily energy expenditure to body mass, equations that have been widely used (see Section 10.11.1). Kendeigh and colleagues used these data to try to account for the limits of distribution of some temperate and tropical passerines and for the presence or absence of migratory behavior (see Section 11.7.1). Unfortunately, these data were often collected under highly artificial laboratory conditions. For example, birds usually were confined in small or intermediate-sized cages and exposed to a range in temperatures and photoperiods. Activity, however, was unlikely to have been similar in type or amount to that found in the field. Schmid (1965c) demonstrated this potential discrepancy. He estimated from the composition and energy density of seeds in the crop that the daily energy expenditure in the field of mourning doves (*Zenaida macroura*) was 297 kJ, which is 4.4 times the value expected from the equations of Kendeigh (1970) at 30°C and 1.5 times the value expected at 0°C.

Some of these difficulties can be avoided by measuring food consumption in the field. As noted, Schmid used this technique on seed-eating doves, but it is most easily accomplished in carnivores because of the particulate nature and defined size of this food supply. Pearson (1964) was one of the first to do this. He examined the predation of carnivores on small rodents. Koplin et al. (1980) and Masman et al. (1988), among others, also used this technique to examine the intake of predatory birds.

10.4.2 Fluctuation in body mass.
Various authors, including Croxall (1982), Grant (1984), and Groscolas (1988), measured fluctuations in body mass to estimate energy budgets in the field in adult seabirds during incubation. Fasting in these species may last from a few days to several months. For example, male emperor penguins (*Aptenodytes forsteri*) incubating an egg in winter may fast for over 100 d during which they lose over 50% of their body mass (Groscolas 1988, Robin et al. 1988). The estimate of rate of metabolism depends critically on the assumptions made on the energy equivalency of mass loss (Groscolas 1988). Early and late stages of starvation generally involve the metabolism of proteins, whereas the (long) intermediate period involves fat metabolism. Because the net energy equivalency of lipids is about 2.2 times that of proteins, knowledge of the stage over which energy expenditure is being estimated is necessary. Some species that have much shorter fasting periods, such as petrels, may never enter the lipid stage of starvation found in emperor penguins. Furthermore, petrels derive oil from the food stored in their stomachs; the oil is metabolized during fasting. These complications make the

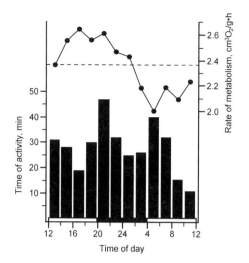

Figure 10.1 Rate of metabolism in the water vole (*Arvicola terrestris*) and time of activity as a function of the time of day. Source: Modified from Drożdż et al. (1978).

estimation of energy expenditure from mass loss difficult. As an added factor, the energy equivalency of mass loss in molting emperor and Adélie penguins is only one-fourth to one-third that in incubating emperor penguins (Groscolas 1988). Although, Helms (1963) tried to use the mass loss of passerines at night to estimate rate of metabolism, estimations from a change in mass are more precise during an extended period when food is not consumed, which is best seen in fasting seabirds or during migration, hibernation, and aestivation.

10.4.3 Oxygen consumption.
Oxygen consumption has been measured continuously in mammals enclosed in chambers supplied with food, nesting material, space for normal activity, and on occasion an activity wheel. These measurements are often referred to as *average daily rates of metabolism*. For example, Figure 10.1 illustrates data collected by Drożdż et al. (1978) on the water vole (*Arvicola terrestris*). Note that a cyclic pattern in rate occurs, in part reflecting a daily rhythm in activity. The mean summed daily rate of metabolism varies with many factors (Figure 10.2), such as ambient temperature, season, and photoperiod. These budgets are more likely to simulate accurately rates of metabolism in small free-living mammals than in small birds because small mammals require a small volume in which to maintain normal activity. This technique is impossible to apply in large vertebrates.

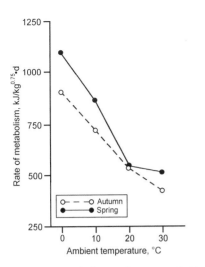

Figure 10.2 Rate of metabolism in the water vole (*Arvicola terrestris*) as a function of ambient temperature and season. Source: Modified from Drożdż et al. (1978).

10.4.4 Doubly labeled water. Nathan Lifson and colleagues (Lifson et al. 1955, Lifson and McClintock 1966) measured the CO_2 production of animals by using doubly labeled water. Their technique, modified for wild animals, was to catch an animal, inject it with $D_2^{18}O$, wait a few minutes for the isotopes to equilibrate with the $H_2^{16}O$ in blood, take an initial blood sample, release the animal, recapture it after an appropriate period, usually a day or two, and take a second blood sample. In comparisons of isotope abundances in the two samples, CO_2 production is estimated by the reduction in ^{18}O, after the decrease is corrected for the amount of ^{18}O lost as water by the decrease in deuterium abundance. The radioactive isotope tritium may replace deuterium.

This technique has been validated repeatedly in the laboratory. Randolph (1980) made simultaneous measurements of oxygen consumption, CO_2 production, and $D_2^{18}O$ turnover in two rodents, *Peromyscus leucopus* and *Tamias striatus*, when they were given food and space sufficient to simulate field energy expenditures. He found that CO_2 production in the laboratory, as estimated by $D_2^{18}O$, was 6.9% less than that measured by gas analysis. He found no significant difference between the amount of CO_2 produced by these rodents when they were confined in large chambers under field conditions (when gas analysis was used) and that produced when they were free-living in the field (when isotopes were used). Reevaluations of

this technique have suggested that it is accurate only to within approximately 10% (Nagy 1989), especially in high-humidity environments.

LeFebvre (1964) first used this method in the field to measure the flight energetics of pigeons, and Mullen (1970, 1971a, 1971b) later used it to measure field expenditures in desert rodents. Nagy (1987) summarized these and other measurements. Doubly labeled water is especially useful for estimating the energy expenditures of poikilotherms in the field because these species potentially show great variability in body temperature as a result of the differential use of behavioral temperature regulation, which makes their energy budgets difficult to estimate from laboratory measurements of oxygen consumption.

10.4.5 Tritium and ^{22}Na. The turnover of tritium alone has been used to estimate fasting rates of metabolism, which is valid as long as the animals remain in water balance and the only source of body water is from the oxidation of food reserves (Nagy 1975). (For an estimate of the error produced by drinking water when energy expenditure is estimated from water flux, see Nagy et al. 1984b, Goldstein and Nagy 1985, and Costa and Trillmich 1988.) It can also be used to estimate field energy expenditures if the only sources of water are from metabolism and the diet (preformed water), but here the composition and water content of the diet must be known (Shoemaker et al. 1976). For example, Shoemaker et al. estimated the field expenditure of jackrabbits (*Lepus californicus*) in this manner, as did Kooyman et al. (1982) and Davis et al. (1983) for penguins. Gallagher et al. (1983) used the radioactive isotope ^{22}Na to measure sodium turnover, which itself is strongly positively correlated with food intake.

10.4.6 Time budgets. Energy budgets estimated from time budgets are constructed from observations of the time spent in various activities, each of which must have a known (or assumed) energy equivalency. Pearson (1954b) wrote the original, classic paper on this relation in Anna's hummingbird (*Calypte anna*). He showed that although the hummingbird spent about 44% of 24h perching (i.e., 82% of the 12-h, 52-min daylight period), only 51% of its energy expenditure occurred during this time (Figure 10.3) under the assump-

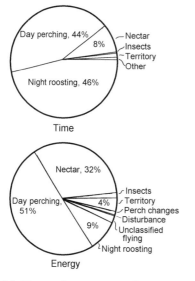

Time

Energy

Figure 10.3 Time and energy expenditures over 24h in the field by Anna's hummingbird (*Calypte anna*). Source: Modified from Pearson (1954b).

tion that the bird became torpid at night. Revised estimates of the cost of hovering (Lasiewski 1963, Wolf and Hainsworth 1971) indicated that *C. anna* spent 66% of the daytime expenditure during perching; that is the cost of flight was not as great as Pearson thought. Stiles (1971) and Carpenter (1976) provided other estimates of the energy expenditure of hummingbirds from time budgets. One reason why hummingbirds spend such a large proportion of the daytime perching may be related to the time required to empty their crops (Karasov et al. 1986a). These observations emphasize that the contribution of a behavior to an energy budget is weighted by the duration and energy intensity of the behavior.

The relation between an energy budget and a time budget is generally given by

$$\dot{E} = \sum_{i=1}^{n} \dot{b}_i \cdot t_i, \qquad (10.2)$$

where $\dot{E}$ is energy expenditure (kJ/d), t_i is the time (h) spent in activity i, $\sum_{i=1}^{n} t_i = 24$h (the time budget), and $\dot{b}_i$ is the energy equivalency (kJ/h) of activity i (McNab 1963b, 1989b; Travis 1982). Such budgets are most widely used in animals whose activities can be easily monitored, especially birds.

The complexities involved in Equation 10.2 are clearly presented in an analysis by Ettinger and King (1980) of the energy budget of the willow fly-catcher (*Empidonax trailli*) during the breeding season in eastern Washington. Based on the suggestions of King (1974) and Ettinger and King (1980), the observations of Walsberg and King (1978), and guesses as to the cost of some activities, Ettinger and King estimated that a male willow flycatcher expends about 56.2 kJ/d during the prenesting phase, when territorial establishment occurs and singing is at its greatest. Austin (1978) provided similar analyses for verdins (*Auriparus flaviceps*); Powell (1979), for fishers (*Martes pennanti*); Belovsky and Jordan (1978), for moose (*Alces alces*); and Collier et al. (1975), for lemmings (*Lemmus sibiricus*).

The problems associated with converting a time budget into an energy budget have been thoroughly discussed (King 1974, Mugaas and King 1981, Williams and Nagy 1984, Goldstein 1988), one of the most obvious problems being the energy equivalencies of various activities. Does perching by a flycatcher during the day in the field (where many visual stimuli including food, partners, competitors, and predators are present) raise the rate of metabolism by 70% compared to the rate when perching at the same temperature at night? If not, by how much? How does this differ in woodpeckers? Or hawks? Is the presumptive difference between woodpeckers and hawks simply due to the difference in body size? Furthermore, an energy budget derived from a time budget usually has no estimates of variance: it simply is the sum of a series of products, although in theory each of the elements of an energy budget, including the time devoted to various activities and their energy equivalencies, has its individual variance. Travis (1982) suggested a method to estimate the variance of the energy budget from knowledge of the variances of its components. Even when energy budgets derived from time budgets agree with expenditures derived from the use of doubly labeled water, the variances around the mean are uncorrelated (Williams and Nagy 1984), so the correspondence between the means may well be fortuitous. Goldstein (1988) recommended that sensitivity analyses be made of the components of a time-energy budget.

The difficulty in converting a time budget into an energy budget can be reduced by direct measurements of the cost of various activities for species with varying morphologies and behaviors. For example, the cost of calling in *Hyla* might exceed 10 times the standard rate (Taigen and Wells 1985, Taigen et al. 1985, Prestwich et al. 1989), and the cost of singing in the Carolina wren (*Thryothorus ludovicianus*), at least 5.2 times the basal rate (Eberhardt 1994). The cost of feather synthesis in birds varies with body mass and with the basal rate of metabolism: birds with high basal rates at a particular mass have a high cost for feather production (Lindström et al. 1993). This cost may be appreciable; the increment is 47% of the basal rate in the bluethroat (*Luscinia svecica*) when averaged throughout the molting period.

10.4.7 Combined methods. Energy budgets estimated from time budgets can be checked by field estimates of food consumption or isotopic dilution. Koplin et al. (1980) compared time-energy budgets with food consumption in American kestrels (*Falco sparvarius*) and white-tailed kites (*Elanus leucurus*). Time-budget estimates were within 5% or deviated as much as 50% with the estimates based on food consumption, depending on the assumptions of the time-budget model. Masman et al. (1988) concluded that their estimates of energy expenditure in European kestrels (*F. tinnunculus*) derived from food intake in the field were about 7% greater than those measured by doubly labeled water. Utter and LeFebvre (1973) made early comparisons of doubly labeled water and time-budget analyses on purple martins (*Progne subis*); Mullen and Chew (1973), on *Perognathus*; and Weathers and Nagy (1980), on phainopeplas (*Phainopepla nitens*). The first two studies gave estimates that were in agreement within 12%, whereas estimates of daily expenditures in the phainopeplas were 39% lower when time budgets were used.

A discrepancy between time-energy and isotopically derived budgets often reflected low estimates derived from a time-budget analysis (see Williams and Nagy 1984). This may occur in several ways. For example, McNab (1978b) estimated field energy expenditures in three-toed sloths (*Bradypus*) from field measurements of body temperature and activity and laboratory measurements of oxygen consumption. These estimates later turned out to be only 71% of the mean value measured by isotopes in the field (Nagy and Montgomery 1980). The principal difficulty in this case was in estimating the cost of activity. The use of appropriately complex measures of the physical environment is also required to obtain an accurate estimate of the cost of temperature regulation (Mugaas and King 1981, Buttemer et al. 1986, Goldstein 1988). Equivalent or standard operative temperatures (see Section 2.9.3) and the incorporation of humidity are preferred to the use of air temperature alone. The discrepancy in estimates of phainopepla energy budgets may have resulted principally from underestimating maintenance metabolism at high ambient temperatures, when the birds were subjected to appreciable solar heat loads.

Weathers et al. (1984) thoroughly examined the difficulties associated with estimating field energy budgets from time budgets in the loggerhead shrike (*Lanius ludovicianus*). These authors measured field expenditure with doubly labeled water and estimated field expenditures from time budgets using various modifications of the Pearson (1954b) method that incorporated laboratory-based estimates of the energy equivalencies of various activities, or simulated field expenditures using the Kendeigh et al. (1977) equations, as modified by Koplin et al. (1980). Various combinations of temperature, convection, and activity costs were used. The Pearson method coupled with the operative temperature and convective exchange gave the estimate most similar to the measurements obtained from doubly labeled water. The accuracy of this estimate is derived from an accurate estimate of thermoregulatory costs and from measured costs for various activities.

In summary, the use of isotopes is the only reasonably accurate way to *measure* energy expenditures in the field at the present time, although field expenditures may be *approximated* by the measurement of oxygen consumption in small mammals in the laboratory when exposed to field conditions. Time budgets may be a reliable means of *estimating* field expenditures, but only if the assumed equivalencies of activities are valid, and if an accurate picture of the physical conditions encountered is used. Such assurances are not available for time-energy budgets, except as obtained by a comparison with isotopic measurements (McNab 1989c).

10.5 FIELD ENERGY EXPENDITURES

The energy expenditures of free-living mammals and birds, as measured by doubly labeled water, have been summarized (Nagy 1987, 1994). Curves for birds, rodents, and lizards are plotted in Figure

Table 10.1 Field energy expenditures (FEE)* of vertebrates determined by doubly labeled water

Group	n	a	b	r^2
Reptiles				
Iguanid lizards	25	0.22	0.799	0.982
Birds				
All birds	50	10.89	0.640	0.907
All birds[†]	28	13.92	0.657	0.966
passerines	26	8.89	0.749	0.899
other than passerines	24	4.80	0.749	0.899
seabirds	15	8.02	0.704	0.911
Mammals				
All mammals	61	5.27	0.732	0.961
Small mammals[‡]	74	7.86	0.621	0.864
All mammals[†]	15	9.69	0.634	0.955
Eutherians	44	4.63	0.762	0.972
Marsupials	17	10.8	0.582	0.978

*Expenditures are in the form $FEE = a \cdot m^b$, where FEE is in the units of kJ/d and body mass is in grams.
Sources: Derived from Nagy (1982, 1987, 1994), [†]Daan et al. (1991), and [‡]Speakman (2000).

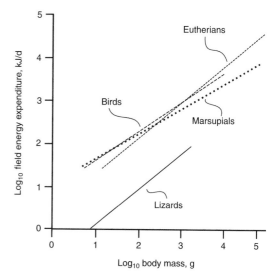

Figure 10.4 Log_{10} field energy expenditures, measured by doubly labeled water, in iguanid lizards, nonpasserine and passerine birds, and rodents as a function of log_{10} body mass. Sources: Modified from Nagy (1982, 1987, 1994).

10.4 as a function of body mass. Field expenditures are proportional to $m^{0.76}$ in eutherian mammals and to $m^{0.73}$ for mammals generally; to $m^{0.64}$ or $m^{0.75}$ in birds, depending on whether birds are treated together or whether they are divided into passerines and nonpasserines; and to $m^{0.80}$ in lizards (Table 10.1). As in standard rates of metabolism, mass-specific rates decrease with an increase in mass. Because few $D_2{}^{18}O$ measurements are available for vertebrates that weigh more than a few hundred grams (due to the cost of the technique), caution should be used in attaching too much confidence in the fitted powers and coefficients describing field expenditures because the accuracy of these estimates depends on the range in body mass.

10.5.1 Mammals. Laboratory measurements of the daily rate of oxygen consumption at 20°C in insectivores and small rodents are proportional (approximately) to $m^{0.50}$ (French et al. 1976, McNab 1980b, Nagy 1987) and similar to or slightly less than field expenditures, a difference that may reflect lower temperatures in the field, often between 4 and 12°C (Figure 10.2). Daily energy expenditures are inversely correlated with ambient temperature in part because time of activity, at least in large herbivores (Belovsky and Slade 1986), increases at low environmental temperatures.

In his most recent survey of data, Nagy (1994) discussed some of the factors that influence field expenditures. Body mass was the most important factor, accounting for 96% of the variation for 61 species of mammals, or 98% of the variation for 44 species of eutherians. Still, an appreciable residual variation existed: in eutherians the maximal field expenditure, corrected for body mass, was 6 times that of the minimal field expenditure. Marsupials had a different scaling relationship than eutherians, and desert eutherians had low field expenditures, but no correlation was found with diet. Somewhat surprisingly, Nagy did not inquire whether the residual variation in field rates correlated with the residual variation in basal rate. When the field expenditures of eutherians (Nagy's Table 2, plus *Bradypus* [Nagy and Montgomery 1980]) are compared to basal rates of metabolism, both expressed as a percentage of the eutherian values expected from mass (Figure 10.5), field rates were strongly correlated with basal rate ($P < 0.0001$; $n = 26$; $r^2 = 0.673$). This correlation was not found in marsupials (Figure 10.5), principally because they have such a small variation in basal rate after the effect of body mass is eliminated (see Section 13.5).

Speakman (2000) analyzed the field energy expenditures of small mammals measured by the doubly labeled water method. He concluded that these expenditures were affected by several factors,

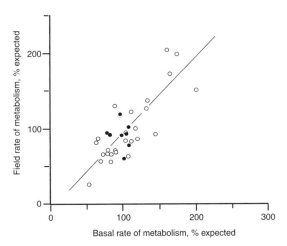

Figure 10.5 Field energy expenditures in eutherians (o) and marsupials (•) expressed relative to the expenditures expected from mass (Nagy 1982), as a function of basal rate of metabolism, expressed relative to the basal rate expected from mass (McNab 1988a). Sources: Derived from Nagy and Montgomery (1980) and Nagy (1994).

including (1) body mass, (2) ambient temperatures, (3) season, (4) latitude, (5) diet, and (6) standard rate of metabolism. Body mass alone accounted for 86.4% of the variation in field rate. Speakman recalculated the correlation several times, dropping "outliers," using reduced major axes, using repeated measurements for many species, and including a "phylogenetic" analysis, but none of these modifications had an appreciable effect on the conclusion that body mass was the most important factor determining field energy expenditures. Field expenditures, as expected, correlated negatively with ambient temperature, which was clearly shown in a pocket-mouse (*Perognathus formosus*) and a kangaroo-rat (*Dipodomys merriami*). Field expenditures were 46% greater in summer than winter, which implies either that thermal acclimatization and behavior effectively reduced the effect of low ambient temperature in winter, or that summer expenditures reflected higher levels of activity or the cost of reproduction, or both. Some dietary classes (grazing, insectivory, carnivory, nectarivory/frugivory) were correlated with high field expenditures, whereas others (folivory, granivory) were associated with low field expenditures. Finally, field expenditures were correlated with "resting" (~basal) rate of metabolism beyond the influences of mass and "phylogeny." A difficulty with this data set, and therefore in this analysis, is that by restricting it to small (<4 kg) mammals, the

ecological diversity of the studied species was small in that most of the species included lived in mid-temperate grasslands or were desert rodents. In any case, this analysis showed the complexity inherent in field estimates of energy expenditure, even given the narrow subset of mammals included. This complexity, as always, is plagued by factor interaction, such as between temperature and season, temperature and latitude, and diet and taxonomy.

The daily field expenditures of mammals according to Nagy (1994) are usually between 2.0 and 2.5 times the basal rate, although this conclusion must be qualified in several ways:

1. Such ratios must incorporate *measured* basal rates, not rates that are assumed to fall on some standard curve. Thus, most of the difference in the field expenditures found between the jackrabbit *Lepus californicus* (Shoemaker et al. 1976) and the sloth *B. variegatus* (Nagy and Montgomery 1980), two mammals that have a similar body mass, is associated with a difference in basal rate: the field expenditure and basal rate of *L. californicus* are 4.0 and 3.3 times, respectively, those of *B. variegatus*.

2. Some of the differences in field rates reflect an active versus a sluggish life-style, which are often correlated with the basal rate of metabolism (because of the correlation of basal rate with muscle mass; see Sections 5.4 and 10.8). Thus, *B. variegatus*, which has a very low basal rate (42% of the expected rate), has a ratio between field and basal expenditures that is only 1.7 (Figure 5.22), so that the mean field expenditure in this sloth is still only 71% of the *basal* rate expected from the all-mammal curve! The cost of activity, therefore, cannot be assumed to be a constant multiple of the basal rate (see McNab 1980b).

3. Because field rates are sensitive to ambient temperature, this ratio is also temperature sensitive. Antarctic fur seal (*Arctocephalus gazella*) females nursing pups on Bird Island, South Georgia, where air temperatures were between −1 and 8.5°C, had field expenditures that were 3.4 times the (expected) basal rate. Northern fur seal (*Callorhinus ursinus*) females nursing pups under similar environmental conditions had similar field expenditures (Costa and Gentry 1986). In contrast, Galápagos fur seal (*A. galapagoensis*) females nursing pups, which encountered air temperatures between 20 and 30°C, had field expenditures that were only 1.1 times the (expected) basal rate (Costa and Trillmich 1988).

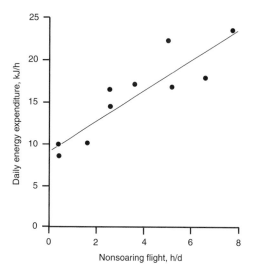

Figure 10.6 Field energy expenditure in the Eurasian kestrel (*Falco tinnunculus*) as a function of the proportion of the day spent in flight. Source: Modified from Masman and Klaassen (1987).

The ratio of field expenditures to "resting" rates of metabolism among small mammals varied from 1.6 to 7.6, with the mean being 3.4, the limited range implying a functional connection between field and "standard" rates of metabolism (Speakman 2000).

10.5.2 Birds. Birds would be expected to have higher field expenditures than mammals because they often have high basal rates and flight is an intensive means of locomotion. For instance, the mean daily field expenditures of Eurasian kestrels is directly correlated with the number of hours per day in which the kestrels use powered flight (Figure 10.6; Masman and Klaassen 1987), and nectar-feeding honeyeaters (Meliphagidae) have much lower field expenditures, corrected for mass, than nectar-feeding hummingbirds (Weathers et al. 1996), reflecting the low cost of foraging by honeyeaters (while perching) and the high cost by hummingbirds (while hovering). Measurements indicate, however, that birds have field expenditures similar to (Nagy 1987) or slightly greater than (King 1974) those of mammals of the same mass (Figure 10.4). Many measurements were made in breeding birds, when the cost of feeding and foraging for the young should increase energy expenditure. At masses greater than 100 g, birds usually have field expenditures that vary between 2.5 and 5.2 times the basal rate. At the upper end of this range, the brown noddy tern (*Anous stolidus*) and wedge-tailed shearwater (*Puffinus pacificus*), both of which fly long distances, expend at a rate between 4.8 and 5.2 times the measured basal rate (Ellis 1985). Some desert partridges, which are sedentary compared to seabirds and have low basal rates, also have very low field expenditures, between 1.5 and 1.9 times the measured basal rates (Goldstein and Nagy 1985, Kam et al. 1987).

The field expenditures of seabirds have been extensively examined. Pelagic species, often characterized by a clutch of one, low nestling growth rates, and an accumulation of large fat reserves in the nestlings, are thought (Ashmole 1963b, Lack 1968b, Harris 1977, Nelson 1977) to face a limited food resource (but see Ricklefs 1979, 1983; Shea and Ricklefs 1985). These species usually have a flight morphology (high aspect ratio and low wing loading) that facilitates a low expenditure for flight foraging at great distances from their nesting sites. The sooty tern (*Sterna fuscata*) has a cost for flight that is only 4.8 times the standard rate because it mixes soaring and flapping flight (see Section 9.6.2) and in the field expends energy only at 3.0 times the (low) measured basal rate (Flint and Nagy 1984), even though it flies far offshore to feed. This species may be so efficient in flight that it can fly indefinitely (Ashmole 1963a). Still lower costs for foraging occur in medium-sized to large albatrosses, as might be expected from their low wing loadings (110–150 N/m²) and high aspect ratios (15.3–15.6; Warham 1977): the Laysan albatross (*Diomedea immutabilis*) has a foraging expenditure that is 2.7 times the (expected) basal rate (Pettit et al. 1988); the gray-headed albatross (*D. chrysostoma*), an expenditure at 2.5 times the (expected) basal rate (Costa and Prince 1987); and the wandering albatross (*D. exulans*), an expenditure at only 1.8 times its (expected) basal rate (Adams et al. 1986). Albatrosses use flapping flight only 6% to 7% of the time flying over the ocean (Pennycuick 1982); they usually rely on dynamic soaring (Adams et al. 1986, Jouventin and Weimerskirch 1990). In the gray-headed albatross the overall field expenditure (including time onshore and at sea) is only 1.8 times the predicted basal rate (Costa and Prince 1987). Associated with this highly efficient flight, wandering albatrosses have been shown through the use of satellites (Jouventin and Weimerskirch 1990) to fly between 3600 and 15,000 km on a foraging trip during the nesting season.

The southern giant-petrel (*Macronectes gigan-teus*), which is the largest (4.5-kg) member of the Procellariidae and half as massive as the largest albatrosses, has a lower aspect ratio (11.9) and a higher wing loading (163 N/m²). Consequently, it uses flapping flight more frequently (24% of the time) when foraging at sea (Pennycuick 1982). Furthermore, giant-petrels feed their young 2 to 3 times more frequently than do albatrosses. Indeed, giant-petrels have a foraging expenditure that is 6.3 times the basal rate (Obst and Nagy 1992), much greater than that found in albatrosses. The high cost of flapping in this species leads to a high field energy budget, 4.6 times the basal rate, which is similar to the field expenditure of its much smaller (0.6-kg) relative, the wedge-tailed shearwater (*Puffinus pacificus*).

Penguins have field expenditures that vary between 2.5 and 3.0 times the basal rate (Ellis 1985), reflecting the lower cost of underwater "flight." Of greater interest are seabirds that fly in air and use their wings for underwater propulsion, namely, auks (Alcidae) in the Northern Hemisphere and diving-petrels (Pelecanoididae) in the Southern Hemisphere: they have exceedingly high field expenditures for their masses (Roby and Ricklefs 1986, Gabrielsen et al. 1991). The dovekie (*Alle alle*), an alcid, had a field expenditure that is 50% greater than expected from body mass, but its ratio of field to basal rate is only 3.9 because this species has a high basal rate (Gabrielsen et al. 1991). High field expenditures in these species also reflect the very high cost of aerial flight as a result of low aspect ratios and high wing loadings. For example, the fairy prion (*Pachyptila turtur*), a procellari-iform that weighs 132 g and hydroplanes across the surface of the ocean, has a wing loading of 39 N/m² and an aspect ratio of 8.6, whereas the common diving-petrel (*Pelecanoides urinatrix*) of similar mass (119 g) has a wing loading of 67 N/m² and an aspect ratio of 7.0 (Warham 1977). Associated with this difference, prions have field ex-penditures that are only 61% to 68% those of diving-petrels (Taylor et al. 1997).

Røskraft et al. (1986) found a further complica-tion: a high rank in the dominance hierarchy of great tits (*Parus major*) and in the pied flycatcher (*Ficedula hypoleuca*) correlated with a high rate of metabolism and large heart mass. Although some concern exists that the correlation between rate and rank might be associated with differential levels of activity during the measurement of metab-olism, the correlation with heart mass suggests that the correlation is real. A correlation of rank with resting or basal rates of metabolism has also been found in willow tits (*P. montanus* [Hogstad 1987]) and dippers (*Cinclus cinclus* [Bryant and Newton 1994]).

Schoener (1969, 1971) described another approach to energy expenditure. He distinguished between species that maximize reproductive success by maximizing energy intake and those in which reproductive success is not limited by energy intake and therefore the time spent feeding is min-imized. Whereas energy maximizers minimize for-aging, time minimizers tend to maximize inactivity or activities other than foraging. Hixon and Carpenter (1988) explored this distinction in two species of hummingbirds, the rufous (*Selasphorus rufus*) and the Costa (*Archilochus costae*). In the Sierra Nevada, California, migrant rufous hum-mingbirds were energy maximizers and nonmi-grant Costa hummingbirds were time minimizers: rufous hummingbirds spent more time foraging and less time sitting and gained mass more rapidly than did Costa hummingbirds. The rufous hum-mingbird gained mass 4 to 8 times as rapidly as the Costa, mainly because the rufous entered torpor at night, thereby reducing energy expenditure and conserving the fat deposits preparatory to migra-tion. Costa hummingbirds rarely entered torpor. Nevertheless, migrant rufous hummingbirds often defended a nectar source against other humming-birds for a few days after the source no longer had nectar sufficient to meet the daily requirements of the hummingbird, because of the great cost associ-ated with finding a better source and displacing its defender (Heinemann 1992).

Bryant and Tatner (1991) summarized the correlates of field energy expenditures in "small" birds. They include brood feeding rates, proportion of time in flight, frequency of activity, body mass, food availability, and (sometimes) ambient temper-ature (although most of the measurements were made during the breeding season when low tem-peratures were not normally encountered). Field expenditures usually were 2 to 4 times the basal rate. Species with expenditures that exceeded 4 times the basal rate not only were breeding, as were almost all species studied, but also fed in flight (swallows, bee-eaters) or under water (dipper, king-fisher). Between 40% and 50% of the species and 16% and 30% of the individuals studied expended energy at rates that exceeded 4 times the (assumed) basal rate. The difficulty in evaluating these data, especially in light of the claim by Drent and Daan (1980) that field expenditures of birds are limited to 4 times the basal rate (see Section 10.6.2), is in knowing the length of time over which this or any other limit applies (Bryant and Tatner 1991). An unnecessary complication is uncertainty as to the basal rate in species that have had field expenditures measured. (Measure basal rate!)

10.5.3 Ectotherms.

Most measurements of field rate of metabolism in ectothermic vertebrates have been obtained from lizards. As expected, they have very low expenditures compared to endotherms, only about 7% of those found in rodents at the same mass (Figure 10.4). Nevertheless, the field expenditures of *Sauromalus*, *Cnemidophorus*, *Callisaurus*, and *Sceloporus*, like those of mammals, are usually 1.3 to 2.5 times the rates measured in the laboratory under standard conditions and field body temperatures (Bennett and Nagy 1977; Nagy 1982, 1983). Of special interest are the field measurements, based on $D_2{}^{18}O$, of the Komodo monitor (*Varanus komodoensis*) obtained by Green et al. (1991). This is the largest lizard: it shows some independence of body temperature from ambient temperature because of its large mass (McNab and Auffenberg 1976), which raises the possibility that this lizard in the field has expenditures somewhat more like those of endotherms than of most lizards. In fact, the few field measurements in Komodo monitors that weighed from 2.2 to 45.2 kg were 1.8 to 2.1 times those expected from the lizard field expenditure equation of Nagy (1982). How these values reflect its thermal biology is unclear.

The predatory style of lizards impacts their field expenditures. In two Kalahari lizards belonging to the genus *Eremias*, the field expenditure was 2.2 times the resting rate in a sit-and-wait species, whereas it was 3.1 times the resting rate in an actively foraging species (Nagy et al. 1984a). This pattern appears to be general: sit-and-wait predatory lizards (such as *Callisaurus* and *Sceloporus*) have lower daily energy expenditures in the field than do pursuing predators (such as *Cnemidophorus*), owing to the greater intensity and amount of activity and the generally higher body temperatures found in pursuing predators (Anderson and Karasov 1981; see also Huey and Pianka 1981, Andrews 1984). The correlation of energy expenditure with behavior occurs within species: the expenditure of *Cnemidophorus hyperythrus* varies as the amount of activity varies with habitat (Karasov and Anderson 1984). That the expenditures of diurnal skinks (*Mabuya*) and nocturnal geckos (*Pachydactylus*) are identical and similar to those of iguanids of the same mass (Nagy and Knight 1989) is surprising because geckos usually operate at lower body temperatures than either skinks or iguanids.

The measurement of the field expenditures of vertebrates has just begun. Many more data are required on both endotherms and ectotherms. The influence of behavior (such as daily and seasonal torpor, feeding mode, migration, territorial defense, and reproduction) on field expenditures and the relation between field and "standard" expenditures need to be examined (some of which are considered in Chapter 11). The principal difficulty with interpreting field expenditures is that they are very complex; scaling field expenditures to body mass leaves much residual variation and does not clearly describe the influence of body mass because so much heterogeneity exists in these data. Furthermore, the scaling curves for field expenditures that Nagy (1987) presented use several values for one species, a decision that is tantamount to weighing some species more than others. Future progress in the analysis of field expenditures will depend on combining them with laboratory measurements under standard conditions and with time-budget analyses. Such a combination will permit a field expenditure to be dissected into its fundamental components and permit the comparison among species of equivalent components of an energy budget (McNab 1989b).

10.6 ENERGETICS OF REPRODUCTION

In addition to the expenditures of energy for maintenance and activity—the two largest components of most daily energy budgets—individuals also expend energy for reproduction. A complication is that the boundaries for the expenditure associated with reproduction are difficult to draw. Is the cost of territorial defense in birds to be included? Or the cost of flight to catch food used to feed altricial young? Here an inclusive definition is used for this topic.

10.6.1 Mammals.

Randolph et al. (1977) described an analysis of the energetics of reproduction in *Sigmodon hispidus*. A female cotton rat weighing 126 g expended some 5472 kJ for maintenance at 22°C over a 38-d period; it spent an additional 2020 kJ during this period when she was pregnant and nursed five young to weaning, a 37% increase in energy expenditure. Of this additional expenditure, 610 kJ was used during pregnancy, which lasted 26 d, and 1410 kJ during lactation, which lasted about 12 d. Consequently, the incremental daily energy expenditure during pregnancy was 23.5 kJ (which is 16% of the maintenance expenditure) and that during lactation was 117.5 kJ (5 times the rate during pregnancy and 82% of the maintenance expenditure). The greater expenditure during lactation is related to the larger mass of nursing young and to the cost of their locomotion and temperature regulation, as well as to the cost of growth itself.

The few data available on other species indicate that pregnant eutherians generally have rates of metabolism that are about 25% above the nonpregnant rate at all body masses, whereas the rate during lactation is approximately double the resting rate at an adult mass of 25 g and 1.5 times the resting rate at a mass of 50 kg. Gittleman and Thompson (1988) estimated that of the total expenditure for reproduction by female eutherians, approximately 20% is used during gestation and 80% during lactation, although they warned that this allocation may be subject to marked variation when the differences among species in gestation period, reproductive strategies, sexual selection, and sex allocation are taken into consideration. They further concluded that some behaviors are reduced so that the increment for reproduction has a smaller impact on the total energy budget. However, some evidence shows that female long-eared bats (*Plecotus auritus*) during lactation have the same energy expenditures as nonreproductive females, which suggests that lactating females make some (unknown) compensatory reduction in other expenditures (McLean and Speakman 1999).

One of the most peculiar patterns of reproduction found in mammals occurs in the East African mole-rat (*Heterocephalus glaber*). It lives in subterranean colonies of up to 300 individuals. Its reproductive behavior has been compared to that of termites, where only one female and one to three males in a colony breed, nonbreeders maintain and defend the colony, and breeding females are much larger than nonbreeding females and males (Jarvis 1981, Lacey and Sherman 1991). If the breeding female dies, some nonbreeding females grow rapidly and compete for reproductive dominance. Litter size usually varies from 5 to 15, although newly reproductive females may have as few as one pup and the maximum known litter size is 27 (Jarvis 1991)! As in other mammals with low basal rates, *H. glaber* shows an increase in rate of metabolism during pregnancy (by 58%) and an even higher increase during lactation (by 150% over the prepregnancy rate; Urison and Buffenstein 1995). The cost of pregnancy increased from one to five young but was independent of litter size up to 17. The large, highly variable litter size may be the ultimate justification for this reproductive system, but it requires a division of labor in the colony (Lacey and Sherman 1991) such that the dominant female is given first access to food returned to the colony and that other costly activities (such as foraging, burrow maintenance, and defense) are divided among the nonreproductive members of the colony (Jarvis 1981, Urison and Buffenstein 1995).

Nicoll and Thompson (1987) presented evidence that species with low basal rates have a greater increase in rate of metabolism during pregnancy (and lactation?) than do species with high basal rates. In other words, pregnancy and lactation require high rates of metabolism (presumably to pay the direct and indirect costs during the "synthesis" of young), and that increment in metabolism may be continuous (as in species with high basal rates) or discontinuous (in species that increase rate of metabolism only during reproduction). This observation is compatible with the view (McNab 1980b) that a high basal rate facilitates a high reproductive output (see Chapter 13).

If lactation is more expensive than pregnancy, does this mean that the cost of producing altricial young from conception to weaning is more expensive than producing precocial young? Or, to take an extreme case, is marsupial reproduction more expensive than eutherian reproduction? Should such comparisons be made in absolute units, such as joules, relative to a sliding scale like basal rate of metabolism, or relative to the probability that young will survive to the age of reproduction? At present these questions go unanswered yet have implications for the theories of ecology and evolution (see Chapter 13). Thompson and Nicoll (1986), at least, suggested that the expenditure for reproduction appears not to be correlated with the form of reproduction but principally with level of basal rate.

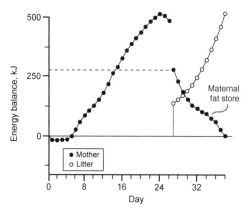

Figure 10.7 Energy balance in *Sigmodon hispidus* before and after the birth of a litter. Source: Modified from Randolph et al. (1977).

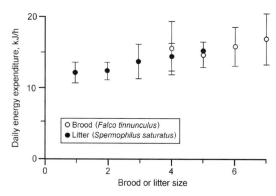

Figure 10.8 Daily field energy expenditures in the ground-squirrel *Spermophilus saturatus* as a function of litter size and in the Eurasian kestrel (*Falco tinnunculus*) as a function of brood size. Sources: Derived from Masman et al. (1989) and Kenagy et al. (1990).

Künkele and Trillmich (1997) examined whether lactating females expend more if young are precocial or altricial, in a comparison of guinea pigs (*Cavia*) and white rats (*Rattus*). They showed that lactating guinea pigs, whose young wean at an age of 23 d but can survive the termination of lactation at the age of 5 d, reached their maximal daily energy expenditures when the young were 6 to 7 d old, whereas the maximal expenditures of lactating white rats occurred at 17 d. Guinea pigs maximally expended energy at 1.9 times the basal rate or 3.9 times the basal rate if milk production is included. The peak sum of energy expenditure and milk production of white rats was 6.2 times the basal rate. The difference between these rodents mainly results from guinea pig young starting to eat solid food from the third day after birth, whereas white rat young only switch over to solid food on day 18. In the face of a food shortage, the early use of solid food by young guinea pigs may permit this species to continue reproduction, whereas rodents with altricial young might be forced to abandon reproduction.

The high cost of lactation requires that energy intake must increase, or that the allocation of energy to other functions be reduced (Gittleman and Thompson 1988). Many mammals time reproduction so that lactation coincides with abundant food supplies, whereas others transfer energy to the period of lactation by withdrawing from an energy store. *Sigmodon* in the laboratory (when food was plentiful) stored approximately 280 kJ as body fat during pregnancy (Randolph et al. 1977), which was used by the end of lactation (Figure 10.7). Similar observations were noted in the ground-

squirrel *Spermophilus undulatus* (Kiell and Millar 1980) and in *Peromyscus leucopus* (Millar 1975). In these cases, energy was transferred from the period of pregnancy (when energy demand is lower) to that of lactation (when energy demand is higher).

The high cost of reproduction in mammals raises the question as to the impact of litter size on energy expenditure. In the golden-mantled ground-squirrel (*S. saturatus*) field measurements by doubly labeled water during the period of peak lactation (between 31 and 35 d after birth) indicated that maternal expenditure increased slightly with an increase in litter size from 1 to 5 (Kenagy et al. 1990; Figure 10.8). The mass of young did not vary with litter size, so that with a normal range of litter sizes (3 to 5), litter mass increased by 67%, whereas the increase in maternal expenditure was only about 12%. At litter sizes of 3 and 5, maternal expenditures were 3.3 to 3.7 times the basal rate, respectively. These measurements, however, do not include the amount of energy transferred in milk to the young. Because the mass of young emerging from the burrows was independent of litter size, Kenagy and coworkers assumed that milk provisioning was proportional to litter size. Therefore, total energy balance (= energy expenditure + lactational transfer) in lactating females showed a 33% increase as the litter size increased from 3 to 5. This increase was correlated with a 21% increase in water loss (in part reflecting the increase in milk transfer) and a decrease in maternal body mass. The daily rates of metabolizable energy exchange varied from 5.2 to 6.9 times the basal rate during the peak of lactation. In white

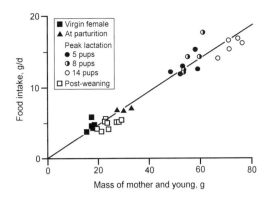

Figure 10.9 Food intake of white mice (*Mus*) as a function of reproductive state, number of pups, and mass of mother and pups. Source: Modified from Hammond and Diamond (1992).

mice, which normally have a litter size of 8 to 10, Hammond and Diamond (1992) showed that food intake of lactating females increased with litter size, and specifically with the mass being provisioned, that is with the sum of that of the mother and young (Figure 10.9). The lactating females had their litters artificially increased from 5 to 26, but the largest litter that was successfully raised was 14 when the sustained rate of metabolism averaged 7.2 times the basal rate. The hypertrophy of the intestine during lactation was sufficient to maintain nutrient uptake with a safety margin.

Food resources have an impact on the energy expenditure of mammals during reproduction. Breeding antarctic fur seals (*Arctocephalus gazella*) responded to a shortage of krill by increasing the duration of foraging trips (1.85 times the duration during a normal krill year) while maintaining the rate of energy expenditure nearly constant (0.93 times the expenditure during a normal year; Costa et al. 1989). The absolute cost of foraging in a krill-shortage year therefore was 1.7 (= 1.85 × 0.93) times that in a year with an abundance of krill. This rate of energy expenditure was 6.7 times the expected basal rate, but only 1.9 times the fasting rate of females nursing their young on land. Under conditions of food shortage, when females are at sea for extended periods, pup mortality increases, mainly as a result of starvation. This response to food shortage contrasts with that of the northern fur seal (*Callorhinus ursinus*), which maintained a constant duration of foraging trips and altered the rate of energy expenditure (Costa and Gentry 1986). Northern fur seals spent 3 times as long resting as did antarctic fur seals, which led Costa

et al. (1989) to conclude that antarctic fur seals may be operating at near their maximal rates. If so, a shortage of food principally has as its consequence a reduced survival of the young.

10.6.2 Birds. Many field measurements (using doubly labeled water) are available on the energetics of reproduction in birds. When feeding young, purple martins (*Progne subis*) expend energy at a rate that is 2.9 to 3.4 times the basal rate, and mockingbirds (*Mimus polyglottus*) expend energy at 2.7 to 3.0 times the basal rate (Utter 1971). House martins (*Delichon urbica*) expend energy at a rate 2.5 to 5.3 times (mean = 3.9) the standard rate when they are feeding young (Hails and Bryant 1979). Diving-petrels (*Pelecanoides*) and the least auklet (*Aethia pusilla*) at this time expend energy at 3.1 to 4.2 times the basal rate (Roby and Ricklefs 1986). In species that forage in flight, such as terns, swallows, and storm-petrels, the cost of foraging is reduced by a morphology that minimizes the cost of flight (Hails 1979, Flint and Nagy 1984, Obst et. al. 1987; Figure 9.16), thereby minimizing field expenditures in spite of extended periods of flight (also see Section 10.5.2). The low basal and daily energy expenditures of soaring birds (Ellis 1985, Wasser 1986) may be a means of evading an undependable food intake. In birds the cost of feeding a brood, thus, may be approximately 2 times that of normal maintenance and activity at moderate environmental temperatures; that is, the maximal sustained expenditure is often stated to be 4 times the standard rate of metabolism (Drent and Daan 1980; but see Section 10.8).

Pennycuick and Bartholomew (1973) examined the consequence of the high cost of reproduction in the lesser flamingo (*Phoeniconaias minor*) in East Africa. It is a specialized filter feeder on blue-green bacteria. Because lake water is filtered at a constant rate, food acquisition is proportional to the density of bacteria in water and the time spent feeding. By assuming that a 1.8-kg flamingo expends 13.5 kJ/h and that an incubating flamingo can use only about 10 h for feeding, the mean concentration of bacteria must be equal to or greater than 0.25 kg/m³ for energy balance to occur. Lower bacterial concentrations would require longer periods for feeding. The abundance of feeding flamingos in the field falls dramatically as bacterial concentrations fall below 0.15 kg/m³: such a concentration would require 16.7 h of feeding to satisfy the assumed energy requirements, an impossible commitment during the breeding season. The principal factor producing the irregular temporal and spatial pattern of breeding in this flamingo may be the local and sporadic occurrence of high bacterial concentrations.

Bryant and Westerterp (1980, 1982, 1983a) provided the most insightful analysis of field expenditures during reproduction. They demonstrated that the daily expenditures of breeding house martins vary with several factors:

1. Large individuals have a greater flying efficiency (glide more?).

2. Daily expenditures increase with the percentage of time spent flying and with the number of times that young are fed (see also Utter and LeFebvre 1973).

3. Daily expenditures increase with an increase in food supply (because more time is spent feeding during good than during poor weather and because flapping flight is used more).

4. Body fat reserves are maintain during good weather, whereas fat reserves diminish during inclement weather.

5. Fat reserves are smallest when foraging distance is greatest.

These factors were used to construct a net energy budget in breeding house martins, which was examined relative to brood size (Bryant and Westerterp 1983b). The net energy balance of a parent, as demonstrated by the increase or decrease in fat stores, equals the amount of energy harvested minus the expenditures by the parent for its own maintenance and activity and the amount contributed to the brood (both sexes contribute approximately equally). The amount of energy expended by parents increased by 10% as the brood increased from 1 to 5 (Figure 10.10), whereas the amount of food fed to the young increased 3.8-fold. The largest, naturally occurring brood is 5 young. When broods were artificially increased to 6 and 7 young, the proportion that fledged fell slightly and body mass of the nestlings decreased, which may have led to a reduced survivorship after fledging. Calculations suggested that a brood size of 5 is the largest that can be raised with a positive energy balance in the parents (Figure 10.10) and thus without an appreciable fat loss by the parents. A second annual brood maximally has 4 young, a reduction that may result

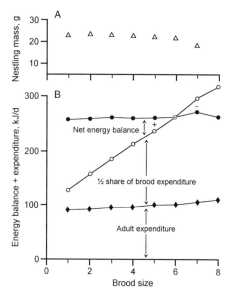

Figure 10.10 (A) Nestling mass and (B) net energy balance of a brooding parent, expenditure of the parent, and one-half of the brood expenditure of house martins (*Delichon urbica*) as a function of brood size. Notice that at brood sizes greater than 6, energy expenditure exceeds income, which corresponds to a reduction in nestling mass. Source: Derived from Bryant and Westerterp (1983b).

from an increase in the daily energy expenditures of parents and a reduction in foraging time produced by a decrease in day length and an increase in inclement weather. The impact of weather on house martins may uniquely apply to species that feed on the wing (e.g., swifts [Koskimies 1950] and swallows), whereas those with radically different habits may have other factors that determine energy expenditure (see Ricklefs 1974).

Field measurements in Eurasian kestrels (*Falco tinnunculus*) indicated that males, which provide most of the food for the offspring and the incubating female, showed no difference in energy expenditure as brood size naturally varied between 4 and 7 (Masman et al. 1989; Figure 10.8). Large broods were provisioned with the same amount of energy per chick because males were able to increase the number of voles caught per hour. When an artificial food shortage was produced (by taking food away from the young), males compensated by increasing food delivery, thereby increasing their daily energy expenditures. If mean brood size was artificially increased from 5.3 to 7.3, then the young grew more slowly and had a higher mortality and the parents lost mass and had a lower (local) survivorship (Dijkstra et al. 1990). If mean brood size was artificially decreased from 5.0 to 2.7, food intake by the young increased. These observations led to the conclusion that kestrels normally operate at expenditures well below the maximum by adjusting clutch size and the timing of reproduction. The factors that limit reproductive output may involve other "costs," including parental survival.

The difficulties encountered by parents in raising successfully a brood under some environmental conditions can be alleviated if other individuals help to feed the young (Emlen 1984). The relevance of nest helpers to energetics is found in two populations of the pied kingfisher (*Ceryle rudis*) in Kenya (Reyer and Westerterp 1985). In both populations the amount of food (in kJ/adult·d) delivered to nestlings increases with the daily expenditures of the adults, although adults cannot increase their daily expenditures without incurring a cost: if daily energy expenditure exceeded 210 kJ, they lost body mass. The highest rate of food delivery to nestlings at Lake Naivasha was 267 kJ/adult·d, which is equivalent to 111.3 kJ/d per nestling, and 101.9 kJ/adult·d at Lake Victoria, or 44.3 kJ/d per nestling. The lower rate at Lake Victoria reflected a lower energy yield per fish, a greater turbidity in the water, and a greater distance between the breeding colony and the fishing grounds. As a consequence, the nestlings at Lake Naivasha *gained* mass at 2.7 g/d, whereas those at Lake Victoria *lost* mass at 5.6 g/d. Under these circumstances, 61% of the nestlings died in nests at Lake Victoria when the parents were unassisted by helpers; at Lake Naivasha only 19% of the young died under similar circumstances.

Pied kingfishers at Lake Victoria usually accepted helpers at the nest. The mortality of young decreased to 22% if the parents were aided by one helper, and to near 0% if two helpers were involved. Nesting pairs at Lake Naivasha, however, usually rejected helpers, and those few that were present made no contribution to a reduction in the mortality of the young. As further evidence of the significance of helpers, the attack or greeting of a helper by the male of a breeding pair of kingfishers can be reversed at both localities by artificially increasing the clutch size at Lake Naivasha or decreasing clutch size at Lake Victoria. Thus, the behavioral response of mated pairs of pied kingfishers was dictated by the balance between reproductive success and limits to the daily energy

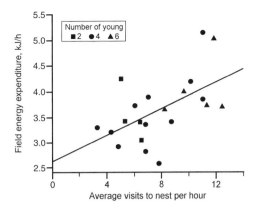

Figure 10.11 Rate of metabolism in savannah sparrows (*Passerculus sandwichensis*) as a function of the rate of nest visits by adults and brood size. Source: Modified from Williams (1987).

expenditures of the parents, a balance that was affected by the quality of the environment in which they lived.

When savannah sparrows (*Passerculus sandwichensis*) raise young, (1) males and females expend energy at approximately 3 times the basal rate, males having a higher power output because they are larger; (2) power output increases with brood size; and (3) power output increases with number of visits to the nest (Figure 10.11; Williams 1987). Williams concluded that small passerines during reproduction expend energy at 3 times the basal rate. This observation and those that some species can successfully raise more than their normal clutch, or that a single parent can successfully raise the entire clutch, imply that some species are normally not expending energy at their maximal rate.

Young must attain independence to ensure the future success of a population. Weathers and Sullivan (1989) examined the cost of this independence in the yellow-eyed junco (*Junco phaeonotus*). They demonstrated that adults, which were neither food nor time limited, have rather low field expenditures during the breeding season (ca. 2.1 times the basal rate), 70% of which was used for maintenance. The expenditure of fledglings increased up to a daily rate that at 10 to 12 weeks was 36% greater than that of the adults. This increase occurred because fledglings spent up to 3 times the period spent by adults foraging for food. Weathers and Sullivan (p. 223) concluded ". . . that energy constraints are the major selective force in

Yellow-eyed Juncos operating not through food limitation among adults but rather through the inefficient foraging of young juncos."

To examine experimentally the consequences for reproduction of variation in the energy expenditure of breeding adults, Lemon (1993) manipulated the rate of energy gain in zebra finches (*Taeniopygia guttatus*). All finches were given the same amount of digestible food, but the proportion of empty seed hulls was varied. With an increase in this proportion, foraging time increased and net energy income decreased. As the net income decreased, interbrood interval increased and brood size decreased, as did juvenile and adult survivorship. The age at the fledging of the young was unaffected by the energy balance. Individuals ". . . that were required to spend a large amount of [time and] energy foraging had less energy available for reproduction and consequently produced smaller broods at a slower rate with lower juvenile survivorship" (p. 960). The diversion in adults of ". . . energy from maintenance to reproduction [may have] caused them to experience earlier reproductive senescence and increased mortality" (p. 960).

Nest attentiveness by birds during incubation also has implications for energetics. In many tropical species and most hole nesters, only one parent incubates the eggs and young. Nest temperatures then do not fall to unacceptable levels when the incubator (usually the female) leaves the nest to feed. In many temperate and polar species, however, both parents incubate and feed the young, thereby diminishing the cooling of eggs and young, as well as reducing the likelihood of nest predation. The periodicity with which parental replacement at the nest occurs depends on many factors in addition to temperature, including the energy requirements of the incubator, the distance from the nest at which feeding occurs, and the difficulty in obtaining sufficient amounts of food. These relations are best known in seabirds, many of which feed at great distances from the nesting sites.

The incubation period of an egg increases with its mass ($m^{0.20}$; Rahn and Ar 1979), but even when egg size is taken into consideration, most species with very long incubation periods are tube-nosed seabirds (Procellariiformes). On Christmas Island in the Pacific Ocean seabirds that feed the farthest offshore, as judged from direct observation and food habits, had the longest periods of continuous incubation (Ashmole and Ashmole 1967). The longer the period, the more likely that an incubat-

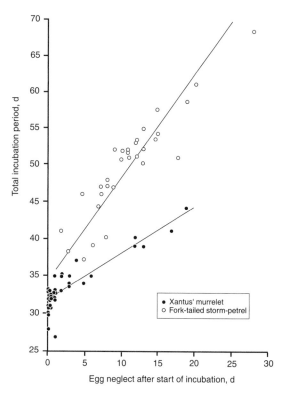

Figure 10.12 Incubation period in the fork-tailed storm-petrel (*Oceanodroma furcata*) and Xantus' murrelet (*Synthliboramphus hypoleuca*) as a function of egg neglect. Sources: Derived from Boersma and Wheelwright (1979) and Murray et al. (1979).

ing adult will neglect eggs to feed at sea (Matthews 1954, Boersma and Wheelwright 1979).

Egg neglect in seabirds extends the incubation period. The incubation period of the fork-tailed storm-petrel (*Oceanodroma furcata*) in Alaska (Figure 10.12) increased by nearly a day and a half for every day of neglect (Boersma and Wheelwright 1979), which presumably compensates for the retarded embryological development produced by low egg temperatures. Egg neglect in Procellariiformes may also entail other penalties: chick mortality in *O. furcata* increases with egg neglect (Boersma and Wheelwright 1979) and the cost of producing an embryo in shearwaters (*Puffinus pacificus*) increases with incubation time (Ackermann et al. 1980). In contrast, incubation period in an alcid, Xantus' murrelet (*Synthliboramphus hypoleuca*), in California increases by 0.62th of a day for each day of neglect (Figure 10.12), which means that some development continues in the absence of the parents (Murray et al. 1979).

The different response to egg neglect in the murrelet and the storm-petrel is related to nest temperatures: in Alaska the storm-petrel eggs cool to 11°C when neglected, whereas in California the murrelet eggs cool to 20°C. The ancient murrelet (*S. antiquus*) in British Columbia compensates for a cool burrow (ca. 10°C) by paying greater attention to the nest, compared to Xantus' murrelet (Sealy 1976). During the breeding season egg neglect increases in the storm-petrel, possibly because food availability is reduced, as indicated by a decrease in adult fat deposits. A long period of egg attentiveness by one parent is found in penguins. Because of the rigorous environment in which they nest and the likelihood of predation by skuas (*Catharacta*) or giant-petrels (*Macronectes*), penguins cannot abandon their eggs and during incubation rely on stored fat (see Section 14.4.1).

The impact of environmental temperature on energetics during reproduction is seen in a comparison of Leach's (*Oceanodroma leucorhoa*) and Wilson's storm-petrels (*Oceanites oceanicus*). Wilson's storm-petrel chicks have higher energy requirements than Leach's storm-petrel chicks (Obst and Nagy 1993), principally because Wilson's storm-petrels face lower environmental temperatures in Antarctica than Leach's storm-petrels do in the Northwest Atlantic. Wilson's storm-petrel adults had field expenditures that were 3.2 times the basal rate when an egg or neonate was present and 3.7 times the basal rate when an older chick was present (Obst et al. 1987). Leach's storm-petrel adults had field expenditures that were higher in Newfoundland (2.7 and 3.1 times the basal rate, respectively) than in New Brunswick (2.1 and 2.5 times the basal rate, respectively; Montevecchi et al. 1992). Field rates of metabolism in the Newfoundland colony increased with the time spent at sea (Figure 10.13). Wilson's storm-petrels met the high energy requirements of their chicks with a high frequency of meal delivery (34% greater than in Leach's storm-petrel) and a high energy density of food (28% greater than in Leach's storm-petrel; Obst and Nagy 1993).

The problem faced by Wilson's storm-petrels in Antarctica is not only the low environmental temperatures but also the short breeding season. This species, like all in the Procellariiformes (except diving-petrels), supplements the food given to nestlings with stomach oil, which is derived from ingested food (Clarke and Prince 1976, Warham et al. 1976, Place et al. 1989). Stomach oil increases

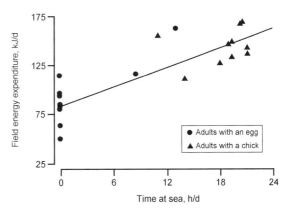

Figure 10.13 Field rate of metabolism in adult Leach's storm-petrel (*Oceanodroma leucorhoa*) as a function of the time spent at sea and whether it had an egg or a chick. Source: Modified from Montevecchi et al. (1992).

the energy density of the food delivered to the nestlings, in this species by 14%. The oil helps to minimize the fledgling period, which otherwise could be attained only if storm-petrels, which feed their young at night, fed their offspring more than once a day, a behavior that would expose the adults to daytime predation. In the antarctic prion (*Pachyptila desolata*) only a few stomachs contained oil, which increased the content's energy density from 5.0 to 9.1 kJ/g (Taylor et al. 1997). The provision of stomach oil in this prion was mainly to younger nestlings, but the long-term energy budgets of nestlings cannot be balanced without the stomach oil: these nestlings expend more energy for maintenance and growth than is taken in as food (Taylor et al. 1997). (Other procellariids depend more on the use of stomach oil than this prion.) In contrast, diving-petrels (*Pelecanoides*) supply no stomach oil to their nestlings. The difference in feeding strategy between adult prions and diving-petrels is correlated with a difference in field energy expenditures: prions have field expenditures (measured by $D_2^{18}O$) that are only about 55% those of diving-petrels. The provision of stomach oil by prions may be permitted by their low field expenditures, or required by their long-distance foraging, whereas the absence of stomach oil in diving-petrels may reflect their higher field expenditures (Roby et al. 1989), or a higher prey density and the pursuit-dive foraging used by diving-petrels (Taylor et al. 1997). The absence of stomach oil in diving-petrels is apparently the derived condition (Taylor et al. 1997).

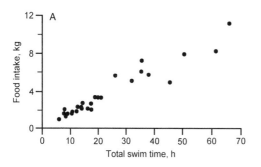

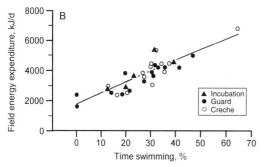

Figure 10.14 A. Food intake in adult Adélie penguins (*Pygoscelis adeliae*) as a function of total swim time during a foraging trip. B. Field rate of metabolism in adult Adélie penguins as a function of the proportion of time during foraging spent swimming at various stages in the breeding cycle. Source: Modified from Chappell et al. (1993).

The energetics of breeding in Antarctica has been explored in the Adélie penguin (*Pygoscelis adeliae* [Chappell et al. 1993]), a small species (4 kg). Adults show a cyclic pattern of feeding and guarding the one or two chicks and of foraging at sea. During incubation (days 1–17 after laying) and when the chicks are small (18–28 d), one parent continuously stays with the young; thereafter the large young group together unattended in creches but are fed by the adults until they are abandoned at the age of 37 to 45 d. The nest relief cycle is longest during incubation (ca. 9.5 d) and shortest during the guarding (40 h) and creche stages (33 h). The food intake of adults increased with the time spent swimming (Figure 10.14A), and energy expenditure increased with the fraction of time spent swimming (Figure 10.14B). The energy expenditure of adults varied with the stage of reproduction: 2.7 times the basal rate during incubation, 3.0 times the basal rate during the guarding stage, and 3.3 times the basal rate during the creche stage—a time-weighted mean of 3.1 times the basal rate. The time required for Adélie adults to supply themselves and their chicks with food

was minimal: 44 min/d during incubation, 58 min/d during the guarding stage, and 73 min/d during the creche stage. Consequently, the reproductive effort in this penguin requires only a modest increase in energy expenditure and in foraging time.

King penguins (*Aptenodytes patagonicus*), which are much larger (13 kg) and feed farther offshore than Adélie penguins, expend energy at a higher rate (4.6 times the standard rate when away from the colony), and the (single) young takes more than 300 d to fledge (Kooyman et al. 1992). The mean field expenditure of king penguins during the breeding season is about 4.3 times the standard rate, whereas those of gentoo (*Pygoscelis papua*), macaroni (*Eudyptes chrysolophus*), and jackass (*Spheniscus demersus*) penguins are 6.0, 7.1, and 6.6 times standard rates, respectively (Davis et al. 1983, 1989; Nagy et al. 1984b). The factors responsible for these differences in the field expenditure are unclear, although Kooyman et al. (1992) hinted that they may be related to size-specific variations in the cost of transport.

Ricklefs (1983) reviewed seabird reproductive energetics. He suggested that the ratio of energy demand to supply was high during both incubation and chick brooding. During incubation it was high because one parent remained continuously with the (single) chick, which means either that the parent went without food or that the chick had to be abandoned periodically. The continuous period of incubation is limited by the body fat stores of the adult to approximately 3 times the length of the interval between incubation periods or to 3 or 4 d, whichever is the shortest. After the chick hatches, brooding is interrupted by the necessity to feed the growing young. Several adjustments by the adult and young facilitate the successful raising of the young. They include the transportation of high-energy-density meals obtained from the consumption of prey with a high content of oils (especially plankton), the partial digestion of the prey, and accumulation of lipids in the parent's proventriculus. Furthermore, the young of many seabirds develop endothermy early, thereby decreasing the amount of incubation required of the parents (Ricklefs et al. 1980, Warham 1990, Ricklefs 1996). Growth rate in seabirds may be limited by the ability of the parents to provide food, which itself is limited by the ability to find food at sea and by a limited meal size. Although Ashmole (1963b, 1971) and Lack (1968b) argued that the low reproductive output of pelagic seabirds results from a variable food availability, Ricklefs (1983) raised the possibility that it may reflect the rate of energy provisioning, that is, the product of meal size and meal energy density (but see Ricklefs and Schew 1994). This difference may reflect ultimate and proximate causes.

In megapodes (brush-turkeys, Megapodiidae), the source of heat for incubation is peculiar: they construct large mounds, often of 3 to 5 tons of organic matter and earth in which to bury their eggs. The heat produced by organic decay, supplemented by solar radiation or geothermal heat, incubates the eggs (Frith 1956). The parental contribution lies originally in constructing the mounds—a cost that is reduced by the repeated use of old mounds—and by opening and closing the mounds, which in the Australian malleefowl (*Leipoa ocellata*) involves approximately 850 kg of mound material, to ensure that the nest temperature remains within an acceptable range. It has an incubation period that lasts 5 to 6 months, and for 2 months the mounds are reworked every day for an average of 5.3 h/d (Frith 1962)! The relative importance of fermentation, solar radiation, and parental activity is illustrated in Figure 10.15 for the malleefowl. A control mound made of soil alone demonstrated the contribution of heat by various physical processes, including solar radiation. An artificial mound included a core of fermenting vegetation, and the natural mound was attended by a pair of birds. Obviously, fermentation makes a significant contribution to the heat content of the nesting mounds early in the nesting cycle, but the opening and closing of the mounds by birds, often on a daily basis, is required to ensure that the heat derived from insolation and fermentation does not overheat the nest. The thermal stability of nesting mounds of the Australian brush-turkey (*Alectura lathami*) increases with size, the largest mounds remaining thermally stable for weeks without intervention of a bird (Seymour and Bradford 1992). Megapodes have large clutches (e.g., 15–24 in the malleefowl and 18–24 in the brush-turkey), and the eggs are among the largest with a very high yolk content (Vleck et al. 1984).

Extensive activity at the mound raises the question of the cost of mound tending. Weathers et al. (1993) tried to estimate the cost of mound-tending behavior in the malleefowl by observing their behavior in the field and by measuring their energy expenditure on a treadmill in the laboratory. Based on the occurrence of gular fluttering, these authors concluded that mound work was equivalent to running at a velocity of 0.57 m/s, which corresponds to a power output that is 3.8 times the basal rate. During the peak of mound tending, malleefowl expended energy at a rate approximately twice that found in birds that use traditional incubation, which raises the question as to why this behavior is used. The answer may be that the "... high investment [in clutch and egg size and yolk content] is possible because the incubation biology of megapodes releases them from physical and temporal constraints that apply to birds [that] ... sit on their eggs and care for their hatchlings" (Vleck et al. 1984, p. 454). Weathers et al. (1993) suggested that mound building permits a higher fecundity (a clutch may be as large as 34 eggs!) and freedom from posthatching parental care. Given the restriction of megapodes to the Andaman Islands, the Philippines, and islands east of Wallace's Line, their presence may reflect the absence of effective mammalian predators (see Dekker 1989).

10.6.3 Ectotherms. Little is known of reproductive energetics in ectothermic vertebrates, and few data are derived from the use of doubly labeled water. Indirect estimates (Andrews 1984) and direct measurements (Anderson and Karasov 1981, Nagy et al. 1984a) suggest that actively foraging lizards allocate larger amounts of energy to reproduction and growth than do sit-and-wait predators, which if correct should mean that *Cnemidophorus* grows faster or has a higher (net) fecundity than *Sceloporus* or *Callisaurus* at the

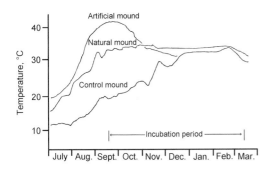

Figure 10.15 Soil temperature in natural nesting mound of a malleefowl (*Leipoa ocellata*), in an artificial mound, and in a mound without organic matter (the "control" mound) as a function of the time of year. Notice that the natural mound has the most constant temperature during the incubation period. Source: Modified from Frith (1956).

same mass. That is, *Cnemidophorus* might have a higher "reproductive effort," which in the opinion of Tinkle and Hadley (1975) is best estimated by the proportion of total energy expenditure allocated to reproduction.

One of the most dramatic differences in reproduction found in reptiles is that some species are oviparous and others are viviparous, a difference found even within some species. Still, little is known of the consequence for energetics of the mode of reproduction. Gravid viviparous lizards belonging to the genus *Sceloporus* have higher rates of metabolism than males or nonreproductive females at the same temperature (Guillette 1982, Beuchat and Vleck 1990), and apparently have higher rates than gravid oviparous lizards, whose rates are not much higher than those of nonreproductive females (Guillette 1982). The higher rate of metabolism in gravid viviparous *S. jarrovi* is confounded by the observation (Beuchat 1986b) that they maintain lower body temperatures in the field in spite of the resulting increase in gestation period (Beuchat 1988). Despite the higher body temperature in males (34.5°C) than in gravid females (32°C), gravid females still have slightly higher rates of metabolism in the field (Beuchat and Vleck 1990). These data can be interpreted to support Guillette's view that the principal difference between oviparous and viviparous lizards is that the former lay eggs at the time when development shifts from a slow (anaerobic?) to a fast (aerobic?) phase, which is when oxygen consumption by embryos greatly increases. (May the selection of a lower body temperature by gravid viviparous *S. jarrovi* reflect a limitation in their capacity to deliver oxygen to the intrauterine young?) *Lacerta vivipera* (Patterson and Davies 1978a) and *S. cyanogenys* (Garrick 1974) also maintain a lower body temperature when pregnant.

Some gravid reptiles maintain a higher body temperature than nonreproductive females; these include *Thamnophis* (Stewart 1965, Gibson and Falls 1979), *Hoplodactylus* (Werner and Whitaker 1978), and *Elgaria* (Stewart 1984), all of which maintain low body temperatures at times other than during reproduction. In these cases, a modest increase in the core temperature of gravid females might be a means of increasing the rate of development without approaching a level of oxygen consumption by the young that would be limited by oxygen transport.

Viviparity in reptiles, which is most common at high latitudes and elevations (Packard et al. 1977, Tinkle and Gibbons 1977), may be a means of shortening gestation period in cool environments through behavioral temperature regulation by gravid females. Yet, no direct evidence is available that this reduction occurs (Tinkle and Gibbons 1977), as logical as this idea might be. Furthermore, viviparous and oviparous reptiles coexist in many environments.

Only spotty information is available on the energetics of reproduction in amphibians. Fitzpatrick (1973) estimated that female lungless salamanders (*Desmognathus ochrophaeus*) use 48% of their annual energy budget for reproduction (65% of which is used to synthesize the eggs and 35% during brooding). This high proportion reflects the low cost of maintenance in plethodontids (but see Wieser 1985). The cost of courtship in *Desmognathus* is less than 1% of ingested energy (Bennett and Houck 1983). In anurans, however, the cost of courtship is quite high: Bucher et al. (1982) measured rates of metabolism in the leptodactylid frog *Physalaemis pustulosus* that were twice the standard rate when calling and 4 times this rate when building a foam nest. In *Ph. pustulosus* calling may last for hours (and occasionally through the night) and building a nest requires from 30 min to 2 h, so these behaviors significantly increase energy expenditure. In *Hyla* the cost of calling at the maximal rate may increase the rate of metabolism to exceed 10 times the resting rate (Taigen and Wells 1985, Taigen et al. 1985, Prestwich et al. 1989).

Even less is known of the energetics of reproduction in other ectothermic vertebrates, although numerous studies have examined changes in body composition during migration and sexual maturation in freshwater, anadromous, and catadromous fishes and lampreys. For example, Beamish (1979) described the energetics of spawning and migration in the anadromous sea lamprey *Petromyzon marinus*. Upon entering freshwater, lampreys stop feeding and rely on fat stored while feeding and maturing at sea. In New Brunswick (Canada) lampreys were collected 140 km upstream from the mouth of a river and compared with other individuals in the same run at a site 60-km upstream from the mouth. From the decrease in body energy content (principally due to a loss in fat), a male was calculated to have spent about 1260 J in swimming

140 km, 42 J in producing gametes, and 1850 J in reproductive activity (mainly for nest construction and locomotor activity), whereas a female spent about 1090 J in migration, 1820 J in gamete formation, and 1760 J in reproductive activity. Thus, the cost of breeding was much greater than the cost of migration in this lamprey, and the combination of these behaviors, during a period when no feeding occurs, leads to a severe depletion of its energy reserves, although doubt exists that the subsequent postspawning death is simply due to this depletion.

10.7 ENERGETICS OF GROWTH

The growth rates of vertebrates have been described repeatedly (e.g., Ricklefs 1968, 1973b, 1979; Frazer and Huggett 1974; Frazer 1977; Case 1978a, 1978b; Andrews 1982; Zullinger et al. 1984). For example, growth rates and their descriptor, the exponential growth constant, vary inversely with body mass in birds (Figure 10.16A) and are influenced by mode of reproduction (Figure 10.16B). A complication is whether the study of the

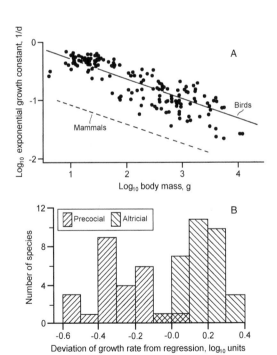

Figure 10.16 A. $\log_{10}$ exponential growth constant in birds and in mammals as a function of $\log_{10}$ body mass. B. Frequency distribution in birds of deviation in growth constant from the constant expected from mass in relation to the precocial or altricial habits of the species. Sources: Modified from Ricklefs (1979) and McNab (1980b).

energetics of growth is best treated as the allocation of energy in growing individuals, which is essentially a biochemical question, or whether it should include the contribution of the parents, as is done here because an ecological and evolutionary approach to the growth of young cannot be divorced from parental behavior.

10.7.1 Mammals. Most detailed descriptions of the energetics of growth in wild mammals concern small species in the laboratory (e.g., Chew and Spencer 1967, Droźdź et al. 1972, Hudson 1974, Gebczynski 1975, McClure and Randolph 1980). The comparative study of McClure and Randolph, and their recalculation of data presented by Chew and Spencer, demonstrated that the proportion of assimilated energy, derived from lactation, used for growth increases with mass: between birth and the attainment of endothermy, *Neotoma* (227 g) uses 37% of energy intake for growth; *Sigmodon* (120 g), 24%; and the pigmy mouse (*Baiomys*, 8 g), only 12%. At small masses, the cost of endothermy, even before its perfection, is high (Figure 5.28), so that the apparently high "efficiency" in diverting assimilated energy into growth in large species may reflect the reduced relative cost of temperature regulation. Large species also increase the transfer of energy to growth by delaying the acquisition of endothermy (McClure and Randolph 1980). Because of the high cost of endothermy, a smaller proportion of assimilated energy is converted into growth by endotherms than by ectotherms (Engelmann 1966, Calow 1977).

During 35 d of lactation, food consumption by a female golden-mantled ground-squirrel (*Spermophilus saturatus*) increased by 2.1-fold; body mass, by 10%; and rate of metabolism, by 20% (Kenagy et al. 1989). If the litter size was 3, body mass of the pups increased from 5.7 g (at birth) to 98.0 g at weaning and to 335.9 g at 100 d. If the litter size was 5, body mass increased from 5.7 g to 76.8 g at weaning and to 297.6 g at 100 d. Estimated milk production in a litter of 3 was 115 kJ/pup·d and in a litter of 5 was 104 kJ/pup·d. A litter size of 5 appeared to force the mother to produce milk at near her physiological limit, the basis of which was unclear.

The energy expenditure devoted to growth in mammals has not been measured in the field, although it has been estimated indirectly in phocid seals that gave birth on ice floes in the Arctic. During lactation most female phocids do not feed and therefore depend principally on their fat stores to fuel their metabolism and to supply milk to their pups. To minimize her dependence on fat reserves for maintenance and to reduce her susceptibility to predation by polar bears, a female produces one young and compresses the time spent lactating by having milk with a very high lipid content (Fedak and Anderson 1982, Ortiz et al. 1984). Female gray seals (*Halichoerus grypus*) transfer to the young about 3 kg of milk per day, which is about 50% fat. The young consume about 84% of the mother's energy stores in 18 d. Approximately 57% of the transferred energy is converted to growth and stored fat by the young; the pups gain 1.8 kg/d (Fedak and Anderson 1982). Harp seals (*Phoca groenlandica*) have further compressed the nursing period to only 9 d, and the pups grow even faster (2.5 kg/d; Stewart and Lavigne 1980), 77% of the mass loss by the mother being transferred into a mass gain by the pup. The extreme to which this trend has been carried is found in hooded seals (*Cystophora cristata*), in which the nursing period is reduced to 4 d (Bowen et al. 1985)! The increase in pup mass during lactation varies between 5.4 and 7.5 kg/d and 76% of the maternal mass loss is stored in the pup.

In temperate and antarctic environments the threat from terrestrial predators is greatly diminished, so the selective pressure to compress lactation in seals there is greatly reduced. Female otariid seals, which generally are smaller and store less energy as body fat (Costa and Trillmich 1988) than female phocids, intersperse suckling with foraging periods, the length of which is limited to 7 d or less by the pups' ability to fast and to gain mass (Gentry and Holt 1986). The rates of milk production in otariids are much lower than those of phocids (Anderson and Fedak 1987). A small temperate phocid, the harbor seal (*Phoca vitulina*), also intersperses feeding with suckling. The female nurses its pup for 24 d and loses 79% of its body energy store, only 50% of which is stored as a mass gain in the pup (Bowen et al. 1992). The combination of a small maternal mass and aquatic habits of the nursing young may account for the low rate of energy storage in the harbor seal pups. In a large temperate phocid, the northern elephant seal (*Mirounga angustirostris*), females transfer 5.5 kg of milk per day to their young over a lactational period of 26.5 d, which involves a loss of 42% of female mass, 55% of which is incorporated into a mass increase by the pup (Costa et al. 1986).

Female arctic walruses (*Odobenus rosmarus*), however, can protect their young from predation by polar bears, which permits them to suckle their young for up to 2 y.

Phocid seal pups channel much of their energy intake during growth to the formation of large deposits of subcutaneous fat: at the termination of nursing over 50% of their mass may be blubber. After weaning, seal pups often have an extended period of fasting, during which they expend energy derived from both the blubber and the core (i.e., from the carcass and viscera, some of which is diffuse fat deposits). The differential use of blubber or core tissues depends on the behavior of the seal pups: if they fast on land, 70% to 80% of the mass loss is derived from the blubber layer, but if they fast in water, blubber contributes only 40% to 50% of the mass loss (Worthy and Lavigne 1990). Land fasters include the gray and elephant seals, whereas water fasters include the harp and hooded seals. The reason for this differential reliance on tissues is that water-fasting pups must maintain a sufficient layer of blubber to act as insulation against heat loss. Because of the high energy density of fat, the principal source of energy in both groups is blubber: approximately 95% of the energy expenditure during fasting in the land fasters and 80% in water fasters.

10.7.2 Birds. Estimates of the cost of growth in birds are influenced by the methods used. In a study of the growth of nestling savannah sparrows (*Passerculus sandwichensis*), the expenditures measured during growth by doubly labeled water progressively surpassed those estimated from measurements of oxygen consumption in the laboratory (Williams and Prints 1986). The principal reason for this discrepancy is that the costs of temperature regulation and activity ("maintenance") become increasingly important as development occurs, and "standard" conditions, when the measurements of oxygen consumption occur, minimize activity.

The energetics of growth in birds not only is correlated with body mass, but also is associated with the type of development, a difference that is clarified when precocial and altricial species have a similar adult mass. A pair of Leach's storm-petrels (*Oceanodroma leucorhoa*, 45 g) expends approximately 1560 kJ over 60 d to raise one semiprecocial young (Ricklefs et al. 1980), whereas a pair of European starlings (*Sturnus vulgaris*, 65 g) expends

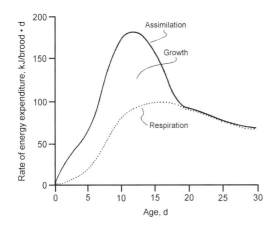

Figure 10.17 Rate of energy expenditure by a brood of house martins (*Delichon urbica*) for growth and respiration as a function of age. Source: Modified from Bryant and Gardiner (1979).

about 1716 kJ over 16 d to raise four altricial young (Westerterp 1973), although noteworthy is that these expenditures occur *during* the growth phase but are not *the* cost of growth. The storm-petrel chick has a low growth rate and the starling young have high growth rates. This difference is typical of the dichotomy between precocial and altricial species: as noted, birds have growth constants that are inversely related to body mass (Figure 10.16A), with precocial species generally falling below the mean curve and altricial species above the curve (Figure 10.16B).

The difference in growth rate between precocial and altricial young reflects the allocation of energy. Precocial young thermoregulate at, or shortly after, hatching, whereas the development of endothermy is usually delayed for 2 weeks or more in altricial young (see Section 5.8.1). The delay of endothermy means that altricial young can channel more assimilated energy into growth, thereby growing rapidly. Hatchling house martins (*Delichon urbica*), for example, direct about 70% of assimilated energy into growth (Bryant and Gardiner 1979). The assimilation of energy by martins is greatest on day 12 after hatching; the amount of energy directed toward growth is maximized on day 10 (Figure 10.17). Growth, thus, is concentrated in the first half of the nestling period and effective endothermy is only attained at about the time growth ceases (Yarbrough 1970, Bryant and Gardiner 1979).

The energy expenditure of nestlings and of their parents, while raising the nestlings, is actually a complicated expenditure for maintenance, activity, and the accumulation of tissue. Only some 13% to 28% of the total expenditures is accumulated in tissue growth (Weathers 1992). The daily energy expenditure of nestlings is not easily described by a power function (Weathers and Siegel 1995): the total expenditure by nestlings from hatching to fledging is proportional to mass at fledging (Weathers 1992). Furthermore, the total expenditure increases as the time to fledging increases, but because this time period itself increases with a decrease in growth rate (Ricklefs 1968), the total energy expenditure of nestlings increases with a decrease in growth rate (Weathers 1992). These correlations were demonstrated in 30 species, including those belonging to 8 orders, including altricial and precocial species, and those with a variety of diets. Hole nesters take longer to fledge than open nesters, as do tropical birds, whereas arctic species tend to have shorter fledging periods (also see Ricklefs 1968, 1996; Klaassen and Drent 1991). This climatic difference may be associated with the latitudinal correlation of basal rate. In any case, "...adaptive modifications in nestling energetics are attained principally through changes in growth rate, which affect total energy expenditure and maximal daily expenditure..." (Weathers 1992, p. 142) in an opposite manner; that is, a high growth rate reduces total expenditure and increases maximal daily energy expenditure.

Most nestling birds deposit fat during growth, but these deposits are largest in species that use foods rich in lipids and (comparatively) poor in proteins (Ricklefs 1979). Such species include petrels, shearwaters, and albatrosses (Procellariiformes), which feed on lipid-rich fish and invertebrates and carry energy-rich lipids in the form of oil from far offshore to feed their young, and the oilbird (*Steatornis*, Caprimulgiformes), which feeds on the nuts and fruits of oil palms. The accumulation of fat might be viewed as a passive energy sink associated with the acquisition of protein (Ricklefs 1979, Ricklefs et al. 1980, Taylor and Konarzewski 1992), but it is more likely an active accumulation of energy preparatory to facing an impending food shortage, either because adults feed on foods that are difficult for naive juveniles to catch, or because the foods are irregularly available (Lack 1968b, Ashmole 1971, Ricklefs 1968, Bryant and Gardiner 1979). Recently, Ricklefs and Schew (1994) suggested a modification of the Lack-Ashmole hypothesis by adding the stochasticity of foraging and provisioning of young by the parents.

In this modification chicks are more likely to survive a variation in feeding frequency if they are overfed (and therefore accumulate fat) than if they are supplied with food at a subsistence level.

The differential use of fat storage in relation to behavior can be seen in gannets and boobies (Sulidae). Tropical boobies (*Sula*) that feed far off-shore, where food is often scarce, have only one young, whereas those that feed inshore and have abundant food supplies have two or more young (Nelson 1977). Booby fledglings do not accumulate large fat stores. In contrast, the northern gannet (*Morus bassanas*), which lives near productive temperate waters, raises only one young, which channels much of the food brought by the parents into a large fat store (Nelson 1977). Unlike boobies, the gannet abandons its young, which plummets into the sea from the nesting cliffs before it can fly and feed; it then swims away from the colony, relying on its fat stores.

Brood size is an important variable influencing the allocation of energy to growth in birds, especially in relation to parental attendance at the nest. In starlings both parents maintained their mass while raising three young, but lost mass if they raised five to seven young, as did a solitary female raising three young (Westerterp et al. 1982). Haartman (1954) noted similar observations in the pied flycatcher (*Ficedula hypoleuca*) and Newton (1966), in the bullfinch (*Pyrrhula pyrrhula*). An optimal brood size maximizes growth rates and minimizes food intake per unit of growth in mass: in starlings the optimum is five young (Figure 10.18). Smaller clutches require brooding because behavioral temperature regulation is ineffective, and larger clutches, through fouling and activity, reduce nest insulation. In large clutches of house martins, fledgling size decreases (Bryant and Gardiner 1979). The occurrence of asynchronous hatching in large clutches may be a means of reducing the peak energy requirements of a brood during growth (Bryant and Gardiner 1979), thereby permitting the parents to raise a larger clutch. This strategy is most marked in raptors, although in this case the last young hatched might not survive sibling competition if food supplies are limited.

10.7.3 Ectotherms. Growth rates in ectotherms, such as reptiles and fish, are only one-tenth to one-thirtieth those of mammals and birds of the same mass (Case 1978b). Yet, field energy expenditures in the lizard *Uta* (Nagy 1983) are strongly corre-

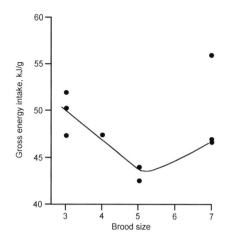

Figure 10.18 Gross mass-specific energy intake by European starlings (*Sturnus vulgaris*) as a function of brood size. Source: Modified from Westerterp et al. (1982).

lated with growth rate. Growth rate and body mass account for 74% of the variation in their field expenditures. Seasonal variations in growth rate in lizards belonging to the genus *Anolis* appear to reflect seasonal variations in the availability of food (Licht 1974) or of water (Stamps 1977, Stamps and Tanaka 1981), as was shown by supplementing the food or water supply. Growth rate in the adder (*Vipera berus*) is greater when faced with a high prey density (Forsman and Lindell 1991). In amphibians growth rates are also correlated with population density and the size of body energy reserves (Crump 1981).

Growth rates in ectotherms are increased through the selection of high body temperatures. Growth rate is maximized in *Bufo boreas* at a body temperature of about 27°C, subject to adequate food availability. A similar pattern of maximizing growth rate has been found in fish (Brett et al. 1969, Brett 1971b, Elliott 1975): the number of meals eaten by brown trout (*Salmo trutta*) in a day increases with ambient temperature and much of this increased energy intake is used for growth. Growth is maximized at 18°C, if three meals per day are available, but if only two meals are available each day, maximal growth occurs at 13°C (Elliott 1975). Clearly, in these ectotherms a balance between growth and maintenance is obtained through the selection of a temperature that is dictated by the amount of food available: a higher food intake permits both a higher growth rate and a higher "maintenance" expenditure because these expenditures are positively coupled.

Parry (1983) cautioned that these two aspects of an organism's existence cannot be separated.

Recent work on energy expenditure and growth in embryonic and larval fish has raised doubts on the connection between rate of metabolism and growth rate. Wieser et al. (1988) started this doubt in a study on the larvae of the roach (*Rutilus rutilus*), a cyprinid, in which growth rate varied with food ration, but rate of metabolism did not. This conclusion was confirmed in larvae of the whitefish (*Coregonus wartmanni*) and the roach (Wieser and Medgysey 1990a, 1990b). Growth rate might be independent of rate of metabolism in rapidly growing fish larvae either because fish larvae channel energy to growth by suppressing other energy-requiring functions or because the cost of growth is not constant throughout life (Wieser 1991).

Rombough (1994) explored the possibility that energy allocation is additive or compensatory during development in a study of Chinook salmon (*Oncorhynchus tshawystscha*). He showed that (1) growth of mass was most rapid at higher temperatures, although the maximal mass attained was greatest at the coolest temperatures (Figure 10.19A); and (2) rate of metabolism increased with mass, although the maximal rate of metabolism was independent of temperature, except at the lowest temperature (Figure 10.19B). No significant correlations were found in larval sockeye salmon between growth rate and rate of metabolism, which led Rombough to suggest that an additive partitioning of energy does not occur in sockeye salmon, at least at early stages of life.

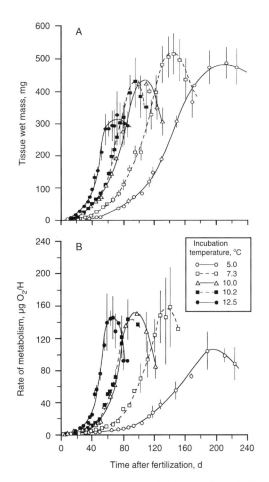

Figure 10.19 (A) Tissue mass and (B) rate of metabolism in embryonic and larval Chinook salmon (*Oncorhynchus tshawystscha*) as a function of time after fertilization and incubation temperature. Source: Modified from Rombough (1994).

10.8 CONSTRAINTS ON ENERGY BUDGETS

The energy expenditures of vertebrates in the field, or for that matter under experimental conditions, may be physiologically or ecologically constrained. The "reality" of constraints, however, is subject to dispute. Narrowly, the dispute in physiology has centered on whether some functional relationship exists between the maximal and standard or basal rate of metabolism. For example, Bennett and Rubin (1979), McNab (1980b), Taylor (1982), and Taigen (1983) maintained or implied that such a proportionality (usually?) exists, often because maximal and standard rates of metabolism scale to body size with identical powers (a shaky basis for concluding a functional connection between maximal and basal rates; see McNab and Eisenberg 1989). In a study of 10 species of passerines, Dutenhoffer and Swanson (1996) went a step further by showing that maximal, cold-induced rate of metabolism was correlated with body mass in a manner parallel to the correlation of basal rate with mass and that the residual variation of maximal rate in the maximal rate–mass correlation was correlated ($r^2 = 0.755$) with residual variation in the basal rate–mass correlation. Koteja (1987), however, concluded that these two rates scale differently in mammals and thus are independent. This "independence" is especially apparent in small marsupials, which have maximal rates of metabolism as high as, or higher than, eutherians in spite of their low basal rates (Dawson and Dawson 1982, Smith and Dawson 1985, Dawson and

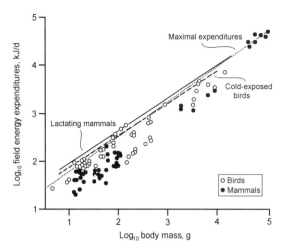

Figure 10.20 Log_{10} field energy expenditures in birds, mammals, lactating mammals, and cold-exposed birds and log_{10} maximal expenditures in birds and mammals as a function of log_{10} body mass. Source: Modified from Weiner (1992).

Olson 1988). Bozinovic (1992) disagreed, pointing out that the use of scaling relationships in which different species of mammals are used to establish maximal and basal rates adds too much heterogeneity to the data: when the same species are used in the study of maximal and basal rates of metabolism, maximal rates are proportional to basal rate. Furthermore, not only were these rates correlated (as they should be because both are correlated with mass), but also the residual variations were correlated, implying a functional connection.

Whether field rates of metabolism are limited and whether the limitation reflects the level of the basal (or standard) rate of metabolism has also been raised. Drent and Daan (1980) maintained that a "prudent predator" during reproduction cannot sustain energy expenditures at rates that exceed 4 times the basal rate. Nagy (1987) and Weiner (1992) pointed out that the maximal rates of metabolism of various birds and mammals (Kirkwood 1983), lactating mammals (Kenagy et al. 1989, Weiner 1989), and cold-exposed birds (Karasov 1990) fall at the upper range of field energy budgets (Figure 10.20), which they interpreted to imply that most mammals and birds are living at rates close to the maximal rates; that is, a limit exists. Furthermore, Koteja (1991) found that field rates of metabolism usually are not correlated with basal rate in breeding birds and marsupials, but that they are in nonreproducing endotherms generally (i.e., birds, mammals, eutherians, and

rodents). A problem contributing to the imprecision of this correlation is the variable conditions under which field measurements are made, not to mention a collective indifference as to whether the basal rate is measured or assumed.

Kersten and Piersma (1987) addressed the factors that produce the high field expenditures of shorebirds. For example, ruddy turnstones (*Arenaria interpes*), black-bellied (grey) plovers (*Pluvialis squatarola*), and Eurasian oystercatchers (*Haematopus ostralegus*) in outdoor cages consumed food at rates that were much greater than expected from mass from the equations of Kendeigh (1970). These rates are 4.1 to 4.4 times the basal rate expected from mass from the non-passerine equation of Aschoff and Pohl (1970). Measured basal rates, however, were 1.2 to 1.5 times the values expected from this relationship. Consequently, the ratio of "existence" expenditure to the measured basal rate was 2.9 to 3.1, well within the usual range in free-ranging birds (Bryant and Tatner 1991). Thus, Kersten and Piersma suggested that the high (simulated) field expenditures of these shorebirds reflected their high basal rates. They (p. 185) went on to argue ". . . that a high [daily energy expenditure], mainly generated by the skeletal muscles, requires a high level of support by the organs in the abdominal cavity, which inevitably results in a high [basal rate of metabolism]." Indeed, seasonal variations in the basal and field energy expenditures of knots (*Calidris canutus*) reflected variation in body composition (Piersma et al. 1996).

Daan et al. (1990) explored the view that the basal rate reflects body composition. They showed that the residual variation in the basal rate of birds, after the effect of body mass was removed, was correlated with the residual variation in the sum of heart and kidney masses. These authors also demonstrated in 26 species of birds that parental field expenditures, measured by doubly labeled water, were correlated with measured basal rates, as were the residuals, the mean ratio between field expenditures and the basal rate being 3.6. Therefore, variation in field energy expenditures in these species was associated with mass-independent variation in body composition. Daan et al. (1990, p. 338) concluded that ". . . the metabolic machinery has . . . been adjusted by natural selection to the energetic requirements during the . . . maximal [daily energy expenditure and] . . . species with a high [parental energy expenditure] . . . display a relatively high [basal rate], [whereas] species with a low [parental energy expenditure]... show a relatively low [basal rate]." Indeed, the low basal rates of tropical birds (Weathers 1979, Hails 1983) may well reflect proportionally small liver, kidney, and heart masses (Rensch and Rensch 1956). Ricklefs et al. (1996), however, cautioned that the kidney and heart of birds are so small that their size alone will unlikely account for the variation in basal rate, and concluded (p. 1064) that "... these organs belong to a suite of characteristics associated with differences in [basal rate of metabolism]," a view similar to that expressed by Kersten and Piersma (1987). Thus, species that have large muscle masses, hearts, kidneys, and guts appear to have high basal rates and field expenditures (Hammond and Diamond 1997, but see Koteja 1996a). The variation in the size of the pectoral muscle masses may be important because they are the principal tissues in which work is accomplished during activity and the generation of heat for temperature regulation in birds (see McNab 1994a). Some variation in basal rate, however, may proximally reflect a variation in thyroid function (Fregly et al. 1963).

Ricklefs et al. (1996) examined the nature of the potential relationship between field energy expenditure and basal rate in birds and mammals. They wanted to know if field expenditures are correlated with basal rates and if so, whether field expenditures represent an amplification of the pathways of metabolism that produce the basal rate (the "shared pathways" model) or whether they represent an expenditure that is the sum of two independent expenditures, one for basal maintenance and one for activity (the "partitioned pathways" model). If the shared model pertained, the correlation coefficient between the daily expenditure and basal rate should be greater than 0, and if the partition model was appropriate, the correlation coefficient between the activity component and basal rate should be greater than 0. They found that the daily energy expenditure in birds was correlated with basal rate in the shared- but not in the partitioned-pathways model. The shared-pathways model correlation was lost, however, when the analysis was modified to include a phylogenetic "correction" (see Garland et al. 1993), although Ricklefs et al. (p. 1055) admitted "... that the phylogeny [must be] known without error...," an optimistic hope—especially given the controversies around the Sibley-Ahlquist phylogeny that was used. In mammals, field expenditures were clearly correlated with basal rates, when both models were accepted, even after a phylogenetic "correction" was made. Ricklefs et al. (p. 1062) concluded that "... a ... fundamental difference [exists between] the energetic physiology of birds and mammals." They suggested that this difference might reflect a more diverse physiology in mammals than birds. A more likely explanation is that an apparently smaller variation in the physiology of birds (say, basal rate) may simply represent an inadequate sampling of the variation present (most studied species being seabirds or temperate passerines, with few tropical land birds included). Should a difference between birds and mammals remain after tropical land birds are studied, the most likely cause for this difference would be the capacity of birds to avoid environmental harshness through flight.

Various factors have been suggested to limit energy expenditure, and most of them have a physiological basis (see Karasov 1986, Weiner 1992). They include limits on foraging rate, rates of digestion and absorption, heat production, mechanical work, and tissue growth. Sometimes these limits are grouped into "central" (usually alimentarily imposed) or "peripheral" (at the sites of substrate use) (Lechner 1978; Wang 1978b; McDevitt and Speakman 1994a, 1994b). For instance, the digestive system repeatedly has been suggested to limit the capacity of a vertebrate to digest food and to assimilate energy and nutrients (Karasov 1986, 1990; Weiner 1987, 1989; Peterson et al. 1990). Evidence for this "central" limit has been difficult to assemble, most of it being indirect: rate of metabolism is correlated with gut volume or surface area, both within species, as a result of acclimation to temperature or fiber content of the diet (Green and Millar 1987, Karasov and Diamond 1988, Woodall and Currie 1989, Hammond and Wunder 1991, Loeb et al. 1991, Koteja 1996b), and among species in association with differences in diet (Chivers and Hladik 1980). Yet, the limitation to energy expenditure in *Peromyscus maniculatus* during cold exposure was greater than during lactation, a pattern that differs from the pattern found in other rodents, which implies that the limitation is not "centrally" located (Koteja 1996a). More direct evidence was found in rufous hummingbirds (*Selaphorus rufus*), which fed 14 to 18 times per hour in the field and took 4 min to clear one-half of the nectar contained in the crop (Karasov et al. 1986a). Between feeding bouts hummingbirds sat on perches, apparently waiting to clear the crop. Indeed, Hammond and Diamond (1997) concluded that the limits to energy expenditure are likely to be "peripheral." The difficulty with these analyses, as always, is the question of cause and effect.

Illustrative of these complexities, Lindström and Kvist (1995) showed that maximal metabolizable energy intake is proportional to basal rate in migrating passerine birds, in terms of both scaling and residual variation. They showed that the mean ratio of maximal metabolizable intake to basal rate was 4.6, whereas the ratio of the daily energy expenditure of reproducing passerines to basal rate was 3.6 (Daan et al. 1990). They (p. 337) concluded ". . . that energy intake rates may not normally limit breeding performance in passerines." Furthermore, the increase in food intake by lactating white mice was correlated with an increase in the intestinal nutrient uptake (Hammond and Diamond 1992). A short-term exposure to cold in *Microtus agrestis* was not limited by the alimentary system (McDevitt and Speakman 1994a), but the length and intensity of the exposure were limited— the maximal expenditure was only 1.7 times the basal rate. But even after this vole was exposed to 5°C for 100 d, there was no evidence that the alimentary tract limited the vole's response to cold (McDevitt and Speakman 1994b). Its principal adjustment to cold was an increased capacity for heat production, apparently associated with an increase in the amount of brown fat and an increase in basal rate. Hammond and Diamond (1997) concluded that limits to performance do exist and they are correlated with standard rates because both depend on energy-supplying organs, which depending on the experience of species, may increase or decrease in size.

A limitation to energy expenditure other than by a physiological limit to the processing of resources might issue from deleterious consequences of high rates of energy expenditure. Evidence for such a limitation would require knowledge of the long-term survival of individuals as a function of energy expenditure. Bryant (1991) showed that free-living house martins that fledged the most young in a lifetime had an intermediate rate of metabolism during the brood-rearing stage (Figure 10.21). A high rate of energy expenditure, even though it permitted a larger brood to be raised in the short term, was associated subsequently with an increased mortal-

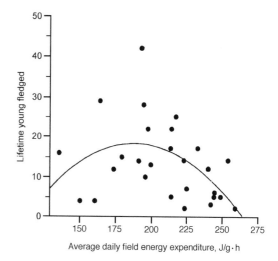

Figure 10.21 Lifetime young fledged by house martins (*Delichon urbica*) as a function of average daily field energy expenditure. Source: Modified from Bryant (1991).

ity, which might have been associated with a reduction in body mass during the breeding effort, and a decreased lifetime fecundity (see also Lemon 1993).

The proposition that a "universal" maximal rate exists is highly unlikely, given the great taxonomic, morphological, and physiological diversity found among vertebrates, and even among mammals and birds. Absolute maxima do not exist, at least because of the influence of body mass, and universal maxima relative to some size-dependent standard, such as the standard or basal rate, probably do not exist. Weiner (1992) argued that the maximal energy expenditure is probably species specific and is more like 7 times the basal rate than the 4 times Drent and Daan (1980) suggested. In a survey of avian field expenditures Bryant and Tatner (1991) noted that 21% of the measurements exceeded the ratio 4:1, and Speakman (2000) showed that 19% of small-mammal field expenditures exceeded the ratio of 4:1.

After an extensive review of the complexities of internal limits to field energy expenditures, Speakman (2000) noted that the capacity for a sustained scope might reflect the supply of resources in the environment (i.e., represented an extrinsic limit). To test this idea, he inquired if the ratio of field to standard rate varied with the diets of small mammals. He found that the correlation was significant and that folivores had low ratios (approximately 2.0), whereas grazers (3.4), insectivores (3.9), and carnivores (4.0) had higher ratios. He

(p. 263) concluded, "[The] extrinsic supply of energy depends on the density of the food, its composition and ease of digestibility, and probably also on the time available for the animals to collect it. Hence small mammals exploiting rich and abundant foods have raised [field expenditures] because the extrinsic supply of energy allows them to sustain this higher expenditure, whereas those exploiting poor resources have suppressed [field expenditures]. The consequence of this extrinsic limitation of [field expenditures] is a link between the ratio of [field] to [standard expenditures] and the diet of the mammals in question, with low ratios linked to poor-quality diets." As noted in Section 5.7.4, these restrictions are carried to an extreme in some arboreal mammals, like tree-sloths, which have low basal rates, small muscle masses, low scope for metabolism, and consequently very low ratios of field to basal rates of energy expenditure, all ultimately derived from the use by sloths of a low-quality (if abundant) food.

A fundamental difficulty with the concept of a limit that is rarely mentioned is that it must include a time period. Whatever the limit, if one exists, it surely can be exceeded for a short-term "emergency," such as evading a predator. Or can an "emergency" include reproduction? Is an hour, a day, a week, or a season appropriate to the definition of a constraint? Nevertheless, the concept that the energy expenditure of vertebrates is always unlimited also is difficult to believe. And that may be the crux of the problem: environmental limitations to energy (and resource) use may be periodic or occasional, which would make them difficult to demonstrate.

10.9 HOME RANGE SIZE AND ENERGY EXPENDITURE

Home range size increases with body mass in terrestrial vertebrates, including endotherms (McNab 1963a, 1983b; Armstrong 1965; Schoener 1968; Milton and May 1976; Harestad and Bunnell 1979; Harvey and Clutton-Brock 1981; Gittleman and Harvey 1982; Mace and Harvey 1983; Mace et al. 1983; Lindstedt et al. 1986) and ectotherms (Turner et al. 1969). The general explanation for this correlation is that large animals, which have higher (total) rates of metabolism and therefore require more food than small species, need larger areas in which to find adequate food supplies. Originally the suggestion was that home range in

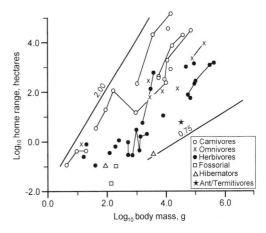

Figure 10.22 Log_{10} home range in mammals as a function of log_{10} body mass and food habits. Source: Modified from Harestad and Bunnell (1979).

mammals should increase proportionally to a functional definition of body size (e.g., basal rate, i.e., $m^{0.75}$) (McNab 1963a). But because field expenditures would be more appropriate, one might expect home range size to increase as $m^{0.64}$ to $m^{0.81}$ (see Section 10.5). However, that is not the case. In birds the exponent of mass that describes the increase in home range falls between 1.16 (Schoener 1968) and 1.23 (Armstrong 1965), and in lizards it is 0.95 (Turner et al. 1969). Harestad and Bunnell (1979) maintained that the exponent in mammals is actually about 1.08, but a close examination of their data on ecologically equivalent species indicates that for some the power may be as high as 1.90 (Figure 10.22). Kelt and Van Vuren (1999), in an analysis of minimal home ranges in mammals, suggested that a mass exists, approximately 100g, at which home range is minimal, but increases at both smaller and larger masses. This minimum was derived from an analysis of Brown et al. (1993), who postulated that this mass was the general energetic optimum for mammals (see Section 3.4).

Why does the exponent describing the size of home range so markedly exceed 0.75? One possibility is that the foods used by large consumers are not as abundant as those used by small consumers so that an increase in consumer mass requires a disproportionately large increase in home range size. Evidence for this explanation is seen in the analyses of Schoener (1968) and Harestad and Bunnell (1979), who demonstrated that the power of mass in carnivorous birds and mammals is much greater than that in herbivores (e.g., 1.39 vs. 0.51 and 1.36 vs. 1.02, respectively; also see Gompper and Gittleman 1991). In four browsing ruminants, however, home range was proportional to $m^{1.38}$ (du Toit 1990).

Lindstedt et al. (1986) made another suggestion for the large exponent of mass. They argued that home range really represents an amount of energy and, therefore, that it should be proportional to the product of rate of energy use (power) and time:

$$\text{area (kJ)} = \text{energy expenditure (kJ/d)} \times \text{time (d)}.$$

They further maintained that the meaningful time unit is "biological" time, which is proportional to $m^{0.25}$ (Lindstedt and Calder 1981). Because total field expenditures in eutherians are approximately proportional to $m^{0.81}$ (see Section 10.5.1), their home ranges might be expected to be proportional to $m^{0.81} \times m^{0.25} = m^{1.06}$, which is what is observed. The time required to traverse home range is approximately proportional to $m^{0.25}$ (Swihart et al. 1988), which gives a concrete expression for "biological" time. Whatever the immediate cause for the increase in home range with body mass, it ultimately reflects an increase in the use of resources at large masses (which is further evidence that total rates are what are ecologically and evolutionarily relevant). In a review of the scaling of home range, Reiss (1988) concluded that our understanding is very limited, but that is undoubtedly related to the complexity of this relationship and to the limited amount of data available.

Factors other than body mass also affect home range size. They include food habits (McNab 1963a, Schoener 1968, Milton and May 1976, Harestad and Bunnell 1979, Swihart et al. 1988), mainly because some foods are less abundant than others: vertebrate eaters have large home ranges, omnivores have intermediate home ranges, and herbivores have small ranges at the same mass (Figure 10.22). Specifically, frugivorous primates have larger home ranges than folivorous primates (Milton and May 1976; Harvey and Clutton-Brock 1981), and among members of the order Carnivora, flesh eaters have the largest home ranges, omnivores and frugivores have intermediate home ranges, and invertebrate-eating specialists have the smallest home ranges. This sequence may reflect the relative abundances of these foods in most environments, but it may also reflect directly the basal rate of metabolism of carnivores: flesh eaters not only use the rarest foods but also have high basal rates, whereas invertebrate eaters use abundant foods and have low basal rates (McNab 1989a).

The relationship between home range size and body mass is further modified by climate and behavior because of their impact on energy availability and expenditure. Grazers and browsers living in arid regions, such as jackrabbits (*Lepus*) and desert bighorn (*Ovis*), have larger home ranges than similar species living in mesic environments, because arid environments have lower rates of plant production than mesic environments (Rosenzweig 1968). Carnivores that live at high latitudes have larger home ranges than those that live at low latitudes (Lindstedt et al. 1986, Gompper and Gittleman 1991). Social primates and social carnivores generally require larger home ranges than solitary mammals of the same summed mass (Milton and May 1976, Harvey and Clutton-Brock 1981, Gittleman and Harvey 1982). A similar trend was shown in social grazers (Kaufmann 1974).

The large home ranges of a group size n of equally sized individuals is expected to have a home range equal, at least, to $[n(1/n)^{0.75}]/[(\Sigma n)^{0.75}]$ times that of a nonsocial individual having a mass equal to Σn (McNab 1983b). For instance, if troop size were 23, the troop home range would be expected to be at least 2.2 times that of a solitary individual, because a given mass of primates will have a higher rate of metabolism if it consists of many small individuals than if it is a few large individuals, a consequence of rate of metabolism being proportional to mass raised to a power less than 1.0 (also see Kleiber [1961], where he compared the mass of cattle to that of rabbits that can be maintained by a given amount of hay). Activity rates may also be much higher in social species (Milton and May 1976), which would further increase energy expenditure and therefore home range size.

10.10 ENERGY BUDGETS IN RESTRICTIVE ENVIRONMENTS: ISLANDS, CAVES, AND BURROWS

Some isolated environments are characterized by low primary production. Such environments include distant oceanic islands, where primary production is limited by island area, and caves, where production is derived principally from detritus transported by water from surface environments. These environments are isolated from productive

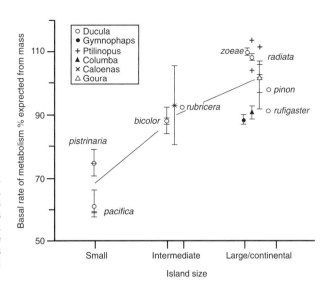

Figure 10.23 Basal rate of metabolism in South Pacific fruit-pigeons belonging to the genus *Ducula*, expressed relative to the values expected from mass in pigeons, as a function of the size of the islands or continents on which they are found. The mean for each species is indicated by a circle, and the mean for each individual in the sample studied is indicated by a horizontal bar. Source: Modified from McNab (2000a).

environments either by distance (as with islands) or by extreme conditions (the absence of light in caves and the adjustments required for a hypogenic existence). Burrows are similar in that some production is derived from the surface by the growth of subterranean tubers, and isolation results from an extreme commitment to a burrowing existence, that is, being fossorial. (Other environments, such as polar communities, also are typified by low primary production, but that occurs on a seasonal basis and they are not as isolated from productive environments as are caves and oceanic islands).

10.10.1 Oceanic islands. Vertebrates facilitate long-term survivorship on oceanic islands by reducing their resource requirements (McNab 1994b). Such reductions occur in several ways:

1. A reduction in body mass. Because a reduction in mass is associated with a reduction in energy expenditure, a small mass facilitates a long-term persistence on small islands, where resources are limited. This reduction is most marked in large mammalian herbivores, such as proboscidians (Sondaar 1977, Roth 1990, Vartanyan et al. 1993), hippopotami (Sondaar 1977, Stuenes 1989), and deer (Sondaar 1977, Lister 1989); edentates (Matthew and de Paulo Couto 1959); and carnivores (Foster 1964, Lomolino 1985). Some endotherms are large on islands (Lomolino 1985), but only if the islands have an unusually large resource base (e.g., bears feeding on salmon on Kodiak Island) or if a normally abundant resource

is not being exploited (e.g., grazing and browsing birds and tortoises in the absence of herbivorous mammals).

2. A reduction in activity. Activity levels of terrestrial birds on islands (e.g., Galápagos) are often greatly reduced in the absence of mammalian predators, a shift that reduces the energy expenditure associated with "flightiness."

3. The evolution of flightlessness in some birds. As seen in Section 9.6.9, some groups of birds repeatedly evolved a flightless condition on ocean islands, which leads to a reduction in energy expenditure (Figure 9.24; McNab 1994a).

4. A reduction in rate of metabolism beyond that produced by factors 1 through 3. Island-specialist pigeons (McNab 2000a), pteropodid bats (McNab and Bonaccorso 2001), and rodents (Arends and McNab 2001) have lower basal rates of metabolism than their relatives that live on large islands or continents (Figure 10.23).

5. Replacement of endotherms by ectotherms: On many islands reptiles have replaced mammals as the principal herbivores (e.g., tortoises on Aldabra and the Galápagos, lizards on many islands) and carnivores (e.g., Komodo monitor on several Lesser Sunda Islands, a crocodile and a python formerly on New Caledonia).

These adjustments, however effective in enhancing survivorship on islands that lack mammalian predators, make them vulnerable to the arrival of humans and their mammalian commensals. As a result, a huge extinction of the faunas of oceanic islands has occurred since the arrival of humans (Steadman 1995).

10.10.2 Caves. Poulson (1964), Poulson and White (1969), and Culver (1988) reviewed the biological characteristics of cave-dwelling specialist (troglobitic) organisms. Among vertebrates, fish and salamanders have evolved a troglobitic existence. Troglobitic fishes have low standard rates of metabolism (Englemann 1909, Poulson 1963, Hüppop 1986). For example, the standard, routine, and active rates of metabolism decreased in a series of amblyopsid fish species with an increased commitment to cave life (Poulson 1963). A similar pattern was seen in the Mexican characin *Astyanax fasciatus*. Surface-dwelling individuals had higher standard and routine rates of metabolism than did individuals found in caves with a low food content; individuals in caves with an abundance of food had intermediate rates of metabolism (Hüppop 1986). Some surface-dwelling individuals had rates of metabolism as low as those found in individuals living in caves with the least amount of food, which suggests how cave populations might have been established. (The inability of Schlagel and Breder [1947] to show a reduction of energy expenditure in cave *Astyanax* is explained by their use of individuals from a food-rich cave inhabited by hybrids between epigean and hypogean fish [Hüppop 1986].)

Associated with the evolution of a reduced energy expenditure has been a reduction in growth rate, an increased longevity, a reduction (or elimination) of eye function and structure, and (often) the elimination of pigment. These changes may well be associated with a reduction in the cost of synthesis and maintenance, as is the case in the evolution of flightlessness in birds on oceanic islands. Energy expenditure has not been measured in troglobitic salamanders, such as some North American species of *Eurycea*, *Gymnophilus*, *Haideotriton*, and *Typhlotriton*. Troglobitic crayfish (Burbanck et al. 1948, Eberly 1960, Caine 1978, Franz 1978, Dickson and Franz 1980) and spiders (Hadley et al. 1981, Anderson and Prestwich 1982) also have low standard rates of metabolism, but apparently not cave amphipods, possibly because the studied species were in caves that were not food limited (Culver and Poulson 1971).

10.10.3 Burrows. Fossorial mammals are characterized by low rates of metabolism (Section 5.4.2), usually explained as an adaptation to

thermal stress in burrows (McNab 1979, 1996), to the composition of burrow atmospheres (Baudinette 1977, Arieli et al. 1977), or to a combination of these factors (Contreras and McNab 1990). Another possibility is that their low rates of metabolism reflect a reduced food supply (Jarvis 1978). An extremist adaptation to a burrowing existence (for a detailed analysis of subterranean mammals, see Nevo [1999]) in many ways parallels an extremist accommodation to a cave existence, including a reduction in vision (*Spalax*) and other structures (legless burrowing reptiles [see Section 9.5.2]), as well as a reduction in rate of metabolism.

In spite of these correlations, no direct observations have been made on the field energy expenditures of cave, burrow, or island specialists to compare them with those of ecologically equivalent species living on the surface or on continents. The conclusion that vertebrates living in environments with restricted resources have low field energy expenditures needs to be demonstrated (for the case in flightless ducks, see Section 9.6.9).

10.11 POPULATION ENERGETICS

Vertebrates usually do not exist as isolated individuals but as functional members of a population, as befits sexually reproducing organisms. The energy expenditure of a population is the collective energy expenditures of the members of that population.

Estimates of the energetics of free-living populations are normally constructed from estimates of individual energy budgets and knowledge of the demographic profile and density of a population, because no practical means exist to measure directly the energetics of populations. Most available estimates of population energy expenditures incorporate laboratory measurements of oxygen consumption (e.g., Mann 1965, Chew and Chew 1970, Grodzinski 1971, Holmes and Sturgis 1973, West and DeWolfe 1974, Burton and Likens 1975, French et al. 1976, Górecki 1977, Grodzinski et al. 1977, Bennett and Gorman 1979) or of food consumption (Buchner 1964, Wiens and Innis 1974, Weiner and Glowacinski 1975, Wiens and Scott 1975). Other estimates may be based on a quantitative model derived from physiological data (Wunder 1978, Kilgore and Armitage 1978). Some population budgets include the costs of reproduction and growth, whereas others do not.

The principal difficulty with estimating population energetics is that the errors associated with estimating population size, demographic parameters, and microclimate are often so great that the variation in individual energy budgets is insignificant. That is, the error in estimating individual energy budgets is probably less than the errors estimating the population parameters incorporated into the overall estimate.

10.11.1 Population maintenance. The information potentially derived from population energetics is illustrated by two analyses of avian energetics by John Wiens. Wiens and Innis (1974) constructed an annual budget for the dickcissel (*Spiza americana*) in South Dakota. By using Kendeigh's (1970) data on food consumption and field data on the proportions of the population that belonged to the adult, juvenile, nestling, and fledgling age classes, these authors converted the population density data (Figure 10.24A) into an estimate of the energy expenditure of dickcissels for each month (Figure

10.24B), the sum of which is the annual population energy budget. This estimate includes the cost of production to some extent, although the inclusion is more implicit (by the inclusion of young and the presumed cost of growth) than explicit. Wiens and Scott (1975) used a similar approach to estimate the population energy expenditure of four seabirds along the Oregon coast (sooty shearwater [*Puffinus griseus*], Leach's storm-petrel [*Oceanodroma leucorhoa*], Brandt's cormorant [*Phalacrocorax penicillatus*], and common murre [*Uria aalge*]). These expenditures in combination with stomach analyses and the energy content of the prey permitted the amounts of food consumed by these species to be estimated. One can easily contest many of the assumptions (e.g.: Are Kendeigh's equations a good estimate of field expenditures in these birds? Are the population numbers much more than a guess? The estimated impact of the resident species would have been increased if the cost of reproduction had been included.). Yet, this analysis suggests that the sooty shearwater, which breeds in New Zealand and appears off the Oregon coast in spring and fall as a migrant, has nearly as great an impact on the food resources along the Oregon coast as the three resident species combined. That impact is selective, however, because of the shearwater's preferential consumption of anchovies.

On occasion, indirect measurements of population energetics through predation have been attempted in the field. In one of the most effective attempts made, Pearson (1964) collected scats of foxes, skunks, and feral cats preying on rodents belonging to the genera *Microtus*, *Reithrodontomys*, and *Mus* on a grassland in California. Predation focused on the *Microtus* population, which reached a high of about 4400 individuals, 86% of which was later found in the scats. The *Microtus* population came close to completely disappearing. In contrast, only 7% of the *Mus* population, which reached a peak of about 7000 individuals, was recovered from scats. These data permitted Pearson to estimate that 753 MJ were transferred from rodents to carnivores over a 6-month period, which is equivalent to supporting four 3-kg carnivores at 1.9 times the basal rate expected from the all-mammal curve.

Because of the difficulty estimating the energetics of populations, few analyses of their energetics have been made. Damuth (1981, 1991) showed that mammals reduce density with an increase in mass

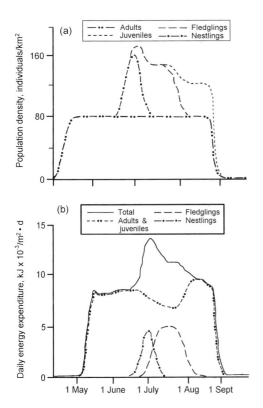

Figure 10.24 (A) Population density and (B) estimated population daily energy expenditure of dickcissels (*Spiza americana*) as a function of time of year. Source: Modified from Wiens and Innis (1974).

and, given that rate of metabolism increases with mass, argued that energy use by mammalian populations is independent of mass, what has been referred to as the "energy equivalence rule." A rather large literature has dealt with this topic. Silva et al. (1997) recently proposed an analysis of these relationships in mammals and birds. They found that population densities tended to decrease with body mass, although they may also decrease at the smallest masses: the highest densities are found in mammals that weigh near 100 g and in birds that weigh 30 g. Furthermore, they observed that population densities and energy expenditures were about 10 times higher in mammals than in birds, which was suggested to reflect the capacity of birds to fly (note that flightless birds were more like mammals, and bats more like birds). Silva et al., however, doubted that population energy expenditure is independent of mass, but an examination of their data indicates such independence, except at the smallest and largest masses (where few data exist).

Few estimates of the energy expenditure of ectothermic vertebrate populations exist, notable exceptions being those of Mann (1965), Burton and Likens (1975), and Bennett and Gorman (1979). Bennett and Gorman, for example, estimated minimal and active energy budgets for several species of lizards on a Caribbean island (Bonaire). Even without including the cost of reproduction and growth, the high population densities and high body temperatures of these lizards led to estimated population energy budgets that were equal to or greater than those of most temperate mammal populations.

In cooler environments, ectothermic vertebrates, such as temperate salamanders, consume a small fraction of community production, but they are exceedingly efficient in converting that intake into growth and reproduction (see Section 4.16). As a consequence, salamander populations in North Temperate forests may have a greater biomass and more protein per hectare than birds at the peak of their breeding season in the same forest (Burton and Likens 1975). These observations led Burton and Likens (1975) and Pough (1980a) to conclude that the secondary production of amphibians and reptiles is as great as that of birds or mammals in terrestrial ecosystems.

10.11.2 Population reproduction and growth.

All stable or expanding populations invest energy in production (i.e., in reproduction and growth).

What factors determine the level at which this investment occurs? Several studies (Slobodkin 1960, Engelmann 1966, McNeill and Lawton 1970, Grodzinski and Wunder 1975, Calow 1977, Humphreys 1979, R. M. May 1979, Townsend and Calow 1981, Lavigne 1982, Wieser 1985) addressed this question.

Engelmann (1966), followed by McNeill and Lawton (1970) and others, showed that the amount of energy invested in production is positively correlated with "respiration" (= rate of metabolism). McNeill and Lawton further indicated that this positive relation is found in poikilotherms and homeotherms, although poikilotherms have a much greater efficiency in converting energy intake to reproduction and growth. This difference exists because of the very high maintenance costs associated with endothermic temperature regulation.

Humphreys (1979) showed that when the $\log_{10}$ of production is plotted as a function of $\log_{10}$ metabolism, a series of parallel curves exist, one for each of the following groups: insectivorous mammals, birds, small-mammal communities, "other" mammals, fish and social insects, invertebrates other than insects, and nonsocial insects (Figure 10.25). In each case the slope of the curves equaled 1.0, which means that the efficiency with which energy intake is converted to production in

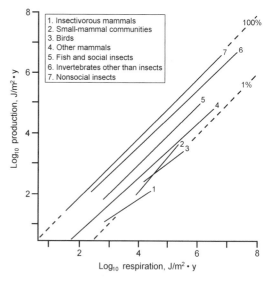

Figure 10.25 $\log_{10}$ production in nonsocial insects, invertebrates other than insects, fish and social insects, "other" mammals, small-mammal communities, birds, and insectivorous mammals as a function of $\log_{10}$ respiration. Source: Modified from Humphreys (1979).

each group is constant. The curves differ in their efficiencies, as represented by the "height" of the curves, which is greatest in nonsocial insects and lowest in insectivorous mammals. R. M. May (1979) pointed out that solitary insects and small mammals represent extremes along a continuum of efficiencies that is bridged by poikilothermic vertebrates and social insects. In general, endotherms convert some 1% to 3% of assimilated energy into production; fish and social insects, about 10%; and other ectotherms, from 21% to 56%.

A positive correlation between production and metabolism is compatible with the view of a trade-off between production and metabolism (Lavigne 1982). Thus, *within* a uniform category (e.g., insectivorous mammals or nonsocial insects) production increases with metabolism. McNab (1980b) proposed a similar view for mammals (see Chapter 13). When, however, species belonging to one category are compared to those belonging to a *different* category, an inverse relation is found between production and metabolism: endotherms have a lower efficiency in transferring energy intake to production than do nonsocial insects. Ultimately, the amount of ingested energy transferred to production depends on the levels of energy intake and metabolism. Lavigne suggested that the rate of production, which has the units of power (= energy/time), is approximately proportional to $m^{0.75}$, irrespective of whether it is measured as growth, milk production, litter mass, or clutch mass, as expected if the ratio of production to metabolism is constant and metabolism is approximately proportional to $m^{0.75}$. This analysis demonstrates the intimate connection that exists between the energetics of individuals and that of populations.

The apparent trade-off between "efficiency" and "power" (see Odum and Pinkerton 1955) found in a comparison of ectotherms and endotherms can also be seen in the different responses of their populations to food shortage. Ectothermic vertebrates, which are characterized by indeterminate growth, grow and reproduce at rates that directly reflect resource availability, whereas endotherms, which have determinant growth (i.e., grow to a fixed adult size), require a fixed minimal amount of food to reach adulthood. If endotherms do not obtain this minimum, they will die or at least will fail to reproduce. However, most poikilotherms with a dwarfed body size will mature even on a restricted food regimen. In the absence of predation and in the presence of a restricted food supply, ectother-

mic populations (e.g., arctic char, *Salvelinus alpinus*) tend to stabilize with numerous small individuals, whereas those of endotherms (hares, arvicolines, moose, grouse) will usually be unstable, often showing cyclic variations (Steen et al. 1991). This difference is often most apparent at high latitudes, which have an appreciable seasonal variation in food resources and requirements (Hansson and Henttonen 1985). Cyclicity of endotherm populations under these conditions issue from "scramble" competition for the limited resource base, combined with a physiology that requires a fixed minimal energy requirement (Steen et al. 1991).

Wieser (1985) provided another view of the difference in the efficiency of production between ecotherms and endotherms. He (p. 509) suggested that the difference between these groups does not reflect biochemical, but rather ecological and evolutionary differences: ". . . whereas the production of an endotherm offspring requires only 2–6% of the metabolizable energy, [a] fish has to spend 35%, [and a] nematode nearly everything it has for this purpose. . . . [T]he difference . . . emerging from this comparison does not prove [ectotherms] to be more efficient than [endotherms] in converting food energy into production. On the contrary: for many ectotherms the cost of reproduction is so high that they are driven to what Calow (1978) called 'reckless reproductive behavior,' that is, face the risk of death by reproduction." Trillmich (1986), however, reminded Wieser that endotherms are not "emancipated" from the cost of reproduction (see Section 10.6), to which Wieser (1986) responded that the "emancipation" is biochemical, not ecological (= "social").

The view Wieser expressed focuses on a fundamental difficulty with efficiencies: an efficiency is a ratio (usually multiplied by 100) of an output divided by an input—and here the question arises, what is the appropriate input? In Wieser's opinion it is not the total energy expenditure, much of which is directed in endotherms for maintenance and activity, but the total expenditure is the appropriate input for Humphrey.

Brett (1986) analyzed the population production of sockeye salmon (*O. nerka*). He examined a population from Babine Lake (British Columbia), Canada. By the time the yearling sockeyes move offshore, 90% of the smolts have died or been consumed, and the remainder weigh 50 to 60 g. By then, the biomass of this cohort has increased by 4%, in spite of the massive loss of small young individuals. This means that the remaining 10% of the population achieves 96% of the production of the cohort at sea, which lasts 1, 2, or 3 y, most of the production occurring in the last year. Smolts that spend 2 or 3 y at sea are much larger and more abundant than the "jacks" that return after only 1 y at sea (at age of 3 y). This cohort produced almost 4000 tons, 2% from 3-y-old individuals (jacks), 41% from 4-y-old individuals, and 57% from 5-y-old individuals. Brett (p. 562) concluded that "... Pacific salmon appear to have evolved a survival and energy efficiency pattern based on saturating their coastal predators in early sea life while themselves feeding heavily and growing rapidly out of the more vulnerable prey size to become a dominant offshore predator."

10.12 CHAPTER SUMMARY

1. An energy budget is a quantitative description of the allocation of energy by a species to various biological functions.

2. Energy budgets have been estimated in many ways, including by using measures of food consumption, fluctuation in body mass, or gas exchange; by using doubly labeled water, other isotopes, or time budgets; and by using a combination of these methods. The only measurements of energy expenditure in the field use isotopes.

3. Field energy expenditures correlate with body mass in lizards, birds, and mammals, with those of birds being somewhat higher than in mammals and those in mammals being 10 to 12 times those of most lizards.

4. Field energy expenditures in mammals and lizards usually are between 1.5 and 2.5 times the standard rate of metabolism; in birds this ratio is somewhat higher, 2.5 to 5.2 times the standard rate.

5. Field expenditures in birds are greatest, independent of body size, in species that spend the most time in flight or have a flight morphology that leads to high expenditures.

6. In mammals rate of metabolism increases during pregnancy by about 25%, independent of body mass, and during lactation by 50% to 100%, depending on mass.

7. Large mammals allocate a larger proportion of energy intake to growth than do small species because of the high cost of maintenance in small species.

8. When feeding young, birds approximately double their rates of energy expenditure, owing mainly to an increase in activity; the increase in energy expenditure has implications for maximal brood size.

9. The growth rate of endotherms is influenced by body size and whether they have altricial or precocial offspring, primarily because of the differential allocation of energy to growth or maintenance.

10. Ectotherms that actively forage for prey obtain more energy and allocate more energy to reproduction and growth than do sit-and-wait predators.

11. Whether energy budgets are constrained by "internal" factors that limit the processing of foods and the performance of work, or by "external" factors that limit the availability of resources is unclear.

12. Energy budgets vary with body composition.

13. Home range is correlated with the level of energy expenditure because it is the area used by an animal to obtain the energy expended, and therefore with the intrinsic and extrinsic factors that influence energy expenditure, namely, body size, food habits, behavior, and climate.

14. Vertebrates that are long-term residents of oceanic islands are generally characterized by low rates of metabolism produced by a small mass, reduced activity, flightlessness, and low standard rates and by the replacement of endotherms by ectotherms.

15. Vertebrates that are specialized to cave or burrow life tend to have reduced rates of metabolism and reduced body structures.

16. Estimates of population energetics are synthetic and difficult to know with precision, but they may point out the relative impacts of different populations, or of subunits of a particular population, on the environment.

17. The production of populations is positively correlated with rate of metabolism in vertebrates having a similar form of maintenance, but a trade-off occurs between production and maintenance when comparing ectotherms with endotherms.

11 Periodicity in the Environment: Balancing Daily and Annual Energy Budgets

11.1 SYNOPSIS

Vertebrates inhabit periodic environments, including those based on solar, lunar, and planetary cycles. The periodicities found in vertebrate behavior, some of which are endogenous, are entrained with these astrophysical periodicities. Endogenous rhythms permit vertebrates to anticipate and prepare for cyclic changes in the environment. Daily energy budgets of vertebrates can be balanced by adjusting the timing of activity, by storing food and body fat, and by entering torpor. The torpors of ectotherms and endotherms are fundamentally different physiological states, but both permit the balance of energy budgets when resources are restricted. In response to an annual cycle in the environment, energy budgets may be balanced in various ways: reproduction, molting, and a high cost of maintenance may be staggered; fat may be stored and used to transfer energy from one period to another; and some species may migrate or enter torpor. Some small mammals may reduce energy expenditure in winter by reducing body mass. A small mass in one sex may compensate for the normal occurrence of a behavior typified by a high rate of energy expenditure. Migration is limited to species with an effective means of long-distance movement. Migration is associated with a reduction in energy expenditure compared to what the expenditure would have been if no migration had occurred; in some shorebirds the reduction in energy expenditure in winter is correlated with a change in body mass, body composition, or both. Seasonal torpor occurs in both ectotherms and endotherms; it depresses energy expenditure. The decreased expenditure in ectothermic torpor may be greater than expected from the decrease in body temperature, whereas that in endotherms is principally derived from the regulation of a body temperature at a level well below that found in normothermia. The principal disadvantage of torpor (at least in endotherms) is a reduction in reproductive output. Aestivation is physiologically similar to hibernation, except that it principally occurs at higher ambient and body temperatures; that is, hibernation is a low-temperature torpor and aestivation is a higher-temperature torpor.

11.2 INTRODUCTION

Most vertebrates face periodic variations in the environment, the few potential exceptions being species that are committed to life in burrows, caves, or deep-sea trenches, and even these environments are unlikely to be absolutely free of all periodic variation in their physical parameters. These periodicities have daily, tidal, lunar, and seasonal rhythms, as demonstrated by variations in light intensity, temperature, rainfall, and sea level. Because these rhythms are fundamentally based on astrophysical and geophysical events, they are reliable and therefore can be used by organisms to anticipate future events, such as an abundance or shortage of food, a warm or cold seasonality, or a wet or dry seasonality. Such predictability focuses attention on unpredictable events, like the local occurrence of rain in deserts, of cyclonic storms in tropical oceans, or of the El Niño/La Niña phenomenon, and how resident vertebrates respond to these challenges. A manifestation of environmental periodicity and predictability is the rhythmic adjustments that vertebrates make in their energy expenditure and behavior. These adjustments are the focus of this chapter.

11.3 PERIODICITY IN THE ENVIRONMENT

Three fundamental periodic phenomena interact in a complex manner on Earth to produce an array of derived local periodicities. These three cyclic phenomena are (1) the daily rotation of Earth on its polar axis, (2) the annual revolution of Earth around the Sun, and (3) the monthly revolution of the Moon around Earth. (For further exploration of these cycles, see Saunders [1976].)

11.3.1 Daily cycle. The alternation of day and night results from the rotation of Earth on its polar axis. Day occurs at a particular site on the surface of Earth when that site faces the Sun, and night occurs when that site faces away from the Sun. The period for complete rotation varies, depending on whether it is measured by a fixed standard (e.g., with reference to distant stars, the so-called sidereal day), or whether it is measured with reference to the Sun (the solar day). The sidereal day is 23 h, 56 min, and 4 s long. The length of the solar day, the relevant biological unit, is not constant because the velocity of Earth's rotation varies through the year: the mean solar day is 24 h, 3 min, and 57 s, although it may vary in length by as much as 16 min. Times of sunrise and sunset vary with longitude, being 1 h later for each 15° westward (24 h/360° = 1 h/15°).

Earth's period of revolution around the Sun has remained constant, but the rotation of Earth on its polar axis has slowed due to the dissipation of rotational energy by tidal forces. This reduction in the period of rotation is estimated to be about 2 s/100,000 y, which means that in the Cambrian, some 570 million years ago, a day was about 21.8 h and a year about 420 d. Wells (1963) estimated from coral growth rings that the middle Devonian had about 400 d/y, which is in agreement with these calculations.

11.3.2 Annual cycle. The annual cycle of Earth's revolution around the Sun with reference to stars is 365 d, 6 h, 9 min, and 9.5 s (the sidereal year). Because the sidereal year is 6 h longer than 365 d, an additional day is added to every fourth year, a leap year. A seasonal variation in climate is produced by this revolution because Earth's polar axis is inclined by 23°27′ with respect to the plane of rotation. Consequently, during part of the year the Northern Hemisphere is inclined toward the Sun, when the Northern Hemisphere has summer and the Southern Hemisphere winter. Later in the cycle of revolution, the opposite conditions pertain.

The progression of seasons has other consequences. One is the apparent movement of the Sun in the sky. In the Northern Hemisphere the Sun reaches its northernmost position on Earth's surface on June 21 or 22, the summer solstice, when the Sun casts a vertical shadow at the tropic of Cancer (23°27′N), and fails to go below the horizon at latitudes above the Arctic Circle (66°30′N). This date is the winter solstice in the Southern Hemisphere. The winter solstice in the Northern Hemisphere and the summer solstice in the Southern Hemisphere occur on December 21 or 22, when the tropic of Capricorn (23°27′S) and Antarctic Circle (66°30′S) are defined. A second consequence of the tilted axis is that day length varies with season at all localities except the equator. Day length increases in both hemispheres from the winter to the summer solstice and decreases from the summer to the winter solstice. From Earth's surface, the Sun appears to cross the equator twice, once on March 20 or 21 and once on September 22 or 23, these dates being defined as the spring and fall equinoxes, depending on the hemisphere in which one lives. These are the only dates on which the entire planet has a 12-h day and night.

The progression of seasons, then, results from the revolution of Earth around the Sun in conjunction with the tilt of Earth's rotational axis and the consequent variation in day length. Although Earth has an elliptical orbit around the Sun, and thus is at times closer to and at other times farther from the Sun, the variation in seasons is not produced by a change in the distance between Earth and the Sun. It is produced because the angle of incident radiation and day length vary with season as a result of the inclination of the polar axis: in summer the incident angle is higher and day length is longer, whereas in winter the opposite is true.

11.3.3 Monthly and tidal cycles. The Moon takes a month (Anglo-Saxon *mona* = moon), by definition, to complete a revolution around Earth. The precise period of this revolution depends on how it is measured: if stars are used to define this period, the (sidereal) month lasts 27 d, 7 h, 43 min, and 11.5 s, but if the Moon's period is measured on Earth at a fixed location from a full moon to the next full moon, this (synodical) month is 29 d, 12 h, 43 min, and 12 s. The difference between these periods is due to the rotation of Earth, and is responsible for the observation that the Moon "rises" slightly later every day—the duration from moonrise to moonrise being about 24.8 h.

The Moon exerts a gravitational pull on Earth's ocean: water tends to pile up on the side of Earth facing the Moon and on the side facing away from the Moon. As the Moon revolves around Earth, these "piles" of water move across the ocean and appear at coastlines as high tides. Generally, two high tides and two low tides occur in each day, subject to strong local modification by coastlines, latitudes, and other factors. High tides, consequently, tend to be 12.4 h apart (i.e., half the period between moonrises), as are low tides. A further complication in tidal height occurs because the Sun also exerts a gravitational pull on the ocean, although it is weak compared to that of the Moon. When the Moon and Sun pull together (i.e., when the Moon and Sun are in alignment with Earth), their gravitational forces are added together, and the subsequent high tides are higher and the low tides lower. The resulting tidal amplitude is greater, and these tides are called *spring tides*. However, when the pull of the Moon is at a right angle to that of the Sun, the tidal amplitude is reduced; these tides are called *neap tides*. The period between successive spring or neap tides is one-half of a synodical month, or 14 d, 18 h, 21 min, and 36 s.

11.4 PERIODICITY IN VERTEBRATES

Most vertebrates show periodicity in their behavior and physiology. They are usually classified as being diurnal (day-active), nocturnal (night-active), or crepuscular (twilight-active). A few, such as shrews and fossorial mammals, are intermittently active throughout the 24-h day. The length and timing of activity are influenced by many factors, such as food habits, body size, the physical environment, and predation (see Kenagy and Vleck 1982). Periodic behavior is entrained (i.e., aligned) with environmental cues, or synchronizers ("zeitgebers"), like photoperiod, rainfall, or temperature. William Rowan first demonstrated that

photoperiodicity in birds is internally generated and related it to their migratory habits (1925, 1926, 1927).

11.4.1 Endogenous circadian rhythms. Many vertebrates maintain a cyclic pattern of activity in continuous darkness or light, even when they are apparently isolated from all environmental factors known to oscillate. These species are said to have a free-running rhythm that is characterized by a distinctive period length. This period varies among species, even among individuals within the same species, but it rarely equals any known, naturally occurring, geophysical period (which indicates its free-running nature). Period length depends on the habits of species (i.e., diurnal or nocturnal) and the intensity of illumination: period length increases with intensity in nocturnal species and decreases with intensity in diurnal species (Aschoff 1960). In contrast to other physiological functions (see Section 4.4), circadian rhythms are almost temperature independent: the Q_{10} of the free-running rhythm in poikilotherms is about 1.02 and in torpid mammals, about 1.1 to 1.3 (Saunders 1976). This temperature independence permits internal rhythms to measure (or reflect) time. The physiological basis for endogenous rhythms is a complex self-timing system, a "biological clock," which may be used in orientation and migration, as well as in timing daily events (Bünning 1973).

The daily activity pattern of an animal in the field results from the interaction of its endogenous rhythm with the periodicity encountered in the environment. The period of activity is entrained with the periodicity of the environment, activity occurring during the day, night, or at twilight hours, depending on the species. The duration of activity, however, may be modified by the length of the day or night. For example, activity in the diurnal, nonhibernating, ground-squirrel *Ammospermophilus leucurus* may be curtailed during the shortest days in winter (Kenagy 1978; Figure 11.1A). In summer, activity in this species does not occupy the entire day; its duration is similar to the free-running period. Similarly, the nocturnal kangaroo-rat *Dipodomys merriami* reduces its duration of activity only in midsummer, when night is shortest, but in winter the long nights are not completely used (Figure 11.1B). The most northernly distributed kangaroo-rat, *D. ordii*, encounters a night that is minimally 7 h in midsummer at the northern limits of distribution (latitude 50°45′N in

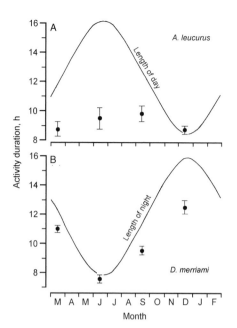

Figure 11.1 Duration of daily activity in (A) antelope ground-squirrels (*Ammospermophilus leucurus*) and (B) Merriam's kangaroo-rat (*Dipodomys merriami*) as a function of the time of year. Source: Modified from Kenagy (1978).

southern British Columbia). Even so, *D. ordii* is active only for 6 h (Kenagy 1976). As Kenagy noted, at latitudes of 34°N, where the duration of night seasonally varies from 9 to 14 h, kangaroo-rats are out of their burrows for only 1 or 2 h, although they may distribute this activity over 10 h or more. Thus, even a restriction of activity to 6 h would not impede feeding, grooming, or social interaction.

At latitudes north of the Arctic Circle vertebrates encounter continuous light in midsummer and continuous dark in midwinter. Faced with these conditions, most vertebrates adjust their activity rhythms. Arctic owls, such as the snowy (*Nyctea scandiaca*), hawk (*Surnia ulula*), and great gray owls (*Strix nebulosa*), are facultatively diurnal in summer. Arctic vertebrates might be expected to show free-running rhythms in summer and in winter, but they generally seem not to do so (Swade and Pittendrigh 1967). The red-backed vole (*Clethrionomys rutilus*) maintained activity during nighttime hours at College, Alaska (65°N), even though in June, when these field observations were made, sunset occurred at 2134 and sunrise was at 0057, the "dark" period being only 103 of the 203 min between sunset and sunrise. When *C.*

rutilus individuals were transferred to Point Barrow, Alaska (71°N), they lost all rhythmicity in their activity, epecially near the summer solstice. An arvicolid rodent native to Point Barrow, *Microtus oeconomus*, was arrhythmic in the laboratory and in the field between early June and mid-August, when no dark period was present, after which it was nocturnal. The arctic freshwater sculpin *Cottus poecilopus* behaves similarly (Andreasson 1973).

In contrast to these species, the brown lemming (*Lemmus sibiricus*) remained rhythmic throughout the summer but simply shifted its activity period in late May from night to day, only to return to nocturnal habits in mid-August. At 69°N latitude in Norway, sand martins (*Riparia riparia*) in summer maintained a "diurnal" activity cycle, stopping their feeding between 2200 and 0700, when aerial insects were least abundant because of the decrease in ambient temperature. In contrast, northern bats (*Eptesicus nilssonii*) fed mainly between 2200 and 0200, which was the period of lowest light intensity (Speakman et al. 2000). The activity of diurnal arctic ground-squirrels (*Spermophilus undulatus*), however, is not greatly affected by continuous light. Diurnal species would be expected to be most affected by continuous dark in the winter north of the Arctic Circle. At that time, ground-squirrels are in hibernation and most birds have migrated. Nevertheless, Irving (1960) showed that some diurnal birds, namely, the snow bunting (*Plectrophenax nivalis*) and the common redpoll (*Carduelis* [*Acanthus*] *flammea*), face prolonged dark periods at the northern limits of their winter distribution. The cold temperatures and food shortage at these latitudes require extended periods of foraging, including some after sunset because their energy requirements cannot be met by restricting foraging to the few daylight hours. Even in midwinter above the Arctic Circle, however, long periods of twilight exist, during which diurnal birds can forage for food (Reinertsen 1983).

A clear light-dark cycle is not necessarily needed to entrain the biological clocks of vertebrates. A slight variation in the daily level of illumination may be sufficient (Swade and Pittendrigh 1967): if the maximal illumination during a 24-h period is equal to or greater than 20 times the minimal illumination, which is similar to what is found at Point Barrow in summer, the activity cycles of most *C. ritulus* voles are entrained to the light cycle.

11.4.2 Endogenous circannual rhythms. Circannual rhythms, as a result of their length, are much harder to study than circadian rhythms. Most examples of endogenous circannual rhythms have been described for birds and mammals (see Pengelley 1974; Gwinner 1986, 1996). The significance of circannual rhythms is that they permit vertebrates to anticipate cyclic changes in the environment and therefore to adjust to climatic conditions, to time migration, and to prepare for hibernation, molting, and reproduction (Klein 1974).

In birds evidence exists for endogenous circannual rhythms in gonadal development, molting, and migratory "restlessness." The willow warbler (*Phylloscopus trochilus*), an Old World warbler (Sylviidae), breeds in central and northern Europe and winters in tropical and southern Africa. When it is exposed to 12 h of light and 12 h of dark in the laboratory, *Ph. trochilus* maintains free-running rhythms in gonadal development, molting, and restlessness for up to 27 months, the periods of these rhythms being about 10 months (Figure 11.2). Similarly, the warblers *Sylvia atricapilla* and *S. borin*, both of which winter in sub-Saharan Africa, demonstrated nine cycles of molt in 8 y of captivity under constant photoperiod (Berthold 1974, 1978). The garden warbler, *S. borin*, also shows a semimonthly cyclic variation in body mass when held at 12 h of light and of dark for up to 30 months; the variation in mass resulted from a variation in food intake, a behavior that may well facilitate refueling during migration (Bairlein 1986). In contrast, the chiff-chaff (*Ph. collybita*), which nests throughout most of Europe but winters in southern Europe, Great Britian, Ireland, and northern Africa, loses any clear rhythms when maintained under the same light-dark regime as *Ph. trochilus* for more than 12 months. The chiff-chaff winters in areas that have a clear photoperiodic rhythm, whereas the other warblers face either no seasonal photoperiodic variation (on the equator) or a reversed rhythm (in southern Africa).

Transequatorial migration raises special problems. Why these migrants do not respond to the increasing day length in the Southern Hemisphere during the northern winter is unclear, but it may relate to a photorefractive period (Marshall 1951) or to a shifting sensitivity to day length. One factor is that species that winter along the equator or in a reversed photoperiod use an internal clock to keep their seasonally dependent behavior congruent with the timing in its breeding range (e.g., see Hamner and Stocking 1970). Furthermore, why do several North American grassland birds, for example, the bobolink (*Dolichonyx oryzivorus*) and the upland sandpiper (*Bartramia longicauda*), winter in central or southern South America, but no South American species winter in North America? A similar asymmetrical pattern exists between Europe and southern Africa. The answer is more likely to be historically than functionally based.

Some mammals show evidence of endogenous circannual rhythms in testicular size, food and water consumption, rate of metabolism, molting, and the propensity to enter torpor. At one extreme, some ground-squirrels (*Spermophilus*) seasonally cycle in and out of a hibernation-like state (Figure 11.3) when maintained at a constant temperature (3°C) and a constant photoperiod (12 or 20 h) for periods up to 4 y (Pengelley and Kelly 1966; Pengelley and Asmudsen 1969, 1974; Heller and Poulson 1970; Mrosovsky 1978). A similar behavior was seen in a pocket-mouse, *Perognathus longimembris* (French 1977a). The free-running period of a complete normothermic-torpor cycle is usually 10 to 11 months when exposed to constant conditions, so rodents under these conditions gradually lose synchrony with their natural environments. (For a summary, see Canguilhem 1985.)

Ground-squirrels, like Old World warblers, have a variable commitment to an endogenous control of their annual cycles. Species with the most

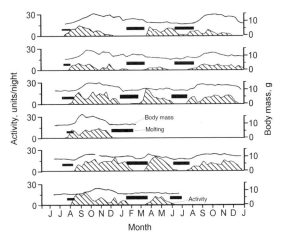

Figure 11.2 Body mass, molting, and restlessness in the willow warbler (*Phylloscopus trochilus*) exposed to 12 h of light and 12 h of dark in the laboratory as a function of the time of year. Source: Modified from Gwinner (1971).

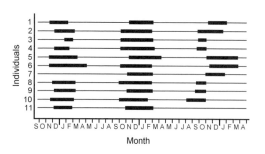

Figure 11.3 Torpid (heavy bars) and normothermic states (thin lines) in the golden-mantled ground-squirrel (*Spermophilus lateralis*) as a function of the time of year. Source: Modified from Pengelley and Kelly (1966).

marked endogenous circannual cycles (e.g., *S. undulatus* in the Arctic and, possibly, *S. lateralis* in the mountains of western North America) live in extreme environments that permit little temporal flexibility for reproduction because of a short summer season produced either by a high latitude or a high altitude. *Spermophilus* species living under less extreme environments, such as *S. beldingi* (Heller and Poulson 1970) and *S. mohavensis*, *S. tereticaudus*, and *S. beecheyi* (Pengelley and Kelly 1966), have abandoned obligatory torpor and an endogenous control of their annual cycles to varying degrees, depending on the environments in which they live. Thus, Michener (1984, p. 81) noted that ground-squirrels ". . . at high [altitudes] and northern latitudes usually have short active seasons terminated by the almost synchronous disappearance of all cohorts, whereas species at low [altitudes] and southern latitudes tend to have long active seasons characterized by asynchronous immergence of cohorts." The dormouse *Glis glis* also does not have an endogenous rhythm for entering torpor but depends directly on the annual cycle in ambient temperature (Jallageas et al. 1989).

In contrast to ground-squirrels, the kangaroo-rat *Dipodomys merriami* shows no fixed circannual reproductive cycle, possibly because this species feeds on a seasonally undependable food resource—desert mast crops—and must time reproduction to take advantage of a sudden increase in available food (Kenagy 1978). To do so, males remain sexually ready, and when food becomes available, this species hoards large quantities of seeds and females rapidly attain a functional reproductive state. A similar responsiveness to the sudden appearance of food is found in some desert birds (Keast and Marshall 1954; Frith 1959;

Keast 1959; Immelmann 1963, 1971). For instance, many birds that reside in the central deserts of Australia, including the zebra finch (*Taeniopygia guttata*), Gouldian finch (*Erythrura gouldiae*), black-faced woodswallow (*Artamus cinereus*), and budgerigar (*Melopsittacus undulatus*), share a series of characteristics that facilitate the initiation of breeding almost immediately after a significant rainfall. These include an increase in clutch number and size, early breeding age, breeding during molt, nomadism, and group cohesion. The proximate cause for the initiation of breeding appears to be rainfall itself. The first eggs may be laid as soon as 7 to 11 d from the beginning of rainfall by zebra finches and within 12 d by the woodswallow (Immelmann 1963).

Many tropical vertebrates have a seasonal pattern in reproduction, but little is known of the environmental cues used by these species to time reproduction (Baker 1947). For example, sooty terns (*Sterna fuscata*) have a 9.7-month reproductive cycle on equatorial Ascension Island (Chapin and Wing 1959), which means that at this locality reproduction can occur potentially in every month. In the tropical Pacific, sooty terns on Christmas Island breed 6 months after an unsuccessful attempt, whereas individuals that successfully breed try again 12 months later (Ashmole 1965). Although many environmental factors may affect the timing of reproduction, internal factors, including the energy demands for feather replacement, may prohibit continuous breeding. Tropical bats, depending as they do on forest production, tend to synchronize the birth of young to coincide with the beginning of the wet season (McNab 1969; Mares and Wilson 1971; Fleming et al. 1972).

A large seasonal change in photoperiod is not required to time events. A neotropical forest bird, the spotted antbird (*Hylophyla naevioides*), responds to an increase in photoperiod as small as 17 min, from 12 h 0 min to 12 h 17 min of light (Hau et al. 1998)! In doing so, both males and females show a growth in gonadal size compared to the gonads in controls that are exposed to a 12-h photoperiod. So photoperiod in the tropics may have enough variation (except exactly on the equator) to permit many (or most) birds to use photoperiod to cue their seasonal behaviors. The stimulation provided by an increase in photoperiod could be overridden by not including crickets in the diet of the antbird (Hau et al. 1998), so the seasonal availability of certain types of food may also contribute to the initiation or timing of seasonal behaviors in the antbird.

Reproduction in some Amazonian fishes is synchronized with wet-dry seasonality (Schwassmann 1976, 1978, 1980). Amazonian fishes can be grouped into those that live in flood-plain lakes and those that live in small streams that do not flood. The former move onto newly flooded plains at high water and shed a large number of eggs at one time, whereas the latter usually breed intermittently, producing many fewer eggs. Flood-plain species may be cued to migration and spawning by a change in water level, but the only experimental evidence for the factors triggering seasonal reproduction was obtained by Kirschbaum (1975, 1979) for the stream-dwelling gymnotid *Eigenmannia virescens*. In this species an increase in water level, rain simulation, or a reduction in water conductivity induces breeding, the latter factor being most important; a reduction in pH does not affect reproduction. The extent to which these cycles of reproduction are endogenous is unclear; an extensive examination of tropical fishes will undoubtedly reveal a great diversity of cues used for the timing of reproduction and in the extent to which the cycles are opportunistic or endogenous. Endogenicity in fish reproductive cycles, as in the hibernation of ground-squirrels, may be most clearly marked when the environment is cyclic and when the penalties for missing environmental cues are harsh. If the environment is unpredictably variable, such as rainfall in a desert, then reproductive (and other) cycles may be opportunistic (for a general review, see Jørgensen 1992). This behavior is marked in Australian amphibians, which are characterized by high growth rates and short larval periods (Main et al. 1959). A similar situation exists in spadefoot toads (*Scaphiopus*) in North American deserts (Mayhew 1965b).

11.4.3 Controversies on the existence of endogenous rhythms.

Some biologists have doubted the existence of endogenous rhythms. Part of this doubt stems from the concern that some of the support for the existence of endogenous rhythms is negative. That is, an "endogenous" rhythm is a rhythm that remains after all environmental cues, such as fluctuations in photoperiod, temperature, gravitational field, sunspot activity, rainfall, and so on, have been eliminated. An exogenous basis for these rhythms might exist either if a particular factor has not been completely eliminated or if some factor other than those already eliminated has not been considered. Another difficulty with the concept of endogenous rhythms is that they tend to persist under constant environmental conditions for limited periods of time, often only for a few cycles. If they are truely endogenous, why do they not persist indefinitely?

Frank A. Brown Jr. searched for external factors dictating rhythms thought by most biologists to be endogenous. Although he (1957, 1960, 1972) claimed that most (all?) "endogenous" circadian rhythms were in fact exogenous, his method of analysis was so complicated, requiring such an elaborate manipulation of data, that most biologists have not accepted his view. The best answer to these doubts remains the positive observation that when vertebrates demonstrate what appear to be circadian or circannual rhythms, they are invariably characterized by distinctive free-running rhythms with a period unequaled by any known geophysical phenomenon.

11.4.4 Significance of endogenous rhythms.

Endogenous daily or annual rhythms are timing devices that permit individuals and populations to anticipate and prepare for cyclic events in the environment. Preparation may involve the deposition of fat for use in winter, hibernation, or migration; molting of plumage or pelage in anticipation of migration or of a seasonal change in weather; or synchronization of a population for reproduction. In a benign environment, such preparation may not be necessary, but in a seasonally harsh environment, its absence may be fatal. The only harsh environments where such preparation does not occur are those in which the fluctuations in conditions are unpredictable. Under these circumstances, the ability to reproduce requires a rapid response to stimuli. Underlying the survival and reproduction of a species in a variable environment is its ability to balance energy budgets on a daily and seasonal time scale.

11.5 BALANCING DAILY ENERGY BUDGETS

Vertebrates have several potential options to balance their daily energy budgets. Because most species are either diurnal or nocturnal, they are usually inactive for approximately 10 to 14 h/d. To balance their energy budgets they must (1) gather and store food during the active period to suffice for the inactive period; (2) store energy, usually as fat, to fuel the inactive period; (3) abandon strict adherence to diurnal or nocturnal activity and feed intermittently throughout the 24-h period; or (4) reduce markedly the energy expenditure during the inactive period. Some vertebrates use each of these behaviors.

11.5.1 Food and fat storage. Daily food storage can occur in two ways: food can be stored in a burrow or other shelter for use during the inactive period or feeding can be enhanced at the end of a period of activity, with the accumuled food in the gut used during the period of inactivity.

Many mammals and birds hoard food for one to a few days, the short term reflecting the perishable nature of much food, for example, animal carcasses and ripe fruit (Vander Wall 1990). Short-term storage has been found among mammals, principally primates and carnivores, and some birds, such as raptors and a variety of passerines. An example of short-term storage is seen in American kestrels (*Falco sparvarius*) in which most storage was during midday and most retrieval occurred at the end of the day (Figure 11.4).

Food storage in the gut probably occurs to some extent in all species but is most effective as a means of balancing the daily energy budget in large species, where the volume of the gut relative to energy demand is greatest, or in mild environments, where energy demands are small. Ruminants often graze or browse extensively before sunset during cold weather, thereby provisioning the rumen with sillage adequate to supply the energy demands of the symbionts through the night. Eurasian kestrels (*F. tinnunculus*) similarly feed just before dark, especially in winter when nighttime is longest (Masman et al. 1986). Mourning doves (*Zenaida macroura*) fill their crop twice daily, once in the morning (recovering from a previous night) and once in the late afternoon (preparing for night; Schmid 1965b).

Some vertebrates might evade the inability to store food by converting food to fat. Lipids are stored as oils in the skeleton of many fishes: in the castor-oil fish (*Ruvettus pretiosus*), 30% of the frontal bone is oil (Bone 1972). Phleger (1988) suggested that oil storage might be more important for temperate species, to tide them over a period of limited food availability, than in tropical species, for which food availability is more dependable. The amount of fat deposited by a bird during the day is inversely related to ambient temperature (King and Farner 1963, 1966; Helms 1968).

11.5.2 Modified activity rhythms and time budgets. The species most likely to abandon strict diurnal or nocturnal activity are small; large vertebrates can readily store adequate amounts of energy during a normal foraging period to meet their requirements during an extended period of inactivity. Shrews (Soricidae) are discontinuously active throughout a 24-h day (Figure 11.5). Hanski (1985) showed among soricine shrews (which do not use torpor) that when food intake is restricted to produce a 5% reduction in body mass, small

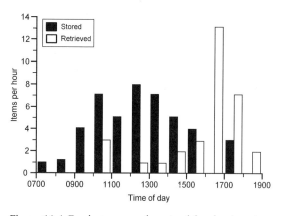

Figure 11.4 Food storage and retrieval by the American kestrel (*Falco sparvarius*) as a function of the time of day. Source: Modified from Collopy (1977).

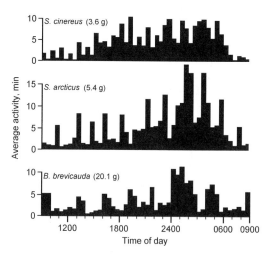

Figure 11.5 Activity in three species of shrews, *Sorex cinereus*, *S. arcticus*, and *Blarina brevicauda*, as a function of the time of day and body mass. Source: Modified from Buchner (1964).

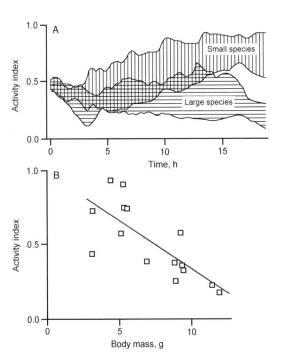

Figure 11.6 Activity index in small (3–5 g) and large (9–12 g) shrews as a function of (A) time and (B) body mass. Source: Modified from Hanski (1985).

shrews, such as *Sorex minutus* (3 g) and *S. cae-cutiens* (5 g), are more active than larger shrews (Figures 11.5 and 11.6A), such as *S. areneus* (9 g) and *S. isodon* (12 g). The amount of activity is inversely related to mass (Figure 11.6B). Hanski noted that the most northernly distributed shrews are small and that a chronically low food availability selects for a small mass and low energy requirements. Activity throughout a 24-h day is also found in some other small mammals, especially those that forage in runways, like arvicolids, or that live in underground burrows, like pocket-gophers and mole-rats, where the threat of diurnal predators is limited.

The potential flexibility of time budgets is seen in the observation that many mammals and birds spend more than 60% of their "active" period at "rest" (Herbers 1981). Furthermore, "... when time budgets of an organism do differ seasonally, geographically, or in relation to prey availability, the proportions of time spent foraging and territorial defense are more likely to change than the amount of time spent resting ..." (Herbers, p. 260). Some of the "resting" period may be required for digestion (e.g., Karasov et al. 1986a) and to avoid predators, but these factors are unlikely to

account for the entire period. Herbers (p. 260) further noted that "... [i]nactivity is the effect of predation efficiency, with the ironic result that extremely efficient predators have more free time and appear lazy whereas inept predators have little free time and thus always appear to be doing something of consequence." Inactivity "... curtails energy expenditures and elongates the interval between feeding periods ... and ... represents buffer time to be utilized when resources are low or metabolic demands are very high" (Herbers, p. 260).

11.5.3 Daily torpor. Many small vertebrates remain committed to a strictly diurnal or nocturnal pattern of activity. Energy budgets can be balanced during the period of inactivity, especially when the animals are faced with harsh environmental conditions or a shortage of food, by suppressing energy expenditure through a decrease in body temperature and rate of metabolism, a physiological state referred to as *torpor*. Broadly defined, torpor includes both the selection of low ambient temperatures by ectotherms and the more striking reduction of body temperature by endotherms. The decrease in body temperature in endotherms when torpid is well beyond the 1 to 4°C decrease found in a circadian cycle (see Reinertsen 1983). The "torpors" of ectotherms and endotherms, although quite different from a physiological perspective, are ecologically similar in that they lead to a reduction in energy expenditure.

In ectotherms the decrease in body temperature during the period of inactivity may or may not be "voluntary." The decrease in body temperature at night in heliothermic lizards occurs because the sun is not available as a heat source for behavioral temperature regulation (see Section 4.8.3). Balancing a daily energy budget is facilitated by the decrease in rate of energy expenditure associated with the fall in body temperature at night, which also has been referred to as *daily torpor* (Hainsworth and Wolf 1978). The daily energy expenditure can be further depressed by selecting lower temperatures when individuals are not feeding (Lillywhite et al. 1973; Elliott 1975). Hessler and Jumars (1974, p. 201) described the marine abyssal megafauna as being "... relatively dormant during the long intermediate periods when no food is available, thereby conserving energy. When the stimulus of food is borne to them by the ubiquitous tidal currents, they become active." The acceptable limits to the use of the terms *torpor* and *dormancy* in ectotherms are fuzzy. Hochachka (1990) described the depression of metabolism, in parallel with the concept of the scope for activity (Fry 1947), as the *scope for survival*, which facilitates survival in an environment with resource limitations.

The clearest use of daily torpor occurs in small birds and mammals. They often find balancing a daily energy budget difficult because of the cost of temperature regulation, small body fat reserves, and low environmental temperatures. The propensity to use daily torpor in the presence of low ambient temperatures (e.g., Ruf and Heldmaier 1992) makes daily torpor in white-toothed shrews (*Crocidura*) and some rodents (*Peromyscus*, *Phodopus*) most common during fall and winter, which tends to blur the difference between a daily and a circannual periodicity. One difference that remains between daily and seasonal torpors, however, is that daily torpor lasts for only a few hours, whereas bouts of hibernation may last for days (see Section 11.8.2).

Daily torpor is widely used by endotherms and appears to have repeatedly evolved. Of 20 or more mammalian orders, at least 7 are known to include species that enter daily torpor (namely, monotremes, didelphid and dasyurid marsupials, insectivores, bats, primates, and rodents). Furthermore, Cade (1964) suggested that torpor evolved independently in at least five lineages of rodents. Birds that enter daily torpor include a dove and some caprimulgids, hummingbirds, swifts, mousebirds, sunbirds, and swallows, and to a lesser extent titmice and manakins (Pearson 1960a, Reinertsen 1983).

Many small endotherms enter torpor, but not because "...their high metabolic rates...need ...a continuously available food supply...," as Bartholomew and Cade (1957, p. 70) suggested. Small endotherms with the highest rates of metabolism, such as soricine shrews, arvicolid rodents, and most temperate passerines, do not use torpor, whereas those with lower rates of metabolism when normothermic, namely, crocidurine shrews, small tenrecs, vespertilionid bats, hummingbirds, sunbirds, and manakins, enter torpor. Why small species with low basal rates of metabolism are most likely to enter torpor (see Section 3.4) is unclear, unless it reflects a correlation of maximal rates of metabolism with basal rate (see Section 10.8): species with low basal rates may not be able to maintain large temperature differentials at low ambient temperatures.

The level of metabolism must be judged against the availability and quality of the foods used. Many of the small endotherms that enter torpor feed on food supplies characterized by a variable availability, such as flying insects (bats, swifts, caprimulgids), mast crops (heteromyid rodents), and nectar (hummingbirds, sunbirds). Small endotherms cannot compensate for a variable food supply by storing fat because of their small mass and high turnover rates (Hainsworth et al. 1977). Juvenile forked-tailed storm-petrels (*Oceanodroma furcata*) permit body temperature to fall to 10 to 25°C, even after they have attained endothermy, if they are not fed, but will recover endothermy after being fed (Boersma 1986). Such a gorge-and-fast pattern occurs in other members of Procellariiformes, swifts (Koskimies 1948), and raptors (Chaplin et al. 1984), among others.

The use of daily torpor is determined not only by the shortage of food. Hummingbirds may preferentially enter torpor at night to conserve body fat for use in migration (Carpenter and Hixon 1988) or to reduce energy expenditure during molting (Hiebert 1992). The rufous hummingbird (*Selasphorus rufus*), which is migratory, stores fat during the day that is conserved at night by entering torpor, whereas Anna's hummingbird (*Calypte anna*), which is nonmigratory, does not accumulate much fat during the day and does not (usually) enter torpor (Beuchat et al. 1979; Hixon and Carpenter 1988). Free-living poorwills (*Phalaenoptilus nuttallii*) enter daily torpor, but the cues used are unclear: entrance was not dictated by a low body mass, a restricted insect availability, or the lunar cycle (Brigham 1992). Furthermore, the entrance by a species (e.g., the common nighthawk [*Chordeiles minor*]) into torpor in the laboratory (Marshall 1955) does not necessarily mean that a species uses torpor in the field (Firman et al. 1993), although (more likely) the cues required for entrance into torpor may not be present in the environment when field studies are made.

The depth and frequency of torpor also reflect the ambient temperature and season (Haftorn 1972; Chaplin 1974, 1976; Reinertsen and Haftorn 1983, 1984; Tannenbaum and Pivorun 1984). Small rodents, such as *Peromyscus leucopus* (Gaertner et al. 1973; Lynch et al. 1978; Vogt and Lynch 1982) and *Phodopus sungorus* (Heldmaier

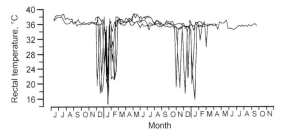

Figure 11.7 Rectal temperature in the Djungarian hamster (*Phodopus sungorus*) as a function of the time of year (when exposed to natural photoperiods). Source: Modified from Heldmaier and Steinlachner (1981a).

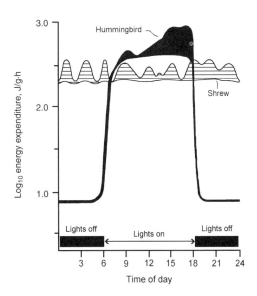

Figure 11.8 Log$_{10}$ energy expenditure in a hummingbird (*Calypte anna*) and a shrew (*Sorex cinereus*) as a function of the time of day. Source: Modified from Bartholomew (1977).

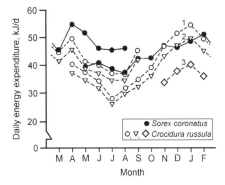

Figure 11.9 Daily energy expenditure in shrews, *Sorex cinereus* (one) and *Crocidura russula* (three), as a function of the time of year. Source: Modified from Genoud (1985).

and Steinlechner 1981a, 1981b), show the maximal propensity to enter torpor in fall and winter (Figure 11.7), a response that seems to reflect both photoperiod and ambient temperature (Heldmaier and Steinlechner 1981a): photoperiod appears to be the most important factor in the hamster and temperature, the most important one in *Peromyscus*. Yet, in *Phodopus* only some members of a population spontaneously enter torpor (Hill 1975; Heldmaier and Steinlechner 1981a), a difference that appears to have a genetic basis (Hill 1975). The differential occurrence of daily torpor in a population indicates that "... torpor ... is primarily aimed to guarantee survival of a fraction of the population for short periods of inaccessibility of food or [the presence of an] extreme cold load (e.g., strong wind, snowstorm) during winter" (Heldmaier and Steinlechner 1981a, p. 270).

The impact on energetics of normothermia and torpor at small masses is readily seen by comparing a 3-g shrew (*Sorex*) that does not enter torpor, with a 3-g hummingbird (*Calypte*) that does. The expenditure of the shrew is nearly twice that of the hummingbird (Figure 11.8): a shrew can sustain a larger expenditure because it can hunt for food in short bursts of activity throughout the day and night, whereas the hummingbird is limited to feeding during the day and must rely on fat stores at night—a precarious behavior at low temperatures for an endotherm of this mass—or it can enter a torpid state. The seasonal impact of daily torpor on the daily energy budget can be demonstrated by a comparison of *Crocidura russula*, which uses daily torpor, with *Sorex coronatus*, which does not (Genoud 1985). In this case *S. coronatus*, which weighs 7 to 9 g, has an estimated energy expendi-

ture that is about 25% greater than that of *C. russula*, which weighs 11 to 13 g, especially when *C. russula* enters torpor for 10 h/d (Figure 11.9). The energy saved by entering torpor can be much greater: the pocket-mouse *Perognathus californicus* at 15°C had a 10-h torpor cycle (2 h in entry, 7.1 h in torpor, and 0.9 h in arousal) during which energy expenditure was only 19% of the cost for continuous maintenance of normothermia (Tucker 1966a).

The intimate connection between the occurrence of torpor in endotherms and the availability of food has been shown repeatedly. In some species

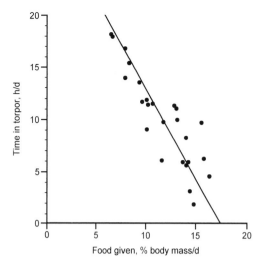

Figure 11.10 Time spent in torpor by the pocket-mouse *Perognathus californicus* as a function of food consumed. Source: Modified from Tucker (1966a).

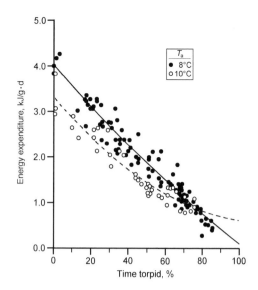

Figure 11.11 Daily energy expenditure in the pocket-mouse *Perognathus longimembris* as a function of the time spent in torpor and environmental temperature. Source: Modified from French (1976).

the shortage of food directly induces torpor. Thus, the proportion of time *P. californicus* spends in torpor increases as the food ration decreases (Figure 11.10), a pattern seen in other rodents (e.g., *Peromyscus* [Tannenbaum and Pivorun 1984]). In *Perognathus longimembris*, this relationship has been translated into an energy budget: daily energy budgets increase with a decrease in the proportion of the day spent in torpor, and at a fixed amount of torpor, the energy budget increases as ambient temperature falls (Figure 11.11). Little or no torpor occurs in the white-toothed shrew (*C. russula*) when well fed, but the time spent in torpor increases as the amount of available food decreases (Vogel et al. 1979). In some mammals (e.g., *Mesocricetus* [Lyman 1954], *Sicista* [Johansen and Krog 1959], *Microdipodops* [Brown and Bartholomew 1969], *P. longimembris* [French 1976]) food must be stored for entrance into torpor. For example, the frequency of torpor in *Microdipodops* increases with the size of the food store (Figure 11.12). The chemical composition of food may also affect the propensity to enter torpor: torpor was most pronounced (i.e., more frequent, longer, and with lower rates of metabolism) in some rodents (*Tamias*, *Peromyscus*) fed a diet containing unsaturated lipids than in individuals fed a diet with saturated lipids (Geiser and Kenagy 1987, Geiser 1991).

Most endotherms minimize the time spent in torpor. *Perognathus californicus* preferentially

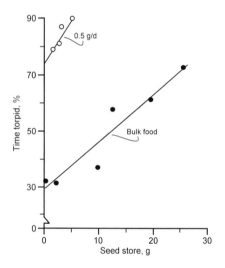

Figure 11.12 Time spent in torpor in the kangaroo mouse (*Microdipodops pallidus*) at 6°C as a function of seed storage when food is unlimited and when food is limited to 0.5 g/d. Source: Modified from Brown and Bartholomew (1969).

selected warm ambient temperatures as long as food was available, but chose low temperatures and entered torpor when the food supply was restricted (Tucker 1966a). The use of torpor to balance an energy budget occurs at the cost of inactivating the animal, thereby reducing social interaction and reproduction.

The term *daily torpor* covers a diversity of responses in endotherms. Although most species that use this behavior are small, some may be relatively large. The diversity is illustrated by several examples:

1. *Hummingbirds.* Because of their small masses and relatively low basal rates (see Section 3.4), all species of hummingbirds studied, including temperate and tropical (Morrison 1962; Hainsworth and Wolf 1970; Bech et al. 1997) and large and small (Lasiewski et al. 1967; Krüger et al. 1982), enter daily torpor. Torpid hummingbirds have body temperatures that are only a few degrees above ambient temperature. Torpor occurs when their body mass (i.e., their energy stores) falls below some minimum (Hainsworth et al. 1977). The decrease in mass at night is reduced as the time spent in torpor increases (Beuchat et al. 1979; Bech et al. 1997). Even incubating hummingbirds may enter torpor when feeding has been diminished by rainy, cold periods at high elevations (Calder and Booser 1973). In compensation, lean, torpid hummingbirds awake early in the morning to feed (also see Carpenter 1974). High altitudes present another problem for hummingbirds: the flowers found there are characterized by low nectar concentrations, which in combination with low ambient temperatures and a small crop volume restricts the ability of hummingbirds to maintain normothermia. Food storage is adequate for 4.2 h of normothermia at 20°C in the lowlands and for 1.4 h at 10°C in the highlands (Hainsworth and Wolf 1972).

The extent to which energy expenditure is reduced by entering torpor is complicated by the high cost of arousal, which raises the possibility that an endotherm might actually expend more energy by entering torpor and immediately arousing than it would by remaining normothermic over the same period. In the rufous hummingbird (*S. rufus*), which readily enters torpor and, like other hummingbirds, prevents body temperature from falling below some minimal level (in this species 9–13°C), the average rate of metabolism during a torpor cycle equals or is less than the average rate during normothermia at the same ambient temperatures (Hiebert 1990). The extent to which rate of metabolism is depressed compared to the rate during normothermia is directly related to the time spent in steady-state torpor (Figure 11.13). This conclusion may not apply to large endotherms that enter torpor because the heat required to increase body temperature during arousal is directly pro-

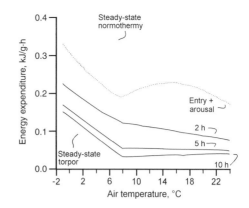

Figure 11.13 Energy expenditure in the rufous hummingbird (*Selasporus rufus*) when in steady-state normothermia, when in steady-state torpor, and when moving between normothermia and torpor of various durations, all as a function of temperature. Source: Modified from Hiebert (1990).

portional to body mass. Then the energy expended to warm the body may exceed the savings attained by entrance into torpor, unless the period spent in torpor is extensive (e.g., note the similarity of energy expenditure of starving and torpid badgers [Harlow 1981a, b]; see Section 11.8.4). In fact, Prothero and Jürgens (1986), based on the high cost of rewarming, asserted that endotherms cannot use daily torpor unless they weigh less than a kilogram, a limit that presumably does not apply to hibernators because their torpid periods are much longer (see Section 11.8.2).

The rufous hummingbird has its greatest reduction in energy expenditure at 8°C (Figure 11.13), which reflects the increased cost of temperature regulation in torpor at lower ambient temperatures. The level at which body temperature in torpor is regulated will influence the amount of energy that is expended and, consequently, the amount of energy that is "saved" by entering torpor. This temperature in hummingbirds is lower in cool environments and higher in warm environments (Wolf and Hainsworth 1972; Hiebert 1990).

2. *Small passerines.* Some small passerines living in cold climates permit body temperature to fall from a daytime high of 40 to 42°C to a nocturnal low in winter of 32 to 36°C at external temperatures of 0 to 10°C (Steen 1958; Haftorn 1972; Chaplin 1974, 1976; Paladino 1986), well above the body temperatures of torpid hummingbirds. This behavior has been observed in *Carduelis*, *Parus*, and *Zonotrichia*. The decrease in body temperature leads to a 10% to 45% reduction in the rate of energy expenditure (Chaplin 1974, 1976;

Reinertsen 1983; Reinertsen and Haftorn 1984, 1986) found in normothermic birds at the same ambient temperature. The propensity of highland willow tits (*P. montanus*) to show a deeper nocturnal hypothermia than lowland tits in spring reflects differences in climate with altitude, whereas in winter and summer no differences are found (Reinertsen 1984). Some small tropical passerines, namely, sunbirds (*Anthrepetes* and *Nectarinia*) and manakins (*Pipra* and *Manacus*), also permit body temperature to fall at night in response to cool temperatures (Cheke 1971, Bucher and Worthington 1982).

3. *Small mammals.* Many small mammals enter daily torpor, most notably some marsupials (Morrison and McNab 1962; Wallis 1979; Geiser 1986), many insectivorous bats, most (all?) crocidurine shrews (Vogel 1974; Vogel et al. 1979; Genoud 1985), at least one soricine shrew (Lindstedt 1980), several tenrecs (Stephenson and Racey 1993a, 1993b, 1994), and some rodents, including *Baiomys taylori* (Hudson 1965), *Perognathus longimembris* (Chew et al. 1967), and *Reithrodontomys megalotis* (Thompson 1985). The propensity to enter torpor is most highly developed in the smaller species. For example, all rodents that have an adult mass less than 10 g enter daily torpor, including the three cited here, which may explain why no arvicolid rodents, which appear to be committed to a rigid form of endothermy (McNab 1992b), have an adult mass much less than 15 to 20 g. Like the rufous hummingbird, the small (27 g) marsupial *Antechinomys laniger* readily enters torpor and regulates body temperature during torpor at a low level (11–17°C). Consequently its rate of metabolism is lowest (Figure 11.14A), torpor is longest (Figure 11.14B), and mass loss is least at intermediate ambient temperatures (Figure 11.14C). Some small nectarivorous flying foxes, including *Syconycteris australis* and *Macroglossus minimus*, enter daily torpor and increase their rate of metabolism at low ambient temperatures when in torpor (Bonaccorso and McNab 1997).

4. *Sedentary endotherms.* Some tropical, arboreal folivorous and frugivorous mammals conserve energy at cool to cold ambient temperatures by greatly reducing thermal conductance without permitting their core body temperatures to fall (see Section 5.5.4). The reduction in rate of metabolism is often so great at low ambient temperatures that it may be below the "basal" or standard rate that

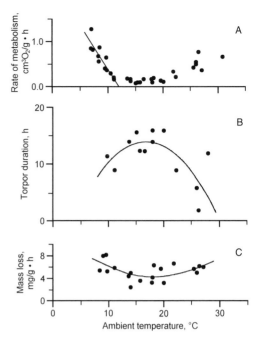

Figure 11.14 (A) Rate of metabolism, (B) torpor duration, and (C) mass loss in the marsupial *Antechinomys laniger* as a function of temperature. Source: Modified from Geiser (1986).

characterizes the zone of thermoneutrality! This behavior was demonstrated in the red panda (*Ailurus fulgens*; Figure 5.15). The panda reduces heat loss by reducing peripheral blood circulation, as shown by low skin temperatures (McNab 1988c). A similar reduction in rate of metabolism at cool temperatures without a fall in core temperature has been found in the viverrids *Arctictis*, *Nandinia*, and *Arctogalidia* (McNab 1995), a tree-kangaroo (*Dendrolagus* [McNab 1988c]), and several lemurs (Daniels 1984; McNab, pers. observ., 1988). A somewhat similar pattern was found in large flying foxes (McNab and Armstrong, 2001).

The reduction in rate of metabolism at low temperatures is effective only for a limited period of time because peripheral tissues cannot be completely cut off from circulation and cannot be permitted to freeze. In the tree-kangaroo reduced rates of metabolism may last up to 40 min at an ambient temperature of 10°C and to 4 h at 14°C (McNab 1988c). This behavior should not be called torpor because the core temperature is maintained at normothermic levels, but it is similar to daily torpor in that it represents a short-term means of reducing energy expenditure associated with a reduction in rate of metabolism and peripheral body temperatures. It is effective only in benign environments and in mammals that have a mass between 0.5 and 15 kg.

11.6 BALANCING ANNUAL ENERGY BUDGETS: INTRODUCTION AND TOLERANCE

The difficulties associated with maintaining a balanced energy budget are clearly demonstrated on an annual basis, especially in environments that show a marked seasonality in physical conditions and food availability. At an extreme, the poles have a continuous dark and cold period in winter, which is replaced by the continuous light and (comparative) warm period in summer. Temperate continental land masses, although somewhat more moderate than polar environments, still have an extremely variable (i.e., intemperate) climate. In many tropical environments the physical environment also fluctuates on a seasonal basis, but these variations are principally in rainfall.

The responses to seasonality in the environment are similar to those used to balance a daily energy budget, but are carried to an extreme because harsh conditions may last for months. Vertebrates can stay in one locality and remain active throughout the year; they can migrate to a more hospitable environment and return only when the conditions are more acceptable; or they can find shelter in a moderate microclimate, where they remain in a state of physiological depression during hostile conditions in the macroenvironment.

11.6.1 Seasonal tolerance: ectotherms. In warm-temperate and tropical environments most ectotherms remain active throughout the year as long as food is available, although their diversity and abundance may be greatly diminished during winter or a dry season. The impact of harsh conditions is usually great on ectotherms. In polar or cold-temperate environments only aquatic ectotherms remain active throughout winter (Patterson and Davies 1978b), and then only as long as water does not completely freeze. These ectotherms are mainly fish but may include some turtles (Gregory 1982).

Few estimates of the seasonal energy budgets of aquatic ectotherms are available. They would be

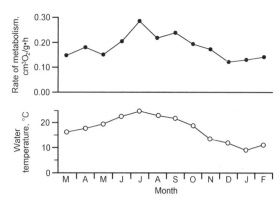

Figure 11.15 Mean rate of metabolism in bluegills (*Leopomis macrochirus*) and water temperature as a function of the time of year. Source: Derived from Wohlschlag and Juliano (1959).

expected to have lower rates of metabolism in winter than summer because of the decrease in water temperature, although seasonal acclimatization of metabolism might partially compensate for the decrease in temperature (see Section 4.5.3). Bluegills (*Lepomis macrochirus*) exposed to a seasonal variation in water temperature of 16.5°C showed little variation in minimal rate of metabolism, but the maximal rate correlated with water temperature (Wohlschlag and Juliano 1959; Figure 11.15).

With respect to the potential impact of energetics on seasonal behavior, Schultz et al. (1991) examined the relationship between body size and the timing of reproduction in a marine teleost, the dwarf surfperch (*Micrometrus minimus*). They showed that small (younger?) females in this viviparous species conceive and bear young later in the spring than do larger (older?) females. Two explanations for this difference were suggested: (1) small energy reserves required small females to reproduce later in the year, when energy reserves were larger; or (2) small females delay reproduction to facilitate their own growth to increase fecundity. Both explanations may apply to the dwarf sunperch.

The response of red-spotted newts (*Notophthalmus viridescens*) to winter depends on where they live. In southern Canada and parts of northern United States newts move onto land to hibernate, whereas in southern localities they remain active in winter. For example, in southern Ohio during summer they have higher rates of aerobic and anaerobic metabolism, which are fueled by feeding, whereas in winter they have a relatively constant, low level of aerobic metabolism, which is fueled by fat stored in the fall. In this region newts store at least 85 mg of fat prior to winter (Jiang and Claussen 1992), which represents enough energy to supply the expenditures of newts for at least 78 d, given that winter temperatures are 1 to 5°C and that the newts have low activity rates. This estimate of the time period increases to 120 d if they have a full stomach at the start of winter (which illustrates the very low rate of metabolism in this species at these temperatures).

Seasonal variations in the activity of terrestrial ectotherms are reflected in energy expenditure. For example, the lizard *Uta stansburiana* had low field expenditures in winter and high expenditures in spring and early summer (Nagy 1983). Adults of this species in winter were found basking between patches of snow at air temperatures near freezing. In winter the rate of food intake, as measured by ^{22}Na, in the scincid lizard *Lampropholis quichenoti* was only about 5% of that in summer (Gallagher et al. 1983). The lizard *Sauromalus obesus*, in contrast, showed a depression in field expenditures in summer when little moist food is available (Nagy and Shoemaker 1975).

11.6.2 Seasonal tolerance: endotherms.

Because endotherms have such high rates of energy expenditure compared to ectotherms, endotherms might be expected to face a shortage of resources during the physically harshest time of year, especially in polar and temperate winters.

The extreme problem faced by endotherms is seen in small, nonmigratory passerines that live throughout the year in subpolar or cold-temperate environments. Two such groups are parids, including tits, titmice, and chickadees, and cardueline finches, including redpolls and goldfinches. For example, black-capped chickadees (*Parus atricapillus*) living near Fairbanks, Alaska (60°N), in winter have a short feeding period, which decreases with day length, light intensity, and low temperature (Kessel 1976). In midwinter chickadees are active only for 6 to 7 h. Furthermore, Alaskan chickadees have not shown any evidence of the mild hypothermia seen in Scandinavian or cold-temperate North American parids (Grossman and West 1977), which raises the question of how they balance their energy budget.

Chickadees and other passerines resident in winter at northern latitudes, including redpolls (*Carduelis flammea*) and goldfinches (C. [*Spinus*]

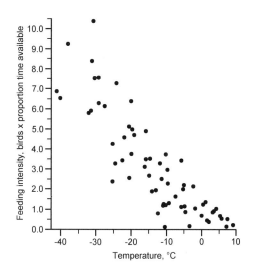

Figure 11.16 Feeding intensities of black-capped chickadees (*Parus atricapillus*) at a feeding station in Fairbanks, Alaska, as a function of temperature. Source: Modified from Kessel (1976).

tristis), make several adjustments to these conditions:

1. Feeding intensity increases with a fall in temperature (Figure 11.16), a pattern that is widespread in birds that reside in cold-temperate environments, for example, diving ducks (Nilsson 1970).

2. The food they eat each day accounts for 30% to 40% of body mass (White and West 1977).

3. They cache seeds and insects in late summer and fall for use in winter (S. Haftorn in Grossman and West 1977).

4. They often store food in esophageal diverticula, which when filled may contain enough food to equal 25% of the daily energy expenditure (White and West 1977).

5. They store body fat with a single peak in early winter (White and West 1977, Carey et al. 1978; Figure 11.17), unlike the bimodal fat cycles in migratory species (see Section 11.7.2).

6. Seasonal fat deposits are sufficient in winter for 23 h of survival at −10°C in a 15-g Michigan goldfinch (Carey et al. 1978).

7. Redpolls store more fat daily than chickadees because redpolls mainly consume lipid-rich seeds, whereas chickadees predominately eat protein-rich insects; this dietary difference is correlated with a higher nighttime metabolism and body temperature in redpolls than in chickadees (Chaplin 1974).

8. Small passerines reduce energy expenditure by seeking shelter, often in dense foliage, tree cav-

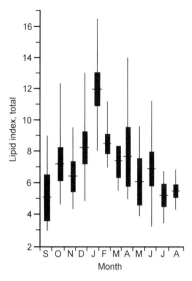

Figure 11.17 Lipid index in common redpolls (*Carduelis flammea*) as a function of the time of year. Source: Modified from White and West (1977).

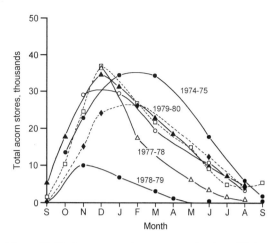

Figure 11.18 Acorn stores by acorn woodpeckers (*Melanerpes formicivorus*) as a function of the year and the time of year. Source: Modified from Koenig and Mumme (1987).

ities (Chaplin 1974), or even snow burrows (Sulkava 1969, Novikov 1972, Lagerstrøm 1979, Korhonen 1981), although the use of snow cavities as shelters appears to be more effective in larger species, such as willow ptarmigan (*Lagopus lagopus*) (Korhonen 1980). (For example, the cost of "free-living" in willow ptarmigan is only about 20% above that in confined individuals exposed to field conditions, and negligible in rock (*L. mutus*) and white-tailed ptarmigan (*L. leucurus*) [Moss 1973].)

Some endotherms hoard appreciable amounts of food to facilitate survivorship during periods of food scarcity in temperate winters and in tropical dry seasons (Vander Wall 1990). Mammals that store food for the long term include shrews, which primarily store insects and some small vertebrates, and a variety of small rodents, which store seeds, nuts, roots, tubers, bulbs, and some vegetation. Birds with these habits include owls, which principally store rodents or (if small) insects; woodpeckers, which store nuts or insects; and passerines, which store nuts, seeds, and insects. In North America four woodpeckers belonging to the genus *Melanerpes* store large amounts of nuts, especially acorns. These include the acorn woodpecker (*M. formicivorus*), Lewis' woodpecker (*M. lewis*), red-headed woodpecker (*M. erythrocephalus*), and red-bellied woodpecker (*M. carolinus*). Acorn

woodpeckers cache large amounts of acorns and other nuts in "granneries," branches of trees in which the acorns are packed into a series of closely spaced holes. The caches are filled by a social group in the fall and early winter and are used in midwinter through spring and summer, reaching a minimum in late summer (Figure 11.18). A similar pattern is seen in the Lewis' woodpecker (Bock 1970).

Carey et al. (1978) concluded that the principal physiological adjustment to winter by goldfinches is an enhanced thermogenic capacity that depends principally on an increased fat supply (see also Yacoe and Dawson 1983). Such an increased capacity permits winter-acclimatized goldfinches to withstand temperatures as low as −70°C for up to 6 to 8 h (Dawson and Carey 1976), a tolerance that varies with season (Figure 11.19). Carey et al. (1978, p. 111) concluded that "... the metabolic machinery necessary to support flight is more than adequate to deal with thermogenic requirements in winter."

Another adjustment made by some small mammals to winter conditions is a reduction in body mass beyond a seasonal loss of body fat. This response is heterogeneous: it may represent the cessation of growth in young, the selective death of old individuals, or the loss of body mass in the old adults that survive winter (Kalela 1957, Fuller et al. 1969, Iverson and Turner 1974). The cessation of growth may be cued by the reduction in photoperiod in fall (Iverson and Turner 1974, Petterborg 1978). The decrease in mass during

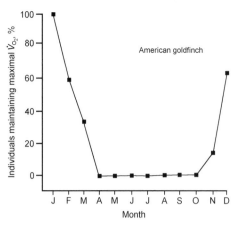

Figure 11.19 Proportion of individual American goldfinches (*Carduelis tristis*) that maintain the maximal capacity to remain normothermic at air temperatures lower than −60°C for more than 3 h as a function of the time of year. Source: Modified from Dawson and Carey (1976).

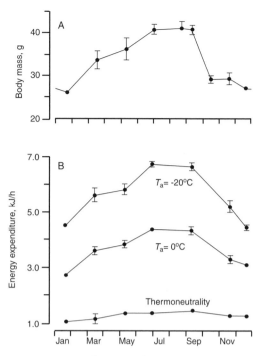

Figure 11.20 (A) Body mass and (B) energy expenditure at various temperatures in the Djungarian hamster (*Phodopus sungorus*) as a function of the time of year. Source: Modified from Heldmaier and Steinlechner (1981b).

winter has been found in *Sorex* (e.g., Dehnel 1949, Mezhzherin 1964, Genoud 1985), *Microtus* (Iverson and Turner 1974), *Clethrionomys* (Sealander 1966, Fuller et al. 1969, Merritt and Zegers 1991), and *Phodopus* (Heldmaier and Steinlechner 1981b). In shrews the reduction in body mass involves a reduction in the skeletal and brain mass (Dehnel 1949, Pucek 1964, Mezhzherin 1964). When dealing with such a fundamental reduction in mass within individuals, this change has been referred to as *Dehnel's phenomenon* (see Section 5.3.3).

The significance of the decrease in mass in winter has been explored in the Djungarian hamster (Figure 11.20A). Its principal consequence is a reduction in energy expenditure (Figure 11.20B; also see Kalela 1957, Iverson and Turner 1974). Such observations emphasize the importance of total units for energy expenditure: preoccupation with mass-specific units may lead to the technically correct, but ecologically meaningless conclusion that "... resting metabolic rate ... showed no definite seasonal changes ..." (Merritt and Zegers 1991). The decrease in mass of *C. gapperi* from 26.2 g (in September) to 21.4 g (in January) implies a decrease of at least 13% ($[21.4/26.2]^{0.71} \times 100 = 87\%$) in resting rate of metabolism.

This Dehnel pattern is exactly opposite to that codified in Bergmann's rule, where an increase in size is correlated with an increase in *latitude* (implying a decrease in temperature). Bergmann's rule describes a change in the body mass of popu-

lations aligned by latitude (but see McNab [1971b] where the occurrence of a positive correlation of mammalian size with latitude in North America occurs only in a minority of species, and even then usually only over a segment of their latitudinal range), whereas Dehnel's phenomenon describes a *seasonal* change of size in individuals. Bergmann's rule implies an *increase*, not a decrease, in energy expenditure with a fall in ambient temperature (see Figure 3.3b), whereas Dehnel's phenomenon implies a *decrease*, not an increase, in energy expenditure with a fall in ambient temperature. Fuller et al. (1969) maintained that a winter decrease in mass is most clearly seen in small mammals that live under snow in cold-temperate and polar environments. In the presence of low temperatures (in subnivean environments) as a result of shallow snow depths, *Clethrionomys* survived less well than *Peromyscus*, which apparently evaded the worst circumstances by entering torpor: "... torpor [may be] superior to continuous activity under these [harsh] conditions" (Fuller et al. 1969, p. 51).

King and Murphy (1985) questioned the assumption that endotherms face a seasonal shortage of resources. They suggested that the principal response of endotherms to a seasonal shortage of resources is a staggering of expenditures (King and Farner 1963, Helms 1968), or that a surplus of resources during one period can be transferred to another period through the storage and use of fat (e.g., the timing of reproduction [Schultz 1991]). In fact, in some mammals (e.g., *Lemmus sibiricus*) molting, once initiated, stops with pregnancy until reproduction has finished, when molting is completed. Another possibility is that endotherms have an energy expenditure that varies with the season. Many endotherms spend much of their "activity" periods resting (Herbers 1981), implying that these species are not normally resource limited. These two responses to seasonality, the first called the *reallocation* hypothesis because energy expenditures are limited and shifted from one function to another and the second called the *increased demand* hypothesis, which implies that energy expenditure is not strictly limited, would be expected to be extremes along a continuum of responses.

Comparatively few endotherms have had their field expenditures measured (or even reliably estimated) throughout the year so that we have little information to judge which pattern is normally followed. What is known in birds is that some species appear to conform to each extreme. Weathers and Sullivan (1993) suggested that granivorous and omnivorous species tend to reallocate expenditures, the total thus remaining fairly constant, whereas carnivorous or insectivorous species had much higher expenditures during the breeding season and thus conformed to the increased demand model. A caution should be given: the data that are available may well be specific to locality and not represent species so much as the circumstances in which the measurements were made. Furthermore, the sexes in one species may conform to different models.

A seasonal reallocation of expenditures has been found at least in juncos (*Junco*), the black-billed magpie (*Pica pica*), and willow ptarmigan (*Lagopus lagopus*). In two species of juncos the energy expenditures measured in the field during the breeding season were only 5% greater than those in winter (Weathers and Sullivan 1993). This similarity resulted from juncos foraging longer in winter than in the breeding season (6.0 vs. 4.5 h/d), which was required because the energy gain from foraging on seeds in winter was lower than when

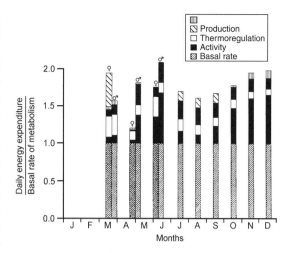

Figure 11.21 Estimated field energy expenditures relative to the expected basal rate of metabolism in the black-billed magpie (*Pica pica*) as a function of the time of year. Here production means egg laying and molting. Source: Modified from Mugaas and King (1981).

foraging on insects in the breeding season (11.8 vs. 15.1 kJ/h). The resulting "saving" in the breeding season was channeled into reproduction. This reallocation (of time and energy), however, may apply in such small birds (mass 18–20 g) only in a moderate winter environment (daytime T_a up to 14°C, nighttime T_a down to −7°C); their expenditures in polar or cold-temperate winters might be much higher than during the breeding season. The estimated energy expenditures of magpies are also similar throughout the year (Figure 11.21), mainly because the higher cost of activity in winter offsets the cost of reproduction during the breeding season and the cost of molting in the fall (Mugaas and King 1981). Estimates of the energy expenditure in a 550-g willow ptarmigan in northern Alaska indicated that winter expenditures averaged only 15% greater than those in summer, even though environmental temperatures fell as low as −35°C (Figure 11.22). Females showed an increase in energy expenditure associated with egg laying, during which time molting was temporarily suspended. Clearly, this ptarmigan timed egg laying and molt, two expensive activities, to coincide with the warmest time of the year (Figure 11.22) and reduced the cost of maintenance in winter through sheltering at night in snow burrows, increasing insulation, and decreasing basal rate (West 1972, also see Moss 1973).

A more common annual pattern, in spite of the tendency to stagger expensive activities during the

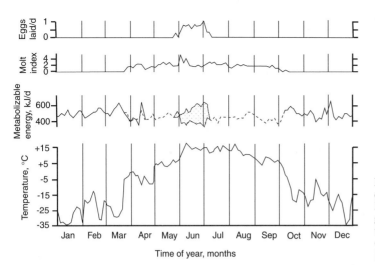

Figure 11.22 Estimated field energy expenditures and the timing of egg laying and molting in the willow ptarmigan (*Lagopus lagopus*) and environmental temperature as a function of the time of year. Source: Modified from West (1968).

year, is that energy expenditure varies with season. This has been seen in several species of birds, including the Cape vulture (*Gyps coprotheres*), long-eared owl (*Asio otus*), European kestrel (*Falco tinnunculus*), and blue penguin (*Eudyptula minor*). The ability to specify details in the annual energy budget reflects the methods used to estimate energy budgets—by far the most detailed (and accurate) estimates coming from studies using the doubly labeled water technique.

Komen and Brown (1993) used the food consumption of nestling and adult Cape vultures to estimate the annual energy expenditure of a reproducing family and of a colony in South Africa. A family consisting of two parents and one nestling expended 1.36 MJ over a nesting period of 136 d, and 3.15 MJ annually. That is, 43% of the annual energy expenditure occurred during reproduction, which occupied 37% of the year. The peak expenditure occurred when the nestling was 80 to 84 d old, an expenditure that was about 2.1 times the minimum. These estimates, coupled with the observation that the colony included 152 breeding pairs and 98 nonbreeding adults, indicated that the colony required 89.4 tons of meat annually. The peak of meat consumption occurred in September and October, when the young had the greatest requirements and when native ungulates and domestic livestock mortalities were the greatest.

Long-eared owls in the Netherlands show higher field expenditures, estimated from food consumption, during the breeding season than in winter (Wijnandts 1984). And although the total expenditures in male and female owls are similar, males

reach the peak in their expenditures when they provide food for themselves, the incubating female, and the developing young. (Note that in this owl, like most others, males are smaller than females.) In winter, long-eared owls appear to survive on a low energy intake, which presumably reflects a low rate of energy expenditure.

Male and female kestrels, like long-eared owls, have similar total field expenditures, even though females are 20% larger. Male expenditures are about 30% greater during the breeding season than in winter, mainly because they too provide food for the females and young (Masman et al. 1988). Female kestrels, however, tend to expend slightly more (7%) energy in winter than during the breeding season, the contrast with the energy budgets of males reflecting sexual role differentiation. In fact, this observation and that in the long-eared owl may be a clue to an ecological significance of sexual dimorphism in endotherms: the smaller sex may well have an added energy expenditure compared to the larger sex, so that when the difference in mass is taken into account, the total expenditures of the sexes are equal, just as the smaller female mass found in some mammals may compensate for the increased expenditures associated with pregnancy and lactation (Powell and Leonard 1983; McNab and Armstrong, 2001).

The energy budget of kestrels was measured by doubly labeled water and by food consumption (Masman et al. 1986). During winter, food intake was concentrated at dusk to compensate for the extended period of darkness. Food intake was highest in females during the egg-laying period, whereas in males it was highest during the nesting phase. Because males started to feed incubating females before the food-capture rate increased, the mass of males decreased by 14% during the breeding season. This imbalance reflects the high cost of hunting: male kestrels can balance energy intake with expenditure only if flight activity is less than 4.6 h/d. During the postreproductive molt kestrels had low rates of metabolism and very low activity levels. The energy saved from a reduction of activity appears to be sufficient to fund feather production. Time allocated to foraging by kestrels, therefore, has a large impact on the energy budget by influencing both energy intake and expenditure.

Unlike these two raptors, the blue penguin showed no difference in energy expenditure, measured by doubly labeled water, between males and females (Gales and Green 1990): the sexes are of equal mass. Rate of energy expenditure during winter, courtship, and early stages of chick rearing was relatively constant in foraging penguins. Low field expenditures (one-half to two-thirds of winter rates) occurred only while fasting during courtship, incubation, and molting, for a total of only 46 d/y. During chick rearing field expenditures progressively increased to rates that were over twice those found in winter. Furthermore, the "expensive" late stages of chick rearing are long, and the cost of foraging increases with the intensity of foraging. The annual expenditure of a nonbreeding penguin was about 448 MJ, whereas that for a breeding penguin was about 533 MJ, or a 19% increase, but that is only half the cost of raising young: a pair would require approximately 1067 MJ/y to raise an average of 1.7 chicks. Gales and Green (1990) calculated that 31% of the annual energy budget of reproducing adults was spent on chick raising, whereas this phase occupies only 16% of the annual cycle.

Krijsveld et al. (1998) explored the impact of body mass on field energy expenditures and its implication for sexual dimorphism. They noted that in some birds females are larger than males, in others the sexes are of the same size, and in still others males are larger than females. Two extreme conditions might be expected to exist: either the ratio of male to female food intake, an approximation of the ratio of male to female energy expenditure, is proportional to the ratio of male to female body mass, or, as was potentially implied by the observations on Eurasian kestrels and long-

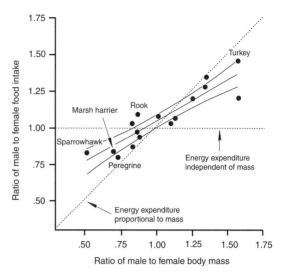

Figure 11.23 Ratio of estimated food intake by males compared to that by females of the same species plotted as a function of the ratio of the body mass of males to that of females. Two curves are indicated, one under the assumption that the ratio of food intake is independent of the ratio of body mass and another under the assumption that the ratio of food intake is proportional to the ratio of body mass. Source: Modified from Krijsveld et al. (1998).

eared owls, the ratio for food intake is constant and independent of the ratio of masses (Figure 11.23). The latter condition would imply that the sexual differences in body mass are due to the consequences for energetics of sexual dimorphism. Under the first condition, the slope of the correlation of food intake to body mass would be 1.0 and under the second, the slope should be 0.0. In fact, the slope was 0.56 (Figure 11.23).

Another interpretation of this correlation can be given. The ratio of food intake (and rate of metabolism) expected from the ratio in body mass should be $(m_{males}/m_{females})^{0.64}$, not $(m_{males}/m_{females})^{1.0}$, because field expenditures of birds are proportional to $m^{0.64}$ (Nagy 1987). Indeed, the slope of 0.56 is much closer to 0.64 than it is to either 1.00 or 0.00 (Figure 11.23). This analysis implies that the ratio of male to female energy expenditures is approximately what is to be expected from the ratio of male to female masses, even though variability about the mean regression line is sufficient to permit the interpretation that in some species a smaller sex is compensatory for a higher intensity of activity, principally during the reproductive season.

Mammals remain active in the face of harsh physical conditions as long as sufficient food

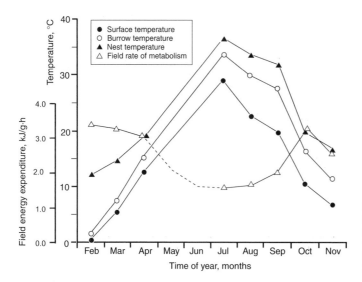

Figure 11.24 Field energy expenditures in the pocket-mouse *Perognathus formosus* measured by doubly labeled water and various environmental temperatures as a function of the time of year. Source: Derived from Mullen and Chew (1973).

remains available, which means that tolerance is restricted to species that feed on grass, browse, immature insects, vertebrates, or (stored) seeds. Small mammals would be expected to show seasonal variation in energy expenditure because they cannot compensate for low ambient temperatures by increasing insulation to any appreciable extent. Thus, the 18-g pocket-mouse *Perognathus formosus* has high daily energy expenditures in winter (Figure 11.24), whereas the jackrabbit (*Lepus californicus*), which weighs 2.3 kg, has a rate of metabolism that is independent of season, except when mean air temperature falls below 10°C (Shoemaker et al. 1976). Merritt and Zegers (1991), however, suggested that the energy expenditure of *Clethrionomys gapperi* is reduced in winter by an increase in insulation and a reduction in body mass. Wunder et al. (1977), Bozinovic et al. (1990), and Wunder (1984) reached similar conclusions, although no doubly labeled water measurements have substantiated them. Seasonal field expenditures are not available for large mammals living at high latitudes.

The performance of small endotherms undoubtedly reflects the microclimate in which they live, which in cold-temperate environments involves the presence or absence of a snow cover (see Sealander 1966). For example, small mammals depend on the presence of a snow cover in cold-temperate regions, as in the tiaga (Pruitt 1957). Under a snow cover of 15 to 20 cm, air temperatures are quite stable and well above those encountered in winter at the snow surface, where the temperature is often more than 20 to 30°C colder. "Before this thickness is

reached activity of shrews and red-backed voles is common on the snow surface; after this thickness is reached surface activity is markedly reduced" (Pruitt 1957, p. 134). The principal advantage of a snow cover to resident endotherms may be the avoidance of exposure to very cold ambient temperatures and high wind velocities, although appreciable rates of heat loss occur in willow ptarmigan (*Lagopus lagopus*) when digging a snow burrow and when in a fresh burrow before the air is warmed (Korhonen 1989). Andreev (1988) estimated that a 1-kg grouse in Siberia consumes in winter about 100 g of browse diet per day, expends 783 kJ/d for basal rate (63% of the total budget), 105 kJ/d for thermoregulation during the 16 h in a snow burrow (13%), and 183 kJ/d for thermoregulation and activity outside the snow burrow (24%). The cost of thermoregulation would have been much higher if the grouse did not shelter in snow burrows. The use of snow burrows by small mammals may only be possible because they construct insulated nests, which may explain why small passerines generally do not use snow burrows (Sulkava 1969).

The arctic fox (*Alopex lagopus*) is the smallest (3–4 kg) terrestrial carnivore that remains active in high arctic winters. On Svalbard the basal rate of metabolism of arctic foxes remains independent of season, although the mass-specific rate decreases in winter because of the deposition of body fat (Prestrud and Nilssen 1992, Fuglei and Øritsland 1999). A similar pattern was found in Svalbard ptarmigan (Mortensen et al. 1983), whereas Svalbard reindeer have low resting rates in winter because of both a small mass and a low rate independent of mass (Nilssen et al. 1984).

The influence of the food resources used by endotherms that remain active throughout the year in cold-temperate and polar environments is seen in a comparison of grouse and lagomorphs (Thomas 1987), a comparison that probably should include arvicolid rodents (see McNab 1992b, Koteja and Weiner 1993). These endotherms share several characteristics: (1) all are obligate herbivores; (2) all (with the possible exception of arvicolids) tolerate extended winter storms that preclude feeding; (3) none show evidence of torpor; and (4) all have small energy reserves. Thomas (1987) suggested that the small fat deposits mainly reflect the poor nutritional quality (as measured by the protein content) of browse and grass in winter. The principal response of these herbivores to avoid starvation during winter is hypertrophy of the gut (Pendergast and Boag 1973, Moss 1974, Gasaway 1976a, Pulliainen 1976, Thomas 1984, Gross et al. 1985, Hammond and Wunder 1991), which permits a more efficient extraction of the nutrients present in low-quality food, an enhanced postgastric fermentation, and a more effective detoxification of plant secondary compounds (see Section 12.7).

In contrast, two grouse appear to have noteworthy fat deposits, the Svalbard rock ptarmigan (*L. mutus hyperboreas*) and the capercaillie (*Tetrao urogallus*). The Svalbard ptarmigan in fall stores in excess of 13% of total mass as fat (this may reach 32%; Mortensen et al. 1983), which decreases to about 1% by April (Grammeltvedt and Steen 1978). This large fat cycle undoubtedly is associated with the near certainty of prolonged snow storms on Svalbard with a diminishing and often unavailable food supply. Why this species puts on fat and other ptarmigan, including the North American rock ptarmigan, and grouse do not is unclear, unless it reflects in Svalbard an inability to move to less stringent environmental conditions. In capercaillie, the largest grouse, males store up to 5.4% of body mass as fat and females, up to 7.5%; females have a mass that is less than one-half that of males (Hissa et al. 1990). Females are more active in winter than males, which leads to the use of their fat stores and requires them to feed. What sets this grouse apart from most other species is also not clear, but it surely faces a harsh climate, consumes poor-quality food, and does not migrate.

The transfer of energy from one season to another through the storage and use of fat deposits is well developed in penguins. In addition to the long fasting periods of emperor and king penguins (see Section 14.3.4), the macaroni penguin annually has two fasting periods, 29 to 32 d during the breeding season and 24 to 25 d during molt (Williams et al. 1992). This penguin loses 31% to 34% and 41% to 47% of their initial masses, respectively, during these fasts. The higher mass loss during molt is due to the high cost of feather replacement. The absence of an increase in plasma urea or uric acid indicates that the fast in the macaroni penguins is fueled by fat metabolism and never lasts long enough to require the burning of protein (unlike the long fasts of emperor penguins).

The preoccupation with adjustments to winter reflects the preponderance of studies in the Northern Hemisphere. In warm-temperate and tropical climates the season that requires adjustments by residents is likely to be a hot, dry summer. Three studies of field energy expenditures in xeric, warm-temperate environments are relevant here:

1. During an unusually dry period, golden bandicoots (*Isoodon auratus*) living on Barrow Island, Western Australia, had poor body condition, a small mass, and low field expenditures (Bradshaw et al. 1994). Five months later body condition improved and body mass increased, even though field energy expenditures remained low and rainfall was minimal, the improvements occurring apparently because the bandicoots shifted from termites to other insects in a mixed diet. After a cyclone dumped 162 mm of rain in 24 h, field energy expenditures of the bandicoots tripled (to 4.0 times the basal rate). Barrow Island mice (*Pseudomys nanus*) had a small mass and very low field expenditures during the extended drought, but both mass and field expenditures increased after the cyclone, the field expenditures by a factor of three.

2. In arid South Africa, aardwolves (*Proteles cristata*) had lower field expenditures in winter (2.1 times the basal rate) than in summer (2.7 times basal rate), much of the seasonal difference being related to a reduced period of activity in winter (4.1 h/d) than in summer (8.8 h/d) (Williams et al. 1997). The especially low winter energy expenditure is correlated with a low basal rate of metabolism (McNab 1984, Anderson et al. 1997).

3. The springbok antelope (*Antidorcas marsupialis*) in the Kalahari Desert had low field expenditures both in the hot, dry summer (55% of the expected value) and in the cold, dry winter (82% of the expected value), the only exception being a short (1-week) rutting season (when field rates were about 155% of the rate expected from mass), which occurred during a hot, wet period in summer (Nagy and Knight 1994).

In these examples, the lowest rates occurred during hot, dry summers, with the exception of the aardwolf, which may have faced a food shortage in winter associated with a shift from feeding (in summer) predominately on an abundant termite (*Trinervitermes*) that is cold sensitive, to a less abundant termite (*Hodotermes*) that is less temperature sensitive in winter.

One of the few studies of seasonal energetics in the tropics, indirect though it may be, was of the seasonal fat cycles of bats in Jamaica (McNab 1976). The amplitude of the annual fat cycle was greatest in insectivorous species, especially those that feed above the forest canopy, and is minimal in frugivores and nectivores. Females have a greater amplitude in their fat cycles than males of the same species. In species with an appreciable fat cycle, fat accumulates during the wet season and is consumed in the dry season.

11.7 BALANCING ANNUAL ENERGY BUDGETS: MIGRATION

The recurrent seasonal movement of a population between two areas, usually one for breeding and the other for residence at a time when conditions on the breeding grounds are inhospitable, is called *migration*. Migration occurs commonly in vertebrates but is limited to species that have an effective means of long-distance movement. It is widespread in birds and is found in some large mammals, bats, and aquatic ectotherms (fishes, turtles). Interest in migration has centered on its timing, means of navigation, and significance (for a compendium, see Baker 1978), but few studies have examined directly the energetics of migration. Two aspects of the energetics of migration are discussed here: its significance for the occurrence of migration, and the energetics of migratory movement itself.

11.7.1 Energy balance and migration. All vertebrates with an effective means of long-distance movement, in theory, could migrate, but they do not. Some birds, for example, are year-long residents in arctic and cold-temperate environments. Migration presumably occurs when the average mortality in individuals wintering on breeding gounds exceeds the average mortality of individuals making two migratory flights (forth and back) and wintering in a mild climate (Lack 1968a). A requirement faced by a species on its wintering grounds, whether they correspond to the breeding range or not, is to balance its energy budget. In temperate and polar environments this may be difficult because of the combination of cold ambient temperatures, which increase energy expenditure in endotherms, short day lengths, which reduce the time for foraging (Kendeigh 1934, 1976; Siebert 1949; Irving 1960; West 1960; Cox 1961; Zimmerman 1965), and a shortage of food. The requirement that energy budgets be balanced does not prevent energy from being transferred from one period (or season) to another through the deposition and use of fat stores.

S. C. Kendeigh and his students extensively studied the significance of energetics for the occurrence of migration. Unlike nonmigratory passerines, migratory species cannot increase sufficiently the rate of food intake in winter to compensate for the decrease in temperature and photoperiod (Siebert 1949). Thus, the maximal energy intake in the migratory dark-eyed junco (*Junco hyemalis*) and white-throated sparrow (*Zonotrichia albicollis*) corresponds to the requirements at the northern limits of their winter distribution (also see Root 1988b), but is less than that required in winter in the breeding range. Resident species, such as the house sparrow (*Passer domesticus*) and bluejay (*Cyanocitta cristata*), increase food intake in a short winter day to exceed the food intake in a long summer day: feeding intensity in nonmigratory species increases with a decrease in ambient temperature (Figure 11.16). Wallgren (1954) similarly analyzed migratory behavior in two species of buntings belonging to the genus *Emberiza*.

The effect of latitude on balancing an energy budget is shown (Figure 11.25) in the calculations of Kendeigh (1976) for the (usually) nonmigratory house sparrow: the expenditure of *P. domesticus* at the northern limits of its distribution (Ft. Churchill, Manitoba) is expected to be 49.0 MJ/y, whereas in

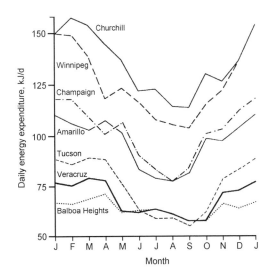

Figure 11.25 Estimated daily energy expenditures of the sedentary house sparrows (*Passer domesticus*) as a function of the time of year and geographic location. Source: Modified from Kendeigh (1976).

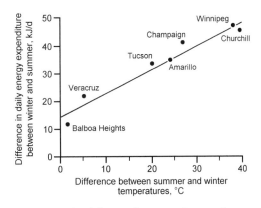

Figure 11.26 The difference between winter and summer in the estimated daily energy expenditure of the house sparrow (*Passer domesticus*) as a function of the latitudinal difference in winter and summer temperatures. Source: Modified from Kendeigh (1976).

Balboa Heights (Panama), where the bird was not then found, the expenditure would be about 23.4 MJ/y. The cost of permanent residency, which was estimated by subtracting the minimal amount of energy expended in summer from the maximal amount expended in winter, is correlated with the difference in mean temperature between the coldest month in winter and the warmest month in summer (Figure 11.26). That is, the premium for migration, or at least fall migration, increases with latitude. In the case of the usually sedentary house sparrow, northern populations migrate in Europe (Summers-Smith 1963) and North America (Broun 1972).

These analyses should be treated with caution because all measurements of energy expenditure were based on food consumption in captive birds, which may not have been similar to the consumption in free-living birds, especially as a result of their confinement in cages. Do birds migrate because they cannot compensate for low temperatures and a short photoperiod, or do they not compensate because they migrate?

Recent work on the energetics of migrating birds has employed heated models and doubly labeled water. Wiersma and Piersma (1994) studied the knot subspecies *Calidris canutus islandica*, which breeds in northeastern Canada and northern Greenland, migrates through Iceland, and winters in northwestern Europe. Other subspecies, includ-

ing the Siberian-breeding *C. c. canutus*, winter in subtropical and tropical environments. Physical models of knots that contained a heating element and were covered with the skin and feathers of knots permitted measurement of the heat input required to maintain a normal body temperature under a variety of environmental conditions. The principal factors determining heat loss (and presumably heat production) were standard operative temperature (see Section 2.9.3), selection of inter-individual distance in flocks, and position relative to wind bearing.

Wiersma and Piersma (1994) estimated the annual cost of maintenance in *C. c. islandica* knots by integrating the cost of maintenance at each locality weighed by the time period the knots spent at each locality (Figure 11.27). With migration to arctic Canada for June and July, energy expenditure increased a little, but by returning to Europe for winter, energy expenditure decreased dramatically compared to what it would have been if the knots had stayed in the Arctic. The total annual expenditure for maintenance for a 130-g knot was 73.0 MJ, 41% of which was estimated to represent basal rate and 59% thermoregulation. If the maximal energy expenditure of the knot is 4.5 times the basal rate, given that the measured basal rate is 3.4 kJ/h, then the maximal rate would be 15.3 kJ/h, which is only 5.1 kJ/h more than the highest cost of maintenance in December and January in the Netherlands. If activity is 1.5 times the basal rate, then knots would have equaled or exceeded the maximal limit at the coldest time of the year in 30 of the last 31 years!

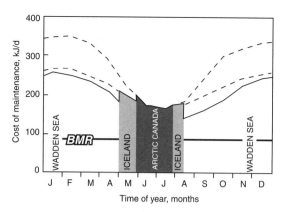

Figure 11.27 Estimated cost of maintenance in knots (*Calidris canutus*) as a function of geographic location and the time of year. The solid curves indicate where the knots are located at a particular time (e.g., Wadden Sea [Netherlands] in winter, Iceland in migration, and Arctic Canada when breeding), and the dashed curves indicate that the knots are elsewhere. BMR, basal rate of metabolism. Source: Modified from Wiersma and Piersma (1994).

This calculation raises the question of why *C. c. islandica* knots winter in the Netherlands. For example, some knots belonging to the Siberian subspecies *C. c. canutus* winter in Mauritania, West Africa. Piersma et al. (1991) estimated the cost of migration between the Netherlands and Mauritania to be approximately 0.3 to 0.5 kJ/km, which given a round-trip distance of 9200 km, would require an expenditure of 2760 to 4600 kJ. Averaged over winter (September to April), this would add 0.47 to 0.79 kJ/h to the expenditure expected in Mauritania (4.97 kJ/h), which is appreciably less than the 9.04 kJ/h required to stay in the Netherlands. Part of the high cost of maintenance during winter in Europe is associated with the larger mass of knots that remain in Europe.

The differential distribution of knot subspecies in winter in relation to body mass has consequences for their energy expenditure, which Piersma et al. (1996) explored. Basal rate of metabolism in knots was correlated with lean body mass, although the composition of the lean mass may be important. Thus, the "nutritional" organs (stomach, intestine, kidneys, and liver) may be the main determinant of basal rate, but changes in the musculature and heart may also contribute to the "effective" lean mass. The size of the lean mass and its composition are adjusted in knots to the climate and food availability on the wintering grounds, factors that lead to a modification of basal rate and therefore the field maintenance costs. Specifically, the subspecies *C. c. islandica*, which, when it winters in the Netherlands, faces a rigorous winter climate, feeds heavily on small molluscs, and has a large stomach, large body mass, high basal rate, and high field expenditure. Individuals belonging to the subspecies *C. c. canutus* that winter in Guinea-Bissau, West Africa, encounter a benign climate. They then have a small mass, low basal rate, and low field expenditure. When they were brought into captivity, lean body masses in individuals belonging to both subspecies decreased, with the greatest reduction occurring in the stomach, intestine, and pectoral muscles. As a result, a marked decrease in basal rate in both subspecies occurred during captivity (also see Piersma et al. 1995). These adjustments reflect the conditions under which the captives were kept, including being fed soft food, which means that many of the differences seen between the two subspecies in the field are the flexible response to the conditions encountered in the field. A large lean mass represents an elevated working capacity, requiring a large supporting machinery. The large lean mass of knots wintering in Europe or breeding in Siberia "... is an adaptive response enabling an appropriate increase in metabolic scope ... without precluding the option of direct energetic savings whenever the enlarged tissues can be dispensed with ..." (Piersma et al. 1996, p. 212).

Doubly labeled water estimates of field energetics were made on the wintering grounds of a related migratory shorebird, the sanderling (*C. alba*). This species breeds in the high Arctic of North America and winters on sandy beaches from temperate North America south to 50°S in South America. One-way migration in this species may be as short as 2000 km or exceed 10,000 km. Castro et al. (1992) found that (1) mean January temperature decreased with latitude; (2) lean body mass increased with latitude; (3) fat deposits increased with latitude; (4) field energy expenditures increased with latitude; and (5) starvation time increased with latitude. These biological characters therefore are correlated with mean January temperature (Figure 11.28). Even though mean January temperature is much warmer in Panama (26.6°C) than in New Jersey (−2.0°C), starvation time is longer in New Jersey (33 h) than in Panama (24 h), mainly due to larger fat stores in New Jersey (278 kJ) than in Panama (95 kJ). This correlation occurs in spite of the observation that the mean field expenditure in New Jersey is 4.2 times the

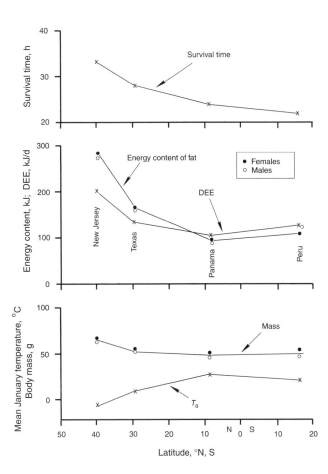

Figure 11.28 Measured daily field energy expenditure (DEE), body mass, body fat content, starvation time, and winter environmental temperatures in the sanderling (*Calidris alba*) as a function of the latitude at the wintering grounds. Source: Modified from Castro et al. (1992).

measured basal rate, whereas that in Panama is 2.1 times the basal rate. Fat storage in this species appears mainly to be a reserve against inclement weather, conditions potentially faced in winter and not associated with the cost of migration.

If movement from the breeding to the wintering range is usually associated with a decrease in the cost of maintenance, what is responsible for the return migration? No consensus exists on the answer to this question. In a study of the dickcissel (*Spiza americana*), Zimmerman (1965) calculated the amount of energy that can be diverted to reproduction and growth (i.e., "productive" energy) for various localities in its winter and summer ranges. A favorable energy balance is greatest in summer at localities from Illinois to Texas, which is the normal breeding range, but it decreases at both higher and lower latitudes. The amount of energy available for production, then, might be increased by northward migration in spring and maintenance costs decreased by southward migration in fall. Tropical finches, however, would have such a small increase in productive

energy as a result of migration that migration would be of little value (Cox 1961).

Whether energy balance is responsible for the patterns of migration that are present is unclear. For example, the cost of maintenance, even in summer, increases with the northward migration of American tree sparrows (*Spizella arborea*) in spring, which means that some factor other than a reduction of maintenance cost must account for the migration (West 1960). The increase in the cost of maintenance may not be important if the increase in food intake associated with a northward spring migration is sufficiently great. Lack (1968a) argued that the breeding range in birds is determined by the amount of food that can be obtained to feed the young. Many migratory arctic and North Temperate birds can be viewed as tropical or warm-temperate species that temporarily move into uninhabited regions to exploit seasonally abundant food resources, principally insects, for reproduction: as befits this suggestion, many of these migratory species spend most of the year on their wintering grounds.

11.7.2 Cost of migration. The energy expenditure for migration and its attendant mortality are integral parts of the balance that determine whether migration occurs or not. The energy expended during migration depends on body size and on whether powered or soaring flight is used. Migratory birds can be separated into those that are time minimizers and those that are energy minimizers (Hedenström 1993). Time minimizers fly as fast as possible without regard to the cost, whereas energy minimizers fly as economically as possible without regard to the time that it takes. These categories, however, grade into each other because by minimizing the time for migration, energy expenditure is reduced (as long as the cost of rapid flight is not too high).

Because the proportion of body mass that can be stored as fat preparatory for migration is described by $m^{-0.27}$ (Lindström 1991), Klaassen (1996) maintained that the maximal flight range of migratory birds decreases with mass. The logic behind this statement is that the flight range based on powered flight is proportional to the fat content (i.e., $m^{0.73}$) and inversely proportional to the rate at which energy is expended in powered flight ($m^{1.17}$; Pennycuick 1972); that is, flight range is proportional to $m^{0.73}/m^{1.17} = m^{-0.44}$. Large birds therefore should be more prone to soar during migration than small birds, at least as long as they are energy minimizers, especially because flight velocity is high in large species. Small birds generally do not soar because of a low flight velocity and a low gliding ratio (see Section 9.6.3). Therefore, small migratory birds generally appear to be time, not energy, minimizers, a correlation Lindström and Alerstam (1992) supported. These authors expected that flight velocity increases along the migratory route. Apparently this increase occurs in willow warblers (*Phylloscopus trochilus*) (Hedenström and Pettersson 1987). Large species, in contrast, can be either time or energy minimizers. If time minimizers, they would use flapping flight, the expense of which can be reduced by increasing the aspect ratio, as is the case in geese and swans. If energy minimizers, they often have low aspect ratios to facilitate landing and takeoff, as in raptors and storks, but depend on thermals, which often means that they must evade expansive areas covered by water and therefore often detour over land (Pennycuick 1972, Hedenström 1993), as do white storks (*Ciconia ciconia*) that migrate from Europe to Africa at the eastern end of the Mediterranean.

The conclusion that powered flight range decreases with mass depends on the cost of flight being greater than $m^{0.73}$. Masman and Klaassen (1987) estimated that rate of metabolism during flight in birds (in the simplest analysis) is proportional to $m^{0.76}$, which if correct suggests that flight range should be proportional to $m^{0.73}/m^{0.76} = m^{-0.03}$, or approximately independent of mass, except that large birds fly at a higher velocity, which may permit large birds to fly farther on a given energy store. Tucker (1970) and Schmidt-Nielsen (1972) concurred that the cost of flight is proportional to mass raised to a power less than 1.0 (see Section 9.6.2).

Much of the direct cost of migration in birds is derived from the catabolism of fat, although recent evidence indicates that some fuel is stored as protein (Jenni-Eiermann and Jenni 1991, Lindström and Peirsma 1993). If 75% of the fuel used for migration is fat and 28% is protein, 22% less energy is available than if all the tissue used was only fat, which implies a 22% reduction in maximal migratory distance (Klaassen 1996). Klaassen et al. (1990) and Klaassen and Biebach (1994) estimated that the fuel used for migration is between 62% and 81% fat, the remainder being protein. The reason for the storage of protein is unclear, unless water production from the metabolism of a mixture of fat and protein is greater than when fat alone is metabolized (Klaassen 1996).

Great variation in the pattern of fat deposition exists among species. Odum et al. (1961) distinguished among migratory species by whether they migrate long or short distances and whether they migrate before or after the peak in fat deposition. These distinctions are related to whether migration is intracontinental or intercontinental (King 1972), that is, whether it entails a long flight over water. Small passerines migrating nonstop to South America from North America fly 1000 to 3500 km at velocities of 30 to 60 km/h (Schnell 1965, Williams et al. 1977), which thus requires 20 to 80 h of continuous flight (Williams et al. 1978). The cost of migration can be reduced by timing migration to coincide with the passage of cold fronts (with attendant tail winds) or by flying at an altitude where head-wind velocity is minimal (Williams et al. 1977). Passerines may double mass through fat deposition preparatory to migration, which involves an increased cost of flight. This increase may be accompanied by hypertrophy of the pectoral muscles (Fry et al. 1972, Marsh and Storer 1981, Dawson et al. 1983, Piersma et al. 1996). Fat deposition often occurs over a short period of time, principally as a result of hyperphagia (King 1972, Dawson et al. 1983).

As evidence of the demands associated with migration, the energy density of fat tissue is greater in migrants and premigrants than in nonmigrants of the same species, and greater in long-distance migrants than in short-distance migrants (Johnston 1970), although the greatest adjustment made by migrants is to increase the size of the fat deposits.

Spring migrants have larger fat deposits than fall migrants (King 1963, King et al. 1963, Johnston 1964), some of which is used after arriving on the breeding grounds (Irving 1960, Johnston 1964). An energy reserve is often needed upon arrival at high-latitude breeding grounds because territorial defense must be initiated immediately to ensure a successful breeding season, but the availability of food is often uncertain due to inclement weather. Autumnal migration often proceeds leisurely without waiting for the accumulation of large fat deposits, especially in species that migrate over land. Autumnal migrants do not require a large energy expenditure immediately upon arriving on the wintering grounds, and fat deposits at the start of autumnal migration can be supplemented en route. This bimodal pattern of fat deposits in migratory species contrasts with the unimodal pattern often seen in nonmigratory species that accumulate fat for use during winter (compare Figures 11.17 and 11.29).

Fat deposition is initiated under the influence of photoperiod (King and Farner 1963, Gifford and Odum 1965), the amount deposited being correlated with environmental temperature (King and Farner 1963, 1966; Helms 1968). Variation in the amount of fat deposited is great within species (e.g., Odum 1958), reflecting the distances that individuals within a species migrate (Odum et al. 1961), which in turn depends on age and sex (Johnston 1966). For example, white-crowned sparrows (*Zonotrichia leucophrys gambelii*) from migratory populations have marked fat deposition in spring and fall (Figure 11.29), whereas individuals from sedentary populations (*Z. l. nuttalli*) show little fat deposition (King and Farner 1963). The African finch *Quelea quelea* deposits fat for migration commensurate with the distance that

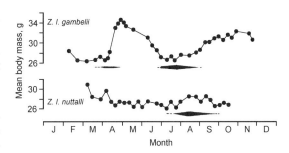

Figure 11.29 Mean body mass and the occurrence of molting (indicated by a bar) in two subspecies of the white-crowned sparrow (*Zonotrichia leucophrys*). The subspecies *Z. l. gambelii* migrates from its breeding grounds in Alaska and Canada to winter in the Great Basin south into Mexico; the subspecies *Z. l. nuttalli* breeds and winters along the Pacific coast in central California. Source: Modified from King and Farner (1963).

populations migrate at the beginning of the wet season, whereas the same populations have nearly identical fat deposits during the dry season (Ward and Jones 1977).

Some indirect estimates of the cost of migration have been made from the change in body mass that occurs during flight. Johnston (1968) compared the fat reserves of palm warblers (*Dendroica palmarum*) in fall migration just before departing across the Gulf of Mexico (when they had 2.66 g of lipids) with those found in migrating warblers collected off the coast of Cuba (with 0.43 g of lipids). Assuming that the birds flew directly from northern Florida to Yucatan (ca. 1040 km) at a velocity of 40 km/h, that lipids have an energy density of 39.3 kJ/g, and that no feeding occurred in flight, the estimated rate of metabolism during flight was 3.37 kJ/h, which is 5.3 times the basal rate of metabolism expected for a 8.7-g (= 8.3 lean mass + 0.4 lipid) passerine from an equation of Aschoff and Pohl (1970). Golden plovers (*Pluvialis dominica*) use 23 g of fat in the direct, non-stop flight from the Aleutian to Hawaiian Islands (Johnston and McFarlane 1967). Given that this distance is about 3800 km, that golden plovers weigh about 120 g, and that they fly at a velocity of 104 km/h, their rate of metabolism during flight would be about 24.7 kJ/h, which is 9.1 times the basal rate expected in nonpasserines from an equation of Aschoff and Pohl (1970). The estimates for linear flight obtained directly by measurement usually fall between about 5 (Bernstein et al. 1973) and 10 (Berger et al. 1970) times the basal rate (see Section 9.6.2).

One of the more remarkable examples of long-distance migration is found in baleen whales. Some feed on plankton in polar seas during summer and migrate to tropical or subtropical waters for winter, where females give birth to young. Little feeding is believed to occur during their residency in tropical waters, in part because plankton concentrations are low. Blubber therefore is stored in summer and consumed during migration and on their wintering "grounds." Brodie (1975) used this pattern to estimate the energy expenditure of fin whales (*Balaenoptera physalus*). An antarctic fin whale weighing 48 metric tons has a maximal change in oil content of 3585 kg, which is stored during the 4-month period of feeding in antarctic waters and is used during the remaining 245 d. If body oil has 36.6 kJ/g, then a fin whale would expend energy at a rate equal to 536,000 kJ/d, which is 108% of the basal rate expected from the all-mammal curve. A similar rate, relative to this standard, is estimated over a 183-d period in a north Pacific fin whale weighing 37 metric tons. These estimates for *Balaenoptera* assume no feeding and obviously include the cost of swimming during migration. Any feeding by these whales in tropical waters means that their energy budgets are underestimated. Clearly, a large mass permits a long period of starvation, mainly because the storage of energy increases with mass more than the rate at which the store is used (see Equation 5.3).

11.8 BALANCING ANNUAL ENERGY BUDGETS: SEASONAL TORPOR

Another solution is available to vertebrates that face seasonally inhospitable environments: they may retreat into shelters on or near the breeding grounds and enter into a state of torpor until the conditions in the external environment return to an acceptable state. When this state occurs in ectotherms, it is sufficiently similar to low-temperature lethargy that torpor has been difficult to define clearly. In endotherms, however, this state is dramatically different from the condition in which a high, regulated body temperature is maintained (= normothermia).

The physiological state found during ecological withdrawal is best referred to as *torpor*, which reflects a reduced responsiveness to external stimuli associated with a reduction in body temperature and rate of metabolism. When torpor occurs over an extended period of time, whether continuous or

not, it is often given an ecological name, depending on the season in which the torpor occurs: *hibernation* in winter and *aestivation* in summer. That is, the terms *hibernation* and *aestivation* refer to seasonal torpors: an animal may show daily torpor (a physiologically depressed state that may recur on a daily basis, see Section 11.5.3), but no animal shows daily hibernation (a term that confuses a daily with a seasonal time period). Hibernation and aestivation are responses to extreme conditions in the environment, normally cold temperatures and a shortage of food in winter and high temperatures and a shortage of food and water in summer, respectively. These states markedly reduce energy expenditure compared with the expenditures in animals that remain active.

11.8.1 Hibernation in ectotherms. The response of ectotherms to extended periods of continuous, or nearly continuous, low temperatures is difficult to define. Some species may enter a cold-induced torpor, which might well be the case in aquatic turtles (Ultsch and Jackson 1982a, 1982b) and which under special circumstances may involve supercooling (Lowe et al. 1971; see also Section 4.7.2) or freezing of extracellular fluids (Storey and Storey 1988a; see Section 4.7.3). Supercooling and freezing, however, are only used under limited conditions because exposure to freezing temperatures may be an important cause of winter mortality in terrestrial ectotherms (see Gregory 1982 and Section 4.7). White and Lasiewski (1971) suggested that snakes that aggregate in dens may burn fat stores to maintain a temperature differential with the environment. No evidence of such behavior exists (Gregory 1982).

Mobile ectothermic vertebrates respond to seasonally low ambient temperatures in several ways. For instance, tropical marine turtles on occasion encounter unexpectedly cold conditions at the temperate limits to their distribution (e.g., the loggerhead *Caretta* in Florida and the green turtle *Chelonia* in the Sea of Cortez). These turtles theoretically could (1) return to tropical waters if the decrease in water temperature were not too abrupt and if the prevailing water currents were appropriate; (2) move into the warm Gulf Stream, which might transport the turtles into even colder water at northern latitudes; or (3) overwinter in a torpid state in or on substrates in shallow water. The latter, at least, occurs in *Chelonia* (Felger et al. 1976) and *Caretta* (Carr et al. 1980).

Because only a few studies have been conducted on hibernation in ectothermic vertebrates, the distribution of this behavior is unknown. Most of the work has been on reptiles (see the excellent reviews by Gregory 1982 and Ultsch 1989). No fishes are known to hibernate (but see Ultsch 1989), probably because thermal acclimatization is an adequate response to cold water, as long as the water does not freeze. Many cold-temperate anurans and salamanders "hibernate": how much of disappearance during winter is cold stupor and how much is seasonal dormancy is unclear, assuming that these states are separable. Some desert amphibians, such as *Scaphiopus*, enter hibernation from an aestivating state that started during a hot, dry summer (Mayhew 1962, 1965b; McClanahan 1967; Ruibal et al. 1969; Seymour 1973a). Among reptiles, hibernation is widespread in temperate and subtropical lizards, snakes, and turtles. It is also found in both species of *Alligator* (Neill 1971, Guggisberg 1972) and in *Crocodylus niloticus* (Guggisberg 1972). *Sphenodon* does not hibernate, contrary to the report by Dawbin (1962), although it is inactive during cold periods in winter.

Hibernation locations (*hibernacula*) may be in water (some anurans, turtles, snakes, and possibly alligators) or on land (under objects, and in loose soil, rock crevasses, or burrows). In cool to cold-temperate environments hibernation usually occurs at temperatures from 2 to 10°C. For example, the temperature of a hibernaculum of *Coluber constrictor* in Utah varied between 3 and 7°C (Brown et al. 1974), and *Crotalus* and *Thamnophis* hibernated in British Columbia and Alberta at body temperatures between 2 and 7°C (Macartney et al. 1989). In warmer climates hibernation may occur at temperatures up to 15°C (Gregory 1982). In cold-temperate localities, where hibernation is required and hibernacula are scarce, hibernation often occurs in congregations (especially among snakes). Under these conditions, some snakes may move remarkable distances to hibernacula (up to 1.8 km for *Coluber* [Brown and Parker 1976] and to 17.7 km for *Thamnophis* [Gregory 1977]).

The ability of any tetrapod to hibernate underwater requires that they obtain sufficient amounts of oxygen from, and dump CO_2 to, the surrounding water. Such exchange is facilitated in ectotherms by low temperatures, which implies low body temperatures and low rates of metabolism but high oxygen tensions in water (see Section 8.9). Given that some turtles have rather high rates of integumental gas exchange (see Section 8.9), some *Chelydra*, *Stenothaerus*, *Chrysemys*, and *Graptemys* hibernate under water (see Ultsch 1989 for a summary). Some adult and larval anurans and larval *Ambystoma* also overwinter in water (see Pinder et al. 1990), but whether this represents "hibernation" in the sense of "torpor" is unclear. More surprising is the observation that some terrestrial and aquatic snakes occasionally (?) hibernate underwater (see Ultsch 1989). For example, Costanzo (1989) showed that *Th. sirtalis* hibernated up to 5.5 months in water covered by ice in an abandoned well in central Wisconsin. Under laboratory simulation of field conditions (5°C, complete darkness) these garter snakes showed a 55% reduction in energy expenditure compared to the expenditure in air at the same temperature.

The physiological state of hibernating ectotherms is characterized by a low body temperature and a low rate of metabolism. The question remains as to whether rate of metabolism is depressed independently of the fall in body temperature, which the measurement on *Th. sirtalis* suggests. Classic ectothermic acclimatization to low temperature involves an *increase* in rate of metabolism at intermediate to low temperatures (see Section 4.5; Figure 4.9). Such an adjustment would accelerate the use of energy stores during hibernation, thereby *shortening* the period over which a hibernator could survive. Thermal acclimatization would be most appropriate only in winter-active species (e.g., *Uta* [Roberts 1968]). This response permits extended periods of growth and reproduction (Lillywhite et al. 1973, Tinkle and Hadley 1973, Ruby 1977) but requires winter feeding. Hibernators, in contrast, would be expected to show either no acclimatization or a *reduction* in rate of metabolism beyond that expected from a decrease in body temperature (the so-called reverse acclimatization).

The response actually found in hibernating ectotherms varies with species. Most show a reduction in rate of metabolism beyond that due to a low body temperature. *Cnemidophorus sexlineatus*, which hibernates, had a high rate of metabolism in summer and a low rate in winter at all temperatures, whereas *Anolis carolinensis*, which does not hibernate, had a high rate in winter (Figure 11.30). A low rate of metabolism in *C. sexlineatus* in winter is similar to that found in the European lizard *Lacerta vivipera*, which also hibernates. The temperatures at which a depression of metabolism occurs often (always?) correspond to the temperatures that are encountered in hibernation (Gregory 1982). Thus, in reptiles (*Dipsosaurus* [Moberly 1963], *Phrynosoma* [Mayhew 1965a]) that are torpid at warm temperatures, a reduction in rate of metabolism occurs only at high temperatures (Figure 11.31). No evidence exists in amphibians of a depression in rate of metabolism beyond that produced by a fall in ambient temperature (Pinder et al. 1990).

Before they enter seasonal torpor, reptiles reduce or eliminate food intake, even in the presence of food. Anorexia is widespread when feeding conflicts with other behaviors that are momentarily of greater importance, such as territorial defense, incubation, and molting (Mrosovsky and Sherry 1980). Anorexia reduces the production of wastes that would have to be eliminated in hibernation by the kidney and gut, and anorexia may be associated with a programmed reduction in body mass. Among reptiles, anorexia has been most clearly shown in the lizard *Phrynosoma* (Mayhew 1965a), but it was also noted in *Alligator*, *Pseudemys*, *Eumeces*, and temperate-zone snakes (Gregory 1982). Similar observations were made on the anurans *Scaphiopus couchii* and *S. hammondii* (Seymour 1973a). In contrast, *Uma* and *Dipsosaurus* continue to feed in the laboratory during the hibernal period. This difference reflects the obligatory hibernation of *Phrynosoma* and *Scaphiopus*, the facultative hibernation of *Dipsosaurus*, and the year-around activity of *Uma*. Many snakes show activity in fall at the entrance of communal hibernacula, a behavior that may permit them to empty their gut preliminary to entering hibernation.

Whether fat is accumulated for use during hibernation in ectotherms is the subject of disagreement. Aleksiuk (1976) stated that little or no fat storage occurs in *Thamnophis*, and Brown et al. (1974) maintained that fat storage in *Coluber* is used for reproduction, not hibernation. Yet, some energy store must be used during hibernation because no

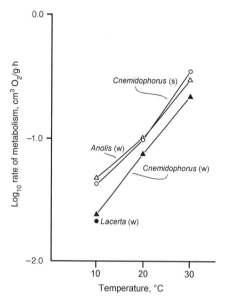

Figure 11.30 Log_{10} rate of metabolism in the lizards *Anolis carolinensis* in winter (w) and *Cnemidophorus sexlineatus* in winter (w) and summer (s) as a function of environmental temperature. *Anolis carolinensis* does not hibernate, whereas *C. sexlineatus* does. A datum from the hibernating European lizard *Lacerta vivipera* is included. Source: Modified from Ragland et al. (1981).

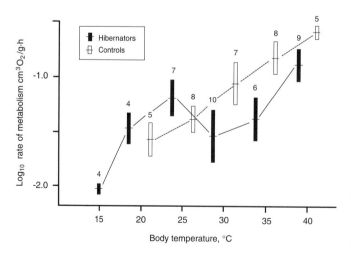

Figure 11.31 Log_{10} rate of metabolism in hibernating and nonhibernating desert iguanas (*Dipsosaurus dorsalis*) as a function of body temperature. Source: Modified from Moberly (1963).

feeding occurs during this period. The storage compounds are mainly lipids, but because the rates of metabolism during hibernation are low (due to low temperatures and possibly to a hibernal depression of metabolism), the amounts used may be difficult to detect (Gregory 1982). These doubts can be resolved only by measuring the amount of fat and other energy-rich compounds throughout the hibernal cycle. Indeed, body lipids are used by *Cnemidophorus tigris* to fuel hibernation, and the lipids contained in the fat bodies are used by females in winter to deposit yolk into their eggs (Gaffney and Fitzpatrick 1973). Approximately half of the energy used during winter dormancy in *Scaphiopus* is derived from body lipids (Seymour 1973a), and they are an important energy source in other anurans as well (Bush 1963, Brenner 1969). In *Scaphiopus* the remaining energy used during hibernation is derived from the fat bodies or body protein (McClanahan 1967, Seymour 1973a). Under severe drought conditions, *Scaphiopus* may remain in torpor a second year, causing males to metabolize their fat bodies and females to degrade their eggs to extend dormancy (Seymour 1973a).

The principal significance of hibernation for ectothermic vertebrates is that it balances a winter energy budget when energy income is very restricted or zero, although hibernation in polar and cold-temperate environments also permits an ectotherm to avoid low environmental temperatures that would be lethal. Hibernation reduces the cost of maintenance during periods when feeding, growth, and reproduction are not possible. The reduction in energy expenditure represented by hibernation is shown in a calculation by Patterson and Davies (1978a): the lizard *Lacerta* spends only 5% of its annual energy budget during the hibernal period, which occupies 44% of the year.

The annual pattern of energy expenditure has been described for two hibernating amphibians. Measurements of oxygen consumption in *S. couchii* when combined with measurements of soil temperature and a factor to convert oxygen consumption into energy expenditure gave estimates (Figure 11.32) of seasonal rates of metabolism (Seymour 1973a). The summed oxygen consumption for the 10 months of dormancy is 1.6 L, which is approximately equal to 800 mg of lipids, assuming that 2.02 L of oxygen is required to oxidize 1 g of fat. This estimate is close to the observed value: lipid content in a 25.5-g toad decreases by 858 mg

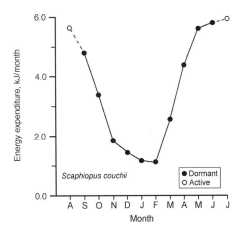

Figure 11.32 Estimated field energy expenditure in the spadefoot toad *Scaphiopus couchii* as a function of the time of year. Source: Modified from Seymour (1973a).

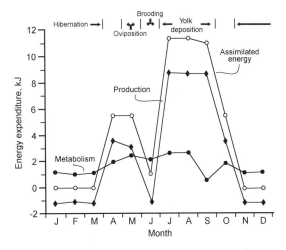

Figure 11.33 Estimated field energy equivalencies of assimilation, production, and metabolism in the plethodontid salamander *Desmognathus ochrophaeus* in relation to the annual behavioral cycle. Source: Derived from Fitzpatrick (1973).

in 10 months of hibernation. Fitzpatrick (1973) divided the annual energy budget of the salamander *Desmognathus ochrophaeus* into various annual activities, including adding yolk to its eggs, oviposition, brooding, and hibernation (Figure 11.33). The very low rate of metabolism in this salamander permits nearly half (46%) of the total annual energy assimilated to be channeled to reproduction.

11.8.2 Hibernation in endotherms. The torpor found in ectotherms and that found in endotherms are physiologically different. These torpors are an evolutionarily convergent means of reducing energy expenditure during periods in which the environment does not permit normal activity owing to either harsh physical conditions or a shortage of food. They differ in that endothermic torpor reflects a state in which body temperature is regulated, albeit at a low level, whereas in ectothermic torpor body temperature is not regulated, except to the extent that ectotherms select hibernacular temperatures. Mammals and birds can normally arouse from torpor, which no true ectotherm can.

Among mammals hibernation is known to occur in an echidna, some marsupials, insectivores, bats, rodents, and carnivores. The North American poorwill (*Phalaenoptilus nuttallii*) is the only bird known to hibernate (see Section 11.8.4), although old reports summarized by McAtee (1947) maintained that some swifts (*Chaetura*) and possibly some swallows (*Progne*) may also hibernate on occasion.

The torpor found in endotherms is characterized by a decrease in body temperature (T_b) with an associated reduction in rate of energy expenditure, the establishment of an equilibrial temperature differential (ΔT), temperature regulation at a low T_b, discontinuity of torpor, and arousal from torpor.

1. Small hibernating endotherms permit T_b to fall to "near" ambient levels. Tucker (1966a) concluded that such a decline in *Perognathus californicus* was passive because he could mathematically simulate the fall by assuming passive heat exchange with the environment. As T_b falls, rate of metabolism decreases following a Q_{10} relation, although the Q_{10} under these conditions is high (often >3.0, and at times up to 6.0; Morrison 1960). The decrease in T_b leads to a decrease in rate of metabolism, which further decreases T_b, so that entrance into torpor can be viewed as a system with positive feedback leading to a high Q_{10}. High Q_{10} values led Morrison to conclude that entrance into torpor was active. Lyman (1963) came to the same conclusion; he noted in ground-squirrels that heart rate and rate of metabolism decrease *before* a decrease in T_b, as did Heldmaier and Ruf (1992). Malan (1988) suggested that some of the decrease in metabolism with entrance into torpor may be associated with respiratory acidosis (i.e., the retention of CO_2). A passive entrance into torpor may be most likely at small masses.

Recent work has not resolved the question of

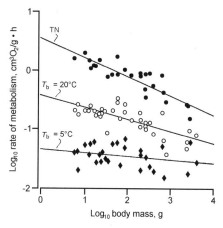

Figure 11.34 Log_{10} rate of metabolism of hibernating mammals when normothermic in thermoneutrality (TN) and when body temperature was 20 and 5°C, as a function of log_{10} body mass. Source: Modified from Geiser (1988).

whether the entrance into torpor is active or passive. Geiser (1988) maintained that the high Q_{10} values for metabolism in small hibernators was evidence for the inhibition of metabolism during their entrance into torpor. However, he did not find high Q_{10} values in large hibernators or in endotherms that entered into daily torpor, the latter being characterized by high-temperature torpor (T_a = 20–25°C). Geiser (1988) rejected many of the reports of high Q_{10} values in large hibernators because before entrance into torpor, they were exposed to low environmental temperatures that required them to have an augmented rate of metabolism. One of the consequences of a high Q_{10} in small, but not in large, hibernators is that the mass-specific rate of metabolism in deep torpor, unlike the rate in normothermia or in shallow or daily torpor, is nearly independent of body mass (Figure 11.34). This means that the total rate of metabolism in these species depends directly on body mass (or nearly so; Kayser 1961).

Snyder and Nestler (1990) provided an entirely different analysis of the entrance into torpor. They argued that Q_{10} values cannot be calculated from the correlation of rate of metabolism with ambient temperature in endotherms entering torpor because this entrance is "regulated" and is not a free fall down a Q_{10} function. Instead, they postulated that body temperature was permitted to decrease through a falling sequence of set points, each of which was characterized by conformation to the Scholander-Irving model represented by Equation 5.1. As a species enters torpor, body temperature, rate of metabolism, and thermal conductance decrease, resulting in a regulated ΔT, until some minimally acceptable body temperature is attained. If ambient temperature falls, rate of metabolism is increased to maintain this body temperature and the increasing ΔT.

With entrance into torpor, Snyder and Nestler (1990) argued that the greatest difference between small and large hibernators is that small species show a greater decrease in thermal conductance. Its reduction with the entrance into torpor means that a small endotherm can maintain a relatively high body temperature in spite of an appreciable decrease in rate of metabolism, a condition that would erroneously be interpreted as a high Q_{10}. These authors (p. 673) maintained that "... Q_{10} is not a valid measure of metabolic state in endotherms and cannot be used. Metabolic rate is the independent variable (Henshaw 1968) and changes in T_b follow from changes in metabolic rate (Wang 1978a), not vice versa as is implied by the proponents of the Q_{10} argument." McNab (1970) expressed a similar view.

Heldmaier and Ruf (1992) suggested that the decrease in rate of metabolism with entrance into torpor resulted from a controlled reduction in ΔT. Rate of metabolism in Djungarian hamsters (*Phodopus sungarus*) and in other mammals that enter torpor is proportional to ΔT at various stages along a torpor bout (compare Figures 5.1 and 11.35). They argued, similar to Lyman (1963), that entrance into torpor was initiated by a "regulated" reduction in rate of metabolism *before* the decrease in ΔT, and maintained that the decrease in thermal

conductance postulated by Snyder and Nessler (1990) was not found.

Whatever the cause of the decrease in rate of metabolism in torpid endotherms, it does not reflect a fall in T_b: these endotherms remain in control of ΔT. French (1985), Snyder and Nessler (1990), and Heldmaier and Ruf (1992) agreed that mass-specific rate of metabolism in torpor is almost independent of body mass, in contrast to that found in normothermia. Given the regulated nature of torpor in endothermic hibernators, the Scholander-Irving relation (Equation 5.1) indicates that independence of mass-specific rates of metabolism from mass reflects either a disproportionate decrease of thermal conductance at small masses, as Snyder and Nessler advocated, or a reduction of ΔT (and therefore T_b), as Heldmaier and Ruf maintained. Both reductions are likely in small hibernators.

Further evidence of the temperature sensitivity of torpor is that the hibernal period in the forest dormouse (*Dryomys nitedula*) increases with latitude (Nevo and Amir 1964), as it does with the European hedgehog (*Erinaceus europeaus*) (Fowler and Racey 1990).

2. An equilibrial temperature differential with the ambient temperature (T_a) is attained in torpor. The size of this differential depends on several factors, including T_a and body mass (McNab 1974b). This conclusion is derived from the observation that when an object is in thermal equilibrium, heat production equals heat loss. Large mammals obviously have larger differentials than small mammals, in part because they have a greater mass, but possibly also because of differences in T_a and Q_{10} among species. Whether bears are "true" hibernators has been questioned (Hock 1960, Morrison 1960, Lyman 1963) because they lack some of the physiological adjustments to low body temperature made by small hibernators, such as the conduction of nerve impulses at low temperatures. The absence of these adjustments, however, is more likely indicative of the large equilibrial differential found in species with a large mass. Large species have a smaller reduction in rate of metabolism during torpor than small species because large species are warmer, unless they select colder microclimates in which to hibernate. The selection of lower environmental temperatures during hibernation by larger species has been shown in hibernating bats among and within caves (McNab 197a). This behavior also occurs in clustering bats, which

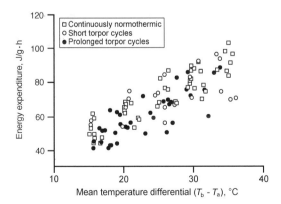

Figure 11.35 Daily energy expenditure in the Djungarian hamster (*Phodopus sungorus*) as a function of the temperature differential maintained between the body and the environment, and the absence or length of torpid periods. Source: Modified from Ruf and Heldmaier (1992).

in terms of heat storage are similar to large bats. Nagel and Nagel (1991) gave a rather different view of hibernaculum selection by cave bats. They implied that selection simply avoided warm and cold temperatures, although cave bats move to altitudes between 700 and 800 m in Germany to ensure cave temperatures between 3 and 7°C.

The equilibrial body temperature in hibernation may be influenced by factors other than ambient temperature and body mass (Geiser and Broome 1993). Among them is the set point for temperature regulation, that is, the lowest body temperature that is regulated in torpor. If ambient temperature falls below the set point, then rate of metabolism increases with a fall in ambient temperature (Figure 11.14A). Thermoregulation at low body temperatures in torpor has been shown in marsupials (Geiser 1986, Gieser and Broome 1993), ground-squirrels (Lyman 1963), a shrew (Lindstedt 1980), and bats (Hock 1951, Bonaccorso and McNab 1997), among others.

One consequence of temperature regulation at low ambient temperatures by hibernators is that the reduction of energy expenditure attained through entrance into torpor is diminished, although the cost of thermoregulation at low ambient temperatures, when maintaining a low body temperature, is much less than when normothermic because ΔT has been reduced (Figure 11.13). As noted, the set point for body temperature may be correlated with environmental conditions (Hiebert 1990): it is higher in endotherms that are torpid at warmer temperatures (MacMillen 1965), and it is usually higher in species that enter daily torpor than in those that enter hibernation (e.g., Geiser 1994). The set point also may decrease with an increase in polysaturated lipids in the diet (Geiser and Kenagy 1987).

3. All endotherms that spontaneously enter torpor arouse periodically from this state as long as conditions in the hibernaculum are within acceptable limits, namely, that the hibernaculum temperatures are not too low. On the other hand, endothermic hibernators often arouse from torpor if ambient temperature approaches 0°C (see Hammel et al. 1968).

Arousal by definition is an active process involving a great increase in rate of heat production (Lyman 1963), a large fraction of which is derived in the early stages of arousal from heat generated by "brown" fat (Hayward and Ball 1966). Although the distribution of brown adipose tissue is widespread among eutherians that enter daily and seasonal torpor (Néchad 1986), controversy about its presence in other vertebrates has existed. Johnston (1971) and Saarela et al. (1989) argued that it was not found in birds, whereas Oliphant (1983) and Olson et al. (1988) described brown fat in some birds. Biochemical evidence from the poorwill, a bird known to enter daily and seasonal torpor, demonstrated that this species does not have brown adipose tissue (Brigham and Trayhurn 1994), which implies that this tissue is unlikely to be found in any bird. Furthermore, an extensive search failed to demonstrate the occurrence of brown fat in any marsupial or monotreme (Hayward and Lisson 1992). Claims that brown fat is widespread (Louden et al. 1985, Rothwell and Stock 1985) are highly exaggerated. Hayward and Lisson (1992), in fact, drew the exact opposite conclusion: brown fat is limited in distribution to eutherians and represents a critical step in their evolution. They suggested that the presence of brown fat may explain why eutherians are the only mammals to dominate cold-temperate and polar environments.

The rate at which a species arouses from torpor depends principally on its mass (Heinrich and Bartholomew 1971, May 1976b, Geiser and Baudinette 1990): large species heat more slowly than small species. Stone and Purvis (1992) agreed with this conclusion, even though they made the argument that large animals "should" heat more rapidly: to the contrary, a reorganization of Equation 2.13 indicates that

$$\frac{dT_b}{dt} = \frac{dQ/dt}{c_p \cdot m} \cong \frac{k \cdot m^{0.75}}{c_p \cdot m^{1.00}} \approx m^{-0.25}.$$

These authors also concluded that warm-up rate varied (negatively) with basal rate and (positively) with the arousal ΔT. Geiser and Baudinette (1990), however, concluded that the warm-up rate correlates positively with basal rate. The period required for warm-up is not constant in a species: it varies with ambient temperature (Geiser 1986, Hiebert 1990), at least in part because of its impact on ΔT (also see Prothero and Jürgens 1986). More difficult to explain, especially if the warm-up rate increases with basal rate, are the observations of Geiser and Baudinette (1990) and Stone and Purvus (1992) that marsupials and eutherians have similar warm-up rates.

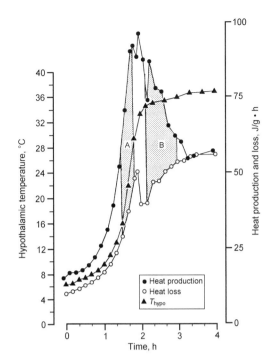

Figure 11.36 Hypothalamic temperature, heat production, and heat loss in the ground-squrrel *Spermophilus lateralis* during arousal from torpor. For a discussion of areas A and B, see text. Source: Modified from Hammel et al. (1968).

The energetics of arousal has been thoroughly examined in the ground-squirrel *Spermophilus lateralis* (Hammel et al. 1968). In this squirrel arousal may be initiated by a decrease in ambient temperature as small as from 2.7 to 2.6°C! At these temperatures the hypothalamic temperature is about 2.86°C. An arousal can be turned off by raising the ambient temperature from 2.6 to 2.7 or 2.8°C, as long as the arousal has not proceeded too far. If, however, the arousal is permitted to proceed, rate of heat production, rate of heat loss, and T_b all markedly increase (Figure 11.36). In time period A (see Figure 11.36), the enthalpy of the squirrel's body (ΔH) increased 1.99 kJ, which given a mass of 43.1 g and the increase in hypothalamic temperature of 13.3°C, means that maximally 29.7% of the mass could have been heated. The initial phase of arousal stopped abruptly after 115 min, when heat production and loss were greatly reduced. Heat loss decreased because innervation of the superficially located pads of brown fat was turned off. In time period B (Figure 11.36), $\Delta H = 3.16$ kJ, only 113 J of which went to increase the temperature of the warmed fraction of the body, the remainder warming the cool tissues,

which if equally distributed, would raise T_b by 8.6°C. Most of this heat production was probably produced by shivering now that the muscles had reached a functional temperature.

The cost of arousal is high, varying with both the increase in body temperature and the arousal frequency. If the ground-squirrel *S. tridecemlineatus* is permitted to arouse spontaneously when hibernating at 11°C, it loses 0.72 g/d, which if fat is equivalent to an expenditure of 27.4 kJ/d (Mrosovsky and Fisher 1970). If, however, the squirrel is repeatedly forced to arise out of torpor, mass loss increases to 1.14 g/d, reflecting a rate of metabolism that is about 60% greater than when arousal is spontaneous.

Given the high cost of arousal, why is hibernation not continuous? No satisfactory answer exists, except for the general suggestion that periodic arousals are required to reestablish a chemical balance (Fisher 1964, Mrosovsky 1971, Galster and Morrison 1975). An often-cited answer is that animals have to eliminate the waste products of metabolism, which may explain why arousal frequency increases with ambient (and therefore body) temperature. The view that body temperature is a better predictor of torpor bout length than rate of metabolism (Geiser and Kenagy 1988) suggests that the buildup of metabolites may only be a partial explanation for the inverse correlation of bout length and ambient temperature (Geiser et al. 1990). Among hibernating bats, the need to restore water lost by evaporation is accomplished by drinking water condensed on the walls of hibernacula, which may be one of the factors most responsible for the arousal of bats (Speakman and Racey 1989, Thomas and Cloutier 1992, Thomas 1995, Thomas and Geiser 1997).

4. Torpor bout length is variable. It depends on ambient temperature (Twente and Twente 1965a, 1965b; Pengelley and Kelly 1966; Brown and Bartholomew 1969; Wang 1973, 1978a; French 1977b, 1982a, 1982b; Twente et al. 1977; Geiser and Kenagy 1988), time of the year (Wang 1973, 1978a; French 1977a, 1982a, 1982b, 1985; Pajunen 1983; Geiser et al. 1990), body temperature (Geiser and Kenagy 1988, Geiser et al. 1990), and body size (French 1985). Bout length may also be influenced by diet: consuming polyunsaturated lipids leads to longer torpor bouts than does consuming saturated lipids (Geiser and Kenagy 1987).

The influence of temperature is complicated. For example, ground-squirrels have longer bouts at

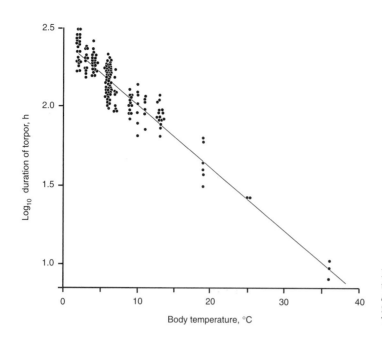

Figure 11.37 Log₁₀ torpor bout length in the ground-squirrel *Spermophilus lateralis* as a function of body temperature. Source: Modified from Twente and Twente (1965b).

colder than warmer temperatures (Figure 11.37). As in daily torpor, exposure to sufficiently low temperatures will lead to a shorter torpor bout because rate of metabolism is elevated (Geiser and Kenagy 1988, Geiser and Broome 1993). As noted, torpor bout may be more closely related to body temperature than to rate of metabolism (Geiser and Kenagy 1988, Geiser et al. 1990), with the exception at low temperatures, which reinforces the centrality of metabolism. Thomas and Geiser (1997) made another suggestion: periodic arousals are more closely correlated with the rate of evaporative water loss than with either body temperature or rate of metabolism. This difference in predictive power was most noticeable at −2°C in the ground-squirrel *S. saturatus*. After arousal and before reentrance into torpor, the ground-squirrel may reestablish its water balance by ingesting condensed water or ice crystals from the wall of the hibernaculum.

All species do not have the same bout length at a given temperature (Pengelley and Kelly 1966, French 1977b). For example, under a regime of 12 h of light and 12 h of darkness, torpor duration in a pocket-mouse (*Perognathus longimembris*) at a fixed ambient temperature of 8°C was polymodal, with torpor bout lengths being less than 24 h, between 24 and 48 h, or between 48 and 72 h (Figure 11.38A). Consequently, torpor lengths could be described as being 1-d, 2-d, 3-d, and so on, bouts. Successive arousals, therefore, tended to

occur at 24 h or multiples of 24 h (Figure 11.38B). The frequencies of torpor bout length and of the intervals between bouts decrease with duration. At an ambient temperature of 8°C, the longest torpor duration was 112 h; at 18°C, it was 45 h. This pattern conntinued in continuous darkness. Then the free-running period of torpor depended on whether the torpor was a 1-d torpor, when it averaged 24.6 h, or a multiday torpor, when the period was 22.4 h. Both of these periodicities ultimately resulted in a loss of synchrony with the external environment. Some individuals kept in continuous darkness for 5 to 6 months shifted from multiday torpor to 1-d torpor (French 1977b).

At a fixed temperature, bout length is shortest in fall and spring and longest in winter (French 1985, Geiser et al. 1990, Geiser and Broome 1993) (Figure 11.39). The increased proportion of time in spring spent in normothermia may increase the ability of a hibernator to assess the environment and to time its emergence with acceptable conditions in the environment, but it pays a price for this behavior in terms of the depletion of fat or food stores.

Views on the relationship expected between torpor bout length and body mass have differed. Large species might be expected to have shorter torpor bouts at a given T_a than smaller species because large species have a larger ΔT and therefore a higher T_b and rate of metabolism. Twente et al. (1977) argued that larger species should have

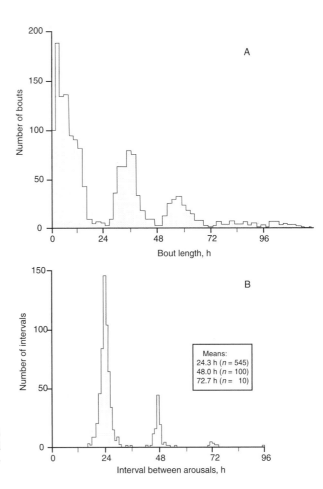

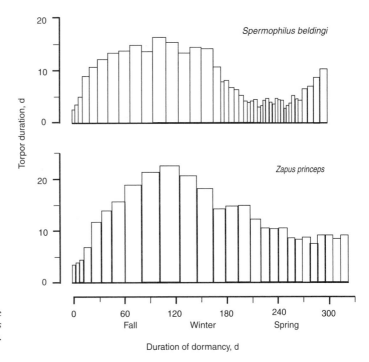

Figure 11.38 Periodicity in the frequency of (A) torpor bout length and (B) torpor bout interval in the pocket-mouse *Perognathus longimembris*. Source: Modified from French (1977b).

Figure 11.39 Torpor bout duration in the rodents *Spermophilus beldingi* and *Zapus princeps* as a function of the time of year. Source: Modified from French (1985).

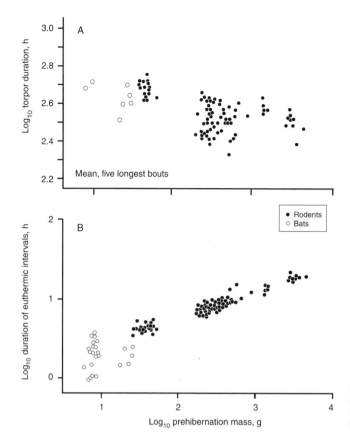

Figure 11.40 (A) $\log_{10}$ torpor duration and (B) $\log_{10}$ euthermic duration in hibernating mammals as a function of $\log_{10}$ body mass. Source: Modified from French (1985).

longer periods because they have lower mass-specific rates of metabolism, which ignores the significance of total rates of metabolism (see Box 3.1). A detailed examination (French 1985) showed that bout length decreases with mass (Figure 11.40A) and that normothermic interval between torpor bouts increases with mass (Figure 11.40B). Consequently, the ratio of time in torpor to time in normothermia varies from about 223:1 at 10g to 26:1 at 1kg. Large bats at a given environmental temperature have shorter bouts than small bats (McNab 1974a). Clearly, energy savings from torpor are greater in small than in large species, and this calculation neglects the time required to enter torpor and the cost of arousal, both of which are greater in large species. Indeed, French et al. (1967, p. 538) found that among coexisting pocket-mice, "... the smaller species, *P. longimembris*, is more responsive to cold weather, and disappears earlier and is inactive longer than [the larger] *P. formosus*." Morrison (1960), however, maintained that large hibernators should be able to hibernate longer than small hibernators because fat deposits, the usual source for the energy expended during hibernation, increase more with mass than does rate of metabolism. The larger fat stores of large hibernators may permit large species to have a hibernal period that is about as long as that in small species in spite of spending more time in normothermia.

5. Endothermic torpor in its various forms can be seen as an extension of sleep (Lyman 1963; Florant et al. 1978; Heller et al. 1978; Walker et al. 1979, 1980). Hibernators enter torpor from the sleeping state, specifically from the slow-wave sleeping pattern, which shows a continuous transition into torpor. Both states involve a decrease in body temperature (usually only 2°C or so in sleep) and a decrease in rate of metabolism. Furthermore, an annual cycle of sleep occurs in mammals, even when they are prevented by high ambient temperatures from entering torpor: the maximal period of sleep occurs in winter (Walker et al. 1980). Walker et al. (1980) suggested that slow-wave sleep was the original energy-conserving resting state in endotherms from which were derived various short- and long-term torpors, such as daily torpor, hibernation, and aestivation, which (compared to

slow-wave sleep) are characterized by lower body temperatures and lower rates of metabolism. Slow-wave sleep does not occur in ectotherms (Walker et al. 1979), an observation that emphasizes the fundamental physiological difference that exists between the torpor and hibernation of endotherms and ectotherms.

11.8.3 The impact of endothermic hibernation. With little doubt, the principal advantage associated with endothermic hibernation is energy conservation. For example, the annual energy budget in the ground-squirrel *Spermophilus richardsoni* is reduced to about 34% of what would be expected of a squirrel maintaining normothermia throughout the year (Wang 1978a) and to 12% of the normothermic value during the hibernal period. One apparent result of this reduction in energy expenditure is an increase in life span (see Section 13.3.4), which has been noted in insectivorous bats (Bourlière 1958, McNab 1982, but see Herreid 1964) and some rodents (French et al. 1967, Hayden and Lindberg 1976, Lyman et al. 1981), and suggested as a possibility in marsupials (Geiser 1986).

An analysis of the energy and time expenditures in the golden-mantled ground-squirrel (*S. saturatus*) indicated that it expended energy at about 0.1 kJ/g·d when hibernating (Figure 11.41); hibernation lasts 7.5 months in Washington State, 63% of the year, during which they expended approximately 17% of their respiratory metabolism

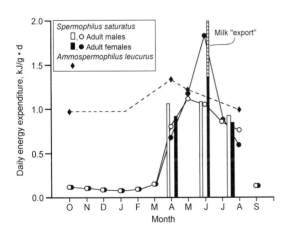

Figure 11.41 Daily field expenditure in the nonhibernating antelope ground-squirrel (*Ammospermophilus leucurus*) and the hibernating golden-mantled ground-squirrel (*Spermophilus saturatus*) as a function of the time of year. Sources: Derived from Karasov (1981) and Kenagy et al. (1989).

(Kenagy et al. 1989). Time-energy budget estimates of energy expenditure were slightly less than doubly labeled water measurements, except in the case of lactating females (Figure 11.41), which were low because this technique measures respiratory exchange and does not account for the transport of milk to the young (except as respiration measures the cost of milk production). When the exported milk is added to the measured respiratory exchange of lactating females, this sum too is slightly greater than the amount estimated by a time-energy budget (Figure 11.41). The great cost of lactation means that female ground-squirrels have appreciably higher rates of energy expenditure than males during the breeding season (Figure 11.41), in spite of the larger mass of males. The small size of females may reflect the great increase in energy expenditure associated with reproduction, that is, mainly lactation (see Section 11.6.2 and Powell and Leonard 1983).

The seasonal pattern of energy expenditure in *S. saturatus*, influenced as as it is by hibernation, is radically different from the pattern seen in the antelope ground-squirrel (*Ammospermophilus leucurus*), a species that does not hibernate. This ground-squirrel has a high field expenditure, as measured by doubly labeled water, throughout the year (Figure 11.41). During a 6-month winter, the antelope ground-squirrel expends over 40% of its annual expenditure.

Although energy expenditure is greatly reduced during hibernation, hibernators still require a source of energy at a time when food acquisition in the environment is minimal or impossible. Some hibernators store food in burrows (e.g., see Lyman 1954, Johansen and Krog 1959, Brown and Bartholomew 1969, Livoreil and Baudoin 1996), but most species use body fat as the principal source of energy. The amount of fat stored undoubtedly varies with several factors, one of which is body mass. For example, fat stores (*F*) generally increase with lean body mass (*m*ᵢ) (Pitts 1962):

$$F(g) = 1.50\,m_i^{0.195},$$

a relationship derived from 177 species of mammals that ranged in mass from 3 g to 130 tons. (Note: Calder [1984] used these data to derive a relationship showing that total body mass [= lean mass + fat] was proportional to $m^{1.20}$.) Factors other than mass that would be expected to influence fat deposition include larger deposits in hibernators (when fat deposits might equal lean mass [Morrison 1960]) than nonhibernators and in populations hibernating at higher latitudes.

How hibernators store body fat, often over a short period of time in preparation for winter, is unclear and may involve an adjustment of the pathways of metabolism. One suggestion (Krzanowski 1961) is that temperate insectivorous bats deposit large quantities of fat in the fall, when ambient temperatures are decreasing, by the selection of cool roosts, which reduces the bat's cost of body maintenance as a result of entering daily torpor. This reduction may permit more energy intake to be channeled into fat synthesis and storage. Speakman and Rowland (1999) found that the behavior of long-eared bats (*Plecotus auritus*) was compatible with this suggestion. This explanation, however, would not apply to species that do not readily enter torpor.

In spite of the contribution of hibernation to balancing a seasonal energy budget, many liabilities are associated with this state. An animal in torpor is helpless, and little or no growth and cell multiplication occur; furthermore, torpor usually permits only one litter per year and requires a short gestation (Hock 1960, Kalabuchov 1960, Lyman 1963). Some of these limitations can be minimized: echidnas (*Tachyglossus aculeatus*) sometimes enter torpor during pregnancy (Geiser and Seymour 1989), as do some marsupials (Morton 1978, Geiser and Masters 1994) and bats (Racey 1973, Audet and Fenton 1988, Kurta et al. 1988, Kurta 1990) during both pregnancy and lactation; bears give birth during the hibernal period; and young ground-squirrels delay entrance into hibernation to extend the period of growth (Kalabuchov 1960, Morrison and Galster 1975, Michener 1983). The usual separation of reproduction and torpor is relevant to the conclusion of Cade (1964, p. 101) that torpor and hibernation are "...not now widely adaptive for rodents..." because so few species use these states. Lyman (1963) agreed with this conclusion, noting that only some 50 per 1000 rodents enter torpor (surely an underestimate because over 50 species of ground-squirrels hibernate), hibernation being an advantage only under specific conditions. He (p. 138) concluded that most rodents find that "...the millstone of a high metabolic rate..." is better than the handicaps associated with hibernation. Maybe the real difference between hibernators and species that do not

enter torpor is found between species "...that spread the reproductive effort and the associated metabolic costs over a long time...[and those] with a short reproductive period and rapid energetically demanding development...[that require being]...locked into rigid homeothermy..." (Geiser and Masters 1994, p. 38). This difference may reflect the dependability of the food supplies used.

Another interpretation Kayser (1961) advocated and Cade (1964) (almost) implied is that the capacity to enter into a torpid state by endotherms is a plesiomorphic character. The diversity of temperature tolerances in torpor and the diversity of stimuli required for entrance into torpor (fat deposition, food accumulation, etc.), as well as the diversity of mammals and birds capable of some kind of daily or seasonal torpor, imply the repeated evolution of these states in endotherms. Augee and Gooden (1992), however, argued that mammalian hibernation is too complex (and presumably too similar in the various species) to have repeatedly evolved; so it is a plesiomorphic character state (but not "primitive"?). Geiser (1994, p. 11), citing the similarity of torpor in birds and mammals, suggested "...that similarities in patterns of torpor do not reflect a common root but a restricted number of physiological options for function at low body temperature and metabolism in endotherms."

The disadvantages associated with hibernation, especially with regard to a reduced rate of reproduction and a reduced r_{max} (see Section 13.4), may explain why migration is preferentially used by birds and why Cade saw a tendency in phylogeny for normothermia to replace hibernation and torpor. In southwestern North America, the ground-squirrel with the widest distribution, *A. leucurus*, does not enter torpor, thereby maximizing the time spent on reproduction and growth (Tucker 1966a), whereas the species with the smallest distribution, *S. mohavensis*, is the most prone to enter torpor (Pengelley and Kelly 1966). The restrictions placed on hibernators may be so great that they cannot compete with nonhibernators unless environmental conditions are so harsh that hibernation is required to survive (Hainsworth and Wolf 1978). Nevertheless, a diversity of mammals use hibernation so that its associated penalties are tolerable. After all, many small mammals hibernate in environments that are not particularly extreme and remain "successful" in the presence of many mammals that never enter torpor.

A most interesting behavior is found in hibernating marmots (*Marmota*), all of which, except for the North American woodchuck (*M. monax*), hibernate in family and social groups. The social impact on hibernation has been shown in *M. marmota* (Arnold 1988, Arnold et al. 1991):

1. A group hibernating in a burrow enters into and arouses from torpor in near synchrony.

2. Although marmots huddling together will presumably conserve energy when normothermic, they are not equally affected by social behavior because individuals in each group tend to maintain their position in the sequence of entrance into and arousal from torpor—and this sequence is not consistently determined by either sex or age.

3. Entrance into torpor was slowed by the presence of normothermic individuals in a group, which however facilitated emergence.

4. An increase in the number of marmots hibernating together delayed from early November to late December the time when burrow temperature fell to 5°C, which is the temperature below which marmots increase rate of metabolism to maintain a fixed body temperature (ca. 7°C).

11.8.4 Marginal hibernation. The distinction between hibernators and nonhibernators is graded, as can be seen in some noteworthy examples of seasonal torpor.

1. *Poorwill*. The poorwill (*Phalaenoptilus nuttallii*) is the only bird known to hibernate. The first to describe this behavior, Jaeger (1948, 1949) noted that some torpid individuals remain in the same place, apparently, for extended periods of time and that preferred locations are used for torpor during several winters. Such behavior is undoubtedly limited to regions where microclimatic temperatures do not fall far below 0°C and only then for short periods of time, that is, in the far southwestern part of the United States and Mexico (Ligon 1970). Poorwills will enter reversible torpor in the laboratory (Marshall 1955; Bartholomew et al. 1957, 1962; Withers 1977), and they enter a torpid state on a daily basis in the spring and fall at the northern limits of their breeding range in southern Canada (Brigham 1992). Recently R. M. Brigham (pers. comm., 2000) showed that torpor in poorwills in winter may last continuously for periods up to several weeks. The poorwill is even known to enter torpor on occasion during incubation and brooding near the northern limits of breeding range in Canada (Kissner and Brigham 1993). These data imply that daily torpor and hibernation are graded, both in the length of an undisturbed period of torpor and in the depth to which body temperature falls. Körtner et al. (2000) demonstrated the graded nature of this difference in the tawny frogmouth (*Podargus strigoides*), which entered daily torpor on 44% of the monitored days during an Australian winter. Because these bouts were daily, this behavior would probably best be called daily torpor, but one can see how difficult (and arbitrary) some of these differences are.

2. *Bears*. Temperate bears spend much of winter in "dormancy." During this period the body temperature of black bears (*Ursus americanus*) falls from an active 37 or 38°C to an inactive 34°C, and occasionally down to 31°C (Hock 1960). The high body temperature in torpor undoubtedly reflects in part a large body mass. Therefore, the claim (e.g., Hock 1960, Morrison 1960) that bears are not "true" hibernators because their body temperatures in torpor are too high ignores the impact of body size on the equilibrial ΔT attained in torpor, and it ignores the ecological context of this torpor. Rate of metabolism during hibernation may decrease by 50% in spite of the continuously high body temperature. In polar bears (*U. maritimus*) only females den in winter, during which body temperature falls to 36°C, but rates of metabolism may decrease by 30% to 50% (Watts and Hansen 1987). The major energy source used by bears in torpor is body fat. Protein metabolism is normally not an important energy source (Nelson et al. 1973), so that nitrogenous waste products do not accumulate, which reduces the necessity for arousal and excretion.

3. *Badger*. Mammalian carnivores other than bears, such as skunks and badgers, have often been reported to enter "winter lethargy." Harlow (1981a, 1981b) studied torpor in the American badger (*Taxidea taxus*). With starvation badgers sometimes entered a shallow torpor, wherein body temperature fell from a normal value of 37 to 28°C. Given the mass of the badgers (ca. 8 kg), they took a long time to cool and to heat: the mean torpor cycle was 29 h (15 h for cooling, 8 h in torpor, 6 h for arousal), which reduced energy expenditure by 27% compared to 29 h of normothermia. The period in which the badger remained in torpor varied from 6 to 18 h. At other

times, starvation did not produce torpor. With 30 d of continuous starvation, body temperature stabilized at 36.4°C and rate of metabolism fell 26%. Thus, torpor does not reduce rate of metabolism more than starvation does, probably because the cost of arousal coupled with the short torpor cycle reduces the energy savings in torpor. Use of starvation, however, is limited by the exhaustion of fat stores and the metabolism of proteins: badgers lost 24% of their original body mass during 30 d of starvation, but they lost as little as 15% during torpor. The maintenance of body temperature during starvation is used at ambient temperatures greater than 10°C. Torpor is used only during the coldest months when burrow temperatures vary between 0 and 5°C. Torpor has also been described in the European badger, *Meles meles* (Slonin 1952, Johansson 1957).

4. *Beaver.* One of the strangest responses to a long winter has been found in the beaver, *Castor canadensis.* Some colonies in Canada at 58°N do not accumulate enough cut branches of alder (*Alnus*), willow (*Salix*), and red osier (*Cornus*) to meet the energy requirements of beaver colonies for the 164 d or so when water is icebound (Novakowski 1967). Furthermore, during winter, young of the year grow, whereas adults and subadults lose mass. Beaver colonies can be found as far north as the delta of the Mackenzie River, Yukon Territories (68°N). Food intake by these beavers decreased by 40% when held in continuous darkness in winter (Aleksiuk and Cowan 1969a, 1969b). Associated with this decrease, thyroid activity and growth also decreased. Most strikingly, these beavers were described as having entered into a paralytic state in which they were nearly immobile. Body temperature within this state, however, maximally fell only 1°C, to 35.7°C. These reactions did not occur in Californian beavers held under identical conditions. Arctic or subarctic beavers in summer have a large food supply, rapid growth, and little body fat. As food availability declines in fall and winter, thyroid activity falls, rate of metabolism decreases, growth stops, and fat deposition occurs. The paralysis of arctic beaver, which needs confirmation, is not torpor in the normal sense (i.e., body temperature is much higher than expected from mass during torpor), but, if confirmed, is ecologically equivalent to hibernation in the reduction of energy expenditure, here obtained principally through a forced reduction in activity.

MacArthur (1989) and Dyck and MacArthur (1992, 1993) gave a rather different analysis of the energetics of beavers. It has a standard relationship between rate of metabolism and ambient temperature, both in the laboratory (MacArthur 1989) and in simulated field conditions in winter (Dyck and MacArthur 1993). Adult and kit beavers showed no evidence of a lowered body temperature, either in the laboratory or in the field (Dyck and MacArthur 1992). (Smith et al. [1991] showed that adult beaver in the field in northern Michigan and Minnesota had maximally a decrease of 2.5°C in core body temperature in winter, which is slightly greater than the circadian variation [1.4°C] in body temperature.) Dyck and MacArthur (1992) agreed with earlier reports that the amount of energy in plants cached at the beginning of winter was often inadequate to fulfill the requirements of beaver colonies for the entire winter. This store may be supplemented by feeding on aquatic vegetation (Dyck and MacArthur 1993).

11.8.5 Aestivation in ectotherms. Aestivation as an ecological and physiological phenomenon is difficult to define. Reno et al. (1972, p. 625) defined it as seasonal "dormancy in response to heat or drought. . . ." It usually refers to torpor during summer in response to a shortage of water or food at times of high temperatures. One difficulty with this definition is that high temperatures per se may have nothing directly to do with aestivation, except that aestivation is a warm-temperature torpor.

Aestivation is most clearly present in vertebrates that live in freshwater subject to seasonal drought. Among fishes such behavior is best known in the African lungfish *Protopterus aethiopicus*, but is apparently also found in other tropical and subtropical fishes (see Section 6.11). During extended periods of aestivation, when no feeding occurs and when the African lungfish is encased in a cocoon made of mucous secretions, rate of metabolism falls (80% from standard measurements [Delaney et al. 1974]). (Note, however, that of the four species of *Protopterus* cocoon formation is known only in two species, *P. annectans* and *P. aethiopicus* [Greenwood 1986]). Although the dry season faced by *Protopterus* is normally only 4 to 6 months long (Janssens 1964), Smith (1935a) was able to keep an individual of this species in aestivation for 3.5 y: by 500 d, body mass had decreased by 38%

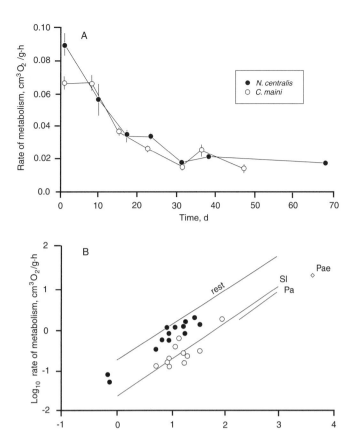

Figure 11.42 A. Rate of metabolism in two anurans, *Neobatrachus centralis* and *Cyclorana maini*, as a function of time from the onset of aestivation. B. Log$_{10}$ rate of metabolism in resting and aestivating anurans and in the aestivating salamander *Siren lacertina* (Sl), aestivating anuran *Pyxicephalus adspersus* (Pa), and the aestivating lungfish *Protopterus aethiopicus* (Pae). The standard curve for resting anurans at 25°C (Gatten et al. 1992) is included. Sources: Modified from Withers (1993) with additional data from Delaney et al. (1974) and Etheridge (1990b).

and rate of metabolism, by 87.5%. Aestivation is originally fueled by the oxidation of body fat, but with time the metabolism of proteins becomes the principal source of energy, which may ultimately limit the period over which aestivation (and starvation) can occur.

Many amphibians also avoid water shortages by entrance into torpor while buried in soil, either with or without a cocoon (see Section 6.11). These amphibians include anurans (e.g., Mayhew 1965b, Balinsky et al. 1967, Lee and Mercer 1967, McClanahan 1967, Ruibal et al. 1969, Seymour 1973a, McClanahan et al. 1976, van Beurden 1980, Loveridge and Withers 1981) and urodeles (Rose 1967, Martof 1969, Reno et al. 1972, Gehlbach et al. 1973). As in *Protopterus*, energy metabolism in amphibians is fueled during torpor principally by fat metabolism (e.g., *Siren*), but in some cases the loss of mass may exceed the fat content of the animal. The extended period that *Phyllomedusa sauvagii* remains inactive during the dry season (Shoemaker et al. 1972) is also aestivation, in this case often at rather high ambient tem-

peratures. The seasonal torpor of *Scaphiopus* may start off as aestivation at the end of summer but continues into winter, when it is associated with low soil temperatures and is best referred to as hibernation. Again, the principal physiological difference between these states may be the temperature at which torpor occurs.

As in hibernation, rate of metabolism in aestivating ectotherms is very low. This has been found in amphibians (*Scaphiopus couchii*, *S. hammondii* [Seymour 1973a], *Siren intermedia* [Gehlbach et al. 1973], *Cyclorana platycephala* [van Beurden 1980], *C. maini* [Withers 1993; see Figure 11.42A], *Neobatrachus* spp. [Withers 1993; see Figure 11.42A], and *Pyxicephalus adspersus* [Loveridge and Withers 1981]) and the lungfish *Protopterus aethiopicus* (Smith 1935a, Delaney et al. 1974). The reduction from standard rates was 60% to 70% in *Pseudobranchus* and *Siren* (Etheridge 1990b), 75% in *Pyxicephalus* (Loveridge and Withers 1981), 80% in *Lepidobatrachus* (McClanahan et al. 1983), and 62% to 86% in *Neobatrachus* and *Cyclorana* (Withers 1995), all of which make cocoons, the decrease reaching a maximum by 30d. *Scaphiopus couchii*, which forms no cocoon, shows an 80% reduction in rate of metabolism during aestivation (Seymour 1973a). The steady-state rate of metabolism during "aestivation" in amphibians and fish is about 20% of the standard rate (Figure 11.42B). The herbivorous chuckwalla (*Sauromalus obesus*) stops feeding, reduces activity, and shows a marked reduction in field rates of metabolism in summer with the onset of high temperatures and the disappearance of moist vegetation (Nagy and Shoemaker 1975).

As a result of the reduction in rate of energy expenditure, the time period for aestivation is greatly extended. For example, a 500-g *Pyxicephalus adspersus* can survive for 260d at 20°C at the dormant rate of metabolism if lipids equal to 3% of body mass are used or 31g of protein is metabolized (Loveridge and Withers 1981). Etheridge (1990b) calculated that a 1125-g *S. lacertina* could aestivate for 2 to 3y without food if it used about 80% of a lipid store and if the store was 48% of body mass. In fact, Martof (1969) had a *S. lacertina* that lived for 5.2y without food and in the process lost 86.5% of its original mass.

The use of aestivation by anurans in extremely hot, dry environments, such as southwestern United States and central Australia, may be the principal response that permits them to tolerate such extreme conditions. As noted, *Scaphiopus* in North America on occasion may remain in aestivation continuously for 2y, if the rains anticipated at the end of nearly a year of withdrawal do not materialize. Some, small proportion of *Cyclorana* may withstand up to 5y of continuous drought in central Australia, only to emerge with rainfall (van Beurden 1980).

11.8.6 Aestivation in endotherms. Some few mammals aestivate, which raises the question of whether the physiological state associated with aestivation in endotherms is different from that found in hibernation. Bartholomew and Cade (1957) and Bartholomew and Hudson (1960) concluded that these torpors are qualitatively similar in being a state in which body temperature is regulated at a low level and rate of metabolism is reduced, whereas they differ quantitatively only in that aestivation occurs at warm temperatures. For example, the cactus mouse (*Peromyscus eremicus*) cannot spontaneously arouse if body temperature falls below 16°C (MacMillen 1965), in which case

it will die within a few days, even if it is artificially warmed. This mouse actively resists entering torpor at ambient temperatures below 15°C. Some other small rodents (*Baiomys taylori* [Hudson 1965], *Perognathus californicus* [Tucker 1966a]) also show high-temperature torpor, as do some tenrecs (*Geogale*, *Microgale* [Stephenson and Racey 1993a, 1993b]).

Aestivation in mammals may be induced either by food or by water shortage (MacMillen 1965), but a water shortage does not necessarily distinguish aestivation from hibernation: Foster et al. (1939) stated that dehydration facilitated entrance into torpor in the hibernator *Spermophilus tridecemlineatus*. Within the genus *Spermophilus*, species hibernate where winters are cold, aestivate in regions of prolonged summer drought, and both hibernate and aestivate in regions where precipitation is restricted to winter and spring. Thus, torpor in its various forms permits *Spermophilus* to exist in the Arctic, in high mountains, and in warm-temperate deserts. In these squirrels torpor can occur at body temperatures from 2 to 30°C. In spite of its flexible use, aestivation, like hibernation, may be a solution of last resort. Some ground-squirrels enter aestivation at the end of summer and then directly shift to hibernation without reappearing above ground. This pattern greatly restricts the time for social interaction: *S. mohavensis* is active for only about 140d (38% of the year), a truly depressed existence. Such behavior undoubtedly reduces the reproductive season and leads to a low r_{max} (see Chapter 13).

11.9 CHAPTER SUMMARY

1. Most vertebrates inhabit periodic environments, the fundamental periodicities being daily, lunar, and annual.

2. Vertebrates show periodicity in their behavior, much of which is imposed by variations in the environment, although some rhythmicity is endogenous.

3. Daily and annual endogenous rhythms anticipate cyclic variations in the environment and permit an organism to prepare for this variation.

4. Many vertebrates compensate for a circadian activity rhythm by storing food or body fat to balance their daily energy budgets.

5. Daily energy budgets can be balanced by the adjustment of activity rhythms, as demonstrated in

some small species that abandon a strict adherence to a diurnal or nocturnal rhythm.

6. The nocturnal decrease in the body temperature and rate of metabolism in many ectothermic vertebrates balances a daily energy budget during periods when food is not available.

7. Some small endotherms use daily torpor to balance daily energy budgets, especially in fall and winter.

8. Three major responses to an annual variation in environmental conditions are seasonal tolerance, migration, and entrance into torpor.

9. Seasonal tolerance to a cold winter in ectotherms is usually restricted to aquatic species and involves a seasonally variable energy expenditure that is normally higher in summer than winter.

10. In endotherms seasonal tolerance involves an energy expenditure that is more or less constant throughout the year (because of a seasonal shift in behavior), is higher in winter (due to a high cost of maintenance), or is higher in summer (due to a high cost of reproduction), although the seasonal amplitude of the expenditure may be diminished by a large mass.

11. In environments with cold winters, small endotherms can remain active throughout winter by caching food, storing fat, and having an enhanced thermogenic capacity.

12. Some small mammals show a reduction in body mass in winter, apparently to reduce resource requirements.

13. Sexual size dimorphism often is compensatory for a seasonally higher rate of energy expenditure in the smaller sex.

14. Migration is limited to vertebrates with a capacity for long-distance movement (e.g., birds, bats, large mammals, fish, and aquatic turtles).

15. Migration to a winter range is associated with a decrease in energy expenditure, thereby facilitating a balanced energy budget. Some of the changes in energy expenditure with respect to latitude are associated with changes in body size and composition.

16. Migratory species exploit a seasonally abundant food resource for reproduction in areas that are nearly uninhabitable in winter.

17. Most of the cost of migration is derived from fat metabolism, the cost depending on the distance migrated, body mass, size of the fat deposits, and the use of storm fronts.

18. Some vertebrates enter into a seasonal torpor in which most physiological functions are depressed; this state may occur in winter (hibernation) or summer (aestivation).

19. Ectothermic vertebrates that hibernate include some temperate and subtropical amphibians, lizards, snakes, turtles, and crocodilians.

20. Hibernation in ectotherms may involve a reduction in physiological functions to rates below those dictated by a decrease in body temperature.

21. Endotherms that hibernate include a variety of small to intermediate-sized mammals, but the only bird known to hibernate is the poorwill.

22. Hibernating endotherms regulate body temperature at low levels and periodically awake to a normothermal state.

23. Whether entrance into torpor is active or passive is unclear, although arousal from torpor clearly is an active process.

24. Endothermic torpor may have evolved from slow-wave sleep as an exaggerated means of conserving energy.

25. Daily or seasonal torpor is used by a minority of endothermic species, apparently because it greatly reduces their reproductive output; it is used only when absolutely required for survival by conditions in the environment.

26. Aestivation is physiologically similar to hibernation, except that it occurs at higher temperatures; it is found in some freshwater fishes, desert amphibians and reptiles, and small mammals.

27. Aestivation in amphibians and fish is accompanied by an 80% reduction in maintenance costs, which thereby extends the period over which dormancy can occur.

12   Diet and Nutrition

12.1 SYNOPSIS

Vertebrates obtain energy and nutrients from food. The foods used vary greatly among fishes, birds, and mammals, but they vary little among living reptiles and even less so among amphibians. Food specialists reduce nutritional requirements by having a small mass or low-intensity metabolism. Mixed diets are widespread, but some combinations of food items are uncommon because they have incompatible digestive requirements. The intestinal morphology of vertebrates reflects the diet and often varies with seasonal variations in the diet. The digestibility of food depends on its chemical composition and the enzymatic capacities of the consumer and in the case of ectotherms, on body temperature. Vegetative parts of plants are difficult to digest and often contain secondary compounds that act as a deterrent to predation. These problems can be solved in part through the use of microbial symbionts. Small herbivores can surmount the poor quality of much vegetation by selectively using high-quality food, but large species need large volumes of food and thus must use abundant, lower-quality food items and require the aid of fermentation. Fermentation occurs in the intestine either before or after the stomach. Foregut fermentation is more efficient in extracting the energy in structural carbohydrates, but it also tends to regulate the rate of passage of food through the gut and therefore limits the amount of food that can be processed, which limits body size. Consequently, large herbivores usually are hindgut fermenters. The palatability of browse is often negatively related to the content of secondary compounds. The abundance and diversity of tropical folivores are positively correlated with the protein content of leaves and negatively correlated with the fiber and tannin contents of the leaves. Tannin in food can be inactivated by the synthesis of tannin-binding proteins in the saliva. The periodic fluctuations in the populations of some consumers may be associated with the induction of secondary compounds in the food as a result of consumption.

12.2 INTRODUCTION

Vertebrates, like all heterotrophs, require food to supply the chemical-potential energy, elements, molecular building blocks (including essential amino acids and vitamins), and much of the water required for existence. They use a great range of organic matter as food, including plant vegetative parts (leaves, roots, bark, cambium, and woody tissue), seeds, fruits, nectar, and pollen; invertebrates and vertebrates; scavenged carcasses; and detritus. Some vertebrates are trophic generalists—taking whatever is available; others are highly specialized. Specialized foods include the blood of endotherms (used by vampire bats) and the scales and fins of teleosts (used by other teleosts). Vertebrates can also be classified broadly as being carnivores, herbivores, and omnivores or can be more finely classified as being invertebrivores, vertebrivores, grazers, browsers, frugivores, nectarivores, and so on.

The amount of food needed by a vertebrate depends not only on its energy (and material) requirements but also on the quality and quantity of food available. Thus, the rate of metabolism of ectotherms (see Section 4.6) and, especially, endotherms (see Section 5.4.1) varies with the kind of food consumed. Some of the influence of food type is associated with what might be called the ecological context in which the food is found. This concept includes seasonal availability, which is more likely to reflect conditions in the environment than to be an intrinsic property of the food, and the means by which a consumer harvests the food, which is as much a function of the consumer as it is of the food used. Other factors influencing the amount of food used stem, at least in part, from an intrinsic property of the food, its molecular composition. The molecular composition of food has an impact through its effect on the presence and availability of nutrients and energy contained in the food and through the presence of molecules that are toxic to the consumer.

This chapter examines some of the morphological and physiological adjustments that vertebrates make to use various food items, as well as some of the adjustments plants apparently make to hinder their consumption. Many of the most radical morphological and physiological modifications in vertebrates to food type are found in species specialized to feed on the vegetative parts of plants, both because plant carbohydrates (and therefore energy) are often not readily usable and because secondary compounds in plants are toxic to vertebrates.

12.3 DIVERSITY IN DIET AND NUTRITION

Food selection couples an organism to its environment. The intensity of the coupling varies with the requirements of the organism for nutrients and energy and with its digestive capacities (McKey et al. 1981). The requirement for specific nutrients and the capacity for digestion are correlated with the food selected. Although many vertebrates are omnivorous, many others show varying degrees of trophic specialization. Because of the correlations of gut length with body mass, of the retention time of food in the gut with gut length, and of the digestion of food with retention time, small vertebrates generally use higher-quality food than do larger vertebrates. Therefore, some diversity in diet may simply reflect the impact of body size (e.g., Demment and Van Soest 1985, but see Foley and Cork 1992).

12.3.1 Diversity in diet. The evolution of trophic diversity has occurred repeatedly in the history of vertebrates. Fishes, for example, show a great diversity in diet. Hyatt (1979) demonstrated

that the diversity in food habits increases at low latitudes in both freshwater and saltwater, reaching its maximum at tropical latitudes (Table 12.1). In temperate and polar zones, most fish are carnivorous (eating either invertebrates, vertebrates, or both), with few having strictly herbivorous habits. Ogden and Lobel (1978), in fact, argued that no strictly herbivorous fish are found in temperate zones, this habit being limited to the tropics. Montgomery (1977), however, demonstrated that at least one eel-blenny living in the cold waters off California is a strict herbivore as an adult. Trophic diversity among fish is most highly developed in association with coral reefs, where specialists feed on corals, echinoderms, and sponges, and in the tropical lowlands, such as Amazonia, where fish feed extensively on fruit, seeds, and leaves (Goulding 1980).

In contrast to fishes, amphibians and reptiles have a much narrower range in food habits. Living amphibians, as adults, are strictly carnivorous, feeding on either invertebrates or vertebrates, although adult sirenid salamanders might be (to some extent) herbivorous, as are most amphibian larvae. Notable exceptions include adult *Hyla truncata* from Brazil, which feeds on fruit (Da Silva et al. 1989), and adult *Rana hexadactyla* (Das and Coe 1994) from south India, which feeds on aquatic vegetation; these observations hint that a greater diversity in anuran food habits may be present in the lowland tropics. Living reptiles are usually carnivorous. Most lizards (in association with a small mass) are insectivorous, but larger lizards tend to feed on vertebrates or on plants (Pough 1973b). Iverson (1982) maintained that the switch from insectivorous habits to herbivory within a species during ontogeny is rare (*Ctenosaura similis* being an example). All snakes are carnivorous, and most preferentially feed on vertebrates, although some of the smallest feed on invertebrates, including insects, other arthropods, and earthworms. A few snakes are egg-eating specialists. All crocodilians are vertebrate-eating specialists. Among turtles, many are omnivorous, some are vertebrate-eating specialists, and tortoises are specialist herbivores. Mortimer (1982) summarized the remarkable range of food specialization found in marine turtles: *Dermochelys* (the leatherback) feeds almost exclusively on jellyfish and tunicates; *Lepidochelys* (the ridley) feeds principally on shrimp and crabs; *Caretta* (the loggerhead), on benthic invertebrates (molluscs,

Table 12.1 Diversity of food habits of fishes in relation to latitude

| | Freshwater | | | | Marine | |
| | Temperate: Canada | Warm Temperate: Lake Ponchartrain, Louisiana | Tropical | | Temperate: Gulf of Maine | Tropical: Marshall Islands |
			Lake Redondo, Amazon, Brazil	Lake Victoria, Africa		
No. of species	222	37	42	100	149	206
Food category (% total)						
Herbivores	9.5	10.8	21.4	14.9	0.7	26.2
Phytoplankton	0.9	5.4	2.4	2.9	0.7	0
Benthic diatoms	3.2	0	4.8	0	0	1.5
Filamentous algae	4.5	5.4	7.1	12.0	0	16.0
Vascular plants	0.9	0	7.1	0	0	8.7
Detritivores	1.4	18.9	9.5	3.9	0.7	3.9
Carnivores	82.4	62.1	40.5	76.0	97.3	61.2
Zooplankton	18.5	5.4	4.8	1.0	16.9	6.3
Benthic invertebrates	43.2	35.1	19.0	34.0	41.2	54.9*
Terrestrial insects	2.7	0	4.8	0	0	0
Fish	18.0	21.6	11.9	41.0	39.2	54.9*
Omnivores	6.8	8.1	28.6	10.0	2.0	8.9

*The consumption of benthic invertebrates and fish was not distinguished.
Source: Condensed from Hyatt (1979).

crustaceans, sponges) and horseshoe crabs; *Eretmochelys* (the hawksbill), on encrusting organisms, including sponges, tunicates, and bryozoans; and *Chelonia* (the green turtle) feeds on seagrasses and algae. The complications inherent with some specialized diets is illustrated by *Eretmochelys*, which feeds principally on sponges: the hawksbill preferentially eats sponges that contain collagen-based skeletons, rather than those using spongin for skeletons (Meylan 1985); it often has a significant intake of siliceous spicules. Even though few living reptiles are herbivorous, many pelycosaurs, therapsids, and saurischian and ornithischian dinosaurs fed exclusively on plant vegetative parts.

Mammals have a great range in food habits, feeding on everything from plankton in the sea to soil fungi and from mayflies to baleen whales. In some cases, the trophic specialization may be extreme, as it is in the koala (*Phascolarctos cinereus*), which feeds on only a few species of *Eucalyptus*; red tree-vole (*Arborimus longicaudus*), which feeds only on the needles of one to four species of conifers, most notably Douglas-fir (*Pseudotsuga menziesii*); and Stephen's wood-rat (*Neotoma stephensi*), which feeds mainly (>90%; Vaughan 1982) on junipers (*Juniperus*). (Notice that the most dramatic examples of specialization on food usually involve the use of plants; see Section 12.6.)

Fleming et al. (1993) showed a temporal and geographic component of trophic specialization in the seasonal behavior of nectarivorous bats in the United States and Mexico. *Leptonycteris curasoae* feeds on the nectar and pollen of Cactaceae and Agavaceae at the northern limits of its summer range in southern New Mexico and Arizona. These plants are characterized by crassulacean acid metabolism (CAM), which incorporates C_3 (Calvin cycle) and C_4 photosynthesis, depending on environmental conditions (Osmond et al. 1973). A fundamental difference between C_3 and C_4 photosynthetic systems is that C_3 photosynthesis usually has a large negative ^{13}C ratio ($\delta^{13}C$), whereas C_4 photosynthesis has a smaller negative $\delta^{13}C$ (Bender 1971). Some of this difference relates to whether CO_2 fixation occurs in the dark (C_4) or in the light (C_3). Under normal environmental conditions CAM plants use C_4 photosynthesis. When animals consume plant matter, they incorporate carbon in the form that they receive it from plants, which in turn reflects the type of photosynthesis. Because *L. curasoae* feeds on cacti and agaves, which are CAM plants, it incorporates ^{13}C into its skeleton with a small negative ratio. This is the case in the sedentary populations in Baja California. In the migratory northern populations, the ratio varies in *L. curasoae* from a low negative ratio during spring, summer, and fall, when they feed on

CAM plants, to highly negative ratios in their winter range in central and southern Mexico, when they feed on C_3 plants (Figure 12.1). These populations have migratory pathways that are timed to coincide with the flowering periods of CAM plants. In contrast, the tropical, sedentary nectarivorous bat *Glossophaga soricina* feeds throughout the year on C_3 plants (Bignoniaceae, Bombacaceae, Leguminosae), as is reflected in its highly negative $\delta^{13}C$ (Figure 12.1).

In an analysis of the contributions of phylogeny and body size to the food habits of New World primates, Ford and Davis (1992) stated that convergences and reversals of mass and food habits occurred repeatedly through history. Although Kay (1984) argued that no primates weighing more than 350 g could be principally insectivorous and none less than 700 g could be strictly folivorous, Ford and Davis found no correlation in small primates between the extent of insectivory and body mass. (Note, however, that many of the smallest neotropical primates, members of the family Callitrichidae, feed on insects and gum; gum may be a high-quality food.) *Cebus* at 3 kg is large for having a highly insectivorous diet. Some of the largest monkeys (e.g., *Alouatta*, *Brachyteles*) are folivorous, although the most folivorous (*Alouatta*) is not the largest. Kinzey (1992) showed that the propensity of monkeys belonging to the subfamily

Pitheciinae to eat hard nuts and seeds increases with body mass.

Birds have nearly as diverse food habits as mammals, except that they are less prone to eat the vegetative parts of plants, with the notable exception of members of the Anseriformes (ducks, geese, and swans), many of which preferentially eat grass or algae. Few birds feed on the leaves of woody plants (with the exception of some grouse, the South American hoatzin [*Opisthocomus hoazin*], and the New Zealand pigeon [*Hemiphaga novaeseelandiae*]), possibly because having a large intestinal storage capacity for food, as is needed when highly indigestible, bulky foods are used, is incompatible with flight (Morton 1978). Although some temperate birds, especially thrushes (Muscicapidae) and waxwings (Bombycillidae), opportunistically consume fruits, most committed frugivores have a tropical distribution and some appear committed to a narrow spectrum of fruit (e.g., trogons feed on Lauraceae [Wheelwright 1983] and the oilbird [*Steatornis caripensis*] feeds on the fruits of some palms).

Diversity in the food habits of vertebrates living in a particular habitat facilitates coexistence. Among sea ducks that inhabit the shorelines of the North Atlantic, scoters (*Melanitta*) tend to feed mainly on bivalves; bufflehead (*Bucephala albeola*), on sand shrimp; common goldeneye (*B. clangula*), on crustaceans; oldsquaw (*Clangula hyemalis*), on a diversity of invertebrates; red-breasted merganser (*Mergus serrator*), on fish; harlequin duck (*Histrionicus histrionicus*), on snails and amphipods; and common eider (*Somateria mollissima*), on sea urchins and bivalves (Stott and Olson 1973, Goudie and Ankney 1986). The smallest and most active species, the bufflehead, selects prey with the smallest proportion of exoskeleton and shell, thereby increasing the usable fraction of the food consumed. Larger sea ducks have greater flexibility in the time spent feeding with respect to environmental conditions, and have relatively larger gizzards and shorter digestive tracts, thereby permitting a specialization on foods with a higher shell content. Most kangaroo-rats feed on seeds, but *Dipodomys microps* eats large quantities of the leaves of *Atriplex* (Kenagy 1973), a specialization that diminishes competition between sympatric kangaroo-rats. A similar pattern is seen in primates from Madagascar (Ganzhorn 1989a), which is at least partially based on differences in leaf chemistry (Ganzhorn 1988, 1989b), and tropical bats (McNab 1971a).

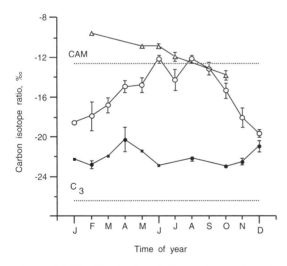

Figure 12.1 The $\delta^{13}C$ in C_3 and CAM plants and in the muscles of the bats *Leptonycteris curasoae* on the mainland (open triangle) and in Baja California (open circle) and in *Glossophaga soricina* (closed circle) as a function of the time of year. Source: Modified from Fleming et al. (1993).

12.3.2 Varied diets. The comparative rareness of highly specialized diets raises a question as to the significance of variation in dietary intake. In an examination of this aspect of diets, Westoby (1978) suggested that the simplest optimal diet would be to use the food that yields the highest nutrient intake per unit of expenditure, especially when no savings in expenditure is attained by feeding on more than one food item. Such a condition would lead to specialization. But various factors contribute to a diverse diet. These include a reduction in the cost of searching for food, a seasonal change in the abundance or nutritional characteristics of foods, and the use of foods that are nutritionally incomplete. Most food specialists are small (e.g., tree-voles, naked mole-rats) or slow-moving (e.g., koala, tree-sloths), have (or require) unusual morphological specializations (e.g., anteaters/termite eaters, woodpeckers), or have an unusually high cost of food collection (e.g., foliage-gleaning birds, badgers, lyrebirds).

Nutritional balancing, in fact, may be one of the more important factors maintaining diversity in the diet. Thus, most birds that eat fruit also eat insects principally because of a shortage of protein in most fruits (Foster 1978, Morton 1973, Herrera 1982, Jordano 1988, Izhaki and Safriel 1989, Bairlein 1996). Brice (1992), however, noted that hummingbirds can survive (at least for a while in captivity) while feeding only on nectar, but not on fruit flies, which suggests that fruit flies may supply protein but insufficient amounts of energy). A similar justification for the mixture of fruit with insects or leaves in the diets of many primates has been used (Kay 1984). Some primates supplement fruits by consuming the seeds and nuts in fruits (Kinzey 1992). As a result, seed-eating frugivorous primates eat fewer insects than folivores that do not eat seeds. Nevertheless, at least some birds are able to feed exclusively on fruits without any appreciable supplement from insects (Levey and Karasov 1989, Karasov and Levey 1990) because they have modified requirements, they feed on protein-rich fruit, or they process large amounts of fruit rapidly.

The variation in diet may be constrained by circumstances, especially in temperate and polar environments. Blue grouse (*Dendragapus obscurus*) shift from eating exclusively conifer needles and buds in winter to eating the leaves of broad-leaved plants, fruits, and flowers in spring and summer (King and Bendell 1982). This shift is most marked in females, which may require a change in nutrition to facilitate reproduction. Rock ptarmigan (*Lagopus mutus*) in some areas of Finland find acceptable food throughout the year, whereas in other areas they have to migrate to find acceptable food in winter (Pulliainen 1970). These ptarmigan preferentially eat birch (*Betula tortuosa*) catkins when available, owing to their high content of nitrogen, digestible protein, and phosphorus (Moss 1967, Pulliainen 1970). Similar observations have been made on willow ptarmigan (*L. lagopus*): they prefer foods high in protein, nitrogen, and phosphorus (Pulliainen and Iivanainen 1981), and consequently in Finland prefer birches (*Betula*) over willow (*Salix*). But the food consumption of ptarmigan is influenced by the presence of other species: in Alaska willow ptarmigan mainly ate willow; rock ptarmigan, mainly birch; and white-tailed ptarmigan (*L. leucurus*), both willow and birch (Moss 1973). Food intake in tetraonids is reduced in winter, when they spend little time foraging and most of the day and night roosting in snow holes, which permits them to subsist on buds and catkins of woody shrubs. As a result, the cost of "free living," that is, the food consumption of individuals in large outdoor pens compared to those held in outdoor cages, is only 20% greater in willow ptarmigan and negligible in the other two species (Moss 1973).

12.3.3 Nutrition. Vertebrates collectively require many factors in their diet, namely, energy, minerals, amino acids, vitamins, and water. Other than energy, the one factor that is most often in short supply is nitrogen in the form of protein. This deficiency is most commonly encountered by leaf and fruit eaters. A protein deficiency may reflect either a low protein content in food, as in many watery, carbohydrate-rich fruits and nectars, or a low protein availability, as in many leaves, when protein is often bound to other molecules, such as hemicellulose. In other words, what is most important is protein availability (Robbins et al. 1987a).

The consequences of low protein contents in many fruits has been explored in frugivorous birds, bats, and primates. Snow (1981) noted that specialized frugivory in birds is limited to species that feed on fruits rich in fat and protein (e.g., Lauraceae, Moraceae, Burseraceae, Palmaceae) that provide a nutritionally complete diet (also see Jordano 1988). These plants usually produce intermediate to large fruits containing a large seed.

Specialized frugivory is limited in distribution to the tropics and is found in such birds as the oilbird (*Steatornis*), cotingas (Cotingidae), manakins (Pipridae), toucans (Ramphastidae), some touracos (Musophagidae), and fruit-pigeons (Columbidae). In temperate environments fruits vary in protein content, depending on the plant species, and with some time for adjustment, blackcap warblers (*Sylvia atricapilla*) can balance their nitrogen budget with an increased fruit intake without resorting to feeding on insects (Bairlein 1996), especially when feeding on fruits rich in protein, such as black elderberries (*Sambucus niger*). Whereas temperate frugivorous birds principally feed insects to their young, some tropical species, including the long-tailed manakin (*Chiroxiphia linearis*), thin-billed euphonia (*Euphonia canirostris*), and the oilbird, feed their young a high proportion of fruits. This is especially the case (cause or effect?) in species that grow slowly (Foster 1978, Morton 1973). The propensity for strict frugivory is most marked in birds other than passerines, and more likely in suboscines than in oscine passerines (Wheelwright et al. 1984).

Not all fruits used have a high protein content: acorn woodpeckers (*Melanerpes formicivorous*) feed heavily on acorns, which have a low protein content. Weathers et al. (1990) argued that this behavior is associated with low growth constants and cooperative breeding (which facilitates an increase in brooding and feeding during an extended nesting period). Any attempt by frugivores to make up for a low protein content in the food by increasing consumption may be limited by the presence of plant secondary compounds that interfere with digestion or other aspects of metabolism (Izhaki and Safriel 1989). The inclusion of "minor" fruits in the diet is correlated with higher protein content than is found in the "major" fruits of the diet (Jordano 1987).

Opportunistic frugivores, in contrast, feed on insects and on small fruits that primarily provide carbohydrates and many small seeds (like the North Temperate *Sambucus*). Some so-called frugivores, such as parrots, in fact are seed predators so that the nutritional characteristics of the pericarp may not be relevant to the nutrition of these species. Monkeys that eat fruits have sometimes been called "messy" because they often tear fruit into pieces and scatter them about. Redford et al. (1984) argued that this behavior is simply a means of sampling fruit for its insect content. They maintained that insects eaten without carbohydrates must be used for energy, but if combined with fruit, can be used for their amino acid content. Nectarivorous birds obtain little protein and free amino acids from nectar (Baker and Baker 1982), but like opportunistic frugivores obtain them from insects.

A special case of protein availability to nectarivores is found in some bats. As noted, nectar per se has little protein, but it has a high content of simple carbohydrates and thus is a good source of energy (but see Section 12.5). The nectarivorous phyllostomid *Leptonycteris curasoae* is able to use the high (20%–40%) protein content of pollen (Howell 1974). The direct digestion of pollen is nearly impossible because of the mechanically and chemically resistant pollen capsule. This bat is able to circumvent this limitation by ingesting pollen and urine, the combination of which in the presence of a low pH withdraws the contents from the capsule, which can then be digested. *Leptonycteris curasoae* cannot maintain body mass on a diet of nectar that lacks protein (Howell 1974). Pollen protein contains significant amounts of proline, an amino acid that is an important constituent of animal collagens, which Howell (1974) suggested is an accommodation by plants to their animal pollinators.

D. W. Thomas (1984) explored the potential consequences for bats of eating fruit with low protein contents. He noted that most frugivorous birds supplement their diet with insects or seeds. Thomas claimed that frugivorous bats in the New World, all of which are members of the Phyllostomidae, supplement fruit with insects, which is sometimes (but not always: see *Artibeus*) the case (Gardner 1977). He suggested that paleotropical fruit bats, members of the family Pteropodidae, increase food intake to ensure an adequate protein intake because they do not supplement fruit with insects. To get sufficient amounts of proteins, pteropodids would have to consume about twice as much energy as phyllostomids of the same mass. What pteropodids do with the hypothesized "excess" energy is unclear; they do not have appreciably higher fat stores, higher rates of carbohydrate loss in feces, an excessive use of brown fat, or higher resting rates of metabolism than do phyllostomids. Thomas suggested that pteropodids may have higher foraging costs than phyllostomids. By the appropriate selection of fruit, however, protein requirements may be met: the

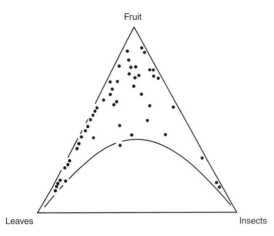

Fruit

Leaves

Insects

Figure 12.2 The mixture of leaves, fruit, and insects in the diets of various primates. Source: Modified from Kay (1984).

phyllostomid *Carollia perspicillata* was able to balance its nitrogen budgets on an array of fruits, but not with *Ficus* (Herbst 1986). Even lactating females were able to balance this budget when feeding on *Piper amalago*, a preferential food of *C. perspicillata*.

Finally, not all mixtures of food seem to be used. Few, if any, vertebrates mix either invertebrates or vertebrates and leaves in a diet, although grass and leaves of woody plants may be combined, as are fruit and insects, fruit and vertebrates, as well as vertebrates and invertebrates (Figure 12.2). The absence of a diet consisting of animals and leaves may represent a digestive incompatibility between the requirements for folivory and those for insectivory (Chivers and Hladik 1980): the initial chemical environment required for the hydrolysis of protein is acidic followed by an alkaline pH, whereas that required by bacteria in the rumen and colon for the hydrolysis of cellulosic carbohydrates is limited to a pH of 6 to 7 (Stevens 1988). Primates may not mix leaves and insects in a diet also because small species (usually <350 g) are normally insectivorous, whereas large species (usually >700 g) usually are folivorous (Kay 1984).

12.4 MORPHOLOGY OF THE GUT IN RELATION TO DIET

All classes of vertebrates that have diverse food habits have a morphologically diverse gastrointestinal tract. This tract generally consists of a buccal cavity, esophagus, stomach, small intestine, large intestine, and rectum, with or without the

presence of one or more caeca (or blind sacs) at the junction of the small and large intestine. This basic pattern is highly modified both quantitatively (by changing the relative lengths of the segments) and qualitatively (by adding or subtracting segments).

12.4.1 Fishes. The gut morphology in fish reflects their diet. Coral-reef herbivores are either grazers, which rasp or suck calcarious substrata, or browsers, which bite or tear benthic macroalgae (Jones 1968, Ogden and Lobel 1978). Browsers have thin-walled stomachs and long intestines, whereas grazers have gizzard-like stomachs and short intestines. The thick stomachs of grazers grind ingested inorganic material, which permits rapid digestion and absorption of the contained organic matter in a short intestine; the longer intestine of browsers facilitates the digestion of plant matter that is not reduced in particle size by the stomach. Within the genus *Acanthurus* (surgeonfishes, Acanthuridae) two stomach types occur: thick-walled in grazers and thin-walled in planktonivores (which have short intestines) and browsers (which feed on macroalgae and have long intestines). Of various carnivorous flatfishes (Pleuronectiformes), those that feed on fish have a simple intestinal loop and heavy gill rakers, those that feed on crustaceans have a long intestinal loop and few gill rakers, and those that feed on polychaetes and molluscs have a complicated intestinal loop and few gill rakers (De Groot 1971). Gill rakers are important in fish eaters to prevent the prey from escaping through the gill slits. Gut length, measured from the buccal cavity through the stomach, is 40% to 54% of tract length in fish eaters, 30% in crustacean eaters, and 20% to 30% in species that eat polychaetes and molluscs. Gut length increases with body mass, the length in herbivores exceeding the length in omnivores, which in turn exceeds that in carnivores (Al-Hussaini 1949). The ratio of intestinal length to body length in fish is less than 1 in carnivores, is 1 to 3 in omnivores, and greater than 3 in herbivores (Sibly 1981). Montgomery (1977) showed that a change in gut length occurs ontogenetically in the eel-blenny *Cebidichthys violaceus* (Stichaeidae) in association with the change in food habits from carnivory to herbivory (Figure 12.3).

12.4.2 Amphibians and reptiles. Little differentiation occurs in gut length and morphology in adult amphibians in accord with their comparatively

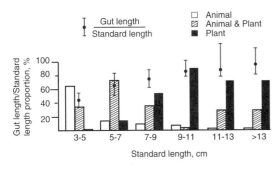

Figure 12.3 Change in gut length and in food habits in an eel-blenny (*Cebidichthys violaceus*) as a function of body size. Source: Modified from Montgomery (1977).

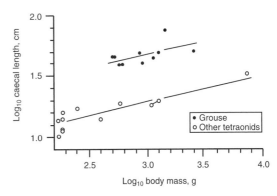

Figure 12.4 Log₁₀ caecal length in grouse and other members of Tetraonldae as a function of log₁₀ body mass. Source: Derived from Leopold (1953).

uniform food habits, but as expected from the herbivory of most tadpoles, they have long intestines. Similarly, reptiles show, relative to fish, birds, and mammals, little differentiation in their gut, mainly because most reptiles are insectivorous or carnivorous. The green turtle (*Chelonia*), however, has a longer intestine and a longer large intestine relative to the small intestine than do carnivorous turtles of similar mass, again reflecting the difficulty in digesting plant carbohydrates (Bjorndal 1985). Herbivorous lizards have a colon that has 1 to 11 transverse partitions, the number depending principally on body size (Iverson 1982). The colon of these lizards contains large numbers of nematodes and bacteria, which contribute to the mechanical and chemical breakdown of plant matter. In fact, the number of nematode species is proportional to the number of colonic valves (and therefore to body size), but no correlation appears to exist between the kinds of dominant bacteria and the size of herbivorous lizards.

12.4.3 Birds. Birds show great diversity in the morphology of their intestinal tract in relation to diet. In many species the esophagus is modified into a crop for food storage; this occurs most commonly in seed and vertebrate eaters, including Falconiformes, Galliformes, Columbiformes, Ploceidae, and Thrincoridae. In the hoatzin (*Opisthocomus*) the crop is stomach-like with digestive glands and powerful musculature; it is functionally equivalent to terms of reticulorumen of ungulates in terms of fermentative digestion of plant carbohydrates (Grajal et al. 1989). The hoatzin is the only bird known to use foregut fermentation.

Birds have variously complex stomachs (Ziswiler and Farmer 1972), including an anterior glandular stomach that secretes proteolytic gastric juices and a posterior muscular stomach (the gizzard) that is used for storage, preliminary proteolytic digestion, and the mechanical breakdown of food. The gizzard may also form pellets for regurgitation of undigestible parts of the food, including bones, hair, and feathers. Species that eat soft foods, such as fish, fruit, and nectar, have distensible stomachs, whereas those that eat hard foods, such as shelled seeds, have muscular stomachs. And if the food is hard and abrasive, like seeds or molluscs, the gizzard is usually lined with a hard, keratinoid layer. In addition, some birds have a pyloric stomach inserted between the muscular stomach and the small intestine. It occurs in many fish eaters, like penguins, grebes, pelicans, cormorants, storks, and herons, but it also is found in hawks, geese, and even cuckoos. The pyloric stomach in grebes contains feathers and in darters contains hairlike processes, both of which may function as filters.

The small intestine of birds is long in herbivores and granivores and short in carnivores, insectivores, and frugivores (but see Karasov and Levey 1990). The large intestine is usually short and straight, but somewhat longer in omnivores. Caeca occur as a pair at the junction of the small and large intestines, although some birds (parrots, kingfishers) have none and caeca are rudimentarily developed in others. Caeca are large in herbivores and some omnivores (cranes, Galliformes, Anseriformes, some members of Charadriiformes, ostrich). Browsing grouse have much longer digestive tracts, especially with much larger caeca, than do quail, partridges, pheasants, and turkeys (Figure 12.4), which eat higher-quality food (Leopold 1953). Caeca in grouse also have a much greater

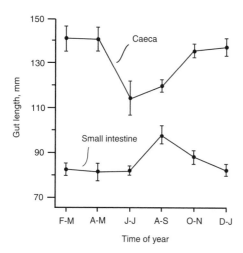

Figure 12.5 Length of the small intestine and caeca in the willow ptarmigan (*Lagopus lagopus*) as a function of the time of year. Source: Modified from Pulliainen and Tunkkari (1983).

diameter than those in other members of Galliformes. Populations of the quail *Callipepla californica* living in mesic environments have a greater intake of green plant matter and have caeca that are 19% longer and intestines 11% longer than populations living in arid environments, where they eat little green matter. The capacity of birds to absorb sugar and amino acids is correlated with the size of the caeca, which may exceed the capacity found in the small intestine (Obst and Diamond 1989).

The flexible nature of the relationship between gut dimensions and diet is seen in some grouse. The length of the caeca and small intestine in the rock ptarmigan varied with season in association with a shift in diet from birch (29% of the intake), berries, and willow in summer and fall to birch alone during winter (Moss 1974). Gasaway (1976a) made similar observations on rock ptarmigan; Pulliainen and Tunkkari (1983), on willow ptarmigan; V. G. Thomas (1984), on willow and rock ptarmigan and on sharp-tailed grouse (*Tympanuchus phasianellus*); and Pendergast and Boag 1973, on spruce grouse (*Canachites canadensis*). For example, the combined caecal length in willow ptarmigan was greater than the length of the small intestine. These segments showed an inverse relationship between length and season (Figure 12.5): the caeca are longest in winter, when the consumed foods were fibrous, whereas the intestine was longest in summer, when berries were heavily eaten but passed through the intestine rapidly (Pulliainen

and Tunkkari 1983). The change in caecal length cannot be completely accounted for by a change in diet, but is also influenced by the increase in food consumption associated with the high cost of maintenance in winter (Pulliainen and Tunkkari 1983, but see Moss 1973). This species also has the largest gizzard of all ptarmigan in association with its consumption of willow stems (V. G. Thomas 1984). When grouse are brought into captivity and fed a high-quality diet, no seasonal variation in gut length occurs (Moss 1972, Pendergast and Boag 1973). A longer small intestine and caecum are found in juvenile and female willow ptarmigan, which may increase the efficiency of digestion in these smaller individuals (Pulliainen 1976).

Few studies of gut dimensions in relation to diet are available in passerines. In a study of European warblers, Jordano (1987) noted that (1) the amount of fruit consumed was inversely correlated with insect intake, (2) fruit use was inversely correlated with fruit size, (3) fruit use increased with intestinal length and gape width, and (4) fruit diameter increased with gape width. That is, small, slender-billed species were principally insectivorous. Frugivorous muscicapids (*Sylvia*), however, have shorter gut passage times and larger intestines than their insectivorous relatives (*Phylloscopus*, *Regulus*). The gut of these warblers showed little correlation with food habits, possibly because of a seasonal shift in food habits associated with a temperate distribution. Dykstra and Karasov (1992) showed that gut length in house wrens (*Troglodytes aeon*) is not fixed: it increased by 14% when food intake increased as a result of cold exposure and forced activity.

12.4.4 Mammals. The morphology of the digestive system in mammals is highly responsive to diet. A part of the diverse morphology—namely, the teeth—is usually neglected because, strictly speaking, they are not part of the gut proper but function in conjunction with the gut. Not only is the dental formula varied in relation to the diet, so too is tooth durability (Janis and Fortelius 1988). Besides grabbing food, teeth are very important in the comminution of food. The comminution of food, rapid digestion, and high rate of metabolism are linked together. This is especially clear in herbivores, when food is mechanically resistant to breakage and the cell walls must be ruptured to release the cell contents and permit bacterial fermentation of the wall fibers. Yet, the teeth of herbivorous mammals must withstand a lifetime of abrasion, most notably in grazers because of the siliceous crystals contained in grasses. Mammals use several methods to strengthen their teeth, including adding hydroxyapatite, increasing the thickness of enamel (*Cebus*, *Cercocebus*, *Pongo*, and *Homo*), increasing cheek tooth size (Elephantidae, †Gomphotheriidae, †Mammutidae, †Stegodontidae; capybara [*Hydrochaerus hydrochaerus*]; and Artiodactyla [Camelidae, Hippopotamidae, Suidae]), increasing tooth height (ungulates, notoungulates, proboscideans, and many rodents), and continuous tooth growth (edentates, some rodents, *Antilocapra*, and *Ovis*).

Diversity in the structure of mammalian guts was demonstrated in several surveys. Amazonian ungulates partition resources based on the food used, their digestive morphology, and habitat segregation (Bodmer 1991). Faunivorous primates have simple, globular stomachs; tortuous small intestines; a short, conical caecum; and a simple, smooth-walled colon (Chivers and Hladik 1980). Frugivores have simple or complex stomachs, the caecum is absent or large, and the large intestine is reduced or enlarged, depending on whether the frugivore supplements a fruit diet with insects or leaves, respectively, to get protein.

Folivores have, as their most conspicuous morphological adaptation to diet, an enlarged chamber in which the fermentation of long-chain, β-linked carbohydrates occurs, in association either with the stomach or with the large intestine. Folivores have a large, coiled caecum—*Dendrohyrax*, the tree hyrax, has three caeca. The most elaborate fermentation chambers are found in association with the stomach, notably in macropodid marsupials, tree-sloths, colobine primates, camels, artiodactyl ruminants, and hippopotami, which clearly demonstrates the convergent evolution of these chambers (see Section 12.7). Vorontsov (1962) and Hansson (1985) showed that northern populations of the Eurasian vole *Clethrionomys glareolis* have a long intestine and large caecum and feed principally on grass and lichens, whereas more southern populations have a short intestine and small caecum and feed mainly on seeds. Similarly, the two wombats have different gut morphologies: *Lasiorhinus* has a long, narrow gut and *Vombatus* has a short, wide gut (Barbosa and Hume 1992a). The proximal colon is the principal site of fermentation in both wombats. The larger distal colon in *Lasiorhinus* may permit greater nitrogen or water

absorption in this wombat, which inhabits drier environments than *Vombatus*.

The morphology of the intestinal tract of Malaysian flying-squirrels varies with diet (Muul and Lim 1978). Flying-squirrels eat seeds, fruit, and the leaves of trees, the larger species being more folivorous; the largest (*Petaurista petaurista*) weighs about 1.5 kg. Because the caecum is where much of the processing of plant matter occurs, large species have larger caeca than smaller species, but those that preferentially eat leaves have, at a given mass, a much larger caecum (Figure 12.6). Thus, small species that feed principally on seeds have a caecum that is only 4 to 5 mm long, whereas the large, leaf-eating *Petaurista* has a caecum that is over 250 mm long!

The food habits of bats may be constrained to some extent by the mechanical requirements for flight, for no bats are truely folivorous (for a slight exception see Kunz and Ingalls [1994], but see Dudley and Vermeij [1994], who noted that these bats are simply compressing the cell contents from the leaves and spitting out the fiber) and none are granivorous. Otherwise, bats are trophically diverse, including eating insects, birds, mammals, fish, frogs, fruit, nectar, pollen, and the blood of endotherms, all of these food habits occurring in the family Phyllostomidae and sporadically in other families. The species with the most generalized food habits have unspecialized stomachs, but those that feed on fluids (nectar, blood) have elongated, tubular stomachs (Forman et al. 1977). The intestines of phyllostomids are generally short, although frugivores have longer intestines than species with other food habits.

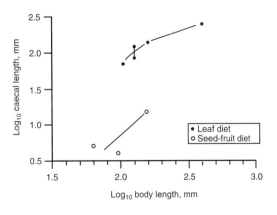

Figure 12.6 Log_{10} caecal length in flying-squirrels as a function of log_{10} body length and food habits. Source: Derived from Muul and Lim (1978).

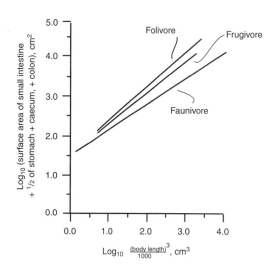

Figure 12.7 $\log_{10}$ absorption area in primates as a function of $\log_{10}$ body size and of food habits. Source: Modified from Chivers and Hladik (1980).

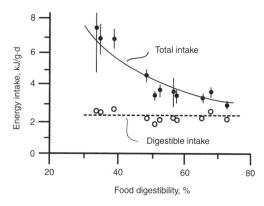

Figure 12.8 Energy intake and digestible intake by arvicolid rodents as a function of food digestibility. Source: Modified from Batzli (1985).

Quantitative differentiation in the gastrointestinal tracts of mammals is appreciable, although such studies should use individuals collected in the field because, as in birds, the diets used in captivity change gut length (e.g., see Hladik 1967). In primates the summed volume of the small intestine and one-half of the stomach, caecum, and colon scales to body mass, as approximated by (body length)^3/1000, in a manner that reflects food habits: at a given mass, faunivores have the smallest volumes; frugivores, larger volumes; and folivores, the largest volumes (Figure 12.7). The large gut volumes found in folivores derive from the use of the caecum and colon, or an enlarged stomach, as fermentation sites. The intermediate position of frugivores occurs because of their use of leaves as a source of protein. The small intestine is the primary site for the absorption of nutrients, but its surface area is distinctly different (i.e., smaller) only in faunivores. Not only do faunivores have small surface areas for absorption, but also their scaling relation has a lower power than is found in association with other diets, which means that the greatest difference between faunivores and other mammals is at large masses.

A large gut size is to be expected mainly when the food is of low quality, which requires a long period for digestion and fermentation to occur, or when an organism has a high nutrient need: grazers and browsers have long intestinal tracts. Thus, *Microtus pennsylvanicus* preferentially eats monocots, principally grasses, and has a larger intestinal tract than *M. ochrogaster*, which mainly feeds on dicots. These species also differ in their propensity to increase tract length in response to a fibrous diet, which is marked in *M. pennsylvanicus* and minimal in *M. ochrogaster* (Batzli et al. 1994). *Microtus ochrogaster* has a longer tract at lower environmental temperatures, especially when food quality is poor (Gross et al. 1985, Hammond and Wunder 1991). An increase in gut size contributes to an increase in nutrient uptake, so that a change in gut size may permit or be required by an increase in rate of energy expenditure. British voles feeding on leaves had larger stomachs than those feeding on seeds, apparently as a response to a larger bulk intake (Lee and Houston 1993). Batzli et al. (1994) and Young Owl and Batzli (1998) described an "integrated processing response" by voles, wherein the digestibility of matter can be maintained constant, in spite of an increased fiber content in the diet (Figure 12.8), through an increased gut size, an increased gut retention time, and an increased absorptive capacity. Such a response results in an increased production of feces (Hammond and Wunder 1991). Some voles, notably lemmings, are more prone to grinding their food into small particles and principally depend on plant cell contents to fuel their nutritional needs. Wunder (1978) and Gross et al. (1985), among others, raised the possibility that the limit to reproduction during periods of high energy demand, such as winter, is not the availability of food itself but the ability to process food, possibly involving a limited gut surface area.

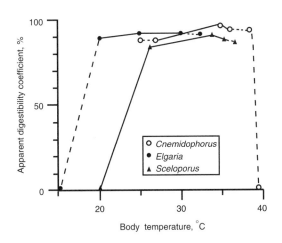

Figure 12.9 Coefficient of apparent digestibility in lizards as a function of body temperature. Source: Derived from Harwood (1979).

12.5 DIGESTION

Many factors, including temperature, food type, and retention of food in the gut, influence the rate at which the complex molecules found in food are digested. The coefficient of apparent digestibility in lizards increases with body temperature (Figure 12.9). This coefficient reaches a maximum at or near the thermal preferendum (Harwood 1979); in the case of *Elgaria*, the coefficient was nearly constant over a wide range in body temperature, just as this lizard has a wide thermal preferendum (Dawson and Templeton 1966). Because of the dependence of digestion on body temperature, the coefficient of digestibility at the thermal preferendum is rather independent of temperature when compared among species (Figure 12.9; see also Woods 1982). The decrease in the digestive coefficient at low temperatures demonstrates the thermal sensitivity of enzymes, although some compensation at moderately low body temperatures may come from an increase in the residency time of food in the gut. Below some "critical" body temperatures digestion will stop. This temperature is about 20°C in *Sceloporus* and 15°C in *Elgaria*. Snakes preparatory to entering hibernation become anorexic and empty their gut (see Section 11.8.1). Black bullheads (*Ictalurus melas*) show no acclimation in peptic digestive capacity and thus have a reduced ability to digest prey at low temperatures (10°C), which requires this fish to metabolize stored fat and glycogen in winter (Nordlie 1966).

Composition of the diet is an exceedingly important factor modifying the capacity to digest food.

Independent of taxon, carnivores have the highest digestive and assimilative efficiencies, 88% and 83%, respectively, whereas herbivores have the lowest efficiencies, 65% and 61%, respectively. Castro et al. (1989) reported somewhat lower values for assimilation efficiency in birds. They also noted that the efficiency increased with the fat content of food and decreased with its protein content (also see Blem 1976, Andreev 1988). As expected, the coefficient of metabolizable energy in both birds and mammals reflects the chemical nature of their diets, although Worthington (1989) maintained that this is not the case within the narrower category of fruits.

The chemical composition of food is not simple. Consider nectar. Baker and Baker (1983) characterized the sugar content of nectar in 765 species of plants by using a ratio of sucrose to the sum of glucose and fructose, the three principal sugars in nectar. When these sugars have equal molar concentrations, the sugar ratio is 0.50. About 55% of all nectars in the survey principally contained glucose and fructose; the remaining 45% had sucrose as the main sugar.

The relative abundance of sugar in a particular plant correlates with the pollinators that visit its flowers (Table 12.2). Hummingbirds (Trochilidae) predominately use nectars that contain sucrose (87% of the plants used by hummingbirds had nectar with a ratio >0.5), whereas perching birds (including New World passerines, sunbirds [Nectariniidae], honeyeaters [Meliphagidae], and honeycreepers [Drepaniidae]) preferentially drink nectars that contain hexose sugars (97% <0.5). New World nectarivorous bats mainly use hexose flowers (100% <0.5), whereas Old World nectarivorous bats are less committed to hexose nectars (57% <0.5). Insects also differentially use hexose-rich nectars (short-tongued bees, flies) or sucrose-rich nectars (long-tongued bees, hawkmoths, moths, and many butterflies). The correlation between nectar composition and pollinator holds even within the plant genus *Erythrina* (Fabaceae): hummingbird flowers secrete predominately sucrose (mean ratio = 1.09; 23 species), whereas those that are pollinated by passerines in the Old and New World secrete hexose (mean ratio = 0.041; 19 species).

A consumer of nectar may well dictate the nectar's sugar composition because, after all, nectar is an enticement to attract pollinators. The Bakers suggested that some consumers of nectar

Table 12.2 Sugar composition of nectars used by various birds

Pollinators	*n*	Sugar ratios*			
		<0.1	0.1–0.499	0.5–0.999	>0.999
Birds					
Hummingbirds (Trochilidae)	140	0.00	0.13	0.32	0.55
New World passerines	12	0.92	0.08	0.00	0.00
Sunbirds (Nectariniidae)	35	0.69	0.26	0.06	0.00
Honeyeaters (Meliphagidae)	22	0.82	0.18	0.00	0.00
Honeycreepers (Fringillidae)	6	0.83	0.17	0.00	0.00
Lorikeets (Psittacidae)	3	0.33	0.67	0.00	0.00
Mammals					
New World (Phyllostomidae)	27	0.33	0.67	0.00	0.00
Old World (Pteropodidae)	7	0.14	0.43	0.29	0.14
Nonvolant mammals	5	0.00	0.40	0.40	0.20

*Sugar ratio = sucrose/(glucose + fructose). Data in columns represent proportions of total ratios encountered.
Source: Derived from Baker and Baker (1983).

characterized by hexose may simply have transferred a taste for fruits, which are dominated by hexoses, to nectar. But that begs the question of why fruits contain much hexose. Another factor is the ability to metabolize sucrose: species that drink glucose-fructose solutions cannot hydrolyze sucrose; that is, they lack sucrase (Martinez del Rio et al. 1988, 1989; Martinez del Rio and Stevens 1989). The occurrence of sucrase in avian families, subfamilies, and genera varies with food habits. If sucrose gets to the colon in birds that lack sucrase, it is rapidly fermented (similar to lactose intolerance in humans), producing lactic acid, reducing pH, and killing cellulolytic bacteria, thereby reducing the absorption of electrolytes and producing diarrhea.

Hummingbirds are distinctive in their preference for nectars that are rich in sucrose (Martinez del Rio 1990a, 1990b). This preference is not due to differences in the processing rate of sucrose over a mixture of hexoses (Martinez del Rio 1990b). Sucrase activity is proportional to maltase activity in birds, and both are much higher (2 to 118 times) in hummingbirds than in passerines (Martinez del Rio 1990a). Why hummingbirds prefer plants with sucrose-rich nectar, however, is unlikely to be simply that they have sucrase, as if that is some historical accident. Could the answer be that the osmotic effect of sugar on the gut is reduced if a given energy content is represented by a disaccharide than by a mixture of monosaccharides? An osmotic load may have its greatest effect in a small mass. Yet, some sunbirds are as small as many hummingbirds, and they prefer hexose-bearing nectars.

The transport of molecules across the intestinal wall is a further complication to nutritional status. Glucose transport is faster in herbivorous fish, lizards, and birds than in carnivorous species (Karasov and Diamond 1983). The variation in rate of amino acid (proline) transport is much less than the variation in glucose transport (Karasov et al. 1985), possibly because all vertebrates need amino acids as building blocks, but glucose is not used as an energy source in carnivores (also see Karazov et al. 1985). In fact, amino acid availability may be limited in plants. Baker and Baker (1982) demonstrated that nectar has a higher concentration of amino acids if the pollinators feed strictly on nectar (e.g., carrion and dung flies that feed on flowers that mimic carrion or dung), but has a lower concentration in vertebrate-pollinated flowers because most vertebrates supplement nectar with insects. The sum of glucose and proline transport across the gut directly correlates with the level of metabolism. Thus, in a comparison of a herbivorous mammal (*Neotoma lepida*) and a herbivorous lizard (*Dipsosaurus dorsalis*) feeding on the same diet (alfalfa pellets), Karasov and Diamond (1983) showed that the mammal had a much greater food consumption (ratio of 8 : 1), higher assimilation efficiency (54% vs. 45%), shorter passage time (1–2 vs. 7 d), and larger small (ratio of 1.8 : 1) and larger (ratio of 6.8 : 1) intestines than the lizard. These differences exist between lizards and nonruminant mammals, in spite of having a similar gut structure

(Karasov et al. 1986b), probably because of the more variable and generally lower body temperature in lizards than in mammals. Mammals clearly eat more food, pass it more rapidly through the gut, and extract a greater proportion of energy from food than lizards.

12.6 SPECIAL PROBLEMS OF HERBIVORY

Two principal problems exist with consuming the vegetative parts of plants: much of the contained energy is in the form of long-chain structural polymers that vertebrates cannot readily hydrolyze, including cellulose, hemicellulose, and lignin, which are collectively referred to as *fiber*. The vegetative parts of plants also often have a wide variety of toxic secondary compounds. These problems either can be faced directly or can be reduced through the selection of high-quality foods.

12.6.1 Processing a herbivorous diet. The energy contained in plant matter is found in the cell constituents and cell walls. Cell constituents can be digested directly by vertebrates, but cell walls normally cannot because they are made of pectin, cellulose, and hemicellulose. The availability of cellulose and hemicellulose can be reduced further by the addition of lignin, which complexes these polysaccharides with cross-linkages. As a consequence, foliage-eating herbivores have been estimated (Karasov 1990) to consume 3 times as much dry matter as insectivores or carnivores to obtain the same amount of usable energy.

Within terrestrial habitats, plant abundance is positively correlated with a high cell wall content (Demment and Van Soest 1985): high-quality forage is rare and low-quality forage is common. Therefore, as the nutrient demands of a herbivore increase, either at a fixed mass or due to an increase in mass, it must use evermore low-quality forage, which requires it to maintain a full gut for continuous processing. Energy and nutritional balance thus may be limited by the capacity for processing in herbivores, rather than by the availability of food. This limit is most pronounced at large masses: the surface area of the gut is proportional to l^2, where l is body length, whereas the nutritional requirements are proportional to $(l^3)^{0.75} = l^{2.3}$. That is, with an increase in mass the increase in surface area for processing food is less than the increase in metabolism (Westoby 1974).

The limit to an increase in body size in a herbi-

vore feeding on abundant food of low nutritional value can be seen in the giant panda (*Ailuropoda melanoleuca*). This panda feeds on the stems, branches, and leaves of bamboo belonging to the genera *Sinarundinaria* and *Fargesia* (Schaller et al. 1985). It possesses a gastrointestinal tract that is similar to the tracts in other members of the order Carnivora and unspecialized for its (nearly) strictly herbivorous diet. Much of the food consists of culms with a diameter greater than 5 cm and leaves, leaves having the lowest fiber content (66% of dry mass) and the highest protein content (13%). Yet, only 19% of the dry matter of bamboo is digested, including nearly none of the celluose and lignin (Dierenfeld et al. 1982). What little digestion occurs depends on crushing food in the mouth, with little chewing, and a very short retention time of food in the gut (ca. 8 h). To obtain enough energy, the panda, which weighs about 85 kg, has a daily field intake of 10 to 18 kg of bamboo (Schaller et al. 1985). This ensures an adequate protein intake given the apparent 90% digestibility of protein. To obtain this amount of food, pandas spend over 90% of the active period feeding. During the spring, when giant pandas feed selectively on the leaves and shoots of *Fargesia*, mass intake may be as high as 38 kg/d! Clearly, an upper limit to body size is set by the increase in requirements for nutrients associated with a large size, combined with the limit on the amount of low-quality food that can be processed imposed by the trade-off between digestive efficiency and passage time.

Some of the limitations associated with a herbivorous diet also can be seen in reptiles. In the green sea turtle (*Chelonia mydas*) a section of the colon (the so-called caecum) is always filled with a green fluid (Bjorndal 1979). This turtle mainly feeds on the seagrass *Thalassia testudinum*: in Nicaragua 78.9% of the food intake is this grass; 9.7% is other seagrasses; 8.2%, algae; 1.8%, bottom matter; and 1.4%, animals (Mortimer 1981). Green turtles are selective feeders in that 56% of the *Th. testudinum* growing in the beds is represented by old leaves, but only 7.8% of the leaves found in stomach contents are old leaves. In part, this selectivity is facilitated by the tendency of green turtles to regraze selected areas (Bjorndal 1980, 1982), thereby ensuring an increase in the quality of the grass; new grass blades have only 50% of the lignin content of old blades and 11% more protein (Bjorndal 1980). Yet, seagrass has a

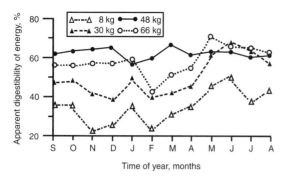

Figure 12.10 Energy digestibility of seagrasses by *Chelonia mydas* as a function of the time of year and body mass. Source: Modified from Bjorndal (1980).

high fiber content, low protein content, and low digestibility. The digestibility of seagrass is further reduced by the presence in grass of phenols and flavenoids that complex with plant protein or the turtle's digestive enzymes (Bjorndal 1985). Digestibility of organic matter in *Chelonia*, however, increases to some extent with an increase in mass (Figure 12.10), possibly because a large mass has higher body temperatures in winter, or because retention time in the gut increases with mass. Green turtles have low growth rates, delayed sexual maturity, and a low energy allocation to reproduction (Bjorndal 1982), which may reflect the low digestibility of seagrass: growth rates of captive green turtles increase if protein is supplemented in the diet.

Eating seagrass may preclude *Chelonia* from extensively eating algae because algae are composed of entirely different structural carbohydrates, which may require a change in the physiology of digestion and the gut microflora. In fact, green turtles in Surinam preferentially feed on algae, and in association with this difference, Surinam turtles have a 2-y, rather than a 3-y, breeding interval and lay more eggs than *Chelonia* from Tortuguero, Costa Rica (Bjorndal 1982), where they feed on seagrass.

Some of the most difficult foods for herbivores to process are the leaves of trees belonging to the genus *Eucalyptus* because they have a low protein content, high cell-wall fiber and tannin contents, and many essential oils. Several marsupials eat these leaves, either in an obligatory manner, such as the koala (*Phascolarctos*) and greater glider (*Schoinobates*), or facultatively, including the ringtail (*Pseudocheirus*) and bushtail (*Trichosurus*) possums. Cork et al. (1983), Chilcott and Hume

(1984a, 1984b), Hume et al. (1984), and Foley (1987) studied the digestive response of marsupials to this food. The most important sources of energy in these leaves are the cell contents. Cell wall digestibility is as low as 25% in the koala and as high as 45% in the ringtail, which is not aided by retention time because of the high lignin content. Only 26% of the lignin is digested by the ringtail and 19%, by the koala. (Note, however, that a question exists whether lignin is ever digested, or whether its *apparent* digestion is a product of the markers used [Bjorndal 1987, Van Soest 1994].) The absorption and excretion of essential oils lead to a poor conversion of digestible to metabolizable energy (Foley 1987). The low rates of metabolism found in these arboreal folivores may well be required to balance an energy budget based on a poorly digested fiber diet (Cork et al. 1983, Chilcott and Hume 1984b). The use of fruit by *Pseudocheirus* in the field may permit higher levels of energy expenditure than can be attained with a strictly folivorous diet (Chilcott and Hume 1984a).

These marsupials also encounter a low nitrogen content in *Eucalyptus* (Degabriele 1981, 1983; Chilcott and Hume 1984a; Hume et al. 1984), as low as 1% to 2%. They can tolerate such low protein contents in their diets because of their low maintenance expenditures. *Pseudocheirus* tolerates lower leaf nitrogen contents than does *Schoinobates* because of the use of caecophagy by *Pseudocheirus* (Hume et al. 1984).

In northern Eurasia, the Alps, and Iceland, the mountain hare (*Lepus timidus*) of necessity eats plants that have a low digestibility (<40%), notably the twigs of birch and willow. It must have a high volume intake for mass maintenance, which occurs only if the digestibility of dry matter is greater than 30% (Pehrson 1979). Otherwise, the volume that must be consumed for mass maintenance would be too great to process. Digestibility can be increased by selecting twigs of small diameter to ensure that the bark-wood ratio is greater than 1.0, because bark has a much higher digestibility than wood. Digestion is also facilitated by the separation in the colon of particles by size, with coarse (wood) particles being eliminated and fine particles being shunted to the caecum.

A 3.5-kg hare would require about 1540 kJ/d for maintenance and, at a minimum, about 96 kJ/d for activity (Pehrson 1983a). The maximal consumption of twigs is about 120 dry g/kg·d, which is equivalent to approximately 2200 kJ of metabolizable energy per day; the minimal expenditure is, thus, about 74% of the maximal intake. Given variations in the physical conditions in the environment and therefore in the cost of thermoregulation, the availability of quality food may limit activity. Large mammals can store fat to use as an energy source during periods of energy shortage, but hares are too small for fat metabolism to be an important means of evading this limitation.

Other nutritional requirements, including nitrogen and protein, calcium, potassium, and magnesium, are met in mountain hares when energy requirements are met (Pehrson 1983b), although some herbivores living in areas characterized by soils that are unusually deficient in essential elements may not be able to maintain elemental balance, even when energy intake is sufficient (Blair-West et al. 1968).

An abundant, but nutritionally limited food is hypogenous fungi. They are widely used by small mammals living in temperate forests (Fogel and Trappe 1978; Maser et al. 1978, 1985; Ovaska and Herman 1986). The ability of a ground-squirrel (*Spermophilus saturatus*) to digest nitrogen and energy contained in a fungus (*Elaphomyces granulatus*) is low (ca. 50%; Cork and Kenagy 1989). Yet, 50% of the digested energy in the ground-squirrel apparently comes from the fungal cell walls, presumably with the aid of microbial fermentation. Ground-squirrels, however, require food with at least 60% to 70% digestibility to maintain body mass, which explains why they switch to new green vegetation in the spring. (Note that the sensitivity of herbivorous mammals to the digestibility of foods is highly variable: *Microtus pennsylvanicus* can maintain mass when the forage intake is 50% cell walls [Keys and Van Soest 1970], whereas *M. ochrogaster* can maintain mass when the digestibility of food is as low as 30% [Batzli and Cole 1979].) Given the poor nutritional quality of fungi, why are they eaten? Cork and Kenagy suggested that their consumption results from their abundance and ease of location. Fungi as a food require only a small nutritional supplement to become part of an adequate diet.

The greatest difficulties with the use of vegetation as a principal food source are found in small species. As with other time periods (see Section 5.3.1), gut retention time is proportional to about $m^{0.25}$ (Demment and Van Soest 1985, Worthington 1989, Karasov 1990, Veloso and Bozinovic 1993, Cork 1994). As a result, the increase in retention

time with body mass facilitates the digestion and fermentation of plant fiber at large masses. Conversely, small species have a short retention time, which requires them to use high-quality foods. This size dependency for the use of fiber led Cork (1994) to characterize mammals that weigh more than 15 kg—which as a limit is too big—as being "fiber-tolerant" and those that are smaller as "fiber-intolerant." The few fiber-intolerant mammals that specialize on folivory are arboreal folivores, leporids, and arvicolid rodents. Leaf eating is rare at masses less than 700 g: most arboreal mammals feed on low-fiber foods, such as fruits, flowers, and insects. As Foley and Cork (1992) noted, small herbivores can partially evade the short gut retention time by separating the soluble from the insoluble fractions of the diet.

Another potential response to a fiber-rich herbivorous diet is to reduce the requirements for energy and nutrients. Mammals that specialize on high-fiber diets usually have low rates of metabolism (McNab 1978b, 1986a; Cork and Foley 1991; Cork 1994), but in the case of arboreal folivores this correlation is complicated by their small muscle masses (Grand 1978) and the correlation of basal rate with muscle mass (McNab 1978b). Veloso and Bozinovic (1993) demonstrated that the basal rate of the rodent *Octodon degu* decreased during acclimation to a high-fiber diet for 27 weeks. They suggested that the degu compensated for the high-fiber diet by increasing food intake, which led to an increase in feces production. Veloso and Bozinovic showed that dry matter intake increased and basal rate decreased as the digestible energy content of consumed food decreased (Figure 12.11). These adjustments permitted the degu to maintain a balanced energy budget when food quality was poor.

The decrease of gut transit time at small masses has several consequences for herbivores:

1. The efficiency of digestion is reduced, markedly in avian frugivores (Levey and Karasov 1989, Worthington 1989, Karasov 1990, Karasov and Levey 1990).
2. Bulk food is rapidly processed.
3. An increased amount of food is processed, which permits nutritional requirements to be met.

Mixed-diet frugivores cannot be maintained in captivity on fruit alone, owing to a nutrient imbalance, mainly because they cannot process sufficient

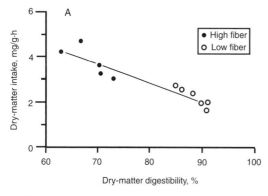

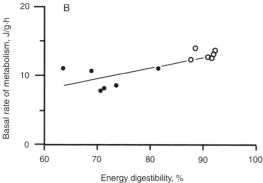

Figure 12.11 A. Dry-matter intake by the rodent *Octodon degu* as a function of dry-matter digestibility and fiber content. B. Basal rate of metabolism in *O. degu* as a function of energy digestibility and fiber content. Source: Modified from Veloso and Bozinovic (1993).

amounts of food to obtain enough protein (Levey and Karasov 1989, Karasov and Levey 1990). In contrast, frugivore specialists can balance their nitrogen budgets because of a high fruit intake, a low nitrogen requirement (or a high nitrogen efficiency), and possibly the consumption of a highly diverse fruit diet (Worthington 1989).

The exclusive use of fruits is limited to the tropics, where fruits are available throughout the year. The occurrence of strict frugivory may also be limited by the occurrence of toxins in fruits (Karasov and Levey 1990). In temperate zones the inclusion of fruits in the diet of birds is limited by the seasonal availability of fruit, which therefore means that birds must switch to other foods when fruits are unavailable. With the shift from fruit to insects, American robins (*Turdus migratorius*) shorten gut retention time (Levey and Karasov 1989).

Tedman and Hall (1985) showed that the larger frugivorous flying foxes *Pteropus alecto* and *Pt.*

poliocephalus have gut transit times that are similar to times of those of the much smaller megachiropteran *Rousettus aegyptiacus*. These short transit times occur in spite of gut lengths that are up to 9 times the body length. Tedman and Hall found no correlation in fruit bats between gut retention time and gut length (as well might also be the case in frugivorous birds; Karasov and Levey 1990) and suggested that a large gut surface area might compensate for the rapid movement through the gut of food that, in the case of megachiropterans, is mainly fruit juice and flesh, the fiber being spit out by the bats after mastication.

12.6.2 Mixed diets. Some herbivores reduce nutritional limits imposed by a particular food by using a mixed diet. Demment (1983) examined the food habits of baboons (*Papio cynocephalus*) in relation to body size. Males are much larger (22.5 kg) than females (13.7 kg). As a consequence, males eat more fibrous food (24.5% vs. 19.0% fiber, respectively). Maximal male size may be limited by the amount of food available and processed. The maximal size of female baboons also may reflect the maximal amount of food that can be processed, but it may correspond to the seasonal increase in rate of metabolism associated with reproduction. If so, the seasonal increase in rate of metabolism expected from the size difference in baboons should be

$$\frac{\text{male}}{\text{female}} \cong \left(\frac{22.5}{13.7}\right)^{0.75} = 1.45.$$

Demment suggested that the maximal rate of metabolism of lactating baboons may be 50% greater than the basal rate, which is close to the difference estimated from body mass (also see Section 11.6.2).

In hyraces, two species, *Procavia johnstoni* and *Heterohyrax brucei*, often have a locally sympatric distribution on kopjes (rocky hills) in Tanzania. In the wet season, when food is abundant, *P. johnstoni* is a grazer and *H. brucei* is a browser (Hoeck 1975). *Procavia johstoni* often feeds on grass in groups, a behavior that prevents the growth of lignified leaves; a similar behavior occurs in wandering herds of ungulates in East Africa (McNaughton 1979). With the onset of the dry season, grass growth stops and the quality of grass declines. At this time the nutritional value of the leaves of some bushes and trees exceeds that of grass, and *P. john-*

stoni switches to feeding on browse until the beginning of the next rainy season, when it returns to feeding on grass, as does *H. brucei* until the first leaves are produced by the bushes and trees that are its usual food.

In the trade-off between digestive efficiency and passage rate of food through the gut, cause is difficult to separate from effect. For example, howler monkeys (*Alouatta palliata*) feed principally on leaves, which they supplement with fruit, and have a long passage time (16–24 h), whereas spider monkeys (*Ateles geoffroyi*) feed principally on fruit, supplemented with young leaves and buds, and have a short passage time (4–5 h). Milton (1981) argued that once a particular food is preferentially used, the attendent morphological, physiological, and behavioral adjustments to this diet tend to preclude the use of other foods, but that concept is difficult to apply except when comparing the most extreme diets. Surely, the preference of *Ateles geoffroyi* for fruit and of *Aloutta palliata* for leaves could be (evolutionarily) reversed under appropriate conditions.

During an arctic winter little high-quality food is available for a herbivore (Batzli 1983). Some arvicolids specialize on grasses (with some mosses, *Lemmus sibricus*), on dicot shrubs (*Dicrostonyx groendlandicus*), or on a mixture of sedges, willows, and herbaceous dicots (*Microtus oeconomus*). The small size of these herbivores minimizes the amount of food that is required and therefore permits a degree of trophic specialization. Arctic ground-squirrels (*Spermophilus undulatus*), however, are much larger and eat a wide variety of plants, especially leguminaceous forbs. The potentially higher energy requirements of *S. undulatus* are avoided in winter because it, unlike arvicolids, hibernates.

Even though digestive efficiency increases with a reduction in passage rate, this does not mean that retention time should always be maximized (Sibly 1981): a zero passage rate obviously cannot be tolerated. Grass rapidly passes through the short gut of barnacle geese (*Branta leucopsis*), but the contained fiber is poorly digested. In contrast, red grouse (*Lagopus lagopus scoticus*) retain digesta for long periods in a gut of large capacity having large caeca. Appreciable proportions of both cellulose (38%) and lignin (44%) are digested by red grouse. (Note again the question as to whether lignin is ever digested.) Although the retention of food in the gut may permit a greater degree of digestion of food, a full gut may, in fact, be a disadvantage: Kenward (1978) showed that woodpigeons (*Columba palumbus*) are more susceptible to predation by hawks at times when the pigeons have full crops.

Jarman (1974) demonstrated the complex nature of food choice. Choice involves the quality and abundance of foods and is associated with the body size, group size, and social structure of consumers, as he showed in East African antelope. Small antelope select high-quality food, which has a limited abundance and thus is capable of sustaining low energy demands (i.e., small masses); these antelope tend to be cryptically colored, solitary, and sedentary and males defend territories. Large antelope, in contrast, use abundant, low-quality foods, which are needed to sustain the high energy requirements of a large mass; they are often conspicuously colored, found in large herds, and highly mobile and do not defend territories. This analysis suggests that "social behavior" is not (completely) separable from "nutritional ecology." Crook and Gartlan (1966) expressed a similar view for primates.

One other adjustment to a low-quality food supply is possible: digestive efficiency can be increased through the use of fermentation produced by symbiotic bacteria. A large mass in herbivorous vertebrates would not be possible without the aid of bacteria. One large (80–100 kg) folivorous mammal that does not depend on fermentation is, as noted, the giant panda, and it spends nearly all of its active hours eating to balance its energy and nutrient budgets. Similarly, two leaf-eating birds from New Zealand also process large volumes of food without any appreciable digestion of complex carbohydrates. One, the kakapo (*Strigops habroptilus*), a flightless parrot, evades the problem of the digestion of complex carbohydrates and lignified fibers by eating the soft shoots of grass, ferns, and sedges; by compressing the juice out of plants; and by regurgitating ingested fiber (Morton 1978). The second, the takahe (*Porphyrio* [*Notornis*] *mantelli*), a flightless gallinule, eats large quantities of grass and produces large volumes of feces (Reid 1974).

12.7 FOREGUT AND HINDGUT FERMENTATION

Fermentation processes plant matter in a few reptiles, a few birds, and many mammals. Through the culturing of anaerobic bacteria, fungi, and protozoa (see Hobson 1988), vertebrates ferment the large, β-linked, structural carbohydrate polymers (especially cellulose, hemicellulose, and possibly lignin) that form the cell wall of plants. Without bacteria vertebrates cannot hydrolyze cellulose because they do not synthesize cellulase. The principal products of fermentation include volatile fatty acids, methane, CO_2, ammonia, and surplus microflora. Volatile fatty acids are the principal source of energy available to vertebrates derived from fermentation. Methane is an energy sink that is passed as a gas. Another product of fermentation is heat: rumen temperature is usually above abdominal temperature and equal to 39 to 41°C. One peculiar apparent exception is found in the naked mole-rat (*Heterocephalus glaber*), in which the temperature of the caecum is 2°C lower than that of the adjacent abdomen (Yahav and Buffenstein 1992). Whether this observation is unique to *Heterocephalus* or is typical of hindgut fermenters is unclear.

The concentration of volatile fatty acids found in the gut reflects the foods used, the retention time of food in the gut, and the differential absorption of volatile fatty acids. Most vertebrates produce the same acids (Table 12.3) listed as follows according to decreasing order of their usual relative abundance: acetate, propionate, butyrate. Evidence from the study of ptarmigan suggests that these abundances reflect the differential absorption of volatile acids, as well as their differential production (Gasaway 1976a, 1976b). Thus, the absorption of butyrate is greater than that of propionate, which is greater than the absorption of acetate. Because absorption is greatest in the fatty acids that have the highest molecular weights, butyrate contributes the greatest amount of energy to the vertebrate depending on fermentation (Gasaway 1976b). The importance of volatile fatty acids is seen in wombats: these acids contribute 30% to 33% of the digestible intake (Barbosa and Hume 1992b), which permits wombats to maintain mass on high-fiber grasses and sedges because of the effectiveness of fermentation and the low-energy requirements of wombats.

An anatomically specialized section of the gastrointestinal tract often is set aside for fermentation to provide the microflora responsible for this process with the appropriate conditions. These conditions include an anaerobic environment, constant chamber temperature, an alkaline or neutral pH, a continuous supply of food, a continuous removal of wastes, and an appropriate diet. Fermentation is located in one of two principal sites: an enlarged foregut, or an enlarged hindgut, either associated with a large caecum or an elaborate colon. Foregut fermentation occurs before gastrointestinal digestion in artiodactyl ruminants (cattle, sheep, goats, antelope, giraffe, deer, and camels) and hippopotami, macropodid marsupials, tree-sloths, and colobine monkeys. Foregut fermentation is also found in one bird, the hoatzin (*Opisthocomus hoazin*). A greater proportion of the energy contained in consumed plants is metabolized in hoatzins than in other herbivorous birds (which use hindgut fermentation), and it is as high as in large mammalian herbivores (Grajal 1995). Hindgut fermentation occurs after gastric digestion and is found in horses, tapirs, rhinoceroses, rabbits and hares, hyraces, elephants, capybara, and some small rodents, although some small rodents have pregastric (Keys et al. 1970) as well as postgastric fermentation. Hindgut fermentation is also found in some iguanid lizards, the green sea turtle, some grouse, and the emu (*Dromaius novaehollandiae*) (Herd and Dawson 1984).

Mammals have been the principal subjects in the study of fermentation. Several summaries (Janis 1976, Parra 1978, Kinnear et al. 1979, Demment and Van Soest 1985, Hofmann 1989) provide an ecological and evolutionary context within which to evaluate the contributions and limitations of fermentation to herbivorous mammals. The few studies of fermentation in reptiles and birds will be compared to that in mammals.

12.7.1 Scaling fermentation in mammals. The use of fermentation by herbivorous mammals is fundamentally influenced by body size (Janis 1976, Parra 1978, Demment and Van Soest 1985, Hofmann 1989). Gut capacity in mammals scales to body mass in an approximately isometric manner (Figure 12.12), irrespective of whether the entire gut (Parra 1978, Demment and Van Soest 1985) or the reticulorumen (Demment 1982) is considered. Therefore, retention time (t), as estimated by the ratio of gut capacity (GC) to rate of metabolism (MR), increases with body mass:

$$t = \left[\frac{GC}{MR}\right] \cong \left[\frac{m^{1.03}}{m^{0.75}}\right] = m^{0.28}. \qquad (12.1)$$

Table 12.3 Volatile fatty acid production by fermenting vertebrates

Species	Location	Total (mM/L)	Acetate	Propionate	Butyrate
			Volatile fatty acids		
			(molar %)		
Fishes					
Acanthurus lineatus	Gut segment IV	18	91	3	6
Acanthurus nigricans	Gut segment V	29	64	27	6
Naso unicornis	Gut segment IV	42	97	1	2
	Gut segment IV	42	97	1	2
Scarus niger	Gut segment IV	10	85	1	2
Centropyge bicolor	Gut segment V	48	74	18	3
Siganus argenteus	Gut segment V	17	86	8	3
Siganus puellus	Gut segment V	23	82	10	2
Reptiles					
Chelonia mydas	Caecum	156	93	2	6
	Anterior colon	191	83	2	15
	Midcolon	207	78	8	14
Iguana iguana	Hindgut		40	9	52
Birds					
Dromaius novaehollandiae	Ileum	14	92	2	7
	Rectum-cloaca	17	87	1	12
Chenonetta jubata	Caeca	62	71	25	5
Lagopus lagopus	Caeca	194	64	23	12
Lagopus mutus	Caeca	265	70	22	8
Opisthocomus hoazin	Crop	115	68	15	7
	Esophagus	70	69	15	9
Mammals					
Vombatus ursinus	Proximal colon (region 1)	87	68	26	6
	Distant colon	47	68	20	6
Lasiorhinus latifrons	Proximal colon (region 1)	104	74	19	7
	Distant colon	88	71	17	7
Thylogale thetis	Stomach	120	68	19	10
	Hindgut	54	72	15	10
Macropus rufogriseus	Stomach	129	65	21	9
	Hindgut	66	70	16	10
Presbytis cristatus	Stomach	130	47	26	18
Dugong dugong	Caecum	183	57	17	25
	Large intestine	236	50	17	32
Sheep	Stomach	99	70	18	10
	Hindgut	50	74	17	6
Cattle	Rumen	110	62	24	17

Sources: Derived from Carroll and Hungate (1954), Bauchop and Martucci (1968), Gasaway (1976a, 1976b), Hume (1977), Murray et al. (1977), Bjorndal (1979), McBee and McBee (1982), Herd and Dawson (1984), Dawson et al. (1989), Grajal et al. (1989), Barbosa and Hume (1992b), and Clements and Choat (1995).

The available data suggest that this analysis is correct (Figure 12.13). Unlike some proposed analyses (Cork and Foley 1991, Justice and Smith 1992), this derivation did not involve the (unnecessary) use of mass-specific rates of metabolism (see Equation 5.4). That is, at small masses retention time decreases (turnover rate of food in the gut increases) and at large masses retention time increases (turnover rate decreases). At a given mass, retention time for food in the gut is greatest in ruminants, followed by hindgut fermenters, omnivores, and carnivores (Figure 12.13; Parra 1978). Small herbivores can compensate for a short retention time by selecting high-quality food (i.e., with a low fiber content). If the quality of food is not high, food intake is inversely related to its digestibility:

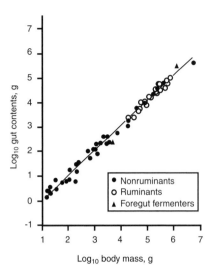

Figure 12.12 Log$_{10}$ gut contents in herbivores as a function of log$_{10}$ body mass. Source: Modified from Demment and Van Soest (1985).

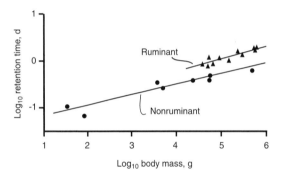

Figure 12.13 Log$_{10}$ retention time of gut contents in ruminants and nonruminants as a function of log$_{10}$ body mass. Source: Modified from Batzli (1985).

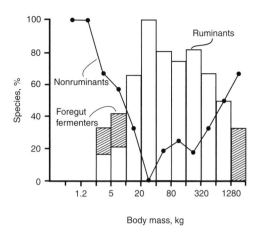

Figure 12.14 Frequency distribution of East African herbivores according to the means by which the food is processed, in relation to log$_2$ body mass. Source: Modified from Demment and Van Soest (1985).

intake is regulated in such a manner that the intake of digestible energy remains constant (Figure 12.8). At large masses, however, herbivores can process food containing a higher fiber content because of the longer retention time of food in the gut.

The interaction of retention time, quality of the diet, and body mass has significance for the distribution of foregut and hindgut fermentation with respect to body size (Demment and Van Soest 1985). Foregut fermenters predominate at intermediate body masses, and hindgut fermenters are most common at small and at very large masses (Figure 12.14). Because of short retention times at small masses, small herbivores use hindgut fermentation, if fermentation is used at all. Effective foregut fermentation can be extended to smaller masses by reducing the maintenance costs of her-

bivores (e.g., kangaroos, tree-sloths) or by having an unusually large chamber for fermentation (e.g., tree-sloths).

Small herbivorous mammals are constrained in their use of fiber as an energy source (Demment and Van Soest 1985, Foley and Cork 1992, Justice and Smith 1992). Small herbivores compensate for the allometric constraint by (1) selectively retaining solutes and fine particles while eliminating coarse fibers, thereby minimizing retention time; (2) using hindgut fermentation, which permits the full extraction of energy and nutrients from the cell contents; (3) increasing colon and caecum size; and (4) using caecophagy (Foley and Cork 1992, Justice and Smith 1992). Justice and Smith developed a mathematical model of fiber use in small herbivores that consistently underestimated the fermentation of fiber at small masses, which may imply the neglect of other adjustments by these species. Hammond and Wunder (1991) noted that fiber digestibilities in *Microtus ochrogaster* increased with dietary fiber content and therefore contributed to a higher proportion of total energy digested. Thus, some small herbivores compensated for a low diet quality and high energy expenditure by increasing gut size, modifying enzymatic digestion, increasing food intake, and increasing foliage preparation by chewing (Hammond and Wunder 1991, Cork and Foley 1991). These adjustments were effective: meadow voles (*Microtus pennsylvanicus*) obtain up to 34% of maintenance energy from fiber (Keys and Van Soest 1970, Foley and Cork 1992) and wood-rats (*Neotoma*)

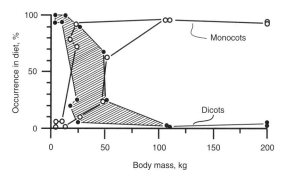

Figure 12.15 The differential use of monocot and dicot plants as food by East African ruminants as a function of body mass. Source: Derived from Sibly (1981).

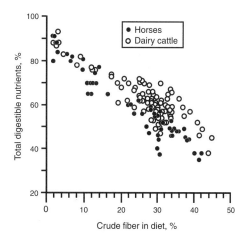

Figure 12.16 Total digestible nutrients attained by horses and cattle as a function of the proportion of the diet that is fiber. Source: Modified from Demment and Van Soest (1985).

obtain 20% of digestible energy from fiber (Justice and Smith 1992).

With an increase in mass, herbivorous mammals switch from the use of foods with low fiber contents, especially dicot leaves, fruit, and seeds, to the use of foods with high fiber contents, especially monocots (e.g., grasses). The switch from dicots to monocots occurs at masses between 20 and 50 kg (Figure 12.15). Dicots are often associated with high processing and absorption rates (Sibly 1981). The switch to eating abundant low-quality monocots from scarce high-quality dicots is required by the increase in total rate of metabolism in large herbivores: enough high-quality food is not available to support a large mass. In a survey (Levin 1976) of 162 dicots, 40 (25%) were toxic, whereas only 13 (11%) of 120 monocots were toxic (see Section 12.8). The toxicity of dicots may well compensate for their higher digestibility and greater nutritional content (Jung 1977). This combination may lead small herbivores to consume large amounts of toxic plants (Jung 1977), which requires the capacity to detoxify the secondary compounds.

The increase in energy expenditure associated with a very large mass cannot be accommodated with a rumen, in spite of its high efficiency, because of the controlled passage of small particles from the rumen (Figure 12.13), so hindgut fermentation is usually used (Figure 12.14). Hindgut fermentation can accommodate the high rate of metabolism associated with a large mass simply by increasing the rate of food intake and passage (compared to foregut fermenters), discarding efficiency, and emphasizing the quantity of food processed. Rumination is less likely to be of value in small foregut fermenters because of the high-quality food that they eat, although it does occur (Figure 12.14) and

has been thought to regulate the rate of fermentation (Hoppe 1977).

Foregut fermentation has some advantages over hindgut fermentation: it aids in the detoxification of secondary compounds (see Section 12.7.3) and is more efficient in extracting energy from structural carbohydrates. Most foregut fermenters are also ruminants; that is, they chew a cud. Ruminants include the artiodactyl Ruminantia (bovids, deer, tragulids) and Tylopoda (camels). Rumination is used to break large particles of food into small particles so that the microflora can ferment more easily the structural carbohydrates of the ingested plant matter. Foregut fermentation coupled with rumination is more efficient in extracting the energy found in plant matter: horses (with hindgut fermentation) digest cellulose with 70% of the efficiency as cows (which are ruminating foregut fermenters), in part because the retention time in cows is 70 to 90 h and that of horses is only 48 h (Sibly 1981). This difference can be seen in Figure 12.16 and generally represents a difference between artiodactyls and perissodactyls (Janis 1976, Demment and Van Soest 1985).

Another advantage of foregut fermentation is that nonprotein nitrogen can be used as a source for nitrogen because of the activity of the microflora. Kinnear et al. (1979) in fact suggested that the fermentation of fiber is an inadequate explanation for the value of foregut fermentation and rumination: they argued that the principal advantages accruing to these phenomena are that they permit the use of foods with lower nutrient levels, and they detoxify secondary compounds—that is, they permit expanded nutritional niches. This expansion is most notable in terms of nitrogen, but it also inserts a trophic level between the food and the herbivore. As a result, energy and nutrients are lost to the microflora, especially in the fermentation of the soluble cell contents of the ingested plants, which can be used directly by the vertebrate herbivores, and in the production of molecules, like methane, which are not used by vertebrates (Kinnear et al. 1979). The nutrient and energy "loss" to microflora is partially recovered by the herbivore's processing of excess microflora. Hindgut fermentation reduces the amount of energy loss by delaying fermentation until after the soluble portion of plant matter is used by the vertebrate.

Many foregut fermenters avoid fermentation of the soluble (nonstructural) fraction of the diet by permitting it to bypass the fermentative portion of the foregut, while retaining the fibrous portion of the food (Cork 1994). Similarly, some caecal and colonic fermenters, especially folivores like the koala (*Phascolarctos cinereus*), greater glider (*Schoinobates volans*), ringtail possum (*Pseudocheirus peregrinus*), and sportive lemur (*Lepilemur mustelinus*), facilitate the elimination of particulate digesta by the retrograde movement of solutes from the colon to the small intestine and caeca, thereby enhancing the processing of the solutes (e.g., Foley 1987, Cork 1994). This separation of the particulate from the soluble digesta permits some (small) species to eat differentially those "feces" that are produced from material derived from the caeca (caecophagy) and to eliminate those feces that are high in fiber (Cork and Foley 1991, Cork 1994). Caecophagy permits hind gut fermenters to use effectively the products of microbial fermentation. Hindgut fermenters that eat food with little fiber (i.e., feed on fruits and seeds and consume few leaves) do not separate the digesta.

When food is not limited, or the food is nutritious, hindgut fermenters (at large and small masses) or nonfermenters (at small masses) tend to dominate foregut fermenters; ruminants dominate principally when resources are limited, or the resource base is of poor quality, because, as noted, foregut fermentation increases the resource base by making a larger fraction of the energy (and nutrients) contained in food available (Kinnear et al. 1979, Sibly 1981). At intermediate masses foregut

fermenters predominate in tropical grasslands, where food is of low quality, whereas hindgut fermenters predominate in tropical forests, especially in the canopy, where food is of a higher quality (Langer 1984, 1986). One of the principal factors encouraging the use of foregut fermentation in grasslands is their magnified environmental cyclicity (compared to forests), which requires on occasion the consumption of food of both low quality and low quantity.

Another specialized group of herbivores are arboreal mammals that feed on the leaves of trees. These folivores depend on microbial fermentation to digest the fibers in plant cell walls. Some are foregut fermenters (e.g., *Bradypus, Colobus, Presbytis, Dendrolagus*) and others are hindgut fermenters (e.g., *Alouatta, Avahi, Lepilemur, Phascolarctos, Schoinobates, Pseudocheirus, Cynocephalus, Dendrohyrax, Petaurista*). Among hindgut fermenters the increase in the size of the fermenting chambers, the caeca and colon, occurs in proportion to the degree of folivory (Cork and Foley 1991; Figures 12.6 and 12.7). The rate of fiber fermentation is quite low, so that these folivores tend to maximize intake, use cell contents and young foliage, and sometimes supplement leaf intake with the consumption of fruits, buds, and flowers (Cork and Foley 1991). Arboreal folivores retain digesta for longer periods than do other mammals of the same mass, and for periods equivalent to those of large herbivores. Foregut fermentation is acceptable in smaller arboreal folivores with their shorter gut retention times only if they have lower nutritional requirements, that is, lower rates of metabolism, as in tree-sloths (Cork and Foley 1991).

12.7.2 Other vertebrate fermenters. Dependence on fermentation also occurs in fishes, reptiles, and birds and may occur in anuran tadpoles. Microbial fermentation of plant matter has been little studied in fishes, although herbivorous fishes are quite diverse in tropical waters. In a survey of 32 tropical marine species, Clements and Choat (1995) found that volatile fatty acids were abundant in the posterior half of the intestine, suggesting that plant fibers were being fermented. Volatile fatty acids were most abundant in fishes that fed principally on macroalgae, followed by some planktivorous acanthurids, with the lowest concentrations in some sediment-eating parrotfish (Scaridae). Some herbivorous fishes have fermentation chambers, but in many with high volatile fatty acids fermentation occurs in the intestine, which implies that the retention of food in the intestine is sufficient to permit fermentation (Clements 1997). One of the distinctive features of herbivorous fishes is that they generally encounter little cellulose (unless they eat aquatic grasses), so that the polysaccharides that they consume include alginic acid, mannans, xylans, and galactans, all from the cell walls of marine algae, as well as laminarin and floridean starch, the storage polysaccharides of algae. This physiological diversity emphasizes the great bias that has constituted fish nutritional physiology: "our knowledge of fish physiology is based largely on studies conducted on carnivorous northern-hemisphere species, and most concern freshwater taxa such as salmonids" (Clements 1997, p. 158). (Such a bias is as prevalent in the study of the physiology of amphibians, reptiles, and birds, as it is for fishes.)

Bjorndal (1997) summarized the data on fermentation in amphibians and reptiles: the digestion of plant matter in the guts of anuran tadpoles may be facilitated by microbes, but is little studied, and the occurrence of gut fermentation in reptiles is limited to a few lizards and to some turtles. The lizard *Iguana iguana* feeds principally on the leaves of tropical trees and vines; it depends on the fermentation of leaves by bacteria in the colon (McBee and McBee 1982). This species obtains 30% to 38% of its daily energy requirements from volatile fatty acids, much of the remainder coming from the soluble cell contents; in this species butyrate may be a more abundant product of fermentation than in other vertebrates (Table 12.3). The green sea turtle (*Chelonia mydas*) has a section of the colon that is specialized to process plant material, the so-called caecum. The principal products of fermentation in this turtle are the same volatile fatty acids that occur in mammalian fermenters in approximately the same quantities (Table 12.3); they contribute approximately 15% of the daily energy requirements of green turtles (Bjorndal 1979). The digestion of food by green turtles increases with mass (Figure 12.10), presumably because thermal stability or body temperature is higher at large masses, but the fermentation of cellulose appears to be independent of mass.

Herbivorous lizards are comparatively rare (ca. 2% of all species [Pough 1973b]). They are principally found in the tropics and hot-temperate deserts and appear to require high, stable body tempera-

tures. Even so, herbivorous lizards are less efficient in digesting food than those with other food habits (Zimmerman and Tracy 1989), in spite of the contribution to digestion of a microflora in the hindgut of most (all?) species. An animal can maximize the digestion of food by maximizing the intake and minimizing the gut transit time (Sibly 1981). In chuckwallas (*Sauromalus obesus*) the effect of low (or variable) body temperatures primarily is on increasing the gut retention time (Zimmerman and Tracy 1989), the low temperature depressing the amount of food that is digested. The principal difficulty faced by herbivorous lizards, then, ". . . seems to be a consequence of a less digestible diet . . ." (Zimmerman and Tracy 1989, p. 383), which may only be circumvented in the warmest climates.

Several birds depend on the fermentative activity of an intestinal microflora. The hoatzin (*Opisthocomus hoazin*), a foregut fermenter from northern South America, has an enlarged crop, which is the fermentative chamber, although Dominguez-Bello et al. (1993) argued that fermentation in the hoatzin is not important for the digestion of cellulose but principally aids in detoxifying plant secondary compounds. The crop and the esophagus constitute 15% of the hoatzin's mass and 75% of its gut mass (Grajal et al. 1989). Fermentation produces volatile fatty acids that are similar in type and abundance to those found in other fermenters (Table 12.3). Other avian fermentation is postgastric, occurring especially in caeca.

Hindgut fermentating birds are most notably grouse, especially ptarmigan (*Lagopus*), spruce grouse (*Canachites*), and capercaille (*Tetrao*). They tend to be large, and like flying-squirrels, the largest species are the most folivorous (Morton 1978). Like other vertebrate fermenters, the volatile fatty acids that occur (in decreasing order of relative abundance) are acetate, propionate, and butyrate (Table 12.4), which undoubtedly means that the microflora and substrates, not the verbrate host, dictate what acids will be produced. Some 7% of the daily energy expenditures of rock ptarmigan (*L. mutus*) is derived from volatile fatty acids, in contrast to 70% being derived from fermentation in artiodactyl foregut fermenters (Gasaway 1976a), although the microflora of foregut fermenters ferment cell contents, which are used directly by hindgut fermenters. As in some hindgut fermenting mammals, willow ptarmigan (*L. lagopus*) and probably other folivorous grouse separate the digesta at the end of the small intestine, with a fluid pulp made of the soluble cell contents and the digested fraction of the cell wall moving into the caeca and the fibrous fraction moving into the colon to form solid droppings (Moss and Parkinson 1972).

The importance of a large body mass for the digestion of fiber is illustrated by the emu (*Dromiceius novaehollandiae*). Herd and Dawson (1984) showed that up to 63% of the standard rate of metabolism and 50% of the maintenance requirements were derived from neutral detergent fiber when the emu was given the highest-fiber diet. This digestion occurred with the aid of symbiotic bacteria, even though the emu lacks a marked differentiation of the gut and has short retention times. The emu survives in arid regions by combining the selection of high-quality foods having a low lignin content with a relatively low rate of metabolism (Calder and Dawson 1978) and a large body size (low turnover rate). Large birds that feed on a highly fibrous, poor-quality diet would be expected to depend on fermentation, including the ostrich (*Struthio camelus*) and rhea (*Rhea americanus*), both of which have a more sacculated hindgut than is found in the emu.

12.7.3 The evolution of foregut and hindgut fermentation.

Janis (1976) described the evolution of ungulates in the context of fermentative herbivory (Figure 12.17). The earliest perissodactyls evolved in the late Paleocene; their molar morphology indicates that they fed principally on a cellulose-rich diet. Janis suggested that artiodactyl "proto-ruminants" lived in tropical forests in the Eocene and ate a low-cellulose diet, presumably including fruits, somewhat similar to the diet of living tragulids (mouse-deer). As climate cooled, the fiber content of plant matter increased. Several responses to the increase in fiber content occurred, including an increase in mass and the development of postgastric fermentation among the Perissodactyla and pregastric fermentation among the Artiodactyla. As seen, an increase in mass increased the tolerance of ungulates to a high fiber content through an increased gut retention time. The development of an appreciable size among the perissodactyls probably required, or resulted from, the development of hindgut fermentation.

Foregut fermentation and rumination developed independently probably 3 times among artiodactyls, once each for bovids, cervids, and camels

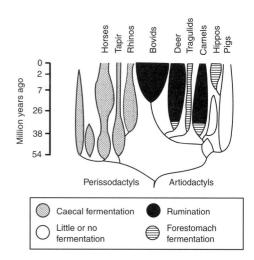

Figure 12.17 The occurrence of fermentation, rumination, and foregut fermentation in relation to the evolution of ungulates. Source: Modified from Janis (1976).

(Figure 12.17). (Note that hippopotami are sometimes included as ruminants, but they are nonruminant foregut fermenters.) They contribute to an increased efficiency of nitrogen utilization and a detoxification of plant secondary compounds, as well as to the digestion of ingested fiber. With the proliferation of artiodactyls in the Oligocene, perissodactyl diversity declined, presumably because the artiodactyls could feed on foods of intermediate fiber content with greater nutritional efficiency. Horses persisted with a large body size by selectively feeding on food of high fiber content. Today zebras (*Equus*), wildebeest (*Connochaetes*), and topi (*Damaliscus*) form mixed herds in East Africa in which zebra consistently select more fibrous parts of plants than do wildebeest. Other living perissodactyls, such as tapirs (*Tapirus*) and the small rhinoceroses (*Rhinoceros sondaicus* and *Dicerorhinus sumatrensis*), ingest food with a lower fiber content than the food ingested by most artiodactyls; they live in tropical forests where the fiber content of the food is less and where few artiodactyls live. Of the large rhinoceroses, two (*R. unicornis* and *Ceratotherium simum*) are grazers and the third, the black rhinoceros (*Diceros bicornis*), preferentially feeds on young browse. Thus, all living perissodactyls are "protected" from artiodactyls, either by the food that they eat or by their large mass. The very large mass of elephants and their mixed diet of grass and browse presumably reduce the ability of ruminant artiodactyls to supplant them. Given that rhinoceroses were derived from taperoids in the Oligocene, a time when the artiodactyls were proliferating, the large body mass of rhinoceroses can be viewed as a response both to an increasing fibrous diet and to competition from foregut, ruminating artiodactyls (Janis 1976). For all of the advantages derived from foregut fermentation, hindgut fermenters collectively have a greater scope for the evolution of diversity (Cork 1994), which is demonstrated at least by their great range in body size and taxonomic diversity.

The separation of herbivores into grazers and browsers may be too simple according to Hofmann (1989), who suggested that a division of herbivores into groups based on food selectivity would be more appropriate. By this classification approximately 25% of mammalian herbivores feed on grass and roughage, depending on plant cell walls for their principal source of energy. Such species include cattle, sheep, mouflon, water buffalo (*Bubalus bubalis*), and banteng (*Bos javanicus*). Another 40% feed principally on plant cell contents, which Hofmann defined as "concentrate selectors," and include roe (*Capreolus capreolus*), white-tailed (*Odocoileus virginianus*), and red deer (*Cervus elaphus*), but no domestic species. The remaining 35% are intermediate in that they have a mixed diet; they include wapiti (*C. elaphus* in North America), impala (*Aepyceros melampus*), domestic goats, caribou (*Rangifer*), and Thomson's gazelle (*Gazella thomsoni*). The relationship of body size to these feeding habits is complex: most smaller (<40 kg) ruminants are concentrate selectors, but oribi (*Ourebia ourebi*, 12–20 kg) and blackbuck (*Antilope cervicapra*, 30–40 kg) are grass and roughage eaters, and many are intermediate feeders. Several concentrate selectors are surprisingly large (greater kudu [*Tragelaphus strepsiceros*], bongo [*T. euryceros*], moose [*Alces alces*], and giraffe [*Giraffa camelopardalis*]), that is, 180 to 1000 kg. Hindgut fermentation is used in concentrate selectors and in those with intermediate diets, although it is less common in species that preferentially feed on grass and roughage (Hofmann 1989).

Owen-Smith and Novellie (1982) and Owen-Smith (1985) described a quantitative model of ungulate herbivores, incorporating such features as rate of movement during foraging, width of food path, bite size and rate, gut capacity and passage rate, and body mass, in an attempt to account for the diversity of ungulates coexisting in Africa. Various predictions were made:

1. Larger browsers maximize nutrient "profit" by including low-quality foods.

2. Grazers do better on grass than browsers confined to grass.

3. Hindgut fermenters should have a broader range of dietary quality than foregut fermenters.

4. Each environment has an optimal ungulate size, given its particular mix of vegetation.

These patterns are found in the African ungulate fauna. Owen-Smith (1985, p. 171) concluded "... that the discrete grading in body sizes [of ungulates] reflects the distribution of leaf sizes and leaf : stem ratios prevailing among woody plants and forbs."

The dependence of mammals on fermentation evolved independently in marsupials, primates, xenarthrans, elephants, and many rodents. It has also evolved several times in birds and reptiles. For example, specialized folivory is found in lizards, including several members of the Iguanidae, the genera *Uromastyx* and *Hydrosaurus* (Agamidae), and the scincid genus *Corucia* (Iverson 1982). Like most living reptiles, living birds that are folivorous have an intermediate to small mass (from an ungulate viewpoint). However, the 12 or so moas (order [†]Dinornithiformes), flightless ratites limited in distribution to New Zealand, occupied browsing niches in the absence of mammalian browsers, weighed from 25 to 240 kg, but were exterminated by Polynesian colonists before the appearance of the first Europeans (Anderson 1989).

The argument has been made that mammals that specialize on the green parts of plants can attain a large mass only if they rely on a symbiotic relationship with bacteria. If that is correct, and nearly all large mammalian herbivores use fermentation, it suggests that large herbivorous pelycosaurs, therapsids, and dinosaurs also depended on fermentation. Maybe the very large masses of herbivorous therapsids and dinosaurs represented a means of attaining the thermal constancy required by a microflora. Yet, elephants are poor digesters of fiber and depend on a high passage rate. Herbivorous dinosaurs too may have had a high passage rate because of their bulk requirements. Norman and Weishampel (1985) suggested that ornithopod dinosaurs used their teeth to grind plant fibers, which if correct may imply that they were preparing food for fermentation.

12.8 THE DISTRIBUTION OF PLANT SECONDARY COMPOUNDS

Many compounds distinctive to green plants are found in concentrations well beyond those required for normal plant metabolism (Levin 1971, Feeny 1975). These compounds include phenols and tannins, terpenes, alkaloids, cyanogenic glycosides, protease inhibitors, lectins, nonprotein amino acids, cardiac glycosides, and oxalates, among others. They include such compounds well known to (and some loved by) humans as strychnine, reserpene, caffeine, cocaine, nicotine, colchicine, mescaline, morphine, cyanide, and urushiols (active ingredients in poison ivy, oak, and sumac). These molecules are toxic to some invertebrates and vertebrates, but as Janzen (1979) pointed out, their toxicity is relative to the organism and is not an intrinsic property of the molecules.

Secondary compounds are not limited in distribution to the vegetative parts of plants but also occur in their reproductive products (Janzen 1977, Herrera 1982, Izhaki and Safriel 1989). For example, tannins are found in acorns (Koenig and Heck 1988) and are abundant in many fruits (Bairlein 1996). Many plants use secondary compounds to protect immature fruits, but the degradation of these compounds is often incomplete with the ripening of the fruits, so some toxic compounds may remain in ripe fruit (Herrera 1982). The occurrence of toxic compounds in fruits reduces their use by birds and mammals, although they also reduce the development of fungal rot (Janzen 1977, Cipollini and Stiles 1993). The antifungal effects are produced by phenols and simple organic acids. They are most prevalent in persistent fall, low-quality fruits, and least well developed in readily eaten, summer, small-seeded fruits. Indeed, Janzen (1977, p. 691) went so far as to suggest that ". . . nutrients and secondary compounds in . . . fruit [are engineered] such that it becomes a desirable food item to a . . . subset of the animal community." The presence of toxic defenses in fruit precludes the consumption of large amounts of fruit (Herrera 1982), so toxic fruits tend to be associated with small crops and lower nutritive rewards. The consumption of fruits may be permitted in birds by an insensitivity to some secondary compounds, such as tannins (Bairlein 1996). Frugivores that have a diverse diet may be required to detoxify a variety of secondary compounds (Jordano 1988).

Environmental conditions influence the occurrence of secondary compounds in woody plants, which has consequences for consumers. For instance, some areas of Amazonia have sterile soils that consist principally of white sand, whereas other areas have soils with a higher organic and nutrient content: in the Guianan shield the abundance and diversity of primates and other mammals are related to the carrying capacity of the soil (Emmons 1984). Two species of titi monkeys (*Callicebus*) are found in Amazonia: one, *C. torquatus*, is restricted to forests growing on the white-sand soils and the other, *C. moloch*, lives in forests growing on richer soils (Kinzey and Gentry 1979). About 70% of the diet of both species is fruit. The remainder of the diet in *C. moloch* is mainly leaves, which because of their abundance permits this monkey to be inactive for appreciable periods of time (or requires inactivity to digest leaves). In contrast, *C. torquatus* supplements its diet of fruit with insects, which Kinzey and Gentry suggested is because the leaves of trees growing on nutrient-poor white sand are precious and therefore are protected by high concentrations of secondary compounds. As a result, this species of titi monkey must spend much of the daylight hours searching for insects to balance its energy and nutrient budgets.

Arboreal folivores encounter more lignin, tannins, and other phenols (i.e., molecules with digestion-reducing capacities) than do fruit, seed, or flower eaters (Cork and Foley 1991). How arboreal folivores handle secondary compounds may have a climatic component: no foregut fermenters live in temperate forests, and what few folivores are found, namely Australian marsupials, are hindgut fermenters (Cork and Foley 1991). Whether the near restriction of arboreal folivores to tropical forests is related to the chemistry of leaves or to historical contingency is unclear. (Yet, a series of temperate folivorous grouse are found in the Northern Hemisphere.) Tropical folivores encounter a wide range of toxins, whereas temperate species principally encounter tannins in evergreen species, tannins not being particularly important in deciduous woody browse (Robbins et al. 1987a). Alkaloids appear to be most common in the tropical lowlands, where herbivory is most highly developed (Levin 1976). Within tropical forests the polyphenol and fiber contents of leaves increase with age (Coley et al. 1985).

Food choice may be determined positively by

nutrient content and negatively by toxicity. The palatability of winter browse to snowshoe hares is not normally correlated with its nutritional content (Bryant 1981a, 1981c; also see Coley et al. 1985). Some secondary compounds, such as oxygenated monoterpenes, are of special interest because they have antimicrobial activity and thus poison the microbial flora that provides the fermentative activity in the caecum of hares. The adventitious shoots of preferred deciduous browse species, including paper birch (*Betula resinifera*), aspen (*Populus tremuloides*), and balsam poplar (*P. balsamifera*), are less palatable than green alder (*Alnus crispa*) or black spruce (*Picea mariana*) in their mature growth stages, even though the hare cannot survive on alder or spruce (Bryant 1981b). The nutritional content of the adventitious shoots is greater, but they often have high concentrations of secondary compounds. Fish feed on algae in proportion to their nutrient content, but blue-green bacteria are generally not eaten, possibly because of high toxicity (Ogden and Lobel 1978). The termites eaten preferentially by the cricetine rodent *Oxymycteris roberti* principally are defended by soldiers, those species using chemical defenses being the least acceptable, a defense similar to the chemically based defenses of plants (Redford 1984).

A geographic pattern in the occurrence of secondary compounds in woody plants was shown by the selective feeding habits of high-latitude hares (Bryant et al. 1994, Swihart et al. 1994). In eastern North America snowshoe hares (*Lepus americanus*) preferred mature to juvenile twigs of northern species of birch (*Betula*) and willow (*Salix*) and juvenile twigs from southern latitudes (Maryland, Connecticut) to juvenile twigs from northern latitudes (Maine). This difference was greater when Alaskan birches and aspen (*Populus*) were compared to those from Maine and Connecticut. Northern snowshoe hares discriminated among twigs more than southern hares, apparently to minimize the cost of detoxification of secondary compounds while maintaining an energy balance. Birches and willows have higher levels of secondary compounds after being browsed (Bryant et al. 1985b) in northern latitudes. Because of the shorter growing period and possibly the greater cyclic variations in hare abundance, the consumption of woody browse at northern latitudes leads to higher levels of secondary compounds. Consequently, free-ranging hares in Alaska and Maine ate less aspen from Alaska than from Maine and Connecticut (Bryant et al. 1994).

The use of woody plants by high-latitude hares also varies with longitude, which has been interpreted as reflecting the differential distribution of browsing mammals (Bryant et al. 1989). Woody plants from regions with few or no browsing mammals would be expected to have fewer secondary compounds and therefore would be more palatable than similar plants from regions with many browsing mammals. Thus, boreal birches and willows from Iceland, which had no native browsing mammals, would be more palatable than those from Alaska, Finland, and Siberia. Indeed, juvenile birch twigs from Iceland were more palatable than those from Finland and Siberia (Bryant et al. 1989). In addition, birch and willow twigs from Finland, where the native mountain hare (*L. timidus*) does not show marked population cycles, are more palatable than those from Alaska and Siberia, where the snowshoe and mountain hares, respectively, have large population fluctuations. Alaskan snowshoe hares ate more heavily defended birches and willows than did Finnish mountain hares, potentially reflecting a difference in their capacities for the detoxification of secondary compounds.

The kinds of chemical defenses have been examined from the viewpoint of available resources (Bryant et al. 1983, Coley et al. 1985, Danell et al. 1985, Bryant and Chapin 1986). Trees and shrubs that are adapted to infertile soils and to low-disturbance environments would be expected to have low-growth rates. Plant parts in these species would be long-lasting, valuable, and difficult to replace, so a chemical defense would be expected. Because carbon would not be limiting under these conditions, carbon-based defenses (e.g., phenols, tannins, and terpenes) would be expected. Plants that live on fertile soils, that live in disturbed environments, and that are fast growing would usually be carbon limited and less likely to invest much energy in chemical defenses. The defenses that are used would likely be based on nitrogen (e.g., alkaloids, cyanogenic glycosides). Of course, long-term, intense browsing will induce even fast-growing species, such as willow, to invest heavily in a chemical defense of adventitious shoots (Bryant et al. 1985b, Tahvanainen et al. 1985).

The correlation between growth rate and the type and amount of chemical defenses by plants led to the suggestion that the cost of chemical defense reduces the amount of energy that they can channel into growth (Feeny 1975, Rhoades and Cates 1976, Coley et al. 1985, Danell et al. 1985). Bryant et al. (1985a), however, argued that low growth rates induced by conditions in the environment (infertile soils, low light intensities, etc.) permit (or require?) the reallocation of carbon to defense. Slow-growing boreal plants have approximately twice as many chemical defenses as fast-growing species (Coley et al. 1985): if boreal evergreens are fertilized with growth-limiting nutrients, the plants increase growth rate and are subject to increased browsing, which implies that fertilization leads to a decrease in the production of secondary compounds.

Plants cannot protect all tissues, so some allocation of chemical defenses occurs (McKey 1974, 1979). One aspect of allocation is the diverse chemical defenses used, producing complex chemical phenotypes that involve different molecules, synthetic pathways, and molecular manipulations and movements. A balance presumably occurs between the benefit that accrues from a defense and the cost of that defense. Some of the important characteristics of secondary compounds influencing that balance are solubility (especially in relation to translocation), molecular composition (which is associated with the cost of synthesis), and molecular size (which may influence the reaction with herbivores). Soluble compounds, such as alkaloids, cyanogenic glycosides, and glucosinolates, can be translocated to rapidly growing tissues (young leaves), which permits them to have some defense at modest cost. Relatively insoluble compounds, like tannins, flavonoids, stilbenes, and tropolones, occur in long-lived tissues (old leaves, wood) (also see Coley et al. 1985). Plants that live in extreme environments may respond to herbivore damage by an unusually diverse spectrum of chemical defenses, which may mean that further herbivore damage would be unacceptably costly to the plant under these conditions (McKey 1979). All woody plants appear to protect juvenile tissues, at least if stimulated to do so by herbivory, whereas only slow-growing species usually protect mature tissues (Bryant et al. 1985a).

Evidence of the cost associated with the use of chemical defenses is found in the observation of Cates (1975) that the palatable (to slugs) morph of a wild ginger, *Asarum caudatum*, produces more seeds and has higher growth rates in the absence of predation than does the unpalatable morph. Furthermore, Janzen (1979) suggested that domesti-cated crop plants have attained a high crop yield in part through having a lower level of chemical protection, compared to their wild relatives; this system can operate only because humans protect domesticated crops from predation. Dimock et al. (1976) showed that the palatability of Douglas-fir (*Pseudotsuga menziesii*) to snowshoe hare and black-tailed deer is based on the genetics of Douglas-fir: genotypes that are more resistant to deer browsing have lower dry-matter and cellulose digestibilities, essential oils with a greater inhibitory effect on rumen flora, and higher phenol contents (Radwan 1972).

12.9 THE RESPONSE OF VERTEBRATES TO SECONDARY COMPOUNDS

All vertebrates that eat green plants encounter secondary compounds, the impact of which is reduced either by selective feeding or by possessing some mechanisms that inactivate these compounds. All possible behaviors seem to occur, including some (e.g., in *Procavia johnstoni*) that avoid plants with the most toxic compounds and others (like in the koala *Phascolarctos*) that specialize on plants that are rich in secondary compounds (essential oils and phenols).

12.9.1 Behavioral responses. The occurrence of secondary compounds may be one of the principal reasons other than nutrition for the diversity of foods in the diet of herbivores (Westoby 1978). For example, browsing adversely affects the growth of trees, so the paper birch (*Betula resinifera*) increases its chemical defenses in the aftermath of browsing. Mature crowns of the paper birch are preferred by snowshoe hares (*Lepus americanus*) in winter (Reichardt et al. 1984), but this food is inaccessible unless the snow is deep. Current-year internodes of mature plants have less resin than juvenile internodes (resin 5% vs. 33% of dry mass, respectively), and the consumption of plant tissues by free-ranging snowshoe hares is negatively correlated with their resin contents (Figure 12.18). Crown branches of *B. resinifera* fed on by moose (*Alces*) were less resinous than those avoided by moose (resin 5% vs. 38% of dry mass). Many resin fractions, however, are not effective repellents, so the total resin content is not necessarily a good index of the chemical protection afforded a plant. Selection by hares is negatively related to the

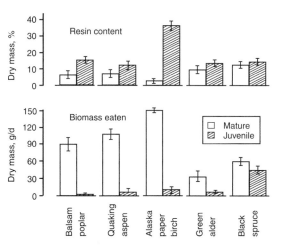

Figure 12.18 The amount of mature and juvenile shoots of five species of trees eaten by snowshoe hares (*Lepus americanus*) in relation to the resin contents of the shoots. Source: Modified from Bryant et al. (1983).

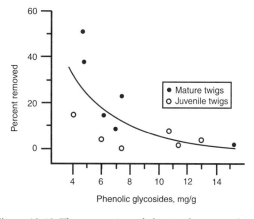

Figure 12.19 The proportion of shoots of various plants eaten by mountain hares (*Lepus timidus*) as a function of the phenolic glycoside content of the plants. Source: Modified from Tahvanainen et al. (1985).

amount of papyriferic acid (Reichardt et al. 1984) and pinosylvin methyl ether produced by alder (*Alnus*) (Bryant et al. 1983).

Browsing preferences represent an interaction of the nutritional and toxic properties of the plants being consumed and the nutritional requirements and capacities for detoxification of the browsers. Under subarctic conditions various birds, including ptarmigan, ruffed grouse, spruce and blue grouse, and capercaille, and mammals, such as snowshoe and mountain hares, moose, and beaver, preferentially consumed foods based negatively on the resin content and not positively on the nutrient content (Bryant and Kuropat 1980). Food preferences were usually greatest for deciduous woody plants, the rank from most to least readily eaten being willow, [aspen], birch, [pine, fur, spruce], *Ledum*, and alder, the bracketed species being used only by some of the listed vertebrates. This ranking is remarkably uniform for this diverse array of birds and mammals. The common denominator in this pattern is the chemical makeup of these plants: the greatest preference is for fast-growing deciduous plants (willow, aspen, birch), and the least preference is for slow-growing evergreens (*Ledum*, spruce) and nitrogen-fixing shrubs and trees (alder).

Mountain hares (*Lepus timidus*) preferentially eat mature to young willow and aspen tissues inversely relative to the amount of phenolic glycosides present. This negative correlation of food holds intraspecifically and interspecifically (Figure 12.19). The pattern of chemically influenced food preferences is complicated by the competitive interactions among browsers: the willow ptarmigan, which is dominant to other sympatric ptarmigan, feeds on the least toxic food, whereas the subordinate rock and white-tailed ptarmigan use more-toxic foods, except in the absence of willow ptarmigan (rock ptarmigan on Iceland and white-tailed ptarmigan in Colorado and southwest Canada), when they feed heavily on willow (Weeden 1969, Bryant and Kuropat 1980). In the absence of competition, food richness is more a positive indicator of food preference for ptarmigan than resin content is a negative factor (Pulliainen and Iivanainen 1981).

Sinclair and Smith (1984) questioned whether snowshoe hares in fact reject food based on the presence of toxic compounds. They noted that the correlation between food preference and resin content is positive, that the correlation between preference and protein-complexing phenols is negative, and that very high resin content in beech buds (ca. 50%) is enough to prevent consumption. As seen by Bryant, snowshoe hares reject juvenile twigs. Sinclair and Smith concluded that no clear pattern exists between the foods used, or rejected, and their chemical composition. A problem with this conclusion is that the chemical composition of food is complicated. Reichardt et al. (1984) pointed out that total phenolics is an inappropriate measure of toxicity: the distribution of specific compounds, such as papyriferic acid, must be examined.

In boreal environments above or beyond timberline, heather (*Calluna vulgaris*) is an abundant plant consumed by ptarmigan and hares. It has high fiber (25%) and low protein (7%) contents (Moss and Parkinson 1972). Red grouse in Scotland can digest only 21% to 30% of heather dry matter, values that are inversely related to intake. In a study of heather grazing by mountain hares, Moss and Hewson (1985) found that juvenile shoots were more nutritious than flowering shoots. Any chemical defense that might have been present in juvenile shoots may have been outweighed by its nutritional content.

An unstudied example of the extreme to which behavior can be used to avoid toxicity imposed by trophic specialization is shown in the arboreal arvicolid from northern California and western Oregon, *Arborimus longicaudus*. This vole lives in the canopy of Douglas-fir (*Ps. menziesii*) forests, often building its nests over 50 m above the ground. In northern California it sometimes builds its nest in redwoods but only in the vicinity of Douglas-firs. Its food is principally the needles of Douglas-fir, although some individuals have been recorded to eat the needles of western hemlock (*Tsuga heterophylla*), grand fir (*Abies grandis*), and Sitka spruce (*Picea sitchensis*) (Taylor 1915, Howell 1926). In captivity *Arborimus longicaudus* will not eat apples, mouse chow, or anything except fresh needles! When eating needles, the tree-vole usually strips and discards the lateral resin ducts and then eats the central vein and attachments (Howell 1926). It often lines its nest with the resin ducts (as an ant repellent?). A close relative, the white-footed vole (*A. albipes*), is a browser on alder (*Aldus*), willow (*Salix*), and Oregon grape (*Berberis*), among other hardwoods and shrubs (Voth et al. 1983). A behavior similar to *A. longicaudus* is found in the chisel-toothed kangaroo-rat (*Dipodomys microps*), which has grooved incisors that aid in stripping hypersaline vesicles in the leaves of *Atriplex* (Kenagy 1973).

Another specialist that feeds nearly exclusively on conifers is Stephen's wood-rat (*Neotoma stephensi*), which preferentially feeds on junipers. This dependency occurs in the southwest desert of the United States, where the year-round activity of this wood-rat requires a dependable water source. It cannot produce a highly concentrated urine and does not evade the harshest conditions in the environment by entering short- or long-term torpor. It tolerates the high concentrations of tannins and terpenoids in the foliage of junipers (Vaughan 1982). Individual junipers are chemically diverse, as demonstrated by the ability of these wood-rats to maintain body mass in feeding trials only when certain junipers were fed to the wood-rats.

Food specialization is seen in arctic arvicolids (Batzli and Jung 1980, Rodgers and Lewis 1985). Brown lemmings (*Lemmus*) feed principally on monocots (grasses); collared lemmings (*Dicrostonyx*), on dicots, especially willow; and the tundra vole (*Microtus oeconomus*), on sedges, willows, and herbaceous dicots. These plants differ in their chemical composition: herbs have the highest concentrations of macronutrients, whereas shrubs and grasses have the highest energy contents (Rodgers and Lewis 1985). Shrubs and some herbs contain the most secondary compounds. Feeding trials showed that sedges (*Carex*) have an extract that is deleterious to *Dicrostonyx*, that *Salix* has an extract that is deleterious to *Lemmus*, and that an evergreen, aromatic shrub, *Ledum*, has an aromatic extract that is deleterious to all three arvicolids (Batzli and Jung 1980). Thus, *Lemmus* and *Dicrostonyx* specialize on particular foods and cannot tolerate the secondary compounds found in the foods used by the other lemming, although both prefer foods with high nitrogen and energy contents (Rodgers and Lewis 1985). *Lemmus*, however, prefers food with a high potassium content and low calcium, magnesium, terpene, glycoside, and tannin contents, whereas *Dicrostonyx* prefers foods with a high magnesium content and a low sodium content, and has a high tolerance of secondary compounds. *Microtus oeconomas*, in contrast, is a generalist herbivore.

As seen among arctic arvicolids, the chemical diversity found in plant communities potentially permits vertebrates feeding on plants to partition these resources based on the diversity in secondary compounds. A similar situation occurs among lemurs in Madagascar (Ganzhorn 1988, 1989b). Lemurs prefer leaves with large amounts of extractable protein, with the exception of *Lepilemur mustelinus*, a lemur that apparently compensates for a low level of protein by having an enlarged caecum in which microbial fermentation occurs. Indriids (*Indri* and *Avahi*) eat leaves with tannins but avoid alkaloids; *Hapalemur* avoids both; and *Lemur fulvus* and *Cheirogaleus major* tolerate both (Ganzhorn 1988). *Lepilemur* also eats leaves with high alkaloid levels. Among African primates, colobines (*Colobus*) do not avoid

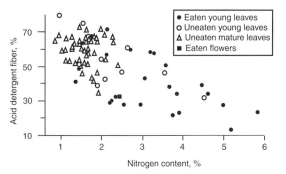

Figure 12.20 Correlation of the fiber content of various plant parts as a function of the amount of nitrogen present and the influence of this correlation on the propensity of the East African colobine monkey *Presbytis rubicunda* to eat these plant parts. Source: Modified from Davies et al. (1988).

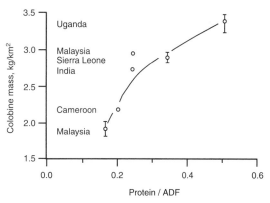

Figure 12.21 Colobine monkey population mass as a function of the ratio of protein to acid detergent fiber (ADF) content in the vegetation. Source: Derived from Waterman et al. (1988) and Oates et al. (1990).

alkaloids, whereas other cercopithecids and hominids do (Ganzhorn 1989b). Clearly, plant secondary chemistry contributes to the ecological separation of sympatric primates, but chemistry alone is not sufficient to account for their diversity in any one locality, other factors being microhabit at selection and food habits (Ganzhorn 1989a).

Colobine monkeys, like ruminants, have an expanded foregut in which microfloral fermentation occurs. Like small artiodactyls, *Colobus satanus* feeds on a high-quality diet, principally leaves and seeds (McKey et al. 1981). It preferentially eats young over mature leaves: young leaves compared to mature leaves are more digestible (less fiber); have higher concentrations of nitrogen (protein), phosphorus, and nonstructural carbohydrates; and have lower concentrations of condensed tannins. Among mature leaves, *C. satanus* prefers species with a high ash content, high digestibility, and low fiber content, although it prefers seeds to most leaves. Seeds have lower contents of phenols and condensed tannins. The leaves that are consumed come from light-demanding, early-successional-stage species that appear to put little investment into defense, but much into growth.

The Bornean colobine *Presbytis rubicunda* lives in forests dominated by members of the family Dipterocarpaceae, whose leaves are characterized by low-protein, high fiber, and high resin contents. Janzen (1974) described these properties as a response to living on low-nutrient, sandy soils that protects valuable leaves from predation. *Presbytis rubicunda* preferentially ate the leaves of rare plants in the forests, especially those of some lianas (Davies et al. 1988). The selected leaves and flowers were characterized by low fiber and high nitrogen contents (Figure 12.20). This combination is characteristic of leguminous trees, which can tolerate poor soils because of the nitrogen-fixing bacterial symbionts found in root nodules and mycorrhizas (Oates et al. 1990).

Waterman et al. (1988) carried this analysis a step further: they argued that mature leaves represent a food resource that colobine monkeys can use when all other foods fail, although the availability of this food would presumably depend on its chemical makeup. In a comparison of five species of colobines (two in Africa, two in southeast Asia, and one in India), their biomass correlated with a ratio that describes the chemical composition of leaves in the forest (i.e., the ratio of protein to acid detergent fiber [ADF]), which incorporates both the attraction of protein (or nitrogen) and the negative contribution of fiber, measured in terms of ADF (Figure 12.21). That this correlation holds so widely is remarkable. Woody plants tend to protect their leaves when these plants live on poor soils. As noted, Kinzey and Gentry (1979) pointed out that the cebid titi monkey *Callicebus torquatus*, a frugivore, consumes significant amounts of insects, a behavior that avoids the use of leaves in regions with nutrient-poor soils that promote the production of leaves with high concentrations of toxic defenses.

In a later analysis of primate biomass and of the abundance of primates on Tiwari Island in Sierra Leone, Oates et al. (1990) found that the protein/ADF ratio accounted for almost all of the variation in colobine biomass, although by incorporating the negative impact of tannins in leaves, for example, in the form protein/(ADF + tannins), the correlation increased slightly. Furthermore, almost all of the variation in an anthropoid biomass in southeastern Asian and African forests was accounted for by the protein/ADF ratio (i.e., $r^2 = 0.997$).

In Tiwari Island, the soil is poor, characterized by a high sand content, low pH, and low concentrations of nutrients. As expected from the suggestion of Janzen (1974), the forest foliage has large amounts of condensed tannins. Nevertheless, this forest supports a high biomass of primates in part because some of the forest has foliage with a high protein-fiber ratio, especially in leguminous trees, and in part because colobines often shift to feeding on seeds, at least at times when young leaves are not available. Colobine abundance often depends on forest gaps because forest gap trees tend to have leaves with less fiber, less phenols, and higher nitrogen contents (Oates et al. 1990). (Note that in the neotropics a favorite food of three-toed tree-sloths [*Bradypus*] are the leaves of *Cecropia*, another forest gap tree, which is protected from predation principally by commensal ants.) Oates et al. went on to state that primate abundance is lowest in the neotropics (usually <700 kg/km²) than in comparable environments in Africa and Asia (often where it is above 1000 kg/km² and up to 3500 kg/km²). They raised the possibility that the low abundance in the neotropics might be due to high abundances of tree-sloths, which in some localities might equal 8000 kg/km².

12.9.2 Physiological responses. Vertebrates respond to plant secondary compounds in a variety of manners (McArthur et al. 1991). These include (1) the formation of inactive complexes, (2) molecular degradation and excretion, and (3) reduction in the requirement for food intake. Experience with toxic compounds in moderate amounts permits a species to adjust its gut flora to detoxify these compounds; for example, Douglas-fir can make up to 50% of the diet of deer (*Odocoileus*) if it has fed on Douglas-fir, but can eat little if it has had no experience with this tree (Oh et al. 1967). This suggests that the presence of toxic compounds in plants would encourage herbivores to treat new foods with caution, learn rapidly, eat several staple food items, and prefer foods with low concentrations of toxic compounds (Freeland and Janzen 1974).

A widespread physiological response of mammals to the chemical composition of ingested plants is the modification of saliva chemistry (Hofmann 1989). These secretions have several functions. In foregut fermenters saliva contains buffers that maintain the rumen pH near 6.5. Saliva also often contains proteins that inactivate secondary compounds. Some of these proteins inactivate the cellulases of the rumen flora, and others, including tannins and other phenols, bind with digestive enzymes in the gut of herbivores. Consequently, salivary proteins that bind with secondary compounds thereby protect gut flora cellulases and gut enzymes.

Tannin-binding proteins occur in the saliva of various browsing mammals, including mule deer (*O. hemionus*), moose (*Alces alces*), and beaver (*Castor canadensis*), and some omnivores, such as black bear (*Usus americanus*) and humans, and are inducible in some mice and rats (Austin et al. 1989, Robbins et al. 1991, Hagerman and Robbins 1993, Foley and McArthur 1994). They are not found or inducible in grazers, namely, cattle and sheep, hamsters, and voles. In deer salivary (parotid) gland sizes are 3 times those of grazers (Robbins et al. 1987b), and deer saliva has higher proline and nitrogen contents than sheep or cattle saliva. That is, grazing mammals encounter low levels of secondary compounds and are less capable of dealing with them than specialist browsers (McArthur et al. 1991). Therefore, the presence of tannins in food has greater effects on grazers than on browsers (Robbins et al. 1987b, McArthur and Sanson 1993a). Mammals that produce salivary tannin-binding proteins lose less nitrogen in the feces and absorb or metabolize fewer tannins than herbivores that do not produce these proteins (Robbins et al. 1991, Foley and McArthur 1994).

Protein structure permits a specificity for the protein-tannin interaction: salivary proteins with a high affinity may bind tannins even in the presence of a high concentration of other proteins, although salivary proteins in herbivorous mammals vary in their ability to bind tannin (Hagerman and Robbins 1993). Thus, dietary generalists, such as bears and mule deer, bind several kinds of tannin, whereas dietary specialists, like moose and beaver, bind only those tannins that are characteristic of their diet. Somewhat surprisingly, arboreal folivores appear not to produce salivary tannin-binding proteins (Robbins et al. 1991, McArthur and Sanson 1993b, Foley and McArthur 1994).

The actions of plant secondary compounds are complex (Karasov et al. 1992). The two principal actions are toxicity (by alkaloids, cyanogenic compounds, nonprotein amino acids) and the binding (especially by tannins) of proteins (possibly digestive enzymes) in the intestine. Toxic molecules generally are smaller and mobile, whereas digestion-reducing molecules are larger and immobile (Foley and McArthur 1994): this distinction may not be sharp, as some tannins may be hydrolyzed, with the resulting smaller molecules acting as toxins. The products of hydrolysis of some condensed tannins, such as quebracho, act as toxins in some ruminants and macropods (Foley and McArthur 1994) but not in (marsupial) folivores (McArthur and Sanson 1993b). In folivores, such as *Trichosurus* and *Pseudocheirus*, most of the ingested tannins are lost in the feces. The preference by snowshoe hares for bitterbush (*Purshia*) as food over blackbrush (*Coleogyne*) may reflect a preference for bitterbush tannins, which may undergo a slower depolymerization and therefore be absorbed more slowly and confer a lower toxicity (Clausen et al. 1990). The microorganisms that ferment plant fibers in foregut fermenters detoxify oxalates, alkaloids, cyanogenic glycosides, and nonprotein amino acids, before they reach the intestine where they would be absorbed (Foley and McArthur 1994). This remains an advantage of foregut over hindgut fermentation, when fermentation occurs posteriorly to the absorption of toxic compounds. Nevertheless, metabolites from poplar and paper birch inhibit the digestion of cellulose at low concentrations in the rumen of the wapiti (*Cervus elaphus*) (Risenhoover et al. 1985).

With some exaggeration, Robbins et al. (1991, p. 480) stated that "... elucidating the effect of tannins on the plant-animal interaction is central to understanding mammalian herbivory...," but their effects surely are widespread and affected by the chemical composition of the food. Dietary tannin does not affect cell wall digestion (Robbins et al. 1987b, although see Hay and Van Hoven 1988) but does reduce protein digestibility with a resulting increased loss of nitrogen in the feces (Dietz et al. 1994). Available protein, not the protein content of food, is what is important (Robbins et al. 1987a), and a particular tannin content in food can be circumvented if the protein content of the food is high enough, at least in some herbivores (e.g., *Microtus ochrogaster* [Lindroth and Batzli 1984]).

In an analysis of the composition of acorns, Koenig (1991) showed that the ability of acorn woodpeckers to metabolize energy (i.e., digested energy minus excreted energy) decreased with an increase in tannin, especially if the tannin is condensed, and further decreased as the lipid content of the acorns increased, even though the addition of lipid alone increases the capacity to metabolize energy. This factor interaction leads to the paradoxical result that acorn woodpeckers are able to maintain body mass in captivity when they are fed acorns from *Quercus lobata*, which has a low tannin content and a low lipid content (7%), but they cannot maintain mass when fed acorns from *Q. agrifolia*, which has a high tannin content and a high lipid content (16.8%). The metabolizable energy coefficient in the two acorns was nearly identical, 60.3% in *Q. lobata* and 64.5% in *Q. agrifolia*.

All tannins are not identical: as noted, condensed tannins, such as quebracho, and hydrolyzed or hydrolyzable tannins, such as tannic acid, do not have the same effect. The protein content of food can compensate for hydrolyzable tannins but apparently not for quebracho, which may inhibit feeding (Lindroth and Batzli 1984, Koenig 1991, McArthur and Sanson 1993a). The effects of secondary compounds, including tannins, may be direct on the consumer or indirect through gut microflora in foregut fermenters (e.g., steenbok [*Raphicerus campestris*]; Hay and Van Hoven 1988) or in hindgut fermenters (e.g., greater glider; Foley 1987).

Herbivores can tolerate toxic secondary compounds if they can be detoxified and excreted. Detoxification involves oxidation, reduction, or hydrolyzation, or any combination of these three processes, or conjugation with glucuronic acid (Lindroth and Batzli 1983). As phenolics increase in the diet, uronic acids in vole urine increase, but decrease with an increase in protein content. The brushtail possum (*Trichosurus vulpecula*) feeds on *Eucalyptus*, which is high in phenolics and leads to high levels of glucuronic acid (see Foley 1987). Up to 40% to 50% of digestible energy from leaves rich in terpenes can be lost in urine, whereas in terpene-free diets only 10% to 15% of digestible energy is lost in the urine (Foley and McArthur 1994). Cork (1981), as reported by Foley and McArthur (1994), estimated that the cost to the koala (*Phascolarctos cinereus*) of excreting glucuronic acid was approximately 20% of fasting glucose production.

If toxic secondary compounds are not detoxified, they may have a variety of consequences for the consuming herbivores. A reduction in the growth rate may occur, as in prairie voles (Lindroth and

Batzli 1984), although it varies with the species and the toxin consumed. A diet of willows resulted in poor growth in *Lemmus*, sedges reduced growth in *Dicrostonyx*, but neither willows nor sedges reduced growth in tundra voles (Jung and Batzli 1981). That is, arvicolid rodents grow slowly when fed plants that are usually avoided and grow rapidly when fed preferred foods (Jung and Batzli 1981, Jean and Bergeron 1986). Toxic compounds may induce renal lesions, as has been found in *M. pennsylvanicus* when fed white clover (*Trifolium repens*) and to a lesser extent other plants (Bergeron et al. 1987). Thomas et al. (1988) showed that processing secondary compounds may have an appreciable cost: meadow voles increased their rate of metabolism by 14% to 23% after being fed a phenol, gallic acid, when it constituted 6% of the dry food intake. If toxic secondary compounds are sufficiently abundant, they may reduce survivorship, at least in arvicolids (Lindroth and Batzli 1984), possibly through the direct action of toxins, but also because of their refusal to eat. Dietz et al. (1994) found that the upper tolerance of tannin by meadow voles is a concentration of about 3% to 5% of food, and then only with a high protein content.

12.9.3 Secondary compounds and the population biology of herbivores.

The interaction among herbivores and the plants that they feed on has been suggested to have profound consequences for the population biology of herbivores. Detailed studies of the arvicolid *Microtus montanus* indicated that the initiation and termination of reproduction are cued by compounds obtained from its food (Berger et al. 1977, 1981; Negus and Berger 1977). These authors showed that the compound 6-methoxybenzoxazolinone (6-MBOA) increases the rate of reproduction in both male and female voles. The immediate precursor of this compound is found in young seedlings. Injury to plant tissue releases enzymes that convert this precursor to 6-MBOA. Plants, then, may integrate environmental conditions to provide a reliable cue for voles to start reproduction. Such a cue may be especially important in environments where photoperiod is unreliable (e.g., compare high and low altitudes at the same latitude). Negus and Berger (1977) tested the importance of this molecule in the field by placing sprouted wheat along *M. montanus* runways. Before placement, no reproduction had occurred, but afterward up to 97% of the population showed evidence of reproduction. This effect was produced both before and after winter solstice, so it could not have been due to photoperiod. Berger et al. (1977) also demonstrated that naturally occurring cinnamic acids and the related vinylphenols derived from food terminated reproduction in *M. montanus*. These compounds are present in late stages of plant growth. Arvicolid rodents are short-lived and tend to live in physically harsh, often unpredictable environments, which may explain why these compounds are used to couple the behavior of these rodents to conditions in the environment.

All arvicolids, however, do not use the same stimulus for reproduction, as Negus and Berger (1998) showed for *Lemmus sibiricus* and *Dicrostonyx groenlandicus* in the high-arctic tundra of the Northwest Territories, Canada. *Lemmus sibiricus*, a specialist on monocots (a low-quality food), has a high volume of food intake and preferentially feeds on plants that have high (although variable) 6-MBOA contents. This compound develops in the spring with the snow melt. In contrast, *D. groenlandicus*, a specialist on dichots (a high-quality food), has a lower volume intake, encounters high concentrations of toxic secondary compounds, and uses an increase in photoperiod as the stimulus to start reproduction in the spring before the snow melt begins. The consumption of monocots and the yearly variation in the abundance of 6-MBOA appear to permit *L. sibiricus* to attain huge population densities, whereas the use of toxic dichots and the reliance on photoperiod as the cue to start reproduction restrict population densities of *D. groenlandicus*.

Food-based chemical stimulation and retardation of reproduction have also been described in birds. Ettinger and King (1981) noted that white-crowned sparrows (*Zonotrichia leucophrys*) in February shifted from feeding on seeds to feeding on green grass, which coincided with a recrudescence of ovarian tissue. In laboratory experiments female, but not male, sparrows showed a growth of gonads when fed wheat seedlings. The growth of the ovaries might have reflected chemical stimulation, or it might have been based on a change in nutritional requirements. California quail (*Callipepla californica*) is widely distributed over western North America, including marginally in some deserts, where it breeds sporadically. Successful breeding in the margins of deserts occurs in this species during wet years. The failure of reproduction is correlated with the synthesis by various annual plants of the phytoestrogens formononetin and genistein during dry years (Leopold et al. 1976). In wet years the concentration of these phytoestrogens in forbs is low, the reproduction of quail is high, and the resulting large seed crop supports a large quail population through winter.

Vaughan and Czaplewski (1985) explored the evolutionary response of a wood-rat (*Neotoma stephensi*) to a specialist diet on tannin- and terpene-rich junipers. They showed that this wood-rat has a low reproductive rate, characterized by a litter of one or two, and normally one or two litters per year. The young, which have a small birth mass, have a low growth rate. Even then, reproductive females cannot assimilate energy at a sufficient rate to pay the cost of maintenance, brooding, and lactation, so they metabolize fat stores. Lactating females in the field may lose up to 40% of their postpartum mass. A large decrease in mass during lactation increases the interval between reproductive periods, which is about 50 d if one young is raised and 70 d if two are raised. Furthermore, reproduction reduces the rigor of females, so that when females are replaced at a den site in the field, it usually occurs after a period of lactation. A low growth rate and a long period of nursing, in conjunction with an extended period of young-mother association, provides for an extended period of dietary training in young wood-rats but leads to a low reproductive rate. Dietary training may lead to an exploitation of the chemical diversity found in junipers (Vaughan and Czaplewski 1985). A similarly low reproductive rate, characterized by small litters, long gestation period, and low growth rate, is a found in another small herbivore that is a specialized feeder on conifers, the red tree-vole (Hamilton 1962).

Freeland (1974) speculated that vole cycles generally reflect the chemical composition of their food supply. He argued that (1) voles have a preference for nontoxic foods; (2) with an increase in vole density, nontoxic foods become increasingly scarce, which requires an increased consumption of toxic foods; and (3) preferred nontoxic plants outcompete toxic plants in the absence of vole predation. Batzli and Pitelka (1970, 1971) generally found these conditions to hold, but they (1975) noted that California voles (*M. californicus*) at high densities do not significantly increase their consumption of toxic plants. Schlesinger (1976), while surveying the literature, also failed to find that voles feed on nontoxic foods, although we now know that the dichotomy between toxic and nontoxic is too simple because what is toxic to one species may be acceptable to another (e.g., Batzli and Jung 1980, Ganzhorn 1988). Freeland maintained that the death of voles at a population high is due to a reduced viability resulting from the consumption of toxic compounds. Phenols in food reduced survival and growth in *M. ochrogaster*, although these toxic effects can be overridden by a diet high in protein (Lindroth and Batzli 1984).

Bryant and his coworkers suggested that the relationship between snowshoe hares and the chemical nature of their food is the principal factor influencing the cyclicity of hare populations (Fox and Bryant 1984, Bryant et al. 1985b). Snowshoe hares show an 8- to 12-y periodicity in population levels, within which the populations may vary from about 20 to 2000/100 ha. For periodic fluctuations to occur in hare populations, a time-delayed, density-dependent mechanism must operate. May (1974, 1976) calculated that a 10-y cycle should have a delay that averages about 2.3 y. Thus, browsing by snowshoe hares in an area of dense population may so deplete the supply of small-diameter twigs that edible plants are forced to produce unpalatable adventitious shoots, which forces the hares to shift to secondary foods, which leads to their starvation and to a crash in their populations. Evidence of an adverse effect by unpalatable browse on snowshoe hares is found in their inability to maintain body mass and in the high concentrations of detoxified phenolics in their urine (Bryant et al. 1985b). Pease et al. (1979) indicated that willow takes about 3 y to recover from intense browsing to produce shoots that have a high palatability, which is approximately the period estimated by May's calculations.

In spite of the many observations that support the view that plant secondary compounds influence the population biology of herbivores, and in spite of the support of some calculations of refractory period length, other observations appear not to be compatible. For example, Oksanen et al. (1987), in a study of *Clethrionomys rufocanus* living on islands and on the Norwegian mainland, noted that (1) these voles preferred shoots of blueberry, *Vaccinium myrtillus*, with a low phenol content and a high nitrogen content; (2) the highest phenol and lowest nitrogen contents were found in blueberries on islands without voles; and (3) the population fluctuations of voles was independent of the chemical composition of blueberry vegetation. These results are generally contrary to what would be expected from a browsing-based induction of secondary compounds. Moss and Hewson (1985) found that juvenile shoots of heather were more nutritious to mountain hares than were flowering shoots. And although Danell et al. (1985) observed that moose preferred to browse on a species of birch that grew rapidly (*Betula pendula*) rather than on one that grew slowly (*B. pubescens*), they found no evidence of induced chemical defenses in juvenile trees. Shoots from trees that had been browsed were more palatable than those from unbrowsed trees.

Another extrinsic "nutritional" factor associated with herbivore population cycles is the storage, release, and exchange of nutrients among the various components of an ecosystem, principally the plants, consumers, feces, and soil (Pitelka and Schultz 1964). In arctic environments the nutrient composition of foliage (e.g., calcium, phosphorus, and protein) varies with the lemming population cycle: it increases with the number of lemmings and falls with their decline—much of the nutrients being tied up in the soil, partially associated with undecomposed feces.

Haukioja et al. (1983) attempted to resolve the conflicting views on "the cause" of herbivore population cycles by examining plant availability, predation, and "self-regulation." They argued that nothing can permit a population outbreak if enough high-quality food is not available. Plant chemical defenses that are rapidly induced by herbivory and rapidly eliminated in association with the reduction of herbivory should stabilize herbivore populations. The length (and possibly the amplitude?) of the cycle would be determined by the lag period for the reduction of the defensive chemicals. These authors suggested that predators may influence the rate of decline in herbivore populations and possibly make the troughs deeper. They further speculated that herbivore population cycles are more common and have a greater amplitude in extreme environments, most notably in boreal regions, because few alternate prey are available for predators. Although Haukioja et al. did not deny the potential importance of intrinsic factors (see Chitty 1960; Krebs 1978, 1979), they concluded that extrinsic factors, especially plant availability (which includes their chemical makeup) and predation, are the principal ones responsible for herbivore cycles.

Yet, not all herbivores respond to extrinsic factors in the same manner. Large, "K-selected" species encounter chemical defenses without large population fluctuations, whereas most small, "r-selected" species have greater fluctuations in density (Haukioja 1980). Consumers with the highest intrinsic population growth constant r_m are more likely to overexploit the resources that maintain the population. This cyclic pattern in population density is the product of natural selection on individuals. "[A]t high densities, selfish tactics emphasize adult survival, dispersal, reservation of resources for own use, etc. more than fecundity. All of them reduce growth rate of the population. Populations show signs of self-regulation without any of the members of the population itself striving for it..." (Haukioja 1980, p. 212).

12.10 CHAPTER SUMMARY

1. All vertebrates obtain energy and nutrients from food, with the amount required depending on the nutritional needs of the vertebrates and the quality and quantity of available food.

2. The diversity of foods used by vertebrates is great, especially among fishes, birds, and mammals. Most adult amphibians feed on invertebrates, whereas most reptiles eat invertebrates or vertebrates, or both. A few lizards and turtles are herbivorous.

3. A diversity in food habits facilitates the co-existence of a diversity of consumers.

4. Most vertebrates have varied diets to reduce the cost of foraging, to compensate for the nutritional inadequacy of particular foods, and to reduce the intake of toxic compounds.

5. The most widespread dietary shortage is in available protein, which is especially marked in fruits and leaves.

6. Most food specialists minimize nutritional requirements by having a small mass or a low-intensity metabolism.

7. Some potential mixtures of food items are not used, apparently because they require different physiological conditions for processing or because they are associated with different body sizes. Few, if any, vertebrates mix large quantities of leaves with animals in their diet.

8. The morphology of the gastrointestinal tract reflects food habits: carnivores have a short, simple tract, whereas herbivores have a long, complicated

tract, often with extensive pregastric or postgastric sacculation.

9. Seasonal variations in the dimensions of the gut are correlated with seasonal variations in the amount of food eaten and its chemical composition.

10. Rate of digestion varies with body temperature in ectotherms, with the optimal rate occurring in the normal range of body temperature during activity.

11. The digestibility of food depends on its chemical composition: animal matter has a high digestibility, whereas some plant matter has the lowest digestibility.

12. The sugar composition of nectars produced by plants is correlated with the type of pollinator consuming the nectar and, ultimately, with the ability of the pollinator to hydrolyze sucrose.

13. Green plant matter is chemically the most complex food resource for vertebrates to exploit because (a) most energy is contained in cellulose, (b) cellulose is often complexed with lignin, (c) the amount of available protein is low, and (d) plants synthesize toxic secondary compounds.

14. Without the aid of bacterial symbionts, few large vertebrates could be specialized herbivores, and those few that do not use fermentation rely on the intake of large quantities of food to compensate for its low digestibility.

15. Folivory is uncommon in small herbivores because the retention time for food in the gut, and therefore the digestion of leaves, decreases with mass.

16. A mixed diet may permit a larger body size than would be obtained by adherence to a specialized diet.

17. Fermentation converts long-chain, ß-linked carbohydrates and other organic compounds into volatile fatty acids, methane, CO_2, ammonia, and surplus bacteria.

18. Fermentation chambers in vertebrates are located before or after the stomach, with foregut fermentation in mammals occurring with greatest frequency at intermediate masses; almost all large herbivorous mammals are hindgut fermenters.

19. A small body mass is associated with a high turnover rate of food in the gut, and thus requires high-quality food; a large mass requires a large volume of food, which of necessity must be of lower quality.

20. Foregut fermentation, especially when coupled with rumination, is more efficient in the extraction of nutrients than is hindgut fermentation, and may permit the detoxification of plant secondary compounds.

21. Perissodactyls persist either because they feed on high-fiber food (usually coupled with a large mass) or because they live in environments that have few hindgut fermenting artiodactyls.

22. Fermentation makes an important contribution to the energetics of some fishes, lizards, turtles, and birds, all being hindgut fermenters, except for one bird, the hoatzin.

23. Many plants synthesize secondary compounds that are toxic to herbivores; these compounds are usually most abundant and diverse in plants that live on infertile soils and in low-disturbance environments, conditions in which plant tissues are long-lasting and nutritionally valuable.

24. The palatability of browse generally is not related to its nutritional content but is often negatively correlated with the presence of specific secondary compounds, including phenols, tannins, terpenes, alkaloids, and cyanogenic glycosides.

25. Plants use soluble, highly toxic compounds in rapidly growing tissues and relatively insoluble compounds in long-lived tissues.

26. Feeding specializations permit herbivores to tolerate particular secondary compounds, but are often associated with intolerance to other compounds, these compounds permitting consumers to partition food resources.

27. The abundance and diversity of folivores in tropical environments are positively correlated with the protein content of leaves and negatively correlated with the fiber and tannin content of leaves.

28. Most browsing mammals produce saliva containing proteins that bind with the tannins encountered in their diet, whereas most grazing mammals do not produce these proteins, mainly because grasses are relatively tannin free.

29. Toxic compounds in the diet, if not detoxified, can reduce the growth rate of young, can cause renal lesions, and can even lead to death.

30. Some compounds found seasonally in green plants appear to cue the seasonal initiation and cessation of reproduction in a few birds and mammals.

31. The periodic fluctuations in arvicolid and snowshoe hare populations may result from the interactions of these mammals with their food supplies, and the levels of secondary compounds induced by browsing.

PART V | Consequences

13 The Significance of Energetics for the Population Ecology of Vertebrates

13.1 SYNOPSIS

The analysis of the influence of energetics on the population ecology of vertebrates is complicated by factor interaction, and this complexity has led to controversies on the relative effects of physiology, ecology, and phylogeny on the parameters of population growth. Eutherian mammals with high basal rates of metabolism usually have short gestation periods, high growth rates, and high fecundities independent of body mass. Life span increases and mortality rates decrease with an increase in body mass; much of the residual variation in mortality, and therefore in life span, is associated with habits that either accelerate mortality, such as a polar distribution, or diminish mortality, such as being volant, arboreal, or fossorial. Eutherians with high basal rates have high population growth constants and population fluctuations, whereas species with low basal rates have low population growth constants and restricted population fluctuations. This dichotomy is related to the r–K continuum. No evidence indicates a correlation of the population growth constant with energy expenditure in marsupials, a condition that may be responsible for the inability of marsupials to outcompete eutherians when eutherians use food resources that permit high rates of metabolism. The independence of population growth from basal rate in marsupials may prevent them from occupying cold-temperate environments. In birds high postnatal growth constants are often associated with high rates of metabolism, and therefore the correlation of basal rate with latitude may explain how birds increase the rate of reproduction to survive in cold-temperate environments with and without the seasonal use of migration. Growth rate is positively correlated with rate of metabolism in ectotherms; some low-energy specialists are characterized by unusually low reproductive rates.

13.2 INTRODUCTION

Physiological ecology generally deals with the response of individuals to the physical conditions faced in the environment. These conditions are described by such parameters as temperature, solar radiation, atmospheric humidity and composition, barometric pressure, water osmolality, and photoperiod. Physiological ecology is simply a modern guise for much of what used to be called *autecology*, which is only one facit of ecology. Others include behavioral, population, and community ecology, these divisions being arbitrary devices used to catalog information and ideas. No sharp distinction exists between "behavior" and "physiology," or between either of these and "ecology," because all interact at the level of the individual. Indeed, one of the great advances in comparative biology during the second half of the twentieth century has been the progressive blurring of the boundaries among these classically defined fields, which raises the question of whether biology can be reduced to one "fundamental" level of organization, for example, the molecular level, a mid-twentieth-century axiom. With a shift from one level of organization to a higher level—from molecules to cells, from cells to tissues and individuals, from individuals to populations, or from populations to communities—some "emergent" characteristics often appear but may not be inherent in the properties at the lower level of organization. (For a discussion of "emergent" properties of organisms, see Mayr [1997].)

Of immediate concern is whether a change in some physiological characters of individuals may have consequences for their populations. For example, one of the fundamental analytic tools in population ecology is the life table, an age-specific account of survivorship and fecundity in a population, or cohort, that permits the population's, or cohort's, mean fecundity to be calculated. Survivorship and fecundity in a population can be combined to estimate its exponential growth constant, r:

$$r = \frac{\ln R_o}{T}, \qquad (13.1)$$

where R_o is mean population fecundity (usually average number of females produced per female-year) and T is mean population generation time (Deevey 1947, Ricklefs 1973a, R. M. May 1976).

A fundamental question is whether energetics has an impact on the parameters of reproduction in vertebrates, that is, on R_o and T. The classic view is one of the "allocation" of energy, which maintains that the amount of energy available to an organism is limited, an organism lives near this limit, and therefore the cost of reproduction is partially or completely derived by subtraction from the expenditure for maintenance (Williams 1966, Gadgil and Bossert 1970). In this case a "trade-off" between reproduction and maintenance occurs in the expenditure of energy: as energy is allocated to reproduction and growth, the cost of maintenance should decrease. Another means by which expenditures might be organized is to have the expenditures for reproduction positively correlated with those for maintenance, a pattern that is most likely to occur if an organism does not normally expend energy at a rate close to its availability in the environment. Then high-maintenance species harvest more energy and expend more on reproduction than low-maintenance species. This pattern has been called a "functional relation" (Konarzewski 1995) and has been claimed to be present in eutherian mammals (McNab 1980b, Derting 1989). A third possibility is that the expenditures for reproduction and maintenance are "independent," a condition that was described for birds (Dunn 1980). For a theoretical description of these alternatives and some of their consequences, see the work by Tuomi et al. (1983) and Section 10.11.2.

13.3 ENERGETICS, GENERATION TIME, AND FECUNDITY IN EUTHERIAN MAMMALS

Several parameters of population reproduction appear to be correlated with basal rate of metabolism in eutherian mammals, which indicates to some extent a functional relation between the costs of reproduction and maintenance. The parameters that are correlated with the basal rate include the length of the gestation period, postnatal growth constant, fecundity, and possibly life span. A major complication, however, is that each of these parameters varies with body mass, so that any correlation of these parameters with energetics must be independent of the influence of body mass. Demonstrating correlations is a difficult task because complete sets of data are available for only a few species and because complex interactions exist among rate of metabolism, body mass, various ecological factors, and taxonomic affiliation.

13.3.1 Gestation period. Gestation period in eutherian mammals increases with body mass (Figure 13.1), the power of mass varying from about 0.15 to 0.19, depending on the taxon and the type of placenta (Kihlström 1972), although the pooled average for mammals is about 0.25 (Sacher and Staffeldt 1974, Millar 1981). Gestation period is longer in species with precocial than with altricial young (Figure 13.1). An appreciable variation in gestation period remains after these factors are considered. Some residual variation is correlated with variation in basal rate independent of the influence of body mass: for example, rodents with low basal rates have longer gestation periods at a given mass than do eutherians with high basal rates; this pertains to both altricial species with basal rates less than about 90% of expected and most precocial species (Figure 13.2). A high basal rate, however, has little effect on the gestation periods of small (15–50 g) altricial species because these periods are already short, although arvicolid rodents have high basal rates and the shortest gestation periods.

In an attempt to examine the impact of metabolism on population ecology while holding taxonomic affiliation constant, N. Vasey and D. T. Rasmussen (pers. comm., 1996) compared 20 pairs of eutherian mammals: each member of the pair had a similar mass and belonged to the same family, but the pair chosen were required to have basal rates that differed by at least 20%. In 15 of the 19 species pairs, the species with the highest basal rate had the shortest gestation period. This distribution is different from a random association of basal rate and gestation period in a sign test at the 0.01 level.

13.3.2 Postnatal growth constant.

Mammals attain adult size as a result of a postnatal period of growth, much of it fueled by maternal lactation. Growth during this period can be characterized by an exponential growth constant (see Ricklefs 1967) that varies inversely with body mass: small species grow proportionally more rapidly than do large species (Figure 13.3). Residual variation in this correlation increases with the basal rate of metabolism in adults, independent of the influence of body mass (Figure 13.4). For example, arvicolids have high growth constants and basal rates; cricetids, intermediate growth constants and basal rates; and heteromyids, low growth constants and basal rates. The correlation of the postnatal growth constant with basal rate occurs over a wide range in body mass. The energy expenditure by mothers for lactation increases with basal rate in *Peromyscus* (Glazier 1985) and *Plecotus* (McLean and

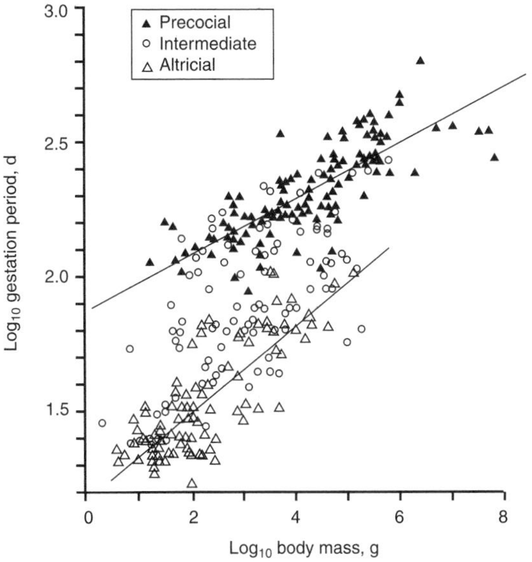

Figure 13.1 Log$_{10}$ gestation period of precocial, intermediate, and altricial eutherian mammals as a function of log$_{10}$ body mass. Source: Modified from Martin and MacLarnon (1985).

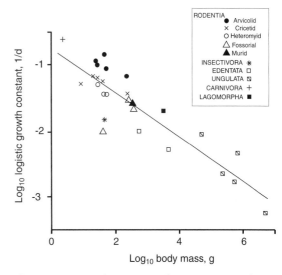

Figure 13.3 Log$_{10}$ logistic growth constant in eutherian mammals as a function of log$_{10}$ body mass. Source: Modified from McNab (1980b).

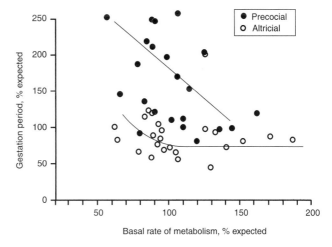

Figure 13.2 Gestation period in precocial and altricial mammals as a function of basal rate of metabolism. Source: Derived from McNab (1980b).

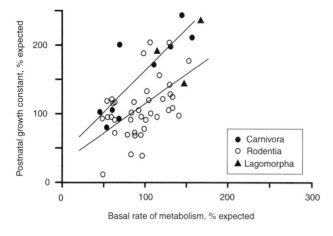

Figure 13.4 Postnatal growth constant in eutherian mammals as a function of basal rate of metabolism. Curves for carnivores and rodents are indicated. Source: Modified from McNab (1980b).

Speakman 2000); the peak energy expenditure of lactating females was greater in rodents and insectivores with high basal rates (Genoud and Vogel 1990); it averages 4–6 times the basal rate (Thompson 1992).

The correlation of the postnatal growth constant with basal rate implies that during postnatal growth young eutherians have high basal rates. Young eutherians often have basal rates that greatly exceed the all-mammal standard during postnatal growth, even in species that have low basal rates as adults (Figure 5.26). High basal rates can occur in young because the mother bears the cost of growth through lactation and because the season for reproduction is synchronized with maximal food availability in the environment. High growth rates in nursing young are advantageous to mothers as high rates shorten the period of dependency of young, which may permit small species to produce more than one litter per year.

Derting (1989) examined whether growth rates increase with basal rate by implanting pellets that contained thyroxine in juvenile cotton rats (*Sigmodon hispidus*). When food supplies were unlimited, basal rate of metabolism, rates of food ingestion, and growth rates increased and sexual maturation developed earlier in individuals with implants than in those that had no implants. When, however, food supplies were restricted, little or no gain in mass occurred even with the thyroxine treatment and its attendant increase in basal rate. Therefore, the presence of a high basal rate permits an increased growth of tissue only if an adequate food supply is available.

13.3.3 Fecundity. The number of female young produced by a female in her lifetime is a measure

of her fecundity, which in mammals equals the product of the number of reproductive periods, the proportion of young that are females, and litter size. The number of reproductive periods is a function of the length of the reproductive life span and age at the time of first reproduction. Small species reproduce more often and have larger litters than large species, so that fecundity decreases with an increase in body size. Western (1979) derived a relation indicating that fecundity is proportional to $m^{-0.28}$. Some of the residual variation in fecundity independent of body mass correlates with variation in basal rate (Figure 13.5). Among small eutherians, arvicolids have high fecundities, large litters, and repeated reproduction in a year, whereas heteromyids have low fecundities, small litters, and a single, annual reproductive period. Cricetids are usually intermediate in terms of both basal rates and fecundities.

An analysis of the energetics of reproduction in 20 pairs of species of eutherians paired by body size and membership in the same family, in which the members of each pair differed in basal rate by at least 20%, demonstrated that in 13 pairs, the species with a higher basal rate had a larger litter size, in 4 pairs the species with the lower basal rate had a larger litter size, and in 3 pairs the species showed no difference in litter size (N. Vasey and D. T. Rasmussen, pers. comm., 1996). This distribution was different from that expected from no effect of basal rate at the 0.025 level. Genoud (1988) showed litter size in shrews to be correlated with basal rate of metabolism, both between the subfamilies Soricinae and Crocidurinae and within the Soricinae (Figure 13.6). Furthermore, in a comparison of the cost of reproduction in *Sorex* and *Crocidura*, Genoud and Vogel (1990) found that

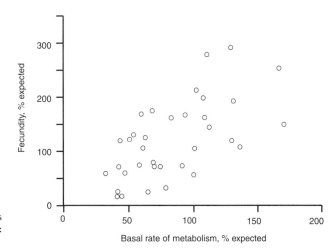

Figure 13.5 Fecundity in eutherian mammals as a function of basal rate of metabolism. Source: Modified from McNab (1980b).

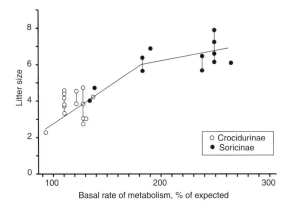

Figure 13.6 Litter size in crocidurine and soricine shrews as a function of basal rate of metabolism. Source: Modified from Genoud (1988).

Sorex, the higher-energy shrews, had shorter gestation periods, less well-developed young, and larger litters than *Crocidura*, the lower-energy shrews. The peak energy expenditure during reproduction increased with litter size and was higher in *Sorex* than in *Crocidura*.

A broader analysis of insectivores indicated that (1) reproductive variables are not correlated with basal rate in the Tenrecidae and (2) a high basal rate is correlated with a short gestation period, a high growth rate, and large litter sizes in the Soricidae (Stephenson and Racey 1995). The difference between these families can be interpreted as demonstrating a phylogenetic distinctiveness, an ecological difference (i.e., a larger size, low basal rate, and the use of seasonal torpor by tenrecs and the minute size, high basal rate, and the commitment to endothermy by shrews), or (more likely)

some combination of an ecological difference embedded in phylogeny.

Rasmussen and Izard (1988) examined the interactions among growth, life-history traits, body mass, brain size, and basal rate in four species of the primate family Lorisidae. They showed that lorisines (lorises and pottos) had lower basal rates of metabolism, longer gestation periods, longer lactation periods, and an older age at the inflection point of the postnatal growth curve than did galacines (galagos). Litter size and the Gompertz growth constant were correlated more closely with basal rate than did with body mass or cranial capacity. Rasmussen and Izard (1988) suggested that the causative factor for the low basal rates in the lorisines is their use of toxic insects as food. This analysis was facilitated "[b]y focusing on a single case of life history divergence, and by selecting species that represented different combinations of body size and metabolic rate…" (p. 363), thereby simplifying the complexity encountered when making broad-scaled, taxonomically diverse allometric analyses.

Mammals with low basal rates are not necessarily forced to have low fecundities because they may accelerate reproduction by increasing their rate of metabolism at a time coinciding with a high resource abundance in the environment. An increase during pregnancy and lactation is most marked in species with low basal rates (Thompson and Nicoll 1986), and occurs to only a modest extent, or not at all, in species with high basal rates (see Section 10.6.1). A seasonal increase in basal rate, however, may not be sufficient to convert a low to a high rate of reproduction because the seasonal "window" in which a high rate of metabolism can be tolerated is short in species using foods that usually require low rates of metabolism. The truncation of the breeding season is especially marked in species that enter torpor (see Section 11.8.3). Some food supplies, such as leaves of woody plants, may not permit eutherians to increase their rate of metabolism, a condition that would lead to a low rate of reproduction. For example, two-toed tree-sloths (*Choloepus*), which have a very low basal rate, show no increase in rate of metabolism during pregnancy (McNab 1978b) and further depress their fecundity by having an extended period of maternal care.

Interspecific studies, as seen, generally indicate that an increase in rate of metabolism is correlated with, or facilitates, an increase in the reproductive rate of mammals, but most intraspecific studies do not indicate such correlations. No connection has been found between the parameters of reproduction and basal rate in cotton rats (Derting and McClure 1989), deer mice (*Peromyscus maniculatus*) (Earle and Lavigne 1990), and house mice (Hayes et al. 1992). Why a difference exists between intraspecific and interspecific studies is unclear, unless the intraspecific differences in basal rate are too small for significant correlations to be shown. Derting and McClure (1989) raised the possibility that the reproductive rate varies with the scope for metabolism, and not the basal rate of metabolism.

13.3.4 Life span. As is typical of biological time periods (see Section 5.3.1), the life span of mammals is proportional to about $m^{0.25}$ (Figure 13.7), although an appreciable variance occurs around the mean curve. Three explanations have been given for the residual variation. One is that life span is correlated with brain size (Sacher and Staffeldt 1974), or at least a combination of body and brain mass (Hofman 1983, 1993). In a sample of 47 mammals, Hofman (1993) showed that 65% of the variation in life span is accounted for by variation in body mass alone; 82%, by brain mass alone; and 90%, by the combination of body and brain masses. The functional basis of the relationship between life span and brain size, assuming that it exists, is not clear: Hofman suggested that individuals after reproduction might contribute to the fitness of their offspring by giving care, which in social species might be associated with larger brains. Bats, however, have long life spans but small brains (Austad and Fischer 1991).

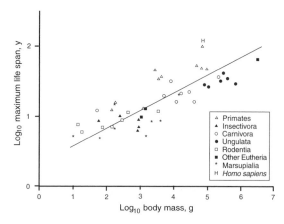

Figure 13.7 Log_{10} maximal life span in mammals as a function of log_{10} body mass. Source: Modified from Hofman (1993).

Another explanation for the variation in life span has been that species with low rates of metabolism have longer life spans (see Hofman 1993). Thus, *Tachyglossus*, *Choloepus*, and bats have long life spans and generally have low basal rates of metabolism. Difficulties exist with this explanation. For example, although temperate bats that enter torpor have longer life spans than tropical bats, tropical bats still have longer life spans than most other mammals (Jürgens and Prothero 1987, Austad and Fischer 1991, Hofman 1993). Marsupials, which usually have lower rates of metabolism than eutherians, have shorter life spans than eutherians (Austad and Fischer 1991). Finally, birds have life spans that are about 2.4 times those of eutherians in spite of the generally higher basal rates of birds. Yet, life span in mammals can be extended by the restriction of energy intake (Weindruch and Walford 1989, Hofman 1993). Hofman concluded that energy metabolism and brain size constrain maximal life span.

The third explanation for the residual variation in life span involves life-history evolution. Long life spans are found in bats and birds (Pomeroy 1990, Austad and Fischer 1991), burrowing mammals (Pomeroy 1990), and arboreal marsupials (Austad and Fischer 1991). The longer life spans of these endotherms reflect life-styles that are typified by reduced environmental hazards and reduced levels of predation because of flight and the use of protected roosts. This reduction in mortality is illustrated by (1) the very low mortality of European swifts, which are so aerial in their habits that they

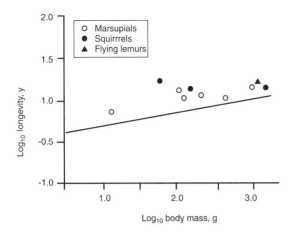

Figure 13.8 Log_{10} longevity in aerial gliding mammals as a function of log_{10} body mass. The regression is for eutherian mammals other than bats. Source: Modified from Holmes and Austad (1994).

are reputed to "roost" in the air (Cramp 1985); (2) the lower mortalities in Afrotropical than in Eurasian land birds, possibly reflecting a greater mortality associated with seasonal migration or residency in a markedly seasonal environment; and (3) the lower mortality in flying aquatic birds than in penguins, potentially correlated with higher levels of marine predation in penguins (Pomeroy 1990). Populations with low mortality rates have low reproductive rates and retarded senescence (Williams 1957, Charlesworth 1980).

Stapp (1992, 1994) observed that the gliding squirrel (*Glaucomys volans*) has nocturnal habits, a low basal rate, low thermal conductance, long period of maternal investment, low growth rates, and a low fecundity compared to other tree-squirrels. He argued that this combination of characteristics reflected a response to the nocturnal microclimate faced by the flying-squirrel and a limited energy availability. These characteristics might also have been derived from a reduced mortality produced by gliding habits, nesting in holes, and nocturnality (Holmes and Austad 1994). An extended life span is typical of gliding mammals (Figure 13.8), including flying lemurs (*Cynocephalus*) and gliding marsupials (*Acrobates, Petaurus, Schoinobates*), as well as flying-squirrels (*Glaucomys, Petaurista*) (Austad and Fischer 1991, Holmes and Austad 1994).

In 1908 Rubner, who noticed (1883) that mass-specific basal rate of metabolism decreased with body mass, observed that life span increased with body mass and suggested that the lifetime energy expenditure of mammals was constant. This conclusion is based on the product of two scaling functions:

mass-specific lifetime energy expenditure (kJ/kg)
= mass-specific energy expenditure (kJ/kg·y)
× lifetime (y)

$$c = c \cdot m^{0.00} = a \cdot m^{-0.25} \cdot b \cdot m^{0.25}.$$

This analysis has its supporters and detractors, much of the discussion being based on whether the scaling function for mass-specific energy expenditure is proportional to $m^{-0.25}$, whether lifetime is proportional to $m^{0.25}$, and therefore whether their product is proportional to $m^{0.00}$. For example, Prinzinger (1993) argued that in birds lifetime mass-specific energy expenditure is proportional to $m^{-0.001}$ (i.e., "independent of mass") and concluded that this result ". . . support[ed] the theory of an absolute metabolic scope during the life cycle" (p. 609). This conclusion implies that total lifetime energy expenditure is proportional to body mass:

total lifetime energy expenditure (kJ)
= mass-specific lifetime expenditure (kJ/kg)
× mass (kg)

$$m^{1.00} = m^{0.00} \times m^{1.00}.$$

Jürgens and Prothero (1991), however, concluded that total lifetime energy expenditure is proportional to mass raised to powers between 0.87 and 0.93, which led to the conclusion that this pattern was incompatible with the Rubner hypothesis. What is unclear is why mass-specific calculations should be used (see Box 3.1). Total lifetime energy expenditure, which is almost directly proportional to body mass, in both mammals and birds, is more profitably considered than a mass-specific lifetime energy expenditure that is independent of body size.

13.4 POPULATION GROWTH IN RELATION TO ENERGETICS IN EUTHERIAN MAMMALS

The growth of populations is described by the population constant r, which is defined in Equation 13.1. In this relationship the parameter r is maximal when R_o is maximal and T is minimal. Most r values in nature are less than r_{max} because the conditions faced by organisms are usually less than optimal, but r_{max} is of intellectual interest because it represents what would be possible if the conditions were optimal; that is, it is a measure of a species's potential.

All factors that influence R_o and T should affect r. Thus, r would be expected to increase with basal rate of metabolism (at a fixed mass) because fecundity appears to increase with basal rate and generation time appears to decrease with basal rate. The impact of basal rate on r, however, should principally be through its effect on T because R_o is incorporated as a natural logarithm, which diminishes the quantitative impact of a numerical change in R_o: in fact, the period from conception to weaning decreases with an increase in basal rate independent of mass (Thompson 1992). If basal rate affects r, it should primarily reflect a correlation of growth rate with basal rate. One consequence of these interactions is shown in the correlation in eutherians of r_{max} with body mass (Figure 13.9A): it is proportional to mass raised to the −0.25 power (Hennemann 1983a). The level at which r_{max} scales

is related to the rate of metabolism: both rate of metabolism and r_{max} are high in endotherms, intermediate in multicellular ectotherms, and low in unicellular ectotherms (compare Figures 3.5 and 13.9B).

So few data are available on the r_{max} of eutherians that its potential dependence on rate of metabolism, independent of the impact of body mass, cannot be directly tested. Hennemann (1983a) estimated r_{max} in mammals from population data on age at the time of first reproduction and annual birthrate of female offspring, in the interactive manner of Cole (1954) and McLaren (1967). The resulting estimates are approximate, but Hennemann concluded that r_{max} indeed increases with the level at which energy is expended, although he maintained that this conclusion did not apply to mammals that live in cold water due to a very high cost of temperature regulation. Schmitz and Lavigne (1984), however, concluded that r_{max} is positively correlated with basal rate in marine mammals. A reexamination of estimated r_{max} values convinced Robinson and Redford (1986) that they were primarily correlated with body mass and taxonomic affiliation, but not with food habits and by implication with basal rate.

Species with a high r_{max} can rebound rapidly from a population low produced by an overexploitation of food resources, predation, or inclement weather. Those with a high r_{max} and, so the argument might go, a high basal rate would be expected to show greater population fluctuations than species with a low r_{max} and a low basal rate. This is exactly what is found when various species are compared in a given environment at the same time (Figure 13.10). For example, among small rodents arvicolids have the highest basal rates and the highest population fluctuations, whereas heteromyids have the lowest basal rates and the lowest fluctuations. Cricetids are intermediate. In a different environment or in the same environment in a different year, the population fluctuations in the same species might be different because conditions have changed, but the comparative amplitudes in the species remain the same (Figure 3.10). At larger masses, the snowshoe hare (*Lepus americanus*) has a high basal rate and large population fluctuations, whereas the arctic hare (*L. arcticus*) has a low basal rate and low fluctuations. Kurta and Ferkin (1991) showed that *Microtus pennsylvanicus*, with a high basal rate, has large population fluctuations, whereas *M. breweri*, a small-island

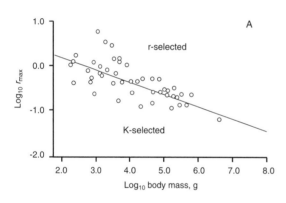

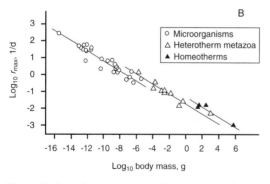

Figure 13.9 A. Log_{10} r_{max} in eutherian mammals as a function of log_{10} body mass. B. Log_{10} r_{max} in microorganisms, ectotherms, and homeotherms as a function of log_{10} body mass. Sources: Modified from Fenchel (1974) and Hennemann (1983a).

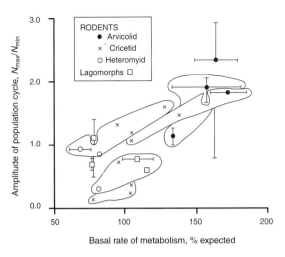

Figure 13.10 Log$_{10}$ amplitude of population density in rodents and lagomorphs as a function of basal rate of metabolism in various environments. Source: Modified from McNab (1980b).

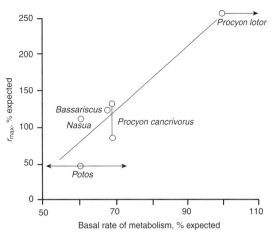

Figure 13.11 The parameter r_{max} as a function of basal rate of metabolism in procyonid carnivores. Source: Derived from Mugaas et al. (1993).

species with a lower basal rate, has population fluctuations that are nearly acyclic.

One of the few groups of mammals that have had their energetics examined in relationship to climate, distribution, and reproduction is the Procyonidae (Mugaas et al. 1993). Species that belong to this family are found in temperate North America south to temperate South America. Tropical species, especially *Potos flavus* and probably *Bassaricyon gabbii*, which are arboreal and highly frugivorous, have low basal rates of metabolism and low r_{max} values, whereas the most temperate species, *Procyon lotor*, has a high basal rate and the highest r_{max}. That is, r_{max} is positively correlated in procyonids with basal rate of metabolism independent of body mass (Figure 13.11). The claim that species that belong to the same genus represent redundant information (e.g., Elgar and Harvey 1987) is obviously wrong here, as *P. cancrivorus* is much more like other tropical procyonids that belong to other genera than it is to the congeneric, but temperate, *P. lotor*. In fact, the high basal rate and high r_{max} in *P. lotor* are undoubtedly connected to its temperate distribution (see Section 14.5.7).

Mammals that have a high r_{max} in association with a high basal rate exploit abundant, often high-quality resources, whereas mammals that have a low r_{max} coupled to a low basal rate use resources that are limited in abundance or quality. That is, those species generally said (in the terminology of MacArthur and Wilson [1967]) to be r-selected

have high basal rates and those said to be K-selected have low basal rates. A high r_{max} can be attained in (at least) two ways: (1) through the evolution of a small body mass, or (2) through an increase in rate of metabolism at a fixed mass. The terms "r-selected" and "K-selected" are relative to a standard that varies with body mass: r-selected species have a higher r_{max} than expected from mass and K-selected species have a lower r_{max} than expected from mass (Figure 13.9A). The mean curve for homeotherms according to Fenchel (1974) is fitted to mammals that are typified by high basal rates (e.g., *Microtus*, *Rattus*, *Bos*), and it is parallel to but above (as expected) Hennemann's fitted mean curve (Figure 13.9B).

The suggested correlation of production with rate of metabolism in eutherians implies that some eutherians have a low r_{max} because they have a low rate of metabolism, which (as seen) may be required by the foods used (see Robinson and Redford [1986] for a different view). Anteaters are K-selected principally because they feed on ants and termites (McNab 1984), not because they are xenarthrans, aardvarks, or pangolins. K-selected mammals have as high a r_{max} as the resources exploited will permit. Other factors may also influence r_{max}: precocial eutherians appear to have a lower r_{max}, at least at small masses (Hennemann 1984b), in part as a result of an extended gestation period, which thereby reduces litter frequency.

BOX 13.1
BOX 13.1
Doubts on the Significance of Energetics for the Parameters of Population Ecology

Hayssen (1984) and Harvey et al. (1991) criticized the concept that variation in the generation time and fecundity of eutherians is correlated with their level of energy expenditure. Hayssen (p. 419) maintained that "...there is no theoretical reason to expect a high correlation between *basal* metabolic rate and a population's maximum rate of increase." Basal rate is not a "special" rate. It is a convenient "standard" rate, and reason does exist to expect a correlation between various population parameters and *some* rate of metabolism. This reasoning maintains that "rate of metabolism" is not simply the rate at which heat is generated, but is a measure of the rate at which an organism functions, including organic synthesis, growth, and reproduction. Besides, the correlations that appear to exist between growth or fecundity and basal rate of metabolism need to be addressed whether or not a "theoretical" framework exists.

The criticism of Harvey et al. (1991) was very different. They suggested that the correlations between population parameters and rate of metabolism proposed by McNab (1980b) do not exist, although they did find a correlation in eutherians between number of offspring per litter and basal rate, which they dismissed as a result of "chance." They claimed that the variation seen in the parameters of reproduction were accounted for in their analysis by variation in body mass and taxonomic affiliation, even though the authors (p. 558) saw "...no reason why our analyses should be systematically biased because they rest on unsatisfactory classifications." Furthermore, Harvey et al. freely used data on metabolism that did not meet standard conditions, so they did not test the correlation of reproductive parameters with basal rate. They also found no correlation of reproduction with field rates of metabolism, which is not surprising considering how variable and nonequivalent field measurements are (but see Figure 10.5).

The correlations proposed by McNab (1980b) and partially corroborated by Hennemann (1983a), Schmitz and Lavigne (1984), Glazier (1985), Thompson and Nicoll (1986), Genoud (1988), Rasmussen and Izard (1988), Genoud and Vogel (1990), Thompson (1992), Mugaas et al. (1993), and Stephenson and Racey (1995) are admittedly very complex and need to be corrected for the impact of body mass. The argument that the remaining residual variation in reproduction is associated only (principally?) with taxonomy does depend on good classifications, as well as on quality data on some standard (i.e., comparable) rate of metabolism. The argument that various characteristics of organisms are "accounted for" by taxonomic affiliation tells us little about the bases of these correlations: close relatives often have similar characteristics because they do the same thing and have a similar size, as well as have a common ancestor. To separate the consequences of similar behavior from those of similar ancestry is very difficult and requires a detailed analysis of the behavior and ecology of the species involved. At present, the tendency to "blame" everything on ancestry is a statement that it is easier to classify an organism than it is to define its behavioral and ecological characteristics. If ancestry is so important in determining the quantitative characteristics of organisms, why should body mass retain its impact independent of phylogeny? The answer may be rather simple: the impact of body mass on rate of metabolism can be shown to act independently of taxonomy because this relationship is easy to analyze, at least in part because it is a quantitative relationship. Furthermore, if rate of metabolism is not functionally important in these relationships, why do they repeatedly appear? That is, why are the correlations with basal rate not random?

The analysis advocated by Harvey et al. (1991), as seen by Westoby et al. (1995), is tantamount to assuming that phylogeny as an explanation has priority over all other explanations. Indeed, the principal difficulty with all analyses of the causes for the occurrence of biological characteristics is character interaction (see Box 5.1). If character interactions are multifaceted and complex, they require large data sets to resolve the interactions more than they require elaborately complex statistical analyses.

13.5 POPULATION ECOLOGY AND ENERGETICS IN MAMMALS OTHER THAN EUTHERIANS

The difference in energetics between marsupials and eutherians (see Section 3.3.1) has implications for population ecology. The basal rates in terrestrial eutherians are high when they feed on vertebrates, on grass, or on a mixed diet of plants and animals, whereas marsupials with these food habits have basal rates that are only 60% to 70% those of eutherians (Table 13.1). However, arboreal folivores, arboreal frugivores, and terrestrial insectivores have low basal rates, irrespective of whether they are marsupials or eutherians. When food habits are associated with high basal rates, high rates occur only in eutherians, and when food or stratum or both are associated with low basal rates in eutherians, low basal rates are also found in marsupials. One interpretation of these data is that some habits *permit* high rates of energy expenditure, a condition that is exploited only by eutherians, whereas some other habits *require* low rates in both marsupials and eutherians (see McNab 1986b, Lillegraven et al. 1987).

No clear correlation of the parameters of population reproduction with rate of metabolism occurs in marsupials and monotremes. For example, gestation period and fecundity are not correlated with basal rate in marsupials (McNab 1986b). At best, the postnatal growth constant is weakly depressed in marsupials with unusually low basal rates. The apparent independence of the parameters of reproduction from basal rate in marsupials may reflect the very restricted range in their basal rate independent of body mass: body mass accounts for 92% of the variation in basal rate among 46 marsupials, whereas it accounts for only 77% of the variation in basal rate among 272 eutherians (McNab 1988a). What is so striking about marsupial basal rates is not that they are low, but that the distribution of basal rates is truncated at the upper end compared to that of eutherians (Figure 13.12): no marsupial has a high basal rate.

Marsupials may have little variation in basal rate *because* reproduction is not coupled to rate of metabolism: the "reason" why carnivorous, grazing, and omnivorous eutherians have a high basal rate may be that they thereby increase their reproductive rate, whereas marsupials with the same food habits do not have high basal rates because such rates do not lead to an increase in reproductive rate (McNab 1986b, Lillegraven et al. 1987). As might be expected, marsupials usually have longer times from conception to weaning than do eutherians (see Section 5.8.1). Marsupials also have lower r_{max} values than eutherians at masses less than 400 g and at masses greater than 10 kg, but at masses between 1.0 and 3.5 kg these groups have similar intrinsic population growth constants, principally because of larger litter sizes in marsupials (Thompson 1987).

An apparent consequence of the coupling between production and basal rate in eutherians and its absence in marsupials is striking. Marsupials coexist with eutherians in the New World when marsupials have food habits that in eutherians are correlated with low rates of metabolism. This coexistence is especially obvious among arboreal frugivore-omnivores, for example, procyonids (eutherians) and didelphids (marsupials). Tropical, nocturnal, arboreal, frugivorous procyonids *Potos*

Table 13.1 The comparative energetics of mammals

| Food habits* | BMR[†]-(n) | | M/E | P[‡] |
	Marsupials (M)	Eutherians (E)		
Terrestrial omnivores	83 (8)	119 (29)	0.70	0.0054
Terrestrial carnivores	89 (9)	151 (9)	0.59	0.0015
Terrestrial insectivores	77 (12)	92 (29)	0.84	0.2731
Terrestrial grazers[§]	95 (5)	146 (19)	0.65	0.0069
Arboreal folivores	75 (5)	82 (6)	0.91	0.6481
Arboreal frugivore/omnivores	84 (4)	73 (3)	1.15	0.1308

*The only habits considered are those found in both marsupials and eutherians, therefore excluding bats, aquatic carnivores, and arboreal seed eaters, among others.
[†]Basal rate of metabolism (BMR) expressed as a percentage of that expected from the all-mammal curve (Equation 3.3).
[‡]Probability that marsupials and eutherians of a food habit have the same basal rates.
[§]Excluding burrowing species, especially those that go into torpor.
Source: Data derived from McNab (1988a).

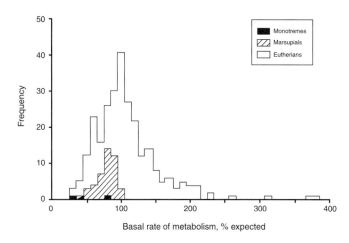

Figure 13.12 Frequency distribution of basal rate of metabolism in monotreme, marsupial, and eutherian mammals. Source: Modified from Lillegraven et al. (1987).

flavus and *Bassaricyon gabbii* coexist with marsupials with similar habits, including *Caluromys* spp., *Caluromysiops irrupta*, and *Glironia venusta*; the basal rate in *Potos* is 86.6% and in *Caluromys* is 89.3% of the values expected from the all-mammal curve (Equation 3.3). Specialist vertebrate-eating marsupials in South America, however, did not survive the connection of South to North America (McNab 1990). The only place where living marsupials have food habits that are associated with high basal rates in eutherians (vertebrate eating and grazing) is in the island refugia of Australia, Tasmania, and New Guinea, where no eutherian carnivores or grazers are found. The survival of marsupials in the presence of ecologically similar eutherians occurs only when food habits require a low basal rate and low r_{max} in eutherians. Otherwise, eutherians displace these marsupials (McNab 1986b); for example, the dingo (*Canis lupus dingo*) displaced the Tasmanian devil (*Sarcophilus harrisii*) and the thylacine (*Thylacinus cynocephalus*) in mainland Australia (Archer 1974). The absence of dingos in Tasmania explains the persistence of these marsupial carnivores.

The ability of eutherians to thrive in polar and cold-temperate environments may be related to the association of a short gestation period, high postnatal growth rate, and high r_{max} with a high rate of metabolism. The weakness of this coupling in marsupials, as indicated by the absence of high basal rates, may explain why marsupials generally do not achieve any appreciable abundance or diversity in these environments, even at high latitudes in South America, where marsupials have had adequate temporal opportunity to exploit cold-temperate conditions (McNab 1986b). For example,

no marsupial is found on Tierra del Fuego and the few marsupials that are found in southern South America, such as *Lestodelphys* and *Dromiciops*, enter seasonal torpor. The only cool-temperate environment that has a diversity of marsupials is Tasmania, which has milder winters than southern South America, again in isolation from potentially competing eutherians.

13.6 POPULATION ECOLOGY AND ENERGETICS IN BIRDS

The correlation of reproduction with energetics in eutherian mammals raises the possibility that the high basal rates found in some birds represent a means of increasing their reproductive output. A major difficulty encountered in evaluating this possibility is the paucity of data available on the standard energetics of an ecologically diverse range of birds. At present, the question as to the dependence of reproduction on energetics in birds cannot be definitively answered, but a tentative answer can be given by comparing birds to eutherians (McNab 1988b).

13.6.1 Postnatal growth constant. As in eutherians, the postnatal growth constant in birds is greatest in species with the highest rates of metabolism. For example:

1. The growth constant decreases with mass by almost the same power (ca. −0.34) as in mammals (Figure 13.13; Ricklefs 1979).
2. The growth constant of birds is about 6 times that of mammals of the same mass, which may reflect higher basal rates in birds.

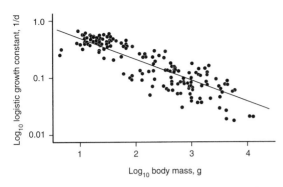

Figure 13.13 Log$_{10}$ logistic growth constant in birds as a function of log$_{10}$ body mass. Source: Modified from Ricklefs (1979).

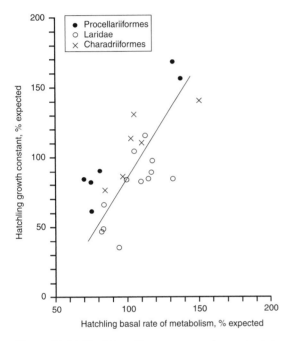

Figure 13.14 Hatchling Gompertz growth constant in Procellariiformes, Charadriiformes, and Laridae as a function of hatchling basal rate of metabolism. Sources: Derived from Drent and Klaassen (1989) and Klaassen and Drent (1991).

3. Some residual variation in the regression of the growth constant on mass is directly correlated with variation in basal rate (Figure 13.14). The principal reason why hummingbirds and frigatebirds have low growth constants, compared to other birds of the same mass (Ricklefs 1968), may be because they have low basal rates.

Padley (1985) found no correlation of postnatal growth constant, or for that matter fecundity, incu-

bation time, and nestling period, with basal rate of metabolism in passerines. The species used were not specified, but undoubtedly were north temperate in distribution, which means that they probably had a rather narrow range in basal rate beyond the influence of body mass, which he found to be a significant correlate of these reproductive parameters. The only hope for finding an appreciable range in passerine standard energetics would be in the study of tropical lowland rainforest species, which have been neglected.

Tropical birds have mean postnatal growth constants that are about 23% lower than temperate and arctic species of the same mass, and have developmental times that are approximately 33% longer than temperate species (Ricklefs 1976). This difference is associated with the predominance of warblers, thrushes, and finches in the North Temperate Zone and of tyrant flycatchers, other suboscines, and tanagers in the neotropics. To a great extent this taxonomic difference reflects a difference in food habits, especially the use of flying insects, nectar, and fruit in the tropics (e.g., swifts, hummingbirds, honeycreepers, euphonias, cotingas, and oilbirds) and immature insects and seeds in temperate zones. As seen, hummingbirds have low basal rates; swifts, which feed on flying insects, have low basal rates; and tropical fruit eaters, including toucans, fruit-pigeons, and some hornbills, have low basal rates by avian standards (McNab 2001). Weathers (1979) also found that basal rate was correlated with latitude (Figure 5.9). A temperate or arctic existence in birds, therefore, may be facilitated by a high rate of metabolism, a high postnatal growth constant, and a rapid attainment of independence by young from parents, which may permit the production of several clutches within a summer to compensate for the annual loss of individuals during migration.

The association of growth rates and basal rate in birds is not uniform: many oceanic birds have comparatively high rates of metabolism but grow slowly. Even among seabirds, the growth constant is correlated with rate of metabolism, but it is set at a lower level than in land birds. Lack (1968b) suggested that these species grow slowly because they face low rates of predation as a result of nesting on islands or on other unusually inaccessible nesting sites. Ricklefs (1979) raised the possibility that the low growth rates reflect energy or nutrient deficiencies in the diet. Some marine birds have a high cost of maintenance because of foraging at great distances from the nesting colonies, or because they live in cold-temperate and polar climates, and these factors may restrict the amount of energy available for the growth of dependent young.

Drent and Klaassen (1989) and Klaassen and Drent (1991) examined in detail the relationships among growth rate, basal rate of metabolism, and latitude. These authors showed that the Gompertz growth constant increased with basal rate, when corrected for body mass, in Procellariiformes, Charadriiformes, and the Laridae (Figure 13.14). During postnatal development in three species of terns (*Sterna*) and one shorebird (*Calidris*), half of the total expenditures were associated with the basal rate, so any change in the basal rate would have an appreciable impact on energy expenditure. A change in basal rate during development was suggested to reflect a change in body composition and therefore represent a rescaling of the machinery of metabolism. The correlation of basal rate with latitude (Figures 5.9 and 13.15), as noted, may be a means of increasing the growth rate and scope of metabolism, which may facilitate survival at northern latitudes. Klaassen and Drent (p. 625) concluded that adjusting ". . . resting metabolic rate is at the core of the adaptation of the organism to its environment."

Konarzewski (1995) questioned this analysis. After "correcting" the data for phylogenetic constraints, he argued that the rate of metabolizable energy is significantly correlated with body mass and latitude, but concluded that either no correlation existed between the costs of production and maintenance or they were negatively correlated with each other (i.e., they corresponded either to the "independence" or the "allocation" models). This conclusion led Konarzewski to reexamine the data of Klaassen and Drent, which showed that a positive correlation existed between the residual variation in chick growth and resting rate of metabolism, even after the data were "corrected" for phylogenetic relationships. However, he proceeded to modify this analysis for the influence of latitude, and then the correlation of growth rate with resting rate of metabolism fell short of statistical significance. Konarzewski (p. 15) remarked, "I feel . . . that this [result] is not persuasive for the existence of the link between metabolic rate and growth rate." (However, reread Westoby et al. [1995] and see Box 5.1.)

A further complication exists in the determination of the postnatal growth constant: precocial young grow more slowly than altricial young (Ricklefs 1968, 1979; Figure 10.16B). This correlation may reflect a trade-off between the amount of energy allocated by the hatchling to maintenance compared to that allocated to growth: if the cost of maintenance is high, as is the case in precocial young that pay the cost of temperature regulation and activity, the amount of energy available for growth is reduced. One of the principal factors determining whether a species is precocial or not is whether the young can feed themselves, that is, whether they use foods, such as algae, plant shoots, fruits, insect larvae, and the like, that are easily captured by naive young (Ricklefs 1979). The combination of body mass, rate of metabolism, and mode of development accounts for most of the variation in postnatal growth constant in birds.

13.6.2 Fecundity. Whether fecundity itself is correlated with basal rate in birds, or whether reproduction is only indirectly correlated with rate of metabolism through postnatal growth rates, is unclear. Fecundity in birds has two components: number of clutches per year (or lifetime) and clutch size. Although many ornithologists, notably David Lack (1947, 1968b), have extensively studied clutch size, especially in relation to latitude, the number of clutches has not been examined relative to rate of metabolism. Clutch size increases in passerines with latitude, which Lack ascribed to the longer day length in temperate summers and therefore the enhanced ability of passerines to forage for food. Ricklefs (1980) followed Ashmole (1963b) in

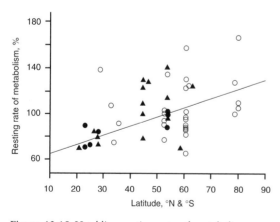

Figure 13.15 Hatchling resting rate of metabolism as a function of latitude. Closed triangles, Procellariiformes; closed dots, Sterninae; open dots, all other species. Source: Modified from Klaassen and Drent (1991).

suggesting that clutch size in birds varies directly with seasonal variations in the productivity of the environment. The exploitation of abundant resources during favorable periods permits a high fecundity in birds, which compensates for the high mortality due to a poor carrying capacity in the environment during unfavorable periods and the mortality associated with migration.

13.7 POPULATION REPRODUCTION AND ENERGETICS IN ECTOTHERMS

The apparent positive correlation of growth rates and reproduction with rates of metabolism in most endotherms may have been a factor contributing to the evolution of their high rates of metabolism, which raises a question of whether such correlations also occur in ectotherms. Ectotherms, by definition, are low-energy organisms, at least in comparison with endotherms, although some ectotherms, most notably salamanders, *Sphenodon*, many turtles, *Xantusia*, and acrochordid snakes, operate at especially low levels of energy expenditure, which may have consequences for their population biology.

In ectotherms r_{max} scales with body mass in a manner similar to that in endotherms, namely, proportional to $m^{-0.28}$, but it is only about 35% that of endotherms of the same mass (Figure 13.9B). In the limited number of reptiles for which growth constants, rates of metabolism, and eccritic temperatures are available, the growth rate correlates with rate of metabolism at the preferred body temperature. For example, the tuatara (*Sphenodon*), acrochordid snakes, and lizards belonging to the genus *Xantusia* have among the lowest known growth rates of all reptiles, and these reptiles have low rates of metabolism (independent of temperature) and low body temperatures. On the other hand, *Cnemidophorus* has a high growth rate (Case 1978a) and a high rate of metabolism, in part because of a high body temperature. Xantusiids have rates of metabolism that are about 50% to 70% of the rates of other lizards of the same mass at the same temperature, and they "... grow very slowly, are late to mature, and have the lowest reproductive potentials known for small lizards ..." (Mautz 1979, p. 583). In the case of acrochordid snakes, low rates of metabolism are coupled with low feeding and reproductive rates, a combination that results in long intervals between reproductive periods (Shine 1986a).

These data suggest that variations in fecundity and postnatal growth rate in reptiles are associated with variations in rate of metabolism. Data on the reproduction of lizards assembled by Tinkle et al. (1970) indicate that generation time may be inversely related to rate of metabolism, in spite of great residual variation in generation time (e.g., in age at the time of first breeding). Again, these correlations are clearest in a family of phlegmatic lizards, the Xantusiidae.

Little is known of these relationships in amphibians and fish, except that fish have very low growth rates compared to other vertebrates (Case 1978a), which in part reflects low body temperatures. Cave-specialist amblyopsid fishes grow slowly and have low fecundities, long life spans, and low rates of metabolism (Poulson 1963; Section 10.10). Some invertebrates also show a correlation between the parameters of reproduction and level of energy expenditure. For example, Anderson (1970, 1978) and Anderson and Prestwich (1982) noted that spiders with very low rates of metabolism, some of which are cave dwelling, produce few eggs, grow slowly, and have long life spans, compared to species with higher rates of metabolism. Cave crayfish grow more slowly, have longer lives, and produce fewer offspring than do surface-dwelling crayfish (Cooper and Cooper 1978), a condition that is correlated with low rates of metabolism in cave crayfish (Eberly 1960).

These correlations, tentative though they may be, raise the question of why, if growth rate, fecundity, and generation time are correlated with rate of metabolism in ectotherms, they do not have still higher rates of metabolism to increase reproductive output even more. Such a query is equivalent to asking why ectotherms are not endotherms. As Pough (1980a, 1983) noted, ectotherms are limited in their level of energy expenditure by a small mass, a large surface-volume ratio, or a highly seasonal food supply (see Section 4.16). The conclusion that some ectotherms show a variation in some demographic parameters in relation to rate of energy expenditure further emphasizes the conclusion that no sharp boundaries exist between ectotherms and endotherms.

13.8 CHAPTER SUMMARY

1. The energy expenditure of vertebrates for reproduction may be subtracted from, independent of, or positively correlated with the expenditure for

maintenance. This system is difficult to dissect because of factor interaction.

2. A short gestation period, high growth constant, and high fecundity are correlated with a high rate of metabolism independent of the influence of body mass in some eutherian mammals, although these correlations often vary with taxonomic affiliation.

3. Life span increases with body mass, a decrease in mortality, and possibly an increase in brain size.

4. Mortality is low in flying and gliding vertebrates and in arboreal and fossorial mammals, that is, in species in which predation is reduced.

5. The maximal growth constant of populations (i.e., the growth constant under ideal environmental conditions) increases with rate of metabolism in eutherians principally because generation time decreases with an increase in rate of metabolism.

6. Eutherian mammals that have high rates of metabolism have larger population fluctuations than do eutherians with low rates of metabolism.

7. The differentiation of eutherians into those with high and low rates of metabolism (at a given mass) is related to the differentiation into r- and K-selected species, respectively.

8. Marsupials do not show a strong correlation of reproduction with energetics, which may explain why no marsupials have high rates of metabolism.

9. The marsupials most likely to coexist with eutherians are those species that have food habits leading to low rates of metabolism in eutherians, which denies eutherians the reproductive benefit of a high rate of metabolism.

10. Eutherians dominate polar and cold-temperate environments because a short-growing season requires mammals to minimize generation time, which requires a high rate of metabolism, requirements that exclude marsupials.

11. Birds have much higher postnatal growth rates than mammals of the same size, which may reflect their higher rates of metabolism.

12. Some of the variation in growth rates in birds is correlated with variation in basal rate of metabolism independent of body mass.

13. In birds the correlation of rate of metabolism with latitude may permit them to accelerate growth rate, thereby permitting repeated reproduction in a short breeding season.

14. Growth rate is positively correlated with rate of metabolism in ectotherms; some ectotherms that have low body temperatures during activity, or have unusually low rates of metabolism at all body temperatures, are characterized by very low growth and reproductive rates.

14 Physiological Limits to the Geographic Distribution of Vertebrates

14.1 SYNOPSIS

Most physiological limits to geographic distribution are likely to result from the interaction of several intertwined factors. Limits associated with thermal and energy exchange, water and salt balance, gas exchange, and food characteristics are complexly determined, and generalizations are difficult to observe. Three generalizations appear to hold. One is that the geographic limit to a breeding range is related to an inability of reproduction to compensate for the mortality encountered at locations beyond the limit to distribution. The second generalization is that when species with similar requirements show competitive interactions, the dominant species is likely to have a greater environmentally based restriction to its distribution, so that the excluded species's survival depends on its greater tolerance of environmental conditions. Finally, climate change potentially has dramatic impacts on the distribution of vertebrates, and even on their survival.

14.2 INTRODUCTION

Many factors interact to limit the distribution of animals. These include various aspects of their morphology, physiology, and behavior, as well as the competitive interactions with other species. Yet, in few cases can we explain why the limits to distribution for a particular species are "here" and not "there." This inability is related in part to the complexity of the interactions between the physical characteristics of the environment and the physiological responses of the organisms. Furthermore, some limits to distribution may reflect historical

factors: many vertebrates that are native to South American rainforests could survive in the rainforests of Southeast Asia. Their absence from Asia may be associated with a physiological or behavioral inability to reach tropical Asia, presumably via Beringia.

This state of affairs has led some biologists to doubt the significance of physiology for setting the limits of distribution, at least in terrestrial vertebrates. For example, Bartholomew (1958, p. 92) stated, ". . . that although the distribution of many marine and aquatic organisms and many terrestrial invertebrates may be explicable in terms of physiological tolerances, no such general statement can at present be made for terrestrial vertebrates." This view reflects the slightly cynical comment that ". . . after much laborious and frustrating effort the investigator of environmental physiology succeeds in proving that the animal in question can actually exist where it lives" (p. 84). (From this view, an astronomer simply shows that a planet, given its mass and distance from a sun, has its particular orbit and velocity, and a chemist. . . .)

In an apparent response to this view and to the conclusion of Dawson (1954) that the desert-dwelling Abert towhee (*Pipilo aberti*) shows a "slight physiological difference" with respect to heat tolerance compared to the coastal California brown towhee (*P. fuscus*) Miller (1963) wrote,

Some . . . investigators . . . have on occasion interpreted their findings as showing that birds have no unusual physiologic adaptations for . . . [desert] conditions. In respect to the conservation of fluids, nothing as striking has come to light as the ability demonstrated in heteromyid rodents

... in which conserving mechanisms are sufficiently perfected to permit metabolic water almost alone to sustain the animal in good condition. Seeming disappointment in not finding like features in birds has led, I believe, to minimizing the significance of the lesser physiologic differentials which were discovered. The significance of these emerges especially when, in the field, we detect mortality in birds confronted with desert conditions and realize that small differences in ability to maintain water balance and to tolerate heat could have been critical to their survival. Furthermore, ... we ... may have expected more spectacular adaptive modifications than the situation actually demands. (p. 666)

An animal species may not be expected to change in evolution any more than it has to in order to meet an environmental situation with reasonable success. (p. 668)

Actually what physiological ecology does is analyze the means by which an "animal can ... exist where it lives," and possibly point out what would be required to extend its range, why (or how) some have done so, and why others have not.

Evolution, of which adaptation is part, usually results from many quantitative changes, among which are changes in physiological functions and their interactions with behavior, important components that potentially contribute to setting limits to geographic distribution. Physiological functions may participate in setting these limits if the physical conditions at the boundaries to distribution exceed the limits of homeostatic regulation, or if the cost in behavior and energy of such regulation is too great. In either case a limit to the distribution of vertebrates represents a limit to adaptation, that is, a limit to the capacity of an organism to respond to changes in the physical or biological characteristics of the environment.

"To do science is to search for repeated patterns, not simply to accumulate facts, and to do the science of geographical ecology is to search for patterns of plant and animal life that can be put on a map" (MacArthur 1972, p. 1). At present, a general analysis of the physiological limits to distribution cannot be given because the diversity of functions potentially involved is great, the number of solutions unknown, and the number of examples examined limited.

Illustrative of the complications encountered is the analysis of Davenport and Sayer (1993) of the factors limiting the distribution of fishes. They concluded that many factors influence distribution, but noted that

... fish are totally excluded [from an area or habitat], by virtue of their physiology, only from a permanently terrestrial existence, from bodies of water with a pH below 4 or above 10 or with a calcium content below 10 mMol/L, from permanently anoxic environments and from hypersaline water bodies with salinities above about [2500 mOsm/kg]. The physiological compromises necessary for less extreme, but still demanding, environments appear to constrain the distribution of certain categories of fish.... [E]xtreme conditions are associated with low-energy life styles, and it is hypothesized that this is because the scope for activity is considerably reduced by the high energetic costs of maintenance and regulation.... (p. 137)

Or, because of a shortage of energy and nutrients.

Many analyses of the physiological limits to distribution have consisted of comparing the limits of distribution with various climatic isopleths, especially minimal temperatures and annual rainfall. As suggestive as these observations might be (see several examples below), they fall short of demonstrating a climatic or a physiological limit to the distribution of species. What also is needed is a demonstration that if a species extended beyond these isopleths, it does not have the reproductive capacity to equal or exceed its mortality and therefore cannnot sustain its population. This is a difficult requirement and has been rarely accomplished.

This chapter explores specific cases in which physiological functions appear to be implicated in setting the limits to distribution. For convenience, these examples are grouped by function. A close examination of these examples suggests that (1) the limits to distribution often involve the inability for reproduction to occur at a rate sufficient to compensate for mortality, and (2) when species interactions are involved in setting the limits to distribution of a species, the dominant species usually has the narrower climatic limit to distribution, whereas the excluded species is more tolerant of climatic factors, which is what permits its survival beyond the borders of tolerance of the dominant species.

This book represents the view that one of the ultimate expressions of physiological ecology is the limits to the distribution of vertebrates.

14.3 THERMAL LIMITS TO THE DISTRIBUTION OF ECTOTHERMS

Because so much effort in the study of physiological ecology of vertebrates has involved temperature relations and energetics, the majority of examples of the physiological limits to distribution fall into this category. But even here few of the cited cases can set a precise limit to geographic distribution by examining any one function; most distributional limits probably reflect an interaction among various physiological, behavioral, and ecological factors.

14.3.1 Ice-fish.
A perfect example of the importance of function interaction that influences geographic distribution is found in ice-fish. In this case the interaction involves temperature, metabolism, and gas exchange. All members of the antarctic teleost family Channichthyidae lack erythrocytes and hemoglobin. This modification reduces the viscosity of blood at low temperatures (see Section 8.4.8). Ice-fish compensate for the resulting reduction in oxygen capacity of blood by being inactive and by having low rates of metabolism and high rates of blood circulation. Hemmingsen and Douglas (1972) showed that rate of metabolism in *Chaenocephalus aceratus* increases with temperature up to 4°C, above which no further increase in rate of metabolism occurs (Figure 14.1), which may well signify that although this fish can tolerate 10°C for a few days, it may not be able to be active for extended periods at these higher temp-

eratures. The extreme adjustment ice-fish make to continuously low temperatures, therefore, is both facilitated and required by the absence of hemoglobin; it appears to limit their ability to tolerate higher temperatures.

Channichthyids are almost exclusively distributed south of the Antarctic Convergence (Figure 14.2), which represents the northern limit of Antarctic Ocean water, the surface temperature of which is about 3°C. Ice-fish living along the shores of Antarctica often encounter water temperatures between −1.0 and −1.4°C. The only ice-fish to be found north of the convergence is *Champsocephalus esox* at the Falkland Islands (Islas Malvinas) and southernmost South America. Unfortunately, little is known of the physiology of this species, except that it too is without hemoglobin. Whether the ice-fish found north of the convergence can withstand higher temperatures than species limited to antarctic waters and if so, how, or whether these species frequent cold microclimates, is unclear and deserves study. For example, *Ch. esox* can be caught at Punta Arenas, Chile, where the following questions could be pursued: What temperatures does this species encounter? Does it have an unusually low rate of metabolism, even for ice-fish? Does it have a higher rate of circulation, or a larger surface area for gas exchange than other species?

14.3.2 Sea snakes.
Sea snakes are limited in their distribution to tropical waters. The hydrophine elaphid *Pelamis platurus* cannot

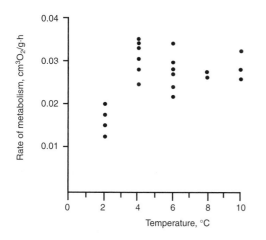

Figure 14.1 Rate of metabolism in the ice-fish *Chaenocephalus aceratus* as a function of water temperature. Source: Modified from Hemmingsen and Douglas (1972).

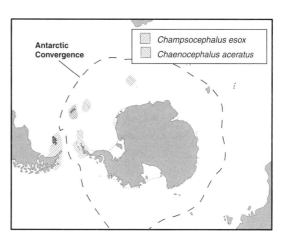

Figure 14.2 Antarctica, the Antarctic Convergence, and the distribution of the ice-fish *Champsocephalus esox* and *Chaenocephalus aceratus*. Source: Modified from Iwami and Kock (1990).

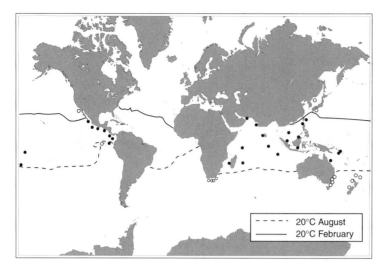

Figure 14.3 Distribution of the sea snake *Pelamis platurus* in relation to the 20°C mean isotherm during the coldest months (August in the Southern Hemisphere and February in the Northern Hemisphere). Breeding records are indicated by ◆, and cold-water strays by ◇. Source: Modified from Dunson and Ehlert (1971).

survive temperatures below 16°C and generally requires temperatures above 20°C (Dunson and Ehlert 1971). Hecht et al. (1974) showed that this species always ceases effective feeding at temperatures below 23°C and usually below 26°C. They concluded that *P. platurus*, which has nearly a circumtropical distribution due to its pelagic habits, is limited by a mean monthly ocean-surface temperature of about 26°C during the coldest month (February in the Northern Hemisphere and August in the Southern Hemisphere). Dunson and Ehlert (1971) interpreted the thermal boundary to be 20°C (Figure 14.3). Individuals collected beyond these boundaries (e.g., in Tasmania, New Zealand, southern Africa, and Japan) are strays that do not constitute members of a breeding population (Hecht et al. 1974). Resident colonies of this sea snake may also be limited to seas with a depth equal to or less than 100 m.

Other sea snakes have restricted geographic ranges, centered in the Indo-Australian region. They too are limited to the tropics, but in some cases these limits are modified by behavior that takes advantage of their littoral habits. For example, whereas *P. platurus* is a rare waif in New Caledonia, *Laticauda colubrina* (a laticaudine elaphid sea snake) is a regular resident in the lagoon at Noumea (Hecht et al. 1974). The waters of this lagoon are above 25°C for only 2 months of the year, which accounts for the absence of *P. platurus*. During winter *L. colubrina* maintains its body temperature between 28 and 30°C by basking on land, presumably to facilitate digestion after it has hunted in the ocean when water temperature

was 21°C. The inability of *P. platurus* to crawl onto land limits its ability to be independent of ocean temperature.

Sea snakes are absent from the tropical waters of the Atlantic Ocean, but they are not absent because environmental conditions are hostile: water temperature in the Caribbean meets the requirements of all sea snakes (Figure 14.3). The only likely explanation appears to be that sea snakes originated in the waters of Southeast Asia and that *P. platurus* moved eastward and arrived at the west coast of tropical America after the Central American bridge had connected North and South America. Most of the west coast of South America is uninhabited by sea snakes owing to the presence of the northward-bound, cold Humboldt Current. *Pelamis platurus* also moved westward through the Indian Ocean to the east coast of Africa, but it has been unable to survive the cool waters of the Cape of Good Hope and of the Benguela Current along southwestern Africa. Here is an example of the interplay of physiological potential with geological opportunity in setting the limits to distribution in a living vertebrate. The proposed construction of a sea-level canal in Panama might permit *P. platurus* to invade the Caribbean and the rest of the tropical Atlantic (Graham et al. 1971).

Hecht et al. (1974) noted that the fossil marine snake †*Pterosphenus*, which was unrelated to living sea snakes but had the lateral compression of living species, has been found in marine deposits of Egypt, Libya, Ecuador, and Alabama. They lived during the early Tertiary and disappeared by the Oligocene. The extinction of these early marine snakes may have been the result of a worldwide mid-Tertiary cooling of ocean temperatures. Only in the late Tertiary were sea temperatures high enough to permit the reinvasion of marine environments, this time by elaphids, but by then it was too late to have access to the tropical Atlantic.

14.3.3 American alligator. Whereas most crocodilians are tropical in distribution, the two species of *Alligator*, one that lives in central China (*A. sinensis*) and the other in the United States (*A. mississippiensis*), are temperate species. The northern limit to the distribution of the American species is northeastern Texas, southern Arkansas, and central Mississippi, Alabama, and Georgia to coastal South and North Carolina. This limit corresponds roughly with the 4°C mean air isotherm in January. Near the northern limits of distribution in South and North Carolina, alligators remain inactive for much of the winter, often in subterranean burrows, which protect them from most freezing temperatures. Sometimes they encounter ice formation on lakes, where they use breathing holes (Brisbin et al. 1982; Hagan et al. 1983). At that time body temperatures may be as low as 4 to 5°C, which is near the alligator's the minimally acceptable temperatures (Brisbin et al. 1982). During warm periods in winter alligators occasionally bask. The inability of crocodiles (*Crocodylus*) to tolerate the warm-temperate environments in which alligators are found may be associated with a differential response to low air temperatures: alligators enter the water when exposed to cold air temperatures, whereas crocodiles leave the water, thereby potentially exposing themselves to freezing temperatures (P. Ross, pers. comm., 1999).

The southern limit to the distribution of *A. mississippiensis* is found along the Gulf Coast from southern Florida to coastal Texas, but it generally avoids salt water. It apparently does not enter Mexico, although it might be expected in the State of Tamaulipas, where it could encounter (and be excluded by?) Moreleti's crocodile (*C. moreletii* [P. Ross, pers. comm., 1999]). Another possibility is that alligators are excluded from Mexico because of their inability to tolerate a tropical climate.

Any difficulties encountered by alligators in a tropical climate might be seen in the Everglades of southern Florida. There alligators in winter rarely come across air temperatures below 10°C and body temperatures do not fall below 15°C, whereas in summer they often have body temperatures above 29°C (S. R. Howarter, pers. comm., 2000). At high summer temperatures, rate of metabolism is high, which requires an increase in food intake. This might be difficult to achieve in the Everglades because the rainy season occurs in summer and with an increase in water level, the prey of alligators, especially fish and turtles, disperse. Consequently, food intake relative to energy expenditure falls in the Everglades during the summer, when most of growth should occur. As a result, alligators are smaller (i.e., shorter and weigh less at a given length) in the Everglades (Jacobson and Kushlan 1989; Dalrymple 1996) than in central Florida (Woodward et al. 1995). The tropical limits to the distribution of alligators may be set by the inability to balance an annual energy budget in a marginal environment (S. R. Howarter, pers. comm., 2000). How then do the characteristics of caimans (*Caiman, Melanosuchus, Paleosuchus*) and crocodiles differ from those of alligators to facilitate life in the tropical lowlands?

14.3.4 Terrestrial ectotherms. Although fishes are found at high latitudes, as long as the water bodies do not freeze, terrestrial ectotherms to varying degrees are limited to lower latitudes, presumably because of the combination of low ambient temperatures in winter, local shortages of shelters to use as hibernacula, and a short reproductive period. In North America, ranids, especially *Rana sylvatica*, have the most northern limits to distribution, going above the Arctic Circle in Alaska and the Yukon Territory; followed progressively by some snakes, principally *Thamnophis sirtalis*; salamanders, especially *Ambystoma*; turtles, which marginally cross the Canadian border; and lizards, which barely cross the Canadian border (Figure 14.4). This sequence is not fixed (Ultsch 1989), for some lizards in Europe cross the Arctic Circle, and a salamander (*Hynobius keyserlingii*) is found above the Arctic Circle in Siberia (see Section 4.7.3). The physiological correlates of these individual limits are poorly understood: the ability to reproduce successfully during a short summer at a rate sufficient to maintain a viable population is likely to be part of the explanation. Why some lizards in Europe and a salamander in Asia cross the Arctic Circle, but none in North America do, is unclear, except as these differences reflect the individual characteristics of species produced by the myopic processes of evolution, and thus reflect the impact of historic contingency.

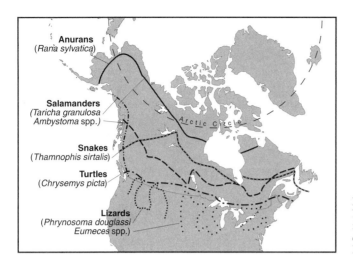

Figure 14.4 Northern limits to distribution in North America of various groups of amphibians and reptiles, as represented by the northernmost species within each group. Source: Derived from Ultsch (1989).

The labels in the figure:

Anurans
(*Rana sylvatica*)

Salamanders
(*Taricha granulosa*
Ambystoma spp.)

Snakes
(*Thamnophis sirtalis*)

Turtles
(*Chrysemys picta*)

Lizards
(*Phrynosoma douglassi*
Eumeces spp.)

Arctic Circle

14.4 THERMAL LIMITS TO THE DISTRIBUTION OF BIRDS

14.4.1 Penguins. Penguins are distinctive birds in a variety of ways, not the least of which is their south polar distribution, although most species actually are found between the Antarctic Circle and the tropic of Capricorn. They are limited in distribution to continental coastlines and islands bathed in cold water (Stonehouse 1967). Penguins breed as far north as the equator in the Galápagos Islands, to 6°30′S along the Peruvian coast in association with the cold Humboldt Current, to 41°S on the east coast of South America in waters cooled by the Falkland Current, and to 25°S along the southeastern coast of Africa along the Benguela Current.

Body size in penguins varies in relation to latitude and water temperature. The largest penguin, the emperor (*Aptenodytes forsteri*, 30 kg) is confined to Antarctica during the winter breeding season. The second largest species, the king (*A. patagonicus*, 15 kg), breeds in summer only on islands that are surrounded by water less than 10°C. The only small penguin breeding on Antarctica per se is the Adélie (*Pygoscelis adeliae*, 5 kg), which, unlike the emperor, breeds in summer.

The correlation of body size with latitude and water temperature is most clearly seen along the west coast of South America. Within the genus *Spheniscus* the largest species (*S. magellanicus*, 5 kg) has a southern (55 to 32°S) distribution and faces water temperatures between 7 and 16°C; *S. humboldti* is intermediate in latitude (33 to 6°S), size (4 kg), and water temperature (15–19°C); and

the equatorial Galápagos penguin (*S. mendiculus*) is the smallest (2 kg) and lives in the warmest sea temperatures (20–25°C). The correlation of penguin body size with water temperature is asymmetrical: small penguins (e.g., 3–6 kg) may live in warm or cold water, but large penguins (>10 kg) do not live in warm water. The largest fossil penguins known (up to 1.2 m tall, compared to the 0.8-m emperor), however, was thought to have lived in seas with a temperature that was 15 to 20°C (Stonehouse 1969).

What is the significance of a large mass in low water temperatures in living species? At one extreme, the emperor penguin breeds in Antarctica during the winter when air temperatures may fall as low as −48°C and wind velocities may be as high as 40 m/s (= 144 km/h) (Le Maho 1977). Male emperors incubate an egg for periods up to 65 d, often at distances of 100 km from open water, which requires that breeding males fast during this period. The entire fast, including both the incubation period and the time for the round-trip travel between the ocean and the nesting site, may last as long as 115 d! During this period males often lose up to 40% of their initial body mass. A large mass (30 kg) is required for an extended period of fasting (Equation 5.4). The Adélie also breeds in Antarctica and can do so with a small mass (5 kg) because its breeding season is in summer, when air temperature may be close to 0°C, and because it lives closer to open water, so that fasting is not nearly so long (ca. 30 d). During fasting this species may still lose about 50% of its original mass. Based on Equation 5.4 and the difference in mass, emperor penguins would be expected to tolerate starvation

1.6 (= $[30/5]^{0.25}$) times as long as Adélie penguins; in fact, emperors can starve 3.9 (= 115/30) times as long, in spite of the colder conditions they face. This increased capacity for starvation is in part due to the huddling of emperors: when huddling is prevented in the field, only 60 d of food deprivation is possible (Le Maho et al. 1976), which is close to the estimate expected from the difference in mass (2.0 vs. 1.6 times).

If a large size is an advantage to penguins in cold climates, what will account for the small size of penguins in warmer climates? The small size of Galápagos or Peruvian penguins is unlikely to be simply the thermal consequence of encountering higher water temperatures, and as noted, many small penguins encounter low water temperatures. A clue for the small size of some warm-water penguins is found in the Galápagos Islands, where breeding penguins are nearly limited in distribution to the north and west coasts of Isabela (Albemarle) and on the shores of the adjacent island of Fernandina (Narborough). In fact, the penguins are concentrated along Bolivar Strait, which separates these two islands. This area of the Galápagos has the coldest water temperatures, which results from the eastward flow of the Cromwell Current (Pak and Zaneveld 1973) and the attendant upwelling of cold water. The upwelling provides nutrients to the surface of the ocean and supports a large plankton crop and rich fish fauna. When this upwelling is diverted by the warm El Niño Current, penguin reproduction is terminated (Boersma 1978). The small mass of the Galápagos penguin may therefore be a means of reducing total food intake in a variable, marginally acceptable environment, conditions that may most strongly affect a flightless bird, which cannot compensate for a local food shortage by exploiting resources at extended distances from nesting colonies: Galápagos penguins rarely stray from the Bolivar Strait. The flightless cormorant (*Phalacrocorax harrisii*) is also limited to the Galápagos Islands and to the Bolivar Strait for similar reasons (Snow 1966), although in terms of mass it is the largest cormorant (3–4 kg).

14.4.2 Giant-petrels and albatrosses. Two families of predominantly Southern Hemisphere seabirds, Diomedidae (albatrosses) and Procellariidae (petrels, shearwaters), differ in foraging strategies, wing morphology, and reproductive cycle, differences that influence their (breeding) distributions, feeding habits, and energy expenditures

(Obst and Nagy 1992). Giant-petrels (*Macronectes*) are the largest members of the family Procellariidae; they are limited in distribution to the high latitudes of the Southern Hemisphere (see Section 10.5.2). There they breed on the Antarctic Peninsula and on some subantarctic islands. The southern giant-petrel (*M. giganteus*) has a mass (4–5 kg) that is similar to those of intermediate albatrosses, but it has a rate of metabolism during foraging that is about 6.3 times its (assumed) basal rate, whereas *Diomedea* albatrosses have rates during foraging that are only 1.8 to 2.7 times the (assumed) basal rates (Obst and Nagy 1992). In giant-petrels this difference is due to a much greater wing loading, a higher wing-flapping frequency, and a frequency in the feeding of chicks that is 2 to 6 times those of albatrosses. Compared to albatrosses, this giant-petrel has a more polar distribution, which requires a faster growth rate and shorter developmental period by the chicks: the developmental period is about 115 d in the giant-petrel but 141 to 281 d in albatrosses. Giant-petrels, therefore, complete their breeding cycle in 4 to 6 months, thereby avoiding an antarctic winter, whereas albatrosses are prevented by their long breeding season from nesting at the highest latitudes. At lower latitudes, giant-petrels breed in lower numbers and spend more time scavenging than pelagic foraging, which is the predominant behavior of the southernmost colonies. The second species, the "northern" giant-petrel (*M. halli*), is smaller (ca. 4 kg) and breeds somewhat north of *M. giganteus* on subantarctic islands, such as Chatham, Bounty, Auckland, and Antipodes, where it encounters many species of breeding albatrosses. The high cost of foraging may make giant-petrels poor competitors with albatrosses.

14.4.3 Anhinga. The anhinga, water turkey, or darter (*Anhinga anhinga*) is a neotropical species that reaches its northern limit to distribution in southeastern United States. Its breeding range extends as far north as Arkansas, eastern Tennessee, and coastal North Carolina. In winter the anhinga is limited to the Gulf Coast and Florida (Figure 14.5). Although many birds have a similar pattern to their distribution, it has thermal significance in the anhinga because, unlike most swimming and diving birds, its feather structure permits water to penetrate to the skin, and they have no oil gland. As a consequence, thermal conductance increases twofold when wet (Hennemann 1982,

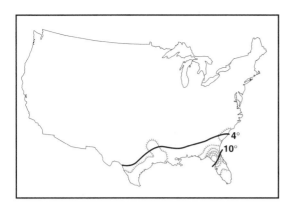

Figure 14.5 The distribution of the anhinga (*Anhinga anhinga*) in winter in the United States in relation to the January mean 4 and 10°C isotherms. Source: Modified from Root (1988c).

1988). The penetration of water to the skin in anhingas, which reduces buoyancy and permits them to capture fish from a stationary position underwater, challenges their ability to regulate body temperature at low air and water temperatures. Thus, its northern limits of distribution in winter, aside from a few strays, coincide with a minimal January atmospheric isotherm of approximately 6°C (Figure 14.5). A complex relationship exists between mean monthly air temperature and the temperature of lakes, depending as it must on their depth and exposure to the sun. The propensity of anhingas to bask in the sun may permit them to tolerate colder water and air temperatures than might otherwise occur, although this tolerance may be modified in relation to cloud cover.

14.4.4 Hummingbirds. The smallest birds belong to the family Trochilidae, which is restricted in distribution to the New World. Most species have a tropical distribution, as might be expected from their use of nectar as food. Yet, at least 14 species enter the United States, 4 of which enter Canada. All that are distributed in Canada and the northern United States migrate to warmer climates for winter. What factors permit some species to have a more northern distribution than others? This question cannot be answered directly at the present time, but at least a comparison can be made of the correlation of physiology and behavior with distribution in some hummingbirds.

Beuchat et al. (1979) compared the energetics of two western North American species, Anna's hummingbird (*Calypte anna*), a permanent resident of the western California lowlands, and the rufous hummingbird (*Selasporus rufus*), which breeds from Alaska to northwestern California and winters in Mexico. These hummingbirds differ in their reaction to temperature: (1) Anna's hummingbird increases food intake with a fall in air temperature, whereas the rufous hummingbird does not, at least at air temperatures less than 20°C; and (2) mass loss during the night in Anna's hummingbird increases at low ambient temperatures, whereas the rufous hummingbird minimizes the loss of mass at night through the use of torpor. These hummingbirds clearly differ in the means by which they balance an energy budget: *C. anna* behaviorally regulates energy gain and *S. rufus* physiologically regulates energy loss. The use of torpor by the rufous hummingbird reflects its smaller mass (3.5 vs. 4.5 g) and its northern distribution, where cool to cold temperatures may occur even in summer.

Several hummingbirds in western North America are limited in distribution to higher elevations in the mountains. These include, remarkably, the smallest species in North America, the calliope (*Stellula calliope*), which weighs only 2.5 g. Unfortunately, most mountain-dwelling species have been little studied. Some reduce energy expenditure by selecting mild microclimates for nesting (see Section 2.5) (Horvath 1964; Calder 1971, 1973a, 1973b, 1974b). Occasionally (2 out of 161 nights) nesting female broad-tailed hummingbirds (*Selasporus platycercus*) enter torpor (Calder and Booser (1973), which raises the question of whether smaller species, such as the calliope, are more prone to enter torpor under the same environmental conditions.

The largest area of high mountains within the range of hummingbirds are the Andes of South America. Some species belonging to the genus *Oreotrochilus* are known to nest at elevations at least up to 4500 m (Smith 1969), where they have been seen to feed on flowers in a snowstorm. The main factor permitting a balanced energy budget in *O. estella* is a reduction in energy expenditure obtained through its (1) propensity to feed from a perch (thereby reducing the cost of feeding by abandoning hovering); (2) roosting and nesting in caves, tunnels, and houses; and (3) use of torpor (Carpenter 1976). Another factor that may be important is that *O. estella* is a rather large species (ca. 8.4 g); indeed, another Andean species, *Patagona gigas*, is even larger (20 g). The principal advantage of a large size is that it permits longer periods of starvation, even though a larger size means an increased cost of maintenance (the Bergmann paradox, see Section 5.3.2). Hummingbirds clearly show a complex array of physiological interactions (among mass, torpor, habitat selection, and energy expenditure) that influences the limits of distribution. They also demonstrate how trade-offs are possible to extend distributions beyond the limits that are expected from a simple view of an organism's responses to the environment.

14.4.5 Starlings and house sparrows. In 1890 European starlings (*Sturnus vulgaris*) were introduced into New York and in 1897 crested mynas (*Acridotheres cristatellus*) were introduced in Vancouver, British Columbia (Johnson and Cowan 1974). Since then, starlings have spread across North America, arriving on the Pacific Coast during the 1950s and in Alaska in the 1970s (Figure 14.6); in contrast, the myna has remained confined to Vancouver, except for an occasional straggler. Why should these two sturnids have reacted so differently to their introductions?

Johnson and Cowan (1974) compared various aspects of the energetics, reproduction, and feeding in the starling and myna at Vancouver in an attempt to answer this question. They showed that (1) the starling has a higher hatching success than the myna (84% in the first and 69% in the second clutch; for the myna it was only 46% and 35%); (2) the hatching success for myna eggs increased to 90% when incubated by starlings, whereas the hatching success for starling eggs fell to 62% when incubated by mynas; (3) mynas had lower nest temperatures that resulted mainly from low nest attentiveness (59% of the time on the nest during the breeding season compared to 77% for the starling); (4) the hatching success of myna eggs incubated by mynas increased to 92% when the nest boxes were artificially heated; and (5) the starling has a posthatching growth constant that is 1.45 times that of the myna.

The common myna (*A. tristis*) in Bengal, India, which is similar in behavior to the crested myna, can maintain an adequate nest temperature with low levels of attentiveness because of high environmental temperatures. Thus, one of the principal problems faced by the crested myna in British Columbia is that it is a tropical species that has not adjusted its behavior to a cool-temperate climate. The crested myna also is characterized by low rates of metabolism at all ages compared to the starling (Johnson and Cowan 1974). A low rate of metabolism in the myna may contribute to its low nest temperature, but it is also implicated in the low postnatal growth rate. As a result of the differences between these species in behavior and physiology, the starling in Vancouver produces 547 young per

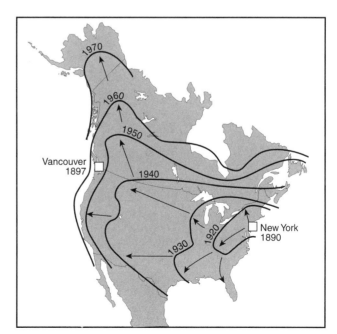

Figure 14.6 The colonization of North America by the European starling (*Sturnus vulgaris*) after its introduction in New York in 1890. The crested myna (*Acridotheres cristatellus*) was introduced in Vancouver, British Columbia, in 1897, where it has remained. Source: Modified from Johnson and Cowan (1974).

100 pairs per year, whereas the myna produces only 238 young per 100 pairs per year.

The myna probably could expand its distribution in North America by adjusting various physiological and behavioral characteristics to a temperate environment. Minor adjustments have been made. For example, crested mynas in Hong Kong have a clutch of 4 eggs, whereas in Vancouver they have a clutch of 4.9 eggs. Mynas in Vancouver, however, persist in feeding their young low-protein fruit, unlike starlings, which feed their young animal matter. The difference between mynas and starlings, then, appears to reflect a tropical versus a temperate origin, respectively. All areas adjacent to Vancouver have a harsher climate, so crested mynas have not been able to establish other breeding populations, thereby expanding their distribution.

No difference in basal rate of metabolism exists between the starling and the myna, although the myna appears to have somewhat poorer insulation (Johnson and Cowan 1975). Johnson and Cowan concluded that the difference in distribution is not caused by a difference in temperature tolerance by adults, but that it is principally due to differences in the temperature sensitivity of their reproduction. This myna probably would have done much better if it had been introduced into southern Florida, as many other tropical birds have. Then, its range would have expanded northward along the Florida peninsula, arriving at some northern limit in central or northern Florida. The functional bases of the northern limits to distribution of tropical exotics should be examined in Florida, a subject that essentially has been ignored.

At present little is known of whether the starling has made any physiological adjustment that facilitates a widespread distribution in North America. Evidence of such an adjustment, however, exists for the house sparrow (*Passer domesticus*), which was released in eastern North America between 1852 and 1860 (Johnston and Selander 1971). Given the short period of time that house sparrows have been in North America (ca. 150 y), they have shown a remarkable flexibility in morphology (Johnston and Selander 1971). Hudson and Kinzey (1966) examined energy expenditure and temperature regulation in house sparrows from populations in Houston (Texas), Boulder (Colorado), Ann Arbor (Michigan), and Syracuse (New York). They found that the Houston population had a basal rate of metabolism that is only 81% of the mean for the other three populations, none of which differed from each other. This low basal rate is correlated with life in a hot, humid climate. Houston birds were able to survive a 2-h exposure to 48.6°C, whereas death occurred in individuals from other populations at lower temperatures. Cold tolerance, plumage mass, and body composition are correlated with latitude in the house sparrow (Blem 1973). This flexibility in basal rate and temperature tolerance may have been an important ingredient in the ability of house sparrows to expand their distribution. In contrast, the Eurasian tree sparrow (*P. montanus*) was introduced into St. Louis, Missouri, in 1870; it has expanded its range only into adjacent areas of central Illinois. This species appears to be morphologically rigid (Barlow 1973), an observation that may be connected with its restricted range. The physiological flexibility of starlings living in various climates in North America and the comparative biology of the house and Eurasian tree sparrows, where they occur together, should be studied.

14.4.6 Finches. In one of the first studies of the physiological correlates of geographic distribution, Salt (1952) examined three species of finches belonging to the genus *Carpodacus*: *C. mexicanus* (house finch), *C. purpureus* (purple finch), and *C. cassinii* (Cassin's finch). In western North America these finches have similar low-altitude distributions in winter. However, in summer the house finch is mainly found in the hot, dry lowlands; the purple finch, in warm, coniferous or mixed forests at low altitudes; and Cassin's finch, in cooler, coniferous forests at high altitudes. Salt measured rates of metabolism as a function of temperature at high (92–95%) and low (12%) relative humidities. He found that the house finch can tolerate high temperatures and low relative humidities, that the purple finch tolerates high relative humidities better than the other two species, and that Cassin's finch is least tolerant of high temperatures, irrespective of humidity. The principal difficulty with this study is that the relationship between rate of metabolism and environmental temperature, as modified by relative humidity, does not agree with what is known from other species: that is, why do his measurements show that a high humidity *depresses* rate of metabolism at high temperatures, when in the experience of other investigators a high humidity at high temperatures retards evaporative water loss, leading to heat storage, with the consequent increase in body temperature and rate of metabolism? Concern for this discrepancy does not deny that a correlation in these *Carpodacus* species may exist between their distribution and their reaction to a combination of relative humidity and temperature. Salt clearly set an excellent example by emphasizing that physical factors are likely to limit distribution when they occur in complex combinations, rather than in isolation.

14.4.7 Magpies. The black-billed magpie (*Pica pica*) is found in North America from Alaska south through the dry mountains and plateaux of western United States, as well as in Europe, Russia, and central Asia, whereas the yellow-billed magpie (*P. nuttalli*) is limited in distribution to the central valleys of California and along the coast of California south of San Francisco. These species face different climates: *P. pica* encounters cold winters and warm, dry summers and *P. nuttalli*, warm winters and hot, humid summers. Correlated with this difference is a greater sensitivity of *P. pica* to an ambient temperature of 40°C (Hayworth and Weathers 1984). Hayworth and Weathers (1984) concluded that the black-billed magpie is limited to dry steppes, but the lack of measurements of temperature regulation at high relative humidities leaves some doubts. The absence of the black-billed magpie from western Oregon and Washington and its presence in coastal Alaska may indicate that the occasional combination of high temperatures and high humidities in summer is enough to exclude this magpie, even though much of the remainder of the year is climatically acceptable. (Is that why the black-billed magpie is not found in eastern North America? Could the yellow-billed magpie tolerate life in the warm, humid southeastern United States?)

14.4.8 The northern limits to bird distribution in winter. Root (1988a) examined the limits to winter distribution in North American terrestrial birds. She found that the most important correlates were vegetation type and minimal January temperature, whereas the eastern and western limits were most strongly correlated with vegetation type and precipitation.

This study was followed by an analysis (Root 1988b) showing that the northern limits to distri-

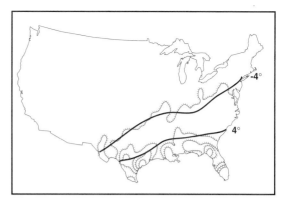

Figure 14.7 The limits to distribution of the eastern phoebe (*Sayornis phoebe*) in winter in the United States. The limit to high abundance is associated with the mean January 4°C isotherm, whereas the absolute northern limit is correlated with the mean January isotherm of –4°C. Source: Modified from Root (1988b).

bution of 51 passerines coincided with a January temperature that corresponded with an estimated energy expenditure that was 2.5 times the basal rate (field expenditures are often 2–3 times basal rate; see Section 10.5.2). For example, the northern limit to distribution of the eastern phoebe (*Sayornis phoebe*) is correlated with the January minimal isotherm equal to –4°C, and the isotherm that corresponds to the northern edge of the area of high abundance is 4°C (Figure 14.7). In this species the isotherms corresponding to high-density populations in winter corresponded to an energy expenditure estimated to be 2.1 times the basal rate. This analysis suggests that many passerines are limited in winter distribution by energy expenditure and, ultimately, energy availability. Root noted that such analyses are often overlooked because most studies have a local scale, which emphasizes species interactions, rather than the influence of broad-scale climatic factors.

Castro (1989) challenged this view by raising questions regarding Root's statistical methods and maintained that even if the northern limits of distribution of passerines were correlated with a rate of metabolism equal to 2.5 times the basal rate, this is not evidence of a constraint. Root (1989) replied by denying that statistical problems existed in her analysis, but appeared to agree that one of the ways for a species to have a more northern limit to distribution in winter would be to have a higher basal rate because the correlation was with a rate that was proportional to basal rate.

14.5 THERMAL LIMITS TO THE DISTRIBUTION OF MAMMALS

14.5.1 Marsupials. One of the most admirable and detailed studies of the influence of energetics on the limits of distribution in a terrestrial endotherm is Brocke's (1970) analysis of the Virginia opossum (*Didelphis virginiana*) in Michigan (at a latitude of ca. 44°N). The critical period of the year for this species in Michigan is December through March, when it often encounters days in which air temperature never rises above 0°C, the so-called freeze-days. The northern limit of distribution in *D. virginiana* is correlated with the iso-pleth for 70 freeze-days (Figure 14.8A). Opossums, thus, are excluded from areas in which 58% (= 70/120) or more of the winter days have temperatures that remain below freezing. Actually, opossums in Michigan are abundant only in areas in which the number of freeze-days is equal to or less than 50. A day in which the maximal temperature never exceeds 0°C prevents opossums from finding sufficient amounts of food in terms of soil invertebrates or carrion.

A limit to geographic distribution is often correlated with more than one variable, so that the causitive factors for the limitation are difficult to determine. In the case of the opossum, its northern

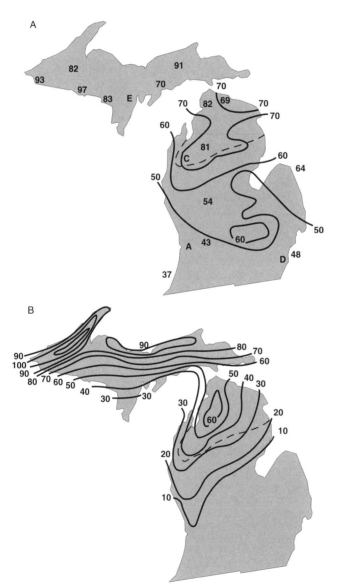

Figure 14.8 The northern limit to distribution of the Virginia opossum (*Didelphis virginiana*) in Michigan (dashed line) in relation to (A) the "freeze-day" isopleths and (B) the snow depth (cm) isopleths. Source: Modified from Brocke (1970).

limit is also correlated with 30 d or more in which the accumulated snow depth is at least 28 cm (Figure 14.8B). These two factors, freeze-days and snow depth, combine to make some areas of central Michigan unacceptable through their restriction on foraging.

When foraging is limited, opossums depend on fat reserves to fuel the cost of maintenance. Field estimates in the 120 d from 1 December to 1 April indicate that adult *Didelphis* males lose about 27% of their original mass, females lose 44%, and juveniles lose 29%. Such large decreases in mass are estimated to represent a 44% decrease in body lipid reserves in heavy individuals and a 72% decrease in light animals. These estimates were obtained within the acceptable geographic range of *Didelphis*. Individuals at or beyond the distributional limit would presumably have even greater fat losses. The apparently smaller mass loss in juveniles is misleading because many die before winter's end. The smallest winter juveniles were collected before 16 February, some of which had essentially no body fat; no opossums of this size were collected later. A cold winter with a heavy snow fall would thus reduce recruitment of young into the breeding population. The high fat loss in females may represent a greater rate of energy expenditure in females than in males with the same level of foraging, or it may represent a similar expenditure at a lower level of foraging. Because copulation does not occur until spring, a cold winter cannot have a direct effect on the fecundity of female opossums (for a contrast, see Section 14.5.2 on armadillos below).

The opossum withstands low environmental temperatures by combining a low rate of metabolism, a variable body temperature, thick insulation, large mass, large autumnal fat stores, and the use of protected shelters. The presence of humans may facilitate survival at low temperatures by providing warmer microclimates and by increasing the food supply. These adjustments, however, are limited, especially when foraging is no longer possible. Unusually harsh winters, typified by very low temperatures or heavy snow falls, have produced a marked decline in the abundance of opossums in Ohio, Wisconsin, and southern Michigan (Brocke 1970).

No marsupial other than *D. virginiana* occurs north of 25°N in North America, which is some 400 km south of the Rio Grande River. The only marsupials found in southern South America are *Lestodelphis halli* in southern Patagonia (ca. 50°S), and *Rhyncholestes raphanurus* and *Dromiciops australis* in south-central Chile (40 to 43°S), each of which has a localized distribution. As noted in Section 13.5, this absence cannot be attributed to a shortage of time to make an adjustment to a cold climate. Both *Dromiciops* and *Lestodelphis* enter hibernation. The general failure of marsupials to invade seasonally cold climates at high latitudes may be related to their low growth rates, especially in environments where the growing season is short. Shrew-opossums, *Caenolestes* spp. and *Lestoros inca*, relatives of *Rhyncholestes*, live in a cold, hostile climate on some of the wet paramos at or above 3000 m in Venezuela, Colombia, Ecuador, and Peru. They, however, are not faced with a short growing season because of their equatorial distribution. The temperate environment with the greatest diversity of marsupials is Tasmania, which has a moderately cold winter. It can be tolerated because of Tasmania's latitude (41 to 43°S) and long season for reproduction and growth.

14.5.2 Armadillos. With few exceptions, armadillos are limited in distribution to the tropics, principally because they combine a low rate of metabolism and a high thermal conductance with a temperature-sensitive food supply (McNab 1980c). They feed mainly on soil invertebrates. Such a food supply is not available when the soil is frozen or covered with snow. A few days of freezing temperatures or snowfall poses no serious problem for armadillos, especially if they are the size of the nine-banded armadillo (*Dasypus novemcinctus*), but as the period of restricted food intake lengthens, the threat of starvation increases because of their high rates of heat loss. A 3.3-kg opossum is estimated to be able to starve for 35.0 consecutive days at 10°C, whereas a 3.3-kg *D. novemcinctus* could only survive about 11.4 d (McNab 1980c). *Dasypus novemcinctus* makes some behavioral adjustments to compensate partially for its temperature sensitivity, including switching its foraging period from night in summer to day in winter, and by burrowing, but these adjustments are only partially effective. As a consequence, *Didelphis virginiana* is found as far north as the Canadian border, whereas *Dasypus novemcinctus* maximally reaches Oklahoma, Arkansas, and the southern portions of Mississippi, Alabama, and Georgia (Humphrey 1974).

Gause (1980) examined in detail the effects of

low temperature on the reproduction of *D. novemcinctus*. Mating and fertilization occur in summer with a peak in July, implantation is delayed for about 4 months, gestation takes about 4.5 months, and the birth of four genetically identical young occurs in March and April. During a normal winter, when the mean January temperature is about 16°C in southern Florida (27°N) and 12°C in northern Florida (30°N), 60% to 80% of the females are pregnant with no evidence of abnormal development or resorption of embryos. A similar condition exists in southern Georgia (31°N) near the northern limits of the armadillo's distribution, where the mean January temperature is about 10°C. During an exceptionally cold winter, when the mean January temperature fell to 13°C in southern Florida, 9°C in northern Florida, and 5°C in southern Georgia, the proportion of pregnant females that have embryos that are being resorbed, or that show abnormal fetal development, greatly increases (32% in southern Florida, 71% in northern Florida, and 85% in southern Georgia). This response means that little recruitment of a new generation occurs in an armadillo population after an unusually cold winter. If such conditions continue for several years in a row, the northern populations decline and the northern limit to distribution retreats to lower latitudes.

Pregnant armadillos make an adjustment to reduce the effect of a cold winter by maximally storing more fat (up to 23% of total mass) than either males (to 14%) or immatures (to 10%), and by leaving burrows to forage soon after a cold front has arrived (Gause 1980). Nevertheless, these behaviors are inadequate during a harsh winter. Immature armadillos are especially vulnerable because they combine a small body mass and a small fat store and must choose in their first fall between using energy intake for growth or for fat storage. The desperate condition faced by immatures is shown by the observation that during a harsh southern Georgia winter mean fat content is only about 1% of total mass; these individuals would not survive winter.

An alternative to facing a temperate winter with a moderate size and large fat stores is potentially available: an armadillo could hibernate. Hibernation is normally associated with a smaller mass than is typical of *D. novemcinctus*. But this is exactly the solution used by the most southerly distributed armadillo in South America: *Zaedyus pichi* (McNab 1980c), which weighs 1.7 kg. Hiber-

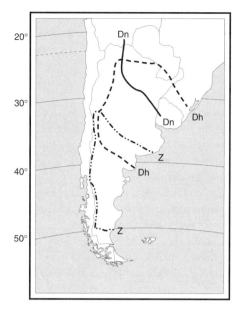

Figure 14.9 The limits to distribution of three species of armadillos, *Zaedyus pichi* (Z), *Dasypus hybridus* (Dh), and *D. novemcinctus* (Dn), in Argentina. Source: Modified from McNab (1980c).

nation has permitted *Z. pichi* to penetrate temperate conditions farther than has the "stoic" behavior of *D. novemcinctus*: the southern distributional limit of *Z. pichi* in Argentina is 17° south of the southern limit of *D. novemcinctus* and 10° south of *D. hybridus*, two armadillos also found in south-central Argentina (Figure 14.9). This pattern gives some comparative measure of the merits of hibernation versus those of the maintenance of body temperature during a cold winter, especially when the food supply is limited and the temperature sensitivity of the animal is high.

14.5.3 Bats. Second only to rodents, bats constitute the most diverse order of mammals in terms of named species, but unlike rodents, bats attain their greatest diversity in the tropical lowlands. Many tropical bats cannot enter temperate climates either because they cannot find appropriate food or because they encounter a hostile climate. The common vampire (*Desmodus rotundus*) in South and North America is limited to regions with a mean minimal temperature of 10°C, or higher, during the coldest month (Figure 14.10). *Desmodus rotundus* roosting at 10°C has a rate of metabolism that is 65% greater than when resting at 20°C (McNab 1973). This means that food consumption must be increased to compensate for the

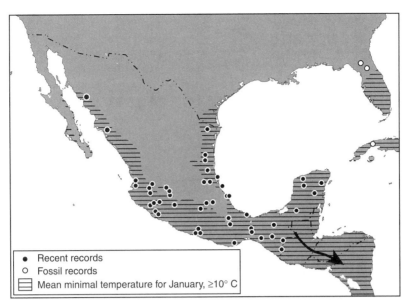

Figure 14.10 Northern limits to distribution of the vampire bat *Desmodus rotundus* in relation to the 10°C January mean minimal isotherm. Source: Modified from McNab (1973).

increased expenditure, but vampires, like all bats, are limited in their capacity to carry a load in flight (see Section 9.6.5). In the Pleistocene *Desmodus* †*magnus* was found in Florida beyond the present 10°C isotherm (Brodkorb 1959; Gut 1959; Olsen 1960), presumably during a warmer period. A limit to meal size represents a limit to the tolerance of cold temperatures, unless torpor is entered or migration is used (neither of which vampires do to any appreciable extent). A limited meal size is especially critical for pregnant females, which presumably have a reduced meal size because they carry a fetus.

Some temperate bats appear to be excluded from tropical environments. This exclusion is clearly seen along the length of peninsular Florida (McNab 1974a), the southern tip of which has a subtropical climate. The principal reason why temperate bats do not adjust to subtropical or tropical environments is that they have attuned reproduction to winter torpor: copulation occurs in the fall or winter and females store sperm. To facilitate sperm storage, females select cave temperatures that are cold enough to enter deep torpor and to reduce arousal.

Two species illustrating this behavior are *Myotis grisescens* and *Pipistrellus subflavus* (McNab 1974a). *Myotis grisescens*, which is large (ca. 13 g) and hibernates in large clusters, requires cold temperatures for hibernation. This species breeds in

caves as far south as northern Florida, although in the Pleistocene it was found 250 km southeast of its present southern limit in north-central Florida (Martin and Webb 1974). During winter most of the north Florida population migrates 450 to 650 km *northward* to Alabama, Tennessee, and North Carolina to find caves sufficiently cold enough to remain in torpor (Tuttle 1976). Migration permits this species to maintain its "temperate" form of reproduction, rather than switch to a "tropical" form, wherein copulation immediately precedes fertilization and implantation. The southern limits to distribution in this species during the breeding season may be determined by a maximally tolerable distance for migration to winter quarters.

Pipistrellus subflavus in many ways behaves similarly to *M. grisescens*, but unlike that species, is small (5 g) and solitary, so it can tolerate higher cave temperatures than *M. grisescens* while remaining torpid. It is the only cave bat that hibernates in Florida. This species too may be committed to fall and winter copulation and sperm storage by the female, behaviors that exclude this species from permanent residency in the tropical lowlands.

Bats, because of their great ecological diversity, are excellent subjects in which to examine the ecological and physiological limits to distribution. Why some species are confined to temperate regions, whereas others apparently of a temperate origin (e.g., *Eptesicus fuscus*, *Lasiurus borealis*, *L. cinereus*) show no such limitation is unknown. Temperate species, at least, must shift their period of copulation from fall to spring upon entrance into a tropical environment, a shift that has been documented in *M. austroriparius* in Florida (Rice 1957). Although few bats other than insectivorous species enter a temperate environment (McNab 1982), some (e.g., molossids, rhinolophids) are much more likely to do so than others (e.g., emballonurids, hipposiderids). The propensity for molossids to enter warm-temperate environments may be associated with their search for insects above and beyond the canopy of tropical forests and their ability to tolerate tropical dry environments. Tropical molossids respond to a marked wet-dry seasonality by the seasonal deposition of body fat (McNab 1976), which may permit them to tolerate the cold-warm seasonality found in warm-temperate environments. Another insectivorous tropical bat that marginally penetrates temperate North America (*Mormoops*) is found in southern Texas and southernmost Arizona and was found in central Florida in the Pleistocene (Ray et al. 1963). Its behavior at marginal localities in winter is worthy of study.

14.5.4 Maximal rates of metabolism and the limits to distribution in rodents. The ability of rodents to increase their rate of metabolism at low ambient temperatures is related to the environments in which they live; that is, it is high in Andean and Arctic species, low in desert and semi-desert grassland species, and intermediate in other species (Bozinovic and Rosenmann 1989). Consequently, the lower lethal temperature, besides being negatively correlated with body mass, is low in Andean and Arctic species, high in tropical species, and intermediate in others. Only rodents that are capable of maintaining a large temperature differential can live at high latitudes and altitudes. Even rodents that enter seasonal torpor in cold climates do not enter torpor because they are incompetent, but do so as a strategy: when normothermic, *Spermophilus*, *Napaeozapus*, and *Tamias* are excellent thermoregulators and maintain large temperature differentials.

14.5.5 Naked mole-rat. The naked mole-rat (*Heterocephalus glaber*) lives as extended family groups in closed burrows in the lowland arid regions of Ethiopia, Somalia, and Kenya. *Hetero-*

cephalus glaber has a very low rate of metabolism, a naked skin, and a small mass, all of which combine to produce marginal endothermy but reduce the likelihood of overheating in warm, humid burrows (McNab 1966b, 1979b). In spite of its poor temperature regulation, the naked mole-rat stores body fat that has a high melting point (McNab 1968), which implies that *H. glaber* never encounters low environmental and body temperatures. The temperature sensitivity of the mole-rat is reflected in the limits to its geographic distribution: the northern and western boundaries are associated with an altitude of approximately 900 m (Figure 14.11) and mean January minimal air temperature of about 17°C with the attendant reduction in soil temperature. The southern, western, and eastern limits to distribution are also correlated with a mean maximal annual rainfall of 50 cm (Figure 14.11). This rodent, then, finds itself trapped in a distribution restricted by the Indian Ocean and the highlands of Kenya and Ethiopia as a result of its extreme adaptation to a hot, dry environment. If the climate in that region were to cool, the naked mole-rat would be unable to survive. Extralimital fossil specimens of *H. glaber* are found in northern Tanzania and eastern Uganda, which may indicate that these localities were warmer and dryer than they are today.

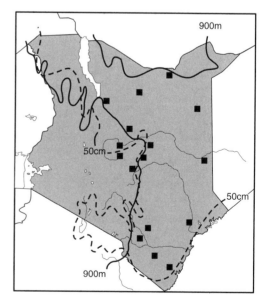

Figure 14.11 Locations (■) of the naked mole-rat (*Heterocephalus glaber*) in Kenya in relation to the 900-m altitude isopleth and the 50-cm rainfall isopleth. Source: Modified from Kingdon (1974).

14.5.6 Flying-squirrels. Two species of flying-squirrels (*Glaucomys*) live in North and Central America. The southern flying-squirrel (*G. volans*) is found in eastern North America from Minnesota, Michigan, southern Ontario, and southern Vermont south to Florida and Texas, and in the mountains of Mexico, Guatemala, and Hondurus. The northern flying-squirrel (*G. sabrinus*) is a boreal species found from Alaska, through Canada, south in the Coast and Cascade Mountains of the west coast to California, in the Rocky Mountains to Utah, and in the Appalachian Mountains to North Carolina. *Glaucomys sabrinus* was found in the lowlands at least as far south as Virginia and Maryland in the late Pleistocene, but with warming both species moved northward, so that *G. sabrinus* appears to have remained beyond the (climatic) reach of *G. volans* at high latitudes and altitudes. Muul (1968) suggested that *G. volans* outcompetes the larger *G. sabrinus* because the southern flying-squirrel breeds earlier and thereby occupies the limited number of breeding sites that are available. The northern flying-squirrel can survive at higher latitudes and altitudes because it is more cold tolerant, the northern limits in *G. volans* being correlated with a mean January temperature of −9.4°C.

A more recent analysis by Stapp et al. (1991) suggested that the northern limit of distribution of *G. volans* is set by the combination of a high expenditure of energy and a progressively decreasing food availability at higher latitudes. They noted that the northern limit in this species coincided with a −15°C January minimal isotherm, which corresponds to an energy expenditure that is 2.5 times the basal rate. At higher latitudes this squirrel could reduce energy expenditure by delaying reproduction, but that would undercut its competitive advantage over *G. sabrinus* (according to Muul [1968]). At high latitudes *G. volans* feeds heavily on the acorns of the red oak (*Quercus rubra*), which reaches its northern limit of distribution at 47 to 49°N, where the January minimal isotherm equals −23°C. At that temperature the nonreproductive energy expenditure of *G. volans* would be approximately 2.7 times the basal rate, and the peak cost of lactation would increase by 15% over that at −15°C. These increased costs, in combination with an increase in the length of winter by 20 to 30 d, means that acorn storage by *G. volans* would have to increase at a latitude where red oak is disappearing. These conflicting latitudinal trends mean that the present limits to a northern distribution in this flying-squirrel cannot be moved northward to any appreciable extent (without a change in climate; see Section 14.9).

14.5.7 Raccoon and other procyonids. Members of the family Procyonidae (narrowly defined, excluding the red panda, *Ailurus fulgens*) are limited in distribution to the New World; most live in tropical Central and South America. *Procyon lotor*, the North American raccoon, is physiologically distinct from tropical procyonids, such as *P. cancrivorus*, *Potos flavus*, *Nasua nasua*, and *N. narica*. Unlike them, it has a high basal rate of metabolism, a broad diet, and a high r_{max} (Figure 13.11). The ringtail (*Bassariscus astutus*), the most temperate in distribution of the remaining procyonids, is intermediate in these characters, and shares with the raccoon a seasonal molt. The estimated r_{max} is high in *Procyon lotor* principally because of large litters and an early age of reproduction. On the opposite extreme is *Potos flavus*, the kinkajou (and probably *Bassaricyon gabbii*, the olingo), which is characterized by a strict tropical distribution, a low basal rate, arboreal habits, and a very low r_{max}.

Mugaas et al. (1993) argued that the evolution of a high basal rate in *Procyon lotor* in association with an increased diversity in food habits (including vertebrates) permitted this species to move out of the tropics and to enter new habitats and climates. They demonstrated that the number of climates in which a procyonid lives is correlated with a "composite" score for procyonid performance (Figure 14.12). This score is the sum of the temperature differential, food diversity, and r_{max}, each normalized for body size in six species of procyonids. The raccoon enters the greatest number of climates (four) "because" it has the highest energy expenditure (per unit of heat loss differential = temperature differential), the greatest diversity in food habits, and the highest r_{max}, which may or may not be independent of each other, whereas the kinkajou has the narrowest climatic range (one) "because" it has a low energy expenditure, a very narrow diet, and a low r_{max}. The rather high climatic diversity in *Bassariscus astutus* (three) is principally associated with a low thermal conductance and a rather high r_{max}.

The Miocene ancestors of living procyonids probably had physiological characteristics similar to present-day *Nasua* and *Procyon cancrivorus*

(Mugaas et al. 1993). From them developed both tropical, arboreal specialists, like *Potos* and *Bassaricyon*, and the temperate specialist *Procyon lotor*. Unlike the assumption that species belonging to the same genus are not independent (e.g., Harvey et al. 1991), "[*P.*] *cancrivorus* shares more in common with other procyonids than it does with [*P.*] *lotor*" (Mugaas et al. 1993, p. 27). "[T]he North American raccoon represents [the] culmination of a divergent evolutionary event that has

given this species the ability to break out of the old procyonid mold [of a low-energy lifestyle in the tropics] and carry the family into new habitats and climates" (Mugaas et al. 1993, p. 28).

Part of the ability of *P. lotor* to tolerate cold climates is related to its flexible body mass. During severe winters raccoons may remain in dens for 2 to 3 months (Mugaas and Seidensticker 1993). Body mass clearly is larger at northern latitudes (Figure 14.13), lean body mass ranging from about 2 to 5 kg. It is even smaller when raccoons are confined to small islands, such as Key Vaca in the Florida Keys, where lean mass is only about 1.5 kg. The deposition of body fat also increases with latitude (Figure 14.13), so that at latitudes near the raccoon's northern limit of distribution in southern Canada (e.g., in Minnesota), up to 50% of its total mass is fat. As with opossums and armadillos near their northern limits of distribution, young raccoons of the year face a conflict in late summer and fall at northern latitudes: should they channel food intake into growth or into fat storage? This choice only applies at the northern limits of distribution; in Florida such a choice is not necessary both because the winters are not harsh and because adult size is relatively small. In Virginia and Minnesota raccoon juveniles do not complete growth until after the first winter, and thus divert much of the energy income in the late summer and early fall to fat storage. For example, Minnesota juveniles in

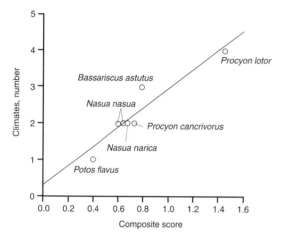

Figure 14.12 The number of climates in which various species of procyonid carnivores are found as a function of a composite score of procyonid performance, which includes basal rate of metabolism, diet breadth, and r_{max}. Source: Modified from Mugaas et al. (1993).

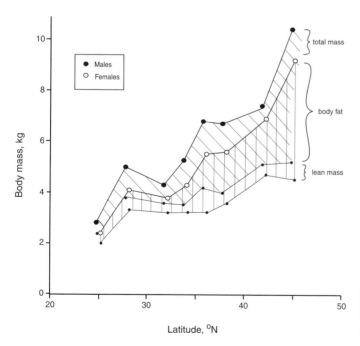

Figure 14.13 Correlation of body mass and body fat content in *Procyon lotor* with latitude. Source: Derived from Mugaas and Seidensticker (1993).

winter have less than half the adult lean body mass, but like adults have 50% of total mass as fat. "Even though [a] large [lean body mass] carries with it a greater absolute energy cost . . . , this is obviously less of a disadvantage in cold climates than the advantage it provides via its attendant increase in fasting endurance" (Mugaas and Seidensticker 1993) (see Section 5.3.2).

14.5.8 Seals. The western Atlantic and Baltic populations of the gray seal (*Halichoerus grypus*), mass of 250 to 280 kg, give birth in the coldest months, namely, mid-January to early March (Hansen and Lavigne 1997). The pups nurse for 3 weeks, during which their mass triples, after which they fast on land or ice for 2 to 4 weeks before they enter the water. The lower limit of thermoneutrality in fasted pups is about −7°C (Nordøy and Blix 1985, Hansen and Lavigne 1997). The northern limit of the breeding distribution in the western Atlantic during January and February is enclosed within the −7.5°C air isotherm. This isotherm also enclosed the breeding populations in the eastern Atlantic (in December and January) and in the Baltic (in March and early April) (Figure 14.14). The increased cost of thermoregulation in fasting pups at air temperatures below −7.5°C appears to be so prohibitively expensive, in light of the fast they are undergoing, that such areas are unacceptable as breeding sites (Hansen and Lavigne 1997).

In contrast to the gray seal, the harbor seal (*Phoca vitulina*) is a small (80–120 kg) species that gives birth in summer, which permits the northern limit of distribution of this species to be farther north than that of the gray seal. The harbor seal probably faces its greatest thermal problem at the southern limits of distribution because it spends much time out of water, where it encounters high temperatures and high rates of solar radiation, factors that may lead to overheating, especially in pups and juveniles. Juvenile harbor seals have high standard rates of metabolism, equal to 2.3 times the rate expected from the all-mammal standard; they have an upper limit to thermoneutrality equal to 25°C (Hansen et al. 1995). At higher ambient temperatures, core body temperature increases. The southern limit of distribution in the harbor seal is completely enclosed by the 25°C air isotherm during July or August, which roughly corresponds to a summer sea isotherm of 20°C (Figure 14.15). Thus, in western North America harbor seals are found as far south as Baja California (17°N), as a result of the southward movement of the cold Japanese Current, but they are found only as far south as North Carolina (35°N) on the Atlantic coast because of the northward movement of the warm Gulf Stream. The combination of air temperature and solar radiation probably is exceedingly important in setting the tropical limits to distribution for many pinnipeds in both the Northern and Southern Hemispheres (Hansen et al. 1995), except for the tropical monk seals (*Monachus*).

14.5.9 Manatees and dugongs. In contrast to pinnipeds, the order Sirenia is presently limited to tropical waters of the New World, Africa, Southeast Asia, and Australia. Living sirenians belong to two families: manatees (Trichechidae) are spottily distributed along the Atlantic coast from Florida to the mouth of the Amazon River, as well as in the Amazon and Orinoco Rivers, and along the Atlantic coast of tropical Africa; and dugongs

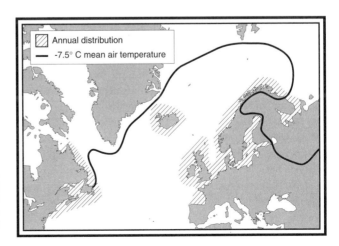

Figure 14.14 The limits of the breeding distribution of the gray seal (*Halichoerus grypus*) in the Atlantic Ocean in relation to the mean January/February air isotherm of −7.5°C. Source: Modified from Hansen and Lavigne (1997).

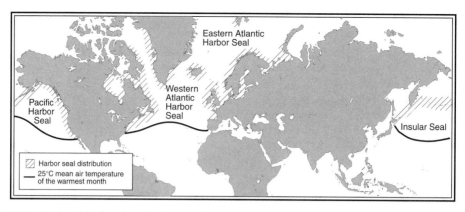

Figure 14.15 The limits of the breeding distribution of the harbor seal (*Phoca vitulina*) in relation to the July/August air isotherm of 25°C. Source: Modified from Hansen et al. (1995).

(Dugongidae) are found in the Indian Ocean and in the tropical western Pacific Ocean. During summer the West Indian manatee (*Trichechus manatus*) is found in Georgia, South Carolina, and on occasion as far north as North Carolina, but in winter they are limited to the warm waters of southern Florida, the Gulf coast of Mexico, and the Caribbean. Unusually cold weather in Florida kills manatees, as occurred in the winter of 1978–1979, in which some 100 manatees were known to have died, at least a third of the deaths being directly related to cold.

This temperature sensitivity is remarkable in a mammal weighing 500 kg, the sensitivity being undoubtedly related to a very low rate of metabolism. Irvine (1983) estimated that *T. manatus* has a basal rate that is only about 34% of the value expected from the all-mammal curve, and Gallivan and Best (1980) found that the basal rate of the Amazonian manatee (*T. inunguis*) is only 45% of the expected value. The low basal rate of manatees probably reflects the poor nutritional characteristics of their vegetarian diet and a tropical distribution.

The thermal limits to manatee distribution in winter occur in Florida. The lower limit of thermoneutrality is difficult to pinpoint accurately because of the variability in rate of metabolism: it may be as high as 24°C or as low as 20°C (Irvine 1983). Irvine suggested that manatees cannot tolerate water temperatures below 20°C for extended periods. This thermal limit translates into northern limits to winter distribution of about 28°N on the Atlantic coast and 27°N on the Gulf coast of Florida. These latitudes correspond to a mean January air temperature of 17 to 18°C. Some man-

atees tolerate higher latitudes in winter by using the warm springs, such as Blue Springs (where winter water temperature is about 22°C) and Crystal River (January water temperature is 23°C), both of which are at 29°N. A few individuals spend winter farther to the north, but only by using the warm effluents of atomic power plants.

The latitudinal limits to distribution of the African manatee (*T. senegalensis*), like the West Indian manatee in Florida, coincides with the 20°C winter atmospheric isotherms on the Atlantic coast, as do the latitudinal limit of dugongs (*Dugong dugong*), in Madagascar, East Africa, and Australia. The only possible exception resides in the northern limits of the dugong in the western Pacific, where winter water temperatures may fall as low as 16°C in the waters off of China and Taiwan. All living sirenians, thus, have rather similar thermal limits to distribution.

The sirenian with the most distinctive thermal distribution was extirpated by humans. In 1741, Georg Wilhelm Steller discovered a dugong, †*Hydrodamalis gigas*, living in the North Pacific and Bering Sea. How Steller's sea-cow could have tolerated such cold water is difficult to understand. Presumably, it too was a vegetarian. Part of its tolerance for cold temperature must have been related to its large size (up to 10 metric tons). The temperature differential in mammals at the lower limit of thermoneutrality is proportional to $m^{0.25}$ (Equation 5.10). Steller's sea-cow was some 20 times the mass of the West Indian manatee, which suggests that its differential should have been 2.1 ($= 20^{0.25}$) times that of the manatee. The West Indian manatee has a temperature differential at the lower limit of thermoneutrality equal approximately to 36.4 − 20.0 = 16.4°C, so that of $^{\dagger}H.$ *gigas* may have been 34.4°C (= 2.1 × 16.4). If the body temperature of the sea-cow was, like the manatee, equal to 36.4°C, the lower limit of thermoneutrality in the sea-cow may have been equal to 2°C, which would have included the water temperatures that were likely to have been encountered. These calculations, as approximate and as assumptive as they are, suggest that the principal adjustment to cold water by $^{\dagger}H.$ *gigas* was associated with a large mass. Steller's sea-cow was only the last of a series of large, cold-water dugongs living in the North Pacific (Domning 1978).

14.6 OSMOTIC LIMITS TO DISTRIBUTION

Water and salt balance are factors undoubtedly important for limiting the distribution of vertebrates, but unfortunately, few demonstrations of these limitations are available, especially in terrestrial environments. Although water balance has been examined in many mammals, especially rodents, the results have rarely been applied to an analysis of their limits to distribution. And in spite of many studies of osmoregulation in fish, and the repeated comments as to the importance of osmoregulation in limiting the distribution of "stenohaline" species, few comparative studies can be found in which the differential penetration of freshwater by marine fish, or of marine water by freshwater fish, has been directly related to the differential capacities of these fish for osmoregulation.

14.6.1 Mullet. Lasserre and Gallis (1975) studied the interaction between the capacity for osmotic regulation in two species of gray mullet and their distribution along an osmotic gradient. A series of artificial ponds in Arcachon Bay, France, had water with concentrations that ranged from 990 to 1000 mOsm/kg near the Atlantic Ocean to freshwater in ponds near the rivers that enter the bay. Seventy-five percent of the individuals found at concentrations greater than 450 mOsm/kg belong to the species *Chelon labrosus*, whereas 90% of the individuals found at concentrations less than 90 mOsm/kg belong to the species *Liza ramada*. Correlated with this difference are the observations that (1) a 5% mortality of *L. ramada* and a 90% mortality of *Ch. labrosus* occurs within 3.5 d following an abrupt transfer to water with a concentration of 15 mOsm/kg, and (2) during longterm, progressive adaptation of these species to freshwa-

ter, *L. ramada* showed no change in plasma osmolality after 150 d, whereas *Ch. labrosus* sustained a 90-mOsm/kg reduction in plasma osmolality in 88 d. *Chelon labrosus*, when forced to stay in freshwater ponds, or to stay in ponds that become fresh as a result of rainfall runoff, die in large numbers.

14.6.2 Pinfish. Carrier and Evans (1976) examined in detail the sodium uptake and efflux in salt water– and freshwater-adapted pinfish (*Lagodon rhomboides*). The kinetics of sodium uptake varied strikingly with acclimation: the Michaelis constant for sodium transport in this marine species in freshwater falls to only 23% of the value in salt water, which is still much higher than is typical of freshwater species. That is, compared to freshwater teleosts, pinfish have a sodium transport system that has a low affinity for sodium, even after acclimation to freshwater. The rate of sodium uptake in an environment having 5 mM sodium fell to 8.4% of the value found in seawater (200 mM sodium), whereas the sodium efflux rate constant in *Lagodon* in freshwater with 5 mM sodium is 18.2% that in seawater. Clearly, the decrease in sodium uptake with the transfer to freshwater is greater than the decrease in sodium loss. Under these conditions pinfish cannot survive long in freshwater. If, however, the freshwater contains 10 mM calcium in addition to 5 mM sodium, the sodium efflux rate constant falls to 6.5% that found in seawater. Then *Lagodon* can survive in freshwater because a sodium transport mechanism with a low affinity can balance the decreased loss of sodium associated with calcium ions.

These adjustments are insufficient to permit *Lagodon* to penetrate freshwater. For example, Odum (1953) found that marine fishes move into freshwater in Florida, although the concept of "freshwater" is complicated to apply because some springs in Florida, such as Homosassa, have high chloride levels, but in no case is *Lagodon* mentioned. Breder (1934) showed that many marine teleosts entered a freshwater lake on Andros Islands in the Bahamas, but they also did not include *Lagodon*. The apparent barrier to *Lagodon* in both areas is the low calcium content of freshwater; spring waters in Florida usually have only 0.5 to 1.7 mM calcium, and the Andros Lake water has 1.0 to 1.5 mM calcium. These levels, although high for freshwater, are too low to permit the penetration of *Lagodon*. Hulet et al. (1967) showed that another marine teleost, the sergeant-major

(*Abudefduf saxatilis*), needs concentrations of calcium between 5 and 15 mM to survive in freshwater. Those marine teleosts that move into freshwaters with calcium concentrations of 0.5 to 1.5 mM may simply have sodium transport mechanisms that have a much higher affinity than is the case in *Lagodon* and apparently *Abudefduf*.

The propensity of marine fish to enter fresh-water appears to be greatest in tropical and subtropical climates (i.e., in warm water). Lasserre and Gallis (1975) showed that the mullet *Chelon* could osmoregulate in seawater down to 7°C (with the plasma maintained at 365 mOsm/kg), but that osmotic regulation broke down at 4°C, when plasma osmolality sharply increased to 435 mOsm/kg. At the lower temperature, mullet could not survive. These observations suggest that a low temperature may turn down metabolism to such an extent that the energy-demanding active transport associated with osmotic regulation cannot balance the diffusional and urinary exchange with the environment (see Figure 6.11). Only at higher temperatures can this balance be maintained, and thus only at higher temperatures can the widespread invasion of freshwater by marine teleosts occur. Yet, salmonids are more prone to anadromy in the cooler regions of their native and transplanted distributions (see Section 14.8.1); these species, of course, have cold-temperate distributions.

14.6.3 Salamanders.

The salamander *Plethodon cinereus* occurs commonly in the soil of the woodlands covering the Blue Ridge Mountains of eastern North America. In the same region, the salamander *P. richmondi* is restricted to isolated talus slopes within the distribution of *P. cinereus*. These distributions at the local level are contiguous but allopatric. The talus environment is much drier than soil. Jaeger (1971) demonstrated that *P. cinereus* has a higher mortality when exposed to dry conditions than *P. richmondi* does. He concluded that *P. cinereus* inhabits moist soil because of its inability to tolerate dry conditions, but that *P. richmondi* is excluded from moist conditions by the presence of *P. cinereus*. The survival of *P. richmondi*, then, depends on its ability to tolerate moderate levels of dehydration. In contrast, although some minor differences existed in the tolerance of two species of *Plethodon* to dehydration in western Oregon, high morality rates kept the populations so low that coexistence occurred with little interaction (Dumas 1956).

14.6.4 Crocodiles and alligator.

Two crocodiles occur in the rivers of northern Australia, *Crocodylus johnstoni*, the freshwater crocodile, and *C. porosus*, the saltwater crocodile. They have a differential distribution along these rivers; *C. johnstoni* is found in freshwater along the upper reaches of the larger rivers, and *C. porosus* is found in estuaries and along ocean beaches (Messel et al. 1979). This pattern can be seen in two rivers, the Daly and the Victoria: an appreciable overlap in the distribution of these species occurs at low salinities (generally <150 mOsm/kg). Short rivers with little freshwater runoff, or long rivers with a tidal influence that penetrates far up the river, may not have any freshwater crocodiles. The boundary between freshwater and salt water fluctuates along the length of the river with rainfall and, thus, with season. For example, the lower reaches of the Adelaide River are much more dilute in July, when *C. johnstoni* is found farther downstream, than in September.

The factors responsible for the distributions of these crocodiles are not understood at the present time: the extent to which these crocodiles are limited in their distribution by the capacity for, or cost of, osmotic regulation or by the competitive interactions between the two species is unclear. Messel et al. (1979) did not encounter freshwater crocodiles along the Tomkinson River, even though the salinity of the river fell to less than 30 mOsm/kg; many *C. porosus* individuals were found at these dilute salinities. On occasion, *C. johnstoni* was found in water with salinities as high as 393 mOsm/kg (Victoria River; Messel et al. 1979). Messel et al. suggested that the saltwater crocodile cannot tolerate hypersaline conditions (>1500 mOsm/kg) for extended periods, and therefore was excluded from some sections of northern Australian rivers.

Both *C. porosus* (Grigg 1977, 1981) and *C. johnstoni* regulate plasma concentration over a wide range of external concentrations (Taplin et al. 1982, 1993). The saltwater crocodile uses lingual salt glands to excrete a hypersaline solution (Taplin and Grigg 1981, Taplin et al. 1982); the freshwater species, however, has lingual salt glands that usually produce a solution much less concentrated than that produced by *C. porosus*. Taplin et al. (1985) described a saltwater population of *C. johnstoni* that excreted sodium at a rate similar to that found in *C. porosus*. In the apparent absence of *C. porosus*, *C. johnstoni* can live in hyperosmotic environments, but whether it needs access to hypoosmotic water is unknown (Taplin et al. 1993). A difference in osmotic regulation may be a factor partially segregating these species, but its importance relative to other factors requires much more information than is presently available.

A similar ecogeographic segregation occurs in southern Florida between the American crocodile (*C. acutus*), which occupies salt water in the Everglades and Keys, and the American alligator (*Alligator mississippiensis*), which lives in freshwater, although again this separation is not absolute, especially in the Everglades. *Alligator mississippiensis* has no lingual salt glands, whereas *C. acutus* has salt glands that produce solutions with the volume and concentration of *C. porosus*. Taplin et al. (1985) argued that Alligatoridae (including *Alligator* and *Caiman*) and the Gavialidae (gavials) evolved in freshwater, whereas the recent radiation within the Crocodylidae (including *Crocodylus* and *Osteolaemus*) was derived from a marine ancestor (Figure 7.19), so that even the most freshwater crocodiles, such as *C. niloticus*, retain the capacity for the production by lingual glands of concentrated solutions at high external concentrations.

14.6.5 Cormorants and anhinga. Among five species of Australian cormorants, one, the black-faced cormorant (*Phalacrocorax fuscescens*), rarely, if ever, enters freshwater. Another species, the pied cormorant (*Ph. varius*), is much more common in salt water than in freshwater. The other three species are commonly found in both environments. An invertebrate-eating cormorant might be prevented from feeding in salt water, unless it had (or could develop) a large nasal gland (to eliminate the excess salt contained in marine invertebrates) or unless it switched to a fish diet (in which case much of the osmotic regulation for the cormorant would be accomplished by the prey). Both of these adjustments are found in the cormorants that tolerate both freshwater and saltwater environments (Thomson and Morley 1966; see Section 7.8.3).

A striking difference is found between the distributions of the double-crested cormorant (*Ph. auritus*) and the anhinga (*Anhinga anhinga*) along the coastal waterways in the southeastern United States. The cormorant is commonly found in saltwater, freshwater, and estuarine environments. Some breeding colonies are located on freshwater lakes. The anhinga is limited to freshwater. Unfortunately, no information is available on the salt

and water balance of the anhinga, although its distribution implies a small, or absent, nasal gland.

The limitation of the anhinga to freshwater could also be related to its foraging style. The cormorant has water-repellant plumage and pursues its prey under water; the cost of the underwater pursuit of fish might be somewhat greater in salt water because of increased buoyancy. The anhinga, however, is a sit-and-wait stabber, which requires the neutral buoyancy attained in freshwater through the loss of the air layer in the plumage. Because the density of salt water is greater than that of freshwater, the anhinga would be positively buoyant in salt water, which would prevent it from using its foraging strategy.

14.6.6 Sandgrouse. Five species of sandgrouse (*Pterocles*) are found in southern Morocco at the northern edge of the Sahara Desert; two are steppe species that reach their southern distributional limits, and the remaining three are desert species that reach their northern distributional limits. The geographic overlap among these species occurs in a region south of the Atlas Mountains, where the mean annual rainfall varies from less than 100 to over 200 mm. Although sandgrouse have long medullary cones compared to other birds, the desert species have shorter cones than the steppe species, mainly because desert species have kidneys that are only about 60% of the mass expected from Equation 7.3 (Hughes 1970b), whereas kidneys in steppe species are about 90% of the expected mass (Thomas and Robin 1977). Thomas and Robin interpreted small kidneys in desert species as a means of reducing water and salt turnover in a dry environment; the sandgrouse do not drink saline water. Sandgrouse males, as has been seen (see Section 7.4.2), are noted for their capacity to transport water adhering to their breast plumage to their young at extended distances from a water hole. Among desert species, at least, a positive correlation of the capacity to transport water occurs with the severity of environmental conditions (Thomas and Robin 1977). Desert species thus appear to conserve water by approaching an anuric condition through a reduction in kidney size. Why an annual rainfall between 100 and 200 mm is limiting to both desert and steppe species, and the extent to which these boundaries are influenced by the presence of other species, are unknown.

14.6.7 Armadillos. The nine-banded armadillo (*Dasypus novemcinctus*), as judged from its kidney structure and inability to maintain its mass on a dry diet, is unable to produce a highly concentrated urine. A hairy armadillo, *Chaetophractus vellerosus*, was found in dry regions of Mendoza, Argentina, and has a kidney with a much greater medulla-cortex ratio, which is a measure of a mammal's capacity to produce a concentrated urine (Greegor 1974; see Section 7.10.3) . The inability of *D. novemcinctus* to tolerate a water shortage is reflected in its distribution in western Texas and southeastern New Mexico. In 1905 a tongue of its distribution extended from Del Rio, Texas, into southeastern New Mexico; in 1972 it no longer lived in New Mexico and west Texas. A decrease in annual rainfall has occurred in this region since 1914 and presently is between 230 and 320 mm; 380 to 400 mm may be the minimally acceptable annual rainfall for this species. Between 1914 and 1972 the number of days in which air temperature did not rise above freezing in this region also decreased; therefore, the retreat of *D. novemcinctus* from New Mexico was unlikely related to temperature. "[T]he moisture barrier [for this species] is moving eastward and the winter barrier northward" (Humphrey 1974, p. 459).

14.6.8 Sewellel. One of the most water-dependent mammals is the sewellel (*Aplodontia rufa*), which is restricted in its distribution to the Coast Range and the Pacific slopes of the Cascade Mountains of British Columbia, Washington, Oregon, and northern California. This rodent has a very high rate of water intake (ca. 33% of body mass/d), mainly because its maximal urine concentration is only 2.2 times the plasma concentration (Nungesser and Pfeiffer 1965). The reasons why *A. rufa* cannot produce a more concentrated urine include that (1) it has a kidney with a thin medulla, (2) it has no renal papillae, and (3) one-third of the nephrons are without a thin segment and only 1% to 2% of the nephrons have loops of Henle long enough to penetrate far into the medulla. Given the poor capacity of the kidneys to produce a concentrated urine, and given the consequently high rate of water intake, *A. rufa* is limited in distribution to areas of heavy rainfall: although the sewellel is found on the western slopes of the Cascades, and therefore has access to the eastern slopes, it does not cross over because the eastern slopes are in a rain shadow. The eastward limits to distribution in this species coincide roughly with the 1000-mm annual rainfall isopleth.

Members of the family Aplodontidae were present as far east as western Montana during the Eocene, and were widely distributed in western Nevada and eastern Oregon in the Miocene and Pliocene (Shotwell 1958). With the elevation of the Cascades, this region was converted into steppe. If these early aplodontids were similar to *A. rufa* in kidney function, they may have disappeared from these inland areas because they could not balance their water budgets. Yet, nagging questions remain: Why should *A. rufa*, or any other mammal for that matter, not respond to a drying environment simply by evolving a kidney capable of producing a concentrated urine? Is the ultimate limit, not the *inability* to produce an effective kidney, but rather a *commitment* to a diet with a high water content, so that the dependence of the food plants themselves on high rainfall limits *A. rufa* to areas of high rainfall? The apparent physiological limitation to distribution in *A. rufa* may ultimately turn out to reflect an ecological or behavioral limitation.

14.7 GAS EXCHANGE AS A LIMIT TO DISTRIBUTION

Considering the number of environments in which oxygen tension may be unacceptably low, such as in stagnant bodies of water, in burrows, and at high altitudes, few examples of the influence of gas exchange on the limits to distribution have been documented in vertebrates.

Gas exchange may be a factor that most likely influences distribution on the local level, rather than having a broad geographic impact. For example, many fishes found in the Amazon basin, such as the piranha (*Serrasalmus*) are limited to "white" waters in which the pH is close to neutral and oxygen tension is relatively high; they may be excluded from "black" waters, where the pH and oxygen tension are low. No detailed examination has been made of the factors limiting the distribution of these fishes. A similar set of physical conditions exists in the black waters of Florida, but they too have been neglected with regard to the distribution of fish. One means of tolerating the low oxygen tensions found in black water, both in Brazil and Florida, is for fish to be air breathers. The geographic impact of water characteristics then may principally affect water-breathing fishes, except when these characteristics permit air-breathing fishes to become a larger component of the fish fauna than normally is the case. The effect of gas exchange on distribution has been less studied in terrestrial vertebrates, which live in an environment where gas exchange usually would be expected to be less important. The three studies included here are indicative of the work that can and should be done.

14.7.1 Darters.

Ultsch et al. (1978) studied six species of the darter genus *Etheostoma* in relation to habitat selection. In winter the distributions of these darters were unaffected by water oxygen tension because of low water temperature (ca. 6°C). In summer water temperature in the streams increased to 20°C and oxygen tension fell. Under these conditions some habitat differentiation appears to be related to gas exchange. For example, the red-lined darter (*E. rufilineatum*) is strictly a fastwater species; it has a high critical oxygen tension (105 mm Hg at 20°C) and a high minimal lethal oxygen tension (mean of 60 mm Hg at 20°C, with some individual values as high as 90 mm Hg). In summer this species is found in water that has an oxygen tension greater than 140 mm Hg. The slack-water darter (*E. boschungi*), in contrast, has a critical oxygen tension equal to 30 mm Hg at 20°C, which permits it to occupy naturally occurring waters with oxygen tensions as low as 55 mm Hg. Its mean lethal tension at 20°C is 14 mm Hg. The difference between these species permits *E. boschungi* to occupy water with a low oxygen tension, but does not require the occupation of these waters. In fact, *E. boschungi* in summer was found in waters with oxygen tensions as high as 139 mm Hg, although it was most abundant at 70 mm Hg. A third species, the eastern swamp darter (*E. fusiforme*), lives in still water. It is able to withstand unusually low oxygen tensions (e.g., at 20°C the lethal tension is 13 mm Hg), yet its critical tension at 20°C is 50 mm Hg. The remaining three species are tolerant of a wide range of oxygen tensions, although they are generally found in fast-running water. Many factors undoubtedly influence the local distribution of these darters, one of them being gas exchange.

14.7.2 Wood-mice.

In a classic study of the limits to distribution posed by low barometric pressure at high altitudes, Kalabuchov (1937) examined highland and lowland subspecies of the wood-mouse *Apodemus sylvaticus*. The most striking difference seen in these mice was their reaction to a change in barometric pressure. Such responses were produced by transplanting lowland mice to the Caucasus Mountains, by transplanting mountain mice to the lowlands, and by experimentally lowering barometric pressure in the laboratory. Mountain *A. sylvaticus* collected at 1500 to 1800 m either maintained the same hemoglobin content and red blood cell count at 150 m as they did at high elevations, or they maintained higher values. The transfer of plains mice to 1475 m produced a 9% increase in the number of red blood cells and in hemoglobin content. A similar increase was found when lowland individuals were exposed in the laboratory to a barometric pressure equivalent to 3000 m. A related species, *A. flavicollis*, showed a 27.2% increase in the number of red blood cells when exposed to this pressure. Other rodents, however, showed a reduction in red blood cells when exposed to 3000 m: a 21.5% reduction in *A. agrarius*, 9.9% in *Clethrionomys glareolus*, and 9.6% in *Cricetulus migratorius*. These responses to a low barometric pressure are correlated with their altitudinal distributions: in the Caucasus Mountains *A. sylvaticus* is found up to 2000 m, whereas *A. agrarius* is only found up to 700 m. A close examination of the distribution of these mice in southern Europe shows that *A. sylvaticus* is found throughout the Alps, but *A. agrarius* is not found at high elevations. Equally, *C. glareolus* and *C. migratorius* are not found in the mountains.

The yellow-necked mouse (*A. flavicollis*) is a particularly interesting case. It shows a marked ability to increase the number of erythrocytes when exposed to a low barometric pressure, but it does not go to unusually high elevations in the mountains; it maximally attains elevations of 1000 to 1200 m. The altitudinal limit of this species, thus, is not related to a physiological inability to increase oxygen-transporting capacity of its blood.

Comparative studies like this one of Kalabuchov should be made in regions where high mountains exist. Especially good areas would be in the Andes of South America and the Himalayas of southern Asia, where the areas of high altitudes are large and where a diverse high-altitude fauna exists. Preliminary observations indicate that high-altitude rodents in the Andes tolerate very low atmospheric oxygen tensions even when exposed to low ambient temperatures (Morrison 1964; Section 8.8.5), but the few available data do not permit an analysis of the altitudinal limits of distribution. Lowland species may be limited by low barometric pressure at high altitudes, but high-altitude species may also face low elevational limits to distribution, which are unlikely due to high barometric pressures. High-altitude limits might be most clearly demonstrated in mammals that live in closed burrows, that is, in species that are doubly denied oxygen (see Lechner 1976).

14.7.3 Fossorial mammals. Mammals that live in closed burrows encounter atmospheres that are isolated from the external atmosphere faced by most mammals. Gas exchange between these two atmospheres depends on diffusion through the soil in which the burrows are found. The rate of this exchange depends on many properties of the burrow and soil, including burrow depth, soil particle size, and water content of the soil (see Section 2.11). Clay soils pose a significant problem for gas exchange because they are made of small particles and have small interstitial spaces, which hold water strongly by capillarity. As a consequence, clay soils are difficult to dry and generally have little air space for the diffusion of gases. As Arieli (1979) observed, burrows in such soils have atmospheres that deviate more from the external atmosphere than burrows in sandy soils. The local distribution of pocket-gophers (*Geomys*) in Florida is correlated with the distribution of sandy soils; gophers avoid clay soils (McNab 1966b). Blind mole-rats (*Spalax*), however, occupy heavy, wet soils (Arieli et al. 1977, Arieli 1979). *Spalax* too must be limited by a sufficiently low soil permeability to gas, unless, like some moles (Olszewski and Skoczén 1965), a significant fraction of gas diffusion occurs through burrow plugs. Moles tolerate heavy, wet soils because the ceilings of their superficial burrows are often broken, creating spaces through which bulk gas exchange with the atmosphere occurs. Investigation of the relations existing among soil composition, the physics of gas exchange, and the microdistribution of fossorial mammals would be very interesting.

14.8 FOOD AND THE LIMITS TO DISTRIBUTION

Food type and quality may limit the distribution of vertebrates. This might be so simple as reflecting the disappearance of a preferred food or the presence or absence of particular chemical structures in the food.

14.8.1 The occurrence of diadromy. The geographic distribution of diadromy, the regular movement of fishes between freshwater and salt water, is complex (McDowell 1987, 1988). It is generally found in "primitive" fishes (lampreys, sturgeons, anguillid eels, and salmonids) and is sporadically represented in other groups. Most interesting is that diadromy has a distinctive geographic distribution (Figure 14.16): anadromous fishes, which breed in freshwater and mature in salt water, are most prominent in the northern subpolar and cool-temperate zones with a small peak in the southern warm-temperate zone; catadromous fishes, which breed in salt water and mature in freshwater, are most common in the warm-temperate and tropical zones; and amphidromous species, in which the movement between freshwater and salt water is not related to reproduction, have a bimodal distribution in the temperate zones.

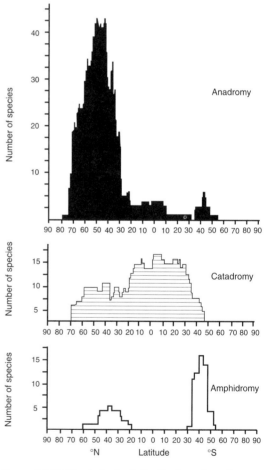

Figure 14.16 The latitudinal distribution of anadromous, catadromous, and amphidromous fishes. Source: Modified from McDowell (1987).

The explanation for this pattern is unclear, in part because of the difficulty of separating the contributions of phylogeny and ecology (McDowell 1987). That an ecological component exists, however, is demonstrated by the observation that several salmonids, both in their native ranges in the Northern Hemisphere and after transplantation to Australia and New Zealand, are more prone to anadromy in the cooler, than in the warmer, segments of their distributions.

Gross (1987) argued that the shift from one behavior to another is based on maximizing fitness through the differential allocation of resources to growth and reproduction as they are influenced by body size (anadromous populations are larger and more fecund) and by production in the aquatic environment. Aquatic production generally is greater in the sea than in freshwater at high latitudes, the very latitudes at which anadromy is most abundant, and is greater in freshwater environments than in the sea at tropical latitudes, where catadromy is most developed (Figure 14.16). "In short, diadromous migrations have evolved to track aquatic production" (Gross 1987, p. 21); that is, diadromous fishes accomplish most of their growth in the most productive waters, in polar seas and tropical freshwaters. A consequence of this geographic pattern is that marine fishes are more likely to move facultatively into freshwater in the tropics than in temperate environments because of the high productivity of tropical freshwaters. Gross argued that amphidromy often is an ancestral stage in the development of anadromy and catadromy, although it may be an evolutionarily stable state if no clear advantage exists for development to be completed in one or another habitat.

14.8.2 Food availability. Food absence may limit the distribution of food specialists. For example, the absence of appropriate fruits may limit the movement of committed frugivores into temperate environments. The only committed frugivorous bird in continental Canada and the United States is the white-crowned pigeon (*Columba leucocephala*), which barely gets into southernmost Florida, and then mainly in the Keys; in all other North American birds frugivory is facultative.

Sherbrooke (1976) noted that the distribution of the heteromyid rodent *Perognathus baileyi* is closely associated with the distribution of the jojoba shrub *Simmondsia chinensis*, which produces a seed that has cyanogenic glucosides and high concentrations of wax. Three species of heteromyids refused to eat the seeds of jojoba in captivity, even though faced with starvation, but *P. baileyi* maintained body mass feeding on jojoba seeds. Although the distributions of *P. baileyi* and *S. chinensis* are not completely congruent, the overlap, even at local levels, is great. Sherbrooke suggested that the toxicity in jojoba seeds and the ability of *P. baileyi* to detoxify the seeds permit the pocket mouse to maintain high population densities.

14.9 CHANGES IN DISTRIBUTION AND ABUNDANCE WITH CHANGES IN THE CLIMATE

Although climate appears rather constant within a human lifetime, the climate on Earth has changed often and sometimes rapidly. Many factors may be responsible for these changes, including the distribution and movement of continents through their effect on currents in the atmosphere and the ocean; the appearance and disappearance of topographic relief on the various landmasses; and changes in the composition of the atmosphere produced by organisms or volcanic activity. Within recorded history, a "Little Ice Age" between 1350 and 1870 greatly affected humans in the Northern Hemisphere and contributed (along with cultural inflexibility) to the demise of the Viking society in Greenland. Evidence indicates that Earth is now going through a warmer period (Figure 2.3A), which may or may not be effected by human activities through a change in the composition of the atmosphere. As climate changes, so too do the abundances, limits of distribution, and even the survival of organisms.

14.9.1 Changes in distribution at the end of the Pleistocene. One of the most dramatic "recent" —in a geological scale—changes in climate and the distribution of vertebrates occurred at the end of the Pleistocene. Pielou (1991) summarized the changes in climate and the distribution of plants and animals, including humans, in North America. About 20,000 years ago the continental ice sheets reached their maximum, after which a warming trend started causing a retreat of the glaciers. The warming trend continued until about 7000 years ago (the "hypsithermal"), when a cooling trend started, although throughout both the cooling and warming trends, many minicooling and warming trends occurred. Two late minicooling trends occurred, one 2800 years ago and the other 600 to 200 years ago (the Little Ice Age). With the melting of the continental ice sheets, much of North America became available for occupation by terrestrial and aquatic vertebrates. During the hypsithermal, a period much warmer than today, many organisms lived farther north than they do at present, including beaver (*Castor*), birch, and spruce on the Seward Peninsula, Alaska. The southern limits of distribution of many species also moved northward: mountain goats (*Oreamnos americanus*) were present in the Sierra Nevada of California during the maximum of the Ice Age but now find that the southern Rocky and Cascade Mountains are unacceptable habitats. The boundary between the paleo-Inuit and paleo-Indian peoples in Labrador has oscillated as the temperatures have increased and decreased (Fitzhugh and Lamb 1986).

14.9.2 Is climate presently changing? Much controversy exists on whether the climate of Earth is changing now, and if so, whether it represents some complex "natural" (i.e., nonhuman mediated) oscillation or is the result of human activities. Countless attempts to collect data and to place them into a reliable mathematical model have resulted in conclusions and predictions that range from (1) no change is occurring to (2) the observed changes are part of a natural oscillation, and (3) marked changes have already occurred and they are only going to become greater in the future. In an extensive review, Mason (1995) showed that (1) atmospheric CO_2 levels have increased by 12.5% between 1958 and 1994 (Figure 2.3B), and 27% since 1765 (before the industrial revolution); (2) the combined air and sea surface temperature increased about 0.7°C between 1910 and 1990 (Figure 2.3A); and (3) sea level in the last 100 years has risen about 10.5 cm, 4 cm from thermal expansion, 4 cm from melting mountain glaciers, and 2.5 cm from the partial melting of the Greenland ice sheet (also see Roemmich 1992). These changes have many secondary effects, including changes in ocean circulation, coastal upwellings, and modification of oceanic productivity (Bakun 1990, Hayward 1997). For example, the zooplankton volume in the California Current has decreased by 80% since 1970 while the sea surface temperature increased by 1.5°C (Roemmich and McGowan 1995).

The causes of these changes are subject to intense

argument, especially given the economic and political consequences of any attempts at correction, if indeed these changes are the result of human activity. According to Mason, a globally coupled atmospheric-deep-ocean model predicts that the average surface air temperature will increase 0.3°C/decade and that sea level will increase 4 cm/decade if the CO_2 concentration in the atmosphere doubles. Such a change will flood many low-lying areas and have a dramatic impact on low oceanic islands, where the endemic native fauna will have no refuge.

14.9.3 Montane amphibians. Some aquatic species of montane amphibians in Brazil (Weygoldt 1989), Costa Rica (Crump et al. 1992, Pounds and Crump 1994, Lips 1998), and Australia (Laurance et al. 1996) have shown dramatic crashs in their populations in the late 1980s. Various of these collapses have been so striking that abundant species have disappeared, most notably the Costa Rican golden toad (*Bufo periglenes*), even to apparent extinction; in Costa Rica 20 of 50 species have disappeared in what appears to be pristine forested communities (Pounds et al. 1996). The explanations for these population declines have included high levels of ultraviolet radiation, acidification of rainwater, the introduction of predators and contaminants, pathogens, and climate change. No one explanation has been accepted as responsible for the population decreases, but the fact that they have occurred at a similar time suggests that they were unlikely to be caused by local events. At Monteverde, Costa Rica, a decrease also occurred in bird and reptile populations, but it was much less in these groups than in amphibians; in birds much of the change was the result of habitat change. Amphibians have several features of their natural history that may make them more susceptible to environmental changes, including a complex life cycle, a diversity of reproductive modes, and a permeable skin (Lips 1998).

Pounds et al. (1999) suggested that the declines in amphibian populations were produced by an increase in the surface temperature of the Pacific Ocean (an El Niño Event), which in the mountains of Costa Rica is expressed as a decrease in mist formation and an increase in temperature. These changes led to the decrease in the mountain amphibian populations and the movement of lowland amphibians, reptiles, and birds to higher elevations. Lowland amphibians, unlike highland species, have an escape from these climatic changes: they can move to higher altitudes. The death of montane amphibians, which may have been ultimately due to climate change, was associated with an outbreak of chytrid fungus in Costa Rica and Australia (Berger et al. 1998, Longcore et al. 1999).

14.9.4 Populations of foraging seabirds. With an increase in the sea surface temperature along the California coast, a dramatic decrease occurred in the zooplankton (Roemmich and McGowan 1995) and a 40% decrease in pelagic seabird abundance (Veit et al. 1996). Sooty shearwater (*Puffinus griseus*) numbers decreased by 90%! Whether that represented a decline in sooty shearwaters, a species that breeds in the Southern Hemisphere, or a relocation of feeding activity to other places in the Pacific basin is unknown. Some local species, such as Brandt's (*Phalacrocorax penicillatus*) and pelagic (*Ph. pelagicus*) cormorants and western gull (*Larus occidentalis*), showed no change in their populations, whereas Xantus' murrelet (*Synthliborhampus hypoleuca*) and Leach's storm-petrel (*Oceanodroma leucorhoa*) increased in abundance. Nevertheless, Ainley et al. (1994) showed that egg laying started later and fewer chicks of seabirds breeding on the Farallon Islands (California) fledged in years when sea surface temperatures were high. Clearly, the increase in sea surface temperature and its correlates have complicated effects on a diverse community of pelagic birds.

14.9.5 Egg laying in passerines. Several recent reports from Europe (Crick et al. 1997, Forchhammer et al. 1998, McCleery and Perrins 1998) and North America (Brown et al. 1999) indicated that egg laying in many passerines started progressively earlier between 1970 and 1995. (A similar pattern was seen in British amphibians [Beebee 1995].) This change is correlated with an increase in temperature associated with the North Atlantic Oscillation, a variation in the circulation of the atmosphere (Forchhammer et al. 1998). The increase in temperature might facilitate an increase in the food supply in the early spring (McCleery and Perrins 1998), thereby permitting earlier nesting. Earlier egg laying in the Mexican jay (*Aphelocoma ultramarina*) was principally correlated with an increase in the monthly minimal temperatures (Brown et al. 1999). Brown and coworkers, noting that if the jays had such a change with only a 0.5°C increase in atmospheric temperature, what

would be the consequence of the 2.8°C rise projected by some models for the middle of the twenty-first century?

14.9.6 Marine mammals in the Arctic. The presence of sea ice in the Arctic is an important structural component of the environment: it provides a substrate for polar bears (*Ursus maritimus*) and pinnipeds; ice provides an environment for the prey of pinnipeds and cetaceans; and the annual formation, movement, and breakup of ice sheets influence the annual migrations of marine mammals (Tynan and DeMaster 1997). Therefore, any appreciable decrease in the amount of ice in the Arctic summer is likely to have profound impacts on marine mammals. Sea ice in the Arctic during the summer has decreased over the last 20 to 30 y coincident with the warming trend in the atmosphere (Maslanik et al. 1996). A continued increase in climatic temperatures will likely lead to a further reduction in summer ice, which will affect seals that use ice for pupping, polar bears that use ice for dispersal and seal hunting, and cetaceans that forage near ice floes. Stirling and Derocher (1993) suggested that (further) global warming would first affect polar bears at the southern limits of distribution in James and Hudson Bays, where today they fast for 4 months with the absence of ice. A longer ice-free period would probably prevent them from occupying these localities. A significant loss of ice would endanger cetaceans that feed at the edge of ice sheets, including the bowhead whale (*Balaena mysticetus*), beluga (*Delphinapterus leucas*), and narwhal (*Monodon monoceros*).

14.10 CHAPTER SUMMARY

1. Few examples can be cited to demonstrate that the limits of geographic distribution are determined simply by the limits to physiological function relative to the physical characteristics in the environment, partly because this topic has attracted little attention.

2. Most limits to distribution probably result from the interaction of several factors.

3. Temperatures limit the distribution of ice-fish, sea snakes, terrestrial ectotherms, penguins, giant-petrels and albatrosses, anhinga, hummingbirds, temperate passerines, opossums, armadillos, bats, rodents, flying-squirrels, procyonids, and manatees and dugongs.

4. Water and salt balance influence the limits to geographic distribution in mullet, pinfish, salamanders, crocodilians, cormorants and anhinga, sandgrouse, armadillos, and sewellel.

5. The importance of gas exchange to the limits of geographic distribution was suggested in darters, wood-mice, and fossorial mammals.

6. Food abundance and quality might limit the distribution of vertebrates, as was suggested for anadromy and catadromy.

7. A geographic limit to distribution is set when a population is unable to reproduce at a rate sufficient to compensate for the mortality determined by the conditions in the environment in regions beyond the limits of distribution.

8. When two species compete for a common resource, the dominant species is (usually?) limited geographically by the physical environment or food resource, whereas the excluded species is found in areas beyond the geographic limits tolerated by the dominant species. That is, the survival of excluded species depends on a greater tolerance of environmental conditions than is present in the dominant species. Possible examples of this interaction were seen in salamanders, albatrosses and giant-petrels, *Carpodacus*, *Pica*, and *Glaucomys*.

9. A change in climate in the past has had profound effects on the distribution of vertebrates, even to the point of leading to their selective extinction. A climate change of unknown proportions is probably occurring, and we already see evidence of changes in the abundance and distribution of vertebrates that may be due to this change.

References

Aalto, M., A. Górecki, R. Meczeva, H. Wallgren, and J. Weiner. 1993. Metabolic rates of the bank voles (*Clethrionomys glareolus*) in Europe along a latitudinal gradient from Lapland to Bulgaria. Ann. Zool. Fenn. 30:233–238.

Abbott, K. D. 1971. Water economy in the canyon mouse *Peromyscus crinitus stephensi*. Comp. Biochem. Physiol. A Physiol. 38:37–52.

Ackermann, R. A., G. C. Whittow, C. V. Paganelli, and T. N. Pettit. 1980. Oxygen consumption, gas exchange, and growth of embryonic wedge-tailed shearwaters (*Puffinus pacificus chlororhynchus*). Physiol. Zool. 53: 210–221.

Adams, N. J. 1992. Embryonic metabolism, energy budgets and cost of production of king *Aptenodytes patagonicus* and gentoo *Pygoscelis papua* penguin eggs. Comp. Biochem. Physiol. A Physiol. 101:497–503.

Adams, N. J., C. R. Brown, and K. A. Nagy. 1986. Energy expenditure of free-ranging wandering albatrosses *Diomedia exulans*. Physiol. Zool. 59:583–591.

Adelman, S., C. R. Taylor, and N. C. Heglund. 1975. Sweating on paws and palms: What is its function? Am. J. Physiol. 229:1400–1402.

Adolph, E. F. 1933. Exchanges of water in the frog. Biol. Rev. 8:224–240.

——. 1949. Quantitative relations in the physiological constitutions of animals. Science 109:579–585.

Afik, D., and B. Pinshow. 1993. Temperature regulation and water economy in desert wolves. J. Arid Environ. 24:197–209.

Ahlberg, P. E. 1995. *Elginerpeton panchei* and the earliest tetrapod clade. Nature 37:420–425.

Ahlberg, P. E., and Z. Johanson. 1998. Osteolepiforms and the ancestry of tetrapods. Nature 395:792–794.

Ahlberg, P. E., and A. R. Milner. 1994. The origin and early diversification of tetrapods. Nature 368:507–514.

Ainley, D. G., W. J. Sydeman, S. A. Hatch, and U. W. Wilson. 1994. Seabird population trends along the west coast of North America: Causes and extent of regional concordance. Stud. Avian Biol. 15:119–133.

Alcocer, I., X. Santacruz, H. Steinbeisser, K. H. Thierauch, and E. M. Del Pino. 1992. Ureotelism as the prevailing mode of nitrogen excretion in larvae of the marsupial frog *Gastrotheca riobambae* (Fowler) (Anura, Hylidae). Comp. Biochem. Physiol. A Physiol. 101:229–231.

Aleksiuk, M. 1976. Reptilian hibernation: Evidence of adaptive strategies in *Thamnophis sirtalis parietalis*. Copeia 1976:170–178.

Aleksiuk, M., and I. M. Cowan. 1969a. Aspects of seasonal energy expenditure in the beaver (*Castor canadensis* Kuhl) at the northern limit of its distribution. Can. J. Zool. 47:471–481.

——. 1969b. The winter metabolic depression in Arctic beavers (*Castor canadensis* Kuhl) with comparisons to California beavers. Can. J. Zool. 47:965–979.

Alexander, R. L. 1995. Evidence of a counter-current heat exchanger in the ray, *Mobula tarapacana* (Chondrichthyes: Elasmobranchii: Batoidea: Myliobatiformes). J. Zool. Lond. 237:377–384.

——. 1996. Evidence of brain-warming in the mobulid rays, *Mobula tarapacana* and *Manta birostris* (Chondrichthyes: Elasmobranchii: Batoidea: Myliobatiformes). Zool. J. Linn. Soc. 118:151–164.

Alexander, R. M. 1965. The lift produced by the heterocercal tails of Selachii. J. Exp. Biol. 43:131–138.

——. 1966. Physical aspects of swimbladder function. Biol. Rev. Camb. Philos. Soc. 41:141–176.

——. 1968. Animal Mechanics. University of Washington Press, Seattle.

——. 1971. Size and Shape. Edward Arnold, London.

——. 1975. The Chordates. Cambridge University Press, Cambridge.

——. 1994. The flight of the pterosaur. Nature 371:12–13.

Alexander, R. M., and A. S. Jayes. 1983. A dynamic similarity hypothesis for the gaits of quadrupedal mammals. J. Zool. Lond. 201:135–152.

Al-Hussaini, A. H. 1949. On the functional morphology of the alimentary tract of some fish in relation to differences in their feeding habits: Anatomy and histology. Q. J. Micrse. Sci. 90:129–139.

Allsbrook, D. B., A. M. Harthoorn, C. P. Luck, and P. G. Wright. 1958. Temperature regulation in the white rhinoceros. J. Physiol. 143:51–52P.

Alt, J. M., H. Stolte, G. M. Eisenbach, and F. Walvig. 1981. Renal electrolyte and fluid excretion in the Atlantic hagfish *Myxine glutinosa*. J. Exp. Biol. 91: 323–330.

Anderson, A. 1989. Prodigious Birds: Moas and Moa-hunting in Prehistoric New Zealand. Cambridge University Press, Cambridge.

Anderson, J. F. 1970. Metabolic rates of spiders. Comp. Biochem. Physiol. 33:51–72.

———. 1978. Energy content of spider eggs. Oecologia 37:41–57.

Anderson, J. F., and K. N. Prestwich. 1982. Respiratory gas exchange in spiders. Physiol. Zool. 55:72–90.

Anderson, M. D., J. B. Williams, and P. R. K. Richardson. 1997. Laboratory metabolism and evaporative water loss of the aardwolf, *Proteles cristatus*. Physiol. Zool. 70:464–469.

Anderson, R. A., and W. H. Karasov. 1981. Contrasts in energy intake and expenditure in sit-and-wait and widely foraging lizards. Oecologia 49:67–72.

Anderson, S. S., and M. A. Fedak. 1987. The energetics of sexual success of grey seals and comparison with the costs of reproduction in other pinnipeds. Symp. Zool. Soc. Lond. 57:319–341.

Andreasson, S. 1973. Seasonal changes in diel activity of *Cottus poecilopus* and *C. gobio* (Pices) at the Arctic Circle. Oikos 24:16–23.

Andreev, A. V. 1988. Ecological energetics of Palaearctic Tetraonidae in relation to chemical composition and digestibility of their winter diets. Can. J. Zool. 66: 1382–1388.

Andrews, R. M. 1982. Patterns of growth in reptiles. *In* C. Gans and F. H. Pough (eds.), Biology of the Reptilia, vol. 13, pp. 273–320. Academic Press, London.

———. 1984. Energetics of sit-and-wait and widely-searching predators. *In* R. A. Seigel (ed.), Vertebrate Ecology and Systematics: A Tribute to Henry S. Fitch, pp. 137–145. Special publ. 10. University of Kansas Museum of Natural History, Lawrence.

Andrews, R. M., and F. H. Pough. 1985. Metabolism of squamate reptiles: Allometric and ecological relationships. Physiol. Zool. 58:214–231.

Ar, A., R. Arieli, and A. Shkolnik. 1977. Blood gas properties and function in the fossorial mole rat under normal and hypoxic-hypercapnic atmospheric conditions. Respir. Physiol. 30:201–218.

Arad, Z., and C. Korine. 1993. Effect of water restriction on energy and water balance and osmoregulation of the fruit bat *Rousettus aegyptiacus*. J. Comp. Physiol. [B] 163:401–405.

Archer, M. 1974. New information about quaternay distribution of the thylacine (Marsupialia, Thylacinidae) in Australia. J. Proc. Roy. Soc. West Aust. 57:43–50.

Arends, A., and B. K. McNab. 2001. The comparative energetics of "caviomorph" rodents. Comp. Biochem. Physiol. A. Physiol. 130:105–122.

Arieli, R. 1979. The atmospheric environment of the fossorial rodent (*Spalax ehrenbergi*): Effects of season, soil texture, rain, temperature and activity. Comp. Biochem. Physiol. A Physiol. 63:569–575.

Arieli, R., and A. Ar. 1981. Heart rate responses of the mole rat (*Spalax ehrenbergi*) in hypercapnic, hypoxic, and cold conditions. Physiol. Zool. 54:14–21.

Arieli, R., A. Ar, and A. Shkolnik. 1977. Metabolic responses of a fossorial rodent (*Spalax ehrenbergi*) to simulated burrow conditions. Physiol. Zool. 50:61–75.

Arieli, R., M. Arieli, G. Heth, and E. Nevo. 1984. Adaptive respiratory variation in 4 chromosomal species of mole rats. Experientia 40:512–514.

Arieli, R., and E. Nevo. 1991. Hypoxic survival differs between two mole rat species (*Spalax ehrenbergi*) of humid and arid habitats. Comp. Biochem. Physiol. A Physiol. 100:543–545.

Armstrong, E. 1983. Brain size and metabolism in mammals. Science 220:1302–1304.

Armstrong, J. T. 1965. Breeding home range in the nighthawk and other birds; its evolutionary and ecological significance. Ecology 46:619–629.

Arnold, W. 1988. Social thermoregulation during hibernation in alpine marmots (*Marmota marmota*). J. Comp. Physiol. [B] 158:151–156.

Arnold, W., G. Heldmaier, S. Ortmann, H. Pohl, T. Ruf, and S. Steinlechner. 1991. Ambient temperatures in hibernacula and their energetic consequences for alpine marmots (*Marmota marmota*). J. Thermal Biol. 16:223–226.

Arrhenius, S. 1912. Theory of Solutions. Yale University Press, New Haven.

Aschoff, J. 1960. Exogenous and endogenous components in circadian rhythms. Cold Springs Harbor Symp. Quant. Biol. 25:11–28.

———. 1981. Der Tagesgäng der Körpertemperatur von Vögeln al Function des Köpergewichtes. J. Ornithol. 122:129–151.

Aschoff, J., and H. Pohl. 1970. Rhythmic variations in energy metabolism. Fed. Proc. 29:1541–1552.

Ashmole, N. P. 1963a. The biology of the wideawake or sooty tern *Sterna fuscata* on Ascension Island. Ibis 103b:297–364.

———. 1963b. The regulation of numbers of tropical oceanic birds. Ibis 103b:458–473.

———. 1965. Adaptive variation in the breeding regime of a tropical sea bird. Proc. Natl. Acad. Sci. U.S.A. 53:311–318.

———. 1968. Body size, prey size, and ecological segregation in five sympatric tropical terns (Aves: Laridae). Syst. Zool. 17:292–304.

———. 1971. Seabird ecology and the marine environment. *In* D. S. Farner and J. R. King (eds.), Avian Biology, vol. I, pp. 223–286. Academic Press, New York.

Ashmole, N. P., and M. J. Ashmole. 1967. Comparative feeding ecology of sea birds of a tropical oceanic island. Bull. Peabody Mus. Nat. Hist. 24:1–13.

Ashton, K. G., M. C. Tracy, and A. deQueiroz. 2000. Is Bergmann's rule valid for mammals? Am. Nat. 156: 390–415.

Asquith, A., and R. Altig. 1986. Osmoregulation of the lesser siren, *Siren intermedia* (Caudata: Amphibia). Comp. Biochem. Physiol. A Physiol. 84:683–685.

Audet, D., and M. B. Fenton. 1988. Heterothermy and the use of torpor by the bat *Eptesicus fuscus* (Chiroptera: Vespertilionidae): A field study. Physiol. Zool. 61:197–204.

Auffenberg, W. 1963. A note on the drinking habits of some land tortoises. Anim. Behav. 11:72–73.

——. 1981. The Behavioral Ecology of the Komodo Monitor. University Presses Florida, Gainesville.

Augee, M. L., and B. A. Gooden. 1992. Monotreme hibernation—some after thoughts. *In* M. L. Augee (ed.), Platypus and Echidnas, pp. 174–176. Royal Zoological Society New South Wales, Sydney.

Austad, S. N., and K. E. Fischer. 1991. Mammalian aging, metabolism, and ecology: Evidence from the bats and marsupials. J. Gerontol. Biol. Sci. 46:B47–B53.

Austin, G. T. 1978. Daily time budget of the postnesting verdin. Auk 95:247–251.

Austin, P. J., L. A. Suchar, C. T. Robbins, and A. E. Hagerman. 1989. Tannin-binding proteins in saliva of deer and their absence in saliva of sheep and cattle. J. Chem. Ecol. 15:1335–1347.

Auth, D. L. 1975. Behavioral ecology of basking in the yellow-bellied turtle, *Chrysemys scripta scripta* (Schoepff). Bull. Fla. State Mus. 20:1–45.

Autumn, K., and D. F. DeNardo. 1995. Behavioral thermoregulation increases growth rate in a nocturnal lizard. J. Herpetol. 29:157–162.

Avery, R. A. 1982. Field studies of body temperature and thermoregulation. *In* C. Gans and F. H. Pough (eds.), Biology of the Reptilia, vol. 12, pp. 93–166. Academic Press, New York.

——. 1984. Physiological aspects of lizard growth: The role of thermoregulation. Symp. Zool. Soc. Lond. 52:407–424.

Avery, R. A., J. D. Bedford, and C. P. Newcombe. 1982. The role of thermoregulation in lizard biology: Predatory efficiency in a temperate diurnal basker. Behav. Ecol. Sociobiol. 11:261–267.

Badgerow, J. P. 1988. An analysis of function in the formation flight of Canada geese. Auk 105:749–755.

Badgerow, J. P., and F. R. Hainsworth. 1981. Energy savings through formation flight? A re-examination of the vee formation. J. Theor. Biol. 93:41–52.

Bairlein, F. 1986. Spontaneous, approximately semi-monthly rhythmic variations of body weight in the migratory garden warbler (*Sylvia borin* Boddaert). J. Comp. Physiol. [B] 156:859–865.

——. 1996. Fruit-eating in birds and its nutritional consequences. Comp. Biochem. Physiol. A Physiol. 113:215–224.

Baker, H. G., and I. Baker. 1982. Chemical constituents of nectar in relation to pollination mechanisms and phylogeny. *In* M. H. Nitecki (ed.), Biochemical Aspects of Evolutionary Biology, pp. 131–171. University of Chicago Press, Chicago.

——. 1983. Floral nectar sugar constituents in relation to pollinator type. *In* C. E. Jones and R. J. Little (eds.), Handbook of Experimental Pollination Biology, pp. 117–141. Scientific and Academic Editions, New York.

Baker, J. R. 1947. The seasons in a tropical rain-forest (New Hebrides). Part 7. Summary and general conclusions. J. Linn. Soc. Lond. 41:248–258.

Baker, R. R. 1978. The Evolutionary Ecology of Animal Migration. Hodder and Staughton, London.

Bakken, G. S. 1976. A heat transfer analysis of animals: Unifying concepts and the application of metabolism chamber data to field ecology. J. Theor. Biol. 60:337–384.

——. 1980. The use of standard operative temperature in the study of the thermal energetics of birds. Physiol. Zool. 53:108–119.

——. 1992. Comment on "standard operative temperatures of two desert rodents, *Gerbillus allenbyi* and *Gerbillus pyramidum*: The effects of morphology, microhabitat and environmental factors." J. Thermal Biol. 17:81–82.

Bakken, G. S., M. T. Murphy, and D. J. Erskine. 1991. The effect of wind and air temperature on metabolism and evaporative water loss rates of dark-eyed juncos, *Junco hyemalis*: A standard operative temperature scale. Physiol. Zool. 64:1023–1049.

Bakker, R. T. 1971. Dinosaur physiology and the origin of mammals. Evolution 25:636–658.

——. 1972a. Anatomical and ecological evidence of endothermy in dinosaurs. Nature 238:81–85.

——. 1972b. Locomotor energetics of lizards and mammals compared. Physiologist 15:76.

——. 1986. The Dinosaur Heresies. William Morrow, New York.

Bakun, A. 1990. Global climate change and intensification of coastal ocean upwelling. Science 247:198–201.

Baldwin, J., and P. W. Hochachka. 1970. Functional significance of isoenzymes in thermal acclimation. Biochem. J. 116:883–887.

Balinsky, J. B., E. L. Choritz, C. G. L. Coe, and G. S. van der Schans. 1967. Amino acid metabolism and urea synthesis in naturally aestivating *Xenopus laevis*. Comp. Biochem. Physiol. 22:59–68.

Ball, E. G., C. F. Strittmatter, and O. Cooper. 1955. Metabolic studies on the gas gland of the swim bladder. Biol. Bull. (Woods Hole) 108:1–17.

Ballantyne, J. S., and T. W. Moon. 1986. Solute effects on mitochondria from an elasmobranch (*Raja erinacea*) and a teleost (*Pseudopleuronectes americanus*). J. Exp. Zool. 239:319–328.

Ballantyne, J. S., C. D. Moyes, and T. W. Moon. 1987. Compatible and counteracting solutes and the evolution of ion and osmoregulation in fishes. Can. J. Zool. 65:1883–1888.

Banavar, J. R., A. Maritan, and A. Rinaldo. 1999. Size and form in efficient transportation networks. Nature 399:130–132.

Bang, B. G. 1964. The nasal organs of the black and turkey vultures; a comparative study of the cathartid species *Coragyps atratus atratus* and *Carthartes aura septentrionalis* (with notes on *Carthartes aura folklandica*, *Pseudogyps bengalensis* and *Nephron percnopterus*). J. Morphol. 115:153–184.

Bang, B. G., and F. B. Bang. 1959. A comparative study of the vertebrate nasal chamber in relation to upper respiratory infections. Bull. Johns Hopkins Hosp. 104:107–149.

Barbosa, P. S., and I. D. Hume. 1992a. Digestive tract morphology and digestion in the wombats (Marsupialia: Vombatidae). J. Comp. Physiol. [B] 162:552–560.

——. 1992b. Hindgut fermentation in the wombats: Two marsupial grazers. J. Comp. Physiol. [B] 162:561–566.

Barham, E. G. 1971. Deep-sea fishes: Lethargy and vertical orientation. *In* G. B. Farquhar (ed.), Proceedings of the International Symposium on Biological Sound Scattering in the Ocean, vol. 5, pp. 100–118. Maury Center for Ocean Science, Washington, D.C.

Barlow, J. C. 1973. Status of the North American population of the European tree sparrow. Ornithol. Monogr. 14:10–23.

Barré, H. 1984. Metabolic and insulative changes in winter- and summer-acclimated king penguin chicks. J. Comp. Physiol. 154:317–324.

Barrell, J. 1916. Influence of Silurian-Devonian climates on the rise of air-breathing vertebrates. Bull. Geol. Soc. Am. 27:387–436.

Barrett, I., and F. J. Hester. 1964. Body temperature of yellowfin and skipjack tunas in relation to sea surface temperature. Nature 203:96–97.

Barrick, R. E., and W. J. Showers. 1994. Thermophysiology of *Tyrannosaurus rex*: Evidence from oxygen isotopes. Science 265:222–224.

Barrick, R. E., W. J. Showers, S. Brande, M. J. Nelson, and M. Turner. 1995. The body temperature of *Tyrannosaurus rex*. Science 267:1667.

Bartels, H. 1964. Comparative physiology of oxygen transport in mammals. Lancet 2:599–604.

Bartels, H., P. Hilpert, K. Barbey, K. Betke, K. Riegel, E. M. Lang, and J. Metcalfe. 1963. Respiratory functions of blood of the yak, llama, camel, Dybowski deer, and African elephant. Am. J. Physiol. 205:331–336.

Bartels, H., R. Schmelzle, and S. Ulrich. 1969. Comparative studies on the respiratory function of mammalian blood. V. Insectivora: Shrew, mole and nonhibernating and hibernating hedgehog. Respir. Physiol. 7:278–286.

Bartholomew, G. A. 1958. The role of physiology in the distribution of terrestrial vertebrates. *In* C. L. Hubbs (ed.), Zoogeography, pp. 81–95. AAAS Symposium 51. American Association for the Advancement of Science (AAAS), Washington, D.C.

——. 1964. The roles of physiology and behaviour in the maintenance of homeostasis in the desert environment. *In* G. M. Hughes (ed.), Homeostasis and Feedback Mechanisms, pp. 7–29. Society for Experimental Biology Symposium 18. Cambridge University Press, Cambridge.

——. 1966. A field study of temperature relations in the Galápagos marine iguana. Copeia 1966:241–250.

——. 1977. Body temperature and energy metabolism. *In* M. S. Gordon (ed.), Animal Physiology: Principles and Adaptations, 3rd ed., pp. 364–449. Macmillan, New York.

Bartholomew, G. A., A. F. Bennett, and W. R. Dawson. 1976. Swimming, diving and lactate production of the marine iguana, *Amblyrhynchus cristatus*. Copeia 1976:709–720.

Bartholomew, G. A., and T. J. Cade. 1957. Temperature regulation, hibernation, and aestivation in the little pocket mouse, *Perognathus longimembris*. J. Mammal. 38:60–72.

——. 1963. The water economy of land birds. Auk 80:504–539.

Bartholomew, G. A., and T. M. Casey. 1977a. Body temperature and oxygen consumption during rest and activity in relation to body size in some tropical beetles. J. Thermal Biol. 2:173–176.

——. 1977b. Endothermy during terrestrial activity in large beetles. Science 195:882–883.

Bartholomew, G. A., and W. R. Dawson. 1979. Thermoregulatory behavior during incubation in Heermann's gulls. Physiol. Zool. 52:422–437.

Bartholomew, G. A., and R. J. Epting. 1975. Allometry of post-flight cooling rates in moths: A comparison with vertebrate homeotherms. J. Exp. Biol. 63:603–613.

Bartholomew, G. A., T. R. Howell, and T. J. Cade. 1957. Torpidity in the white-throated swift, Anna hummingbird, and poor-will. Condor 59:145–155.

Bartholomew, G. A., and J. W. Hudson. 1960. Aestivation in the Mohave ground squirrel, *Citellus mohavensis*. Bull. Mus. Comp. Zool. 124:193–208.

Bartholomew, G. A., J. W. Hudson, and T. R. Howell. 1962. Body temperature, oxygen consumption, evaporative water loss, and heart rate in the poor-will. Condor 64:117–125.

Bartholomew, G. A., and R. C. Lasiewski. 1965. Heating and cooling rates, heart rate and simulated diving in the Galapagos marine iguana. Comp. Biochem. Physiol. 16:573–582.

Bartholomew, G. A., R. C. Lasiewski, and E. C. Crawford. 1968. Patterns of panting and gular flutter in cormorants, pelicans, owls, and doves. Condor 70:31–34.

Bartholomew, G. A., P. Leitner, and J. L. Nelson. 1964. Body temperature, oxygen consumption, and heart rate in three species of Australian flying foxes. Physiol. Zool. 37:179–198.

Bartholomew, G. A., and J. R. B. Lighton. 1986. Oxygen consumption during hover-feeding in free-ranging Anna hummingbirds. J. Exp. Biol. 123:191–199.

Bartholomew, G. A., and R. E. MacMillen. 1960. The water requirements of mourning doves and their use of sea water and NaCl solutions. Physiol. Zool. 33:171–178.

——. 1961. Water economy of the California quail and its use of sea water. Auk 78:505–514.

Bartholomew, G. A., and M. Rainey. 1971. Regulation of body temperature in the rock hyrax, *Heterohyrax brucei*. J. Mammal. 52:81–95.

Bartholomew, G. A., and V. A. Tucker. 1963. Control of changes in body temperature, metabolism, and circulation by the agamid lizard, *Amphibolurus barbatus*. Physiol. Zool. 36:199–218.

——. 1964. Size, body temperature, thermal conductance, oxygen consumption, and heart rate in Australian varanid lizards. Physiol. Zool. 37:341–354.

Bartholomew, G. A., V. A. Tucker, and A. K. Lee. 1965. Oxygen consumption, thermal conductance, and heart rate in the Australian skink *Tiliqua scincoides*. Copeia 1965:169–173.

Barton, M., and A. C. Barton. 1987. Effects of salinity on oxygen consumption of *Cyprinodon variegatus*. Copeia 1987:230–232.

Bassett, J. E. 1982. Habitat aridity and intraspecific differences in the urine concentrating ability of insectivorous bats. Comp. Biochem. Physiol. A Physiol. 72:703–708.

——. 1986. Habitat aridity and urine concentrating ability of nearctic, insectivorous bats. Comp. Biochem. Physiol. A Physiol. 83:125–131.

Bassett, J. E., and J. E. Wiebers. 1979. Subspecific differences in the urine concentrating ability of *Myotis lucifugus*. J. Mammal. 60:395–397.

Batzli, G. O. 1983. Responses of arctic rodent populations to nutritional factors. Oikos 40:396–406.

——. 1985. Nutrition. *In* R. H. Tamarin (ed.), Biology of New World *Microtus*, pp. 779–811. Special Publ. 8. American Society of Mammalogy, Lawrence, Kan.

Batzli, G. O., A. D. Broussard, and R. J. Oliver. 1994. The integrated processing response in herbivorous small mammals. *In* D. J. Chivers and P. Langer (eds.), The Digestive System in Mammals: Food, Form and Function, pp. 324–336. Cambridge University Press, Cambridge.

Batzli, G. O., and F. R. Cole. 1979. Nutritional ecology of microtine rodents: Digestibility of forage. J. Mammal. 60:740–750.

Batzli, G. O., and H.-J. G. Jung. 1980. Nutritional ecology of microtine rodents: Resource utilization near Atkasook, Alaska. Arct. Alp. Res. 12:483–499.

Batzli, G. O., and F. A. Pitelka. 1970. Influence of meadow mouse populations on California grassland. Ecology 51:1027–1039.

——. 1971. Condition and diet of cycling populations of the California vole, *Microtus californicus*. J. Mammal. 52:141–163.

——. 1975. Vole cycles: Test of another hypothesis. Am. Nat. 109:482–487.

Bauchop, T., and R. W. Martucci. 1968. Ruminant-like digestion of the langur monkey. Science 161:698–700.

Baudinette, R. V. 1972. Energy metabolism and evaporative water loss in the California ground squirrel. J. Comp. Physiol. 81:57–72.

——. 1991. The energetics and cardiorespiratory correlates of mammalian terrrestrial locomotion. J. Exp. Biol. 160:209–231.

——. 1994. Locomotion in macropodid marsupials: Gaits, energetics and heat balance. Aust. J. Zool. 42:103–123.

Baudinette, R. V., and P. Gill. 1985. The energetics of "flying" and "paddling" in water: Locomotion in penguins and ducks. J. Comp. Physiol. [B] 155:373–380.

Baudinette, R. V., K. Nagle, and R. A. D. Scott. 1976. Locomotor energetics in dasyurid marsupials. J. Comp. Physiol. 109:159–168.

Baudinette, R. V., and K. Schmidt-Nielsen. 1974. Energy cost of gliding flight in herring gulls. Nature 248:83–84.

Baum, D. A., and A. Larson. 1991. Adaptation reviewed: A phylogenetic methodology for studying character macroevolution. Syst. Zool. 40:1–18.

Baverstock, P. R. 1976. Water balance and kidney function in four species of *Rattus* from ecologically diverse environments. Aust. J. Zool. 24:7–17.

Beachy, C. K., and R. C. Bruce. 1992. Lunglessness in plethodontid salamanders is consistent with the hypothesis of a mountain stream origin: A response to Ruben and Boucot. Am. Nat. 139:839–847.

Beamish, F. W. H. 1964. Respiration of fishes with special emphasis on standard oxygen consumption. III. Influence of oxygen. Can. J. Zool. 42:355–366.

——. 1978. Swimming capacity. *In* W. S. Hoar and D. J. Randall (eds.), Fish Physiology, vol. VII, pp. 101–187. Academic Press, New York.

——. 1979. Migration and spawning energetics of the anadromous sea lamprey, *Petromyzon marinus*. Environ. Biol. Fish. 4:3–7.

Bech, C., A. S. Abe, J. F. Steffensen, M. Berger, and J. E. P. W. Bicudo. 1997. Torpor in three species of Brazilian hummingbirds under semi-natural conditions. Condor 99:780–788.

Beckenbach, A. T. 1975. Influence of body size and temperature on the critical oxygen tension of some plethodontid salamanders. Physiol. Zool. 48:338–347.

Becker, E. L., R. Bird, J. W. Kelly, J. Schilling, S. Solomon, and N. Young. 1958. Physiology of marine teleosts. I. Ionic composition of tissue. Physiol. Zool. 31:224–227.

Bedford, J. J. 1983. The effect of reduced salinity on tissue and plasma composition of the dogfish, *Squalus acanthias*. Comp. Biochem. Physiol. A Physiol. 76:81–84.

Beebee, T. J. C. 1995. Amphibian breeding and climate. Nature 374:219–220.

Beiwener, A. A. 1983. Allometry of quadrupedal locomotion: The scaling of duty factor, bone curvature and limb orientation to body size. J. Exp. Biol. 105:147–171.

Belkin, D. A. 1963. Anoxia: Tolerance in reptiles. Science 139:492–493.

——. 1964. Variations in heart rate during voluntary diving in the turtle *Pseudemys concinna*. Copeia 1964:321–330.

——. 1968. Aquatic respiration and underwater survival of two freshwater turtle species. Respir. Physiol. 4:1–14.

Bell, B. D. 1991. Recent avifaunal changes and the history of ornithology in New Zealand. Acta XX Congr. Int. Ornithol. 20:195–230.

Bell, G. P., G. A. Bartholomew, and K. A. Nagy. 1986. The roles of energetics, water economy, foraging behavior, and geothermal refugia in the distribution of the bat, *Macrotus californicus*. J. Comp. Physiol. [B] 156:441–450.

Bellamy, D., and I. Chester Jones. 1961. Studies on *Myxine glutinosa*—I. The chemical composition of the tissues. Comp. Biochem. Physiol. 3:175–183.

Belovsky, G. E., and P. A. Jordan. 1978. The time-energy budget of a moose. Theor. Popul. Biol. 14:76–104.

Belovsky, G. E., and J. B. Slade. 1986. Time budgets of grassland herbivores: Body size similarities. Oecologia 70:53–62.

Bender, M. M. 1971. Variations in the $^{13}C/^{12}C$ ratios of plants in relation to the pathway of photosynthetic carbon dioxide fixation. Phytochemistry 10:1239–1244.

Benedict, F. G. 1915. Factors affecting basal metabolism. J. Biol. Chem. 20:263–299.

———. 1938. Vital Energetics: A Study in Comparative Basal Metabolism. Publ. 425. Carnegie Institute, Washington, D.C.

Bennett, A. F. 1972. Ventilation in two species of lizards during rest and activity. Comp. Biochem. Physiol. A Physiol. 46:653–671.

———. 1978. Activity metabolism of the lower vertebrates. Annu. Rev. Physiol. 40:447–469.

———. 1980. The metabolic foundations of vertebrate behavior. Bioscience 30:452–456.

———. 1991. The evolution of activity capacity. J. Exp. Biol. 160:1–23.

Bennett, A. F., and B. Dalzell. 1973. Dinosaur physiology: A critique. Evolution 27:170–174.

Bennett, A. F., and W. R. Dawson. 1972. Aerobic and anaerobic metabolism during activity in the lizard Dipsosaurus dorsalis. J. Comp. Physiol. 81:289–299.

———. 1976. Metabolism. In C. Gans and W. R. Dawson (eds.), Biology of the Reptilia, vol. 5, pp. 127–223. Academic Press, New York.

Bennett, A. F., W. R. Dawson, and G. A. Bartholomew. 1975. Effects of activity and temperature on aerobic and anaerobic metabolism in the Galapagos marine iguana. J. Comp. Physiol. 100:317–329.

Bennett, A. F., and T. T. Gleeson. 1976. Activity metabolism in the lizard Sceloporus occidentalis. Physiol. Zool. 49:65–76.

Bennett, A. F., T. T. Gleeson, and G. C. Gorman. 1981. Anaerobic metabolism in a lizard (Anolis bonairensis) under natural conditions. Physiol. Zool. 54:237–241.

Bennett, A. F., and G. C. Gorman. 1979. Population density and energetics of lizards on a tropical island. Oecologia 42:339–358.

Bennett, A. F., and L. D. Houck. 1983. The energetic cost of courtship and aggression in a plethodontid salamander. Ecology 64:979–983.

Bennett, A. F., R. Huey, H. John-Alder, and K. Nagy. 1984. The parasol tail and thermoregulatory behavior of the cape ground squirrel Xerus inauris. Physiol. Zool. 57:57–62.

Bennett, A. F., and H. B. John-Alder. 1984. The effect of body temperature on the locomotory energetics of lizards. J. Comp. Physiol. [B] 155:21–27.

Bennett, A. F., and P. Licht. 1972. Anaerobic metabolism during activity in lizards. J. Comp. Physiol. 81:277–288.

———. 1973. Relative contributions of anaerobic and aerobic energy production during activity in Amphibia. J. Comp. Physiol. 87:351–360.

———. 1974. Anaerobic metabolism during activity in amphibians. Comp. Biochem. Physiol. A Physiol. 48:319–327.

———. 1975. Evaporative water loss in scaleless snakes. Comp. Biochem. Physiol. A Physiol. 52:213–215.

Bennett, A. F., and K. A. Nagy. 1977. Energy expenditure in free-ranging lizards. Ecology 58:697–700.

Bennett, A. F., and J. A. Ruben. 1979. Endothermy and activity in vertebrates. Science 206:649–654.

Bennett, P. M., and P. H. Harvey. 1987. Active and resting metabolism in birds: Allometry, phylogeny and ecology. J. Zool. Lond. 213:327–363.

Bentley, P. J. 1959. Studies on the water and electrolyte metabolism of the lizard Tachysaurus rugosus (Gray). J. Physiol. 145:37–47.

———. 1966. The physiology of urinary bladder of Amphibia. Biol. Rev. Camb. Philos. Soc. 41:275–316.

Bentley, P. J., and W. F. C. Blumer. 1962. Uptake of water by the lizard, Moloch horridus. Nature 194:699–700.

Bentley, P. J., A. K. Lee, and A. R. Main. 1958. Comparison of dehydration and hydration of two genera of frogs (Heleioporus and Neobatrachus) that live in areas of varying aridity. J. Exp. Biol. 35:677–684.

Bentley, P. J., and K. Schmidt-Nielsen. 1966. Cutaneous water loss in reptiles. Science 151:1547–1549.

Berger, L., R. Speare, P. Daszak, D. E. Green, A. A. Cunningham, C. L. Goggin, R. Slocombe, M. A. Ragan, A. D. Hyatt, K. R. McDonald, H. B. Hines, K. R. Lips, G. Marantelli, and H. Parkes. 1998. Chytridiomycosis causes amphibian mortality associated with population declines in the rain forests of Australia and Central America. Proc. Natl. Acad. Sci. U.S.A. 95:9031–9036.

Berger, M., and J. S. Hart. 1972. Die Atmung beim Kolibri Amazilia fimbriata während des Schwirrfluges bei verschiedenen Umgebungstemperaturen. J. Comp. Physiol. 81:363–379.

Berger, M., J. S. Hart, and O. Z. Roy. 1970. Respiration, oxygen consumption and heart rate in some birds during rest and flight. Z. Vgl. Physiol. 66:201–214.

Berger, P. J., N. Negus, E. H. Sanders, and P. D. Gardner. 1981. Chemical triggering of reproduction in Microtus montanus. Science 214:69–70.

Berger, P. J., E. H. Sanders, P. D. Gardner, and N. C. Negus. 1977. Phenotic plant compounds functioning as reproductive inhibitors in Microtus montanus. Science 195:575–577.

Bergeron, J.-M., L. Jodoin, and Y. Jean. 1987. Pathology of voles (Microtus pennsylvanicus) fed with plant extracts. J. Mammal. 68:73–79.

Bergmann, C. 1847. Ueber die Verhältnisse der Wärmeökonomie der Tiere zu ihrer Grösse. Gottinger Stud. 3:595–708.

Bergmann, C., and R. Leuckart. 1852. Anatomisch-physiologische Übersicht des Tierreichs. J. B. Mueller, Stuttgart.

Berliner, R. W., N. G. Levinsky, D. G. Davidson, and M. Eden. 1958. Dilution and concentration of the urine and the action of antidiuretic hormone. Am. J. Med. 24:730–744.

Berman, D. I., A. N. Leirikh, and E. I. Mikailova. 1984. Winter hibernation of the Siberian salamander Hynobius keyserlingi. J. Evol. Biochem. Physiol. 20:323–327. (In Russian)

Bernstein, M. H. 1971. Cutaneous water loss in small birds. Condor 73:468–469.

Bernstein, M. H., S. P. Thomas, and K. Schmidt-Nielsen. 1973. Power input during flight of the fish crow, Corvus ossifragus. J. Exp. Biol. 58:401–410.

Berthold, P. 1974. Circannial rhythms in birds with different migratory habits. In E. T. Pengelley (ed.), Circannual Clocks, pp. 55–94. Academic Press, New York.

———. 1978. Endogenous control as a possible basis for varying migratory habits in different bird populations. Experientia 34:1451.

Bethke, R. W., and V. G. Thomas. 1988. Differences in flight and heart muscle mass among geese, dabbling ducks, and diving ducks relative to habitat use. Can. J. Zool. 66:2024–2028.

Beuchat, C. A. 1986a. Phylogenetic distribution of the urinary bladder in lizards. Copeia 1986:512–517.

———. 1986b. Reproductive influences on the thermoregulatory behavior of a live-bearing lizard. Copeia 1986:971–979.

———. 1988. Temperature effects during gestation in a viviparous lizard. J. Thermal Biol. 13:135–142.

———. 1990. Body size, medullary thickness, and urine concentrating ability in mammals. Am. J. Physiol. 258:R298–R308.

Beuchat, C. A., S. B. Chaplin, and M. L. Morton. 1979. Ambient temperature and daily energetics of two species of hummingbirds, *Calypte anna* and *Selasphorus rufus*. Physiol. Zool. 52:280–295.

Beuchat, C. A., and S. Ellner. 1986. A quantitative test of life history theory: Thermoregulation by a viviparous lizard. Ecol. Monogr. 56:45–60.

Beuchat, C. A., F. H. Pough, and M. M. Stewart. 1984. Response to simultaneous dehydration and thermal stress in three species of Puerto Rican frogs. J. Comp. Physiol. 154:579–585.

Beuchat, C. A., and D. Vleck. 1990. Metabolic consequences of viviparity in a lizard, *Sceloporus jarrovi*. Physiol. Zool. 63:555–570.

Beuchat, C. A., D. Vleck, and E. J. Braun. 1986. Role of the urinary bladder in osmotic regulation of neonatal lizards. Physiol. Zool. 59:539–551.

Bickler, P. E. 1984. Blood acid-base status of an awake heterothermic rodent, *Spermophilus tereticaudus*. Respir. Physiol. 57:307–316.

Bicudo, J. E. P. W., and K. Johansen. 1979. Respiratory gas exchange in the airbreathing fish, *Synbranchus marmoratus*. Environ. Biol. Fish. 4:55–64.

Bilinski, E. 1974. Biochemical aspects of fish swimming. Biochem. Biophys. Perspect. Mar. Biol. 1:239–288.

Birchard, G. F., D. L. Kilgore Jr., and D. L. Boggs. 1984. Respiratory gas concentrations and temperatures within the burrows of three species of burrow-nesting birds. Wilson Bull. 96:451–456.

Bjorndal, K. A. 1979. Cellulose digestion and volatile fatty acid production in the green turtle, *Chelonia mydas*. Comp. Biochem. Physiol. A Physiol. 63:127–133.

———. 1980. Nutrition and grazing behavior of the green turtle *Chelonia mydas*. Mar. Biol. 56:147–154.

———. 1982. The consequences of herbivory for the life history pattern of the Caribbean green turtle, *Chelonia mydas*. In K. A. Bjorndal (ed.), Biology and Conservation of Sea Turtles, pp. 111–116. Smithsonian Institution Press, Washington, D.C.

———. 1985. Nutritional ecology of sea turtles. Copeia 1985:736–751.

———. 1987. Digestive efficiency in a temperate herbivorous reptile *Gopherus polyphemus*. Copeia 1987:714–720.

———. 1997. Fermentation in reptiles and amphibians. In R. I. Mackie and B. A. White (eds.), Gastrointestinal Microbiology, vol. I. Gastrointestinal Ecosystems and Fermentations, pp. 199–230. Chapman and Hall, New York.

Black, C. P., and S. M. Tenney. 1980. Oxygen transport during progressive hypoxia in high-altitude and sea-level waterfowl. Respir. Physiol. 39:217–239.

Black, E. C. 1940. The transport of oxygen by the blood of freshwater fish. Biol. Bull. Camb. Philos. Soc. 79:215–229.

———. 1958. Hyperactivity as a lethal factor in fish. J. Fish. Res. Board Can. 15:573–586.

Blackburn, T. M., and K. J. Gaston. 1996. Spatial patterns in body sizes of bird species in the New World. Oikos 77:436–446.

Blackburn, T. M., K. J. Gaston, and N. Loder. 1999. Geographic gradients in body size: A clarification of Bergmann's rule. Diversity Distrib. 5:165–174.

Blair-West, J. R., J. P. Coghlan, D. A. Denton, J. F. Nelson, E. Orchard, B. A. Scoggins, R. D. Wright, K. Myers, and C. L. Junqueira. 1968. Physiological, morphological and behavioural adaptation to a sodium deficient environment by wild native Australian and introduced species of animals. Nature 217:922–928.

Blake, B. H. 1977. The effects of kidney structure and the annual cycle of water requirements in gold-mantled ground squirrels and chipmunks. Comp. Biochem. Physiol. A Physiol. 58:413–419.

Blaxter, J. H. S., C. S. Wardle, and B. L. Roberts. 1971. Aspects of the circulatory physiology and muscle systems of deep-sea fish. J. Mar. Biol. Assoc. U.K. 51:991–1006.

Blaylock, L. A., R. Ruibal, and K. Platt-Aloia. 1976. Skin structure and wiping behavior of phyllomedusine frogs. Copeia 1976:283–295.

Blem, C. R. 1973. Geographic variation in the bioenergetics of the house sparrow. Ornithol. Monogr. 14:96–121.

———. 1976. Efficiency of energy utilization of the house sparrow, *Passer domesticus*. Oecologia 25:257–264.

Bligh, J., and A. M. Harthoorn. 1965. Continuous radiotelemetric records of the deep body temperature of some unrestrained African mammals under near-natural conditions. J. Physiol. 176:145–162.

Block, B. A. 1986. Structure of the brain and eye heater tissue in marlins, sailfish, and spearfishes. J. Morphol. 190:169–189.

Block, B. A., and F. G. Carey. 1985. Warm brain and eye temperatures in sharks. J. Comp. Physiol. 156:229–236.

Block, B. A., and J. R. Finnerty. 1994. Endothermy in fishes: A phylogenetic analysis of constraints, predispositions, and selection pressures. Environ. Biol. Fish. 40:283–302.

Block, B. A., J. R. Finnerty, A. F. R. Stewart, and J. Kidd. 1993. Evolution of endothermy in fish: Mapping physiological traits on a molecular phylogeny. Science 260:210–214.

Blum, J. J. 1977. On the geometry of four dimensions and the relationship between metabolism and body mass. J. Theor. Biol. 64:599–601.

Bock, C. E. 1970. The ecology and behavior of the Lewis' woodpecker (*Asyndesmus lewis*). Univ. Calif. Publ. Zool. 92:1–91.

Bodmer, R. E. 1991. Influence of digestive morphology on resource partitioning in Amazonian ungulates. Oecologia 85:361–365.

Boersma, P. D. 1978. Breeding patterns of Galapagos penguins as an indicator of oceanographic conditions. Science 200:1481–1483.

———. 1986. Body temperature, torpor, and growth in chicks of fork-tailed storm-petrels (*Oceanodroma furcata*). Physiol. Zool. 59:10–19.

Boersma, P. D., and N. T. Wheelwright. 1979. Egg neglect in the Procellariiformes: Reproductive adaptations in the fork-tailed storm-petrel. Condor 81:157–165.

Bogert, C. M. 1949a. Thermoregulation in reptiles: A factor in evolution. Evolution 3:195–211.

———. 1949b. Thermoregulation and eccritic body temperatures in Mexican lizards of the genus *Sceloporus*. Ann. Inst. Biol. Mex. 20:415–426.

———. 1953. Body temperatures of the tuatara under natural conditions. Zoologica (NY) 38:63–64.

———. 1959. How reptiles regulate their body temperature. Sci. Am. 200:105–120.

Bogert, C. M., and R. B. Cowles. 1947. Moisture loss in relation to habitat selection in some Floridian reptiles. Am. Mus. Novit. 1358:1–34.

Boggs, D. F., D. L. Kilgore Jr., and G. F. Birchard. 1984. Respiratory physiology of burrowing mammals and birds. Comp. Biochem. Physiol. A Physiol. 77: 1–7.

Bonaccorso, F. J., A. Arends, M. Genoud, D. Cantoni, and T. Morton. 1992. Thermal ecology of mustached and ghost-faced bats (Mormoopidae) in Venezuela. J. Mammal. 73:365–378.

Bonaccorso, F. J., and B. K. McNab. 1997. Plasticity of basal metabolism and temperature regulation in blossom bats (Pteropodidae): Impact on distribution. J. Mammal. 78:1073–1088.

Bonaventura, J., C. Bonaventura, and B. Sullivan. 1974. Urea tolerance as a molecular adaptation of elasmobranch hemoglobins. Science 186:57–59.

Bone, Q. 1972. Buoyancy and hydrodynamic function of integument in the castor-oil fish, *Ruvettus pretiosus* (Pisces: Gempylidae). Copeia 1972:78–87.

Bone, Q., and A. D. Chubb. 1983. The retial system of the locomotor muscles in the thresher shark. J. Mar. Biol. Assoc. U.K. 63:239–241.

Booth, D. T., and M. E. Feder. 1991. Formation of hypoxic boundary layers and their biological implications in a skin-breathing aquatic salamander, *Desmognathus quadramaculatus*. Physiol. Zool. 64:1307–1321.

Boucot, A. J., and C. Janis. 1983. Environment of the early Palaeozoic vertebrates. Palaeogeogr. Palaeoclim. Palaeoecol. 41:251–287.

Bourlière. F. 1958. Comparative biology of aging. J. Gerontol. Suppl. I 13:16–24.

Bowen, W. D., O. T. Oftedal, and D. J. Boness. 1985. Birth to weaning in 4 days: Remarkable growth in the hooded seal, *Cystophora cristata*. Can. J. Zool. 63:2841–2846.

———. 1992. Mass and energy transfer during lactation in a small phocid, the harbor seal (*Phoca vitulina*). Physiol. Zool. 65:844–866.

Boyce, M. S. 1978. Climate variability and body size variation in the muskrats (*Ondatra zibethicus*) of North America. Oecologia (Berl.) 36:1–19.

Boyer, D. R. 1965. Ecology of the basking habit in turtles. Ecology 46:99–118.

Bozinovic, F. 1992. Scaling basal and maximum metabolic rate in rodents and the aerobic capacity model for the evolution of endothermy. Physiol. Zool. 65: 921–932.

Bozinovic, F., F. F. Novoa, and C. Veloso. 1990. Seasonal changes in energy expenditure and digestive tract of *Abrothrix andinus* (Cricetidae) in the Andes Range. Physiol. Zool. 63:1216–1231.

Bozinovic, F., and M. Rosenmann. 1989. Maximum metabolic rate of rodents: Physiological and ecological consequences on distributional limits. Funct. Ecol. 3:173–181.

Bozinovic, F., M. Rosenmann, F. F. Novoa, and R. G. Medel. 1995. Mediterranean type of climatic adaptation in the physiological ecology of rodent species. *In* M. T. Kalin Arroyo, P. H. Zedler, and M. D. Fox (eds.), Ecology and Biogeography of Mediterranean Ecosystems in Chile, California, and Australia, pp. 347–362. Springer-Verlag, New York.

Bradbury, J. W., and S. L. Vehrencamp. 1976. Social organization and foraging in emballonurid bats. I. Field studies. Behav. Ecol. Sociobiol. 1:337–381.

Bradford, D. F. 1974. Water stress of free-living *Peromyscus truei*. Ecology 55:1407–1414.

Bradley, S. R., and D. R. Deavers. 1980. A reexamination of the relationship between thermal conductance and body weight in mammals. Comp. Biochem. Physiol. A Physiol. 65:465–476.

Bradshaw, S. D. 1970. Seasonal changes in the water and electrolyte metabolism of *Amphibolurus* lizards in the field. Comp. Biochem. Physiol. 36:689–718.

Bradshaw, S. D., and A. R. Main. 1968. Behavioural attitudes and regulation of temperature in *Amphibolurus* lizards. J. Zool. Lond. 154:193–221.

Bradshaw, S. D., K. D. Norris, C. R. Dickman, P. C. Withers, and D. Murphy. 1994. Field metabolism and turnover in the golden bandicoot (*Isoodon auratus*) and other small mammals from Barrow Island, Western Australia. Aust. J. Zool. 42:29–41.

Bradshaw, S. D., and V. H. Shoemaker. 1967. Aspects of water and electrolyte changes in a field population of *Amphibolurus* lizards. Comp. Biochem. Physiol. 20:855–865.

Brame, A. H., and D. B. Wake. 1963. The salamanders of South America. Los Ang. Cty. Mus. Contrib. Sci. 69:1–72.

Bramwell, C. D., and D. B. Fellgett. 1973. Thermal regulation in sail lizards. Nature 242:203–205.

Bramwell, C. D., and C. R. Whitfield. 1974. Biomechanics of *Pteranodon*. Philos. Trans. R. Soc. Lond. B Biol. Sci. 267:503–581.

Brattstrom, B. H. 1963. A preliminary review of the thermal requirements of amphibians. Ecology 44:238–255.

——. 1965. Body temperatures of reptiles. Am. Midl. Nat. 73:376–422.

——. 1968. Thermal acclimation in anuran amphibians as a function of latitude and altitude. Comp. Biochem. Physiol. 24:93–111.

——. 1973. Rate of heat loss by large Australian monitor lizards. Bull. South. Calif. Acad. Sci. 72: 52–54.

Brauer, R. W., V. G. Sidelyova, M. B. Dail, G. I. Galazii, and R. D. Roer. 1984. Physiological adaptation of cottoid fishes of Lake Baikal to abyssal depths. Comp. Biochem. Physiol. A Physiol. 77:699–705.

Braun, E. J. 1978. Renal response of the starling (*Sturnus vulgaris*) to an intravenous salt load. Am. J. Physiol. 234:F270–F278.

Braun, E. J., and W. H. Dantzler. 1972. Function of mammalian-type and reptilian-type nephrons in kidney of desert quail. Am. J. Physiol. 222:617–629.

Bray, A. A. 1985. The evolution of the terrestrial vertebrates: Environmental and physiological considerations. Philos. Trans. R. Soc. Lond. B Biol. Sci. 309: 289–322.

Braysher, M. L. 1971. The structure and function of the nasal gland from the Australian sleepy lizard *Trachydosaurus* (formally *Tiliqua*) *rugosus*: Family Scincidae. Physiol. Zool. 44:129–136.

——. 1976. The excretion of hyperosmotic urine and other aspects of the electrolyte balance of the lizard *Amphibolurus maculosus*. Comp. Biochem. Physiol. A Physiol. 54:341–345.

Breder, C. M. 1934. Ecology of an oceanic fresh-water lake, Andros Island, Bahamas, with special reference to its fishes. Zoologica (NY) 18:57–88.

Brenner, B. M., C. Baylis, and W. M. Deen. 1976. Transport of molecules across renal glomerular capillaries. Physiol. Rev. 56:502–534.

Brenner, F. J. 1969. The role of temperature and fat deposition in hibernation and reproduction in two species of frogs. Herpetologica 25:105–113.

Brett, J. R. 1944. Some lethal temperature relations of Algonquin Park fishes. Univ. Toronto Stud. Biol. 52:1–49.

——. 1956. Some principles in the thermal requirements of fishes. Q. Rev. Biol. 31:75–87.

——. 1964. The respiratory metabolism and swimming performance of young sockeye salmon. J. Fish. Res. Board Can. 21:1183–1226.

——. 1965. The relation of size to rate of oxygen consumption and sustained swimming speed of sockeye salmon (*Oncorhynchus nerka*). J. Fish. Res. Board Can. 22:1491–1501.

——. 1970. Temperature: Fishes. *In* O. Kinne (ed.), Marine Ecology, pp. 513–616. Wiley-Interscience, New York.

——. 1971a. Energetic responses of salmon to temperature. A study of some thermal relations in the physiology and freshwater ecology of sockeye salmon (*Oncorhynchus nerka*). Am. Zool. 11:99–113.

——. 1971b. Satiation time, appetite, and maximum food intake of sockeye salmon *Oncorhynchus nerka*. J. Fish. Res. Board Can. 28:409–415.

——. 1986. Production energetics of a population of sockeye salmon, *Oncorhynchus nerka*. Can. J. Zool. 64:555–564.

Brett, J. R., J. E. Shelbourn, and C. T. Shoop. 1969. Growth rate and body composition of fingerling sockeye salmon, *Oncorhynchus nerka*, in relation to temperature and ration size. J. Fish. Res. Board Can. 26:2363–2394.

Bretz, W. L., and K. Schmidt-Nielsen. 1972. The movement of gas in the respiratory system of the duck. J. Exp. Biol. 56:57–65.

Brice, A. T. 1992. The essentiality of nectar and arthropods in the diet of the Anna's hummingbird (*Calypte anna*). Comp. Biochem. Physiol. A Physiol. 101:151–155.

Brigham, R. M. 1992. Daily torpor in a free-ranging goatsucker, the common poorwill (*Phalaenoptilus nuttallii*). Physiol. Zool. 65:457–472.

Brigham, R. M., and P. Trayhurn. 1994. Brown fat in birds? A test for the mammalian bat-specific mitochondrial uncoupling protein in common poorwills. Condor 96:208–211.

Brill, R. W., D. L. Guernsey, and E. D. Stevens. 1978. Body surface and gill heat loss rates in restrained skipjack tuna. *In* G. D. Sharp and A. E. Dizon (eds.), The Physiological Ecology of Tunas, pp. 261–276. Academic Press, New York.

Brioli, F. 1938. Beobachtungen an *Pterodactylus*. Sitzungsbericht. Bayer. Akad. Wissenschaft. Abteil. 16:139–154.

——. 1941. Haare bei Reptilien. Anat. Anz. 92:62–68.

Brisbin, I. L. Jr., E. A. Standora, and M. J. Vargo. 1982. Body temperatures and behavior of American alligators during cold winter weather. Am. Midl. Nat. 107:209–218.

Brocke, R. H. 1970. The winter ecology and bioenergetics of the opossum, *Didelphis marsupialis*, as distributional factors in Michigan. Ph.D. dissertation, Michigan State University, East Lansing.

Brodie, P. F. 1975. Cetacean energetics, an overview of intraspecific size variation. Ecology 56:152–161.

Brodkorb, P. 1959. The Pleistocene avifauna of Arredondo, Florida. Bull. Fla. State Mus. 4:269–291.

Brody, S. 1945. Bioenergetics and Growth. Reinhold, Baltimore.

Brody, S., and R. C. Procter. 1932. Relation between basal metabolism and mature body weight in different species of mammals and birds. Univ. Missouri Agr. Exp. Station Res. Bull. 116:89–101.

Broun, M. 1972. Apparent migratory behavior in the house sparrow. Auk 89:187–189.

Brower, J. C. 1983. The aerodynamics of *Pteranodon* and *Nyctosaurus*, two large pterosaurs from the Upper Cretaceous of Kansas. J. Vert. Paleontol. 3:84–124.

Brower, J. C., and J. Veinus. 1981. Allometry in pterosaurs. Paleontol. Contrib. Univ. Kansas 105:1–32.

Brown, C. R., and G. G. Foster. 1992. The thermal and energetic significance of clustering in the speckled mousebird, *Colius striatus*. J. Comp. Physiol. 162:658–664.

Brown, F. A. Jr. 1957. Response of a living organism under "constant conditions" including pressure, to a barometric-pressure-correlated, cyclic, external variable. Biol. Bull. (Woods Hole) 112:288–304.

——. 1960. Response to pervasive geophysical factors and the biological clock problem. Cold Spring Harbor Symp. Quant. Biol. 25:57–71.

——. 1972. The "clocks" timing biological rhythms. Am. Sci. 60:756–766.

Brown, G. W., Jr., J. James, R. J. Henderson, W. N. Thomas, R. O. Robinson, A. L. Thomson, E. Brown, and S. G. Brown. 1966. Urocolytic enzymes in liver of the Dipnoan *Protopterus aethiopicus*. Science 153:1653–1654.

Brown, J. A., S. M. Taylor, and C. J. Gray. 1983. Glomerular ultrastructure of the trout, *Salmo gairdneri*: Glomerular capillary epithelium and the effects of environmental salinity. Cell Tissue Res. 230:205–218.

Brown, J. H., and G. A. Bartholomew. 1969. Periodicity and energetics of torpor in the kangaroo mouse, *Microdipodops pallidus*. Ecology 50:705–709.

Brown, J. H., and C. R. Feldmeth. 1971. Evolution in constant and fluctuating environments: Thermal tolerances of desert pupfish (*Cyprinodon*). Evolution 25:390–398.

Brown, J. H., and R. C. Lasiewski. 1972. Metabolism of weasels: The cost of being long and thin. Ecology 53:939–943.

Brown, J. H., and A. K. Lee. 1969. Bergmann's rule and climatic adaptation in woodrats (*Neotoma*). Evolution 23:329–338.

Brown, J. H., P. A. Marquet, and M. L. Taper. 1993. Evolution of body size: Consequences of an energetic definition of fitness. Am. Nat. 142:573–584.

Brown, J. L., S.-H. Li, and N. Bhagabati. 1999. Long-term trend toward earlier breeding in an American bird: A response to global warming? Proc. Natl. Acad. Sci. U.S.A. 96:5565–5569.

Brown, W. L. 1958. General adaptation and evolution. Syst. Zool. 7:157–168.

Brown, W. S., and W. S. Parker. 1976. Movement ecology of *Coluber constrictor* near communal hibernacula. Copeia 1976:225–242.

Brown, W. S., W. S. Parker, and J. A. Elder. 1974. Thermal and spatial relationships of two species of colubrid snakes during hibernation. Herpetologica 30:32–38.

Brownfield, M. S., and B. A. Wunder. 1976. Relative medullary area: A new structural index for estimating urinary concentration capacity of mammals. Comp. Biochem. Physiol. A Physiol. 55:69–75.

Bruton, M. N. 1979. The survival of habitat desiccation by air breathing clariid catfishes. Environ. Biol. Fish. 4:273–280.

Bryant, D. M. 1991. Constraints on energy expenditure by birds. Acta XX Congr. Int. Ornithol. 4:1989–2001.

Bryant, D. M., and A. Gardiner. 1979. Energetics of growth in house martins (*Delichon urbica*). J. Zool. Lond. 189:275–304.

Bryant, D. M., and A. V. Newton. 1994. Metabolic costs of dominance in dippers, *Cinclus cinclus*. Anim. Behav. 48:447–455.

Bryant, D. M., and P. Tatner. 1991. Intraspecies variation in avian energy expenditure: Correlates and constraints. Ibis 133:236–245.

Bryant, D. M., and K. R. Westerterp. 1980. The energy budget of the house martin (*Delichon urbica*). Ardea 68:91–102.

——. 1982. Evidence for individual differences in foraging efficiency amongst breeding birds: A study of house martins *Delichon urbica* using the doubly-labelled water technique. Ibis 124:187–192.

——. 1983a. Short-term variability in energy turnover by breeding house martins *Delichon urbica*: A study using doubly-labelled water ($D_2{}^{18}O$). J. Anim. Ecol. 52:525–543.

——. 1983b. Time and energy limits to brood size in house martins (*Delichon urbica*). J. Anim. Ecol. 52:905–925.

Bryant, J. P. 1981a. Hare Trigger. Nat. Hist. 90:46–52.

——. 1981b. Phytochemical deterrence of snowshoe hare browsing by adventitious shoots of four Alaskan trees. Science 213:889–890.

——. 1981c. The regulation of snowshoe hare feeding behaviour during winter by plant antiherbivore chemistry. *In* Proceedings of the 1st International Lagomorph Conference, pp. 720–731. University of Guelph Press, Ontario.

Bryant, J. P., and F. S. Chapin III. 1986. Browsing-woody plant interactions during boreal forest plant succession. Ecol. Stud. Anal. Synth. 57:213–255.

Bryant, J. P., F. S. Chapin, P. Reichardt, and T. Clausen. 1985a. Adaptation to resource availability as a determinant of chemical defense strategies in wood plants. Recent Adv. Phytochem. 19:219–237.

Bryant, J. P., and P. J. Kuropat. 1980. Selection of winter forage in subarctic browsing vertebrates: The role of plant chemistry. Annu. Rev. Ecol. Syst. 11:261–285.

Bryant, J. P., R. K. Swihart, P. B. Reichardt, and L. Newton. 1994. Biogeography of woody plant chemical defense against snowshoe hare browsing: Comparison of Alaska and eastern North America. Oikos 70:385–395.

Bryant, J. P., J. Tahvanainen, M. Sulkinoja, R. Julkunen-Tiitto, P. Reichardt, and T. Green. 1989. Biogeographic evidence for the evolution of chemical defense by boreal birch and willow against mammalian browsing. Am. Nat. 134:20–34.

Bryant, J. P., G. D. Wieland, T. Clausen, and P. Kuropat. 1985b. Interactions of snowshoe hare and feltleaf willow in Alaska. Ecology 66:1564–1573.

Bryant, J. P., G. D. Wieland, P. B. Reichardt, V. E. Lewis, and M. C. McCarthy. 1983. Pinosylvin methyl ether deters snowshoe hare feeding on green alder. Science 222:1023–1025.

Bucher, T. L., M. J. Ryan, and G. A. Bartholomew. 1982. Oxygen consumption during resting, calling, and nest building in the frog *Physalaemus pustulosus*. Physiol. Zool. 55:10–22.

Bucher, T. L., and A. Worthington. 1982. Nocturnal hypothermia and oxygen consumption in manakins. Condor 84:327–331.

Buchner, C. H. 1964. Metabolism, food capacity, and feeding behavior in four species of shrews. Can. J. Zool. 42:259–279.

Buffenstein, R., and S. Yahav. 1991. Is the naked mole-rat *Heterocephalus glaber* an endothermic yet poikilothermic mammal? J. Thermal Biol. 16:227–232.

Bullard, R. W., C. Broumand, and F. R. Meyer. 1966. Blood characteristics and volume in two rodents native to high altitude. J. Appl. Physiol. 21:994–998.

Bullock, T. H. 1955. Compensation for temperature in the metabolism and activity of poikilotherms. Biol. Rev. Camb. Philos. Soc. 30:311–342.

Bünning, E. 1973. The Physiological Clock: Circadian Rhythms and Biological Chronometry, 3rd ed.Springer-Verlag, New York.

Burbanck, W. D., J. P. Edwards, and M. P. Burbanck. 1948. Toleration of lowland oxygen tension by cave and stream crayfish. Ecology 29:360–367.

Burger, J. W. 1962. Further studies on the function of the rectal gland in the spiny dogfish. Physiol. Zool. 35:205–217.

Burger, J. W., and W. N. Hess. 1960. Function of the rectal gland in the spiny dogfish. Science 131:670–671.

Burggren, W. W. 1982. "Airgulping" improves blood oxygen transport during aquatic hypoxia in the goldfish, *Carassius auratus*. Physiol. Zool. 55:327–334.

Burke, J. D. 1966. Vertebrate blood oxygen capacity and body weight. Nature 212:46–48.

Burness, G. P., R. C. Ydenberg, and P. W. Hochachka. 1998. Interindividual variability in body composition and resting oxygen consumption rate in breeding tree swallow, *Tachycineta bicolor*. Physiol. Zool. 71:247–256.

Burton, A. C. 1934. The application of the theory of heat flow to the study of energy metabolism. J. Nutr. 7:497–533.

Burton, T. M., and G. E. Likens. 1975. Energy flow and nutrient cycling in salamander populations in the Hubbard Brook Experimental Forest, New Hampshire. Ecology 56:1068–1080.

Bury, R. B., and T. G. Balgooyen. 1976. Temperature selectivity in the legless lizard *Anniella pulchra*. Copeia 1976:152–155.

Busch, C. 1987. Haematological correlates of burrowing in *Ctenomys*. Comp. Biochem. Physiol. A Physiol. 86:461–463.

——. 1988. Consumption of blood, renal function and utilization of free water by the vampire bat, *Desmodus rotundus*. Comp. Biochem. Physiol. A Physiol. 90:141–146.

Bush, F. M. 1963. Effects of light and temperature on the gross composition of the toad, *Bufo fowleri*. J. Exp. Zool. 153:1–13.

Butler, P. J. 1991. Exercise in birds. J. Exp. Biol. 160:233–262.

Butler, P. J., and A. J. Woakes. 1979. Changes in heart rate and respiratory frequency during natural behaviour of ducks, with particular reference to diving. J. Exp. Biol. 79:283–300.

Buttemer, W. A., A. M. Hayworth, W. W. Weathers, and K. A. Nagy. 1986. Time-budget estimates of avian energy expenditure: Physiological and meteorological considerations. Physiol. Zool. 59:131–149.

Buxton, P. A. 1923. Animal Life in Deserts. Edward Arnold, London (reprinted 1955).

Cade, T. J. 1964. The evolution of torpidity in rodents. Ann. Acad. Sci. Fenn. 71:79–112.

Cade, T. J., and G. A. Bartholomew. 1959. Sea-water and salt utilization by savannah sparrows. Physiol. Zool. 32:230–238.

Cade, T. J., and J. A. Dybas. 1962. Water economy of the budgerygah. Auk 79:345–364.

Cade, T. J., and L. Greenwald. 1966. Nasal salt excretion in falconiform birds. Condor 68:338–350.

Cade, T. J., and G. L. Maclean 1967. Transport of water by adult sandgrouse to their young. Condor 69:323–343.

Cade, T. J., C. A. Tobin, and A. Gold. 1965. Water economy and metabolism of two estrildine finches. Physiol. Zool. 38:9–33.

Calder, W. A. 1969. Temperature relations and underwater endurance of the smallest homeothermic diver, the water shrew. Comp. Biochem. Physiol. 30:1075–1082.

——. 1971. Temperature relationships and nesting of the Calliope hummingbird. Condor 73:314–321.

——. 1973a. Microhabitat selection during nesting of hummingbirds in the Rocky Mountains. Ecology 54:127–134.

——. 1973b. An estimate of the heat balance of a nesting hummingbird in a chilling climate. Comp. Biochem. Physiol. A Physiol. 46:291–300.

——. 1974a. Consequences of body size for avian energetics. *In* R. A. Paynter Jr. (ed.), Avian Energetics, pp. 86–144. Publ. 15. Nuttall Ornithology Club, Cambridge, Mass.

——. 1974b. The thermal and radiant environment of a winter hummingbird nest. Condor 76:268–273.

——. 1984. Size, Function, and Life History. Harvard University Press, Cambridge.

Calder, W. A., and J. Booser. 1973. Hypothermia of broad-tailed hummingbirds during incubation in nature with ecological correlations. Science 180:751–753.

Calder, W. A., and E. J. Braun. 1983. Scaling of osmotic regulation in mammals and birds. Am. J. Physiol. 244:R601–R606.

Calder, W. A., and T. J. Dawson. 1978. Resting metabolic rates of ratite birds: The kiwis and the emu. Comp. Biochem. Physiol. A Physiol. 60:479–481.

Calder, W. A., and J. R. King. 1972. Body weight and the energetics of temperature regulation: A re-examination. J. Exp. Biol. 56:775–780.

Calder, W. A., and K. Schmidt-Nielsen. 1966. Evaporative cooling and respiratory alkalosis in the pigeon. Proc. Nat. Acad. Sci. U.S.A. 55:750–756.

Calloway, N. O. 1976. Body temperature: Thermodynamics of homeothermism. J. Theor. Biol. 8:331–344.

Calow, P. 1977. Conversion efficiencies in heterotrophic organisms. Biol. Rev. Camb. Philos. Soc. 52:385–409.

——. 1978. Life Cycles. Chapman and Hall, London.

Campbell, J. D. 1985. The influence of metabolic cost upon the level and precision of behavioral thermoregulation in an eurythermic lizard. Comp. Biochem. Physiol. A Physiol. 81:597–601.

Campbell, J. W., J. E. Vorhaben, and D. D. Smith Jr. 1987. Uricoteley: Its nature and origin during the evolution of tetrapod vertebrates. J. Exp. Zool. 243: 349–363.

Canals, M., M. Rosenmann, and F. Bozinovic. 1989. Energetics and geometry of huddling in small mammals. J. Theor. Biol. 141:181–189.

Canguilhem, B. 1985. Rythmes circannuels chez les mammifères hibernants sauvages. Can. J. Zool. 63:453–463.

Cannon, B., J. Nedergaard, L. Romert, U. Sundin, and J. Svartengren. 1978. The biochemical mechanism of thermogenesis in brown adipose tissue. In L. C. H. Wang and J. W. Hudson (eds.), Strategies in Cold: Natural Torpidity and Thermogenesis, pp. 567–594. Academic Press, New York.

Caple, G., R. P. Balda, and W. R. Willis. 1983. The physics of leaping animals and the evolution of preflight. Am. Nat. 121:455–467.

Carey, C., W. R. Dawson, L. C. Maxwell, and J. A. Faulkner. 1978. Seasonal acclimatization to temperature in Cardueline finches. J. Comp. Physiol. 125: 101–113.

Carey, F. G. 1982. A brain heater in the swordfish. Science 216:1327–1329.

Carey, F. G., G. Gabrielson, J. W. Kanwisher, O. Brazier, J. G. Casey, and H. L. Pratt Jr. 1982. The white shark, Carcharodon carcharias, is warm-bodied. Copeia 1982:254–260.

Carey, F. G., and Q. H. Gibson. 1983. Heat and oxygen exchange in the rete mirabile of the bluefin tuna, Thunnus thynnus. Comp. Biochem. Physiol. A Physiol. 74:333–342.

——. 1987. Blood flow in the muscle of free-swimming fish. Physiol. Zool. 60:138–148.

Carey, F. G., and J. V. Scharold. 1990. Movements of blue sharks (Prionace glauca) in depth and course. Mar. Biol. 106:329–342.

Carey, F. G., and K. Schmidt-Nielsen. 1962. Secretion of iodide by the nasal glands of birds. Science 137: 866–867.

Carey, F. G., and J. M. Teal. 1969a. Mako and porbeagle: Warm-bodied sharks. Comp. Biochem. Physiol. 28:199–204.

——. 1969b. Regulation of body temperature by the bluefin tuna. Comp. Biochem. Physiol. 28:205–213.

Carey, F. G., J. M. Teal, and J. W. Kanwisher. 1981. The visceral temperatures of mackerel sharks (Lamnidae). Physiol. Zool. 54:334–344.

Carey, F. G., J. M. Teal, J. W. Kanwisher, K. D. Lawson, and J. S. Beckett. 1971. Warm-bodied fish. Am. Zool. 11:137–145.

Carlson, K. J. 1968. The skull morphology and estivation burrows of the Permian lungfish, Gnathorhiza serrata. J. Geol. 76:641–663.

Carmi, N., B. Pinshow, W. P. Porter, and J. Jaeger. 1992. Water and energy limitations on flight duration in small migratory birds. Auk 109:268–276.

——. 1995. Reply to Klaassen's commentary concerning water and energy limitations on flight range. Auk 112:263.

Carpenter, F. L. 1974. Torpor in an Andean hummingbird: Its ecological significance. Science 183:545–547.

——. 1975. Bird hematocrits: Effects of high altitude and strength of flight. Comp. Biochem. Physiol. A Physiol. 50:415–417.

——. 1976. Ecology and evolution in an Andean hummingbird. Univ. Calif. Publ. Zool. 106:1–75.

Carpenter, F. L., and M. A. Hixon. 1988. A new function for torpor: Fat conservation in a wild migrant hummingbird. Condor 90:373–378.

Carpenter, R. E. 1969. Structure and function of the kidney and the water balance of desert bats. Physiol. Zool. 42:288–302.

——. 1985. Flight physiology of flying foxes, Pteropus poliocephalus. J. Exp. Biol. 114:619–647.

——. 1986. Flight physiology of intermediate-sized fruit bats (Pteropodidae). J. Exp. Biol. 120:79–103.

Carpenter, R. E., and M. A. Stafford. 1970. The secretory rates and chemical stimulus for secretion of the nasal salt gland in the Rallidae. Condor 72:316–324.

Carr, A., L. Ogren, and C. McVea. 1980. Apparent hibernation by the Atlantic loggerhead turtle Caretta caretta off Cape Canaveral, Florida. Biol. Conserv. 19:7–14.

Carrier, J. C. 1974. Physiological effects of environmental calcium on freshwater survival and sodium kinetics in the stenohaline marine teleost, Lagodon rhomboides. Ph.D. dissertation, University of Miami, Coral Gables, Fla.

Carrier, J. C., and D. H. Evans. 1976. The role of environmental calcium in freshwater survival of the marine teleost, Lagodon rhomboides. J. Exp. Biol. 5:529–538.

Carroll, E. J., and R. E. Hungate. 1954. The magnitude of the microbial fermentation in the bovine rumen. Appl. Microbiol. 2:205–214.

Carroll, R. L. 1965. Lungfish burrows from the Michigan coal basin. Science 148:963–964.

——. 1988. Vertebrate Paleontology and Evolution. Freeman, New York.

Carroll, R. L., and P. Gaskell. 1978. The order Microsauria. Mem. Am. Philos. Soc. 126:1–211.

Carter, G. S. 1931. Aquatic and aerial respiration in animals. Biol. Rev. Camb. Philos. Soc. 6:1–35.

——. 1935a. Results of the Cambridge expedition to British Guiana, 1933. The fresh waters of the rainforest areas of British Guiana. Zool. J. Linn. Soc. 39:147–186.

——. 1935b. Reports of the Cambridge expedition to British Guiana, 1933. Respiratory adaptations of the fishes of the forest waters, with descriptions of the accessory respiratory organs of Electrophorus electricus (Linn.)(= Gymnotus electricus auctt.) and Plecostomus plecostomus (Linn.). Zool. J. Linn. Soc. 39:219–233.

Carter, G. S., and L. C. Beadle. 1930. Notes on the habits and development of Lepidosiren paradoxa. Zool. J. Linn. Soc. 37:197–203.

——. 1931. The fauna of the swamps of the Paraquayan Chaco in relation to its environment—II. Respiratory adaptations in the fishes. Zool. J. Linn. Soc. 37: 327–368.

Case, T. J. 1978a. Speculations on the growth rate and reproduction of some dinosaurs. Paleobiology 4:320–328.

——. 1978b. On the evolution and adaptive significance of postnatal growth rates in the terrestrial vertebrates. Q. Rev. Biol. 53:243–282.

Castro, G. 1989. Energy costs and avian distributions: Limitations or chance?—a comment. Ecology 70:1181–1182.

Castro, G., C. Carey, J. Whittembury, and C. Monge. 1985. Comparative responses of sea level and montane rufous-collared sparrows, *Zonotrichia capensis*, to hypoxia and cold. Comp. Biochem. Physiol. A Physiol. 82:847–850.

Castro, G., J. P. Myers, and R. E. Ricklefs. 1992. Ecology and energetics of sanderlings migrating to four latitudes. Ecology 73:833–844.

Castro, G., N. Stoyan, and J. P. Myers. 1989. Assimilation efficiency in birds: A function of taxon or food type? Comp. Biochem. Physiol. A Physiol. 92:271–278.

Cates, R. G. 1975. The interface between slugs and wild ginger: Some evolutionary aspects. Ecology 56:391–400.

Cavagna, G. A., N. C. Heglund, and R. C. Taylor. 1977. Mechanical work in terrestrial locomotion: Two basic mechanisms for minimizing energy expenditure. Am. J. Physiol. 233:R243–R261.

Chaffee, R. R. J., W. W. Mayhew, M. Drebin, and Y. Cassuto. 1963. Studies on the thermogenesis in cold-acclimated birds. Can. J. Biochem. Physiol. 41:2215–2220.

Chaffee, R. R. J., and J. C. Roberts. 1971. Temperature acclimation in birds and mammals. Annu. Rev. Physiol. 33:155–202.

Chamberlin, T. C. 1900. On the habitat of the early vertebrates. J. Geol. 8:400–412.

Chapin, J. P., and L. W. Wing, 1959. The wideawake calendar, 1953 to 1958. Auk 76:153–158.

Chaplin, S. B. 1974. Daily energetics of the black-capped chickadee, *Parus atricapillus*, in winter. J. Comp. Physiol. 89:321–330.

——. 1976. The physiology of hypothermia in the black-capped chickadee, *Parus atricapillus*. J. Comp. Physiol. 112:335–344.

——. 1982. The energetic significance of huddling behavior in common bushtits (*Psaltriparus minimus*). Auk 99:424–430.

Chaplin, S. B., D. A. Diesel, and J. A. Kasparie. 1984. Body temperature regulation in red-tailed hawks and great horned owls: Responses to air temperature and food deprivation. Condor 86:175–181.

Chapman, L. J., C. A. Chapman, D. A. Brazeau, B. McLaughlin, and M. Jordan. 1999. Papyrus swamps, hypoxia, and faunal diversification: Variation among populations of *Barbus neumayeri*. J. Fish. Biol. 54:310–327.

Chapman, L. J., F. Galis, and J. Shinn. 2000. Phenotypic plasticity and the possible role of genetic assimilation: Hypoxia-induced trade-offs in the morphological traits of an African cichlid. Ecol. Lett. 3:387–393.

Chapman, L. J., and K. G. Hulen. 2001. Implications of hypoxia for the brain size and gill morphometry of mormyrid fishes. J. Zool. Lond.

Chapman, L. J., L. Kaufman, and C. A. Chapman. 1994a. Why swim upside down? A comparative study of two mochokid catfishes. Copeia 1994:130–135.

Chapman, L. J., L. S. Kaufman, C. A. Chapman, and F. E. McKenzie. 1994b. Hypoxia tolerance in twelve species of East African cichlids: Potential for low oxygen refugia in Lake Victoria. Conserv. Biol. 9:1274–1288.

Chapman, L. J., and K. F. Liem. 1995. Papyrus swamps and the respiratory ecology of *Barbus neumayeri*. Environ. Biol. Fish. 44:183–197.

Chappell, M. A., and G. A. Bartholomew. 1981. Standard operative temperatures and thermal energetics of the antelope ground squirrel *Ammospermophilus leucurus*. Physiol. Zool. 54:81–93.

Chappell, M. A., and T. L. Bucher. 1987. Effects of temperature and altitude on ventilation and gas exchange in chukars (*Alectoris chukar*). J. Comp. Physiol. [B] 157:129–136.

Chappell, M. A., and T. M. Ellis. 1987. Resting metabolic rates in boid snakes: Allometric relationships and temperature effects. J. Comp. Physiol. [B] 157:227–235.

Chappell, M. A., K. R. Morgan, S. L. Souza, and T. L. Bucher. 1989. Convection and thermoregulation in two Antarctic seabirds. J. Comp. Physiol. 159:313–322.

Chappell, M. A., V. H. Shoemaker, D. N. James, S. K. Maloney, and T. L. Bucher. 1993. Energetics of foraging in breeding Adélie penguins. Ecology 74:2450–2461.

Chappell, M. A., and L. R. G. Snyder. 1984. Biochemical and physiological correlates of deer mouse a-chain hemoglobin polymorphisms. Proc. Natl. Acad. Sci. U.S.A. 81:5484–5488.

Charles, N., R. Field, and R. Shine. 1985. Notes on the reproductive biology of Australian pythons, genera *Aspidites*, *Liasis*, and *Morelia*. Herpetol. Rev. 16:45–48.

Charlesworth, B. 1980. Evolution in Age-Structured Populations. Cambridge University Press, Cambridge.

Charnov, E. L. 1993. Life History Invariants: Some Explorations of Symmetry in Evolutionary Ecology. Oxford University Press, Oxford.

Chassin, P. S., C. R. Taylor, N. C. Heglund, and H. J. Seeherman. 1976. Locomotion in lions: Energetic cost and maximum aerobic capacity. Physiol. Zool. 49:1–10.

Cheeke, P. R., and J. W. Amberg. 1973. Comparative calcium excretion by rats and rabbits. J. Anim. Sci. 37:450–454.

Cheke, R. A. 1971. Temperature rhythms in African montane sunbirds. Ibis 113:500–506.

Chew, R. M. 1955. The skin and respiratory water losses of *Peromyscus maniculatus sonoriensis*. Ecology 36:463–467.

Chew, R. M., and A. E. Chew. 1970. Energy relationships of the mammals of a desert shrub (*Larrea tridentata*) community. Ecol. Monogr. 40:1–21.

Chew, R. M., R. G. Lindberg, and P. Hayden. 1967. Temperature regulation in the little pocket mouse,

Perognathus longimembris. Comp. Biochem. Physiol. 21:487–505.

Chew, R. M., and E. Spencer. 1967. Development of metabolic response to cold in young mice of four species. Comp. Biochem. Physiol. 22:873–888.

Chilcott, M. J., and I. D. Hume. 1984a. Digestion of *Eucalyptus andrewsii* foliage by the common ringtail possum, *Pseudocheirus peregrinus.* Aust. J. Zool. 32:605–613.

———. 1984b. Nitrogen and urea metabolism and nitrogen requirements of the common ringtail possum, *Pseudocheirus peregrinus,* fed *Eucalyptus andrewsii* foliage. Aust. J. Zool. 32:615–622.

Childress, J. J., and M. H. Nygaard. 1973. The chemical composition of midwater fishes as a function of depth of occurrence off southern California. Deep-Sea Res. 20:1093–1109.

Chinsamy, A., L. M. Chiappe, and P. Dodson. 1994. Growth rings in Mesozoic birds. Nature 368:196–197.

Chitty, D. 1960. Population processes in the vole and their relevance to general theory. Can. J. Zool. 38:99–113.

Chivers, D. J., and C. M. Hladik. 1980. Morphology of the gastrointestinal tract in primates: Comparisons with other mammals in relation to diet. J. Morphol. 166:337–386.

Cholette, C., A. Gagnon, and P. Germain. 1970. Isosmotic adaptation in *Myxine glutinosa* L. 1. Variations of some parameters and role of amino acid pool of the muscle cells. Comp. Biochem. Physiol. 33:333–346.

Christensen, C. U. 1974. Adaptations in the water economy of some anuran Amphibia. Comp. Biochem. Physiol. A Physiol. 47:1035–1049.

Christian, K. A., and C. R. Tracy. 1981. The effect of the thermal environment on the ability of hatchling Galápagos land iguanas to avoid predation during dispersal. Oecologia 49:218–223.

Christian, K. A., C. R. Tracy, and W. P. Porter. 1983. Seasonal shifts in body temperature and use of microhabitats by Galápagos land iguanas (*Conolophus pallidus*). Ecology 64:463–468.

Christman, S. P. 1974. Geographic variation for salt water tolerance in the frog *Rana sphenocephala.* Copeia 1974:773–778.

Church, N. S. 1960. Heat loss and the body temperature of flying insects. II. Heat conduction within the body and its loss by radiation and convection. J. Exp. Biol. 37:186–212.

Churchill, T. A., and K. B. Storey. 1992a. Responses to freezing exposure by hatchling turtles *Pseudemys scripta elegans*: Factors influencing the development of freeze tolerance by reptiles. J. Exp. Biol. 167:221–233.

———. 1992b. Freezing survival of the garter snake *Thamnophis sirtalis parietalis.* Can. J. Zool. 70:99–105.

Cipollini, M. L., and E. W. Stiles. 1993. Fruit rot, antifungal defense, and palatability of fleshy fruits for frugivorous birds. Ecology 74:751–762.

Clark, H. 1955. Metabolism of the black snake embryo. I. Nitrogen excretion. J. Exp. Biol. 30:492–501.

Clarke, A. 1993. Seasonal acclimatization and latitudinal compensation in metabolism: Do they exist? Funct. Ecol. 7:139–149.

Clarke, A., and N. M. Johnston. 1999. Scaling of metabolic rate with body mass and temperature in teleost fish. J. Anim. Ecol. 68:893–905.

Clarke, A., and P. A. Prince. 1976. The origin of stomach oil in marine birds: Analyses of the stomach oil from six species of subantarctic procellariiform birds. J. Exp. Mar. Biol. Ecol. 23:15–20.

Clarke, F. W. 1924. The Data of Geochemistry. Bull. U.S. Geol. Surv. 770:1–841.

Clausen, T. P., F. D. Provenza, E. A. Burritt, P. B. Reichardt, and J. P. Bryant. 1990. Ecological implications of condensed tannin structure: A case study. J. Chem. Ecol. 16:2381–2392.

Claussen, D. L. 1967. Studies of water loss in two species of lizards. Comp. Biochem. Physiol. 20:115–130.

———. 1969. Studies on water loss and rehydration in anurans. Physiol. Zool. 42:1–14.

Claussen, D. L., M. D. Townsley, and R. G. Bausch. 1990. Supercooling and freeze-tolerance in the European wall lizard, *Podarcis muralis,* with a revisional history of the discovery of freeze-tolerance in vertebrates. J. Comp. Physiol. [B] 160:137–143.

Clemens, D. T. 1988. Ventilation and oxygen consumption in rosy finches and house finches at sea level and high altitude. J. Comp. Physiol. [B] 158:57–66.

———. 1990. Interspecific variation and effects of altitude on blood properties of rosy finches (*Leucosticte arctoa*) and house finches (*Carpodacus mexicanus*). Physiol. Zool. 63:288–307.

Clemens, W. A., and L. G. Nelms. 1993. Paleoecological implications of Alaskan terrestrial vertebrate fauna in latest Cretaceous time at high paleolatitudes. Geology 21:503–506.

Clements, K. D. 1997. Fermentation and gastrointestinal microorganisms in fishes. *In* R. I. Mackie and B. A. White (eds.), Gastrointestinal Microbiology, vol. I. Gastrointestinal Ecosystems and Fermentations, pp. 156–198. Chapman and Hall, New York.

Clements, K. D., and J. H. Choat. 1995. Fermentation in tropical marine herbivorous fishes. Physiol. Zool. 68:355–378.

Cogger, H. G., and A. Holmes. 1960. Thermoregulatory behaviour in a specimen of *Morelia spilotes variegata* Gray (Serpentes: Boidae). Proc. Linn. Soc. N.S.W. 85:328–333.

Cohen, N. W. 1952. Comparative rates of dehydration and hydration in some California salamanders. Ecology 33:462–479.

Cole, L. C. 1943. Experiments on toleration of high temperature in lizards with reference to adaptive coloration. Ecology 24:94–108.

———. 1954. The population consequences of life history phenomena. Q. Rev. Biol. 29:103–137.

Coley, P. D., J. P. Bryant, and F. S. Chapin III. 1985. Resource availability and plant antiherbivore defense. Science 230:895–899.

Collier, B. D., N. C. Stenseth, S. Barkley, and R. Osborn. 1975. A simulation model of energy acquisition and utilization by the brown lemming *Lemmus trimucronatus* at Barrow, Alaska. Oikos 26:276–294.

Collopy, M. W. 1977. Food caching by female American Kestrels in winter. Condor 79:63–68.

Congleton, J. L. 1974. The respiratory response to asphyxia of *Typhlogobius californiensis* (Teleostei: Gobiidae) and some related gobies. Biol. Bull. 146:186–205.

Conte, F. P. 1965. Effects of ionizing radiation on osmoregulation in fish *Oncorhynchus kisutch*. Comp. Biochem. Physiol. 15:292–302.

Contreras, L. C. 1986. Bioenergetics and distribution of fossorial *Spalacopus cyanus* (Rodentia): Thermal stress, or cost of burrowing? Physiol. Zool. 59:20–28.

Contreras, L. C., and B. K. McNab. 1990. Thermoregulation and energetics of subterranean mammals. *In* E. Nevo and O. A. Reig (eds.), Evolution of Subterranean Mammals at the Organismal and Molecular Levels, pp. 231–250. Wiley-Liss, New York.

Cooch, F. G. 1964. A preliminary study of the survival value of a functional salt gland in prairie Anatidae. Auk 81:380–393.

Cooper, J. E., and M. R. Cooper. 1978. Growth, longevity, and reproductive strategies in Shelta Cave crayfishes. Natl. Speleol. Soc. Bull. 40:97. (Abstract)

Cope, E. D. 1898. The crocodilians, lizards and snakes of North America. Reports U.S. Nat. Mus. 1900: 155–1270.

Cork, S. J. 1981. Digestion and metabolism in the koala (*Phascolarctos cinereus* Goldfuss): An arboreal folivore. PhD. thesis, University of New South Wales.

——. 1994. Digestive constraints on dietary scope in small and moderately-small mammals: How much do we really understand? *In* D. J. Chivers and P. Langer (eds.), The Digestive System in Mammals: Food, Form and Function, pp. 337–369. Cambridge University Press, Cambridge.

Cork, S. J., and W. J. Foley. 1991. Digestive and metabolic strategies of arboreal mammalian folivores in relation to chemical defenses in temperate and tropical forests. *In* R. T. Palo and C. T. Robbins (eds.), Plant Defenses against Mammalian Herbivory, pp. 133–166. CRC Press, Boca Raton.

Cork, S. J., I. D. Hume, and T. J. Dawson. 1983. Digestion and metabolism of a natural foliar diet (*Eucalyptus punctata*) by an arboreal marsupial, the koala (*Phascolarctos cinereus*). J. Comp. Physiol. 153:181–190.

Cork, S. J., and G. J. Kenagy. 1989. Nutritional value of hypogeous fungus for a forest-dwelling ground squirrel. Ecology 70:577–586.

Corner, E. D. S., E. J. Denton, and G. R. Forster. 1969. On the buoyancy of some deep-sea sharks. Proc. R. Soc. Lond. 171B:415–429.

Cortés, A., M. Rosenmann, and C. Báez. 1990. Función del riñon y del pasaje nasal en la conservación de agua corporal en roedores simpátridos de Chile central. Rev. Chil. Hist. Nat. 63:279–291.

Cortés, A., C. Zuleta, and M. Rosenmann. 1988. Comparative water economy of sympatric rodents in a Chilean semi-arid habitat. Comp. Biochem. Physiol. A Physiol. 91:711–714.

Costa, D. P. 1988. Methods for studying the energetics of freely diving animals. Can. J. Zool. 66:45–52.

Costa, D. P., J. P. Croxall, and C. D. Duck. 1989. Foraging energetics of antarctic fur seals in relation to changes in prey availability. Ecology 70:596–606.

Costa, D. P., and R. L. Gentry. 1986. Reproductive energetics of the northern fur seal. *In* R. L. Gentry and G. L. Kooyman (eds.), Fur Seals: Maternal Strategies on Land and at Sea, pp. 79–101. Princeton University Press, Princeton, N.J.

Costa, D. P., B. J. LeBoeuf, A. C. Huntley, and C. L. Ortiz. 1986. The energetics of lactation in the northern elephant seal, *Mirounga angustirostris*. J. Zool. Lond. 209:21–23.

Costa, D. P., and P. A. Prince. 1987. Foraging energetics of grey-headed albatrosses *Diomedea chrysostoma* at Bird Island, South Georgia. Ibis 129:149–158.

Costa, D. P., and F. Trillmich. 1988. Mass changes and metabolism during the perinatal fast: A comparison between antarctic (*Arctocephalus gazella*) and Galápagos fur seals (*Arctocephalus galapagoensis*). Physiol. Zool. 61:160–169.

Costanzo, J. P. 1985. The bioenergetics of hibernation in the eastern garter snake *Thamnophis sirtalis sirtalis*. Physiol. Zool. 58:682–692.

——. 1989. A physiological basis for prolonged submergence in hibernating garter snakes *Thamnophis sirtalis*: Evidence for an energy-sparing adaptation. Physiol. Zool. 62:580–592.

Costanzo, J. P., and D. L. Claussen. 1990. Natural freeze tolerance in the terrestrial turtle, *Terrapene carolina*. J. Exp. Zool. 254:228–232.

Costanzo, J. P., D. L. Claussen, and R. E. Lee Jr. 1988. Natural freeze tolerance in a reptile. Cryo-Lett. 9:380–385.

Coulter, G. W. 1967. Low apparent oxygen requirements of deep-water fishes in Lake Tanganyika. Nature 215:317–318.

Cowles, R. B. 1962. Semantics in biothermal studies. Science 135:670.

——. 1967. Black pigmentation: Adaptation for concealment or heat conservation? Science 158:1340–1341.

Cowles, R. B., and C. M. Bogert. 1944. A preliminary study of the thermal requirements of desert reptiles. Bull. Am. Mus. Nat. Hist. 83:267–296.

Cox, C. B. 1967. Cutaneous respiration and the origin of the modern Amphibia. Proc. Linn. Soc. Lond. 178:37–47.

Cox, G. W. 1961. The relation of energy requirements of tropical finches to distribution and migration. Ecology 42:253–266.

Cragg, M. M., J. B. Balinsky, and E. Baldwin. 1961. A comparative study of nitrogen excretion in some Amphibia and reptiles. Comp. Biochem. Physiol. 3:227–235.

Cramp, S. 1985. Birds of Europe, the Middle East and North Africa, vol. 4. Oxford University Press, Oxford.

Crawford, E. C., and G. Kampe. 1971. Physiological responses of the lizard *Sauromalus obesus* to changes in ambient temperature. Am. J. Physiol. 220:1256–1260.

Crawford, E. C., and R. C. Lasiewski. 1968. Oxygen consumption and respiratory evaporation of the emu and rhea. Condor 70:333–339.

Crawshaw, L. I. 1977. Physiological and behavioral reactions of fishes to temperature change. J. Fish. Res. Board Can. 34:730–734.

Crawshaw, L. I., and H. T. Hammel. 1971. Behavioral thermoregulation in two species of Antarctic fish. Life Sci. 10:1009–1020.

——. 1973. Behavioral temperature regulation in the California horn shark, *Heterodontus francisci*. Brain Behav. Evol. 7:447–452.

——. 1974. Behavioral regulation of internal temperature in the brown bullhead, *Ictalurus nebulosus*. Comp. Biochem. Physiol. A Physiol. 47:51–60.

Crawshaw, L. I., H. T. Hammel, and W. F. Garey. 1973. Brainstem temperature affects gill ventilation in the California scorpionfish. Science 181:579–581.

Crick, H. Q. P., C. Dudley, D. E. Glue, and D. L. Thomson. 1997. UK birds are laying eggs earlier. Nature 388:526.

Crompton, A. W., C. R. Taylor, and J. A. Jagger. 1978. Evolution of homeothermy in mammals. Nature 272:333–336.

Crook, J. H., and J. S. Gartlan. 1966. Evolution of primate societies. Nature 210:1200–1203.

Crowder, W. C., M. Nie, and G. R. Ultsch. 1998. Oxygen uptake in bullfrog tadpoles (*Rana catesbeiana*). J. Exp. Biol. 250:121–134.

Crowley, S. R. 1985. Thermal sensitivity of sprint-running in the lizard *Sceloporus undulatus*: Support for a conservative view of thermal physiology. Oecologia 66:219–225.

Croxall, J. P. 1982. Energy costs of incubation and moulting in petrels and penguins. J. Anim. Ecol. 51:177–194.

Crump, M. L. 1981. Energy accumulation and amphibian metamorphosis. Oecologia 38:235–247.

Crump, M. L., F. R. Hensley, and K. L. Clark. 1992. Apparent decline of the golden toad: Underground or extinct? Copeia 1992:415–420.

Culik, B. 1992. Diving heart rates in Adélie penguins (*Pygoscelis adeliae*). Comp. Biochem. Physiol. A Physiol. 102:487–490.

Culik, B., and R. P. Wilson. 1991. Energetics of underwater swimming in Adélie penguins (*Pygoscelis adeliae*). J. Comp. Physiol. [B] 161:285–291.

Culver, D. C. 1982. Cave Life: Evolution and Ecology. Harvard University Press, Cambridge.

Culver, D. C., and T. L. Poulson. 1971. Oxygen consumption and activity in closely related amphipod populations from cave and surface habitats. Am. Midl. Nat. 85:74–84.

Cunningham, J. T., and D. M. Reid. 1932. Experimental researches on the emission of oxygen by the pelvic filaments of the male *Lepidosiren* with some experiments on *Symbranchus marmoratus*. Proc. R. Soc. Lond. 110:234–248.

Curtis, B. J., and C. M. Wood. 1991. The function of the urinary bladder *in vivo* in the freshwater rainbow trout. J. Exp. Biol. 155:567–583.

Cutts, C. J., and J. R. Speakman. 1994. Energy savings in formation flight of pink-footed geese. J. Exp. Biol. 189:251–261.

Cuyler, L. C., and N. A. Øritsland. 1993. Metabolic strategies for winter survival by Svalbard reindeer. Can. J. Zool. 71:1787–1792.

Czopek, J. 1965. Quantitative studies on the morphology of respiratory surfaces in amphibians. Acta Anat. 62:296–323.

Daan, S., D. Masman, and A. Groenewold. 1990. Avian basal metabolic rates: Their association with body composition and energy expenditure in nature. Am. J. Physiol. 259:R333–R340.

Daan, S., D. Masman, A. M. Strijkstra, and G. J. Kenagy. 1991. Daily energy turnover during reproduction in birds and mammals: Its relationship to basal metabolic rate. Acta XX Congr. Int. Ornithol. 4:1976–1987.

Daan, S., D. Masman, A. Strijkstra, and S. Verhulst. 1989. Intraspecific allometry of basal metabolic rate: Relations with body size, temperature, composition, and circadian phase in the kestrel, *Falco tinnunculus*. J. Biol. Rhythms 4:267–283.

Daeschler, E. B., N. H. Shubin, K. S. Thomson, and W. W. Amaral. 1994. A Devonian tetrapod from North America. Science 265:639–642.

Dafré, A. L., and D. Wilhelm Fihlo. 1989. Root effect hemoglobins in marine fish. Comp. Biochem. Physiol. A Physiol. 92:467–471.

Dahlberg, M. L., D. L. Shumway, and P. Doudoroff. 1968. Influence of dissolved oxygen and carbon dioxide on swimming performance of largemouth bass and coho salmon. J. Fish. Res. Board Can. 25:49–70.

Dall, W., and N. E. Milward. 1969. Water intake, gut absorption and sodium fluxes in amphibious and aquatic fishes. Comp. Biochem. Physiol. 30:247–260.

Dalrymple, G. H. 1996. Growth of the American alligator in the Shark Valley region of Everglades National Park. Copeia 1996:212–216.

Damuth, J. 1981. Population density and body size in mammals. Nature 290:699–700.

——. 1991. Of size and abundance. Nature 351:268–269.

Danell, K., K. Huss-Danell, and R. Bergström. 1985. Interactions between browsing mouse and two species of birch in Sweden. Ecology 66:1867–1878.

Daniels, C. B., and J. Pratt. 1992. Breathing in long necked dinosaurs: Did the sauropods have bird lungs? Comp. Biochem. Physiol. A Physiol. 101:43–46.

Daniels, F., and R. A. Alberty. 1975. Physical Chemistry, 4th ed. Wiley, New York.

Daniels, H. L. 1984. Oxygen consumption in *Lemur fulvus*: Deviation from the ideal model. J. Mammal. 65:584–592.

Dantzler, W. H. 1970. Kidney function in desert vertebrates. *In* G. K. Benson and J. G. Phillips (eds.), Hormones and the Environment, pp. 157–189. Cambridge University Press, Cambridge.

Dantzler, W. H., and B. Schmidt-Nielsen. 1966. Excretion in fresh-water turtle (*Pseudemys scripta*) and desert tortoise (*Gopherus agassizii*). Am. J. Physiol. 210:198–210.

Darden, T. R. 1972. Respiratory adaptations of a fossorial mammal, the pocket gopher (*Thomomys bottae*). J. Comp. Physiol. 78:121–137.

Das, B. K. 1927. The bionomics of certain air-breathing fishes of India, together with an account of the development of their air-breathing organs. Philos. Trans. R. Soc. B Biol. Sci. 216:183–219.

Das, I., and M. Coe. 1994. Dental morphology and diet in anuran amphibians from south India. J. Zool. Lond. 233:417–427.

Da Silva, H. R., M. C. De Britto-Pereira, and U. Caramaschi. 1989. Frugivory and seed dispersal by *Hyla truncata*, a neotropical treefrog. Copeia 1989:781–783.

Davenport, J., and M. D. J. Sayer. 1993. Physiological determinant of distribution in fish. J. Fish Biol. 43 (Suppl. A):121–145.

Davies, A. G., E. L. Bennett, and P. G. Waterman. 1988. Food selection by two south-east Asian colobine monkeys (*Presbytis rubicunda* and *Presbytis melalophos*) in relation to plant chemistry. Biol. J. Linn. Soc. 34:33–56.

Davis, R. E. 1955. Heat transfer in the goldfish, *Carassius auratus*. Copeia 1955:207–209.

Davis, R. W., J. P. Croxall, and M. J. O'Connell. 1989. The reproductive energetics of the gentoo (*Pygoscelis papua*) and macaroni (*Eudyptes chrysolophus*) penguins at South Georgia. J. Anim. Ecol. 58:59–74.

Davis, R. W., G. L. Kooyman, and J. P. Croxall. 1983. Water flux and estimated metabolism of free-ranging gentoo and macaroni penguins at South Georgia. Polar Biol. 2:41–46.

Dawbin, W. H. 1962. The tuatara in its natural habitat. Endeavour 21:16–24.

Dawson, A. B. 1951. Functional, and degenerate or rudimentary glomeruli in the kidney of two species of Australian frog, *Cyclorama* (*Chiroleptes*) *platycephalus* and *alboguttatus* (Gunther). Anat. Rec. 109:417–429.

Dawson, T. J. 1972. Likely effects of standing and lying on the radiant heat load experienced by a resting kangaroo on a summer day. Aust. J. Zool. 20:17–22.

Dawson, T. J., and W. R. Dawson. 1982. Metabolic scope and conductance in response to cold of some dasyurid marsupials and Australian rodents. Comp. Biochem. Physiol. A Physiol. 71:59–64.

Dawson, T. J., and M. J. S. Denny. 1969a. A bioclimatological comparison of the summer day microenvironments of two species of arid zone kangaroos. Ecology 50:328–332.

——. 1969b. Seasonal variation in the plasma and urine electrolyte concentration of the arid zone kangaroos *Megaleia rufa* and *Macropus robustus*. Aust. J. Zool. 17:777–784.

Dawson, T. J., and A. J. Hulbert. 1970. Standard metabolism, body temperature, and surface areas of Australian marsupials. Am. J. Physiol. 218:1233–1238.

Dawson, T. J., A. B. Johns, and A. M. Beal. 1989. Digestion in the Australian wood duck (*Chenonetta jubata*): A small avian herbivore showing selective digestion of the hemicellulose component of fiber. Physiol. Zool. 62:522–540.

Dawson, T. J., and J. M. Olson. 1988. Thermogenic capabilities of the opossum *Monodelphis domestica* when warm and cold acclimated: Similarities between American and Australian marsupials. Comp. Biochem. Physiol. A Physiol. 89:85–91.

Dawson, T. J., D. Robertshaw, and C. R. Taylor. 1974. Sweating in the kangaroo: A cooling mechanism during excercise, but not in heat. Am. J. Physiol. 227:494–498.

Dawson, T. J., and K. Schmidt-Nielsen. 1966. Effect of thermal conductance on water economy in the antelope jack rabbit, *Lepus alleni*. J. Cell. Physiol. 67:463–472.

Dawson, W. R. 1954. Temperature regulation and water requirements of the brown and Abert towhees, *Pipilo fascus* and *Pipilo aberti*. Univ. Calif. Publ. Zool. 59:81–124.

——. 1975. On the physiological significance of the preferred body temperatures of reptiles. *In* D. M. Gates and R. B. Schmerl (eds.), Perspectives in Biophysical Ecology, pp. 443–473. Springer-Verlag, New York.

Dawson, W. R., and G. A. Bartholomew. 1958. Metabolic and cardiac responses to temperature in the lizard *Dipsosaurus dorsalis*. Physiol. Zool. 31:100–111.

Dawson, W. R., G. A. Bartholomew, and A. F. Bennett. 1977. A reappraisal of the aquatic specializations of the Galapagos marine iguana (*Amblyrhynchus cristatus*). Evolution 31:891–897.

Dawson, W. R., and A. F. Bennett. 1973. Roles of metabolic level and temperature regulation in the adjustment of western plumed pigeons (*Lophophops ferruginea*) to desert conditions. Comp. Biochem. Physiol. A Physiol. 44:249–266.

Dawson, W. R., and C. Carey. 1976. Seasonal acclimatization to temperature in cardueline finches. J. Comp. Physiol. 112:317–333.

Dawson, W. R., and F. C. Evans. 1957. Relation of growth and development to temperature regulation in nestling field and chipping sparrows. Physiol. Zool. 30:315–327.

Dawson, W. R., R. L. Marsh, and M. E. Yacoe. 1983. Metabolic adjustments of small passerine birds for migration and cold. Am. J. Physiol. 245:R755–R767.

Dawson, W. R., and T. L. Poulson. 1962. Oxygen capacity of lizard bloods. Am. Midl. Nat. 68:154–164.

Dawson, W. R., and J. R. Templeton. 1966. Physiological responses to temperature in the alligator lizard, *Gerrhonotus multicarinatus*. Ecology 47:759–765.

Dayan, T., D. Simberloff, and E. Tchernov. 1994. Morphological change in Quaternary mammals: A note for species interactions? *In* R. A. Martin and A. D. Barnosky (eds.), Morphological Changes in Quaternary Mammals of North America, pp. 71–83. Cambridge University Press, New York.

Dayan, T., D. Simberloff, E. Tchernov, and Y. Yom-Tov. 1989. Inter- and intraspecific character displacement in mustelids. Ecology 70:1526–1539.

——. 1991. Calibrating the paleothermometer: Climate, communities, and the evolution of size. Paleobiology 17:189–199.

DeCosta, J., M. Alonso-Bedate, and A. Fraile. 1981. Temperature acclimation in amphibians: Changes in lactate dehydrogenase activities and isoenzyme patterns in several tissues from adult *Discoglossus pictus pictus* (Otth). Comp. Biochem. Physiol. B Biochem. Mol. Biol. 70:331–339.

Deevey, E. S. 1947. Life tables for natural populations of animals. Q. Rev. Biol. 22:283–314.

Degabriele, R. 1981. A relative shortage of nitrogenous food in the ecology of the koala (*Phascolarctos cinereus*). Aust. J. Ecol. 6:139–141.

Degabriele, R. 1983. Nitrogen and the koala (*Phascolarctos cinereus*): Some indirect evidence. Aust. J. Ecol. 8:75–76.

Degani, G. 1981. Salinity tolerance and osmoregulation in *Salamandra salamandra* (L.) from different populations. J. Comp. Physiol. [B] 145:133–137.

Degani, G., N. Silanikove, and A. Shkolnik. 1984. Adaptation of green toad (*Bufo viridis*) to terrestrial life by urea accumulation. Comp. Biochem. Physiol. A Physiol. 77:585–587.

De Groot, S. J. 1971. On the interrelationships between morphology of the alimentary tract, food and feeding behaviour of flatfishes (Pisces: Pleuronectiformes). Neth. J. Sea Res. 5:121–196.

Dehadrai, P. V. 1962. Respiratory function of the swimbladder of *Notopterus* (Lacepede). Proc. Zool. Soc. Lond 139:341–357.

Dehadrai, P. V., and S. D. Tripathi. 1976. Environment and ecology of freshwater air-breathing teleosts. *In* G. M. Hughes (ed.), Respiration of Amphibious Vertebrates, pp. 39–72. Academic Press, London.

Dehnel, A. 1949. Studies on the genus *Sorex* L. Ann. Univ. Marie Curie-Sklodowska, Sect. C Biol. 4:17–102.

DeJong, A. A. 1976. The influence of simulated solar radiation on the metabolic rate of white-crowned sparrows. Condor 78:174–179.

Dejours, P. 1975. Principles of Comparative Respiratory Physiology. American Elsevier, New York.

Dekker, R. W. R. J. 1989. Predation and the western limits of megapode distribution (Megapodidae; Aves). J. Biogeograph. 16:317–321.

Delaney, R. G., S. Lahiri, and A. P. Fishman. 1974. Aestivation of the African lungfish *Prototperus aethiopicus*: Cardiovascular and respiratory functions. J. Exp. Biol. 61:111–128.

Delson, J., and W. G. Whitford. 1973. Adaptation of the tiger salamander, *Ambystoma tigrinum*, to arid habitats. Comp. Biochem. Physiol. A Physiol. 46:631–638.

Demment, D. M. 1982. The scaling of ruminoreticulum size with body weight in East African ungulates. Afr. J. Ecol. 20:43–47.

——. 1983. Feeding ecology and the evolution of body size of baboons. Afr. J. Ecol. 21:219–233.

Demment, D. M., and P. J. Van Soest. 1985. A nutritional explanation for body-size patterns of ruminant and nonruminant herbivores. Am. Nat. 125:641–672.

Denison, R. H. 1941. The soft anatomy of *Bothriolepis*. J. Paleontol. 15:553–561.

——. 1956. A review of the habitat of the earliest vertebrates. Fieldiana, Geol. 11:359–457.

Densmore, L. D., and H. C. Dessauer. 1982. Low protein divergence of species within the circumtropical genus *Crocodylus*: Result of a post-Pliocene transoceanic dispersal and radiation? Fed. Proc. 41:1004.

Denton, D. A., M. Reich, and F. J. R. Hird. 1963. Ureotelism of echidna and platypus. Science 139:1225.

Denton, E. J., and N. B. Marshall. 1958. The buoyancy of bathypelagic fishes without a gas-filled swimbladder. J. Mar. Biol. Assoc. U.K. 37:753–767.

de Ricqlès, A. J. 1974. Evolution of endothermy: Histological evidence. Evol. Theory 1:51–80.

——. 1980. Tissue structures of dinosaur bone: Functional significance and possible relation to dinosaur physiology. *In* R. D. K. Thomas and E. C. Olson (eds.), A Cold Look at the Warm-Blooded Dinosaurs, pp. 103–139. AAAS Symposium 28. American Association for the Advancement of Science (AAAS), Washington, D.C.

Derrickson, E. M. 1989. The comparative method of Elgar and Harvey: Silent ammunition for McNab. Funct. Ecol. 3:123–127.

Derting, T. L. 1989. Metabolism and food availability as regulators of production in juvenile cotton rats. Ecology 70:587–595.

Derting, T. L., and P. A. McClure. 1989. Intraspecific variation in metabolic rate and its relationship with productivity in the cotton rat, *Sigmodon hispidus*. J. Mammal. 70:520–531.

de Vlaming, V. L., and M. Sage. 1973. Osmoregulation in the euryhaline elasmobranch, *Dasyatis sabina*. Comp. Biochem. Physiol. 45:31–44.

DeVries, A. L. 1970. Freezing resistance in Antarctic fishes. *In* M. W. Holdgate (ed.), Antarctic Ecology, vol. 1, pp. 320–328. Academic Press, New York.

——. 1971a. Glycoproteins as biological antifreeze agents in Antarctic fishes. Science 172:1152–1155.

——. 1971b. Freezing resistance in fishes. *In* J. S. Hoar and D. J. Randall (eds.), Fish Physiology, vol. VI, pp. 157–190. Academic Press, New York.

——. 1980. Biological antifreezes and survival in freezing environment. *In* R. Giles (ed.), Animals and Environmental Fitness, pp. 583–607. Pergamon Press, Oxford.

——. 1982. Biological antifreeze agents in coldwater fishes. Comp. Biochem. Physiol. A Physiol. 73:627–640.

——. 1984. Role of glycopeptides and peptides in inhibition of crystallization of water in polar fishes. Philos. Trans. R. Soc. Lond. B Biol. Sci. 304:575–588.

DeVries, A. L., and J. T. Eastman. 1981. Physiology and ecology of notothenioid fishes of the Ross Sea. J. R. Soc. N.Z. 11:329–340.

DeVries, A. L., S. K. Komatsu, and R. E. Feeney. 1970. Chemical and physical properties of freezing point-depressing glycoproteins from Antarctic fishes. J. Biol. Chem. 245:2901–2913.

DeVries, A. L., and D. E. Wohlschlag. 1969. Freezing resistence in some Antarctic fishes. Science 163:1074–1075.

DeWitt, C. B. 1967. Precision of thermoregulation and its relation to environmental factors in the desert iguana, *Dipsosaurus dorsalis*. Physiol. Zool. 40:49–66.

Dhindsa, D. S., A. S. Hoversland, and J. Metcalfe. 1971. Comparative studies of the respiratory functions of mammalian blood. VII. Armadillo (*Dasypus novemcinctus*). Respir. Physiol. 13:198–208.

Dial, B. E., and L. L. Grismer. 1992. A phylogenetic analysis of physiological-ecological character evolution in the lizard genus *Coleonyx* and its implications for historical biogeographic reconstruction. Syst. Biol. 41:178–195.

Dickson, G. W., and R. Franz. 1980. Respiration rates, ATP turnover and adenylate energy charge in excised gills of surface and cave crayfish. Comp. Biochem. Physiol. A Physiol. 65:375–379.

Dierenfeld, E. S., H. F. Hintz, J. B. Robertson, P. J. van Soest, and O. T. Oftedahl. 1982. Utilization of bamboo by the giant panda. J. Nutr. 112:636–641.

Dietz, B. A., A. E. Hagerman, and G. W. Barrett. 1994. Role of condensed tannin on salivary tannin-binding proteins, bioenergetics, and nitrogen digestibility in *Microtus pennsylvanicus*. J. Mammal. 75:880–889.

Dijkstra, C., A. Bult, S. Bijlsma, S. Daan, T. Meijer, and M. Zijlstra. 1990. Brood size manipulations in the Kestrel (*Falco tinnunculus*): Effects on offspring and parent survival. J. Anim. Ecol. 599:269–285.

Dimock, E. J., R. R. Silen, and V. E. Allen. 1976. Genetic resistance in Douglas-fir to damage by snowshoe hare and black-tailed deer. Forestry Sci. 22:106–121.

Dizon, A. E., and R. W. Brill. 1979. Thermoregulation in tunas. Am. Zool. 19:249–265.

Dmi'el, R., and D. Tel-Tzur. 1985. Heat balance of two starling species (*Sturnis vulgaris* and *Onychognathus tristrami*) from temperate and desert habitats. J. Comp. Physiol. [B] 155:395–402.

Dobbs, G. H., and A. L. DeVries. 1975a. Renal function in Antarctic teleost fishes: Serum and urine composition. Mar. Biol. 29:59–70.

———. 1975b. The aglomerular nephron of antarctic teleosts: A light and electron microscopic study. Tissue Cell 7:159–170.

Dobbs, G. H., Y. Lin, and A. L. DeVries. 1974. Aglomerularism in Antarctic fish. Science 185:793–794.

Dol'nik, V. R. 1982. Time and energy budgets in free-living birds. Acad. Sci. U.S.S.R. Proc. Zool. Inst. 113:1–37.

Dolph, C. J., H. A. Braun, and E. W. Pfeiffer. 1962. The effect of vasopressin upon urine concentration in *Aplodontia rufa* (Sewellel) and the rabbit. Physiol. Zool. 35:263–269.

Dominguez-Bello, M. G., M. C. Ruiz, and F. Michelangeli. 1993. Evolutionary significance of foregut fermentation in the hoatzin (*Opisthocomus hoazin*; Aves: Opisthocomidae). J. Comp. Physiol. [B] 163:594–601.

Domning, D. P. 1978. Sirenian evolution in the North Pacific Ocean. Univ. Calif. Publ. Geol. Sci. 118:1–176.

Doncaster, C. P., E. Dumonteil, H. Barré, and P. Jouventin. 1990. Temperature regulation of young coypus (*Myocastor coypus*) in air and water. Am. J. Physiol. 259:R1220–R1227.

Dorst, J. 1970. A Field Guide to the Larger Mammals of Africa. Houghton Mifflin, Boston.

Douglas, E. L., K. S. Peterson, J. R. Gysi, and D. J. Chapman. 1985. Myoglobin in the heart tissue of fishes lacking hemoglobin. Comp. Biochem. Physiol. A Physiol. 81:885–888.

Drent, R. H., and S. Daan. 1980. The prudent parent: Energetic adjustments in avian breeding. Ardea 68:225–252.

Drent, R. H., and M. Klaassen. 1989. Energetics of avian growth: The causal link with BMR and metabolic scope. *In* C. Bech and R. E. Reinertsen (eds.), Physiology of Cold Adaptation in Birds, pp. 349–359. Plenum, New York.

Drewes, R. C., S. S. Hillman, R. W. Putnam, and O. M. Sokol. 1977. Water, nitrogen and ion balance in the African tree frog *Chiromantis petersi* Boulanger (Anura: Rhacophoridae), with comments on the structure of the integument. J. Comp. Physiol. 116:257–267.

Driedzic, W. R., and P. W. Hochachka. 1978. Energy metabolism during exercise. *In* W. S. Hoar and D. J. Randall (eds.), Fish Physiology, vol. 7, pp. 503–543. Academic Press, New York.

Drożdż, A., A. Gorecki, W. Grodzinski, and J. Pelikan. 1978. Bioenergetics of water voles (*Arvicola terrestris* L.) from southern Moravia. Ann. Zool. Fenn. 8:97–103.

Drożdż, A., A. Gorecki, and K. Sawicka-Kapusta. 1972. Bioenergetics of growth in common voles. Acta Theriol. 17:245–257.

Dubale, M. S. 1951. A comparative study of the extent of gill surface in some representative Indian fishes, and its bearing on the origin of the air-breathing habit. J. Univ. Bombay 19:90–101.

Dudley, R., and G. J. Vermeij. 1992. Do the power requirements of flapping flight constrain folivory in flying animals? Funct. Ecol. 6:101–104.

———. 1994. Energetic constraints of folivory: Leaf fractionation by frugivorous bats. Funct. Ecol. 8:668.

Duellman, W. E., and L. Trueb. 1986. Biology of Amphibians. McGraw-Hill, New York.

Duke, J. T., and G. R. Ultsch. 1990. Metabolic oxygen regulation and conformity during submergence in the salamanders *Siren lacertina*, *Amphiuma means*, and *Amphiuma tridactylum*, and a comparison with other giant salamanders. Oecologia 84:16–23.

Duman, J. G., and A. L. DeVries. 1975. The role of macromolecular antifreezes in cold water fishes. Comp. Biochem. Physiol. A Physiol. 52:193–199.

Dumas, P. C. 1956. The ecological relations of sympatry in *Plethodon dunni* and *Plethodon vehiculum*. Ecology 37:484–495.

———. 1966. Studies of the *Rana* species complex in the Pacific Northwest. Copeia 1966:60–74.

Dunbrack, R. L., and M. A. Ramsay. 1993. The allometry of mammalian adaptations to seasonal environments: A critique of the fasting endurance hypothesis. Oikos 66:336–342.

Duncker, H.-R. 1991. Constructional and ecological prerequisites for the evolution of homeothermy. *In* N. Schmidt-Kittler and K. Vogel (eds.), Constructional Morphology and Evolution, pp. 331–357. Springer-Verlag, Berlin.

Dunitz, J. D., and S. A. Benner. 1986. Body temperature and the specific heat of water. Nature 324:418.

Dunn, E. H. 1980. On the variability in energy allocation of nestling birds. Auk 97:19–27.

Dunn, J. F. 1988. Low-temperature adaptation of oxidative energy production in cold-water fishes. Can. J. Zool. 66:1098–1104.

Dunson, W. A. 1960. Aquatic respiration in *Trionyx spinifer asper*. Herpetologica 16:277–283.

——. 1966. A new site of sodium uptake in fresh-water turtles. Am. Zool. 6:320. (Abstract)

——. 1969. Electrolyte excretion by the salt gland of the Galápagos marine iguana. Am. J. Physiol. 216:995–1002.

——. 1970. Some aspects of electrolyte and water balance in three estuarine reptiles, the diamondback terrapin, American and "salt water" crocodiles. Comp. Biochem. Physiol. 32:161–174.

——. 1974. Salt gland excretion in a mangrove monitor lizard. Comp. Biochem. Physiol. A Physiol. 47:1245–1255.

——. 1975. Salt and water balance in sea snakes. *In* W. A. Dunson (ed.), The Biology of Sea Snakes, pp. 329–353. University Park Press, Baltimore.

——. 1978. Role of the skin in sodium and water exchange of aquatic snakes placed in seawater. Am. J. Physiol. 235:R151–R159.

——. 1982. Salinity relations of crocodiles in Florida Bay. Copeia 1982:374–385.

Dunson, W. A., and M. K. Dunson. 1973. Convergent evolution of sublingual glands in the marine file snake and the true sea snakes. J. Comp. Physiol. 86:193–208.

——. 1974. Interspecific differences in fluid concentration and secretion rate of sea snake salt glands. Am. J. Physiol. 227:430–438.

Dunson, W. A., and G. W. Ehlert. 1971. Effects of temperature, salinity, and surface water flow on distribution of the sea snake *Pelamis*. Limnol. Oceanogr. 16:845–853.

Dunson, W. A., and F. J. Mazzotti. 1988. Some aspects of water and sodium exchange of freshwater crocodilians in fresh water and sea-water: Role of the integument. Comp. Biochem. Physiol. A Physiol. 90:391–396.

Dunson, W. A., R. K. Packer, and M. K. Dunson. 1971. Sea snakes: An unusual salt gland under the tongue. Science 173:437–441.

Dunson, W. A., and A. M. Taub. 1967. Extrarenal salt excretion in sea snakes (*Laticauda*). Am. J. Physiol. 213:975–982.

Dutenhoffer, M. S., and D. L. Swanson. 1996. Relationship of basal to summit metabolic rate in passerine birds and the aerobic capacity model for the evolution of endothermy. Physiol. Zool. 69:1232–1254.

du Toit, J. T. 1990. Home range—body mass relations: A field study on African browsing ruminants. Oecologia 85:301–303.

du Toit, J. T., J. U. M. Jarvis, and G. N. Louw. 1985. Nutrition and burrowing energetics of the Cape mole-rat *Georychus capensis*. Oecologia 66:81–87.

Dyck, A. P., and R. A. MacArthur. 1992. Seasonal patterns of body temperature and activity in free-ranging beaver (*Castor canadensis*). Can. J. Zool. 70:1668–1672.

——. 1993. Daily energy requirements of beaver (*Castor canadensis*) in a simulated winter microhabitat. Can. J. Zool. 71:2131–2135.

Dyktsra, C. R., and W. H. Karasov. 1992. Changes in gut structure and function of house wrens (*Troglodytes aedon*) in response to increased energy demands. Physiol. Zool. 65:422–442.

Earle, M., and D. M. Lavigne. 1990. Intraspecific variation in body size, metabolic rate, and reproduction of deer mice (*Peromyscus maniculatus*). Can. J. Zool. 68:381–388.

Eastman, J. T. 1988. Lipid storage systems and the biology of two neutrally buoyant antarctic notothenioid fishes. Comp. Biochem. Physiol. B Biochem. Mol. Biol. 90:529–537.

Eastman, J. T., and A. L. DeVries. 1981. Buoyancy adaptations in a swim-bladderless antarctic fish. J. Morphol. 167:91–102.

Eastman, J. T., A. L. DeVries, R. E. Coalson, R. E. Nordquist, and R. B. Boyd. 1979. Renal conservation of antifreeze peptide in Antarctic eelpout, *Rhigophila dearborni*. Nature 282:217–218.

Eberhardt, L. S. 1994. Oxygen consumption during singing by male Carolina Wrens (*Thryothorus ludovicianus*). Auk 111:124–130.

Eberly, W. R. 1960. Competition and evolution in cave crayfishes of southern Indiana. Syst. Zool. 9:29–32.

Eckert, S. A., K. L. Eckert, P. Ponganis, and G. L. Kooyman. 1989. Diving and foraging behavior of leatherback sea turtles (*Dermochelys coriacea*). Can. J. Zool. 67:2834–2840.

Economos, A. C. 1979. On structural theories of basal metabolic rate. J. Theor. Biol. 80:445–450.

Edwards, N. A. 1975. Scaling of renal functions in mammals. Comp. Biochem. Physiol. A Physiol. 52:63–66.

Edwards, R. M., and H. Haines. 1978. Effects of ambient water vapor pressure and temperature on evaporative water loss in *Peromyscus maniculatus* and *Mus musculus*. J. Comp. Physiol. 128:177–184.

Edwards, R. R. C., J. H. S. Blaxter, U. K. Gopalan, and C. V. Mathew. 1970. A comparison of standard oxygen consumption of temperate and tropical bottom-dwelling marine fish. Comp. Biochem. Physiol. 34:491–495.

Egginton, S., and B. D. Sidell. 1989. Thermal acclimation induces adaptive changes in subcellular structure of fish skeletal muscle. Am. J. Physiol. 256:R1–R9.

Eldon, G. A. 1979. Breeding, growth and aestivation of the Canterbury mudfish, *Neochanna burrowsius* (Salmoniformes: Galaxiidae). N.Z. J. Mar. Freshw. Res. 13:331–346.

Elgar, M. A., and P. H. Harvey. 1987. Basal metabolic rates in mammals: Allometry, phylogeny and ecology. Funct. Ecol. 1:25–36.

Elick, G. E., and J. A. Sealander. 1972. Comparative water loss in relation to habitat selection in small colubrid snakes. Am. Midl. Nat. 88:429–439.

Ellington, C. P. 1991. Limitations on animal flight performance. J. Exp. Biol. 160:71–91.

Elliott, A. B., and L. Karunakaran. 1974. Diet of *Rana cancrivora* in fresh water and brackish water environments. J. Zool. Lond. 174:203–215.

Elliott, J. M. 1975. Number of meals in a day, maximum weight of food consumed in a day and maximum rate of feeding for brown trout, *Salmo trutta* L. Freshw. Biol. 5:287–303.

——. 1981. Some aspects of thermal stress on freshwater teleosts. *In* A. D. Pickering (ed.), Stress and Fish, pp. 209–245. Academic Press, London.

Ellis, H. I. 1976. Thermoregulation in four species of nesting herons. Ph.D. dissertation, University of Florida, Gainesville.

——. 1980. Metabolism and solar radiation in dark and white herons in hot climates. Physiol. Zool. 53:358–372.

——. 1985. Energetics of free-ranging sea birds. *In* G. C. Whittow and H. Rahn (eds.), Seabird Energetics, pp. 203–234. Plenum, New York.

Ellis, H. I., and J. Frey. 1984. Energetics of thermoregulation in Heermann's gull and a test of a latitudinal gradient of metabolism. Am. Zool. 24:138A.

Ellis, T. M., and M. A. Chappell. 1987. Metabolism, temperature relations, maternal behavior, and reproductive energetics in the ball python (*Python regius*). J. Comp. Physiol. 157:393–402.

Else, P. L., and A. J. Hulbert. 1981. Comparison of the "mammal machine" and the "reptile machine": Energy production. Am. J. Physiol. 240:R3–R9.

——. 1985a. Mammals: An allometric study of metabolism at tissue and mitochondrial level. Am. J. Physiol. 248:R415–R421.

——. 1985b. An allometric comparison of the mitochondria of mammalian and reptilian tissues: The implications for the evolution of endothermy. J. Comp. Physiol. [B] 156:3–11.

——. 1987. Evolution of mammalian endothermic metabolism: "Leaky" membranes as a source of heat. Am. J. Physiol. 253:R1–R7.

Elton, C. 1927. Animal Ecology. Sidgwick and Jackson, London.

Ely, C. A. 1944. Development of *Bufo marinus* larvae in dilute sea water. Copeia 1944:256.

Emery, N., T. L. Poulson, and W. B. Kinter. 1972. Production of concentrated urine by avian kidneys. Am. J. Physiol. 223:180–187.

Emlen, S. T. 1984. Cooperative breeding in birds and mammals. *In* J. R. Krebs and N. B. Davies (eds.), Behavioral Ecology: An Evolutionary Approach, 2nd ed., pp. 305–339. Blackwell, Oxford.

Emmons, L. H. 1984. Geographic variation in densities and diversities of non-flying mammals in Amazonia. Biotropica 16:210–222.

Engelmann, M. D. 1966. Energetics, terrestrial field studies and animal productivity. *In* J. B. Cragg (ed.), Advances in Ecological Research, vol. III, pp. 73–115. Academic Press, New York.

Enig, M. J., J. Ramsey, and D. Eby. 1976. Effect of temperature on pyruvate metabolism in the frog: The role of lactate dehydrogenase isozymes. Comp. Biochem. Physiol. B Physiol. 53:145–148.

Epting, R. J. 1980. Functional dependence of the power for hovering on wing disc loading in hummingbirds. Physiol. Zool. 53:347–357.

Epting, R. J., and T. M. Casey. 1973. Power output and wing disc loading in hovering hummingbirds. Am. Nat. 107:761–765.

Erlinge, S. 1987. Why do European stoats *Mustela erminea* not follow Bergmann's rule? Holarct. Ecol. 10:33–39.

Erskine, D. J., and J. R. Spotila. 1977. Heat-energy-budget analysis and heat transfer in the large mouth blackbass (*Micropterus salmoides*). Physiol. Zool. 50:157–169.

Esch, H. 1960. Ueber die Koerpertemperaturen und den Waermehaushet von *Apis mellifica*. Z. Vgl. Physiol. 43:305–335.

Etheridge, K. 1990a. Water balance in estivating sirenid salamanders (*Siren lacertina*). Herpetologica 46:400–406.

——. 1990b. The energetics of estivating sirenid salamanders (*Siren lacertina* and *Pseudobranchus striatus*). Herpetologica 46:407–414.

Ettinger, A. O., and J. R. King. 1980. Time and energy budgets of the willow flycatcher (*Empidonax trailli*) during the breeding season. Auk 97:533–546.

——. 1981. Consumption of green wheat enhances photostimulated ovarian growth in white-crowned sparrows. Auk 98:832–834.

Evans, D. H. 1969. Studies on the permeability to water of selected marine, freshwater and euryhaline teleosts. J. Exp. Biol. 50:689–703.

——. 1975. Ionic exchange mechanisms in fish gills. Comp. Biochem. Physiol. A Physiol. 51:491–495.

——. 1980. Osmotic and ionic regulation by freshwater and marine fishes. *In* M. A. Ali (ed.), Environmental Physiology of Fishes, pp. 93–122. Plenum, New York.

——. 1984. The role of gill permeability and transport mechanisms in euryhalinity. *In* W. S. Hoar and D. J. Randall (eds.), Fish Physiology, vol. 10, pp. 239–283. Academic Press, London.

——. 1993. Osmotic and ionic regulation. *In* D. H. Evans (ed.), The Physiology of Fishes, pp. 315–341. CRC Press, Boca Raton.

Everson, I., and R. Ralph. 1968. Blood analyses of some antarctic fish. Br. Antarctic Surv. Bull. 15:59–62.

Fair, J. W. 1970. Comparative rates of rehydration from soil in two species of toads, *Bufo boreas* and *Bufo punctatus*. Comp. Biochem. Physiol. 34:281–287.

Fänge, R. 1963. Structure and function of the excretory organs of myxinoids. *In* A. Brodal and R. Fänge (eds.), The Biology of the Myxine, pp. 516–529. Universitetforlaget, Oslo.

——. 1966. Physiology of the swimbladder. Physiol. Rev. 46:299–322.

Fänge, R., and K. Fugelli. 1962. Osmoregulation in chimaeroid fishes. Nature 196:689.

——. 1963. The rectal gland of elasmobranchs, and osmoregulation in chimeroid fishes. Sarsia 10:27–34.

Fänge, R., K. Schmidt-Nielsen, and M. Robinson. 1958. Control of secretion from the avian salt gland. Am. J. Physiol. 195:321–326.

Farlow, J. O. 1990. Dinosaur energetics and thermal biology. *In* D. B. Weishampel, P. Dodson, and H. Osmólska (eds.), The Dinosauria, pp. 43–55. University of California Press, Berkeley.

Farlow, J. O., P. Dodson, and A. Chinsamy. 1995.

Dinosaur biology. Annu. Rev. Ecol. System. 26:445–471.

Farlow, J. O., C. V. Thompson, and D. E. Rosner. 1974. Plates of the dinosaur *Stegosaurus*: Forced convection heat loss fins? Science 192:1123–1125.

Farmer, M., H. J. Fyhn, U. E. H. Fyhn, and R. W. Noble. 1979. Occurrence of Root effect hemoglobins in Amazonian fishes. Comp. Biochem. Physiol. A Physiol. 62:115–124.

Fedak, M. A., and S. S. Anderson. 1982. The energetics of lactation: Accurate measurements from a large wild mammal, the grey seal (*Halichoerus grypus*). J. Zool. Lond. 198:473–479.

Fedak, M. A., B. Pinshow, and K. Schmidt-Nielsen. 1974. Energy cost of bipedal running. Am. J. Physiol. 227:1038–1044.

Fedak, M. A., and H. J. Seeherman. 1979. Reappraisal of energetics of locomotion shows identical cost in bipeds and quadrupeds including ostrich and horse. Nature 282:713–716.

Feder, M. E. 1976a. Lunglessness, body size, and metabolic rate in salamanders. Physiol. Zool. 49:398–406.

——. 1976b. Oxygen consumption and body temperature in neotropical and temperate zone lungless salamanders (Amphibia: Plethodontidae). J. Comp. Physiol. 110:197–208.

——. 1978. Environmental variability and thermal acclimation in neotropical and temperate zone salamanders. Physiol. Zool. 51:7–16.

——. 1983a. Integrating the ecology and physiology of plethodontid salamanders. Herpetologica 39:291–310.

——. 1983b. Metabolic and biochemical correlates of thermal acclimation in the rough-skinned newt *Taricha granulosa*. Physiol. Zool. 56:513–521.

——. 1983c. The relation of air breathing and locomotion to predation on tadpoles, *Rana berlandieri*, by turtles. Physiol. Zool. 56:522–531.

Feder, M. E., A. F. Bennett, W. W. Burggren, and R. B. Huey. 1987. New Directions in Ecological Physiology. Cambridge University Press, New York.

Feder, M. E., A. G. Gibbs, G. A. Griffith, and J. Tsuji. 1984. Thermal acclimation of metabolism in salamanders: Fact or artifact? J. Thermal Biol. 9:255–260.

Feduccia, A. 1973. Dinosaurs as reptiles. Evolution 27:166–169.

——. 1974. Endothermy, dinosaurs and *Archaeopteryx*. Evolution 28:503–504.

——. 1996. The Origin and Evolution of Birds. Yale University Press, New Haven.

Feeny, P. 1975. Biochemical evolution between plants and their insect herbivores. *In* L. E. Gilbert and P. H. Raven (eds.), Coevolution of Animals and Plants, pp. 3–19. University of Texas Press, Austin.

Feinsinger, P., and S. B. Chaplin. 1975. On the relationship between wing disc loading and foraging strategy in hummingbirds. Am. Nat. 109:217–224.

Feinsinger, P., R. K. Colwell, J. Terborgh, and S. B. Chaplin. 1979. Elevation and the morphology, flight energetics, and foraging ecology of tropical hummingbirds. Am. Nat. 113:481–497.

Feist, D. D., and P. R. Morrison. 1981. Seasonal changes in metabolic capacity and norepinephrine thermogenesis in the Alaskan red-backed vole: Environmental cues and annual differences. Comp. Biochem. Physiol. A Physiol. 69:697–700.

Feldman, H. A., and T. A. MacMahon. 1983. The 3/4 mass exponent for energy metabolism is not a statistical artifact. Respir. Physiol. 52:149–163.

Felger, R. S., K. Cliffton, and P. J. Regal. 1976. Winter dormancy in sea turtles: Independent discovery and exploitation in the Gulf of California by two local cultures. Science 192:283–285.

Felsenstein, J. 1985. Phylogenies and the comparative method. Am. Nat. 125:1–15.

Fenchel, T. 1974. Intrinsic rate of natural increase: The relationship with body size. Oecologia 14:317–326.

Ferguson, R. G. 1958. The preferred temperature of fish and their midsummer distribution in temperate lakes and streams. J. Fish. Res. Board Can. 15:607–624.

Finstad, B., K. J. Nilssen, and O. A. Gulseth. 1989. Sea-water tolerance in freshwater-resident arctic char (*Salvelinus alpinus*). Comp. Biochem. Physiol. A Physiol. 92:599–600.

Firman, M. C., R. M. Brigham, and R. M. R. Barclay. 1993. Do free-ranging common nighthawks enter torpor? Condor 95:157–162.

Fish, F. E. 1992. Aquatic locomotion. *In* T. E. Tomasi and T. H. Horton (eds.), Mammalian Energetics, pp. 34–63. Cornell University Press, Ithaca, N.Y.

——. 1994. Influence of hydrodynamic design and propulsive mode on mammalian swimming energetics. Aust. J. Zool. 42:79–101.

Fisher, C. P., E. Lindgren, and W. R. Dawson. 1972. Drinking patterns and behavior of Australian desert birds in relation to their ecology and abundance. Condor 74:111–136.

Fisher, K. C. 1964. On the mechanism of periodic arousal in the hibernating ground squirrel. Ann. Acad. Sci. Fenn. Ser. A 71:143–156.

Fitzhugh, W. W., and H. F. Lamb. 1986. Vegetation history and culture change in Labrador prehistory. Arct. Alp. Res. 17:357–370.

Fitzpatrick, L. C. 1973. Energy allocation in the Allegheny mountain salamander, *Desmognathus ochrophaeus*. Ecol. Monogr. 43:43–58.

Fleming, C. A. 1982. George Edward Lodge: The Unpublished New Zealand Bird Paintings. Nova Pacifica, Wellington.

Fleming, T. H., E. T. Hooper, and D. E. Wilson. 1972. Three Central American bat communities: Structure, reproductive cycles, and movement patterns. Ecology 53:555–569.

Fleming, T. H., R. A. Nuñez, and L. da S. Lobo Sternberg. 1993. Seasonal changes in the diets of migrant and non-migrant nectarivorous bats as revealed by carbon stable isotope analysis. Oecologia 94:72–75.

Fletcher, G. L., M. H. Kao, and R. M. Fourney. 1986. Antifreeze peptides confer freezing resistance to fish. Can. J. Zool. 64:1897–1901.

Fletcher, G. L., M. J. King, and M. H. Kao. 1987. Low temperature regulation of antifreeze glycopeptide levels in Atlantic cod (*Gadus morhua*). Can. J. Zool. 65:227–233.

Flint, E. N., and K. A. Nagy. 1984. Flight energetics of free-living sooty terns. Auk 101:288–294.

Florant, G. L., B. M. Turner, and H. C. Heller. 1978. Temperature regulation during awakefulness, sleep, and hibernation in marmots. Am. J. Physiol. 235: R82–R88.

Fogel, R., and J. M. Trappe. 1978. Fungus consumption (mycophagy) by small mammals. Northwest Sci. 52:1–30.

Foley, W. J. 1987. Digestion and energy metabolism in a small arboreal marsupial, the greater glider (*Petauroides volans*), fed high-terpene *Eucalyptus* foliage. J. Comp. Physiol. [B] 157:355–362.

Foley, W. J., and S. J. Cork. 1992. Use of fibrous diets by small herbivores: How far can the rules be "bent"? Trends Ecol. Evol. 7:159–162.

Foley, W. J., and C. McArthur. 1994. The effects and costs of allelochemicals for mammalian herbivores: An ecological perspective. *In* D. J. Chivers and P. Langer (eds.), The Digestive System in Mammals: Food, Form and Function, pp. 370–391. Cambridge University Press, Cambridge.

Fontaine, M. 1930. Recherche sur le milieu intérieur de la lamproi marine (*Petromyzon marinus*). Ses variations en fonction de celles du milieu estérieur. C. R. Acad. Sci. Paris 191:680–682.

Forchhammer, M. C., E. Post, and N. Chr. Stenseth. 1998. Breeding phenology and climate. Nature 391: 29–30.

Ford, J., and E. H. Sedgewick. 1967. Bird distribution in the Nullarbor plain and Great Victoria Desert region, Western Australia. Emu 67:99–124.

Ford, S. M., and L. C. Davis. 1992. Systematics and body size: Implications for feeding adaptations in New World monkeys. Am. J. Phys. Anthropol. 88:415–468.

Forey, P., and P. Janvier. 1994. Evolution of the early vertebrates. Am. Sci. 82:554–565.

Forman, G. L., C. J. Phillips, and C. S. Rouk. 1977. Alimentary tract. *In* R. J. Baker, J. K. Jones Jr., and D. C. Carter (eds.), Biology of Bats on the New World Family Phyllostomatidae, Part III, pp. 205–227. Special Publ. 16. Museum Texas Technological University, Lubbock.

Forsman, A., and L. E. Lindell. 1991. Trade-off between growth and energy storage in male *Vipera beras* (L.) under different prey densities. Funct. Ecol. 5:717–723.

Forster, M. E. 1990. Confirmation of the low metabolic rate of hagfish. Comp. Biochem. Physiol. A Physiol. 96:113–116.

Forster, R. P. 1942. The nature of the glucose reabsorptive process in the frog renal tubule. Evidence for intermittency of glomerular function in the intact animal. J. Cell. Comp. Physiol. 20:55–69.

Foster, J. B. 1964. Evolution of mammals on islands. Nature 202:234–235.

Foster, M. A., R. C. Foster, and R. K. Meyer. 1939. Hibernation and the endocrines. Endocrinology 24: 603–613.

Foster, M. S. 1978. Total frugivory in tropical passerines: A reappraisal. Trop. Ecol. 19:131–154.

Fox, J. F., and J. P. Bryant. 1984. Instability of the snowshoe hare and wood plant interaction. Oecologia 63:128–135.

Foxon, G. E. H. 1933. Pelvic fins of the *Lepidosiren*. Nature 131:732–733.

——. 1955. Problems of the double circulation in vertebrates. Biol. Rev. Camb. Philos. Soc. 30:196–228.

Fowler, P. A., and P. A. Racey. 1990. Daily and seasonal cycles of body temperatures and aspects of heterothermy in the hedgehog *Erinaceus europaeus*. J. Comp. Physiol. [B] 160:299–307.

Frair, W., R. G. Ackman, and N. Mrosovsky. 1972. Body temperature of *Dermochelys coriacea*: Warm turtle from cold water. Science 177:791–793.

Frazer, J. F. D. 1977. Growth of young vertebrates in the egg or uterus. J. Zool. Lond. 183:189–201.

Frazer, J. F. D., and A. St. G. Huggett. 1974. Species variations in the foetal growth rates of eutherian mammals. J. Zool. Lond. 174:481–509.

Freda, J., and D. G. McDonald. 1988. Physiological correlates of interspecific variation in acid tolerance in fish. J. Exp. Biol. 136:243–258.

Freeland, W. J. 1974. Vole cycles: Another hypothesis. Am. Nat. 108:238–245.

Freeland, W. J., and D. H. Janzen. 1974. Strategies in herbivory by mammals: The role of plant secondary compounds. Am. Nat. 108:269–289.

Fregly, M. J., K. M. Cook, and A. B. Otis. 1963. Effect of hypothyroidism on tolerance of rats to heat. Am. J. Physiol. 204:1039–1044.

French, A. R. 1976. Selection of high temperatures for hibernation by the pocket mouse, *Perognathus longimembris*: Ecological advantages and energetic consequences. Ecology 57:185–191.

——. 1977a. Circannual rhythmicity and entrainment of surface activity in the hibernator, *Perognathus longimembris*. J. Mammal. 58:37–43.

——. 1977b. Periodicity of recurrent hypothermia during hibernation in the pocket mouse, *Perognathus longimembris*. J. Comp. Physiol. 115:87–100.

——. 1982a. Interspecific differences in the pattern of hibernation in the ground squirrel *Spermophilus beldingi*. J. Comp. Physiol. 148:83–91.

——. 1982b. Effects of temperature on the duration of arousal episodes during hibernation. J. Appl. Physiol. 52:216–220.

——. 1985. Allometries of the durations of torpid and euthermic intervals during mammalian hibernation: A test of the theory of metabolic control of the timing of changes in body temperature. J. Comp. Physiol. [B] 156:13–19.

French, N. R., W. E. Grant, W. Grodzinski, and D. M. Swift. 1976. Small mammal energetics in grassland ecosystems. Ecol. Monogr. 46:201–220.

French, N. R., B. G. Maza, and A. P. Aschwanden. 1967. Life spans of *Dipodomys* and *Perognathus* in the Mojave desert. J. Mammal. 48:537–548.

Fried, A. N. 1980. An adaptive advantage of basking behavior in an anuran amphibian. Physiol. Zool. 53: 433–444.

Frith, H. J. 1956. Temperature regulation in the nesting

mounds of the mallee-fowl, *Leipoa ocellata* Gould. Aust. CSIRO 1:79–95.

——. 1959. Ecology of wild ducks in inland Australia. *In* J. A. Keast, R. L. Crocker, and C. S. Christian (eds.), Biogeography and Ecology in Australia, pp. 383–395. Junk, The Hague.

——. 1962. The Mallee-fowl, The Bird That Builds an Incubator. Angus and Robertson, Sydney.

Fromm, P. O. 1963. Studies on renal and extra-renal excretion in a freshwater teleost, *Salmo gairdneri*. Comp. Biochem. Physiol. 10:121–128.

Frumkin, R., B. Pinshow, and Y. Weinstein. 1986. Metabolic heat production and evaporative heat loss in desert Phasianids: Chuckar and sand partridge. Physiol. Zool. 59:592–605.

Fry, C. H., I. J. Ferguson-Lees, and R. J. Dowsett. 1972. Flight muscle hypertrophy and ecophysiological variation of yellow wagtail *Motacilla flava* races at Lake Chad. J. Zool. Lond. 167:293–306.

Fry, F. E. J. 1947. Effects of the environment on animal activity. Univ. Toronto Stud. Biol. Ser. 55:1–62.

Fuentes, E. R., and F. M. Jaksic. 1979. Latitudinal size variation of Chilean foxes: Tests of alternative hypotheses. Ecology 60:43–47.

Fuglei, E., and N. A. Oritsland. 1999. Seasonal trends in body mass, food intake and resting metabolic rate, and induction of metabolic depression in arctic foxes (*Alopex lagopus*) at Svalbard. J. Comp. Physiol. 169:361–369.

Full, R. J. 1986. Locomotion without lungs: Energetics and performance of a lungless salamander. Am. J. Physiol. 251:R775–R780.

Fuller, W. A., L. L. Stebbins, and G. B. Dyke. 1969. Overwintering of small mammals near Great Slave Lake, northern Canada. Arctic 22:35–55.

Fulton, W. 1986. The Tasmanian mudfish *Galaxius cleaveri*. Fishes Sahul 4:150–151.

Gabrielsen, G. W., J. R. E. Taylor, M. Konarzewski, and F. Mehlum. 1991. Field and laboratory metabolism and thermoregulation in Dovekies (*Alle alle*). Auk 108:71–78.

Gadgil, M., and W. H. Bossert. 1970. Life history consequences of natural selection. Am. Nat. 104:1–24.

Gaertner, R. A., J. S. Hart, and O. Z. Roy. 1973. Seasonal spontaneous torpor in the white-footed mouse, *Peromyscus leucopus*. Comp. Biochem. Physiol. A Physiol. 45:169–181.

Gaffney, F. G., and L. C. Fitzpatrick. 1973. Energetics and lipid cycles in the lizard, *Cnemidophorus tigris*. Copeia 1973:446–452.

Gage, S. H., and S. P. Gage. 1886. Aquatic respiration in soft-shelled turtles: A contribution to the physiology of respiration in vertebrates. Am. Nat. 20:233–236.

Gagge, A. P. 1937. A new physiological variable associated with sensible and insensible prespiration. Am. J. Physiol. 120:277–287.

——. 1940. Standard operative temperature: A generalized temperature scale, applicable to direct and partional calorimetry. Am. J. Physiol. 131:93–103.

Gagge, A. P., and J. D. Hardy. 1967. Thermal radiation: Exchange of the human by partitional calorimetry. J. Appl. Physiol. 23:248–258.

Gales, R., and B. Green. 1990. The annual energetics cycle of little penguins (*Eudyptila minor*). Ecology 71:2297–2312.

Gallagher, K. J., D. A. Morrison, R. Shine, and G. C. Grigg. 1983. Validation and use of ^{22}Na turnover to measure food intake in free-ranging lizards. Oecologia 60:76–82.

Gallivan, G. J., and R. C. Best. 1980. Metabolism and respiration of the Amazonian manatee (*Trichechus inunguis*). Physiol. Zool. 53:245–253.

Galloway, T. 1933. The osmotic pressure and saline content of the blood of *Petromyzon fluviatilis*. J. Exp. Biol. 10:313–316.

Galster, W., and P. R. Morrison. 1975. Gluconeogenesis in arctic ground squirrels. Am. J. Physiol. 228:325–330.

Gans, C. 1970a. Strategy and sequence in the evolution of the external gas exchangers of ectothermal vertebrates. Forma Functio 3:61–104.

——. 1970b. Respiration in early tetrapods—the frog is a red herring. Evolution 24:723–734.

——. 1973. Uropeltid snakes—survivors in a changing world. Endeavour 28:146–151.

——. 1975. Tetrapod limblessness: Evolution and functional correlaries. Am. Zool. 15:455–467.

Gans, C., T. Krakauer, and C. V. Paganelli. 1968. Water loss in snakes: Interspecific and intraspecific variability. Comp. Biochem. Physiol. 27:747–761.

Ganzhorn, J. U. 1988. Food partitioning among Malagasy primates. Oecologia 75:436–450.

——. 1989a. Niche separation of seven lemur species in the eastern rainforest of Madagascar. Oecologia 79:279–286.

——. 1989b. Primate species separation in relation to secondary plant chemicals. Hum. Evol. 4:125–132.

Gardner, A. L. 1977. Feeding habits. *In* R. J. Baker, J. Knox Jones Jr., and D. C. Carter (eds.), Biology of Bats of the New World Family Phyllostomatidae, Part II, pp. 293–350. Special Publ. 13. Museum Texas Technological University, Lubbock.

Garey, W. F. 1962. Cardiac responses of fishes in asphyxic environments. Biol. Bull. (Woods Hole) 122:362–368.

——. 1970. Cardiac output of the carp (*Cyprinus capio*). Comp. Biochem. Physiol. 33:181–189.

Garland, T., Jr. 1983. The relation between maximal running speed and body mass in terrestrial mammals. J. Zool. Lond. 199:157–170.

Garland, T., Jr., A. W. Dickerman, C. M. Janis, and J. A. Jones. 1993. Phylogenetic analysis of covariance by computer-simulation. Syst. Biol. 42:265–292.

Garland, T., Jr., P. H. Harvey, and A. R. Ives. 1992. Procedures for the analysis of comparative data using phylogenetically independent contrasts. Syst. Biol. 41:18–32.

Garland, T., Jr., R. B. Huey, and A. F. Bennett. 1991. Phylogeny and coadaptation of thermal physiology in lizards: A reanalysis. Evolution 45:1969–1975.

Garrick, L. D. 1974. Reproductive influences on behavioral thermoregulation in the lizard, *Sceloporus cyanogenys*. Physiol. Behav. 12:85–91.

Gasaway, W. C. 1976a. Seasonal variation in diet, volatile

fatty acid production and size of cecum of rock ptarmigan. Comp. Biochem. Physiol. 53:109–114.

——. 1976b. Volatile fatty acids and metabolizable energy derived from cecal fermentation in the willow ptarmigan. Comp. Biochem. Physiol. 53:115–122.

Gates, D. M. 1962. Energy Exchange in the Biosphere. Harper and Row, New York.

——. 1980. Biophysical Ecology. Springer-Verlag, New York.

Gatten, R. E., Jr. 1974a. Effects of temperature and activity on aerobic and anaerobic metabolism and heart rate in the turtles *Pseudemys scripta* and *Terrapene ornata*. Comp. Biochem. Physiol. A Physiol. 48:619–648.

——. 1974b. Effect of nutritional status on the preferred body temperature of the turtles *Pseudemys scripta* and *Terrapene ornata*. Copeia 1974:912–917.

——. 1980. Aerial and aquatic oxygen uptake by freely-diving snapping turtles (*Chelydra serpentina*). Oecologia 46:266–271.

——. 1987. Activity metabolism of anuran amphibians: Tolerance to dehydration. Physiol. Zool. 60:576–585.

Gatten, R. E., Jr., and R. M. McClung. 1981. Thermal selection by an amphisbaenian. J. Thermal Biol. 6:49–51.

Gatten, R. E., Jr., K. Miller, and R. J. Full. 1992. Energetics at rest and during locomotion. *In* M. E. Feder and W. W. Burggren (eds.), Environmental Physiology of the Amphibians, pp. 314–377. University of Chicago Press, Chicago.

Gaunt, A. S., and C. Gans. 1969. Diving bradycardia and withdrawal bradycardia in *Caiman crocodilus*. Nature 223:207–208.

Gaunt, A. S., R. S. Hikida, J. R. Jehl Jr., and L. Fenbert. 1990. Rapid atrophy and hypertrophy of an avian flight muscle. Auk 107:649–659.

Gause, G. E. 1980. Physiological and morphometric responses of the nine-banded armadillo (*Dasypus novemcinctus*) to environmental factors. Ph.D. dissertation, University of Florida, Gainesville.

Gebczynski, M. 1975. Heat economy and the energy cost of growth in the bank vole during the first month of postnatal life. Acta Theriol. 20:379–434.

Gee, J. H. 1980. Respiratory patterns and antipredator responses in the central mudminnow, *Umbra limi*, a continuous, facultative, air-breathing fish. Can. J. Zool. 58:819–827.

Gee, J. H., and P. A. Gee. 1991. Reaction of gobioid fishes to hypoxia: Buoyancy control and aquatic surface respiration. Copeia 1991:17–28.

Gee, J. H., and J. B. Graham. 1978. Respiratory and hydrostatic functions of the intestine of the catfishes *Hoplosternum thoracatum* and *Brochis splendens* (Callichthyidae). J. Exp. Biol. 74:1–16.

Gee, J. H., and T. G. Northcote. 1963. Comparative ecology of two sympatric species of dace (*Rhinichthys*) in the Fraser River system, British Columbia. J. Fish. Res. Board Can. 20:105–118.

Gehlbach, F. R., R. Gordon, and J. B. Jordan. 1973. Aestivation of the salamander, *Siren intermedia*. Am. Midl. Nat. 89:455–463.

Gehlbach, F. R., J. R. Kimmel, and W. A. Weems. 1969. Aggregations and body relations in tiger salamanders (*Ambystoma tigrinum*) from the Grand Canyon rims, Arizona. Physiol. Zool. 42:173–182.

Geise, W., and K. E. Linsenmair. 1986. Adaptations of the reed frog *Hyperolius Viridiflavus* (Amphibia, Anura, Hyperoliidae) to its arid environment. II. Some aspects of the water economy of *Hyperolius viridiflavus nitidulus* under wet and dry season conditions. Oecologia 68:542–548.

——. 1988. Adaptations of the reed frog *Hyperolius viridiflavus* (Amphibia, Anura, Hyperoliidae) to its arid environment. IV. Ecological significance of water economy with comments on thermoregulation and energy allocation. Oecologia 77:327–338.

Geiser, F. 1986. Thermoregulation and torpor in the Kultarr, *Antechinomys laniger* (Marsupialia: Dasyuridae). J. Comp. Physiol. [B] 156:751–757.

——. 1988. Reduction of metabolism during hibernation and daily torpor in mammals and birds: Temperature effect or physiological inhibition? J. Comp. Physiol. [B] 158:25–37.

——. 1991. The effect of unsaturated and saturated dietary lipids on the pattern of daily torpor and the fatty acid composition of tissues and membranes of the deer mouse *Peromyscus maniculatus*. J. Comp. Physiol. [B] 161:590–597.

——. 1994. Hibernation and torpor in marsupials: A review. Aust. J. Zool. 42:1–16.

Geiser, F., and R. V. Baudinette. 1990. The relationship between body mass and rate of rewarming from hibernation and daily torpor in mammals. J. Exp. Biol. 151:349–359.

Geiser, F., and L. S. Broome. 1993. The effect of temperature on the pattern of torpor in a marsupial hibernator. J. Comp. Physiol. [B] 163:133–137.

Geiser, F., S. Hiebert, and G. J. Kenagy. 1990. Torpor bout duration during the hibernation season of two sciurid rodents: Interrelations with temperature and metabolism. Physiol. Zool. 63:489–503.

Geiser, F., and G. J. Kenagy. 1987. Polyunsaturated lipid diet lengthens torpor and reduces body temperature in a hibernator. Am. J. Physiol. 252:R897–R901.

——. 1988. Torpor duration in relation to temperature and metabolism in hibernating ground squirrels. Physiol. Zool. 61:442–449.

Geiser, F., and P. Masters. 1994. Torpor in relation to reproduction in the Mulgara, *Dasycercus cristicauda* (Dasyuridae: Marsupialia). J. Thermal Biol. 19:33–40.

Geiser, F., and R. S. Seymour. 1989. Torpor in a pregnant echidna, *Tachyglossus aculeatus* (Monotremata: Tachyglossidae). Aust. Mammal. 12:81–82.

Geist, V. 1987. Bergmann's rule is invalid. Can. J. Zool. 65:1035–1038.

——. 1990. Bergmann's rule is invalid: A reply to J. D. Paterson. Can. J. Zool. 68:1613–1615.

Geluso, K. N. 1975. Urine concentration cycles of insectivorous bats in the laboratory. J. Comp. Physiol. 99:309–319.

——. 1978. Urine concentrating ability and renal structure of insectivorous bats. J. Mammal. 59:312–323.

Genoud, M. 1985. Ecological energetics of two European

shrews: *Crocidura russula* and *Sorex coronatus* (Soricidae: Mammalia). J. Zool. Lond. 207:63–85.

———. 1988. Energetic strategies of shrews: Ecological constraints and evolutionary implications. Mammal Rev. 18:173–193.

———. 1993. Temperature regulation in subtropical tree bats. Comp. Biochem. Physiol. A Physiol. 104:321–331.

Genoud, M., and F. J. Bonaccorso. 1986. Temperature regulation, rate of metabolism, and roost temperature in the greater white-lined bat, *Saccopteryx bilineata* (Emballonuridae). Physiol. Zool. 59:49–64.

Genoud, M., F. J. Bonaccorso, and A. Arends. 1990. Rate of metabolism and temperature regulation in two small tropical insectivorous bats (*Peropteryx macrotis* and *Natalus tumidirostris*). Comp. Biochem. Physiol. A Physiol. 97:229–324.

Genoud, M., and P. Vogel. 1990. Energy requirements during reproduction and reproductive effort in shrews (Soricidae). J. Zool. Lond. 220:41–60.

Gentry, R. L., and J. R. Holt. 1986. Attendance behavior of northern fur seals. *In* R. L. Gentry and G. L. Kooyman (eds.), Fur Seals: Maternal Strategies at Land and Sea, pp. 41–60. Princeton University Press, Princeton.

Gessaman, J. A. 1972. Bioenergetics of the snowy owl. Arct. Alp. Res. 4:223–238.

Gettinger, R. D. 1975. Metabolism and thermoregulation of a fossorial rodent, the northern pocket gopher (*Thomomys talpoides*). Physiol. Zool. 48:311–322.

Gibbons, A. 1996. New feathered fossil brings dinosaurs and birds closer together. Science 274:720–721.

Gibson, A. R., and J. B. Falls. 1979. Thermal biology of the common garter snake *Thamnophis sirtalis* (L.). I. Temporal variation, environmental effects and sex differences. Oecologia 43:79–97.

Gifford, C. E., and E. P. Odum. 1965. Bioenergetics of lipid deposition in the bobolink, a trans-equatorial migrant. Condor 67:383–403.

Girgis, S. 1961. Aquatic respiration in the common Nile turtle, *Trionyx triunguis* (Forskal). Comp. Biochem. Physiol. 3:206–217.

Gittleman, J. L., and P. H. Harvey. 1982. Carnivore home range, metabolic needs and ecology. Behav. Ecol. Sociobiol. 10:57–63.

Gittleman, J. L., and S. D. Thompson. 1988. Energy allocation in mammalian reproduction. Am. Zool. 28:863–875.

Glazier, D. S. 1985. Relationship between metabolic rate and energy expenditure for lactation in *Peromyscus*. Comp. Biochem. Physiol. A Physiol. 80:587–590.

Gleeson, T. T., G. S. Mitchell, and A. F. Bennett. 1980. Cardiovascular responses to graded activity in the lizards *Varanus* and *Iguana*. Am. J. Physiol. 239:R174–R179.

Glenn, M. E. 1970. Water relations in three species of deer mice (*Peromyscus*). Comp. Biochem. Physiol. 33:231–248.

Goff, G. P., and G. B. Stenson. 1988. Brown adipose tissue in leatherback sea turtles: A thermogenic organ in an endothermic reptile? Copeia 1988:1071–1075.

Goldstein, D. L. 1983. Effect of wind on avian metabolic rate with particular reference to Gambel's quail. Physiol. Zool. 56:485–492.

———. 1988. Estimates of daily energy expenditure in birds: The time-energy budget as an integrator of laboratory and field studies. Am. Zool. 28:829–844.

Goldstein, D. L., and E. J. Braun. 1988. Contributions of the kidneys and intestines to water conservation, and plasma levels of antidiuretic hormone, during dehydration in house sparrows (*Passer domesticus*). J. Comp. Physiol. 158:353–361.

Goldstein, D. L., and K. A. Nagy. 1985. Resource utilization by desert quail: Time and energy, food and water. Ecology 66:378–387.

Goldstein, D. L., J. B. Williams, and E. J. Braun. 1990. Osmoregulation in the field by salt-marsh savannah sparrows *Passerculus sandwichensis beldingi*. Physiol. Zool. 63:669–682.

Goldstein, L., and R. P. Forster. 1971. The role of uricolysis in the production of urea by fishes and other aquatic vertebrates. Comp. Biochem. Physiol. 14:567–576.

Golightly, R. J., Jr., and R. D. Ohmart. 1983. Metabolism and body temperature of two desert canids, coyote and kit-fox. J. Mammal. 64:624–635.

Gompper, M. E., and J. L. Gittleman. 1991. Home range scaling: Interspecific and comparative trends. Oecologia 87:343–348.

Gonzalez, R. J., and D. G. McDonald. 1994. The relationship between oxygen uptake and ion loss in fish from diverse habitats. J. Exp. Biol. 190:95–108.

Goodfriend, W., D. Ward, and A. Subach. 1991. Standard operative temperatures of two desert rodents, *Gerbillus allenbyi* and *Gerbillus pyramidum*: The effects of morphology, microhabitat and environmental factors. J. Thermal Biol. 16:157–166.

Goodman, D. E. 1971. Thermoregulation in the brown water snake, *Natrix taxispilota*, with discussion of the ecological significance of thermal preferenda in the order Squamata. Ph.D. dissertation, University of Florida, Gainesville.

Goolish, E. M. 1991. Anaerobic swimming metabolism of fish: Sit-and-wait versus active forager. Physiol. Zool. 64:485–501.

Gordon, M. S. 1959. Osmotic and ionic regulation in Scottish brown trout and sea trout (*Salmo trutta* L.). J. Exp. Biol. 36:253–260.

———. 1962. Osmotic regulation in the green toad. J. Exp. Biol. 39:261–270.

Gordon, M. S., B. W. Belman, and P. H. Chow. 1976. Comparative studies on the metabolism of shallow water and deep-sea marine fishes. IV. Patterns of aerobic metabolism in the mesopelagic deep-sea fangtooth fish *Anoplogaster cornuta*. Mar. Biol. 35:287–293.

Gordon, M. S., I. Boetius, D. H. Evans, R. McCarthy, and L. C. Oglesby. 1969. Aspects of the physiology of terrestrial life in amphibious fishes. I. The mudskipper, *Periophthalmus sobrinus*. J. Exp. Biol. 50:141–149.

Gordon, M. S., S. Fischer, and E. T. Tarifeno. 1970. Aspects of the physiology of terrestrial life in amphibious fishes. II. The Chilean clingfish, *Sisyases sanguineus*. J. Exp. Biol. 53:559–572.

Gordon, M. S., W.-S. Ng, and A. W.-S. Yip. 1978. Aspects of the physiology of terrestrial life in amphibious fishes. III. The Chinese mudskipper *Periophthalmus cantonensis*. J. Exp. Biol. 72:57–75.

Gordon, M. S., K. Schmidt-Nielsen, and H. M. Kelly. 1961. Osmotic regulation in the crab-eating frog (*Rana cancrivora*). J. Exp. Biol. 38:659–678.

Gordon, M. S., and V. A. Tucker. 1965. Osmotic regulation in the tadpoles of the crab-eating frog (*Rana cancrivora*). J. Exp. Biol. 42:437–445.

———. 1968. Further observations on the physiology of salinity adaptation in the crab-eating frog (*Rana cancrivora*). J. Exp. Biol. 49:185–193.

Górecki, A. 1977. Energy flow through the common hamster population. Acta Theriol. 22:25–66.

Gottschalk, C. W., W. E. Lassiter, M. Mylle, K. J. Ullrich, B. Schmidt-Nielsen, R. O'Dell, and G. Pehling. 1963. Micropuncture study of composition of loop of Henle fluid in desert rodents. Am. J. Physiol. 204:532–535.

Gottschalk, C. W., and M. Mylle. 1959. Micropuncture study of the mammalian urinary concentrating mechanism: Evidence for the countercurrent hypothesis. Am. J. Physiol. 196:927–936.

Goudie, R. I., and C. D. Ankney. 1986. Body size, activity budgets, and diets of sea ducks wintering in Newfoundland. Ecology 67:1475–1482.

Gould, L. L., and F. Heppner. 1974. The vee formation of Canada geese. Ank 91:494–506.

Gould, S. J., and R. C. Lewontin. 1979. The spandrels of San Marco and the Panglossian paradigm: A critique of the adaptationist programme. Proc. R. Soc. Lond. B 205:581–598.

Gould, S. J., and E. S. Vrba. 1982. Exaptation—a missing term in the science of form. Paleobiology 8:4–15.

Goulding, M. 1980. The Fishes and The Forest. University of California Press, Berkeley.

Goyal, S. P., P. K. Ghosh, and I. Prakash. 1981. Significance of body fat in relation to basal metabolic rate in some Indian desert rodents. J. Arid Environ. 4:59–62.

Graham, J. B. 1974. Aquatic respiration in the sea snake *Pelamis platurus*. Respir. Physiol. 21:1–7.

———. 1976. Respiratory adaptations of marine air-breathing fishes. *In* G. M. Hughes (ed.), Respiration of Amphibious Vertebrates, pp. 165–187. Academic Press, New York.

———. 1997. Air-Breathing Fishes. Academic Press, San Diego.

Graham, J. B., R. H. Rosenblatt, and C. Gans. 1978. Vertebrate air breathing arose in fresh waters and not in the oceans. Evolution 32:459–463.

Graham, J. B., I. Rubinoff, and H. K. Hecht. 1971. Temperature physiology of the sea snake *Pelamis platurus*: An index of its colonization potential in the Atlantic Ocean. Proc. Natl. Acad. Sci. U.S.A. 68:1360–1363.

Grajal, A. 1995. Digestive efficiency of the hoatzin, *Opisthocomus hoazin*: A folivorous bird with foregut fermentation. Ibis 137:383–388.

Grajal, A., S. D. Strahl, R. Parra, M. G. Domingues, and A. Neher. 1989. Foregut fermentation in the hoatzin, a neotropical leaf-eating bird. Science 243:1236–1238.

Grammeltvedt, R. 1978. Atrophy of a breast muscle in a single-fibre type (*M. pectoralis*) in fasting willow grouse, *Lagopus lagopus* (L.). J. Exp. Zool. 205:195–204.

Grammeltvedt, R., and J. B. Steen. 1978. Fat deposition in Spitzbergen ptarmigan (*Lagopus mutus hyperboreas*). Arctic 31:496–498.

Grand, T. I. 1977. Body weight: Its relation to tissue composition, segment distribution, and motor function. I. Interspecific comparison. Am. J. Physiol. Anthropol. 47:211–240.

———. 1978. Adaptations of tissue and limb segments to facilitate moving and feeding in arboreal folivores. *In* G. G. Montgomery (ed.), The Ecology of Arboreal Folivores, pp. 231–241. Smithsonian Institution Press, Washington, D.C.

———. 1983. Body weight: Its relationship to tissue composition, segmental distribution of mass, and motor function. III. The Didelphidae of French Guyana. Aust. J. Zool. 31:299–312.

Grant, G. S. 1984. Energy cost of incubation to the parent seabird. *In* G. C. Whittow and H. Rahn (eds.), Seabird Energetics, pp. 59–71. Plenum, New York.

Graves, J. E., and G. N. Somero. 1982. Electrophoretic and functional enzymatic evolution in four species of eastern Pacific barracudas from different thermal environments. Evolution 36:97–106.

Gray, I. E. 1954. Comparative study of the gill area of marine fishes. Biol. Bull. (Woods Hole) 107:219–225.

Greegor, D. H. 1974. Comparative ecology and distribution of two species of armadillos, *Chaetophractus vellerosus* and *Dasypus novemcinctus*. Ph.D. dissertation, University of Arizona, Tucson.

———. 1975. Renal capacities of an Argentine desert armadillo. J. Mammal. 56:626–632.

Green, B., D. King, M. Braysher, and A. Saim. 1991. Thermoregulation, water turnover and energetics of free-living Komodo dragons, *Varanus komodoensis*. Comp. Biochem. Physiol. A Physiol. 99:97–101.

Green, D. A., and J. S. Millar. 1987. Changes in gut dimensions and capacity of *Peromyscus maniculatus* relative to diet quality and energy needs. Can. J. Zool. 65:2159–2162.

Greene, C. W. 1926. The physiology of the spawning migration. Physiol. Rev. 6:201–241.

Greenewalt, C. H. 1975. The flight of birds. Trans. Am. Philos. Soc. 65:5–67.

———. 1976. Could pterosaurs fly? Science 188:676.

Greenwald, L. 1971. Sodium balance in the leopard frog (*Rana pipiens*). Physiol. Zool. 44:149–161.

———. 1972. Sodium balance in amphibians from different habitats. Physiol. Zool. 45:229–237.

Greenwald, O. E. 1974. Thermal dependence of striking and prey capture by gopher snakes. Copeia 1974:141–148.

Greenwood, P. H. 1986. The natural history of African lungfishes. J. Morphol. Suppl. 1:163–179.

Greer, A. E., J. D. Lazelle, and R. M. Wright. 1973. Anatomical evidence for a countercurrent heat exchanger in the leatherback turtle (*Dermochelys coriaceae*). Nature 244:181.

Gregory, P. T. 1977. Life-history parameters of the red-sided garter snake (*Thamnophis sirtalis parietalis*) in an extreme environment, the Interlake region of Manitoba. Nat. Mus. Nat. Sci. (Ottawa) Publ. Zool. 13:1–44.

——. 1982. Reptilian hibernation. *In* C. Gans and F. H. Pough (eds.), Biology of the Reptilia, vol. 13, pp. 53–154. Academic Press, New York.

Griffith, R. W. 1981. Composition of the blood serum of deep-sea fishes. Biol. Bull. 160:250–264.

——. 1985. Habitat, phylogeny and the evolution of osmoregulatory strategies in primitive fishes. *In* R. E. Foreman, A. Gorbman, J. M. Dodd, and R. Olsson (eds.), Evolutionary Biology of Primitive Fishes, pp. 69–80. Plenum, New York.

——. 1987. Freshwater or marine origin of the vertebrates? Comp. Biochem. Physiol. A Physiol. 87:523–531.

——. 1994. The life of the first vertebrates. Bioscience 44:408–417.

Griffith, R. W., and P. K. T. Pang. 1979. Mechanisms of osmoregulation in the coelacanth: Evolutionary implications. Occ. Pap. Calif. Acad. Sci. 134:79–93.

Griffith, R. W., P. K. T. Pang, A. K. Srivastava, and G. E. Pickford. 1973. Serum composition of freshwater stingrays (Potamotrygonidae) adapted to fresh and dilute sea water. Biol. Bull. (Woods Hole) 144:304–320.

Griffith, R. W., B. L. Umminger, B. F. Grant, P. K. T. Pang, and G. E. Pickford. 1974. Serum composition of the coelacanth, *Latimeria chalumnae* Smith. J. Exp. Zool. 187:87–102.

Grigg, G. C. 1965. Studies on the Queensland lungfish, *Neoceratodus forsteri* (Krefft). III. Aerial respiration in relation to habits. Aust. J. Zool. 13:413–421.

——. 1967. Some respiratory properties of the blood of four species of Antarctic fishes. Comp. Biochem. Physiol. 23:139–148.

——. 1969. Temperature-induced changes in the oxygen equilibrium curve of the blood of the brown bullhead *Ictalurus nebulosus*. Comp. Biochem. Physiol. 28:1203–1223.

——. 1977. Physiological adaptations of *Crocodylus porosus*. *In* H. Messel and H. T. Butler (eds.), Australian Animals and Their Environment, pp. 335–354. Shakespeare Head Press, Sydney.

——. 1981. Plasma homeostasis and cloacal urine composition in *Crocodylus porosus* caught along a saline gradient. J. Comp. Physiol. 144:261–270.

Grigg, G. C., and J. Alchin. 1976. The role of cardiovascular system in thermoregulation of *Crocodylus johnstoni*. Physiol. Zool. 49:24–36.

Grigg, G. C., L. A. Beard, T. Moutton, M. T. Queirol Melo, and L. E. Taplin. 1998. Osmoregulation by the broad-snouted caiman, *Caiman latirostris*, in estuarine habitat in southern Brazil. J. Comp. Physiol. [B] 168:445–452.

Grismer, L. L. 1988. Phylogeny, taxonomy, classification, and biogeography of eublepharid geckos. *In* R. Estes and G. Pregill (eds.), Phylogenetic Relationships of the Lizard Families, pp. 369–469. Stanford University Press, Stanford.

Grodzinski, W. 1971. Energy flow through populations of small mammals in the Alaskan taiga forest. Acta Theriol. 16:231–275.

Grodzinski, W., M. Makomaska, R. Tertil, and J. Weiner. 1977. Bioenergetics and total impact of vole populations. Oikos 29:494–510.

Grodzinski, W., and B. A. Wunder. 1975. Ecological energetics of small mammals. *In* F. B. Golley, K. Petrusewicz, and L. Ryszkowski (eds.), Small Mammals: Their Productivity and Population Dynamics, pp. 173–204. Cambridge University Press, Cambridge.

Groscolas, R. 1988. The use of body mass loss to estimate metabolic rate in fasting sea birds: A critical examination based on emperor penguins (*Aptenodytes forsteri*). Comp. Biochem. Physiol. A Physiol. 90:361–366.

Gross, J. E., Z. Wang, and B. A. Wunder. 1985. Effects of food quality and energy needs: Changes in gut morphology and capacity of *Microtus ochrogaster*. J. Mammal. 66:661–667.

Gross, M. R. 1987. Evolution of diadromy in fishes. Symp. Am. Fish. Soc. 1:14–25.

Grossman, A. F., and G. C. West. 1977. Metabolic rate and temperature regulation of winter acclimatized black-capped chickadees *Parus atricapillus* of interior Alaska. Ornis Scand. 8:127–138.

Guderley, H., and P. Blier. 1988. Thermal acclimation in fish: Conservative and labile properties of swimming muscle. Can. J. Zool. 66:1105–1115.

Guggisberg, C. A. W. 1972. Crocodiles: Their Natural History, Folklore and Conservation. David and Charles, Newton Abbott, Devon, England.

Guiliano, W. M., R. Patiño, and R. S. Lutz. 1998. Comparative reproductive and physiological responses of northern bobwhite and scaled quail to water deprivation. Comp. Biochem. Physiol. A Physiol. 119:781–786.

Guillette, L. J. 1982. Effects of gravity on the metabolism of the reproductively bimodal lizard, *Sceloporus aeneus*. J. Exp. Zool. 223:33–36.

Guimond, R. W., and V. H. Hutchison. 1972. Pulmonary branchial and cutaneous gas exchange in the mud puppy, *Necturus maculosus maculosus* (Rafinesque). Comp. Biochem. Physiol. A Physiol. 42:367–392.

——. 1973a. Trimodal gas exchange in the large aquatic salamander, *Siren lacertina* (Linnaeus). Comp. Biochem. Physiol. A Physiol. 46:249–268.

——. 1973b. Aquatic respiration: An unusual strategy in the hellbender *Cryptobranchus alleganiensis alleganiensis* (Daudin). Science 182:1263–1265.

——. 1974. Aerial and aquatic respiration in the congo eel *Amphiuma means means* (Garden). Respir. Physiol. 20:147–159.

Gut, H. J. 1959. A Pleistocene vampire bat from Florida. J. Mammal. 40:534–538.

Gwinner, E. 1971. A comparative study of circannual rhythms in warblers. *In* M. Menaker (ed.), Biochronometry, pp. 405–427. U.S. National Academy of Sciences, Washington, D.C.

——. 1986. Circannual Rhythms. Springer-Verlag, Berlin.

———. 1996. Circadian and circannual programmes in avian migration. J. Exp. Biol. 199:39–48.

Haartman, L. von. 1954. Der Trauerfliegenschnaepper—III. Die Nahrungsbiologie. Acta Zool. Fenn. 83:1–96.

Hadley, N. F., G. A. Ahearn, and F. G. Howarth. 1981. Water and metabolic relations of cave-adapted and epigean lycosid spiders in Hawaii. J. Arachnol. 9:215–222.

Haftorn, S. 1972. Hypothermia of tits in winter. Ornis Scand. 3:153–166.

Hagan, J. M., P. C. Smithson, and P. D. Doerr. 1983. Behavioral response of the American alligator to freezing weather. J. Herpetol. 17:402–404.

Hagerman, A. E., and C. T. Robbins. 1993. Specificity of tannin-binding salivary proteins relative to diet selection by mammals. Can. J. Zool. 71:628–633.

Hailey, A., and P. M. C. Davies. 1986. Life style, latitude and activity metabolism of natricine snakes. J. Zool. Lond. [A] 209:461–476.

Hailey, A., C. Gaitanaki, and N. S. Loumbourdis. 1987. Metabolic recovery from exhaustive activity by a small lizard. Comp. Biochem. Physiol. A Physiol. 88:683–689.

Hails, C. J. 1979. A comparison of flight energetics in hirundines and other birds. Comp. Biochem. Physiol. A Physiol. 63:581–585.

———. 1983. The metabolic rate of tropical birds. Condor 85:61–65.

Hails, C. J., and D. M. Bryant. 1979. Reproductive energetics of a free-living bird. J. Anim. Ecol. 48:471–482.

Haim, A., N. Fairall, and P. W. Prinsloo. 1985c. The ecological significance of calcium bicarbonate in the urine of subterranean rodents: Testing a hypothesis. Comp. Biochem. Physiol. A Physiol. 82:867–869.

Haim, A., G. Heth, and E. Nevo. 1985a. Urine analysis of three rodent species with emphasis on calcium and magnesium bicarbonate. Comp. Biochem. Physiol. A Physiol. 80:503–506.

Haim, A., G. Heth, E. Nevo, N. Gruener, and T. Goldstein. 1985b. Urine analysis of three rodent species with emphasis on calcium and magnesium bicarbonate. Comp. Biochem. Physiol. A Physiol. 80:503–506.

Haim, A., and I. Izhaki. 1993. The ecological significance of resting metabolic rate and non-shivering thermogenesis for rodents. J. Thermal Biol. 18:71–81.

Haim, A., E. van der Straeten, and W. M. Cooreman. 1987. Urine analysis of European moles *Talpa europaea* and white rats *Rattus norvegicus* kept on a carnivore's diet. Comp. Biochem. Physiol. A Physiol. 88:179–181.

Hainsworth, F. R. 1987. Precision and dynamics of positioning by Canada geese in formation. J. Exp. Biol. 128:445–462.

———. 1988. Induced drag savings from ground effect and formation flight in brown pelicans. J. Exp. Biol. 135:431–444.

Hainsworth, F. R., B. G. Collins, and L. L. Wolf. 1977. The function of torpor in hummingbirds. Physiol. Zool. 50:215–222.

Hainsworth, F. R., and L. L. Wolf. 1970. Regulation of oxygen consumption and body temperature during torpor in a hummingbird, *Eulampis jugularis*. Science 168:368–369.

———. 1972. Crop volume, nectar concentration and hummingbird energetics. Comp. Biochem. Physiol. A Physiol. 42:359–366.

———. 1978. Economics of temperature regulation and torpor in non-mammalian organisms. *In* L. Wang and J. W. Hudson (eds.), Strategies in Cold: Natural Torpidity and Thermogenesis, pp. 147–184. Academic Press, New York.

Hall, F. G. 1924. The functions of the swimbladder of fishes. Biol. Bull. (Woods Hole) 47:79–117.

———. 1937. Adaptations of mammals to high altitudes. J. Mammal. 18:468–472.

———. 1965. Hemoglobin and oxygen: Affinities in seven species of Sciuridae. Science 148:1350–1351.

———. 1966. Minimal utilizable oxygen and the oxygen dissociation curve of blood of rodents. J. Appl. Physiol. 21:375–378.

Hall, F. G., D. B. Hill, and E. S. Guzman Barron. 1936. Comparative physiology in high altitudes. J. Cell. Comp. Physiol. 8:301–313.

Halstead, L. B. 1985. The vertebrate invasion of fresh water. Philos. Trans. R. Soc. Lond. B Biol. Sci. 309:243–258.

Hamilton, T. H. 1961. The adaptive significance of intraspecific trends of variation in wing length and body size among bird species. Evolution 15:180–195.

Hamilton, W. J., III. 1962. Reproductive adaptations of the red tree mouse. J. Mammal. 43:486–504.

———. 1973. Life's Color Code. McGraw-Hill, New York.

Hamilton, W. J., III, and F. Heppner. 1967. Radiant solar energy and the function of black homeotherm pigmentation: An hypothesis. Science 155:196–197.

Hammel, H. T. 1955. Thermal properties of fur. Am. J. Physiol. 182:369–376.

———. 1956. Infrared emissivities of some Arctic fauna. J. Mammal. 37:375–378.

———. 1976a. Colligative properties of a solution. Science 192:748–756.

———. 1976b. On the origin of endothermy in mammals. Isr. J. Med. Sci. 12:905–915.

Hammel, H. T., L. I. Crawshaw, and H. P. Cabanac. 1973. The activation of behavioral responses in the regulation of body temperature in vertebrates. *In* P. Lomax and E. Schoenbaum (eds.), The Pharmacology of Thermoregulation, pp. 124–141. Karger, Basal.

Hammel, H. T., T. J. Dawson, R. M. Abrams, and H. T. Andersen. 1968. Total calorimetric measurements on *Citellus lateralis* in hibernation. Physiol. Zool. 41:341–357.

Hammond, K. A., and J. Diamond. 1992. An experimental test for a ceiling on sustained metabolic rate in lactating mice. Physiol. Zool. 65:952–977.

———. 1997. Maximal sustained energy budgets in humans and animals. Nature 386:457–462.

Hammond, K. A., and B. A. Wunder. 1991. The role of diet quality and energy need in the nutritional ecology of a small herbivore, *Microtus ochrogaster*. Physiol. Zool. 64:541–567.

Hamner, W. M. 1971. On seeking an alternative to the

endogenous reproductive rhythm hypothesis in birds. *In* M. Menaker (ed.), Biochronometry, pp. 448–462. National Academy of Sciences, Washington, D.C.

Hamner, W. M., and J. Stocking. 1970. Why don't bobolinks breed in Brazil? Ecology 51:743–751.

Hand, S. C., and G. N. Somero. 1982. Urea and methylamine effects on rabbit muscle phosphofructokinase. J. Biol. Chem. 257:734–741.

Hanegan, J. L., and J. E. Heath. 1970a. Activity patterns and energetics of the moth, *Hyalophora cecropia*. J. Exp. Biol. 53:611–627.

——. 1970b. Mechanisms for the control of body temperature in the moth, *Hyalophora cecropia*. J. Exp. Biol. 53:349–362.

Haney, D. C., and F. G. Nordlie. 1997. Influence of environmental salinity on routine metabolic rate and critical oxygen tension of *Cyprinodon variegatus*. Physiol. Zool. 70:511–518.

Hansen, S., and D. M. Lavigne. 1997. Temperature effects on the breeding distribution of grey seals (*Halichoerus grypus*). Physiol. Zool. 70:436–443.

Hansen, S., D. M. Lavigne, and S. Innes. 1995. Energy metabolism and thermoregulation in juvenile harbor seals (*Phoca vitulina*) in air. Physiol. Zool. 68:290–315.

Hanski, I. 1985. What does a shrew do in an energy crisis? *In* R. M. Sibly and R. H. Smith (eds.), Behavioural Ecology: Ecological Consequences of Adaptive Behaviour, pp. 247–252. Blackwell Science, Oxford.

Hansson, L. 1985. Geographic differences in bank voles *Clethrionomys glareolus* in relation to ecogeographical rules and possible demographic and nutritive strategies. Ann. Zool. Fenn. 22:319–328.

Hansson, L., and H. Henttonen. 1985. Gradients in density variations of small rodents: The importance of latitude and snow cover. Oecologia 67:394–402.

Happold, D. C. D., and M. Happold. 1988. Renal form and function in relation to the ecology of bats (Chiroptera) from Malawi, Central Africa. J. Zool. Lond. 215:629–655.

Hardy, A. 1960. Was man more aquatic in the past? New Sci. 7:642–654.

Harestad, A. S., and F. L. Bunnell. 1979. Home range and body weight—a reevaluation. Ecology 60:389–402.

Hargitay, B., and W. Kuhn. 1951. Das Multiplikationsprinzip als Grundlage der Harnkonzentrierung in der Niere. Z. Electrochem. 55:539–558.

Harkness, D. R. 1972. The regulation of hemoglobin oxygenation. *In* G. H. Stollerman (ed.), Advances in Internal Medicine, vol. 17, pp. 189–214. Year Book Medical, Chicago.

Harlow, H. J. 1981a. Torpor and other physiological adaptations of the badger (*Taxidea taxus*) to cold environments. Physiol. Zool. 54:267–275.

——. 1981b. Metabolic adaptations to prolonged food deprivation by the American badger *Taxidea taxus*. Physiol. Zool. 54:276–284.

Harlow, P., and G. C. Grigg. 1984. Shivering thermogenesis in a brooding diamond python. Copeia 1984:959–965.

Harris, M. P. 1977. Comparative ecology of seabirds in the Galápagos archipelago. *In* B. Stonehouse and C. M. Perrins (eds.), Evolutionary Ecology, pp. 65–76. MacMillan, London.

Hart, J. S. 1956. Seasonal changes in insulation of the fur. Can. J. Zool. 34:53–57.

——. 1957. Climatic and temperature induced changes in the energetics of homeotherms. Rev. Can. Biol. 16:133–174.

——. 1962. Seasonal acclimation in four species of small wild birds. Physiol. Zool. 35:224–236.

——. 1971. Rodents. *In* G. C. Whittow (ed.), Comparative Physiology of Thermoregulation, vol. II, pp. 1–149. Academic Press, New York.

Hart, J. S., and O. Heroux. 1953. A comparison of some seasonal and temperature-induced changes in *Peromyscus*: cold resistance, metabolism, and pelage insulation. Can. J. Zool. 31:528–534.

——. 1955. Exercise and temperature regulation in lemmings and rabbits. Can. J. Biochem. Physiol. 33:428–435.

Hart, J. S., H. Pohl, and J. S. Tener. 1965. Seasonal acclimatization in varying hare (*Lepus americanus*). Can. J. Zool. 43:731–744.

Hartman, F. A. 1961. Locomotor mechanisms of birds. Smithson. Misc. Collect. 143:1–91.

——. 1963. Some flight mechanisms of bats. Ohio J. Sci. 63:59–65.

Harvey, E. N. 1928. The oxygen consumption of luminous bacteria. J. Gen. Physiol. 11:469–475.

Harvey, P. H., and P. M. Bennett. 1983. Brain size, energetics, ecology, and life history patterns. Nature 306:314–315.

Harvey, P. H., and T. H. Clutton-Brock. 1981. Primate home-range size and metabolic needs. Behav. Ecol. Sociobiol. 8:151–155.

Harvey, P. H., and M. D. Pagel. 1991. The Comparative Method in Evolutionary Biology. Oxford University Press, Oxford.

Harvey, P. H., M. D. Pagel, and J. A. Rees. 1991. Mammalian metabolism and life histories. Am. Nat. 137:556–566.

Harwood, R. H. 1979. The effect of temperature on the digestive efficiency of three species of lizards, *Cnemidophorus tigris*, *Gerrhonotus multicarinatus*, and *Sceloporus occidentalis*. Comp. Biochem. Physiol. A Physiol. 63:417–433.

Hau, M., M. Wikelski, and J. C. Wingfield. 1998. A neotropical forest bird can measure the slight changes in tropical photoperiod. Proc. R. Soc. Lond. B 265:89–95.

Haukioja, E. 1980. On the role of plant defences in the fluctuation of herbivore populations. Oikos 35:202–213.

Haukioja, E., K. Kapiainen, P. Niemelä, and J. Tuomi. 1983. Plant availability hypothesis and other explanations of herbivore cycles: Complementary or exclusive alternatives? Oikos 40:419–432.

Hay, L., and W. Van Hoven. 1988. Tannins and digestibility in the steenbok (*Raphicerus campestris*). Comp. Biochem. Physiol. A Physiol. 91:509–511.

Hayden, P., and R. G. Lindberg. 1976. Survival of laboratory-reared pocket mice, *Perognathus longimembris*. J. Mammal. 57:266–272.

Hayes, J. P. 1989a. Field and maximal metabolic rates of deer mice (*Peromyscus maniculatus*) at low and high altitudes. Physiol. Zool. 62:732–744.

——. 1989b. Altitudinal and seasonal effects on aerobic metabolism of deer mice. J. Comp. Physiol. [B] 159: 453–459.

Hayes, J. P., T. Garland, and M. R. Dohm. 1992. Individual variation in metabolism and reproduction of *Mus*: Are energetics and life history linked? Funct. Ecol. 6:5–14.

Hayssen, V. 1984. Basal metabolic rate and the intrinsic rate of increase: An empirical and theoretical reexamination. Oecologia 64:419–421.

Hayssen, V., and R. C. Lacy. 1985. Basal metabolic rates in mammals: Taxonomic differences in the allometry of BMR and body mass. Comp. Biochem. Physiol. A Physiol. 81:741–754.

Hayward, J. S., and E. G. Ball. 1966. Quantitative aspects of brown adipose tissue thermogenesis during arousal from hibernation. Biol. Bull. (Woods Hole) 131:94–103.

Hayward, J. S., and P. A. Lisson. 1992. Evolution of brown fat: Its absence in marsupials and monotremes. Can. J. Zool. 70:171–179.

Hayward, T. L. 1997. Pacific Ocean climate change: Atmospheric forcing, ocean circulation and ecosystem response. Trends Ecol. Erol. 12:150–154.

Hayworth, A. M., and W. W. Weathers. 1984. Temperature regulation and climatic adaptation in black-billed and yellow-billed magpies. Condor 86:19–26.

Hazlehurst, G. A., and J. M. V. Rayner. 1992. Flight characteristics of Triassic and Jurassic Pterosauria: An appraisal based on wing shape. Paleobiology 18:447–463.

Heath, J. E. 1964a. Reptilian thermoregulation: Evaluation of field studies. Science 146:784–785.

——. 1964b. Head-body temperature differences in horned lizards. Physiol. Zool. 37:273–279.

——. 1965. Temperature regulation and diurnal activity in horned lizards. Univ. Calif. Publ. Zool. 64:97–136.

——. 1968. The origins of thermoregulation. *In* E. T. Drake (ed.), Evolution and Environment, pp. 259–278. Yale University Press, New Haven.

Heath, J. E., and P. A. Adams. 1967. Regulation of heat production by large moths. J. Exp. Biol. 47:21–33.

Heatwole, H. 1977a. Heart rate during breathing and apnea in marine snakes (Reptilia, Serpentes). J. Herpetol. 11:67–76.

——. 1977b. Voluntary submergence time and breathing rhythm in the homalopsine snake, *Cerberus rhynchops*. Aust. Zool. 19:155–166.

Heatwole, H., and R. S. Seymour. 1975. Pulmonary and cutaneous oxygen uptake in sea snakes and a file snake. Comp. Biochem. Physiol. A Physiol. 51:399–405.

——. 1978. Cutaneous oxygen uptake in three groups of aquatic snakes. Aust. J. Zool. 26:481–486.

Heatwole, H., F. Torres, S. Blasini de Austin, and A. Heatwole. 1969. Studies on anuran water balance—I. Dynamics of evaporative water loss by the coqui, *Eleutherodactylus portoricensis*. Comp. Biochem. Physiol. 28:245–269.

Hecht, M. K., C. Kropach, and B. M. Hecht. 1974. Distribution of the yellow-bellied sea snake, *Pelamis platurus*, and its significance in relation to the fossil record. Herpetologica 30:387–396.

Hedenström, A. 1993. Migration by soaring or flapping flight in birds: The relative importance of energy cost and speed. Philos. Trans. R. Soc. Lond. B Biol. Sci. 342:353–361.

Hedenström, A., and J. Pettersson. 1987. Migration routes and wintering areas of willow warblers *Phylloscopus trochilus* (L.) ringed in Fennoscandia. Ornis Fenn. 64:137–143.

Heglund, N. C., C. R. Taylor, and T. A. McMahon. 1974. Scaling stride frequency and gait to animal size: Mice to horses. Science 186:1112–1113.

Heinemann, D. 1992. Resource use, energetic profitability, and behavioral decisions in migrant rufous hummingbirds. Oecologia 90:137–149.

Heinrich, B. 1972a. Energetics of temperature regulation and foraging in a bumblebee, *Bombus terricola* Kirby. J. Comp. Physiol. 77:49–64.

——. 1972b. Patterns of endothermy in bumblebee queens, drones and workers. J. Comp. Physiol. 77:65–79.

——. 1974. Thermoregulation in endothermic insects. Science 185:747–756.

——. 1977. Why have some animals evolved to regulate a high body temperature? Am. Nat. 111:623–640.

——. 1981a. The mechanisms and energetics of honeybee swarm temperature regulation. J. Exp. Biol. 91:25–55.

——. 1981b. Ecological and evolutionary perspectives. *In* B. Heinrich (ed.), Insect Thermoregulation, pp. 236–302. Wiley, New York.

——. 1984. Strategies of thermoregulation and foraging in two vespid wasps, *Dolichovespula maculata* and *Vespula vulgaris*. J. Comp. Physiol. 154:175–180.

——. 1987. Thermoregulation by winter-flying endothermic moths. J. Exp. Biol. 127:313–332.

——. 1994. Untitled book review. Am. Sci. 82:486–487.

Heinrich, B., and G. A. Bartholomew. 1971. An analysis of pre-flight warm-up in the sphinx moth, *Manduca sexta*. J. Exp. Biol. 55:223–239.

Heinrich, B., and T. M. Casey. 1973. Metabolic rate and endothermy in sphinx moths. J. Comp. Physiol. 82: 195–206.

——. 1978. Heat transfer in dragonflies: "Fliers" and "perchers." J. Exp. Biol. 74:17–36.

Heisinger, J. F., and R. P. Breitenbach. 1969. Renal structural characteristics as indexes of renal adaptation for water conservation in the genus *Sylvilagus*. Physiol. Zool. 42:160–172.

Heisinger, J. F., T. S. King, H. W. Halling, and B. L. Fields. 1973. Renal adaptations to macro- and micro-habitats in the family Cricetidae. Comp. Biochem. Physiol. A Physiol. 44:767–774.

Heldmaier, G., and T. Ruf. 1992. Body temperature and metabolic rate during natural hypothermia in endotherms. J. Comp. Physiol. [B] 162:696–706.

Heldmaier, G., and S. Steinlechner. 1981a. Seasonal pattern and energetics of short daily torpor in the

Djungarian hamster, *Phodopus sungorus*. Oecologia 48:265–270.

———. 1981b. Seasonal control of energy requirements for thermoregulation in the Djungarian hamster (*Phodopus sungorus*), living in natural photoperiod. J. Comp. Physiol. [B] 142:429–437.

Heller, H. C. 1972. Measurements of convective and radiative heat transfer in small mammals. J. Mammal. 53:289–295.

Heller, H. C., and T. L. Poulson. 1970. Circannian rhythms—II. Endogenous and exogenous factors controlling reproduction and hibernation in chipmunks (*Eutamias*) and ground squirrels (*Spermophilus*). Comp. Biochem. Physiol. 33:357–383.

———. 1972. Altitudinal zonation of chipmunks (*Eutamias*): Adaptations to aridity and high temperatures. Am. Midl. Nat. 87:296–313.

Heller, H. C., J. M. Walker, G. L. Florant, S. F. Glotebach, and R. J. Berger. 1978. Sleep and hibernation: Electrophysiological and thermoregulatory homologies. *In* L. C.-H. Wang and J. W. Hudson (eds.), Strategies in Cold: Natural Torpidity and Thermogenesis, pp. 225–265. Academic Press, New York.

Helms, C. W. 1963. Tentative field estimates of metabolism in buntings. Auk 80:318–334.

———. 1968. Food, fat, and feathers. Am. Zool. 8:151–167.

Hemmingsen, A. M. 1960. Energy metabolism as related to body size and respiratory surfaces, and its evolution. Rep. Steno Mem. Hosp. Nord. Insulin Lab. 9:1–110.

Hemmingsen, E. A., and E. L. Douglas. 1970. Respiratory characteristics of the hemoglobin-free fish *Chaenocephalus aceratus*. Comp. Biochem. Physiol. 33:733–744.

———. 1972. Respiratory and circulatory responses in a hemoglobin-free fish, *Chaenocephalus aceratus*, to changes in temperature and oxygen tension. Comp. Biochem. Physiol. A Physiol. 43:1031–1043.

Hemmingsen, E. A., E. L. Douglas, and G. C. Grigg. 1969. Oxygen consumption in an Antarctic hemoglobin-free fish, *Pagetopsis macropterus*, and in three species of *Notothenia*. Comp. Biochem. Physiol. 29:467–470.

Hemmingsen, E. A., E. L. Douglas, K. Johansen, and R. W. Millard. 1972. Aortic blood flow and cardiac output in the hemoglobin-free fish *Chaenocephalus aceratus*. Comp. Biochem. Physiol. A Physiol. 43:1045–1051.

Hemmingsen, E. A., E. L. Douglas, and J. B. Stewart. 1973. Cardiovascular and respiratory studies on hemoglobin-free ice fishes. Antarct. J. U.S. 8:203–204.

Hennemann, W. W. 1982. Energetics and spread-winged behavior of anhingas in Florida. Condor 84:91–96.

———. 1983a. Relation among body mass, metabolic rate and the intrinsic rate of natural increase in mammals. Oecologia 56:104–108.

———. 1983b. Environmental influences on the energetics and behavior of anhingas and double-crested cormorants. Physiol. Zool. 56:201–216.

———. 1984a. Spread-winged behaviour of double-crested and flightless cormorants *Phalacocorax auritus* and

P. harrisi: Wing drying or thermoregulation? Ibis 126: 230–239.

———. 1984b. Intrinsic rates of natural increase of altricial and precocial eutherian mammals: The potential price of precociality. Oikos 93:363–368.

———. 1988. Energetics and spread-winged behavior in anhingas and double-crested cormorants: The risks of generalization. Am. Zool. 28:845–851.

Henshaw, R. E. 1968. Thermoregulation during hibernation: Application of Newton's law of cooling. J. Theor. Biol. 20:79–90.

Heppner, F. 1970. The metabolic significance of differential absorption of radiant energy by black and white birds. Condor 72:50–59.

Herbers, J. M. 1981. Time resources and laziness in animals. Oecologia 49:252–262.

Herbst, L. H. 1986. The role of nitrogen from fruit pulp in the nutrition of the frugivorous bat *Carollia perspicillata*. Biotropica 18:39–44.

Herd, R. M., and T. J. Dawson. 1984. Fiber digestion in the emu, *Dromaius novaehollandiae*, a large bird with a simple gut and high rates of passage. Physiol. Zool. 57:70–84.

Herreid, C. F., II. 1963. Temperature regulation and metabolism in Mexican freetail bats. Science 142: 1573–1574.

———. 1964. Bat longevity and metabolic rate. Exp. Gerontol. 1:1–9.

Herreid, C. F., II, and B. Kessel. 1967. Thermal conductance in birds and mammals. Comp. Biochem. Physiol. 21:405–414.

Herrera, C. M. 1982. Defense of ripe fruit from pests: Its significance in relation to plant-disperser interactions. Am. Nat. 120:218–241.

Hertz, P. E., and R. B. Huey. 1981. Compensation for altitudinal changes in the thermal environment by some *Anolis* lizards on Hispaniola. Ecology 62:515–521.

Hertz, P. E., R. B. Huey, and E. Nevo. 1983. Homage to Santa Anita: Thermal sensitivity of sprint speed in agamid lizards. Evolution 37:1075–1084.

Hesse, R., W. C. Allee, and K. P. Schmidt. 1951. Ecological Animal Geography. Wiley, New York.

Hessler, R. R., and P. A. Jumars. 1974. Abyssal community analysis from replicate box cores in the central North Pacific. Deep-Sea Res. 21:185–209.

Heusner, A. A. 1982a. Energy metabolism and body size. I. Is the 0.75 mass exponent of Kleiber's equation a statistical artifact? Respir. Physiol. 48:1–12.

———. 1982b. Energy metabolism and body size. II. Dimensional analysis and energetic non-similarity. Respir. Physiol. 48:13–25.

———. 1991. Size and power in mammals. J. Exp. Biol. 160:25–54.

Hew, C. L., D. Slaughter, G. L. Fletcher, and S. B. Joshi. 1981. Antifreeze glycoproteins in the plasma of Newfoundland Atlantic cod (*Gadus morhua*). Can. J. Zool. 59:2186–2192.

Hickman, C. P., Jr. 1968. Ingestion, intestinal absorption, and elimination of seawater and salts in the southern flounder, *Paralichthys lethostigma*. Can. J. Zool. 46: 457–466.

Hickman, C. P., Jr., and B. F. Trump. 1969. The kidney. *In* W. S. Hoar and D. J. Randall (eds.), Fish Physiology, pp. 91–239. Academic Press, New York.

Hiebert, S. M. 1990. Energy costs and temporal organization of torpor in the rufous hummingbird (*Selasphorus rufus*). Physiol. Zool. 63:1082–1097.

——. 1992. Time-dependent thresholds for torpor initiation in the rufous hummingbird (*Selasphorus rufus*). J. Comp. Physiol. [B] 162:249–255.

Hildebrand, M. 1965. Symmetrical gaits of horses. Science 150:701–708.

Hill, L., and W. H. Dawbin. 1969. Nitrogen excretion in the tuatara. *Sphenodon punctatus*. Comp. Biochem. Physiol. 31:453–468.

Hill, R. D., R. C. Schneider, G. C. Liggins, A. H. Schuette, R. L. Elliott, M. Guppy, P. W. Hochachka, J. Qvist, K. J. Falke, and W. M. Zapol. 1987. Heart rate and body temperature during free diving of Weddell seals. Am. J. Physiol. 253:R344–R351.

Hill, R. W. 1975. Daily torpor in *Peromyscus leucopus* on an adequate diet. Comp. Biochem. Physiol. A Physiol. 51:413–423.

Hillenius, W. J. 1992. The evolution of nasal turbinates and mammalian endothermy. Paleobiology 18:17–29.

——. 1994. Turbinates in therapsids: Evidence for late Permian origins of mammalian endothermy. Evolution 48:207–229.

Hillman, P. E. 1969. Habitat specificity in three sympatric species of *Ameiva* (Reptilia: Teiidae). Ecology 50:476–481.

Hillman, S. S. 1976. Cardiovascular correlates of maximal oxygen consumption rates in anuran amphibians. J. Comp. Physiol. 109:199–207.

Hillman, S. S., G. C. Gorman, and R. Thomas. 1979a. Water loss in *Anolis* lizards: Evidence for acclimation and intraspecific differences along a habitat gradient. Comp. Biochem. Physiol. A Physiol. 62:491–494.

Hillman, S. S., V. H. Shoemaker, R. Pitman, and P. C. Withers. 1979b. Reassessment of aerobic metabolism in amphibians during activity. J. Comp. Physiol. 129:309–313.

Hillman, S. S., and P. C. Withers. 1979. An analysis of respiratory surface area as a limit to activity metabolism in anurans. Can. J. Physiol. 57:2100–2105.

Hillyard, S. D. 1981. Energy metabolism and osmoregulation in desert fishes. *In* R. J. Naiman and D. L. Soltz (eds.), Fishes in North American Deserts, pp. 385–409. Wiley, New York.

Himms-Hagen, J. 1978. Biochemical aspects of nonshivering thermogenesis. *In* L. C. H. Wang and J. W. Hudson (eds.), Strategies in Cold: Natural Torpidity and Thermogenesis, pp. 595–617. Academic Press, New York.

Hinds, D. S. 1973. Acclimatization of thermoregulation in the desert cottontail, *Sylvilagus audubonii*. J. Mammal. 54:708–728.

——. 1977. Acclimatization of thermoregulation in desert-inhabiting jackrabbits (*Lepus alleni* and *Lepus californicus*). Ecology 58:246–264.

Hinds, D. S., and R. E. MacMillen. 1985. Scaling of energy metabolism and evaporative water loss in heteromyid rodents. Physiol. Zool. 58:282–298.

Hinsley, S. A., P. N. Ferns, D. H. Thomas, and B. Pinshow. 1993. Black-bellied Sandgrouse (*Pterocles orientalis*) and pin-tailed sandgrouse (*Pterocles alchata*): Closely related species with differing bioenergetic adaptations to arid zones. Physiol. Zool. 66:20–42.

Hirano, T., D. W. Johnson, H. A. Bern, and S. Utida. 1973. Studies on water and ion movements in the isolated urinary bladder of selected freshwater, marine and euryhaline teleosts. Comp. Biochem. Physiol. A Physiol. 45:529–540.

Hirshfield, M. F., C. R. Feldmeth, and D. L. Soltz. 1980. Genetic differences in physiological tolerances of Amargosa pupfish (*Cyprinodon nevadensis*) populations. Science 207:999–1001.

Hissa, R., and R. Palokangas. 1970. Thermoregulation in the titmouse. Comp. Biochem. Physiol. 33:941–953.

Hissa, R., H. Rintamäki, P. Virtanen, H. Lindén, and V. Vihko. 1990. Energy reserves of the capercaillie *Tetrao urogallus* in Finland. Comp. Biochem. Physiol. A Physiol. 97:345–351.

Hixon, M. A., and F. L. Carpenter. 1988. Distinguishing energy maximizers from time minimizers: A comparative study of two hummingbird species. Am. Zool. 28:913–925.

Hladik, C. M. 1967. Surface relative du tractus digestif de quelques Primates. Morphologie des villosites intestinales et correlations avec le regime alimentaire. Mammalia 31:120–147.

Hobson, P. N. 1988. Rumen Microbial Ecosystem. Elsevier Science, New York.

Hochachka, P. W. 1988. Channels and pump-determinants of metabolic cold adaptation strategies. Comp. Biochem. Physiol. B Biochem. Mol. Biol. 90:515–519.

——. 1990. Scope for survival: A conceptual "mirror" to Fry's scope for activity. Trans. Am. Fish. Soc. 119:622–628.

Hochachka, P. W., and G. N. Somero. 1984. Biochemical Adaptation. Princeton University Press, Princeton.

Hock, R. J. 1951. The metabolic rates and body temperatures of bats. Biol. Bull. (Woods Hole) 101:289–299.

——. 1960. Seasonal variations in physiologic functions of Arctic ground squirrels and black bears. Bull. Mus. Comp. Zool. 124:155–171.

——. 1964. Physiological responses of deer mice to various native altitudes. *In* W. H. Weihe (ed.), The Physiological Responses to High Altitude, pp. 59–72. Macmillan, New York.

Hodgman, C. D., R. C. Weast, C. W. Wallace, and S. M. Selby (eds.). 1954. Handbook of Chemistry and Physics. Chemical Rubber Publishing, Cleveland, Ohio.

Hoeck, H. N. 1975. Differential feeding behaviour of the sympatric hyrax *Procavia johnstoni* and *Heterohyrax brucei*. Oecologia 22:15–47.

Hoffmann, C. K. 1890. Reptilien. *In* Dr. H. G. Bronn's Klassen und Ordnungen des Tier-Reichs, vol. 6, pp. 1401–2089. Leipzig.

Hofman, M. A. 1983. Energy metabolism, brain size and longevity in mammals. Q. Rev. Biol. 58:495–512.

——. 1993. Encephalization and the evolution of longevity in mammals. J. Evol. Biol. 6:209–227.

Hofmann, R. R. 1989. Evolutionary steps of ecophysiological adaptation and diversification of ruminants: A comparative view of their digestive system. Oecologia 78:443–457.

Hogstad, O. 1987. It is expensive to be dominant. Auk 104:333–336.

Hokkanen, J. E. I. 1990. Temperature regulation of marine mammals. J. Theor. Biol. 145:465–485.

Holeton, G. F. 1970. Oxygen uptake and circulation by a hemoglobinless Antarctic fish (Chaenocephalus aceratus Lonnberg) compared with three red-blooded Antarctic fish. Comp. Biochem. Physiol. 34:457–471.

——. 1974. Metabolic cold adaptation of polar fish: Fact or artifact? Physiol. Zool. 47:137–152.

Holeton, G. F., and D. J. Randall. 1967. The effect of hypoxia upon the partial pressure of gases in the blood and water afferent and efferent to the gills of rainbow trout. J. Exp. Biol. 46:317–327.

Holeton, G. F., and E. D. Stevens. 1978. Swimming energetics of an Amazonian characin in "black" and "white" water. Can. J. Zool. 56:983–987.

Holland, K. N., R. W. Brill, R. K. C. Chang, J. R. Silbert, and D. A. Fournier. 1992. Physiological and behavioural thermoregulation in bigeye tuna (Thunnus obesus). Nature 358:410–412.

Holmes, D. J., and S. N. Austad. 1994. Fly now, die later: Life-history correlates of gliding and flying in mammals. J. Mammal. 75:224–226.

Holmes, E. B. 1985. Are lungfishes the sister group of tetrapods? Biol. J. Linn. Soc. 25:379–397.

Holmes, R. T., and F. W. Sturgis. 1973. Annual energy expenditure by the avifauna of a northern hardwoods ecosystem. Oikos 24:24–29.

Holt, J. P., and E. A. Rhode. 1976. Similarity of renal glomerular hemodynamics in mammals. Am. Heart J. 92:465–472.

Hoppe, P. P. 1977. Comparison of voluntary food and water consumption and digestion in Kirk's dikdik and suni. East Afr. Wildl. J. 15:41–48.

Hopson, J. A. 1969. The origin and adaptive radiation of mammal-like reptiles and nontherian mammals. Ann. N.Y. Acad. Sci. 167:199–216.

——. 1973. Endothermy, small size, and the origin of mammalian reproduction. Am. Nat. 107:446–452.

Hopson, J. A., and H. R. Barghusen. 1986. An analysis of therapsid relationships. In N. Hotton III, P. D. MacLean, J. J. Roth, and E. C. Roth (eds.), The Ecology and Biology of Mammal-like Reptiles, pp. 83–106. Smithsonian Institution Press, Washington, D.C.

Hopson, J. A., and A. W. Crompton. 1969. Origin of mammals. In T. Dobzhansky, M. K. Hecht, and W. C. Steere (eds.), Evolutionary Biology, vol. 3, pp. 15–72. Appleton-Century-Crofts, New York.

Horn, M. H., and K. C. Riegle. 1981. Evaporative water loss and intertidal vertical distribution in relation to body size and morphology of stichaeoid fishes from California. J. Exp. Mar. Biol. Ecol. 50:273–288.

Horner, J. R. 1982. Evidence of colonial nesting and "site fidelity" among ornithischian dinosaurs. Nature 297:675–676.

Horvath, O. 1964. Seasonal differences in rufous hummingbird nest height and their relation to nest climate. Ecology 45:235–241.

Houck, M. A., J. A. Gauthier, and R. E. Strauss. 1990. Allometric scaling in the earliest fossil bird, Archeopteryx lithographica. Science 247:195–198.

Houston, A. H., and J. A. Madden. 1968. Environmental temperature and plasma electrolyte regulation in the carp Cyprinus carpio. Nature 217:969–970.

Howe, S., D. L. Kilgore Jr., and C. Colby. 1987. Respiratory gas concentrations and temperatures within nest cavities of the northern flicker (Colaptes auratus). Can. J. Zool. 65:1541–1547.

Howell, A. B. 1926. Voles of the genus Phenacomys. II. Life history of the red tree mouse Phenacomys longicaudus. North Am. Fauna 48:39–64.

Howell, B. J., F. W. Baumgardner, K. Bondi, and H. Rahn. 1970. Acid-base balance in cold-blooded vertebrates as a function of body temperature. Am. J. Physiol. 218:600–606.

Howell, D. J. 1974. Bats and pollen: Physiological aspects of the syndrome of chiropterophily. Comp. Biochem. Physiol. A Physiol. 48:263–276.

Hoyt, D. F., and C. R. Taylor. 1981. Gait and energetics of locomotion in horses. Nature 292:239–240.

Hoy-Thomas, J. A., and R. S. Miles. 1971. Palaeozoic Fishes. Saunders, Philadelphia.

Huber, G. C. 1917. On the morphology of the renal tubules of vertebrates. Anat. Rec. 13:305–339.

Hudson, J. W. 1962. The role of water in the biology of the antelope ground squirrel (Citellus leucurus). Univ. Calif. Publ. Zool. 64:1–56.

——. 1965. Temperature regulation and torpidity in the pygmy mouse, Baiomys taylori. Physiol. Zool. 38:243–254.

——. 1974. The estrous cycle, reproduction, growth, and development of temperature regulation in the pigmy mouse, Baiomys taylori. J. Mammal. 55:572–588.

Hudson, J. W., W. R. Dawson, and R. W. Hill. 1974. Growth and development of temperature regulation in nestling cattle egrets. Comp. Biochem. Physiol. A Physiol. 49:717–741.

Hudson, J. W., and S. L. Kinzey. 1966. Temperature regulation and metabolic rhythms in populations of the house sparrow, Passer domesticus. Comp. Biochem. Physiol. 17:203–217.

Hudson, J. W., and L. C.-H. Wang. 1968. Thyroid function in desert ground squirrels. In C. C. Hoff and M. L. Riedesel (eds.), Physiological Systems in Semiarid Environments, pp. 17–33. University of New Mexico Press, Albuquerque.

Huey, R. B. 1974. Behavioral thermoregulation in lizards: Importance of associated costs. Science 184:1001–1003.

Huey, R. B., and A. F. Bennett. 1987. Phylogenetic studies of coadaptations: Preferred temperatures versus optimal performance temperatures of lizards. Evolution 41:1098–1115.

Huey, R. B., and J. G. Kingsolver. 1989. Evolution of

thermal sensitivity of ectotherm performance. Tree 4:131–135.

Huey, R. B., and E. R. Pianka. 1981. Ecological consequences of foraging mode. Ecology 62:991–999.

Huey, R. B., and M. Slatkin. 1976. Cost and benefits of lizard thermoregulation. Q. Rev. Biol. 51:363–384.

Huey, R. B., and T. P. Webster. 1976. Thermal biology of *Anolis* lizards in a complex fauna: The *cristatellus* group on Puerto Rico. Ecology 57:985–994.

Huggins, A. K., G. Skutch, and E. Baldwin. 1969. Ornithine-urea cycle enzymes in teleostean fish. Comp. Biochem. Physiol. 28:587–602.

Hughes, G. M. 1966. The dimensions of fish gills in relation to their function. J. Exp. Biol. 45:177–195.

——. 1972a. Distribution of oxygen tension in the blood and water along the secondary lamella of the icefish gill. J. Exp. Biol. 56:481–492.

——. 1972b. Gills of a living coelacanth, *Latimeria chalumnae*. Experientia 28:1301–1302.

——. 1973. Respiratory responses to hypoxia in fish. Am. Zool. 13:475–489.

——. 1976. On the respiration of *Latimeria chalumnae*. Zool. J. Linn. Soc. 59:195–208.

——. 1977. Dimensions and the respiration of lower vertebrates. *In* T. J. Pedley (ed.), Scaling Effects in Animal Locomotion, pp. 57–81. Academic Press, London.

Hughes, G. M., S. C. Dube, and J. S. Datta Munshi. 1974. Surface area of the respiratory organs of the climbing perch, *Anabas testudineus* (Pisces: Anabatidae). J. Zool. Lond. 170:227–243.

Hughes, G. M., and M. Morgan. 1973. The structure of fish gills in relation to their respiratory function. Biol. Rev. Camb. Philos. Soc. 48:419–475.

Hughes, M. R. 1968. Renal and extrarenal sodium excretion in the common tern *Sterna hirundo*. Physiol. Zool. 41:210–219.

——. 1970a. Cloacal and salt-gland ion excretion in the seagull, *Larus glaucescens*, acclimated to increasing concentrations of sea water. Comp. Biochem. Physiol. 32:315–325.

——. 1970b. Relative kidney size in nonpasserine birds with functional salt glands. Condor 72:164–168.

——. 1975. Salt gland secretion produced by the gull, *Larus glaucescens*, in response to stomach loads of different sodium and potassium concentrations. Comp. Biochem. Physiol. A Physiol. 51:909–913.

Hulbert, A. J. 1980. Evolution from ectothermia towards endothermia. *In* Z. Szeleny and M. Szekely (eds.), Satellite 28th International Congress of Physiological Sciences, pp. 237–247. Kiado, Budapest.

Hulbert, A. J., and T. J. Dawson. 1974. Thermoregulation in perameloid marsupials from different environments. Comp. Biochem. Physiol. A Physiol. 47:591–616.

Hulbert, A. J., and P. L. Else. 1981. Comparison of the "mammal machine" and the "reptile machine": Energy use and thyroid activity. Am. J. Physiol. 241:R350–R356.

Hulet, W. H., S. J. Masel, L. H. Jodrey, and R. G. Wehr. 1967. The role of calcium in the survival of marine teleosts in dilute sea water. Bull. Mar. Sci. 17:677–688.

Hume, I. D. 1977. Production of volatile fatty acids in two species of wallaby and in sheep. Comp. Biochem. Physiol. A Physiol. 56:299–304.

——. 1989. Reading the entrails of evolution. New Sci. 15:43–47.

Hume, I. D., W. J. Foley, and M. J. Chilcott. 1984. Physiological mechanisms of foliage digestion in the greater glider and ringtail possum (Marsupialia: Pseudocheiridae). *In* A. P. Smith and I. D. Hume (eds.), Possums and Gliders, pp. 247–251. Australian Mammal Society, Sydney.

Hummel, D., and M. Beukenberg. 1989. Aerodynamische Interferenzeffekte beim formationsflug von Vögeln. J. Ornithol. 130:15–24.

Humphrey, S. R. 1974. Zoogeography of the nine-banded armadillo (*Dasypus novemcinctus*) in the United States. Bioscience 24:457–462.

Humphreys, W. F. 1979. Production and respiration in animal populations. J. Anim. Ecol. 48:427–453.

Hüppop, K. 1986. Oxygen consumption of *Astyanax fasciatus* (Characidae: Pisces): A comparison of epigean and hypogean populations. Environ. Biol. Fish. 17:299–308.

Hureau, J.-C. 1966. Biologie de *Chaenichthys rhinoceratus* Richardson, et probleme du sang incolore de Chaenichthyidae, poissons de mers australes. Bull. Soc. Zool. Fr. 91:735–751.

Hutchinson, J. C. D. 1955. Evaporative cooling in fowls. J. Agric. Sci. 45:48–59.

Hutchinson, J. C. D., and G. D. Brown. 1969. Penetrance of cattle coats by radiation. J. Appl. Physiol. 26:454–464.

Hutchison, V. H. 1961. Critical thermal maxima in salamanders. Physiol. Zool. 34:92–125.

Hutchison, V. H., H. G. Dowling, and A. Vinegar. 1966. Thermoregulation in a brooding female Indian python, *Python molurus bivittatus*. Science 151:694–696.

Hutchison, V. H., and R. K. Dupré. 1992. Thermoregulation. *In* M. E. Feder and W. W. Burggren (eds.), Environmental Physiology of the Amphibians, pp. 206–249. University of Chicago Press, Chicago.

Hutchison, V. H., H. B. Haines, and G. Engbretson. 1976. Aquatic life at high altitude: Respiratory adaptations in the Lake Titicaca frog, *Telmatobius culeus*. Respir. Physiol. 27:115–129.

Hutchison, V. H., and J. Larimer. 1960. Reflectivity of the integuments of some lizards from different habitats. Ecology 41:199–204.

Hutchison, V. H., W. G. Whitford, and M. Kohl. 1968. Relation of body size and surface area to gas exchange in anurans. Physiol. Zool. 41:65–85.

Huxley, J. 1932. On Relative Growth. Methuen, London.

Hyatt, K. D. 1979. Feeding strategies. *In* W. S. Hoar, D. J. Randall, and J. R. Brett (eds.), Fish Physiology, vol. 8, pp. 71–119. Academic Press, New York.

Hyvärinen, H. 1984. Winter strategy of voles and shrews in Finland. *In* J. F. Merritt (ed.), Winter Ecology of Small Mammals, pp. 139–148. Special Publ. Carnegie Mus. Nat. Hist. 10:1–380.

Idler, D. R., and I. Bitners. 1959. Biochemical studies on sockeye salmon during spawning migration. V.

Cholesterol, fat, protein and water in the body of the standard fish. J. Fish. Res. Board Can. 16:235–241.

Immelmann, K. 1963. Drought adaptations in Australian desert birds. Proc. XIII Int. Ornithol. Congr. 2:649–657.

——. 1971. Ecological aspects of periodic reproduction. In D. S. Farner and J. R. King (eds.), Avian Biology, vol. I, pp. 341–389. Academic Press, New York.

Ireland, M. P., and I. M. Simons. 1977. Adaptation of the axolotl (Ambystoma mexicanum) to a hyperosmotic medium. Comp. Biochem. Physiol. A Physiol. 56:415–417.

Iriarte, J. A., W. L. Franklin, W. E. Johnston, and K. H. Redford. 1990. Biogeographic variation of food habits and body size of the American puma. Oecologia 85:185–190.

Irvine, B. A. 1983. Manatee metabolism and its influence on distribution in Florida. Biol. Conserv. 25:315–334.

Irving, L. 1956. Physiological insulation of swine as bare-skinned mammals. J. Appl. Physiol. 9:414–420.

——. 1957. The usefulness of Scholander's views on adaptive insulation of animals. Evolution 11:257–259.

——. 1960. Birds of Anaktuvuk Pass, Kobuk, and Old Crow. Bull. U.S. Nat. Mus. 217:1–409.

Irving, L., E. C. Black, and V. Safford. 1941a. The influence of temperature upon the combination of oxygen with the blood of trout. Biol. Bull. (Woods Hole) 80:1–17.

Irving, L., and J. S. Hart. 1954. The metabolism and insulation of seals as bare-skinned mammals in cold water. Can. J. Zool. 35:497–511.

Irving, L., and J. Krog. 1954. The body temperature of arctic and subarctic birds and mammals. J. Appl. Physiol. 6:667–680.

——. 1955. Temperature of skin in the Arctic as a regulator of heat. J. Appl. Physiol. 7:355–364.

Irving, L., J. Krog, and M. Monson. 1955. The metabolism of some Alaskan animals in winter and summer. Physiol. Zool. 28:173–185.

Irving, L., L. J. Peyton, and M. Monson. 1956. Metabolism and insulation of swine as bare-skinned mammals. J. Appl. Physiol. 9:421–426.

Irving, L., K. Schmidt-Nielsen, and N. S. B. Abrahamsen. 1957. On the melting points of animal fats in cold climates. Physiol. Zool. 30:93–105.

Irving, L., P. F. Scholander, and S. W. Grinnell. 1941b. The respiration of the porpoise, Tursiops truncatus. J. Cell. Comp. Physiol. 17:145–168.

Iverson, J. B. 1982. Adaptations to herbivory in iguanine lizards. In G. M. Burghardt and A. S. Rand (eds.), Iguanas of the World, pp. 60–76. Noyes, Park Ridge, N.J.

Iverson, S. L., and B. N. Turner. 1974. Winter weight dynamics in Microtus pennsylvanicus. Ecology 55:1030–1041.

Iwami, T., and K.-H. Kock. 1990. Channichthyidae. In O. Gon and P. C. Heemstra (eds.), Fishes of the Southern Ocean, pp. 381–399. J. L. B. Smith Institute of Ichthyology, Grahamstown, Union of South Africa.

Izhaki, I., and U. N. Safrel. 1989. Why are there so few exclusively frugivorous birds? Experiments on fruit digestibility. Oikos 54:23–32.

Jackson, D. C., J. Allen, and P. K. Strupp. 1976. The contribution of non-pulmonary surfaces to CO_2 loss in 6 species of turtles at 20°C. Comp. Biochem. Physiol. A Physiol. 55:243–246.

Jacob, F. 1977. Evolution and tinkering. Science 196:1161–1166.

Jacob, J. S., and H. S. McDonald. 1976. Diving bradycardia in four species of North American aquatic snakes. Comp. Biochem. Physiol. A Physiol. 53:69–72.

Jacobson, T., and J. A. Kushlan. 1989. Growth dynamics in the American alligator. J. Zool. Lond. 219:309–328.

Jaeger, E. C. 1948. Does the poor-will "hibernate"? Condor 50:45–46.

——. 1949. Further observations on the hibernation of the poor-will. Condor 51:105–109.

Jaeger, R. G. 1971. Moisture as a factor influencing the distribution of two species of terrestrial salamanders. Oecologia 6:191–207.

Jallageas, M., N. Mas, and I. Assenmacher. 1989. Further demonstration of the ambient temperature dependence of the annual biological cycles in the edible dormouse, Glis glis. J. Comp. Physiol. [B] 159:333–338.

James, F. C. 1970. Geographic size variation in birds and its relationship to climate. Ecology 51:365–389.

Jamison, R. L., and C. R. Robertson. 1979. Recent formulations of the urinary concentrating mechanism: A status report. Kidney Int. 16:537–545.

Janis, C. M. 1976. The evolutionary strategy of the Equidae and the origins of rumen and cecal digestion. Evolution 30:757–774.

Janis, C. M., and M. Fortelius. 1988. On the means whereby mammals achieve increased functional durability of their dentitions, with special reference to limiting factors. Biol. Rev. 63:197–230.

Jansky, L. 1962. Maximal steady state metabolism and organ thermogenesis in mammals. In J. T. Hannon and E. Viereck (eds.), Comparative Physiology of Temperature Regulation, pp. 175–201. Arctic Aeromedical Laboratory, Fort Wainwright, Ala.

——. 1973. Non-shivering thermogenesis and its thermoregulatory significance. Biol. Rev. Camb. Philos. Soc. 48:85–132.

Janssens, P. A. 1964. The metabolism of the aestivating African lungfish. Comp. Biochem. Physiol. 11:105–117.

Janssens, P. A., and P. P. Cohen. 1966. Ornithine-urea cycles enzymes in the African lungfish Protopterus aetiopicus. Science 152:358–359.

Janvier, P. 1985. Environmental framework of the diversification of the Osteostraci during the Silurian and Devonian. Philos. Trans. R. Soc. Lond. Biol. Sci. 309:259–272.

Janzen, D. H. 1974. Tropical blackwater rivers, animals, and mast fruiting by the Dipterocarpaceae. Biotropica 6:69–103.

——. 1977. Why fruits rot, seeds mold, and meat spoils. Am. Nat. 111:691–713.

——. 1979. New horizons in the biology of plant defenses. In G. A. Rosenthal and D. H. Janzen (eds.),

Herbivores: Their Interaction with Secondary Plant Metabolites, pp. 331–350. Academic Press, New York.

Jarman, P. J. 1974. The social organization of antelope in relation to their ecology. Behaviour 48:215–267.

Jarvik, E. 1952. On the fish-like tail in the ichthyostegid stegocephalians. Medd. Gronl. 114:5–90.

——. 1964. Specializations in early vertebrates. Ann. Soc. R. Zool. Belg. 94:11–95.

——. 1968. Aspects of vertebrate phylogeny. Nobel Symp. 4:497–527.

——. 1981. Lungfishes, tetrapods, paleontology, and plesiomorphy. Syst. Zool. 30:378–384.

Jarvis, J. U. M. 1981. Eusociality in a mammal: Cooperative breeding in naked mole-rat colonies. Science 212:571–573.

——. 1991. Reproduction of naked mole-rats. In P. W. Sherman, J. U. M. Jarvis, and R. D. Alexander (eds.), The Biology of the Naked Mole-Rat, pp. 384–425. Princeton University Press, Princeton.

Jean, Y., and J.-M. Bergeron. 1986. Can voles (Microtus pennsylvanicus) be poisoned by secondary metabolites of commonly eaten foods? Can. J. Zool. 64:158–162.

Jehl, J. R., Jr. 1997. Fat loads and flightlessness in Wilson's Phalaropes. Condor 99:538–543.

Jenni-Eiermann, S., and L. Jenni. 1991. Metabolic responses to flight and fasting in night-migrating passerines. J. Comp. Physiol. 161:465–474.

Jiang, S., and D. L. Claussen. 1992. A bioenergetic budget for overwintering newts (Notophthalmus viridescens) from southern Ohio: Their fat reserves and aerobic metabolic rates in water. Comp. Biochem. Physiol. A Physiol. 101:743–750.

Johansen, K. 1961. Temperature regulation in the nine-banded armadillo (Dasypus novemcinctus mexicanus). Physiol. Zool. 34:126–144.

——. 1966. Air breathing in the teleost Symbranchus marmoratus. Comp. Biochem. Physiol. 18:383–395.

Johansen, K., and A. S. F. Ditadi. 1966. Double circulation in the giant toad, Bufo paracnemis. Physiol. Zool. 39:140–150.

Johansen, K., D. Hanson, and C. Lenfant. 1970. Respiration in a primitive air breather, Amia calva. Respir. Physiol. 9:162–174.

Johansen, K., and J. Krog. 1959. Diurnal body temperature variations and hibernation in the birchmouse Sicista betulina. Am. J. Physiol. 196:1200–1204.

Johansen, K., and C. Lenfant. 1972. A comparative approach to the adaptability of O_2-HB affinity. In M. Rorth and P. Astrup (eds.), Oxygen Affinity of Hemoglobin and Red Cell Acid Base Status, pp. 750–783. Munksgaard, Copenhagen.

Johansen, K., C. Lenfant, and D. Hanson. 1968. Cardiovascular dynamics in the lungfishes. Z. Vgl. Physiol. 59:157–186.

Johansen, K., G. Lykkeboe, R. E. Weber, and G. M. O. Maloiy. 1976. Blood respiratory properties of the naked mole rat Heterocephalus glaber, a mammal of low body temperature. Respir. Physiol. 28:303–314.

Johansen, K., C. P. Mangum, and G. Lykkeboe. 1978b. Respiratory properties of the blood of Amazon fishes. Can. J. Zool. 56:898–906.

Johansen, K., C. P. Mangum, and R. E. Weber. 1978a. Reduced blood O_2 affinity associated with air breathing in osteoglossid fishes. Can. J. Zool. 56:891–897.

Johansson, B. 1957. Some biochemical and electrocardiographic data on the badger. Acta Zool. (Stockh.) 38:205–218.

John-Alder, H. B., T. Garland Jr., and A. F. Bennett. 1986. Locomotory capacities, oxygen consumption, and the cost of locomotion of the shingle-back lizard (Trachydosaurus rugosus). Physiol. Zool. 59:523–531.

John-Alder, H. B., P. J. Morin, and S. Lawler. 1988. Thermal physiology, phenology, and distribution of tree frogs. Am. Nat. 132:506–520.

Johnson, O. W. 1974. Relative thickness of the renal medulla in birds. J. Morphol. 142:277–284.

Johnson, O. W., and J. N. Mugaas. 1970a. Some histological features of avian kidneys. Am. J. Anat. 127:423–436.

——. 1970b. Quantitative and organizational features of the avian renal medulla. Condor 72:288–292.

Johnson, O. W., and R. D. Ohmart. 1973a. Some features of water economy and kidney microstructure in the large-billed savannah sparrow (Passerculus sandwichensis rostratus). Physiol. Zool. 46:276–284.

——. 1973b. The renal medulla and water economy in vesper sparrows (Pooecetes gramineus). Comp. Biochem. Physiol. A Physiol. 44:655–661.

Johnson, S. R., and I. M. Cowan. 1974. Thermal adaptations as a factor affecting colonizing success of introduced Sturnidae (Aves) in North America. Can. J. Zool. 52:1559–1576.

——. 1975. The energy cycle and thermal tolerance of the starlings (Aves, Sturnidae) in North America. Can. J. Zool. 53:55–68.

Johnson, S. R., and G. C. West. 1975. Growth and development of heat regulation in nestlings and metabolism in adult common murre and thick-billed murre. Ornis Scand. 6:109–115.

Johnston, D. W. 1963. Heart weights of some Alaskan birds. Wilson Bull. 75:435–446.

——. 1964. Ecologic aspects of lipid deposition in some postbreeding arctic birds. Ecology 45:848–852.

——. 1966. A review of the vernal fat deposition picture in overland migrant birds. Bird-Banding 37:172–183.

——. 1968. Body characteristics of palm warblers following an overwater flight. Auk 85:13–18.

——. 1970. Caloric density of avian adipose tissue. Comp. Biochem. Physiol. 34:827–832.

——. 1971. The absence of brown adipose tissue in birds. Comp. Biochem. Physiol. A Physiol. 40:1107–1108.

Johnston, D. W., and R. W. McFarlane. 1967. Migration and bioenergetics of flight in the Pacific golden plover. Condor 69:156–168.

Johnston, I. A., A. Clarke, and P. Ward. 1991. Temperature and metabolic rate in sedentary fish from the Antarctic, North Sea, and Indo-West Pacific Ocean. Mar. Biol. 109:191–195.

Johnston, I. A., and A. Wokoma. 1986. Effects of temperature and thermal acclimation of contractile properties and metabolism of skeletal muscle in the flounder (Platichthys flesus L.). J. Exp. Biol. 120:119–130.

Johnston, R. F., and R. K. Selander. 1971. Evolution in

the house sparrow. 2. Adaptive differentiation in North American populations. Evolution 25:1–28.

Jones, D. N., R. W. R. J. Dekker, and C. S. Roselaar. 1995. The Megapodes. Oxford University Press, Oxford.

Jones, F. R. H., and N. B. Marshall. 1953. The structure and functions of the teleostean bladder. Biol. Rev. 28:16–83.

Jones, P. L., and B. D. Sidell. 1982. Metabolic responses of striped bass (*Morone saxatilis*) to temperature acclimation. II. Alterations in metabolic carbon sources and distributions of fiber types in locomotory muscle. J. Exp. Zool. 219:163–171.

Jones, R. M. 1980a. Nitrogen excretion by *Scaphiopus* tadpoles in ephemeral ponds. Physiol. Zool. 53:26–31.

——. 1980b. Metabolic consequences of accelerated urea synthesis during seasonal dormancy of spadefoot toads, *Scaphiopus couchi* and *Scaphiopus multiplicatus*. J. Exp. Zool. 212:255–267.

Jones, R. M., and S. S. Hillman. 1978. Salinity adaptation in the salamander *Batrachoseps*. J. Exp. Biol. 76:1–10.

Jones, R. S. 1968. Ecological relationships in Hawaiian and Johnston Island Acanthuridae (sturgeonfishes). Micronesica 4:309–361.

Jordan, F., D. C. Haney, and F. G. Nordlie. 1993. Plasma osmotic regulation and routine metabolism in the Eustis pupfish, *Cyprinodon variegatus hubbsi* (Teleostei: Cyprinodontidae). Copeia 1993:784–789.

Jordano, P. 1987. Frugivory, external morphology and digestive system in Mediterranean sylviid warblers *Sylvia* spp. Ibis 129:175–189.

——. 1988. Diet, fruit choice and variation in body condition of frugivorous warblers in Mediterranean scrubland. Ardea 76:193–209.

Jørgensen, C. B. 1991. Water and salt balance at low temperature in a cold temperate zone anuran, the toad *Bufo bufo*. Comp. Biochem. Physiol. A Physiol. 106:377–384.

——. 1992. Growth and reproduction. *In* M. E. Feder and W. W. Burggren (eds.), Environmental Physiology of the Amphibians, pp. 439–466. University of Chicago Press, Chicago.

Jouventin, P., and H. Weimerskirch. 1990. Satellite tracking of wandering albatrosses. Nature 343:746–748.

Jung, H.-J. G. 1977. Responses of mammalian herbivores to secondary plant compounds. Biologist 59:123–136.

Jung, H.-J. G., and G. O. Batzli. 1981. Nutritional ecology of microtine rodents: Effects of plant extracts on the growth of arctic microtines. J. Mammal. 62:286–292.

Junqueira, L. C. U., G. Malnic, and C. Monge. 1966. Reabsorptive function of the ophidian cloaca and large intestine. Physiol. Zool. 39:151–159.

Jürgens, K. D., H. Bartels, and R. Bartels. 1981. Blood oxygen transport and organ weights of small bats and small non-flying mammals. Respir. Physiol. 45:243–260.

Jürgens, K. D., M. Pietschmann, K. Yamaguchi, and T. Kleinschmidt. 1988. Oxygen binding properties, capillary densities and heart weights in high altitude camels. J. Comp. Physiol. [B] 158:469–477.

Jürgens, K. D., and J. Prothero. 1987. Scaling of maximal lifespan in bats. Comp. Biochem. Physiol. A Physiol. 88:361–367.

——. 1991. Lifetime energy budgets in mammals and birds. Comp. Biochem. Physiol. A Physiol. 100:703–709.

Justice, K. E., and F. A. Smith. 1992. A model of dietary fiber utilization by small mammalian herbivores, with empirical results for *Neotoma*. Am. Nat. 139:398–416.

Kahl, M. P., Jr. 1963. Thermoregulation in the wood stork, with special reference to the role of the legs. Physiol. Zool. 36:141–151.

Kalabuchov, N. I. 1937. Some physiological adaptations of the mountain and plain forms of the wood-mouse (*Apodemus sylvaticus*) and other species of mouse-like rodents. J. Anim. Ecol. 6:254–272.

——. 1960. Comparative ecology of hibernating rodents. Bull. Mus. Comp. Zool. 124:45–74.

Kalela, O. 1957. Regulation of reproduction rate in subarctic populations of the vole *Clethrionomys rufocanus* (Sund.). Ann. Acad. Sci. Fenn. Sect. IV Biologica Ser. A 34:1–60.

Kam, M., A. A. Degan, and K. A. Nagy. 1987. Seasonal energy, water and food consumption of free-living chukars (*Alectoris chukar*) and sand partridges (*Ammoperdix heyi*) in the Negev Desert. Ecology 68:1029–1037.

Kamel, S., and R. E. Gatten Jr. 1983. Aerobic and anaerobic activity metabolism of limbless and fossorial reptiles. Physiol. Zool. 56:419–429.

Kamel, S., J. E. Marsden, and F. H. Pough. 1985. Diploid and tetraploid gray treefrogs (*Hyla chrysoscelis* and *H. versicolor*) have similar metabolic rates. Comp. Biochem. Physiol. A Physiol. 82:217–220.

Kanwisher, J., G. Gabrielsen, and N. Kanwisher. 1981. Free and forced diving in birds. Science 211:717–719.

Kao, M. H., G. L. Fletcher, N. C. Wang, and C. L. Hew. 1986. The relationship between molecular weight and antifreeze polypeptide activity in marine fish. Can. J. Zool. 64:578–582.

Karasov, W. H. 1981. Daily energy expenditure and the cost of activity in a free-living mammal. Oecologia 51:253–259.

——. 1986. Energetics, physiology and vertebrate ecology. Trends Ecol. Evol. 1:101–104.

——. 1990. Digestion in birds: Chemical and physiological determinants and ecological implications. Stud. Avian Biol. 13:391–415.

Karasov, W. H., and R. A. Anderson. 1984. Interhabitat differences in energy acquisition and expenditure in a lizard. Ecology 65:235–247.

Karasov, W. H., R. K. Buddington, and J. M. Diamond. 1985. Adaptation of intestinal and amino acid transport in vertebrate evolution. *In* R. Gilles and M. Gilles-Baillien (eds.), Transport Processes, Iono- and Osmoregulation, pp. 227–239. Springer-Verlag, Berlin.

Karasov, W. H., and J. M. Diamond. 1983. Adaptive regulation of sugar and amino acid transport by vertebrate intestine. Am. J. Physiol. 245:G443-G462.

——. 1988. Interplay between physiology and ecology in digestion: Intestinal nutrient transporters vary within and between species. Bioscience 38:602–611.

Karasov, W. H., and D. J. Levey. 1990. Digestive system trade-offs and adaptations of frugivorous passerine birds. Physiol. Zool. 63:1248–1270.

Karasov, W. H., M. W. Meyer, and B. W. Darken. 1992. Tannic acid inhibition of amino acid and sugar absorption by mouse and vole intestine: Tests following acute and subchronic exposure. J. Chem. Ecol. 18:719–736.

Karasov, W. H., E. Petrossian, L. Rosenberg, and J. M. Diamond. 1986b. How do food passage rate and assimilation differ between herbivorous lizards and nonruminant mammals? J. Comp. Physiol. 156:599–609.

Karasov, W. H., D. Phan, J. M. Diamond, and F. L. Carpenter. 1986a. Food passage and intestinal nutrient absorption in hummingbirds. Auk 103:453–464.

Karnaky, K. J., Jr. 1980. Ion-secreting epithelia: Chloride cells in the head region of *Fundulus heteroclitus*. Am. J. Physiol. 238:Rl85–Rl98.

Karnaky, K. J., Jr., S. A. Ernst, and C. W. Philpott. 1976. Teleost chloride cell. I. Response of pupfish *Cyprinodon variegatus* gill Na, K-ATPase and chloride cell fine structure to various high salinity environments. J. Cell Biol. 70:144–156.

Katz, U. 1989. Strategies of adaptation to osmotic stress in anuran amphibia under salt and burrowing conditions. Comp. Biochem. Physiol. A Physiol. 93:499–503.

Kaufmann, J. H. 1974. Social ethology of the whiptail wallaby, *Macropus parryi*, in northeastern New South Wales. Anim. Behav. 22:281–369.

Kaul, R., and V. H. Shoemaker. 1989. Control of thermoregulatory evaporation in the waterproof treefrog *Chiromantis xerampelina*. J. Comp. Physiol. [B] 158:643–649.

Kay, F. R. 1977. 2,3-Diphosphoglycerate, blood oxygen dissociation and the biology of mammals. Comp. Biochem. Physiol. A Physiol. 57:309–316.

Kay, R. F. 1984. On the use of anatomical features to infer foraging behavior in extinct primates. *In* P. S. Rodman and J. G. H. Cant (eds.), Adaptations for Foraging in Nonhuman Primates, pp. 21–53. Columbia University Press, New York.

Kayser, C. 1961. The Physiology of Natural Hibernation. Pergamon Press, New York.

Keast, [J.] A. 1959. Australian birds: Their zoogeography and adaptations to an arid environment. *In* J. A. Keast, R. L. Crocker, and C. S. Christian (eds.), Biogeography and Ecology in Australia, pp. 89–114. Junk, The Hague.

Keast, J. A., and A. J. Marshall. 1954. The influence of drought and rainfall on reproduction in Australian desert birds. Proc. Zool. Soc. Lond. 124:493–499.

Kellner, A. W. A. 1996. Reinterpretation of a remarkably well preserved pterosaur soft tissue from the Early Cretaceous of Brazil. J. Vert. Paleontol. 16:718–722.

Kelt, D. A., and D. Van Vuren. 1999. Energetic constraints and the relationship between body size and home range area in mammals. Ecology 80:337–340.

Kempton, R. T. 1953. Studies on the elasmobranch kidney. II. Reabsorption of urea by the smooth dogfish, *Mustelus canis*. Biol. Bull. (Woods Hole) 104:45–56.

Kenagy, G. J. 1972. Saltbush leaves: Excision of hypersaline tissue by a kangaroo rat. Science 178:1094–1096.

———. 1973. Adaptation for leaf eating in the Great Basin kangaroo rat, *Dipodomys microps*. Oecologia 12:383–412.

———. 1976. The periodicity of daily activity and its seasonal changes in free-ranging and captive kangaroo rats. Oecologia 24:105–140.

———. 1978. Seasonality of endogenous circadian rhythms in a diurnal rodent *Ammospermophilus leucurus* and a nocturnal rodent *Dipodomys merriami*. J. Comp. Physiol. 128:21–36.

Kenagy, G. J., D. Masman, S. M. Sharbaugh, and K. A. Nagy. 1990. Energy expenditure during lactation in relation to litter size in free-living golden-mantled ground squirrels. J. Anim. Ecol. 59:73–88.

Kenagy, G. J., S. M. Sharbaugh, and K. A. Nagy. 1989. Annual cycle of energy and time expenditure in a golden-mantled ground squirrel population. Oecologia 78:269–282.

Kenagy, G. J., R. D. Stevenson, and D. Masman. 1989. Energy requirements for lactation and postnatal growth in captive golden-mantles ground squirrels. Physiol. Zool. 62:470–487.

Kenagy, G. J., and D. Vleck. 1982. Daily temporal organization of metabolism in small mammals: Adaptation and diversity. *In* J. Achoff, S. Daan, and G. Groos (eds.), Vertebrate Circadian Systems, pp. 322–338. Springer-Verlag, Berlin.

Kendall, M. D., P. Ward, and S. Bacchus. 1973. A protein reserve in the pectoralis major flight muscle of *Quelea quelea*. Ibis 115:600–601.

Kendeigh, S. C. 1934. The role of environment in the life of birds. Ecol. Monogr. 4:299–417.

———. 1949. Effect of temperature and season on energy resources of the English sparrow. Auk 66:113–127.

———. 1969a. Tolerance of cold and Bergmann's rule. Auk 86:13–25.

———. 1969b. Energy responses of birds to their thermal environments. Wilson Bull. 81:441–449.

———. 1970. Energy requirements for existence in relation to size of bird. Condor 72:60–65.

———. 1976. Latitudinal trends in the metabolic adjustments of the house sparrow. Ecology 57:509–519.

Kendeigh, S. C., V. R. Dol'nik, and V. M. Gavrilov. 1977. Avian energetics. *In* J. Pinowski and S. C. Kendeigh (eds.), Granivorous Birds in Ecosystems, pp. 127–204. Cambridge University Press, New York.

Kennerly, T. E. 1964. Microenvironmental conditions of the pocket gopher burrow. Texas J. Sci. 16:395–441.

Kenward, R. E. 1978. Hawks and doves: Factors affecting success and selection in goshawk attacks on woodpigeons. J. Anim. Ecol. 47:449–460.

Kerr, J. G. 1898. Notes on the dry-season habits of *Lepidosiren*, communicated to him in a letter by Mr. R. J. Hunt, of Paraguay. Proc. Zool. Soc. Lond. 1898:41–44.

Kersten, M., and T. Piersma. 1987. High levels of energy expenditure in shorebirds: Metabolic adaptations to an energetically expensive way of life. Ardea 75:175–187.

Kessel, B. 1976. Winter activity patterns of black-capped chickadees in interior Alaska. Wilson Bull. 88:36–61.

Keys, A., and J. B. Bateman. 1932. Branchial responses to adrenaline and to pitressin in the eel. Biol. Bull. (Woods Hole) 63:327–336.

Keys, J. E., Jr., and P. J. Van Soest. 1970. Digestibility of forages by the meadow vole (*Microtus pennsylvanicus*). J. Dairy Sci. 53:1502–1508.

Keys, J. E., Jr., P. J. Van Soest, and E. P. Young. 1970. The effect of increasing dietary and wall content on the digestibility of hemicellulose and cellulose in swine and rats. J. Anim. Sci. 31:1172–1177.

Khalil, F. 1948a. Excretion in reptiles—II. Nitrogen constituents of the urinary concentrations of the oviparous snake *Zamenis diadema*, Schlegel. J. Biol. Chem. 172:101–103.

——. 1948b. Excretion in reptiles—III. Nitrogen constituents of the urinary concentrations of the viviparous snake *Eryx thebaicus*, Reuss. J. Biol. Chem. 172:105–106.

——. 1951. Excretion in reptiles—IV. Nitrogen constituents of the excreta of lizards. J. Biol. Chem. 189:443–445.

Khalil, F., and G. Haggag. 1955. Ureotelism and uricotelism in tortoises. J. Exp. Zool. 130:423–432.

——. 1958. Nitrogenous excretion in crocodiles. J. Exp. Biol. 35:552–555.

Kiell, D. J., and J. S. Millar. 1980. Reproduction and nutrient reserves of arctic ground squirrels. Can. J. Zool. 58:416–421.

Kihlström, J. E. 1972. Period of gestation and body weight in some placental mammals. Comp. Biochem. Physiol. A Physiol. 43:673–679.

Kilgore, D. L., and K. B. Armitage. 1978. Energetics of yellow-bellied marmot populations. Ecology 59:78–88.

King, D. 1980. The thermal biology of free-living sand goannas (*Varanus gouldii*) in Southeastern Australia. Copeia 1980:755–767.

King, J. R. 1963. Autumnal migratory-fat deposition in the white-crowned sparrow. Proc. Int. Ornithol. Congr. 13:940–949.

——. 1964. Oxygen consumption and body temperature in relation to ambient temperature in the white-crowned sparrow. Comp. Biochem. Physiol. 12:13–24.

——. 1972. Adaptive periodic fat storage by birds. Proc Int. Ornithol. Congr. 15:200–217.

——. 1974. Seasonal allocation of time and energy resources in birds. *In* R. A. Paynter (ed.), Avian Energetics, pp. 4–70. Publ. 15. Nuttall Ornithological Club, Cambridge, Mass.

King, J. R., S. Barker, and D. S. Farner. 1963. A comparison of energy reserves during the autumnal and vernal migratory periods in the white-crowned sparrow, *Zonotrichia leucophrys gambelii*. Ecology 44:513–521.

King, J. R., and D. S. Farner. 1961. Energy metabolism, thermoregulation and body temperatures. *In* A. J. Marshall (ed.), Biology and Comparative Physiology of Birds, pp. 215–288. Academic Press, New York.

——. 1963. The relationship of fat deposition to zugunruhe and migration. Condor 65:200–223.

——. 1964. Terrestrial animal in humid heat: Birds. *In* D. B. Dill (ed.), Handbook of Physiology, section 4: Adaptation to the Environment, pp. 603–624. American Physiological Society, Washington, D.C.

——. 1966. The adaptive role of winter fattening in the white-crowned sparrow with comments on its regulation. Am. Nat. 100:403–418.

King, J. R., and M. E. Murphy. 1985. Periods of nutritional stress in the annual cycles of endotherms: Fact or fiction? Am. Nat. 25:955–964.

King, R. D., and J. F. Bendell. 1982. Foods selected by blue grouse (*Dendragapus obscurus fuliginosus*). Can. J. Zool. 60:3268–3281.

Kingdon, J. 1974. East African Mammals, vol. II, part B. Academic Press, New York.

Kinnear, J. E., A. Cockson, P. Christensen, and A. R. Main. 1979. The nutritional biology of the ruminants and ruminant-like mammals—a new approach. Comp. Biochem. Physiol. A Physiol. 64:357–365.

Kinzey, W. G. 1992. Diet and dental adaptations in the Pitheciinae. Am. J. Phys. Anthropol. 88:499–514.

Kinzey, W. G., and A. H. Gentry. 1979. Habitat utilization in two species of *Callicebus*. *In* R. W. Sussman (ed.), Primate Ecology: Problem-Oriented Field Studies, pp. 89–100. Wiley, New York.

Kirkwood, J. K. 1983. A limit to metabolisable energy intake in mammals and birds. Comp. Biochem. Physiol. A Physiol. 75:1–3.

Kirsch, J. A. W., T. F. Flannery, M. S. Springer, and F.-J. Lapointe. 1995. Phylogeny of the Pteropodidae (Mammalia: Chiroptera) based on DNA hybridisation, with evidence for bat monophyly. Aust. J. Zool. 43:395–428.

Kirschbaum, F. 1975. Environmental factors control the periodical reproduction of tropical electric fish. Experientia 31:1159–1160.

——. 1979. Reproduction of the weakly electric fish *Eigenmannia virescens* (Rhamphichthyidae, Teleostei) in captivity. Behav. Ecol. Sociobiol. 4:331–355.

Kirschner, L. B. 1967. Comparative physiology: Invertebrate excretory organs. Annu. Rev. Physiol. 29:169–196.

——. 1993. The energetics of osmotic regulation in ureotelic and hypoosmotic fishes. J. Exp. Zool. 267:19–26.

——. 1995. Energetics of osmoregulation in fresh water vertebrates. J. Exp. Zool. 271:243–252.

Kissner, K. J., and R. M. Brigham. 1993. Evidence for the use of torpor by incubating and brooding common poorwills *Phalaenoptilus nuttallii*. Ornis Scand. 24:333–334.

Kitchell, J. F. 1969. Thermophilic and thermophobic responses of snakes in a thermal gradient. Copeia 1969:189–191.

Kitzan, S. M., and P. R. Sweeny. 1968. A light and electron microscope study of the structure of *Protopterus annectens* epidermis. I. Mucus production. Can. J. Zool. 46:767–772.

Klaassen, M. 1995. Water and energy limitations on flight range. Auk 112:260–262.

——. 1996. Metabolic constraints on long-distance migration in birds. J. Exp. Biol. 199:57–64.

Klaassen, M., and H. Biebach. 1994. Energetics of fattening and starvation in the long-distance migratory garden warbler, *Sylvia borin*, during the migratory phase. J. Comp. Physiol. [B] 164:362–371.

Klaassen, M., and R. Drent. 1991. An analysis of hatchling resting metabolism: In search of ecological correlates that explain deviations from allometric relations. Condor 93:612–629.

Klaassen, M., M. Kersten, and B. J. Ens. 1990. Energetic requirements for maintenance and premigratory body mass gain of waders wintering in Africa. Ardea 78:209–220.

Klauber, L. M. 1956. Rattlesnakes: Their Habits, Life Histories, and Influence on Mankind, 2 vols. University of California Press, Berkeley.

Klaus, S., G. Heldmaier, and D. Ricquier. 1988. Seasonal acclimation of bank voles and wood mice: Nonshivering thermogenesis and thermogenic properties of brown adipose tissue mitochondria. J. Comp. Physiol. [B] 158:157–164.

Kleckner, N. W., and B. D. Sidell. 1985. Comparison of maximal activities of enzymes from tissues of thermally acclimated and naturally acclimatized chain pickerel (*Esox niger*). Physiol. Zool. 58:18–28.

Kleiber, M. 1932. Body size and metabolism. Hilgardia 6:315–353.

——. 1947. Body size and metabolic rate. Physiol. Rev. 27:511–541.

——. 1961. Fire of Life: An Introduction to Animal Energetics. Wiley, New York.

——. 1970. Conductivity, conductance and transfer constant for animal heat. Fed. Proc. Fed. Am. Soc. Exp. Biol. 29:660.

——. 1972. A new Newton's law of cooling? Science 178:1283–1285.

——. 1973. Perspectives on linear heat transfer. Science 181:186.

——. 1975. Metabolic turnover rate: A physiological meaning of the metabolic rate per unit body weight. J. Theor. Biol. 53:199–204.

Kleiber, M., and J. F. Dougherty. 1934. The influence of environmental temperature on the utilization of food energy in baby chicks. J. Gen. Physiol. 17:701–706.

Klein, H. 1974. The adaptational value of the internal annual clocks in birds. *In* E. T. Pengelley (ed.), Circannual Clocks, pp. 347–391. Academic Press, New York.

Klein, R. G. 1986. Carnivore size and quaternary climate change in southern Africa. Quat. Res. 26:153–170.

Klein, R. G., and K. Scott. 1989. Glacial/interglacial size variation in fossil spotted hyaenas (*Crocuta crocuta*) from Britain. Quat. Res. 32:88–95.

Klinger, S. A., J. J. Magnuson, and G. W. Gallepp. 1982. Survival mechanisms of the central mudminnow (*Umbra limi*), flathead minnow (*Pimephales promelas*) and brook stickleback (*Culaea inconstans*) for low oxygen in winter. Environ. Biol. Fish. 7:113–120.

Klir, J. J., J. E. Heath, and N. Bennan. 1990. An infrared thermographic study of surface temperature in relation to external thermal stress in the Mongolian gerbil, *Meriones unguiculatus*. Comp. Biochem. Physiol. A Physiol. 96:141–146.

Knutson, R. M. 1974. Heat production and temperature regulation in eastern skunk cabbage. Science 186:746–747.

Koban, M., and D. D. Feist. 1982. The effect of cold on norepinephrine turnover in tissues of seasonally acclimated redpolls, *Carduelis flammea*. J. Comp. Physiol. 146:137–144.

Kobelt, F., and K. E. Linsenmair. 1986. Adaptations of the reed frog *Hyperolius viridiflavus* (Amphibia, Anura, Hyperoliidae) to its arid environment. Oecologia 68:533–541.

Koenig, W. D. 1991. The effects of tannins and lipids on digestion of acorns by acorn woodpeckers. Auk 108:79–88.

Koenig, W. D., and M. K. Heck. 1988. Ability of two species of oak woodland birds to subsist on acorns. Condor 90:705–708.

Koenig, W. D., and R. L. Mumme. 1987. Population Ecology of the Cooperatively Breeding Acorn Woodpecker. Princeton University Press, Princeton.

Koford, C. B. 1968. Peruvian desert mice: Water independence, competition, and breeding cycle near the Equator. Science 160:552–553.

Kokko, J. P., and F. C. Rector Jr. 1972. Countercurrent multiplication system without active transport in inner medulla. Kidney Int. 2:214–223.

Kokshaysky, N. V. 1973. Functional aspects of some details of bird wing configuration. Syst. Zool. 22:442–450.

——. 1977. Some scale dependent problems in aerial animal locomtion. *In* T. J. Pedley (ed.), Scale Effects in Animal Locomotion, pp. 421–435. Academic Press, London.

Komen, J., and C. J. Brown. 1993. Food requirements and the timing of breeding of a Cape vulture colony. Ostrich 64:86–92.

Konarzewski, M. 1995. Allocation of energy to growth and respiration in avian postembyronic development. Ecology 76:8–19.

Konarzewski, M., and J. Diamond. 1995. Evolution of basal metabolic rate and organ masses in laboratory mice. Evolution 49:1239–1248.

Kooijman, S. A. L. M. 2000. Dynamic Energy and Mass Budgets in Biological Systems, 2nd ed. Cambridge University Press, Cambridge.

Kooyman, G. L. 1963. Erythrocyte analysis of some Antarctic fishes. Copeia 1963:457–458.

Kooyman, G. L., M. A. Castellini, and R. W. Davis. 1981. Physiology of diving in marine mammals. Annu. Rev. Physiol. 43:343–356.

Kooyman, G. L., Y. Cherel, Y. Le Maho, J. P. Croxall, P. H. Thorson, V. Ridoux, and C. A. Kooyman. 1992. Diving behavior and energetics during foraging cycles in king penguins. Ecol. Monogr. 62:143–163.

Kooyman, G. L., R. W. Davis, J. P. Croxall, and D. L. Costa. 1982. Diving depths and energy requirements of king penguins. Science 217:726–727.

Koplin, J. R., M. W. Collopy, A. R. Baumann, and H. Levenson. 1980. Energetics of two wintering raptors. Auk 97:795–806.

Korhonen, K. 1980. Microclimate in the snow burrows of willow grouse (*Lapopus lagopus*). Ann. Zool. Fenn. 17:5–9.

———. 1981. Temperature in the nocturnal shelters of the redpoll (*Acanthus flammea* L.) and the Siberian tit (*Parus cinctus* Budd.) in winter. Ann. Zool. Fenn. 18:165–168.

———. 1989. Heat loss of willow grouse (*Lagopus l. lagopus* L.) in a snow environment. J. Thermal Biol. 14:27–31.

Körtner, G., R. M. Brigham, and F. Geiser. 2000. Winter torpor in a large bird. Nature 467:318.

Koskimies, J. 1948. On temperature regulation and metabolism in the swift, *Micropus a. apus* L., during fasting. Experientia 4:274–276.

———. 1950. The life of the swift, *Micropus apus* (L.), in relation to the weather. Ann. Acad. Sci. Fenn. Biol. 15:1–151.

———. 1962. Ontogeny of thermoregulation and energy metabolism in some gallinaceous birds. Trans. Congr. Int. Union Game Biol. Bologna 5:149–160.

Koskimies, J., and L. Lahti. 1964. Cold-hardiness of the newly hatched young in relation to ecology and distribution in ten species of European ducks. Auk 81:281–307.

Koteja, P. 1987. On the relation between basal and maximal metabolic rate in mammals. Comp. Biochem. Physiol. A Physiol. 87:205–208.

———. 1991. On the relation between basal and field metabolic rates in birds and mammals. Funct. Ecol. 5:56–64.

———. 1996a. Limits to the energy budget in a rodent, *Peromyscus maniculatus*: The central limitation hypothesis. Physiol. Zool. 69:981–993.

———. 1996b. Limits to the energy budget in a rodent, *Peromyscus maniculatus*: Does gut capacity set the limit? Physiol. Zool. 69:994–1020.

Koteja, P., and J. Weiner. 1993. Mice, voles and hamsters: Metabolic rates and adaptive strategies in muroid rodents. Oikos 66:505–514.

Kozłowski, J., and J. Weiner. 1997. Interspecific allometries are by-products of body size optimization. Am. Nat. 149:352–380.

Krakauer, T., C. Gans, and C. V. Paganelli. 1968. Ecological correlation of water loss in burrowing reptiles. Nature 218:659–660.

Kramer, D. L. 1983. The evolutionary ecology of respiratory mode in fishes: An analysis based on the costs of breathing. Environ. Biol. Fish. 9:145–158.

Kramer, D. L., C. C. Lindsey, G. E. E. Moodie, and E. D. Stevens. 1978. The fishes and the aquatic environment of the central Amazon basin, with particular reference to respiratory patterns. Can. J. Zool. 56:717–729.

Kramer, D. L., D. Manley, and R. Burgeois. 1983. The effect of respiratory mode and oxygen concentration on the risk of aerial predation in fishes. Can. J. Zool. 61:653–665.

Krebs, C. J. 1978. A review of the Chitty hypothesis of population regulation. Can. J. Zool. 56:2463–2480.

———. 1979. Dispersal, spacing behaviour and genetics in relation to population fluctuations in the vole *Microtus townsendii*. Fortschr. Zool. 25:61–77.

Krijgsveld, K. L., C. Dijkstra, G. H. Visser, and S. Daan. 1998. Energy requirements for growth in relation to sexual size dimorphism in marsh harrier *Circus aeruginosus* nestlings. Physiol. Zool. 71:693–702.

Krogh, A. 1904. Some experiments on the cutaneous respiration of vertebrate animals. Skand. Arch. Physiol. 16:348–357.

Krogh, A., and J. Leitch. 1919. The respiratory function of the blood in fishes. J. Physiol. 52:288–300.

Krüger, K., R. Prinzinger, and K.-L. Schuchmann. 1982. Torpor and metabolism in hummingbirds. Comp. Biochem. Physiol. A Physiol. 73:679–689.

Krzanowski, A. 1961. Weight dynamics of bats wintering in a cave at Pulawy (Poland). Acta Theriol. 4:242–264.

Kubb, R. N., J. R. Spotila, and D. R. Pendergrast. 1980. Mechanisms of heat transfer and time-dependent modeling of body temperatures in the largemouth bass (*Micropterus salmoides*). Physiol. Zool. 53:222–239.

Kuhn, W., A. Ramel, H. J. Kuhn, and E. Marti. 1963. The filling mechanism of the swimbladder. Experientia 19:497–552.

Kuhnen, G. 1986. O_2 and CO_2 concentrations in burrows of euternmic and hibernating golden hamsters. Comp. Biochem. Physiol. A Physiol. 84:517–522.

Künkele, J., and F. Trillmich. 1997. Are precocial young cheaper? Lactation energetics in the guinea pig. Physiol. Zool. 70:589–596.

Kunz, T. H., and K. A. Ingalls. 1994. Folivory in bats: An adaptation derived from frugivory. Funct. Ecol. 8:665–668.

Kurta, A. 1990. Torpor patterns in food-deprived *Myotis lucifugus* (Chiroptera: Vespertilionidae) under simulated roost conditions. Can. J. Zool. 69:255–257.

Kurta, A., G. P. Bell, K. A. Nagy, and T. H. Kunz. 1988. Energetics of pregnancy and lactation in free-ranging little brown bats (*Myotis lucifugus*). Physiol. Zool. 62:804–818.

Kurta, A., and M. Ferkin. 1991. The correlation between demography and metabolic rate: A test using the beach vole (*Microtus breweri*) and the meadow vole (*Microtus pennsylvanicus*). Oecologia 87:102–105.

LaBrie, S. J., and I. D. W. Sutherland. 1962. Renal function in water snakes. Am. J. Physiol. 203:995–1000.

Lacey, E. A., and P. W. Sherman. 1991. Social organization of naked mole-rat colonies: Evidence for divisions of labour. *In* P. W. Sherman, J. U. M. Jarvis, and R. D. Alexander (eds.), The Biology of the Naked Mole-Rat, pp. 273–336. Princeton University Press, Princeton.

Lack, D. 1947. The significance of clutch size. Ibis 89:302–352.

———. 1968a. Bird migration and natural selection. Oikos 19:1–9.

———. 1968b. Ecological Adaptation for Breeding in Birds. Methuen, London.

Lagerstrøm, M. 1979. Goldcrests (*Regulus regulus*) roosting in the snow. Ornis Fenn. 56:170–171.

Lahiri, S. 1975. Blood oxygen affinity and alveolar ventilation in relation to body weight in mammals. Am. J. Physiol. 229:529–536.

Lambert, R., and G. Teissier. 1927. Théorie de la similitude biologique. Ann. Physiol. 3:212–246.

Lambert, W. D. 1991. Altriciality and its implications for

dinosaur thermoenergetic physiology. N. Jb. Geol. Palaont. Abh. 182:73–84.

Lamprey, H. F. 1963. Ecological separation of the large mammal species in the Tarangire Game Reserve, Tanganyika. East Afr. Wildl. J. 1:63–92.

Langer, P. 1984. Anatomical and nutritional adaptations in wild herbivores. In F. M. C. Gilchrist and R. I. Mackie (eds.), Herbivore Nutrition in the Subtropics and Tropics, pp. 185–203. Science Press, Craighill, South Africa.

——. 1986. Large mammalian herbivores in tropical forests with either hindgut- or forestomach-fermentation. Z. Saugeteirkd. 51:173–187.

Langille, B. L., and B. Crisp. 1980. Temperature dependence of blood viscosity in frogs and turtles: Effect on heat exchange with environment. Am. J. Physiol. 239:R248–R253.

Langvatn, R., and S. D. Albon. 1986. Geographic clines in body weight of Norwegian red deer: A novel explanation of Bergmann's rule? Holarct. Ecol. 9:285–293.

Lasiewski, R. C. 1963. Oxygen consumption of torpid, resting, active, and flying hummingbirds. Physiol. Zool. 36:122–140.

——. 1969. Physiological responses to heat stress in the poor-will. Am. J. Physiol. 217:1504–1509.

Lasiewski, R. C., and W. A. Calder. 1971. A preliminary allometric analysis of respiratory variables in resting birds. Respir. Physiol. 11:152–166.

Lasiewski, R. C., A. L. Costa, and M. H. Bernstein. 1966a. Evaporative water loss in birds—I. Characteristics of the open flow method of determination, and their relation to estimates of thermoregulatory ability. Comp. Biochem. Physiol. 19:445–457.

——. 1966b. Evaporative water loss in birds—II. A modified method for determination by direct weighing. Comp. Biochem. Physiol. 19:459–470.

Lasiewski, R. C., and W. R. Dawson. 1967. A re-examination of the relation between standard metabolic rate and body weight in birds. Condor 69:13–23.

Lasiewski, R. C., and R. J. Lasiewski. 1967. Physiological responses of the blue-throated and Rivoli's hummingbirds. Auk 84:34–48.

Lasiewski, R. C., and R. S. Seymour. 1972. Thermoregulatory responses to heat stress in four species of birds weighing approximately 40 grams. Physiol. Zool. 45:106–118.

Lasiewski, R. C., W. W. Weathers, and M. H. Bernstein. 1967. Physiological responses of the giant hummingbird, Patagona gigas. Comp. Biochem. Physiol. 23:797–813.

Lasserre, P., and J.-L. Gallis. 1975. Osmoregulation and differential penetration of two grey mullets, Chelon labrosus (Risso) and Liza ramada (Risso) in estuarine fish ponds. Aquaculture 5:323–344.

Laurance, W. F., K. R. McDonald, and R. Speare. 1996. Epidemic disease and the catastrophic decline of Australian rain forest frogs. Conserv. Biol. 10:406–413.

Lauren, D. L. 1985. The effect of chronic saline exposure on the electrolyte balance, nitrogen metabolism, and corticosterone titer in the American alligator, Alligator mississippiensis. Comp. Biochem. Physiol. A Physiol. 81:217–223.

Lavigne, D. M. 1982. Similarity in energy budgets of animal populations. J. Anim. Ecol. 51:195–206.

Lavigne, D. M., S. Innes, G. A. J. Worthy, K. M. Kovacs, O. J. Schmitz, and J. P. Hickie. 1985. Metabolic rates of seals and whales. Can. J. Zool. 64:279–284.

Lawson, D. A. 1975. Pterosaur from the latest Cretaceous of West Texas: Discovery of the largest flying creature. Science 187:947–948.

Layne, J. R., Jr., and R. E. Lee Jr. 1989. Seasonal variation in freeze tolerance and ice content of the tree frog Hyla versicolor. J. Exp. Zool. 249:133–137.

LeBoeuf, B. J., Y. Naito, A. C. Huntley, and T. Asaga. 1989. Prolonged, continuous, deep diving by northern elephant seals. Can. J. Zool. 67:2514–2519.

Lechner, A. J. 1976. Respiratory adaptations in burrowing pocket gophers from sea level and high altitude. J. Appl. Physiol. 41:168–173.

——. 1978. The scaling of maximal oxygen consumption and pulmonary dimensions in small mammals. Respir. Physiol. 34:29–44.

Lee, A. K. and J. A. Badham. 1963. Body temperature, activity, and behavior of the agamid lizard, Amphibolurus barbatus. Copeia 1963:387–394.

Lee, A. K., and E. H. Mercer. 1967. Cocoon surrounding desert-dwelling frogs. Science 157:87–88.

Lee, W. B., and D. C. Houston. 1993. The effect of diet quality on gut anatomy of British voles (Microtinae). J. Comp. Physiol. [B] 163:337–339.

LeFebvre, E. A. 1964. The use of D_2O^{18} for measuring energy metabolism in Columba livia at rest and in flight. Auk 81:403–416.

Lehtola, K. A. 1973. Ordovician vertebrates from Ontario. Contrib. Mus. Paleontol. Univ. Mich. 24:23–30.

Leim, A. H., and W. B. Scott. 1966. Fishes of the Atlantic coast of Canada. Bull. Fish. Res. Board Can. 155:1–485.

Le Maho, Y. 1977. The emperor penguin: A strategy to live and breed in the cold. Am. Sci. 65:680–693.

Le Maho, Y., P. Delclitte, and J. Chatonnet. 1976. Thermoregulation in fasting emperor penguins under natural conditions. Am. J. Physiol. 231:913–922.

Lemaire, M., M. Goffart, J. Closon, and R. Winand. 1969. La fonction thyroidienne chez l'unau (Choloepus hoffmanni Peters). Gen. Comp. Endocrinol. 12:181–199.

Lemon, W. C. 1993. The energetics of lifetime reproductive success in the zebra finch (Taeniopygia guttata). Physiol. Zool. 66:946–963.

Lemons, D. E., and L. I. Crawshaw. 1985. Behavioral and metabolic adjustments to low temperatures in the large mouth bass (Micropterus salmoides). Physiol. Zool. 58:175–180.

Lenfant, C., and K. Johansen. 1967. Respiratory adaptations in selected amphibians. Respir. Physiol. 2:247–260.

——. 1968. Respiration in the African lungfish Protopterus aethiopicus. I. Respiratory properties of blood and normal patterns of gas exchange. J. Exp. Biol. 49:437–452.

Lenfant, C., K. Johansen, and G. C. Grigg. 1966. Respiratory properties of blood and pattern of gas exchange in the lungfish *Neoceratodus forsteri* (Krefft). Respir. Physiol. 2:1–21.

Lenfant, C., K. Johansen, and J. D. Torrance. 1970. Gas transport and oxygen storage capacity in some pinnipeds and the sea otter. Respir. Physiol. 9:277–286.

Leonard, W. R., and M. L. Robertson. 1997. Rethinking the energetics of bipedality. Current Anthropol. 38: 304–309.

Leopold, A. S. 1953. Intestinal morphology of gallinaceous birds in relation to food habits. J. Wildl. Manage. 17:197–203.

Leopold, A. S., M. Erwin, J. Oh, and B. Browning. 1976. Phytoestrogens: Adverse effects on reproduction in California quail. Science 191:98–100.

Levey, D. J., and W. H. Karasov. 1989. Digestive responses of temperate birds switched to fruit or insect diets. Auk 106:675–686.

Levin, D. A. 1971. Plant phenolics: An ecological perspective. Am. Nat. 105:157–181.

——. 1976. Alkaloid-bearing plants: An ecogeographic perspective. Am. Nat. 110:261–284.

Lewis, W. M. 1970. Morphological adaptations of cyprinodontoids for inhabiting oxygen deficient waters. Copeia 1970:319–326.

Licht, P. 1974. Response of *Anolis* lizards to food supplementation in nature. Copeia 1974:215–221.

Licht, P., and A. F. Bennett. 1972. A scaleless snake: Tests of the role of reptilian scales in water loss and heat transfer. Copeia 1972:702–707.

Licht, P., and A. G. Brown. 1967. Behavioral thermoregulation and its role in the ecology of the red-bellied newt, *Taricha rivularis*. Ecology 48:598–611.

Licht, P., W. R. Dawson, V. H. Shoemaker, and A. R. Main. 1966. Observations on the thermal relations of Western Australian lizards. Copeia 1966:97–110.

Licht, P., M. E. Feder, and S. Bledsoe. 1975. Salinity tolerance and osmoregulation in the salamander *Batrachoseps*. J. Comp. Physiol. 102:123–134.

Lifson, N., G. B. Gordon, and R. McClintock. 1955. Measurement of total carbon dioxide production by means of D_2O^{18}. J. Appl. Physiol. 7:704–710.

Lifson, N., and R. McClintock. 1966. Theory of use of the turnover rates of body water for measuring energy and material balance. J. Theor. Biol. 12:46–74.

Liggins, G. W., and G. C. Grigg. 1985. Osmoregulation of the cane toad, *Bufo marinus*, in salt water. Comp. Biochem. Physiol. A Physiol. 82:613–619.

Lighthill, M. J. 1977. Introduction to the scaling of aerial locomotion. *In* T. J. Pedley (ed.), Scale Effects in Animal Locomotion, pp. 365–404. Academic Press, London.

Ligon, J. D. 1970. Still more responses of the poor-will to low temperatures. Condor 72:496–498.

Lillegraven, J. A. 1976. Biological considerations of the marsupial-placental dichotomy. Evolution 29:707–722.

Lillegraven, J. A., S. D. Thompson, B. K. McNab, and J. L. Patton. 1987. The origin of eutherian mammals. Biol. J. Linn. Soc. 32:281–336.

Lillywhite, H. B. 1970. Behavioral temperature regulation in the bullfrog, *Rana catesbeiana*. Copeia 1970:158–168.

——. 1971. Thermal modulation of cutaneous mucus discharge as a determinant of evaporative water loss in the frog, *Rana catesbeiana*. Z. Vgl. Physiol. 73:84–104.

——. 1975. Physiological correlates of basking in amphibians. Comp. Biochem. Physiol. A Physiol. 52:323–330.

——. 1980. Behavioral thermoregulation in Australian elaphid snakes. Copeia 1980:452–458.

——. 1987. Temperature, energetics, and physiological ecology. *In* R. A. Seigel, J. T. Collins, and S. S. Novak (eds.), Snakes: Ecology and Evolutionary Biology, pp. 422–471. Macmillan, New York.

Lillywhite, H. B., and T. M. Ellis. 1994. Ecophysiological aspects of the coastal-estuarine distribution of acrochordid snakes. Estuaries 17:53–61.

Lillywhite, H. B., and P. Licht. 1974. Movement of water over toad skin: Functional role of epidermal sculpturing. Copeia 1974:165–171.

——. 1975. A comparative study of integumentary mucous secretions in amphibians. Comp. Biochem. Physiol. A Physiol. 51:937–941.

Lillywhite, H. B., P. Licht, and P. Chelgren. 1973. The role of behavioral thermoregulation in the growth energetics of the toad, *Bufo boreas*. Ecology 54: 375–383.

Lillywhite, H. B., A. K. Mittal, T. K. Garg, and N. Agrawal. 1997. Wiping behavior and its ecophysiological significance in the Indian tree frog *Polypedates maculatus*. Copeia 1997:88–100.

Lindeborg, R. G. 1952. Water requirements of certain rodents from xeric and mesic habitats. Contrib. Lab. Vert. Biol. Univ. Mich. 58:1–32.

Lindroth, R. L., and G. O. Batzli. 1983. Detoxification of some naturally occurring phenolics by prairie voles: A rapid assay of glucuronidation metabolism. Biochem. Syst. Ecol. 11:405–409.

——. 1984. Plant phenolics as chemical defenses: Effects of natural phenolics on survival and growth of prairie voles (*Microtus ochrogaster*). J. Chem. Ecol. 10:229–244.

Lindsey, C. C. 1966. Body sizes of poikilotherm vertebrates at different latitudes. Evolution 20:456–465.

Lindstedt, S. L. 1980. Regulated hypothermia in the desert shrew. J. Comp. Physiol. 137:173–176.

Lindstedt, S. L., and M. S. Boyce. 1985. Seasonality, fasting endurance, and body size in mammals. Am. Nat. 125:873–878.

Lindstedt, S. L., and W. A. Calder. 1981. Body size, physiological time, and longevity of homeothermic animals. Q. Rev. Biol. 56:1–16.

Lindstedt, S. L., B. J. Miller, and S. W. Buskirk. 1986. Home range, time, and body size in mammals. Ecology 67:413–418.

Lindström, Å. 1991. Maximum fat deposition rates in migrating birds. Ornis Scand. 22:12–19.

Lindström, Å., and T. Alerstam. 1992. Optimal fat loads in migrating birds: A test of the time-minimization hypothesis. Am. Nat. 140:477–491.

Lindström, Å., and A. Kvist. 1995. Maximum energy

intake is proportional to basal metabolic rate in passerine birds. Proc. R. Soc. Lond. B 261:337–343.

Lindström, Å., and T. Piersma. 1993. Mass changes in migrating birds: The evidence for fat and protein storage re-examined. Ibis 135:70–78.

Lindström, Å., G. H. Visser, and S. Daan. 1993. The energetic cost of feather synthesis is proportional to basal metabolic rate. Physiol. Zool. 66:490–510.

Linthicum, D. S., and F. G. Carey. 1972. Regulation of brain and eye temperatures by the bluefin tuna. Comp. Biochem. Physiol. A Physiol. 43:425–433.

Lips, K. R. 1998. Decline of a tropical montane amphibian fauna. Conserv. Biol. 12:106–111.

Lissaman, P. B. S., and C. Schollenberger. 1970. Formation flight of birds. Science 168:1003–1005.

Lister, A. M. 1989. Rapid dwarfing of red deer on Jersey in the Last Interglacial. Nature 342:539–542.

Littleford, R. A., W. F. Keller, and N. W. Phillips. 1947. Studies on the vital limits of water loss in the plethodontid salamanders. Ecology 28:440–447.

Livezey, B. C. 1989. Flightlessness in grebes (Aves: Podicipedidae): Its independent evolution in three genera. Evolution 43:29–54.

——. 1992a. Morphological corollaries and ecological implications of flightlessness in the kakapo (Psittaciformes: Strigops habroptilus). J. Morphol. 213:105–145.

——. 1992b. Flightlessness in the Galápagos cormorant (Compsohalieus [Nannopterum] harrisi): Heterochrony, gi[g]antism, and specialization. Zool. J. Linn. Soc. 105:155–224.

Livingstone, D. L. 1963. Chemical composition of rivers and lakes. Geol. Surv. Prof. Paper 440-G:1–64.

Livoreil, B., and C. Baudoin. 1996. Differences in food hoarding behaviour in two species of ground squirrels Spermophilus tridecemlineqtus and Spermophilus spilosoma. Ethol. Ecol. Evol. 8:199–205.

Loeb, S. C., R. G. Schwab, and M. W. Demment. 1991. The responses of pocket gophers (Thomomys bottae) to changes in diet quality. Oecologia 86:542–551.

Lomolino, M. V. 1985. Body size of mammals on islands: The island rule reexamined. Am. Nat. 125:310–316.

Longcore, J. E., A. P. Pessier, and D. K. Nichols. 1999. Batrachochytrium dendrobatidis gen. et sp. nov., a chytrid pathogenic to amphibians. Mycologia 91:219–227.

Lonnberg, E. 1899. Salamanders with and without lungs. Zool. Anz. 22:545–548.

Lonsdale, K., and D. J. Suter. 1971. Uric acid and dihydrate in bird urine. Science 172:958–959.

Louden, A., N. Rothwell, and M. Stock. 1985. Brown fat, thermogenesis, and physiological birth in a marsupial. Comp. Biochem. Physiol. A Physiol. 81:815–819.

Louw, E., G. N. Louw, and C. P. Retief. 1972. Thermolability, heat tolerance and renal function in the dassie or hyrax, Procavia capensis. Zool. Afr. 7:451–469.

Lovegrove, B. G. 1989. The cost of burrowing by the social mole rats (Bathyergidae) Crytomys damarensis and Heterocephalus glaber: The role of soil moisture. Physiol. Zool. 62:449–469.

Lovegrove, B. G., G. Heldmaier, and M. Knight. 1991a.

Seasonal and circadian energetic parameters in an arboreal rodent, Thallomys paedulcus, and a burrow-dwelling rodent, Aethomys namaquensis, from the Kalahari Deset. J. Thermal Biol. 16:199–209.

Lovegrove, B. G., G. Heldmaier, and T. Ruf. 1991b. Perspectives of endothermy revisited: The endothermic temperature range. J. Thermal Biol. 16:185–197.

Loveridge, J. P. 1970. Observations on nitrogenous excretion and water relations of Chiromantis xerampelina (Amphibia, Anura). Arnoldia (Rhod.) 5:1–6.

——. 1976. Strategies of water conservation in southern African frogs. Zool. Afr. 11:319–333.

Loveridge, J. P., and P. C. Withers. 1981. Metabolism and water balance of active and cocooned African bullfrogs Pyxicepahlus adspersus. Physiol. Zool. 54:203–214.

Lowe, C. H., and W. G. Heath. 1969. Behavioral and physiological responses to temperature in the desert pupfish Cyprinodon macularis. Physiol. Zool. 42:53–59.

Lowe, C. H., P. J. Lardner, and E. A. Halpern. 1971. Supercooling in reptiles and other vertebrates. Comp. Biochem. Physiol. A Physiol. 39:125–135.

Lundgren, B. O., and K.-H. Kiessling. 1985. Seasonal variation in catalytic activities in breast muscle of some migratory birds. Oecologia 66:468–471.

——. 1988. Comparative aspects of fibre types, areas, and capillary supply in the pectoralis muscle of some passerine birds with differing migratory behaviour. J. Comp. Physiol. [B] 158:165–173.

Lustick, S. 1969. Bird energetics: Effects of artificial radiation. Science 163:387–390.

——. 1971. Plumage color and energetics. Condor 73:121–122.

Lustick, S., M. Adam, and A. Hinko. 1980. Interaction between posture, color, and the radiative heat load in birds. Science 208:1052–1053.

Lustick, S., B. Battersby, and M. Kelty. 1978. Behavioral thermoregulation: Orientation toward the sun in herring gulls. Science 200:81–83.

Lustick, S., S. Talbolt, and E. L. Fox. 1970. Absorption of radiant energy in redwinged blackbirds (Agelaius phoeniceus). Condor 72:471–473.

Lutcavage, M. E., P. G. Bushnell, and D. R. Jones. 1992. Oxygen stores and aerobic metabolism in the leatherback turtle. Can. J. Zool. 70:348–351.

Lutz, P. L., I. S. Longmuir, and K. Schmidt-Nielsen. 1974. Oxygen affinity of bird blood. Respir. Physiol. 20:325–330.

Lutz, P. L., and J. D. Robertson. 1971. Osmotic constituents of the coelacanth Latimeria chalumnae Smith. Biol. Bull. (Woods Hole) 141:553–560.

Lutz, P. L., and K. Schmidt-Nielsen. 1977. Effect of simulated altitude on blood gas transport in the pigeon. Respir. Physiol. 30:383–388.

Lyman, C. P. 1954. Activity, food consumption and hoarding in hibernators. J. Mammal. 35:545–552.

——. 1963. Hibernation in mammals and birds. Am. Sci. 51:127–138.

Lyman, C. P., R. C. O'Brien, G. C. Green, and E. D. Papafraugos. 1981. Hibernation and longevity in the Turkish hamster, Mesocricetus brandti. Science 212:668–670.

Lynch, G. R., F. D. Vogt, and H. R. Smith. 1978. Seasonal study of spontaneous daily torpor in the white-footed mouse, *Peromyscus leucopus*. Physiol. Zool. 51:289–299.

Macari, M., D. L. Ingram, and M. J. Dauncey. 1983. Influence of thermal and nutritional acclimation on body temperatures and metabolic rate. Comp. Biochem. Physiol. A Physiol. 74:549–553.

MacArthur, D. L., and J. W. T. Dandy. 1982. Physiological aspects of overwintering in the boreal chorus frog (*Pseudacris triseriata maculata*). Comp. Biochem. Physiol. A Physiol. 72:137–141.

MacArthur, R. A. 1989. Energy metabolism and thermoregulation of beaver (*Castor canadensis*). Can. J. Zool. 68:2409–2416.

MacArthur, R. H. 1972. Geographical Ecology. Harper and Row, New York.

MacArthur, R. H, and E. O. Wilson. 1967. The Theory of Island Biogeography. Princeton University Press, Princeton.

Macartney, J. M., K. W. Larsen, and P. T. Gregory. 1989. Body temperatures and movements of hibernating snakes (*Crotalus* and *Thamnophis*) and thermal gradients of natural hibernacula. Can. J. Zool. 67:108–114.

Macdonald, A. G., I. Gilchrist, and C. S. Wardle. 1987. Effects of hydrostatic pressure on the motor activity of fish from shallow water and 900 m depths; some results of Challenger cruise 6B/85. Comp. Biochem. Physiol. A Physiol. 88:543–547.

Macdonald, J. A., and R. M. G. Wells. 1991. Viscosity of body fluids from antarctic notothenioid fish. *In* G. diPrisco, B. Maresca, and B. Tota (eds.), Biology of Antarctic Fish, pp. 163–178. Springer-Verlag, Berlin.

Mace, G. M., and P. H. Harvey. 1983. Energetic constraints on home-range size. Am. Nat. 122:120–132.

Mace, G. M, P. H. Harvey, and T. H. Clutton-Brock. 1983. Vertebrate home-range size and energetic requirements. *In* I. R. Swingland and P. J. Greenwood (eds.), The Ecology of Animal Movement, pp. 32–55. Clarendon Press, Oxford.

Macfarlane, W. V., B. Howard, H. Haines, P. J. Kennedy, and C. M. Sharpe. 1971. Heirarchy of water and energy turnover of desert mammals. Nature 234:483–484.

MacKay, R. S. 1964. Galapagos tortoise and marine iguana deep body temperatures measured by radio telemetry. Nature 204:355–358.

MacMahon, T. 1973. Size and shape in biology. Science 179:1201–1204.

——. 1975. Allometry and biomechanics: Limb bones in adult ungulates. Am. Nat. 109:547–563.

MacMillen, R. E. 1964. Population ecology, water relations, and social behavior of s southern California semidesert rodent fauna. Univ. Calif. Publ. Zool. 71:1–66.

——. 1965. Aestivation in the cactus mouse, *Peromyscus eremicus*. Comp. Biochem. Physiol. 16:227–248.

——. 1972. Water economy of nocturnal desert rodents. Symp. Zool. Soc. Lond. 31:147–174.

MacMillen, R. E., and E. A. Christopher. 1975. The water relations of two populations of noncaptive desert rodents. *In* N. F. Hadley (ed.), Environmental Physiology of Desert Organisms, pp. 117–137. Dowden, Hutchinson, and Ross, Stroudsberg, Penn.

MacMillen, R. E., and D. E. Grubbs. 1976. Water metabolism in rodents. *In* D. H. Johnson (ed.), Progress in Animal Biometerology, vol. I, pp. 63–69. Swetz and Zeitlinger, Lisse, Netherlands.

MacMillen, R. E., and D. S. Hinds. 1983. Water regulatory efficiency in heteromyid rodents: A model and its application. Ecology 64:152–164.

MacMillen, R. E., and A. K. Lee. 1967. Australian desert mice: Independence of exogenous water. Science 158:383–385.

——. 1969. Water metabolism of Australian hopping mice. Comp. Biochem. Physiol. 28:493–514.

——. 1970. Energy metabolism and pulmocutaneous water loss of Australian hopping mice. Comp. Biochem. Physiol. 35:355–369.

MacMillen, R. E., and J. E. Nelson. 1969. Bioenergetics and body size in dasyurid marsupials. Am. J. Physiol. 217:1246–1251.

MacMillen, R. E., and C. H. Trost. 1966. Water economy and salt balance in white-winged and Inca doves. Auk 83:441–456.

Maddox, P. A., W. M. Deen, and B. M. Brenner. 1974. Dynamics of glomerular ultrafiltration. VI. Studies in the primate. Kidney Int. 5:271–278.

Magnin, E. 1962. Recherches sur la systématique et de la biologie des acepenserides *Acipenser sturio*, *Acipenser oxyrhynchus*, et *Acipenser fulvescens*. Ann. Stat. Centr. Hydrob. Appl. 9:170–242.

Magnuson, J. J. 1970. Hydrostatic equilibrium of *Euthynnus affinus*, a pelagic teleost without a gas bladder. Copeia 1970:56–85.

Magnuson, J. J., A. L. Beckel, K. Mills, and S. B. Brandt. 1985. Surviving winter hypoxia: Behavioral adaptations of fishes in a northern Wisconsin winterkill lake. Environ. Biol. Fish. 14:241–250.

Magnuson, J. J., J. W. Keller, A. L. Beckel, and G. W. Gallepp. 1983. Breathing gas mixtures different from air: An adaptation for survival under the ice of a facultative air-breathing fish. Science 220:312–314.

Maher, W. J. 1964. Growth rate and development of endothermy in the snow bunting (*Plectophenax nivalis*) and Lapland longspur (*Calcarius lapponicus*) at Barrow, Alaska. Ecology 45:520–528.

Mahoney, S. A., L. Fairchild, and R. E. Shea. 1985. Temperature regulation in great frigate birds *Fregata minor*. Physiol. Zool. 58:138–148.

Mahoney, S. A., and J. R. King. 1977. The use of the equivalent black-body temperature in the thermal energetics of small birds. J. Thermal Biol. 2:115–120.

Main, A. R., and P. J. Bentley. 1964. Water relations of Australian burrowing frogs and tree frogs. Ecology 45:379–382.

Main, A. R., M. J. Littlejohn, and A. K. Lee. 1959. Ecology of Australian frogs. *In* J. A. Keast, R. L. Crocker, and C. S. Christian (eds.), Biogeography and Ecology in Australia, pp. 398–411. Junk, The Hague.

Maina, J. N., A. S. King, and D. Z. King. 1982. A morphometric analysis of the lung of a species of bat. Respir. Physiol. 50:1–11.

Maina, J. N., A. S. King, and G. Settle. 1989. An allometric study of pulmonary morphometric parameters in birds, with mammalian comparisons. Philos. Trans. R. Soc. Lond. 326:1–57.

Malan, A. 1988. pH and hypometabolism in mammalian hibernation. Can. J. Zool. 66:95–98.

Mann, K. H. 1965. Energy transformations by a population of fish in the River Thames. J. Anim. Ecol. 34:253–275.

Marden, J. H. 1987. Maximum lift production during takeoff in flying animals. J. Exp. Biol. 130:235–258.

——. 1990. Maximum load-lifting and induced power output of Harris' hawks are general functions of flight muscle mass. J. Exp. Biol. 149:511–514.

Marder, J. 1973. Body temperature regulation in the brown-necked raven (*Corvus corax ruficollis*)—II. Thermal changes in the plumage of ravens exposed to solar radiation. Comp. Biochem. Physiol. A Physiol. 45:431–440.

——. 1983. Cutaneous water evaporation—II. Survival of birds under extreme thermal stress. Comp. Biochem. Physiol. A Physiol. 75:433–439.

Marder, J., and J. Ben-Asher. 1983. Cutaneous water evaporation—I. Its significance in heat-stressed birds. Comp. Biochem. Physiol. A Physiol. 75:425–431.

Mares, M. A. 1977. Water independence in a South American non-desert rodent. J. Mammal. 58:653–656.

Mares, M. A., R. A. Ojeda, C. E. Borghi, S. M. Giannoni, G. B. Diaz, and J. K. Braun. 1997. How desert rodents overcome halophytic plant defenses. Bioscience 47:699–704.

Mares, M. A., and D. E. Wilson. 1971. Bat reproduction during the Costa Rican dry season. Bioscience 21:471–477.

Margaria, R. 1938. Sulla fisiologia e specialmente sul consumo energetico della marcia e della corsa a varie velocita ed inclinazioni del terreno. Atti Acad. Naz. Lincei Mem. Cl. Sci. 7:299–368.

Marks, J. S. 1993. Molt of bristle-thighed curlews in the northwestern Hawaiian Islands. Auk 110:573–587.

Marples, B. J. 1932. The structure and development of the nasal glands of birds. Proc. Zool. Soc. Lond. 1932:829–844.

Marquet, P. A., J. C. Ortiz, F. Bozinovic, and F. M. Jaksic. 1989. Ecological aspects of thermoregulation at high altitudes: The case of Andean *Liolaemus* lizards in northern Chile. Oecologia 81:16–20.

Marsh, R. L. 1984. Adaptations of the gray catbird *Dumetella carolinesis* to long-distance migration: Flight muscle hypertrophy associated with elevated body mass. Physiol. Zool. 57:105–117.

Marsh, R. L., and R. W. Storer. 1981. Correlation of flight-muscle size and body mass in Cooper's hawks: A natural analogue of power training. J. Exp. Biol. 91:363–368.

Marshall, A. J. 1951. The refractory period of testis rhythm in birds and its possible bearing on breeding and migration. Wilson Bull. 63:238–261.

Marshall, E. K. 1934. The comparative physiology of the kidney in relation to theories of renal secretion. Physiol. Rev. 14:133–159.

Marshall, E. K., and H. W. Smith. 1930. The glomerular development of the vertebrate kidney in relation to habitat. Biol. Bull. (Woods Hole) 59:135–153.

Marshall, J. T. 1955. Hibernation in captive goatsuckers. Condor 57:129–134.

Marshall, N. B. 1971. Explorations in the Life of Fishes. Harvard University Press, Cambridge, Mass.

Martill, D. M., and D. M. Unwin. 1989. Exceptionally well preserved pterosaur wing membrane from the Cretaceous of Brazil. Nature 340:138–140.

Martin, C. J. 1902. Thermal adjustment and respiratory exchange in monotremes and marsupials—a study in the development of homoeothermism. Philos. Trans. R. Soc. Lond. 195:1–37.

Martin, K. L. M. 1991. Facultative aerial respiration in an intertidal sculpin, *Clinocottus analis* (Scorpaeniformes: Cottidae). Physiol. Zool. 64:1341–1355.

——. 1995. Time and tide wait for no fish: Intertidal fishes out of water. Environ. Biol. Fish. 44:165–181.

Martin, R. A., and S. D. Webb. 1974. Late Pleistocene mammals from the Devil's Den Fauna, Levy County. *In* S. D. Webb (ed.), Pleistocene Mammals of Florida. pp. 114–145. University Presses of Florida, Gaineville.

Martin, R. D. 1981. Relative brain size and basal metabolic rate in terrestrial vertebrates. Nature 293:57–60.

Martin, R. D., and A. M. MacLarnon. 1985. Gestation period, neonatal size and maternal investment in placental mammals. Nature 313:220–223.

Martinez del Rio, C. 1990a. Dietary, phylogenetic, and ecological correlates of intestinal sucrase and maltase activity in birds. Physiol. Zool. 63:987–1011.

——. 1990b. Sugar preferences in hummingbirds: The influence of subtle chemical differences on food choice. Condor 92:1022–1030.

Martinez del Rio, C., W. H. Karasov, and D. J. Levey. 1989. Physiological basis and ecological consequences of sugar preferences in cedar waxwings. Auk 106:64–71.

Martinez del Rio, C., and B. R. Stevens. 1989. Physiological constraint on feeding behavior: Intestinal membrane disaccharides of the starling. Science 243:794–796.

Martinez del Rio, C., B. R. Stevens, D. E. Dankeke, and P. T. Andreadis. 1988. Physiological correlates of preference and aversion for sugars in three species of birds. Physiol. Zool. 61:222–229.

Martof, B. S. 1969. Prolonged inanition in *Siren lacertina*. Copeia 1969:285–289.

Maser, C., J. M. Trappe, and R. A. Nussbaum. 1978. Fungal-small mammal interrelationships with emphasis on Oregon coniferous forests. Ecology 59:799–809.

Maser, Z., C. Maser, and J. M. Trappe. 1985. Food habits of the northern flying squirrel (*Glaucomys sabrinus*) in Oregon. Can. J. Zool. 63:1084–1088.

Maslanik, J. A., M. C. Serreze, and R. G. Barry. 1996. Recent decreases in Arctic summer ice cover and linkages to atmospheric circulation anomalies. Geophys. Res. Lett. 23:1677–1680.

Masman, D., S. Daan, and H. J. A. Beldhuis. 1988.

Ecological energetics of the kestrel: Daily energy expenditure throughout the year based on time-energy budget, food intake and doubly labeled water methods. Ardea 76:64–81.

Masman, D., C. Dijkstra, S. Daan, and A. Bult. 1989. Energetic limitation of avian parental effort: Field experiments in the Kestrel (*Falco tinnunculus*). J. Evol. Biol. 2:435–455.

Masman, D., M. Gordijn, S. Daan, and C. Dijkstra. 1986. Ecological energetics of the kestrel: Field estimates of energy intake throughout the year. Ardea 74:24–39.

Masman, D., and M. Klaassen. 1987. Energy expenditure during free flight in trained and free-living Eurasian kestrels (*Falco tinnunculus*). Auk 104:603–616.

Mason, B. J. 1995. Predictions of climate changes caused by man-made emissions of greenhouse gases: A critical assessment. Contemp. Physics 36:299–319.

Massman, W. H., and A. L. Pachecho. 1957. Disappearance of young Atlantic croakers from the York River, Virginia. Trans. Am. Fish. Soc. 89:154–159.

Matter, R. M. 1966. Studies of the nitrogen metabolism of *Amia calva* with special reference to ammonia and urea production and elimination. Ph. D. dissertation, University of Missouri, Columbia.

Matthew, W. D., and C. de Paulo Couto. 1959. The Cuban edentates. Bull. Am. Mus. Nat. Hist. 117:1–56.

Matthews, G. V. T. 1954. Some aspects of incubation in the Manx shearwater, *Procellaria puffinus*, with particular reference to chilling resistance in the embryo. Ibis 96:432–440.

Mautz, W. J. 1979. The metabolism of reclusive lizards, the Xantusiidae. Copeia 1979:577–584.

May, M. L. 1976a. Thermoregulation and adaptation to temperature in dragonflies (Odonata: Anisoptera). Ecol. Monogr. 46:1–32.

——. 1976b. Warming rates as a function of body size in periodic endotherms. J. Comp. Physiol. 111:55–70.

——. 1979a. Energy metabolism of dragonflies (Odonata: Anisoptera) at rest and during endothermic warm up. J. Exp. Biol. 83:79–94.

——. 1979b. Insect thermoregulation. Annu. Rev. Entomol. 24:313–349.

May, M. L., and T. M. Casey. 1983. Thermoregulation and heat exchange in euglossine bees. Physiol. Zool. 56:541–551.

May, R. M. 1974. Stability and Complexity in Model Ecosystems. Princeton University Press, Princeton.

——. 1976. Models for single populations. *In* R. M. May (ed.), Theoretical Ecology, Principles and Applications, pp. 4–25. Saunders, Philadelphia.

——. 1979. Production and respiration in animal communities. Nature 282:443–444.

Mayhew, W. W. 1962. *Scaphiopus couchi* in California's Colorado Desert. Herpetologica 18:153–161.

——. 1965a. Hibernation in the horned lizard, *Phrynosoma m'calli*. Comp. Biochem. Physiol. 16:103–119.

——. 1965b. Adaptations of the amphibian, *Scaphiopus couchi*, to desert conditions. Am. Midl. Nat. 74:95–109.

Mayr, E. 1956. Geographical character gradients and climatic adaptation. Evolution 10:105–108.

——. 1997. This Is Biology. Harvard University Press, Cambridge.

Mayr, E., and W. B. Provine, eds. 1980. The Evolutionary Synthesis: Perspectives on the Unification of Biology. Harvard University Press, Cambridge.

Mazzotti, F. J., and W. A. Dunson. 1989. Osmoregulation in crocodilians. Am. Zool. 29:903–920.

McArthur, A. J., and J. A. Clark. 1988. Body temperature of homeotherms and the conservation of energy and water. J. Thermal Biol. 13:9–13.

McArthur, C., A. E. Hagerman, and C. T. Robbins. 1991. Physiological strategies of mammalian herbivores against plant defenses. *In* R. T. Palo and C. T. Robbins (eds.), Plant Defenses against Mammalian Herbivory, pp. 103–114. CRC Press, Boca Raton.

McArthur, C., and G. D. Sanson. 1993a. Nutritional effects and costs of a tannin in a grazing and a browsing macropodid marsupial herbivore. Funct. Ecol. 7:690–696.

——. 1993b. Nutritional effects and costs of a tannin in two marsupial arboreal folivores. Funct. Ecol. 7:697–703.

McAtee, W. L. 1947. Torpidity in birds. Am. Midl. Nat. 38:191–206.

McBean, R. L., and L. Goldstein. 1970. Accelerated synthesis of urea in *Xenopus laevis* during osmotic stress. Am. J. Physiol. 219:1124–1130.

McBee, R. H., and V. H. McBee. 1982. The hindgut fermentation in the green iguana, *Iguana iguana*. *In* G. M. Burghardt and A. S. Rand (eds.), Iguanas of the World, pp. 77–83. Noyes, Park Ridge, N.J.

McClanahan, L. 1967. Adaptations of the spadefoot toad, *Scaphiopus couchii*, to desert environments. Comp. Biochem. Physiol. 20:73–79.

McClanahan, L., and R. Baldwin. 1969. Rate of water uptake through the integument of the desert toad, *Bufo punctatus*. Comp. Biochem. Physiol. 28:381–389.

McClanahan, L., R. Ruibal, and V. H. Shoemaker. 1983. Rate of cocoon formation and its physiological correlates in a ceratophryd frog. Physiol. Zool. 56:430–435.

McClanahan, L., V. H. Shoemaker, and R. Ruibal. 1976. Structure and function of the cocoon of a ceratophryd frog. Copeia 1976:179–185.

McCleery, R. H., and C. M. Perrins. 1998. Temperature and egg-laying trends. Nature 391:30–31.

McClure, P. A., and J. C. Randolph. 1980. Relative allocation of energy to growth and development of homeothermy in the eastern wood rat (*Neotoma floridana*) and hispid cotton rat (*Sigmodon hispidus*). Ecol. Monogr. 50:199–219.

McCutcheon, F. H. 1936. Hemoglobin function during the life history of the bullfrog. J. Cell. Comp. Physiol. 8:63–81.

McCutcheon, F. H., and F. G. Hall. 1937. Hemoglobin in the Amphibia. J. Cell. Comp. Physiol. 9:191–197.

McDevitt, R. M., and J. R. Speakman. 1994a. Limits to sustainable metabolic rate during transient exposure to low temperatures in short-tailed field voles (*Microtus agrestis*). Physiol. Zool. 67:1103–1116.

——. 1994b. Central limits to sustainable metabolic rate have no role in cold acclimation of the short-tailed field vole (*Microtus agrestis*). Physiol. Zool. 67:1117–1139.

McDiarmid, R. W., and M. S. Foster. 1987. Cocoon formation in another hylid frog, *Smilisca baudinii*. J. Herpetol. 21:352–355.

McDowell, R. M. 1987. Evolution and importance of diadromy. *In* M. J. Dadswell, R. J. Klauda, C. M. Moffitt, R. L. Saunders, R. A. Rulifson, and J. E. Cooper (eds.), Common Strategies of Andromous Fishes, pp. 1–13. Symposium. 1. American Fisheries Society, Bethesda, Md.

——. 1988. Diadromy in Fishes. Croom Helm, London.

McFarland, W. N., and F. W. Munz. 1965. Regulation of body weight and serum composition by hagfish in various media. Comp. Biochem. Physiol. 14:383–398.

McFarland, W. N., and W. A. Wimsatt. 1969. Renal function and its relation to the ecology of the vampire bat, *Desmodus rotundus*. Comp. Biochem. Physiol. 28:985–1006.

McGahan, J. 1973. Gliding flight of the Andean condor in nature. J. Exp. Biol. 58:225–237.

McGinnis, S. M., and L. L. Dickson. 1967. Thermoregulation in the desert iguana *Dipsosaurus dorsalis*. Science 156:1757–1759.

McKey, D. B. 1974. Adaptive patterns in alkaloid physiology. Am. Nat. 108:305–320.

——. 1979. The distribution of secondary compounds within plants. *In* G. A. Rosenthal and D. H. Janzen (eds.), Herbivores: Their Interaction with Secondary Plant Metabolites, pp. 55–133. Academic Press, New York.

McKey, D. B., J. S. Gartlan, P. G. Waterman, and G. M. Choo. 1981. Food selection by black colobus monkeys (*Colobus satanas*) in relation to plant chemistry. Biol. J. Linn. Soc. 16:115–146.

McLaren, I. A. 1967. Seals and group selection. Ecology 48:104–110.

McLean, J. A., and J. R. Speakman. 1999. Energy budgets of lactating and non-reproductive brown long-eared bats (*Plecotus auritus*) suggest females use compensation lactation. Funct. Ecol. 13:360–372.

——. 2000. Effects of body mass and reproduction on basal metabolic rate of brown long-eared bats (*Plecotus auritus*). Physiol. Biochem. Zool. 73:112–121.

McMasters, J. H. 1976. Aerodynamics of the long pterosaur wing. Science 191:899.

McNab, B. K. 1963a. Bioenergetics and the determination of home range size. Am. Nat. 97:133–140.

——. 1963b. A model of the energy budget of a wild mouse. Ecology 44:521–532.

——. 1966a. An analysis of the body temperatures of birds. Condor 68:47–55.

——. 1966b. The metabolism of fossorial rodents: A study of convergence. Ecology 47:712–733.

——. 1968. The influence of fat deposits on the basal rate of metabolism in desert homoiotherms. Comp. Biochem. Physiol. 26:337–343.

——. 1969. The economics of temperature regulation in Neotropical bats. Comp. Biochem. Physiol. 31:227–268.

——. 1970. Body weight and the energetics of temperature regulation. J. Exp. Biol. 53:329–348.

——. 1971a. The structure of tropical bat faunas. Ecology 52:352–358.

——. 1971b. On the ecological significance of Bergmann's rule. Ecology 52:845–854.

——. 1973. Energetics and the distribution of vampires. J. Mammal. 54:131–144.

——. 1974a. The behavior of temperate cave bats in a subtropical environment. Ecology 55:943–958.

——. 1974b. The energetics of endotherms. Ohio J. Sci. 74:370–380.

——. 1976. Seasonal fat reserves of bats in two tropical environments. Ecology 57:332–338.

——. 1978a. The evolution of endothermy in the phylogeny of mammals. Am. Nat. 112:1–21.

——. 1978b. Energetics of arboreal folivores: Physiological problems and ecological consequences of feeding on an ubiquitous food supply. *In* G. G. Montgomery (ed.), The Ecology of Arboreal Folivores, pp. 153–162. Smithsonian Institution Press, Washington, D.C.

——. 1978c. The comparative energetics of marsupials. J. Comp. Physiol. 125:113–128.

——. 1979a. Climatic adaptation in the energetics of heteromyid rodents. Comp. Biochem. Physiol. A Physiol. 62:813–820.

——. 1979b. The influence of body size on the energetics and distribution of fossorial and burrowing mammals. Ecology 60:1010–1021.

——. 1980a. On estimating thermal conductance in endotherms. Physiol. Zool. 53:145–156.

——. 1980b. Food habits, energetics, and the population biology of mammals. Am. Nat. 116:106–124.

——. 1980c. Energetics and the limits to a temperate distribution in armadillos. J. Mammal. 61:606–627.

——. 1982. Evolutionary alternatives in the physiological ecology of bats. *In* T. H. Kunz (ed.), The Ecology of Bats, pp. 151–200. Plenum, New York.

——. 1983a. Energetics, body size, and the limits to endothermy. J. Zool. Lond. 199:1–29.

——. 1983b. Ecological and behavioral consequences of adaptation to various food resources. *In* J. F. Eisenberg and D. G. Kleiman (eds.), Advances in the Study of Mammalian Behavior, pp. 664–697. Special Publ. 7. American Society of Mammalogy, Lawrence, Kans.

——. 1984. Physiological convergence amongst ant-eating and termite-eating mammals. J. Zool. Lond. 203:485–510.

——. 1986a. The influence of food habits on the energetics of eutherian mammals. Ecol. Monogr. 56:1–19.

——. 1986b. Food habits, energetics, and the reproduction of marsupials. J. Zool. Lond. 208:595–614.

——. 1988a. Complications inherent in scaling basal rate of metabolism in mammals. Q. Rev. Biol. 63:25–54.

——. 1988b. Food habits and the basal rate of metabolism in birds. Oecologia 77:343–349.

——. 1988c. Energy conservation in a tree-kangaroo (*Dendrolagus matschiei*) and the red panda (*Ailurus fulgens*). Physiol. Zool. 61:280–292.

——. 1989a. Basal rate of metabolism, body size, and food habits in the order Carnivora. *In* J. L. Gittleman (ed.), Carnivore Behavior, Ecology, and Evolution, pp. 335–354. Cornell University Press, Ithaca, N.Y.

——. 1989b. Body mass, food habits, and the use of torpor in birds. *In* C. Bech and R. E. Reinertsen (eds.), Physiology of Cold Adaptation in Birds, pp. 283–291. Plenum, New York.

——. 1989c. Laboratory and field studies of the energy expenditure of endotherms: A comparison. Trends Ecol. Evol. 4:111–112.

——. 1989d. Temperature regulation and rate of metabolism in three Bornean bats. J. Mammal. 70:153.

——. 1990. On the selective persistence of mammals in South America. *In* K. H. Redford and J. F. Eisenberg (eds.), Advances in Tropical Mammalogy, pp. 605–614. Sandhill Crane Press, Gainesville, Fla.

——. 1991. The energy expenditure of shrews. *In* J. S. Findley and T. L. Yates (eds.), The Biology of the Soricidae, pp. 35–45. Museum of Southwestern Biology, University of New Mexico, Albuquerque.

——. 1992a. A statistical analysis of mammalian rates of metabolism. Funct. Ecol. 6:672–679.

——. 1992b. The comparative energetics of rigid endothermy: The Arvicolidae. J. Zool. Lond. 227:585.

——. 1992c. Energy expenditure: A short history. *In* T. E. Tomasi and T. H. Horton (eds.), Mammalian Energetics, pp. 1–15. Cornell University Press, Ithaca.

——. 1994a. Energy conservation and the evolution of flightlessness in birds. Am. Nat. 144:628–642.

——. 1994b. Resource use and the survival of land and freshwater vertebrates on oceanic islands. Am. Nat. 144:643–660.

——. 1995. Energy expenditure and conservation in frugivorous and mixed-diet carnivorans. J. Mammal. 76:206–222.

——. 1997. On the utility of uniformity in the definition of basal rate of metabolism. Physiol. Zool. 70:718–720.

——. 1999. On the comparative ecological and evolutionary significance of total and mass-specific rates of metabolism. Physiol. Biochem. Zool. 72:642–644.

——. 2000a. The influence of body mass, climate, and distribution on the energetics of South Pacific pigeons. Comp. Biochem. Physiol. A Physiol. 127:309–329.

——. 2000b. The standard energetics of mammalian carnivores: Felidae and Hyaenidae. Can. J. Zool. 78:2227–2239.

——. 2001. Energetics of toucans, a barbet, and a hornbill: Implications for avian folivory. Auk 118.

McNab, B. K., and W. Auffenberg. 1976. The effect of large body size on the temperature regulation of the Komodo dragon, *Varanus komodoensis*. Comp. Biochem. Physiol. A Physiol. 55:345–350.

McNab, B. K., and F. J. Bonaccorso. 1995a. The energetics of Australasian swifts, frogmouths, and nightjars. Physiol. Zool. 68:245–261.

——. 1995b. The energetics of pteropodid bats. *In* P. A. Racey and S. M. Swift (eds.), Ecology, Evolution and Behaviour of Bats, pp. 111–112. Symposium 67. Zoology Society of London, Clarendon Press, Oxford.

——. 2001. The metabolism of New Guinean pteropodid bats. J. Comp. Physiol. [B] 171:201–214.

McNab, B. K., and J. F. Eisenberg. 1989. Brain size and its relation to the rate of metabolism in mammals. Am. Nat. 133:157–167.

McNab, B. K., and P. R. Morrison. 1963. Body temperature and metabolism in subspecies of *Peromyscus* from arid and mesic environments. Ecol. Monogr. 33:63–82.

McNabb, F. M. A. 1969. A comparative study of water balance in three species of quail—II. Utilization of saline drinking solutions. Comp. Biochem. Physiol. 28:1059–1074.

McNabb, F. M. A., and R. A. McNabb. 1975a. Proportions of ammonia, urea, urate and total nitrogen in avian urine and quantitative methods for their analysis on a single urine sample. Poult. Sci. 54:1498–1505.

McNabb, R. A., and F. M. A. McNabb. 1975b. Urate excretion by the avian kidney. Comp. Biochem. Physiol. A Physiol. 51:253–258.

McNaughton, S. J. 1979. Grazing as an optimization process: Grass-ungulate relationships in the Serengeti. Am. Nat. 113:691–703.

McNeill, S., and J. H. Lawton. 1970. Annual production and respiration in animal populations. Nature 225:472–474.

McVicar, A. J., and J. C. Rankin. 1985. Dynamics of glomerular filtration in the river lamprey, *Lampetra fluviatilis* L. Am. J. Physiol. 249:F132–F138.

Meek, R. P., and J. J. Childress. 1973. Respiration and the effect of pressure in the mesopelagic fish *Anoplogaster cornuta* (Beryciformes). Deep-Sea Res. 20:1111–1118.

Mendes, E. G. 1945. Contribuição para a fisiologia dos sistemas respiratorio e circulatório de *Siphonops annulatus* (Amphíbia—Gymnophonia). Zoologia (Sao Paulo) 9:25–67.

Merritt, J. F., and D. A. Zegers. 1991. Seasonal thermogenesis and body-mass dynamics of *Clethrionomys gapperi*. Can. J. Zool. 69:2771–2777.

Meserve, P. L. 1978. Water dependence in some Chilean arid zone rodents. J. Mammal. 59:217–219.

Messel, H., A. G. Wells, and W. J. Green. 1979. Surveys of tidal river systems in the Northern Territory of Australia and their crocodile populations. Monogr. 7. Pergammon Press, Sydney.

Meylan, A. 1985. The role of sponge collagens in the diet of the hawksbill turtle (*Eretmochelys imbricata*). *In* A. Bairati and R. Garrone (eds.), Biology of Invertebrate and Lower Vertebrate Collagens, pp. 191–196. Plenum, New York.

Mezhzherin, V. A. 1964. Dehnel's phenomenon and its possible explanation. Acta Theriol. 8:95–114.

Mezhzherin, V. A., and G. L. Melnikova. 1966. Adaptive importance of seasonal changes in some morpho-physiological indices in shrews. Acta Theriol. 11:503–521.

Michener, G. R. 1983. Kin selection, matriarchies, and the evolution of sociality in ground-dwelling sciurids. *In* J. F. Eisenberg and D. G. Kleiman (eds.), Advances in the Study of Mammalian Behavior, pp. 528–572. Special Publ. 7. American Society of Mammalogy, Lawrence, Kan.

——. 1984. Age, sex, and species differences in the annual cycles of ground-dwelling sciurids: Implications for sociality. *In* J. O. Murie and G. R. Michener (eds.), Biology of Ground-dwelling Squirrels: Annual Cycles, Behavioral Ecology, and Sociality, pp. 81–107. University of Nebraska Press, Lincoln.

Midtgård, U. 1989. A morphometric study of structures important for cold resistance in the arctic Iceland gull compared to herring gulls. Comp. Biochem. Physiol. A Physiol. 93:399–402.

Millar, J. S. 1975. Tactics of energy partitioning in breeding *Peromyscus*. Can. J. Zool. 53:967–976.

——. 1981. Post partum reproductive characteristics of eutherian mammals. Evolution 35:1149–1163.

Millar, J. S., and G. J. Hickling. 1990. Fasting endurance and the evolution of mammalian body size. Funct. Ecol. 4:5–12.

Millard, A. R. 1995. The body temperature of *Tyrannosaurus rex*. Science 267:1666.

Miller, A. H. 1963. Desert adaptations in birds. Proc. Int. Ornithol. Congr. 13:666–674.

Miller, A. H., and R. C. Stebbins. 1964. The Lives of Desert Animals in Joshua Tree National Monument. University of California Press, Berkeley.

Miller, M. A. 1957. Burrows of the Sacramento Valley pocket gopher in flood-irrigated alfalfa fields. Hilgardia 26:431–452.

Milton, K. 1981. Food choice and digestive strategies of two sympatric primate species. Am. Nat. 117:496–505.

Milton, K., and M. L. May. 1976. Body weight, diet and home range area in primates. Nature 259:459–462.

Milton, P. 1971. Oxygen consumption and osmoregulation in the shanny, *Blennius pholis*. J. Mar. Biol. Assoc. U.K. 51:247–265.

Minnich, J. E. 1970. Water and electrolyte balance of the desert iguana, *Dipsosaurus dorsalis*, in its natural habitat. Comp. Biochem. Physiol. 35:921–933.

——. 1972. Excretion of urate salts by reptiles. Comp. Biochem. Physiol. A Physiol. 41:535–549.

Moberly, W. R. 1963. Hibernation in the desert iguana, *Dipsosaurus dorsalis*. Physiol. Zool. 36:152–160.

——. 1968. The metabolic responses of the common iguana, *Iguana iguana*, to walking and diving. Comp. Biochem. Physiol. 27:21–32.

Moldenhauer, R. R., and J. A. Wiens. 1970. The water economy of the sage sparrow, *Amphipiza belli nevadensis*. Condor 72:265–275.

Molnar, R. E., and J. Wiffen. 1994. A late Cretaceous polar dinosaur fauna from New Zealand. Cretac. Res. 15:689–706.

Mommsen, T. P., and P. J. Walsh. 1989. Evolution of urea synthesis in vertebrates: The piscine connection. Science 243:72–75.

——. 1992. Biochemical and environmental perspectives on nitrogen metabolism in fishes. Experientia 48:583–593.

Montevecchi, W. A., V. L. Birt-Friesen, and D. K. Cairns. 1992. Reproductive energetics and prey harvest of Leach's storm-petrels in the northwest Atlantic. Ecology 73:823–832.

Montgomery, W. L. 1977. Diet and gut morphology in fishes, with special reference to the monkeyface prickleback, *Cebidichthys violaceus* (Stichaeidae: Blennioidei). Copeia 1977:178–182.

Moore, J. A. 1939. Temperature tolerance and rates of development in the eggs of Amphibia. Ecology 20:459–478.

——. 1949. Patterns of evolution in the genus *Rana*. *In* G. L. Jepsen, E. Mayr, and G. G. Simpson (eds.), Genetics, Paleontology, and Evolution, pp. 315–338. Princeton University Press, Princeton.

Morgan, E. 1982. The Aquatic Ape. Souvenir, London.

Morhardt, S. S., and D. M. Gates. 1974. Energy-exchange analysis of the Belding ground squirrel and its habitat. Ecol. Monogr. 44:17–44.

Morris, R. 1956. The osmoregulatory ability of the lampern (*Lampetra fluviatilis* L.) in sea water during the course of its spawning migration. J. Exp. Biol. 33:235–248.

——. 1958. The mechanism of marine osmoregulation in the lampern (*Lampetra fluviatilis* L.) and the causes of its breakdown during the spawning migration. J. Exp. Biol. 35:649–665.

——. 1960. General problems of osmoregulation with special reference to cyclostomes. Symp. Zool. Soc. Lond. 1:1–14.

——. 1965. Studies on salt and water balance in *Myxine glutinosa* (L.). J. Exp. Biol. 42:359–371.

Morrison, P. R. 1960. Some interrelations between weight and hibernation function. Bull. Mus. Comp. Zool. 124:75–91.

——. 1962. Modification of body temperature by activity in Brazilian hummingbirds. Condor 64:315–323.

——. 1964. Wild animals at high altitudes. Symp. Zool. Soc. Lond. 13:49–55.

Morrison, P. R., and W. Galster. 1975. Patterns of hibernation in the arctic ground squirrel. Can. J. Zool. 53:1345–1355.

Morrison, P. R., K. Kerst, C. Reynafarje, and J. Ramos. 1963b. Hematocrit and hemoglobin levels in some Peruvian rodents from high and low altitude. Int. J. Biometeorol. 7:45–50.

Morrison, P. R., K. Kerst, and M. Rosenmann. 1963a. Hematocrit and hemoglobin levels in some Chilean rodents from high and low altitude. Int. J. Biometeorol. 7:45–50.

Morrison, P. R., and B. K. McNab. 1962. Daily torpor in a Brazilian murine opossum. Comp. Biochem. Physiol. 6:57–68.

Morrison, P. R., and J. H. Petajan. 1962. The development of temperature regulation in the opossum, *Didelphis marsupialis virginiana*. Physiol. Zool. 35:52–65.

Morrison, P. R., and F. A. Ryser. 1951. Temperature and metabolism in some Wisconsin mammals. Fed. Proc. 10:93–94.

——. 1952. Weight and body temperature in mammals. Science 116:231–232.

Morrison, P. R., F. A. Ryser, and R. L. Strecker. 1954. Growth and the development of temperature regulation in the tundra redback vole. J. Mammal. 35:376–386.

Morrison, P. R., and W. J. Tietz. 1957. Cooling and

thermal conductivity in three small Alaskan mammals. J. Mammal. 38:78–86.

Mortensen, A., and A. S. Blix. 1986. Seasonal changes in resting metabolic rate and mass-specific conductance in Svalbard ptarmigan, Norwegian rock ptarmigan and Norwegian willow ptarmigan. Ornis Scand. 17:8–13.

Mortensen, A., S. Unander, M. Kolstad, and A. S. Blix. 1983. Seasonal changes in body composition and crop content of Spitzbergen ptarmigan *Lagopus mutus hyperboreas*. Ornis Scand. 14:144–148.

Mortimer, J. A. 1981. The feeding ecology of the West Caribbean green turtle (*Chelonia mydas*) in Nicaragua. Biotropica 13:49–58.

——. 1982. Feeding ecology of sea turtles. *In* K. A. Bjorndal (ed.), Biology and Conservation of Sea Turtles, pp. 103–109. Smithsonian Institution Press, Washington, D.C.

Morton, E. S. 1973. On the evolutionary advantages and disadvantages of fruit eating in tropical birds. Am. Nat. 107:8–22.

——. 1978. Avian arboreal folivores: Why not? *In* G. G. Montgomery (ed.), The Ecology of Arboreal Folivores, pp. 123–130. Smithsonian Institution Press, Washington, D.C.

Morton, S. R. 1978. Torpor and nest-sharing in free-living *Sminthopsis crassicaudata* (Marsupialia) and *Mus musculus* (Rodentia). J. Mammal. 59:569–575.

Moss, R. 1967. Food selection and nutrition in ptarmigan (*Lagopus mutus*). Symp. Zool. Soc. Lond. 21:207–216.

——. 1972. Effects of captivity on gut lengths in red grouse. J. Wildl. Manage. 36:99–104.

——. 1973. The digestion and intake of winter foods by wild ptarmigan in Alaska. Condor 75:293–300.

——. 1974. Winter diets, gut lengths, and interspecific competition in Alaskan ptarmigan. Auk 91:737–746.

Moss, R., and R. Hewson. 1985. Effects on heather of heavy grazing by mountain hares. Holarct. Ecol. 8:280–284.

Moss, R., and J. A. Parkinson. 1972. The digestion of heather (*Calluna vulgaris*) by red grouse (*Lagopus lagopus scoticus*). Br. J. Nutr. 27:285–298.

Motais, R., J. Isaia, J. C., Rankin, and J. Maetz. 1969. Adaptive changes of the water permeability of the teleostean gill epithelium in relation to external salinity. J. Exp. Biol. 51:529–546.

Moyle, V. 1949. Nitrogenous excretion in chelonian reptiles. Biochem. J. 44:581–584.

Mrosovsky, N. 1971. Hibernation and the Hypothalamus. Appleton Century Crofts, New York.

——. 1978. Circannual cycles in hibernators. *In* L. C. Wang and J. W. Hudson (eds.), Strategies in Cold, pp. 21–65. Academic Press, New York.

——. 1980. Thermal biology of sea turtles. Am. Zool. 20:531–547.

Mrosovsky, N., and K. C. Fisher. 1970. Sliding set points for body weight in ground squirrels during the hibernation season. Can. J. Zool. 48:241–247.

Mrosovsky, N., and P. C. H. Pritchard. 1971. Body temperatures of *Dermochelys coriacea* and other sea turtles. Copeia 1971:624–631.

Mrosovsky, N., and D. F. Sherry. 1980. Animal anorexias. Science 207:837–842.

Muchlinski, A. E., J. M. Hogan, and R. J. Stoutenburgh. 1990. Body temperature regulation in a desert lizard, *Sauromalus obesus*, under undisturbed field conditions. Comp. Biochem. Physiol. A Physiol. 95:579–583.

Mugaas, J. N., and J. R. King. 1981. Annual variation of daily energy expenditure by the black-billed magpie: A study of thermal and behavioral energetics. Stud. Avian Biol. 5:1–78.

Mugaas, J. N., and J. Seidensticker. 1993. Geographic variation of lean body mass and a model of its effect on the capacity of the raccoon to fatten and fast. Bull. Fl. Mus. Nat. Hist. 36:85–107.

Mugaas, J. N., J. Seidensticker, and K. P. Mahlke-Johnson. 1993. Metabolic adaptation to climate and distribution of the raccoon *Procyon lotor* and other Procyonidae. Smithson. Contrib. Zool. 542:1–34.

Muir, B. S. 1969. Gill dimensions as a function of fish size. J. Fish. Res. Board Can. 26:165–170.

Muir, B. S., and G. M. Hughes. 1969. Gill dimensions for three species of tunny. J. Exp. Biol. 51:271–285.

Mullen, R. K. 1970. Respiratory metabolism and body water turnover rates of *Perognathus formosus* in its natural environment. Comp. Biochem. Physiol. 32:259–265.

——. 1971a. Energy metabolism and body water turnover rates of two species of free-living kangaroo rats, *Dipodomys merriami* and *Dipodomys microps*. Comp. Biochem. Physiol. A Physiol. 39:379–390.

——. 1971b. Energy metabolism of *Peromyscus crinitus* in its natural environment. J. Mammal. 52:633–635.

Mullen, R. K., and R. M. Chew. 1973. Estimating the energy metabolism of free-living *Perognathus formosus*: A comparison of direct and indirect methods. Ecology 54:633–637.

Mullen, T. L., and R. H. Alvarado. 1976. Osmotic and ionic regulation in amphibians. Physiol. Zool. 49:11–23.

Munsey, L. D. 1972a. Salinity tolerance of the African pipid frog, *Xenopus laevis*. Copeia 1972:584–586.

——. 1972b. Water loss in five species of lizards. Comp. Biochem. Physiol. A Physiol. 43:781–794.

Munz, F. W., and W. N. McFarland. 1964. Regulatory function of a primitive vertebrate kidney. Comp. Biochem. Physiol. 13:381–400.

Munz, F. W., and R. W. Morris. 1965. Metabolic rate of the hagfish, *Eptatretus stoutii* (Lockington) 1878. Comp. Biochem. Physiol. 16:1–6.

Murray, K. G., K. Winnett-Murray, and G. L. Hunt Jr. 1979. Egg neglect in Xantus' murrelet. Proc. Colonial Waterbird Group 3:186–195.

Murray, R. M., H. Marsh, G. E. Heinsohn, and A. V. Spain. 1977. The role of the midgut caecum and large intestine in the digestion of sea grasses by the dugong. Comp. Biochem. Physiol. A Physiol. 56:7–10.

Murrish, D. E., and K. Schmidt-Nielsen. 1970. Water transport in the cloaca of lizards: Active or passive? Science 170:324–326.

Musacchia, X. J. 1959. The viability of *Chrysemys picta*

submerged at various temperatures. Physiol. Zool. 32:47–50.

Muul, I. 1968. Behavioral and physiological influences on the distribution of the flying squirrel, *Glaucomys volans*. Misc. Publ. Mus. Zool. Univ. Michigan 134:1–66.

Muul, I., and B. L. Lim. 1978. Comparative morphology, food habits, and ecology of some Malaysian arboreal rodents. *In* G. G. Montgomery (ed.), The Ecology of Arboreal Folivores, pp. 361–368. Smithsonian Institution Press, Washington, D.C.

Myers, B. C., and M. M. Eells. 1968. Thermal aggregation in *Boa* constrictor. Herpetologica 24:61–66.

Myers, G. S. 1949. Usage of anadromous, catadromous and allied terms for migratory fishes. Copeia 1949: 89–97.

Nagel, A., and R. Nagel. 1991. How do bats choose optimal temperatures for hibernation? Comp. Biochem. Physiol. A Physiol. 99:323–326.

Nagy, K. A. 1972. Water and electrolyte budgets of a free-living desert lizard. J. Comp. Physiol. 79:39–62.

———. 1975. Water and energy budgets of free-living animals: Measurement using isotopically labelled water. *In* N. F. Hadley (ed.), Environmental Physiology of Desert Organisms, pp. 227–245. Dowden, Hutchinson and Ross, Stoudsberg, Penns.

———. 1982. Energy requirements of free-living iguanid lizards. *In* G. M. Burghardt and A. S. Rand (eds.), Iguanas of the World, pp. 49–59. Noyes, Pine Ridge, N.J.

———. 1983. Ecological energetics. *In* R. B. Huey, E. R. Pianka, and T. W. Schoener (eds.), Lizard Ecology, pp. 24–54. Harvard University Press, Cambridge.

———. 1987. Field metabolic rate and food requirement scaling in mammals and birds. Ecol. Monogr. 57: 111–128.

———. 1989. Field bioenergetics: Accuracy of models and methods. Physiol. Zool. 62:237–252.

———. 1994. Field bioenergetics of mammals: What determines field metabolic rates? Aust. J. Zool. 42:43–53.

Nagy, K. A., R. B. Huey, and A. F. Bennett. 1984a. Field energetics and foraging mode of Kalahari lacertid lizards. Ecology 65:588–596.

Nagy, K. A., and M. H. Knight. 1989. Comparative field energetics of a Kalahari skink (*Mabuya striata*) and gecko (*Pachydactylus bibroni*). Copeia 1989:13–17.

———. 1994. Energy, water, and food use by springbok antelope (*Antidorcas marsupialis*) in the Kalahari Desert. J. Mammal. 75:860–872.

Nagy, K. A., and G. G. Montgomery. 1980. Field metabolic rate, water flux, and food consumption in three-toed sloths (*Bradypus variegatus*). J. Mammal. 61:465–472.

Nagy, K. A., D. Odell, and R. S. Seymour. 1972. Temperature regulation by the inflorescence of *Philodendron*. Science 178:1195–1197.

Nagy, K. A., and C. C. Peterson. 1988. Scaling of water flux rate in animals. Univ. Calif. Publ. Zool. 120:1–172.

Nagy, K. A., R. S. Seymour, A. K. Lee, and R. Braithwaite. 1978. Energy and water budgets in free-living

Antechinus stuartii (Marsupialia: Dasyuridae). J. Mammal. 59:60–68.

Nagy, K. A., and V. H. Shoemaker. 1975. Energy and nitrogen budgets of the free-living desert lizard *Sauromalus obesus*. Physiol. Zool. 48:252–262.

Nagy, K. A., W. R. Siegfried, and R. P. Wilson. 1984b. Energy utilization by free-ranging jackass penguins, *Spheniscus demersus*. Ecology 165:1648–1655.

Naiman, R. J., S. D. Gerking, and R. E. Stuart. 1976. Osmoregulation in the Death Valley pupfish *Cyprinodon milleri* (Pisces: Cyprinodontidae). Copeia 1976:807–810.

Nash, J. 1931. The number and size of glomeruli in the kidneys of fishes, with observations on the morphology of the renal tubules of fishes. Am. J. Anat. 47:425–439.

Navas, C. A. 1996. Metabolic physiology, locomoter performance, and thermal niche breadth in neotropical anurans. Physiol. Zool. 69:1481–1501.

Néchad, M. 1986. Structure and development of brown adipose tissue. *In* P. Trayhurn and D. G. Nichols (eds.), Brown Adipose Tissue, pp. 1–30. Edward Arnold, London.

Needham, J. 1931. Chemical Embryology. Cambridge University, Cambridge.

Negus, N. C., and P. J. Berger. 1977. Experimental triggering of reproduction in a natural population of *Microtus montanus*. Science 196:1230–1231.

———. 1998. Reproductive strategies of *Dicrostonyx groenlandicus* and *Lemmus sibiricus* in high arctic tundra. Can. J. Zool. 76:391–400.

Neill, W. H., R. K. C. Chang, and A. E. Dizon. 1976. Magnitude and ecological implications of thermal inertia in skipjack tuna, *Katsumonus pelamis* (Linnaeus). Environ. Biol. Fish. 1:61–80.

Neill, W. H., and E. D. Stevens. 1974. Thermal inertia versus thermoregulation in "warm" turtles and tunas. Science 184:1008–1010.

Neill, W. T. 1950. An estivating bowfin. Copeia 1950: 240.

———. 1957. Historical biogeography of present day Florida. Bull. Fla. State Mus. 2:175–220.

———. 1958. The occurrence of amphibians and reptiles in saltwater areas, and a bibliography. Bull. Mar. Sci. Gulf Caribb. 8:1–97.

———. 1971. The Last of the Ruling Reptiles: Alligators, Crocodiles, and Their Kin. Columbia University Press, New York.

Nelson, E. M. 1961. The comparative morphology of the definitive swim bladder in the Catastomidae. Am. Midl. Nat. 65:101–110.

Nelson, J. B. 1977. Some relationships between food and breeding in the marine Pelecaniformes. *In* B. Stonehouse and C. Perrins (eds.), Evolutionary Ecology, pp. 77–87. University Park Press, Baltimore.

Nelson, J. S. 1994. Fishes of the World, 3rd ed. J. Wiley and Sons, New York.

Nelson, R. A., H. W. Wahner, J. D. Jones, R. D. Ellefson, and P. E. Zollman. 1973. Metabolism of bears before, during, and after winter sleep. Am. J. Physiol. 224: 491–496.

Nevo, E. 1979. Adaptive convergence and divergence in

subterranean mammals. Annu. Rev. Ecol. Syst. 10: 269–308.

——. 1999. Mosaic Evolution of Subterranean Mammals: Regression, Progression and Global Convergence. Oxford University Press, Oxford.

Nevo, E., and E. Amir. 1964. Geographic variation in reproduction and hibernation patterns of the forest dormouse. J. Mammal. 45:69–87.

Nevo, E., A. Beiles, G. Heth, and S. Simson. 1986. Adaptive differentiation of body size in speciating mole rats. Oecologia 69:327–333.

Nevo, E., and A. Shkolnik. 1974. Adaptive metabolic variation of chromosome forms of mole rats, *Spalax*. Experientia 30:724–726.

Newell, R. C. 1966. The effect of temperature on the metabolism of poikilotherms. Nature 212:426–428.

Newell, R. C., and V. I. Pye. 1970. The influence of thermal acclimation on the relation between oxygen consumption and temperature in *Littorina littorea* (L.) and *Mytilus edulis* L. Comp. Biochem. Physiol. 34: 385–397.

Newman, M. T. 1956. Adaptation of man to cold climates. Evolution 10:101–105.

Newsome, A. E. 1965. The distribution of red kangaroos, *Megaleia rufa* (Desmarest), about sources of persistent food and water in central Australia. Aust. J. Zool. 13:289–299.

Newton, I. 1701. Scala graduum caloris. Calorum descriptiones & signa. Philos. Trans. Lond. 22:824–829.

——. 1966. Fluctuations in the weights of bullfinches. Br. Birds 59:89–100.

Nichols, J. W., and L. J. Weber. 1989. Comparative oxygen affinity of fish and mammalian myoglobins. J. Comp. Physiol. [B] 159:205–209.

Nicol, S. C. 1978. Rates of water turnover in marsupials and eutherians: A comparative review with new data on the Tasmanian devil. Aust. J. Zool. 26:465–473.

Nicoll, M. E., and S. D. Thompson. 1987. Basal metabolic rates and energetics of reproduction in therian mammals: Marsupials and placentals compared. Symp. Zool. Soc. Lond. 57:7–27.

Nilssen, K. J., J. A. Sundsfjord, and A. S. Blix. 1984. Regulation of metabolic rate in Svalbard and Norwegian reindeer. Am. J. Physiol. 247:R837–R841.

Nilsson, L. 1970. Food-seeking activity of south Swedish diving ducks in the non-breeding season. Oikos 21: 145–154.

Noble, G. K. 1925. The integumentary, pulmonary, and cardiac modifications correlated with increased cutaneous respiration in the Amphibia: A solution of the 'hairy frog' problem. J. Morphol. Physiol. 40:341–416.

——. 1929. The adaptive modifications of the arboreal tadpoles of *Hoplophryne* and the torrent tadpoles of *Staurois*. Bull. Am. Mus. Nat. Hist. 58:291–334.

Noll-Banholzer, U. 1979. Body temperature, oxygen consumption, evaporative water loss and heart rate in the fennec. Comp. Biochem. Physiol. A Physiol. 62:585–592.

Norberg, U. M., and J. M. V. Rayner. 1987. Ecological morphology and flight in bats (Mammalia: Chiroptera): Wing adaptations, flight performance, foraging strategy and echolocation. Philos. Trans. R. Soc. Lond. B Biol. Sci. 316:335–427.

Nordlie, F. G. 1966. Thermal acclimation and peptic digestive capacity in the black bullhead, *Ictalurus melas* (Raf.). Am. Midl. Nat. 75:416–424.

——. 1978. The influence of environmental salinity on respiratory oxygen demands in the euryhaline teleost, *Ambassis interrupta* Bleeker. Comp. Biochem. Physiol. A Physiol. 59:271–274.

Nordlie, F. G., D. C. Haney, and S. J. Walsh. 1992. Comparisons of salinity tolerances and osmotic regulatory capabilities in populations of sailfin molly (*Poecilia latipinna*) from brackish and fresh waters. Copeia 1992:741–746.

Nordlie, F. G., and C. W. Leffler. 1975. Ionic regulation and the energetics of osmoregulation in *Mugil cephalus* Lin. Comp. Biochem. Physiol. A Physiol. 51:125–131.

Nordlie, F. G., and S. J. Walsh. 1989. Adaptive radiation in osmotic regulatory patterns among three species of cyprinodontids (Teleostei: Atherinomorpha). Physiol. Zool. 62:1203–1218.

Nordlie, F. G., S. J. Walsh, D. C. Haney, and T. F. Nordlie. 1991. The influence of ambient salinity on routine metabolism in the teleost *Cyprinodon variegatus* Lacepède. J. Fish Biol. 38:115–122.

Nordøy, E. S., and E. S. Blix. 1985. Energy sources in fasting grey seal pups evaluated with computed tomography. Am. J. Physiol. 249:R471–R476.

Norman, D. B., and D. B. Weishampel. 1985. Ornithopod feeding mechanisms: Their bearing on the evolution of herbivory. Am. Nat. 126:151–164.

Norris, K. S. 1953. The ecology of the desert iguana *Dipsosaurus dorsalis*. Ecology 34:265–287.

——. 1967. Color adaptation in desert reptiles and its thermal relationships. *In* W. W. Milstead (ed.), Lizard Ecology: A Symposium, pp. 162–229. University of Missouri Press, Columbia.

Norris, K. S., and W. R. Dawson. 1964. Observations on the water economy and electrolyte excretion of chuckwallas (Lacertilia, *Sauromalus*). Copeia 1964: 638–646.

Norris, R. A., and F. S. L. Williamson. 1955. Variation in relative heart size of certain passerines with increase in altitude. Wilson Bull. 67:78–83.

Novakowski, N. S. 1967. The winter bioenergetics of a beaver population in northern latitudes. Can. J. Zool. 45:1107–1110.

Novikov, G. A. 1972. The use of under-snow refuges among small birds: The sparrow family. Aquilo Ser. Zool. 13:95–97.

Novoa, F. F., M. Rosenmann, and F. Bozinovic. 1991. Physiological responses of four passerine species to simulated altitudes. Comp. Biochem. Physiol. A Physiol. 99:179–183.

Nungesser, W. C., and E. W. Pfeiffer. 1965. Water balance and maximum concentrating capacity in the primitive rodent, *Aplodontia rufa*. Comp. Biochem. Physiol. 14:289–297.

Oates, J. F., G. H. Whitesides, A. G. Davies, P. G. Waterman, S. M. Green, G. L. Dasilva, and S. Mole. 1990. Determinants of variation in tropical forest primate biomass: New evidence from West Africa. Ecology 71:328–343.

Obst, B. S., and J. M. Diamond. 1989. Interspecific variation in sugar and amino acid transport by the avian cecum. J. Exp. Zool. Suppl. 3:117–126.

Obst, B. S., and K. A. Nagy. 1992. Field energy expenditures of the southern giant-petrel. Condor 94:801–810.

——. 1993. Stomach oil and the energy budget of Wilson's storm-petrel nestlings. Condor 95:792–805.

Obst, B. S., K. A. Nagy, and R. E. Ricklefs. 1987. Energy utilization by Wilson's storm-petrel (*Oceanites oceanicus*). Physiol. Zool. 60:200–210.

Ochoa, H. 1999. Energetics of northern phocid seals: The influence of seasonality on food intake and energy expenditure. Ph.D. dissertation, University of Florida, Gainesville.

Odum, E. P. 1958. The fat deposition picture in the white-throated sparrow in comparison with that in long-range migrants. Bird-Banding 29:105–108.

Odum, E. P., and C. E. Connell. 1956. Lipid levels in migrating birds. Science 123:892–894.

Odum, E. P., C. E. Connell, and H. L. Stoddard. 1961. Flight energy and estimated flight ranges of some migratory birds. Auk 78:515–527.

Odum, H. T. 1953. Factors controlling marine invasion into Florida fresh waters. Bull. Mar. Sci. Gulf Caribb. 3:134–156.

Odum, H. T., and R. C. Pinkerton. 1955. Time's speed regulator: The optimum efficiency for maximum power output in physical and biological systems. Am. Sci. 43:331–343.

Ogden, J. C., and P. S. Lobel. 1978. The role of herbivorous fishes and urchins in coral reef communities. Environ. Biol. Fish. 3:49–63.

Oguri, M. 1964. Rectal glands of marine and freshwater sharks: Comparative histology. Science 144:1151–1152.

Oh, H. K., M. B. Jones, and W. M. Longhurst. 1967. Comparison of rumen microbial inhibition resulting from various essential oils insolated from relatively unpalatable plant species. Appl. Microbiol. 16:39–44.

Ohmart, R. D., and R. C. Lasiewski. 1971. Roadrunners: Energy conservation by hypothermia and absorption of sunlight. Science 172:67–69.

Ohmart, R. D., L. Z. McFarland, and J. T. Morgan. 1970. Urographic evidence that urine enters the rectum and ceca of the roadrunner (*Geococcyx californianus*). Comp. Biochem. Physiol. 35:487–489.

Ohmart, R. D., and E. L. Smith. 1970. Use of sodium chloride solutions by Brewer's sparrow and tree sparrow. Auk 87:329–341.

——. 1971. Water deprivation and use of sodium chloride solutions by vesper sparrows (*Pooecetes gramineus*). Condor 73:364–366.

Oksanen, L., T. Oksanen, A. Lukkari, and S. Sirén. 1987. The role of phenol-based inducible defense in the interaction between tundra populations of the vole *Clethrionomys rufocanus* and the dwarf shrub *Vaccinium myrtillus*. Oikos 50:371–380.

Oliphant, I. W. 1983. First observations of brown fat in birds. Condor 85:350–354.

Olowo, J. P., and L. J. Chapman. 1996. Papyrus swamps and variation in the respiratory behaviour of the African fish *Barbus neumayeri*. Afr. J. Ecol. 34:211–222.

Olsen, S. J. 1960. Additional remains of Florida's Pleistocene vampire. J. Mammal. 41:458–462.

Olson, E. C. 1971. Vertebrate Paleozoology. Wiley, New York.

Olson, J. M., W. R. Dawson, and J. J. Camilliere. 1988. Fat from black-capped chickadees: Avian brown adipose tissue? Condor 90:529–537.

Olson, K. R., and P. O. Fromm. 1971. Excretion of urea by two teleosts exposed to different concentrations of ambient ammonia. Comp. Biochem. Physiol. A Physiol. 40:999–1007.

Olson, S. L. 1973. Evolution of the rails of the South Atlantic Islands (Aves: Rallidae). Smithson. Contrib. Zool. 152:1–53.

Olszewski, J. L., and S. Skoczén 1965. The airing of burrows of the mole, *Talpa europaea* Linnaeus, 1758. Acta Theriol. 10:181–193.

Øritsland, N. A. 1970. Energetic significance of absorption of solar radiation in polar homeotherms. *In* H. W. Holdgate (ed.), Antarctic Ecology, vol. I, pp. 464–470. Academic Press, London.

Ortiz, C. L., B. J. Le Boeuf, and D. P. Costa. 1984. Milk intake of elephant seal pups: An index of parental investment. Am. Nat. 124:416–422.

Osmólska, H. 1979. Nasal salt gland in dinosaurs. Acta Palaeontol. Pol. 24:205–215.

Osmond, C. B., W. G. Allaway, B. G. Sutton, J. H. Troughton, O. Queiroz, U. Lüttge, and K. Winter. 1973. Carbon isotope discrimination in photosynthesis of CAM plants. Nature 246:41–42.

Ostrom, J. H. 1969. Osteology of *Deinonychus antirrhopus*, an unusual theropod from the lower Cretaceous of Montana. Bull. Peabody Mus. Nat. Hist. 30:1–165.

——. 1974. *Archaeopteryx* and the origin of flight. Q. Rev. Biol. 49:27–47.

——. 1976. *Archaeopteryx* and the origin of birds. Biol. J. Linn. Soc. 8:91–182.

Osuga, D. T., and R. E. Feeney. 1978. Antifreeze glycoproteins from Arctic fish. J. Biol. Chem. 253:5338–5341.

Ott, M. E., N. Heisler, and G. R. Ultsch. 1980. A re-evaluation of the relationship between temperature and the critical oxygen tension in freshwater fishes. Comp. Biochem. Physiol. A Physiol. 67:337–340.

Ovaska, K., and T. B. Herman. 1986. Fungal consumption by six species of small mammals in Nova Scotia. J. Mammal. 67:208–211.

Owen-Smith, N. 1985. Niche separation among African ungulates. *In* E. S. Vrba (ed.), Species and Speciation, pp. 167–171. Monograph 4. Transvaal Museum, Pretoria.

Owen-Smith, N., and P. Novellie. 1982. What should a clever ungulate eat? Am. Nat. 119:151–178.

Packard, G. C. 1968. Oxygen consumption of *Microtus montanus* in relation to ambient temperature. J. Mammal. 49:215–220.

——. 1974. The evolution of air-breathing in Paleozoic gnathostome fishes. Evolution 28:320–325.

——. 1976. Devonian amphibians: Did they excrete carbon dioxide via skin, gills, or lungs? Evolution 30:270–280.

Packard, G. C., and M. J. Packard. 1990. Patterns of survival at subzero temperatures by hatchling painted turtles and snapping turtles. J. Exp. Zool. 254: 233–236.

——. 1993. Hatchling painted turtles (*Chrysemys picta*) survive exposure to subzero temperatures during hibernation by avoiding freezing. J. Comp. Physiol. [B] 163:147–152.

——. 1995. The basis for cold tolerance in hatchling painted turtles (*Chrysemys picta*). Physiol. Zool. 68: 129–148.

Packard, G. C., C. R. Tracy, and J. J. Roth. 1977. The physiological ecology of reptilian eggs and embryos, and the evolution of viviparity within the class Reptilia. Biol. Rev. Camb. Philos. Soc. 52:71–105.

Padian, K. 1983. A functional analysis of flying and walking in pterosaurs. Paleobiology 9:218–239.

Padian, K., and J. M. V. Rayner. 1993. Structural fibers of the pterosaur wing: Anatomy and aerodynamics. Naturwissenschaften 80:361–364.

Padley, D. 1985. Do the life history parameters of passerines scale to metabolic rate independently of body mass? Oikos 45:285–287.

Pajunen, I. 1983. Ambient temperature dependence of the body temperature and of the duration of the hibernation periods in the garden dormouse, *Eliomys quercinus* L. Cryobiology 20:690–697.

Pak, H., and J. R. V. Zaneveld. 1973. The Cromwell Current on the east side of the Galapagos Islands. J. Geophys. Res. 78:7845–7859.

Paladino, F. V. 1985. Temperature effects on locomotion and activity bioenergetics of amphibians, reptiles, and birds. Am. Zool. 25:965–972.

——. 1986. Transient nocturnal hypothermia in white-crowned sparrows. Ornis Scand. 17:78–80.

Paladino, F. V., and J. R. King. 1979. Energetic cost of terrestrial locomotion: Biped and quadruped runners compared. Rev. Can. Biol. 38:321–323.

——. 1984. Thermoregulation and oxygen consumption during terrestrial locomotion by white-crowned sparrows *Zonotrichia leucophrys gambelii*. Physiol. Zool. 57:226–236.

Paladino, F. V., M. P. O'Connor, and J. R. Spotila. 1990. Metabolism of leatherback turtles, gigantothermy, and thermoregulation of dinosaurs. Nature 344:859–860.

Palzenberger, M., and H. Pohla. 1992. Gill surface area of water-breathing freshwater fish. Rev. Fish Biol. Fisheries 2:187–216.

Pang, P. K. T., R. W. Griffith, and J. W. Atz. 1977. Osmoregulation in elasmobranchs. Am. Zool. 17:365–377.

Parer, J. T., and J. Metcalfe. 1967a. Respiratory studies of monotremes. 1. Blood of platypus (*Ornithorhynchus anatinus*). Respir. Physiol. 3:136–142.

——. 1967b. Respiratory studies of monotremes. 2. Blood of echidna (*Tachyglossus setosus*). Respir. Physiol. 3:143–150.

Parra, R. 1978. Comparison of foregut and hindgut fermentation in herbivores. *In* G. G. Montgomery (ed.), The Ecology of Arboreal Folivores, pp. 205–229. Smithsonian Institution Press, Washington, D.C.

Parry, G. 1961. Osmotic and ionic changes in blood and muscle of migrating salmonids. J. Exp. Biol. 38:411–427.

Parry, G. D. 1983. The influence of the cost of growth on ectotherm metabolism. J. Theor. Biol. 101:453–477.

Parsons, P. E., and C. R. Taylor. 1977. Energetics of brachiation versus walking: A comparison of a suspended and an inverted pendulum mechanism. Physiol. Zool. 50:182–188.

Paterson, J. D. 1990. Comment—Bergmann's rule is invalid: A reply to V. Geist. Can. J. Zool. 68: 1610–1612.

Patterson, J. W., and P. M. C. Davies. 1978a. Energy expenditure and metabolic adaptation during winter dormancy in the lizard *Lacerta vivipara*, Jacquin. J. Thermal Biol. 3:183–186.

——. 1978b. Thermal acclimation in temperate lizards. Nature 275:646–647.

Paul, J. 1986. Body temperature and the specific heat of water. Nature 323:300.

Payan, P., L. Goldstein, and P. R. Forster. 1973. Gills and kidneys in ureosmotic regulation in euryhaline sharks. Am. J. Physiol. 224:367–372.

Payne, H. J., and J. D. Burke. 1964. Blood oxygen capacity in turtles. Am. Midl. Nat. 71:460–465.

Paynter, R. A. 1971. Nasal glands in *Cinclodes nigrofumosus*, a maritime passerine. Bull. Br. Ornithol. Club 91:11–12.

Peaker, M., and J. L. Linzell. 1977. Salt glands in birds and reptiles. Cambridge University Press, Cambridge.

Pearson, O. P. 1950. The metabolism of hummingbirds. Condor 52:145–152.

——. 1954a. Habits of the lizard *Liolaemus multiformis multiformis* at high altitudes in southern Peru. Copeia 1954:111–116.

——. 1954b. The daily energy requirements of a wild Anna hummingbird. Condor 56:317–322.

——. 1960a. Torpidity in birds. Bull. Mus. Comp. Zool. 124:93–103.

——. 1960b. The oxygen consumption and bioenergetics of harvest mice. Physiol. Zool. 33:152–160.

——. 1964. Carnivore-mouse predation: An example of its intensity and bioenergetics. J. Mammal. 45:177–188.

——. 1977. The effect of substrate and of skin color on thermoregulation of a lizard. Comp. Biochem. Physiol. A Physiol. 58:353–358.

Pearson, O. P., and D. F. Bradford. 1976. Thermoregulation of lizards and toads at high altitudes in Peru. Copeia 1976:155–170.

Pease, J. L., R. H. Vowles, and L. B. Keith. 1979. Interaction of snowshoe hares and wood vegetation. J. Wildl. Manage. 43:43–60.

Pehrson, A. 1979. Winter food consumption and digestibility in caged mountain hares. *In* K. Myers and C. D. MacInnes (eds.), Proceedings of the World Lagomorph Conference, pp. 732–742. University of Guelph Press, Ontario.

——. 1983a. Maximal winter browse intake in captive mountain hares. Finn. Game Res. 41:45–55.

——. 1983b. Digestibility and retention of food components in caged mountain hares *Lepus timidus* during winter. Holarct. Ecol. 6:395–403.

Pelster, B., C. R. Bridges, and M. K. Grieshaber. 1988. Respiratory adaptations of the burrowing marine teleost *Lumpenus lampretaeformis* (Walbaum). II. Metabolic adaptations. J. Exp. Mar. Biol. Ecol. 124: 43–55.

Pendergast, B. A., and D. A. Boag. 1973. Seasonal changes in the internal anatomy of spruce grouse in Alberta. Auk 90:307–317.

Pengelley, E. T. (ed.) 1974. Circannual Clocks. Annual Biological Rhythms. Academic Press, New York.

Pengelley, E. T., and S. J. Asmudsen. 1969. Free-running periods of endogenous circannian rhythms in the golden-mantled ground squirrel, *Citellus lateralis*. Comp. Biochem. Physiol. 30:177–183.

——. 1974. Circannual rhythmicity in hibernating mammals. *In* E. T. Pengelley (ed.), Circannual Clocks: Annual Biological Rhythms, pp. 95–160. Academic Press, New York.

Pengelley, E. T., and K. H. Kelly. 1966. A "circannian" rhythm in hibernating species of the genus *Citellus* with observations on their physiological evolution. Comp. Biochem. Physiol. 19:603–617.

Penman, H. L. 1940a. Gas and vapour movements in the soil. I. The diffusion of vapours through porous solids. J. Agr. Sci. 30:437–462.

——. 1940b. Gas and vapour movements in the soil. II. The diffusion of carbon dioxide through porous solids. J. Agr. Sci. 30:570–581.

Penn, M. R., and B. S. Campbell. 1981. Some aspects of thermoregulation in three species of southern African tortoise. S. Afr. J. Zool. 16:35–43.

Pennycuick, C. J. 1969. The mechanics of bird migration. Ibis 111:525–556.

——. 1971a. Gliding flight of the white-backed vulture *Gryps africanus*. J. Exp. Biol. 55:13–38.

——. 1971b. Control of gliding angle in Ruppell's griffon vulture *Gyps ruppellii*. J. Exp. Biol. 55:39–46.

——. 1972. Soaring behaviour and performance of some East African birds observed from a motor glider. Ibis 114:178–218.

——. 1975. On the running of the gnu (*Connochaetes taurinus*) and other animals. J. Exp. Biol. 63:775–799.

——. 1982. The flight of petrels and albatrosses (Procellariiformes) observed in South Georgia and its vicinity. Philos. Trans. R. Soc. Lond. B Biol. Sci. 300:75–106.

——. 1986. Mechanical constraints on the evolution of flight. *In* K. Padian (ed.), The Origin of Birds and the Evolution of Flight, pp. 83–98. Mem. 8. California Academy Sciences, San Francisco.

——. 1987. Flight of seabirds. *In* J. P. Croxall (ed.), Seabirds: Feeding Biology and Role in Marine Ecosystems, pp. 43–62. Cambridge University Press, Cambridge.

——. 1988. On the reconstruction of pterosaurs and their manner of flight, with notes on vortex wakes. Biol. Rev. Camb. Philos. Soc. 63:299–331.

Pennycuick, C. J., and G. A. Bartholomew. 1973. Energy budget of the lesser flamingo (*Phoeniconaias minor* Geoffroy). East Afr. Wildl. J. 11:199–207.

Perrin, M. R., and B. S. Campbell. 1981. Some aspects of thermoregulation in three species of southern African tortoise. S. Afr. J. Zool. 16:35–43.

Peters, R. H. 1983. The Ecological Implications of Body Size. Cambridge University Press, Cambridge.

Peterson, C. C. 1996. Anhomeostasis: Seasonal water and solute relations in two populations of the desert tortoise (*Gopherus agassizii*) during chronic drought. Physiol. Zool. 69:1324–1358.

Peterson, C. C., K. A. Nagy, and J. Diamond. 1990. Sustained metabolic scope. Proc. Natl. Acad. Sci. U.S.A. 87:2324–2328.

Peterson, C. R. 1987. Daily variation in the body temperatures of free-ranging garter snakes. Ecology 68:160–169.

Petterborg, L. J. 1978. Effect of photoperiod on body weight in the vole, *Microtus montanus*. Can. J. Zool. 56:431–435.

Pettigrew, J. D. 1986. Flying primates? Megabats have the advanced pathway from eye to midbrain. Science 231:1304–1306.

——. 1995. Flying primates: Crashed, or crashed through? Symp. Zool. Soc. Lond. 67:3–26.

Pettit, M. J., and T. L. Beitinger. 1985. Oxygen acquisition of the reedfish, *Erpetoichthys calabaricus*. J. Exp. Biol. 114:289–306.

Pettit, T. N., K. A. Nagy, H. I. Ellis, and G. C. Whittow. 1988. Incubation energetics of the Laysan alabatross. Oecologia 74:546–550.

Pfeiffer, E. W. 1968. Comparative anatomical observations of the mammalian renal pelvis and medulla. J. Anat. 102:321–331.

Pfeiffer, E. W., W. C. Nungesser, D. A. Iverson, and J. F. Wallerius. 1960. The renal anatomy of the primitive rodent, *Aplodontia rufa*, and a consideration of its functional significance. Anat. Rec. 137:227–235.

Phillips, C. J., R. P. Coppinger, and D. S. Schimel. 1981. Hyperthermia in running sled dogs. J. Appl. Physiol. 51:135–142.

Phillips, P. K., and J. E. Heath. 1992. Heat exchange by the pinna of the African elephant (*Loxodonta africana*). Comp. Biochem. Physiol. A Physiol. 101: 693–699.

Phleger, C. F. 1988. The importance of skull lipid as an energy reserve during starvation in the ocean sturgeon, *Acanthurus bahianus*. Comp. Biochem. Physiol. A Physiol. 91:97–100.

Pianka, E. R. 1970. On *r*- and *K*-selection. Am. Nat. 104:592–597.

Pickering, A. D., and R. Morris. 1970. Osmoregulation of *Lampetra fluviatilis* L. and *Petromyzon marinus* (Cyclostomata) in hyperosmotic solutions. J. Exp. Biol. 53:231–243.

Pickford, G. E., and F. B. Grant. 1967. Serum osmolality in the coelacanth, *Latimeria chalumnae*: Urea retention and ion regulation. Science 155:568–570.

Pidcock, S., L. E. Taplin, and G. C. Grigg. 1997. Differences in renal-cloacal function between *Crocodylus porosus* and *Alligator mississippiensis* have implications for crocodilian evolution. J. Comp. Physiol. [B] 167:153–158.

Pielou, E. C. 1991. After the Ice Age: The Return of Life to Glaciated North America. University of Chicago Press, Chicago.

Piermarini, P. M., and D. H. Evans. 1998. Osmoregulation of the Atlantic stingray (*Dasyatis sabrina*) from the freshwater Lake Jesup of the St. John's River, Florida. Physiol. Zool. 71:553–560.

Piersma, T. 1988. Breast muscle atrophy and constraints on foraging during the flightless period of wing moulting great crested grebes. Ardea 76:96–106.

Piersma, T., L. Bruinzeel, R. Drent, M. Kersten, J. Van der Meer, and P. Wiersma. 1996. Variability in basal metabolic rate of a long-distance migrant shorebird (red knot, *Calidris canutus*) reflects shifts in organ sizes. Physiol. Zool. 69:191–217.

Piersma, T., N. Cadée, and S. Daan. 1995. Seasonality in basal metabolic rate and thermal conductance in a long-distance migrant shorebird, the knot (*Calidris canutus*). J. Comp. Physiol. [B] 165:37–45.

Piersma, T., R. Drent, and P. Wiersma. 1991. Temperate versus tropical wintering in the world's northernmost breeder, the knot: Metabolic scope and resource levels restrict subspecific options. Acta XX Congr. Int. Ornithol. 2:761–772.

Pinder, A. W., K. B. Storey, and G. R. Ultsch. 1990. Estivation and hibernation. *In* M. E. Feder and W. W. Burggren (eds.), Environmental Physiology of the Amphibians, pp. 250–274. University of Chicago Press, Chicago.

Pitelka, F. A., and A. M. Schultz. 1964. The nutrient-recovery hypothesis for arctic microtine cycles. *In* D. J. Crisp (ed.), Grazing in Terrestrial and Marine Environments, pp. 55–68. Blackwell Science, Oxford.

Place, A. R., and D. A. Powers. 1979. Genetic variation and relative catalytic efficiencies: Lactate dehydrogenase B allozymes of *Fundulus heteroclitus*. Proc. Natl. Acad. Sci. U.S.A. 76:2354–2358.

Place, A. R., N. C. Stoyan, R. E. Ricklefs, and R. G. Butler. 1989. Physiological basis of stomach oil formation in Leach's storm-petrel (*Oceanodroma leucorhoa*). Auk 106:687–699.

Plakke, R. K., and E. W. Pfeiffer. 1964. Blood vessels of the mammalian renal medulla. Science 146:1683–1685.

Platt, T., and W. Silvert. 1981. Ecology, physiology, allometry and dimensionality. J. Theor. Biol. 93:855–860.

Poe, S. 1996. Data set in congruence and the phylogeny of crocodilians. Syst. Biol. 45:393–414.

Pohl, H. 1965. Temperature regulation and cold acclimation in the golden hamster. J. Appl. Physiol. 20:405–410.

Pohl, H., and J. S. Hart. 1965. Thermoregulation and cold acclimation in a hibenator, *Citellus tridecemlineatus*. J. Appl. Physiol. 20:398–404.

Pohl, H., and G. West. 1973. Daily and seasonal variation in metabolic response to cold during rest and forced exercise in the common redpoll. Comp. Biochem. Physiol. A Physiol. 45:851–867.

Pomeroy, D. 1990. Why fly? The possible benefits for lower mortality. Biol. J. Linn. Soc. 40:53–65.

Pond, C. M., and M. A. Ramsay. 1992. Allometry of the distribution of adipose tissue in Carnivora. Can. J. Zool. 70:342–347.

Porter, W. P., and D. M. Gates. 1969. Thermodynamic equilibria of animals with environment. Ecol. Monogr. 39:245–270.

Porter, W. P., J. W. Mitchell, W. A. Beckman, and C. B. DeWitt. 1973. Behavioral implications of mechanistic ecology. Oecologia 13:1–54.

Post, E., and N. C. Stenseth. 1999. Climatic variability, plant phenology, and northern ungulates. Ecology 80:1322–1339.

Potts, W. T. W. 1954. The energetics of osmotic regulation in brackish- and fresh-water animals. J. Exp. Biol. 31:618–630.

———. 1985. Discussion. Philos. Trans. R. Soc. Lond. B Biol. Sci. 309:319–320.

Potts, W. T. W., and G. Parry. 1964. Osmotic and Ionic Regulation in Animals. Pergamon Press, Oxford.

Potts, W. T. W., and P. P. Rudy. 1972. Aspects of osmotic and ionic regulation in the sturgeon. J. Exp. Biol. 56:703–715.

Pough, F. H. 1969. Physiological aspects of the burrowing of sand lizards (*Uma*, Iguanidae) and other lizards. Comp. Biochem. Physiol. 31:869–884.

———. 1973a. Heart rate, breathing and voluntary diving of the elephant trunk snake, *Acrochordus javanicus*. Comp. Biochem. Physiol. A Physiol. 44:183–189.

———. 1973b. Lizard energetics and diet. Ecology 54:837–844.

———. 1976. The effect of temperature on oxygen capacity of reptile blood. Physiol. Zool. 49:141–151.

———. 1977a. The relationship between body size and blood oxygen affinity in snakes. Physiol. Zool. 50:77–87.

———. 1977b. Ontogenetic change in molecular and functional properties of blood of garter snakes, *Thamnophis sirtalis*. J. Exp. Zool. 201:47–56.

———. 1980a. The advantages of ectothermy for tetrapods. Am. Nat. 115:92–112.

———. 1980b. Blood oxygen transport and delivery in reptiles. Am. Zool. 20:173–185.

———. 1980c. Environmental adaptations in the blood of lizards. Comp. Biochem. Physiol. 31:885–901.

———. 1983. Amphibians and reptiles as low-energy systems. *In* W. P. Aspey and S. I. Lustick (eds.), Behavioral Energetics: The Cost of Survival in Vertebrates, pp. 141–188. Ohio State University Press, Columbus.

Pough, F. H., and R. M. Andrews. 1984. Individual and sibling-group variation in metabolism of lizards: The aerobic capacity model for the origin of endothermy. Comp. Biochem. Physiol. A Physiol. 79:415–419.

———. 1985. Use of anaerobic metabolism by free-ranging lizards. Physiol. Zool. 58:205–213.

Pough, F. H., and C. Gans. 1982. The vocabulary of reptilian thermoregulation. *In* C. Gans and H. F. Pough (eds.), Biology of the Reptilia, vol. 12, pp. 17–23. Academic Press, New York.

Pough, F. H., and T. L. Taigen. 1990. Metabolic correlates of the foraging and social behaviour of dart-poison frogs. Anim. Behav. 39:145–155.

Poulson, T. L. 1963. Cave adaptation in amblyopsid fisheries. Am. Midl. Nat. 70:257–290.

——. 1964. Animals in aquatic environments: Animals in caves. *In* D. B. Dill (ed.), Adaptation to the Environment, Section 4, Handbook of Physiology. American Physiological Society, Washington, D.C.

——. 1965. Countercurrent multipliers in avian kidneys. Science 148:389–391.

Poulson, T. L., and G. A. Bartholomew. 1962a. Salt balance in the savannah sparrow. Physiol. Zool. 35:109–119.

——. 1962b. Salt utilization in the house finch. Condor 64:245–252.

Poulson, T. L., and W. B. White. 1969. The cave environment. Science 165:971–981.

Pounds, J. A., and M. L. Crump. 1994. Amphibian declines and climate disturbance: The case of the golden toad and the harlequin frog. Conserv. Biol. 8:72–85.

Pounds, J. A., M. P. L. Fogden, and J. H. Campbell. 1999. Biological response to climate change on a tropical mountain. Nature 398:611–615.

Pounds, J. A., M. P. L. Fogden, J. M. Savage, and G. C. Gorman. 1996. Tests of null models for amphibian declines on a tropical mountain. Conserv. Biol. 11:1307–1322.

Powell, R. A. 1979. Ecological energetics and foraging strategies of the fisher (*Martes pennanti*). J. Anim. Ecol. 48:195–212.

Powell, R. A., and R. D. Leonard. 1983. Sexual dimorphism and energy expenditure for reproduction in female fisher *Martes pennanti*. Oikos 40:166–174.

Powers, D. A. 1972. Hemoglobin adaptation for fast and slow water habitats in sympatric catostomid fishes. Science 177:360–362.

——. 1980. Molecular ecology of teleost fish hemoglobins: Strategies for adapting to changing environments. Am. Zool. 20:139–162.

Powers, D. A., H. J. Fyhn, U. E. H. Fyhn, J. P. Martin, R. L. Garlick, and S. C. Wood. 1979. A comparative study of the oxygen equilibria of blood from 40 genera of Amazonian fishes. Comp. Biochem. Physiol. A Physiol. 62:67–85.

Powers, D. A., G. S. Greaney, and A. R. Place. 1979. Physiological correlation between lactate dehydrogenase genotype and haemoglobin function in killifish. Nature 277:240–241.

Powers, D. A., and D. Powers. 1975. Predicting gene frequencies in natural populations: A testable hypothesis. *In* C. Markert (ed.), The Isozymes, Genetics and Evolution, vol. 4, pp. 63–84. Academic Press, New York.

Powers, D. R. 1992. Effect of temperature and humidity on evaporative water loss in Anna's hummingbird (*Calypte anna*). J. Comp. Physiol. [B] 162:74–84.

Prance, G. T., and J. R. Arias. 1975. A study of the floral biology of *Victoria amazonica* (Poepp.) Sowerby (Nymphaeaceae). Acta Amazonica 5:109–139.

Prange, H. D., and K. Schmidt-Nielsen. 1969. Evaporative water loss in snakes. Comp. Biochem. Physiol. 28:973–975.

——. 1970. The metabolic cost of swimming in ducks. J. Exp. Biol. 53:763–777.

Precht, H. 1958. Theory of temperature adaptation in cold-blooded animals. *In* C. L. Prosser (ed.), Physiological Adaptation, pp. 50–78. American Physiological Society, Washington, D.C.

Preest, M. R., and C. A. Beuchat. 1994. Nitrogen excretion in hummingbirds: Ammonotely in a bird? Am. Zool. 37:A53.

Preest, M. R., and F. H. Pough. 1989. Interaction of temperature and hydration on locomotion of toads. Funct. Ecol. 3:693–699.

Prestrud, P., and K. Nilssen. 1992. Fat deposition and seasonal variation in body composition of arctic foxes in Svalbard. J. Wildl. Manage. 56:221–233.

Prestwich, K. N., K. E. Brugger, and M. Topping. 1989. Energy and communication in three species of hylid frogs: Power input, power output and efficiency. J. Exp. Biol. 144:53–80.

Price, K. S., Jr. 1967. Fluctuations in two osmoregulatory components, urea and sodium chloride, of the clearnose skate, *Raja eglanteria* BOSC 1802—II. Upon natural variation of the salinity of the external medium. Comp. Biochem. Physiol. 23:77–82.

Price, K. S., Jr., and E. P. Creaser Jr. 1967. Fluctuations in two osmoregulatory components, urea and sodium chloride, of the clearnose skate, *Raja eglanteria* BOSC 1802—I. Upon laboratory modification of external salinities. Comp. Biochem. Physiol. 23:65–76.

Prinzinger, R. 1988. Energy metabolism, body-temperature and breathing parameters in nontorpid blue-naped mousebirds *Urocolius macrourus*. J. Comp. Physiol. [B] 157:801–806.

——. 1993. Life span in birds and the ageing theory of absolute metabolic scope. Comp. Biochem. Physiol. A Physiol. 105:609–615.

Prinzinger, R., and I. Hänssler. 1980. Metabolism-weight relationship in some small nonpasserine birds. Experientia 36:1299–1300.

Prinzinger, R., A. Preßmar, and E. Schleucher. 1991. Body temperature in birds. Comp. Biochem. Physiol. A Physiol. 99:499–506.

Pritchard, P. C. H., and W. F. Greenhood. 1968. The sun and the turtle. J. Int. Turtle Tort. Soc. 2:20–25, 34.

Prothero, J. 1992. Scaling of bodily proportions in adult terrestrial mammals. Am. J. Physiol. 262:R492–R503.

Prothero, J., and K. D. Jürgens. 1986. An energetic model of daily torpor in endotherms. J. Theor. Biol. 121:403–415.

Pruitt, W. O., Jr. 1957. Observations on the bioclimate of some taiga mammals. Arctic 10:131–138.

Pucek, Z. 1963. Seasonal changes in the brain case of some representatives of the genus *Sorex* from the Palearctic. J. Mammal. 44:523–536.

——. 1964. Morphological changes in shrews kept in captivity. Acta Theriol. 8:137–166.

Pulliainen, E. 1970. Winter nutrition of the rock ptarmigan, *Lagopus mutus* (Montin), in northern Finland. Ann. Zool. Fenn. 7:295–302.

——. 1976. Small intestine and caeca lengths in the willow grouse (*Lagopus lagopus*) in Finnish Lapland. Ann. Zool. Fenn. 13:195–199.

Pulliainen, E., and J. Iivanainen. 1981. Winter nutrition of the willow grouse (*Lagopus lagopus* L.) in the extreme north of Finland. Ann. Zool. Fenn. 18: 263–269.

Pulliainen, E., and P. Tunkkari. 1983. Seasonal changes in the gut length of the willow grouse (*Lagopus lagopus*) in Finnish Lapland. Ann. Zool. Fenn. 20: 53–56.

Pullin, R. S. V., D. J. Morris, C. R. Bridges, and R. J. A. Atkinson. 1980. Aspects of the respiratory physiology of the burrowing fish *Cepola rubescens* L. Comp. Biochem. Physiol. A Physiol. 66:35–42.

Purdue, J. P. 1989. Changes during the Holocene in the size of white-tailed deer (*Odocoileus virginianus*) from central Illinois. Quat. Res. 32:307–316.

Pusey, B. J. 1986. The effect of starvation on oxygen consumption and nitrogen excretion in *Lepidogalaxias salamandroides* (Mees). J. Comp. Physiol. [B] 156: 701–705.

——. 1989. Aestivation in the teleost fish *Lepidogalaxias salamandroides* (Mees). Comp. Biochem. Physiol. A Physiol. 92:137–138.

Quilliam, T. A., J. A. Clarke, and A. J. Salsbury. 1971. The ecological significance of certain new hematological findings in the mole and hedgehog. Comp. Biochem. Physiol. A Physiol. 40:89–102.

Quinn, T., and D. E. Schneider. 1991. Respiration of the teleost fish *Ammodytes hexapterus* in relation to its burrowing behavior. Comp. Biochem. Physiol. A Physiol. 98:71–75.

Racey, P. A. 1973. Environmental factors affecting the length of gestation in heterothermic bats. J. Reprod. Fertil. Suppl. 19:175–189.

Radwan, M. A. 1972. Differences between Douglas-fir genotypes in relation to browsing preference by black-tailed deer. Can. J. For. Res. 2:250–255.

Ragland, I. M., L. C. Wit, and J. C. Sellers. 1981. Temperature acclimation in the lizards *Cnemidophorus sexlineatus* and *Anolis carolinensis*. Comp. Biochem. Physiol. A Physiol. 70:33–36.

Rahn, H. 1966a. Aquatic gas exchange: Theory. Respir. Physiol. 1:1–12.

——. 1966b. Gas transport from the external environment to the cell. *In* A. V. S. de Reuck and R. Porter (eds.), Development of the Lung, pp. 3–23. Ciba Foundation Symposium, Churchill, London.

Rahn, H., and A. Ar. 1979. The avian egg: Incubation time and water loss. Condor 76:147–152.

Rahn, H., and B. J. Howell. 1976. Bimodal gas exchange. *In* G. M. Hughes (ed.), Respiration of Amphibious Vertebrates, pp. 271–285. Academic Press, New York.

Rahn, H., K. B. Rahn, B. J. Howell, C. Gans, and S. M. Tenney. 1971b. Air breathing of the garfish (*Lepisosteus osseus*). Respir. Physiol. 11:285–307.

Rahn, H., O. D. Wangensteen, and L. E. Farhi. 1971a. Convection and diffusion gas exchange in air or water. Respir. Physiol. 12:1–16.

Ralph, R., and I. Emerson. 1968. The respiratory metabolism of some antarctic fish. Comp. Biochem. Physiol. 27:299–307.

Ramaswamy, M., and T. G. Reddy. 1983. Ammonia and urea excretion in three species of air-breathing fish subjected to aerial exposure. Proc. Indian Acad. Sci. (Anim. Sci.) 92:293–297.

Ramsey, J. A. 1935. Methods of measuring the evaporation of water from animals. J. Exp. Biol. 12:355–372.

Randall, D. J., C. M. Wood, S. F. Perry, H. Bergman, G. M. O. Maloiy, T. P. Mommsen, and P. A. Wright. 1989. Urea excretion as a strategy in a very alkaline environment. Nature 337:165–166.

Randolph, J. C. 1980. Daily energy metabolism of two rodents (*Peromyscus leucopus* and *Tamias striatus*) in their natural environment. Physiol. Zool. 53:70–81.

Randolph, P. A., J. C. Randolph, K. Mattingly, and M. M. Foster. 1977. Energy costs of reproduction in the cotton rat, *Sigmodon hispidus*. Ecology 58:31–45.

Rao, K. P., and T. H. Bullock. 1954. Q_{10} as a function of size and habitat temperature in poikilotherms. Am. Nat. 88:33–44.

Rashevsky, N. 1960. Mathematical Biophysics, 3rd ed., vol. 2. Dover, New York.

Rasmussen, D. T., and M. K. Izard. 1988. Scaling of growth and life history traits relative to body size, brain size, and metabolic rate in lorises and galagos (Lorisidae, Primates). Am. J. Phys. Anthropol. 75: 357–367.

Ray, C. 1958. Vital limits and rates of dessication in salamanders. Ecology 39:75–83.

——. 1960. The application of Bergmann's and Allen's rules to the poikilotherms. J. Morphol. 106:85–108.

Ray, C. E., S. J. Olsen, and H. J. Gut. 1963. Three mammals new to the Pleistocene fauna of Florida, and a reconsideration of five earlier records. J. Mammal. 44:373–395.

Rayner, J. M. V. 1979. A vortex theory of animal flight. Part 2. The forward flight of birds. J. Fluid Mech. 91:731–763.

——. 1988. Form and function in avian flight. Curr. Ornithol. 5:1–66.

——. 1989. Mechanics and physiology of flight in fossil vertebrates. Trans. R. Soc. Edinb. Earth Sci. 80: 311–320.

Read, L. J. 1971a. Chemical composition of body fluids and urine of the holocephalan *Hydrolagus colliei*. Comp. Biochem. Physiol. A Physiol. 39:185–192.

——. 1971b. The presence of high ornithine-urea cycle enzyme activity in the teleost *Opsanus tau*. Comp. Biochem. Physiol. B Biochem. Mol. Biol. 39:409–413.

Reagan, N. L., and P. A. Verrell. 1991. The evolution of plethodontid salamanders: Did terrestrial mating facilitate lunglessness? Am. Nat. 138:1307–1313.

Redford, K. H. 1984. Mammalian predation on termites: Tests with the burrowing mouse (*Oxymycterus robertii*) and its prey. Oecologia 65:145–152.

Redford, K. H., G. A. Bouchardet da Fonseca, and T. E. Lachner. 1984. The relationship between frugivory and insectivory in primates. Primates 25:433–440.

Reeves, R. B. 1976. Temperature-induced changes in

blood acid-base status: pH and pCO_2 in a binary buffer. J. Appl. Physiol. 40:752–761.

Regal, P. J. 1966. Thermophilic response following feeding in certain reptiles. Copeia 1966:588–590.

———. 1967. Voluntary hypothermia in reptiles. Science 155:1551–1553.

Reichardt, P. B., J. P. Bryant, T. P. Clausen, and G. D. Wieland. 1984. Defense of winter-dormant Alaska paper birch against snowshoe hares. Oecologia 65:58–69.

Reid, B. 1974. Faeces of takahe (*Notornis mantelli*): A general discussion relating the quantity of faeces to the type of food and the estimated energy requirements of the bird. Notornis 21:306–311.

Reid, R. E. H. 1987. Bone and dinosaurian "endothermy." Modern Geol. 11:133–154.

Reinertsen, R. E. 1983. Nocturnal hypothermia and its energetic significance for small birds living in the arctic and subarctic regions. A review. Polar Res. 1:269–284.

———. 1984. Seasonal and local variation of nocturnal hypothermia between lowland and highland willow tits in central Norway. Fauna Norv. Ser. C Cinclus 7:70–74.

Reinertsen, R. E., and S. Haftorn. 1983. Nocturnal hypothermia and metabolism in the willow tit *Parus montanus* at 63°N. J. Comp. Physiol. 151:109–118.

———. 1984. The effect of short-time fasting on metabolism and nocturnal hypothermia in the willow tit *Parus montanus*. J. Comp. Physiol. [B] 154:23–28.

———. 1986. Different metabolic strategies of northern birds for nocturnal survival. J. Comp. Physiol. 156:655–663.

Reiss, M. 1988. Scaling of home range size: Body size, metabolic needs and ecology. Trends Ecol. Evol. 3:85–86.

Reno, H. W., F. R. Gehlbach, and R. A. Turner. 1972. Skin and aestivational cocoon of the aquatic amphibian, *Siren intermedia* Le Conte. Copeia 1972:625–631.

Rensch, I., and B. Rensch. 1956. Relative Organmasse bei tropischen Warmblutern. Zool. Anz. 156:106–124.

Reyer, H.-U., and K. Westerterp. 1985. Parental energy expenditure: A proximate cause of helper recruitment in the pied kingfisher (*Cerlye rudis*). Behav. Ecol. Sociobiol. 17:363–369.

Reynafarje, B., and P. R. Morrison. 1962. Myoglobin levels in some tissues from wild Peruvian rodents native to high altitude. J. Biol. Chem. 237:2861–2864.

Reynolds, P. S., and R. M. Lee III. 1996. Phylogenetic analysis of avian energetics: Passerines and nonpasserines do not differ. Am. Nat. 147:735–759.

Rhoades, D. F., and R. G. Cates. 1976. Toward a general theory of plant antiherbivore chemistry. *In* J. W. Wallace and R. L. Mansell (eds.), Biochemical Interactions between Plants and Insects, pp. 168–213. Plenum, New York.

Rice, D. W. 1957. Life history and ecology of *Myotis austroriparius* in Florida. J. Mammal. 38:15–32.

Rich, T. H. V., and P. V. Rich. 1989. Polar dinosaurs and biotas of the early Cretaceous of southeastern Australia. Nat. Geogr. Res. 5:15–53.

Richet, C. 1885. Récherches de calorimetre. Arch. Physiol. 6:237–291.

———. 1889. La chaleur animale. Paris.

Ricklefs, R. E. 1967. A graphical method of fitting equations to growth curves. Ecology 48:978–983.

———. 1968. Patterns of growth in birds. Ibis 110:419–451.

———. 1973a. Ecology. Chiron Press, Newton, Mass.

———. 1973b. Patterns of growth in birds. II. Growth rate and mode of development. Ibis 115:177–201.

———. 1974. Energetics of reproduction in birds. *In* R. A. Paynter (ed.), Avian Energetics, pp. 152–292. Publ. 15. Nuttall Ornithological Club, Cambridge, Mass.

———. 1976. Growth rates of birds in the humid New World tropics. Ibis 118:179–207.

———. 1979. Adaptation, constraint, and compromise in avian postnatal development. Biol. Rev. Camb. Philos. Soc. 54:269–290.

———. 1980. Geographical variation in clutch size among passerine birds: Ashmole's hypothesis. Auk 97:38–49.

———. 1983. Some considerations on the reproductive energetics of pelagic seabirds. Stud. Avian Biol. 8:84–94.

———. 1996. Avian energetics, ecology, and evolution. *In* C. Carey (ed.), Avian Energetics and Nutritional Ecology, pp. 1–30. Chapman and Hall, New York.

Ricklefs, R. E., and F. R. Hainsworth. 1968. Temperature dependent behavior of the cactus wren. Ecology 49:227–233.

Ricklefs, R. E., M. Konarzewski, and S. Daan. 1996. The relationship between basal metabolic rate and daily energy expenditure in birds and mammals. Am. Nat. 147:1047–1071.

Ricklefs, R. E., and W. A. Schew. 1994. Foraging stochasticity and lipid accumulation by nestling petrels. Funct. Ecol. 8:159–170.

Ricklefs, R. E., S. C. White, and J. Cullen. 1980. Energetics of postnatal growth in Leach's storm-petrel. Auk 97:566–575.

Riegel, J. A. 1978. Factors affecting glomerular function in the Pacific hagfish, *Eptatretus stoutii*. J. Exp. Biol. 73:261–277.

Riegel, K., H. Bartels, I. O. Buss, P. G. Wright, E. Kleihauer, C. P. Luck, J. T. Parer, and J. Metcalfe. 1967. Comparative studies of the respiratory functions of mammalian blood. IV. Fetal and adult African elephant blood. Respir. Physiol. 2:182–195.

Riggs, A. 1979. Studies of the hemoglobins of Amazonian fishes: An overview. Comp. Biochem. Physiol. A Physiol. 62:257–271.

Risenhoover, K. L., L. A. Renecker, and L. E. Morgantini. 1985. Effects of secondary metabolites from balsam poplar and paper birch on cellulose digestion. J. Range Manage. 38:370–371.

Robbins, C. T., A. E. Hagerman, P. J. Austin, C. McArthur, and T. A. Hanley. 1991. Variation in mammalian physiological responses to a condensed tannin and its ecological implications. J. Mammal. 72:480–486.

Robbins, C. T., T. A. Hanley, A. E. Hagerman, O. Hjeijord, D. L. Baker, C. C. Schwartz, and W. W. Mautz. 1987a. Role of tannins in defending plants against ruminants: Reduction in protein availability. Ecology 68:98–107.

Robbins, C. T., S. Mole, E. A. Hagerman, and T. A. Hanley. 1987b. Role of tannins in defending plants against ruminants: Reduction in dry matter digestion? Ecology 68:1606–1615.

Roberts, J. S., and B. Schmidt-Nielsen. 1966. Renal ultrastructure and excretion of salt and water by three terrestrial lizards. Am. J. Physiol. 211:476–486.

Roberts, L. A. 1968. Oxygen consumption in the lizard *Uta stansburiana*. Ecology 49:809–819.

Robertshaw, D., and C. R. Taylor. 1969. A comparison of sweat gland activity in eight species of East African bovids. J. Physiol. 203:135–143.

Robertson, I. C., and P. J. Weatherhead. 1992. The role of temperature in microhabitat selection by northern water snakes (*Nerodia sipedon*). Can. J. Zool. 70: 417–422.

Robertson, J. D. 1954. The chemical composition of the blood of some aquatic chordates including members of the *Tunicata*, *Cyclostomata* and *Osteichthyes*. J. Exp. Biol. 31:424–442.

———. 1957. The habitat of the early vertebrates. Biol. Rev. Camb. Philos. Soc. 32:156–187.

Robertson, S. L., and E. N. Smith. 1979. Thermal implications of cutaneous blood flow in the American alligator. Comp. Biochem. Physiol. A Physiol. 62: 569–572.

———. 1981. Thermal conductance and its relation to thermal time constants. J. Thermal Biol. 6:129–143.

Robin, J. P., M. Frain, C. Sardet, R. Groscolas, and Y. Le Maho. 1988. Protein and lipid utilization during long-term fasting in emperor penguins. Am. J. Physiol. 254:R61–R68.

Robinson, D. E., G. S. Campbell, and J. R. King. 1976. An evaluation of heat exchange in small birds. J. Comp. Physiol. 105:153–166.

Robinson, J. G., and K. H. Redford. 1986. Intrinsic rate of natural increase in neotropical forest mammals: Relationships to phylogeny and diet. Oecologia 68:516–520.

Robinson, P. L. 1971. A problem of faunal replacement on Permo-Triassic continents. Palaeontology 14:131–153.

Roby, D. E., K. L. Brink, and A. R. Place. 1989. Relative passage rates of lipid and aqueous digesta in the formation of stomach oils. Auk 106:303–313.

Roby, D. E., and R. E. Ricklefs. 1986. Energy expenditure in adult least auklets and diving petrels during the chick-rearing period. Physiol. Zool. 59:661–678.

Rodgers, A. R., and M. C. Lewis. 1985. Diet selection in Arctic lemmings (*Lemmus sibericus* and *Dicrostonyx groenlandicus*): Food preferences. Can. J. Zool. 63: 1161–1173.

Roemmich, D. 1992. Ocean warming and sea level rise along the Southwest U.S. coast. Science 257:373–375.

Roemmich, D., and J. McGowan. 1995. Climatic warming and the decline of zooplankton in the California Current. Science 267:1324–1326.

Rogowitz, G. L., and J. A. Gessaman. 1990. Influence of air temperature, wind and irradiance on metabolism of white-tailed jackrabbits. J. Thermal Biol. 15:125–131.

Rombough, P. J. 1994. Energy partitioning during fish development: Additive or compensatory allocation of energy to support growth? Funct. Ecol. 8:178–186.

Rombough, P. J., and B. M. Moroz. 1990. The scaling and potential importance of cutaneous and branchial surfaces in respiratory gas exchange in young chinook salmon (*Oncorhynchus tshawytscha*). J. Exp. Biol. 154:1–12.

Romer, A. S. 1947. Review of the Labyrinthodontia. Bull. Mus. Comp. Zool. Harvard 99:1–368.

———. 1956. Osteology of Reptiles. University of Chicago Press, Chicago.

———. 1966. Vertebrate Paleontology, 3rd ed. University of Chicago Press, Chicago.

———. 1972. Skin breathing—primary or secondary? Respir. Physiol. 14:183–192.

Romer, A. S., and B. H. Grove. 1935. Environment of the early vertebrates. Am. Midl. Nat. 16:805–856.

Romer, A. S., and E. C. Olson. 1954. Aestivation in a Permian lungfish. Brevoria 30:1–8.

Root, R. W. 1931. The respiratory function of the blood of marine fishes. Biol. Bull. (Woods Hole) 61:427–456.

———. 1949. Aquatic respiration in the musk turtle. Physiol. Zool. 172–178.

Root, T. 1988a. Environmental factors associated with avian distributional boundaries. J. Biogeogr. 15:489–505.

———. 1988b. Energy constraints on avian distributions and abundances. Ecology 69:330–339.

———. 1988c. Atlas of Wintering North American Birds. University of Chicago Press, Chicago.

———. 1989. Energy constraints on avian distributions: A reply to Castro. Ecology 70:1183–1185.

Root, T., T. P. O'Connor, and W. R. Dawson. 1991. Standard metabolic level and insulative characteristics of eastern house finches, *Carpodacus mexicanus* (Muller). Physiol. Zool. 64:1279–1295.

Rose, F. L. 1967. Seasonal changes in lipid levels of the salamander *Amphiuma means*. Copeia 1967:662–666.

Rosen, D. E., P. L. Forey, B. G. Gardiner, and C. Patterson. 1981. Lungfishes, tetrapods, paleontology, and plesiomorphy. Bull. Am. Mus. Nat. Hist. 167: 159–276.

Rosenberger, A. E., and L. J. Chapman. 2000. Respiratory characters of three species of haplochromine cichlids: Implications for use of wetland refugia. J. Fish Biol. 57:483–501.

Rosenmann, M., and P. R. Morrison. 1974. Physiological responses to hypoxia in the tundra vole. Am. J. Physiol. 227:734–739.

———. 1975. Metabolic responses of highland and lowland rodents to simulated high altitudes and cold. Comp. Biochem. Physiol. A Physiol. 51:523–530.

Rosenzweig, M. L. 1968. Net primary production of terrestrial communities: Prediction from climatological data. Am. Nat. 102:67–74.

Røskraft, E., T. Järvi, M. Bakken, C. Bech, and R. E. Reinertsen. 1986. The relationship between social status and resting metabolic rate in great tits (*Parus major*) and pied flycatchers (*Ficedula hypoleuca*). Anim. Behav. 34:838–842.

Rosser, B. W., and J. C. George. 1986a. The avian pectoralis: Histochemical characterization and distribution of muscle fibers. Can. J. Zool. 64:1174–1185.

——. 1986b. Slow muscle fibers in the pectoralis of the turkey vulture (*Cathartes aura*): An adaptation for soaring flight. Zool. Anz. 217:252–258.

Roth, J. J. 1973. Vascular supply to the ventral pelvic regions of anurans as related to water balance. J. Morphol. 140:443–460.

Roth, V. L. 1990. Insular dwarf elephants: A case study in body mass estimation and ecological inference. *In* J. Damath and B. J. MacFadden (eds.), Body Size in Mammalian Paleobiology, pp. 151–179. Cambridge University Press, Cambridge.

Rothwell, N. J., and M. J. Stock. 1985. Biological distribution and significance of brown adipose tissue. Comp. Biochem. Physiol. A Physiol. 82:745–751.

Rounsevell, D. 1970. Salt excretion in the Australian pipit, *Anthus novaeseelandiae* (Aves: Motacillidae). Aust. J. Zool. 18:373–377.

Rovedatti, M. G., P. M. Castañé, A. Salibián, and S. Espina. 1988. Studies on the urinary waste products in South American anurans from different habitats. Comp. Biochem. Physiol. A Physiol. 90:249–252.

Rowan, W. 1925. Relation of light to bird migration and developmental changes. Nature 115:494–495.

——. 1926. On photoperiodism, reproductive periodicity and the annual migrations of birds and certain fishes. Proc. Boston Soc. Nat. Hist. 38:147–189.

——. 1927. Migration and reproductive rhythm in birds. Nature 119:351–352.

Rowe, M. F. 1999. Physiological responses of African elephants to a cold environment: How does the elephant keep its heat? Master's thesis, Louisiana State University, Shreveport.

Ruben, J. A. 1976a. Aerobic and anaerobic metabolism during activity in snakes. J. Comp. Physiol. 109:147–157.

——. 1976b. Correlation of enzymatic activity, muscle myoglobin concentration and lung morphology with activity metabolism in snakes. J. Exp. Zool. 197:313–320.

——. 1977. Some correlates of cranial and cervical morphology with predatory modes in snakes. J. Morphol. 152:89–100.

——. 1991. Reptilian physiology and the flight capacity of *Archaeopteryx*. Evolution 45:1–17.

——. 1995. The evolution of endothermy in mammals and birds: From physiology to fossils. Annu. Rev. Physiol. 57:69–95.

Ruben, J. A., and A. F. Bennett. 1980. Antiquity of the vertebrate pattern of activity metabolism and its possible relation to vertebrate origins. Nature 286:886–888.

Ruben, J. A., and A. J. Boucot. 1989. The origin of the lungless salamanders (Amphibia: Plethodontidae). Am. Nat. 134:161–169.

Ruben, J. A., W. J. Hillenius, N. R. Geist, A. Leitch, T. D. Jones, P. J. Currie, J. R. Horner, and G. Espe III. 1996. The metabolic status of some late Cretaceous dinosaurs. Science 273:1204–1207.

Ruben, J. A., and J. K. Parrish. 1990. Antiquity of the

chordate pattern of exercise metabolism. Paleobiology 16:355–359.

Rubner, M. 1883. Ueber den Einfluss der Körpergrosse auf Stoff- und Kraftwechsel. Z. Biol. 19:535–562.

——. 1908. Das Problem der Lebensdauer und seine Beziehungen zum Wachstum und zur Ernährung. Oldenbourg, München.

Ruby, D. E. 1977. Winter activity in Yarrow's spiny lizard, *Sceloporus yarrovi*. Herpetologica 33:322–333.

Rudd, J. T. 1954. Vertebrates without erythrocytes and blood pigment. Nature 173:848–850.

——. 1958. Vertebrates without blood pigment: A study of the fish family Chaenichthyidae. Int. Congr. Zool. 15:526–528.

Ruf, T., and G. Heldmaier. 1992. The impact of daily torpor on energy requirements in the Djungarian hamster, *Phodopus sungorus*. Physiol. Zool. 65:994–1010.

Ruibal, R. 1959. The ecology of a brackish water population of *Rana pipiens*. Copeia 1959:315–322.

——. 1961. Thermal relations of five species of tropical lizards. Evolution 15:98–111.

——. 1962a. Osmoregulation in amphibians from heterosaline habitats. Physiol. Zool. 35:133–147.

——. 1962b. The adaptive value of bladder water in the toad, *Bufo cognatus*. Physiol. Zool. 35:218–223.

Ruibal, R., and S. Hillman. 1981. Cocoon structure and function in the burrowing hylid frog, *Pternohyla fodiens*. J. Herpetol. 15:403–408.

Ruibal, R., and R. Philibosian. 1970. Eurythermy and niche expansion in lizards. Copeia 1970:645–653.

Ruibal, R., L. Tevis, and V. Roig. 1969. The terrestrial ecology of the spadefoot toad *Scaphiopus hammondii*. Copeia 1969:571–584.

Russell, D. A. 1967. Systematics and morphology of American mosasaurs. Bull. Peabody Mus. Nat. Hist. 23:1–240.

Ryser, F. A., and P. R. Morrison. 1954. Cold resistance in the young ring-necked pheasant. Auk 71:253–266.

Rytand, D. A. 1938. The number and size of mammalian glomeruli as related to kidney and to body weight, with methods for their enumeration and measurement. Am. J. Anat. 62:507–520.

Saarela, S., R. Hissa, A. Pyornila, R. Harjula, M. Ojanen, and M. Orell. 1989. Do birds possess brown adipose tissue? Comp. Biochem. Physiol. A Physiol. 92:219–228.

Sacca, R., and W. Burggren. 1982. Oxygen uptake in air and water in the air-breathing reedfish *Calamoichthys calabaricus*: Role of skin, gills and lungs. J. Exp. Biol. 97:179–186.

Sacher, G. A., and E. F. Saffeldt. 1974. Relation of gestation time to brain weight for placental mammals: Implications for the theory of vertebrate growth. Am. Nat. 108:593–615.

Saha, N., and B. K. Ratha. 1987. Active ureogenesis in a freshwater air-breathing teleost, *Heteropneustes fossilis*. J. Exp. Zool. 241:137–141.

——. 1989. Comparative study of ureogenesis in freshwater, air-breathing teleosts. J. Exp. Zool. 252:1–8.

Saint-Paul, U. 1984. Physiological adaptation to hypoxia

of a neotropical characoid fish *Colossoma macropomum*, Serrasalmidae. Envir. Biol. Fish. 11:53–62.

Salt, G. W. 1952. The relation of metabolism to climate and distribution in three finches of the genus *Carpodacus*. Ecol. Monogr. 22:121–152.

——. 1964. Respiratory evaporation in birds. Biol. Rev. Camb. Philos. Soc. 39:113–136.

Salthe, S. S. 1965. Comparative catalytic studies of lactic dehydrogenases in the Amphibia: Environmental and physiological correlations. Comp. Biochem. Physiol. 16:393–408.

Sand, H., G. Cederland, and K. Danell. 1995. Geographical and latitudinal variation in growth patterns and adult body size of Swedish moose (*Alces alces*). Oecologia 102:433–442.

Saramago, J. 1991. The Year of the Death of Ricardo Reis. Harcourt Brace, San Diego.

Sarrus, F., and J. F. Rameaux. 1839. Application des sciences accessoires et principalement des mathématiques à la physiologie générale. Bull. Acad. Med. Paris 3: 1094–1100.

Saunders, D. S. 1976. Insect Clocks. Pergamon, Oxford.

Sawaya, P. 1946. Sobre a biologia de alguns peixes de respiração aérea. Zoologia (Sao Paulo) 11:255–286.

——. 1947. Metabolismo respiratório de anfíbio gymnophiona, *Typhlonectes compressicauda*. Zoologia (Sao Paulo) 12:51–56.

Schaller, G. B., Hu Jinchu, Pan Wenshi, and Zhu Jing. 1985. The Giant Pandas of Wolong. University of Chicago Press, Chicago.

Scheid, P. 1979. Mechanisms of gas exchange in bird lungs. Rev. Physiol. Biochem. Pharmacol. 86:137–186.

Schildmacher, H. 1932. Ueber den Einfluss des Salzwassers auf die Entwicklung der Nasendruesen. J. Ornithol. 80:293–299.

Schindelmeiser, J., and H. Greven. 1981. Nitrogen excretion in intra- and extrauterine larvae of the ovovivparous salamander, *Salamandra salamandra* (L.) (Amphibia, Urodela). Comp. Biochem. Physiol. A Physiol. 70:563–565.

Schlagel, S. R., and C. M. Breder, Jr. 1947. A study of the oxygen consumption of blind and eyed cave characins in light and in darkness. Zoologica 32:17–27.

Schlesinger, W. H. 1976. Toxic foods and vole cycles: Additional data. Am. Nat. 110:315–317.

Schmid, W. D. 1965a. Some aspects of the water economies of nine species of amphibians. Ecology 46:261–269.

——. 1965b. High temperature tolerances of *Bufo hemiophrys* and *Bufo cognatus*. Ecology 46:559–560.

——. 1965c. Energy intake of the mourning dove, *Zenaidura macroura marginella*. Science 150:1171–1172.

——. 1968. Natural variations in nitrogen excretion of amphibians from different habitats. Ecology 49:180–185.

——. 1982. Survival of frogs in low temperature. Science 215:687–698.

Schmid, W. D., and R. E. Barden. 1965. Water permeability and lipid content of amphibian skin. Comp. Biochem. Physiol. 15:423–427.

Schmidt-Nielsen, B. 1958. Urea excretion in mammals. Physiol. Rev. 38:139–168.

Schmidt-Nielsen, B., and R. P. Forster. 1954. The effect of dehydration and low temperature on renal function in the bullfrog. J. Cell. Comp. Physiol. 44:233–246.

Schmidt-Nielsen, B., and R. O'Dell. 1961. Structure and concentrating mechanism in the mammalian kidney. Am. J. Physiol. 200:1119–1124.

Schmidt-Nielsen, B., R. O'Dell, and H. Osaki. 1961. Interdependence of urea and electrolytes in production of a concentrated urine. Am. J. Physiol. 200:1125–1132.

Schmidt-Nielsen, B., and K. Schmidt-Nielsen. 1950. Evaporative water loss in desert rodents in their natural habitat. Ecology 31:75–85.

Schmidt-Nielsen, B., K. Schmidt-Nielsen, T. R. Houpt, and S. A. Jarnum. 1956. Water balance of the camel. Am. J. Physiol. 185:185–194.

Schmidt-Nielsen, B., and K. Skadhauge. 1967. Function of the excretory system of the crocodile (*Crocodylus acutus*). Am. J. Physiol. 212:973–980.

Schmidt-Nielsen, K. 1960. The salt-excreting gland of marine birds. Circulation 21:955–967.

——. 1964. Desert Animals. Oxford University Press, London.

——. 1972. Locomotion: Energy cost of swimming, flying, and running. Science 177:222–228.

——. 1984. Why Is Animal Size So Important? Cambridge University Press, New York.

Schmidt-Nielsen, K., A. Borut, P. Lee, and E. C. Crawford. 1963. Nasal salt excretion and the possible function of the cloaca in water conservation. Science 142:1300–1301.

Schmidt-Nielsen, K., E. C. Crawford, A. E. Newsome, K. S. Rawson, and H. T. Hammel. 1967. Metabolic rate of camels: Effect of body temperature and dehydration. Am. J. Physiol. 212:341–346.

Schmidt-Nielsen, K., T. J. Dawson, H. T. Hammel, D. Hinds, and D. C. Jackson. 1965. The jack rabbit—a study in desert survival. Hvalradets Skr. 48:125–142.

Schmidt-Nielsen, K., and R. Fänge. 1958. The function of the salt gland in the brown pelican. Auk 75: 282–289.

Schmidt-Nielsen, K., F. R. Hainesworth, and D. E. Murrish. 1970. Counter-current heat exchange in the respiratory passages: Effect on water and heat balance. Respir. Physiol. 9:263–276.

Schmidt-Nielsen, K., S. A. Jarnum, and T. R. Houpt. 1957. Body temperature of the camel and its relation to water economy. Am. J. Physiol. 188:103–112.

Schmidt-Nielsen, K., C. B. Jorgensen, and H. Osaki. 1958. Extrarenal salt excretion in birds. Am. J. Physiol. 193:101–107.

Schmidt-Nielsen, K., and Y. T. Kim. 1964. The effect of salt intake on the size and function of the salt gland of ducks. Auk 81:160–172.

Schmidt-Nielsen, K., and J. L. Larimer. 1958. Oxygen dissociation in curves of mammalian blood in relation to body size. Am. J. Physiol. 195:424–428.

Schmidt-Nielsen, K., and P. Lee. 1962. Kidney function in the crab-eating frog (*Rana cancrivora*). J. Exp. Biol. 39:167–177.

Schmidt-Nielsen, K., and W. J. L. Sladen. 1958. Nasal secretion in the Humboldt penguin. Nature 181: 1217–1218.

Schmidt-Nielsen, K., and C. R. Taylor. 1968. Red blood cells: Why or why not? Science 162:274–275.

Schmitz, O. J., and D. M. Lavigne. 1984. Intrinsic rate of increase, body size, and specific metabolic rate in marine mammals. Oecologia 62:305–309.

Schmuck, R., F. Kobelt, and K. E. Linsenmair. 1988. Adaptations of the reed frog Hyperolius viridiflavus (Amphibia, Anura, Hyperoliidae) to its arid environment. V. Iridophores and nitrogen metabolism. J. Comp. Physiol. [B] 158:537–546.

Schnell, G. D. 1965. Recording the flight-speed of birds by Doppler radar. Living Bird 4:79–87.

Schnell, G. D., and J. J. Hellack. 1979. Bird flight speeds in nature: Optimized or a compromise? Am. Nat. 113: 53–66.

Schoener, T. W. 1968. Sizes of feeding territories among birds. Ecology 49:123–141.

——. 1969. Optimal size and specialization in constant and fluctuating environments: An energy-time approach. Brookhaven Symp. Biol. 22:103–114.

——. 1971. Theory of feeding strategies. Annu. Rev. Ecol. Syst. 2:369–404.

Scholander, P. F. 1940. Experimental investigations on the respiratory function in diving mammals and birds. Hvalradets Skr. 22:1–131.

——. 1954. Secretion of gases against high pressures in the swimbladder of deep sea fishes. II. The rete mirabile. Biol. Bull. (Woods Hole) 107:260–277.

——. 1955. Evolution of climatic adaptation in homeotherms. Evolution 9:15–26.

——. 1956. Climatic rules. Evolution 10:339–340.

——. 1966. The role of solvent pressure in osmotic systems. Proc. Natl. Acad. Sci. U.S.A. 55:1407–1414.

——. 1967. Osmotic mechanism and negative pressure. Science 156:67–69.

——. 1971. State of water in osmotic processes. Microvasc. Res. 3:215–232.

Scholander, P. F., W. Flagg, V. Walters, and L. Irving. 1953. Climatic adaptation in arctic and tropical poikilotherms. Physiol. Zool. 26:67–92.

Scholander, P. F., H. T. Hammel, E. D. Bradstreet, and E. A. Hemmingsen. 1965. Sap pressure in vascular plants. Science 148:339–346.

Scholander, P. F., A. R. Hargens, and S. L. Miller. 1968. Negative pressure in the interstitial fluid of animals. Science 161:321–328.

Scholander, P. F., R. Hock, V. Walters, and L. Irving. 1950c. Adaptation to cold in arctic and tropical mammals and birds in relation to body temperature, insulation, and basal metabolic rate. Biol. Bull. (Woods Hole) 99:259–271.

Scholander, P. F., R. Hock, V. Walters, F. Johnston, and L. Irving. 1950b. Heat regulation in some arctic and tropical mammals and birds. Biol. Bull. (Woods Hole) 99:225–236.

Scholander, P. F., and L. van Dam. 1954. Secretion of gases against high pressures in the swimbladder of deep sea fishes. I. Oxygen dissociation in blood. Biol. Bull. (Woods Hole) 107:247–259.

——. 1957. The concentration of hemoglobin in some cold water arctic fishes. J. Cell. Comp. Physiol. 49:1–4.

Scholander, P. F., L. van Dam, J. W. Kanwisher, H. T. Hammel, and M. S. Gordon. 1957. Supercooling and osmoregulation in arctic fish. J. Cell. Comp. Physiol. 49:5–24.

Scholander, P. F., V. Walters, R. Hock, and L. Irving. 1950a. Body insulation of some arctic and tropical mammals and birds. Biol. Bull. (Woods Hole) 99: 225–236.

Schottle, E. 1932. Morphologie und Physiologie der Atmung bei Wasser-, Schlamm-, und landlebenden Gobiiformen. Z. Wissenschftl. Zool. 140:1–114.

Schroeder, C. A. 1978. Temperature elevation in palm inflorescences. Principes 22:26–29.

Schuchmann, K.-L., and R. Prinzinger. 1988. Energy metabolism, nocturnal torpor and respiratory frequency in a green hermit (Phaetornis guy). J. Ornithol. 129:469–472.

Schulz, A. R. 1988. Energy metabolism in the whole animal revisited. Respir. Physiol. 73:11–20.

Schultz, E. T. 1991. The effect of energy reserves on breeding schedule: Is there a saturation point? Funct. Ecol. 5:819–824.

Schultz, E. T., L. M. Clifton, and R. R. Warner. 1991. Energetic constraints and size-based tactics: The adaptive significance of breeding-schedule variation in a marine fish (Embiotocidae: Micrometrus minimus). Am. Nat. 138:1408–1430.

Schwartz, G. G., and L. A. Rosenblum. 1981. Allometry of primate hair density and the evolution of human hairlessness. Am. J. Phys. Anthropol. 55:9–12.

Schwassmann, H. O. 1976. Ecology and taxonomic status of different geographic populations of Gymnorhamphichthys hypostomus Ellis (Pisces, Cypriniformes, Gymnotoidei). Biotropica 8:25–40.

——. 1978. Times of annual spawning and reproductive strategies in Amazonian fishes. In J. E. Thorpe (ed.), Rhythmic Activity of Fishes, pp. 187–200. Academic Press, London.

——. 1980. Biological rhythms: Their adaptive significance. In M. A. Ali (ed.), Environmental Physiology of Fishes, pp. 613–630. Plenum, New York.

Scott, I., and P. R. Evans. 1992. The metabolic output of avian (Sturnus vulgaris, Calidris alpina) adipose tissue liver and skeletal muscle: Implications for BMR/body mass relationships. Comp. Biochem. Physiol. A Physiol. 103:329–332.

Scott, I. M., M. K. Yousef, and W. G. Bradley. 1972. Body fat content and metabolic rate of rodents: Desert and mountain. Proc. Soc. Exp. Biol. Med. 141:818–821.

Scott, J. R., C. R. Tracy, and D. Pettus. 1982. A biophysical analysis of daily and seasonal utilization of climate space by a montane snake. Ecology 63:482–493.

Sealander, J. A. 1964. The influence of body size, season, sex, age and other factors upon some blood parameters in small mammals. J. Mammal. 45:598–616.

——. 1966. Seasonal variations in hemoglobin and hematocrit values in the northern red-backed mouse, *Clethrionomys rutilus dawsoni* (Merriam), in interior Alaska. Can. J. Zool. 44:213–224.

Sealy, S. G. 1976. Biology of nesting ancient murrelets. Condor 78:294–306.

Searcy, W. A. 1979. Sexual selection and body size in male red-winged blackbirds. Evolution 33:649–661.

——. 1980. Optimum body sizes at different ambient temperatures: An energetics explanation of Bergmann's rule. J. Theor. Biol. 83:579–593.

Seymour, R. S. 1973a. Energy metabolism of dormant spadefoot toads (*Scaphiopus*). Copeia 1973:435–445.

——. 1973b. Gas exchange in spadefoot toads beneath the ground. Copeia 1973:452–460.

——. 1974. How sea snakes avoid the bends. Nature 250:489–490.

Seymour, R. S., M. C. Barnhart, and G. A. Bartholomew. 1984. Respiratory gas exchange during thermogenesis in *Philodendron selloum* Koch. Planta 161:229–232.

Seymour, R. S., and D. F. Bradford. 1992. Temperature regulation in the incubation mounds of the Australian brush-turkey. Condor 94:134–150.

Seymour, R. S., G. P. Dobson, and J. Baldwin. 1981. Respiratory and cardiovascular physiology of the aquatic snake, *Acrochordus arafurae*. J. Comp. Physiol. 144:215–227.

Shafland, P. L., and J. M. Pestrak. 1982. Lower lethal temperatures for fourteen non-native fishes in Florida. Environ. Biol. Fish. 7:149–156.

Shaklee, J. B., J. A. Christiansen, B. D. Sidell, C. L. Prosser, and G. S. Whitt. 1977. Molecular aspects of temperature acclimation in fish: Contributions of changes in enzyme activities and isozyme patterns to metabolic reorganization in the green sunfish. J. Exp. Zool. 201:1–20.

Sharov, A. G. 1971. New flying reptiles from the Mesozoic deposits of Kazakhstan and Kirgizia. Tr. Paleo. Inst. Acad. Nauk SSSR 130:104–113. (In Russian)

Sharratt, B. A., I. C. Jones, and D. Bellamy. 1964. Water and electrolyte composition of the body and renal function of the eel (*Anguilla anguilla* L.). Comp. Biochem. Physiol. 11:9–18.

Shea, R. E., and R. E. Ricklefs. 1985. An experimental test of the idea that food supply limits growth rate in a tropical pelagic seabird. Am. Nat. 126:116–122.

Sherbrooke, W. C. 1976. Differential acceptance of toxic jojoba seed (*Simmondsia chinensis*) by four Sonoran Desert heteromyid rodents. Ecology 57:596–602.

Sherman, E., and S. G. Stadlen. 1986. The effect of dehydration on rehydration and metabolic rate in a lunged and a lungless salamander. Comp. Biochem. Physiol. A Physiol. 85:483–487.

Shield, J. 1972. Acclimation and energy metabolism of the dingo, *Canis dingo*, and the coyote, *Canis latrans*. J. Zool. Lond. 168:483–501.

Shine, R. 1985. The evolution of viviparity in reptiles: An ecological analysis. *In* B. C. Gans and F. Billett (eds.), Biology of the Reptilia, vol. 15, pp. 605–694. J Wiley, New York.

——. 1986a. Ecology of a low-energy specialist: Food habits and reproductive biology of the Arafura filesnake (Acrochordidae). Copeia 1986:424–457.

——. 1986b. Evolutionary advantages of limblessness: Evidence from the pygopodid lizards. Copeia 1986:525–529.

Shirley, E. K., and K. Schmidt-Nielsen. 1967. Oxalate metabolism in the pack rat, sand rat, hamster and white rat. J. Nutr. 91:496–502.

Shkolnik, A., and A. Borut. 1969. Temperature and water relations in two species of spiny mice (*Acomys*). J. Mammal. 50:245–255.

Shkolnik, A., and K. Schmidt-Nielsen. 1976. Temperature regulation in hedgehogs from temperate and desert environments. Physiol. Zool. 49:56–64.

Shoemaker, V. H. 1964. The effects of dehydration on electrolyte concentrations in a toad, *Bufo marinus*. Comp. Biochem. Physiol. 13:261–271.

——. 1972. Osmoregulation and excretion in birds. *In* D. S. Farner and J. R. King (eds.), Avian Biology, vol. II, pp. 527–574. Academic Press, New York.

Shoemaker, V. H., M. A. Baker, and J. P. Loveridge. 1989. Effect of water balance on thermoregulation in waterproof frogs (*Chiromantis* and *Phyllomedusa*). Physiol. Zool. 62:133–146.

Shoemaker, V. H., D. Balding, R. Ruibal, and L. L. McClanahan. 1972. Uricotelism and low evaporative water loss in a South American frog. Science 175:1018–1020.

Shoemaker, V. H., and P. E. Bickler. 1979. Kidney and bladder function in a uricotelic treefrog (*Phyllomedusa sauvagei*). J. Comp. Physiol. [B] 133:211–218.

Shoemaker, V. H., P. Licht, and W. R. Dawson. 1966. Effects of temperature on kidney function in the lizard *Tiliqua rugosa*. Physiol. Zool. 39:244–252.

Shoemaker, V. H., and L. L. McClanahan. 1973. Nitrogen excretion in the larvae of a land-nesting frog (*Leptodactylus bufonius*). Comp. Biochem. Physiol. A Physiol. 44:1149–1156.

——. 1975. Evaporative water loss, nitrogen excretion and osmoregulation in phyllomedusine frogs. J. Comp. Physiol. 100:331–345.

——. 1980. Nitrogen excretion and water balance in amphibians of Borneo. Copeia 1980:446–451.

Shoemaker, V. H., L. L. McClanahan, and R. Ruibal. 1969. Seasonal changes in body fluids in a field population of spadefoot toads. Copeia 1969:585–591.

Shoemaker, V. H., L. L. McClanahan, P. C. Withers, S. S. Hillman, and R. C. Drewes. 1987. Thermoregulatory response to heat in the waterproof frogs *Phyllomedusa* and *Chiromantis*. Physiol. Zool. 60:365–372.

Shoemaker, V. H., K. A. Nagy, and W. R. Costa. 1976. Energy utilization and temperature regulation by jackrabbits (*Lepus californicus*) in the Mojave desert. Physiol. Zool. 49:364–375.

Shoemaker, V. H., and C. Sigurdson. 1989. Brain cooling via evaporation from the eyes in a waterproof frog. Am. Zool. 29:106A.

Shotwell, J. A. 1958. Evolution and biogeography of the aplodontid and mylagaulid rodents. Evolution 12:451–484.

Sibley, C., and J. E. Ahlquist. 1990. Phylogeny and

Classification of Birds: A Study in Molecular Evolution. Yale University Press, New Haven.

Sibly, R. M. 1981. Strategies of digestion and defecation. *In* C. R. Townsend and P. Calow (eds.), Physiological Ecology: An Evolutionary Approach to Resource Use, pp. 109–139. Sinauer, Sunderland, Mass.

Sibly, R. M., and P. Calow. 1986. Physiological Ecology of Animals: An Evolutionary Approach. Blackwell, Oxford.

Sidell, B. D. 1977. Turnover of cytochrome C in skeletal muscle of green sunfish (*Lepomis cyanellus*, R.) during thermal acclimation. J. Exp. Zool. 199:233–250.

Sidell, B. D., and I. A. Johnston. 1985. Thermal sensitivity of contractile function in chain pickerel, *Esox niger*. Can. J. Zool. 63:811–816.

Siebert, H. C. 1949. Differences between migrant and non-migrant birds in food and water intake at various temperatures and photoperiods. Auk 66:128–153.

Silva, A. 1956. The relation of coloration in mammals to low temperature. J. Mammal. 37:378–381.

Silva, M., J. H. Brown, and J. A. Downing. 1997. Differences in population density and energy use between birds and mammals: A macroecological perspective. J. Anim. Ecol. 66:327–340.

Silver, H., N. F. Colovos, J. B. Holter, and H. H. Hayes. 1969. Fasting metabolism of white-tailed deer. J. Wildl. Manage. 33:490–498.

Sinclair, A. R. E., and J. N. M. Smith. 1984. Do plant secondary compounds determine feeding preferences of snowshoe hares? Oecologia 61:403–410.

Singh, B. N., and G. M. Hughes. 1971. Respiration of an air-breathing catfish *Clarias batrachus* (Linn.). J. Exp. Biol. 55:421–434.

Sinsch, U. 1984. Thermal influences on the habitat preference and the diurnal activity in three European *Rana* species. Oecologia 64:125–131.

——. 1989. Behavioural thermoregulation of the Andean toad (*Bufo spinulosus*) at high altitudes. Oecologia Berl. 80:32–38.

——. 1991. Cold acclimation in frogs (*Rana*): Microhabitat choice, osmoregulation, and hydromineral balance. Comp. Biochem. Physiol. A Physiol. 98:469–477.

Skadhauge, E. 1974a. Renal concentrating ability in selected West Australian birds. J. Exp. Biol. 61:269–276.

——. 1974b. Cloacal resorption of salt and water in the galah (*Cacatua roseicapilla*). J. Physiol. 240:763–773.

——. 1976. Cloacal absorption of urine in birds. Comp. Biochem. Physiol. A Physiol. 55:93–98.

——. 1981. Osmoregulation in Birds. Springer-Verlag, Berlin.

Skadhauge, E., and S. D. Bradshaw. 1974. Saline drinking and cloacal excretion of salt and water in the zebra finch. Am. J. Physiol. 227:1263–1267.

Skadhauge, E., S. K. Mahoney, and T. J. Dawson. 1991. Osmotic adaptation of the emu (*Dromaius novaehollandiae*). J. Comp. Physiol. [B] 161:173–178.

Skadhauge, E., and B. Schmidt-Nielsen. 1967. Renal medullary electrolyte and urea gradient in chickens and turkeys. Am. J. Physiol. 212:1313–1318.

Slip, D. J., and R. Shine. 1988a. Thermophilic response to feeding of the diamond python, *Morelia s. spilota* (Serpentes: Boidae). Comp. Biochem. Physiol. A Physiol. 89:645–650.

——. 1988b. Reptilian endothermy: A field study of thermoregulation by brooding diamond pythons. J. Zool. Lond. 216:367–378.

——. 1988c. Thermoregulation of free-ranging diamon pythons, *Morelia spilota* (Serpentes: Boidae). Copeia 1988:984–995.

Sloan, R. E., J. K. Rigby Jr., L. M. Van Valen, and D. Gabriel. 1986. Gradual dinosaur extinction and simultaneous ungulate radiation in the Hell Creek Formation. Science 232:629–633.

Slobodkin, L. B. 1960. Ecological energy relationships at the population level. Am. Nat. 94:213–236.

——. 1964. The strategy of evolution. Am. Sci. 52:342–357.

Slonin, A. D. 1952. Animal Heat and Its Regulation in the Mammalian Organism. Academy of Sciences, Leningrad and Moscow. (In Russian)

Smith, B. K., and T. J. Dawson. 1985. Use of helium-oxygen to examine the effect of cold adaptation on the summit metabolism of a marsupial, *Dasyuroides byrnei*. Comp. Biochem. Physiol. A Physiol. 81:445–449.

Smith, C. T. C. 1969. A high altitude hummingbird on the volcano Cotopaxi. Ibis 111:17–22.

Smith, D. W., R. O. Peterson, T. D. Drummer, and D. S. Sheputis. 1991. Over-winter activity and body temperature patterns in northern beavers. Can. J. Zool. 69:2178–2182.

Smith, E. N. 1976. Heating and cooling rates of the American alligator, *Alligator mississippiensis*. Physiol. Zool. 49:37–48.

——. 1979. Behavioral and physiological thermoregulation of crocodilians. Am. Zool. 19:239–247.

Smith, E. N., R. D. Allison, and W. E. Crowder. 1974. Bradycardia in a free ranging American alligator. Copeia 1974:770–772.

Smith, E. N., S. L. Robertson, and S. R. Adams. 1981. Thermoregulation of the spiny soft-shelled turtle *Trionyx spinifer*. Physiol. Zool. 54:74–80.

Smith, E. N., and E. W. Tobey. 1983. Heart rate response to forced and voluntary diving in swamp rabbits *Sylvilagus aquaticus*. Physiol. Zool. 56:632–638.

Smith, H. W. 1929. The composition of the body fluids of the goose fish (*Lophius piscatorius*). J. Biol. Chem. 82:71–75.

——. 1930. Metabolism of the lung-fish, *Protopterus aethiopicus*. J. Biol. Chem. 88:97–130.

——. 1931a. Observations on the African lung-fish, *Protopterus aethiopicus*, and on evolution from water to land environments. Ecology 12:164–181.

——. 1931b. The absorption and excretion of water and salts by elasmobranch fishes. II. Marine elasmobranchs. Am. J. Physiol. 98:296–310.

——. 1932. Water regulation and its evolution in the fishes. Q. Rev. Biol. 7:1–26.

——. 1935a. The metabolism of lung-fish. I. General considerations of the fasting metabolism in the active fish. J. Cell. Comp. Physiol. 6:43–67.

——. 1935b. The metabolism of lung-fish. II. Effect of feeding meat on metabolic rate. J. Cell. Comp. Physiol. 6:335–349.

——. 1936. The retention and physiological role of urea in the Elasmobranchii. Biol. Rev. Camb. Philos. Soc. 11:49–82.

——. 1959. From Fish to Philosopher. Little, Brown, Boston.

Smith, H. W., and C. G. Smith. 1931. The absorption and excretion of water and salts by elasmobranch fishes. I. Fresh water elasmobranchs. Am. J. Physiol. 98:279–295.

Smith, K. L., and R. R. Hessler. 1974. Respiration of benthopelagic fishes: In situ measurements at 1230 meters. Science 184:72–73.

Smith, R. J. 1984. Allometric scaling in comparative biology: Problems of concept and method. Am. J. Physiol. 246:R152–R160.

Smith, R. L., and D. Rhodes. 1983. Body temperature of the salmon shark, *Lamna ditropis*. J. Mar. Biol. Assoc. U.K. 63:243–244.

Smocovitis, V. B. 1996. Unifying Biology: The Evolutionary Synthesis and Evolutionary Biology. Princeton University Press, Princeton.

Smyth, M., and G. A. Bartholomew. 1966a. The water economy of the black-throated sparrow and the rock wren. Condor 68:447–458.

——. 1966b. Effects of water deprivation and sodium chloride on the blood and urine of the mourning dove. Auk 83:597–602.

Smyth, M., and H. N. Coulombe. 1971. Notes on the use of desert springs by birds in California. Condor 73:240–243.

Snell, R. R., and K. M. Cunnison. 1983. Relation of geographic variation in the skull of *Microtus pennsylvanicus* to Climate. Can. J. Ecol. 61:1232–1241.

Snow, B. K. 1966. Observations on the behaviour and ecology of the flightless cormorant *Nannopterum harrisi*. Ibis 108:265–280.

Snow, D. W. 1981. Tropical frugivorous birds and their food plants: A world survey. Biotropica 13:1–14.

Snyder, G. K. 1971. Influence of temperature and hematocrit on blood viscosity. Am. J. Physiol. 220:1667–1672.

——. 1976. Respiratory characteristics of whole blood and selected aspects of circulatory physiology in the common short-nosed fruit bat, *Cynopterus brachyotis*. Respir. Physiol. 28:239–247.

Snyder, G. K., and J. R. Nestler. 1990. Relationships between body temperature, thermal conductance, Q_{10} and energy metabolism during daily torpor and hibernation in rodents. J. Comp. Physiol. [B] 159:667–675.

Snyder, G. K., and W. W. Weathers. 1975. Temperature adaptations in amphibians. Am. Nat. 109:93–101.

Snyder, L. R. G., J. P. Hayes, and M. A. Chappell. 1988. Alpha-chain hemoglobin polymorphisms are correlated with altitude in the deer mouse, *Peromyscus maniculatus*. Evolution 42:689–697.

Somero, G. N. 1975. The role of isozymes in adaptation to varying temperatures. *In* C. L. Markert (ed.), Isozymes II: Physiological Function, pp. 221–234. Academic Press, New York.

——. 1982. Physiological and biochemical adaptation of deep-sea fishes: Adaptive responses to the physical and biological characteristics of the abyss. *In* W. G. Ernst and J. G. Morin (eds.), The Environment of the Deep Sea, pp. 256–278. Prentice-Hall, Englewood Cliffs, N.J.

Somero, G. N., A. C. Giese, and D. E. Wohlschlag. 1968. Cold adaptation of the antarctic fish *Trematomus bernacchii*. Comp. Biochem. Physiol. 26:223–233.

Somero, G. N., and J. F. Siebenaller. 1979. Inefficient lactate dehydrogenases of deep-sea fishes. Nature 282:100–102.

Sondaar, P. Y. 1977. Insularity and its effect on mammal evolution. *In* M. K. Hecht, P. C. Goody, and B. M. Hecht (eds.), Major Patterns in Vertebrate Evolution, pp. 671–707. Plenum, New York.

Southwick, E. E. 1983. The honey bee cluster as a homothermic superorganism. Comp. Biochem. Physiol. A Physiol. 75:641–645.

Spaargaren, D. H. 1992. Transport function of branching structures and the "surface law" for basic metabolic rate. J. Theor. Biol. 154:495–504.

Sparti, A. 1990. Comparative temperature regulation of African and European shrews. Comp. Biochem. Physiol. A Physiol. 97:391–397.

Spatz, H.-Ch. 1991. Circulation, metabolic rate, and body size in mammals. J. Comp. Physiol. [B] 161:231–236.

Speakman, J. R. 1990. On Blum's four-dimensional geometric explanation for the 0.75 exponent in metabolic allometry. J. Theor. Biol. 144:139–141.

——. 2000. The cost of living: Field metabolic rates of small mammals. Adv. Ecol. Res. 30:177–297.

Speakman, J. R., and D. Banks. 1998. The function of flight formations in greylag geese *Anser anser*; energy saving or orientation? Ibis 140:280–287.

Speakman, J. R., and P. A. Racey. 1989. Hibernal ecology of the pipistrelle bat: Energy expenditure, water requirements and mass loss—implications for survival and the function of winter emergence flights. J. Anim. Ecol. 58:797–813.

Speakman, J. R., and A. Rowland. 1999. Preparing for inactivity: How insectivorous bats deposit a fat store for hibernation. Proc. Nutr. Soc. 58:123–131.

Speakman, J. R., J. Rydell, P. I. Webb, J. P. Hayes, G. C. Hays, I. A. R. Hulbert, and R. M. McDevitt. 2000. Activity patterns of insectivorous bats and birds in northern Scandinavia (69°N), during continuous midsummer daylight. Oikos 88:75–86.

Sperber, I. 1944. Studies on the mammalian kidney. Zool. Bidr. Upps. 22:249–431.

Spight, T. M. 1967. The water economy of salamanders: Water uptake after dehydration. Comp. Biochem. Physiol. 20:767–771.

——. 1968. The water economy of salamanders: Evaporative water loss. Physiol. Zool. 41:195–203.

Spotila, J. R. 1980. Constraints of body size and environment on the temperature regulation of dinosaurs. *In* R. D. K. Thomas and E. C. Olson (eds.), A Cold Look at the Warm Blooded Dinosaurs, pp. 233–252.

AAAS Selected Symposium 28. American Academy for the Advancement of Sciences (AAAS), Washington, D.C.

Spotila, J. R., P. W. Lommen, G. S. Bakken, and D. M. Gates. 1973. A mathematical model for body temperatures of large reptiles: Implications for dinosaur ecology. Am. Nat. 107:391–404.

Spotila, J. R., M. P. O'Connor, P. Dodson, and F. V. Paladino. 1991. Hot and cold running dinosaurs: Body size, metabolism and migration. Modern Geol. 16:203–227.

Spray, D. C., and M. L. May. 1972. Heating and cooling rates in four species of turtles. Comp. Biochem. Physiol. A Physiol. 41:507–522.

Staaland, H. 1967a. Anatomical and physiological adaptations of the nasal glands in Charadriiformes birds. Comp. Biochem. Physiol. 23:933–944.

——. 1967b. Temperature sensitivity of the avian salt gland. Comp. Biochem. Physiol. 23:991–993.

Stahl, W. R. 1965. Organ weights in primates and other mammals. Science 150:1039–1042.

——. 1967. Scaling of respiratory variables in mammals. J. Appl. Physiol. 22:453–460.

Stamps, J. A. 1977. Rainfall, moisture and dry season growth rates in Anolis aeneus. Copeia 1977:415–419.

Stamps, J. A., and S. Tanaka. 1981. The influence of food and water on growth rates in a tropical lizard (Anolis aeneus). Ecology 62:33–40.

Standaert, T., and K. Johansen. 1974. Cutaneous gas exchange in snakes. J. Comp. Physiol. 89:313–320.

Standora, E. A., J. R. Spotila, and R. E. Foley. 1982. Regional endothermy in the sea turtle, Chelonia mydas. J. Thermal Biol. 7:159–165.

Stapp, P. 1992. Energetic influences on the life history of Glaucomys volans. J. Mammal. 73:914–920.

——. 1994. Can predation explain life-history strategies in mammalian gliders? J. Mammal. 75:227–228.

Stapp, P., P. J. Perkins, and W. W. Mautz. 1991. Winter energy expenditure and the distribution of southern flying squirrels. Can. J. Zool. 69:2548–2555.

Starling, E. H. 1899. The glomerular functions of the kidney. J. Physiol. 24:317–330.

Steadman, D. W. 1995. Prehistoric extinctions of Pacific island birds: Biodiversity meets zooarcheology. Science 267:1123–1131.

Stebbins, R. C. 1954. Amphibians and Reptiles of Western North America. McGraw-Hill, New York.

Steen, I., and J. B. Steen. 1965. The importance of legs in the thermoregulation of birds. Acta Physiol. Scand. 63:285–291.

Steen, J. B. 1958. Climatic adaptation in some small northern birds. Ecology 39:625–629.

——. 1963. The physiology of the swimbladder in the eel Anguilla vulgaris. III. The mechanism of gas secretion. Acta Physiol. Scand. 59:221–241.

——. 1970. The swimbladder as a hydrostatic organ. In W. S. Hoar and D. J. Randall (eds.), Fish Physiology, vol. 4, pp. 413–443. Academic Press, New York.

——. 1971. Comparative Physiology of Respiratory Mechanisms. Academic Press, London.

Steen, J. B., and T. Berg. 1966. The gills of two species of haemoglobin-free fishes compared to those of other teleosts—with a note on severe anaemia in an eel. Comp. Biochem. Physiol. 18:517–526.

Steen, J. B., and P. S. Enger. 1957. Muscular heat production in pigeons during exposure to cold. Am. J. Physiol. 191:157–158.

Steen, J. B., and A. Kruysse. 1964. The respiratory function of teleostean gills. Comp. Biochem. Physiol. 12:127–142.

Steen, J. B., H. Steen, and N. Ch. Stenseth. 1991a. Population dynamics of poikilotherms and homeotherm vertebrates: Effects of food shortage. Oikos 60:269–272.

Steen, J. B., Ø. Tøien, and P. Fiske. 1991b. Metabolic adaptations to hypothermia in snipe hatchlings (Gallinago media). J. Comp. Physiol. [B] 161:155–158.

Steen, W. B. 1929. On the permeability of the frog's bladder to water. Anat. Rec. 43:215–220.

Stefanski, M., R. E. Gatten Jr., and F. H. Pough. 1989. Activity metabolism of salamanders: Tolerance to dehydration. J. Herpetol. 23:45–50.

Stein, R. S. 1975. Dynamic analysis of Pteranodon ingens: A reptilian adaptation to flight. J. Paleontol. 49:534–548.

Stensiö, E. A. 1927. The Downtownian and Devonian vertebrates of Spitzbergen. I. Family Cephalaspidae. Skr. Svalbard Nordhishavet. 12:1–391.

——. 1968. The cyclostomes with special reference to the diphyletic origin of the Petromyzontida and Myxinoidea. Nobel Symp. 4:13–71.

Stephenson, P. J., and P. A. Racey. 1993a. Reproductive energetics of the Tenrecidae (Mammalia: Insectivora). I. The large-eared tenrec, Geogale aurita. Physiol. Zool. 66:643–663.

——. 1993b. Reproductive energetics of the Tenrecidae (Mammalia: Insectivora). II. The shrew-tenrecs, Microgale spp. Physiol. Zool. 66:664–685.

——. 1994. Seasonal variation in resting metabolic rate and body temperature of streaked tenrecs, Hemicentetes nigriceps and H. semispinosus (Insectivora: Tenrecidae). J. Zool. Lond. 232:285–294.

——. 1995. Resting metabolic rate and reproduction in the Insectivora. Comp. Biochem. Physiol. A Physiol. 112:215–223.

Steudel, K., W. P. Porter, and D. Sher. 1994. The biophysics of Bergmann's rule: A comparison of the effects of pelage and body size variation on metabolic rate. Can. J. Zool. 72:70–77.

Steunes, S. 1989. Taxonomy, habits, and relationships of the subfossil Madagascan hippopotami Hippopotamus lemerlei and H. madagascariensis. J. Vert. Paleontol. 9:241–268.

Stevens, C. E. 1988. Comparative Physiology of the Vertebrate Digestive System. Cambridge University Press, New York.

Stevens, E. D. 1973. The evolution of endothermy. J. Theor. Biol. 38:597–611.

Stevens, E. D., and F. E. J. Fry. 1970. The rate of thermal exchange in a teleost, Tilapia mossambica. Can. J. Zool. 48:221–226.

——. 1971. Brain and muscle temperatures in ocean caught and captive skipjack tuna. Comp. Biochem. Physiol. A Physiol. 38:203–211.

——. 1974. Heat transfer and body temperature in non-thermoregulatory teleosts. Can. J. Zool. 52:1137–1143.

Stevens, E. D., and G. F. Holeton. 1978. The partitioning of oxygen intake from air and from water by erythrinids. Can. J. Zool. 56:965–969.

Stevens, E. D., and A. M. Sutterlin. 1976. Heat transfer between fish and ambient water. J. Exp. Biol. 65:131–145.

Stevenson, R. D. 1985. Body size and limits to the daily range of body temperature in terrestrial ectotherms. Am. Nat. 125:102–117.

Stevenson, R. D., C. R. Peterson, and J. S. Tsuji. 1985. The thermal dependence of locomotion, tongue flicking, digestion, and oxygen consumption in the wandering garter snake. Physiol. Zool. 58:46–57.

Stewart, G. R. 1965. Thermal ecology of the garter snakes *Thamnophis sirtalis* (Hallowell) *concinnus* and *Thamnophis ordinoides* (Baird and Girard). Herpetologica 21:81–102.

Stewart, J. R. 1984. Thermal biology of the live bearing lizard *Gerrhonotus coeruleus*. Herpetologica 40:349–355.

Stewart, R. E. A., and D. M. Lavigne. 1980. Neonatal growth of northwest Atlantic harp seals, *Pageophilus groenlandicus*. J. Mammal. 61:670–680.

Stiassny, M. L., L. R. Parenti, and G. D. Johnson. 1996. Interrelationships of Fishes. Academic Press, San Diego.

Stiffler, D. F., and R. H. Alvarado. 1974. Renal function in response to reduced osmotic load in larval salamanders. Am. J. Physiol. 226:1243–1249.

——. 1980. Renal regulation of electrolytes in two species of ambystomatid salamanders. Copeia 1980:918–921.

Stiffler, D. F., M. L. DeRuyter, and C. R. Talbot. 1990. Osmotic and ionic regulation in the aquatic caecilian *Typhlonectes compressicauda* and the terrestrial caecilian *Ichthyophis kohtaoensis*. Physiol. Zool. 63:649–668.

Stiles, F. G. 1971. Time, energy, and territoriality of the Anna hummingbird (*Calypte anna*). Science 173:818–821.

Stinner, J. N., and V. H. Shoemaker. 1987. Cutaneous gas exchange and low evaporative water loss in the frogs *Phyllomedusa sauvagei* and *Chiromantis xerampelina*. J. Comp. Physiol. [B] 157:423–427.

Stirling, I., and A. E. Derocher. 1993. Possible impacts of climatic warming on polar bears. Arctic 46:240–245.

Stoicovici, F., and E. A. Pora. 1951. Comportarea la variatiuni de salinitate. Nota XXX. Stud. Cercet. Stiint., Acad. Rep. Pop. Romane, Fil. Cluj. 2:159–219.

Stone, G. N., and A. Purvis. 1992. Warm-up rates during arousal from torpor in heterothermic mammals: Physiological correlates and a comparison with heterothermic insects. J. Comp. Physiol. [B] 162:284–295.

Stone, P. A., J. L. Dobie, and R. P. Henry. 1992. Cutaneous surface area and bimodal respiration in softshelled (*Trionyx spiniferus*), stinkpot (*Sternotherus odoratus*), and mud turtles (*Kinosternon subrubrum*). Physiol. Zool. 65:311–330.

Stonehouse, B. 1967. The general biology and thermal balances of penguins. *In* J. B. Cragg (ed.), Advances in Ecological Research, pp. 131–196. Academic Press, London.

——. 1969. Environmental temperatures of Tertiary penguins. Science 163:673–675.

Storer, R. W. 1966. Sexual dimorphism and food habits in three North American accipiters. Auk 83:423–436.

Storey, J. M., and K. B. Storey. 1985a. Adaptations of metabolism for freeze tolerance in the gray tree frog, *Hyla versicolor*. Can. J. Zool. 63:49–54.

——. 1985b. Triggering of cryoprotectant synthesis by the initiation of ice nucleation in the freeze tolerant frog, *Rana sylvatica*. J. Comp. Physiol. [B] 156:191–195.

Storey, K. B. 1987. Glycolysis and the regulation of cryoprotectant synthesis in liver of the freeze tolerant wood frog. J. Comp. Physiol. [B] 157:373–380.

——. 1990. Life in a frozen state: Adaptive strategies for natural freeze tolerance in amphibians and reptiles. Am. J. Physiol. 258:R559–R568.

Storey, K. B., J. R. Layne Jr., M. M. Cutwa, T. A. Churchill, and J. M. Storey. 1993. Freezing survival and metabolism of box turtles, *Terrapene carolina*. Copeia 1993:628–634.

Storey, K. B., and J. M. Storey. 1986. Freeze tolerance and intolerance as strategies of winter survival in terrestrially-hibernating amphibians. Comp. Biochem. Physiol. A Physiol. 83:613–617.

——. 1988a. Freeze tolerance in animals. Physiol. Rev. 68:27–84.

——. 1988b. Freeze tolerance: Constraining forces, adaptive mechanisms. Can. J. Zool. 66:1122–1127.

——. 1992. Natural freeze tolerance in ectothermic vertebrates. Annu. Rev. Physiol. 54:619–637.

Storey, K. B., J. M. Storey, S. P. J. Brooks, T. A. Churchill, and R. J. Brooks. 1988. Hatching turtles survive freezing during winter hibernation. Proc. Natl. Acad. Sci. U.S.A. 85:8350–8354.

Stott, R. S., and D. P. Olson. 1973. Food-habitat relationship of sea ducks on the New Hampshire coastline. Ecology 54:996–1007.

Strahan, R. 1962. Survival of the hag, *Paramyxine atami* Dean, in diluted sea water. Copeia 1962:471–473.

Strunk, T. H. 1971. Heat loss from a Newtonian animal. J. Theor. Biol. 33:35–61.

——. 1973. Perspectives on linear heat transfer. Science 181:184–185.

Studier, E. H., and T. P. Baca. 1968. Atmospheric conditions in artificial rodent burrows. Southwest. Nat. 13:401–410.

Studier, E. H., and J. W. Procter. 1971. Respiratory gases in burrows of *Spermophilus tridecemlineatus*. J. Mammal. 52:631–633.

Studier, E. H., and D. E. Wilson. 1983. Natural urine concentrations and composition in neotropical bats. Comp. Biochem. Physiol. A Physiol. 75:509–515.

Studier, E. H., S. J. Wisniewski, A. T. Feldman, R. W. Dapson, B. C. Boyd, and D. E. Wilson. 1983. Kidney structure in neotropical bats. J. Mammal. 64:445–452.

Stuenes, S. 1989. Taxonomy, habits, and relationships of

the subfossil Madagascaran hippopotami *Hippopota-mus lemerlei* and *H. madagascariensis*. J. Vert. Pale-ontol. 9:241–268.

Styczynska-Jurewicz, E. 1970. Bioenergetics of osmoreg-ulation in aquatic animals. Polish Arch. Hydrobiol. 17:295–302.

Suhr, J. D. 1976. Nitrogen excretion, urea cycle activity and osmoregulation in *Amphiuma tridactylum*, *Siren intermedia*, *Cryptobranchus bishopi*, and *Necturus maculosus*. Ph.D. dissertation, University of Nebraska, Lincoln.

Sulkava, S. 1969. On small birds spending the night in the snow. Aquilo Ser. Zool. 7:33–37.

Summers-Smith, D. 1963. The House Sparrow. Collins, London.

Sverdrup, H. U., M. W. Johnson, and R. H. Fleming. 1942. The Oceans. Prentice-Hall, New York.

Swade, R. H., and C. S. Pittendrigh. 1967. Circadian locomotor rhythms of rodents in the Arctic. Am. Nat. 101:431–466.

Swain, S. D. 1992. Flight muscle catabolism during overnight fasting in a passerine bird, *Eremophila alpestris*. J. Comp. Physiol. [B] 162:383–392.

Swanson, D. L. 1991a. Seasonal adjustments in metabo-lism and insulation in the dark-eyed junco. Condor 93:538–545.

———. 1991b. Substrate metabolism under cold stress in seasonally acclimatized dark-eyed juncos. Physiol. Zool. 64:1578–1592.

———. 1993. Cold tolerance and thermogenic capacity in dark-eyed juncos in winter: Geographic variation and comparison with American tree sparrows. J. Thermal Biol. 18:275–281.

Sweeney, T. E., and C. A. Beuchat. 1993. Limitations of methods of osmometry: Measuring the osmolality of biological fluids. Am. J. Physiol. 264:R469–R480.

Swihart, R. K., J. P. Bryant, and L. Newton. 1994. Lati-tudinal patterns in consumption of woody plants by snowshoe hares in the eastern United States. Oikos 70:427–434.

Swihart, R. K., N. A. Slade, and B. J. Bergstrom. 1988. Relating body size to the rate of home range use in mammals. Ecology 69:393–399.

Szarski, H. 1964. The structure of respiratory organs in relation to body size in Amphibia. Evolution 18:118–126.

Tahvanainen, J., E. Helle, R. Julkunen-Tiitto, and A. Lavola. 1985. Phenolic compounds of willow bark as deterrents against feeding by mountain hare. Oecolo-gia 65:319–323.

Taigen, T. L. 1983. Activity metabolism of anuran amphibians: Implications for the origin of endothermy. Am. Nat. 121:94–109.

Taigen, T. L., and C. A. Beuchat. 1984. Anaerobic thresh-old of anuran amphibians. Physiol. Zool. 57:641–647.

Taigen, T. L., S. B. Emerson, and F. H. Pough. 1982. Ecological correlates of anuran exercise physiology. Oecologia 52:49–56.

Taigen, T. L., and F. H. Pough. 1983. Prey preference, foraging behavior, and metabolic characteristics of frogs. Am. Nat. 122:509–520.

Taigen, T. L., and K. D. Wells. 1985. Energetics of vocal-ization by an anuran amphibian (*Hyla versicolor*). J. Comp. Physiol. 155:163–170.

Taigen, T. L., K. D. Wells, and R. L. Marsh. 1985. The enzymatic basis of high metabolic rates in calling frogs. Physiol. Zool. 58:719–726.

Tannenbaum, M. G., and E. B. Pivorun. 1984. Differ-ences in daily torpor patterns among three southeast-ern species of *Peromyscus*. J. Comp. Physiol. [B] 154:233–236.

Taplin, L. E. 1984a. Evolution and zoogeography of the crocodilians: A new look at an ancient order. *In* M. Archer and G. Clayton (eds.), Vertebrate Zoogeogra-phy and Evolution in Australasia, pp. 361–370. Hesperian Press, Perth.

———. 1984b. Drinking of freshwater but not sea water by the estuarine crocodile (*Crocodylus porosus*). Comp. Biochem. Physiol. A Physiol. 77:763–767.

———. 1985. Sodium and water budgets of the fasted estuarine crocodile, *Crocodylus porosus*, in sea water. J. Comp. Physiol. [B] 155:501–513.

Taplin, L. E., and G. C. Grigg. 1981. Salt glands in the tongue of the estuarine crocodile *Crocodylus porosus*. Science 212:1045–1047.

———. 1989. Historical zoogeography of the eusuchian crocodilians: A physiological perspective. Am. Zool. 29:885–901.

Taplin, L. E., G. C. Grigg, and L. Beard. 1985. Salt gland function in fresh water crocodiles: Evidence for a marine phase in eusuchian evolution? *In* G. C. Grigg, R. Shine, and H. Ehmann (eds.), Biology of Australasian Frogs and Reptiles, pp. 403–410. Royal Society New South Wales, Sydney.

———. 1993. Osmoregulation of the Australian fresh-water crocodile, *Crocodylus johnstoni*, in fresh and saline waters. J. Comp. Physiol. [B] 163:70–77.

Taplin, L. E., G. C. Grigg, P. Harlow, T. M. Ellis, and W. A. Dunson. 1982. Lingual salt glands in *Crocody-lus acutus* and *C. johnstoni* and their absence from *Alligator mississippiensis* and *Caiman crocodilus*. J. Comp. Physiol. 149:43–47.

Taplin, L. E., and J. P. Loveridge. 1988. Nile crocodiles, *Crocodylus niloticus*, and estuarine crocodiles, *Croco-dylus porosus*, show similar osmoregulatory responses on exposure to seawater. Comp. Biochem. Physiol. A Physiol. 89:443–448.

Taylor, C. R. 1968. The minimum water requirements of some East African bovids. Symp. Zool. Soc. Lond. 21:195–206.

———. 1969. Metabolism, respiratory changes, and water balance of an antelope, the eland. Am. J. Physiol. 217:317–320.

———. 1970a. Strategies of temperature regulation: Effect on evaporation in East African ungulates. Am. J. Physiol. 219:1131–1135.

———. 1970b. Dehydration and heat: Effects on tempera-ture regulation in East African ungulates. Am. J. Physiol. 219:1136–1139.

———. 1977. The energetics of terrestrial locomotion and body size in vertebrates. *In* T. J. Pedley (ed.), Scale Effects in Animal Locomotion, pp. 127–141. Academic Press, London.

———. 1980. Mechanical efficiency of terrestrial locomotion: A useful concept? *In* H. Y. Elder and E. R. Trueman (ed.), Aspects of Animal Movement, pp. 235–244. Cambridge University Press, Cambridge.

——. 1981. Evolution of mammalian homeothermy: A two-step process? *In* K. Schmidt-Nielsen, L. Bolis, and C. R. Taylor (eds.), Comparative Physiology: Primitive Mammals, pp. 103–111. Cambridge University Press, Cambridge.

——. 1982. Scaling limits of metabolism to body size: Implications for animal design. *In* C. R. Taylor, K. Johansen, and L. Bolis (eds.), A Companion to Animal Physiology, pp. 161–170. Cambridge University Press, Cambridge.

Taylor, C. R., S. L. Caldwell, and V. J. Roundtree. 1972. Running up and down hills: Some consequences of size. Science 178:1096–1097.

Taylor, C. R., R. Dmi'el, M. Fedak, and K. Schmidt-Nielsen. 1971b. Energetic cost of running and heat balance in a large bird, the rhea. Am. J. Physiol. 221:597–601.

Taylor, C. R., N. C. Heglund, and G. M. O. Maloiy. 1982. Energetics and mechanics of terrestrial locomotion. I. Metabolic energy consumption as a function of speed and body size in birds and mammals. J. Exp. Biol. 97:1–21.

Taylor, C. R., and C. P. Lyman. 1967. A comparative study of the environmental physiology of an East African antelope, the eland, and the hereford steer. Physiol. Zool. 40:280–295.

——. 1972. Heat storage in running antelopes: Independence of brain and body temperatures. Am. J. Physiol. 222:114–117.

Taylor, C. R., D. Robertshaw, and R. Hofmann. 1969. Thermal panting: A comparison of wildebeest and zebu cattle. Am. J. Physiol. 217:907–910.

Taylor, C. R., and V. J. Roundtree. 1973a. Running on two or on four legs: Which consumes more energy? Science 179:186–187.

——. 1973b. Temperature regulation and heat balance in running cheetahs: A strategy for sprinters? Am. J. Physiol. 224:848–851.

Taylor, C. R., K. Schmidt-Nielsen, R. Dmi'el, and M. Fedak. 1971a. Effect of hyperthermia on heat balance during running in the African hunting dog. Am. J. Physiol. 220:823–827.

Taylor, C. R., K. Schmidt-Nielsen, and J. L. Raab. 1970. Scaling of energetic cost of running to body size in mammals. Am. J. Physiol. 219:1104–1107.

Taylor, C. R., S. Shkolnik, R. Dmi'el, D. Baharav, and A. Borut. 1974. Running in cheetahs, gazelles, and goats: Energy cost and limb configuration. Am. J. Physiol. 227:848–850.

Taylor, J. M., S. C. Smith, and J. H. Calaby. 1985. Altitudinal distribution and body size among New Guinean *Rattus* (Rodentia: Muridae). J. Mammal. 66:353–358.

Taylor, J. R. E. 1985. Ontogeny of thermoregulation and energy metabolism in pygoscelid penguin chicks. J. Comp. Physiol. [B] 155:615–627.

Taylor, J. R. E., and M. Konarzewski. 1992. Budget of elements in the little auk (*Alle alle*) chicks. Funct. Ecol. 6:137–144.

Taylor, J. R. E., A. Place, and D. D. Roby. 1997. Stomach oil and reproductive energetics in Antarctic prions, *Pachyptila desolata*. Can. J. Zool. 75:490–500.

Taylor, W. P. 1915. Description of a new subgenus (*Arborimus*) of *Phenacomys*, with a contribution to knowledge of the habits and distribution of *Phenacomys longicaudus* True. Proc. Calif. Acad. Sci. 5:111–161.

Tedman, R. A., and L. S. Hall. 1985. The morphology of the gastrointestinal tract and food transit time in the fruit bats *Pteropus alecto* and *P. poliocephalus* (Megachiroptera). Aust. J. Zool. 33:625–640.

Templeman, W. 1965. Mass mortalities of marine fishes in the Newfoundand area presumably due to low temperature. Int. Comm. Northwest Atl. Fish. Spec. Publ. 6:137–147.

Templeton, J. R. 1964. Nasal salt excretion in terrestrial lizards. Comp. Biochem. Physiol. 11:223–229.

——. 1967. Nasal salt gland excretion and adjustment to sodium loading in the lizard, *Ctenosaura pectinata*. Copeia 1967:136–140.

Tenney, S. M., and J. B. Tenney. 1970. Quantitative morphology of cold-blooded lungs: Amphibia and Reptilia. Respir. Physiol. 9:197–215.

Thomas, D. H., and G. L. Maclean. 1981. Comparison of physiological and behavioural thermoregulation and osmoregulation in two sympatric sandgrouse species (Aves: Pteroclididae). J. Arid Environ. 4:335–358.

Thomas, D. H., and A. P. Robin. 1977. Comparative studies of thermoregulatory and osmoregulatory behaviour and physiology of five species of sandgrouse (Aves: Pterocliidae) in Morocco. J. Zool. Lond. 183:229–249.

Thomas, D. W. 1984. Fruit intake and energy budgets of frugivorous bats. Physiol. Zool. 57:457–467.

——. 1995. The physiological ecology of hibernation in vespertilionid bats. Symp. Zool. Soc. Lond. 67:233.

Thomas, D. W., and D. Cloutier. 1992. Evaporative water loss by hibernating little brown bats, *Myotis lucifugus*. Physiol. Zool. 65:443–456.

Thomas, D. W., and F. Geiser. 1997. Periodic arousals in hibernating mammals: Is evaporative water loss involved? Funct. Ecol. 11:585–591.

Thomas, D. W., C. Samson, and J.-M. Bergeron. 1988. Metabolic costs associated with the ingestion of plant phenolics by *Microtus pennsylvanicus*. J. Mammal. 69:512–515.

Thomas, R. D. K., and E. C. Olson. 1980. A Cold Look at the Warm-Blooded Dinosaurs. AAAS Select. Symp. 28. American Association for the Advancement of Science (AAAS), Washington, D.C.

Thomas, S. P. 1975. Metabolism during flight in two species of bats, *Phyllostomus hastatus* and *Pteropus gouldii*. J. Exp. Biol. 63:273–293.

Thomas, S. P., and R. A. Suthers. 1972. The physiology and energetics of bat flight. J. Exp. Biol. 57:317.

Thomas, V. G. 1984. Winter diet and intestinal proportions of rock and willow ptarmigan and sharp-tailed grouse in Ontario. Can. J. Zool. 62:2258–2263.

——. 1987. Similar winter energy strategies of grouse, hares and rabbits in northern biomes. Oikos 50:206.

Thomerson, J. E., T. B. Thorson, and R. L. Hempel.

1977. The bull shark, *Carcharhinus leucas*, from the upper Mississippi River near Alton, Ilinois. Copeia 1977:166–168.

Thompson, D'A. W. 1917. On Growth and Form. Cambridge University Press, Cambridge.

Thompson, G. G., and P. C. Withers. 1997. Standard and maximal metabolic rates of goannas (Squamata: Varanidae). Physiol. Zool. 70:307–323.

Thompson, S. D. 1985. Subspecific differences in metabolism, thermoregulation, and torpor in the western harvest mouse *Reithrodontomys megalotis*. Physiol. Zool. 58:430–444.

——. 1987. Body size, duration of parental care, and the intrinsic rate of natural increase in eutherian and metatherian mammals. Oecologia 71:201–209.

Thompson, S. D., R. E. MacMillen, E. M. Burke, and C. R. Taylor. 1980. The energetic cost of bipedal hopping in small mammals. Nature 287:223–224.

Thompson, S. D., and M. E. Nicoll. 1986. Basal metabolic rate and energetics of reproduction in therian mammals. Nature 321:690–693.

——. 1992. Gestation and lactation in small mammals: Basal metabolic rate and the limits of energy use. *In* T. E. Tomasi and T. H. Horton (eds.), Mammalian Energetics, pp. 213–259. Cornell University Press, Ithaca.

Thomson, J. A., and N. W. Morley. 1966. Physiological correlates of habitat selection in Australian cormorants. Emu 66:17–26.

Thomson, K. S. 1967. Notes on the relationship of the rhipidistian fishes and the ancestry of the tetrapods. J. Paleontol. 41:660–674.

——. 1968. A critical review of certain aspects of the diphyletic theory of tetrapod relationships. Nobel Symp. 4:285–305.

——. 1969. The biology of the lobe-finned fishes. Biol. Rev. Camb. Philos. Soc. 44:91–154.

——. 1971. The adaptation and evolution of early fishes. Q. Rev. Biol. 46:139–166.

——. 1980. The ecology of Devonian lobe-finned fishes. *In* A. L. Panchen (ed.), The Terrestrial Environment and the Origin of Land Vertebrates, pp. 1–10. Academic Press, London.

Thorson, T. B. 1955. The relationship of water economy to terrestrialism in amphibians. Ecology 36:100.

——. 1970. Freshwater stingrays, *Potamotrygon* spp.: Failure to concentrate urea when exposed to saline medium. Life Sci. 9:893–900.

——. 1971. Movement of bull sharks, *Carcharhinus leucas*, between Caribbean Sea and Lake Nicaragua demonstrated by tagging. Copeia 1971:336–338.

——. 1972. The status of the bull shark, *Carcharhinus leucas*, in the Amazon River. Copeia 1972:601–605.

Thorson, T. B., C. M. Cowan, and D. E. Watson. 1966b. Sharks and sawfish in the Lake Izabal-Rio Dulce system, Guatemala. Copeia 1966:620–622.

——. 1967. *Potamotrygon* spp.: Elasmobranchs with low urea content. Science 158:375–377.

——. 1973. Body fluid solutes of juveniles and adults of the euryhaline bull shark *Carcharhinus leucas* from freshwater and saline environments. Physiol. Zool. 46:29–42.

Thorson, T. B., and A. Svihla. 1943. Correlation of the habitats of amphibians with their ability to survive the loss of body water. Ecology 24:374–381.

Thorson, T. B., and D. E. Watson. 1975. Reassignment of the African freshwater stingray, *Potamotrygon garouaensis*, to the genus *Dasyatis*, on physiologic and morphologic grounds. Copeia 1975:701–712.

Thorson, T. B., D. E. Watson, and C. M. Cowan. 1966a. The status of the freshwater shark of Lake Nicaragua. Copeia 1966:385–402.

Tinkle, D. W., and J. W. Gibbons. 1977. The distribution and evolution of viviparity in reptiles. Misc. Publ. Mus. Zool. Univ. Mich. 154:1–55.

Tinkle, D. W., and N. F. Hadley. 1973. Reproductive effort and winter activity in the viviparous montane lizard *Sceloporus jarrovi*. Copeia 1973:272–277.

——. 1975. Lizard reproductive effort: Caloric estimates and comments on its evolution. Ecology 56:427–434.

Tinkle, D. W., H. M. Wilbur, and S. G. Tilley. 1970. Evolutionary strategies in lizard reproduction. Evolution 24:55–74.

Todd, E. S. 1973. Positive buoyancy and air-breathing: A new piscine gas bladder function. Copeia 1973: 461–464.

Tornberg, R., M. Mönkkönen, and M. Pahkala. 1999. Changes in diet and morphology of Finnish goshawks from 1960s to 1990s. Oecologia 121:369–376.

Torre-Bueno, J. R., and L. Larochelle. 1978. The metabolic cost of flight in unrestrained birds. J. Exp. Biol. 75:223–229.

Torres, J. J., B. W. Belman, and J. J. Childress. 1979. Oxygen consumption rates of midwater fishes off California. Deep-Sea Res. 26A:185–197.

Torres, J. J., and G. N. Somero. 1988. Vertical distribution and metabolism in antarctic mesopelagic fishes. Comp. Biochem. Physiol. B Biochem. Mol. Biol. 90:521–528.

Townsend, C. R., and P. Calow. 1981. Physiological Ecology: An Evolutionary Approach to Resource Use. Blackwell Science, Oxford.

Tracy, C. R. 1972. Newton's law: Its application for expressing heat losses from homeotherms. Bioscience 22:656–659.

——. 1973. Perspectives on linear heat transfer. Science 181:185–186.

——. 1976. A model of the dynamic exchanges of water and energy between a terrestrial amphibian and its environment. Ecol. Monogr. 46:293–326.

Travis, J. 1982. A method for the statistical analysis of time-energy budgets. Ecology 63:19–25.

Tregear, R. T. 1965. Hair density, wind speed, and heat loss in mammals. J. Appl. Physiol. 20:796–801.

Tribe, M. A., and K. Bauler. 1968. Temperature dependence of "standard metabolic rate" in a poikilotherm. Comp. Biochem. Physiol. 25:427–436.

Trillmich, F. 1986. Are endotherms emancipated? Some considerations on the cost of reproduction. Oecologia 69:631–633.

Trojan, M. 1977. Water balance and renal adaptations in four Palaearctic hamsters. Naturwissenschaften 64:591–592.

——. 1979. Vergleichende Untersuchungen über den

Wasseraushalt und die Nierenfunktion der paläarktischen Hamster *Cricetus cricetus* (Leske, 1779), *Mesocricetus auratus* (Waterhouse, 1839), *Cricetulus griseus* (Milne-Edwards, 1867) and *Phodopus sungorus* (Pallas, 1770). Zool. Jb. Physiol. 83:192–223.

Tsugawa, K. 1976. Direct adaptation of cells to temperature: Similar changes of LDH isozyme patterns by *in vitro* and *in situ* adaptations in *Xenopus laevis*. Comp. Biochem. Physiol. B Biochem. Mol. Biol. 55:259–263.

———. 1980. Thermal dependence in kinetic properties of lactate dehydrogenase from the African clawed toad, *Xenopus laevis*. Comp. Biochem. Physiol. B Biochem. Mol. Biol. 66:459–466.

Tsuji, J. S. 1988a. Seasonal profiles of standard metabolic rate of lizards (*Sceloporus occidentalis*) in relation to latitude. Physiol. Zool. 61:230–240.

———. 1988b. Thermal acclimation of metabolism in *Sceloporus* lizards from different latitudes. Physiol. Zool. 61:241–253.

Tucker, V. A. 1966a. Diurnal torpor and its relation to food consumption and weight changes in the California pocket mouse *Perognathus californicus*. Ecology 47:245–252.

———. 1966b. Oxygen consumption of a flying bird. Science 154:150–151.

———. 1968a. Respiratory physiology of house sparrows in relation to high altitude flight. J. Exp. Biol. 48:55–66.

———. 1968b. Respiratory exchange and evaporative water loss in the flying budgerigar. J. Exp. Biol. 48:67–87.

———. 1970. Energetic cost of locomotion in animals. Comp. Biochem. Physiol. 34:841–846.

———. 1972. Metabolism during flight in the laughing gull. Am. J. Physiol. 222:237–245.

———. 1973. Bird metabolism during flight: Evaluation of a theory. J. Exp. Biol. 58:689–709.

———. 1975. The energetic cost of moving about. Am. Sci. 63:413–419.

Tuomi, J., T. Hakala, and E. Haukioja. 1983. Alternative concepts of reproductive effort, costs of reproduction, and selection in life-history evolution. Am. Zool. 23:25–34.

Turner, F. B., R. I. Jennrich, and J. D. Weintraub. 1969. Home ranges and body size in lizards. Ecology 50:1076–1081.

Turner, J. S., and C. R. Tracy. 1983. Blood flow to appendages and control of heat exchange in American alligators. Physiol. Zool. 56:195–200.

Turney, L. D., and V. H. Hutchison. 1974. Metabolic scope, oxygen debt and diurnal oxygen consumption cycle of the leopard frog, *Rana pipiens*. Comp. Biochem. Physiol. A Physiol. 49:583–601.

Tuttle, M. D. 1976. Population ecology of the gray bat (*Myotis grisescens*): Philopatry, timing, and patterns of movement, weight loss during migration, and seasonal adaptive strategies. Occas. Papers Mus. Nat. His. Univ. Kansas 54:1–38.

Twente, J. W., and J. A. Twente. 1965a. Effects of core temperature upon duration of hibernation of *Citellus lateralis*. J. Appl. Physiol. 20:411–416.

———. 1965b. Regulation of hibernating periods by temperature. Proc. Natl. Acad. Sci. U.S.A. 54:1058–1061.

Twente, J. W., J. A. Twente, and R. M. Moy. 1977. Regulation of arousal from hibernation by temperature in three species of *Citellus*. J. Appl. Physiol. 42:191–195.

Tynan, C. T., and D. P. DeMaster. 1997. Observations and predictions of arctic climatic change: Potential effects on marine mammals. Arctic 50:308–322.

Ullrich, K. J., and K. H. Jarausch. 1956. Untersuchungen zum problem der Harn konzentrierung und harnverdunnung. Pflugers Arch. Gesamte Physiol. 262:537–550.

Ullrich, K. J., B. Schmidt-Nielsen, R. O'Dell, G. Pheling, C. W. Gottschalk, W. E. Lassiter, and M. Mylle. 1963. Micropuncture study of composition of proximal and distal tubular fluid in rat kidney. Am. J. Physiol. 204:527–531.

Ultsch, G. R. 1973a. A theoretical and experimental investigation of the relationships between metabolic rate, body size, and oxygen exchange capacity. Respir. Physiol. 18:143–160.

———. 1973b. The effects of water hyacinths (*Eichhornia crassipes*) on the microenvironment of aquatic communities. Arch. Hydrobiol. 72:460–473.

———. 1974a. Gas exchange and metabolism in the Sirenidae (Amphibia: Caudata)—I. Oxygen consumption of submerged sirenids as a function of body size and respiratory surface area. Comp. Biochem. Physiol. A Physiol. 47:485–498.

———. 1974b. The allometric relationship between metabolic rate and body size: The role of the skeleton. Am. Midl. Nat. 92:500–504.

———. 1974c. In vivo permeability coefficient to oxygen of the skin of *Siren intermedia*. Am. J. Physiol. 226:1219–1220.

———. 1976a. Respiratory surface area as a factor controlling the standard rate of O_2 consumption of aquatic salamanders. Respir. Physiol. 26:357–369.

———. 1976b. Eco-physiological studies of some metabolic and respiratory adaptations of sirenid salamanders. *In* G. M. Hughes (ed.), Respiration of Amphibious Vertebrates, pp. 287–312. Academic Press, London.

———. 1987. The potential role of hypercarbia in the transition from water-breathing to air-breathing in vertebrates. Evolution 41:442–445.

———. 1989. Ecology and physiology of hibernation and overwintering among freshwater fishes, turtles, and snakes. Biol. Rev. 64:435–516.

———. 1995. On adjusting metabolic rates for body size. Florida Sci. 58:270–273.

———. 1996. Gas exchange, hypercarbia and acid-base balance, paleoecology, and the evolutionary transition from water-breathing to air-breathing among vertebrates. Palaeogeogr. Palaeoclimat. Palaeoecol. 123:1–27.

Ultsch, G. R., H. Boschung, and M. J. Ross. 1978. Metabolism, critical oxygen tension, and habitat selection in darters (*Etheostoma*). Ecology 59:99–107.

Ultsch, G. R., and D. C. Jackson. 1982a. Long-term submergence at 3°C of the turtle, *Chrysemys picta belli*, in normoxic and severely hypoxic water. I. Survival, gas exchange and acid-base status. J. Exp. Biol. 96:11–28.

———. 1982b. Long-term submergence at 3°C of the turtle,

Chrysemys picta belli, in normoxic and severly, hypoxic water. III. Effects of changes in ambient P_{O_2} and subsequent air breathing. J. Exp. Biol. 97:87–99.

——. 1996. pH and temperature in ectothermic vertebrates. Bull. Alabama Mus. Nat. Hist. 18:1–41.

Ultsch, G. R., D. C. Jackson, and R. Moalli. 1981. Metabolic oxygen conformity among lower vertebrates: The toadfish revisited. J. Comp. Physiol. 142:439–443.

Umminger, B. L. 1969. Physiological studies on supercooled killifish (*Fundulus heteroclitus*). 2. Serum organic constituents and the problem of supercooling. J. Exp. Zool. 172:409–423.

Unwin, D. M., and N. N. Bakhurina. 1994. *Sordes pilosus* and the nature of the pterosaur flight apparatus. Nature 371:62–64.

Urison, N. T., and R. B. Buffenstein. 1994. Kidney concentrating ability of a subterranean xeric rodent, the naked mole-rat (*Heterocephalus glaber*). J. Comp. Physiol. [B] 163:676–681.

Urison, N. T., and R. B. Buffenstein. 1995. Metabolic and body temperature changes during pregnancy and lactation in the naked mole rat (*Heterocephalus glaber*). Physiol. Zool. 68:402–420.

Urist, M. R., and K. A. Van de Putte. 1967. Comparative biochemistry of the blood of fishes. *In* P. W. Gilbert, R. F. Mathewson, and D. P. Rall (eds.), Sharks, Skates and Rays, pp. 271–292. Johns Hopkins Press, Baltimore, Md.

Utter, J. M. 1971. Daily energy expenditures of free-living purple martins (*Progne subis*) and mockingbirds (*Mimus polyglottos*) with comparison of two northern populations of mockingbirds. Ph.D. dissertation, Rutgers University, New Brunswick, NJ.

Utter, J. M., and E. A. LeFebrve. 1970. Energy expenditure for free flight by the purple martin (*Progne subis*). Comp. Biochem. Physiol. 35:713–719.

——. 1973. Daily energy expenditure of purple martins (*Progne subis*) during the breeding season: Estimates using D_2O^{18} and time budget methods. Ecology 54:597–604.

Val, A. L., and V. M. F. de Almeida-Val. 1995. Fishes of the Amazon and Their Environment. Springer-Verlag, Berlin.

van Berkum, F. H. 1986. Evolutionary patterns of the thermal sensitivity of sprint speed in *Anolis* lizards. Evolution 40:594–604.

van Berkum, F. H., F. H. Pough, M. M. Stewart, and P. F. Brussard. 1982. Altitudinal and interspecific differences in the rehydration abilities of Puerto Rican frogs (*Eleutherodactylus*). Physiol. Zool. 55:130–136.

van Beurden, E. K. 1980. Energy metabolism of dormant Australian water-holding frogs (*Cyclorana platycephalus*). Copeia 1980:787–799.

Van Damme, R., D. Bauwens, and R. F. Verheyen. 1990. Evolutionary rigidity of thermal physiology: The case of the cool temperate lizard *Lacerta vivipara*. Oikos 57:61–67.

Vander Wall, S. B. 1990. Food Hoarding in Animals. University of Chicago Press, Chicago.

Van Mierop, L. H. S., and S. M. Barnard. 1978. Further observations on thermoregulation in the brooding female *Python molurus bivittatus* (Serpentes: Boidae). Copeia 1978:615–621.

Van Mierop, L. H. S., T. Walsh, and D. L. Marcellini. 1983. Reproduction of *Chondropython viridis* (Reptilia, Serpentes, Boidae). *In* D. L. Marcellini (ed.), The Sixth Annual Reptile Symposium on Captive Propagation and Husbandry, pp. 265–274. Zoological Consortium, Thurmont, Md.

Van Soest, P. J. 1994. Nutritional Ecology of the Ruminant. Comstock, Ithaca, N.Y.

Vanstone, W. E., E. Roberts, and H. Tsuyuki. 1964. Changes in the multiple hemoglobin patterns of some Pacific salmon, genus *Oncorhynchus* during the parr-smolt transformation. Can. J. Physiol. Pharmacol. 42:697–703.

van Voorhies, W. V., J. A. Raymond, and A. L. DeVries. 1978. Glycoproteins as biological antifreeze agents in the cod, *Gadus ogac* (Richardson). Physiol. Zool. 51:347–353.

Vartanyan, S. L., V. E. Garutt, and A. V. Sher. 1993. Holocene dwarf mammoths from Wrangel Island in the Siberian Arctic. Nature 362:337–340.

Vaughan, T. A. 1982. Stephen's woodrat, a dietary specialist. J. Mammal. 63:53–62.

Vaughan, T. A., and N. J. Czaplewski. 1985. Reproduction in Stephen's woodrat: The wages of folivory. J. Mammal. 66:429–443.

Veghte, J. H. 1964. Thermal and metabolic responses of the gray jay to cold stress. Physiol. Zool. 37:316–328.

Veghte, J. H., and C. F. Herreid. 1965. Radiometric determination of feather insulation and metabolism of arctic birds. Physiol. Zool. 38:267–275.

Veit, R. R., P. Pyle, and J. A. McGowan. 1996. Ocean warming and long-term change in pelagic bird abundance with the California current system. Mar. Ecol. Prog. Ser. 139:11–18.

Veloso, C., and F. Bozinovic. 1993. Dietary and digestive constraints on basal energy metabolism in a small herbivorous rodent. Ecology 74:2003–2010.

Verheyen, E., R. Blust, and W. Decleir. 1994. Metabolic rate, hypoxia tolerance and aquatic surface respiration of some lacustrine and riverine African cichlid fishes (Pisces: Cichlidae). Comp. Biochem. Physiol. A Physiol. 107:403–411.

Verma, L. R. 1970. A comparative study of temperature regulation in *Apis mellifera* L. and *Apis indica* F. Am. Bee J. 110:390–391.

Videler, J. J., and B. A. Nolet. 1990. Costs of swimming measured at optimum speed: Scale effects, differences between swimming styles, taxonomic groups and submerged and surface swimming. Comp. Biochem. Physiol. A Physiol. 97:91–99.

Vinegar, A., and V. H. Hutchison. 1965. Pulmonary and cutaneous gas exchange in the green frog, *Rana clamitans*. Zoologia (N.Y.) 50:47–53.

Vinegar, A., V. H. Hutchison, and H. G. Dowling. 1970. Metabolism, energetics, and thermoregulation during brooding of snakes of the genus *Python* (Reptilia: Boidae). Zoologia (N.Y.) 55:19–48.

Vleck, C. M., D. F. Holt, and D. Vleck. 1979. Metabolism of avian embryos: Patterns in altricial and precocial birds. Physiol. Zool. 52:363–377.

Vleck, D. 1979. The energy cost of burrowing by the pocket gopher *Thomomys bottae*. Physiol. Zool. 52:122–136.

——. 1981. Burrow structure and foraging costs in the fossorial rodent *Thomomys bottae*. Oecologia 49:391–396.

Vleck, D., C. M. Vleck, and R. S. Seymour. 1984. Energetics of embryonic development in the megapode birds, mallee fowl *Leipoa ocellata* and brush turkey *Alectura lathami*. Physiol. Zool. 57:444–456.

Vogel, P. 1974. Kälteresistenz and reversible Hypothermie der Etruskerspitz maus (*Suncus etruscus*, Soricidae, Insectivora). Z. Saugetierk. 39:78–88.

Vogel, P., M. Burgener, J.-P. Lardet, M. Genoud, and H. Frey. 1979. Influence de la temperature et de la nourriture disponible sur la torpeur chez la Musaraigne musette (*Crocidura russula*) en captivity. Bull. Soc. Vaudoise Sci. Nat. 74:325–332.

Vogel, S., and W. L. Bretz. 1972. Interficial organisms: Passive ventilation in the velocity gradients near surfaces. Science 175:210–211.

Vogel, S., C. P. Ellington, and D. L. Kilgore. 1973. Wind-induced ventilation of the burrow of the prairie-dog, *Cynomys ludovicianus*. J. Comp. Physiol. 85:1–14.

Vogt, F. D., and G. R. Lynch. 1982. Influence of ambient temperature, nest availability, huddling, and daily torpor on energy expenditure in the white-footed mouse *Peromyscus leucopus*. Physiol. Zool. 55:56.

Voigt, C. C., and Y. Winter. 1999. Energetic cost of hovering flight in nectar-feeding bats (Phyllostomidae: Glossophaginae) and its scaling in moths, birds and bats. J. Comp. Physiol. [B] 169:38–48.

von Hoesslin, H. 1888. Ueber die Ursache der scheinbaren Abhängigkeit des Umsatzes von der Grösse der Körperoberfläche. Du Bois-Reymond Arch. Anat. Physiol. 11:323–379.

Vorobyeva, E., and H.-P. Schultze. 1991. Description and systematics of panderichthyid fishes with comments on their relationship to tetrapods. *In* H.-P. Schultze and L. Trueb (eds.), Origins of the Higher Groups of Tetrapods: Controversy and Consensus, pp. 68–109. Cornell University Press, Ithaca, N.Y.

Vorontsov, N. N. 1962. The ways of food specialization and evolution of the alimentary system in Muroidea. Symp. Theriol. Brno 4:360–377.

Voth, E. H., C. Maser, and M. L. Johnson. 1983. Food habits of *Arborimus albipes*, the white-footed vole, in Oregon. Northwest Sci. 57:1–7.

Wake, D. B. 1970. The abundance and diversity of tropical salamanders. Am. Nat. 104:211–213.

Wake, D. B., and J. F. Lynch. 1976. The distribution, ecology, and evolutionary history of plethodontid salamanders in tropical America. Nat. Hist. Mus. Los Ang. Cty. Sci. Bull. 25:1–65.

Wakeman, J. M., and G. R. Ultsch. 1975. The effects of dissolved O_2 and CO_2 on metabolism and gas-exchange partitioning in aquatic salamanders. Physiol. Zool. 48:348–359.

Waldschmidt, A., and E. F. Müller. 1988. A comparison of postnatal thermal physiology and energetics in an altricial (*Gerbillus perpallidus*) and a precocial (*Acomys cahirinus*) rodent species. Comp. Biochem. Physiol. A Physiol. 90:169–181.

Walker, J. M., A. Garber, R. J. Berger, and H. C. Heller. 1979. Sleep and estivation (shallow torpor): Continuous processes of energy conservation. Science 204:1098–1100.

Walker, J. M, E. H. Haskell, R. J. Berger, and H. C. Heller. 1980. Hibernation and circannual rhythms of sleep. Physiol. Zool. 53:8–11.

Wallgren, H. 1954. Energy metabolism of two species of the genus *Emberiza* as correlated with distribution and migration. Acta Zool. Fenn. 84:1–110.

Wallis, R. L. 1979. Responses to low temperature in small marsupial mammals. J. Thermal Biol. 4:105–111.

Walsberg, G. E. 1988. Consequences of skin color and fur properties for solar heat gain an ultraviolet irradiance in two mammals. J. Comp. Physiol. [B] 158:213–221.

Walsberg, G. E., G. S. Campbell, and J. R. King. 1978. Animal coat color and radiative heat gain: A re-evaluation. J. Comp. Physiol. 126:211–222.

Walsberg, G. E., and J. R. King. 1978. The energetic consequences of incubation for two passerine species. Auk 95:644–655.

Walsh, P. J., E. Danulat, and T. P. Mommsen. 1990. Variation in urea excretion in the gulf toadfish *Opsanus beta*. Mar. Biol. 106:323–328.

Walton, B. M. 1993. Physiology and phylogeny: The evolution of locomotor energetics in hylid frogs. Am. Nat. 141:26–50.

Walvig, F. 1960. The integument of the icefish *Chaenocephalus aceratus* (Lonnberg). Nytt Mag. Zool. (Oslo) 9:31–36.

Wang, L. C. H. 1973. Radiotelemetric study of hibernation under natural and laboratory conditions. Am. J. Physiol. 224:673–677.

——. 1978a. Energetic and field aspects of mammalian torpor: The Richardson's ground squirrel. *In* L. C. H. Wang and J. W. Hudson (eds.), Strategies in Cold: Natural Torpidity and Thermogenesis, pp. 109–145. Academic Press, New York.

——. 1978b. Factors limiting maximum cold induced heat production. Life Sci. 23:2089–2098.

Wang, Z.-X., N.-Z. Sun, and W.-F. Sheng. 1989. Aquatic respiration in soft-shelled turtles, *Trionyx sinensis*. Comp. Biochem. Physiol. A Physiol. 92:593–598.

Warburg, M. R. 1965. Studies on the water economy of some Australian frogs. Aust. J. Zool. 13:317–330.

——. 1971. On the water economy of Israel amphibians: The anurans. Comp. Biochem. Physiol. 40:911–924.

Ward, P., and P. J. Jones. 1977. Pre-migratory fattening in three races of the red-billed quelea *Quelea quelea* (Aves: Ploceidae), an intra-tropical migrant. J. Zool. Lond. 181:43–56.

Warham, J. 1977. Wing loadings, wing shapes, and flight capabilities of Procellariiformes. N.Z. J. Zool. 4:73.

——. 1990. The Petrels. Their Ecologic and Breeding Systems. Academic Press, London.

Warham, J., R. Watts, and R. J. Dainty. 1976. The composition, energy content and function of the stomach oils of petrels (order Procellariiformes). J. Exp. Mar. Biol. Ecol. 23:1–13.

Wasser, J. S. 1986. The relationship of energetics of falconiform birds to body mass and climate. Condor 88:57–62.

Waterman, P. G., J. A. M. Ross, E. L. Bennett, and A. G. Davies. 1988. A comparison of the floristics and leaf chemistry of the tree flora in two Malaysian rain forests and the influence of leaf chemistry on populations of colobine monkeys in the Old World. Biol. J. Linn. Soc. 34:1–32.

Wathen, P. M., J. W. Mitchell, and W. P. Porter. 1971. Theoretical and experimental studies of energy exchange from jackrabbit ears and cylindrically shaped appendages. Biophys. J. 11:1030–1047.

Watt, W. B. 1968. A captive significance of pigment polymorphism in *Colius* butterflies. I. Variation of melanin pigment in relation to thermoregulation. Evolution 22:437–458.

Watts, D. C., and R. L. Watts. 1966. Carbomoyl phosphate synthetase in the Elasmobranchii: Osmoregulatory function and evolutionary implications. Comp. Biochem. Physiol. 17:785–798.

Watts, P. D., and S. E. Hansen. 1987. Cyclic starvation as a reproductive strategy in the polar bear. Symp. Zool. Soc. Lond. 57:305–318.

Weathers, W. W. 1970. Physiological thermoregulation in the lizard *Dipsosaurus dorsalis*. Copeia 1970:549–557.

——. 1979. Climatic adaptation in avian standard metabolic rate. Oecologia 42:81–89.

——. 1992. Scaling nestling energy requirements. Ibis 134:142–153.

Weathers, W. W., W. A. Buttemer, A. M. Hayworth, and K. A. Nagy. 1984. An evaluation of time-budget estimates of daily energy expenditure in birds. Auk 101:459–472.

Weathers, W. W., W. D. Koenig, and M. T. Stanback. 1990. Breeding energetics and thermal ecology of the acorn woodpecker in central coastal California. Condor 92:341–359.

Weathers, W. W., and J. J. McGrath. 1972. Acclimation to simulated altitude in the lizard *Dipsosaurus dorsalis*. Comp. Biochem. Physiol. A Physiol. 42:263–268.

Weathers, W. W., and K. A. Nagy. 1980. Simultaneous doubly labeled water ($^3HH^{18}O$) and time-budget estimates of daily energy expenditures in *Phainopepla nitens*. Auk 97:861–867.

Weathers, W. W., D. C. Paton, and R. S. Seymour. 1996. Field metabolic rate and water flux of nectarivorous honeyeaters. Aust. J. Zool. 44:445–460.

Weathers, W. W., R. S. Seymour, and R. V. Baudinette. 1993. Energetics of mound-tending behaviour in the malleefowl, *Leipoa ocellata* (Megapodidae). Anim. Behav. 45:333–341.

Weathers, W. W., and R. B. Siegel. 1995. Body size establishes the scaling of avian postnatal metabolic rate: An interspecific analysis using phylogenetically independent contrasts. Ibis 137:532–542.

Weathers, W. W., and K. A. Sullivan. 1989. Juvenile foraging proficiency, parental effort, and avian reproductive success. Ecol. Monogr. 59:223–246.

——. 1993. Seasonal patterns of time and energy allocation by birds. Physiol. Zool. 66:511–536.

Weathers, W. W., and F. N. White. 1971. Physiological thermoregulation in turtles. Am. J. Physiol. 221:706–710.

Weaver, J. C. 1983. The improbable endotherm: The energetics of the sauropod dinosaur *Brachiosaurus*. Paleobiology 9:173–182.

Weber, R. E. 1982. Intraspecific adaptation of hemoglobin function in fish to oxygen availability. *In* A. D. F. Addink and N. Spronk (eds.), Exogenous and Endogenous Influences on Metabolic and Neural Control, pp. 87–102. Pergamon Press, Oxford.

Weber, R. E., M. A. Heath, and F. N. White. 1986. Oxygen binding functions of blood and hemoglobin from the Chinese pangolin, *Manis pentadactyla*: Possible implications of burrowing and low body temperature. Respir. Physiol. 64:103–112.

Webster, M. D., G. S. Campbell, and J. R. King. 1985. Cutaneous resistance to water-vapor diffusion in pigeons and the role of the plumage. Physiol. Zool. 58:58–70.

Webster, M. D., and W. W. Weathers. 1988. Effect of wind and air temperature on metabolic rate in verdins, *Auriparus flaviceps*. Physiol. Zool. 61:543–554.

Weeden, R. B. 1969. Foods of rock and willow ptarmigan in central Alaska with comments on interspecific competition. Auk 86:271–281.

Weigmann, R. 1929. Die Wirkung starker Abkühlung auf Amphibien und Reptilien. Z. Wiss. Zool. 134:641–692.

Weindruch, R., and R. L. Walford. 1989. The Retardation of Aging and Disease by Dietary Restriction. Charles C Thomas, Springfield, Ill.

Weiner, J. 1977. Energy metabolism of the roe deer. Acta Theriol. 22:3–24.

——. 1987. Maximum energy assimilation rates in the Djungarian hamster (*Phodopus sungorus*). Oecologia 72:297–302.

——. 1989. Metabolic constraints to mammalian energy budgets. Acta Theriol. 34:3–35.

——. 1992. Physiological limits to sustainable energy budgets in birds and mammals: Ecological implications. Trends Ecol. Evol. 7:384–388.

Weiner, J., and Z. Glowacinski. 1975. Energy flow through a bird community in a deciduous forest in southern Poland. Condor 77:233–242.

Weir, J. S. 1972. Spatial distribution of elephants in an African national park in relation to environmental sodium. Oikos 23:1–13.

Weis-Fogh, T. 1952. Weight economy of flying insects. Trans. Ninth Int. Congr. Entomol. 1:341–347.

Wellnhofer, P. 1987. Die Flughaut von *Pterodactylus* (Reptilia, Pterosauria) am Beispiel des Wiener Exemplares von *Pterodactylus kochi* (Wagner). Ann. Naturhist. Mus. Wien 88A:149–162.

——. 1988. Terrestrial locomotion in pterosaurs. Hist. Biol. 1:3–16.

——. 1996. The Illustrated Encyclopedia of Prehistoric Flying Reptiles. Barnes and Noble, New York.

Wells, J. W. 1963. Coral growth and geochronometry. Nature 197:948–950.

Wells, R. M. G. 1986. Cutaneous oxygen uptake in the antarctic icequab, *Rhigophila dearborni* (Pisces: Zoarchidae). Polar Biol. 5:175–180.

——. 1987. Respiration of antarctic fish from McMurdo Sound. Comp. Biochem. Physiol. A Physiol. 88:417–424.

Wells, R. M. G., G. C. Grigg, L. A. Beard, and G. Summers. 1989. Hypoxic responses in a fish from a stable environment: Blood oxygen transport in the antarctic fish *Pagothenia borchgrevinki*. J. Exp. Biol. 141:97–111.

Werner, Y. L., and A. H. Whitaker. 1978. Observations and comments on the body temperatures of some New Zealand reptiles. N.Z. J. Zool. 5:375–393.

West, G. B., J. H. Brown, and B. J. Enquist. 1997. A general model for the origin of allometric scaling laws in biology. Science 276:122–126.

——. 1999. The fourth dimension of life: Fractal geometry and allometric scaling of organisms. Science 284:1677–1679.

West, G. C. 1960. Seasonal variation in the energy balance of the tree sparrow in relation to migration. Auk 77:306–329.

——. 1965. Shivering and heat production in wild birds. Physiol. Zool. 38:111–120.

——. 1968. Bioenergetics of captive willow ptarmigan under natural conditions. Ecology 49:1035–1045.

——. 1972. Seasonal differences in resting metabolic rate of Alaskan ptarmigan. Comp. Biochem. Physiol. A Physiol. 42:867–876.

West, G. C., and B. B. DeWolfe. 1974. Populations and energetics of taiga birds near Fairbanks, Alaska. Auk 91:757–775.

West, G. C., E. R. R. Funke, and J. S. Hart. 1968. Power spectral density and probability analysis of electromyograms in shivering birds. Can. J. Physiol. Pharmacol. 46:703–706.

West, G. C., and J. S. Hart. 1966. Metabolic responses of evening grosbeaks to constant and to fluctuating temperatures. Physiol. Zool. 39:171–184.

Western, D. 1979. Size, life history and ecology in mammals. Afr. J. Ecol. 17:185–204.

Westerterp, K. 1973. The energy budget of the nestling starling, *Sturnus vulgaris*. A field study. Ardea 61: 137–158.

Westerterp, K., W. Gortmaker, and H. Wijngaarden. 1982. An energetic optimum in brood-raising in the starling *Sturnus vulgaris*: An experimental study. Ardea 70:153–162.

Westoby, M. 1974. An analysis of diet selection by large generalist herbivores. Am. Nat. 108:290–304.

——. 1978. What are the biological bases of varied diets? Am. Nat. 112:627–631.

Westoby, M., M. R. Leishman, and J. M. Lord. 1995. On misinterpreting the "phylogenetic correction." J. Ecol. 83:531–534.

Weygoldt, P. 1989. Changes in the composition of mountain stream frog communities in the Atlantic Mountains of Brazil: Frogs as indicators of environmental deteriorations? Stud. Neotrop. Fauna Environ. 24: 249–255.

Wheeler, P. E. 1984. The evolution of bipedality and loss of functional body hair in hominids. J. Human Evol. 13:91–98.

——. 1985. The loss of functional body hair in man: The influence of thermal environment, body form and bipedality. J. Human Evol. 14:23–28.

Wheelwright, N. T. 1983. Fruits and the ecology of the resplendent quetzals. Auk 100:286–301.

Wheelwright, N. T., W. A. Haber, K. G. Murray, and C. Guindon. 1984. Tropical fruit-eating birds and their food plants: A survey of a Costa Rican lower montane forest. Biotropica 16:173–192.

White, C. M., and G. C. West. 1977. The annual lipid cycle and feeding behavior of Alaskan redpolls. Oecologia 27:227–238.

White, E. I. 1958. Original environment of the craniates. *In* T. S. Westoll (ed.), Studies on Fossil Vertebrates, pp. 212–234. Athone Press, London.

White, F. N., G. A. Bartholomew, and J. L. Kinney. 1978. Physiological and ecological correlates of tunnel nesting in the European bee-eater, *Merops apiaster*. Physiol. Zool. 51:140–154.

White, F. N., J. Kinney, W. R. Siegfried, and A. C. Kemp. 1984. Thermal and gaseous conditions of hornbill nests. Res. Rept. Natl. Geogr. Soc. 17:931–936.

White, F. N., and R. C. Lasiewski. 1971. Rattlesnake denning: Theoretical considerations on winter temperatures. J. Theor. Biol. 30:553–557.

White, H. L. 1929. Observations on the nature of glomerular activity. Am. J. Physiol. 90:689–704.

Whitford, W. G., and V. H. Hutchison. 1963. Cutaneous and pulmonary gas exchange in the spotted salamander, *Ambystoma maculatum*. Biol. Bull. 124:344–354.

——. 1965. Gas exchange in salamanders. Physiol. Zool. 38:228–242.

——. 1967. Body size and metabolic rate in salamanders. Physiol. Zool. 40:127–133.

Whittembury, J., R. Lozano, C. Torres, and C. Monge. 1968. Blood viscosity in high altitude polycythemia. Acta Physiol. Latoam. 18:355–359.

Whybrow, P. J. 1981. Evidence for the presence of nasal salt glands in Hadrosauridae (Ornithischia). J. Arid Environ. 4:43–57.

Wickler, S. J., and R. L. Marsh. 1981. Effects of nestling age and burrow depth on CO_2 concentrations in the burrows of bank swallows (*Riparia riparia*). Physiol. Zool. 54:132–136.

Wiens, J. A., and G. S. Innis. 1974. Estimation of energy flow in bird communities: A population bioenergetics model. Ecology 55:730–746.

Wiens, J. A., and J. M. Scott. 1975. Model estimation of energy flow in Oregon coastal seabird populations. Condor 77:439–452.

Wiersma, P., and T. Piersma. 1994. Effects of microhabitat, flocking, climate and migratory goal on energy expenditure in the annual cycle of red knots. Condor 96:257–279.

Wieser, W. 1985. A new look at energy conversion in ecothermic and endothermic animals. Oecologia 66: 506–510.

——. 1986. More on energy conversion in ectotherms and endotherms: Biochemical versus social costs. Oecologia 69:634.

——. 1991. Limitations of energy acquisition and energy use in small poikilotherms: Evolutionary implications. Funct. Ecol. 5:234–240.

Wieser, W., H. Forstner, N. Medgyesy, and S. Hinterleit-
ner. 1988. To switch or not to switch: Partitioning of
energy between growth and activity in larval cyprinids
(Cyprinidae: Teleostei). Funct. Ecol. 2:499–507.

Wieser, W., and N. Medgysey. 1990a. Aerobic maximum
for growth in the larvae and juveniles of a cyprinid fish,
Rutilus rutilus (L.): Implications for energy budgeting
in small poikilotherms. Funct. Ecol. 4:233–242.

———. 1990b. Cost and efficiency of growth in the larvae
of two species of fish with widely differing metabolic
rates. Proc. R. Soc. Lond. B 242:51–56.

Wigginton, J. D., and F. S. Dobson. 1999. Environmental influences on geographic variation in body size of
western bobcats. Can. J. Zool. 77:802–813.

Wijnandts, H. 1984. Ecological energetics of the long-
eared owl (*Asio otus*). Ardea 72:1–92.

Wilder, I. W., and E. R. Dunn. 1920. The correlation of
lunglessness in salamanders with a mountain brook
habitat. Copeia 1920:63–68.

Wilhelm Filho, D., G. J. Eble, G. Kassner, F. X. Caprario,
A. L. Dafré, and M. Ohira. 1992. Comparative hema-
tology in marine fish. Comp. Biochem. Physiol. A
Physiol. 102:311–321.

Wilhelm Filho, D., and E. Reischl. 1981. Heterogeneity
and functional properties of hemoglobins from south
Brazilian freshwater fish. Comp. Biochem. Physiol. B
Biochem. Mol. Biol. 69:463–470.

Williams, G. C. 1957. Pleiotropy, natural selection and
the evolution of senescence. Evolution 11:398–411.

———. 1966. Adaptation and Natural Selection. Princeton
University Press, Princeton.

Williams, J. B. 1987. Field metabolism and food consumption of savannah sparrows during the breeding
season. Auk 104:277–289.

Williams, J. B., M. D. Anderson, and P. R. K. Richardson. 1997. Seasonal differences in field metabolism,
water requirements, and foraging behavior of free-
living aardwolves. Ecology 78:2588–2602.

Williams, J. B., and K. A. Nagy. 1984. Daily energy
expenditure of savannah sparrows: Comparison of
time-energy budget and doubly-labeled water estimates. Auk 101:221–229.

Williams, J. B., M. M. Pacelli, and E. J. Braun. 1991. The
effect of water deprivation on renal function in con-
scious unrestrained Gambel's quail (*Callipepla gam-
bellii*). Physiol. Zool. 64:1200–1216.

Williams, J. B., and A. Prints. 1986. Energetics of growth
in nestling savannah sparrows: A comparison of
doubly labeled water and laboratory estimates. Condor
88:74–83.

Williams, T. C., T. J. Klonowski, and P. Berkeley. 1976.
Angle of Canada goose V flight formation measured by
radar. Auk 93:554–559.

Williams, T. C., J. M. Williams, L. C. Ireland, and J. M.
Teal. 1977. Autumnal bird migration over the western
North Atlantic Ocean. Am. Birds 31:251–267.

———. 1978. Estimated flight time for transatlantic autumnal migrants. Am. Birds 32:275–280.

Williams, T. D., K. Gihebremeskel, G. Williams, and
M. A. Crawford. 1992. Breeding and moulting fasts in
macaroni penguins: Do birds exhaust their fat reserves?
Comp. Biochem. Physiol. A Physiol. 703:783–785.

Willmer, E. N. 1934. Some observations on the respiration of certain tropical fresh-water fishes. J. Exp. Biol.
11:283–306.

Willoughby, E. J. 1966. Water requirements of the
ground dove. Condor 68:243–248.

———. 1968. Water economy of the Stark's lark and grey-
backed finch-lark from the Namib desert of South West
Africa. Comp. Biochem. Physiol. 27:723–745.

Wilson, F. R., G. Somero, and C. L. Prosser. 1974.
Temperature-metabolism relations of two species of
Sebastes from different thermal environments. Comp.
Biochem. Physiol. B Biochem. Mol. Biol. 47:485–491.

Wilson, K. J. 1974. The relationship of oxygen supply for
activity to body temperature in four species of lizards.
Copeia 1974:920–934.

Wilson, K. J., and D. L. Kilgore. 1978. The effects of
location and design on the diffusion of respiratory
gases in mammal burrows. J. Theor. Biol. 71:73–101.

Wilson, M. A., and A. C. Echternacht. 1987. Geographic
variation in the critical thermal minimum of the green
anole, *Anolis carolinensis* (Sauria:Iguanidae), along a
latitudinal gradient. Comp. Biochem. Physiol. A
Physiol. 87:757–760.

Winberg, G. G. 1956. Rate of metabolism and food
requirements of fishes. J. Fish. Res. Board Can. Transl.
Ser. 194(1960):1–202.

———. 1961. New information on metabolic rate in fishes.
J. Fish. Res. Board Can. Transl. Ser. 362(1961):1–11.

Winslow, C. E. A., L. P. Herrington, and A. P. Gagge.
1937. Physiological reactions of the human body to
varying environmental temperatures. Am. J. Physiol.
120:1–22.

Winslow, R. M., C. C. Monge, N. J. Statham, C. G.
Gibson, S. Charache, J. Whittembury, O. Moran, and
R. L. Berger. 1981. Variability of oxygen affinity of
blood: Human subjects native to high altitude. J. Appl.
Physiol. 51:1411–1416.

Winter, Y. 1998. Energetic cost of hovering flight in
a nectar-feeling bat measured with fast-response
respirometry. J. Comp. Physiol. [B] 168:434–444.

Winter, Y., and O. von Helversen. 1998. The energy cost
of flight: Do small bats fly more cheaply than birds?
J. Comp. Physiol. [B] 168:105–111.

Wirz, H. 1953. Der osmotische Druck de Blutes in der
Nieren Papille. Helv. Physiol. Acta 11:20–29.

Withers, P. C. 1977. Respiration, metabolism, and heat
exchange of euthermic and torpid poorwills and hummingbirds. Physiol. Zool. 50:43–52.

———. 1978. Models of diffusion mediated gas exchange
in animal burrows. Am. Nat. 112:1101–1112.

———. 1981. An aerodynamic analysis of bird wings as
fixed aerofoils. J. Exp. Biol. 90:143–162.

———. 1982. Effect of diet and assimilation efficiency on
water balance for two desert rodents. J. Arid Environ.
5:375–384.

———. 1993. Metabolic depression during aestivation in
the Australian frogs, *Neobatrachus* and *Cyclorana*.
Aust. J. Zool. 41:467–473.

———. 1995. Cocoon formation and structure in the
aestivating Australian desert frogs, *Neobatrachus* and
Cyclorana. Aust. J. Zool. 43:429–441.

Withers, P. C., and J. D. Campbell. 1985. Effects of

environmental cost on thermoregulation in the desert iguana. Physiol. Zool. 58:329–339.

Withers, P. C., S. S. Hillman, and R. C. Drewes. 1984. Evaporative water loss and skin lipids of anuran amphibians. J. Exp. Zool. 232:11–17.

Withers, P. C., S. S. Hillman, R. C. Drewes, and O. M. Sokal. 1982b. Water loss and nitrogen excretion in sharp-nosed reed frogs (*Hyperolius nasutus*: Anura, Hyperoliidae). J. Exp. Biol. 97:335–343.

Withers, P. C., G. Louw, and S. Nicolson. 1982a. Water loss, oxygen consumption and colour changes in "waterproof" reed frogs (*Hyperolius*). S. Afr. J. Sci. 78:30–32.

Withers, P. C., and J. B. Williams. 1990. Metabolic and respiratory physiology of an arid-adapted Australian bird, the spinifex pigeon. Condor 92:961–969.

Wittenberg, J. B., and R. A. Haedrich. 1974. The choroid rete mirabile of the fish eye. II. Distribution and relation to the pseudobranch and to the swimbladder rete mirabile. Biol. Bull. (Woods Hole) 146:137–156.

Wittenberg, J. B., and B. A. Wittenberg. 1962. Active secretion of oxygen into the eye of fish. Nature 194:106–107.

——. 1974. The choroid rete mirabile of the fish eye. I. Oxygen secretion and structure: Comparison with the swimbladder rete mirabile. Biol. Bull. (Woods Hole) 146:116–136.

Wohlschlag, D. E. 1960. Metabolism of an antarctic fish and the phenomenon of cold adaptation. Ecology 41:287–292.

——. 1962. Antarctic fish growth and metabolic differences related to sex. Ecology 43:589–597.

——. 1963. An antarctic fish with unusually low metabolism. Ecology 44:557–564.

——. 1964. Respiratory metabolism and ecological characteristics of some fishes in McMurdo Sound, Antarctica. *In* M. O. Lee (ed.), Biology of the Antarctic Seas, pp. 33–62. Publ. 1190. American Geophysical Union, Washington, D.C.

Wohlschlag, D. E., and R. O. Juliano. 1959. Seasonal changes in bluegill metabolism. Limnol. Oceanogr. 4:195–209.

Wolf, L. L., and F. R. Hainsworth. 1971. Time and energy budgets of territorial hummingbirds. Ecology 52:980–988.

——. 1972. Environmental influence on regulated body temperature in torpid hummingbirds. Comp. Biochem. Physiol. A Physiol. 41:167–173.

Wolf, N. G., and D. L. Kramer. 1987. Use of cover and the need to breathe: The effects of hypoxia on vulnerability of dwarf gouramis to predatory snakeheads. Oecologia 73:127–132.

Wolf, N. G., P. R. Swift, and F. G. Carey. 1988. Swimming muscle helps warm the brain of lamnid sharks. J. Comp. Physiol. [B] 157:709–715.

Wood, C. M., S. F. Perry, P. A. Wright, H. L. Bergman, and D. J. Randall. 1989. Ammonia and urea dynamics in the Lake Magadi tilapia, a ureotelic teleost fish adapted to an extremely alkaline environment. Respir. Physiol. 77:1–20.

Wood, J. T., and S. I. Lustick. 1989. The effects of artificial solar radiation on wind-stressed tufted titmice (*Parus bicolor*) and Carolina chickadees (*Parus carolinesis*) at low temperatures. Comp. Biochem. Physiol. A Physiol. 92:473–477.

Wood, S. C., R. E. Weber, G. M. O. Maloiy, and K. Johansen. 1975. Oxygen uptake and blood respiratory properties of the caecilian *Boulengerula taitanus*. Respir. Physiol. 24:355–363.

Wood, S. C., R. E. Weber, and D. A. Powers. 1979. Respiratory properties of blood and hemoglobin solutions from the piranha. Comp. Biochem. Physiol. A Physiol. 62:163–167.

Woodall, P. F., and G. J. Currie. 1989. Food consumption, assimilation and rate of food passage in the Cape rock elephant shrew, *Elephantalus edwardii* (Macroscelidea: Macroscelidinae). Comp. Biochem. Physiol. A Physiol. 92:75–79.

Woodhead, P. M. J., and A. D. Woodhead. 1959. The effects of low temperatures on the physiology and distribution of the cod, *Gadus morhua* L. in the Barents Sea. Proc. Zool. Soc. Lond. 133:181–199.

Woods, P. E. 1982. Vertebrate digestive and assimilation efficiencies: Taxonomic and trophic comparisons. Biologist 64:58–77.

Woodward, A. R., J. H. White, and S. B. Linda. 1995. Maximum size of the alligator (*Alligator mississippiensis*). J. Herpetol. 29:507–513.

Worthington, A. H. 1989. Adaptations for avian frugivory: Assimilation efficiency and gut transit time of *Manacus vitellinus* and *Pipra mentalis*. Oecologia 80:381–389.

Worthy, G. A. J., and D. M. Lavigne. 1990. Mass loss, metabolic rate, and energy utilization by harp and gray seal pups during postweaning fast. Physiol. Zool. 60:352–364.

Wourms, J. P. 1967. Annual fishes. *In* F. H. Wilt and N. K. Wessells (eds.), Methods in Developmental Biology, pp. 123–137. T. Y. Crowell, New York.

Wright, P. G. 1984. Why do elephants flap their ears? S. Afr. J. Zool. 19:266–269.

Wunder, B. A. 1978. Implications of a conceptual model for the allocation of energy resources by small mammals. *In* D. P. Snyder (ed.), Populations of Small Mammals under Natural Conditions, pp. 68–75. Pymatuning Symp. Ecol. No. 5. University of Pittsburgh, Pittsburgh.

——. 1984. Strategies for and environmental cueing mechanisms of seasonal changes in thermoregulatory parameters of small mammals. *In* J. F. Merritt (ed.), Winter Ecology of Small Mammals, pp. 165–172. Special Publ. 10. Carnegie Museum of Natural History Pittsburgh, Penn.

——. 1992. Morphophysiological indicators of the energy state of small mammals. *In* T. E. Tomasi and T. H. Horton (eds.), Mammalian Energetics, pp. 83–104. Cornell University Press, Ithaca.

Wunder, B. A., D. S. Dobkin, and R. D. Gettinger. 1977. Shifts of thermogenesis in the prairie vole (*Microtus ochrogaster*). Oecologia 29:11–26.

Wunder, B. A., and P. R. Morrison. 1974. Red squirrel metabolism during incline running. Comp. Biochem. Physiol. A Physiol. 48:153–161.

Wygoda, M. L. 1984. Low cutaneous evaporative water loss in arboreal frogs. Physiol. Zool. 57:329–337.

——. 1988. Adaptive control of water loss resistance in an arboreal frog. Herpetologica 44:251–257.

Yacoe, M. E., and W. R. Dawson. 1983. Seasonal acclimatization in American goldfinches: The role of the pectoralis muscle. Am. J. Physiol. 245:R265–R271.

Yahav, S., and R. Buffenstein. 1992. Caecal function provides the energy of fermentation without liberating heat in the poikilothermic mammal, *Heterocephalus glaber*. J. Comp. Physiol. [B] 162:216–218.

Yancey, P. H., M. E. Clark, S. C. Hand, R. D. Bowlus, and G. N. Somero. 1982. Living with water stress: Evolution of osmolyte systems. Science 217:1214–1222.

Yancey, P. H., and G. N. Somero. 1978. Urea-requiring lactate dehydrogenases of marine elasmobranch fishes. J. Comp. Physiol. 125:135–141.

——. 1979. Counteraction of urea destabilization of protein structure by methylamine osmoregulatory compounds of elasmobranch fishes. Biochem. J. 183:317–323.

——. 1980. Methylamine osmoregulatory solutes of elasmobranch fishes counteract urea inhibition of enzymes. J. Exp. Zool. 212:205–213.

Yarbrough, C. G. 1970. The development of endothermy in nestling gray-crowned rosy finches, *Leucosticte tephrocotis griseonucha*. Comp. Biochem. Physiol. 34:917–925.

——. 1971. The influence of distribution and ecology on the thermoregulation of small birds. Comp. Biochem. Physiol. A Physiol. 39:235–266.

Young Owl, M., and G. O. Batzli. 1998. The integrated processing response of voles to fibre content of natural diets. Funct. Ecol. 12:4–13.

Yousef, M. K., and H. D. Johnson. 1975. Thyroid activity in desert rodents: A mechanism for lowered metabolic rate. Am. J. Physiol. 229:427–431.

Yousef, M. K., H. D. Johnson, W. G. Bradley, and S. M. Seif. 1974. Triated water-turnover rate in rodents: Desert and mountain. Physiol. Zool. 47:153–162.

Zamachowski, W. 1977a. The water economy in some European species of anuran amphibians during the annual cycle. I. Water content of the organism. Acta Biol. Cracoviensia Zool. Ser. 20:181–189.

——. 1977b. The water economy in some European species of anuran amphibians during the annual cycle. II. Skin permeability in vitro. Acta Biol. Cracoviensia Zool. Ser. 20:191–206.

——. 1977c. The water economy in some European species of anuran amphibians during the annual cycle. III. Resistance to water storage. Acta Biol. Cracoviensia Zool. Ser. 20:207–228.

Zerba, E., and G. E. Walsberg. 1992. Exercise-generated heat contributes to thermoregulation by Gambel's quail in the cold. J. Exp. Biol. 171:409–422.

Zervanos, S. M. 1975. Seasonal effects of temperature on the respiratory metabolism of the collared peccary (*Tayassu tajacu*). Comp. Biochem. Physiol. A Physiol. 50:365–371.

Zimmerman, J. L. 1965. Bioenergetics of the dickcissel, *Spiza americana*. Physiol. Zool. 38:370–389.

Zimmerman, L. C., and C. R. Tracy. 1989. Interactions between the environment and ectothermy and herbivory in reptiles. Physiol. Zool. 62:374–409.

Ziswiler, V., and D. S. Farmer. 1972. Digestion and the digestive system. *In* D. S. Farner and J. R. King (eds.), Avian Biology, vol. II, pp. 343–430. Academic Press, New York.

Zullinger, E. M., R. E. Ricklefs, K. H. Redford, and G. M. Mace. 1984. Fitting sigmoidal equations to mammalian growth curves. J. Mammal. 65:607–636.

Taxonomic Index

Manacus 356
Manidae 181, 439
Manis 104, 108, 256, 257
Manta 120, 140
Marmota 42, 256, 257, 385
Marsupialia 99, 100, 108, 175, 298,
   312, 356, 378, 436, 441, 442, 458
Martes 91, 310
Masticophis 278, 279
Megachiroptera 291, 406
Megapodiidae 116, 325
Megacrex 295
Megaptera 104
Melanerpes 359, 395, 423
Melanitta 117, 393
Melanosuchus 451
Meles 386
Meliphaga 192
Meliphagidae 192, 314, 401, 402
Melopsittacus 176, 191, 282, 348
Menuridae 394
Mergus 117, 393
Meriones 105
Mesitornis 293
Mesocricetus 110, 111, 212, 256, 354
Metatheria 35, 38
Microchiroptera 291
Microdipodops 40, 354
Microgale 388
Micrometrus 357
Micronisus 198
Micropterus 51, 271, 272
Microtus 93, 210, 211, 258, 334, 339,
   346, 360, 400, 405, 407, 410, 420,
   423, 424, 425, 438, 439
Mimus 319
Mirounga 260, 328
Misgurnus 239
Mobula 120
Mobulidae 120
Moloch 177
Molossidae 100, 291, 462
Monachus 97, 465
Monias 293
Monodon 475
Monopeltis 179
Monopterus 143, 226, 239
Monotremata 298, 352, 441, 442
Moraceae 394
Morelia 67, 122, 123
Mormoopidae 95
Mormoops 462
Mormyridae 236
Morone 54
Morus 330
Motacilla 294
Mugil 148, 149
Muridae 211, 213, 216, 433
Mus 259, 298, 319, 339, 437
Muscicapidae 393, 443
Musophagidae 395
Mustela 91, 103, 204, 302
Mustelidae 40, 94, 97
Mycteria 105
Mycterioperca 138
Myotis 212, 461
Myoxocephalus 50, 59
Myrmecobius 100
Myrmecophagidae 72, 104, 108, 394

Myxine 136, 146
Myxinoidea 135, 136, 151, 152, 154

Nandinia 106, 356
Napaeozapus 462
Naso 409
Nasua 439, 463, 464
Natrix 52, 278
Nectarinia 356
Nectariniidae 352, 356, 401, 402
Necturus 150, 161, 206, 207, 245, 246,
   247
Neobatrachus 164, 169, 170, 387
Neoceratodus 144, 226, 239, 241, 242,
   251
Neochanna 172
Neocirrhites 50
Neophema 177, 192
Neotoma 89, 103, 212, 214, 215, 216,
   327, 392, 402, 410, 420, 425
Nerodia 66, 67, 260
Noctuidae 118
Notaden 164
Notomys 183, 204, 209, 213
Notophthalmus 356
Notopterus 240
Notornis 407
Notothenia 50, 234
Nototheniidae 144, 233, 234
Nototheniodia 271
Numenius 294
Nyctea 16, 346
Nycticebus 113, 296, 299

Oceanites 323
Oceanodroma 198, 200, 322, 323, 328,
   339, 352, 475
Octodon 213, 405, 406
Octodontidae 213, 216
Odobenus 261, 328
Odocoileus 90, 92, 415, 422
Ogcocephalus 226
Oikopleura 273
Oncorhynchus 49, 61, 68, 138, 143,
   144, 145, 150, 223, 228, 271, 272,
   273, 331, 341
Ondatra 92, 97
Onychomys 212, 214
Ophioderma 273
Ophiosaurus 64, 276, 278
Opisthocomus 293, 393, 397, 408, 409,
   413
Opsanus 143, 227, 228, 230, 235
Oreamnos 474
Oreochromis 143
Oreotrochilus 454
Ornithorhynchus 203, 223
Orycteropus 104, 181, 257, 439
Oryx 177
Osphronemidae 239
Osphronemus 239
Osteichthyes 136, 142, 152, 223
Osteoglossidae 239
Osteoglossum 237, 238, 241
Osteolaemus 202, 469
Osteopilus 167, 274
Otomys 257
Ourebia 415
Ovibos 92, 103

Ovis 92, 336, 399
Oxymycteris 417

Pachydactylus 316
Pachymedusa 156, 162, 167, 169
Pachyptila 293, 315, 323
Pagetopsis 234
Pagothenia 52, 233
Paleosuchus 451
Palmaceae 394
Pan 299
Panthera 91, 296, 298
Papio 406
Parachirrhites 50
Paraechinus 126
Paramyxine 136
Paridae 112, 352
Parulinae 443
Parus 19, 20, 22, 112, 254, 315, 355,
   358
Passer 190, 192, 196, 259, 366, 367,
   456
Passerculus 191, 192, 193, 194, 294,
   321, 328
Passeriformes 38, 41, 175, 182, 194,
   197, 287, 290, 295, 312, 352, 355,
   358, 395, 402, 475
Patagona 454
Pelamis 198, 263, 449, 450
Pelecaniformes 295
Pelecanoides 315, 319, 323
Pelecanoididae 285, 286, 287, 289,
   315
Pelecaniformes 198
Pelecanus 182, 198, 199, 287
Percidae 49
Periophthalmus 143, 237, 239, 241
Perissodactyla 411
Perodicticus 113, 296
Perognathus 40, 117, 214, 215, 216,
   311, 313, 347, 353, 354, 356, 364,
   376, 380, 381, 382, 388, 473
Peromyscus 33, 98, 99, 104, 113, 117,
   180, 210, 211, 212, 216, 254, 258,
   259, 309, 318, 333, 352, 353, 354,
   360, 388, 433, 437
Petaurista 293, 399, 412, 437
Petauroides 293
Petaurus 293, 437
Petromyzon 136, 137, 326
Petromyzontia 136, 472
Phaeton 291
Phainopepla 311
Phalacrocorax 18, 114, 182, 198, 199,
   200, 201, 295, 339, 453, 469, 475
Phalaenoptilus 181, 352, 376, 385
Phalaropus 294
Phaps 176
Phascolarctos 392, 394, 404, 412, 418,
   423
Phasianidae 192
Phenacomys 86, 87
Phoca 105, 328, 465, 466
Phocidae 261
Phodopus 98, 212, 352, 353, 360, 377
Phoeniconaias 319
Phrynosoma 55, 63, 64, 69, 74, 83,
   156, 374, 452
Phrynosomatidae 64

Subject Index